IMMUNO BIOLOGY

THE IMMUNE SYSTEM IN HEALTH AND DISEASE

FOURTH EDITION

P.327 — Antibody transfer to newborn.

IMMUNO BIOLOGY

THE IMMUNE SYSTEM IN HEALTH AND DISEASE

FOURTH EDITION

Charles A. Janeway, Jr.
Yale University Medical School

Paul Travers
Anthony Nolan Research Institute, London

Mark Walport
Imperial College School of Medicine, London

J. Donald Capra
Oklahoma Medical Research Foundation

CB
CURRENT BIOLOGY PUBLICATIONS

GARLAND PUBLISHING
ALERE FLAMMAM
Taylor & Francis Group

Text Editors: Penelope Austin, Eleanor Lawrence
Editorial Assistant: Richard Woof
Copyeditor: Bruce Goatly
Production Editor: Emma Hunt
Indexer: Liza Weinkove
Illustration and Layout: Blink Studio, London

Distributors:

Inside North America: Garland Publishing, 19 Union Square West, New York, NY 10003, US.
Inside Japan: Nankodo Co. Ltd., 42-6, Hongo 3-Chome, Bunkyo-ku, Tokyo 113, Japan.
Outside North America and Japan: Churchill Livingstone, Robert Stevenson House, 1–3 Baxter's Place, Leith Walk, Edinburgh, EH1 3AF.

ISBN 0 8153 3217 3 (paperback) Garland
ISBN 0 4430 6275 7 (paperback) Churchill Livingstone
ISBN 0 4430 6274 9 (paperback) International Student Edition

A catalog record for this book is available from the British Library.

Library of Congress Cataloging-in-Publication Data
Janeway, Charles.
Immunobiology: the immune system in health and disease /
Charles A. Janeway, Jr., Paul Travers, Mark Walport;
with the assistance of J. Donald Capra.—4th ed.
p.cm.
ISBN 0-8153-3217-3 (pbk.)
1. Immunity.I. Travers, Paul, 1956–II. Walport, Mark.
III. Title.
QR181.J37 1999
616.07'9- - dc21
98-30316
CIP

This book was produced using QuarkXpress 3.32 and Adobe Illustrator 7.0.

Printed in United States of America.

Published by Current Biology Publications, part of Elsevier Science London
Middlesex House, 34–42 Cleveland Street, London W1P 6LB, UK
and Garland Publishing, a member of the Taylor & Francis Group,
19 Union Square West, New York, NY 10003, US.

Preface for the fourth edition

Immunology continues to accelerate its rate of progress, not only in the areas of the basic biology of adaptive immunity, but also in the study of host responses to various infectious agents, the understanding of allergy, rapid progress towards new vaccines, and other areas too numerous to mention. As we were preparing this fourth edition of our book, we again called on many individuals to read chapters or parts of chapters in which they were greater experts than we are or could ever claim to be. They are acknowledged on page ix. As in the three previous editions, we would like to thank them for their earnest efforts, and to absolve them of responsibility for the final product, which is ours alone.

For this edition, we again have a new author, J. Donald Capra of Oklahoma Medical Research Foundation, who took responsibility for the chapters that deal with B cell antigen recognition and B cell development. Mark Walport undertook further overhaul of the chapters on the more clinical subjects of immunology. Paul Travers and Charlie Janeway remain as lead authors, but the help of these two people has made our work far easier. A new chapter on signal transduction by lymphocyte antigen receptors is included for the first time, inserted between the chapters on antigen recognition and the chapters on lymphocyte development, so that the signals that drive lymphocyte development and survival can be placed in context.

As for the first three editions of this book, we have again been helped by a very talented and tireless team of editors, illustrators, and publishers. Miranda Robertson oversaw the work as a whole, aided by Penny Austin and Eleanor Lawrence. Matthew McClements has continued his wonderful work making good pictures from bad sketches. Sarah Gibbs, Emma Hunt, Sheila Wand and Richard Woof provided publishing and organizational support throughout the process of putting the book together. Again, we want to thank our publishers for their continuing support of our effort to remain current: Peter Newmark of Current Biology (Elsevier) and Libby Borden, recently retired President of Garland Publishing (a member of the Taylor & Francis Group).

Now that we are beginning to understand the immune system and how it works, future challenges will come more than ever from trying to use immunity in its broadest sense to promote human health world-wide. Nowhere is this more apparent than in the battle against AIDS. Recent discoveries of new receptors for this vicious virus give hope for novel treatments based on inhibitors of these receptors, while drug therapy promises a respite from the deadly progress of this disease. But these are stopgap measures at best, and it is our belief that a vaccine should remain the top priority. Vaccines for many other debilitating diseases are also being developed, including discussion of vaccinating against autoimmune diseases and cancer. We are confident that this will eventually be achieved, just as Jenner showed the world over two hundred years ago that he could prevent small pox by vaccinating a boy with cow pox. The only real question is when.

Finally, we would like to thank our long-suffering families and the members of our respective laboratories for the hours of neglect we have inflicted on them over the years. Charlie Janeway again wants to thank Kim Bottomly for her unflagging support, and also his children, Katie, Hannah, and Megan for being at least understanding of the preoccupation of their father with something as abstruse as modern immunobiology. Paul Travers continues to be indebted to Rose Zamoyska for her support and encouragment, and for reminding him of the data when speculations run too wild. Mark Walport thanks Julia for her loving support, and his children, Louise, Robert, Emily, and Fiona for coping with their father's abstraction by matters immunological. Don Capra thanks the staff at the Oklahoma Medical Research Foundation who thought that OMRF had a new full time president, not realizing that several trips to London, days rewriting chapters, and countless hours searching for reprints were now a common element of their new president's responsibilities. He also thanks the immunologists at OMRF for helping with many of the chapters, especially Carol Webb, Linda Thompson, Paul Kincade, Kendra White, Nadine Tuaillon, Dee Dee Stafford, Moe Reichlin and Roger van der Heijden.

Charles A. Janeway Jr.
Yale University School of Medicine
Howard Hughes Medical Institute

Paul Travers
The Anthony Nolan
Research Institute

Mark Walport
Imperial College School of Medicine

J. Donald Capra
Oklahoma Medical Research
Foundation

Preface for the first edition

This book is intended as an introductory text for use in immunology courses for medical students, advanced undergraduate biology students, and graduate students. It attempts to present the field of immunology from a consistent viewpoint, that of the host's interaction with an environment containing myriad species of potentially harmful microbes. The justification for this particular approach is that the absence of components of the immune system is virtually always made clinically manifest by an increased susceptibility to infection. Thus, first and foremost, the immune system exists to protect the host from infection, and its evolutionary history must have been shaped largely by this challenge. Other aspects of immunology, such as allergy, autoimmunity, graft rejection, and immunity to tumors are treated as variations on this basic protective function in which the nature of the antigen is the major variable.

We have attempted to structure the book logically, focusing mainly on the adaptive immune response mediated by antigen-specific lymphocytes operating by clonal selection. The first part of the book summarizes our understanding of immunology in conceptual terms and introduces the main players—the cells, tissues, and molecules of the immune system. It also contains a 'toolbox' of experiments and techniques that form the experimental basis of immunology. The middle three parts of the book deal with three main aspects of adaptive immunity: how the immune system recognizes and discriminates among different molecules; how individual cells develop so that each bears a unique receptor directed at foreign, and not at self, molecules; and how these cells are activated when they encounter microbes whose molecular components bind their receptors, and the effector mechanisms that are used to eliminate these microbes from the body.

Having described the major features of lymphocytes and of adaptive immune responses, the last part of the book integrates this material at the level of the intact organism, examining when, how, and where immune responses occur, and how they fail in some instances. In this part of the book we also consider those aspects of host defense that do not involve clonal selection of lymphocytes, known as innate immunity or natural resistance. We then look at the role of the immune system in causing rather than preventing disease, focusing on allergy, auto-immunity, and graft rejection as examples. Finally, we consider how the immune system can be manipulated to the benefit of the host, emphasizing endogenous regulatory mechanisms and the possibility of vaccinating not only against infection, but also against cancer and immunological diseases. The book also contains a glossary of key terms, biographical notes on some immunologists, and summary tables of key molecules.

In preparing this textbook, we have striven for a coherent overall presentation of concepts supported by experiment and observation. We have had the chapters read by experts (see page ix) who have helped us to eliminate errors of fact and conclusion, to improve presentation, and to achieve better balance; we are sincerely grateful to them for their hours of hard work. Any shortcomings of the book are not their fault, but ours. We plan to keep this material current by revising the book each year, again relying on a panel of experts for each chapter. But the greatest help in improving the value of the book will come from its readers; we welcome your comments and criticisms, and we will work to incorporate your ideas into each new update. Thus, we plan to make this the first current textbook: we are attempting to do this because we believe that the field of immunology is advancing so rapidly at present that annual updating is essential.

There is no doubt that we have both omitted and included too much; the field of immunology covers such a broad area of interest that one of the most daunting challenges in writing this book has been in deciding what has to be discussed and what can be ignored. The judgements we have made are personal and so will not be shared exactly by anyone else. Again, we welcome your input and suggestions about places where this judgement is in clear error.

Finally, we want to thank the many people who have worked so hard to make this book possible. Our illustrator, Celia Welcomme, has brought her extraordinary talents to bear on the figures, while the book itself was edited by three highly skilled and knowledgeable individuals, Miranda Robertson, Rebecca Ward, and Eleanor Lawrence; led by Miranda, they questioned every word, sentence, comma and figure in the book, and made us tear our hair out in trying to be clear and accurate at the same time. If we have failed, it is not their fault. Peter Newmark and Vitek Tracz have provided intelligent and even inspirational guidance, motivated in no small part by the memory of our original publisher, the remarkable Gavin Borden, who sadly died young and never saw this book, into which he put so much, in print. Nothing would have been achieved without the diligence of Becky Palmer, who kept all of us organized over a long period, with the help of Emma Dorey, Sylvia Purnell, Gary Brown, and many other people at Current Biology Ltd. Charlie Janeway wants to thank several patient secretaries, especially Liza Cluggish, Anne Brancheau, Susan Morin, and Kara McCarthy for all their help. Finally, our families have suffered more than we have, from neglect, absence and fits of ill temper. Thus, we thank Kim Bottomly, Katie, Hannah and Megan Janeway, and Rose Zamoyska for their forbearance and support.

Charles A. Janeway Jr.
Yale University School of Medicine
Howard Hughes Medical Institute

Paul Travers
Birkbeck College
University of London

April 1994

Acknowledgements

Text

We would like to thank the following experts who read parts or the whole of the chapters indicated and provided us with invaluable advice.

Chapter 2: Leonard A. Herzenberg, Stanford University Medical School; Leonore A. Herzenberg, Stanford University Medical School; I. Lefkovitz, University of Oxford.

Chapter 3: A.B. Edmundson, Oklahoma Medical Research Foundation; E.W. Voss Jr., University of Illinois, Urbana; E.A. Padlan, National Institute of Diabetes and Digestive and Kidney Diseases, Bethesda; D. Stollar, Tufts University School of Medicine, Boston; E.S. Ward, University of Texas; I.A. Wilson, Scripps Research Institute, La Jolla; E.E. Max, Food and Drug Administration, Bethesda U. Storb, University of Chicago; H. Kratzin, Max-Planck-Institute, Goettingen; J. Borst, The Netherlands Cancer Institute, Amsterdam; D.T. Fearon, University of Cambridge School of Clinical Medicine, UK; T.F. Tedder, Duke University Medical Center, Durham, US; A. DeFranco, Walter and Eliza Hall Institute, Melbourne.

Chapter 4: E. Palmer, Basel Institute for Immunology; H. Ploegh, Massachusetts Institute of Technology; P. Cresswell, Howard Hughes Medical Institute, Yale University School of Medicine; I.A. Wilson, Scripps Research Institute, La Jolla; J. Trowsdale, University of Cambridge; A. Weiss, Howard Hughes Medical Institute, University of California, San Francisco; A. DeFranco, The Walter and Eliza Hall Institute of Medical Research, Melbourne; D. Cantrell, Imperial Cancer Research Fund, London; J. Borst, The Netherlands Cancer Institute, Amsterdam; H.J. Gueze, University of Utrecht.

Chapter 5: A. Pawson, Samuel Lunenfeld Research Institute, Toronto; C.C. Goodnow, The John Curtain School of Medical Research, Canberra; D.T. Fearon, University of Cambridge School of Clinical Medicine.

Chapter 6: C.C. Goodnow, The John Curtain School of Medical Research, Canberra; K. Rajewsky, University of Cologne; F. Alt, Howard Hughes Medical Institute, Children's Hospital, Boston; M. Cooper, Howard Hughes Medical Institute, University of Alabama; P.W. Kincade, Oklahoma Medical Research Foundation; E.E. Max, Food and Drug Administration, Bethesda; D. Namazee, National Jewish Center for Immunology and Respiratory Medicine; F. Melchers, Basel Institute Immunology; P.J. Gearhart, National Institute on Aging, Baltimore; Y. Liu, DNAX Research Institute, Palo Alto; A. Solomon, University of Tennessee Medical Center, Knoxville; J.C. Weill, Faculte de Medecine Necker, Paris; R.G. Mage, National Institute of Allergy and Infectious Diseases, Bethesda; P. Casali, Cornell University Medical College, New York; G. Kelsoe, University of Maryland School of Medicine, Baltimore; I. Maclennan, University of Birmingham Medical School, UK; S. Korsmeyer, Washington University School of Medicine, St. Louis; T.H. Rabbits, Medical Research Council, Cambridge, UK.

Chapter 7: W. van Ewijk, Erasmus University of Rotterdam; A. Hayday, Yale University; M. Merkenschlager, Medical Research Council, Clinical Sciences Center, London; E. Robey, University of Califonia, Berkeley; M.J. Bevan, Howard Hughes Medical Institute, University of Washington, Seattle; R. Zamoyska, National Institute for Medical Research, London.

Chapter 8: Tim Springer, Harvard Medical School; J.P. Allison, University of California, Berkeley; K.M. Murphy, Washington University School of Medicine, St. Louis; G. Griffiths, Sir William Dunn School of Pathology, Oxford; D.M. Paulnock, University of Wisconsin Medical School, Madison.

Chapter 9: M.M. Frank, Duke University Medical Center, Durham US; K. Reid, Medical Research Council, University of Oxford; J. Ravetch, The Rockefeller University, New York; J.G.J. van de Winkel, University Hospital Utrecht; S.J. Galli, Beth Israel Deaconess Medical Center-East, Boston; K. Rajewsky, University of Cologne.

Chapter 10: L. Picker, University of Texas; M.M. Frank, Duke University Medical Center, Durham US; S. Gordon, Sir William Dunn School of Pathology, Oxford; D.M. Paulnock, University of Wisconsin, Madison; R. Modlin, University of California, Los Angeles; P.C. Doherty, St. Jude Children's Research Hospital, Memphis; J.J. Oppenheim, Frederick Cancer Research and Development Center; P.M. Murphy, National Institute of Allergy and Infectious Diseases, Bethesda; J. Sprent, Scripps Research Institute, La Jolla; A. Hayday, Yale University School of Medicine; C.A. Biron, Brown University, Providence; E.O. Long, National Institute of Allergy and Infectious Diseases, Rockville; R.H. Hardy, Fox Chase Cancer Center, Philadelphia.

Chapter 11: F.S. Rosen, The Center for Blood Research, Boston; A.J. McMichael, University of Oxford; R.A. Weiss, Chester Beatty Research Institute, London; C. Kinnon, Institute of Child Health, London.

Chapter 12: A.B. Kay, Imperial College School of Medicine, London; J. Ravetch, The Rockefeller University, New York; J.G.J. van de Winkel, University Hospital Utrecht; S.J. Galli, Beth Israel Deaconess Medical Center-East, Boston.

Chapter 13: C.C. Goodnow, The John Curtain School of Medical Research, Canberra; H. McDevitt, Stanford University School of Medicine; D.C. Wraith, University of Bristol; J. Miller, The Walter and Eliza Hall Institute of Medical Research, Melbourne; R. Lechler, Imperial College School of Medicine, London.

Chapter 14: S. Cobbold, Sir William Dunn School of Pathology, Oxford; D.C. Wraith, University of Bristol; H.J. Stauss, Imperial College School of Medicine, London; T. Boon, Ludwig Institute for Cancer Research, Brussels; G. Dougan, Imperial College of Science, Technology and Medicine, London; R. Modlin, University of California, Los Angeles; G.R. Crabtree, Howard Hughes Medical Institute, Stanford.

Appendix I: N.A. Barclay, Sir William Dunn School of Pathology, Oxford.

Appendix II: J.J. Oppenheim, Frederick Cancer Research and Development Center; R.D. Schreiber, Washington University School of Medicine, St. Louis; K. Moore, DNAX Research Institute, Palo Alto.

Appendix III: A. Zlotnik, DNAX Research Institute, Palo Alto.

Photographs

The following photographs have been reproduced with the kind permission of the journal in which they originally appeared.

Chapter 1
Fig. 1.9 from the *Journal of Experimental Medicine* 1972, **135**:200-219, by copyright permission of the Rockefeller University Press.

Chapter 3
Fig. 3.1 panel a from *Nature* 1992, **360**:369-372. © 1992 MacMillan Magazines Ltd.
Fig. 3.5 top left panel from *Advances in Immunology* 1969, **11**:1-30. © 1969 Academic Press.
Fig. 3.9 top panel from *Science* 1990, **248**:712-719. © 1990 by the AAAS. Middle panel, from *Structure* 1993, **1**:83-93 © 1993 Current Biology.
Fig. 3.11 from *Science* 1986, **233**:747-753. © 1986 by the AAAS.
Fig. 3.22 top panel from the *European Journal of Immunology* 1988, **18**:1001-1008.

Chapter 4
Fig. 4.12 from *Science* 1995, **268**:533-539. © 1995 by the AAAS.
Fig. 4.15 from *Cell* 1996, **84**:505-507. © 1996 Cell Press
Fig. 4.31 from *Science* 1996, **274**:209-219. © 1996 by the AAAS.
Fig. 4.35 from *Science* 1996, **274**:209-219. © 1996 by the AAAS.
Fig. 4.36 from *Nature* 1996, **384**:188-192. © 1994 MacMillan Magazines.
Fig. 4.40 from *Nature* 1997, **387**:630-634. © 1997 MacMillan Magazines.

Chapter 6
Fig. 6.3 right panel from the *European Journal of Immunology* 1987, **17**:1473-1484.

Chapter 7
Fig. 7.5 from *Nature* 1994, **372**:100-103. © 1994 MacMillan Magazines.
Fig. 7.19 from *International Immunology* 1996, **8**:1537-1548. © 1996 Oxford University Press.

Chapter 8
Fig. 8.28 panel c from *Second International Workshop on Cell Mediated Cytotoxicity*. Eds. P.A. Henkart, and E. Martz. New York, Plenum Press 1985, 99-119.
Fig. 8.36 panels a and b from *Second International Workshop on Cell Mediated Cytotoxicity* . Eds. P.A. Henkart, and E. Martz. New York, Plenum Press 1985, 99-119, panel c from *Immunology Today* 1985, **6**:21-27.

Chapter 9
Fig. 9.18 from *Nature* 1994, **372**:336-343. © 1994 MacMillan Magazines.
Fig. 9.34 right panel from *Essays in Biochemistry* 1986, **22**:27-68. © 1986, Academic Press.
Fig. 9.35 planar conformation from the *European Journal of Immunology* 1988, **18**:1001-1003.
Fig. 9.49 from *Blut* 1990, **60**:309-318. © 1994 MacMillan Magazines.

Chapter 10
Fig. 10.15 top left panel from *Nature* 1994, **367**:338-345. © 1994 MacMillan Magazines. Bottom left panel from the *Journal of Immunolgy* 1990, **144**:2287-2294. © 1990 The American Association of Immunologists.
Fig. 10.35 left panel from the *Journal of Immunolgy* 1985, **134**:1349-1359. © 1985 The American Association of Immunologists. Middle and right panels from *Annual Reviews in Immunolgy* 1989, **7**:91-109. © 1989 Annual Reviews Inc.

Chapter 11
Fig. 11.6 top left panel from *International Reviews in Experimental Pathology* 1986, **28**:45-78. © 1986, Academic Press.
Fig. 11.24 from *Cell* 1998, **93**:665-671. © 1998 Cell Press Ltd.
Fig. 11.25 from the *Nature* 1995, **373**:117-122. © 1995 MacMillan Magazines.

Chapter 13
Fig. 13.31 from the *Journal of Experimental Medicine* 1992, **176**:1355-1364, by copyright permission of the Rockefeller University Press.

Chapter 14
Fig. 14.15 from *Mechanisms of Cytotoxicity by Natural Killer Cells.* R.B. Herberman (Ed.) New York, Academic Press 1985, 195. © 1985, Academic Press.

CONTENTS

List of Headings

Part II THE RECOGNITION OF ANTIGEN

Chapter 3: Structure of the Antibody Molecule and Immunoglobulin Genes

The structure of a typical antibody molecule.

The interaction of the antibody molecule with specific antigen.

The generation of diversity in the humoral immune response.

Structural variation in immunoglobulin constant regions.

Chapter 4: Antigen Recognition by T Lymphocytes

The generation of T-cell ligands.

Part V THE IMMUNE SYSTEM IN HEALTH AND DISEASE

Chapter 10: Host Defense Against Infection

Infection and innate immunity.

Non-adaptive host responses to infection.

Adaptive immunity to infection.

Immunological memory.

Chapter 11: Failures of Host Defense Mechanisms

Pathogens have evolved various means of evading or subverting normal host defenses.

Inherited immunodeficiency diseases.

PART I

AN INTRODUCTION TO IMMUNOBIOLOGY

Basic Concepts in Immunology

1

Immunology is a relatively new science. Its origin is usually attributed to Edward Jenner (Fig. 1.1), who discovered in 1796 that cowpox, or vaccinia, induced protection against human smallpox, an often fatal disease. Jenner called his procedure vaccination, and this term is still used to describe the inoculation of healthy individuals with weakened or attenuated strains of disease-causing agents to provide protection from disease. Although Jenner's bold experiment was successful, it took almost two centuries for smallpox vaccination to become universal, an advance that enabled the World Health Organization to announce in 1979 that smallpox had been eradicated (Fig. 1.2), arguably the greatest triumph of modern medicine.

Fig. 1.1 Edward Jenner. Portrait by John Raphael Smith. Reproduced courtesy of the Yale Historical Medical Library.

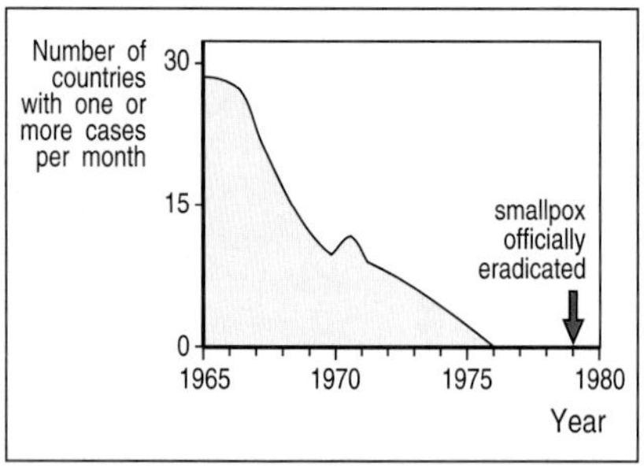

Fig. 1.2 The eradication of smallpox by vaccination. After a period of 3 years in which no cases of smallpox were recorded, The World Health Organization was able to announce in 1979 that smallpox had been eradicated.

When Jenner introduced vaccination he knew nothing of the infectious agents that cause disease: it was not until late in the 19th century that Robert Koch proved that infectious diseases are caused by **microorganisms**, each one responsible for a particular disease, or **pathology**. We now recognize four broad categories of disease-causing microorganisms, or **pathogens**: these are **viruses**; **bacteria**; pathogenic **fungi**; and other relatively large and complex eukaryotic organisms collectively termed **parasites**.

The discoveries of Koch and other great 19th century microbiologists stimulated the extension of Jenner's strategy of vaccination to other diseases. In the 1880s, Louis Pasteur devised a vaccine against cholera in chickens, and developed a rabies vaccine that proved a spectacular success upon its first trial use in a boy bitten by a rabid dog. These practical triumphs led to a search for the mechanism of protection and to the development of the science of immunology. In 1890, Emil von Behring and Shibasaburo Kitasato discovered that the serum of vaccinated individuals contained substances—which they called **antibodies**—that specifically bound to the relevant pathogen.

A specific **immune response**, such as the production of antibodies against a particular pathogen, is known as an **adaptive immune response**, because it occurs during the lifetime of an individual as an adaptation to infection with that pathogen. In many cases, an adaptive immune response confers life-long **protective immunity** to re-infection with the same pathogen. This distinguishes such responses from **innate immunity**, which, at the time that von Behring and Kitasato discovered antibodies, was known chiefly through the work of the great Russian immunologist Elie Metchnikoff. Metchnikoff discovered that many microorganisms could be engulfed and digested by **phagocytic cells**, which he called **macrophages**. These cells are immediately available to combat a wide range of pathogens without requiring prior exposure, and act in the same way in all normal individuals, and so are considered innate. Antibodies, by contrast, are produced only in response to specific infections, and the antibodies present in a given individual directly reflect the infections to which he or she has been exposed.

Indeed, it quickly became clear that specific antibodies can be induced against a vast range of substances. Such substances are known as **antigens** because they can stimulate the generation of antibodies. We shall see, however, that not all adaptive immune responses entail the production of antibodies, and the term antigen is now used in a rather broader sense to describe any substance capable of being recognized by the adaptive immune system.

Both innate immunity and adaptive immune responses depend upon the activities of white blood cells, or **leukocytes**. Innate immunity largely involves granulocytes and macrophages. Granulocytes, also called polymorphonuclear leukocytes, are a diverse collection of white blood cells whose prominent granules give them their characteristic staining patterns; they include the neutrophils, which are phagocytic. The macrophages of humans and other vertebrates are presumed to be the direct evolutionary descendants of the phagocytic cells present in simpler animals, such as those that Metchnikoff observed in sea stars. Adaptive immune responses depend upon **lymphocytes**, which provide the life-long immunity that can follow exposure to disease or vaccination. The innate and adaptive immune systems together provide a remarkably effective defense system that ensures that, although we spend our lives surrounded by potentially pathogenic microorganisms, we become ill only relatively rarely. Many infections are handled successfully by the innate immune system and cause no pathology, whereas those that trigger adaptive immunity are usually met successfully and are followed by lasting immunity.

The main focus of this book will be on the diverse mechanisms of adaptive immunity, whereby specialized classes of lymphocytes recognize and target pathogenic microorganisms or the cells infected with them. We shall see, however, that all the cells involved in innate immune responses also participate in adaptive immune responses. Indeed, most of the effector actions that the adaptive immune system uses to destroy invading microorganisms depend upon them.

In this chapter, we first introduce the cells of the immune system, and the tissues in which they develop and through which they circulate or migrate. In later sections, we outline the specialized functions of the different types of cells and the mechanisms whereby they eliminate infection.

The components of the immune system.

The cells of the immune system originate in the **bone marrow**, where many of them also mature. They then migrate to patrol the tissues, circulating in the blood and in a specialized system of vessels called the **lymphatic system**.

1-1 The white blood cells of the immune system derive from precursors in the bone marrow.

All the cellular elements of blood, including the red blood cells that transport oxygen, the platelets that trigger blood clotting in damaged tissues, and the white blood cells of the immune system, derive ultimately from the same **progenitor** or precursor cells, the **hematopoietic stem cells** in the bone marrow. As these stem cells can give rise to all of the different types of blood cells, they are often known as pluripotent hematopoietic stem cells. Initially they give rise to stem cells of more limited potential, which are the immediate progenitors of red blood cells, platelets, and the two main categories of white blood cells. The different types of blood cell and their lineage relationships are summarized in Fig. 1.3. We shall be concerned here with all the cells derived from the common lymphoid progenitor and the myeloid progenitor, apart from the megakaryocytes and red blood cells.

The **myeloid progenitor** is the precursor of the granulocytes, macrophages, and mast cells of the immune system. Macrophages are one of the two types of phagocyte in the immune system and are distributed widely in the body tissues, where they play a critical part in innate immunity. They are the mature form of **monocytes**, which circulate in the blood and differentiate continuously into macrophages upon migration into the tissues. **Mast cells**, whose blood-borne precursors are not well defined, also differentiate in the tissues. They reside mainly near small blood vessels and release substances that affect vascular permeability when activated. Although best known for their role in orchestrating allergic responses, they are believed to play a part in protecting mucosal surfaces against pathogens.

The **granulocytes** are so called because they have densely staining granules in their cytoplasm; they are also sometimes called **polymorphonuclear leukocytes** because of their oddly shaped nuclei. There are three types of granulocyte, all of which are relatively short-lived and are produced in larger numbers during immune responses, when they leave the blood to migrate to

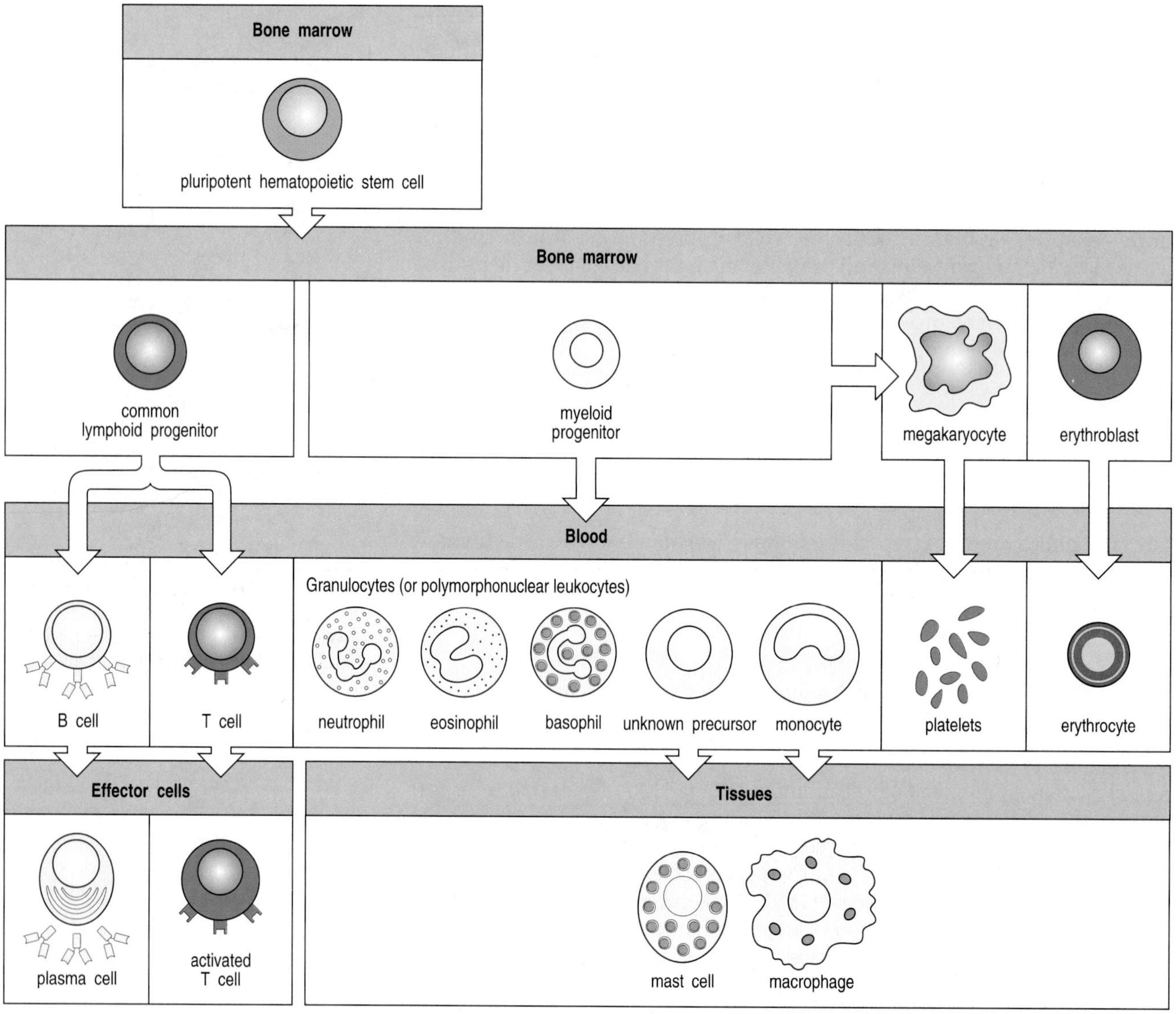

Fig. 1.3 All the cellular elements of blood, including the lymphocytes of the adaptive immune system, arise from hematopoietic stem cells in the bone marrow. These pluripotent cells divide to produce two more specialized types of stem cells, a lymphoid stem cell (lymphoid progenitor), which gives rise to T and B lymphocytes, and a myeloid stem cell (myeloid progenitor), which gives rise to leukocytes, erythrocytes (red blood cells that carry oxygen), and the megakaryocytes that produce platelets, which are important in blood clotting. Although we have illustrated only one progenitor cell for the T and B lymphocytes, an alternative that has not been ruled out is that both T- and B-cell lineages arise directly from the pluripotent stem cell. The T and B lymphocytes are distinguished by their sites of differentiation—T cells in the thymus and B cells in the bone marrow—and by their antigen receptors. B lymphocytes differentiate on activation into antibody-secreting plasma cells, and T lymphocytes differentiate into cells that can kill infected cells or activate other cells of the immune system. The leukocytes that derive from the myeloid stem cell are the monocytes, and the basophils, eosinophils, and neutrophils, which are collectively termed either granulocytes, because of the cytoplasmic granules whose characteristic staining gives them a distinctive appearance in blood smears, or polymorphonuclear leukocytes, because of their irregularly shaped nuclei. Monocytes differentiate into macrophages in the tissues and these are the main tissue phagocytic cell of the immune system. Neutrophils, the most important phagocytic cells, have functions similar to those of macrophages but remain in the bloodstream until recruited to sites of infection; eosinophils are blood-borne cells that are involved mainly in inflammation and the defense against parasites, whereas basophils are found in blood and are similar in some ways to eosinophils, with which they share a common precursor, and to mast cells, which arise from a separate lineage. Mast cells also arise from precursors in bone marrow but complete their maturation in tissues; they are important in allergic responses.

sites of infection or inflammation. **Neutrophils**, which are the other phagocytic cell of the immune system, are the most numerous and most important cellular component of the innate immune response: hereditary deficiencies in neutrophil function lead to overwhelming bacterial infection, which is

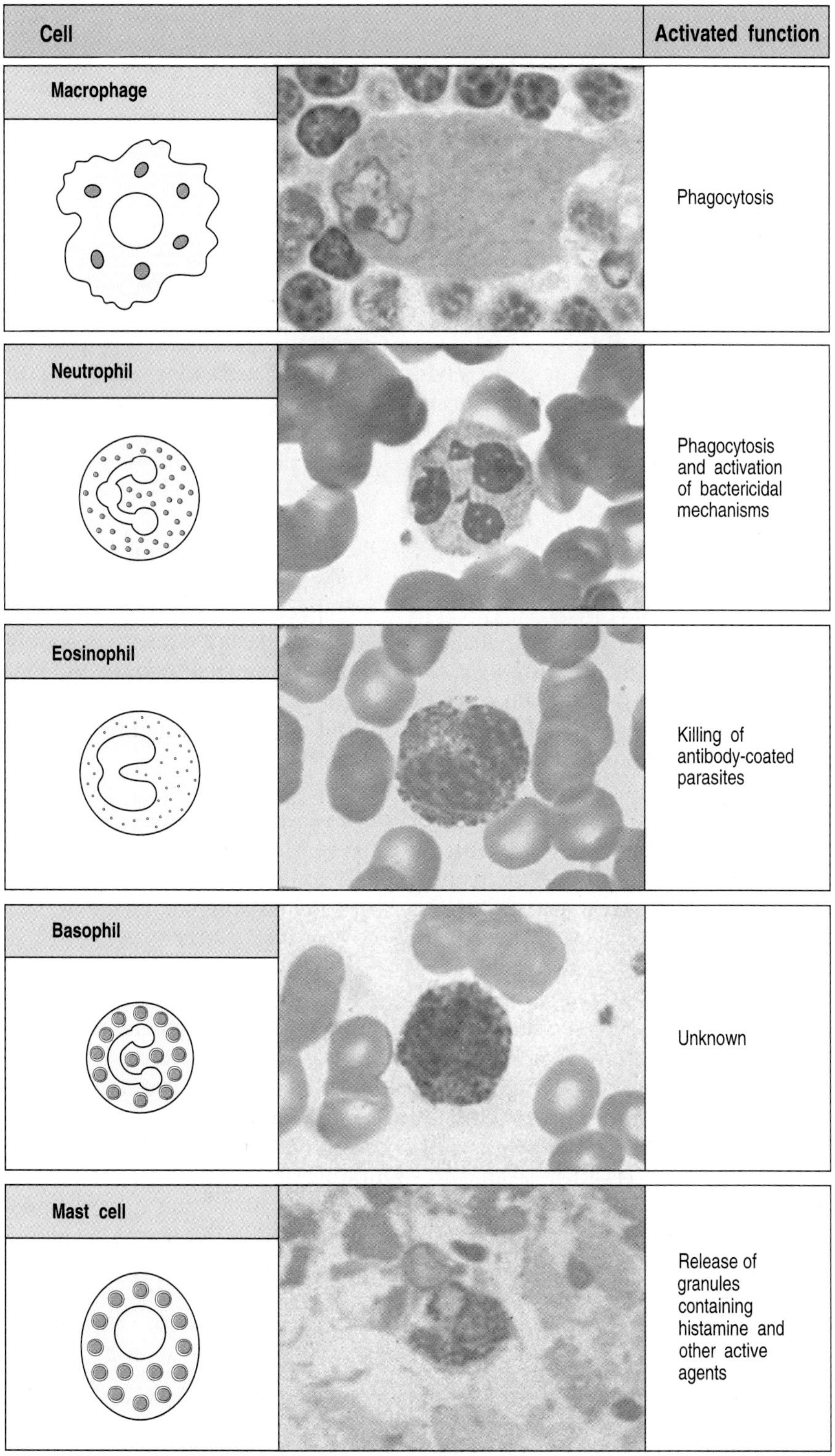

Cell
Activated function
Macrophage
Phagocytosis
Neutrophil
Phagocytosis and activation of bactericidal mechanisms
Eosinophil
Killing of antibody-coated parasites
Basophil
Unknown
Mast cell
Release of granules containing histamine and other active agents

Fig. 1.4 Myeloid cells in innate and adaptive immunity. Several different cell types of the myeloid lineage perform important functions in the immune response; these cells are shown schematically in the left column in the form in which they will be represented throughout the rest of the book. A photomicrograph of each cell type is shown in the center column. Macrophages and neutrophils are primarily phagocytic cells that engulf IgG antibody-coated pathogens, which they destroy in intracellular vesicles after pathogen uptake. The other myeloid accessory effector cells are primarily secretory cells, which release the contents of their prominent granules upon binding to antibody-coated particles. Eosinophils are thought to be involved in attacking large parasites such as worms, whereas the function of basophils is unclear. Mast cells are tissue cells that trigger a local inflammatory response by releasing substances that act on local blood vessels. They bind the IgE class of antibody and are activated by antigen binding to IgE. Photographs courtesy of N Rooney and B Smith.

fatal if untreated. **Eosinophils** are thought to be important chiefly in defense against parasitic infections. The function of **basophils** is probably similar and complementary to that of eosinophils and mast cells; we shall discuss the functions of these cells in Chapter 9 and their role in allergic inflammation in Chapter 12. The cells of the myeloid lineage are shown in Fig. 1.4.

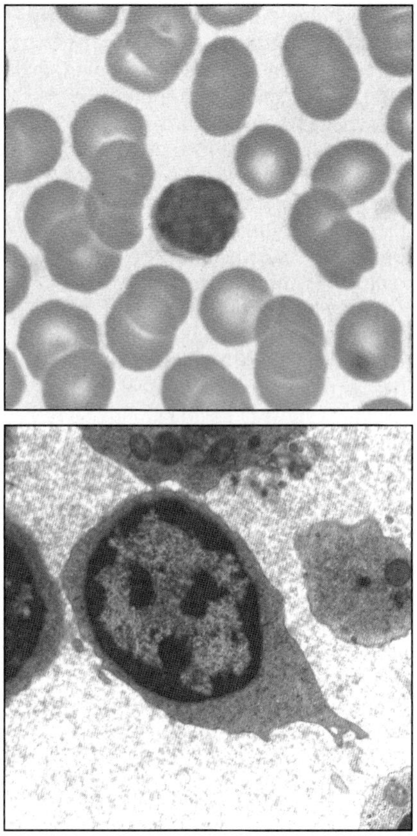

Fig. 1.5 Lymphocytes are mostly small and inactive cells. The top panel shows a light micrograph of a small lymphocyte surrounded by red blood cells. Note the condensed chromatin of the nucleus, indicating little transcriptional activity, the relative absence of cytoplasm, and the small size. The bottom panel shows a transmission electron micrograph of a small lymphocyte. Note the condensed chromatin, the scanty cytoplasm and the absence of rough endoplasmic reticulum and other evidence of functional activity. Photographs courtesy of N Rooney.

The **common lymphoid progenitor** gives rise to the lymphocytes, with which most of this book will be concerned. There are two major types of lymphocyte: **B lymphocytes** or **B cells**, which when activated differentiate into **plasma cells** that secrete antibodies; and **T lymphocytes** or **T cells**, of which there are two main classes. One class differentiates on activation into **cytotoxic T cells**, which kill cells infected with viruses, whereas the second class of T cells differentiates into cells that activate other cells such as B cells and macrophages.

Most lymphocytes are small, featureless cells with few cytoplasmic organelles and much of the nuclear chromatin inactive, as shown by its condensed state (Fig. 1.5). This appearance is typical of inactive cells and it is not surprising that, as recently as the early 1960s, textbooks could describe these cells, now the central focus of immunology, as having no known function. Indeed, lymphocytes have no functional activity until they encounter antigen, which is necessary to trigger their proliferation and the differentiation of their specialized functional characteristics.

Both T and B lymphocytes bear highly diverse receptors on their surface that allow them to recognize antigen. Collectively, these receptors are highly diverse in their antigen specificity, but an individual lymphocyte is equipped with receptors that will recognize only one particular antigen. Each lymphocyte therefore recognizes a different antigen. Together, the receptors of all the different lymphocytes are capable of recognizing a very wide diversity of antigens. The **B-cell antigen receptor (BCR)** is a membrane-bound form of the antibody that they will secrete when activated. Antibody molecules as a class are now generally known as **immunoglobulins**, usually shortened to **Ig**, and the antigen receptor of B lymphocytes is known as **surface immunoglobulin**. Immunoglobulin molecules will be discussed in detail in Chapter 3, and the development of B lymphocytes will be described in Chapter 6. The **T-cell antigen receptor (TCR)** is related to immunoglobulin but is quite distinct from it, and we shall describe the T-cell receptor in detail in Chapter 4 and T-cell development in Chapter 7. A third class of lymphoid cell, called a **natural killer cell** or **NK cell** (Fig. 1.6), lacks antigen-specific receptors and so is part of the innate immune system.

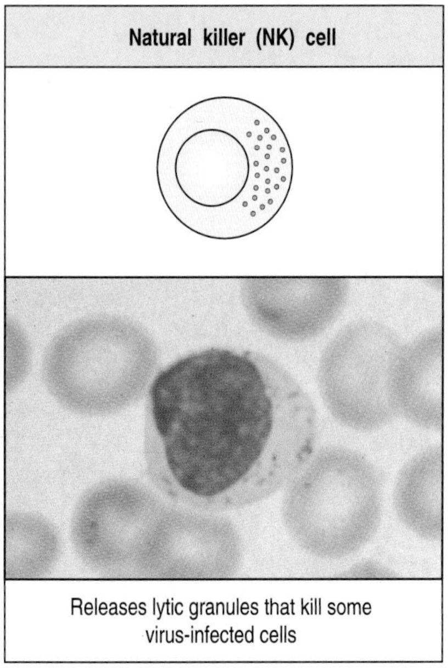

Natural killer (NK) cell
Releases lytic granules that kill some virus-infected cells

Fig. 1.6 Natural killer (NK) cells. These are large granular lymphocyte-like cells with important functions in innate immunity. Although lacking antigen-specific receptors, they can detect and attack certain virus-infected cells. Photograph courtesy of N Rooney and B Smith.

1-2 Lymphocytes mature in the bone marrow or the thymus.

The **lymphoid organs** are organized tissues where lymphocytes make interactions with non-lymphoid cells that are important either to their development or to the initiation of adaptive immune responses. They can be divided broadly into **primary** or **central lymphoid organs**, where lymphocytes are generated, and **secondary** or **peripheral lymphoid organs**, where adaptive immune responses are initiated. The central lymphoid organs are the bone marrow and the **thymus**, a large organ in the upper chest: the location of the thymus, with the other lymphoid organs, is shown schematically in Fig. 1.7.

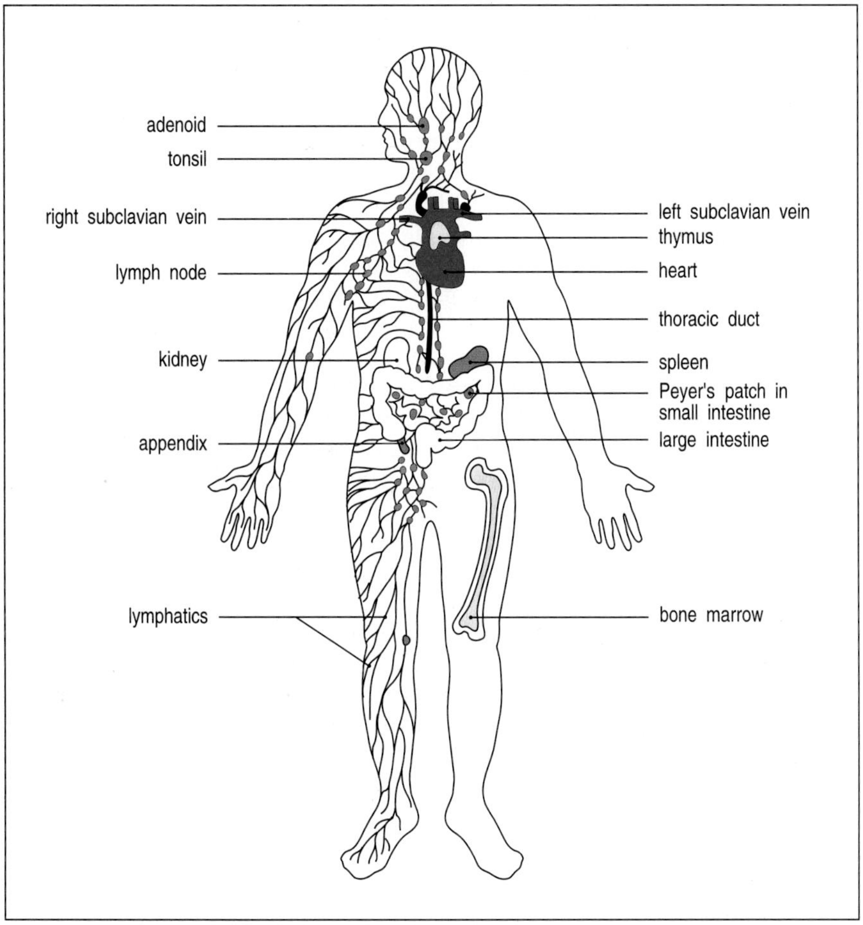

Fig. 1.7 The distribution of lymphoid tissues in the body. Lymphocytes arise from stem cells in bone marrow, and differentiate in the central lymphoid organs (yellow): B cells in bone marrow and T cells in the thymus. They migrate from these tissues through the bloodstream to the peripheral lymphoid tissues (blue), the lymph nodes, spleen, and lymphoid tissues associated with mucosa, like the gut-associated lymphoid tissues such as tonsils, Peyer's patches, and appendix. These are the sites of lymphocyte activation by antigen. Lymphatics drain extracellular fluid as lymph through the lymph nodes and into the thoracic duct, which returns the lymph to the bloodstream by emptying into the left subclavian vein. Lymphocytes that circulate in the bloodstream enter the peripheral lymphoid organs, and are eventually carried by lymph to the thoracic duct where they re-enter the bloodstream. Lymphoid tissue is also associated with other mucosa such as the bronchial linings (not shown).

Both B and T lymphocytes originate in the bone marrow but only B lymphocytes mature there: T lymphocytes migrate to the thymus to undergo maturation. Thus B lymphocytes are so called because they are bone marrow derived, and T lymphocytes because they are thymus derived. Once they have completed their maturation, both types of lymphocyte enter the bloodstream, from which they migrate to the peripheral lymphoid organs.

1-3 The peripheral lymphoid organs are specialized to trap antigen and allow the initiation of adaptive immune responses.

Pathogens can enter the body by many routes and set up infections anywhere, but antigen and lymphocytes will eventually encounter each other in the peripheral lymphoid organs—the lymph nodes, the spleen, and mucosal lymphoid tissues (see Fig. 1.7). Lymphocytes are continually recirculating through these tissues, to which antigen is also carried from all sites of infection and where it is displayed by specialized cells.

The **lymph nodes** are highly organized lymphoid structures that are the sites of convergence of an extensive system of vessels that collect the extracellular fluid from the tissues and return it to the blood. This extracellular fluid is produced continuously by filtration from the blood, and is called **lymph**. The vessels that carry it are called **lymphatic vessels**, or sometimes just **lymphatics** (see Fig. 1.7). The **afferent lymphatic vessels**, which drain fluid from the tissues, also carry cells bearing antigens from sites of infection in most parts of the body to the lymph nodes, where they are trapped. In the lymph nodes, B lymphocytes are localized in **follicles**, with T cells more diffusely distributed in surrounding **paracortical areas**, also referred to as T-cell zones. Some of the B-cell follicles include **germinal centers**, where B cells are undergoing intense proliferation after encountering their specific antigen and their cooperating T cells (Fig. 1.8). B and T lymphocytes are segregated in a similar fashion in the other peripheral lymphoid tissues, and we shall see that this organization promotes the crucial interactions that occur between B and T cells upon encountering antigen.

The **spleen** is a fist-sized organ just behind the stomach (see Fig. 1.7) that collects antigen from the blood. It also collects and disposes of senescent red blood cells. Its organization is shown schematically in Fig. 1.9. The bulk of the spleen is composed of **red pulp**, which is the site of red blood cell disposal. The lymphocytes surround the arterioles entering the organ, forming areas of **white pulp**, the inner region of which is divided into a **periarteriolar lymphoid sheath** (PALS), containing mainly T cells, and a flanking **B-cell corona**.

The **gut-associated lymphoid tissues** (GALT), which include the **tonsils**, **adenoids**, and **appendix**, and specialized structures called **Peyer's patches** in the small intestine, collect antigen from the epithelial surfaces of the gastrointestinal tract. In Peyer's patches, which are the most important and highly organized of these tissues, the antigen is collected by specialized epithelial cells called **M cells**. The lymphocytes form a follicle consisting of a large central dome of B lymphocytes surrounded by smaller numbers of

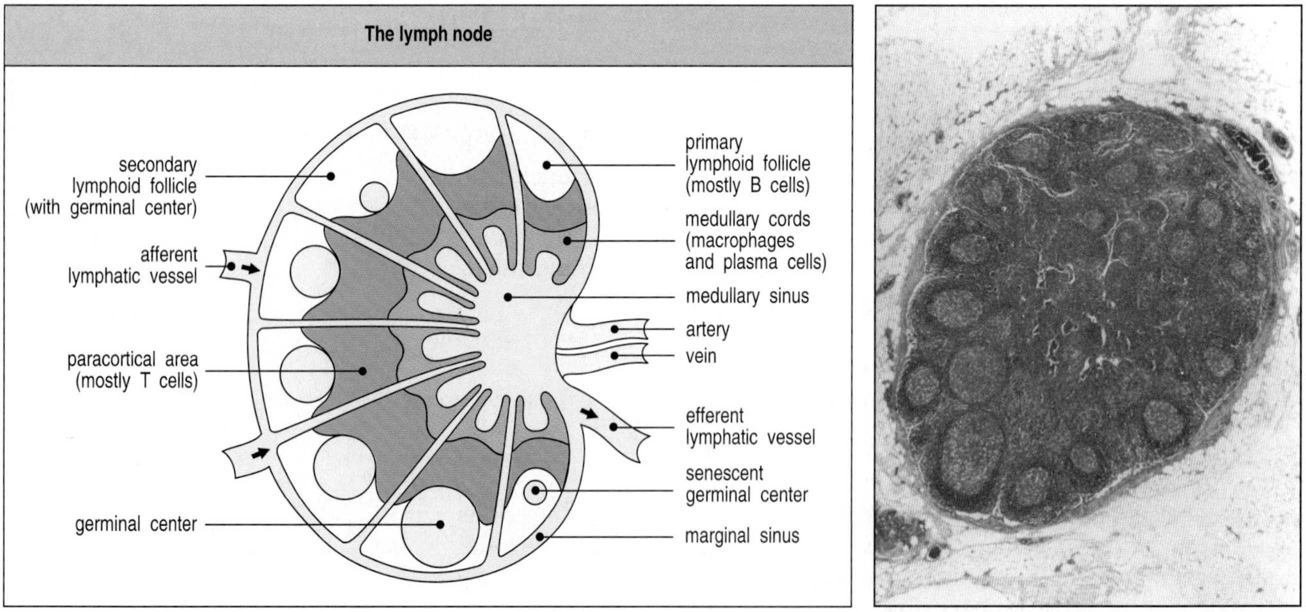

Fig. 1.8 Organization of a lymph node. As shown in the diagram on the left, a lymph node consists of an outermost cortex and an inner medulla. The cortex is composed of an outer cortex of B lymphocytes organized into lymphoid follicles, and deep, or paracortical, areas made up mainly of T lymphocytes and specialized cells known as dendritic cells. Some of the B-cell follicles contain central areas of intense B-cell proliferation called germinal centers. These follicles are known as secondary lymphoid follicles. Lymph draining from the extracellular spaces of the body carries antigens from the tissues to the lymph node via the afferent lymphatics. Lymph leaves by the efferent lymphatic in the medulla. The medulla consists of strings of macrophages and antibody-secreting plasma cells known as the medullary cords. Naive lymphocytes enter the node from the bloodstream through specialized post-capillary venules (not shown) and leave with the lymph through the efferent lymphatic. The light micrograph shows a section through a lymph node, with prominent follicles containing germinal centers. Magnification x 7. Photograph courtesy of N Rooney.

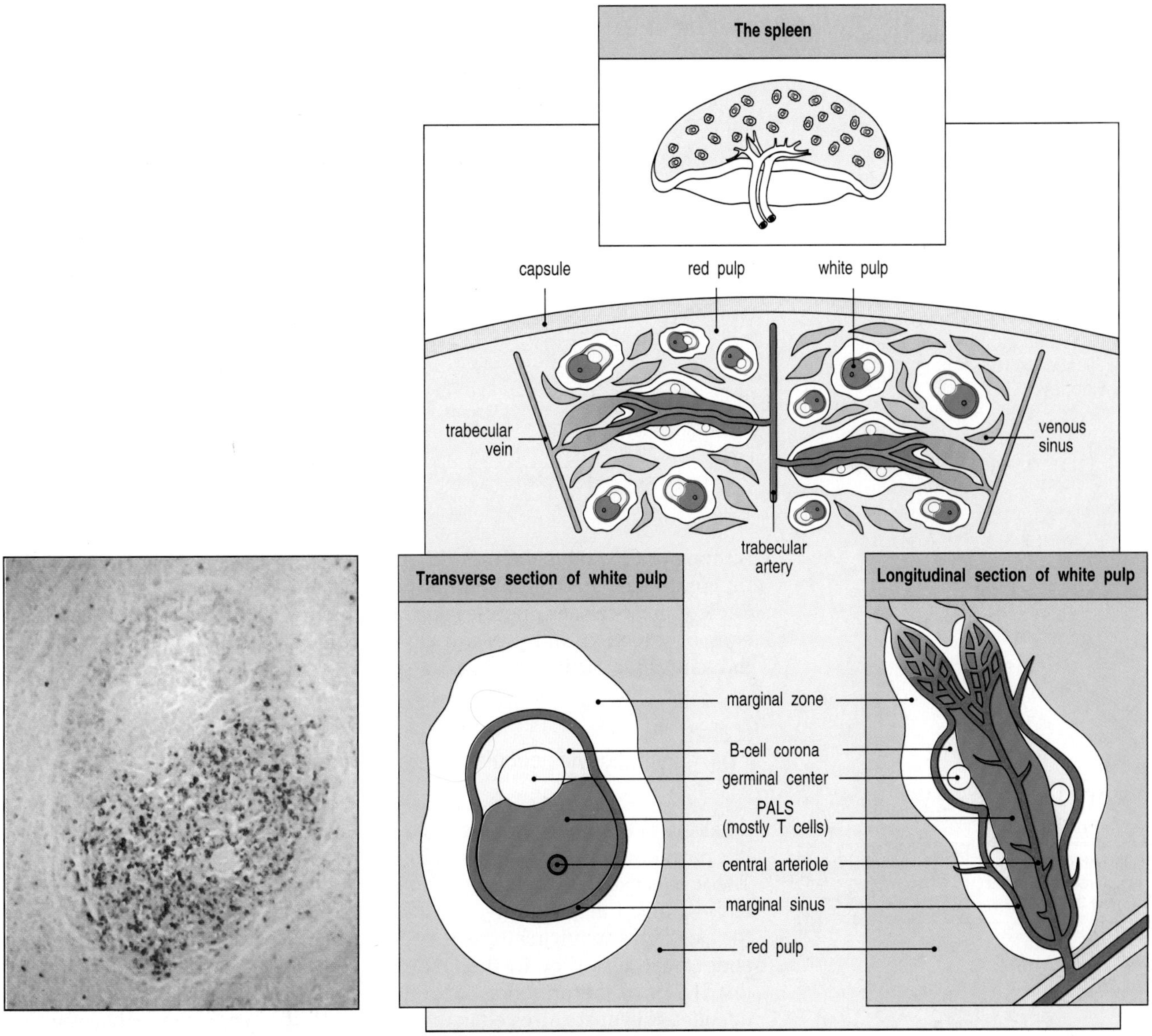

Fig. 1.9 Organization of the lymphoid tissues of the spleen. The spleen consists of red pulp (pink areas in the top panel), which is a site of red blood cell destruction, interspersed with lymphoid white pulp (yellow and blue areas in the lower panels). The center panel shows an enlargement of a small section of the spleen showing the arrangement of discrete areas of white pulp around central arterioles. Most of the white pulp is shown in transverse section, with one portion shown in longitudinal section. The bottom two diagrams show enlargements of a transverse section (lower left) and longitudinal section (lower right) of white pulp. In each area of white pulp, blood carrying lymphocytes and antigen flows from a trabecular artery into a central arteriole. Cells and antigen then pass into a marginal sinus and drain into a trabecular vein. The marginal sinus is surrounded by a marginal zone of lymphocytes. Within the marginal sinus and surrounding the central arteriole is the periarteriolar lymphoid sheath (PALS), made up of T cells (stained darkly in the light micrograph, which shows a transverse section of white pulp). The lymphoid follicles consist mainly of B cells (lightly stained), including germinal centers (the unstained cells lying between the B- and T-cell areas in the micrograph). Although the organization of the spleen is similar to that of a lymph node, antigen enters the spleen from the blood rather than from the lymph. Section stained with hematoxylin and eosin. Photograph courtesy of J C Howard.

T lymphocytes (Fig. 1.10). Similar but more diffusely organized aggregates of lymphocytes protect the respiratory epithelium, where they are known as **bronchial-associated lymphoid tissue** (BALT), and other mucosa, where they are known simply as **mucosal-associated lymphoid tissue** (MALT).

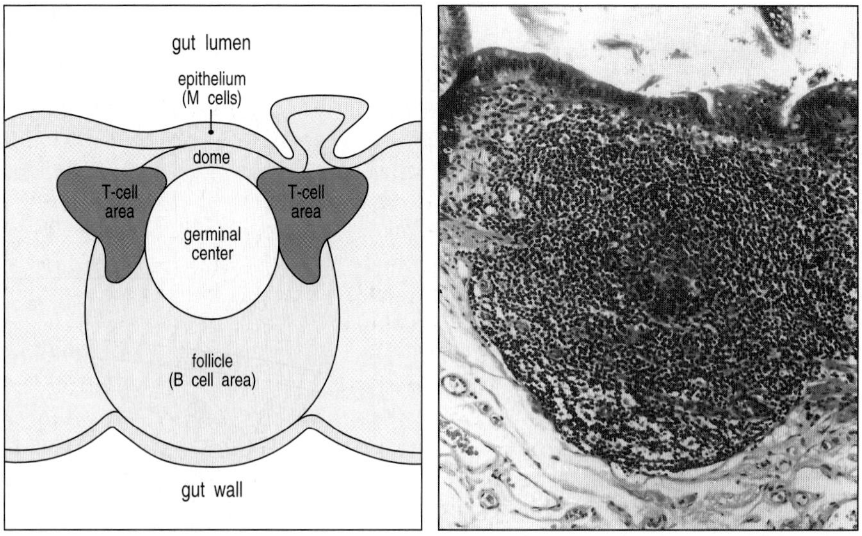

Fig. 1.10 Organization of typical gut-associated lymphoid tissue. As the diagram on the left shows, the bulk of the lymphoid tissue is B cells, organized in a large and highly active domed follicle. T cells occupy the areas between follicles. The antigen enters across a specialized epithelium made up of so-called M cells. Although this tissue looks very different from other lymphoid organs, the basic divisions are maintained. The light micrograph shows a section through the gut wall. The dome of gut-associated lymphoid tissue can be seen lying beneath the epithelial tissues. Magnification x 16. Photograph courtesy of N Rooney.

Although remarkably different in appearance, the lymph nodes, spleen, and mucosal-associated lymphoid tissues all share the same basic architecture. Each of these tissues operates on the same principle, trapping antigen from sites of infection and presenting it to migratory small lymphocytes, thus inducing adaptive immune responses.

1-4 Lymphocytes circulate between blood and lymph.

Small T and B lymphocytes that have matured in the bone marrow and thymus but have not yet encountered antigen are referred to as **naive lymphocytes**. These cells circulate continually from the blood into the peripheral lymphoid tissues, which they enter by means of specialized adhesive interactions with the capillaries supplying these tissues that allow them to squeeze between the endothelial cells. They are then returned to the blood via the lymphatic vessels (Fig. 1.11). In the presence of an infection, lymphocytes that recognize the infectious agent are arrested in the lymphoid tissue where they proliferate and differentiate into **effector cells** capable of combating the infection.

When an infection occurs in the periphery, for example, large amounts of antigen are taken up by phagocytic cells, which then travel from the site of infection through the afferent lymphatic vessels into the lymph nodes (see Fig. 1.11). In the lymph nodes, these cells display the antigen to recirculating lymphocytes, which they also help to activate. Once these specific lymphocytes have undergone a period of proliferation and differentiation, they leave the lymph nodes as effector cells through the **efferent lymphatic vessel** (see Fig. 1.8).

All the lymphoid tissues trap cells and antigen arriving from sites of infection so that antigen can be presented to naive lymphocytes migrating in from the blood and an adaptive immune response stimulated. The lymphoid tissues are thus not static structures but vary quite dramatically depending upon whether or not infection is present. The diffuse mucosal lymphoid tissues can appear and disappear in response to infection, whereas the architecture of the more organized tissues changes in a defined way during an infection. For example, the B-cell follicles of the lymph nodes expand as B lymphocytes proliferate to form germinal centers (see Fig. 1.8), and the entire lymph node enlarges, a phenomenon familiarly known as swollen glands.

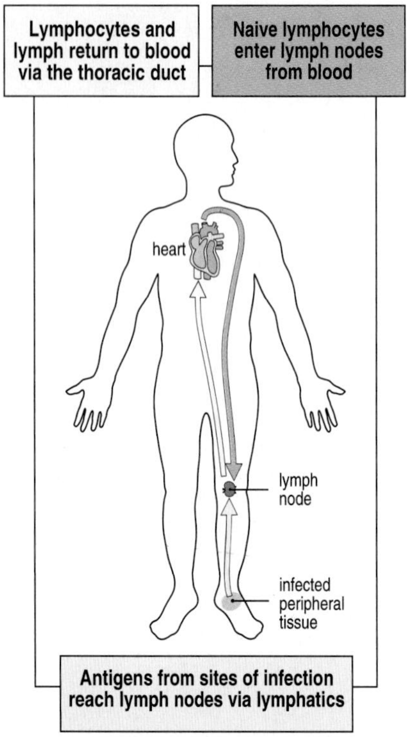

Fig. 1.11 Circulating lymphocytes encounter antigen in peripheral lymphoid tissues.

Summary.

Immune responses are mediated by leukocytes, which derive from precursors in the bone marrow. These give rise to the polymorphonuclear leukocytes and the macrophages of the innate immune system, and also to the lymphocytes of the adaptive immune system. There are two major types of lymphocyte: B lymphocytes, which mature in the bone marrow; and T lymphocytes, which mature in the thymus. The bone marrow and thymus are thus known as the central lymphoid organs. Monocytes and mast-cell precursors migrate to the body tissues where they mature, whereas all the other cells of the immune system circulate in the blood. Lymphocytes recirculate continually from the bloodstream through the peripheral lymphoid organs, where cells displaying the antigen are trapped, returning to the bloodstream through the lymphatic vessels. The three major types of peripheral lymphoid tissue are: the spleen, which collects antigens from the blood; the lymph nodes, which collect antigen from sites of infection in the tissues; and the mucosal-associated lymphoid tissues (MALT), which collect antigens from the epithelial surfaces of the body. Adaptive immune responses are initiated in these peripheral lymphoid tissues.

Principles of innate and adaptive immunity.

The phagocytes of the innate immune system provide a first line of defense against many common microorganisms and are essential for the control of common bacterial infections. However, they cannot always eliminate infectious organisms, and there are many pathogens that they cannot recognize. The lymphocytes of the adaptive immune system have evolved to provide a more versatile means of defense that, in addition, provides an increased level of protection from a subsequent re-infection with the same pathogen. The cells of the innate immune system play a crucial part in the initiation and subsequent direction of adaptive immune responses. Moreover, because there is a delay of 4–7 days before the initial adaptive immune response takes effect, the innate immune response has a critical role in controlling infections during this period.

1-5 Many bacteria activate phagocytes and trigger inflammatory responses.

Macrophages and neutrophils have surface receptors that have evolved to recognize and bind common constituents of many bacterial surfaces. Bacterial molecules binding to these receptors trigger the cells to engulf the bacterium and also induce the secretion of biologically active molecules by these phagocytes. These molecules include **cytokines**, which are defined as proteins released by cells that affect the behavior of other cells. The cytokines released by phagocytes in response to bacterial constituents have a range of effects that are collectively known as **inflammation**. Inflammation is traditionally defined by the four Latin words *calor*, *dolor*, *rubor*, and *tumor*, meaning heat, pain, redness, and swelling, all of which reflect the effects of cytokines on the local blood vessels (Fig. 1.12). Dilation and increased permeability of the blood vessels during inflammation lead to increased local

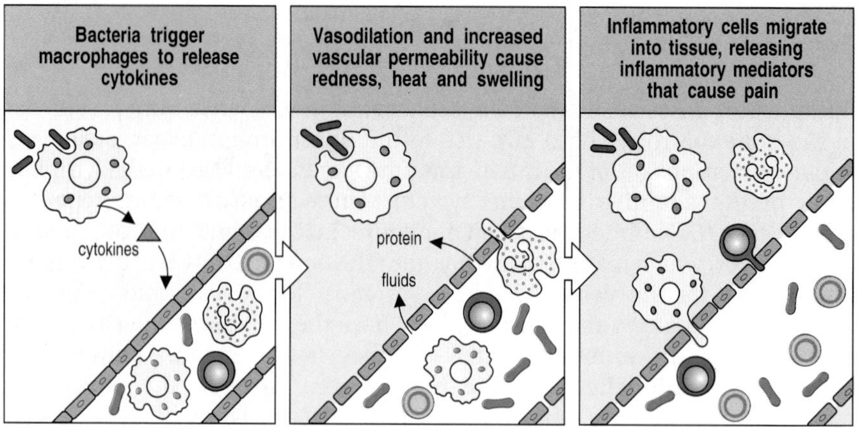
Bacteria trigger macrophages to release cytokines
cytokines
Vasodilation and increased vascular permeability cause redness, heat and swelling
protein
fluids
Inflammatory cells migrate into tissue, releasing inflammatory mediators that cause pain

Fig. 1.12 Bacterial infection triggers an inflammatory response. Macrophages encountering bacteria in the tissues are triggered to release cytokines that increase the permeability of blood vessels, allowing fluid and proteins to pass into the tissues. The stickiness of the endothelial cells of the blood vessels is also changed, so that cells adhere to the blood vessel wall and are able to crawl through it; first neutrophils and then macrophages are shown entering the tissue from a blood vessel. The accumulation of fluid and cells at the site of infection causes the redness, swelling, heat, and pain, known collectively as inflammation. Neutrophils and macrophages are the principal inflammatory cells. Later in an immune response, activated lymphocytes may also contribute to inflammation.

blood flow and the leakage of fluid, and account for the heat, redness, and swelling. Cytokines also have important effects on the adhesive properties of the endothelium, causing circulating leukocytes to stick to the endothelial cells of the blood vessel wall and migrate between them to the site of infection, to which they are attracted by yet other cytokines. The migration of cells into the tissue and their local actions account for the pain. The main cell types seen in an inflammatory response in its initial phases are neutrophils, followed by macrophages; these are therefore known as **inflammatory cells.**

Inflammatory responses later in an infection also involve the lymphocytes of the adaptive immune response, which have meanwhile been activated by antigen that has drained from the site of infection via the afferent lymphatics. The activation of lymphocytes depends critically on interactions with phagocytic cells; bacterial constituents induce changes in the surface molecules expressed by the phagocytic cells that are crucial to the central part they play in the induction of adaptive immune responses; we shall discuss this in detail in Chapter 8.

1-6 Lymphocytes are activated by antigen to give rise to clones of antigen-specific cells that mediate adaptive immunity.

The defense systems of innate immunity are effective in combating many pathogens but they can only recognize microorganisms bearing surface molecules that are common to many pathogens and that have remained unchanged in the course of evolution. Such highly conserved molecules can be recognized by the neutrophils and macrophages of vertebrates. Not surprisingly, many bacteria have evolved a protective capsule that enables them to conceal these molecules and thereby avoid provoking phagocytic cells. Viruses carry no such unvarying molecules and are rarely recognized by phagocytic cells. Moreover, the surface molecules of pathogens evolve much faster than could any ordinary vertebrate recognition system. The recognition mechanism used by the lymphocytes of the adaptive immune response has evolved to overcome these problems.

Instead of bearing several different receptors, each recognizing a different surface molecule of a pathogen, each naive lymphocyte entering the bloodstream bears antigen receptors of only a single specificity. However, the specificity of these receptors is determined by a unique genetic mechanism that operates during the development of lymphocytes in the bone marrow

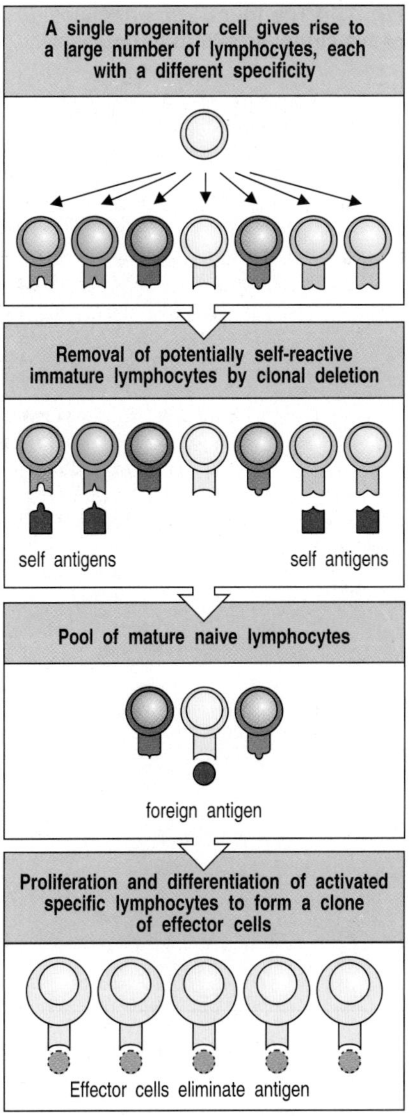

Fig. 1.13 The clonal selection hypothesis. During its normal course of development, each lymphocyte progenitor is able to give rise to many lymphocytes, each bearing a distinct antigen receptor. Lymphocytes with receptors that bind ubiquitous self antigens are eliminated early in development, before they become able to respond, ensuring tolerance to such self antigens. When antigen interacts with the receptor on a mature lymphocyte, that cell is activated to become a lymphoblast and then starts to divide. It gives rise to a clone of identical progeny, all of whose receptors bind the same antigen. Antigen specificity is thus maintained as the progeny proliferate and differentiate into effector cells. Once antigen has been eliminated by these effector cells, the immune response ceases.

and thymus to generate thousands of different variants of the genes encoding the receptor molecules. Thus, although an individual lymphocyte carries receptors of only one specificity, the specificity of each lymphocyte is different. Along with the other mechanisms that diversify receptor specificity, this ensures that the millions of lymphocytes in the body collectively carry millions of different antigen receptor specificities (Fig. 1.13). This constitutes the **receptor repertoire** of the individual. During the lifetime of the individual these lymphocytes undergo a process akin to natural selection: only those lymphocytes that encounter an antigen to which their receptor binds will be activated to proliferate and differentiate into effector cells.

This selective mechanism was first proposed in the 1950s by F. McFarlane Burnet to explain why antibodies, which can be induced in response to virtually any antigen, are produced in each individual only to those antigens to which he or she is exposed. He postulated the pre-existence in the body of many different potential antibody-producing cells, each having the ability to make antibody of a different specificity and displaying on its surface a membrane-bound version of the antibody serving as a receptor for antigen. On binding antigen, the cell is activated to proliferate and produce many identical progeny, known as a **clone**; these cells can now secrete antibodies with a specificity identical to that of the surface receptor. McFarlane Burnet called this the **clonal selection theory**.

1-7 Clonal selection of lymphocytes is the central principle of adaptive immunity.

Remarkably, at the time that McFarlane Burnet formulated his theory, nothing was known of the antigen receptors of lymphocytes; indeed the function of lymphocytes themselves was still obscure. Lymphocytes did not take center stage until the early 1960s, when James Gowans discovered that removal of the small lymphocytes from rats resulted in the loss of all known adaptive immune responses. These immune responses were restored when the lymphocytes were replaced. This led to the realization that lymphocytes must be the units of clonal selection, and their biology became the focus of the new field of **cellular immunology**.

Clonal selection of lymphocytes with diverse receptors elegantly explained adaptive immunity but it raised one significant intellectual problem. If the antigen receptors of lymphocytes are generated randomly during the lifetime of an individual, how are lymphocytes prevented from recognizing antigens on the tissues of the body and attacking them? Peter Medawar had shown, in 1953, that if exposed to foreign tissues during embryonic development, animals become immunologically **tolerant** to these tissues and will not subsequently make immune responses to them. Burnet proposed that

Fig. 1.14 The four basic principles of the clonal selection hypothesis.

Postulates of the clonal selection hypothesis
Each lymphocyte bears a single type of receptor with a unique specificity
Interaction between a foreign molecule and a lymphocyte receptor capable of binding that molecule with high affinity leads to lymphocyte activation
The differentiated effector cells derived from an activated lymphocyte will bear receptors of identical specificity to those of the parental cell from which that lymphocyte was derived
Lymphocytes bearing receptors specific for ubiquitous self molecules are deleted at an early stage in lymphoid cell development and are therefore absent from the repertoire of mature lymphocytes

developing lymphocytes that are potentially self-reactive are removed before they can mature. He has since been proved right in this too, although the mechanisms of tolerance are still being worked out, as we shall see when we discuss the development of lymphocytes in Chapters 6 and 7.

Clonal selection of lymphocytes is the single most important principle in adaptive immunity. It is shown schematically in Fig. 1.13, and its four basic postulates are listed in Fig. 1.14. The last of the problems posed by the clonal selection theory—that of how the diversity of lymphocyte antigen receptors is generated—was solved in the 1970s when advances in molecular biology made it possible to clone the genes encoding antibody molecules.

1-8 The structure of antibody molecules illustrates the problem of lymphocyte antigen receptor diversity.

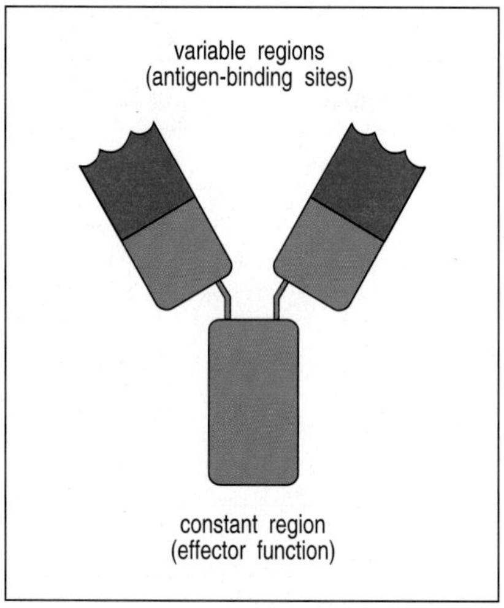

Fig. 1.15 Structure of an antibody molecule. The two arms of the Y-shaped antibody molecule contain the variable regions that form the two identical antigen-binding sites. The stem can take one of only a limited number of forms and is known as the constant region. It is the region that engages the effector mechanisms that antibodies activate to eliminate pathogens.

Antibodies, as discussed above, are the secreted form of the B-cell antigen receptor. Because they are produced in very large quantities in response to antigen, they can be studied by traditional biochemical techniques; indeed their structure was understood long before recombinant DNA technology made it possible to study the membrane-bound antigen receptors of lymphocytes. The startling feature that emerged from the biochemical studies was that an antibody molecule is composed of two distinct regions: a **constant region** that can take one of only four or five biochemically distinguishable forms; and a **variable region** that can take an apparently infinite variety of subtly different forms that allow it to bind specifically to an equally vast variety of different antigens.

This division is illustrated in the simple schematic diagram in Fig. 1.15, where the antibody is depicted as a Y-shaped molecule, with the constant region shown in blue and the variable region in red. The two variable regions, which are identical in any one antibody molecule, determine the antigen-binding specificity of the antibody, and the constant region determines how the antibody disposes of the pathogen once it is bound.

Each antibody molecule has a two-fold axis of symmetry and is composed of two identical heavy chains and two identical light chains (Fig. 1.16). Both heavy and light chains have variable as well as constant regions; the variable regions combine to form each antigen-binding site so that both chains contribute to the antigen-binding specificity of the antibody molecule. The structure of antibody molecules will be described in detail in Chapter 3, where we shall also discuss the structural and genetic basis for the different functional properties of antibodies conferred by their constant regions. For the time being we are concerned only with the properties of immunoglobulin molecules as antigen receptors, and the central issue of how the diversity of the variable regions is generated.

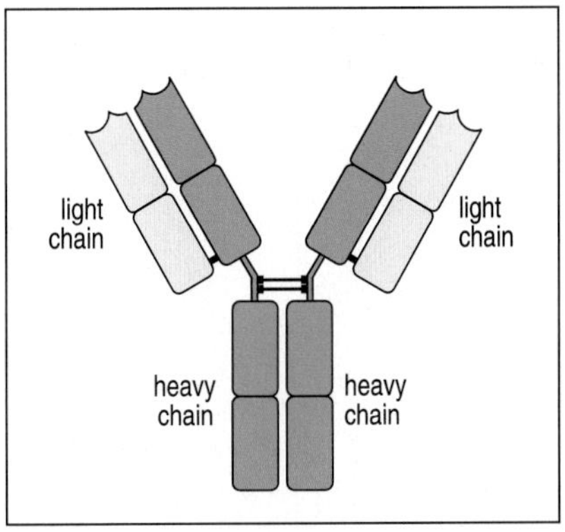

Fig. 1.16 Antibodies are made up of four protein chains. There are two types of chain in an antibody molecule: a larger chain called the heavy chain (green), and a smaller one called the light chain (yellow). Each chain has both a variable and a constant region, and there are two identical light chains and two identical heavy chains in each antibody molecule.

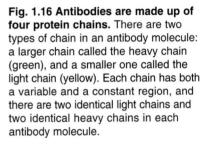

1-9 Each developing lymphocyte generates a unique antigen receptor by rearranging its receptor genes.

How are antigen receptors with an almost infinite range of specificities encoded by a finite number of genes? This question was answered in 1976, when Susumu Tonegawa discovered that the genes for immunoglobulin variable regions are inherited as sets of **gene segments**, each encoding a part of the variable region of one of the polypeptide chains that make up an immunoglobulin molecule (Fig. 1.17). During B-cell development in the bone marrow, these gene segments are joined by DNA recombination to form a stretch of DNA that codes for an entire variable region. Because there are many different segments in each set, and different gene segments are joined together in different cells, each cell generates unique genes for the variable regions of the heavy and light chains of the immunoglobulin molecule.

This mechanism has three important consequences. First, it enables a limited number of gene segments to generate a very diverse set of proteins. Second, as each cell assembles a different set of gene segments to encode its antigen receptor, each cell expresses a unique receptor specificity. Third, as gene rearrangement involves an irreversible change in a cell's DNA, all the progeny of that cell will inherit genes encoding the same receptor specificity. This general scheme was later confirmed for the genes encoding the antigen receptor on T lymphocytes. The main distinctions between B and T lymphocyte receptors are that the surface immunoglobulin that serves as the B-cell receptor has two identical antigen recognition sites and can be secreted, whereas the **T-cell receptor** has a single antigen recognition site and is always a cell-surface molecule. We shall see later that these receptors also recognize antigen in very different ways.

The potential diversity of lymphocyte receptors generated in this way is enormous. Just a few hundred different gene segments can combine in different ways to generate thousands of different receptor chains. The diversity of lymphocyte receptors is further amplified by junctional diversity, created by adding or subtracting nucleotides in the process of joining gene segments, and by the fact that each receptor is made by pairing two different

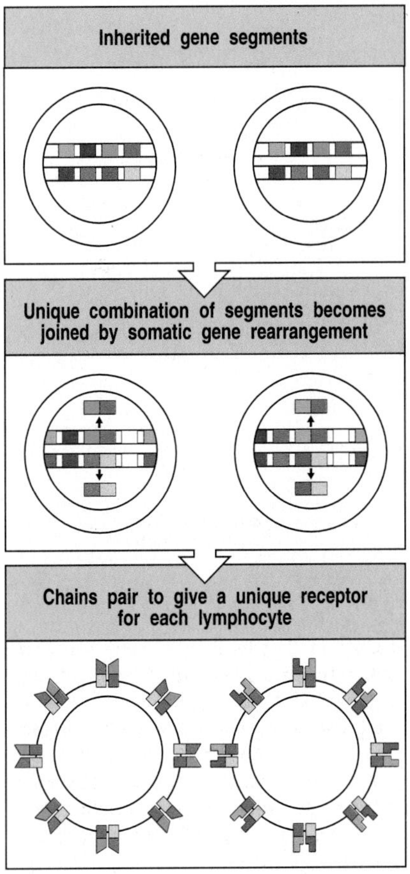

Fig. 1.17 The diversity of lymphocyte antigen receptors is generated by somatic gene rearrangements. Different parts of the variable region of antigen receptors are encoded by sets of gene segments. During a lymphocyte's development, one member of each set of gene segments is joined randomly to the others by an irreversible process of DNA recombination. The juxtaposed gene segments make up a complete gene encoding the variable part of one chain of the receptor, which is unique to that cell. This random process is repeated for the other set of gene segments, giving rise to the other chain. The expressed rearranged genes produce the two types of polypeptide chain that come together to form the unique antigen receptor on the lymphocyte surface. Once the required recombination events have occurred, further gene rearrangement is prohibited. Thus, the receptor specificity of a lymphocyte cannot change once it has been determined, and the lymphocyte can express only one receptor specificity. Each lymphocyte bears many copies of its unique receptor.

variable chains, each encoded in distinct sets of gene segments. A thousand different chains of each type could generate 10^6 distinct antigen receptors through this **combinatorial diversity**. Thus a small amount of genetic material can encode a truly staggering diversity of receptors; there are lymphocytes of at least 10^8 different specificities in an individual at any one time. Once gene rearrangement is complete, the antigen receptor is expressed on the surface of the developing lymphocyte, which is now ready to interact with antigen.

1-10 Lymphocytes proliferate in response to antigen in peripheral lymphoid tissue.

Because each lymphocyte has a different antigen-binding specificity, the fraction of lymphocytes that can bind and respond to any given antigen is very small. To generate sufficient specific effector lymphocytes to fight an infection, an activated lymphocyte must proliferate before its progeny finally differentiate into effector cells. This **clonal expansion** is a feature of all adaptive immune responses.

Lymphocyte activation and proliferation is initiated in the draining lymphoid tissues, where phagocytic cells carrying antigen are located. These display the antigen to the naive recirculating lymphocytes as they migrate through the lymphoid tissue before returning to the bloodstream via the efferent lymph. On recognizing its specific antigen, the small lymphocyte stops migrating and enlarges. The chromatin in its nucleus becomes less dense, nucleoli appear, the volume of cytoplasm increases, and the synthesis of new RNA and protein is induced. Within a few hours, the cell looks completely different, and activated cells at this stage are called **lymphoblasts** (Fig. 1.18).

The antigen-specific cells now begin to divide, normally duplicating two to four times every 24 hours for 3–5 days, so that one naive lymphocyte gives rise to a clone of around 1000 daughter cells of identical specificity. These then differentiate into effector cells that are able, in the case of B cells, to secrete antibody, or in the case of T cells are able to destroy infected cells or activate other cells of the immune system. These changes also affect the recirculation of lymphocytes—because of changes in the expression of specialized adhesion molecules on their surface, they cease circulating through the blood and lymph and instead migrate between the endothelial cells at sites of infection, signaled by the cytokines released by inflammatory cells at these sites.

After a naive lymphocyte has been activated, it takes 4–5 days before clonal expansion is complete and the lymphocytes have differentiated into effector cells. That is why adaptive immune responses occur only after a delay of several days. Effector cells have only a limited lifespan and, once antigen is removed, most of the antigen-specific cells generated by the clonal expansion of small lymphocytes undergo **programmed cell death**, or **apoptosis**. However, some persist after the antigen has been eliminated. This is the basis of **immunological memory**, which ensures a more rapid and effective response on a second encounter with a pathogen and thereby provides lasting protective immunity.

The characteristics of immunological memory are observed readily by comparing the antibody response of an individual to a first or **primary immunization** with the response elicited in the same individual by a **secondary** or **booster immunization** with the same antigen. As detailed in Fig. 1.19, the **secondary antibody response** occurs after a shorter lag phase,

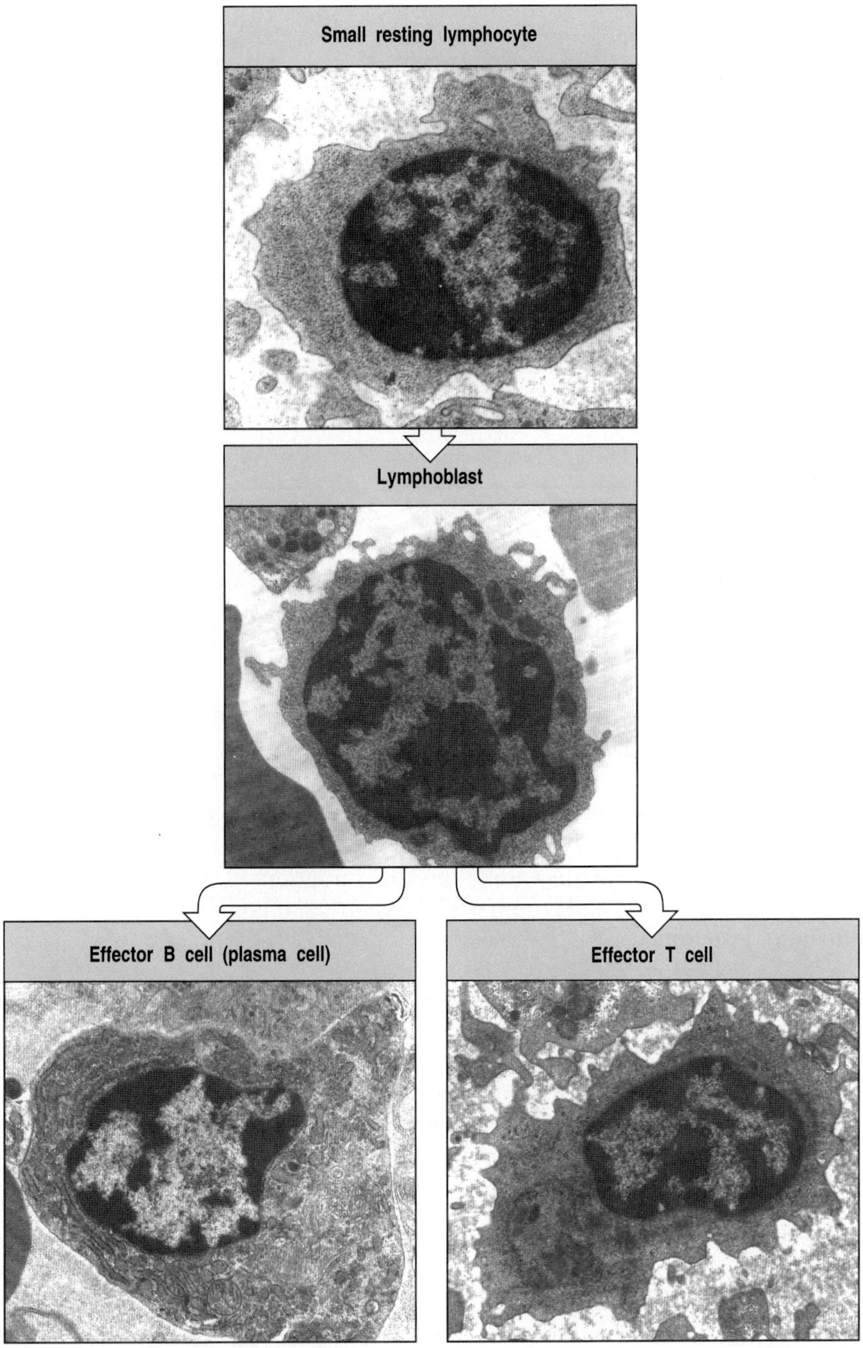

Fig. 1.18 Transmission electron micrographs of lymphocytes at various stages of activation to effector function. Small resting lymphocytes (upper panel) have not yet encountered antigen. Note the small amount of cytoplasm, the absence of rough endoplasmic reticulum, and the condensed chromatin, all indicative of an inactive cell. This could be either a T cell or a B cell. Small circulating lymphocytes are trapped in lymph nodes when their receptors encounter antigen on antigen-presenting cells. Stimulation by antigen induces the lymphocyte to become an active lymphoblast. This cell undergoes clonal expansion by repeated division, which is followed by differentiation to effector function. The central micrograph shows an activated lymphoblast responding to antigen. Note the large size, the nucleoli, the enlarged nucleus with diffuse chromatin, and the active cytoplasm; again, T and B cells are similar in appearance. The lower panels show effector T and B lymphocytes. Note the large amount of cytoplasm, the nucleus with prominent nucleoli, abundant mitochondria and the presence of rough endoplasmic reticulum, all hallmarks of active cells. The rough endoplasmic reticulum is especially prominent in antibody-secreting B cells, usually called plasma cells, which are synthesizing and secreting very large amounts of protein in the form of antibody. Photographs courtesy of N Rooney.

achieves a markedly higher plateau level, and produces antibodies of higher affinity. We shall describe the mechanisms of these remarkable changes in Chapters 9 and 10. The cellular basis of immunological memory is the clonal expansion and clonal differentiation of cells specific for the eliciting antigen, and it is therefore entirely antigen specific.

It is immunological memory that allows successful vaccination and prevents re-infection with pathogens that have been repelled successfully by an adaptive immune response. Immunological memory is the most important biological consequence of the development of adaptive immunity based on clonal selection, although the cellular and molecular basis of immunological memory is still not fully understood, as we shall see in Chapter 10.

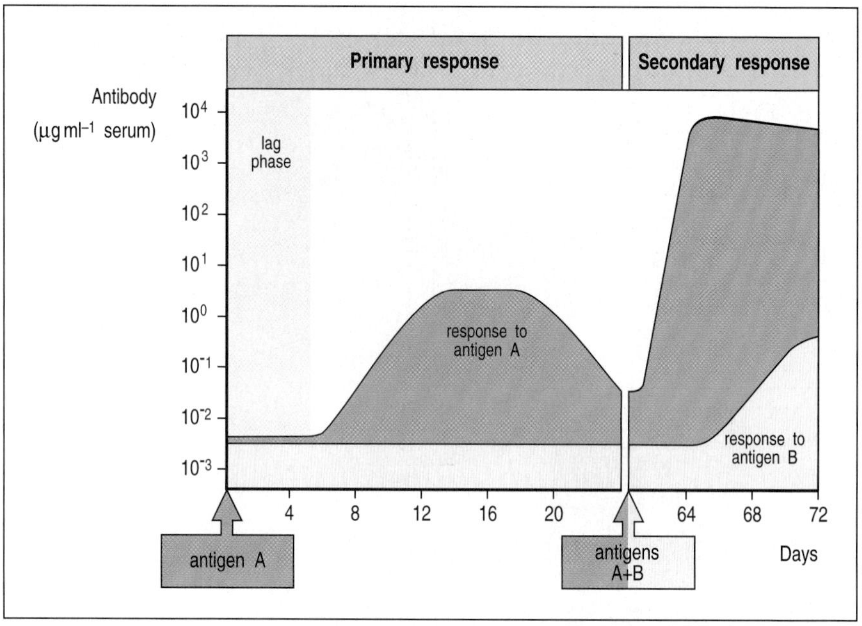

Fig. 1.19 The course of a typical antibody response. Antigen A introduced at time zero encounters little specific antibody in the serum. After a lag phase, antibody against antigen A (blue) appears; its concentration rises to a plateau, and then declines. When the serum is tested for antibody against another antigen, B (yellow), there is none present, demonstrating the specificity of the antibody response. When the animal is later challenged with a mixture of antigens A and B, a very rapid and intense response to A occurs. This illustrates immunological memory, the ability of the immune system to make a second response to the same antigen more efficiently and effectively, providing the host with a specific defense against infection. This is the main reason for boosting immunity after vaccination. Note that the response to B resembles the initial or primary response to A, as this is the first encounter of the host with antigen B.

1-11 Interaction with other cells as well as with antigen is necessary for lymphocyte activation.

Peripheral lymphoid tissues are specialized not only to trap phagocytic cells that ingest antigen (see Section 1-3) but also to promote the interactions between cells that are necessary for the initiation of adaptive immune responses. The spleen and lymph nodes in particular are highly organized for the latter function.

All lymphocyte responses to antigen require not only the signal that results from antigen binding, but also a second signal, which is delivered by another cell. For most B-cell responses, the second signal comes from a T cell (Fig. 1.20, left panel); for T cells (see Fig. 1.20, right panel), the second signal can be delivered by any of three cell types: dendritic cells, macrophages, or B cells. Mature **dendritic cells** are cells with a distinctive branched morphology found exclusively in the T-cell areas of lymphoid tissue. The precursors of

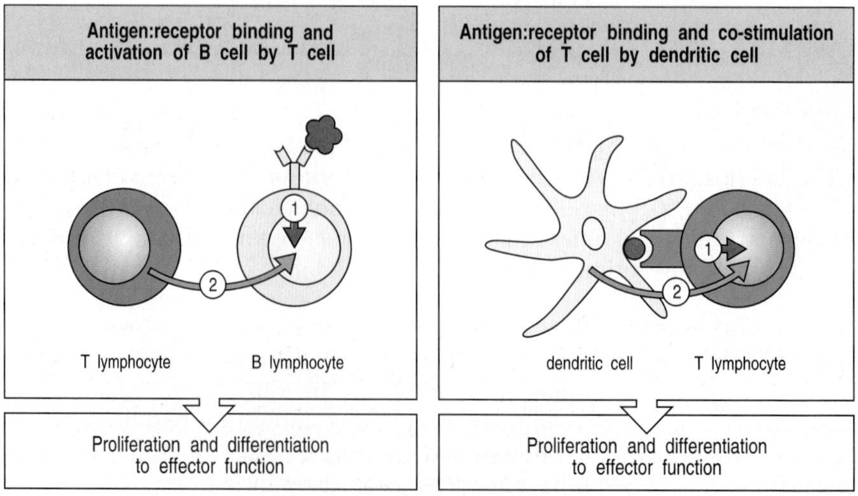

Fig. 1.20 Two signals are required for lymphocyte activation. As well as receiving a signal through their antigen receptor, mature lymphocytes must also receive a second signal to become activated. For B cells (left panel), the second signal is usually delivered by a T cell. For T cells (right panel) it is delivered by a professional antigen-presenting cell, shown here as a dendritic cell.

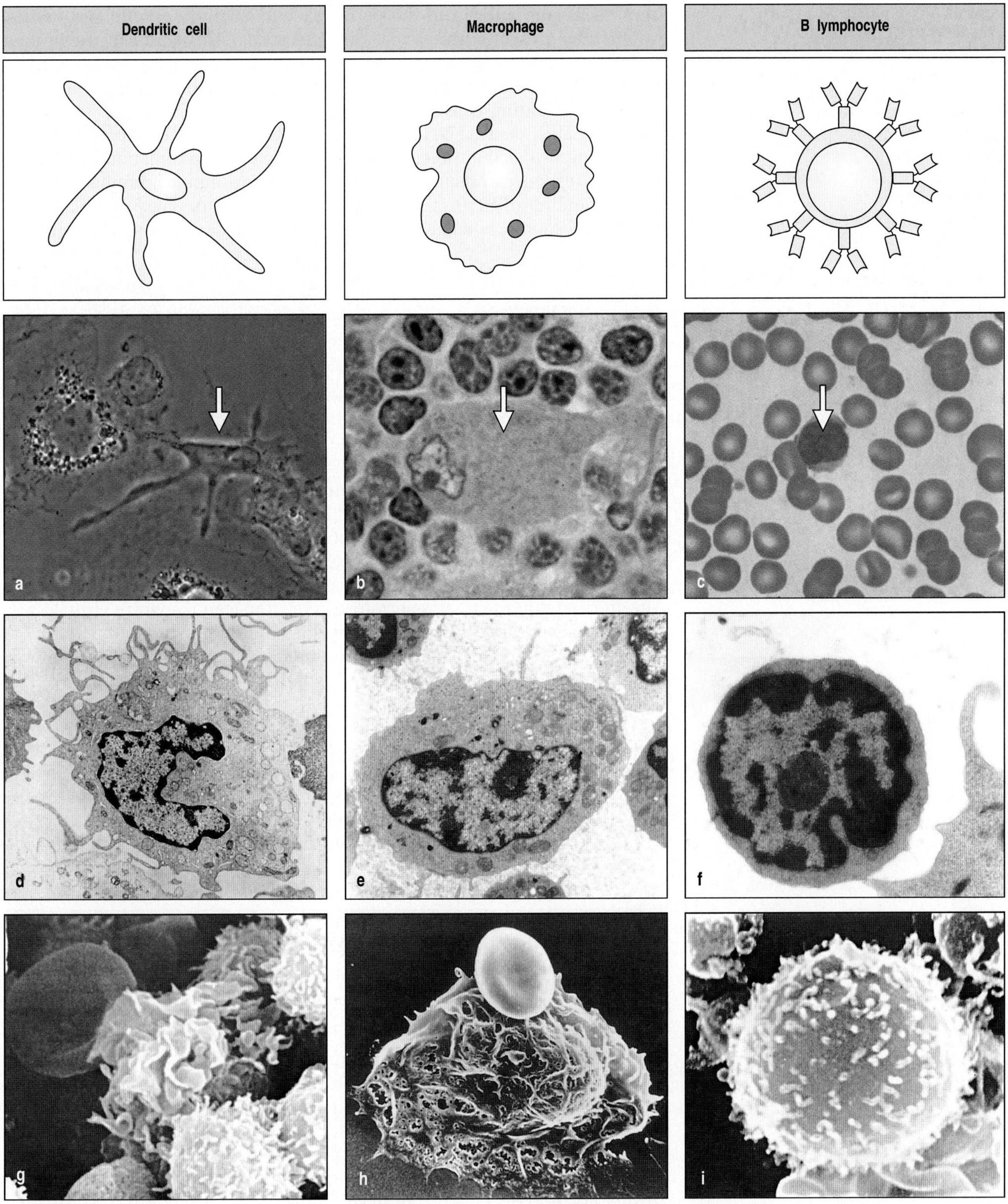

Fig. 1.21 The professional antigen-presenting cells. The three types of professional antigen-presenting cell are shown in the form in which they will be depicted throughout this book (top row), as they appear in the light microscope (second row), by transmission electron microscopy (third row) and by scanning electron microscopy (bottom row). Dendritic cells are found in lymphoid tissues and are thought to have a critical role in immunity to many antigens. Macrophages are specialized to internalize and present particulate antigens, and B cells have antigen-specific receptors that allow them to internalize large amounts of specific antigen and present it to T cells. Photographs courtesy of R M Steinman (a); N Rooney (b, c, e, f); S Knight (d, g); P F Heap (h, i).

these cells trap antigen in the periphery and migrate to lymphoid tissues, where they present antigens to T cells. Because of this, and because of their ability to deliver activating signals, dendritic cells, along with macrophages and B cells, are known as **professional antigen-presenting cells**, or often just **antigen-presenting cells**. The three cell types that can present antigen to T cells are illustrated in Fig. 1.21. Dendritic cells are the most important antigen-presenting cell of the three, with a central role in the initiation of adaptive immune responses. Macrophages can also mediate innate immune responses directly and make a crucial contribution to the effector phase of the adaptive immune response, whereas B cells contribute mainly to adaptive immunity.

Summary.

The early innate systems of defense, which depend on invariant receptors recognizing common features of pathogens, are crucially important, but they cannot confer protection from novel types of pathogen and do not lead to immunological memory. These are the unique features of adaptive immunity, which is based on clonal selection of lymphocytes bearing antigen-specific receptors. The clonal selection of lymphocytes provides a theoretical framework for understanding all the key features of adaptive immunity. Each lymphocyte carries cell-surface receptors of a single specificity, generated by random recombination of variable receptor gene segments and the pairing of different variable chains. This produces lymphocytes each bearing a distinct receptor, so that the total repertoire of receptors can recognize virtually any antigen. If the receptor on a lymphocyte is specific for a ubiquitous self antigen, the cell is eliminated by encountering the antigen early in development. When a recirculating lymphocyte encounters foreign antigen in lymphoid tissues, it is induced to proliferate and its progeny then differentiate into effector cells that can eliminate a specific infectious agent. A subset of these proliferating lymphocytes differentiates into memory cells, ready to respond rapidly to the same pathogen if it is encountered again. The details of these processes of recognition, development, and differentiation form the main material of the middle three parts of this book.

The recognition and effector mechanisms of adaptive immunity.

Clonal selection describes the basic operating principle of the adaptive immune response but not its mechanisms. In the last part of this chapter, we shall outline the mechanisms by which pathogens are detected by lymphocytes and eventually destroyed in a successful adaptive immune response. The distinct lifestyles of different pathogens require different response mechanisms not only to ensure the destruction of pathogens but also for their detection and recognition (Fig. 1.22). We have already seen that there are two different kinds of antigen receptor: the surface immunoglobulin of B cells, and the smaller antigen receptor of T cells. These surface receptors are adapted to recognize antigen in two different ways: B cells recognize antigen that is present outside the cells of the body, where, for example, most bacteria are found; T cells, by contrast, can detect antigens generated inside host cells, for example by viruses.

The immune system protects against four classes of pathogen		
Type of pathogen	**Examples**	**Diseases**
Extracellular bacteria, parasites, fungi	*Streptococcus pneumoniae* *Clostridium tetani* *Trypanosoma brucei*	Pneumonia Tetanus Sleeping sickness
Intracellular bacteria, parasites	*Mycobacterium leprae* *Leishmania donovani* *Plasmodium falciparum*	Leprosy Leishmaniasis Malaria
Viruses (intracellular)	Variola Influenza Varicella	Smallpox Flu Chickenpox
Parasitic worms (extracellular)	*Ascaris* *Schistosoma*	Ascariasis Schistosomiasis

Fig. 1.22 The major pathogen types confronting the immune system and some of the diseases that they cause.

The **effector mechanisms** that operate to eliminate pathogens in an adaptive immune response are essentially identical to those of innate immunity. Indeed, it seems likely that specific recognition by clonally distributed receptors evolved as a late addition to existing innate effector mechanisms to produce the present-day adaptive immune response. We begin by outlining the effector actions of antibodies, which depend almost entirely on recruiting cells and molecules of the innate immune system.

1-12 Extracellular pathogens and their toxins are eliminated by antibodies.

Antibodies, which were the first specific product of the immune response to be identified, are found in the fluid component of blood, or **plasma**, and in extracellular fluids. Because body fluids were once known as humors, immunity mediated by antibody is known as **humoral immunity**.

As we have seen in Fig. 1.15, antibodies are Y-shaped molecules whose arms form two identical antigen-binding sites that are highly variable from one molecule to another, providing the diversity required for specific antigen recognition. The stem of the Y, which defines the **class** of the antibody and determines its functional properties, takes one of only five major forms, or **isotypes**. Each of the five antibody classes engages a distinct set of effector mechanisms for disposing of antigen once it is recognized. We shall describe the isotypes and their actions in detail in Chapters 3 and 9.

The simplest and most direct way in which antibodies can protect from pathogens or their toxic products is by binding to them and thereby blocking their access to cells that they may infect or destroy (Fig. 1.23, left panels). This is known as **neutralization** and is important for protection against bacterial toxins and against pathogens such as viruses, which can thus be prevented from entering cells and replicating.

Binding by antibodies, however, is not sufficient on its own to arrest the replication of bacteria that multiply outside cells. In this case, one role of antibody is to enable a phagocytic cell to ingest and destroy the bacterium. This is important for the many bacteria that are resistant to direct recognition by phagocytes; instead, the phagocytes recognize the constant region of the antibodies coating the bacterium (see Fig. 1.23, middle panels). The coating of pathogens and foreign particles in this way is known as **opsonization**.

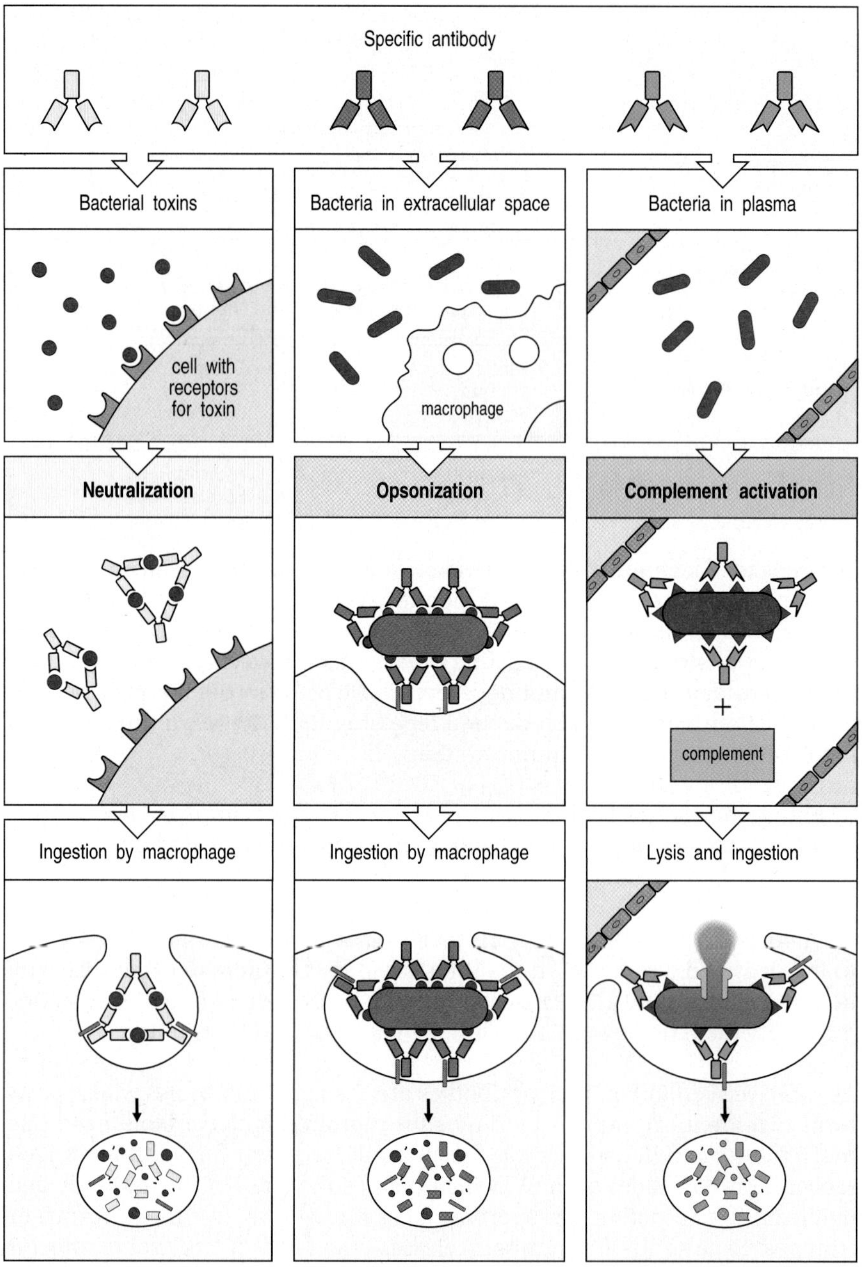

Fig. 1.23 Antibodies can participate in host defense in three main ways. The left panels show antibodies binding to and neutralizing a bacterial toxin, preventing it from interacting with host cells and causing pathology. Unbound toxin can react with receptors on the host cell, whereas the toxin:antibody complex cannot. Antibodies also neutralize complete virus particles and bacterial cells by binding to them and inactivating them. The antigen:antibody complex is eventually scavenged and degraded by macrophages. Antibodies coating an antigen render it recognizable as foreign by phagocytes (macrophages and neutrophils), which then ingest and destroy it; this is called opsonization. The middle panels show opsonization and phagocytosis of a bacterial cell. The right panels show activation of the complement system by antibodies coating a bacterial cell. Bound antibodies form a receptor for the first protein of the complement system, which eventually forms a protein complex on the surface of the bacterium that, in some cases, can kill the bacterium directly but more generally favors the uptake and destruction of the bacterium by phagocytes. Thus, antibodies target pathogens and their products for disposal by phagocytes.

The third function of antibodies is to activate a system of plasma proteins known as **complement**. The complement system, which we shall discuss in detail in Chapter 9, can directly destroy bacteria, and this is important in a few bacterial infections (see Fig. 1.23, right panels). However, the main function of complement is, like that of the antibodies themselves, to coat the surface of pathogens and enable phagocytes to engulf and destroy bacteria that they would otherwise not recognize. Complement also enhances the bactericidal actions of phagocytes; indeed it is so called because it 'complements' the activities of antibodies.

Antibodies of different isotypes are found in different compartments of the body and differ in the effector mechanisms that they recruit, but all pathogens and particles bound by antibody are eventually delivered to phagocytes for ingestion, degradation, and removal from the body (see Fig. 1.23, bottom panels).

The complement system and the phagocytes that antibodies recruit are not themselves antigen-specific; they depend upon antibody molecules to mark the particles as foreign. Antibodies are the sole contribution of B cells to the adaptive immune response. T cells, by contrast, have a variety of effector actions.

1-13 T cells are needed to control intracellular pathogens and to activate B-cell responses to most antigens.

Pathogens are accessible to antibodies only in the blood and the extracellular spaces. However, some bacterial pathogens and parasites, and all viruses, replicate inside cells, where they cannot be detected by antibodies. The destruction of these invaders is the function of the T lymphocytes, or T cells, which are responsible for the **cell-mediated immune responses** of adaptive immunity.

Cell-mediated reactions depend on direct interactions between T lymphocytes and cells bearing the antigen that the T cells recognize. The actions of cytotoxic T cells are the most direct. These cells recognize body cells infected with viruses, which replicate inside cells by using the synthetic machinery of the cell itself. The replicating virus eventually kills the cell, releasing the new virus particles. Antigens derived from the replicating virus, however, are meanwhile displayed on the surface of infected cells, where they are recognized by cytotoxic T cells. These cells can then control the infection by killing the infected cell before viral replication is complete (Fig. 1.24).

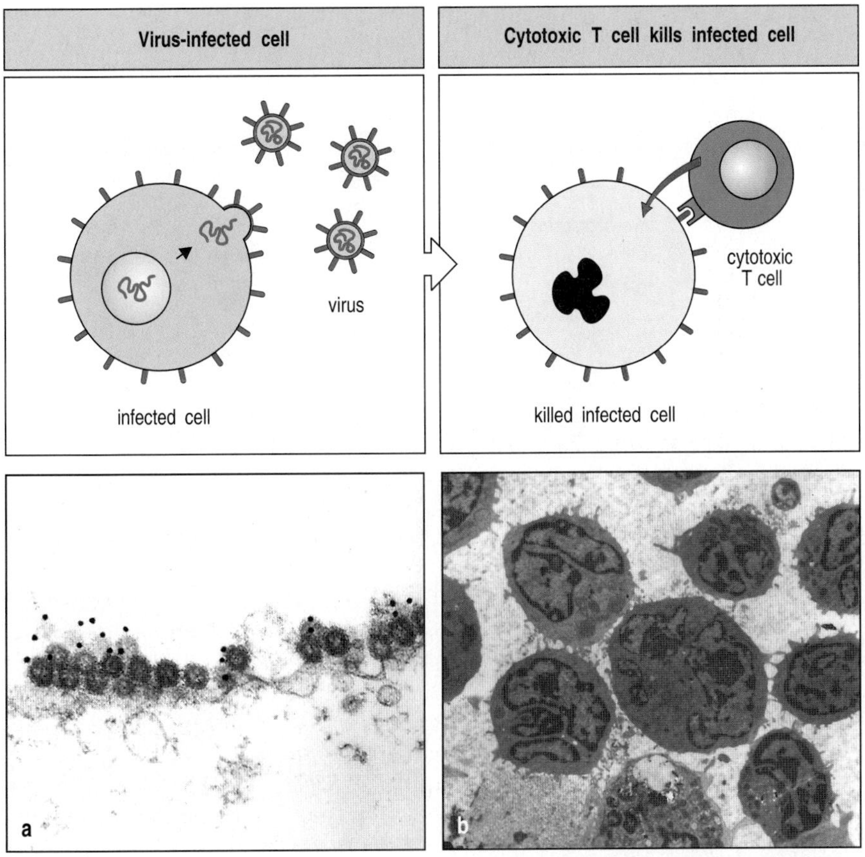

Fig. 1.24 Mechanism of host defense against intracellular infection by viruses. Cells infected by viruses are recognized by specialized T cells called cytotoxic T cells, which kill the infected cells directly. The killing mechanism involves the activation of nucleases in the infected cell, which cleave host and viral DNA. Panel a is a transmission electron micrograph showing the plasma membrane of a Chinese hamster ovary (CHO) cell infected with influenza virus. Many virus particles can be seen budding from the cell surface. Some of these have been labeled with a monoclonal antibody that is specific for a viral protein and is coupled to gold particles, which appear as the solid black dots in the micrograph. Panel b is a transmission electron micrograph of a virus-infected cell surrounded by reactive T lymphocytes. Note the close apposition of the membranes of the virus-infected cell and the T cell in the upper left corner of the micrograph, and the clustering of the cytoplasmic organelles in the T cell between the nucleus and the point of contact with the infected cell. Panel a courtesy of M Bui and A Helenius; panel b courtesy of N Rooney.

T lymphocytes are also important in the control of intracellular bacterial infections. Some bacteria grow only in the vesicles of macrophages; important examples are *Mycobacterium tuberculosis* and *M. leprae*, the pathogens that cause tuberculosis and leprosy (Fig. 1.25). Bacteria entering macrophages are usually destroyed in the lysosomes, which contain a variety of enzymes and antimicrobial substances. Intracellular bacteria survive because the vesicles they occupy do not fuse with the lysosomes. These infections can be controlled by a second subset of T cells, known as a **T_H1 cells**, which activate macrophages, inducing the fusion of their lysosomes with the vesicles containing the bacteria and at the same time stimulating other antibacterial mechanisms of the phagocyte. T_H1 cells also release cytokines that attract macrophages to the site of infection.

T cells destroy intracellular pathogens by killing infected cells and by activating macrophages but they also have a central role in the destruction of extracellular pathogens by activating B cells. This is the specialized role of a third subset of T cells, called **helper T cells** consisting mainly of **T_H2 cells**. We shall see in Chapter 9, when we discuss the humoral immune response in detail, that only a few antigens with special properties are capable of activating naive B lymphocytes on their own. Most antigens require an accompanying signal from helper T cells before they can stimulate B cells to proliferate and differentiate into cells capable of secreting antibody (see Fig. 1.20). The ability of T cells to activate B cells was discovered long before it was recognized that a functionally distinct class of T cells activates macrophages, and the term helper T cell was originally coined to describe T cells that activate B cells. Although the designation 'helper' was later extended to T cells that activate macrophages (hence the H in T_H1), we consider this usage confusing and we will, in the remainder of this book, reserve the term helper T cells for all T cells that activate B cells.

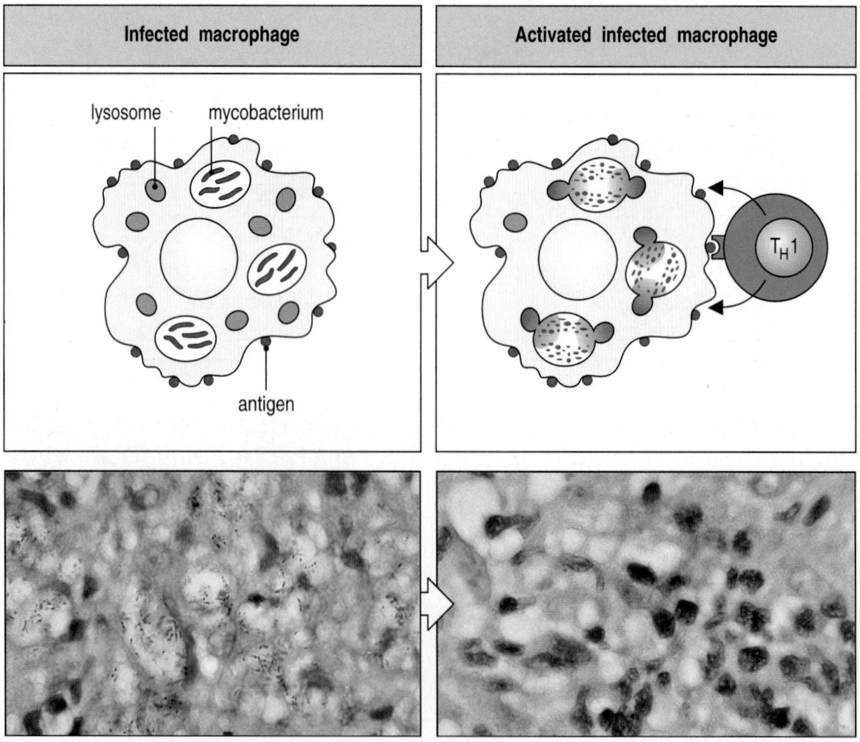

Fig. 1.25 Mechanism of host defense against intracellular infection by mycobacteria. Mycobacteria infecting macrophages live in cytoplasmic vesicles that resist fusion with lysosomes and consequent destruction of the bacteria by macrophage bactericidal activity. However, when a specific T_H1 cell recognizes an infected macrophage, it releases macrophage-activating molecules or cytokines that induce lysosomal fusion and the activation of macrophage bactericidal activities. The elimination of mycobacteria from the vesicles of activated macrophages can be seen in the light micrographs (bottom row) of resting (left) and activated (right) macrophages infected with mycobacteria. The cells have been stained with an acid-fast red dye to reveal the presence of the mycobacteria, which are prominent as red-staining rods in the resting macrophages but have been eliminated from the activated macrophages. Photographs courtesy of G Kaplan.

1-14 T cells are specialized to recognize foreign antigens as peptide fragments bound to proteins of the major histocompatibility complex.

All the effects of T lymphocytes depend upon interactions with cells containing foreign proteins. In cytotoxic T cells and T_H1 cells, the proteins are produced by pathogens infecting the target cell or having been ingested by it. Helper T cells, in contrast, recognize and interact with B cells that have bound and internalized foreign antigen via their surface immunoglobulin. In all cases, T cells recognize their targets by detecting peptide fragments derived from these foreign proteins and bound to specialized cell-surface molecules on the infected host cells, on phagocytes, or on B cells. The molecules that display peptide antigen to T cells are membrane glycoproteins encoded in a cluster of genes bearing the cumbersome name **major histocompatibility complex**, abbreviated to **MHC**.

The human **MHC molecules** were first discovered as the result of attempts to use skin grafts from donors to repair badly burned pilots and bomb victims during World War II. The patients rejected the grafts, and eventually genetic experiments on inbred mice led to the identification of genes causing the rapid rejection of skin grafts between mice that differed only at these genetic loci and at no other. Because they control the compatibility of tissue grafts, these genes became known as 'histocompatibility genes'. Later, it was found that several closely linked genes specify histocompatibility, which led to the term major histocompatibility complex. The central role of the MHC in antigen recognition by T cells, which we shall discuss in Chapter 4, revealed the true physiological function of the proteins encoded by the MHC. This, in turn, led to an explanation for the major effect on tissue compatibility for which they were named. We shall discuss these diverse functions of MHC molecules in Chapters 4, 7, and 13.

1-15 Two major types of T cell recognize peptides bound by two different classes of MHC molecule.

There are two types of MHC molecule, called MHC class I and MHC class II, which differ in subtle ways but share most of their major structural features. The most important of these structural features is the two outer extracellular domains that form a long cleft in which a single peptide fragment is trapped during the synthesis and assembly of the MHC molecule inside the cell. The MHC molecule bearing its cargo of peptide is then transported to the cell surface (Fig. 1.26) where it displays the peptide to T cells. The antigen receptors of T lymphocytes are specialized to recognize a foreign antigenic peptide fragment bound to an MHC molecule.

The most important differences between the two classes of MHC molecule lie not in their structure but in the source of the peptides that they trap and carry to the cell surface. **MHC class I molecules** collect peptides derived from proteins

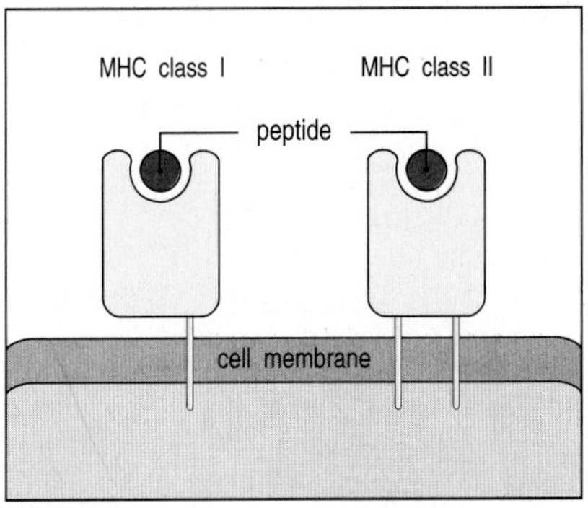

Fig. 1.26 MHC molecules display peptide fragments of antigens on the surface of cells. MHC molecules are membrane proteins whose outer extracellular domains form a cleft in which a peptide fragment is bound. These fragments, which are derived from proteins degraded inside the cell, including foreign protein antigens, are bound by the newly synthesized MHC molecule before it reaches the surface. There are two kinds of MHC molecule—MHC class I and MHC class II—with related but distinct structures and functions.

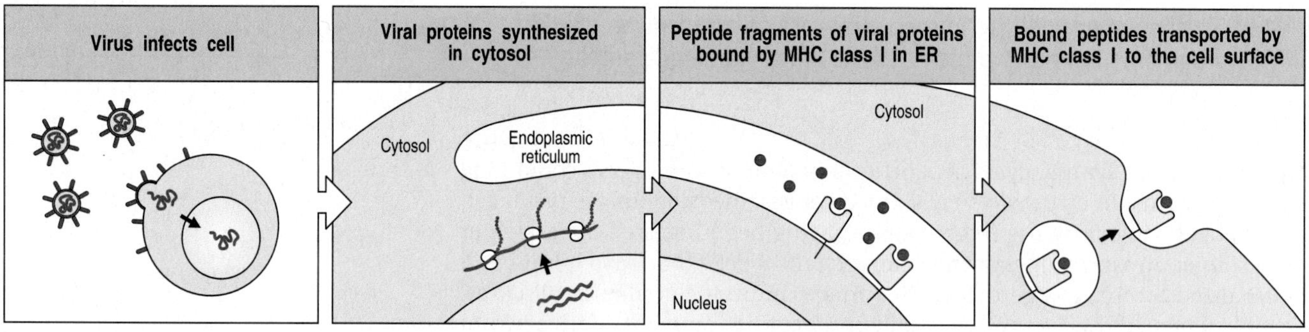

Fig. 1.27 MHC class I molecules present antigen derived from proteins in the cytosol. In cells infected with viruses, viral proteins are synthesized in the cytosol. Peptide fragments of viral proteins are transported into the endoplasmic reticulum where they are bound by MHC class I molecules, which then deliver the peptides to the cell surface.

synthesized in the cytosol, and are thus able to display fragments of viral proteins on the cell surface (Fig. 1.27). **MHC class II molecules** bind peptides derived from proteins in intracellular membrane-bound vesicles, and thus display peptides derived from pathogens living in macrophage vesicles or internalized by phagocytic cells or B cells (Fig. 1.28). We shall see in Chapter 4 exactly how peptides from these different sources are made differentially accessible to the two types of MHC molecule.

Once they reach the cell surface with their cargo of antigenic peptides, the two classes of MHC molecule are recognized by different functional classes of T cell. MHC class I molecules bearing viral peptides are recognized by cytotoxic T cells, which kill the infected cell (Fig. 1.29); MHC class II molecules bearing peptides derived from pathogens living in the vesicles of macrophages or taken up by B cells are recognized by the T_H1 or T_H2 cells (Fig. 1.30). How the ability to recognize the appropriate type of MHC

Fig. 1.28 MHC class II molecules present antigen originating in intracellular vesicles. Some bacteria infect cells and grow in intracellular vesicles. Peptides derived from such bacteria are bound by MHC class II molecules and transported to the cell surface (top row). MHC class II molecules also bind and transport peptides derived from antigen that has been bound and internalized by B-cell antigen receptor-mediated endocytosis into intracellular vesicles (bottom row).

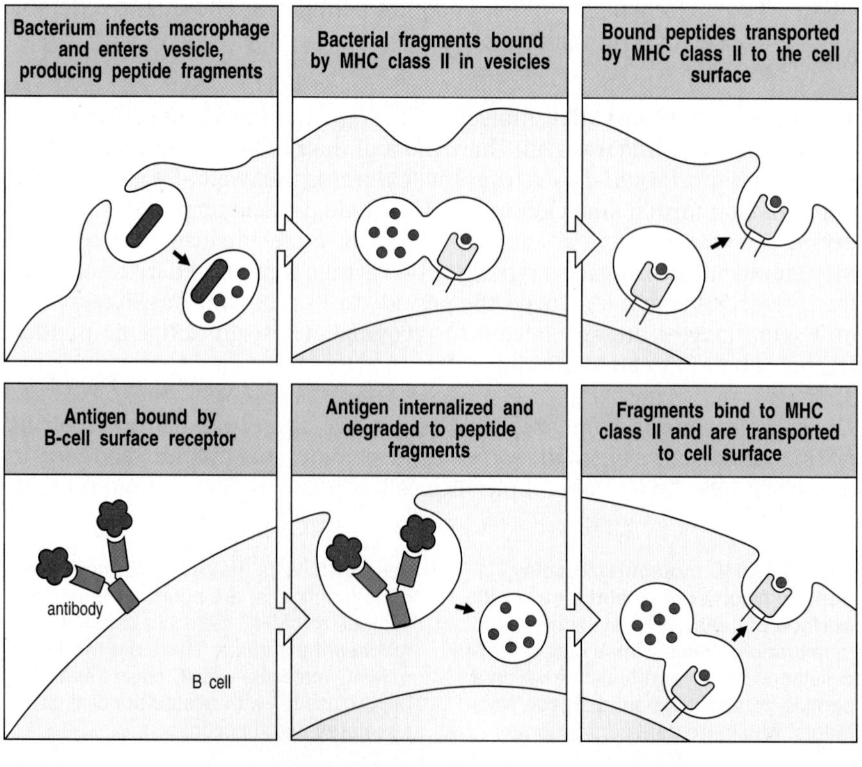

molecule becomes coupled during T-cell development to the distinct functions of the different classes of T cells is a central question in immunology and is a major topic of Chapter 7.

On recognizing their targets, the three types of T cell are stimulated to release different sets of effector molecules. These can directly affect their target cells or help to recruit other effector cells in ways we shall discuss in Chapter 8. These effector molecules include many cytokines, which have a crucial role in the clonal expansion of lymphocytes as well as in innate immune responses and in the effector actions of most immune cells; thus, understanding the actions of cytokines is central to understanding the various behaviors of the immune system.

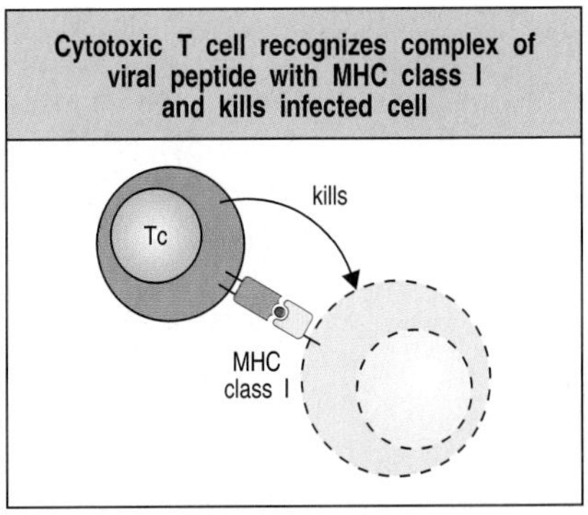

Fig. 1.29 Cytotoxic T cells recognize antigen presented by MHC class I molecules and kill the cell. The peptide:MHC class I complex on virus-infected cells is detected by antigen-specific cytotoxic T cells. Cytotoxic T cells are pre-programmed to kill cells.

1-16 Defects in the immune system result in increased susceptibility to infectious disease.

We tend to take for granted the ability of our immune systems to free our bodies of infection and prevent its recurrence. In some people, however, parts of the immune system fail. In the most severe of these **immunodeficiency diseases**, adaptive immunity is completely absent, and death occurs in infancy from overwhelming infection unless heroic measures are taken. Other less catastrophic failures lead to recurrent infections with particular types of pathogen, depending on the particular deficiency, and much has been learned about the functions of the different components of the human immune system through the study of these immunodeficiencies, as will be discussed in Chapter 11.

Twenty years ago, a devastating form of immunodeficiency appeared, the **acquired immune deficiency syndrome**, or **AIDS**, which is itself caused by an infectious agent. This disease destroys the T cells that activate macrophages, leading to infections caused by intracellular bacteria and other pathogens normally controlled by these T cells. Such infections are the major cause of death from this increasingly prevalent immunodeficiency disease.

AIDS is caused by a virus, the **human immunodeficiency virus**, or **HIV**, that has evolved several strategies by which it not only evades but also subverts the protective mechanisms of the adaptive immune response. In Chapter 11 we shall also discuss how other viruses, and bacteria and parasites as well, have evolved ways of avoiding destruction by the immune system. The conquest of many of the world's leading causes of disease, including malaria and the various diarrheal diseases (the leading killers of children), as well as of the more recent threat from AIDS, might depend upon a better understanding of the interactions of these pathogens with the cells of the immune system.

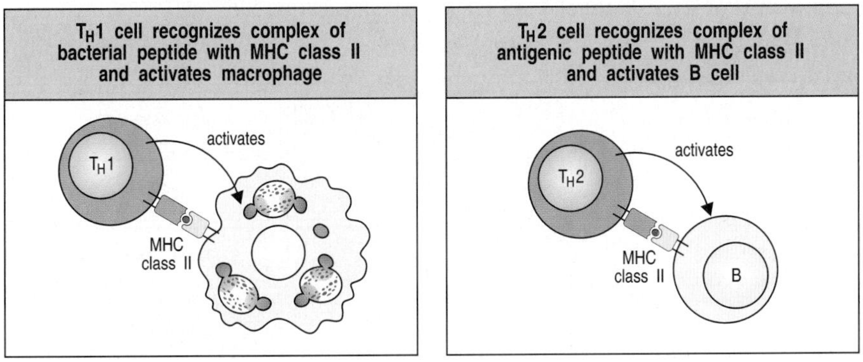

Fig. 1.30 T_H1 and T_H2 cells recognize antigen presented by MHC class II molecules. T_H1 and T_H2 cells both recognize peptides bound to MHC class II molecules. On recognition of their specific antigen on infected macrophages, T_H1 cells activate the macrophage, leading to the destruction of the intracellular bacteria (left panel). When T_H2 cells recognize antigen on B cells, helper T cells activate these cells to proliferate and differentiate into antibody-producing plasma cells (right panel).

1-17 Understanding adaptive immune responses is important for the control of allergies, autoimmune disease, and organ graft rejection, and also for vaccination.

Many medically important diseases are associated with a normal immune response directed against an inappropriate antigen, often in the absence of infectious disease. Immune responses in the absence of infection occur in **allergy**, where the antigen is an innocuous foreign substance, in **autoimmune disease**, where the response is to a self antigen, and in **graft rejection**, where the antigen is borne by a foreign cell. What we call a successful immune response or a failure, and whether the response is considered harmful or beneficial to the host, depends not on the response itself but rather on the nature of the antigen (Fig. 1.31).

Allergic diseases, which include asthma, are an increasingly common cause of disability in the developed world, and many other important diseases are now recognized as autoimmune. An autoimmune response directed against pancreatic β cells is the leading cause of diabetes in the young. In allergies and autoimmune diseases, the powerful protective mechanisms of the adaptive immune response are the cause of serious damage to the host.

Immune responses to harmless antigens, to body tissues, or to organ grafts are, like all other immune responses, highly specific. At present, the usual way to treat these responses is with **immunosuppressive drugs**, which inhibit all immune responses, desirable or undesirable. If instead it were possible to suppress only those lymphocyte clones responsible for the unwanted response, the disease could be cured or the grafted organ protected without impeding protective immune responses. Although antigen-specific suppression of immune responses can be induced experimentally, the molecular basis of suppression is unknown. If one could achieve control of the immune response, then the dream of antigen-specific **immunoregulation** to control unwanted immune responses could become a reality. We shall see in Chapter 10 how the mechanisms of immune regulation are beginning to emerge from a better understanding of the functional subsets of lymphocytes and the cytokines that control them, and we shall discuss the present state of understanding of allergies, autoimmune disease, graft rejection, and immunosuppressive drugs in Chapters 12, 13, and 14.

Fig. 1.31 Immune responses can be beneficial or harmful depending on the nature of the antigen. Beneficial responses are shown in white, harmful responses in shaded boxes. Where the response is beneficial, its absence is harmful.

Antigen	Effect of response to antigen	
	Normal response	Deficient response
Infectious agent	Protective immunity	Recurrent infection
Innocuous substance	Allergy	No response
Grafted organ	Rejection	Acceptance
Self organ	Autoimmunity	Self tolerance
Tumor	Tumor immunity	Cancer

1-18 Stimulation of an adaptive immune response is the most effective way to prevent infectious disease.

Although the specific suppression of immune responses must await advances in basic research on immune regulation and its application, the deliberate stimulation of an immune response by immunization, or vaccination, has achieved many successes in the two centuries since Jenner's pioneering experiment.

Mass immunization programs have led to the virtual eradication of several diseases that used to be associated with significant morbidity (illness) and mortality (Fig. 1.32). Immunization is considered so safe and so important that most states in the USA require children to be immunized against up to seven common childhood diseases. Impressive as these accomplishments are, there are still many diseases for which we lack effective vaccines. And even where a vaccine for diseases such as measles or polio can be used effectively in developed countries, technical and economic problems can prevent its widespread use in developing countries, where mortality from these diseases is still high. The tools of modern immunology and molecular biology are being applied to develop new vaccines and improve old ones, and

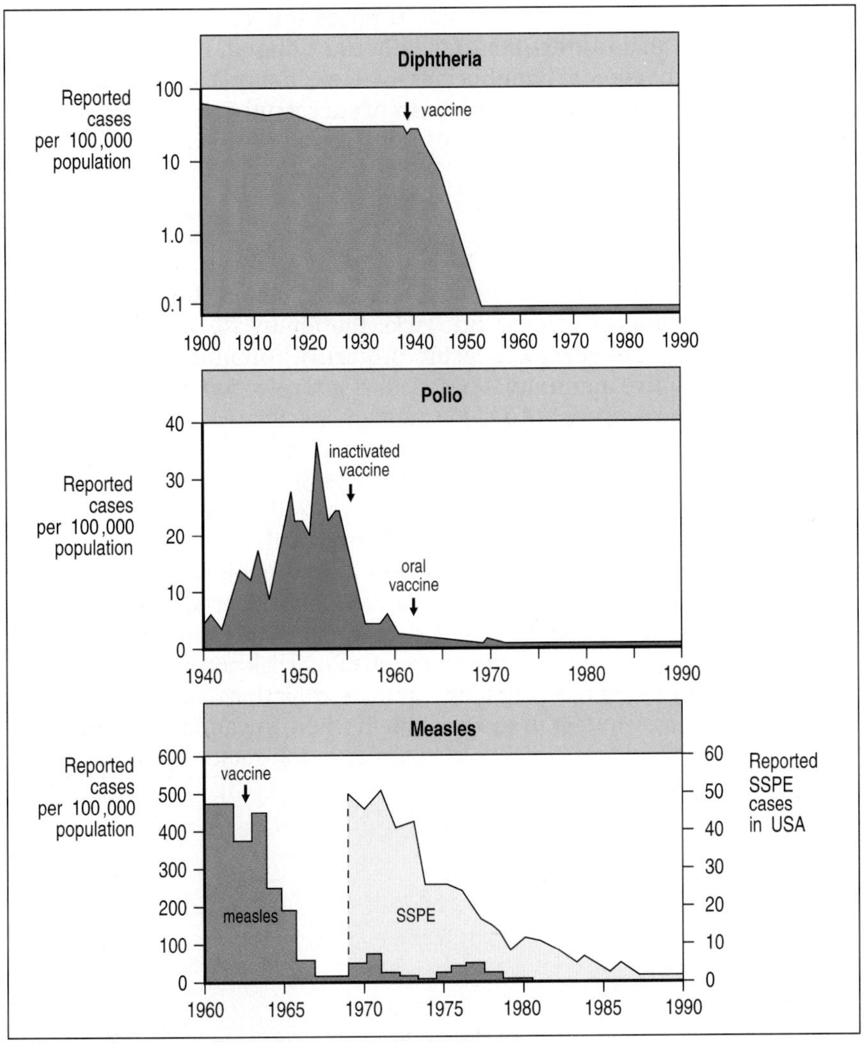

Fig. 1.32 Successful vaccination campaigns. Diphtheria, polio, and measles and its consequences have been virtually eliminated in the USA, as shown in these three graphs. SSPE stands for subacute sclerosing panencephalitis, a brain disease that is a late consequence of measles infection in a few patients. When measles was prevented, SSPE disappeared 10–15 years later. However, as these diseases have not been eradicated worldwide, immunization must be maintained in a very high percentage of the population to prevent their reappearance.

we shall discuss these advances in Chapter 14. The prospect of controlling these important diseases is tremendously exciting. The guarantee of good health is a critical step towards population control and economic development. At a cost of pennies per person, great hardship and suffering can be alleviated.

Summary.

Lymphocytes have two distinct recognition systems specialized for detection of extracellular and intracellular pathogens. B cells have cell-surface immunoglobulin molecules as receptors for antigen and, upon activation, secrete the immunoglobulin as soluble antibody that provides defense against pathogens in the extracellular spaces of the body. T cells have receptors that recognize peptide fragments of intracellular pathogens transported to the cell surface by the glycoproteins of the major histocompatibility complex (MHC). Two classes of MHC molecule transport peptides from different intracellular compartments to present them to distinct types of effector T cell:cytotoxic T cells that kill infected target cells, and T_H1 and T_H2 cells that mainly activate macrophages and B cells. Thus, T cells are crucially important for both the humoral and cell-mediated responses of adaptive immunity. The adaptive immune response seems to have engrafted specific antigen recognition by highly diversified receptors onto innate defense systems. These have a central role in the effector actions of both B and T lymphocytes. The antigen-specific suppression of adaptive immune responses is the goal of treatment for important human diseases involving inappropriate activation of lymphocytes, whereas the specific stimulation of an adaptive immune response is the basis of successful vaccination.

Summary to Chapter 1.

The immune system defends the host against infection. Innate immunity serves as a first line of defense but lacks the ability to recognize certain pathogens and to provide the specific protective immunity that prevents re-infection. Adaptive immunity is based on clonal selection from a repertoire of lymphocytes bearing highly diverse antigen-specific receptors that enable the immune system to recognize any foreign antigen. In the adaptive immune response, antigen-specific lymphocytes proliferate and differentiate into effector cells that eliminate pathogens. Host defense requires different recognition systems and a wide variety of effector mechanisms to seek out and destroy the wide variety of pathogens in their various habitats within the body and at its surface. Not only can the adaptive immune response eliminate a pathogen but, in the process, it also generates increased numbers of differentiated memory lymphocytes through clonal selection and this allows a more rapid and effective response upon re-infection. The regulation of immune responses, whether to suppress them when unwanted or to stimulate them in the prevention of infectious disease, is the major medical goal of research in immunology.

General references.

Historical background

Burnet, F.M.: *The Clonal Selection Theory of Acquired Immunity.* London, Cambridge University Press, 1959.

Gowans, J.L.: **The Lymphocyte—a disgraceful gap in medical knowledge.** *Immunol. Today* 1996, **17**:288-291.

Landsteiner, K.: *The Specificity of Serological Reactions,* 3rd edn. Boston, Harvard University Press, 1964.

Metchnikoff, E.: *Immunity in the Infectious Diseases,* 1st edn. New York, Macmillan Press, 1905.

Silverstein, A.M.: *History of Immunology,* 1st edn. London, Academic Press, 1989.

Biological background

Alberts, B., Bray, D., Lewis, J., Raff, M., Roberts, K., and Watson, J.D.: *Molecular Biology of the Cell,* 3rd edn. New York, Garland Publishing, 1994.

Ryan, K.J., (ed): *Medical Microbiology,* 3rd edn. East Norwalk, CT, Appleton-Lange, 1994.

Stryer, L.: *Biochemistry,* 4th edn. New York, Freeman, 1995.

Primary journals devoted solely or primarily to immunology

Autoimmunity
Clinical and Experimental Immunology
Comparative and Developmental Immunology
European Journal of Immunology
Immunity
Immunogenetics
Immunology
Infection and Immunity
International Immunology
Journal of Experimental Medicine
Journal of Immunology
Regional Immunology
Thymus

Primary journals with frequent papers in immunology

Cell
Current Biology
EMBO Journal
Journal of Biological Chemistry
Journal of Cell Biology
Journal of Clinical Investigation
Molecular Cell Biology
Nature
Proceedings of the National Academy of Sciences, USA
Science

Review journals in immunology

Annual Reviews in Immunology
Contemporary Topics in Microbiology and Immunology
Current Opinion in Immunology
Immunological Reviews
Immunology Today
Proceedings of the International Congress of Immunology: Progress in Immunology, **1–8**, 1971–1992.
Research in Immunology
Seminars in Immunology
The Immunologist
The Immunologist, **3**: Proceedings Issue, 9th International Congress of Immunology

Advanced textbooks in immunology, compendia, etc.

Lachmann, P.J., Peters, D.K., Rosen, F.S., Walport, M.J. (eds): *Clinical Aspects of Immunology,* 5th edn. Oxford, Blackwell Scientific Publications, 1993.

Mak, T.W. and Simard, J.J.L.: *Handbook of Immune Response Genes.* New York, Plenum Press, 1998.

Paul, W.E. (ed): *Fundamental Immunology,* 4th edn. New York, Raven Press, 1998.

Roitt, I.M., and Delves, P.J. (eds): *Encyclopedia of Immunology,* 3rd edn. London/San Diego, Academic Press, 1992.

Rosen, F.S., Geha, R.S.: *Case Studies in Immunology: A Clinical Companion,* 2nd edn. New York, Garland Publishing/Current Biology, 1999.

The Induction, Measurement, and Manipulation of the Immune Response

2

Before you read further, a word from the authors about this chapter. We have written it to be read when it is needed to understand a particular method; later chapters reference Chapter 2 as appropriate. Although it is also written so that it can be read from start to finish, most students will want to dip into this toolbox of methods when they encounter a reference to it in later chapters, rather than tackling it now. To make the sections relevant to later chapters easy to identify, the edges of the paper in this chapter are colored, and the most useful methods for any part of the book are color coded to the part in which they are needed. However, we recommend that you read the first six sections of Chapter 2 before continuing with the rest of the book.

The description of the immune system outlined in Chapter 1 is drawn from the results of many different kinds of experiment and from the study of human disease. Immunologists have devised a wide variety of techniques for inducing, measuring, and characterizing immune responses, and for altering the immune system through cellular, molecular, and genetic manipulation. Before we examine the cellular and molecular basis of host defense in the remainder of this book, we shall look at how the immune system is studied and introduce the specialized language of immunology. In this chapter, we describe many basic immunological phenomena that experimental immunologists seek to explain in terms of the cellular and molecular features of the immune system. Since modern molecular genetics has an important role in the analysis of the immune system and of human disease, genetic analysis of the immune system by using recently developed techniques for genetic manipulation is discussed. We also describe clinical tests used to assess immune function in patients with immunological disorders.

Immunological techniques are widely applied in many other areas of biology and medicine. The use of antibodies to detect specific molecules in complex mixtures and in tissues is of particular importance. We therefore devote an entire section of this chapter to the antibody-based methods used by immunologists, by basic scientists in many other biological disciplines, and by clinicians. These methods illustrate the specificity and utility of antibodies, whose structure and generation are an important theme of subsequent parts of this book.

The induction and detection of immune responses.

Most of this book focuses on **adaptive immunity**, that is, on the antigen-specific immune responses of lymphocytes to foreign materials. These responses are normally directed at antigens borne by pathogenic micro-organisms. However, the immune system can also be induced to respond to simple non-living antigens, and experimental immunologists have mainly focused on the responses to these simple antigens in developing our understanding of the immune response. Thus, we shall begin by discussing how such adaptive immune responses are induced and detected.

The deliberate induction of an immune response is known as **immunization**. Experimental immunizations are routinely carried out by injecting the test antigen into the animal or human subject, and we shall see that the route, dose, and form in which antigen is administered can profoundly affect whether a response occurs and the type of response that is produced. To determine whether an immune response has occurred and to follow its course, the immunized individual is monitored for the appearance of immune reactants directed at the specific antigen. Immune responses to most antigens elicit the production of both specific antibodies and specific effector T cells. Monitoring the antibody response usually involves the analysis of relatively crude preparations of **antiserum** (plural: **antisera**). The **serum** is the fluid phase of clotted blood, which, if taken from an immunized individual, is called antiserum because it contains specific antibodies against the immunizing antigen as well as other soluble serum proteins. To study immune responses mediated by T cells, blood lymphocytes or cells from lymphoid organs such as the spleen are tested; T-cell responses are more commonly studied in experimental animals than in humans.

Any substance that can elicit an immune response is said to be **immunogenic** and is called an **immunogen**. There is a clear operational distinction between an immunogen and an antigen. An **antigen** is defined as any substance that can bind to a specific antibody. All antigens therefore have the potential to elicit specific antibodies but some need to be attached to an immunogen in order to do so. This means that although all immunogens are antigens, not all antigens are immunogenic.

The following sections describe some of the most commonly used techniques for inducing, detecting, and measuring adaptive immune responses. These techniques are used to address many questions in immunology. What determines whether a particular substance will be immunogenic or not? How does one raise antibodies against substances that are not by themselves immunogenic? And what determines which type of response will be provoked by a particular immunization? We shall first examine the nature of antigens and the features that make a substance immunogenic, before turning to a general consideration of how the response is detected.

2-1 Antibodies can be produced against almost any substance.

When antibodies were first discovered as the agents of resistance to infection, it was thought likely that their ability to bind pathogens had been selected over evolutionary time. However, Karl Landsteiner soon showed that antibodies could be elicited against a virtually limitless range of molecules, including synthetic chemicals never found in the natural environment. This demonstrated unequivocally that the repertoire of possible antibodies in any individual is essentially unlimited, and that the genes encoding individual antibodies could not have been selected for their action against pathogens. Landsteiner's work radically changed the way that immunologists thought about the antibody response, forcing them to the conclusion that evolution must have selected not for specific antibody structures but rather for the ability to generate an open repertoire of antibodies of diverse structure, a subject we shall focus on in Chapter 3. The ability to generate such a large diversity of antibodies means that no two individuals are likely to make the same response to any antigen. It also alerted immunologists to the potential utility of antibodies for detecting and measuring almost any substance, even when present in a complex mixture of other molecules.

In order to determine the range of antibodies that could be produced, Landsteiner studied the immune response to small organic molecules, such as phenyl arsonates and nitrophenyls. Although these simple structures do not provoke antibodies when injected by themselves, Landsteiner found that antibodies could be raised against them if the molecule was attached covalently to a protein carrier. He therefore termed them **haptens** (from the Greek *haptein*, to fasten). Animals immunized with a hapten–carrier conjugate produced three distinct sets of antibodies (Fig. 2.1). One set comprised hapten-specific antibodies that reacted with the same hapten on any carrier, as well as with free hapten. The second set of antibodies was specific for the carrier protein, as shown by their ability to bind both the hapten-modified and unmodified carrier protein. Finally, some antibodies reacted only with the specific conjugate of hapten and carrier used for immunization. Landsteiner studied mainly the antibody response to the hapten, as these small molecules could be synthesized in many closely related forms. He observed that antibodies raised against a particular hapten bind that hapten but, in general, fail to bind even very closely related chemical structures. The binding of haptens by anti-hapten antibodies has played an important part in defining the precision of antigen binding by antibody molecules. Anti-hapten antibodies are also important medically as they mediate allergic reactions to penicillin and other compounds that elicit antibody responses when they attach to self proteins (see Section 12-11).

Antisera contain many different antibody molecules that bind to the immunogen in slightly different ways (see Fig. 2.1). Some of the antibodies in an antiserum are cross-reactive. A **cross-reaction** is defined as the binding of an antibody to an antigen other than the immunogen; most cross-react with closely related molecules but, on occasion, some bind antigens having no clear relationship to the immunogen. These cross-reacting antibodies can create problems when the antiserum is used for the detection of a specific antigen by using the techniques outlined in the next part of this chapter. They can be removed from an antiserum by **absorption** with the cross-reactive antigen, leaving behind the antibodies that bind only to the immunogen. Absorption can be performed by affinity chromatography using immobilized antigen, a technique that is also used for purification of antibodies or antigens (see Section 2-13). However, most problems of cross-reactivity can be avoided by making monoclonal antibodies. Monoclonal antibodies are homogeneous antibodies derived from a single antibody-producing cell (see Section 2-10). They can be selected for a lack of cross-reactivity and their binding properties can be reliably defined.

The antigens used most frequently in experimental immunology are proteins, and antibodies to proteins are of enormous utility in experimental biology and medicine. Therefore, in this chapter we shall focus on the production and use of anti-protein antibodies. Although antibodies can also be made to

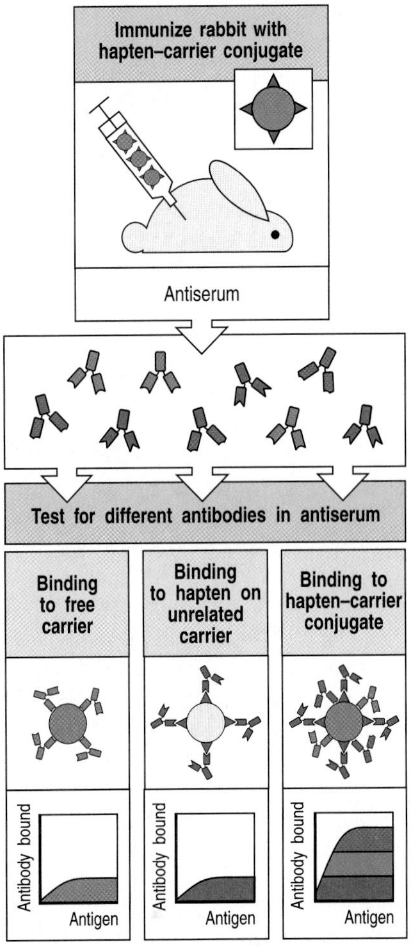

Fig. 2.1 Antibodies can be elicited by small chemical groups called haptens only when the hapten is linked to an immunogenic protein carrier. Three types of antibodies are produced. One set (blue) binds the carrier protein alone and is called carrier-specific. One set (red) binds to the hapten on any carrier or to free hapten in solution and is called hapten-specific. One set (purple) only binds the specific conjugate of hapten and carrier used for immunization, apparently binding to sites at which the hapten joins the carrier, and is called conjugate-specific. The amount of antibody of each type in this serum is shown schematically in the graphs at the bottom; note that the original antigen binds more antibody than the sum of anti-hapten and anti-carrier owing to the additional binding of conjugate-specific antibody.

carbohydrates, to nucleic acids, and to other structural classes of antigen (see Chapter 9), their induction generally requires the attachment of the antigen to a protein carrier. Thus, the immunogenicity of protein antigens determines the outcome of virtually every immune response.

2-2 The immunogenicity of a protein depends on its presentation to T cells.

Although any structure can be recognized by antibody as an antigen, usually only proteins elicit fully developed adaptive immune responses. This is because proteins have the ability to engage T cells, which contribute to inducing most antibody responses and are required for immunological memory. Proteins engage T cells because the T cells recognize antigens as peptide fragments of proteins bound to major histocompatibility complex (MHC) molecules (see Section 1-14). An adaptive immune response that includes immunological memory can be induced by other classes of antigen only when they are attached to a protein carrier that can engage the necessary T cells. Immunological memory is produced as a result of the initial or **primary immunization**. This is also known as **priming**, as the animal or person is now 'primed' like a pump to mount a more potent response to subsequent challenges with the same antigen. The response to each immunization is increasingly intense, so that **secondary**, **tertiary**, and subsequent responses are of increasing magnitude. Repetitive challenge with antigen to achieve a heightened state of immunity is known as **hyperimmunization**.

Certain properties of a protein that favor the priming of an adaptive immune response have been defined by studying antibody responses to simple natural proteins like hen egg-white lysozyme and to synthetic polypeptide antigens (Fig. 2.2). The larger and more complex a protein, and the more distant its

Fig. 2.2 Intrinsic properties and extrinsic factors that affect the immunogenicity of proteins.

Factors that influence the immunogenicity of proteins		
Parameter	**Increased immunogenicity**	**Decreased immunogenicity**
Size	Large	Small (MW<2500)
Dose	Intermediate	High or low
Route	Subcutaneous > intraperitoneal > intravenous or intragastric	
Composition	Complex	Simple
Form	Particulate	Soluble
	Denatured	Native
Similarity to self protein	Multiple differences	Few differences
Adjuvants	Slow release	Rapid release
	Bacteria	No bacteria
Interaction with host MHC	Effective	Ineffective

relationship to self proteins, the more likely it is to elicit a response. This is because such responses depend on the protein's being degraded into peptides that can bind to MHC molecules, and on the subsequent recognition of these peptide:MHC complexes by T cells that have survived a process of selection against responsiveness to self. The larger and more distinct a protein antigen is, the more likely it is to contain such peptides. Particulate or aggregated antigens are more immunogenic because they are taken up more efficiently by the specialized antigen-presenting cells responsible for initiating a response (see Section 1-11); indeed small soluble proteins are unable to induce a response unless they are made to aggregate in some way. Many vaccines, for example, use aggregated protein antigens to potentiate the immune response. As we shall see in Sections 2-3 and 2-4, the way in which protein antigens are administered can greatly influence the induction and character of the immune response. For the purposes of this chapter we are focusing on empirical observations that have emerged from the practice of immunization. A clearer understanding of how these various parameters determine immunogenicity will become evident when the priming of T cells is described in Chapter 8.

2-3 The response to a protein antigen is influenced by dose, form, and route of administration.

The magnitude of the immune response depends on the dose of immunogen administered. Below a certain threshold dose, most proteins do not elicit any immune response. Above the threshold dose, there is a gradual increase in the response as the dose of antigen is increased, until a broad plateau level is reached, followed by a decline at very high antigen doses (Fig. 2.4). As most infectious agents enter the body in small numbers, immune responses are generally elicited only by pathogens that multiply to a level sufficient to exceed the antigen dose threshold. The broad response optimum allows the system to respond to infectious agents across a wide range of doses. At very high antigen doses the immune response is inhibited, which may be important in maintaining tolerance to abundant self proteins such as plasma proteins. In general, secondary and subsequent immune responses occur at lower antigen doses and achieve higher plateau values, which is a sign of immunological memory. However, under some conditions, very low or very high doses of antigen may induce specific unresponsive states, known respectively as acquired **low-zone** or **high-zone tolerance**.

The route by which antigen is administered also affects both the magnitude and the type of response obtained. Antigens injected subcutaneously generally elicit the strongest responses, whereas antigens injected or transfused directly into the bloodstream tend to induce unresponsiveness or tolerance unless they bind to host cells or contain aggregates that are readily taken up by antigen-presenting cells. Antigens administered solely to the gastro intestinal tract have distinctive effects, frequently eliciting a local antibody response in the intestinal lamina propria, while producing a systemic state of tolerance that manifests as a diminished response to the same antigen if subsequently administered in immunogenic form elsewhere in the body. This 'split tolerance' may be important in avoiding allergy to antigens in food, as the local response prevents food antigens from entering the body, while the inhibition of systemic immunity helps to prevent the formation of IgE antibodies, which are the cause of such allergies (see Chapter 12). In contrast, protein antigens that enter the body through the respiratory epithelium tend to elicit allergic responses for reasons that are not clear.

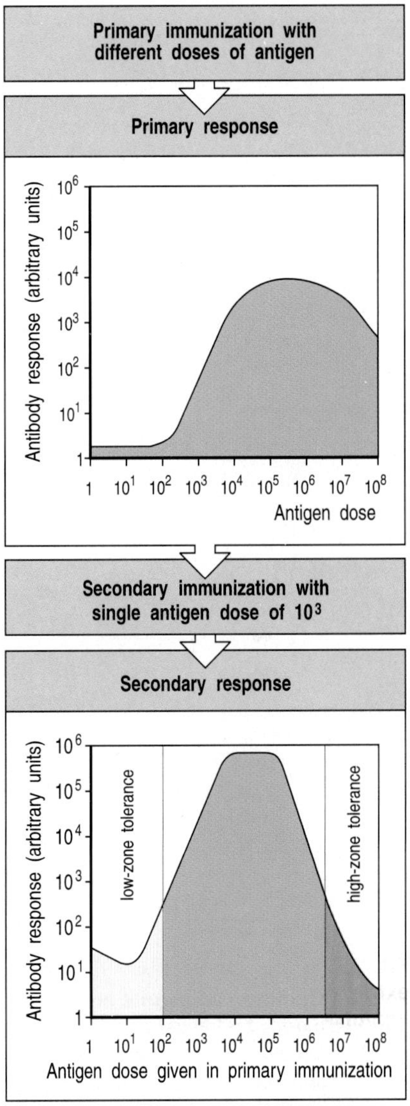

Fig. 2.3 The dose of antigen used in an initial immunization affects the primary and secondary antibody response. The typical antigen dose–response curve shown here illustrates the influence of dose on both a primary antibody response (amounts of antibody produced expressed in arbitrary units) and the effect of the dose used for priming on a secondary antibody response elicited by a dose of antigen of 10^3 arbitrary mass units. Very low doses of antigen do not cause an immune response at all. Slightly higher doses appear to inhibit specific antibody production, an effect known as low-zone tolerance. Above these doses there is a steady increase in the response with antigen dose to reach a broad optimum. Very high doses of antigen also inhibit immune responsiveness to a subsequent challenge, a phenomenon known as high-zone tolerance.

2-4 Immunogenicity can be enhanced by administration of proteins in adjuvants.

Most proteins are poorly immunogenic or non-immunogenic when administered by themselves. Strong adaptive immune responses to protein antigens almost always require that the antigen be injected in a mixture known as an **adjuvant**. An adjuvant is any substance that enhances the immunogenicity of substances mixed with it. Adjuvants differ from protein carriers in that they do not form stable linkages with the immunogen. Furthermore, adjuvants are needed primarily for initial immunizations, whereas carriers are required to elicit not only primary but also subsequent responses to haptens. Commonly used adjuvants are listed in Fig. 2.4.

Adjuvants can enhance immunogenicity in two different ways. First, adjuvants convert soluble protein antigens into particulate material, which is more readily ingested by antigen-presenting cells such as macrophages. For example, the antigen can be adsorbed on particles of the adjuvant (such as alum), or made particulate by emulsification in mineral oils. This enhances immunogenicity somewhat, but such adjuvants are relatively weak unless they also contain bacteria or bacterial products. Such microbial constituents are the second means by which adjuvants enhance immunogenicity, and although their exact contribution to enhancing immunogenicity is unknown, they are clearly the more important component of an adjuvant. Microbial products may signal macrophages or dendritic cells to become more effective antigen-presenting cells, and their role is considered in more detail in Chapter 8. One of their effects is to induce the production of inflammatory cytokines and potent local inflammatory responses; this effect is probably intrinsic to their activity in enhancing responses, but precludes their use in humans. Nevertheless, some human vaccines contain microbial antigens

Fig. 2.4 Common adjuvants and their use. Adjuvants are mixed with the antigen and usually render it particulate, which helps to retain the antigen in the body and promotes uptake by macrophages. Most adjuvants include bacteria or bacterial components that stimulate macrophages, aiding in the induction of the immune response. ISCOMs (immune stimulatory complexes) are small micelles of the detergent Quil A; when viral proteins are placed in these micelles, they apparently fuse with the antigen-presenting cell, allowing the antigen to enter the cytosol. Thus, the antigen-presenting cell can stimulate a response to the viral protein, much as a virus infecting these cells would stimulate an anti-viral response.

Adjuvants that enhance immune responses

Adjuvant name	Composition	Mechanism of action
Incomplete Freund's adjuvant	Oil-in-water emulsion	Delayed release of antigen; enhanced uptake by macrophages
Complete Freund's adjuvant	Oil-in-water emulsion with dead mycobacteria	Delayed release of antigen; enhanced uptake by macrophages; induction of co-stimulators in macrophages
Freund's adjuvant with MDP	Oil-in-water emulsion with muramyldipeptide (MDP), a constituent of mycobacteria	Similar to complete Freund's adjuvant
Alum (aluminum hydroxide)	Aluminum hydroxide gel	Delayed release of antigen; enhanced macrophage uptake
Alum plus *Bordetella pertussis*	Aluminum hydroxide gel with killed *B. pertussis*	Delayed release of antigen; enhanced uptake by macrophages; induction of co-stimulators
Immune stimulatory complexes (ISCOMs)	Matrix of Quil A containing viral proteins	Delivers antigen to cytosol; allows induction of cytotoxic T cells

that can also act as effective adjuvants. For example, purified constituents of the bacterium *Bordetella pertussis*, which is the causal agent of whooping cough, are used as both antigen and adjuvant in the triplex DPT (diphtheria, pertussis, tetanus) vaccine against these diseases.

2-5 B-cell responses are detected by antibody production.

B cells contribute to adaptive immunity by secreting antibodies, and the response of B cells to an injected immunogen is usually measured by analyzing the specific antibody produced in a **humoral immune response**. This is most conveniently achieved by assaying the antibody that accumulates in the fluid phase of the blood or **plasma**; such antibodies are known as circulating antibodies. Circulating antibody is usually measured by collecting blood, allowing it to clot, and then isolating the serum from the clotted blood. The amount and characteristics of the antibody in the resulting antiserum are then determined using the assays we shall describe in the next part of this chapter.

The most important characteristics of an antibody response are the specificity, amount, isotype or class, and affinity of the antibodies produced. The specificity determines the ability of the antibody to distinguish the immunogen from other antigens. The amount of antibody can be determined in many different ways and is a function of the number of responding B cells, their rate of antibody synthesis, and the persistence of the antibody after production. The persistence of an antibody in the plasma and extracellular fluid bathing the tissues is determined mainly by its isotype (see Chapter 3); each isotype has a different half-life *in vivo*. The isotypic composition of an antibody response also determines the biological functions these antibodies can perform and the sites in which antibody will be found. Finally, the strength of binding of the antibody to its antigen is termed its **affinity**. Binding strength is important, since the higher the affinity of the antibody for its antigen, the less antibody is required to eliminate the antigen, as antibodies with higher affinity will bind at lower antigen concentrations. All these parameters of the humoral immune response help to determine the capacity of that response to protect the host from infection.

2-6 T-cell responses are detected by their effects on other cells or by the cytokines they produce.

The measurement of antibody responses in humoral immunity is fairly simple; by contrast, immunity that is mediated by T cells, called **cell-mediated immunity**, is technically far more difficult to measure. This is principally because T cells do not make a secreted antigen-binding product, so there is no simple binding assay for their antigen-specific responses. T-cell activity can be divided into an induction phase, in which T cells are activated to divide and differentiate, and an effector phase, in which their function is expressed. Both phases require that the T cell interacts with another cell and that it recognizes specific antigen displayed in the form of peptide:MHC complexes on the surface of this interacting cell. In the induction phase, the interaction must be with an antigen-presenting cell able to deliver co-stimulatory signals, whereas, in the effector phase, the nature of the target cell depends on the type of effector T cell that has been activated. Most commonly, the presence of T cells that have responded to a specific antigen is detected by their subsequent *in vitro* proliferation when re-exposed to the same antigen.

However, T-cell proliferation indicates only that cells able to recognize that antigen have been activated previously; it does not reveal what effector function they mediate. The effector function of a T cell is assayed by its effect on an appropriate target cell. As we learned in Chapter 1, several basic effector functions have been defined for T cells. Cytotoxic CD8 T cells can kill infected target cells, thus preventing further replication of obligate intracellular pathogens, whereas CD4 T cells can activate either B cells or macrophages in ways that are determined largely by the cytokines they produce (see Section 8-17). The different T-cell effector responses that can be elicited by immunization determine the functional outcome of an immune response. However, no general principles that allow one to predict accurately the type of immune response produced by a particular immunization regimen have emerged. The ability to control the type of immune response produced remains a central goal of immunology, as we shall see in Chapters 10–14.

Summary.

Adaptive immunity is studied by eliciting a response through deliberate infection or, more commonly, by injection of antigens in an immunogenic form, and by measuring the outcome in terms of humoral and cell-mediated immunity. Intrinsic properties of the antigen determine its immunogenic potential. However, the elicitation of an immune response is heavily influenced by the dose and route of antigen administration and by the adjuvants used to administer it. The main parameters of the antibody response are the amount, affinity, isotype, and specificity of the antibody produced, the isotype determining the functional capabilities of the humoral immune response to a given antigen. The main parameters of the cell-mediated immune response are the numbers of T cells able to respond and their functional properties.

The measurement and use of antibodies.

Antibody molecules are highly specific for their corresponding antigen, being able to detect one molecule of a protein antigen out of more than 10^8 similar molecules. This makes antibodies both easy to isolate and study, and invaluable as probes of biological processes. Whereas standard chemistry would have great difficulty in distinguishing between two such closely related proteins as human and pig insulin, or two such closely related structures as *ortho*- and *para*-nitrophenyl, antibodies can be made that discriminate between these two structures absolutely. The value of antibodies as molecular probes has stimulated the development of many sensitive and highly specific techniques to measure their presence, to determine their specificity and affinity for a range of antigens, and to ascertain their functional capabilities. Many standard techniques used throughout biology exploit the specificity and stability of antigen binding by antibodies. Comprehensive guides to the conduct of these antibody assays are available in many books on immunological methodology; we shall illustrate here only the most important techniques, especially those used in studying the immune response itself. These examples also illustrate the unique properties of antibody molecules that are explained by their structure and genetic origin, as we shall see in Chapter 3.

2-7 The amount and specificity of an antibody can be measured by its direct binding to antigen.

The presence of specific antibody can be detected by using many different assays. Some measure the direct binding of the antibody to its antigen. Such assays are based on **primary interactions** and we shall describe several in this section. Others determine the amount of antibody present by the changes it induces in the physical state of the antigen, such as the precipitation of soluble antigen or the clumping of antigenic particles; these are called **secondary interactions** and will be described in the next section. Both types of assay can be used to measure the amount and specificity of the antibodies produced after immunization, and both can be applied to a wide range of other biological problems. Here, we shall describe several of these assays that are commonly used in immunology, biology, and medicine. As such assays were originally conducted with sera from immune individuals, or antisera, they are commonly referred to as **serological assays**, and the use of antibodies is often called **serology**. The amount of antibody is usually determined by antigen-binding assays after titration of the antiserum by serial dilution, and the point at which binding falls to 50% of the maximum is usually referred to as the **titer** of an antiserum.

Two commonly used direct binding assays are **radioimmunoassay** (**RIA**) and **enzyme-linked immunosorbent assay** (**ELISA**). For these one needs a pure preparation of a known antigen or antibody, or both. In a radioimmunoassay, a pure component (antigen or antibody) is radioactively labeled, usually with ^{125}I. For the ELISA, an enzyme is linked chemically to the antibody or antigen. The unlabeled component (again either antigen or antibody) is attached to a solid support, such as the wells of a plastic multiwell plate, which will adsorb a certain amount of any protein. Most commonly, the antigen is attached to the solid support and the binding of labeled antibody is assayed. The labeled antibody is allowed to bind to the unlabeled antigen, under conditions where non-specific adsorption is blocked, and any unbound antibody and other proteins are washed away. Antibody binding is measured directly in terms of the amount of radioactivity retained by the coated wells in radioimmunoassay, whereas in ELISA, binding is detected by a reaction that converts a colorless substrate into a colored reaction product (Fig. 2.5). The color change can be read directly in the reaction tray, making data collection very easy, and ELISA also avoids the hazards of radioactivity. This makes ELISA the preferred method for most direct-binding assays.

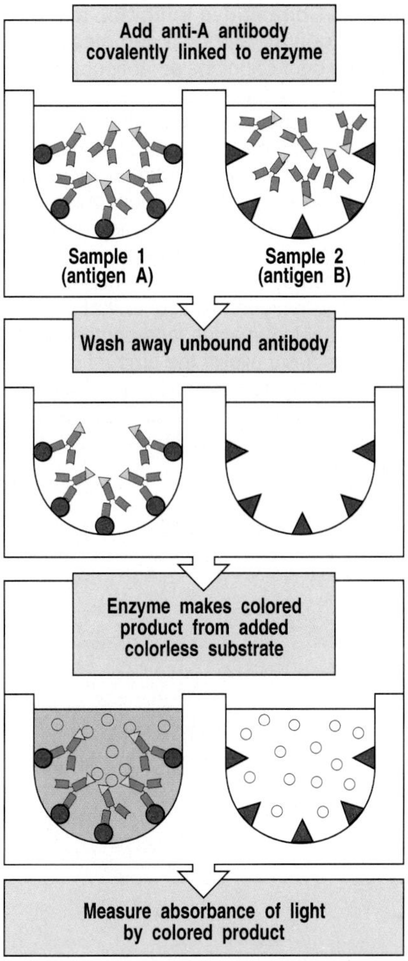

Fig. 2.5 The principle of the enzyme-linked immunosorbent assay (ELISA). To detect antigen A, purified antibody specific for antigen A is linked chemically to an enzyme. The samples to be tested are coated onto the surface of plastic wells to which they bind non-specifically; residual sticky sites on the plastic are blocked by adding irrelevant proteins (not shown). The labeled antibody is then added to the wells under conditions where non-specific binding is prevented, so that only binding to antigen A causes the labeled antibody to be retained on the surface. Unbound labeled antibody is removed from all wells by washing, and bound antibody is detected by an enzyme-dependent color-change reaction. This assay allows arrays of wells known as microtiter plates to be read in fiberoptic multichannel spectrometers, greatly speeding the assay. Modifications of this basic assay allow antibody or antigen in unknown samples to be measured as shown in Figs 2.6 and 2.29 (see also Section 2-9).

These assays illustrate two crucial aspects of all serological assays. First, at least one of the reagents must be available in a pure, detectable form in order to obtain quantitative information. Second, there must be a means of separating the bound fraction of the labeled reagent from the unbound, free fraction so that the percentage of specific binding can be determined. Normally, this separation is achieved by having the unlabeled partner trapped on a solid support. Labeled molecules that do not bind can then be washed away, leaving just the labeled partner that has bound. In Fig. 2.5, the unlabeled antigen is attached to the well and the labeled antibody is trapped by binding to it. The separation of bound from free is an essential step in every assay that uses antibodies.

These assays do not allow one to measure directly the amount of antigen or antibody in a sample of unknown composition, as both depend on the binding of a pure labeled antigen or antibody. There are various ways around this problem, one of which is to use a **competitive inhibition assay**, as shown in Fig. 2.6. In this type of assay, the presence and amount of a particular antigen in an unknown sample is determined by its ability to compete with a labeled reference antigen for binding to an antibody attached to a plastic well. A standard

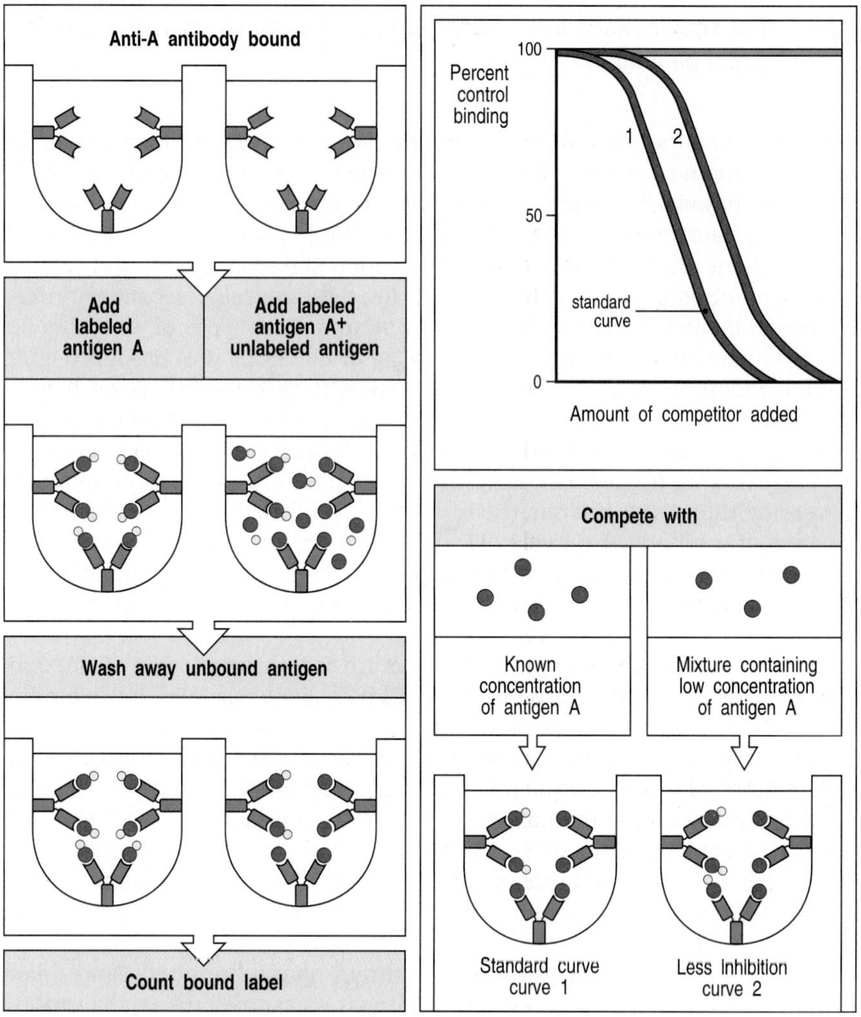

Fig. 2.6 Competitive inhibition assay for antigen in unknown samples. A fixed amount of unlabeled antibody is attached to a set of wells, and a standard reference preparation of a labeled antigen is bound to it. Unlabeled standard or test samples are then added in various amounts and the displacement of labeled antigen is measured, generating characteristic inhibition curves. A standard curve is obtained by using known amounts of unlabeled antigen identical to that used as the labeled species, and comparison with this curve allows the amount of antigen in unknown samples to be calculated. The green line on the graph represents a sample lacking any substance that reacts with anti-A antibodies.

curve is first constructed by adding varying amounts of a known, unlabeled standard preparation; the assay can then measure the amount of antigen in unknown samples by comparison with the standard. The competitive binding assay can also be used for measuring antibody in a sample of unknown composition by attaching the appropriate antigen to the plate and measuring the ability of the test sample to inhibit the binding of a labeled specific antibody.

All the assays described so far rely on pure preparations of an antibody. However, antibody to any one antigen makes up a very small percentage of the total protein in an antiserum, even after repeated immunizations. Therefore, antibody must be purified before it can be labeled. Specific antibody can be isolated from an antiserum by using **affinity chromatography**, which exploits the specific binding of antibody to antigen held on a solid matrix. Antigen is bound covalently to small, chemically reactive beads, which are loaded into a column, and the antiserum is allowed to pass over the beads. The specific antibodies bind, while all the other proteins in the serum, including antibodies to other substances, can be washed away. The specific antibodies are then eluted, typically by lowering the pH to 2.5 or raising it to greater than 11. This demonstrates that antibodies bind stably under physiological conditions of salt concentration, temperature, and pH, but that the bonds are non-covalent because the binding is reversible. Affinity chromatography can also be used to purify antigens from complex mixtures by coating the beads with specific antibody. The technique is known as affinity chromatography because it separates molecules on the basis of their affinity for one another.

2-8 Antibody binding can be detected by changes in the physical state of the antigen.

The direct measurement of antibody binding to antigen is used in most quantitative serological assays. However, some important assays are based on the ability of antibody binding to alter the physical state of the antigen it binds to. These secondary interactions can be detected in a variety of ways. For instance, when the antigen is displayed on the surface of a large particle such as a bacterium, antibodies can cause the bacteria to clump or **agglutinate**. The same principle applies to the reactions used in blood typing, only here the target antigens are on the surface of red blood cells and the clumping reaction caused by antibodies against them is called **hemagglutination** (from the Greek *haima*, blood).

Hemagglutination is used to determine the ABO blood group of blood donors and transfusion recipients. Clumping or agglutination is induced by antibodies or agglutinins called anti-A or anti-B that bind to the A or B blood group substances respectively (Fig. 2.7). These blood-group antigens are arrayed in many copies on the surface of the red blood cell, allowing the cells to agglutinate when crosslinked by antibodies. Because hemagglutination involves the crosslinking of blood cells by the simultaneous binding of antibody molecules to identical antigens on different cells, this reaction demonstrates that each antibody molecule has at least two identical antigen-binding sites.

When sufficient quantities of antibody are mixed with soluble macromolecular antigens, a visible precipitate consisting of large aggregates of antigen crosslinked by antibody molecules can form. The amount of precipitate depends on the amounts of antigen and antibody, and on the ratio between them (Fig. 2.8). This **precipitin reaction** provided the first quantitative assay for antibody but is now seldom used in immunology. However, it is important

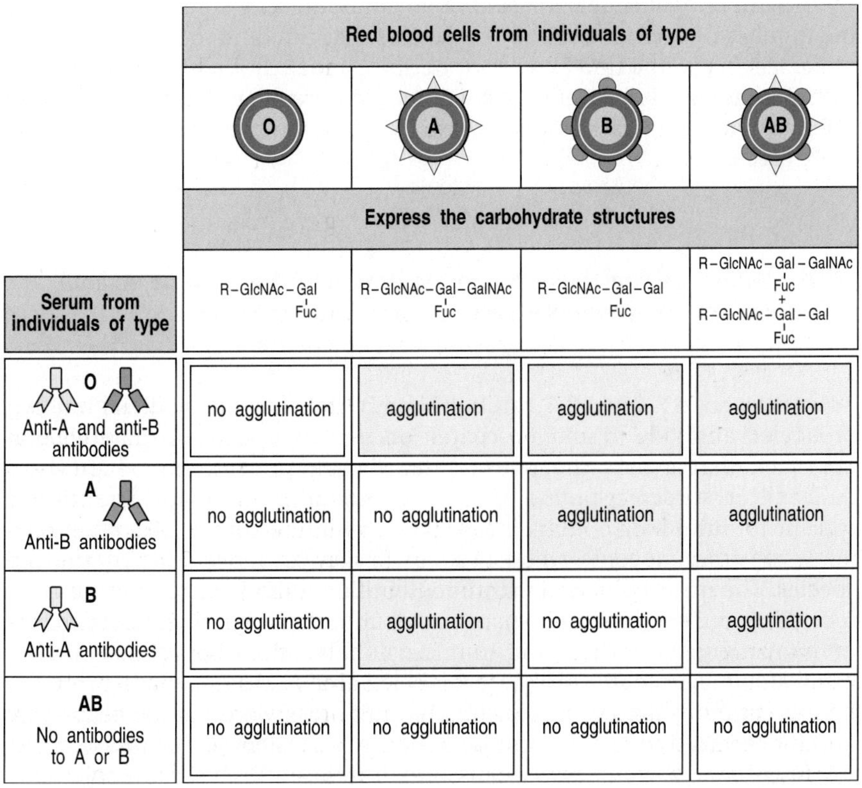

Red blood cells from individuals of type
O
A
B
AB
Express the carbohydrate structures
R–GlcNAc–Gal–Fuc
R–GlcNAc–Gal–GalNAc–Fuc
R–GlcNAc–Gal–Gal–Fuc
R–GlcNAc–Gal–GalNAc–Fuc + R–GlcNAc–Gal–Gal–Fuc
Serum from individuals of type
O Anti-A and anti-B antibodies: no agglutination | agglutination | agglutination | agglutination
A Anti-B antibodies: no agglutination | no agglutination | agglutination | agglutination
B Anti-A antibodies: no agglutination | agglutination | no agglutination | agglutination
AB No antibodies to A or B: no agglutination | no agglutination | no agglutination | no agglutination

Fig. 2.7 Hemagglutination is used to type blood groups and match compatible donors and recipients for blood transfusion. Common gut bacteria bear antigens that are similar or identical to blood group antigens, and these stimulate the formation of antibodies to these antigens in individuals who do not bear the corresponding antigen on their own red blood cells (left column); thus, type O individuals, who lack A and B, have both anti-A and anti-B antibodies, while type AB individuals have neither. The pattern of agglutination of the red blood cells of a transfusion donor or recipient with anti-A and anti-B antibodies reveals the individual's ABO blood group. Before transfusion, the serum of the recipient is also tested for antibodies that agglutinate the red blood cells of the donor, and vice versa, a procedure called a cross-match, which may detect potentially harmful antibodies to other blood groups that are not part of the ABO system.

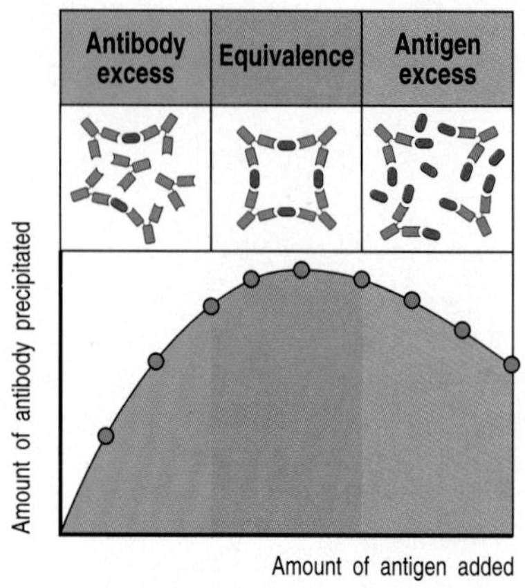

Fig. 2.8 Antibody can precipitate soluble antigen. Analysis of the precipitate can generate a precipitin curve. Different amounts of antigen are added to a fixed amount of antibody, and precipitates form by antibody crosslinking of antigen molecules. The precipitate is recovered and the amount of precipitated antibody measured; the supernatant is tested for residual antigen or antibody. This defines zones of antibody excess, equivalence, and antigen excess. At equivalence, the largest antigen:antibody complexes form. In the zone of antigen excess, some of the immune complexes are too small to precipitate. These soluble immune complexes can cause pathological damage to small blood vessels when they form *in vivo* (see Chapter 13).

to understand the interaction of antigen with antibody that leads to this reaction, as the production of **antigen:antibody complexes**, also known as **immune complexes**, *in vivo* occurs in almost all immune responses and occasionally can cause significant pathology (see Chapters 12 and 13).

In the precipitin reaction, various amounts of soluble antigen are added to a fixed amount of serum containing antibody. As the amount of antigen added increases, the amount of precipitate generated also increases up to a maximum and then declines (see Fig. 2.8). When small amounts of antigen are added, antigen:antibody complexes are formed under conditions of antibody excess so that each molecule of antigen is bound extensively by antibody and crosslinked to other molecules of antigen. When large amounts of antigen are added, only small antigen:antibody complexes can form and these are often soluble in this zone of antigen excess. Between these two zones, all of the antigen and antibody is found in the precipitate, generating a zone of equivalence. At equivalence, very large lattices of antigen and antibody are formed by crosslinking. While all antigen:antibody complexes can potentially produce disease, the small, soluble immune complexes formed in the zone of antigen excess may persist and cause pathology *in vivo*.

The precipitin reaction is affected by the number of binding sites that each antibody has for antigen, and by the maximum number of antibodies that can be bound by an antigen molecule or particle at any one time. These quantities are defined as the **valence** of each antibody and the valence of the antigen: the valence of both the antibodies and the antigen must be two or greater before any precipitation can occur. The valence of an antibody depends on its structural class (see Section 3-20).

Antigen will be precipitated only if it has several antibody-binding sites. This condition is usually satisfied in macromolecular antigens as they have a complex surface to which antibodies of many different specificities can bind. The site on an antigen to which each distinct antibody molecule binds is called an **antigenic determinant** or an **epitope**. Steric considerations limit the number of distinct antibody molecules that can bind to a single antigen molecule at any one time, however, because antibody molecules binding to epitopes that partially overlap will compete for binding. For this reason, the valence of an antigen is almost always less than the number of epitopes on the antigen (Fig. 2.9).

2-9 Anti-immunoglobulin antibodies are a useful tool for detecting bound antibody molecules.

As we learned in Section 2-7, antibody can be detected by the direct binding of labeled antibody to antigen coated on plastic surfaces. A more general approach that avoids the need to label each preparation of antibody molecules is to detect bound, unlabeled antibody with a labeled antibody specific for immunoglobulins themselves. Immunoglobulins, like other proteins, are immunogenic when used to immunize individuals of another species. The majority of **anti-immunoglobulin antibodies** raised in this way recognize conserved features shared by all immunoglobulin molecules of the immunizing species. These anti-immunoglobulin antibodies can be purified using affinity chromatography, then labeled and used as a general probe for bound antibody. Anti-immunoglobulin antibodies were first developed by Robin Coombs to detect the antibodies that cause **hemolytic disease of the newborn**, or **erythroblastosis fetalis**, and the test for this disease is still called the Coombs test. Hemolytic disease of the newborn occurs when a mother makes IgG antibodies specific for the **Rhesus** or **Rh blood group antigen** expressed on the red blood cells of her fetus. Rh-negative mothers make these antibodies when they are exposed to Rh-positive fetal red blood cells bearing the paternally inherited Rh antigen. Maternal IgG antibodies are normally transported across the placenta to the fetus where they protect the newborn infant against infection. However, IgG anti-Rh antibodies coat the fetal red blood cells which are then destroyed by phagocytic cells in the liver, causing a hemolytic anemia in the fetus and newborn infant.

Since the Rh antigens are widely spaced on the red blood cell surface, the IgG anti-Rh antibodies cannot fix complement and cause lysis of red blood cells *in vitro*. Furthermore, for reasons that are not fully understood, antibodies to Rh blood group antigens do not agglutinate red blood cells as do antibodies to the ABO blood group antigens. Thus detecting these antibodies was difficult until anti-human immunoglobulin antibodies were developed. With these, maternal IgG antibodies bound to the fetal red blood cells can be detected after washing the cells to remove unbound immunoglobulin that is present in the fetal serum. Adding anti-human immunoglobulin antibodies to the washed fetal red blood cells agglutinates any cells to which maternal antibodies are bound. This is the **direct Coombs test** (Fig. 2.10), so called because it directly detects antibody bound to the surface of the fetal red blood cells. An **indirect Coombs test** is used to detect non-agglutinating anti-Rh antibody in maternal serum; the serum is first incubated with Rh-positive red blood cells, which bind the anti-Rh antibody, after which the antibody-coated cells are washed to remove unbound immunoglobulin and are then agglutinated with anti-immunoglobulin antibody (see Fig. 2.10). The indirect Coombs test allows Rh incompatibilities that might lead to hemolytic disease of the newborn to be detected and this knowledge allows the disease to be prevented, as we shall see in Chapter 10. The Coombs test is also commonly employed to detect anti bodies to drugs that bind to red blood cells and cause hemolytic anemia.

Anti-immunoglobulin antisera have found many uses in clinical medicine and biological research since their introduction. Labeled anti-immunoglobulin antibodies can be used in radioimmunoassay or ELISA to detect binding of unlabeled antibody to antigen-coated plates. The ability of anti-immunoglobulins to react with antibodies of all specificities demonstrates that antibody molecules have constant features recognizable by the anti-immunoglobulin, in addition to the variability required for antibodies to discriminate between a myriad of antigens. The presence of both constant and variable features in one protein posed a genetic puzzle for immunologists, the solution to which is described in Chapter 3. Some anti-immunoglobulin antibodies made in rabbits react with only a subset of human immunoglobulin molecules. It was this property of anti-immunoglobulin antibodies that led to the discovery that several distinct sets of antibodies, the immunoglobulin classes or isotypes, are present in human serum (see Chapters 3 and 11).

Anti-immunoglobulins specific for each isotype can be produced by immunizing an animal of a different species with a pure preparation of one isotype and then removing those antibodies that cross-react with immunoglobulins of other isotypes by using affinity chromatography. Anti-isotype antibodies can be used to measure how much antibody of a particular isotype in an antiserum reacts with a given antigen. This reaction is particularly important for detecting small amounts of specific antibodies of the IgE isotype, which are responsible for most allergies. The presence in an individual's serum of IgE binding to an antigen correlates with allergic reactions to that antigen.

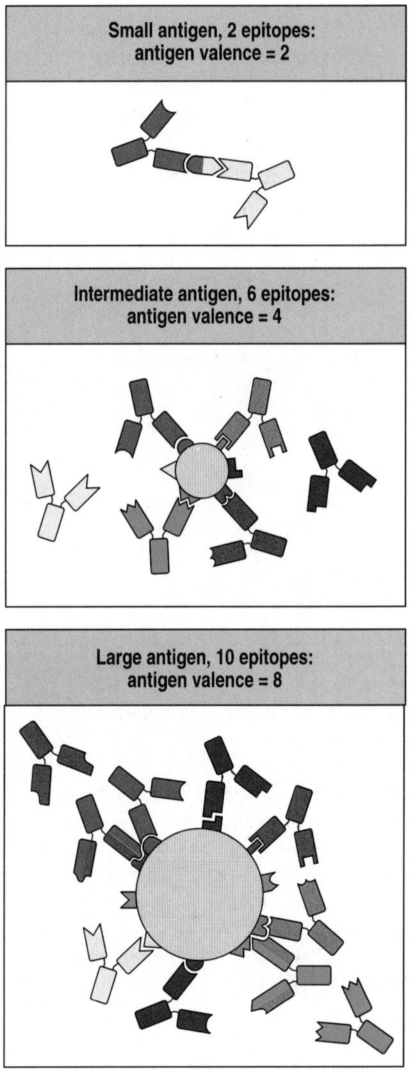

Fig. 2.9 Different antibodies bind to distinct epitopes on an antigen molecule. The surface of an antigen possesses many potential antigenic determinants or epitopes, distinct sites to which an antibody can bind. The number of antibody molecules that can bind to a molecule of antigen at one time defines the antigen's valence. Steric considerations can limit the number of different antibodies that bind to the surface of an antigen at any one time (center and bottom panels) so that the number of epitopes on an antigen is always greater than or equal to its valence.

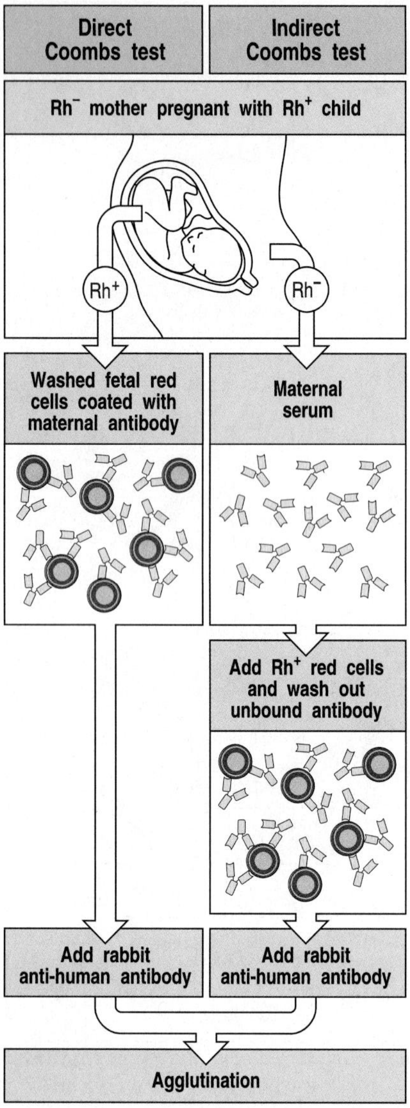

Fig. 2.10 The Coombs direct and indirect anti-globulin tests for antibody to red blood cell antigens. A Rh^- mother of a Rh^+ fetus can become immunized to fetal red blood cells that enter the maternal circulation at the time of delivery. In a subsequent pregnancy with a Rh^+ fetus, IgG anti-Rh antibodies can cross the placenta and damage the fetal red blood cells. In contrast to anti-Rh antibodies, maternal anti-ABO antibodies are of the IgM isotype and cannot cross the placenta, and so do not cause harm. Anti-Rh antibodies do not agglutinate red blood cells but their presence on the fetal red cell surface can be shown by washing away unbound immunoglobulin and then adding antibody to human immunoglobulin, which agglutinates the antibody-coated cells. Anti-Rh antibodies can be detected in the mother's serum in an indirect Coombs test; the serum is incubated with Rh^+ red blood cells, and once the antibody binds, the red cells are treated as in the direct Coombs test.

An alternative approach to detecting bound antibodies exploits bacterial proteins that bind to immunoglobulins with high affinity and specificity. One of these, **Protein A** from the bacterium *Staphylococcus aureus*, has been exploited widely in immunology for the affinity purification of immunoglobulin and for the detection of bound antibody. The use of standard second reagents such as labeled anti-immunoglobulin or Protein A to detect antibody bound specifically to its antigen allows great savings in reagent labeling costs, and also provides a standard detection system so that results in different assays can be compared directly.

2-10 Antisera contain heterogeneous populations of antibody molecules, while monoclonal antibodies are homogeneous molecules having a single specificity.

The antibodies generated in a natural immune response or after immunization in the laboratory are a mixture of molecules of different specificities and affinities. Some of this heterogeneity results from the production of antibodies that bind to different epitopes on the immunizing antigen, but even antibodies directed at a single antigenic determinant such as a hapten can be markedly heterogeneous, as shown by **isoelectric focusing**. In this technique, proteins are separated on the basis of their isoelectric point, the pH at which their net charge is zero. By electrophoresing proteins in a pH gradient for long enough, each molecule migrates along the pH gradient until it reaches the pH at which it is neutral and is thus concentrated (focused) at that point. When antiserum containing anti-hapten antibodies is treated in this way and then transferred to a solid support such as nitrocellulose paper, the anti-hapten antibodies can be detected by their ability to bind labeled hapten. The binding of antibodies of various isoelectric points to the hapten shows that even antibodies that bind the same antigenic determinant can be heterogeneous.

Antisera are valuable for many biological purposes but they have certain inherent disadvantages that relate to the heterogeneity of the antibodies they contain. First, each antiserum is different from all other antisera, even if raised in a genetically identical animal by using the identical preparation of antigen and the same immunization protocol. Second, antisera can be produced in only limited volumes, and thus it is impossible to use the identical serological reagent in a long or complex series of experiments or clinical tests. Finally, even antibodies purified by affinity chromatography (see Section 2-13) may include minor populations of antibodies that give unexpected cross-reactions, which confound the analysis of experiments. To avoid these problems, and to harness the full potential of antibodies, it was necessary to develop a way of making an unlimited supply of antibody molecules of homogeneous structure and known specificity. This has been achieved through the production of monoclonal antibodies from hybrid antibody-forming cells or, more recently, by genetic engineering, as we shall see next.

Biochemists in search of a homogeneous preparation of antibody that they could subject to detailed chemical analysis turned early to proteins produced by patients with multiple myeloma, a common tumor of plasma cells. It was known that antibodies are normally produced by plasma cells and since this disease is associated with the presence of large amounts of a homogeneous gamma globulin called a **myeloma protein** in the patient's serum, it seemed likely that myeloma proteins would serve as models for normal antibody molecules. Thus, much of the early knowledge of antibody structure came from studies on myeloma proteins. These studies showed that monoclonal antibodies could be obtained from immortalized plasma cells. However, the antigen specificity of most myeloma proteins was unknown, which limited their usefulness as objects of study or as immunological tools.

This problem was solved by Georges Köhler and Cesar Milstein, who devised a technique for producing a homogeneous population of antibodies of known antigenic specificity. They did this by fusing spleen cells from an immunized mouse to cells of a mouse myeloma to produce hybrid cells that both proliferated indefinitely and secreted antibody specific for the antigen used to immunize the spleen cell donor. The spleen cell provides the ability to make specific antibody, while the myeloma cell provides the ability to grow indefinitely in culture and secrete immunoglobulin continuously. By using a myeloma cell partner that produces no antibody proteins itself, the antibody produced by the hybrid cells comes only from the immune spleen cell partner. After fusion, the hybrid cells are selected using drugs that kill the myeloma parental cell, while the unfused parental spleen cells have a limited lifespan and soon die, so that only hybrid myeloma cell lines or **hybridomas** survive. Those hybridomas producing antibody of the desired specificity are then identified and cloned by regrowing the cultures from single cells (Fig. 2.11). Since each hybridoma is a **clone** derived from fusion with a single B cell, all the antibody molecules it produces are identical in structure, including their antigen-binding site and isotype. Such antibodies are therefore called **monoclonal antibodies**. This technology has revolutionized the use of antibodies by providing a limitless supply of antibody of a single and

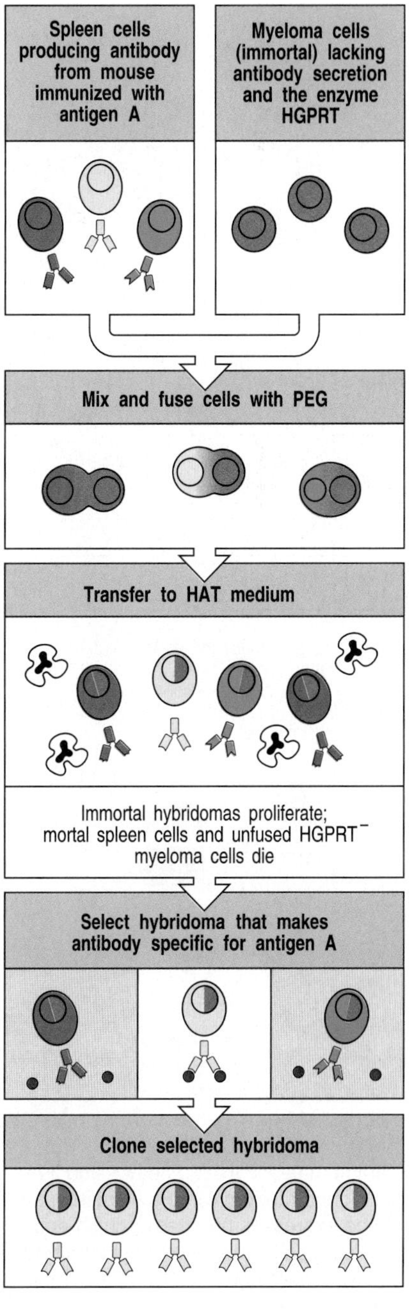

Fig. 2.11 The production of monoclonal antibodies. Mice are immunized with antigen A and given an intravenous booster immunization three days before they are killed, in order to produce a large population of spleen cells secreting specific antibody. Spleen cells die after a few days in culture. In order to produce a continuous source of antibody they are fused with immortal myeloma cells by using polyethylene glycol (PEG) to produce a hybrid cell line called a hybridoma. The myeloma cells are selected beforehand to ensure that they are not secreting antibody themselves and that they are sensitive to the hypoxanthine-aminopterin-thymidine (HAT) medium that is used to select hybrid cells because they lack the enzyme hypoxanthine: guanine phosphoribosyl transferase (HGPRT). The HGPRT gene contributed by the spleen cell allows hybrid cells to survive in the HAT medium, and only hybrid cells can grow continuously in culture because of the malignant potential contributed by the myeloma cells. Therefore, unfused myeloma cells and unfused spleen cells die in the HAT medium, as shown here by cells with dark, irregular nuclei. Individual hybridomas are then screened for antibody production, and cells that make antibody of the desired specificity are cloned by growing them up from a single antibody-producing cell. The cloned hybridoma cells are grown in bulk culture to produce large amounts of antibody. As each hybridoma is descended from a single cell, all the cells of a hybridoma cell line make the same antibody molecule, which is thus called a monoclonal antibody.

known specificity. Monoclonal antibodies are now used in most serological assays, as diagnostic probes, and as therapeutic agents. So far, however, only mouse monoclonals are routinely produced and efforts to use this same approach to make human monoclonal antibodies have met with very limited success.

Recently, a novel technique for producing antibody-like molecules has been introduced. Gene segments encoding the antigen-binding variable or V domains of antibodies are fused to genes encoding the coat protein of a bacteriophage. Bacteriophage containing such gene fusions are used to infect bacteria, and the resulting phage particles have coats that express the antibody-like fusion protein, with the antigen-binding domain displayed on the outside of the bacteriophage. A collection of recombinant phage, each displaying a different antigen-binding domain on its surface, is known as a **phage display library**. In much the same way that antibodies specific for a particular antigen can be isolated from a complex mixture by affinity chromatography (see Section 2-13), phage expressing antigen-binding domains specific for a particular antigen can be isolated by selecting the phage in the library for binding to that antigen. The phage particles that bind are recovered and used to infect fresh bacteria. Each phage isolated in this way will produce a monoclonal antigen-binding particle analogous to a monoclonal antibody (Fig. 2.12). The genes encoding the antigen-binding site, which are unique to each phage, can then be recovered from the phage DNA and used to construct genes for a complete antibody molecule by joining them to parts of immunoglobulin genes that encode the invariant parts of an antibody. When these reconstructed antibody genes are introduced into a suitable host cell line, such as the non-antibody-producing myeloma cells used for hybridomas, the transfected cells can secrete antibodies with all the desirable characteristics of monoclonal antibodies produced from hybridomas.

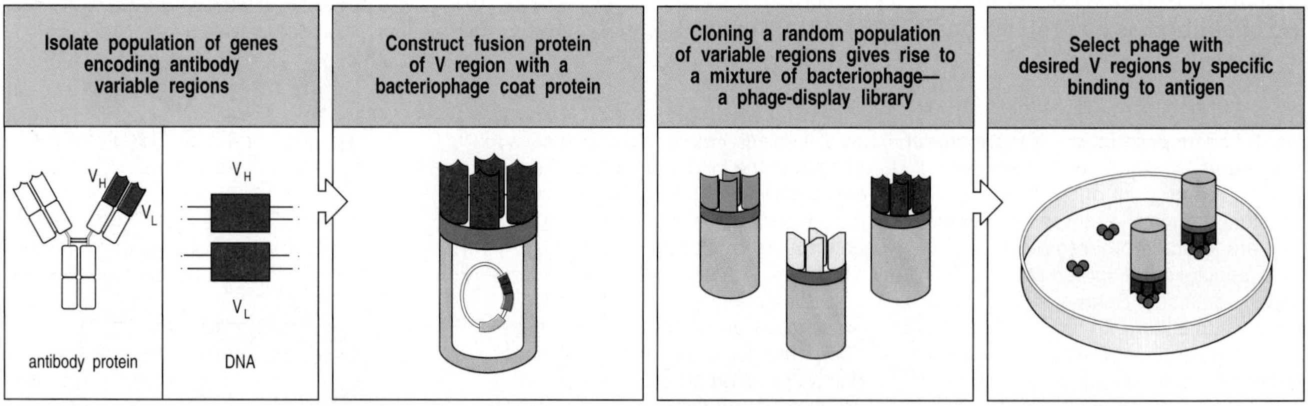
Isolate population of genes encoding antibody variable regions
Construct fusion protein of V region with a bacteriophage coat protein
Cloning a random population of variable regions gives rise to a mixture of bacteriophage—a phage-display library
Select phage with desired V regions by specific binding to antigen
V_H
V_L
antibody protein
DNA

Fig. 2.12 The production of antibodies by genetic engineering. Short primers to consensus sequences in heavy- and light-chain variable (V) regions of immunoglobulin genes are used to generate a library of heavy- and light-chain V-region DNAs by the polymerase chain reaction (see Fig. 2.30), with spleen DNA as the starting material. These heavy- and light-chain V-region genes are cloned randomly into a filamentous phage such that each phage expresses one heavy-chain and one light-chain V region as a surface fusion protein with antibody-like properties. The resulting phage display library is multiplied in bacteria, and the phage are then bound to a surface coated with antigen. The unbound phage are washed away; the bound phage are recovered, multiplied in bacteria, and again bound to antigen. After a few cycles, only specific high-affinity antigen-binding phage are left. These can be used like antibody molecules, or their V genes can be recovered and engineered into antibody genes to produce genetically engineered antibody molecules (not shown). This technology may replace the hybridoma technology for producing monoclonal antibodies and has the advantage that humans can be used as the source of DNA.

2-11 The affinity of an antibody can be determined directly by measuring binding to small monovalent ligands.

The **affinity** of an antibody is the strength of binding of a monovalent ligand to a single antigen-binding site. The affinity of an antibody that binds small antigens, such as haptens that can diffuse freely across a dialysis membrane, can be determined directly by the technique of **equilibrium dialysis**. A known amount of antibody, whose molecules are too large to cross a dialysis membrane, is placed in a dialysis bag and offered various amounts of antigen. Molecules of antigen that bind to the antibody are no longer free to diffuse across the dialysis membrane, so only the unbound molecules of antigen equilibrate across it. By measuring the concentration of antigen inside the bag and in the surrounding fluid, one can determine the amount of the antigen that is bound as well as the amount that is free when equilibrium has been achieved. Given that the amount of antibody present is known, the affinity of the antibody and the number of specific binding sites for the antigen per molecule of antibody can be determined from this information. The data are usually analyzed using **Scatchard analysis** (Fig. 2.13); such analyses were used to demonstrate that a molecule of IgG antibody has two identical antigen-binding sites.

Whereas affinity measures the strength of binding of an antigenic determinant to a single antigen-binding site, an antibody reacting with an antigen that has multiple identical epitopes or with the surface of a pathogen will

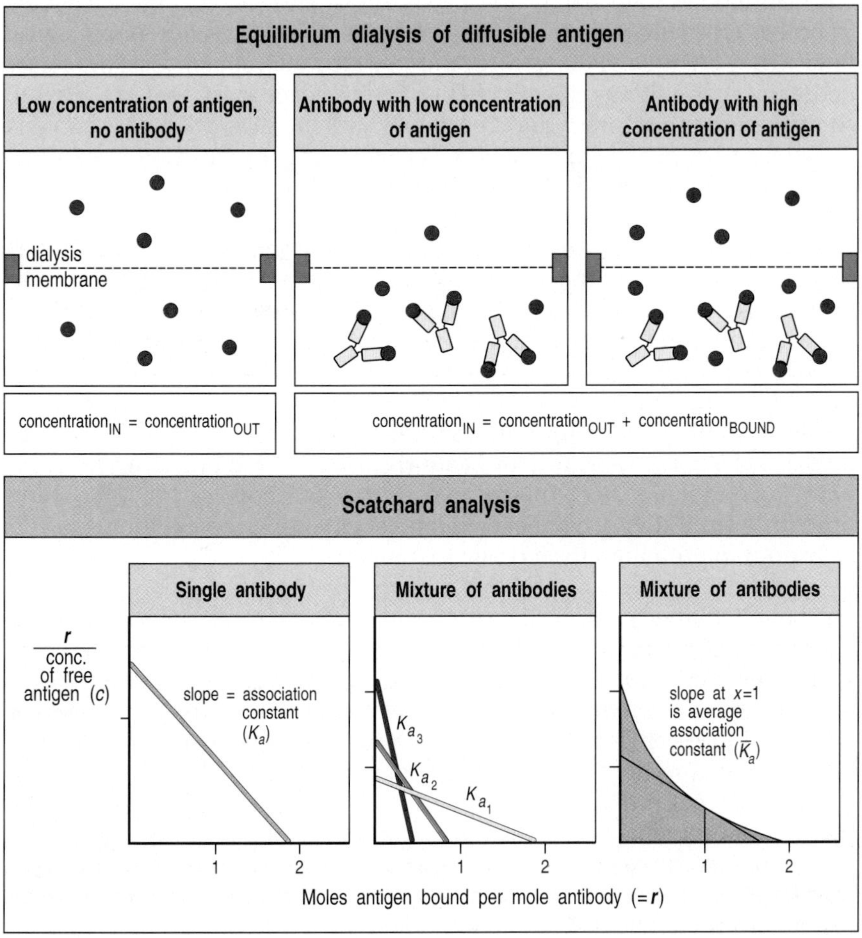

Fig. 2.13 The affinity and valence of an antibody can be determined by equilibrium dialysis. A known amount of antibody is placed in the bottom half of a dialysis chamber and exposed to different amounts of a diffusible monovalent antigen, such as a hapten. At equilibrium, the concentration of free antigen will be the same on each side of the membrane, so that at each concentration of antigen added, the fraction of the antigen bound is determined from the difference in concentration of total antigen in the top and bottom chambers. This information can be transformed into a Scatchard plot as shown here. In Scatchard analysis, the ratio r/c (where r = moles of antigen bound per mole of antibody and c = molar concentration of free antigen) is plotted against r. The number of binding sites per antibody molecule can be determined from the value of r at infinite free-antigen concentration, where r/free = 0, in other words at the x-axis intercept. The analysis of a monoclonal IgG antibody molecule, in which there are two identical antigen-binding sites per molecule, is shown in the left panel. The slope of the line is determined by the affinity of the antibody molecule for its antigen; if all the antibody molecules in a preparation are identical, as for this monoclonal antibody, then a straight line is obtained whose slope is equal to $-K_a$, where K_a is the association (or affinity) constant and the dissociation constant $K_d = 1/K_a$. However, antiserum raised even against a simple antigenic determinant such as a hapten contains a heterogeneous population of antibody molecules (see Section 2-10). Each antibody molecule would, if isolated, make up part of the total and give a straight line whose x-axis intercept is less than two, as this antibody molecule contains only a fraction of the total binding sites in the population (middle panel). As a mixture, they give curved lines with an x-axis intercept of two for which an average affinity ($\bar{K}_a$) can be determined from the slope of this line at a concentration of antigen where 50% of the sites are bound, or at $x = 1$ (right panel). The association constant determines the equilibrium state of the reaction Ag + Ab = Ag:Ab, where antigen = Ag and antibody = Ab, and K_a = [Ag:Ab]/[Ag][Ab]. This constant reflects the 'on' and 'off' rates for antigen binding to the antibody; with small antigens such as haptens, binding is usually as rapid as diffusion allows, whereas differences in off rates determine the affinity constant. However, with larger antigens the 'on' rate may also vary as the interaction becomes more complex.

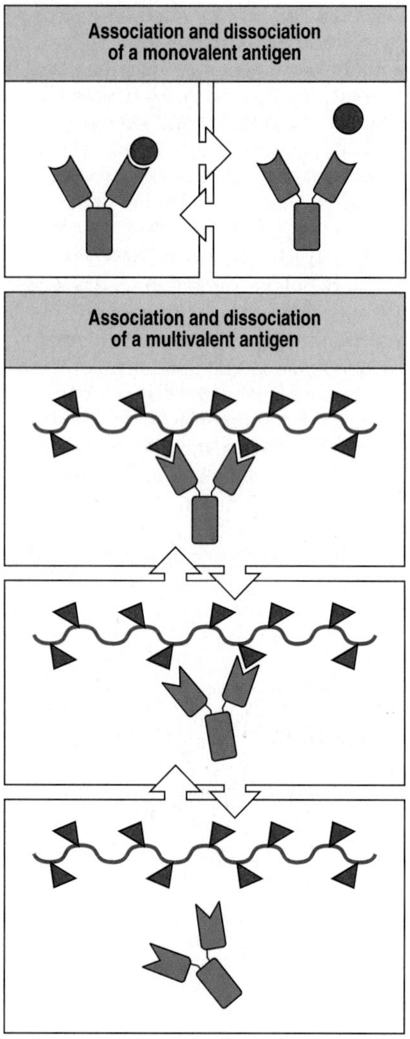

Fig. 2.14 The avidity of an antibody is its strength of binding to intact antigen. When an IgG antibody binds a ligand with multiple identical epitopes, both binding sites can bind the same molecule or particle. The overall strength of binding, called avidity, is greater than the affinity, the strength of binding of a single site, since both binding sites must dissociate at the same time for the antibody to release the antigen. This property is very important in the binding of antibody to bacteria, which usually have multiple identical epitopes on their surfaces.

often bind the same molecule or particle with both of its antigen-binding sites. This increases the apparent strength of binding, since both binding sites must release at the same time in order for the two molecules to dissociate. This is often referred to as **cooperativity** in binding, but it should not be confused with the cooperative binding found in a protein such as hemoglobin in which binding of ligand at one site enhances the affinity of a second binding site for its ligand. The overall strength of binding of an antibody molecule to an antigen or particle is called its **avidity** (Fig. 2.14). For IgG antibodies, bivalent binding can significantly increase avidity; in IgM antibodies, which have ten identical antigen-binding sites, the affinity of each site for a monovalent antigen is usually quite low, but the avidity of binding of the whole antibody to a surface such as a bacterium that displays multiple identical epitopes can be very high.

2-12 Antibodies can be used to identify antigen in cells, tissues, and complex mixtures of substances.

Since antibodies bind stably and specifically to antigen, they are invaluable as probes for identifying a particular molecule in cells, tissues, or biological fluids. They are used in this way to study a wide range of biological processes and clinical conditions. In this section, a few techniques that are used to study the immune system, and in cell biology generally, will be described; a complete treatment of this subject can be found in any of the excellent methodology books available (see list at end of Chapter 2).

Antibody molecules can be used to locate their target molecules accurately in single cells or in tissue sections by a variety of different labeling techniques. As in all serological tests, the antibody binds stably to its antigen, allowing unbound antibody to be removed by thorough washing. As antibodies to proteins recognize the surface features of the native, folded protein, the native structure of the protein being sought usually needs to be preserved, either by using only the most gentle chemical fixation techniques or by using frozen tissue sections that are fixed only after the antibody reaction has been performed. Some antibodies, however, bind proteins even if they are denatured, and such antibodies will bind specifically even to protein in fixed tissue sections.

The bound antibody can be visualized with a variety of sensitive techniques, and the specificity of antibody binding coupled to sensitive detection provides remarkable detail about the structure of cells. One very powerful technique for identifying antibody-bound molecules in cells or tissue sections is **immunofluorescence**, in which a fluorescent dye is covalently attached directly to the specific antibody. More commonly, bound antibody is detected by fluorescent anti-immunoglobulin, a technique known as **indirect immunofluorescence**. The dyes chosen for immunofluorescence are excited by light of one wavelength, usually blue or green, and emit light of a different wavelength in the visible spectrum. By using selective filters, only the light coming from the dye or fluorochrome used is detected in the fluorescence

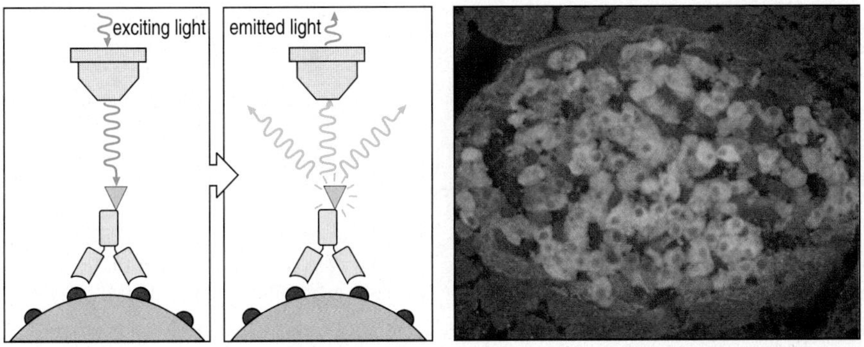

Fig. 2.15 Immunofluorescence microscopy. Antibodies labeled with a fluorescent dye such as fluorescein (green triangle) are used to reveal the presence of their corresponding antigens in cells or tissues. The stained cells are examined in a microscope that exposes them to blue or green light to excite the fluorescent dye. The excited dye emits light at a characteristic wavelength, which is captured by viewing the sample through a selective filter. This technique is applied widely in biology to determine the location of molecules in cells and tissues. Different antigens can be detected in tissue sections by labeling antibodies with dyes of distinctive color. Here, antibodies to the protein glutamic acid decarboxylase (GAD) coupled to a green dye are shown to stain the β cells of pancreatic islets of Langerhans. The α cells do not make this enzyme and are labeled with antibodies to the hormone glucagon coupled with an orange fluorescent dye. GAD is an important antigen in diabetes, a disease in which the insulin-secreting β cells of the islets of Langerhans are destroyed by an immune attack on self tissues (see Chapter 13). Photograph courtesy of M Solimena and P De Camilli.

microscope (Fig. 2.15). Although Albert Coons first devised this technique to identify the plasma cell as the source of antibody, it can be used to detect the distribution of any protein. By attaching different dyes to different antibodies, the distribution of two or more molecules can be determined in the same cell or tissue section (see Fig. 2.15).

An alternative method of detecting a protein in tissue sections is to use **immunohistochemistry**, in which the antibody is chemically coupled to an enzyme that converts a colorless substrate into a colored reaction product whose deposition can be directly observed under a light microscope. This technique is analogous to the ELISA assay described in Section 2-7.

The recent development of the confocal fluorescent microscope, which uses computer-aided techniques to produce an ultrathin optical section of a cell or tissue, gives very high resolution immunofluorescence microscopy without the need for elaborate sample preparation. The resolution of the confocal microscope can be further increased using low intensity illumination so that two photons are required to excite the fluorochrome. A pulsed laser beam is used and only when it is focused into the focal plane of the microscope is the intensity sufficient to excite fluorescence. In this way the fluorescence emission itself can be restricted to the optical section. To examine cells at even higher resolution, antibodies labeled with gold particles can be applied to ultrathin sections examined in the transmission electron microscope. Antibodies labeled with gold particles of distinctive diameters enable two or more proteins to be studied simultaneously. The difficulty with this technique is in staining the ultrathin section, as few molecules of antigen will be present in each section.

In order to raise antibodies to membrane proteins and other cellular structures that are difficult to purify, mice are often immunized with whole cells or crude cell extracts. Antibodies to the individual molecules are then obtained by preparing monoclonal antibodies that bind to the cell used for immunization. To characterize the molecules identified by these antibodies, cells of the same type are labeled with radioisotopes and dissolved in non-ionic detergents that disrupt cell membranes but do not interfere with antigen:antibody interactions. This allows the labeled protein to be isolated by binding to the antibody. The antibody is usually attached to a solid support, such as the beads that are used in affinity chromatography (see Section 2-13) or to protein A. Cells can be labeled in two main ways for this **immunoprecipitation analysis.** All of the proteins in a cell can be labeled metabolically by growing the cell in radioactive amino acids that are incorporated into cellular protein (Fig. 2.16). Alternatively, one can label only the cell-surface proteins by radioiodination under conditions that prevent iodine from crossing the plasma membrane and labeling proteins inside the cell, or by a reaction that labels only membrane proteins with biotin, a small molecule that is detected readily by labeled avidin, a protein found in egg whites that binds biotin with very high affinity.

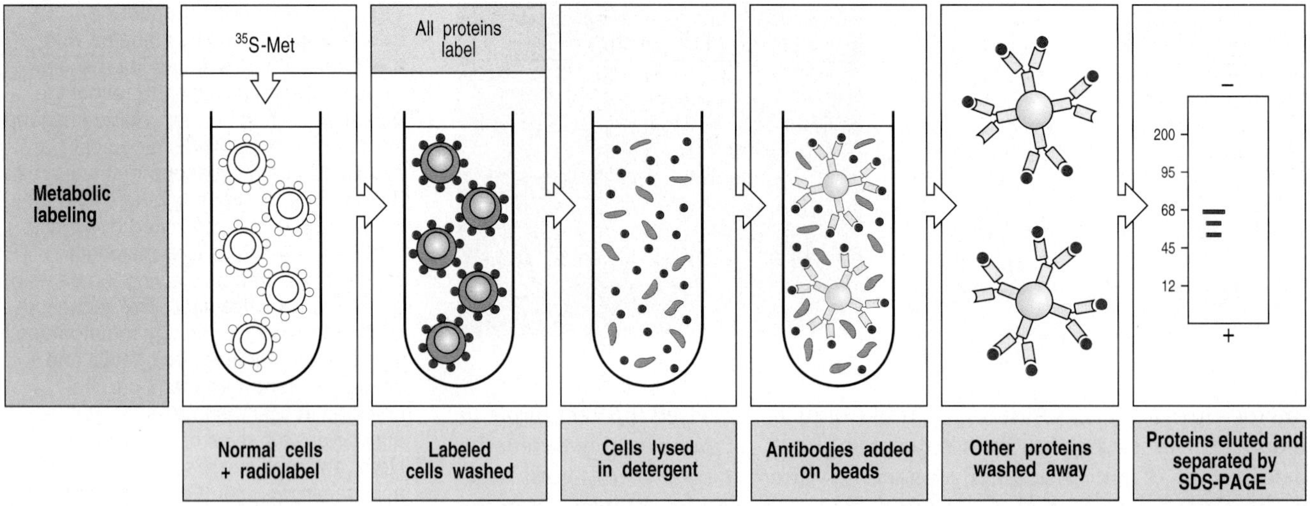

Fig. 2.16 Cellular proteins reacting with an antibody can be characterized by immunoprecipitation of labeled cell lysates. All actively synthesized cellular proteins can be labeled metabolically by incubating cells with radioactive amino acids (shown here for methionine) or one can label just the cell-surface proteins by using radioactive iodine in a form that cannot cross the cell membrane or by a reaction with the small molecule biotin, detected by its reaction with labeled avidin (not shown). Cells are lysed with detergent and individual labeled cell-associated proteins can be precipitated with a monoclonal antibody attached to beads. After unbound proteins have been washed away, the bound protein is eluted in the detergent sodium dodecyl sulfate (SDS), which dissociates it from the antibody and also coats the protein with a strong negative charge, allowing it to migrate according to its size in polyacrylamide gel electrophoresis (PAGE). The positions of the labeled proteins are determined by autoradiography using X-ray film. This technique of SDS-PAGE can be used to determine the molecular weight and subunit composition of a protein. Patterns of protein bands observed with metabolic labeling are usually more complex than those revealed by radioiodination, owing to the presence of precursor forms of the protein (right panel). The mature form of a surface protein can be identified as being the same size as that detected by surface iodination or biotinylation (not shown).

Once the labeled proteins have been isolated by the antibody, they can be characterized in several ways. The most common is polyacrylamide gel electrophoresis (PAGE) of the proteins after dissociating them from antibody in the strong ionic detergent, sodium dodecyl sulfate (SDS), a technique generally abbreviated as **SDS-PAGE**. SDS binds relatively homogeneously to proteins, conferring a charge that allows the electrophoretic field to drive protein migration through the gel. The rate of migration is controlled mainly by protein size (see Fig. 2.16). Proteins of differing charges can be separated using isoelectric focusing (see Section 2-10). This technique can be combined with SDS-PAGE in a procedure known as **two-dimensional gel electrophoresis**. For this, the immunoprecipitated protein is eluted in urea, a non-ionic solubilizing agent, and run on an isoelectric focusing gel in a narrow tube of polyacrylamide. This first-dimensional isoelectric focusing gel is then placed across the top of an SDS-PAGE slab gel, which is then run vertically to separate the proteins by molecular weight (Fig. 2.17). Two-dimensional gel

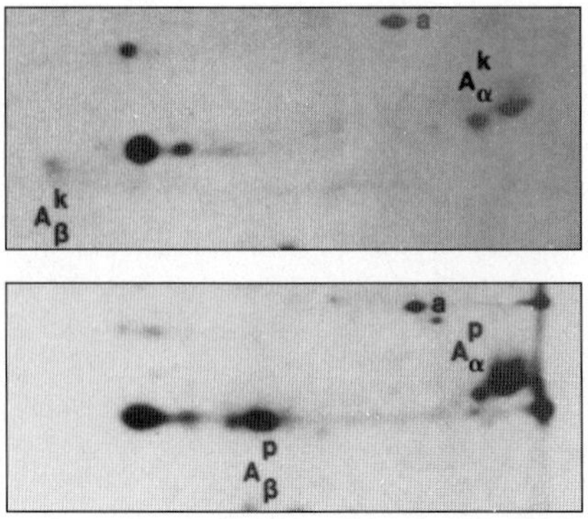

Fig. 2.17 Two-dimensional gel electrophoresis of MHC class II molecules. Proteins in mouse spleen cells have been labeled metabolically (see Fig. 2.16), precipitated with a monoclonal antibody against the mouse MHC class II molecule H2-A, and separated by isoelectric focusing in one direction and SDS-PAGE in a second direction at right angles to the first; hence the term two-dimensional gel electrophoresis. This allows one to distinguish molecules of the same molecular weight on the basis of their charge. The separated proteins are detected by autoradiography. The MHC class II molecules are composed of two chains, α and β, and in the different MHC class II molecules these have different isoelectric points (compare upper and lower panels). The MHC genotype of mice is indicated by lower case superscripts (k,p). Actin, a common contaminant, is marked a. Photographs courtesy of J F Babich.

Fig. 2.18 Western blotting is used to identify antibodies to the human immunodeficiency virus (HIV) in serum from infected individuals. The virus is dissociated into its constituent proteins by treatment with the detergent SDS, and its proteins are separated using SDS-PAGE. The separated proteins are transferred to a nitrocellulose sheet and reacted with the test serum. Anti-HIV antibodies in the serum bind to the various HIV proteins and are detected with enzyme-linked anti-human immunoglobulin, which deposits colored material from a colorless substrate. This general methodology will detect any combination of antibody and antigen and is used widely, although the denaturing effect of SDS means that the technique works most reliably with antibodies that recognize the antigen when it is denatured.

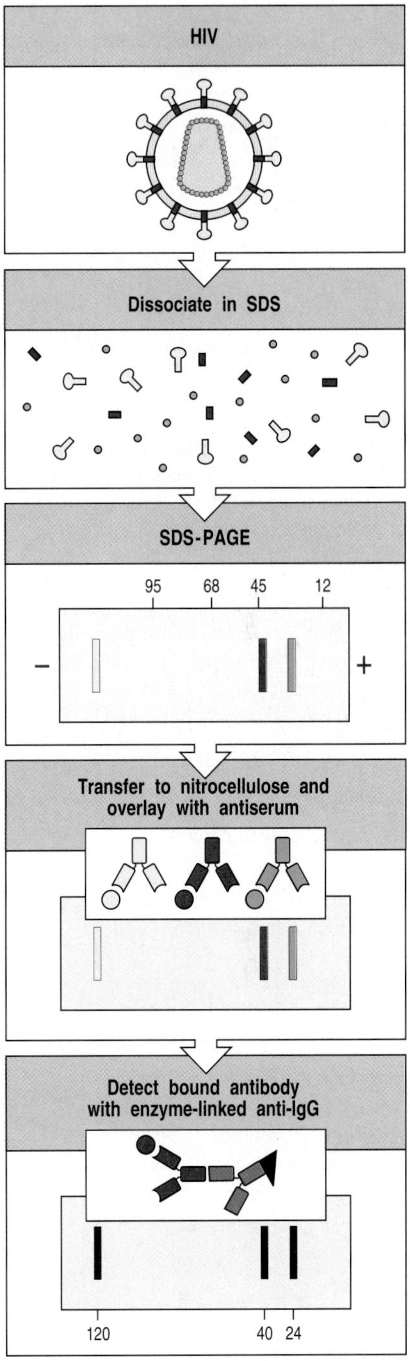

electrophoresis is a powerful technique that allows many hundreds of proteins in a complex mixture to be distinguished from one another.

An alternative approach that avoids the problem of radiolabeling cells is to solubilize all cellular proteins by placing unlabeled cells directly in detergent and running the lysate on SDS-PAGE. The size-separated proteins are then transferred from the gel to a stable support such as nitrocellulose paper. Specific proteins are detected by antibodies able to react with SDS-solubilized proteins (mainly those that react with denatured sequences) and their positions revealed by anti-immunoglobulin that is labeled with radioisotopes or an enzyme. This procedure is called **immunoblotting** or **Western blotting**. (The latter term arose because the comparable technique for detecting specific DNA sequences is known as Southern blotting, after Ed Southern who devised it, which in turn provoked the name Northern for blots of size-separated RNA, and Western for blots of size-separated proteins.) Western blots are used in many applications in basic research and clinical diagnosis. They are often used to test sera for the presence of antibodies to specific proteins, for example to detect antibodies to different constituents of the human immunodeficiency virus, HIV (Fig. 2.18).

2-13 Antibodies can be used to identify genes and their products.

Immunoprecipitation and immunoblotting are useful for determining the molecular weight and isoelectric point of a protein as well as its abundance, distribution, and whether, for example, it undergoes changes in molecular weight and isoelectric point as a result of processing within the cell. However, these techniques do not provide a definitive characterization of the protein. More and more commonly, the full description of a protein is derived from the DNA that encodes the protein. Antibodies specific for the protein can be used to isolate the purified protein using affinity chromatography (Fig. 2.19). Small amounts of amino acid sequence can then be obtained from the protein's amino-terminal end or from peptide fragments generated by proteolysis. The information in these amino acid sequences is used to construct a set of synthetic oligonucleotides corresponding to the possible DNA sequences, which are then used as probes to isolate the gene encoding the protein from either a library of DNA sequences complementary to mRNA (a cDNA library) or a genomic DNA library.

An alternative approach uses antibodies directly to identify and isolate a gene encoding a cell-surface protein. A specific antibody is used to detect the expression of the protein on the surface of a cell type that does not normally express it, after the cell has been transfected with the gene in the form of a cDNA. A suitable cDNA library is prepared from total mRNA isolated from a cell type known

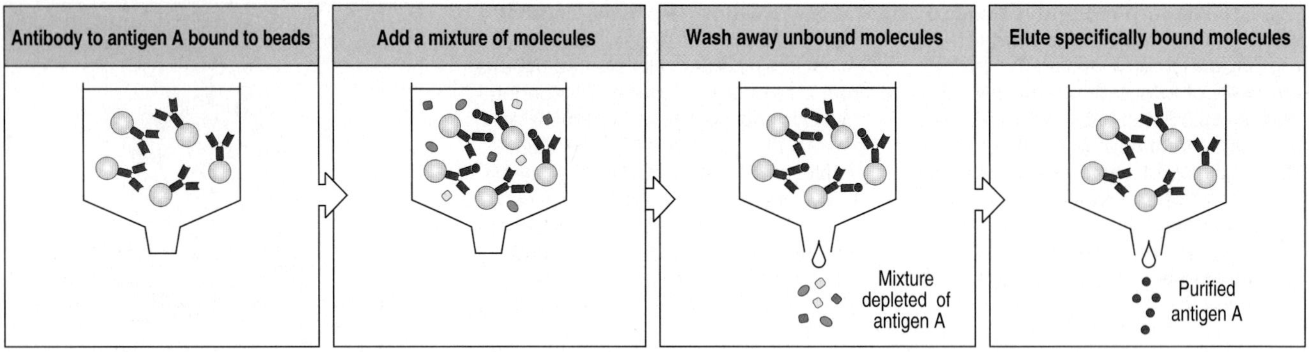
Antibody to antigen A bound to beads
Add a mixture of molecules
Wash away unbound molecules
Elute specifically bound molecules
Mixture depleted of antigen A
Purified antigen A

Fig. 2.19 Affinity chromatography uses antigen:antibody binding to purify antigens or antibodies. To purify a specific antigen from a complex mixture of molecules, a monoclonal antibody is attached to an insoluble matrix, such as chromatography beads, and the mixture of molecules is passed over the matrix. The specific antibody binds the antigen of interest; other molecules are washed away. Specific antigen is then eluted by altering the pH, which can usually disrupt antibody:antigen bonds. Antibodies can be purified in the same way on beads coupled to antigen (not shown).

to express the protein. The cDNA library is then cloned into special vectors, called expression vectors, which are constructed to allow the genes they carry to be expressed upon transfection into cultured mammalian cells. These vectors drive expression of the gene in the transfected cells without integrating into the host cell DNA. Cells expressing the protein are isolated by binding to antibody (see Section 2-15), and the vectors are recovered by lysing the cells (Fig. 2.20).

The recovered vectors are then introduced into bacterial cells where they replicate rapidly, and these amplified vectors are used in a second round of transfection in mammalian cells. After several cycles of transfection, isolation, and amplification in bacteria, single colonies of bacteria are picked and the vectors prepared from cultures of each colony are used in a final transfection to identify a cloned vector carrying the cDNA of interest, which is then isolated and characterized. This methodology has been used to isolate many genes encoding cell-surface molecules.

The full amino acid sequence of the protein can be deduced from the nucleotide sequence of its cDNA, and this often gives clues to the nature of the protein and its biological properties. The nucleotide sequence of the gene and

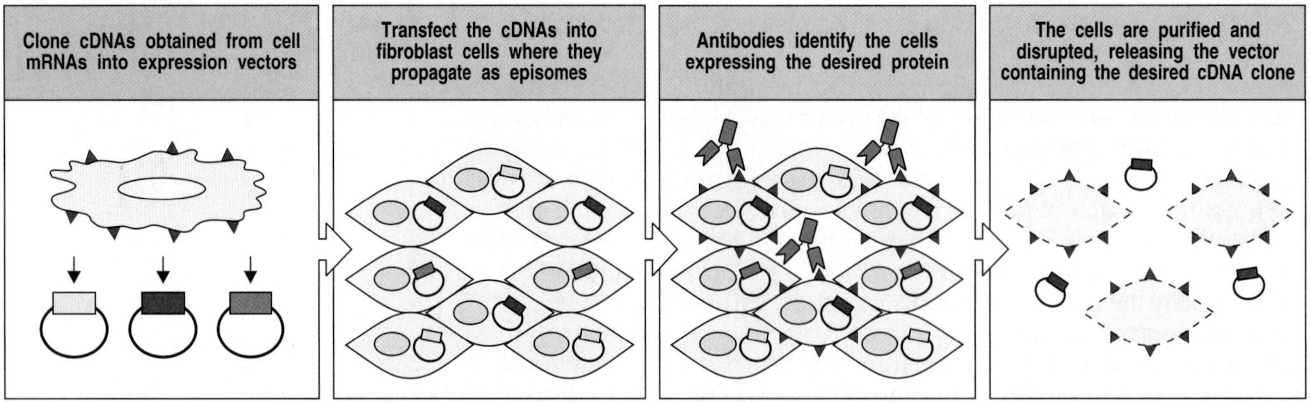
Clone cDNAs obtained from cell mRNAs into expression vectors
Transfect the cDNAs into fibroblast cells where they propagate as episomes
Antibodies identify the cells expressing the desired protein
The cells are purified and disrupted, releasing the vector containing the desired cDNA clone

Fig. 2.20 The gene encoding a cell-surface molecule can be isolated by expressing it in fibroblasts and detecting its protein product with monoclonal antibodies. Total mRNA from a cell line or tissue expressing the protein is isolated, converted into cDNA, and cloned as cDNAs in a vector designed to direct expression of the cDNA in fibroblasts. The entire cDNA library is used to transfect cultured fibroblasts. Fibroblasts that have taken up cDNA encoding a cell-surface protein express the protein on their surface; they can be isolated by binding a monoclonal antibody against that protein. The vector containing the gene is isolated from the cells that express the antigen and used for more rounds of transfection and re-isolation until uniform positive expression is obtained, ensuring that the correct gene has been isolated. The cDNA insert can then be sequenced to determine the sequence of the protein it encodes and can also be used as the source of material for large-scale expression of the protein for analysis of its structure and function. The method illustrated is limited to cloning genes for single-chain proteins (that is, those encoded by only one gene) that can be expressed in fibroblasts. It has been used to clone many genes of immunological interest such as that for CD4.

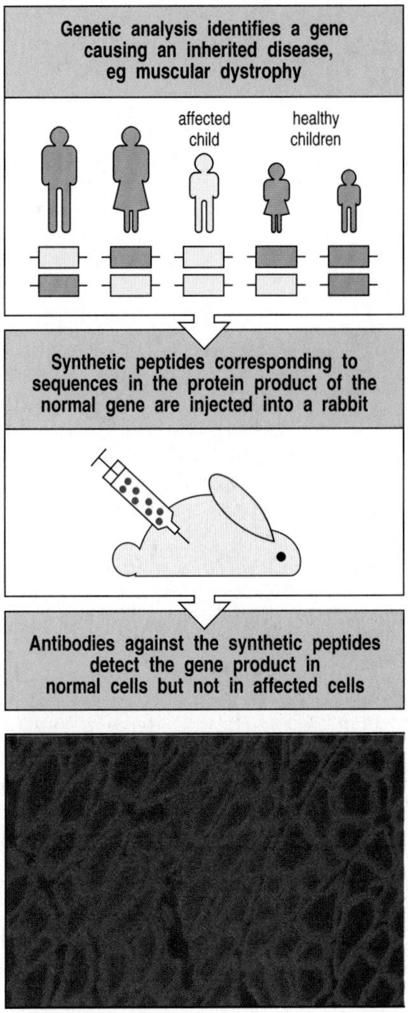

Fig. 2.21 The use of antibodies to detect the unknown protein product of a known gene is called reverse genetics. When the gene responsible for a genetic disorder such as Duchenne muscular dystrophy is isolated, the amino acid sequence of the unknown protein product of the gene can be deduced from the nucleotide sequence of the gene, and synthetic peptides representing parts of this sequence can be made. Antibodies are raised against these peptides and purified from the antiserum by affinity chromatography on a peptide column (see Fig. 2.19). Labeled antibody is used to stain tissue from individuals with the disease and from unaffected individuals to determine differences in the presence, amount, and distribution of the normal gene product. The product of the dystrophin gene is present in normal mouse skeletal muscle cells, as shown in the bottom panel (red fluorescence), but is missing from the cells of mice bearing the mutation *mdx*, the mouse equivalent of Duchenne muscular dystrophy (not shown). Photograph (x15) courtesy of H G W Lidov and L Kunkel.

its regulatory regions can be determined from genomic DNA clones. The gene can be manipulated and introduced into cells by transfection for larger-scale production and functional studies. This approach has been used to characterize many immunologically important proteins, such as the MHC glycoproteins.

The converse approach is taken to identify the unknown protein product of a cloned gene. The gene sequence is used to construct synthetic peptides of 10–20 amino acids that are identical to part of the deduced protein sequence, and antibodies are then raised against these peptides by coupling them to carrier proteins; the peptides behave as haptens. These anti-peptide antibodies often bind the native protein and so can be used to identify its distribution in cells and tissues and to try to ascertain its function (Fig. 2.21). This approach to identifying the function of a gene is often called 'reverse genetics' as it works from gene to phenotype rather than from phenotype to gene, which is the classical genetic approach. The great advantage of reverse genetics over the classical approach is that it does not require a detectable phenotypic genetic trait in order to identify a gene.

Summary.

The interaction of an antibody molecule with its ligand serves as the paradigm for immunological specificity, an essential concept in immunology. This is best understood by studying the binding of antibodies to antigens, which illustrates their tremendous power to discriminate between related antigens and their high affinity of binding to particular structures. The behavior of antibodies in serological assays shows that antibody molecules are highly diverse, symmetrically bivalent and have both constant and variable structural features. How the immune system produces the millions of different antibody molecules found in plasma while maintaining their overall structural identity that allows anti-immunoglobulin antibodies to detect any antibody molecule is the main subject of Chapter 3. In Chapter 9, we will learn about the production of antibody by B cells and why the amount, specificity, isotype, and affinity of antibody molecules are important in humoral immunity; here, we have learned how these attributes can be measured in a wide variety of assays, each giving its own type of information about the antibody response. Since antibodies can be raised to any structure, can bind it with high affinity and specificity, and can be made in unlimited amounts through monoclonal antibody production, they are particularly powerful tools of investigation. Many different techniques using antibodies have been devised and have a central role in both clinical medicine and biological research.

The study of lymphocytes.

The analysis of immunological specificity has focused largely on the antibody molecule because it is the most accessible agent of adaptive immunity. However, all adaptive immune responses are mediated by lymphocytes, so an understanding of immunobiology must be based on an understanding of lymphocyte behavior. To study and assay lymphocyte behavior, the cells must be isolated and the distinct functional lymphocyte subpopulations identified and separated. This section emphasizes studies on T lymphocytes, as the only known effector function of B lymphocytes is to produce antibodies, the subject of the preceding part of this chapter.

2-14 Lymphocytes can be isolated from blood, bone marrow, lymphoid organs, epithelia, and sites of inflammation.

The first step in studying lymphocytes is to isolate them so that their behavior can be analyzed *in vitro*. Human lymphocytes can be isolated most readily from peripheral blood by density centrifugation over a step gradient consisting of a mixture of the carbohydrate polymer Ficoll™ and the dense iodine-containing compound metrizamide. This yields a population of mononuclear cells at the interface that has been depleted of red blood cells and most polymorphonuclear leukocytes or granulocytes (Fig. 2.22). The resulting population, called **peripheral blood mononuclear cells**, consists mainly of lymphocytes and monocytes. Although this population is readily accessible, it is not necessarily representative of the lymphoid system, as only recirculating lymphocytes can be isolated from blood. In experimental animals, and occasionally in humans, lymphocytes can be isolated from lymphoid organs, such as spleen, thymus, bone marrow, lymph nodes, or mucosal-associated lymphoid tissues, most commonly the palatine tonsils in humans (see Fig. 1.7). A specialized population of lymphocytes resides in surface epithelia; these cells are isolated by fractionating the epithelial layer after its detachment from the basement membrane. Finally, in situations where local immune responses are prominent, lymphocytes can be isolated from the site of the response itself. For example, in order to study the autoimmune reaction that is thought to be responsible for rheumatoid arthritis, an inflammatory response in joints, lymphocytes are isolated from the fluid aspirated from the inflamed joint space.

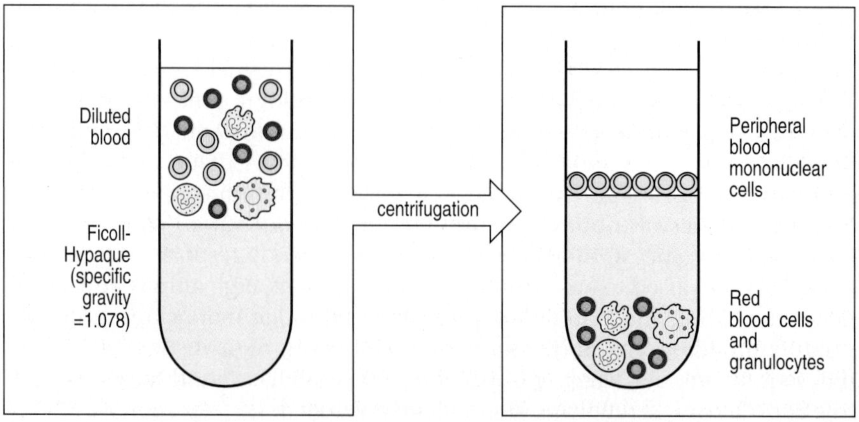

Fig. 2.22 Peripheral blood mononuclear cells can be isolated from whole blood by Ficoll-Hypaque™ centrifugation. Diluted anti-coagulated blood (left panel) is layered over Ficoll-Hypaque and centrifuged. Red blood cells and polymorphonuclear leukocytes or granulocytes are more dense and centrifuge through the Ficoll-Hypaque, while mononuclear cells consisting of lymphocytes together with some monocytes band over it and can be recovered at the interface (right panel).

2-15 Lymphocyte populations can be purified and characterized by antibodies specific for cell-surface molecules.

Resting lymphocytes present a deceptively uniform appearance, all being small round cells with a dense nucleus and little cytoplasm (see Fig. 1.5). However, these cells comprise many functional subpopulations, which are usually identified and distinguished from each other on the basis of their differential expression of cell-surface proteins, which can be detected using specific antibodies (Fig. 2.23). B and T lymphocytes, for example, are identified unambiguously and separated from each other by antibodies to the constant regions of the B- and T-cell antigen receptors. T cells are further subdivided on the basis of expression of the co-receptor proteins CD4 and CD8.

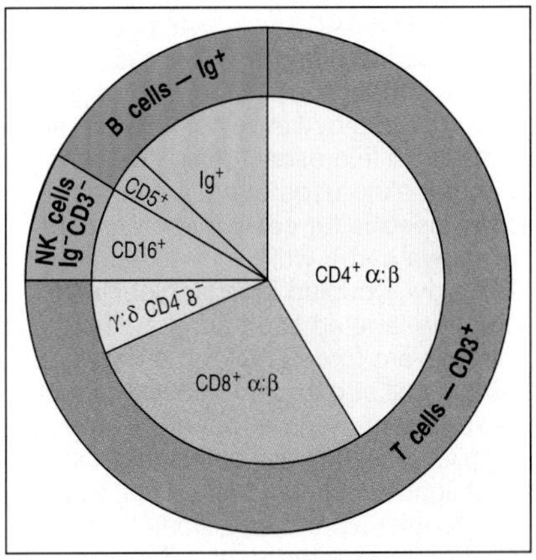

Fig. 2.23 The distribution of lymphocyte subpopulations in human peripheral blood. As shown on the outside of the circle, lymphocytes can be divided into T cells bearing T-cell receptors (detected with anti-CD3 antibodies), B cells bearing immunoglobulin receptors (detected with anti-immunoglobulin), and null cells including natural killer (NK) cells, that label with neither. Further divisions of the T-cell and B-cell populations are shown inside. Using anti-CD4 and anti-CD8 antibodies, α:β T cells can be subdivided into two populations, whereas γ:δ T cells are identified with antibodies against the γ:δ T-cell receptor and mainly lack CD4 and CD8. A minority population of B cells express CD5 on their surface (see Section 6-13).

An immensely powerful tool for defining and enumerating lymphocytes is the **flow cytometer**, which detects and counts individual cells passing in a stream through a laser beam. A flow cytometer equipped to separate the identified cells is called a **fluorescence-activated cell sorter** (**FACS**). These instruments are used to study the properties of cell subsets identified using monoclonal antibodies to cell-surface proteins. Individual cells within a mixed population are first tagged by treatment with specific monoclonal antibodies labeled with fluorescent dyes, or by specific antibodies followed by labeled anti-immunoglobulin. The mixture of labeled cells is then forced with a much larger volume of saline through a nozzle, creating a fine stream of liquid containing cells spaced singly at intervals. As each cell passes through a laser beam it scatters the laser light, and any dye molecules bound to the cell will be excited and will fluoresce. Sensitive photomultiplier tubes detect both the scattered light, which gives information on the size and granularity of the cell, and the fluorescence emissions, which give information on the binding of the labeled monoclonal antibodies and hence on the expression of cell-surface proteins by each cell (Fig. 2.24).

In the cell sorter, the signals passed back to the computer are used to generate an electric charge, which is passed from the nozzle through the liquid stream at the precise time that the stream breaks up into droplets, each containing no more than a single cell; droplets containing a charge can then be deflected from the main stream of droplets as they pass between plates of opposite charge, so that positively charged droplets are attracted to a negatively charged plate, and vice versa. In this way, specific subpopulations of cells, distinguished by the binding of the labeled antibody, can be purified from a mixed population of cells. Alternatively, to deplete a population of cells, the same fluorochrome can be used to label different antibodies directed at marker proteins expressed by the various undesired cell types. The cell sorter can be used to direct labeled cells to a waste channel, retaining only the unlabeled cells.

When cells are labeled with a single fluorescent antibody, the data from a flow cytometer are usually displayed in the form of a histogram of fluorescence intensity versus cell numbers. If two or more antibodies are used, each coupled to different fluorescent dyes, then the data are more usually displayed in the form of a two-dimensional scatter diagram or as a contour diagram, where the fluorescence of one dye-labeled antibody is plotted against that of a second, with the result that a population of cells labeling with one antibody can be further subdivided by its labeling with the second antibody (see Fig. 2.24). By examining large numbers of cells, flow cytometry can give quantitative data on the percentage of cells bearing different molecules, such as surface immunoglobulin, which characterizes B cells, the T-cell receptor-associated

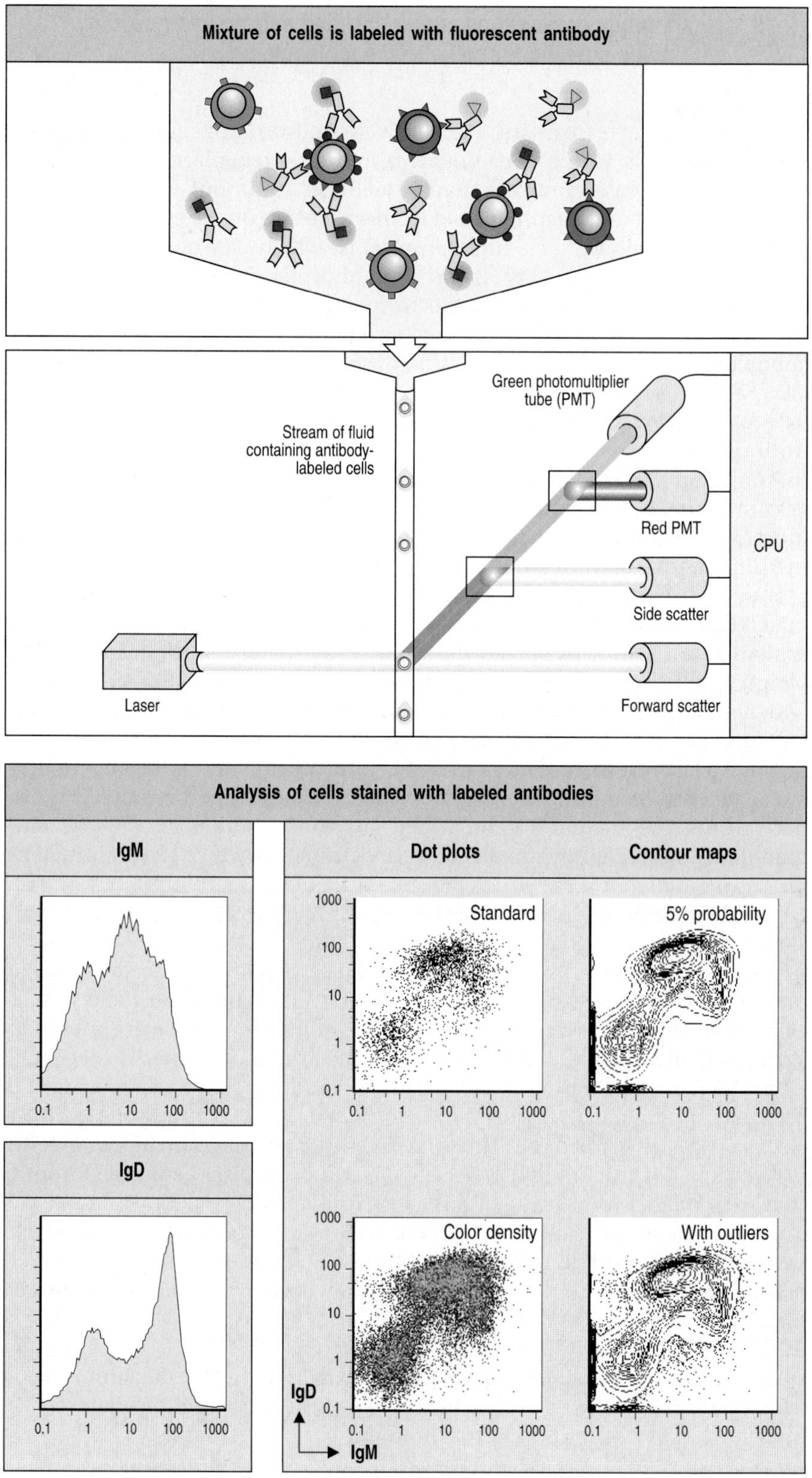

Fig. 2.24 The FACS™ allows individual cells to be identified by their cell-surface antigens and to be sorted. Cells to be analyzed by flow cytometry are first labeled with fluorescent dyes (top panel). Direct labeling uses dye-coupled antibodies specific for cell-surface antigens (as shown here), while indirect labeling uses a dye-coupled immunoglobulin to detect un-labeled cell-bound antibody. The cells are forced through a nozzle in a single-cell stream that passes through a laser beam (second panel). Photomultiplier tubes (PMTs) detect the scattering of light, which is a sign of cell size and granularity, and emissions from the different fluorescent dyes. This information is analyzed by computer (CPU). By examining many cells in this way, the number of cells with a specific set of characteristics can be counted and levels of expression of various molecules on these cells can be measured. The lower part of the figure shows how this data can be represented, using the expression of two surface immunoglobulins, IgM and IgD, on a sample of B cells from a mouse spleen. The two immunoglobulins have been labeled with different colored dyes. When the expression of just one type of molecule is to be analyzed (IgM or IgD), the data is usually displayed as a histogram, as in the left-hand panels. Histograms display the distribution of cells expressing a single measured parameter (for example, size, granularity, fluorescence color). When two or more parameters are measured for each cell (IgM and IgD), various types of two-color plots can be used to display the data, as shown in the right-hand panel. All four plots represent the same data. The horizontal axis represents intensity of IgM fluorescence and the vertical axis the intensity of IgD fluorescence. Two-color plots provide more information than histograms; they allow recognition, for example, of cells that are 'bright' for both colors, 'dull' for one and bright for the other, dull for both, negative for both, and so on. For example, the cluster of dots in the extreme lower left portions of the plots represents cells that do not express either immunoglobulin, and are mostly T cells. The standard dot plot (upper left) places a single dot for each cell whose fluorescence is measured. It is good for picking up cells that lie outside the main groups but tends to saturate in areas containing a large number of cells of the same type. A second means of presenting these data is the color dot plot (lower left), which uses color density to indicate high-density areas. A contour plot (upper right) draws 5% 'probability' contours, with 5% of the cells lying between each contour providing the best monochrome visualization of regions of high and low density. The lower right plot is a 5% probability contour map which also shows outlying cells as dots.

molecules known as CD3, and the CD4 and CD8 co-receptor proteins that distinguish the major T-cell subsets. Likewise, FACS analysis has been instrumental in defining stages in the early development of B and T cells. As the power of the FACS technology has grown, progressively more antibodies labeled with distinct fluorescent dyes can be used at the same time. Three-,

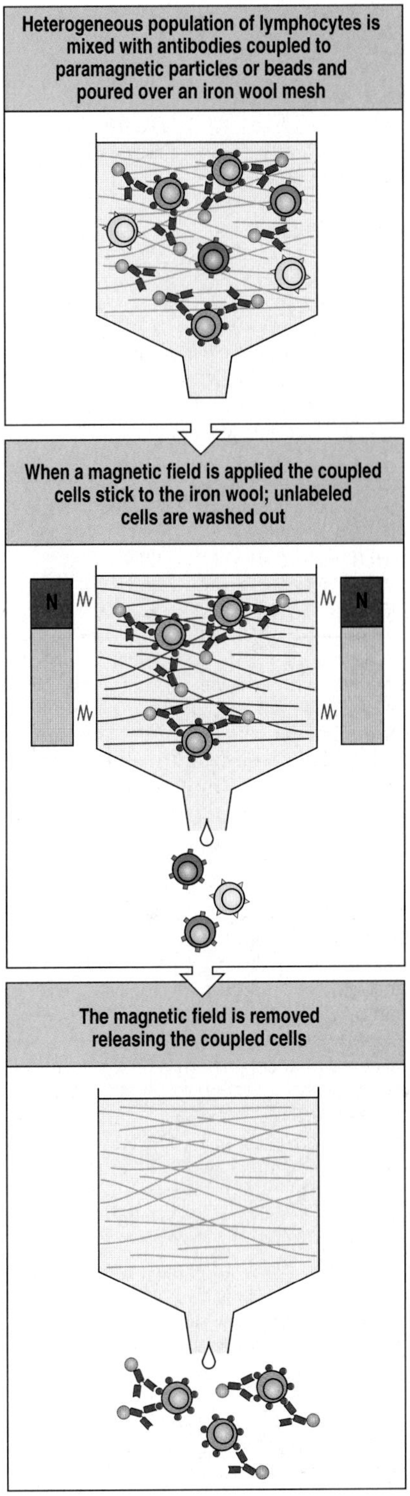

Heterogeneous population of lymphocytes is mixed with antibodies coupled to paramagnetic particles or beads and poured over an iron wool mesh
When a magnetic field is applied the coupled cells stick to the iron wool; unlabeled cells are washed out
N
N
The magnetic field is removed releasing the coupled cells

Fig. 2.25 Lymphocyte subpopulations can be separated physically by using antibodies coupled to paramagnetic particles or beads. A mouse monoclonal antibody specific for a particular cell-surface molecule is coupled to paramagnetic particles or beads. It is mixed with a heterogeneous population of lymphocytes and poured over an iron mesh in a column. A magnetic field is applied so that the antibody-bound cells stick to the iron wool while cells which have not bound antibody are washed out; these cells are said to be negatively selected for lack of the molecule in question. The bound cells are released by removing the magnetic field; they are said to be positively selected for presence of the antigen recognized by the antibody.

four- and even five-color analyses can now be handled by very powerful machines. FACS analysis has been applied to a broad range of problems in immunology; indeed, it played a vital role in the early identification of AIDS as a disease in which T cells bearing CD4 are depleted selectively (see Chapter 11).

Although the FACS is superb for isolating small numbers of cells in pure form, when large numbers of lymphocytes must be prepared quickly, mechanical means of separating cells are preferable. A powerful and efficient way of isolating lymphocyte populations is to couple paramagnetic beads to monoclonal antibodies that recognize distinguishing cell-surface molecules. These antibody-coated beads are mixed with the cells to be separated, and run through a column containing material that attracts the paramagnetic beads when the column is placed in a strong magnetic field. Cells binding the magnetically labeled antibodies are retained; cells lacking the appropriate surface molecule can be washed away (Fig. 2.25). The bound cells are positively selected for expression of the particular cell-surface molecule, while the unbound cells are negatively selected for its absence.

A particular cell population can also be isolated by binding to antibody-coated plastic surfaces, a technique known as **panning**, or by removing unwanted cells by killing them by treatment with specific antibody and complement. Cells can also be passed over columns of antibody-coated, nylon-coated steel wool and different populations differentially eluted. This technique extends affinity chromatography to cells, and is now a very popular way to separate cells. All these techniques can also be used as a pre-purification step prior to sorting out highly purified populations by FACS.

The main conclusion reached from studies on isolated lymphocyte populations is that lymphocytes bearing particular combinations of cell-surface proteins represent distinct developmental stages that have particular functions, which suggested that these proteins must be involved directly in the function of the cell. For this reason, such surface molecules were originally called **differentiation antigens**. When groups of monoclonal antibodies were found to recognize the same differentiation antigen, they were said to define a **cluster of differentiation**, abbreviated to **CD**, followed by an arbitrarily assigned number. This is the origin of the CD nomenclature for lymphocyte cell-surface antigens. The known CD antigens are listed in Appendix I.

2-16 Lymphocytes can be stimulated to grow by polyclonal mitogens or by specific antigen.

To function in adaptive immunity, rare antigen-specific lymphocytes must proliferate extensively before they differentiate into functional effector cells in order to generate sufficient numbers of effector cells of a particular specificity. Thus, the analysis of induced lymphocyte proliferation is a central issue

Mitogen	Responding cells
Phytohemagglutinin (PHA) (red kidney bean)	T cells
Concanavalin (ConA) (Jack bean)	T cells
Pokeweed mitogen (PWM) (Pokeweed)	T and B cells
Lipopolysaccharide (LPS) (*Escherichia coli*)	B cells (mouse)

Fig. 2.26 Polyclonal mitogens, many of plant origin, stimulate lymphocyte proliferation in tissue culture. Many of these mitogens are used to test the ability of lymphocytes in human peripheral blood to proliferate.

in their study. However, it is difficult to detect the proliferation of normal lymphocytes in response to specific antigen because only a minute proportion of cells will be stimulated to divide. Enormous impetus was given to the field of lymphocyte culture by the finding that certain substances induce many or all lymphocytes of a given type to proliferate. These substances are referred to collectively as **polyclonal mitogens** because they induce mitosis in lympho cytes of many different specificities or clonal origins. T and B lymphocytes are stimulated by different polyclonal mitogens (Fig. 2.26). Polyclonal mitogens seem to trigger essentially the same growth response mechanisms as antigen. Lymphocytes normally exist as resting cells in the G_0 phase of the cell cycle. When stimulated with polyclonal mitogens, they rapidly enter the G_1 phase and progress through the cell cycle. In most studies, lymphocyte proliferation is most simply measured by the incorporation of ^{3}H-thymidine into DNA. This assay is used clinically for assessing the ability of lymphocytes from patients with suspected immunodeficiencies to proliferate in response to a non-specific stimulus (see Section 2-22).

Once lymphocyte culture had been optimized using the proliferative response to polyclonal mitogens as an assay, it became possible to detect antigen-specific T-cell proliferation in culture by measuring ^{3}H-thymidine uptake in response to an antigen to which the T-cell donor had been previously immunized (Fig. 2.27). This is the assay most commonly used for assessing T-cell responses after immunization, but it reveals little about the functional capabilities of the responding T cells. These must be ascertained by functional assays, as described in the next section.

2-17 T-cell effector functions can be measured in four ways—target-cell killing, macrophage activation, B-cell activation, or cytokine production.

As we learned in Section 2-6, effector T cells are detected by their effects on target cells displaying antigen or by the secretion of specific cytokines that act on such target cells. Measurement of these effector functions is the basis of T-cell bioassays used to assess both T-cell specificity for antigen and T-cell effector function.

Activated CD8 T cells generally kill any cells that display the specific peptide:MHC class I complex they recognize. Therefore, CD8 T-cell function can be determined using the simplest and most rapid T-cell bioassay—the killing of a target cell by a cytotoxic T cell. This is usually detected in a ^{51}Cr-release assay. Live cells will take up, but do not spontaneously release, radioactively labeled sodium chromate, $Na_2{}^{51}CrO_4$. When these labeled cells are killed, the

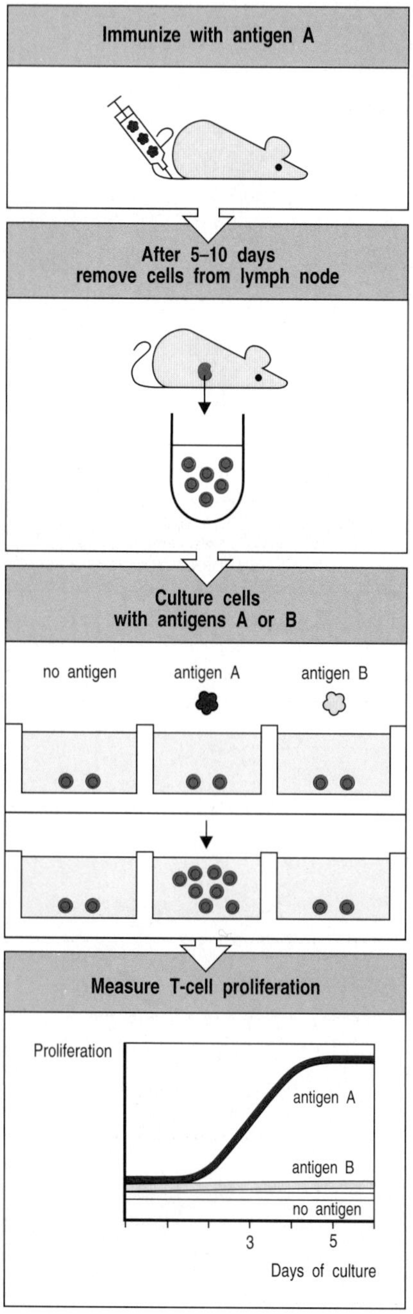

Fig. 2.27 Antigen-specific T-cell proliferation is used frequently as an assay for T-cell responses. T cells from mice or humans that have been immunized with an antigen (A) proliferate when they are exposed to antigen A and antigen-presenting cells but not when cultured with unrelated antigens to which they have not been immunized (antigen B). Proliferation can be measured by incorporation of ^{3}H-thymidine into the DNA of actively dividing cells. Antigen-specific proliferation is a hallmark of specific CD4 T-cell immunity.

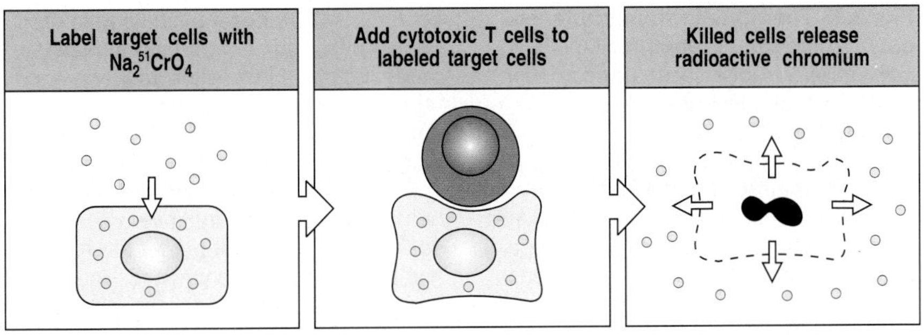

Fig. 2.28 Cytotoxic T-cell activity is often assessed by chromium release from labeled target cells. Target cells are labeled with radioactive chromium as $Na_2{}^{51}CrO_4$, washed to remove excess radioactivity and exposed to cytotoxic T cells. Cell destruction is measured by the release of radioactive chromium into the medium, detectable within 4 hours of mixing target cells with T cells.

radioactive chromate is released and its presence in the supernatant of mixtures of target cells and cytotoxic T cells can be measured (Fig. 2.28). In a similar assay, proliferating target cells such as tumor cells can be labeled with ^3H-thymidine, which is incorporated into the replicating DNA. On attack by a cytotoxic T cell, the DNA of the target cells is rapidly fragmented and released into the supernatant, and one can measure either the release of these fragments or the retention of ^3H-thymidine in chromosomal DNA. These assays provide a rapid, sensitive, and specific measure of the activity of cytotoxic T cells.

CD4 T-cell functions usually involve the activation rather than the killing of cells bearing specific antigen, which for CD4 cells is a specific peptide:MHC class II complex. The activating effects of CD4 T cells on B cells or macrophages are mediated in large part by non-specific mediator proteins called cytokines, which are released by the T cell when it recognizes antigen (see Chapter 8). Thus, CD4 T-cell function is usually studied by measuring the type and amount of these released proteins. As different effector T cells release different amounts and types of cytokines, one can learn about the effector potential of that T cell by measuring the proteins it produces.

Cytokines can be detected by their activity in biological assays of cell growth, where they serve either as growth factors or growth inhibitors. A more specific assay is a modification of ELISA known as a **capture** or **sandwich ELISA**. In this assay, the cytokine is characterized by its ability to bridge between two monoclonal antibodies reacting with different epitopes on the cytokine molecule (Fig. 2.29). Sandwich ELISA can also be carried out by placing the cells themselves on a surface coated with antibody to a cytokine. After a short incubation, the cytokine released by each cell is trapped on the antibody coat and the presence of cytokine-secreting cells can be revealed when the cells are washed off and a labeled second anti-cytokine antibody is added. The position of each cell releasing cytokine is marked by a distinct spot in this assay, which is therefore known as an **ELISPOT assay**. ELISPOT can also be used to detect specific antibody secretion by B cells, in this case by using antigen-coated surfaces to trap specific antibody and labeled anti-immunoglobulin to detect the bound antibody.

Sandwich ELISA avoids a major problem of cytokine bioassays, the ability of different cytokines to stimulate the same response in a bioassay. Bioassays

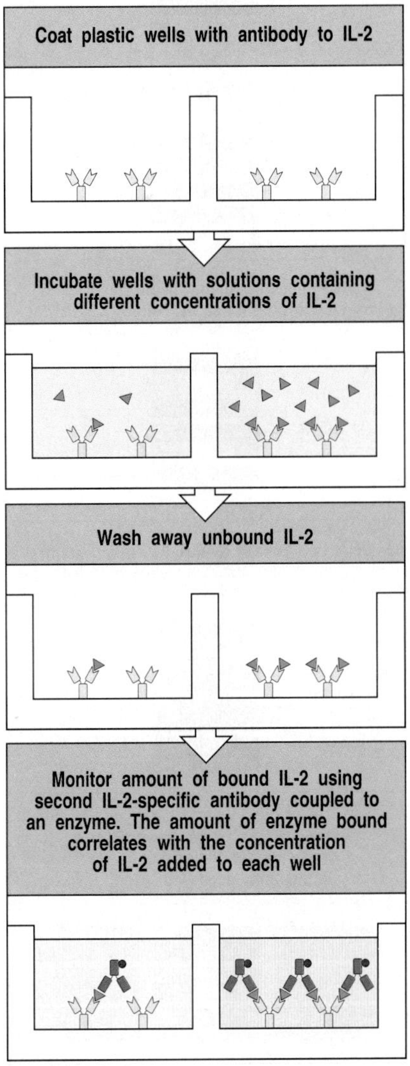

Coat plastic wells with antibody to IL-2
Incubate wells with solutions containing different concentrations of IL-2
Wash away unbound IL-2
Monitor amount of bound IL-2 using second IL-2-specific antibody coupled to an enzyme. The amount of enzyme bound correlates with the concentration of IL-2 added to each well

Fig. 2.29 Measurement of interleukin-2 (IL-2) production by sandwich ELISA. When T cells are activated with a mitogen or antigen they usually release the T-cell growth factor IL-2. In this assay, one unlabeled anti-IL-2 antibody is attached to the plastic, and then the IL-2-containing fluid is added. After washing, bound IL-2 is detected by binding a second, labeled anti-IL-2 antibody directed at a different epitope. This assay is highly specific because cytokines and other antigens that cross-react with one antibody are very unlikely to cross-react with the other. It can detect and quantify IL-2 and many other cytokines with great sensitivity and precision.

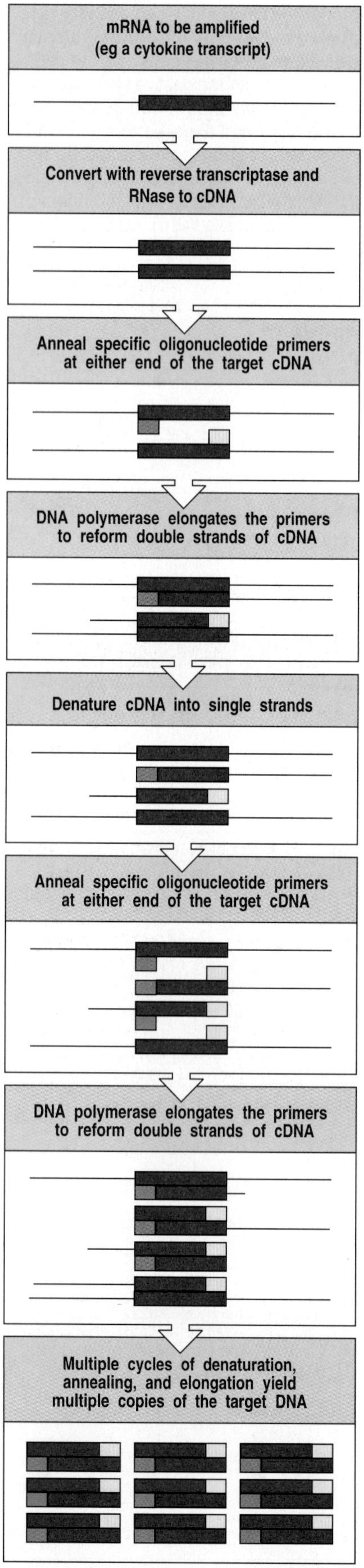

Fig. 2.30 The reverse transcriptase–polymerase chain reaction (RT–PCR). Total mRNA from a tissue or population of cells is first converted into cDNA by using reverse transcriptase. For the polymerase chain reaction, synthetic oligonucleotide primers complementary to the nucleotide sequence of a given cytokine transcript are made. The cDNA is separated into single strands in the presence of an excess of these two primers so that on reannealing, the primers bind their complementary sequence in the cDNA. The DNA is then replicated using the DNA polymerase Taq from the bacterium *Thermus aquaticus*, which is stable at the high temperatures used to separate the DNA strands between each replication cycle. Replication starts from each primer, using the cDNA as a template. The replicated double-stranded DNA is separated into single strands by heating and the mixture is then cooled so that a new cycle of primer annealing and replication can commence. The first products are random in length, but in subsequent cycles only the sequence delimited by the primers is replicated; many copies of this DNA accumulate.

must always be confirmed by inhibition of the response with neutralizing monoclonal antibodies specific for the cytokine. Another way of identifying cells actively producing a given cytokine is to stain them with a fluorescently tagged anti-cytokine monoclonal antibody and identify and count them by FACS.

A quite different approach to detecting cytokine production is to determine the presence and amount of the relevant cytokine mRNA in stimulated T cells. This can be done for single cells by *in situ* hybridization and for cell populations by **reverse transcriptase–polymerase chain reaction (RT–PCR)**. Reverse transcriptase is an enzyme used by certain RNA viruses, such as the human immunodeficiency virus (HIV-1) that causes AIDS, to convert an RNA genome into a DNA copy, or cDNA. In RT–PCR, mRNA is isolated from cells and cDNA copies made using reverse transcriptase. The desired cDNA is then selectively amplified by the polymerase chain reaction using sequence-specific primers (Fig. 2.30). When the products of the reaction are subjected to electrophoresis on an agarose gel, the amplified DNA can be visualized as a band of a specific size. The amount of amplified cDNA sequence will be proportional to its representation in the mRNA; stimulated T cells actively producing a particular cytokine will produce large amounts of that particular mRNA and thus give correspondingly large amounts of the selected cDNA on RT–PCR. The level of cytokine mRNA in the original tissue is usually determined by comparison with the outcome of RT–PCR on the mRNA produced by a so-called 'housekeeping gene' expressed by all cells.

2-18 Homogeneous T lymphocytes can be obtained as T-cell hybrids, cloned T-cell lines, T-cell tumors, or as primary clones obtained by limiting dilution.

Just as the analysis of antibody specificity and structure has been aided greatly by the development of hybridomas making monoclonal antibodies, the analysis of specificity and effector function in T cells has depended heavily on monoclonal populations of T lymphocytes. These can be obtained in four distinct ways. First, as for B-cell hybridomas (see Section 2-10), normal T cells proliferating in response to specific antigen can be fused to malignant T-cell lymphoma lines to generate **T-cell hybrids**. The hybrids express the receptor of the normal T cell, but proliferate indefinitely owing to the cancerous state of the lymphoma parent. T-cell hybrids can be cloned to yield a population of cells all having the same T-cell receptor. When stimulated by their specific antigen these cells release cytokines such as the T-cell growth factor interleukin-2 (IL-2), and the production of cytokines is used as an assay to assess the antigen specificity of the T-cell hybrid.

T-cell hybrids are excellent tools for the analysis of T-cell specificity, as they grow readily in suspension culture. However, they cannot be used to analyze the regulation of specific T-cell proliferation in response to antigen because they are continually dividing. T-cell hybrids also cannot be transferred into an animal to test for function *in vivo* because they would give rise to tumors. Functional analysis of T-cell hybrids is also confounded by the fact that the malignant partner cell affects their behavior in functional assays. Therefore, the regulation of T-cell growth and the effector functions of T cells must be studied using **T-cell clones**. These are clonal cell lines of a single T-cell type and antigen specificity, which are derived from cultures of heterogeneous T cells, called **T-cell lines**, whose growth is dependent on periodic restimulation with specific antigen and, frequently, on the addition of T-cell growth factors (Fig. 2.31). T-cell clones also require periodic restimulation with antigen and are more tedious to grow than T-cell hybrids but, because their growth depends on specific antigen recognition, they maintain antigen specificity, which is often lost in T-cell hybrids. Cloned T-cell lines can be used for studies of effector function both *in vitro* and *in vivo*. In addition, the proliferation of T cells, a critical aspect of clonal selection, can be characterized only in cloned T-cell lines, where such growth is dependent on antigen recognition. Thus, both types of monoclonal T-cell line have valuable applications in experimental studies.

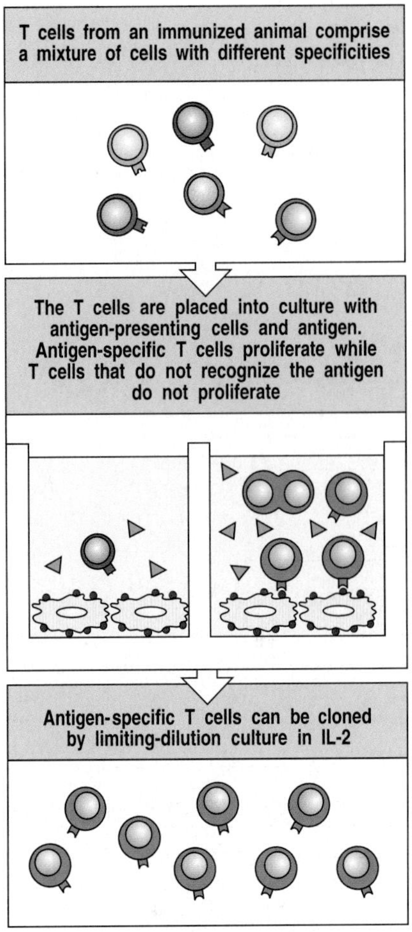

Fig. 2.31 Production of cloned T-cell lines. T cells from an immune donor, comprising a mixture of cells with different specificities, are activated with antigen and antigen-presenting cells. Single responding cells are cultured by limiting dilution in the T-cell growth factor IL-2, which selectively stimulates the responding cells to proliferate. From these single cells, cloned lines specific for antigen are identified and can be propagated by culture with antigen, antigen-presenting cells, and IL-2.

Studies of human T cells have relied largely on T-cell clones because a suitable fusion partner for making T-cell hybrids has not been identified. However, a human T-cell lymphoma line, called Jurkat, has been characterized extensively because it secretes IL-2 when its antigen receptor is crosslinked with anti-receptor monoclonal antibodies. This simple assay system has yielded much information about signal transduction in T cells. One of the Jurkat cell line's most interesting features, shared with T-cell hybrids, is that it stops growing when its antigen receptor is crosslinked. This has allowed mutants lacking the receptor or having defects in signal transduction pathways to be selected simply by culturing the cells with anti-receptor antibody and selecting those that continue to grow. Thus, T-cell tumors, T-cell hybrids, and cloned T-cell lines all have valuable applications in experimental immunology.

Finally, primary T cells from any source can be isolated as single, antigen-specific cells by limiting dilution rather than by first establishing a mixed population of T cells in culture as a T-cell line and then deriving clonal subpopulations. During the growth of T-cell lines, particular T-cell clones can come to dominate the cultures and give a false picture of the number and specificities in the original sample. Direct cloning of primary T cells avoids this artifact.

The response of a lymphocyte population is a measure of the overall response, but the frequency of lymphocytes able to respond to a given antigen can be determined only by limiting dilution culture. This assay makes use of the Poisson distribution, a statistical function that describes how objects are distributed at random. For instance, when a sample of heterogeneous T cells is distributed equally into a series of culture wells, some wells will receive no T cells specific for a given antigen, some will receive one specific T cell, some two, and so on. The T cells in the wells are activated with specific antigen, antigen-presenting cells, and growth factors. After allowing several days for their growth and differentiation, the cells in each well are tested for a response to antigen, such as cytokine release or the ability to kill specific target cells. The assay is replicated with different numbers of T cells in the samples. The logarithm of the proportion of wells in which there is no response is plotted against the number of cells initially added to each well. If cells of one type, typically antigen-specific T cells because of their rarity, are the only limiting factor for obtaining a response, then a straight line is obtained. From the Poisson distribution, it is known that there is, on average, one antigen-specific cell per well when the proportion of negative wells is 37%. Thus, the frequency of antigen-specific cells in the population equals the reciprocal of the number of cells added to each well when 37% of the wells are negative. After priming, the frequency of specific cells goes up substantially, reflecting the antigen-driven proliferation of antigen-specific cells. The limiting dilution assay can also be used to measure the frequency of B cells that can make antibody to a given antigen.

Summary.

The cellular basis of adaptive immunity is the clonal selection of lymphocytes by antigen. Therefore, to study adaptive immune responses, one must isolate lymphocytes and characterize them. Lymphocytes can be divided into subpopulations by using antibodies that detect cell-surface molecules expressed selectively on cells of a given type. Subsets defined in this way also differ functionally, suggesting that the cell-surface molecules detected are important for the function of that cell. Antibodies to cell-surface antigens can be used to separate lymphocytes physically, using magnetic beads, or by fluorescence-activated cell sorting, which also allows quantitative analysis of cell subpopulations. The functional capabilities of these isolated populations can then be tested *in vitro* and *in vivo.* Both B and T cells proliferate in response to specific antigen. The only known effector function of B cells is to secrete antibody whereas T lymphocytes can be functionally subdivided and their effector functions measured in four different ways—target-cell killing, macrophage activation, B-cell activation and cytokine secretion. Individual T cells can also be cloned, either as T-cell hybridomas or as continuously growing lines of normal T cells, which are valuable for analyzing the specificity, function, and signaling properties of T cells.

Analyzing immune responses in intact or manipulated organisms.

One of the ultimate goals of immunobiology is to understand the immune response *in vivo* and to control it. To do so, techniques to study immunity in live animals and in human patients are essential. The following sections describe how immunity is measured and characterized in the intact organism, be it a mouse or a human being. From these observations much is learned about the functioning of the intact immune system. The cellular and molecular basis for these observed functions is the subject of much of this book. Experimental animals, and in particular inbred mice, can also be manipulated by various means for the purposes of studying immune functions. This can be achieved by transferring lymphocytes or antibodies from one animal to another, or by altering the genome, either by inserting new genes to create transgenic animals, or deleting genes by using gene knock-out techniques.

2-19 Protective immunity can be assessed by challenge with infectious agents.

An adaptive immune response against a pathogen often confers long-lasting immunity against infection with that pathogen; successful vaccination achieves the same end. The very first experiment in immunology, Jenner's successful vaccination against smallpox, is still the model for assessing the presence of such protective immunity. The assessment of protective immunity conferred by vaccination has three essential steps. First, an immune response is elicited by immunization with a candidate vaccine. Second, the immunized individuals, along with unimmunized controls, are challenged with the infectious agent (Fig. 2.32). Finally, the prevalence and severity of infection in the immunized individual is compared with the course of the disease in the unimmunized controls. For obvious reasons, such experiments are usually carried out first in animals, if a suitable animal model for the infection exists. However, eventually a trial must be carried out in humans. In this case, the infectious challenge is usually provided naturally by carrying out the trial in a region where the disease is prevalent. The efficacy of the vaccine is determined by assessing the prevalence and severity of new infections in the immunized and control populations. Such studies necessarily give less precise results than a direct experiment but, for most diseases, they are the only way of assessing a vaccine's ability to induce protective immunity in humans.

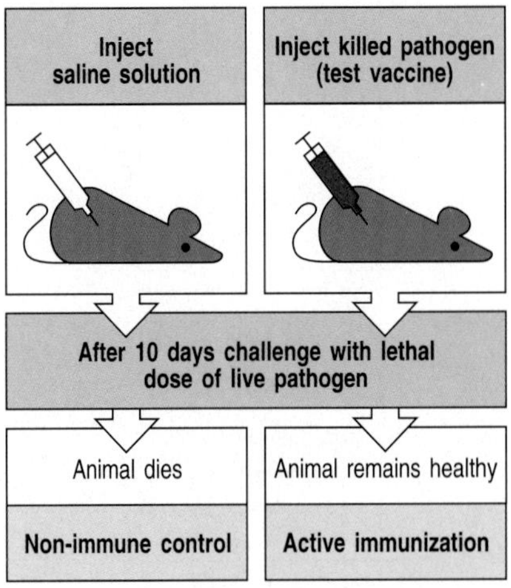

Fig. 2.32 *In vivo* assay for the presence of protective immunity after vaccination in animals. Mice are injected with the test vaccine or a control such as saline solution. Different groups are then challenged with lethal or pathogenic doses of the test pathogen or with an unrelated pathogen as a specificity control (not shown). Unimmunized animals die or become severely infected. Successful vaccination is seen as specific protection of immunized mice against infection with the test pathogen. This is called active immunity and the process is called active immunization.

2-20 Immunity can be transferred by antibodies or by lymphocytes.

The tests described in the previous section show that protective immunity has been established, but cannot show whether it involves humoral immunity, cell-mediated immunity, or both. When these studies are carried out in inbred mice, the nature of protective immunity can be determined by transferring serum or lymphoid cells from an immunized donor animal to an unimmunized syngeneic recipient (that is, a genetically identical animal of the same inbred strain) (Fig. 2.33). If protection against infection can be conferred by the transfer of serum, the immunity is provided by circulating antibodies and is called **humoral immunity**. Transfer of immunity by antiserum or purified antibodies provides immediate protection against many pathogens and against toxins such as those of tetanus and snake venom. However, although protection is immediate, it is temporary, lasting only so long as the transferred antibodies remain active in the recipient's body. This type of transfer is therefore called **passive immunization**. Only **active immunization** with antigen can provide lasting immunity. Moreover, the recipient may become immunized to the antiserum used to transfer immunity. Horse or sheep sera are the usual sources of anti-snake venoms used in humans, and repeated administration can lead either to serum sickness (see Section 12-17) or, if the recipient becomes allergic to the foreign serum, to anaphylaxis (see Section 12-11).

Protection against many diseases cannot be transferred with serum but can be transferred by lymphoid cells from immunized donors. The transfer of lymphoid cells from an immune donor to a normal syngeneic recipient is called **adoptive transfer** or **adoptive immunization**, and the immunity transferred is called **adoptive immunity**. Immunity that can be transferred only with lymphoid cells is called **cell-mediated immunity**. Such cell transfers must be between genetically identical donors and recipients, such as members of the same inbred strain of mouse, so that the donor lymphocytes are not rejected by the recipient and do not attack the recipient's tissues. Adoptive transfer of

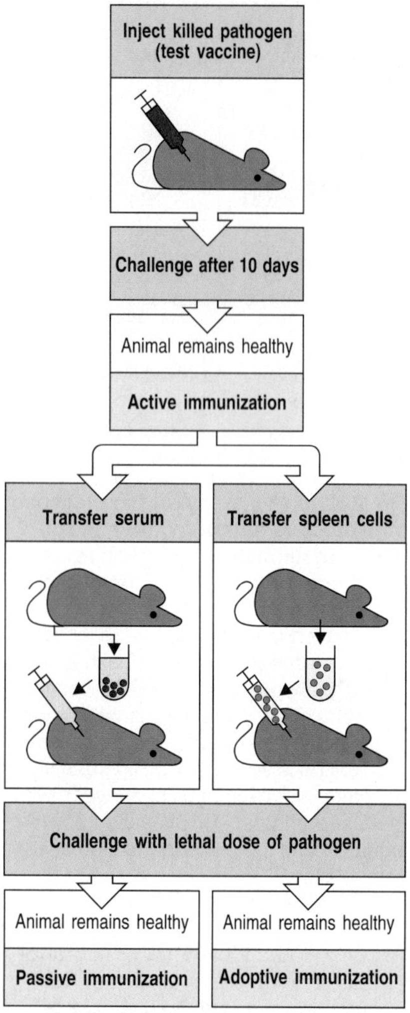

Fig. 2.33 Immunity can be transferred by antibodies or by lymphocytes. Successful vaccination leads to a long-lived state of protection against the specific immunizing pathogen. If this immune protection can be transferred to a normal syngeneic recipient with serum from an immune donor, then immunity is mediated by antibodies; such immunity is called humoral immunity and the process is called passive immunization. If immunity can only be transferred by infusing lymphoid cells from the immune donor into a normal syngeneic recipient, then the immunity is called cell-mediated immunity and the transfer process is called adoptive transfer or adoptive immunization. Passive immunity is short-lived, as antibody is eventually catabolized, but adoptively transferred immunity is mediated by immune cells, which can survive and provide longer-lasting immunity.

immunity is used clinically in humans in experimental approaches to cancer therapy or as an adjunct to bone marrow transplantation; in these cases, the patient's own T cells, or the T cells of the bone marrow donor, are given.

2-21 Local responses to antigen can indicate the presence of active immunity.

Active immunity is often studied *in vivo*, especially in humans, by injecting antigens locally in the skin. If a reaction appears, this indicates the presence of antibodies or immune lymphocytes that are specific for that antigen; the **tuberculin test** is an example of this. When people have had tuberculosis they develop cell-mediated immunity that can be detected as a local response when their skin is injected with a small amount of tuberculin, an extract of *Mycobacterium tuberculosis*, the pathogen that causes tuberculosis. The response typically appears a day or two after the injection and consists of a raised, red, and hard (or indurated) area in the skin, which then disappears as the antigen is degraded.

The immune system can also make less desirable responses, such as the hypersensitivity reactions responsible for allergies (see Chapter 12). Local intracutaneous injections of minute doses of the antigens that cause allergies are used to determine which antigen triggers a patient's allergic reactions. Local responses that happen in the first few minutes after antigen injection in immune recipients are called **immediate hypersensitivity reactions**, and they can be of several forms, one of which is the wheal and flare response described in Chapter 12. Immediate hypersensitivity reactions are mediated by specific antibodies of the IgE class formed as a result of earlier exposures to the antigen. Responses that take hours to days to develop, such as the tuberculin test, are referred to as **delayed-type hypersensitivity** responses and are caused by pre-existing immune T cells. This latter type of response was observed by Jenner when he tested vaccinated individuals with a local injection of vaccinia virus.

These tests work because the local deposit of antigen remains concentrated in the initial site of injection, eliciting responses in local tissues. They do not cause generalized reactions if sufficiently small doses of antigen are used. However, local tests carry a risk of systemic allergic reactions, and they should be used with caution in people with a history of hypersensitivity.

2-22 The assessment of immune responses and immunological competence in humans.

The methods used for testing immune function in humans are necessarily more limited than those used in experimental animals, but many different tests are available, some of which have been mentioned already. They fall into several groups depending on the reason the patient is being studied.

Assessment of protective immunity in humans generally relies on tests conducted *in vitro*. To assess humoral immunity, specific antibody levels in the patient's serum are assayed using the test microorganism or a purified microbial product as antigen. To test for humoral immunity against viruses, antibody production is often measured by the ability of serum to neutralize the infectivity of live virus for tissue culture cells. In addition to providing information about protective immunity, the presence of antibody to a particular pathogen indicates that the patient has been exposed to it, making such tests of crucial importance in epidemiology. At present, testing for antibody to HIV is the main screening test for infection with this virus, critical both for the patient and in blood banking, where blood from infected donors must be excluded from the supply. Essentially similar tests are used in investigating allergy, where allergens are used as the antigens in tests for specific IgE antibody by ELISA or radioimmunoassay (see Section 2-7), which may be used to confirm the results of skin tests.

Cell-mediated immunity to infectious agents can be tested either by skin test with extracts of the pathogen, as in the tuberculin test (see Section 2-21), or by the ability of the pathogen or an extract from it to stimulate T-cell proliferative responses *in vitro* (Section 2-16). These tests provide information about the exposure of the patient to the disease and also about their ability to mount an adaptive immune response to it.

Patients with immune deficiency (see Chapter 11) are usually detected clinically by a history of recurrent infection. To determine the competence of the immune system in such patients, a battery of tests is usually conducted (Fig. 2.34); these focus with increasing precision as the nature of the defect is narrowed down to a single element. The presence of the various cell types in blood is determined by routine hematology, often followed by FACS analysis (see Section 2-15) of lymphocyte subsets, and the measurement of serum immunoglobulins. The phagocytic competence of freshly isolated polymorphonuclear leukocytes and monocytes is tested, and the efficiency of the complement system (see Chapter 9) is determined by testing the dilution of serum required for lysis of 50% of antibody-coated red blood cells (this is denoted the CH_{50}).

Evaluation of the cellular components of the human immune system			
	B cells	T cells	Phagocytes
Normal numbers ($\times 10^9$ per liter of blood)	Approximately 0.3	Total 1.0–2.5 CD4 0.5–1.6 CD8 0.3–0.9	Monocytes 0.15–0.6 Polymorphonuclear leukocytes Neutrophils 3.00–5.5 Eosinophils 0.05–0.25 Basophils 0.02
Measurement of function *in vivo*	Serum Ig levels Specific antibody levels	Skin test	—
Measurement of function *in vitro*	Induced antibody production in response to pokeweed mitogen	T-cell proliferation in response to phytohemagglutinin or to tetanus toxoid	Phagocytosis Nitro blue tetrazolium uptake Intracellular killing of bacteria
Specific defects	See Fig. 11.8	See Fig. 11.8	See Fig. 11.8

Evaluation of the humoral components of the human immune system					
	Immunoglobulins				Complement
Component	IgG	IgM	IgA	IgE	
Normal levels	600–1400 mg dl^{-1}	40–345 mg dl^{-1}	60–380 mg dl^{-1}	0–200 IU ml^{-1}	CH_{50} of 125–300 IU ml^{-1}

Fig. 2.34 The assessment of immunological competence in humans. Both humoral and cell-mediated aspects of host defense can be checked, usually in a prescribed sequence, to identify the presence of an immune response or the causes of immunological incompetence. The initial screen consists of measuring levels of immunoglobulin and complement, and counting lymphocytes and phagocytic cells. (IgE is present, if at all, at very low levels and is measured in international units (IU) per ml; the CH_{50} of complement is the dilution at which 50% of antibody-coated red blood cells are lysed.) This initial screen usually indicates whether a defect in humoral or T-cell mediated immunity is present, and also whether it affects the induction or mediation of a response. In Chapter 11, defects in host defense known as immunodeficiency diseases are described in detail.

In general, if such tests reveal a defect in one of these broad compartments of immune function, more specialized testing is then needed to determine the precise nature of the defect. Tests of lymphocyte function are often valuable, starting with the ability of polyclonal mitogens to induce T-cell proliferation and B-cell secretion of immunoglobulin in tissue culture (see Section 2-16). These tests can eventually pinpoint the cellular defect in immunodeficiency.

In patients with autoimmune diseases (see Chapter 13), the same parameters are usually analyzed to determine whether there is a gross abnormality in the immune system. However, most patients with such diseases show few abnormalities in general immune function. To determine whether a patient is producing antibody against their own cellular antigens, the most informative test is to react their serum with tissue sections, which are then examined for bound antibody by indirect immunofluorescence using anti-human immunoglobulin labeled with fluorescent dye (see Section 2-12). Most autoimmune diseases are associated with the production of broadly characteristic patterns of autoantibodies directed at self tissues. These patterns aid in the diagnosis of the disease and help to distinguish autoimmunity from tissue inflammation due to infectious causes.

2-23 Irradiation kills lymphoid cells, allowing the study of immune function by adoptive transfer and the study of lymphocyte development in bone marrow chimeras.

Ionizing radiation from X-ray or γ-ray sources kills lymphoid cells at doses that spare the other tissues of the body. This makes it possible to eliminate immune function in a recipient animal before attempting to restore immune function by adoptive transfer, and allows the effect of the adoptively transferred cells to be studied in the absence of other lymphoid cells. James Gowans originally used this technique to prove the role of the lymphocyte in immune responses. He showed that all active immune responses could be transferred to irradiated recipients by small lymphocytes from immunized donors. This technique can be refined by transferring only certain lymphocyte subpopulations, such as B cells, CD4 T cells, and so on. Even cloned T-cell lines have been tested for their ability to transfer immune function, and have been shown to confer adoptive immunity to their specific antigen. Such adoptive transfer studies are a cornerstone in the study of the intact immune system, as they can be carried out rapidly, simply, and in any strain of mouse.

Somewhat higher doses of radiation eliminate all cells of hematopoietic origin, allowing replacement of the entire hematopoietic system, including lymphocytes, from donor bone marrow stem cells. The resulting animals are called **radiation bone marrow chimeras** from the Greek word *chimera*, a mythical animal that had the head of a lion, the tail of a serpent and the body of a goat. This technique is used to examine the development of lymphocytes as opposed to their effector functioning, and it has been particularly important in studying T-cell development, as we shall see in Chapter 7. Essentially the same technique is used in humans to replace bone marrow when it fails, as in aplastic anemia or after nuclear accidents, or to eradicate the bone marrow and replace it with normal marrow in the treatment of certain cancers.

2-24 Genetic defects can prevent the development of all lymphocytes.

There are several inherited immunodeficiencies in humans that are described as severe combined immune deficiency, or SCID, because they are characterized by defects in both humoral and cell-mediated immunity (we shall describe these various syndromes further in Chapter 11). Patients with these disorders suffer from a lack of lymphocytes, or lymphocyte function, and are remarkably susceptible to infection with a wide range of agents: most can survive only if completely isolated from their surroundings. Some SCID patients can be treated by bone marrow transplantation. SCID individuals are a dramatic illustration of the importance of lymphocytes in host defense and of the origin of all lymphocytes from a bone marrow progenitor.

In the mouse, a recessive mutation called *scid* prevents lymphocyte differentiation (see Chapter 11). Such mice have normal microenvironments for both B- and T-lymphocyte differentiation from stem cells, so grafting normal bone marrow into homozygous *scid/scid* mice can generate an intact immune system. Individual components of the mature immune system can also be transferred to *scid/scid* mice to generate animals expressing only the functions of particular subpopulations of lymphocytes. *Scid* mice are useful for distinguishing those immune functions that are innate (see Chapter 10) as opposed to those that require adaptive immunity mediated by specific lymphocytes. More recent studies use mice mutant in the *RAG-1* or *RAG-2* genes. These mice are completely devoid of functional T and B cells, whereas *scid* mice produce some lymphocytes as they age.

2-25 T cells can be eliminated selectively by removal of the thymus or by the *nude* mutation, while B cells are absent in agammaglobulinemic humans and genetically manipulated mice.

The importance of T-cell function *in vivo* can be ascertained in mice with no T cells of their own. Under these conditions, the effect of a lack of T cells can be studied, and T-cell subpopulations can be restored selectively to analyze their specialized functions. T lymphocytes originate in the thymus, and neonatal **thymectomy**, the surgical removal of the thymus of a mouse at birth, prevents T-cell development from occurring because the export of most functionally mature T cells only occurs after birth in the mouse. Alternatively, adult mice can be thymectomized and then irradiated and reconstituted with bone marrow; such mice will develop all hematopoietic cell types except mature T cells.

The recessive *nude* mutation in mice is caused by a mutation in the gene for the transcription factor Wnt and in homozygous form causes hairlessness and absence of the thymus. Consequently, these animals fail to develop T cells from bone marrow progenitors. Grafting thymectomized or *nude/nude* mice with thymic epithelial elements depleted of lymphocytes allows the graft recipients to develop normal mature T cells. This procedure allows the role of the non-lymphoid thymic stroma to be examined; it has been crucial in determining the role of thymic stromal cells in T-cell development (see Chapter 7).

There is no single site of B-cell development in mice, so techniques such as thymectomy cannot be applied to the study of B-cell function and development in rodents. However, **bursectomy**, the surgical removal of the **Bursa of Fabricius** in birds, can inhibit the development of B cells in these species. In fact, it was the effect of thymectomy versus bursectomy that led to the naming of T cells for thymus-derived lymphocytes and B cells for bursal-derived lymphocytes. There are no known spontaneous mutations (analogous to the *nude* mutation) in mice that produce animals with T cells but no B cells. However, such mutations exist in humans, leading to a failure to mount humoral immune responses or make antibody. The diseases produced by such mutations are called **agammaglobulinemias** because they were originally detected as the absence of gamma globulins. The genetic basis for one form of this disease in humans has now been established (see Chapter 11), and some features of the disease can be reproduced in mice by targeted disruption of the corresponding gene (see Section 2-27). Several different mutations in crucial regions of immunoglobulin genes have already been produced by gene targeting and have provided mice lacking B cells.

2-26 Individual genes can be introduced into mice by transgenesis.

The function of genes has traditionally been studied by observing the effects of spontaneous mutations in whole organisms and, more recently, by analyzing the effects of targeted mutations in cultured cells. The advent of gene cloning and *in vitro* mutagenesis now make it possible to produce specific mutations in whole animals. Mice with extra copies or altered copies of a gene in their genome can be generated by **transgenesis**, which is now a well established procedure. To produce **transgenic mice**, a cloned gene is introduced into the mouse genome by microinjection into the male pronucleus of a fertilized egg, which is then implanted into the uterus of a pseudopregnant female mouse. In some of the eggs, the injected DNA becomes integrated randomly into the genome, giving rise to a mouse that has an extra genetic element of known structure, the **transgene** (Fig. 2.35).

The transgene, to be studied in detail, needs to be introduced onto a stable, well-characterized genetic background. However, it is difficult to prepare transgenic embryos successfully in inbred strains of mice, and transgenic mice are routinely prepared in F2 embryos (that is, the embryo formed after the mating of two F1 animals). The transgene must then be bred onto a well-characterized genetic background; this requires 10 generations of back-crossing with an inbred strain to assure that the integrated transgene is largely (>99%) free of heterogeneous genes from the founder mouse of the transgenic mouse line (Fig. 2.36).

This technique allows one to study the impact of a newly discovered gene on development, to identify the regulatory regions of a gene required for its normal tissue-specific expression, to determine the effects of its overexpression or

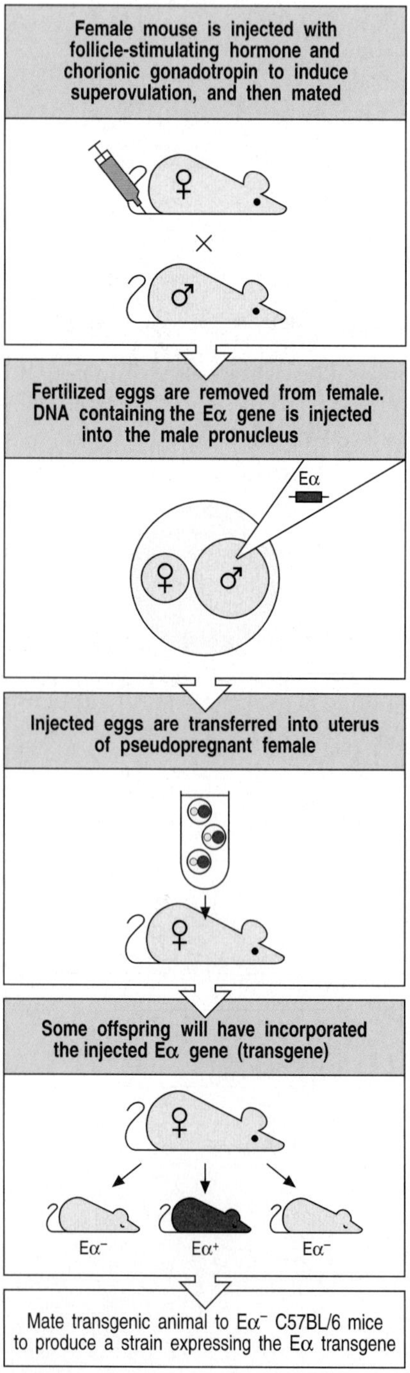
Female mouse is injected with follicle-stimulating hormone and chorionic gonadotropin to induce superovulation, and then mated
♀
×
♂
Fertilized eggs are removed from female. DNA containing the Eα gene is injected into the male pronucleus
Eα
♀
♂
Injected eggs are transferred into uterus of pseudopregnant female
♀
Some offspring will have incorporated the injected Eα gene (transgene)
♀
Eα⁻
Eα⁺
Eα⁻
Mate transgenic animal to Eα⁻ C57BL/6 mice to produce a strain expressing the Eα transgene

Fig. 2.35 The function and expression of genes can be studied *in vivo* by using transgenic mice. DNA encoding a protein of interest, here the mouse MHC class II protein Eα, is purified and micro-injected into the male pronuclei of fertilized eggs. The eggs are then implanted into pseudopregnant female mice. The resulting offspring are screened for the presence of the transgene in their cells, and positive mice are used as founders that transmit the transgene to their offspring, establishing a line of transgenic mice that carry one or more extra genes. The function of the Eα gene used here is tested by breeding the transgene into C57BL/6 mice that carry an inactivating mutation in their endogenous Eα gene.

Fig. 2.36 The breeding of transgenic co-isogeneic or congenic mouse strains. Transgenic mouse strains are routinely made in F2 mice. To produce mice on an inbred background, the transgene is introgressively backcrossed onto a standard strain, usually C57BL/6 (B6). The presence of the transgene is tracked by carrying out PCR on genomic DNA extracted from the tail of young mice. After 10 generations of backcrossing, mice are >99% genetically identical, so that any differences observed between the mice are likely to be due to the transgene itself. The same technique can be used to breed a gene knock-out into a standard strain of mice, as most gene knock-outs are made in the 129 strain of mice (see Fig. 2.38). The mice are then intercrossed and homozygous knock-out mice detected by an absence of an intact copy of the gene of interest (as determined by PCR).

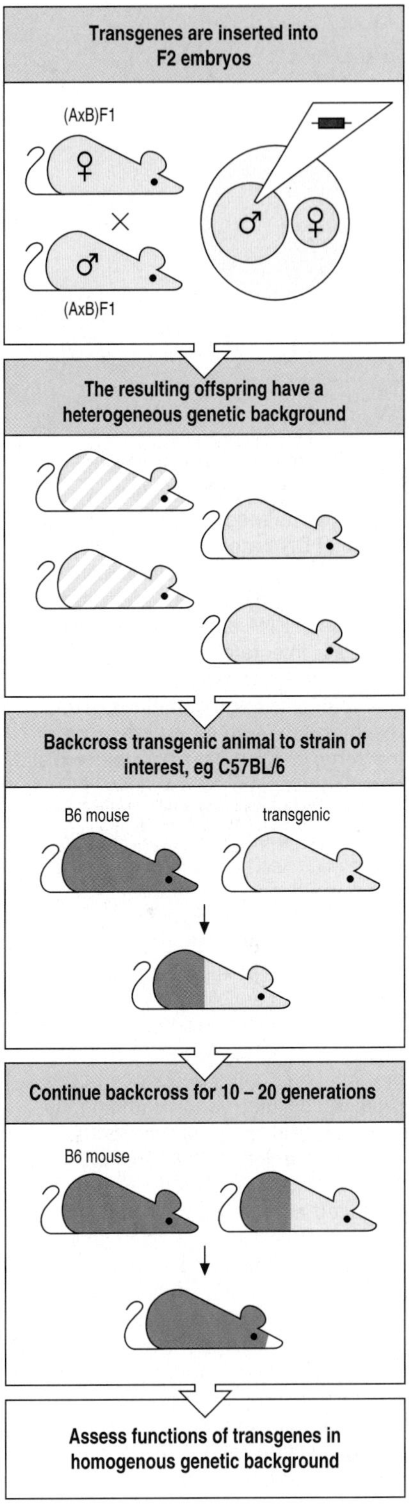

expression in inappropriate tissues, and to find out the impact of mutations on gene function. Transgenic mice have been particularly useful in studying the role of T-cell and B-cell receptors in lymphocyte development, as will be described in Chapters 6 and 7.

2-27 The role of individual genes can be studied *in vivo* by gene knock-out.

In many cases, the functions of a particular gene can be fully understood only if a mutant animal that does not express the gene can be obtained. Whereas genes used to be discovered through identification of mutant phenotypes, it is now far more common to discover and isolate the normal gene and then determine its function by replacing it *in vivo* with a defective copy. This procedure is known as **gene knock-out**, and it has been made possible by two fairly recent developments: a powerful strategy to select for targeted mutation by homologous recombination, and the development of continuously growing lines of **embryonic stem cells** (**ES cells**). These are embryonic cells which, on implantation into a blastocyst, can give rise to all cell lineages in a chimeric mouse.

The technique of **gene targeting** takes advantage of the phenomenon known as **homologous recombination** (Fig. 2.37). Cloned copies of the target gene are altered to make them non-functional and are then introduced into the ES cell where they recombine with the homologous gene in the cell's genome, replacing the normal gene with a non-functional copy. Homologous recombination is a rare event in mammalian cells, and thus a powerful selection strategy is required to detect those cells in which it has occurred. Most commonly, the introduced gene construct has its sequence disrupted by an inserted antibiotic-resistance gene such as that for neomycin resistance. If this construct undergoes homologous recombination with the endogenous copy of the gene, the endogenous gene is disrupted but the antibiotic-resistance gene remains functional, allowing cells that have incorporated the gene to be selected in culture for resistance to the neomycin-like drug G418. However, antibiotic resistance on its own shows only that the cells have taken up and integrated the neomycin-resistance gene. To be able to select for those cells in which homologous recombination has occurred, the ends of the construct usually carry the thymidine kinase gene from the herpes simplex virus (HSV-tk). Cells that incorporate DNA randomly usually retain the entire DNA construct including HSV-tk, whereas homologous recombination between the construct and cellular DNA, the desired result, involves the exchange of homologous DNA sequences so that the non-homologous HSV-tk genes at the ends of the construct are eliminated. Cells carrying HSV-tk are killed by the anti-viral drug ganciclovir, and so cells with homologous recombinations have the unique feature of being resistant to both neomycin and ganciclovir, allowing them to be selected efficiently when these drugs are added to the cultures (see Fig. 2.37).

This technique can be used to produce homozygous mutant cells in which the effects of knocking-out a specific gene can be analyzed. Diploid cells in which both copies of a gene have been mutated by homologous recombination can be selected after transfection with a mixture of constructs in which the gene to be targeted has been disrupted by one or other of two different antibiotic-resistance genes. Having obtained a mutant cell with a functional defect, the defect can be ascribed definitively to the mutated gene if the mutant phenotype can be reverted with a copy of the normal gene transfected into the mutant cell. Restoration of function means that the defect in the mutant gene has been complemented by the normal gene's function. This technique is very powerful as it allows the gene that is being transferred to be mutated in precise ways to determine which parts of the protein are required for function.

To knock out a gene *in vivo*, it is only necessary to disrupt one copy of the cellular gene in an ES cell. ES cells carrying the mutant gene are produced by targeted mutation (as in Fig. 2.37), and injected into a blastocyst which is re-implanted into the uterus. The cells carrying the disrupted gene become incorporated into the developing embryo and contribute to all

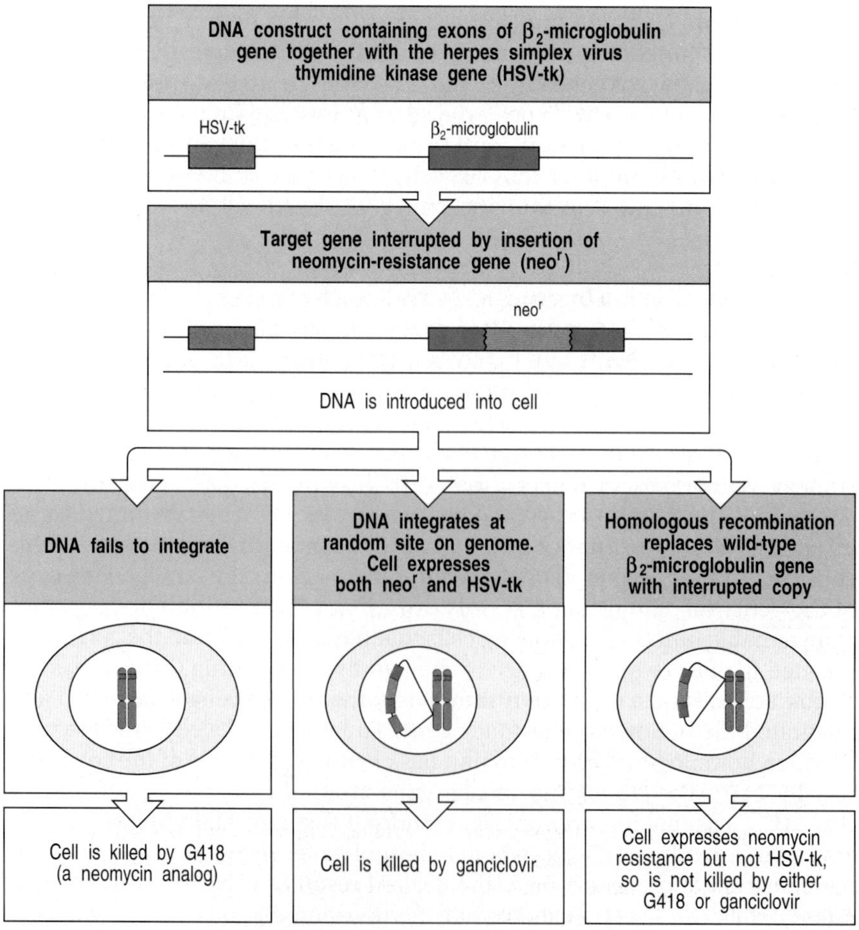

Fig. 2.37 The deletion of specific genes can be accomplished by homologous recombination. When pieces of DNA are introduced into cells, they can integrate into cellular DNA in two different ways. If they randomly insert into sites of DNA breaks, the whole piece is usually integrated, often in several copies. However, extrachromosomal DNA can also undergo homologous recombination with the cellular copy of the gene, in which case only the central, homologous region is incorporated into cellular DNA. Inserting a selectable marker gene such as resistance to neomycin (*neo^r*) into the coding region of a gene does not prevent homologous recombination, and it achieves two goals. First, any cell that has integrated the injected DNA is protected from the neomycin-like antibiotic G418. Second, when the gene recombines with homologous cellular DNA, the *neo^r* gene disrupts the coding sequence of the modified cellular gene. Homologous recombinants can be discriminated from random insertions if the gene for herpes simplex virus thymidine kinase (HSV-tk) is placed at one or both ends of the DNA construct, which is often known as a 'targeting construct' because it targets the cellular gene. In random DNA integrations, HSV-tk is retained. HSV-tk renders the cell sensitive to the anti-viral agent ganciclovir. However, as HSV-tk is not homologous to the target DNA, it is lost from homologous recombinants. Thus, cells that have undergone homologous recombination are uniquely both G418- and ganciclovir-resistant, and survive in a mixture of the two antibiotics. The presence of the disrupted gene has to be confirmed by Southern blotting or by PCR using primers in the *neo^r* gene and in cellular DNA lying outside the region used in the targeting construct. By using two different resistance genes one can disrupt the two cellular copies of a gene, making a deletion mutant (not shown).

tissues of the resulting chimeric offspring, including those of the germline. The mutated gene can therefore be transmitted to some of the offspring of the original chimera, and further breeding of the mutant gene to homozygosity produces mice that completely lack the expression of that particular gene product (Fig. 2.38). The effects of the absence of the gene's function can then be studied. In addition, the parts of the gene that are essential for its function can be identified by determining whether function can be restored by introducing different mutated copies of the gene back into the genome by transgenesis. The manipulation of the mouse genome by gene knock-out and transgenesis is revolutionizing our understanding of the role of individual genes in lymphocyte development and function, as we shall see throughout this book.

Because the most commonly used ES cells are derived from a poorly characterized strain of mice known as strain 129, the analysis of the function of a gene knock-out often requires extensive backcrossing to another strain, just as in transgenic mice (see Fig. 2.36). One can track the presence of the mutant copy of the gene by the presence of the *neo*^{r} gene. After sufficient backcrossing, the mice are intercrossed to produce mutants on a stable genetic background.

A problem with gene knock-outs arises when the function of the gene is essential for the survival of the animal; in such cases the gene is termed a **recessive lethal gene** and homozygous animals cannot be produced. However, by making chimeras with mice that are deficient in B and T cells, it is possible to analyze the function of recessive lethal genes in lymphoid cells. To do this, ES cells with homozygous lethal loss-of-function mutations are injected into blastocysts of mice lacking the ability to rearrange their antigen receptor genes because of a mutation in their recombinase-activating genes (*RAG* knock-out mice). As these chimeric embryos develop, the *RAG*-deficient cells can compensate for any developmental failure resulting from the gene knock-out in the ES cells in all except the lymphoid lineage. So long as the mutated ES cells can develop into hematopoietic progenitors in the bone marrow, the embryos will survive and all of the lymphocytes in the resulting chimeric mouse will be derived from the mutant ES cells (Fig. 2.39).

A second powerful technique achieves tissue-specific or developmentally regulated gene deletion by employing the DNA sequences and enzymes used by bacteriophage P1 to excise itself from a host cell's genome. Integrated bacteriophage P1 DNA is flanked by recombination signal sequences called

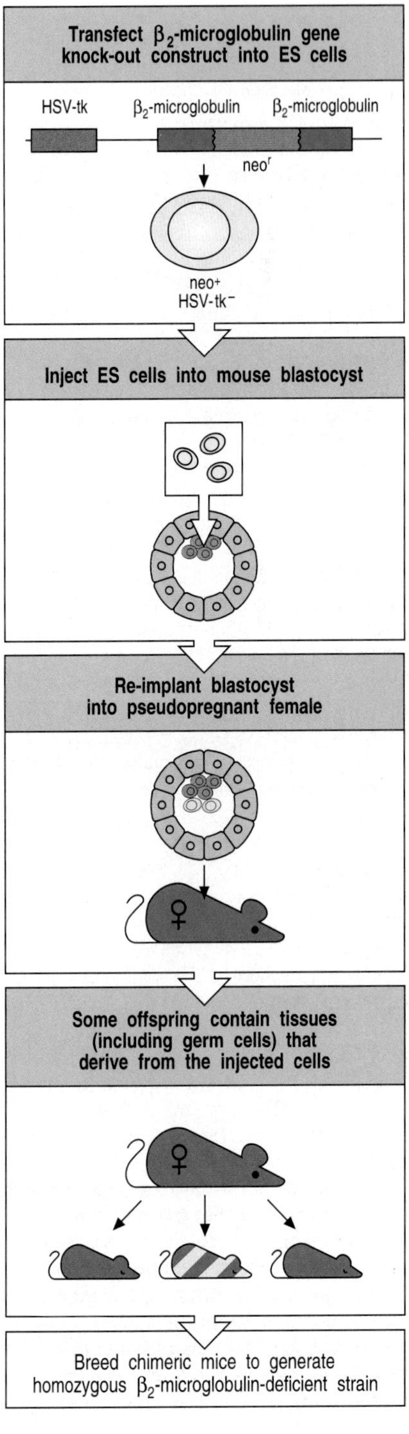

Fig. 2.38 Gene knock-out in embryonic stem cells enables mutant mice to be produced. Specific genes can be inactivated by homologous recombination in cultures of embryonic stem cells (ES cells). Homologous recombination is carried out as described in Fig. 2.37. In this example, the gene for β_2-microglobulin in ES cells is disrupted by homologous recombination with a targeting construct. Only a single copy of the gene needs to be disrupted. ES cells in which homologous recombination has taken place are injected into mouse blastocysts. If the mutant ES cells give rise to germ cells in the resulting chimeric mice (striped in the figure), then the mutant gene can be transferred to their offspring. By breeding the mutant gene to homozygosity, a mutant phenotype is generated. These mutant mice are usually of the 129 strain as gene knock-out is generally conducted in ES cells derived from the 129 strain of mice. In this case, the homozygous mutant mice lack MHC class I molecules on their cells, as MHC class I molecules have to pair with β_2-microglobulin for surface expression. The β_2-microglobulin-deficient mice can then be bred with mice transgenic for subtler mutants of the deleted gene, allowing the effect of such mutants to be tested *in vivo*.

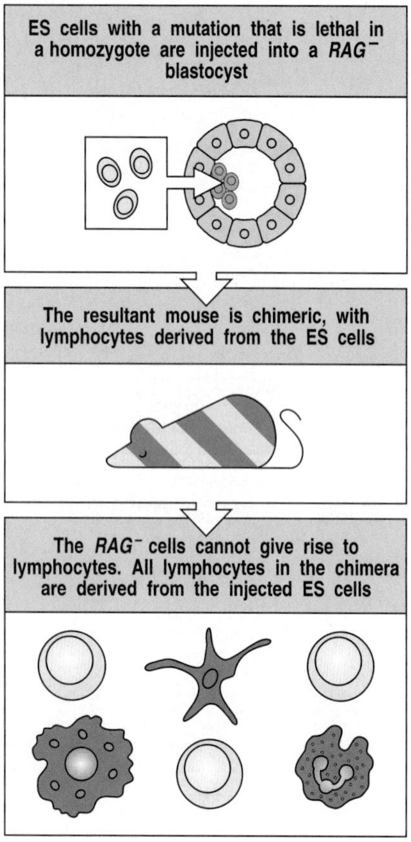

ES cells with a mutation that is lethal in a homozygote are injected into a RAG^- blastocyst
The resultant mouse is chimeric, with lymphocytes derived from the ES cells
The RAG^- cells cannot give rise to lymphocytes. All lymphocytes in the chimera are derived from the injected ES cells

Fig. 2.39 The role of recessive lethal genes in lymphocyte function can be studied using RAG-deficient chimeric mice. ES cells homozygous for the lethal mutation are injected into a RAG-deficient blastocyst (top panel). The RAG-deficient cells can give rise to all the tissues of a normal mouse except lymphocytes, and so can compensate for any deficiency in the developmental potential of the mutant ES cells (middle panel). If the mutant ES cells are capable of differentiating into hematopoietic stem cells, that is, if the gene function that has been deleted is not essential for this developmental pathway, then all the lymphocytes in the chimeric mouse will be derived from the ES cells (bottom panel), as RAG-deficient mice cannot make lymphocytes of their own.

loxP sites. A recombinase, Cre, recognizes these sites, cuts the DNA and joins the two ends, thus excising the intervening DNA in the form of a circle. This mechanism can be adapted to allow the deletion of specific genes in a transgenic animal only in certain tissues or at certain times in development. First, *loxP* sites flanking a gene, or perhaps just a single exon, are introduced by homologous recombination (Fig. 2.40). Usually, the introduction of these sequences into flanking or intronic DNA does not disrupt the normal function of the gene. Mice containing such *loxP* mutant genes are then mated with mice made transgenic for the Cre recombinase, under the control of a tissue-specific or inducible promoter. When the Cre recombinase is active, either in the appropriate tissue or when induced, it excises the DNA between the inserted *loxP* sites, thus inactivating the gene or exon. Thus, for example, using a T-cell specific promoter to drive expression of the Cre recombinase, a gene can be deleted only in T cells, while remaining functional in all other cells of the animal. This is an extremely powerful genetic technique that is still in its infancy and is certain to yield exciting results in the future.

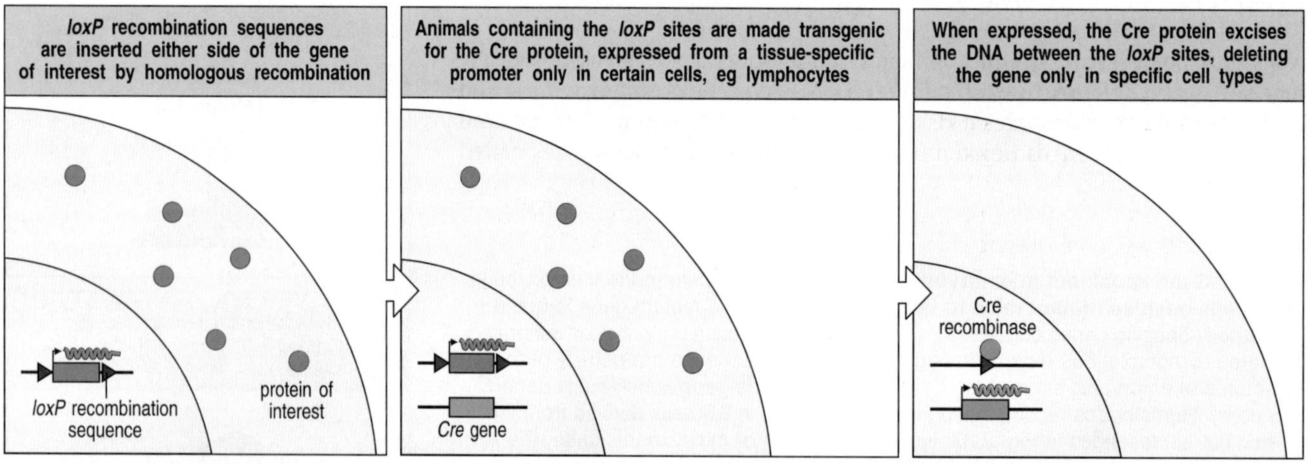

Fig. 2.40 The P1 bacteriophage recombination system can be used to eliminate genes in particular cell lineages. The P1 bacteriophage protein Cre excises DNA that is bounded by recombination signal sequences called *loxP* sequences. These sequences can be introduced at either end of a gene by homologous recombination (left panel). Animals carrying genes flanked by *loxP* can also be made transgenic for the gene for the Cre protein, which is placed under the control of a tissue-specific promoter so that it is expressed only in certain cells or only at certain times during development (middle panel). In the cells in which the Cre protein is expressed, it recognizes the *loxP* sequences and excises the DNA lying between them (right panel). Thus, individual genes can be deleted only in certain cell types or only at certain times. In this way, genes that are essential for the normal development of a mouse can be analyzed for their function in the developed animal and/or in specific cell types. Genes are shown as boxes, RNA as squiggles, and proteins as colored balls.

Summary.

The measurement of immune function in intact organisms is essential to a full understanding of the immune system in health and disease. The ability of an immunized individual to resist infection is still the standard assay for protective immunity conferred by infection or vaccination. Local reactions to antigens injected into the skin can provide information about antibody and T-cell responses to the antigen, a procedure that is particularly important in testing for allergic reactions. Finally, many assays *in vitro* such as the analysis of specific antibody in serum and the proliferative responses of T cells to mitogens and specific antigen are used to assess immune function in human patients. Manipulation of the immune system *in vivo* reveals the need for each of its components. The use of irradiation or mutation to eliminate lymphocytes or particular lymphocyte lineages, and then adoptively transferring mature lymphocytes, isolated subpopulations, cloned T-cell lines, or bone marrow stem cells allows one to study the functions and development of immune cells in an *in vivo* setting. The role of individual genes in lymphocyte development and function can be studied *in vivo* by manipulating the mouse genome, adding genes by transgenesis or eliminating them through gene knock-out. These two techniques can be combined to give detailed information about structure–function relationships in genes and their protein products, either in cultured cells or *in vivo*. These powerful techniques are increasing our understanding of immunobiology at an astonishing rate. The use of mutant mice in the study of host defenses to specific pathogens should provide a new understanding of these highly complex processes.

Summary to Chapter 2.

The immune system is very complex. To analyze it properly, it must be broken down into its individual components, and these must be studied both in isolation and in the context of the larger system. In this chapter we have described how immune responses are induced and measured, and how the immune system can be manipulated experimentally. An appreciation of the methodologies and findings described in this chapter is essential for a full understanding of immunobiology, and many of the techniques included here are used routinely in the experiments described in subsequent chapters. Some, especially the use of monoclonal antibodies for identifying molecules in cells and tissues and the manipulation of the mouse genome, also have general applications in biology.

General method references.

Ausubel, M.: *Current Protocols in Molecular Biology*, 1st edn. New York, Greene Publishing Associates and Wiley Interscience, 1987. Continuous updates added.

Coligan, J.E.: *Current Protocols in Immunology*, 1st edn. New York, Greene Publishing Associates and Wiley Interscience, 1991. Continuous updates added.

Green, M.C. (ed): *Genetic Variant and Strains of the Laboratory Mouse*, 1st edn. New York, Gustav Fischer Verlag, 1981.

Harlow, E. and Lane, D.: *Antibodies: a Laboratory Manual.* Cold Spring Harbor, NY, Cold Spring Harbor Laboratory Press, 1988.

Rose, N.R., Conway de Macario, E., Fahey, J.L., Friedman, H., Penn, G.M. (eds): *Manual of Clinical Laboratory Immunology*, 4th edn. Washington DC, American Society of Microbiology, 1992.

Sambrook, J., Fritsch, E.F., Maniatis, T.: *Molecular Cloning: A Laboratory Manual*, 2nd edn. Cold Spring Harbor, NY, Cold Spring Harbor Laboratory Press, 1989.

Weir, D. (ed): *The Handbook of Experimental Immunology, vol 1.*, 5th edn. Oxford, Blackwell Scientific Publications, 1996.

Journals and series.

Cytometry
Journal of Immunological Methods
Methods in Enzymology

Specific techniques.

Chen, J., Shinkai, Y., Young, F., Alt, F.W.: **Probing immune functions in RAG-deficient mice.** *Curr. Opin. Immunol.* 1994, **6**:313-319.

Mueller, R., Sarvetnick, N.: **Transgenic/knockout mice—tools to study autommunity.** *Curr. Opin. Immunol.* 1995, **7**:799-803.

Radbruch, A., Recktenwald, D.: **Detection and isolation of rare cells.** *Curr. Opin. Immunol.* 1995, **7**:270-273.

Yeung, R.S.M., Penninger, J., Mak, T.W.: **T-cell development and function in gene-knockout mice.** *Curr. Opin. Immunol.* 1994, **6**:298-307.

PART II

THE RECOGNITION OF ANTIGEN

Structure of the Antibody Molecule and the Immunoglobulin Genes

3

Antibodies are the antigen-specific products of B cells, and the production of antibody in response to infection is the main contribution of B cells to adaptive immunity. Antibodies were the first of the molecules that participate in specific immune recognition to be characterized and are still the best understood. Collectively, antibodies form a family of plasma proteins known as the immunoglobulins, whose basic building block, the immunoglobulin domain, is used in various forms in many molecules of both the immune system and other biological recognition systems.

The antibody molecule has two separable functions: one is to bind specifically to molecules from the pathogen that elicited the immune response; the other is to recruit various cells and molecules to destroy the pathogen once the antibody is bound to it. These functions are structurally separated in the antibody molecule, one part of which specifically recognizes antigen and the other engages the effector mechanisms that will dispose of it. The antigen-binding region varies extensively between antibody molecules and is thus known as the **variable region** or **V region**. The variability of antibody molecules allows each molecule to recognize a particular antigen, and the total repertoire of antibodies made by a single individual is large enough to ensure that virtually any structure can be bound. The region of the antibody molecule that engages the effector functions of the immune system does not vary in the same way and is thus known as the **constant region** or **C region**, although it has in fact five main forms, or isotypes, which are specialized for activating different immune effector mechanisms.

The remarkable diversity of antibody molecules is the consequence of a highly specialized mechanism by which the antibody genes expressed in any given cell are assembled by DNA rearrangements that join together two or three different gene segments to form a V-region gene during the development of the B cell. Subsequent DNA rearrangement can attach the assembled V-region gene to any C-region gene and thus produce antibodies of any of the five isotypes.

B cells do not secrete antibody until they have been stimulated by specific antigen, which they recognize by means of membrane-bound immunoglobulin molecules that serve as their antigen receptors. The binding of antigen to these surface receptors is a crucial step in inducing the B cell to proliferate and differentiate into an antibody-secreting cell. In this chapter, we shall describe the structural and functional properties of antibody molecules and explain the specialized genetic processes that generate their diversity and functional versatility. The way in which surface immunoglobulin molecules act as receptors that transmit activating signals to B cells will be described in Chapter 5; the assembly and expression of immunoglobulin genes during B-cell development will be described in Chapter 6; and the functioning of the humoral immune response will be discussed in Chapter 9.

The structure of a typical antibody molecule.

Antibody molecules are roughly Y-shaped molecules consisting of three equal-sized segments, loosely connected by a flexible tether. Three schematic representations of this structure, determined by X-ray crystallography, are shown in Fig. 3.1. The aim of this section is to explain how this structure is formed and how it allows antibody molecules to carry out their dual tasks: binding on the one hand to a wide variety of antigens, and on the other hand, to a limited number of effector molecules and cells. As we shall see, each of these tasks is carried out by separate parts of the molecule; two ends are variable between different antibody molecules and are involved in antigen binding, whereas the third end is conserved and interacts with effector molecules.

All antibodies are constructed in the same way from paired heavy and light polypeptide chains, and the generic term **immunoglobulin** (**Ig**) is used for all such proteins. Within this general category, however, five classes of immunoglobulin—IgM, IgD, IgG, IgA, and IgE—can be distinguished biochemically as well as functionally, whereas more subtle differences confined to the variable region account for the specificity of antigen binding. We shall describe the general structural features of immunoglobulin molecules, using the IgG molecule as an example.

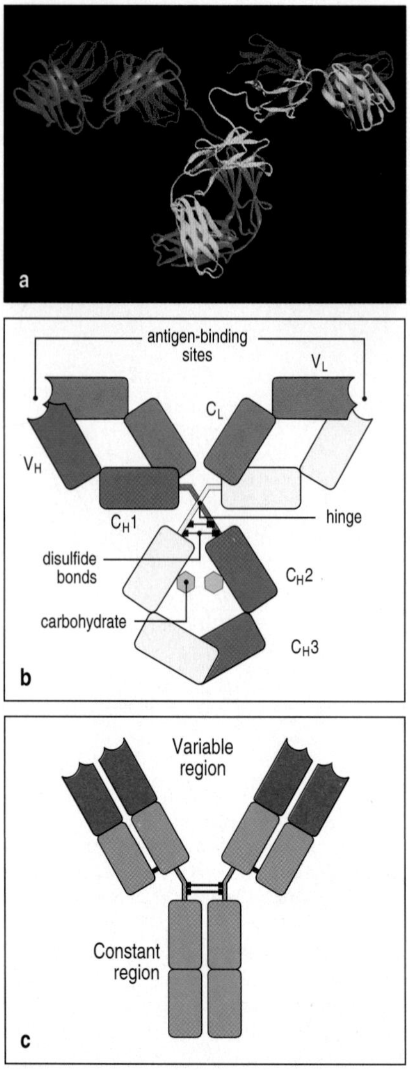

Fig. 3.1 Structure of an antibody molecule. Panel a: a ribbon diagram based on the X-ray crystal structure of an IgG antibody shows the course of the backbone polypeptide chains. Three globular regions form a Y. The two antigen-binding sites are at the tips of the arms, which are tethered to the trunk of the Y by a flexible hinge region. Panel b: a schematic representation of the structure shown in a, illustrating the four-chain composition and the separate domains comprising each chain. Panel c: a simplified schematic representation that will be used throughout this book. Photograph courtesy of A McPherson and L Harris.

3-1 IgG antibodies consist of four polypeptide chains.

IgG antibodies are large molecules (with a molecular weight of approximately 150 kDa) composed of two different polypeptide chains. One, of approximately 50 kDa, is termed the **heavy** or **H chain**, and the other, of 25 kDa, is termed the **light** or **L chain** (Fig. 3.2). The two chains are present in an equimolar ratio, and each IgG molecule contains two heavy chains and two light chains. The two heavy chains are linked to each other by disulfide bonds and each heavy chain is linked to a light chain by a disulfide bond. In any one immunoglobulin molecule, the two heavy chains and the two light chains are identical.

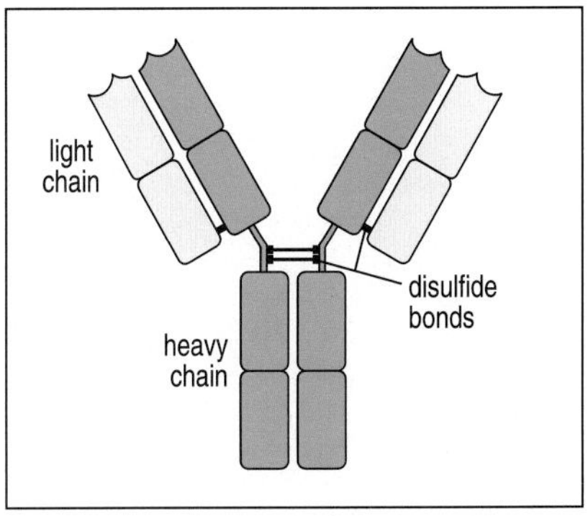

Fig. 3.2 Immunoglobulin molecules are composed of two types of chain: heavy chains and light chains. Each immunoglobulin molecule is made up of two heavy chains (green) and two light chains (yellow) joined by disulfide bonds so that each heavy chain is linked to a light chain and the two heavy chains are linked together.

There are two types of light chain found in antibodies, which are termed lambda (λ) and kappa (κ) chains. No functional difference has been found between antibodies having λ or κ light chains, and either type of light chain may be found in antibodies of any of the five major classes. The ratio of the two types of light chain varies from species to species. In mice, the average κ to λ ratio is 20:1, whereas in humans it is 2:1 and in cattle it is 1:20. The reason for this variation is unknown. Distortions of this ratio can sometimes be used to detect immune system abnormalities: for example, an excess of λ light chains in a human might indicate the presence of a B-cell tumor producing λ chains.

By contrast, the class of an antibody is defined by the structure of its heavy chain. There are five main **heavy-chain classes** or **isotypes**, some of which have several subtypes, and these determine the functional activity of an antibody molecule. The five major classes of immunoglobulin are **immunoglobulin M** (**IgM**), **immunoglobulin D** (**IgD**), **immunoglobulin G** (**IgG**), **immunoglobulin A** (**IgA**), and **immunoglobulin E** (**IgE**), and their heavy chains are denoted by the corresponding lower case Greek letter (μ, δ, γ, α, and ε, respectively). Their distinctive functional properties are conferred by the carboxy-terminal part of the heavy chain, where it is not associated with the light chain. We shall describe the distinct heavy-chain isotypes in more detail later. The general structural features of all the isotypes are similar and we shall consider IgG, the most abundant isotype in blood plasma, as a typical antibody molecule.

3-2 The heavy and light chains are composed of constant and variable regions.

The amino acid sequences of many immunoglobulin heavy and light chains have been determined and reveal two important features of antibody molecules. First, each chain consists of a series of similar, although not identical, amino acid sequences, each about 110 amino acids in length. The light chain comprises two such sequences, whereas the heavy chain of the IgG antibody contains four. This suggests that the immunoglobulin chains have evolved by repeated duplication of an ancestral gene corresponding to one folded domain of the protein. As we shall see, these sequences do in fact correspond to separate structural domains in the folded protein (see Fig. 3.1).

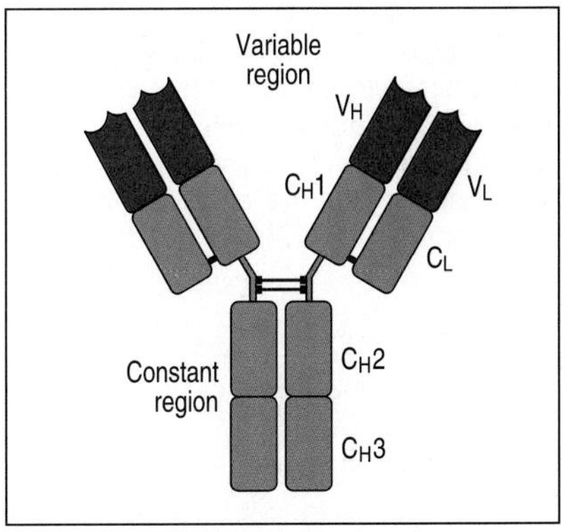

Fig. 3.3 The heavy and light chains of an immunoglobulin can be divided into domains on the basis of sequence similarity. The amino-terminal domain (red) of each chain is variable in sequence when several antibodies are compared; the remaining domains are constant (blue). The two domains of the light chains are termed V_L and C_L. IgG has four domains in the heavy chain, which are termed V_H, C_H1, C_H2 and C_H3.

The second important feature revealed by sequence comparisons is that the amino-terminal sequences of both the heavy and light chains vary greatly between different antibodies. The variability in sequence is limited to approximately the first 110 amino acids, corresponding to the first domain, whereas the carboxy-terminal sequences are constant between immunoglobulin chains, either light or heavy, of the same isotype (Fig. 3.3). The variable domains (**V domains**) make up the V region of the antibody, and the constant domains of the heavy chains (**C domains**) make up the C region (see Fig. 3.3).

3-3 The antibody molecule can readily be cleaved into functionally distinct fragments.

The antibody molecule comprises three equal-sized globular portions joined by a flexible stretch of polypeptide chain known as the **hinge region** to form a crude 'Y' shape (see Fig. 3.1). Each arm of the Y is formed by the association of a light chain with the amino-terminal half of a heavy chain, whereas the trunk of the Y is formed by the pairing of the carboxy-terminal halves of the two heavy chains. The association of the heavy and light chains is such that the V_H and V_L domains are paired, as are the C_H1 and C_L domains. The C_H3 domains pair with each other but the C_H2 domains do not interact; carbohydrate side chains attached to the C_H2 domains lie between the two heavy chains. The two antigen-binding sites are formed by the paired V_H and V_L domains at the end of the two arms of the Y (see Fig. 3.1, center panel).

Proteolytic enzymes (proteases) that cleave polypeptide sequences have been used to dissect the structure of antibody molecules and to determine which parts of the molecule are responsible for its various functions. Limited digestion with the protease papain cleaves antibody molecules into three fragments (Fig. 3.4). Two fragments are identical and contain the antigen-

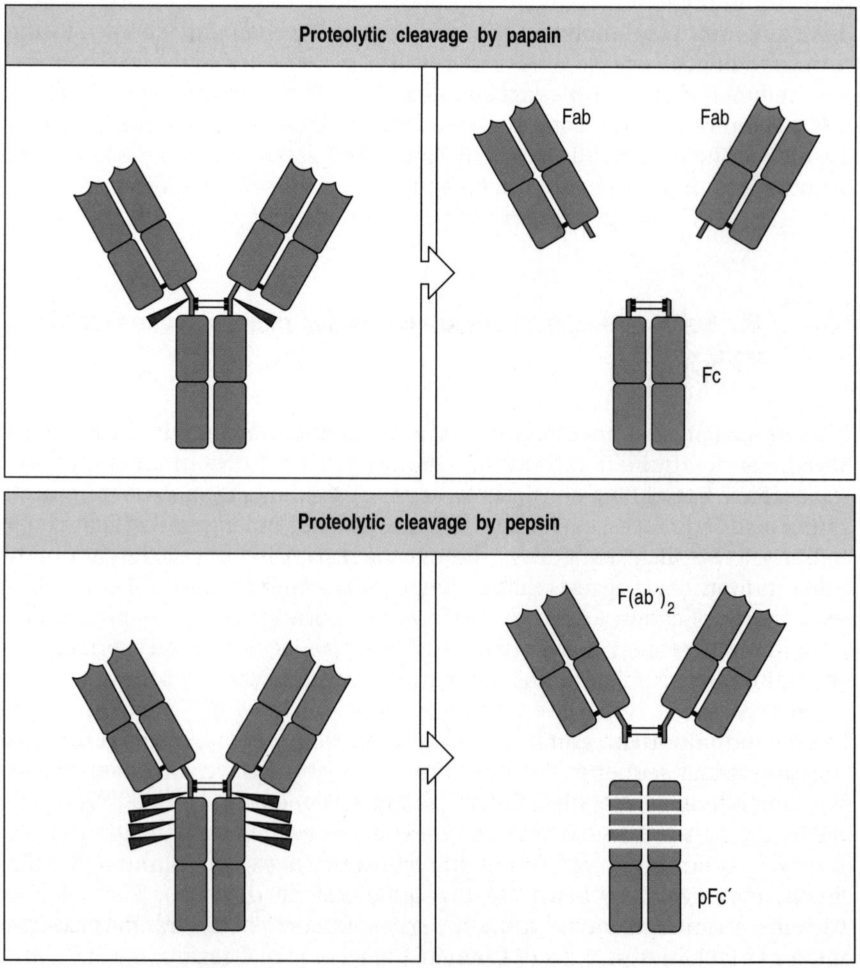

Fig. 3.4 The Y-shaped immunoglobulin molecule can be dissected by partial digestion with proteases. Papain cleaves the immunoglobulin molecule into three pieces, two Fab fragments and one Fc fragment (upper panels). The Fab binds antigens. The Fc is crystallizable and contains C regions. Pepsin cleaves an immuno-globulin to yield one $F(ab')_2$ fragment and many small pieces of the Fc fragment, the largest of which is called the pFc′ fragment (lower panels). $F(ab')_2$ is written with a prime because it contains a few more amino acids than Fab, including the cysteines that are necessary for the disulfide bonds.

binding activity, and these are termed the **Fab fragments**, for Fragment antigen binding. The Fab fragments correspond to the arms of the antibody molecule, which contain the complete light chains paired with the V_H and C_H1 domains of the heavy chains. The other fragment contains no antigen-binding activity but was originally observed to crystallize readily, and for this reason was named the **Fc fragment**, for Fragment crystallizable. This fragment corresponds to the paired C_H2 and C_H3 domains and is the part of the antibody molecule that interacts with effector molecules and cells.

The exact pattern of fragments obtained after proteolysis depends on where the protease cleaves the antibody molecule in relation to the disulfide bonds that link the two heavy chains. These lie in the hinge region between the C_H1 and C_H2 domains, and, as illustrated in Fig. 3.4, papain cleaves the antibody molecule on the amino-terminal side of the disulfide bonds, releasing the two arms of the antibody as separate Fab fragments, whereas in the Fc fragment the carboxy-terminal halves of the heavy chains remain linked.

A second protease, pepsin, cleaves in the same general region of the antibody molecule as papain but on the carboxy-terminal side of the disulfide bonds (see Fig. 3.4), producing a fragment, the $F(ab')_2$ fragment, in which the two arms of the antibody molecule remain linked. In this case the remaining part of the heavy chain is cut into several small fragments. The $F(ab')_2$ fragment has exactly the same antigen-binding characteristics as the original antibody but is unable to interact with any effector molecule and thus is of potential value in therapeutic applications of antibodies as well as in research into the role of the Fc portion.

Genetic engineering techniques also now permit the construction of a truncated Fab comprising only the V region of a heavy chain linked by a stretch of synthetic peptide to a V region of a light chain. This is called **single-chain Fv**, named from Fragment variable. Fv molecules may become valuable therapeutic agents because of their small size, allowing ready tissue penetration. They may be coupled to protein toxins to yield immunotoxins with potential application, for example, in tumor therapy.

3-4 The immunoglobulin molecule is flexible, especially at the hinge region.

The hinge region that links the Fc and Fab portions of the antibody molecule is in reality a flexible tether, allowing independent movement of the two Fab arms, rather than a rigid hinge. For example, electron microscopy of antibody complexes with a bivalent hapten capable of crosslinking two antigen-binding sites demonstrates that the angle between the two Fab arms can vary (Fig. 3.5). Some flexibility is also found at the junction between the V and C domains, allowing bending and rotation of the V domain relative to the C domain—for example, in the crystal structure of the antibody molecule shown in Fig. 3.1 (top panel), not only are the two hinge regions clearly different, but the angle between the V and C domains in each of the two Fab arms is also different. This range of motion has led to the junction between the V and C domains being referred to as a 'molecular ball-and-socket joint'. Flexibility at both the hinge and V–C junction enables the binding of both arms of the antibody molecule to sites that are different distances apart, for instance sites on bacterial cell-wall polysaccharides. Flexibility of the hinge also permits the interaction of antibodies with the antibody-binding proteins that mediate immune effector mechanisms, as will be described in Chapter 9.

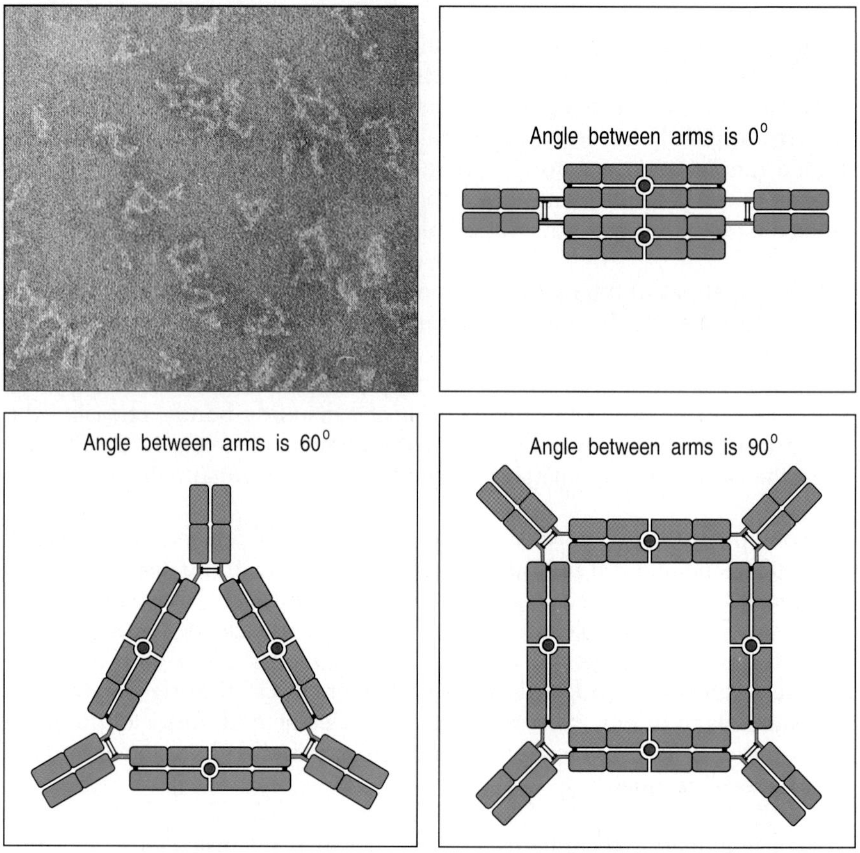
Angle between arms is 0°
Angle between arms is 60°
Angle between arms is 90°

Fig. 3.5 Antibody arms are joined by a flexible hinge. A small bifunctional hapten (red ball in diagrams) that can crosslink two antigen-binding sites is used to create antigen:antibody complexes, which can be seen in the electron micrograph. Linear, triangular, and square forms are seen, with short projections or spikes. Limited pepsin digestion removes these spikes, which therefore correspond to the Fc portion of the antibody; the $F(ab')_2$ pieces remain crosslinked by antigen. The interpretation of the complexes is shown in the diagrams. The angle between the arms of the antibody molecules varies, from 0° in the antibody dimers, through 60° in the triangular forms, to 90° in the square forms, showing that the connections between the arms are flexible. Photograph (x 300,000) courtesy of N M Green.

3-5 Each domain of an immunoglobulin molecule has a similar structure.

The discrete globular domains of immunoglobulin chains fall into two distinct structural categories, corresponding to V and C domains. The similarities and differences between these two domains can be seen in the diagram of a light chain in Fig. 3.6. Both domains are constructed from two sheets formed by adjacent strands of the polypeptide chain that pack together and are linked by a disulfide bridge, forming a roughly cylindrical shape. The sheets that form this structure are known as **β sheets** and the structure they form is known as a **β barrel** or **β sandwich**. β barrels can be formed in many different ways in different proteins: the β barrel that is formed by immunoglobulin chains is known as the **immunoglobulin fold**. In Sections 3-11 and 3-22 we shall see how each domain is encoded by a discrete exon, separated by an intron from the next domain exon.

Both the essential similarity of the V and C domains and the critical difference between them are most clearly seen in the bottom panels of Fig. 3.6, where the cylindrical domains are opened out to reveal how the polypeptide chain folds to create each layer. The main difference between the V and C domain structures is that the V domain is larger and has an extra loop of polypeptide chain. We shall see in the next section that the flexible loops of the V-region domain form the antigen-binding site of the antibody molecule.

Many of the amino acids that are common to all the domains of the heavy and light chains lie in the core of the immunoglobulin fold and are critical to the stability of its structure. For that reason, other proteins having sequences homologous to those of immunoglobulins are believed to have domains with

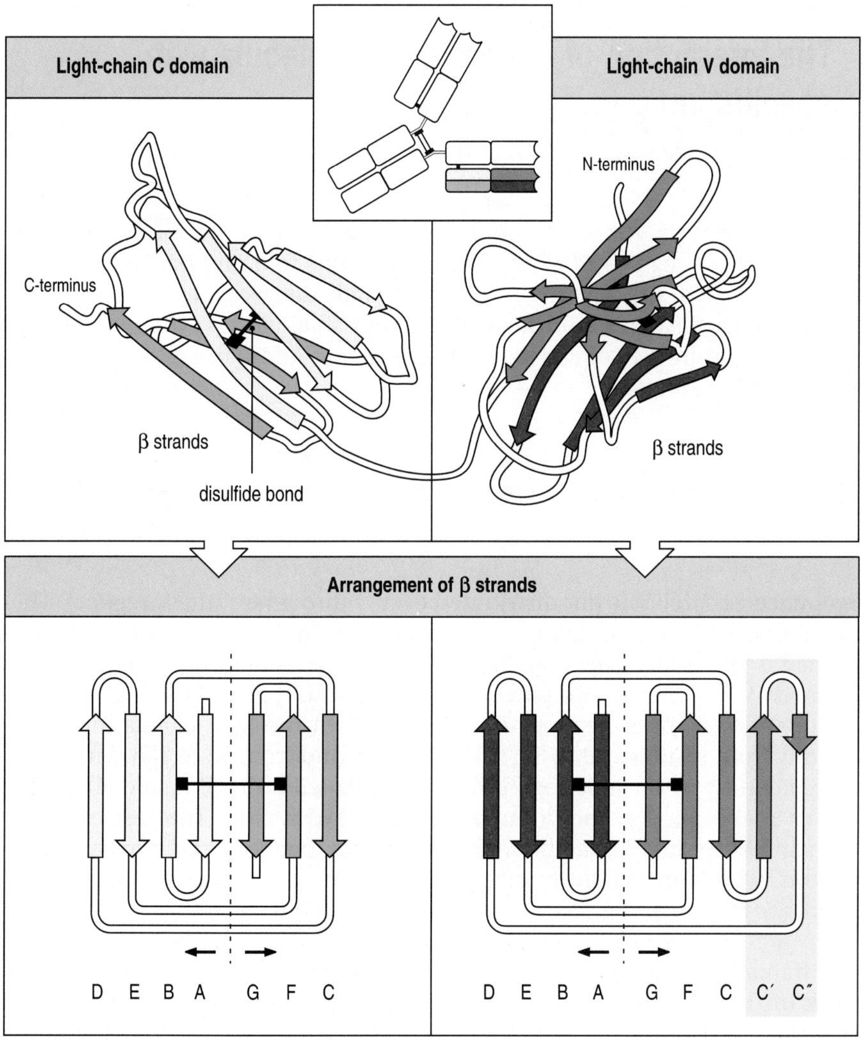

Fig. 3.6 The structure of immunoglobulin constant and variable domains. The upper panels show schematically the folding pattern of the C and V domains of an immunoglobulin light chain. Each domain is a globular structure in which several strands of polypeptide chain come together to form two antiparallel β sheets that are held together by a disulfide bond. The strands in each sheet are shown in distinct colors in the folded structure but their arrangement can be seen more clearly when the sheets are opened out, as shown in the lower panels. The β strands are lettered sequentially with respect to their occurrence in the amino acid sequence of the domains; the order in each β sheet is characteristic of immunoglobulin domains. The β strands C′ and C″ that are found in the V domains but not in the C domains are indicated by a shaded background. The characteristic 4-strand plus 3-strand (C-region type domain) or 4-strand plus 5-strand (V-region type domain) are typical immunoglobulin superfamily domain building blocks, found in a whole range of other proteins as well as antibodies.

a similar structure. These **immunoglobulin domains** are found in many proteins of the immune and nervous systems, as well as in other proteins thought to be involved in cell–cell recognition.

Summary.

IgG antibodies are made up of four polypeptide chains, comprising two identical light chains and two identical heavy chains, and can be thought of as forming a flexible Y-shaped structure. Each of the four chains has a variable (V) region at its amino terminus, which contributes to the antigen-binding site, and a constant (C) region, which in the heavy chain determines the isotype and hence the functional properties of the antibody. The light chains are bonded to the heavy chains by many non-covalent interactions and by disulfide bonds, and the V regions of the heavy and light chains pair to generate two identical antigen-binding sites, which lie at the tips of the arms of the Y. The possession of two antigen-binding sites allows antibody molecules to crosslink antigens. The trunk of the Y, or Fc fragment, is composed of the two carboxy-terminal domains of the two heavy chains. Joining the arms of the Y to the trunk are the flexible hinge regions. The Fc fragment and hinge regions differ in antibodies of different isotypes, thus determining their functional properties. However, the overall organization of the domains is similar in all isotypes.

The interaction of the antibody molecule with specific antigen.

In the previous section we described the structure of the antibody molecule and how the V regions of heavy and light chains fold and pair to form the antigen-binding site. In this section we shall discuss the different ways in which antigens can bind to antibody and shall address the question of how variation in the sequences of the antibody V domains determines the specificity for antigen.

3-6 Localized regions of hypervariable sequence form the antigen-binding site.

Sequence variability is not distributed evenly throughout the V regions. The distribution of variable amino acids can be seen clearly by using what is termed a **variability plot** (Fig. 3.7), where the sequences of many different antibody V regions are compared. Three regions of particular variability can be identified in both heavy and light chains. In the light chains these are roughly from residues 28 to 35, from 49 to 59, and from 92 to 103. These are designated **hypervariable regions** and are denoted HV1, HV2, and HV3. The most variable part of the domain is in the HV3 region. The rest of the V domain shows less variability and the regions between the hypervariable regions, which are relatively invariant, are termed the **framework regions**. There are four such regions, designated FR1, FR2, FR3, and FR4.

The framework regions form the β sheets that provide the structural framework of the domain, whereas the hypervariable sequences correspond to three loops at one edge of each sheet that are juxtaposed in the folded protein

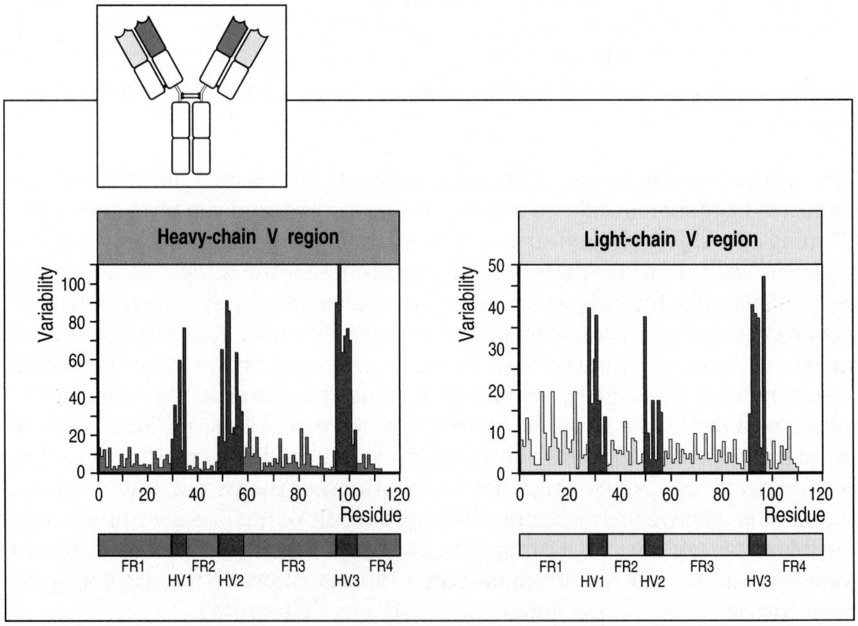

Fig. 3.7 There are discrete regions of hypervariability in V domains. The figure shows an analysis of a sequence comparison of several dozen heavy- and light-chain V regions. At each amino acid position the degree of variability is the ratio of the number of different amino acids seen in all of the sequences together to the frequency of the most common amino acid. Three hypervariable regions (HV1, HV2, and HV3) are indicated in red and correspond to CDR1, CDR2, and CDR3. They are flanked by less variable framework regions (FR1, FR2, FR3, and FR4, shown in blue or yellow).

(Fig. 3.8). Thus, not only is sequence diversity focused on particular parts of the V regions but it is localized to a particular part of the surface of the molecule. Moreover, when the V_H and V_L domains pair in the antibody molecule, the hypervariable loops from each domain are brought together, creating a single hypervariable site at the tip of the Fab fragment that forms the binding site for antigens, the **antigen-binding site** or **antibody-combining site**. As the three hypervariable loops constitute the binding site for antigen and determine specificity by forming a surface complementary to the antigen, they are more commonly termed the **complementarity-determining regions**, or **CDRs**, and are denoted **CDR1**, **CDR2**, and **CDR3**. One consequence of the contribution of CDRs from both V_H and V_L domains to the antigen-binding site is that it is the combination of the heavy and the light chain that determines the final antigen specificity. Thus, one way in which the immune system is able to generate antibodies of different specificities is by generating different combinations of heavy- and light-chain V regions. This means of producing variability is known as **combinatorial diversity**; we will encounter a second form of combinatorial diversity when we come to consider how the genes encoding the heavy- and light-chain V regions are created from smaller segments of DNA (see Section 3-11).

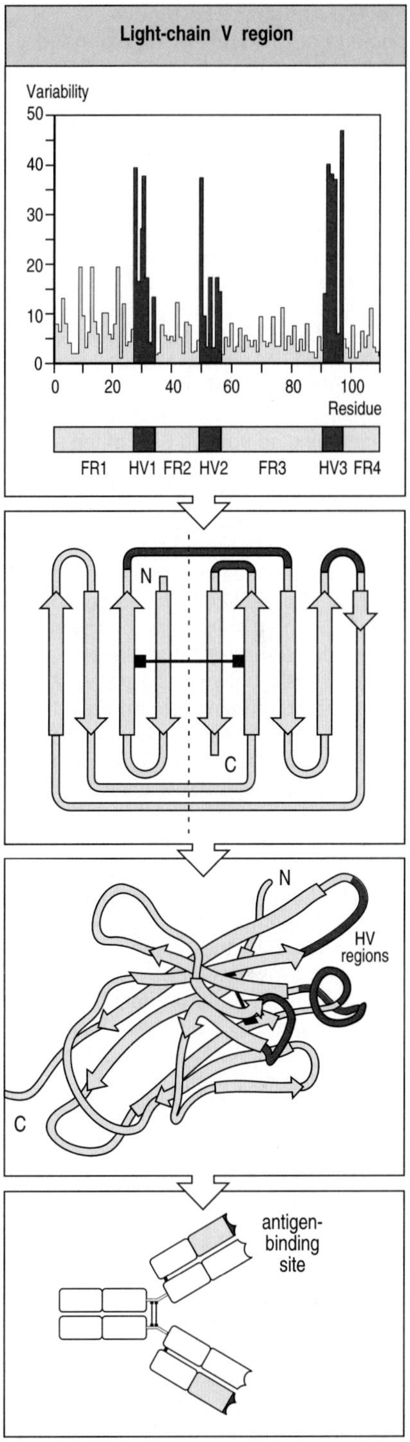

Fig. 3.8 The hypervariable regions lie in discrete loops of the folded structure. When the hypervariable regions are positioned on the structure of a V domain it can be seen that they lie in loops that are brought together in the folded structure. In the antibody molecule, the pairing of a heavy and a light chain brings together the hypervariable loops from each chain to create a single hypervariable surface, which forms the antigen-binding site at the tip of each arm.

3-7 Small molecules bind to clefts between the heavy- and light-chain V domains.

In early investigations of antigen binding, the only available sources of single species of antibody molecules were tumors of antibody-secreting cells, which produce homogeneous antibody molecules. The specificities of the tumor-derived antibodies were unknown, so that large numbers of compounds had to be screened to identify ligands that could be used in the analysis of antigen binding. In general, the substances found to bind to the antibodies were small chemical compounds, or haptens, such as phosphorylcholine or vitamin K_1. Structural analysis of antigen:antibody complexes between antibodies and their hapten ligands provided the first direct evidence that the hypervariable regions form the antigen-binding site, and established the structural basis for antigen specificity. Subsequently, with the discovery of monoclonal antibodies (see Section 2-10), it became possible to make large amounts of pure antibody specific for many different sorts of antigen. From these it has been possible to obtain a more general picture of how antibodies react with their antigens, confirming and extending the view of antibody:antigen interactions derived from the study of haptens.

The surface of the antibody molecule formed by the juxtaposition of the CDRs of the heavy and light chains creates the site to which antigens bind. Clearly, as the sequences of the CDRs are different in different antibodies, so are the shapes of the surfaces created by these CDRs. As a general principle, antibodies bind ligands whose surfaces are complementary to that of the antibody. For a small antigen, such as a hapten or a short peptide, the surface to which it binds may be formed by a pocket or groove lying between the heavy- and light-chain V domains (Fig. 3.9, top and center panels). Other antigens, such as protein molecules, can be of the same size as, or larger than, antibodies themselves, and cannot fit into such grooves or pockets. In these cases the interface between the two molecules is often an extended surface involving all of the CDRs and, in some cases, can include some of the framework regions of the antibody (see Fig. 3.9, bottom panel). This surface need not be concave, but can be flat, undulating or even convex.

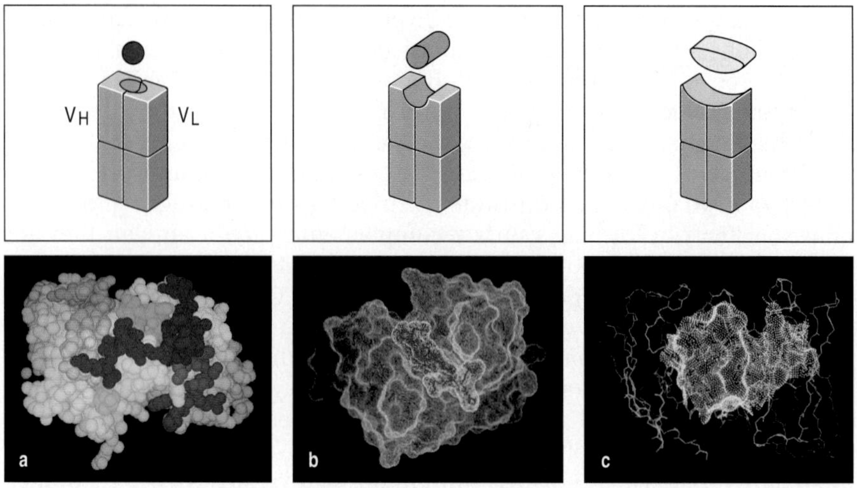

Fig. 3.9 Antigens can bind in pockets or grooves, or on extended surfaces in the binding sites of antibodies. The panels in the top row show schematic representations of the different types of binding site in a Fab fragment of an antibody. Left, pocket; center, groove; right, extended surface. Below are examples of each type. Panel a: space-filling representation of the interaction of a small peptide antigen with the complementarity determining regions (CDRs) of a Fab fragment as viewed looking into the antigen-binding site. Seven amino acid residues of the antigen, shown in red, are bound in the antigen-binding pocket. Five of the six CDRs (H1, H2, H3, L1, and L3) interact with the peptide, whereas L2 does not. The CDR loops are colored as follows: L2, magenta; L3, green; H1, blue; H2, pale purple; H3, yellow. Panel b: in a complex of an antibody with a peptide from the human immunodeficiency virus, the peptide (orange) binds along a groove formed between the heavy- and light-chain V domains (green). Panel c: complex between hen egg-white lysozyme and the Fab fragment of its corresponding antibody (HyHel5). Two extended surfaces come into contact, as can be seen from this computer-generated image, where the surface contour of the lysozyme molecule (yellow dots) is superimposed on the antigen-binding site. Residues in the antibody that make contact with the lysozyme are shown in full (red); for the rest of the Fab fragment only the peptide backbone is shown (blue). All six CDRs of the antibody are involved in the binding. Photographs a and b courtesy of A I Wilson and R L Stanfield. Photograph c courtesy of S Sheriff.

3-8 Antibodies bind to extended sites on the surfaces of native protein antigens.

The biological function of antibodies is to bind to pathogens and their products, and to facilitate their removal from the body. Some of the most important pathogens have polysaccharide coats, and antibodies that recognize the carbohydrate moieties of these are essential in providing immune protection from such pathogens. In many cases, however, the antigens that provoke an immune response are proteins. For example, protective antibodies against viruses recognize viral coat proteins. In such cases, the antibody binds to the native conformation of the protein and the determinants recognized are therefore areas on the surface of the protein. Regions of a molecule that are recognized specifically by antibodies are called **antigenic determinants** or **epitopes**. Such sites on a protein surface are likely to be composed of amino acids from different parts of the sequence that have been brought together by protein folding. Epitopes of this kind are known as **conformational** or **discontinuous epitopes** because the site is composed of segments of the protein that are discontinuous in the primary sequence but are contiguous in the three-dimensional structure. In contrast, an epitope composed of a single segment of polypeptide chain is termed a **continuous** or **linear epitope**. Although antibodies raised against native proteins usually recognize discontinuous epitopes, some can be found to bind peptide fragments of the protein. Conversely, antibodies raised against peptide fragments of a protein or against synthetic peptides corresponding to part of its sequence are occasionally found to bind to the native protein. This makes it possible, in some cases, to use synthetic peptides in vaccines aimed at raising antibodies against an intact protein of a pathogen.

3-9 Antigen:antibody interactions involve a variety of forces.

The interaction between an antibody and its antigen can be disrupted by high salt concentrations, extremes of pH, detergents, and sometimes by competition with high concentrations of the pure epitope itself. The binding is therefore a reversible non-covalent interaction. The forces, or bonds, involved in these non-covalent interactions are outlined in Fig. 3.10.

Non-covalent forces	Origin	
Electrostatic forces	Attraction between opposite charges	$-\overset{+}{N}H_3 \quad \overset{-}{O}OC-$
Hydrogen bonds	Hydrogen shared between electronegative atoms (N,O)	$>N — H - - O=C<$; δ^- δ^+ δ^-
Van der Waals forces	Fluctuations in electron clouds around molecules oppositely polarize neighboring atoms	$\delta^+ \rightleftharpoons \delta^-$; $\delta^- \rightleftharpoons \delta^+$
Hydrophobic forces	Hydrophobic groups interact unfavorably with water and tend to pack together to exclude water molecules. The attraction also involves van der Waals forces	H_2O ; δ^+ ; δ^- ; H_2O ; δ^- ; δ^+ ; H_2O ; H_2O

Fig. 3.10 The non-covalent forces that hold together the antigen:antibody complex. Partial charges found in electric dipoles are shown as δ^+ or δ^-. Electrostatic forces diminish as the inverse square of the distance separating the charges, whereas van der Waals forces, which are more numerous in most antigen–antibody contacts, fall off as the sixth power of the separation and therefore operate only over very short ranges. Covalent bonds rarely occur between antigens and antibodies.

The non-covalent forces in antigen:antibody binding can involve electrostatic interactions, either between charged amino acid side chains, as in salt bridges, or between electric dipoles, as in hydrogen bonds and short-range van der Waals forces. High salt concentrations and extremes of pH disrupt antigen:antibody binding by weakening electrostatic interactions. This principle is employed in the purification of antigens by using affinity columns of immobilized antibodies or vice versa (see Section 2-7).

Hydrophobic interactions occur when two hydrophobic surfaces come together to exclude water. The strength of hydrophobic interactions is proportional to the surface area that is hidden from water. For some antigens, hydrophobic interactions probably account for most of the binding energy, although this is hard to quantify experimentally. In some cases, water molecules are trapped in pockets in the interface between antigen and antibody. These trapped water molecules may also contribute to binding, especially between polar residues.

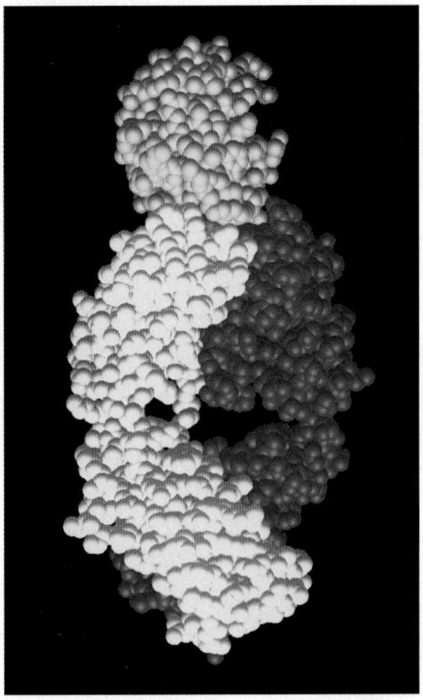

Fig. 3.11 The complex of lysozyme with the antibody D1.3. The interaction of the Fab fragment of D1.3 with hen egg-white lysozyme is shown, with the lysozyme in blue, the heavy chain in purple and the light chain in yellow. A glutamine residue of lysozyme, shown in red, protrudes between the two V domains of the antigen-binding site and makes hydrogen bonds important to the antigen:antibody binding. Original photograph courtesy of R J Poljak.

The contribution of each of these forces to the overall interaction depends on the specific antibody and antigen involved. A striking difference from other protein–protein interactions is that antibodies possess many aromatic amino acids in their antigen-binding sites; these amino acids participate mainly in van der Waals and hydrophobic interactions, and sometimes in hydrogen bonds. In general, the hydrophobic and van der Waals forces operate over very short ranges and serve to pull together two surfaces that are complementary in shape: hills on one surface must fit into valleys on the other for good binding to occur. In contrast, electrostatic interactions between charged side chains, and hydrogen bonds bridging oxygen and/or nitrogen atoms, accommodate specific features or reactive groups while strengthening the interaction overall. For example, in the complex of hen egg-white lysozyme with the antibody D1.3 (Fig. 3.11), strong hydrogen bonds are formed between the antibody and a particular glutamine in the lysozyme molecule that protrudes between the V_H and V_L domains. Lysozymes from partridge and turkey have another amino acid in place of the glutamine and do not bind to the antibody. In the high-affinity complex of hen egg-white lysozyme with another antibody, HyHel5 (see Fig. 3.9, bottom panel), two salt bridges between two basic arginines on the surface of the lysozyme interact with two glutamic acids, one each from the V_H CDR1 and CDR2 loops. Again, lysozymes that lack one of the two arginine residues show a 1000-fold decrease in affinity. Thus, overall surface complementarity, together with specific electrostatic and hydrogen-bonding interactions, appear to determine antibody specificity. Genetic engineering by site-directed mutagenesis can tailor an antibody-binding site to its complementary antigenic epitope; this process has its natural counterpart in maturation to higher affinity by the process of somatic hypermutation as an antibody response progresses, which we shall discuss later (see Section 3-18).

Summary.

X-ray crystallographic analysis of antigen:antibody complexes has demonstrated that the hypervariable loops of immunoglobulin V regions determine the specificity of antibodies. With protein antigens, the antibody molecule contacts the antigen over a broad area of its surface that is complementary to the surface recognized on the antigen. Electrostatic interactions, hydrogen bonds, van der Waals forces, and hydrophobic interactions can all contribute to binding. Amino acid side chains in most or all of the hypervariable loops make contact with antigen and determine both the specificity and the affinity of the interaction. Other parts of the V region play little part in the direct contact with the antigen but provide a stable structural framework for the hypervariable loops and help determine their position and conformation. Antibodies raised against native proteins usually bind to the surface of the protein and make contact with residues that are discontinuous in the primary structure of the molecule; however, they may occasionally bind peptide fragments of the protein, and antibodies raised against peptides derived from a protein can often be used to detect the native protein molecule. Peptides binding to antibodies usually bind in the cleft between the V regions of the heavy and light chains, where they make specific contact with some, but not necessarily all, of the hypervariable loops. This is also the usual mode of binding for carbohydrate antigens and small molecules such as haptens.

The generation of diversity in the humoral immune response.

Virtually any substance can elicit an antibody response. Furthermore, the response even to a simple antigen is diverse, comprising many different antibody molecules each with a unique affinity and fine specificity. The complete collection of antibody specificities available within an individual is known as the **antibody repertoire** and in humans consists of as many as 10^{11} different antibody molecules, and perhaps many more. Before it was possible to examine the immunoglobulin genes directly, there were two main hypotheses for the origin of this diversity. According to one, the **germline theory**, there is a separate gene for each different antibody chain and the antibody repertoire is largely inherited. By contrast, **somatic diversification theories** proposed that a limited number of inherited V-region sequences undergo alteration within B cells during the lifetime of an individual to generate the observed repertoire. The cloning of the genes that encode immunoglobulins showed that the antibody repertoire is, in fact, generated during B-cell development by DNA rearrangements. These combine and assemble different V-region gene segments from a relatively small group of inherited V-region sequences at each locus; diversity is further enhanced by a process of somatic hypermutation in mature B cells. Thus both theories were partially correct.

3-10 Immunoglobulin genes are rearranged in antibody-producing cells.

The DNA sequences encoding the V and C regions of immunoglobulin chains are separated by some considerable distance in the genome in all cells except for lymphocytes of the B lineage, in which rearrangements of the DNA occur early in ontogeny to bring them together. This was originally discovered 20 years ago when it first became possible to study the arrangement of the immunoglobulin genes by using restriction enzyme analysis and Southern blotting. Chromosomal DNA is first cut with a restriction enzyme, and the DNA fragments containing particular V- and C-region sequences are then identified by hybridization with radiolabeled DNA probes specific for the relevant V- and C-region sequences. In germline DNA, the V- and C-region sequences are on separate DNA fragments. However, in DNA from the antibody-producing B cell that was the source of the V- and C-region DNA probes, the probes hybridize to the same DNA fragment, showing that the V- and C-region sequences are adjacent in these cells. A typical experiment using human DNA is shown in Fig. 3.12.

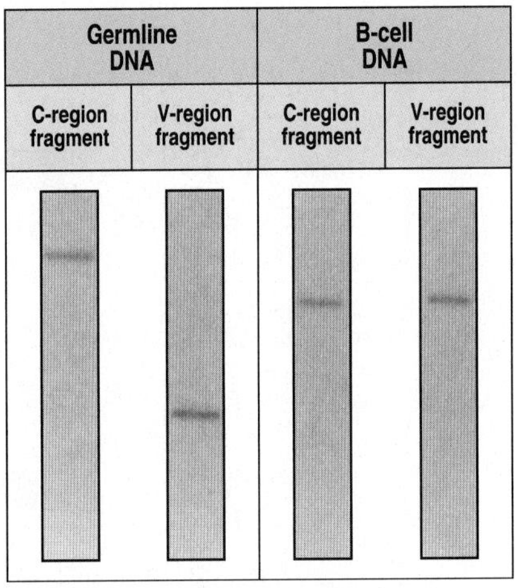

Fig. 3.12 Immunoglobulin genes are rearranged in B cells. The two photographs on the left (germline DNA) show a Southern blot of a restriction enzyme digest of DNA of a normal person. The locations of immunoglobulin DNA sequences are identified by hybridization with V- and C-region probes. The V and C regions are found in distinct DNA fragments. The two photographs on the right (B-cell DNA) are of a similar restriction digest of DNA from peripheral blood lymphocytes from a patient with chronic lymphocytic leukemia (see Chapter 6), in which a particular clone of B cells is greatly expanded. The malignant B cells express the V region from which the V-region probe was obtained, and a unique rearrangement is visible. In this DNA, the V and C regions are found in the same fragment, which is a different size from either the C- or the V-region germline fragments. A population of normal B lymphocytes has many different rearranged genes, so they yield a smear of DNA fragment sizes, normally not visible as a crisp band. Photograph courtesy of S Wagner and L Luzzatto.

This simple experiment shows that segments of genomic DNA are rearranged in somatic cells of the B-lymphocyte lineage. This process of rearrangement is known as **somatic recombination**, to distinguish it from the meiotic recombination that takes place during the production of gametes.

3-11 Complete V regions are generated by the somatic recombination of separate gene segments.

It is now known that the DNA rearrangements that bring together the DNA encoding the V and C regions of the immunoglobulin chain actually join separate segments of the V-region DNA, one of which is adjacent to the DNA encoding the C region. In the light chain, each V domain is encoded in two separate DNA segments. The first segment encodes the first 95–101 amino acids of the light chain and is termed a **V gene segment** because it makes up most of the V domain. The second segment encodes the remainder of the V domain (up to 13 amino acids) and is termed a **joining** or **J gene segment**.

The process of rearrangement that leads to the production of an immunoglobulin light-chain gene is shown in Fig. 3.13 (center panel). The joining of a V and a J gene segment creates a continuous piece of DNA encoding the whole of the light-chain V region. The J gene segments are separated from the C region (C) genes only by non-coding DNA, and are joined to them by RNA splicing after transcription, not by DNA recombination. In the experiment shown in Fig. 3.12 therefore, the germline DNA fragment identified by the 'V-region probe' contains the V gene segment, and that identified by the 'C-region probe' actually contains both the J and C gene segments.

The heavy-chain V regions are encoded in three gene segments. In addition to the V and J gene segments (denoted V_H and J_H to distinguish them from the light-chain gene segments), there is a third gene segment called the **diversity** or **D_H gene segment**, which lies between the V_H and J_H gene segments. The process of recombination that generates a complete heavy-chain V region is shown in Fig. 3.13 (right panel), and occurs in two separate stages. In the first, a D_H gene segment is joined to a J_H gene segment; then a V_H gene segment rearranges to DJ_H to complete a heavy-chain V-region exon. As with the light chains, RNA splicing joins the assembled V-region sequence to the C-region coding sequence found downstream.

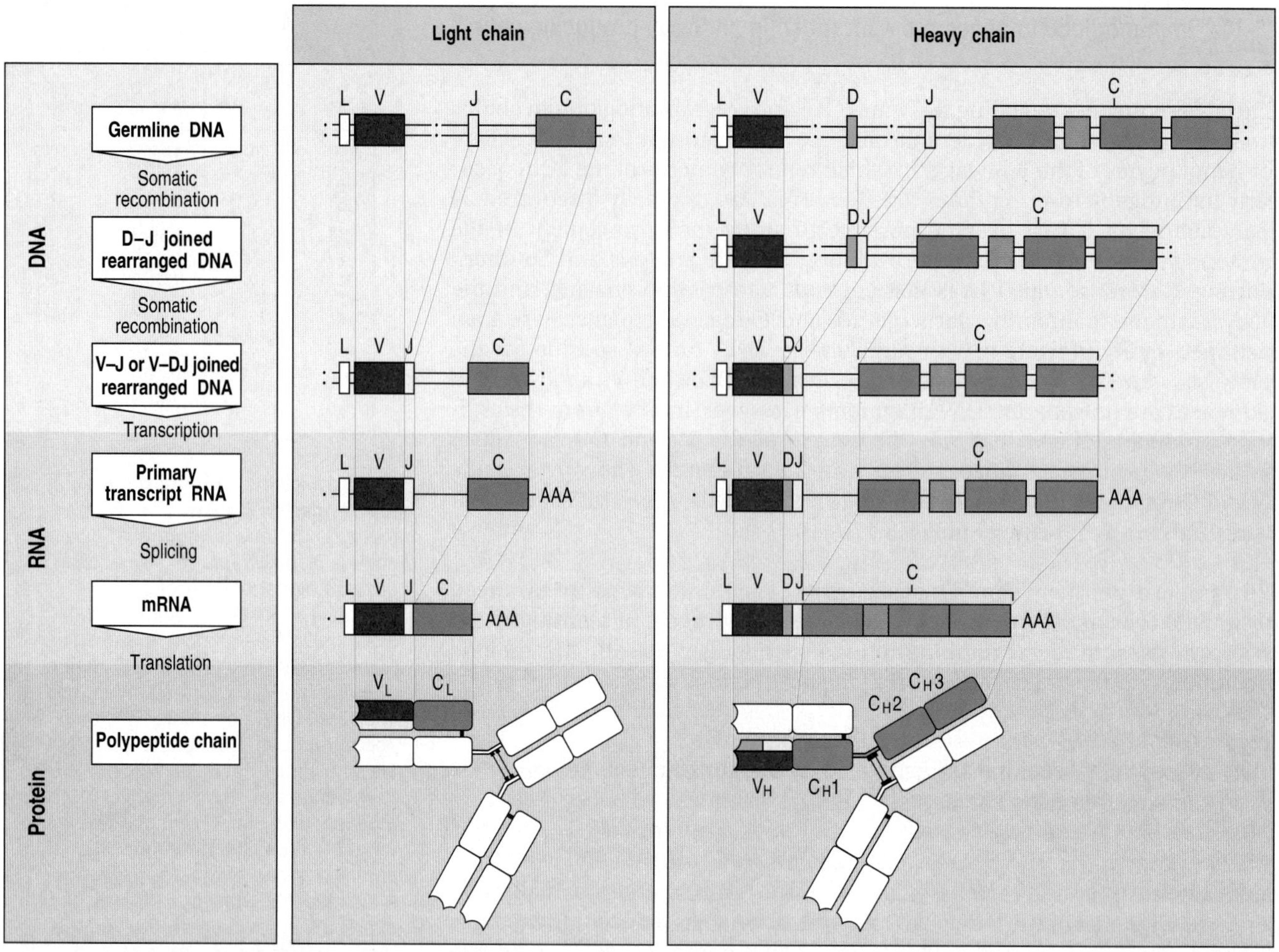

Fig. 3.13 V-region genes are constructed from gene segments. Light-chain V-region genes are constructed from two segments (left panel). A variable (V) and a joining (J) gene segment in the genomic DNA are joined to form a complete light-chain V-region exon. The C region is encoded in a separate exon and is joined to the V-region exon by splicing of the light-chain RNA to remove the L to V and the J to C introns. Immunoglobulin chains are extracellular proteins and the V gene segment is preceded by an exon encoding a leader peptide (L), which directs the protein into the cell's secretory pathways and is then cleaved. Heavy-chain V regions are constructed from three gene segments (right panel). First the diversity (D) and J gene segments join, then the V gene segment joins to the combined DJ sequence, forming a complete V_H exon. The heavy-chain C-region genes (only one is shown here for simplicity) are each encoded by several exons: note the separate exon encoding the hinge domain (purple). The C-region exons, together with the leader sequence, are spliced to the V-domain sequence during processing of the heavy-chain RNA transcript. The leader sequence is removed and carbohydrate moieties are attached after translation.

3-12 V-region gene segments are present in multiple copies.

For simplicity, we have discussed the formation of a complete immunoglobulin V-region gene as though there were only single copies of each gene segment. In fact, there are multiple copies of all of the gene segments in germline DNA. Their numbers can be estimated by probing restriction-enzyme digests of germline DNA: in humans, the numbers of functional gene segments of each type have also been determined by gene cloning and sequencing, and are shown in Fig. 3.14. These numbers can vary between individuals because of insertion or deletion of gene segments by meiotic recombination, or because of mutations that transform a functional gene into a non-functional pseudogene.

Fig. 3.14 The numbers of functional gene segments for the V regions of human heavy and light chains. These numbers are derived from exhaustive cloning and sequencing of DNA from one individual and exclude all pseudogenes (mutated and non-functional versions of a gene sequence). Owing to genetic polymorphism, the numbers will not be the same for all humans.

Number of functional gene segments in human immunoglobulin loci			
Segment	Light chains		Heavy chain
	κ	λ	H
Variable (V)	40	30	65
Diversity (D)	0	0	27
Joining (J)	5	4	6

The functional gene segments are organized into three clusters: the κ, λ, and heavy-chain genes. These clusters are located on different chromosomes and each is organized slightly differently, as shown in Fig. 3.15. For the λ light-chain genes, located on chromosome 22, there is a cluster of V_λ gene segments, followed by pairs of J_λ gene segments and C_λ genes. In the κ light-chain genes, on chromosome 2, the cluster of V_κ gene segments is followed by a cluster of J_κ gene segments, then by a single C_κ gene. Finally, the organization of the heavy-chain genes, on chromosome 14, resembles that of the κ genes, with separate clusters of V_H, D_H, and J_H gene segments and of C_H genes.

The human V gene segments can be grouped into families on the basis of similarity of DNA sequence; both the heavy-chain and κ-chain V gene segments can be subdivided into seven families whose members are more than 80% homologous, whereas there are eight families of V_λ gene segments. The families can be further grouped into clans, made up of families that are more similar to each other than to families in other clans. Human V_H gene segments fall into three such clans. All of the V_H gene segments identified from amphibians, reptiles and mammals also fall into the same three clans, suggesting that these clans existed in a common ancestor of these modern species.

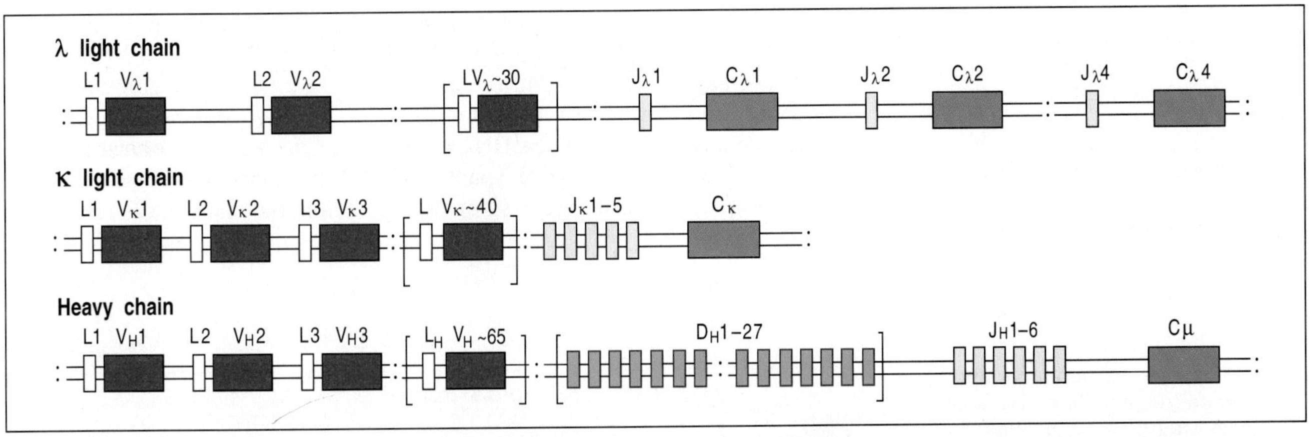

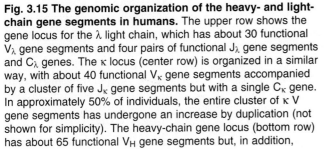

Fig. 3.15 The genomic organization of the heavy- and light-chain gene segments in humans. The upper row shows the gene locus for the λ light chain, which has about 30 functional V_λ gene segments and four pairs of functional J_λ gene segments and C_λ genes. The κ locus (center row) is organized in a similar way, with about 40 functional V_κ gene segments accompanied by a cluster of five J_κ gene segments but with a single C_κ gene. In approximately 50% of individuals, the entire cluster of κ V gene segments has undergone an increase by duplication (not shown for simplicity). The heavy-chain gene locus (bottom row) has about 65 functional V_H gene segments but, in addition, there is a cluster of around 27 D segments lying between the V_H gene segments and the six J_H gene segments. The heavy-chain locus also contains a large cluster of C_H genes that will be described in Fig. 3.24. For simplicity we have shown only a single C_H gene in this diagram without illustrating its separate exons, have omitted pseudogenes, and have shown all V regions in the same orientation. L, leader sequence. This diagram is not to scale: the total length of the heavy chain cluster is over 2 megabases (2 million bases), whereas some of the D segments are only six bases long.

3-13 Rearrangement of V, D, and J gene segments is guided by flanking sequences in DNA.

When the non-coding regions flanking the different heavy and light chain V, D, and J gene segments are compared, conserved sequences are found adjacent to the points at which recombination takes place. The sequences consist of a conserved block of seven nucleotides (the **heptamer** 5′CACAGTG3′), which is always contiguous with the coding sequence, followed by a spacer of roughly 12 or 23 base pairs (bp), followed by a second conserved block of nine nucleotides (the **nonamer** 5′ACAAAAACC3′) (Fig. 3.16). The spacer varies in sequence but its length is conserved and corresponds to one or two turns of the DNA double helix. This would bring the heptamer and nonamer sequences to one side of the DNA helix, where they can be bound by the protein complex that catalyzes recombination. The heptamer–spacer–nonamer is often called a **recombination signal sequence** or RSS.

VDJ recombination occurs only between gene segments located on the same chromosome and the process follows another rule, that recombination can link only a gene segment flanked by a 12mer-spaced RSS to one with a 23mer-spaced RSS (**the 12/23 rule**). Thus, for the heavy chain, a D_H gene segment can be joined to a J_H gene segment and a V_H gene segment to a D_H gene segment, but V_H gene segments cannot be joined to J_H gene segments directly, as both V_H and J_H gene segments are flanked by 23 bp spacers and the D_H gene segments have 12 bp spacers on both sides (see Fig. 3.16). It is now apparent that a D–D fusion also occurs in most species. In humans this is seen in approximately 5% of antibodies. These fusions break the 12/23 rule but add substantially to the diversity of the antibody repertoire. Similar D–D fusions occur in the T-cell receptor locus as well but, as we shall see in Chapter 4, the T-cell receptor recombination signal sequences are arranged so fusions can occur without breaking the 12/23 rule. D–D fusion is generally the major mechanism accounting for the extra length of the CDR3 loops found in the heavy chains of some antibodies.

The mechanism of gene segment rearrangement is similar for heavy- and light-chain genes, although only one joining event is needed for light-chain genes whereas two are needed to generate a complete heavy-chain V exon. The commonest mode of rearrangement (Fig. 3.17, left panels) involves the looping-out and deletion of the DNA intervening between two gene segments. The 12mer-spaced and 23mer-spaced recombination signal sequences are brought together by interactions between proteins that specifically recognize

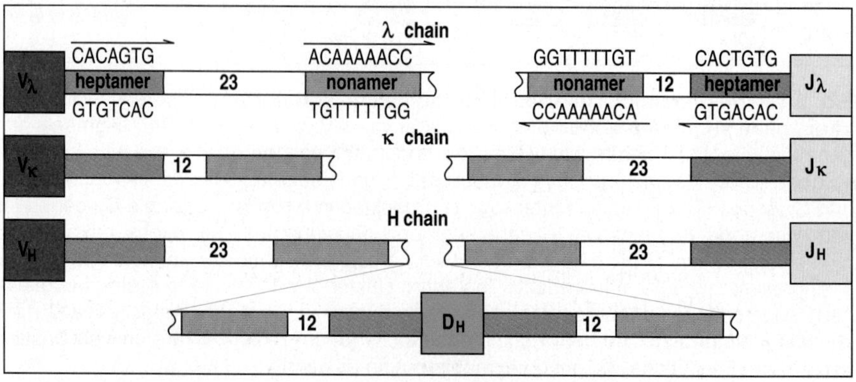

Fig. 3.16 Conserved heptamer and nonamer sequences flank the gene segments encoding the V regions of heavy (H) and light (λ and κ) chains. The spacing between the heptamer and nonamer sequences is always either approximately 12 bp or approximately 23 bp, and joining almost always involves a 12 bp and a 23 bp recombination signal sequence.

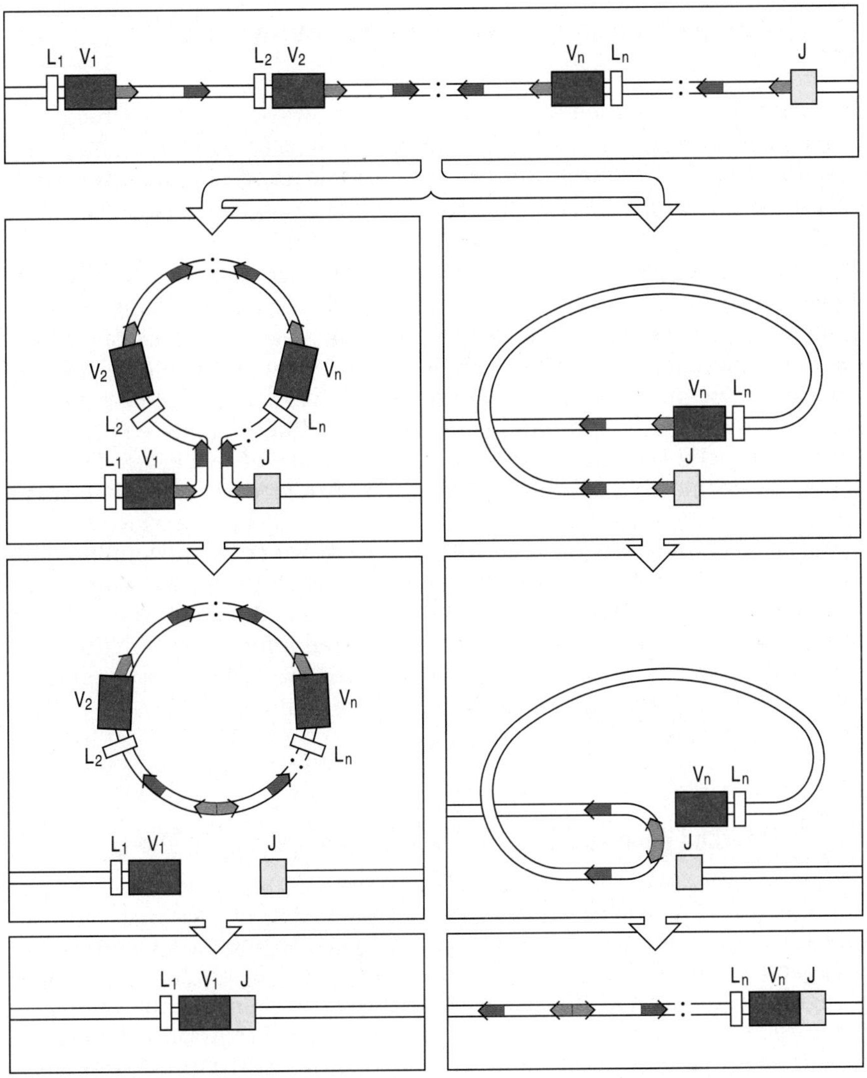

Fig. 3.17 V-region gene segments are joined by recombination. In every V-region recombination event, the signals flanking the gene segments are brought together to allow recombination to take place. For simplicity, the recombination of light chains is illustrated; for heavy chains, two separate recombination events are required to generate a functional V region. In some cases, as shown in the left panels, the V and J gene segments have the same transcriptional orientation. Juxtaposition of the recombination signal sequences results in the looping out of the intervening DNA. Heptamers are shown in orange, nonamers in purple and the arrows represent the directions of the recombination heptamer and nonamer signals (see Fig. 3.16). Recombination occurs at the ends of the heptamer sequences, creating a signal joint and releasing the intervening DNA in the form of a closed circle. Subsequently, the joining of the V and J gene segments creates the coding joint. In other cases, illustrated in the right panels, the V and J gene segments are initially oriented in opposite transcriptional directions. Bringing together the signal sequences in this case requires a more complex looping of the DNA. Joining the ends of the two heptamer sequences now results in the inversion of the intervening DNA. Again, the joining of the V and J segments creates a functional V-region exon.

the length of spacer between the heptamer and nonamer signals and thus enforce the 12/23 rule for recombination. The two DNA molecules are then broken and re-ligated. The ends of the heptamer sequences are joined precisely in a head-to-head fashion to form a **signal joint** in a circular piece of DNA, which is then lost from the genome when the cell divides. The joining of the V and J gene segments, to form what is called the **coding joint**, is imprecise, and consequently generates much additional variability.

A second mode of recombination can occur between two gene segments that have opposite transcriptional orientations. This mode of recombination is less common, although such rearrangements account for about half of all V_κ to J_κ joins; in roughly 50% of individuals, the transcriptional orientation of half of their V_κ gene segments is opposite to that of the J_κ gene segments. The mechanism of recombination is essentially the same but the DNA that lies between the two gene segments meets a different fate (see Fig. 3.17, right panels). When the recombination signals in such cases are brought together and recombination takes place, the intervening DNA is not lost from the chromosome but is retained in an inverted orientation.

3-14 Antibody diversity is generated by four main processes.

Antibody diversity is generated in four main ways, three of which are consequences of the process of recombination used to create complete immunoglobulin V regions, whereas the fourth is a mutational process that occurs later, acting only on rearranged DNA encoding the V regions.

First, there are multiple different copies of each type of gene segment that make up an immunoglobulin V region, and different combinations of gene segments can be used in different rearrangement events. This is responsible for a substantial part of the diversity of the heavy- and light-chain V regions. A second source of combinatorial diversity arises through the pairing of different combinations of heavy- and light-chain V regions to form the antigen-binding site. With these two means of generating diversity alone, approximately 2.5×10^6 different antibody molecules could, in theory, be made (see Section 3-15). Third, additional diversity is introduced at the joints between the different gene segments as a result of the recombination process. Finally, somatic hypermutation introduces point mutations into rearranged V-region genes and is the only means by which the antigen specificity of an immunoglobulin can be altered after recombination has created functional heavy- and light-chain genes. We will discuss these four mechanisms at greater length in the following sections.

3-15 Inherited gene segments are used in different combinations.

There are multiple copies of the V, D, and J gene segments, each of which is capable of contributing to an immunoglobulin V region. Many different V regions can therefore be made by selecting different combinations of these segments. For human κ light chains, there are approximately 40 functional V_κ gene segments and 5 J_κ gene segments, and thus potentially 200 different V_κ regions. For λ light chains there are approximately 30 functional V_λ gene segments and 4 J_λ gene segments, yielding 120 possible V_λ regions. So in all, 320 different light chains can be made. For the heavy chains of humans, there are 65 functional V_H gene segments, approximately 27 D_H gene segments and 6 J_H gene segments, and thus around 11,000 different possible V_H regions ($65 \times 27 \times 6 \approx 11{,}000$). As both the heavy- and the light-chain V regions contribute to antibody specificity, each of the 320 different light chains could be combined with each of the approximately 11,000 heavy chains to give around 3.5×10^6 different antibody specificities. The ability to create many different specificities by making many different combinations of a small number of gene segments is known as combinatorial diversity. For each of these loci, we have given the number of germline V gene segments contributing to functional antibodies; the total number of V gene segments is larger but the additional gene segments do not appear in expressed immunoglobulin molecules and are pseudogenes.

In practice, combinatorial diversity is likely to be less than we might expect from the theoretical calculations above. Not all V gene segments are used at the same frequency; some are common in antibodies; others are found only rarely. Moreover, not all V_H regions pair successfully with all V_L regions. However, two further processes add greatly to repertoire diversity: imprecise joining of V, D, and J gene segments and somatic hypermutation.

3-16 Variable addition and subtraction of nucleotides at the junctions between the gene segments encoding the V region contributes to diversity in the third hypervariable region.

Of the three hypervariable loops in the protein chains of immunoglobulins, two are encoded within the V gene segment DNA. The third (CDR3) falls at the junction between the V gene segment and the J gene segment, and in the heavy chain is partially encoded by the D gene segment. In both heavy and light chains, the diversity of the third hypervariable region is significantly increased by the addition and deletion of nucleotides at two steps in the formation of the junctions between gene segments. This is known as **junctional diversity**. The added nucleotides are known as **P-nucleotides** and **N-nucleotides** and their addition is illustrated schematically in Fig. 3.18.

P-nucleotides are so called because they make up palindromic sequences added to the ends of the gene segments. They are thought to occur in the following way. When the two heptamers of the recombination signal sequence are brought together in the course of DNA rearrangement, the DNA is cleaved precisely between the heptamer and the coding sequence of the gene segments to be joined (see Fig. 3.18, top panel). The two heptamers are then joined to remove the intervening DNA (see Fig. 3.17), but the cleaved ends of the coding segments are not directly ligated to one an other. Instead, the cleaved ends are sealed to form hairpins (see Fig. 3.18, second panel) and a single-stranded cleavage subsequently occurs at a random point within the coding sequence so that a single-stranded tail is formed from a few nucleotides of the coding sequence plus the complementary nucleotides from the other DNA strand (see Fig. 3.18, third and fourth panels). In most light-chain gene rearrangements, DNA repair enzymes then fill in complementary nucleotides on the single-stranded tails and the two double-stranded ends are then rejoined, leaving short palindromic sequences at the joint. In heavy-chain gene rearrangements and in some human light-chain genes, however, N-nucleotides are first added by a quite different mechanism.

N-nucleotides are so called because they are non-template-encoded. They are added by an enzyme called **terminal deoxynucleotidyl transferase (TdT)** to single-stranded ends of the coding DNA after hairpin cleavage. After the addition of up to 20 nucleotides by this enzyme, the two single-stranded stretches at the ends of the gene segments form base pairs over a short region. Repair enzymes then trim off any non-matching bases, synthesize complementary bases to fill in the remaining single-stranded DNA, and ligate them to the P-nucleotides (see Fig. 3.18, last three panels). N-nucleotides are less common in light-chain genes because TdT is expressed only for a short period in B-cell development, during the assembly of the heavy-chain genes, which undergo rearrangement before the light-chain genes are assembled.

As well as the addition of nucleotides at the V–J junction (light chain), and especially the V–D and D–J junctions of the heavy chain, nucleotides can also be deleted. This is accomplished by a variety of exonucleases. Thus, the length of heavy-chain CDR3 can be even shorter than the smallest D segment. In some instances it is difficult, if not impossible, to recognize the D segment that contributed to CDR3 formation because of the excision of most of its nucleotides.

As the total number of nucleotides added by these processes is random, the added nucleotides often disrupt the reading frame of the coding sequences beyond the joint. Such frameshifts will normally lead to a non-functional

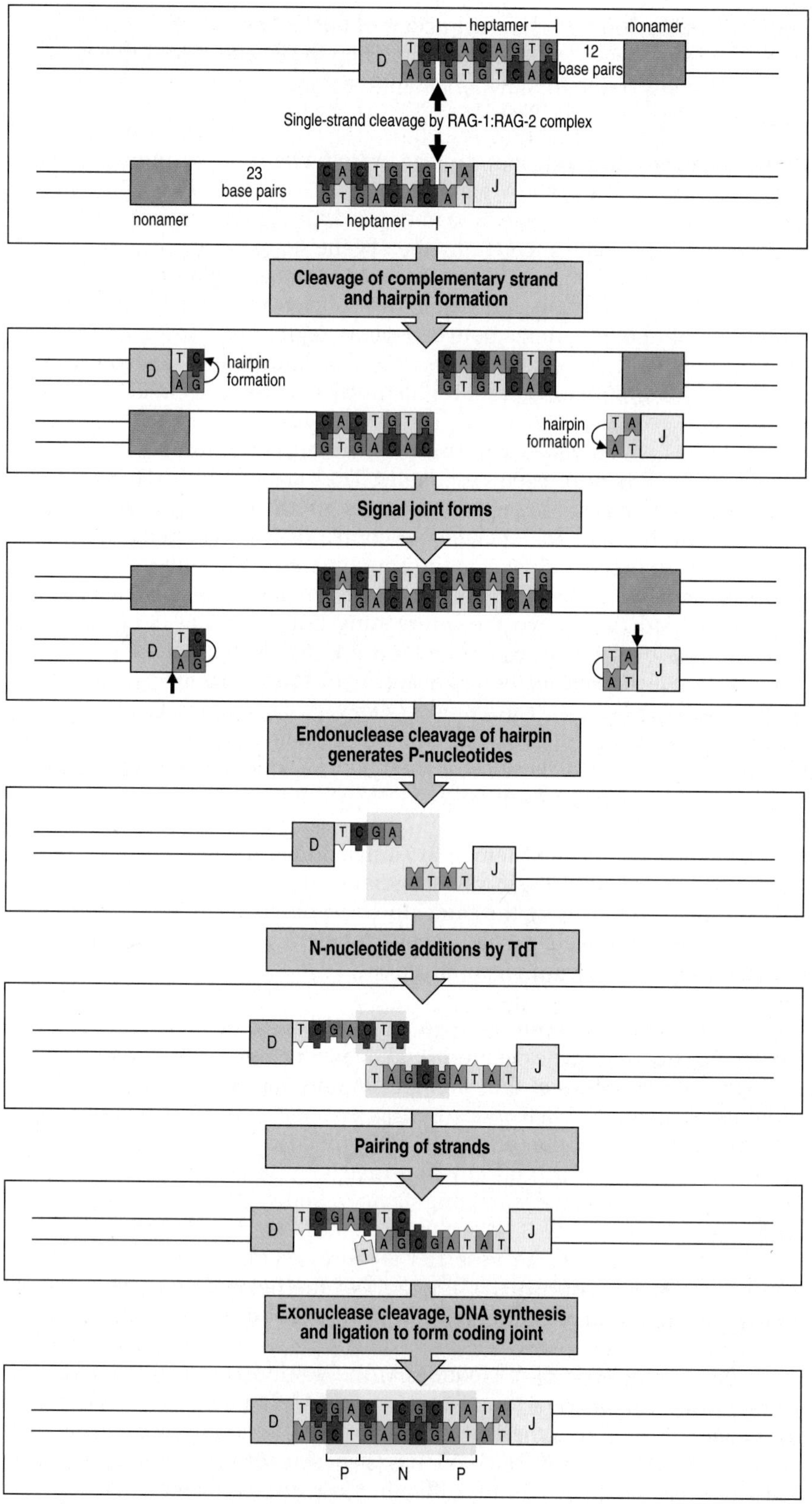

Fig. 3.18 Enzymatic steps in the rearrangement of immunoglobulin gene segments. The process is illustrated for a D_H to J_H rearrangement, showing the full sequence of the heptamer signal sequence and the first two nucleotides of the gene segments to be joined; however, the same steps occur in V_H to D_H and in V_L to J_L rearrangements. Rearrangement begins with the action of the RAG-1:RAG-2 complex, which recognizes the recombination signal sequences and cuts one strand of the double-stranded DNA precisely at the end of the heptamer sequences (top panel). The 5′ cut end of this DNA strand then reacts with the complementary uncut strand, breaking it to leave a double-stranded break at the end of the heptamer sequence, and forming a hairpin by joining to the cut end of its complementary strand on the other side of the break (second panel). Subsequently, the two heptamer sequences are ligated to form the signal joint, while an endonuclease cleaves the DNA hairpin at a random site (third panel) to yield a single-stranded DNA end (fourth panel). Depending on the site of cleavage, this single-stranded DNA may contain nucleotides that were originally complementary in the double-stranded DNA and which therefore form short DNA palindromes, as indicated by the shaded box in the fourth panel. Such stretches of nucleotides that originate from the complementary strand are known as P-nucleotides. For example, the sequence GA at the end of the D segment shown is complementary to the preceding sequence TC. Where the enzyme terminal deoxynucleotidyl transferase (TdT) is present, nucleotides are added at random to the ends of the single-stranded segments (fifth panel), indicated by the shaded box surrounding these non-templated, or N, nucleotides. The two single-stranded ends then pair (sixth panel). Exonuclease trimming of unpaired nucleotides and repair of the coding joint by DNA synthesis and ligation leaves both the P- and N-nucleotides present in the final coding joint (indicated by shading in the bottom panel). The randomness of insertion of P- and N-nucleotides makes an individual P–N region a valuable marker for following an individual B-cell clone as it develops, for instance in hypermutation studies (see Fig. 3.19).

protein, and DNA rearrangements leading to such disruptions are known as **non-productive rearrangements.** As roughly two in every three rearrangements will be non-productive, many B cells never succeed in producing functional immunoglobulin molecules, and junctional diversity is therefore achieved only at the expense of considerable wastage. We shall discuss this further when we describe the development of B cells in Chapter 6. The rearrangement of immunoglobulin genes is tightly regulated to ensure that each B cell expresses only one rearranged heavy-chain gene and one rearranged light-chain gene (see Sections 6-7 and 6-8).

3-17 Specialized enzymes are required for somatic recombination of V gene segments.

The complex of several enzymes that act in concert to effect somatic V-region gene recombination is termed the **'V(D)J recombinase.'** This complex is mostly made up of cleavage and repair enzymes present in all cell types and required for the normal maintenance of nuclear DNA. The first cleavage step, however, requires an additional specialized heterodimeric endonuclease formed from the products of two genes called ***RAG-1*** and ***RAG-2***, for **recombination-activating genes.** *RAG-1* has sequence similarities to a yeast gene, *HRP-1*, and to bacterial **topoisomerases**, which catalyze the breakage and rejoining of DNA. *RAG-1* and *RAG-2* are normally expressed together only in developing lymphocytes. If they are artificially expressed in cells in culture that do not make antibodies, they can cause the rearrangement of artificially introduced unrearranged immunoglobulin genes. Mice in which either of the *RAG* genes is knocked out suffer a complete block in primary lymphocyte development at the gene rearrangement stage. A second specialized component of the V(D)J recombinase complex is TdT, discussed in the previous section.

The other components of the recombinase complex that have been identified are proteins with a role in repairing double-stranded breaks in DNA. These include the enzyme DNA ligase IV, the enzyme DNA-dependent protein kinase (DNA-PK), and Ku, a well-known autoantigen, which is a heterodimer (Ku 70; Ku 86) that associates tightly with DNA-PK. Mutant mice in which DNA-PK is defective cannot join DNA at the junctions between the gene segments encoding the V region, and so can make only trivial amounts of immunoglobulins or T-cell receptors. Such mice suffer from a *s*evere *c*ombined *i*mmune *d*eficiency—hence the name for this mutation, ***scid***.

3-18 Rearranged V genes are further diversified by somatic hypermutation.

The mechanisms for generating diversity described so far all take place during the rearrangement of gene segments in the initial development of B cells in primary lymphoid organs. There is an additional mechanism that generates diversity throughout the V region and which operates on B cells in secondary lymphoid organs after functional antibody genes have been assembled. This process, known as **somatic hypermutation**, introduces point mutations into the V regions of the rearranged heavy- and light-chain genes at a very high rate, giving rise to mutant immunoglobulin molecules on the surface of the B cells (Fig. 3.19). Some of the mutant immunoglobulin molecules bind antigen better than the original surface immunoglobulin, and B cells expressing them are thus preferentially selected to mature into antibody-secreting cells. This gives rise to a phenomenon called **affinity maturation** of the antibody population, which we will discuss in more detail in Chapters 9 and 10.

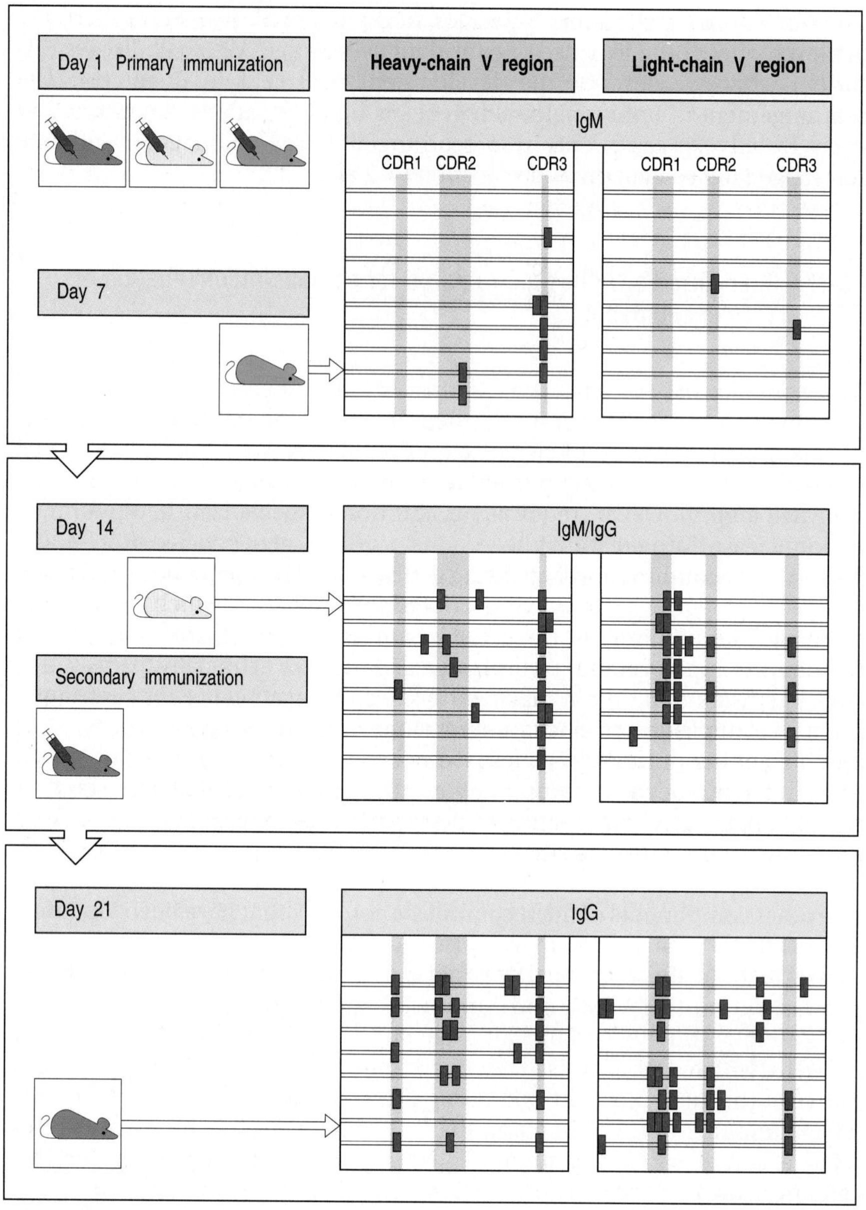

Fig. 3.19 Somatic hypermutation introduces diversity into expressed immunoglobulin genes. This figure illustrates an experiment in which somatic hypermutation in immunoglobulin V regions was demonstrated by sequencing heavy- and light-chain V regions of immunoglobulins specific for the same antigen at different times after immunization. Three groups of mice were immunized, in this case with a small hapten, oxazolone, for which the majority of antibodies produced use a single V_H and V_L. Seven days after immunization, B cells were taken from one group of animals and oxazolone-specific hybridomas were produced from them. The hybridomas secreted predominantly IgM antibodies and showed little sequence variation in the V regions. Amino acid positions that differ from the prototypic V-region sequences are shown as red bars on the lines representing the sequences. At day 7 most of the differences lie in the junctional regions, that is, in CDR3. After a further 7 days the oxazolone-specific antibodies from the second group of mice were analyzed. Both IgG and IgM antibodies were present and all six CDRs were affected. A third group of mice were given a secondary immunization and were analyzed after a further 7 days. At this time, most of the antibodies were IgG and all showed extensive changes in the V regions. Base changes that change amino acid sequence (productive changes) and silent changes were both seen throughout the V region but productive changes were concentrated in CDRs.

Somatic hypermutation occurs when B cells respond to antigen. The immunoglobulin C-region genes, and other genes expressed in the B cells, are not affected, whereas all rearranged V-region genes are mutated even if they are the result of non-productive rearrangements and are not expressed. The pattern of nucleotide base changes in non-productive V-region genes illustrates the result of somatic hypermutation without selection for enhanced binding to antigen. The base changes are distributed widely through the V region, but not completely randomly: there are certain 'hotspots' of mutation that indicate a preference for characteristic short motifs of four to five nucleotides, and perhaps also certain ill-defined secondary structural features. The pattern of base changes in the expressed V-region genes is different. The net result of selection for enhanced binding to antigen is that base changes that alter amino acid sequences tend to be clustered in the CDR regions, whereas silent mutations that preserve amino acid sequence and do not alter structure are scattered throughout the sequence.

Most of the antibody diversity in an adult individual derives from somatic alterations acquired during the lifetime of the individual. The combination of heritable and acquired components of diversity that we have described operates in several mammalian immune systems. Other species achieve a mix of inherited and acquired diversity by different means: birds do not use somatic recombination or hypermutation to create diversity, but create their antibody repertoires by gene conversion from germline pseudogenes. Overall, it would appear that there is strong selective pressure to generate sufficient diversity in the immune system to protect the organism from common pathogens, and that several different mechanisms have evolved towards this end.

Summary.

Diversity within the antibody repertoire is achieved by several means. Many V-region gene segments are present in the genome of an individual, and thus provide a heritable source of diversity. Additional diversity results from the formation of a complete V-region exon by the random recombination of separate V, D, and J gene segments. Variability of the junctions between segments is increased by the insertion of random numbers of P- and N-nucleotides and by variable deletion of nucleotides at the ends of some coding sequences. The association of different light- and heavy-chain V regions to form the antigen-binding site contributes further diversity. Finally, after an immunoglobulin has been expressed, the coding sequences for its V regions are modified by somatic hypermutation upon stimulation of the B cell by antigen. The combination of all these sources of diversity creates a vast repertoire of antibody specificities from a relatively limited number of genes.

Structural variation in immunoglobulin constant regions.

So far we have focused on the structural versatility of the V region of the antibody molecule, discussing only the general structural features of the C region as illustrated by IgG, the most abundant type of antibody in plasma. We now turn to the structural features that distinguish the heavy-chain C regions of antibodies of the five major isotypes and confer on them their specialized functional properties. A given V_H region may be expressed with any of the different C_H regions through a mechanism known as **isotype switching**, which involves further DNA rearrangements.

3-19 The principal immunoglobulin isotypes are distinguished by the structure of their heavy-chain constant regions.

The five main isotypes of immunoglobulin are IgM, IgD, IgG, IgE, and IgA. In humans, IgG antibodies can be further subdivided into four subclasses (IgG1, IgG2, IgG3, and IgG4), whereas IgA antibodies are found as two subclasses (IgA1 and IgA2). The IgG isotypes in humans are named in order of their abundance in serum, with IgG1 being the most abundant. The heavy chains that define these isotypes are designated by the lower-case Greek letters μ, δ,

Fig. 3.20 The properties of the human immunoglobulin isotypes. IgM is so called because of its size: although monomeric IgM is only 190 kDa, it normally forms pentamers, known as macroglobulin (hence the M), of very large molecular weight (see Fig. 3.22). IgA dimerizes to give a molecular weight of around 390 kDa in secretions. IgE antibody is associated with immediate-type hypersensitivity. When fixed to tissue mast cells IgE has a much longer half-life than in plasma (shown here). The activation of the alternative pathway of complement by IgA1 is caused not by its Fc portion but by its Fab portion.

	Immunoglobulin								
	IgG1	IgG2	IgG3	IgG4	IgM	IgA1	IgA2	IgD	IgE
Heavy chain	γ_1	γ_2	γ_3	γ_4	μ	α_1	α_2	δ	ε
Molecular weight (kDa)	146	146	165	146	970	160	160	184	188
Serum level (mean adult mg ml^{-1})	9	3	1	0.5	1.5	3.0	0.5	0.03	5×10^{-5}
Half-life in serum (days)	21	20	7	21	10	6	6	3	2
Classical pathway of complement activation	++	+	+++	−	+++	−	−	−	−
Alternative pathway of complement activation	−	−	−	−	−	+	−	−	−
Placental transfer	+++	+	++	−/+	−	−	−	−	−
Binding to macrophages and other phagocytes	+	−	+	−	−	+	+	−	+
High-affinity binding to mast cells and basophils	−	−	−	−	−	−	−	−	+++
Reactivity with staphylococcal Protein A	+	+	−/+	+	−	−	−	−	−

γ, ε, and α, as shown in Fig. 3.20, which also lists the major physical properties of the different human isotypes. IgM forms pentamers in serum, which accounts for its high molecular weight.

Sequence differences between immunoglobulin heavy chains cause the various isotypes to differ in several characteristic respects. These include the number and location of interchain disulfide bonds, the number of attached oligosaccharide moieties, the number of C domains, and the length of the hinge region (Fig. 3.21). IgM and IgE heavy chains contain an extra C domain that replaces the hinge region found in γ, δ, and α chains. The absence of the

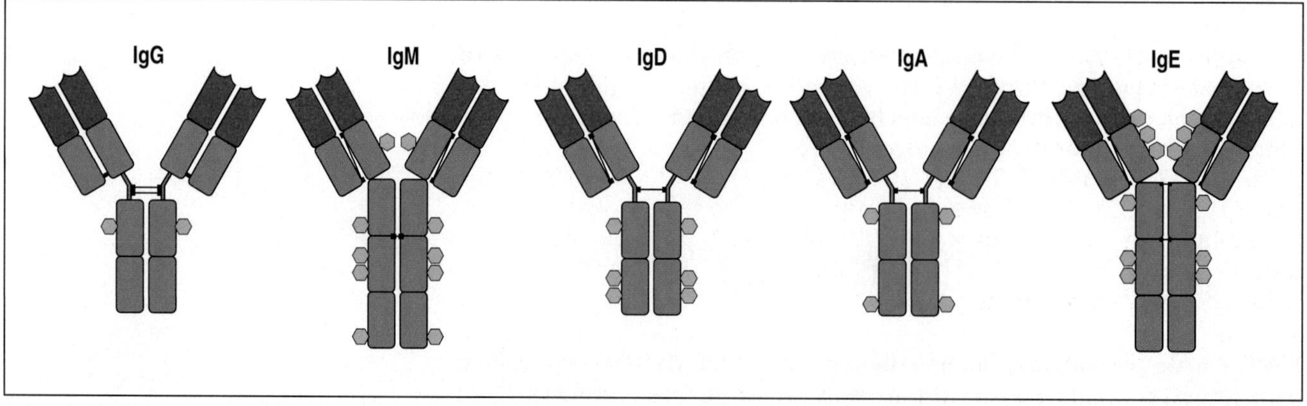

Fig. 3.21 The structural organization of the main human immunoglobulin isotype monomers. In particular, note the differences in the numbers and locations of the disulfide bonds linking the chains. Both IgM and IgE lack a hinge region but each contains an extra heavy-chain domain. The isotypes also differ in the distribution of N-linked carbohydrate groups, as shown in turquoise.

hinge region does not imply that IgM and IgE molecules lack flexibility; electron micrographs of IgM molecules binding to ligands show that the Fab arms can bend relative to the Fc portion. However, such a difference in structure may have functional consequences that are not yet characterized.

3-20 IgM and IgA can form polymers.

Although all immunoglobulin molecules are constructed from a basic unit of two heavy and two light chains, both IgM and IgA can form multimers (Fig. 3.22). IgM and IgA C regions contain a 'tailpiece' of 18 amino acids that contains a cysteine residue essential for polymerization. IgM molecules are found as pentamers, and occasionally hexamers, in plasma, whereas IgA in mucous secretions, but not in plasma, is mainly found as a dimer (see Fig. 3.22). An additional separate 15 kDa polypeptide chain called the J chain promotes polymerization by linking to the cysteines of the tailpiece, which is found only in the secreted forms of the μ and α chains (see Section 3-24). (This J chain should not be confused with the J gene segment; see Section 3-11.) In the case of IgA, polymerization is required for transport through epithelia, as we shall discuss further in Chapter 9.

The polymerization of immunoglobulin molecules is thought to be important in antibody binding to repetitive epitopes. The dissociation rate of one individual epitope from one individual binding site influences the strength of binding, or **affinity**, of that site: the lower the dissociation rate, the higher the affinity (see Section 2-11). An antibody molecule has two or more identical antigen-binding sites, and if it attaches to two or more repeating epitopes on a single target antigen, it will only dissociate when all sites dissociate. The dissociation rate of the whole antibody from the whole antigen will therefore be much slower than the rate for the individual binding sites, giving a greater effective binding strength, or **avidity**. This consideration is particularly relevant for pentameric IgM, which has 10 antigen-binding sites. IgM antibodies frequently recognize repetitive epitopes such as those expressed by bacterial cell-wall polysaccharides, but the binding of individual sites is often of low affinity because IgM is made early in immune responses, before affinity maturation. Multi-site binding makes up for this, dramatically improving the overall functional binding strength.

3-21 Immunoglobulin C regions confer functional specialization.

The secreted immunoglobulins protect the body in a variety of ways, as we have outlined briefly here and will discuss further in Chapter 9. In some cases it is enough for the immunoglobulin to bind antigen. For instance, by binding tightly to a toxin or virus, an antibody can prevent it from recognizing its receptor on a host cell. The V regions on their own are sufficient for this. The C region is essential, however, for recruiting the help of other cells and molecules to destroy and dispose of pathogens, and it confers functionally distinct properties on each of the various isotypes. Immunoglobulin has three main effector functions.

First, the Fc portions of different isotypes are recognized by specialized receptors expressed by immune effector cells. The Fc portions of IgG1 and IgG3 antibodies are recognized by Fc receptors expressed on the surface of phagocytic cells such as macrophages and neutrophils, which can thereby bind and engulf pathogens coated with antibodies of these isotypes. The Fc portion of IgE binds Fc receptors on mast cells, basophils, and activated eosinophils,

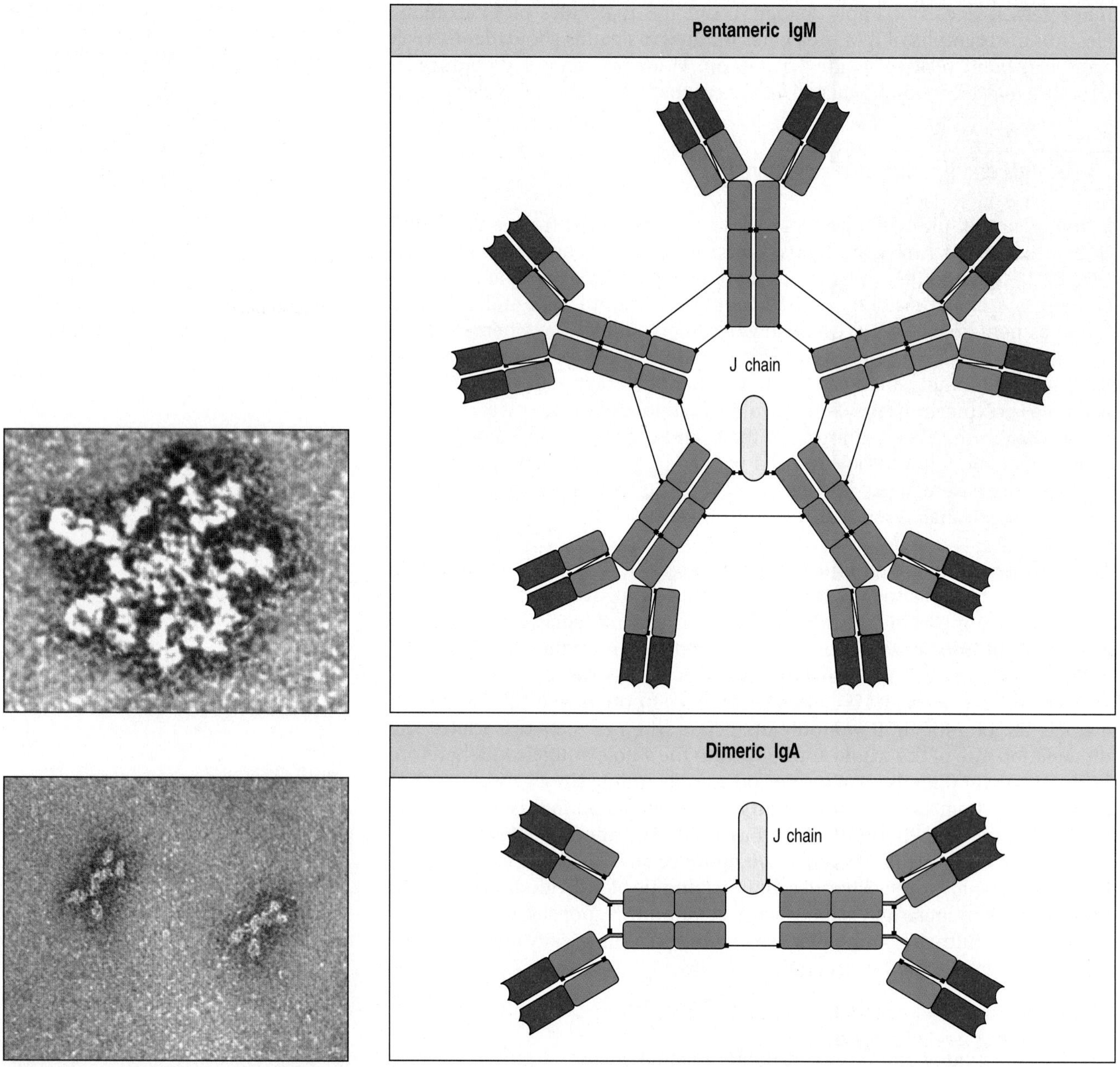

Fig. 3.22 The IgM and IgA molecules can form multimers. IgM and IgA are usually synthesized as multimers in association with an additional polypeptide chain, the J chain. In pentameric IgM, the monomers are crosslinked by disulfide bonds to each other and to the J chain. The top left panel shows an electron micrograph of an IgM pentamer, showing the arrangement of the monomers in a flat disc. IgM can also form hexamers that lack a J chain but are more efficient in complement activation. In dimeric IgA, the monomers have disulfide bonds to the J chain as well as to each other. The bottom left panel shows an electron micrograph of dimeric IgA. Photographs (x 900,000) courtesy of K H Roux and J M Schiff.

enabling these cells to respond to the binding of specific antigen by releasing inflammatory mediators. Second, the Fc portions of antigen:antibody complexes can bind to complement (see Fig. 1.23) and initiate the complement cascade, which helps to recruit and activate phagocytes, can aid the engulfment of microbes by phagocytes, and can also directly destroy pathogens (see Chapter 9). Third, the Fc portion can deliver antibodies to places they would not reach without active transport. These include the mucous secretions, tears, and milk (IgA), and the fetal blood circulation by transfer from the pregnant mother.

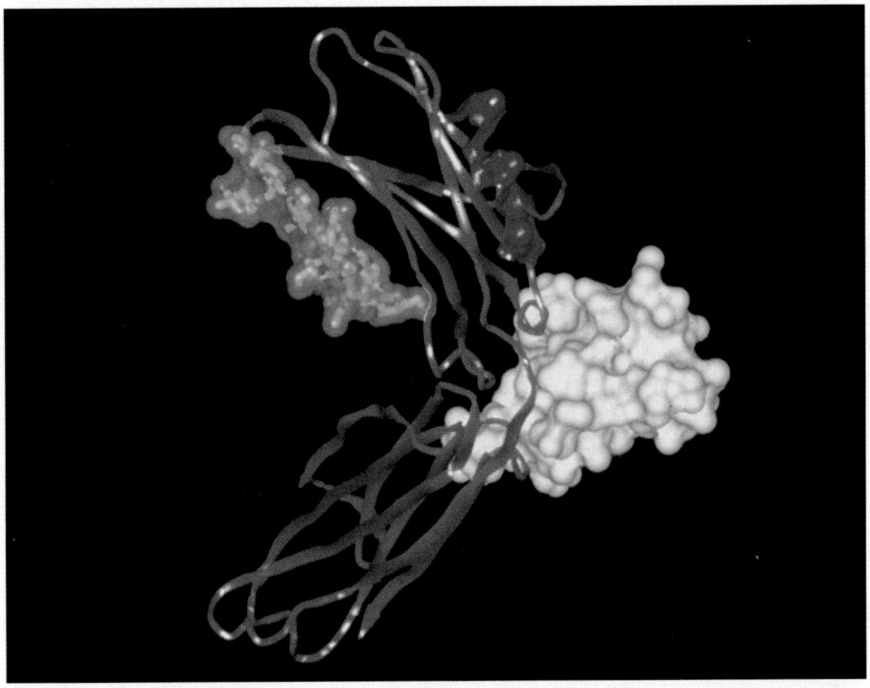

Fig. 3.23 Protein A of *Staphylococcus aureus* bound to a fragment of the Fc region of IgG. A fragment of the Fc portion of a single IgG heavy chain is complexed with a fragment of the immunoglobulin-binding Protein A from *S. aureus*. The Fc fragment has two domains, C_H2 and C_H3, shown in magenta. A carbohydrate chain is attached to an asparagine residue in the C_H2 domain: all the atoms are shown and the surface is outlined in green. The fragment of Protein A (white) is bound between the two domains of the Fc fragment. The amino acids that bind to the complement component C1q (red) lie in the C_H2 domain. Photograph courtesy of C Thorpe.

The role of the Fc portion in these effector functions can be demonstrated by studying enzymatically treated immunoglobulins that have had one or other domain of the Fc cleaved off (see Section 3-3) or, more recently, by genetic engineering, which permits detailed mapping of the exact amino-acid residues within the Fc that are needed. Many kinds of microorganism seem to have responded to the destructive potential of the Fc portion by manufacturing proteins that either bind to it or proteolytically cleave it, and so prevent the Fc piece from working. Examples of these are Protein A and Protein G made by *Staphylococcus* spp. (Fig. 3.23), and protein D of *Haemophilus* spp. Researchers can exploit these to help to map the Fc and as immunological reagents (see Section 2-9).

3-22 The same V_H exon can associate with different C_H genes in the course of an immune response.

The V-region exons expressed by any given B cell are determined during its early differentiation in the bone marrow and, although they may subsequently be modified by somatic hypermutation, no further V-gene segment recombination occurs. All the progeny of that B cell will therefore express the same assembled V genes. By contrast, the C-region genes expressed in a B cell change in its progeny as they mature and proliferate in the course of an immune response. Every B cell begins by expressing IgM, and the first antibody produced in an immune response is always IgM. Later in the immune response, however, the same assembled V region may be expressed in IgG, IgA, or IgE antibodies. This change is known as the **isotype switch** and is stimulated in the course of an immune response by cytokines released by T cells. How cytokines influence B-cell responses will be discussed in detail in Chapter 9. Here we are concerned with the molecular basis of the isotype switch.

The immunoglobulin C-region genes form a large cluster spanning about 200 kilobases (kb) to the 3′ side of the J_H gene segments (Fig. 3.24): each C-region gene is split into separate exons (not shown in the figure)

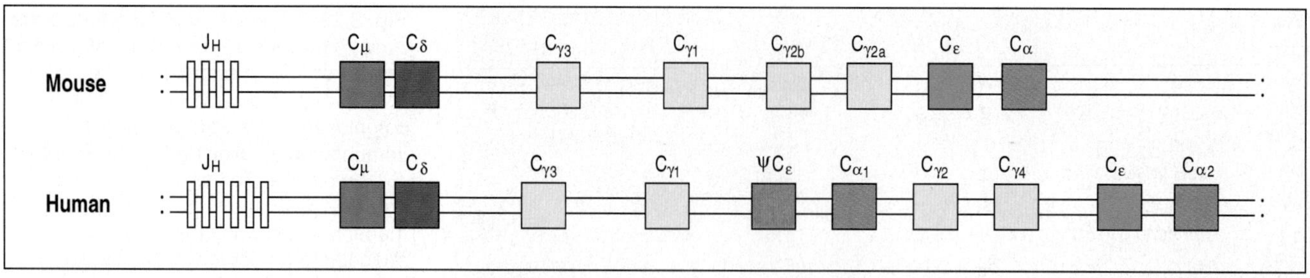

Fig. 3.24 The organization of the immunoglobulin heavy-chain C-region genes in mice and humans (not to scale). In humans, the cluster shows evidence of evolutionary duplication of a unit consisting of two γ genes, an ε gene and an α gene. One of the ε genes has become inactivated and is now a pseudogene (ψ); hence only one subtype of IgE is expressed. For simplicity, other pseudogenes are not illustrated, and the exon details within each C gene are not shown. The classes of immunoglobulins found in mice are called IgM, IgD, IgG1, IgG2a, IgG2b, IgG3, IgA, and IgE.

corresponding to the separate domains of the folded protein. The C-region gene encoding the μ chain lies closest to the J_H gene segments, and therefore closest to the assembled V-region exon after DNA rearrangement. In the absence of isotype switching a complete μ heavy-chain transcript is produced from the rearranged gene. Any J_H gene segments remaining between the assembled V gene and the Cμ gene are removed during RNA processing to generate the mature mRNA. μ heavy chains are therefore the first to be expressed and IgM is the first immunoglobulin isotype to be expressed during B cell development.

Immediately 3′ to the μ gene lies the δ gene encoding the C region of the IgD heavy chain. IgD is co-expressed with IgM on the surface of almost all B cells, although this isotype is secreted in only small amounts and its function is unknown. Indeed, mice in which the delta exons have been deleted by homologous recombination seem to have essentially normal immune systems. The possibility that changes in the ratio of surface IgD to IgM may be associated with unresponsiveness or with memory in B cells will be discussed in Chapter 6. B cells expressing IgM and IgD have not undergone isotype switching, which, as we shall see shortly, entails an irreversible change in the DNA. Instead, these cells produce a long primary transcript that is differentially cleaved and spliced to yield two distinct mRNA molecules. In one of these, the VDJ exon is linked to the C_μ exons to encode a μ heavy chain, and in the other the VDJ exon is linked to the C_δ exons to encode a δ heavy chain (Fig. 3.25).

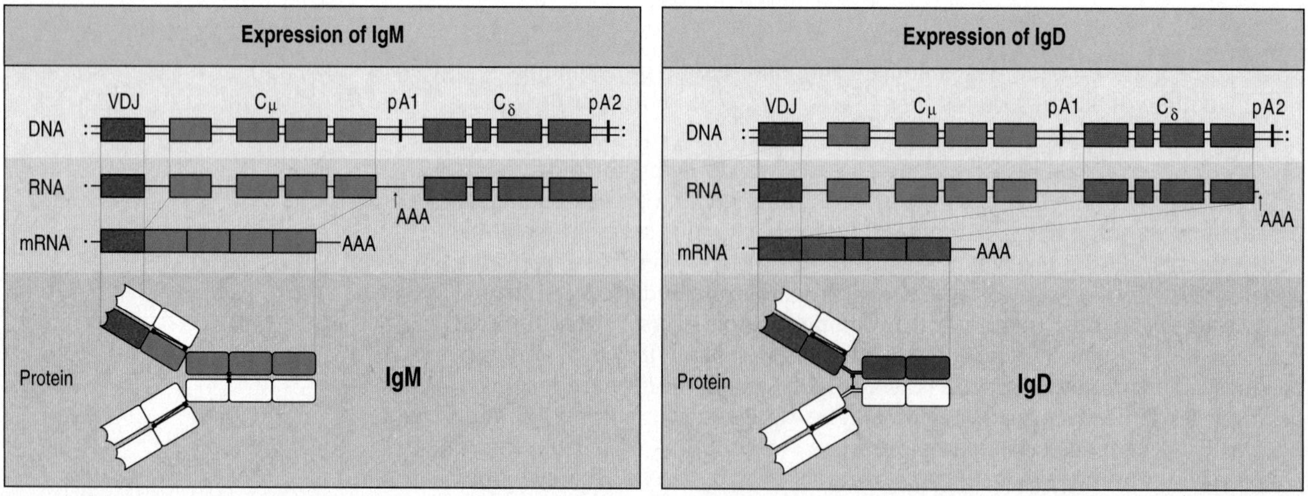

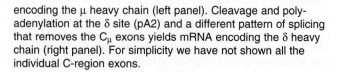

Fig. 3.25 Co-expression of IgD and IgM is regulated by RNA processing. In mature B cells, transcription initiated at the V_H promoter extends through both C_μ and C_δ exons. This long primary transcript is then processed by cleavage and polyadenylation, and by splicing. Cleavage and polyadenylation at the μ site (pA1) and splicing between C_μ exons yields an mRNA encoding the μ heavy chain (left panel). Cleavage and polyadenylation at the δ site (pA2) and a different pattern of splicing that removes the C_μ exons yields mRNA encoding the δ heavy chain (right panel). For simplicity we have not shown all the individual C-region exons.

Switching to other isotypes occurs only after B cells have been stimulated by antigen. It occurs through a specialized mechanism guided by stretches of repetitive DNA known as **switch regions**. Switch regions lie in the intron between the rearranged V-region exon and the μ gene, and at equivalent sites upstream of the C genes encoding each of the other heavy-chain isotypes, with the exception of the δ gene (Fig. 3.26, top panel). The μ switch region (S_μ) consists of about 150 repeats of the sequence [(GAGCT)$_n$(GGGGGT)], where n is usually three but can be as many as seven. The sequences of the other switch regions (S_γ, S_α, and S_ε) differ in detail but all contain repeats of the GAGCT and GGGGGT sequences.

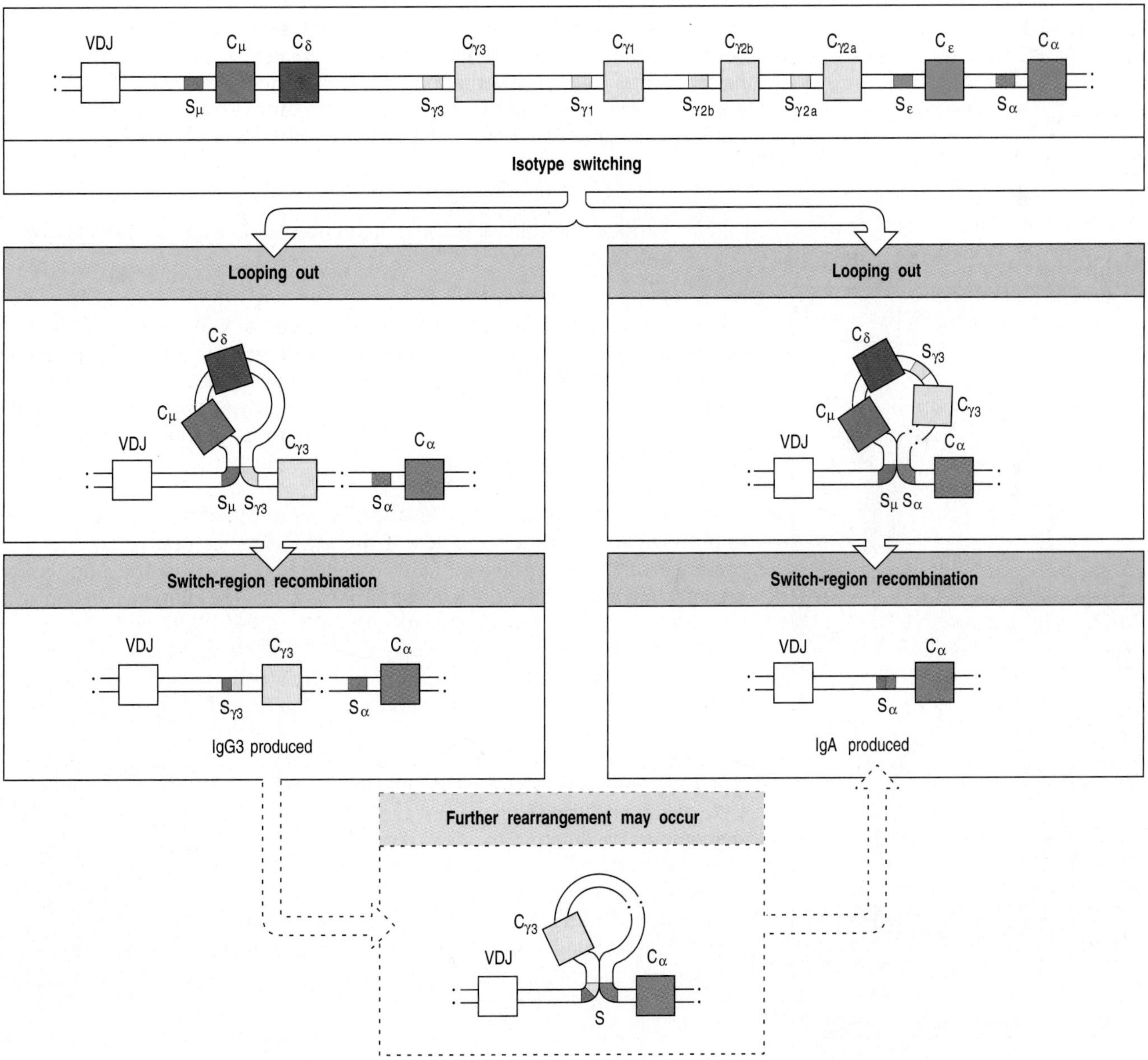

Fig. 3.26 Isotype switching involves recombination between specific signals. Repetitive DNA sequences that guide isotype switching are found upstream of each of the immunoglobulin C-region genes, with the exception of the δ gene. Switching occurs by recombination between these repetitive sequences, or switch signals, with deletion of the intervening DNA. The initial switching event takes place from the μ switch region; switching to other isotypes can take place subsequently from the recombinant switch region formed after μ switching. S, switch region.

When a B cell switches from co-expression of the IgM and IgD to express IgG, DNA recombination occurs between S_μ and S_γ, the C_μ and C_δ coding regions are deleted, and γ heavy-chain transcripts are made from the recombined gene (illustrated for switching to γ3 in Fig. 3.26, left panels). Some of the progeny of this IgG-producing cell may subsequently undergo a further switching event to produce a different isotype, for example IgA, as shown in the bottom panel of Fig. 3.26. Alternatively, as shown in the right panels of Fig. 3.26, the switch recombination may occur between S_μ and one of the switch regions downstream of the γ genes so that the cell switches from IgM to IgA or IgE (illustrated for IgA only). All switch recombination events produce genes that can encode a functional protein because the switch sequences lie in introns and therefore cannot cause frameshifts.

Switch recombination is unlike V-region gene segment recombination in several ways. First, all isotype switch recombination is productive; second, it uses different recombination signal sequences and enzymes; third, it happens after antigen stimulation and not during B-cell development in the bone marrow (see Chapter 6); and fourth, the switching process is not random but is regulated by T cells, as will be discussed in Chapter 9.

3-23 Various differences between immunoglobulins can be detected by antibodies.

When an immunoglobulin is used as an antigen, it will be treated like any other foreign protein and will elicit an antibody response. Anti-immunoglobulin antibodies can be made that recognize the amino acids that characterize the isotype of the injected antibody. Such anti-isotypic antibodies recognize all immunoglobulins of the same isotype in all members of the species from which the injected antibody came.

It is also possible to raise antibodies that recognize differences in immunoglobulins from members of the same species that are due to the presence of multiple alleles of the individual C genes in the population (genetic polymorphism). Such allelic variants are called **allotypes**. In contrast to anti-isotypic antibodies, anti-allotypic antibodies will recognize immunoglobulin of a particular isotype only in some members of a species. Finally, as individual antibodies differ in their V regions, one can raise antibodies against unique sequence variants, which are called **idiotypes**.

A schematic picture of the differences between idiotypes, allotypes, and isotypes is given in Fig. 3.27. Historically, the main features of immunoglobulins were defined by using isotypic and allotypic genetic markers. The independent segregation of allotypic markers revealed the existence of separate heavy-chain, κ, and λ genes. Rabbits, which are unique in having heavy-chain V-region allotypes, provided the earliest evidence that the V and C regions were encoded by distinct gene segments. The finding that V-region allotypes could be associated with different isotypes of immunoglobulin confounded the dogma that one gene encoded one polypeptide chain, and presaged the discovery of somatic recombination.

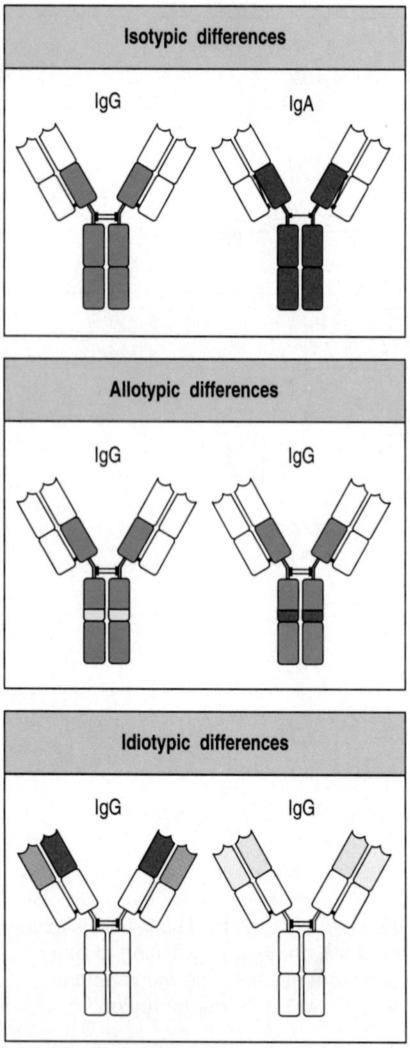

Fig. 3.27 Different types of variation between immunoglobulins. Differences between constant regions due to usage of a different C-region gene are called isotypes; differences due to different alleles of the same C gene are called allotypes; differences due to particular rearranged V_H and V_L genes are called idiotypes.

3-24 Transmembrane and secreted forms of immunoglobulin are generated from alternative heavy-chain transcripts.

Immunoglobulins of all heavy-chain isotypes can be produced either in secreted form or as a membrane-bound receptor. All B cells initially express the transmembrane form of IgM; after antigen stimulation some of their progeny differentiate into plasma cells producing the secreted form of IgM whereas others undergo isotype switching to express transmembrane immunoglobulins of a different isotype before switching to the production of secreted antibody.

The membrane forms of all isotypes are monomers: IgM and IgA polymerize only when they are secreted. In its membrane-bound form the immunoglobulin molecule has a hydrophobic transmembrane domain of about 25 amino acid residues, which anchors it to the surface of the B lymphocyte. This transmembrane domain is absent from the secreted form, whose carboxy terminus is a secretory tail. The two different carboxy termini of the transmembrane and secreted forms of immunoglobulin heavy chains are encoded in separate exons and production of the two forms is achieved by alternative RNA processing (Fig. 3.28). The last two exons of the heavy-chain C-region genes contain the sequence encoding the transmembrane and cytoplasmic regions; if the primary transcript is cleaved and polyadenylated at a site downstream of these exons, the sequence encoding the carboxy terminus of the secreted form is removed by splicing and the cell-surface form of immunoglobulin is produced. Alternatively, if the primary transcript is cleaved at the polyadenylation site located before the last two exons, only the secreted molecule can be produced. This differential RNA processing is illustrated for the μ heavy-chain gene in Fig. 3.28, but occurs in the same way for all isotypes.

Although the production of membrane and secreted versions of the heavy chain is achieved by similar mechanisms to those that allow the co-expression of surface IgM and IgD (see Fig. 3.25), these two instances of alternative RNA processing act at different stages in the life of the B cell, and on different primary transcripts. B cells make a long heavy-chain transcript that can be processed to give either transmembrane IgM or IgD before they are stimulated by antigen; a B cell that is activated ceases to co-express IgD with IgM, either because μ and δ sequences have been removed as a consequence of an isotype switch or, in IgM-secreting plasma cells, because transcription from the V_H promoter no longer extends through the C_δ exons.

Summary.

The isotypes of immunoglobulins are defined by their heavy-chain C regions, each isotype being encoded by a separate C-region gene. The heavy-chain C-region genes lie in a cluster 3′ to the V-region genes. A productively rearranged V-region exon is initially expressed in μ and δ heavy chains, but the same V-region exon can subsequently be associated with any one of the other isotypes by the process of isotype switching in which the DNA is rearranged to place the V region 5′ to different C-region genes. Unlike VDJ recombination, isotype switching occurs only in specifically activated B cells and is always productive. The immunological functions of the various isotypes differ; thus isotype switching varies the response to the same antigen at different times or under different conditions. Immunoglobulin RNA can be

Fig. 3.28 Transmembrane and secreted forms of immunoglobulins are derived from the same heavy-chain sequence by alternative RNA processing. Each heavy-chain C gene has two exons (MC, yellow) that encode the transmembrane region and cytoplasmic tail of the transmembrane form, and an SC sequence (orange) encoding the carboxy terminus of the secreted form. In the case of IgD, the SC sequence is present on a separate exon, but for the other isotypes, including IgM as shown here, the SC sequence is contiguous with the last C-domain exon. The events that dictate whether a heavy-chain RNA will result in a secreted or transmembrane immunoglobulin occur during processing of the initial transcript. Each heavy-chain C gene has two potential polyadenylation sites (shown as pA_s and pA_m). In the upper panel, the transcript is cleaved and polyadenylated at the second site (pA_m). Splicing between a site located between the $C_\mu4$ exon and the SC sequence, and a second site at the 5′ end of the MC exons, results in removal of the SC sequence and joining of the MC exons to the $C_\mu4$ sequence. This generates the transmembrane form of the heavy chain. In the lower panel, the primary transcript is cleaved and polyadenylated at the first site (pA_s), eliminating the MC exons and giving rise to the secreted form of the heavy chain.

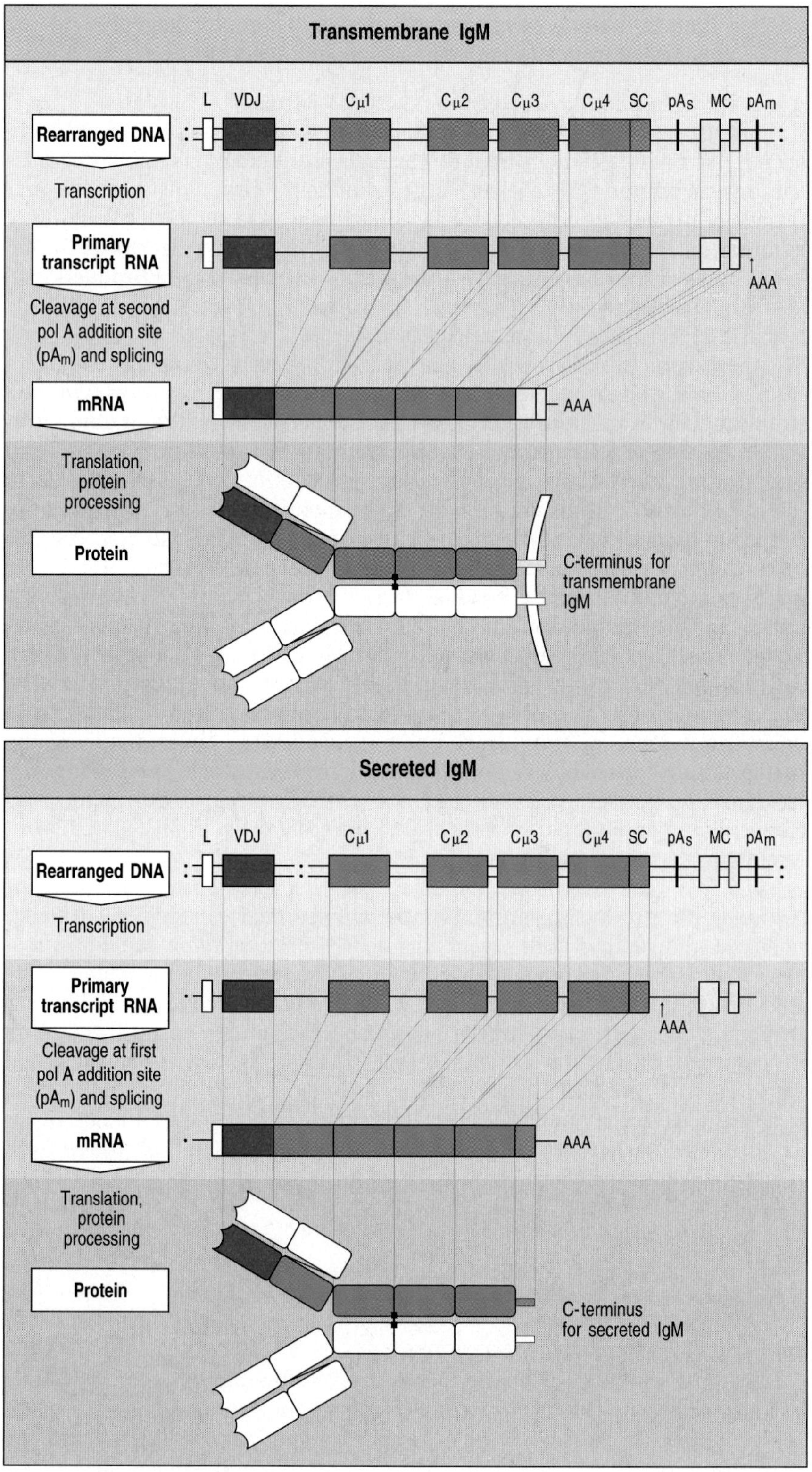

processed in two different ways to produce either membrane-bound immunoglobulin, which acts as the B-cell receptor for antigen, or secreted antibody. In this way, the B-cell antigen receptor has the same specificity as the antibody that the B cell secretes upon activation.

Summary to Chapter 3.

This chapter has provided an overview of the antibody molecule. In Chapter 1 we introduced the clonal selection hypothesis and suggested that the principles of clonal selection provide a framework for understanding the adaptive immune system. The clonal selection hypothesis requires that there be a broad repertoire of antibody specificities and that these specificities be distributed clonally. The organization of the antibody molecule and of the genes encoding the antibody molecule allows diversity to be generated within the antigen-binding site. Specialized mechanisms of somatic recombination and somatic mutation are responsible for generating virtually unlimited diversity from a relatively small number of inherited gene segments. Moreover, the process of creating this diversity ensures that the specificity of each receptor is unique to the B cell in which it occurs, and hence is distributed clonally. A second principle of the clonal selection hypothesis requires that the effector antibody molecules secreted by a B cell have the same specificity as the immunoglobulin on the cell surface. Alternative processing of heavy-chain RNA transcripts allows variation in the carboxy-terminal sequences of the heavy chains and production of transmembrane and secreted immunoglobulins, and ensures that the antigen specificity remains the same for both. Membrane-bound immunoglobulin can therefore serve as an antigen-specific receptor that signals the B cell to proliferate and secrete antibody in response to specific antigen. Different heavy-chain C-region genes determine the different effector functions of distinct antibody isotypes. Moving an active heavy-chain V-region exon to a new expression site adjacent to a different C-region gene switches the isotype and ensures that the same antigen specificity is retained in antibodies with different functional properties.

General references.

Ager, A., Callard, R., Ezine, S., Gerard, C., and Lopez-Botet, M.: **Immune receptor supplement.** *Immunol. Today* 1996, **17**.

Casali, P., and Silberstein, L.E.S. (eds): **Immunoglobulin gene expression in development and disease.** *Ann. N. Y. Acad. Sci.* 1995, **764**.

Davies, D.R., and Chacko, S.: **Antibody structure.** *Acc. Chem. Res.* 1993, **26**:421-427.

DeFranco, A.L., Blum, J.H., Stevens, T.L., Law, D.A., Chan, V.W.F., Foy, S.P., Datta, S.K., and Matsuuchi, L.: **Structure and function of the B-cell antigen receptor.** *Chem. Immunol.* 1994, **59**:156-172.

Fearon, D.T., and Carter, R.H.: **The CD19/CR2/TAPA-1 complex of B lymphocyteslinking natural to acquired immunity.** *Annu. Rev. Immunol.* 1995, **13**:127-149.

Frazer, K., and Capra, J.D.: Immunoglobulins: structure and function, in Paul W.E. (ed): *Fundamental Immunology*, 4th edn. New York, Raven Press, 1998.

Hames, B.D., and Glover, D.M.: Molecular Immunology, in *Frontiers in Molecular Biology*, 2nd edn. Oxford, IRL Press, 1996.

Honjo, T., and Alt, F.W. (eds): *Immunoglobulin Genes*, 2nd edn. London, Academic Press, 1996.

Max, E.E.: *Immunoglobulins: molecular genetics*, in Paul W.E. (ed): *Fundamental Immunology*, 3rd edn. New York, Raven Press, 1993.

Poljak, R.J.: **Structure of antibodies and their complexes with antigens.** *Mol. Immunol.* 1991, **28**:1341-1345.

Reth, M.: **The B-cell antigen receptor complex and co-receptors.** *Immunol. Today* 1995, **16**:310-313.

Schatz, D.G., Oettinger, M.A., and Schlissel, M.S.: **V(D)J recombination molecular biology and regulation.** *Annu. Rev. Immunol.* 1992, **10**:359-383.

Tuaillon, N., and Capra, J.D.: **Use of D gene segments with irregular spacers in terminal deoxynucleotidyltransferase (TDT)+/+ and TdT-/- mice carrying a human heavy chain transgenic minilocus.** *Proc. Natl. Acad. Sci. USA* 1998, **95**:1703-1708.

Wagner, S.D., and Neuberger, M.S.: **Somatic hypermutation of immunoglobulin genes.** *Annu. Rev. Immunol.* 1996, **14**:441-457.

Wilson, P.C., de Bouteiller, O., Liu, Y-J., Potter, K., Banchereau, J., Capra, J.D., and Pascual, V.: **Somatic hypermutation introduces insertions and deletions into immunoglobulin V genes.** *J. Exp. Med.* 1998, **187**:59-70.

Section references.

3-1 IgG antibodies consist of four polypeptide chains.

Edelman, G.M.: **Antibody structure and molecular immunology.** *Scand. J. Immunol.* 1991, **34**:4-22.

Faber, C., Shan, L., Fan, Z-c., Guddat, L.W., Furebring, C., Ohlin, M., Borrebaeck, C.A.K., and Edmundson, A.B.: **Three-dimensional structure of a human Fab with high affinity for tetanus toxoid.** *Immunotech.* 1998, **3**:253-270.

Harris, L.J., Larson, S.B., Hasel, K.W., Day, J., Greenwood, A., and McPherson, A.: **The 3-dimensional structure of an intact monoclonal antibody for canine lymphoma.** *Nature* 1992, **360**:369-372.

3-2 The heavy and light chains are composed of constant and variable regions.

Han, W.H., Mou, J.X., Sheng, J., Yang, J., and Shao, Z.F.: **Cryo-atomic force microscopy—a new approach for biological imaging at high resolution.** *Biochem.* 1995, **34**:8215-8220.

3-3 The antibody molecule can readily be cleaved into functionally distinct fragments.

Porter, R.R.: **Structural studies of immunoglobulins.** *Scand. J. Immunol.* 1991, **34**:382-389.

Yamaguchi, Y., Kim, H., Kato, K., Masuda, K., Shimada, I., and Arata, Y.: **Proteolytic fragmentation with high specificity of mouse IgG—mapping of proteolytic cleavage sites in the hinge region.** *J. Immunol. Methods* 1995, **181**:259-267.

3-4 The immunoglobulin molecule is flexible, especially at the hinge region.

Gerstein, M., Lesk, A.M., and Chothia, C.: **Structural mechanisms for domain movements in proteins.** *Biochem.* 1994, **33**:6739-6749.

Kim, J.K., Tsen, M.F., Ghetie, V., and Ward, E.S.: **Evidence that the hinge region plays a role in maintaining serum levels of the murine IgG1 molecule.** *Mol. Immunol.* 1995, **32**:467-475.

3-5 Each domain of an immunoglobulin molecule has a similar structure.

Barclay, A.N., Brown, M.H., Law, S.K., McKnight, A.J., Tomlinson, M.G., and van der Merwe, P.A. (eds): *The Leukocyte Antigen Factsbook,* 2nd edn. London, Academic Press, 1997.

Hsu, E., and Steiner, L.A.: **Primary structure of immunoglobulin through evolution.** *Curr. Opin. Struct. Biol.* 1992, **2**:422-430.

3-6 Localized regions of hypervariable sequence form the antigen-binding site.

Chitarra, V., Alzari, P.M., Bentley, G.A., Bhat, T.N., Eisele, J.L., Houdusse, A., Lescar, J., Souchon, H., and Poljak, R.J.: **3-dimensional structure of a heteroclitic antigen–antibody cross reaction complex.** *Proc. Nat. Acad. Sci. USA* 1993, **90**:7711-7715.

Gilliland, L.K., Norris, N.A., Marquardt, H., Tsu, T.T., Hayden, M.S., Neubauer, M.G., Yelton, D.E., Mittler, R.S., and Ledbetter, J.A.: **Rapid and reliable cloning of antibody variable regions and generation of recombinant single-chain antibody fragments.** *Tissue Antigens* 1996, **47**:1-20.

3-7 Small molecules bind to clefts between the heavy- and light-chain V domains.

Padlan, E.A.: **Anatomy of the antibody molecule.** *Mol. Immunol.* 1994, **31**:169-217.

3-8 Antibodies bind to extended sites on the surfaces of native protein antigens.

Davies, D.R., and Cohen, G.H.: **Interactions of protein antigens with antibodies.** *Proc. Natl. Acad. Sci. USA* 1996, **93**:7-12.

Stanfield, R.L., and Wilson, I.A.: **Protein–peptide interactions.** *Curr. Opin. Struct. Biol.* 1995, **5**:103-113.

Wilson, I.A., and Stanfield, R.L.: **Antibody-antigen interactions—new structures and new conformational changes.** *Curr. Opin. Struct. Biol.* 1994, **4**:857-867.

3-9 Antigen:antibody interactions involve a variety of forces.

Braden, B.C., and Poljak, R.J.: **Structural features of the reactions between antibodies and protein antigens.** *Faseb J.* 1995, **9**:9-16.

Hanin, V., Déry, O., Boquet, D., Sagot, M-A., Créminon, C., Couraud, J-Y., and Grassi, J.: **Importance of hydropathic complementarity for the binding of the neuropeptide substance p to a monclonal antibody: equilibrium and kinetic studies.** *Mol. Immunol.* 1997, **34**:829-838.

Ros, R., Schwesinger, F., Anselmetti, D., Kubon, M., Schäfer, R., Plückthun, A., and Tiefenauer, L.: **Antigen binding forces of individually addressed single-chain Fv antibody molecules.** *Proc. Natl. Acad. Sci. USA* 1998, **95**:7402-7405.

3-10 Immunoglobulin genes are rearranged in antibody-producing cells.

Tonegawa, S.: **Somatic generation of immune diversity.** *Scand. J. Immunol.* 1993, **38**:305-317.

Waldmann, T.A.: **The arrangement of immunoglobulin and T-cell receptor genes in human lymphoproliferative disorders.** *Adv. Immunol.* 1987, **40**:247-321.

3-11 Complete V regions are generated by the somatic recombination of separate gene segments.

Lansford, R., Okada, A., Chen, J., Oltz, E.M., Blackwell, T.K., Alt, F.W., and Rathbun, G.: Mechanism and control of immunoglobulin gene rearrangement, in *Molecular Immunology,* 2nd edn. Oxford, IRL Press, 1995.

3-12 V-region gene segments are present in multiple copies.

Cook, G.P., and Tomlinson, I.M.: **The human immunoglobulin V-H repertoire.** *Immunol. Today* 1995, **16**:237-242.

Kofler, R., Geley, S., Kofler, H., and Helmberg, A.: **Mouse variable-region gene families—complexity, polymorphism, and use in nonautoimmune responses.** *Immunol. Rev.* 1992, **128**:5-21.

Matsuda, F., and Honjo, T.: **Organization of the human immunoglobulin heavy-chain locus.** *Adv. Immunol.* 1996, **62**:1-29.

3-13 Rearrangement of V, D, and J gene segments is guided by flanking sequences in DNA.

Agrawal, A., and Schatz, D.G.: **RAG1 and RAG2 form a stable postcleavage synaptic complex with DNA containing signal ends in V(D)J recombination.** *Cell* 1997, **89**:43-53.

Grawunder, U., West, R.B., and Lieber, M.R.: **Antigen receptor gene rearrangement.** *Curr. Opin. Immunol.* 1998, **10**:172-180.

Nadel, B., Tang, A., Escuro, G., Lugo, G., and Feeney, A.J.: **Sequence of the spacer in the recombination signal sequence affects V(D)J rearrangement frequency and correlates with nonrandom V_κ usage *in vivo*.** *J. Exp. Med.* 1998, **187**:1495-1503.

3-14 Antibody diversity is generated by four main processes.

Fanning, L.J., Connor, A.M., and Wu, G.E.: **Development of the immunoglobulin repertoire.** *Clin. Immunol. and Immunopath.* 1996, **79**:1-14.

Stewart, A.K., and Schwartz, R.S.: **Immunoglobulin V regions and the B cell.** *Blood* 1994, **83**:1717-1730.

3-15 Inherited gene segments are used in different combinations.

Lee, A., Desravines, S., and Hsu, E.: **IgH diversity in an individual with only one million B lymphocytes.** *Develop. Immunol.* 1993, **3**:211-222.

Radic, M.Z., and Weigert, M.: **Genetic and structural evidence for antigen selection of anti-DNA antibodies.** *Annu. Rev. Immunol.* 1994, **12**:487-520.

3-16 Variable addition and subtraction of nucleotides at the junctions between the gene segments encoding the V region contributes to diversity in the third hypervariable region.

Gauss, G.H., and Lieber, M.R.: **Mechanistic constraints on diversity in human V(D)J recombination.** *Mol. Cell. Biol.* 1996, **16**:258-269.

Lewis, S.M.: **The mechanism of V(D)J joining—lessons from molecular, immunological, and comparative analyses.** *Adv. Immunol.* 1994, **56**:27-150.

3-17 Specialized enzymes are required for somatic recombination of V gene segments.

Blunt, T., Finnie, N.J., Taccioli, G.E., Smith, G.C.M., Demengeot, J., Gottlieb, T.M., Mizuta, R., Varghese, A.J., Alt, F.W., Jeggo, P.A., and Jackson, S.P.: **Defective DNA-dependent protein kinase activity is linked to V(D)J recombination and DNA-repair defects associated with the murine–scid mutation.** *Cell* 1995, **80**:813-823.

Gu, Z., Jin, S., Gao, Y., Weaver, D.T., and Alt, F.W.: **Ku70-deficient embryonic stem cells have increased ionizing radiosensitivity, defective DNA end-binding activity, and inability to support V(D)J recombination.** *Proc. Natl. Acad. Sci. USA* 1997, **94**:8076-8081.

Lin, W.C., and Desiderio, S.: **V(D)J recombination and the cell-cycle.** *Immunol. Today* 1995, **16**:279-289.

Li, Z.Y., Otevrel, T., Gao, Y.J., Cheng, H.L., Seed, B., Stamato, T.D., Taccioli, G.E., and Alt, F.W.: **The XRCC4 gene encodes a novel protein involved in DNA double-strand break repair and V(D)J recombination.** *Cell* 1995, **83**:1079-1089.

3-18 Rearranged V genes are further diversified by somatic hypermutation.

Cascalho, M., Wong, J., Steinberg, C., and Wabl, M.: **Mismatch repair coopted by hypermutation.** *Science* 1998, **279**:1207-1210.

Klotz, E.L., Hackett, J., and Storb, U.: **Somatic hypermutation of an artificial test substrate within an Igκ transgene.** *J. Immunol.* 1998, **161**:782-790.

Neuberger, M.S., and Milstein, C.: **Somatic hypermutation.** *Curr. Opin. Immunol.* 1995, **7**:248-254.

Tomlinson, I.M., Walter, G., Jones, P.T., Dear, P.H., Sonnhammer, E.L.L., and Winter, G.: **The imprint of somatic hypermutation on the repertoire of human germline V genes.** *J. Mol. Biol.* 1996, **256**:813-817.

3-19 The principal immunoglobulin isotypes are distinguished by the structure of their heavy-chain constant regions.

Davies, D.R., and Metzger, H.: **Structural basis of antibody function.** *Annu. Rev. Immunol.* 1983, **1**:87-117.

3-20 IgM and IgA can form polymers.

Hendrickson, B.A., Conner, D.A., Ladd, D.J., Kendall, D., Casanova, J.E., Corthesy, B., Max, E.E., Neutra, M.R., Seidman, C.E., and Seidman, J.G.: **Altered hepatic transport of IgA in mice lacking the J chain.** *J. Exp. Med.* 1995, **182**:1905-1911.

Niles, M.J., Matsuuchi, L., and Koshland, M.E.: **Polymer IgM assembly and secretion in lymphoid and nonlymphoid cell-lines—evidence that J chain is required for pentamer IgM synthesis.** *Proc. Nat. Acad. Sci. USA* 1995, **92**:2884-2888.

3-21 Immunoglobulin C regions confer functional specialization.

Helm, B.A., Sayers, I., Higginbottom, A., Machado, D.C., Ling, Y., Ahmad, K., Padlan, E.A., and Wilson, A.P.M.: **Identification of the high affinity receptor binding region in human IgE.** *J. Biol. Chem.* 1996, **271**:7494-7500.

Jefferis, R., Lund, J., and Goodall, M.: **Recognition sites on human IgG for Fcγ receptors—the role of glycosylation.** *Immunol. Letters* 1995, **44**:111-117.

Sensel, M.G., Kane, L.M., and Morrison, S.L.: **Amino acid differences in the N-terminus of C_H2 influence the relative abilities of IgG2 and IgG3 to activate complement.** *Mol. Immunol.* **34**:1019-1029.

3-22 The same V_H exon can associate with different C_H genes in the course of an immune response.

Morrison, S.L., Porter, S.B., Trinh, K.R., Wims, L.A., Denham, J., and Oi, V.T.: **Variable region domain exchange influences the functional properties of IgG.** *J. Immunol.* 1998, **160**:2802-2808.

Stavnezer, J.: **Immunoglobulin class switching.** *Curr. Opin. Immunol.* 1996, **8**:199-205.

3-23 Various differences between immunoglobulins can be detected by antibodies.

Fields, B.A., Goldbaum, F.A., Ysern, X., Poljak, R.J., and Mariuzza, R.A.: **Molecular basis of antigen mimicry by an anti idiotope.** *Nature* 1995, **374**:739-742.

3-24 Transmembrane and secreted forms of immunoglobulin are generated from alternative heavy-chain transcripts.

Caldwell, J., McElhone, P., Brokaw, J., Anker, R., and Pollok, B.A.: **Coexpression of full length and truncated Igμ chains in human B lymphocytes results from alternative splicing of a single primary RNA transcript.** *J. Immunol.* 1991, **146**:4344-4351.

Lyczak, J.B., Zhang, K., Saxon, A., and Morrison, S.L.: **Expression of novel secreted isoforms of human IgE proteins.** *J. Biol. Chem.* 1996, **271**:3428-3436.

Antigen Recognition by T Lymphocytes

4

In an adaptive immune response, antigen is recognized by two distinct sets of highly variable receptor molecules—the immunoglobulins that serve as antigen receptors on B cells and the antigen-specific receptors of T cells. As we saw in Chapter 3, immunoglobulins are secreted as antibodies by activated B cells and bind pathogens or their toxic products in the extracellular spaces of the body. Binding by antibody neutralizes viruses and marks pathogens for destruction by phagocytes and complement; these mechanisms will be discussed in Chapter 9. T cells, by contrast, recognize only antigens that are displayed on cell surfaces. These antigens may derive from pathogens, such as viruses or intracellular bacteria, that replicate within cells, or from pathogens or their products that cells internalize by endocytosis from the extracellular fluid.

T cells can detect the presence of intracellular pathogens because infected cells display on their surface peptide fragments derived from the pathogens' proteins. These foreign peptides are delivered to the cell surface by specialized host-cell glycoproteins encoded in a large cluster of genes that were first identified by their potent effects on the immune response to transplanted tissues. For that reason, the gene complex was termed the **major histocompatibility complex** (**MHC**), and the peptide-binding glycoproteins are still called **MHC molecules**. The recognition of antigen as a small peptide fragment bound to an MHC molecule and displayed at the cell surface is one of the most distinctive features of T cells, and will be the central focus of this chapter.

We shall begin by discussing the mechanisms of antigen processing and presentation, whereby protein antigens are degraded into peptides inside cells and the peptides are then carried to the cell surface stably bound to MHC molecules. We shall see that there are two different classes of MHC molecule, known as MHC class I and MHC class II, that deliver peptides from different cellular compartments to the surface of the infected cell. Peptides from the cytosol are bound to MHC class I molecules and recognized by CD8 T cells, whereas peptides generated in vesicles are bound to MHC class II molecules and recognized by CD4 T cells. The two functional subsets of T cells are thereby activated to initiate the destruction of pathogens resident in these two different cellular compartments. CD4 T cells may also activate B cells that have internalized specific antigen, and thus stimulate the production of antibodies to extracellular pathogens and their products.

In the second part of this chapter, we shall see that there are several genes for each class of MHC molecule: that is, the MHC is polygenic. Each of these genes has many alleles: that is, the MHC is also highly polymorphic. Indeed, the most remarkable feature of the classical MHC genes is their genetic variability. MHC polymorphism has a profound effect on antigen recognition by T cells, and the combination of polygeny and polymorphism greatly extends the range of peptides that can be presented to T cells by each individual and by each population at risk from an infectious pathogen.

Finally, we shall describe the **T-cell receptors** themselves. As might be expected from their function as highly variable antigen-recognition structures, T-cell receptors are closely related to antibody molecules in structure and, like immunoglobulins, are encoded in V, D, and J gene segments, which rearrange to form a complete variable-region exon. There are, however, important differences between T-cell receptors and immunoglobulins that reflect the unique features of antigen recognition by the T-cell receptor.

The generation of T-cell ligands.

The protective function of T cells depends on their ability to recognize cells that are harboring pathogens or that have internalized pathogens or their products. T cells do this by recognizing peptide fragments of pathogen-derived proteins in the form of complexes of peptide and MHC molecules on the surface of these cells. Because the generation of peptides from an intact antigen involves modification of the native protein, it is commonly referred to as **antigen processing**, whereas the display of the peptide at the cell surface by the MHC molecule is referred to as **antigen presentation**. In this section, we shall see how the structure and intracellular transport of the two classes of MHC molecule enables them to bind to a wide range of peptides derived from pathogens present in the cytosol or in the vesicular compartment of cells, and to present them for recognition by the appropriate functional type of T cell.

4-1 T cells recognize a complex of a peptide fragment bound to an MHC molecule.

Antigen recognition by T cells clearly differs from that employed by B-cell receptors and their secreted counterparts, antibody molecules. Antigen recognition by B cells involves direct binding to the native protein structure, and it can be shown by X-ray crystallography that antibodies typically bind to the surface of the protein, contacting amino acids that are discontinuous in the primary structure but are brought together in the folded protein. T cells, on the other hand, were found to respond to contiguous short amino acid sequences in proteins. These sequences were often buried within the native structure of the protein and could not be recognized directly by T-cell receptors unless some kind of unfolding or 'processing' of the protein antigen into peptide fragments had occurred.

The significance of this observation became clear only with the realization that the peptide fragments that stimulate T cells are recognized only when bound to an appropriate MHC molecule. The ligand recognized by the T cell is thus a complex of peptide and MHC molecule. The evidence for the role of the MHC was at first indirect and has only relatively recently been shown conclusively by stimulating T cells with purified peptide:MHC complexes. As we shall see later in the chapter, the T-cell receptor interacts with its ligand by making contacts both with the MHC molecule and with the peptide fragment of antigen.

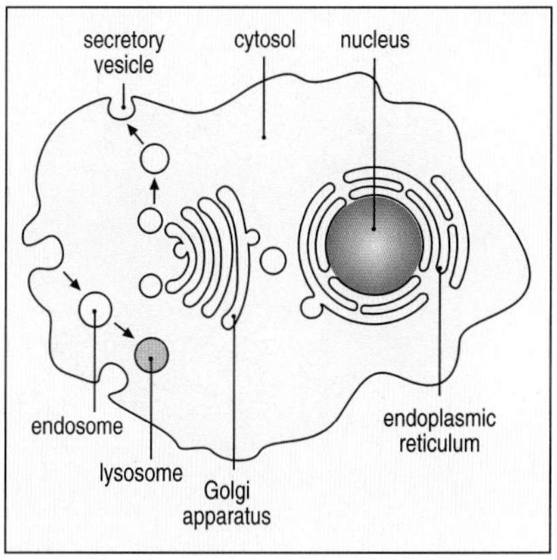

Fig. 4.1 There are two major compartments within cells, separated by membranes. The first is the cytosol, which is contiguous with the nucleus via the nuclear pores in the nuclear membrane. The second is the vesicular system, which comprises the endoplasmic reticulum, Golgi apparatus, endosomes, lysosomes, and other intracellular vesicles. The vesicular system can be thought of as contiguous with the extracellular fluid, as secretory vesicles bud off from the endoplasmic reticulum and are transported via the Golgi membranes to move vesicular contents out of the cell, whereas endosomes take up extracellular material into the vesicular system.

4-2 T cells with different functions recognize peptides produced in two distinct intracellular compartments.

Infectious agents can replicate in either of two distinct intracellular compartments (Fig. 4.1). Viruses and certain bacteria replicate in the cytosol or in the contiguous nuclear compartment (Fig. 4.2, left panel), whereas many pathogenic bacteria and some eukaryotic parasites replicate in the endosomes and lysosomes that form part of the vesicular system (Fig. 4.2, center panel). The immune system has different strategies for eliminating infections from these two sites. Cells infected with viruses or with bacteria that live in the cytosol are eliminated by cytotoxic T cells; these T cells are distinguished by the cell-surface molecule **CD8**. The function of CD8 T cells is to kill infected cells; this is an important means of eliminating sources of new viral particles and obligate cytosolic bacteria, thus freeing the host of infection.

Pathogens and their products in the vesicular compartments of cells are detected by a different class of T cell, distinguished by surface expression of the molecule **CD4**. CD4 T cells are specialized to activate other cells and fall into two functional classes: T_H1 cells (sometimes known as inflammatory T cells), which activate macrophages to kill the intravesicular pathogens they harbor, and T_H2 cells or helper T cells, which activate B cells to make antibody. Microbial antigens may enter the vesicular compartment in either of two ways. Some bacteria, including the mycobacteria that cause tuberculosis and leprosy, invade macrophages and flourish in intracellular vesicles. Other bacteria proliferate outside cells, where they cause pathology by secreting toxins and other proteins. These bacteria and their toxic products can be internalized by endocytosis, entering intracellular vesicles in this way. In particular, B cells take up and internalize specific antigen by receptor-mediated endocytosis of antigen bound to their surface immunoglobulin receptor (Fig. 4.2, right panel).

To produce an appropriate response to infectious microorganisms, T cells need to be able to detect the presence of intracellular pathogens and to

	Cytosolic pathogens	Intravesicular pathogens	Extracellular pathogens and toxins

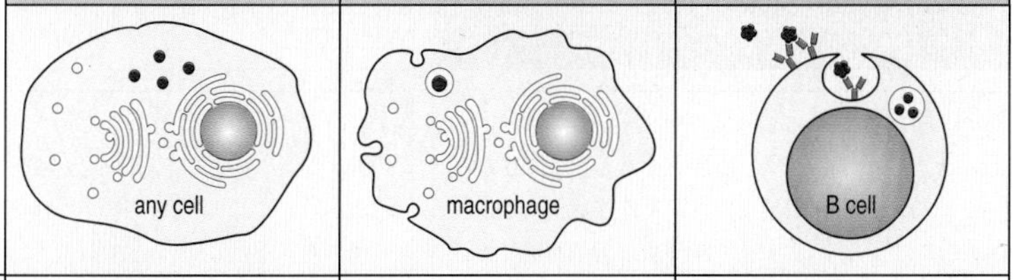

	Cytosolic pathogens	Intravesicular pathogens	Extracellular pathogens and toxins
Degraded in	Cytosol	Endocytic vesicles (low pH)	Endocytic vesicles (low pH)
Peptides bind to	MHC class I	MHC class II	MHC class II
Presented to	CD8 T cells	CD4 T cells	CD4 T cells
Effect on presenting cell	Cell death	Activation to kill intravesicular bacteria and parasites	Activation of B cells to secrete Ig to eliminate extracellular bacteria/toxins

Fig. 4.2 Pathogens and their products can be found in either the cytosolic or the vesicular compartment of cells. Left panel: all viruses and some bacteria replicate in the cytosolic compartment. Their antigens are presented by MHC class I molecules to CD8 T cells. Center panel: other bacteria and some parasites are engulfed into endosomes, usually by phagocytic cells such as macrophages, and are able to proliferate within the endocytic vesicles. Their antigens are presented by MHC class II molecules to CD4 T cells. Right panel: proteins derived from extracellular pathogens may enter the vesicular system of cells by binding to surface molecules followed by endocytosis. This is illustrated for proteins bound by surface immunoglobulin of B cells, which thereby present antigens to CD4 helper T cells stimulating the B cells to produce soluble antibody. (The endoplasmic reticulum and Golgi apparatus have been omitted for simplicity.) Other types of cell may also internalize antigens in this way and be able to activate T cells.

distinguish between foreign material coming from the cytosolic and vesicular compartments. This is achieved through the delivery of peptides to the cell surface from each of these intracellular compartments by a different class of MHC molecule. MHC class I molecules deliver peptides originating in the cytosol to the cell surface, where they are recognized by CD8 T cells. MHC class II molecules deliver peptides originating in the vesicular system to the cell surface, where they are recognized by CD4 T cells (see Fig. 4.2). We shall see later, when we discuss the recognition of MHC molecules by the T-cell receptor, how the molecules CD8 and CD4 help in the differential recognition of MHC class I and MHC class II molecules by the two major subsets of T cells.

4-3 The two classes of MHC molecule have distinct subunit structures but similar three-dimensional structures.

The **MHC class I** and **MHC class II** molecules are cell-surface glycoproteins closely related in overall structure and function, although they have different subunit structures. Both molecules have two domains that resemble immunoglobulin domains, and two domains that fold together to create a long cleft that is the site where peptides bind. However, differences in their structures allow them to serve distinct functions in antigen presentation,

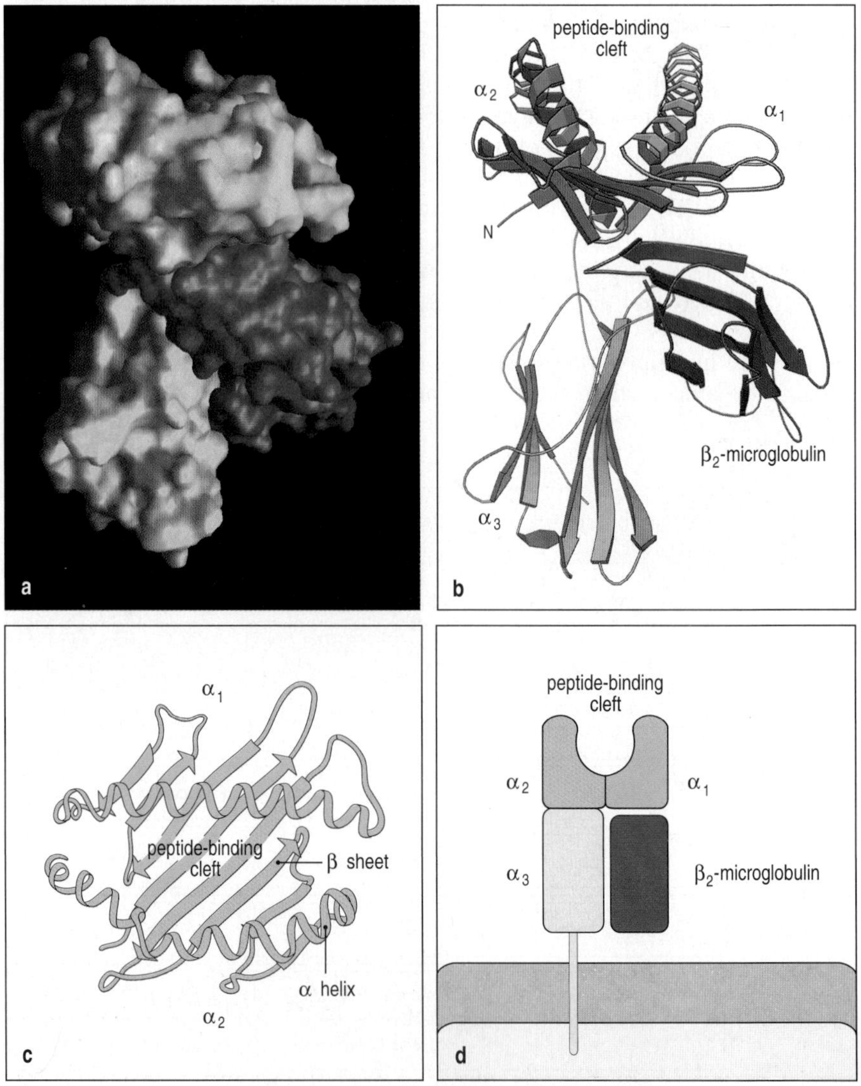

Fig. 4.3 The structure of an MHC class I molecule, determined by X-ray crystallography. Panel a shows a computer graphic representation of a human MHC class I molecule, HLA-A2, which has been cleaved from the cell surface by the enzyme papain. The surface of the molecule is shown, colored according to the domains described below. Panel b shows a ribbon diagram of that structure. Shown schematically in panel d, the MHC class I molecule is a heterodimer of a membrane-spanning α chain (molecular weight 43,000 Da), non-covalently associated with β_2-microglobulin (12,000 Da), which does not span the membrane. The α chain folds into three domains: α_1, α_2, and α_3. The α_3 domain and β_2-microglobulin show similarities in amino acid sequence to immunoglobulin constant domains and have similar folded structures, whereas the α_1 and α_2 domains fold together into a single structure consisting of two segmented α helices lying on a sheet of eight antiparallel β strands. The folding of the α_1 and α_2 domains creates a long cleft or groove, which is the site at which peptide antigens bind to the MHC molecules. The transmembrane region and the short stretch of peptide that connects the external domains to the cell surface are not seen in panels a and b as they have been removed by the papain digestion. As can be seen in panel c, looking down on the molecule from above, the sides of the cleft are formed from the inner faces of the two α helices; the β-pleated sheet formed by the pairing of the α_1 and α_2 domains creates the floor of the cleft. We shall use the schematic representation in panel d throughout this text.

binding peptides from different intracellular sites and activating different subsets of T cells. Purified peptide:MHC class I and peptide:MHC class II complexes have been characterized structurally, allowing us to describe in detail both the MHC molecules themselves and the way in which they bind peptides.

MHC class I structure is outlined in Fig. 4.3. MHC class I molecules consist of two polypeptide chains, an α or heavy chain encoded in the MHC, and a smaller non-covalently associated chain, β_2-microglobulin, which is not encoded in the MHC. Only the class I α chain spans the membrane. The molecule has four domains, three formed from the MHC-encoded α chain, and one contributed by β_2-microglobulin. The α_3 domain and β_2-microglobulin have a folded structure that closely resembles that of an immunoglobulin domain (see Section 3-5). The most remarkable feature of MHC class I molecules is the structure of the α_1 and α_2 domains, which pair to generate a cleft on the surface of the molecule that is the site of peptide binding.

MHC class II molecules consist of a non-covalent complex of two chains, α and β, both of which span the membrane (Fig. 4.4). The crystal structure of the MHC class II molecule shows that it is folded very much like the MHC class I molecule. The major differences lie at the ends of the peptide-binding cleft, which are more open in MHC class II molecules. The main consequence

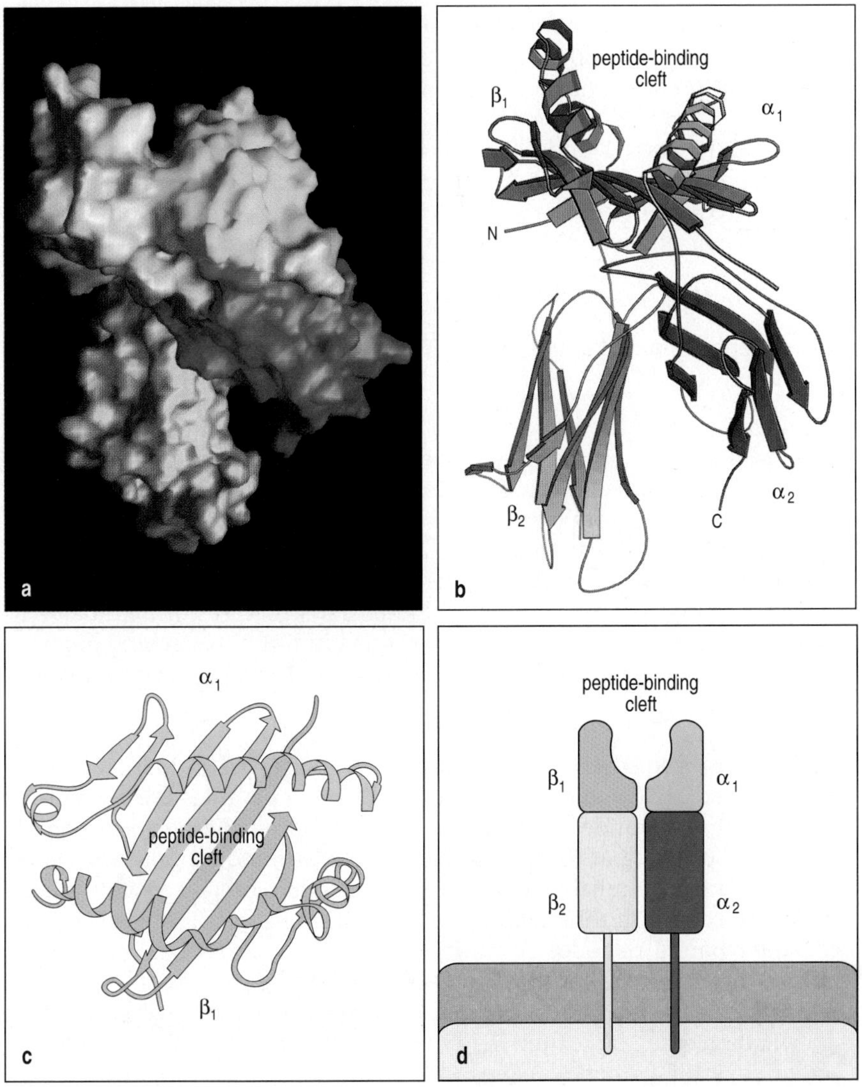

Fig. 4.4 MHC class II molecules resemble MHC class I molecules in structure. The MHC class II molecule is composed of two transmembrane glycoprotein chains, α (34,000 Da) and β (29,000 Da), as shown schematically in panel d. Each chain has two domains, and the two chains together form a compact four-domain structure similar to that of the MHC class I molecule (compare with panel d of Fig. 4.3). Panel a shows a computer graphic representation of the surface of the MHC class II molecule, in this case the human protein HLA-DR1, and panel b shows the equivalent ribbon diagram. The α_2 and β_2 domains, like the α_3 and β_2-microglobulin domains of the MHC class I molecule, have amino acid sequence and structural similarities to immunoglobulin constant domains; in the MHC class II molecule, the two domains forming the peptide-binding cleft are contributed by different chains and are therefore not joined by a covalent bond (see panels c and d). Another important difference, not apparent in this diagram, is that the peptide-binding groove of the MHC class II molecule is open at both ends.

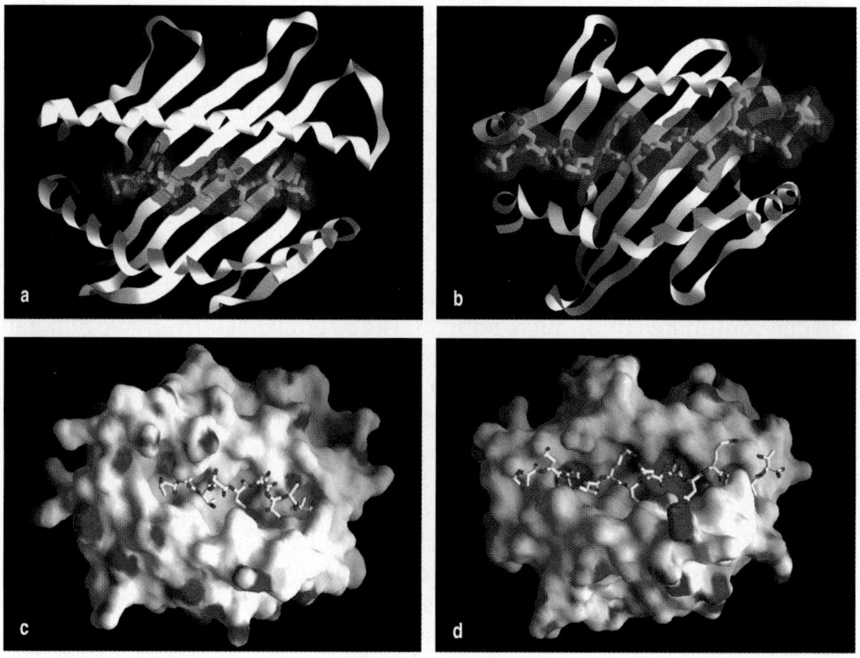

Fig. 4.5 MHC molecules bind peptides tightly within the cleft. The original crystal structure of an MHC class I molecule contained a mixture of naturally occurring peptide antigens, and details of the peptide:MHC interaction could not be discerned. When MHC molecules are crystallized with a single synthetic peptide antigen bound to their cleft, the details of peptide binding are revealed. In MHC class I molecules (panels a and c) the peptide is bound in an elongated conformation with both ends tightly bound at either end of the cleft. In the case of MHC class II molecules (panels b and d), the peptide is also bound in an elongated conformation but the ends of the peptide are not tightly bound and the peptide extends beyond the cleft. The upper surface of the peptide:MHC complex is recognized by T cells, and is composed of residues of the MHC molecule and the peptide. In representations c and d, the electrostatic potential of the MHC molecule surface is shown, with blue areas indicating a positive potential and red a negative potential.

of this is that the ends of a peptide bound to an MHC class I molecule are substantially buried within the molecule, whereas the ends of peptides bound to MHC class II molecules are not.

In both MHC class I and MHC class II molecules, bound peptides are sandwiched between two α-helical segments of the MHC molecule (Fig. 4.5). The T-cell receptor interacts with this ligand, making contacts both with the MHC molecule and with the peptide fragment of antigen.

4-4 Peptides are stably bound to MHC molecules.

An individual may be infected by a wide variety of different pathogens whose proteins will not necessarily have peptide sequences in common. If T cells are to be alerted to all possible intracellular infections the MHC molecules on each cell (both class I and class II) must be able to bind stably to many different peptides. This behavior is quite distinct from that of other peptide-binding receptors, such as those for peptide hormones, which usually bind only a single type of peptide. The crystal structures of peptide:MHC complexes have helped to show how a single binding site can bind peptides with high affinity, while retaining the ability to bind a wide variety of different peptides.

MHC molecules bind peptide ligands as an integral part of the MHC molecular structure, and MHC molecules are unstable when peptides are not bound. The stability of peptide binding is important because otherwise, peptide exchanges occurring at the cell surface would prevent peptide:MHC complexes from being reliable indicators of infection or of specific antigen uptake. As a result of this stability, when MHC molecules are purified from cells, their bound peptides co-purify with them, and this has enabled the peptides bound by specific MHC molecules to be analyzed. For this purpose, the peptides are eluted from the MHC molecules by denaturing the complex in acid to release the bound peptides, which can then be purified and sequenced. Pure synthetic versions of these peptides can also be incorporated into previously empty MHC molecules and the structure of the complex determined, revealing details of the contacts between the MHC molecule and the peptide. From an analysis of the sequences of peptides bound to specific MHC molecules, and a structural analysis of the peptide:MHC complex, a detailed picture of the binding interactions has been built up. We shall first discuss the peptide-binding properties of MHC class I molecules.

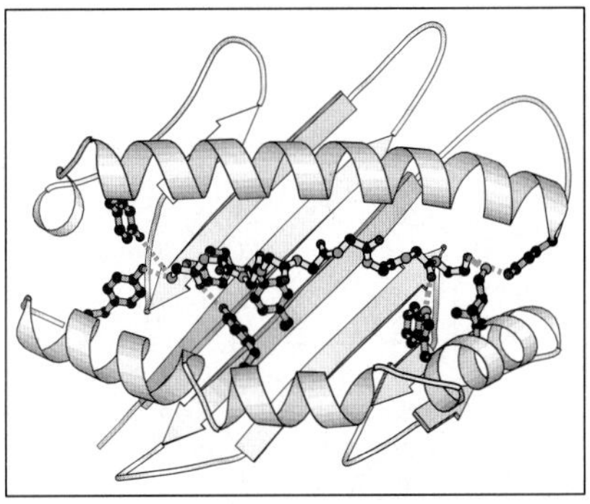

Fig. 4.6 Peptides are bound to MHC class I molecules by their ends. MHC class I molecules interact with the backbone of a bound peptide (shown in yellow) through a series of hydrogen bonds and ionic interactions (shown as dotted blue lines) at each end of the peptide. (The amino terminus of the peptide is to the left; the carboxy terminus to the right.) Black circles are carbon atoms; red are oxygen; blue are nitrogen. The amino acid residues in the MHC molecule that form these bonds are common to all MHC class I molecules and their side chains are shown in full (in gray) upon a ribbon diagram of the MHC class I binding groove. A cluster of tyrosine residues common to all MHC class I molecules forms hydrogen bonds to the amino terminus of the bound peptide, while a second cluster of residues forms hydrogen bonds and ionic interactions with the peptide backbone at the carboxy terminus and with the carboxy terminus itself.

4-5 MHC class I molecules bind short peptides of 8–10 amino acids by both ends.

Peptides that bind to MHC class I molecules are usually 8–10 amino acids long. The binding of the peptide is stabilized at its two ends by contacts between atoms in the free amino and carboxy termini and invariant sites that are found at each end of the peptide-binding groove of all MHC class I molecules (Fig. 4.6). These contacts are thought to be the main stabilizing contacts for peptide:MHC class I complexes because synthetic peptide analogs lacking terminal amino and carboxyl groups fail to bind stably to MHC class I molecules. The peptide lies in an elongated conformation along the groove; variations in peptide length appear to be accommodated, in most cases, by a kinking in the peptide backbone. However, two examples of MHC class I molecules where the peptide is able to extend out of the groove at the carboxy terminus suggest that some length variation may also be accommodated in this way.

These interactions provide broad peptide:MHC binding specificity. MHC molecules are highly polymorphic at certain sites in the peptide-binding cleft, however, and interactions between the polymorphic amino acid side chains and the peptide mean that different allelic variants of MHC molecules bind different peptides preferentially. Peptides binding to a given allelic variant of an MHC molecule have been shown to have the same or very similar amino acid residues at two or three specific positions along the peptide sequence. The amino acid side chains at these positions insert into pockets in the MHC molecule that are lined by the polymorphic amino acids. Because the binding of these side chains anchors the peptide to the MHC molecule, the peptide residues involved have been called **anchor residues**. Both the position and identity of these anchor residues can vary depending on the particular MHC class I allele that is binding the peptide. However, most peptides that bind to MHC class I molecules have an anchor residue at the carboxy terminus that is usually hydrophobic or in some instances basic (Fig. 4.7). Changing any anchor residue can prevent the peptide from binding and, conversely, most synthetic peptides of suitable length that contain these anchor residues will bind the appropriate MHC class I molecule, in most cases irrespective of the sequence of the peptide at other positions. These features of peptide binding allow MHC class I molecules to bind a wide variety of different peptides.

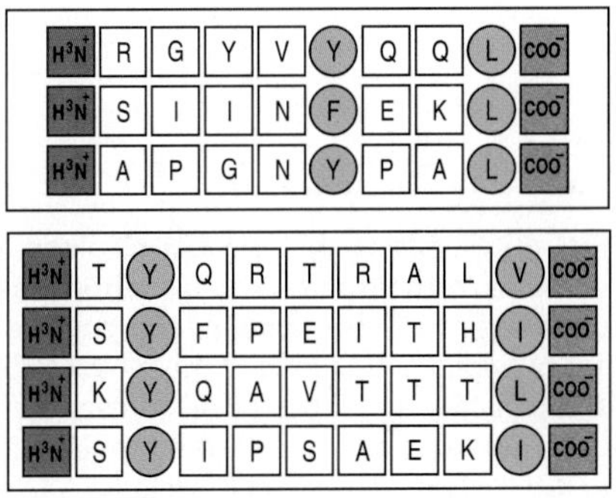

Fig. 4.7 Peptides bind to MHC molecules through structurally related anchor residues. Peptides eluted from two different MHC class I molecules are shown. The anchor residues (green) differ for peptides binding different alleles of class I MHC molecules but are similar for all peptides binding to the same MHC molecule. The upper and lower panels show peptides that bind to two different alleles of MHC class I molecules respectively. The anchor residues binding a particular MHC molecule need not be identical but are always related (for example, phenylalanine (F) and tyrosine (Y) are both aromatic amino acids, whereas valine (V), leucine (L) and isoleucine (I) are all large hydrophobic amino acids). Peptides also bind to MHC class I molecules through their amino (blue) and carboxyl (red) termini.

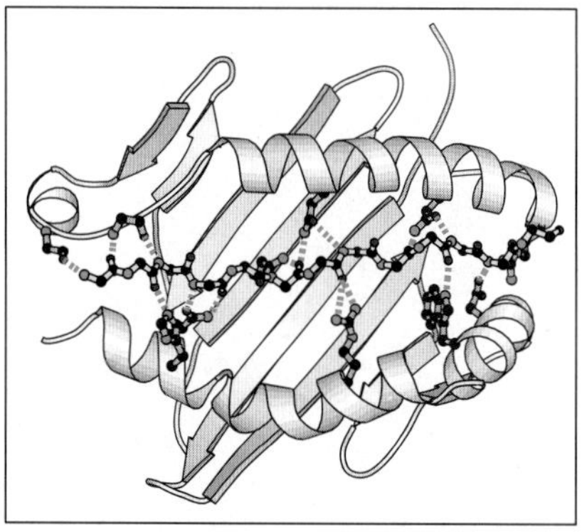

Fig. 4.8 Peptides bind to MHC class II molecules by interactions along the length of the binding groove. A peptide (yellow; shown as the peptide backbone only, with the amino terminus to the left and the carboxy terminus to the right), is bound by an MHC class II molecule through a series of hydrogen bonds (dotted blue lines) that are distributed along the length of the peptide. The hydrogen bonds towards the amino terminus of the peptide are made with the backbone of the class II polypeptide chain, whereas throughout the peptide's length are bonds made with residues that are highly conserved in MHC class II molecules. The side chains of these residues are shown in gray upon the ribbon diagram of the MHC class II binding groove.

4-6 The length of the peptides bound by MHC class II molecules is not constrained.

Peptide binding to MHC class II molecules has also been analyzed by elution of bound peptides and by X-ray crystallography, and is different in several ways from peptide binding to MHC class I molecules. Peptides that bind to MHC class II molecules are at least 13 amino acids long and can be much longer. The clusters of conserved residues that in MHC class I molecules bind the two ends of a peptide are not found in MHC class II molecules and the ends of the peptide are not bound. Instead, the peptide lies in an extended conformation along the MHC class II peptide-binding groove. It is held in this groove both by peptide side chains that protrude into shallow and deep pockets lined by residues that vary between MHC class II molecules, and by interactions between the peptide backbone and side chains of conserved MHC class II residues that line all MHC class II peptide-binding grooves (Fig. 4.8). Although there are fewer crystal structures of MHC class II-bound peptides than of MHC class I, the available data show that amino acid side chains at residues 1, 4, 6, and 9 of a minimal MHC class II-bound peptide are held in these binding pockets.

The binding pockets of MHC class II molecules are more permissive in their accommodation of different amino acid side chains than are those of the MHC class I molecule, making it more difficult to define anchor residues and predict which peptides will be able to bind particular MHC class II molecules (Fig. 4.9). Nevertheless, it is usually possible to detect a pattern of binding with different alleles of MHC class II molecules, and to associate this binding

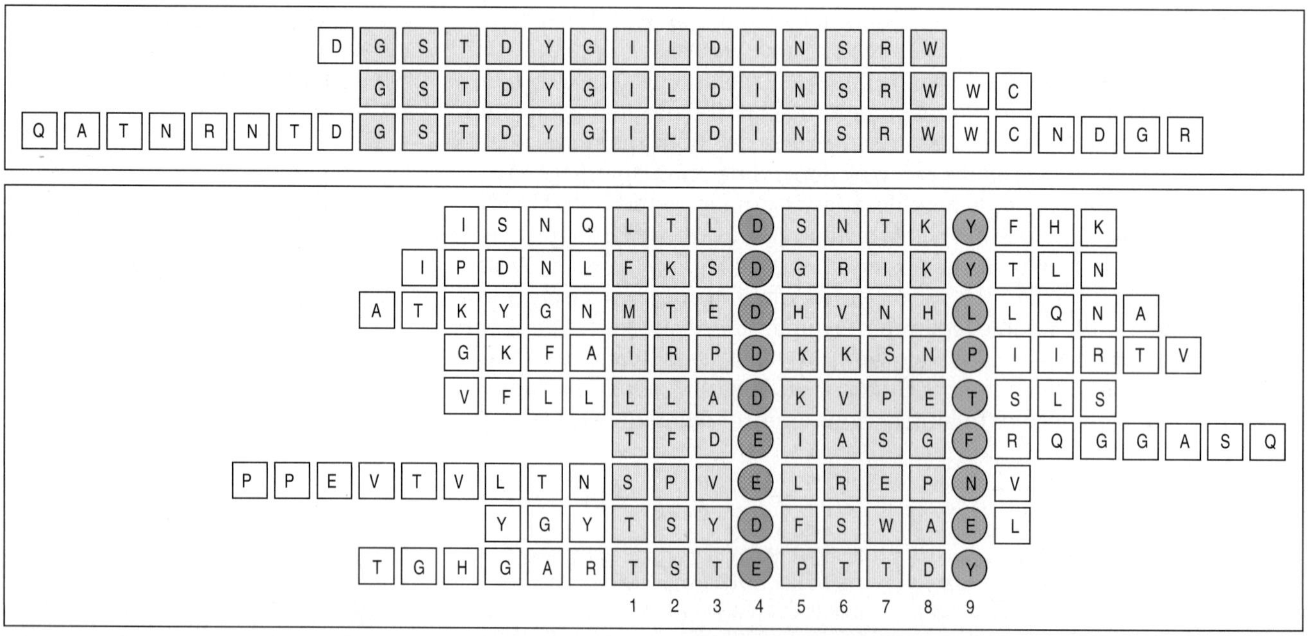

Fig. 4.9 Peptides that bind MHC class II molecules are variable in length and their anchor residues lie at various distances from the ends of the peptide. The sequences of a set of peptides that bind to the mouse MHC class II A^k allele are shown in the upper panel. All contain the same core sequence but differ in length. In the lower panel, different peptides binding to the human MHC class II allele HLA-DR3 are shown. The lengths of these peptides can vary, and so by convention the first anchor residue is denoted as residue 1. Note that all of the peptides share a negatively charged residue (aspartic acid (D) or glutamic acid (E)) in the P4 position (blue) and tend to have a hydrophobic residue (for example, tyrosine (Y), leucine (L), proline (P), phenylalanine (F)) in the P9 position (green).

motif with the amino acids that make up the groove of the MHC class II allele. Because the peptide is bound by its backbone and allowed to emerge from both ends of the binding groove there is, in principle, no upper limit to the length of peptides binding to MHC class II molecules. However, it appears that longer peptides bound to MHC class II molecules are trimmed by peptidases to peptides of 13–17 amino acids in most cases. Like MHC class I molecules, MHC class II molecules that lack bound peptide are unstable, but the critical stabilizing interactions that the peptide makes with the MHC class II molecule are not yet known.

4-7 The two classes of MHC molecule are expressed differentially on cells.

MHC class I and MHC class II molecules have a distinct distribution among cells that reflects the different effector functions of the T cells that recognize them (Fig. 4.10). MHC class I molecules, as we saw in Section 4-2, present peptides from pathogens in the cytosol, commonly viruses, to CD8 cytotoxic T cells, which are specialized to kill any cell that they specifically recognize. As viruses can infect any nucleated cell, almost all such cells express MHC class I molecules, although the level of constitutive expression varies from one cell type to the next. For example, cells of the immune system express abundant MHC class I on their surface, whereas liver cells (hepatocytes) express relatively low levels (see Fig. 4.10). Non-nucleated cells, such as mammalian red blood cells, express little or no MHC class I, and thus the interior of red

Tissue	MHC class I	MHC class II
Lymphoid tissues		
T cells	+++	+*
B cells	+++	+++
Macrophages	+++	++
Other antigen-presenting cells (eg Langerhans' cells)	+++	+++
Epithelial cells of the thymus	+	+++
Other nucleated cells		
Neutrophils	+++	–
Hepatocytes	+	–
Kidney	+	–
Brain	+	–†
Non-nucleated cells		
Red blood cells	–	–

Fig. 4.10 The expression of MHC molecules differs between tissues. MHC class I molecules are expressed on all nucleated cells, although they are most highly expressed in hematopoietic cells. MHC class II molecules are normally expressed only by a subset of hematopoietic cells and by thymic stromal cells, although they may be expressed by other cell types on exposure to the inflammatory cytokine interferon-γ. * In humans, activated T cells express MHC class II molecules, whereas in mice, all T cells are MHC class II-negative. † In the brain, most cell types are MHC class II-negative but microglia, which are related to macrophages, are MHC class II-positive.

blood cells is a site in which an infection can go undetected by cytotoxic T cells. As red blood cells cannot support viral replication, this is of no great consequence for viral infection, but it may be the absence of MHC class I that allows the *Plasmodium* species that cause malaria to live in this privileged site.

In contrast, the main function of the CD4 T cells that recognize MHC class II molecules is to activate other effector cells of the immune system. Thus MHC class II molecules are normally found on B lymphocytes and macrophages—cells that participate in immune responses—but not on other tissue cells (see Fig. 4.10). When CD4 T cells recognize peptides bound to MHC class II molecules on B cells, they stimulate the B cells to produce antibody. Likewise, CD4 T cells recognizing peptides bound to MHC class II molecules on macrophages activate these cells to destroy the pathogens in their vesicles. We shall see in Chapter 8 that MHC class II molecules are also expressed on specialized antigen-presenting cells in lymphoid tissues where naive T cells encounter antigen and are first activated. The expression of both MHC class I and MHC class II molecules is regulated by cytokines, in particular interferons, released in the course of immune responses. Interferon-γ (IFN-γ), for example, increases the expression of MHC class I and MHC class II molecules, and can induce the expression of MHC class II molecules on certain cell types that do not normally express them. Interferons also enhance the antigen-presenting function of MHC class I molecules by inducing the expression of key components of the intracellular machinery that enables peptides to be loaded onto the MHC molecules. We now turn to this machinery.

4-8 Peptides that bind to MHC class I molecules are actively transported from the cytosol to the endoplasmic reticulum.

The antigen fragments that bind to MHC class I molecules for presentation to CD8 T cells are typically derived from viruses that take over the cell's biosynthetic mechanisms to make their own proteins. All proteins are made in the cytosol. Proteins destined for the cell surface, including both classes of MHC molecule, are translocated during their synthesis into the lumen of the endoplasmic reticulum, where they must fold correctly before they can be transported to the cell surface. Because the peptide-binding site of the MHC class I molecule is formed in the lumen of the endoplasmic reticulum and is never exposed to the cytosol, how are peptides derived from viral proteins in the cytosol able to bind to MHC class I molecules for delivery to the cell surface?

The answer to this question was first suggested by the behavior of mutant cells with a defect in antigen presentation by MHC class I molecules. Although both chains of MHC class I molecules are synthesized normally in these cells, the MHC class I proteins are expressed only at very low levels on the cell surface. The defect in these cells can be corrected by the addition of synthetic peptides, suggesting both that the mutation affects the supply of peptides to MHC class I molecules and that peptide is required for their normal cell-surface expression. This was the first indication that MHC molecules are unstable in the absence of bound peptide.

Analysis of the DNA in the mutant cells showed that two genes encoding members of the ATP-binding cassette, or ABC, family of proteins are mutant or absent in these cells. These two genes map within the MHC itself (see Section 4-15). ATP-binding cassette proteins are associated with membranes in many cells, including bacterial cells, and they mediate ATP-dependent transport of ions, sugars, amino acids, or peptides across the membrane.

The two ATP-binding cassette proteins deleted in the mutant cells are associated with the endoplasmic reticulum membrane. They are encoded in the MHC and are inducible by interferon. Transfection of the mutant cells with both genes restores presentation of cytosolic peptides by the cell's MHC class I molecules. These proteins are now called **Transporters associated with Antigen Processing-1** and **-2** (**TAP-1** and **TAP-2**). The two TAP proteins form a heterodimer (Fig. 4.11) and mutations in either TAP gene can prevent antigen presentation by MHC class I molecules.

In assays *in vitro* using microsomal vesicles that mimic the endoplasmic reticulum, vesicles from normal cells will internalize peptides, which then bind to MHC class I molecules already present in the microsomal lumen. Vesicles from TAP-1 or TAP-2 mutant cells do not transport peptides. Peptide transport into the normal microsomes requires ATP hydrolysis, proving that the TAP-1:TAP-2 complex is an ATP-dependent peptide transporter that selectively loads peptides into the lumen of the endoplasmic reticulum. Such experiments have also shown that the TAP transporter has some specificity for the peptides it will transport. The TAP-1:TAP-2 transporter prefers peptides of eight or more amino acids with hydrophobic or basic residues at the carboxy terminus—the exact features of peptides that bind MHC class I molecules.

Schematic diagram of the TAP transporter

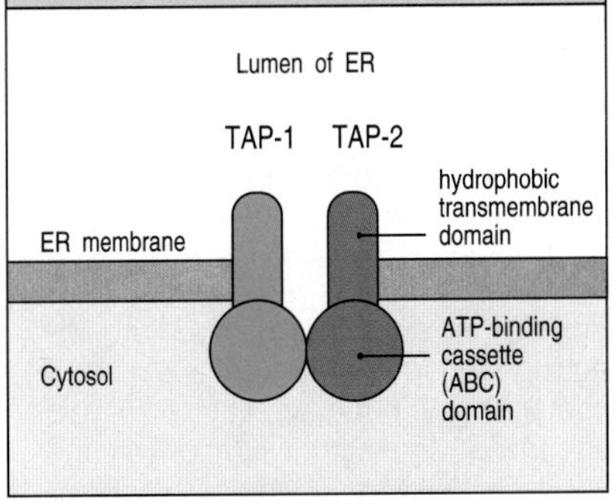

Fig. 4.11 The TAP-1 and TAP-2 transporter molecules form a heterodimer in the endoplasmic reticulum membrane. All protein molecules that belong to the ATP-binding cassette (ABC) transporter family have four domains, two complex transmembrane domains that each have multiple transmembrane regions, and two ATP-binding domains. Both TAP-1 and TAP-2 encode one hydrophobic and one ATP-binding domain and assemble into a heterodimer to form a four-domain transporter. On the basis of similarities between the TAP molecules and other members of the ABC-transporter family, it is believed that the ATP-binding domains lie within the cytoplasm of the cell, whereas the hydrophobic domains project through the membrane into the lumen of the endoplasmic reticulum (ER).

4-9 Peptides of cytosolic proteins are generated in the cytosol before transport into the endoplasmic reticulum.

Proteins in cells are continually being degraded and replaced with newly synthesized proteins. A major part in cytosolic protein degradation is played by a large, multicatalytic protease complex called the **proteasome** (Fig. 4.12). The proteasome is a large cylindrical complex of some 28 subunits, arranged as four stacked rings, each of seven subunits, and it has a hollow core lined by the active sites of the proteolytic subunits of the proteasome. Proteins to be degraded are introduced into the core of the proteasome and are there broken down into short peptide fragments.

Various lines of evidence implicate the proteasome in the production of peptide ligands for MHC class I molecules. For example, the proteasome takes part in the ubiquitin-dependent degradation pathway for cytosolic proteins; experimentally tagging proteins with ubiquitin also results in more efficient presentation of their peptides by MHC class I molecules. Moreover, inhibitors of the proteolytic activity of the proteasome also inhibit antigen presentation by MHC class I molecules. Whether the proteasome is the only cytosolic protease capable of generating peptides for transport into the endoplasmic reticulum is not known.

Two subunits of the proteasome, called LMP2 and LMP7, are encoded within the MHC near the TAP-1 and TAP-2 genes (see Section 4-15) and their expression is induced by interferon, as is that of the MHC class I and TAP molecules. LMP2 and LMP7 substitute for two constitutive subunits of the proteasome, which they displace in cells in which they are expressed. A third subunit, MECL-1, which is not encoded within the MHC, is also induced by interferon treatment and also displaces a constitutive proteasome subunit. These three inducible subunits and their constitutive counterparts are thought to be the active proteases of the proteasome. The replacement of the constitutive components by their interferon-inducible counterparts seems to change the specificity of the proteasome: in interferon-treated cells, there is increased cleavage of polypeptides after hydrophobic and

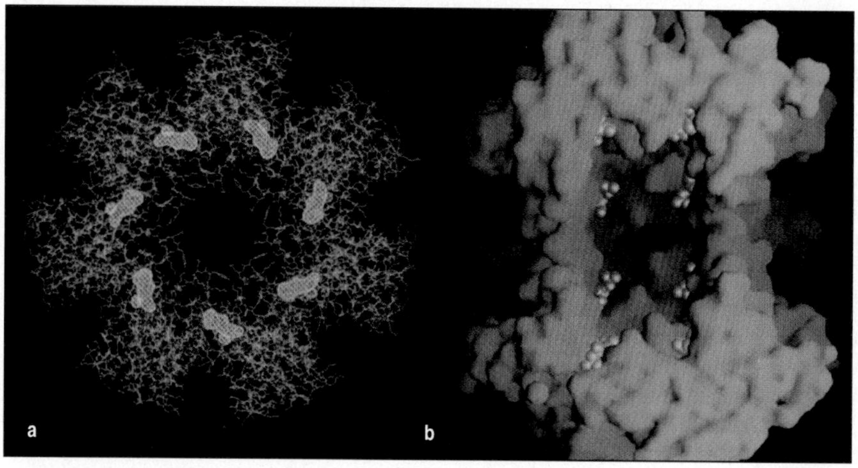

Fig. 4.12 The structure of the proteasome. The digestion of cytosolic proteins is carried out by a large protease complex, the proteasome. Proteasomes are ubiquitous in eukaryotes and archaebacteria and are conserved both in structure and function; the structure of an archaebacterial proteasome has been solved and allows us to visualize how these functions are carried out. The proteasome contains 28 subunits arranged to form a cylindrical structure composed of four rings, each of seven subunits. Panel a shows a horizontal cross-section through the proteasome, depicting the arrangement of the seven subunits that comprise each ring; panel b shows a longitudinal section, in which the surface of the proteasome can be seen. The subunits that form the two central rings of the archaebacterial proteasome contain the protease activity, and the active sites of these proteases are indicated in green in panel a and in gold in panel b. It is not known exactly how the mammalian proteasome degrades cytosolic proteins but it has only six proteolytic sites, three in each of the two central rings, and these sites lie in the center of the cylinder. Thus, it is likely that proteins have to unfold and pass through the center of a cylindrical structure like that shown here for degradation to occur. Photographs (× 667,000) courtesy of W Baumeister.

basic residues, and reduced cleavage after acidic residues. This produces peptides with carboxy-terminal residues that are preferred anchor residues for peptide binding to most MHC class I molecules and are also the preferred structures for transport by TAP.

As well as peptides from cytosolic proteins, MHC class I molecules present peptides derived from membrane and secreted proteins, for example the glycoproteins of viral envelopes. Membrane and secreted proteins are normally translocated into the lumen of the endoplasmic reticulum during their biosynthesis. Yet the peptides bound by MHC class I molecules bear evidence that such proteins are degraded in the cytosol: asparagine-linked carbohydrate moieties commonly present on membrane-bound or secreted proteins can be removed in the cytosol by an enzyme reaction that changes the asparagine residue into aspartic acid, and this diagnostic sequence change can be seen in some peptides presented by MHC class I molecules. It now appears that endoplasmic reticulum proteins can be returned to the cytosol via the same translocation system by which they were initially transported across the endoplasmic reticulum membrane. This newly discovered mechanism, known as retrograde translocation, may be the normal mechanism by which proteins in the endoplasmic reticulum are turned over, and by which misfolded proteins in the endoplasmic reticulum are removed and degraded. Once in the cytosol, the polypeptides are degraded by the proteasome, and the resulting peptides may then be transported back into the lumen of the endoplasmic reticulum via the TAP transporter and loaded onto MHC class I molecules.

4-10 Newly synthesized MHC class I molecules are retained in the endoplasmic reticulum by binding a TAP-1-associated protein until they bind peptide.

The binding of peptide is an important step in the assembly of stable MHC class I molecules. When the supply of peptides into the endoplasmic reticulum is disrupted, as in the TAP mutant cells, newly synthesized MHC class I molecules are held in the endoplasmic reticulum in a partially folded state. (This explains why cells with mutations in TAP-1 or TAP-2 fail to express MHC class I molecules at the cell surface.) As we shall see, the folding of an MHC class I molecule depends on its association first with β_2-microglobulin and then with peptide, and this process involves a number of accessory proteins with a chaperone-like function. Only when peptide is bound is the MHC class I molecule released from the endoplasmic reticulum and allowed to reach the surface of the cell.

In humans, newly synthesized MHC class I α chains bind to a chaperone protein, **calnexin**, which retains the MHC class I molecule in a partially folded state in the endoplasmic reticulum. Calnexin is also known to associate with partially folded T-cell receptors, immunoglobulins, and MHC class II molecules, and so has a central role in the assembly of many molecules important in immunology. When β_2-microglobulin binds to the α chain, the partially folded α:β_2-microglobulin heterodimer dissociates from calnexin and now binds to a complex of proteins, one of which, calreticulin, is similar to calnexin and probably carries out a similar chaperone function. A second component of the complex is the TAP-1-associated protein **tapasin**, also encoded by a gene that lies within the MHC. Tapasin forms a bridge between class I MHC molecules and the TAP-1 subunit of the transporter, allowing the partially folded α:β_2-microglobulin heterodimer to await the transport of a suitable peptide from the cytosol. Finally, the binding of a peptide to the partially folded heterodimer releases it from the complex of TAP:tapasin: calreticulin and allows the now fully folded MHC class I molecule to leave the endoplasmic reticulum and be transported to the cell surface (Fig. 4.13).

Most of the peptides transported by TAP will not bind to the MHC molecules in that cell and are rapidly cleared out of the endoplasmic reticulum; there is evidence that they are transported back into the cytosol by an ATP-dependent transport mechanism distinct from the TAP transporter. It is not yet clear whether the TAP transporter plays a direct role in loading MHC class I

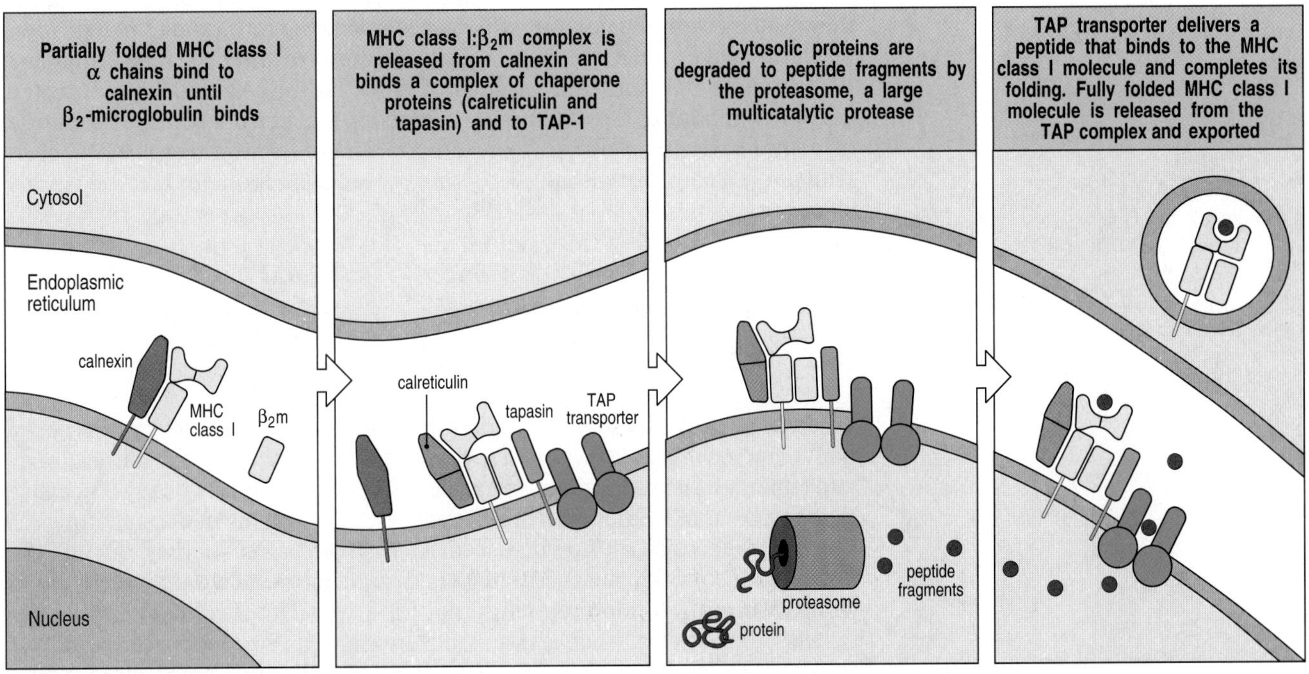
Partially folded MHC class I α chains bind to calnexin until β2-microglobulin binds
MHC class I:β2m complex is released from calnexin and binds a complex of chaperone proteins (calreticulin and tapasin) and to TAP-1
Cytosolic proteins are degraded to peptide fragments by the proteasome, a large multicatalytic protease
TAP transporter delivers a peptide that binds to the MHC class I molecule and completes its folding. Fully folded MHC class I molecule is released from the TAP complex and exported
Cytosol
Endoplasmic reticulum
calnexin
MHC class I
β2m
Nucleus
calreticulin
tapasin
TAP transporter
proteasome
peptide fragments
protein

Fig. 4.13 MHC class I molecules do not leave the endoplasmic reticulum unless they bind peptides. MHC class I α chains assemble in the endoplasmic reticulum with a membrane-bound protein, calnexin. When this complex binds β_2-microglobulin (β_2m) it is released from calnexin, and the partially folded MHC class I molecule then binds to the TAP-1 subunit of the TAP transporter by interacting with one molecule of the TAP-associated protein tapasin. It is retained within the endoplasmic reticulum until released by the binding of a peptide, which completes the folding of the MHC class I molecule. Peptides generated by the degradation of proteins in the cytoplasm are transported into the lumen of the endoplasmic reticulum by the TAP transporter. Once peptide has bound to the MHC molecule, the peptide: MHC complex is transported through the Golgi complex to the cell surface.

molecules with peptide or whether binding to the TAP transporter merely allows the MHC class I molecule to scan the transported peptides before they diffuse through the lumen of the endoplasmic reticulum and are transported back into the cytosol.

In cells with mutant TAP genes, the MHC class I molecules are unstable and are eventually translocated back into the cytosol of the cell, where they are degraded, indicating that the MHC class I molecule must bind a peptide to complete its folding and be transported onwards from the endoplasmic reticulum. Even in normal cells, MHC class I molecules are retained in the endoplasmic reticulum for some time, suggesting that they are usually present in excess of peptide. This is very important for the function of MHC class I molecules because they must be immediately available to transport viral peptides to the cell surface at any time that the cell becomes infected. In uninfected cells, peptides derived from self proteins fill the peptide-binding cleft of the mature MHC class I molecules present at the cell surface. When a cell is infected by a virus, the presence of excess MHC class I molecules in the endoplasmic reticulum allows the rapid presentation at the cell surface of peptides derived from the pathogen.

Because the presentation of viral peptides by MHC class I molecules signals CD8 T cells to kill the infected cell, some viruses have evolved mechanisms to evade recognition by interfering with this pathway. The herpes simplex virus, for example, prevents the transport of viral peptides into the endoplasmic reticulum by producing a protein that binds to and inhibits the TAP transporter. A second mechanism is that used by adenoviruses, which encode a protein that binds to MHC class I molecules and retains them in the endoplasmic reticulum, again preventing the appearance at the cell surface of MHC class I molecules loaded with viral peptides. A third mechanism is that used by cytomegalovirus, which accelerates the retrograde translocation of MHC class I molecules back into the cytosol of the cell, where they are degraded. The advantage to a virus of blocking the recognition of infected cells is so great that it would not be surprising if other steps in the formation of peptide:MHC complexes, for example the association of the MHC class I:chaperone complex with the TAP transporter, were found to be inhibited by some viruses.

4-11 Peptides presented by MHC class II molecules are generated in acidified endocytic vesicles.

Whereas viruses and some bacteria replicate in the cytosol, several classes of pathogen, including *Leishmania* spp. and the mycobacteria that cause leprosy and tuberculosis, replicate in intracellular vesicles in macrophages. Because they reside in membrane-enclosed vesicles, the proteins of these pathogens are not accessible to proteasomes. Instead, after activation of the macrophage, proteins in these sites are degraded by endosomal or lysosomal proteases into peptide fragments that bind to MHC class II molecules for delivery to the cell surface (Fig. 4.14). Here they can be recognized by CD4 T cells, which also recognize peptide fragments derived from extracellular pathogens and proteins that are internalized into similar intracellular vesicles.

Most of what we know about the processing of proteins in the endocytic pathway has come from experiments in which simple proteins are fed to macrophages; in this way the processing of added antigen can be quantified. Proteins that bind to surface immunoglobulin on B cells and are internalized by receptor-mediated endocytosis are processed by the same pathway.

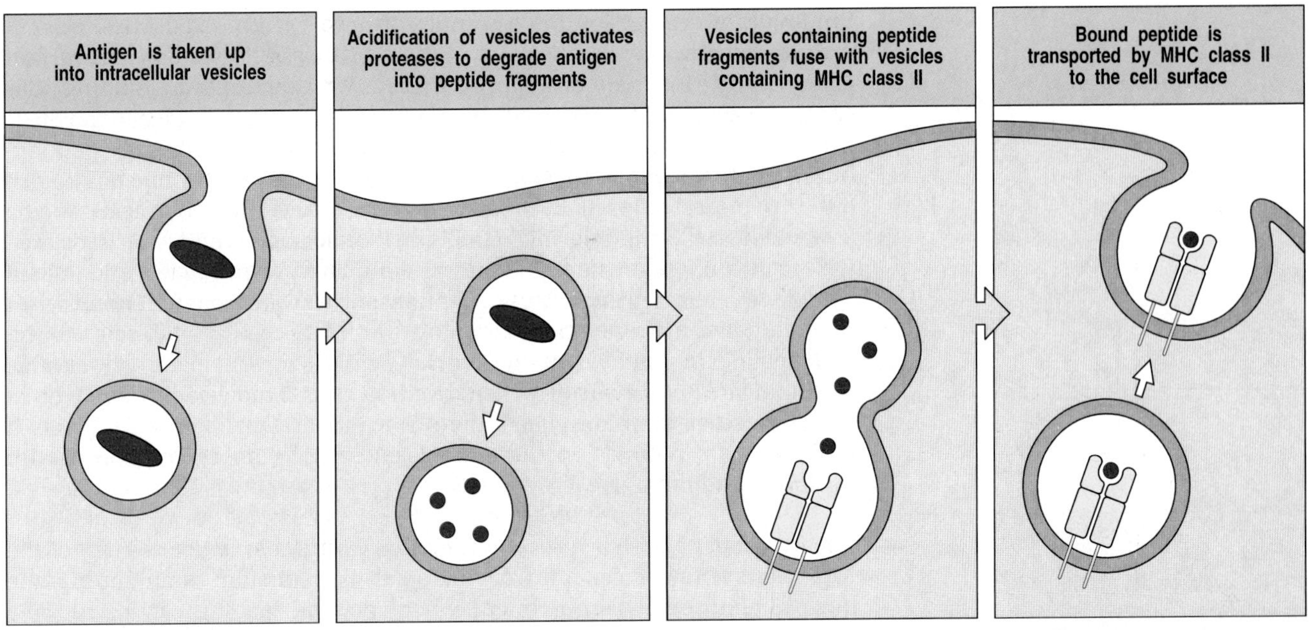
Antigen is taken up into intracellular vesicles
Acidification of vesicles activates proteases to degrade antigen into peptide fragments
Vesicles containing peptide fragments fuse with vesicles containing MHC class II
Bound peptide is transported by MHC class II to the cell surface

Fig. 4.14 Antigens that bind to MHC class II molecules are degraded in acidified endocytic vesicles. In some cases, the source of the antigen may be bacteria or parasites that have invaded the cell to replicate in intracellular vesicles. In other cases, as illustrated here, microorganisms or foreign proteins are taken up by phagocytic cells or endocytosed by other professional antigen-presenting cells. The pH of the endocytic vesicles containing the engulfed pathogens progressively decreases, activating proteases that reside within the vesicle to degrade the engulfed material. At some point on their pathway to the cell surface, newly synthesized MHC class II molecules pass through such acidified vesicles and bind peptide fragments of the antigen, transporting the peptides to the cell surface.

Internalized protein antigens that enter cells through endocytosis become enclosed in vesicles known as endosomes that become increasingly acidic as they progress into the interior of the cell. The vesicles of the endosomal pathway contain proteases, known as acid proteases, that are activated at low pH and eventually degrade proteins contained in the vesicles. Proteins of pathogens growing in intracellular vesicles are also handled by this pathway of antigen processing.

Drugs, such as chloroquine, that raise the pH of endocytic vesicles inhibit the presentation of antigens that enter the cell in this way, suggesting that acid proteases are responsible for the processing of internalized antigen. Among these acid proteases are the cysteine proteases cathepsins B, D, S, and L, the last of which is the most active enzyme in this family of related proteases. Antigen processing can be mimicked to some extent by digestion of proteins with these enzymes *in vitro* at acid pH. Of these, cathepsins S and L may be the predominant proteases involved in the processing of vesicular antigens; mice engineered to lack expression of cathepsin B or cathepsin D show normal antigen processing, whereas mice with mutations in cathepsin S and cathepsin L are phenotypically defective in antigen processing.

4-12 The invariant chain directs newly synthesized MHC class II molecules to acidified intracellular vesicles.

The function of MHC class II molecules is to present peptides generated in the intracellular vesicles of B cells, macrophages, and other antigen-presenting cells to CD4 T cells. However, the biosynthetic pathway for MHC class II molecules, like that of other cell-surface glycoproteins, starts with their translocation into the endoplasmic reticulum, and they must therefore be prevented from binding prematurely to peptides transported into the endoplasmic reticulum lumen by the TAP transporter, or to the cell's own newly synthesized polypeptides. As the endoplasmic reticulum is richly endowed with unfolded and partially folded polypeptide chains, a general mechanism is needed to prevent their binding in the open-ended MHC class II peptide-binding groove.

Binding is prevented by the assembly of newly synthesized MHC class II molecules with a protein known as the MHC class II-associated **invariant chain (Ii)**. The invariant chain forms trimers, with each subunit binding non-covalently to an MHC class II α:β heterodimer (Fig. 4.15). Ii binds to the MHC class II molecule with part of its polypeptide chain lying within the peptide-binding groove, thus blocking the groove and preventing the binding of either peptides or partially folded proteins. While this complex is being assembled in the endoplasmic reticulum, its component parts are associated with calnexin. Only when assembly is completed to produce a nine-chain complex is it released from calnexin for transport onward from the endoplasmic reticulum. In this nine-chain complex, the MHC class II molecule cannot bind peptides or unfolded proteins, so that peptides present in the endoplasmic reticulum are not usually presented by MHC class II molecules. Moreover, in the absence of invariant chains there is evidence that many MHC class II molecules are retained in the endoplasmic reticulum as complexes with misfolded proteins.

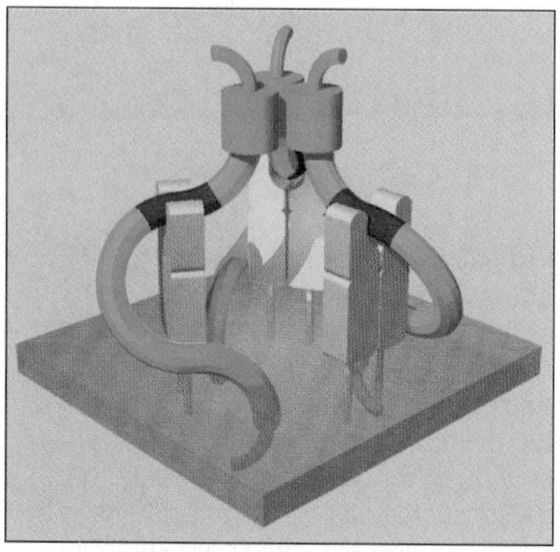

The invariant chain has a second function, which is to target delivery of the MHC class II molecules from the endoplasmic reticulum to an appropriate low-pH endosomal compartment where peptide loading can occur. The complex of MHC class II α:β heterodimers with invariant chain is retained for 2–4 hours in this compartment. During this time, the invariant chain is cleaved by proteases, such as cathepsin S or cathepsin L, in several steps, as shown in Fig. 4.16. The initial cleavage events generate a truncated form of the invariant chain that remains bound to the MHC class II molecule and retains it within the proteolytic compartment. A subsequent cleavage releases the MHC class II molecule from the membrane-associated fragment of Ii, leaving a short fragment of Ii, called **CLIP** (for **class II-associated invariant-chain peptide**)

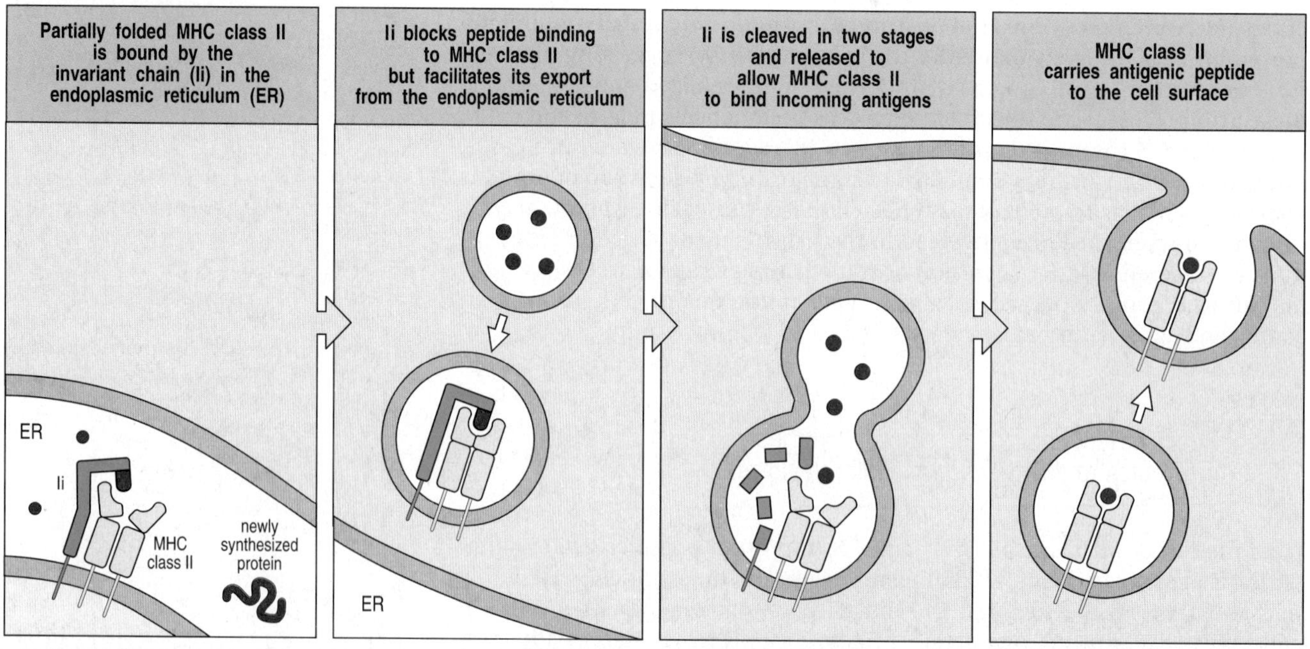
Partially folded MHC class II is bound by the invariant chain (Ii) in the endoplasmic reticulum (ER)
Ii blocks peptide binding to MHC class II but facilitates its export from the endoplasmic reticulum
Ii is cleaved in two stages and released to allow MHC class II to bind incoming antigens
MHC class II carries antigenic peptide to the cell surface
ER
Ii
MHC class II
newly synthesized protein
ER

Fig. 4.15 The MHC class II-associated invariant chain delays peptide binding and targets MHC class II molecules to the endosomes. The invariant chain (Ii) assembles with newly synthesized MHC class II molecules in the endoplasmic reticulum, where it prevents the MHC class II molecule from binding intracellular peptides and partially folded proteins present in the lumen, and directs its export through the Golgi apparatus to acidified endosomes containing peptides of resident bacteria or engulfed extracellular proteins. Here the invariant chain is cleaved in stages, and the MHC class II molecule binds antigenic peptide and is transported to the cell surface. A model for the trimeric invariant chain bound to αβ heterodimers is shown in the upper left. The CLIP portion is shown in red, the rest of the invariant chain in green, and the α:β MHC class II heterodimer in yellow. Model structure courtesy of P Cresswell.

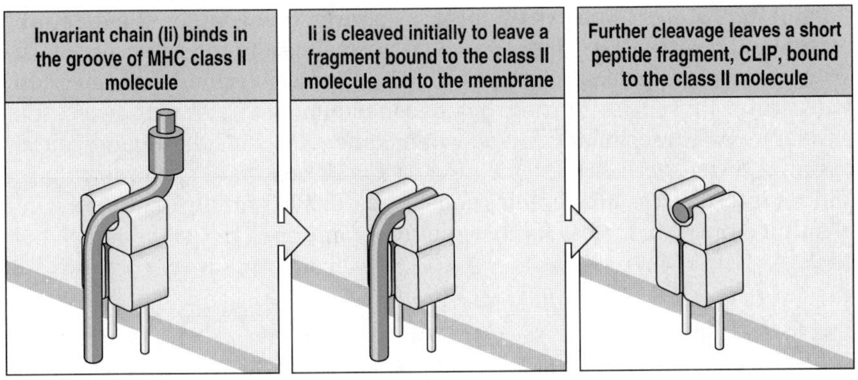

Invariant chain (Ii) binds in the groove of MHC class II molecule
Ii is cleaved initially to leave a fragment bound to the class II molecule and to the membrane
Further cleavage leaves a short peptide fragment, CLIP, bound to the class II molecule

Fig. 4.16 The invariant chain is cleaved to leave a peptide fragment, CLIP, bound to the MHC class II molecule. The invariant chain (Ii) binds to MHC class II molecules with a section of its polypeptide chain lying along the peptide-binding groove (left panel). In an acidified endocytic vesicle Ii is cleaved, first at one side of the class II molecule (center panel). The remaining portion of Ii (known as the Leupeptin Induced Peptide or LIP fragment) retains the transmembrane and cytoplasmic segments and hence the signals that target Ii:MHC class II complexes to the endosomal pathway. Subsequent cleavage (right panel) of LIP leaves only a short peptide still bound by the class II molecule; this peptide is the CLIP fragment.

still bound to the MHC class II molecule. MHC class II molecules that have CLIP associated with them still cannot bind other peptides, and CLIP must either dissociate or be displaced to allow peptides to bind and be delivered to the cell surface. Cathepsin S carries out the cleavage of Ii in most class II-positive cells, including antigen-presenting cells, whereas cathepsin L appears to substitute for cathepsin S in thymic cortical epithelial cells. As we shall see in Chapter 7, thymic cortical epithelial cells may express a distinct spectrum of peptide:MHC class II complexes that play a role in selecting the CD4 T-cell receptor repertoire.

The exact pathway to, and characteristics of, the endocytic compartment in which invariant chain is cleaved and MHC class II molecules encounter peptides is not clearly defined. Most newly synthesized MHC class II molecules are brought towards the cell surface in vesicles, which at some point fuse with incoming endosomes. However, there is also evidence that some MHC class II:Ii complexes are first transported to the cell surface and then re-internalized into endosomes. In either case, MHC class II:Ii complexes enter the endocytic pathway and are there exposed to an acidic, proteolytic environment in which the invariant chain is cleaved and pathogens and their proteins are broken down into peptides available to bind to the MHC class II molecule. Electron microscopy studies using antibodies tagged with gold particles to localize Ii and MHC class II molecules within the cell suggest that the site where Ii is cleaved and where peptides bind to MHC class II molecules is a specific endocytic compartment, called the **MIIC** (**MHC class II compartment**), late in the endosomal pathway (Fig. 4.17).

As with MHC class I molecules, MHC class II molecules in uninfected cells bind peptides derived from self proteins. 'Empty' MHC class II molecules that do not bind peptide after dissociation from the invariant chain are unstable, leading to aggregation and rapid degradation at the acidic pH of the endosomal compartment. It is therefore not surprising that peptides derived from MHC

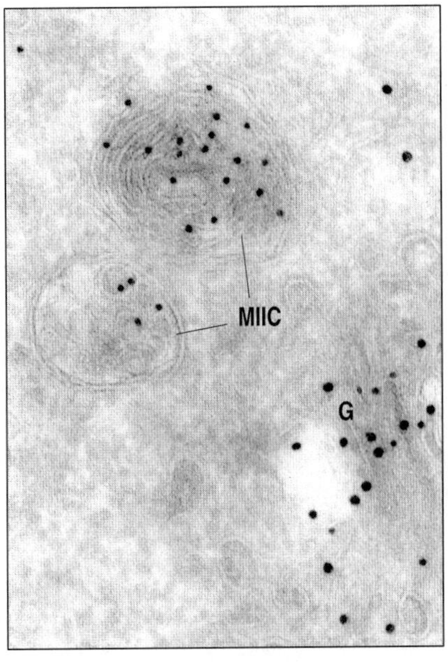

Fig. 4.17 MHC class II molecules are loaded with peptide in specialized vesicles. MHC class II molecules are transported from the Golgi (labeled G in this electron micrograph of an ultrathin section of a B cell) to the cell surface via specialized intracellular vesicles called the MHC class II compartment (MIIC) (indicated by the arrows). These have a complex morphology showing internal vesicles and sheets of membrane. Antibodies labeled with different-sized gold particles identify the presence of both MHC class II molecules (small gold particles) and the invariant chain (large gold particles) in the Golgi, whereas the MIIC shows the presence only of MHC class II molecules. Magnification × 135,000. Photograph courtesy of Hans J Geuze.

class II molecules themselves form a large proportion of the peptides presented by MHC class II molecules in normal cells. This suggests that, as for MHC class I molecules, excess MHC class II molecules are generated. Thus, when a cell is infected by mycobacteria or other pathogens that proliferate in intracellular vesicles, when a phagocyte engulfs a pathogen, or when pathogen-derived proteins bind to a B cell's surface immunoglobulin and are internalized, the peptides generated from the pathogen proteins find plentiful empty MHC class II molecules to bind.

4-13 A specialized MHC class II-like molecule catalyzes loading of MHC class II molecules with endogenously processed peptides.

Just as mutant cell lines first led to an appreciation of the role of the TAP transporters in peptide loading on MHC class I molecules, observations on mutant human B-cell lines revealed an unsuspected component of the vesicular antigen-processing pathway. MHC class II molecules in these mutant cell lines assemble correctly with the invariant chain and seem to follow the normal vesicular route, but fail to bind peptides derived from internalized proteins and often arrive at the cell surface with the CLIP peptide still bound.

The defect in these mutant cells lies in a MHC class II-like molecule called **HLA-DM** in humans (H-2M in mice). The HLA-DM locus is found near the TAP and LMP genes in the class II region of the MHC (see Fig. 4.19), and it encodes an α chain and a β chain that closely resemble those of other MHC class II molecules although, unlike other class II molecules, HLA-DM does not appear to require peptide for stabilization. The DM molecule is not expressed at the cell surface but rather is found predominantly in the MIIC compartment. HLA-DM binds to and stabilizes empty MHC class II molecules that would otherwise aggregate; in addition, it catalyzes both the release of the CLIP fragment from MHC class II:CLIP complexes and the binding of other peptides to the resulting empty MHC class II molecule (Fig. 4.18). HLA-DM is also able to catalyze the release of peptides other than CLIP from MHC class II molecules, depending on the stability of the MHC:peptide complex. In the presence of a mixture of peptides capable of binding to MHC class II molecules, as occurs in the MIIC itself, HLA-DM will continuously bind and rebind to peptide:MHC class II complexes, removing unstably bound peptides and allowing other peptides to replace them. As we will see later (in Section 8-6), antigens presented by MHC class II molecules may have to persist on the surface of antigen-presenting cells for some days before encountering T cells able to recognize them. The ability of HLA-DM to remove unstably bound peptides, sometimes called 'peptide editing,' ensures that the peptide:MHC class II complexes displayed on the surface of the antigen-presenting cell will survive long enough to be able to stimulate the appropriate CD4 cells.

In a subset of cells, namely in the epithelial cells of the thymus and in B cells, a second atypical MHC class II molecule is expressed, called HLA-DO (or, in mice, H-2O). This molecule is a heterodimer of the HLA-DNα chain and the HLA-DOβ chain. DO resembles DM in that it is not expressed at the cell surface, being found only in intracellular vesicles, and it does not appear to bind peptides. Instead, the HLA-DO molecule is a negative regulator of HLA-DM, binding to it inside the cell and inhibiting both the HLA-DM-catalyzed release of CLIP from, and the binding of other peptides to, MHC class II molecules. Expression of the HLA-DOβ chain is not increased by IFN-γ, whereas the expression of HLA-DM, like that of other MHC class II molecules, is increased. Thus, during inflammatory responses, the increased

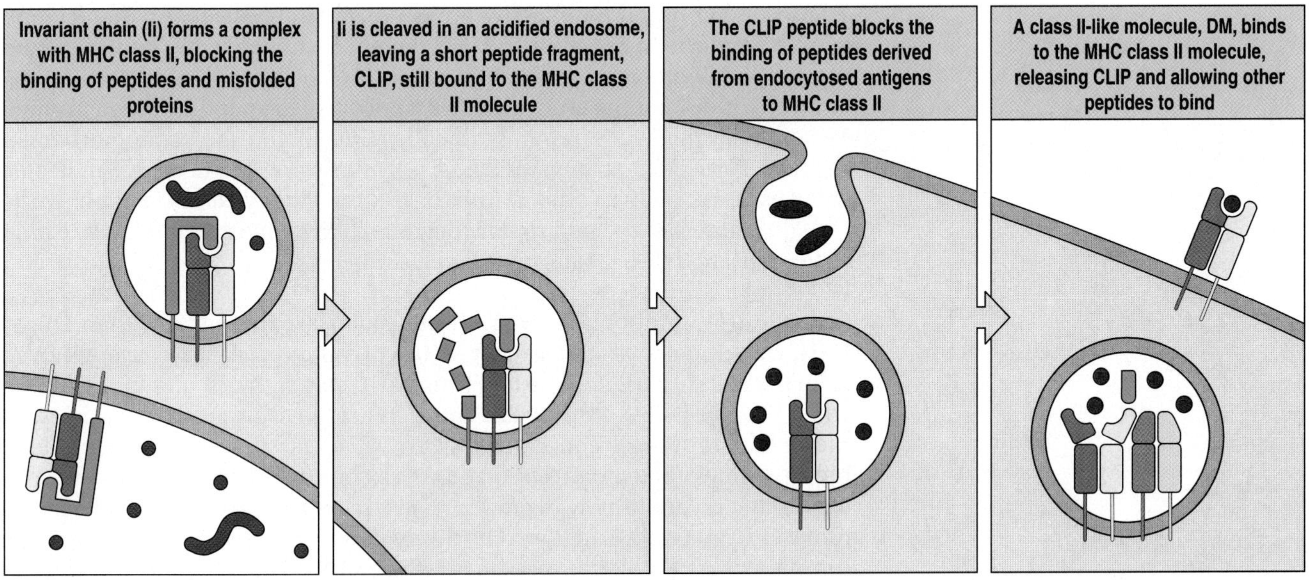

Invariant chain (Ii) forms a complex with MHC class II, blocking the binding of peptides and misfolded proteins
Ii is cleaved in an acidified endosome, leaving a short peptide fragment, CLIP, still bound to the MHC class II molecule
The CLIP peptide blocks the binding of peptides derived from endocytosed antigens to MHC class II
A class II-like molecule, DM, binds to the MHC class II molecule, releasing CLIP and allowing other peptides to bind

Fig. 4.18 HLA-DM facilitates the loading of antigenic peptides onto class II molecules. The invariant chain binds to newly synthesized class II molecules and blocks the binding of peptides and unfolded proteins in the endoplasmic reticulum and during the transport of the class II molecule into acidified endocytic vesicles (first panel). In such vesicles, proteases cleave the invariant chain, leaving the CLIP peptide bound to the class II molecule (second panel). Pathogens and their proteins are broken down into peptides within acidified endocytic vesicles, but these peptides cannot bind to MHC class II molecules that are occupied by CLIP (third panel). The class II-like molecule, DM, binds to MHC class II:CLIP complexes, catalyzing the release of CLIP and the binding of antigenic peptides (fourth panel).

expression of HLA-DM is able to overcome the inhibitory effects of HLA-DO. Why the antigen-presenting ability of thymic epithelial cells and of B cells should be regulated in this way is not known; in the thymic epithelial cells the purpose may be to select developing CD4 T cells with a repertoire of self peptides distinct from that to which they will be exposed as mature T cells (see Section 7-14).

The role of the DM molecule in facilitating the binding of peptides to MHC class II molecules appears to parallel the role of the TAP molecules in facilitating the binding of peptides to MHC class I molecules. Thus it seems likely that a specialized mechanism to deliver peptides efficiently has co-evolved with the MHC molecules themselves. It is also likely that pathogens have evolved strategies to inhibit this peptide-loading process, much as viruses have found ways to subvert the process of antigen processing and presentation through MHC class I molecules.

4-14 Stable peptide binding by MHC molecules allows effective antigen presentation at the cell surface.

To enable MHC molecules to perform their essential function of signaling intracellular infection, it is important that the peptide:MHC complex should be stable at the cell surface. If the complex were to dissociate too readily, the pathogen in the infected cell could escape detection. Conversely, MHC molecules on uninfected cells could pick up peptides released by MHC molecules on infected cells and falsely signal to cytotoxic T cells that a healthy cell is infected, triggering its unwarranted destruction. The stable binding of peptide by MHC molecules makes both these undesirable outcomes unlikely.

The ability to stimulate T cells can be used to measure the persistence of antigen:MHC complexes on cell surfaces. In this way it can be shown that peptide:MHC complexes expressed on live cells are lost at the same rate as the MHC molecule itself, indicating that the binding of peptide is essentially irreversible. This stability of binding permits even rare peptides to be transported efficiently to the cell surface by MHC molecules, and allows long-term display of these complexes on the surface of the infected cell, thus fulfilling the first of the requirements for effective antigen presentation.

The second criterion for effective antigen presentation is that if dissociation of a peptide from a cell-surface MHC molecule should occur, peptides from the surrounding extracellular fluid would not be able to bind to the now empty peptide-binding groove. In fact, removal of the peptide from a purified MHC class I molecule requires denaturation of the molecule, and when peptide dissociates at the surface of a live cell, the MHC class I molecule changes conformation, the β_2-microglobulin moiety dissociates, and the α chain is internalized and rapidly degraded. Thus, most empty MHC class I molecules are quickly lost from the cell surface.

At neutral pH, empty MHC class II molecules are more stable than empty MHC class I molecules, yet empty MHC class II molecules are also removed from the cell surface. Peptide loss from MHC class II molecules is most likely when the molecules are recycled through acidified intracellular vesicles. At the acidic pH of these endocytic vesicles, MHC class II molecules are able to bind peptides that are present in the vesicles, but those that fail to do so aggregate and are rapidly degraded. Thus, both the MHC class I and class II molecules on a cell surface are effectively prevented from acquiring peptides from the surrounding extracellular fluid. This ensures that T cells act selectively on infected cells or on cells specialized for antigen uptake and display, while sparing surrounding healthy cells.

Summary.

The most distinctive feature of antigen recognition by T cells is the form of the ligand recognized by the T-cell receptor. This comprises a peptide derived from the foreign antigen bound to an MHC molecule. MHC molecules are cell-surface glycoproteins with a peptide-binding groove that can bind a wide variety of different peptides. The MHC molecule binds the peptide in an intracellular location and delivers it to the cell surface, where the combined ligand can be recognized by a T cell. There are two classes of MHC molecule, MHC class I and MHC class II, which deliver peptides from proteins degraded in different intracellular sites. MHC class I molecules bind peptides from proteins degraded in the cytoplasm of the cell. Multicatalytic protease complexes in the cytoplasm, known as proteasomes, degrade both cytosolic proteins and misfolded or excess membrane and secretory proteins that have been transported back into the cytoplasm from the endoplasmic reticulum. Peptides produced by proteasomes are transported into the endoplasmic reticulum by a heterodimeric ATP-binding protein called TAP (Transporter Associated with Antigen Processing), and they are then available for binding by partially folded MHC class I molecules that are held tethered to TAP. Peptide binding is an integral part of MHC class I assembly, and must occur before the MHC class I molecule can complete its folding and leave the endoplasmic reticulum for the surface of the cell. By binding stably to peptides from proteins degraded in the cytosol, MHC class I molecules serve to display peptides from viruses or other cytosolic pathogens on the surface of infected cells. In contrast, MHC class II molecules are prevented from binding to peptides in the endoplasmic reticulum by their early association with the invariant chain, which fills and blocks their peptide-binding groove. Instead, they are targeted by the invariant chain to an acidic endocytic compartment where, in the presence of active proteases, in particular the cathepsins S or L, and with the help of a specialized MHC class II-like molecule that catalyzes peptide loading, the invariant chain is released and other peptides are bound. In this way, MHC class II molecules bind peptides from proteins that are degraded in endocytic vesicles, thereby capturing peptides from pathogens that enter the vesicular system of macrophages, or peptides from the specific antigens internalized by the immunoglobulin receptors of B cells. Different types of T cell are activated on recognizing foreign peptides presented by the different classes of MHC molecule. The CD8 T cells that recognize MHC class I:peptide complexes are specialized to kill any cells displaying foreign peptides and so rid the body of cells infected with viruses and other cytosolic pathogens. The CD4 T cells that recognize MHC class II:peptide complexes are specialized to activate other effector cells of the immune system; macrophages, for example, are activated to kill the intra-vesicular pathogens they harbor and B cells to secrete immunoglobulins against foreign molecules. Thus, the two classes of MHC molecule deliver peptides from different cellular compartments to the cell surface, where they are recognized by T cells mediating distinct and appropriate effector functions.

The major histocompatibility complex of genes: organization and polymorphism.

The function of the MHC molecules is to bind peptide fragments derived from pathogens and display them on the cell surface for recognition by the appropriate T cells. The consequences of such presentation are almost always deleterious to the pathogen: virus-infected cells are killed, macrophages are activated to kill bacteria in intracellular vesicles, and B cells are activated to produce antibody molecules capable of eliminating or neutralizing extracellular pathogens. Thus, there is strong selective pressure in favor of any pathogen that can mutate its structural genes to escape presentation by an MHC molecule.

Two separate properties of the MHC make it difficult for pathogens to evade immune responses in this way. First, the MHC is **polygenic**: there are several MHC class I and MHC class II genes, encoding proteins with different ranges of peptide-binding specificities. Second, the MHC is highly **polymorphic**: there are multiple alleles of each gene. The MHC genes are, in fact, the most polymorphic genes known. In this section, we shall describe the organization of the genes in the MHC and discuss how the allelic variation in MHC molecules arises. We shall also see how the effect of polygeny and polymorphism on the range of peptides bound contributes to the ability of the immune system to respond to a multitude of different and rapidly evolving pathogens.

4-15 Many proteins involved in antigen processing and presentation are encoded by genes within the major histocompatibility complex.

The major histocompatibility complex extends over approximately 4 centimorgans of DNA, or about 4×10^6 base pairs, and contains more than 200 genes in humans. As work continues to define the genes within and around the MHC, both its extent and the numbers of genes involved are likely to grow; in fact, recent studies suggest that the MHC may span at least 7×10^6 base pairs. The genes encoding the α chains of MHC class I molecules and the α and β chains of MHC class II molecules are linked within the complex; the genes for β_2-microglobulin and the invariant chain lie on separate chromosomes. Figure 4.19 shows the general organization of these genes in the MHC of humans and of the mouse. The particular combination of MHC alleles found on an individual chromosome is known as an **MHC haplotype**.

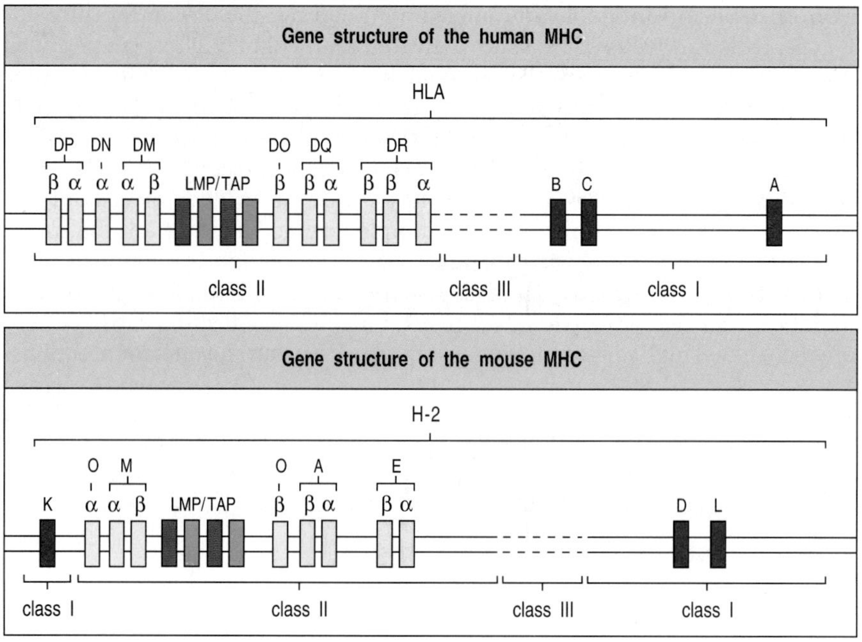

Fig. 4.19 The genetic organization of the major histocompatibility complex (MHC) in humans and the mouse. The organization of the principal MHC genes is shown for both human (where the MHC is called HLA and is on chromosome 6) and mouse (in which the MHC is called H-2 and is on chromosome 17). The organization of the MHC genes is similar in both species. There are separate regions of MHC class I and of MHC class II genes, although in the mouse an MHC class I gene appears to have translocated relative to the human MHC so that, in mice, the class I region is split in two. In both species there are three main class I genes, which are called HLA-A, -B, and -C in humans, and H2-K, -D, and -L in the mouse. The gene for β_2-microglobulin, although it encodes part of the MHC class I molecule, is located on a different chromosome, chromosome 15 in humans and chromosome 2 in the mouse. The class II region includes the genes for the α and β chains of the antigen-presenting class II MHC molecules HLA-DR, -DP, and -DQ (H-2, A and E in the mouse). In addition, the genes for the TAP-1:TAP-2 peptide transporter, the LMP genes that encode proteasome subunits, the genes encoding the DMα and DMβ chains, and the genes encoding the α and β chains of the DO molecule (DNα and DOβ, respectively) are also in the MHC class II region. The so-called class III genes encode various other proteins with functions in immunity (see Fig. 4.20).

There are three class I α-chain genes in humans, called HLA-A, -B, and -C. There are also three pairs of MHC class II α- and β-chain genes, called HLA-DR, -DP, and -DQ. However, in many haplotypes the HLA-DR cluster contains an extra β-chain gene whose product can pair with the DRα chain. This means that the three sets of genes give rise to four types of MHC class II molecule. All the MHC class I and class II molecules are capable of presenting antigens to T cells and, because each protein binds a different range of peptides, the presence of several loci means that any one individual is equipped to present a much broader range of different peptides than if only one MHC protein of each class were expressed at the cell surface.

The two TAP genes lie in the MHC class II region, in close association with the LMP genes that encode components of the proteasome, whereas the gene for tapasin, which binds to both TAP and empty MHC class I molecules, lies at the centromeric edge of the MHC. The genetic linkage of the MHC class I genes, whose products deliver cytosolic peptides to the cell surface, with the TAP, tapasin, and proteasome genes, which encode the molecules that generate peptides in the cytosol and transport them into the endoplasmic reticulum, suggests that the entire MHC has been selected during evolution for antigen processing and presentation.

Moreover, when cells are treated with the cytokines IFN-α, -β, or -γ, there is a marked increase in transcription of MHC class I α chain and β_2-microglobulin, and in the MHC-linked proteasome, tapasin, and TAP genes. Interferons are produced early in viral infections as part of the innate immune response, as described in more detail in Chapter 11, and this effect of interferons increases the ability of cells to process viral proteins and present the resulting peptides at the cell surface, thus helping to activate T cells and initiate the later phases of the adaptive immune response. The coordinated regulation of the genes encoding these components may be facilitated by the linkage of many of them in the MHC.

The DM genes, whose function is to catalyze peptide binding to MHC class II molecules, are clearly related to the MHC class II genes. The DNα and DOβ genes, which encode the DO molecule, a negative regulator of DM, are also clearly related to the class II genes. The DMα, β, and DNα genes, but not the DOβ gene, are coordinately regulated with the genes encoding other MHC class II molecules and the invariant chain; expression of all of these is induced by IFN-γ (but not by IFN-α or -β), via the production of a transcriptional activator known as **MHC class II transactivator** (**CIITA**). The absence of CIITA in patients with the bare lymphocyte syndrome causes severe immunodeficiency as described later, in Chapter 11.

4-16 A variety of genes with specialized functions in immunity are also encoded in the MHC.

Although the most important known function of the gene products of the MHC is the processing and presentation of antigens to T cells, many other genes map within this region; some of these are known to have other roles in the immune system, but many have yet to be characterized functionally. Figure 4.20 shows the detailed organization of the human MHC.

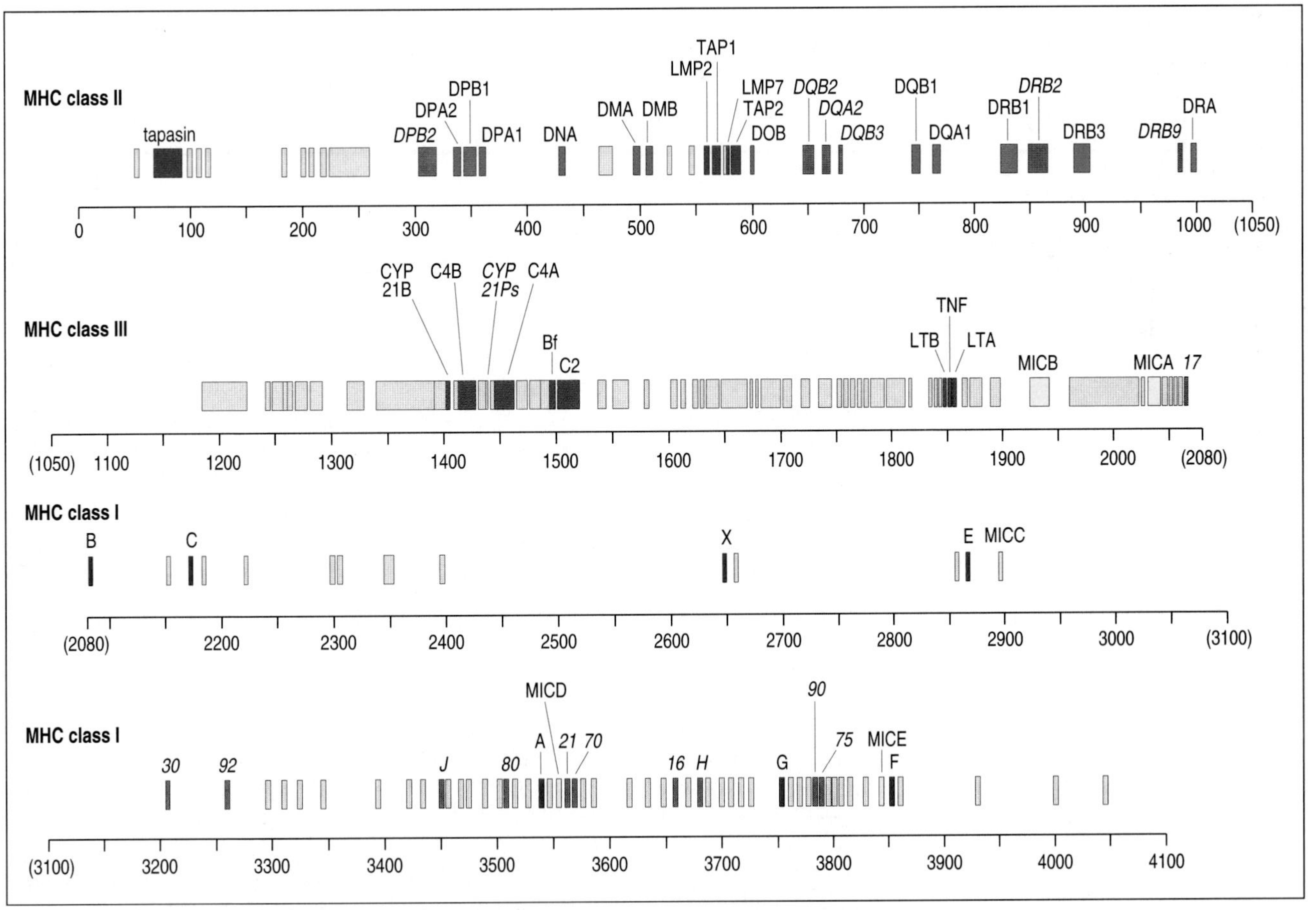

Fig. 4.20 Detailed map of the human MHC region. The organizations of the class I, class II, and class III regions of the MHC are shown, with approximate genetic distances given in thousands of base pairs (kb). Most of the genes in the class I and class II regions are mentioned in the text; the additional genes indicated in the class I region (for example, E, F, G, H, J, and X) are class I-like genes, encoding class IB molecules; the additional class II genes are pseudogenes. The genes shown in the class III region encode the complement proteins C4 (two genes, shown as C4A and C4B), C2 and Factor B (shown as Bf) as well as genes that encode the cytokines tumor necrosis factor α (TNF) and lymphotoxin (LTA, LTB). Closely linked to the C4 genes is the gene encoding 21-hydroxylase (shown as CYP 21B), an enzyme involved in steroid synthesis. Genes shown in gray and named in italic are pseudogenes.

In addition to the highly polymorphic MHC class I and class II genes, there are many MHC class I genes encoding variants of these proteins that show little polymorphism, most of which have yet to be assigned a function. These genes are linked to the class I region of the MHC and the exact number of genes varies greatly between species and even between members of the same species. These genes have been termed **MHC class IB** genes; like MHC class I genes, they encode β_2-microglobulin-associated cell-surface molecules. Their expression on cells is variable, both in the amount expressed at the cell surface and in the tissue distribution.

In mice, one of these molecules, H2-M3, can present peptides with N-formylated amino termini, which is of interest because all prokaryotes initiate protein synthesis with N-formylmethionine. Cells infected with cytosolic bacteria can be killed by CD8 T cells that recognize N-formylated bacterial peptides bound to this MHC class IB molecule. Whether an equivalent class IB molecule exists in humans is not known.

The large number of MHC class IB genes (50 or more in the mouse) means that many different class IB molecules can exist in a single animal. These may, like the protein that presents N-formylmethionyl peptides, have specialized roles in antigen presentation. Some class IB genes, for instance the members of the **MIC** gene family, are under a different regulatory control from the classical MHC class I genes and are induced in response to cellular stress (such as heat shock). These MHC class IB genes are expressed in fibroblasts and epithelial cells and may play a part in innate immunity or in the induction of immune responses in circumstances where interferons are not produced.

Other MHC class IB genes may function to inhibit cell killing by NK cells, a role apparently played by several class I MHC molecules, and one we shall discuss in more detail in Chapter 11. Such a role has been suggested for the MHC class I molecule HLA-G, which is expressed on fetus-derived placental cells that migrate into the uterine wall. These cells express no classical MHC class I molecules and cannot be recognized by CD8 T cells but, unlike other cells lacking classical MHC class I molecules, they are not killed by NK cells. The combination of the lack of classical MHC class I molecules and the expression of HLA-G may protect the fetus from attack by either CD8 T cells or NK cells. Likewise, HLA-E has a specialized role in recognition by NK cells. HLA-E binds a very restricted subset of peptides, derived from the leader peptides of other HLA class I molecules, and it is known that these peptide: HLA-E complexes can bind to CD94, one of the receptors present on NK cells. The consequence of HLA-E recognition by NK cells blocks killing. Because recognition of HLA-E can inform NK cells that class I molecules are being synthesized within the cell, it is possible that an imbalance between recognition of HLA-E and recognition of other class I molecules could signal NK cells that a virus infection was sequestering class I molecules within the cell (see Section 4-10). In this way viruses that seek to evade immune recognition by decreasing class I antigen presentation could be thwarted.

Some MHC class I-like genes map outside the MHC region. One family of such genes, called CD1, also functions in antigen presentation to T cells, although it does not present peptide antigens. Instead, the CD1 molecule is able to bind and present the mycobacterial membrane components mycolic acid and lipoarabinomannan. These can be derived either from internalized mycobacteria or from the uptake of the lipoarabinomannans by the mannose receptor expressed by many phagocytic cells (see Section 10-5). These ligands will thus be delivered into the endocytic pathway, where antigens are normally delivered to MHC class II molecules, as we have discussed earlier. The CD1 molecule, although similar to MHC class I molecules in its subunit organization and association with β_2-microglobulin, behaves like a MHC class II molecule in not being retained within the endoplasmic reticulum by association with the TAP complex but being targeted to endocytic vesicles where it binds its lipid ligands. It appears that the CD1 genes have evolved as a separate lineage of antigen-presenting molecules able to present microbial lipids and glycolipids.

Still other MHC class IB genes have functions that appear unrelated to the immune system. The HLA-HFe gene lies some 3 million base pairs from HLA-A; its product, expressed on cells in the intestinal tract, has a function in iron metabolism. Individuals defective for this gene have an iron storage disease, hemochromatosis, in which an abnormally high level of iron is retained in the liver and other organs. Mice lacking β_2-microglobulin, and hence defective in the expression of all class I molecules, show a similar iron overload. Exactly how this gene product regulates the levels of iron within the body is not known, but it is unlikely to involve an immunological mechanism.

Of the other genes that map within the MHC, some have products such as the complement components C2, Factor B and C4, or the cytokines tumor necrosis factor-α and tumor necrosis factor-β (lymphotoxin), that have important functions in immunity. These have been termed MHC class III genes, and are shown in Fig. 4.20. The functions of these genes will be discussed in Chapters 9 and 10.

As we shall see in Chapter 13, many studies have established associations between susceptibility to certain diseases and particular allelic variants of genes in the MHC. Although most of these diseases are known or suspected to have an immune etiology, this is not true of all of them, and it is important to remember that there are many genes lying within the MHC that have no known or suspected immunological function. One of these is the enzyme 21-hydroxylase, which causes congenital adrenal hyperplasia and, in severe cases, salt-wasting syndrome when it is deficient. Even where a disease-related gene is clearly homologous to immune system genes, as is the case with HLA-HFe, the disease mechanism may not be immune-related. Disease associations mapping to the MHC must therefore be interpreted with caution, in the light of a detailed understanding of its genetic structure and the functions of its individual genes. Much remains to be learned about the latter and about the significance of all the genetic variation localized within the MHC. For instance, the complement component C4 genes are highly polymorphic, but the adaptive significance of this genetic variability is not well understood. In contrast we now have considerable insight into how polymorphism in the MHC class I and MHC class II genes can affect resistance or susceptibility to disease.

4-17 The protein products of MHC class I and class II genes are highly polymorphic.

Because there are three genes encoding MHC class I molecules and four possible sets of MHC class II molecules, every individual will express at least three different MHC class I proteins and four MHC class II proteins on his or her cells. In fact, the number of different MHC proteins expressed on the cells of most individuals is greater because of the extreme polymorphism of the MHC and the co-dominant expression of MHC genes.

The term **polymorphism** comes from the Greek *poly*, meaning many, and *morph*, meaning shape or structure. As used here, it means variation at a single genetic locus and its product within a species; the individual variant

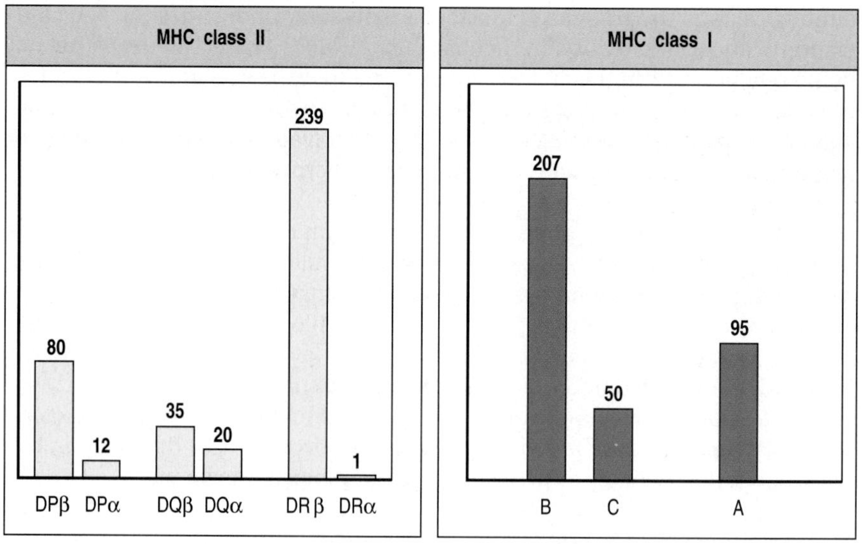

Fig. 4.21 Human MHC genes are highly polymorphic. With the notable exception of the DRα locus, which is functionally monomorphic, each locus has many alleles. The number of different alleles is shown in this figure by the height of the bars and is obtained from studies mainly of Caucasoid populations. In other populations, such as Amerindian or Oriental populations, new alleles are found, so that the total diversity in these loci is greater than represented here. In fact, it is impossible to determine the total amount of variability at these loci without detailed worldwide studies.

genes are termed **alleles**. There are more than 200 alleles of some MHC class I and class II loci (Fig. 4.21), each allele being present at a relatively high frequency in the population. For these reasons, the chance that the corresponding MHC locus on both chromosomes of an individual will encode the same allele is small; most individuals will be **heterozygous** at these loci. Both alleles are expressed in the cell, so expression is said to be co-dominant, and both products function in presenting antigens to T cells (Fig. 4.22). The extensive polymorphism at each locus has the potential to double the number of distinct MHC molecules expressed in an individual and thereby increases the diversity already available through polygeny, the existence of multiple functionally equivalent loci (Fig. 4.23).

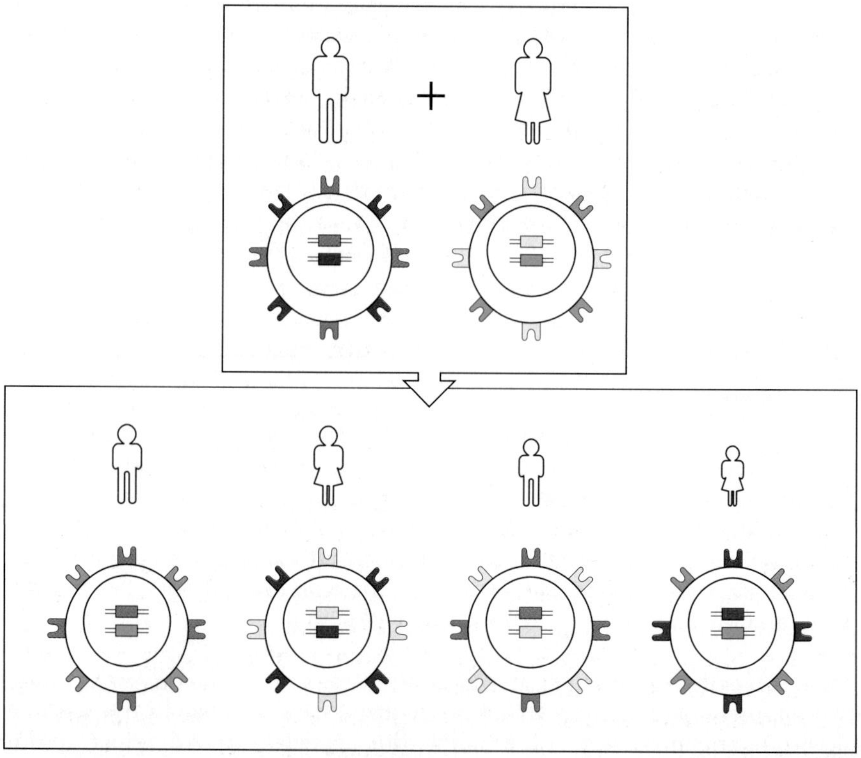

Fig. 4.22 Expression of MHC alleles is co-dominant. The MHC is so polymorphic that most individuals are likely to be heterozygous at each locus. Alleles are expressed from both MHC haplotypes in any one individual, and the products of all alleles are found on all expressing cells. In any mating, there are four possible combinations of haplotypes that can be found in the offspring; thus siblings are also likely to differ in the MHC alleles they express, there being one chance in four that an individual will share both haplotypes with a sibling. One consequence of this is the difficulty of finding suitable donors for tissue transplantation.

Thus, with three MHC class I genes and four potential sets of MHC class II genes on each chromosome, a human typically expresses six different MHC class I molecules and eight different MHC class II molecules on his or her cells. For the MHC class II genes, the number of different products may be increased still further by the combination of α and β chains from different chromosomes (so that two α chains and two β chains can give rise to four different products). It has been shown that, in mice, not all combinations of α and β chains can pair to form stable dimers and so, in practice, the exact number of different MHC class II molecules expressed depends on which alleles are present on each chromosome.

All MHC products are polymorphic to a greater or lesser extent, with the exception of the DRα chain and its homolog Eα in the mouse. These chains do not vary in sequence between different individuals and are said to be **monomorphic**. This might indicate a functional constraint that prevents variation in the DRα and Eα proteins, but no such special function has yet been found. Many mice, both domestic and wild, have a mutation in the Eα gene that prevents synthesis of the Eα protein; they thus lack cell-surface E molecules, so if E molecules have a special function it is unlikely to be an essential one. All other MHC class I and class II genes are polymorphic.

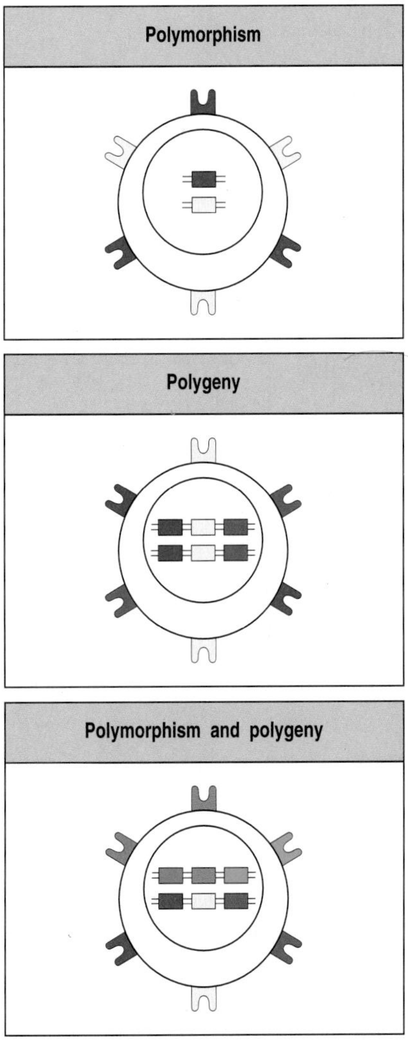

Fig. 4.23 Polymorphism and polygeny contribute to the diversity of MHC molecules expressed by an individual. The MHC genes are highly polymorphic, so that each individual is likely to be heterozygous, that is, to express two different, allelic MHC molecules from each locus. However, no matter how polymorphic the genes, no individual can express more than two alleles of a given gene. The duplication of the MHC genes, leading to polygeny, overcomes this limitation. Polymorphism and polygeny combine to produce the diversity in MHC molecules seen both within an individual and in the population at large.

4-18 MHC polymorphism determines which peptides can be bound and presented to T cells.

The products of individual MHC alleles can differ from one another by up to 20 amino acids, making each allele quite distinct. Most of these differences are localized to exposed surfaces of the outer domain of the molecule, and to the peptide-binding groove in particular (Fig. 4.24). The polymorphic residues that line the peptide-binding groove determine the peptide-binding properties of the different MHC molecules.

We have seen that peptides bind to MHC class I molecules through specific anchor residues (see Section 4-5), and that the amino acid side chains of these residues anchor the peptide by binding in pockets that line the peptide-binding groove. Polymorphism in MHC class I molecules affects the amino acids lining these pockets and thus their binding specificity. In consequence, the anchor residues of peptides that bind to each allelic variant are different. The set of anchor residues that allows binding to a given MHC class I molecule is called a **sequence motif**. These sequence motifs make it possible to identify peptides within a protein that can potentially bind the appropriate MHC molecule, which may be very important in designing peptide vaccines (see Chapter 14). Different allelic variants of MHC class II molecules also bind different peptides but the more open structure of the MHC class II peptide-binding groove, and the greater length of the peptides bound in it, allow greater flexibility in peptide binding. It is therefore more difficult to predict which peptides will bind to MHC class II molecules.

In rare cases, a protein will have no peptides with a suitable motif for binding to any of the MHC molecules expressed on the cells of an individual. When this happens, the individual fails to respond to the antigen. Such failures in responsiveness to simple antigens were first reported in inbred animals, where they were called **immune response (Ir) gene defects**. These defects were identified and mapped to genes within the MHC long before the function of MHC molecules was understood. Indeed, they were the first clue to the antigen-presenting function of MHC molecules, although it was only much later that the Ir genes were shown to encode MHC class II molecules. Ir

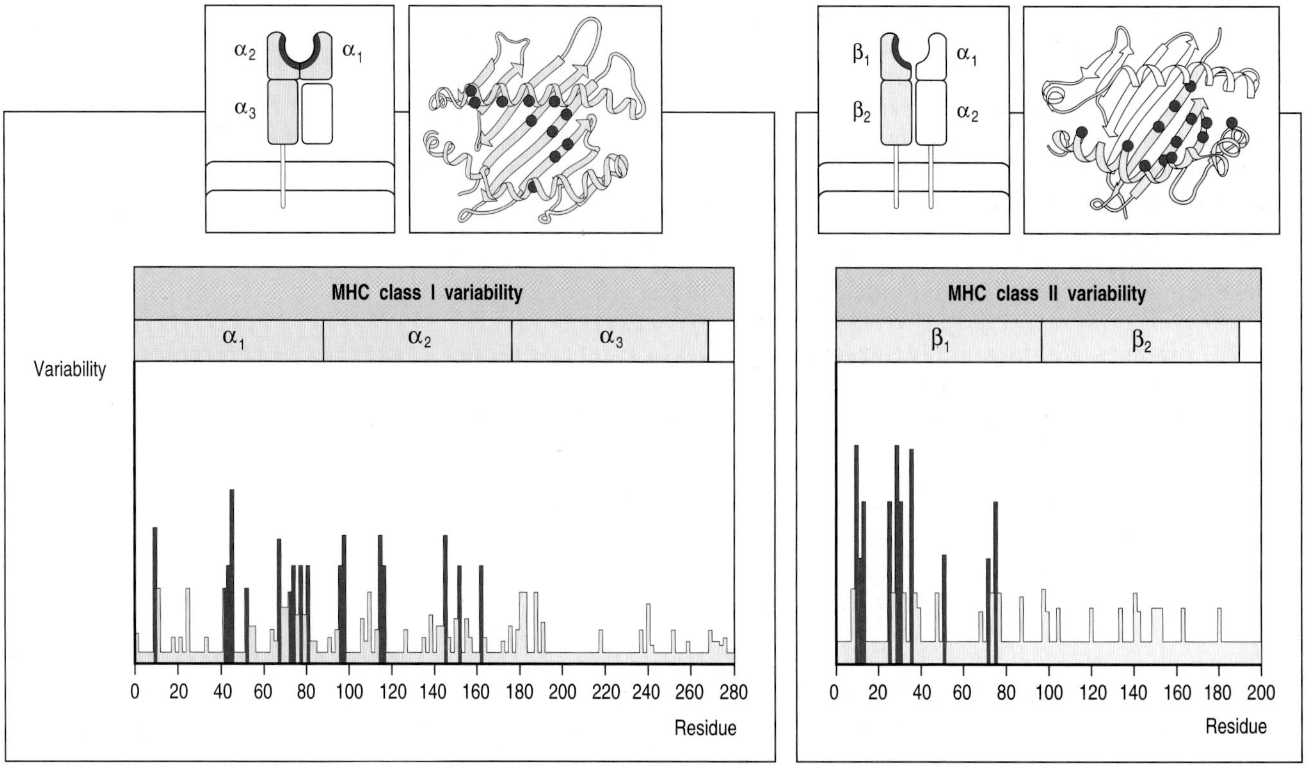

Fig. 4.24 Allelic variation occurs at specific sites within the MHC molecules. Variability plots of the MHC molecules show that the variation arising from polymorphism in the MHC molecules is restricted to the amino-terminal domains (α_1 and α_2 domains of class I molecules, and β_1 and α_1 domains of MHC class II molecules), the domains that form the peptide-binding cleft. Moreover, allelic variability is clustered in specific sites within the amino-terminal domains, lying in positions that line the peptide-binding cleft, either on the floor of the groove or directed inwards from the walls.

gene defects are common in inbred strains of mice because the mice are homozygous for all their MHC genes and thus express only one allelic variant from each gene locus. Ordinarily, the polymorphism of MHC molecules guarantees a sufficient number of different MHC molecules in a single individual to make this type of non-responsiveness unlikely, even to relatively simple antigens such as small toxins. This has obvious importance for host defense.

4-19 MHC polymorphism affects antigen recognition by T cells, both indirectly, by controlling peptide binding, and directly, through contacts between the T-cell receptor and the MHC molecule itself.

Ir gene defects, which are a dramatic example of the effects of MHC polymorphism on the immune response, were identified in experimental animals (guinea pigs and subsequently mice) through the failure to mount an immune response to specific foreign antigens. Initially, the only evidence linking the defect to the MHC was genetic—mice of one MHC genotype could make antibody in response to a particular antigen, whereas mice of a different MHC genotype, but otherwise genetically identical, could not. The MHC genotype was somehow controlling the ability of the immune system to detect or respond to specific antigens, but it was not clear at that time that direct recognition of MHC molecules was involved.

Later experiments showed that the antigen specificity of T-cell recognition was controlled by MHC molecules. The immune responses affected by the Ir genes were known to be dependent on T cells, and this led to a series of experiments aimed at ascertaining how MHC polymorphism might control the responses of T cells. The earliest of these experiments showed that T cells could be activated only by macrophages or B cells that shared MHC alleles with the mouse in which the T cells originated; this provided the first evidence that antigen recognition by T cells is dependent on the presence of specific MHC molecules in the antigen-presenting cell. The clearest example of this feature of T-cell recognition came, however, from studies of virus-specific cytotoxic T cells for which Peter Doherty and Rolf Zinkernagel were awarded the Nobel Prize in 1996.

When mice are infected with a virus, they generate cytotoxic T cells that kill self cells infected with the virus, while sparing uninfected cells or cells infected with unrelated viruses. The cytotoxic T cells are thus virus-specific. A particularly striking outcome of these experiments, however, was that the specificity of the cytotoxic T cells was also affected by allelic polymorphism in MHC molecules: cytotoxic T cells induced by viral infection in mice of MHC genotype a (MHC^a) would kill any MHC^a cell infected with that virus but would not kill cells of MHC genotype b, or c, and so on, even if they were infected with the same virus. Because the MHC genotype restricts the antigen specificity of T cells, this effect is called **MHC restriction**. Together with the earlier studies on both B cells and macrophages, this showed that MHC restriction is a critical feature of antigen recognition by all functional classes of T cells.

Because different MHC molecules bind different peptides, MHC restriction in responses to viruses and other complex antigens could be explained solely on this indirect basis. However, it can be seen from Fig. 4.24 that some of the polymorphic amino acids on MHC molecules are located on the α helices flanking the peptide-binding cleft in such a way that they would be exposed on the outer surface of the peptide:MHC complex and can be directly contacted by the T-cell receptor. It is therefore not surprising that, when T cells are tested for their ability to recognize the same peptide bound to different MHC molecules, they readily distinguish this peptide bound to MHC^a from the same peptide bound to MHC^b. Thus, specificity in a T-cell receptor is defined both by the peptide and by the MHC molecule binding it (Fig. 4.25). This restricted recognition may sometimes be caused by differences in the conformation of the bound peptide imposed by the different

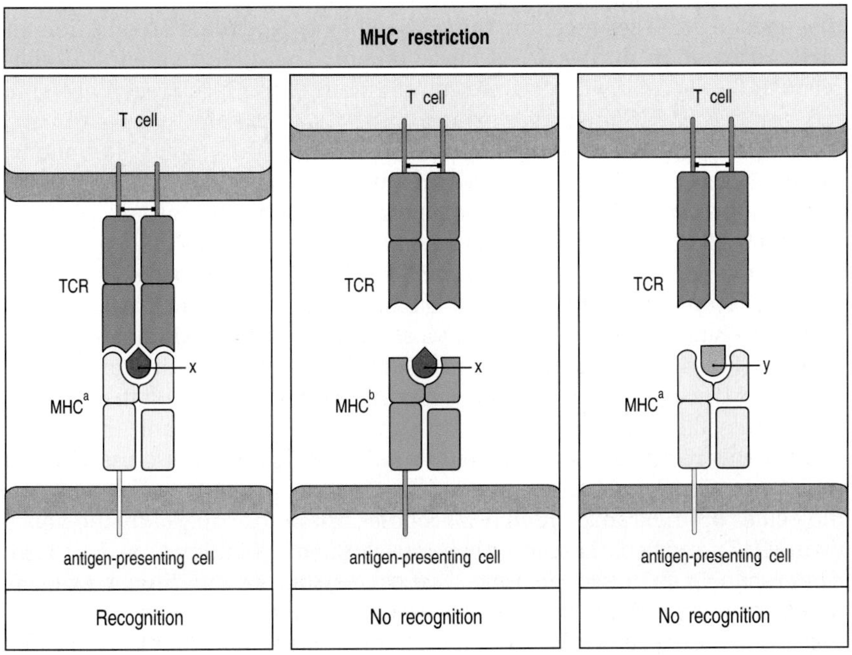

Fig. 4.25 T-cell recognition of antigens is MHC-restricted. The antigen-specific T-cell receptor (TCR) recognizes a complex of antigenic peptide and MHC. One consequence of this is that a T cell specific for peptide x and a particular MHC allele, MHC^a (left panel), will not recognize the complex of peptide x with a different MHC allele, MHC^b (center panel), or the complex of peptide y with MHC^a (right panel). The co-recognition of peptide and MHC molecule is known as MHC restriction because the MHC molecule is said to restrict the ability of the T cell to recognize antigen. This restriction may either result from direct contact between MHC molecule and T-cell receptor or be an indirect effect of MHC polymorphism on the peptides that bind or on their bound conformation.

MHC molecules rather than by direct recognition of polymorphic amino acids on the MHC molecule itself. Thus MHC restriction in antigen recognition reflects the combined effect of differences in peptide binding, and of direct contact between the MHC molecule and the T-cell receptor.

4-20 Non-self MHC molecules are recognized by 1–10% of T cells.

The discovery of MHC restriction, in revealing the physiological function of the MHC molecules, also helped explain the otherwise puzzling phenomenon of non-self MHC recognition in graft rejection. Transplanted tissues or organs from donors bearing MHC molecules that differ from those of the recipient—even by as little as one amino acid—are reliably rejected. The rapid and very potent cell-mediated immune response to the transplanted tissue results from the presence in any individual of large numbers of T cells that are specifically reactive to particular non-self or **allogeneic MHC molecules**. Early studies on T-cell responses to allogeneic MHC molecules used the **mixed lymphocyte reaction**. In a mixed lymphocyte reaction, T cells from one individual are mixed with lymphocytes from a second individual; usually, these latter lymphocytes are prevented from dividing by irradiation or treatment with the cytostatic drug mitomycin C. Such studies have shown that roughly 1–10% of all T cells in an individual will respond to stimulation by cells from any allogeneic individual. This type of T-cell response is called **alloreactivity** because it represents recognition of allelic polymorphism on allogeneic MHC molecules.

Before the role of the MHC molecules in antigen presentation was understood, it was a mystery why so many T cells should recognize non-self MHC molecules, as there is no reason why the immune system should have evolved a defense against tissue transplants. However, once it was appreciated that T-cell receptors have evolved to recognize foreign peptides in combination with polymorphic MHC molecules, alloreactivity became easier to explain. From experiments in which T cells from animals lacking MHC class I and class II molecules have been artificially driven to mature, it has now been shown that the ability to recognize MHC molecules is inherent in the genes that encode the T-cell receptor, rather than being dependent on selection for MHC recognition during T-cell development. The high frequency of alloreactive T cells clearly reflects the commitment of the T-cell receptor to the recognition of MHC molecules in general.

As we shall see in Chapter 7, however, mature T cells have survived a stringent selection process for the ability to respond to foreign, but not self, peptides bound to self MHC molecules. It is therefore thought that the alloreactivity of mature T cells reflects the cross-reactivity of T-cell receptors normally specific for a variety of foreign peptides bound by self MHC molecules. Given a T-cell receptor that is normally specific for a self MHC molecule binding a foreign peptide (Fig. 4.26, left panel), there are two ways in which it may bind to non-self MHC molecules. In some cases, the peptide bound by the non-self MHC molecule interacts strongly with the T-cell receptor, and the T-cells bearing this receptor are stimulated to respond. This type of cross-reactive recognition arises because the spectrum of peptides bound by non-self MHC molecules on the transplanted tissues differ from those bound by the host's own MHC, and it is known as peptide-dominant binding (see Fig. 4.26, center panel). In a second type of cross-reactive recognition, known as MHC-dominant binding, alloreactive T cells respond because of direct binding of the T-cell receptor to distinctive features of the non-self MHC molecule

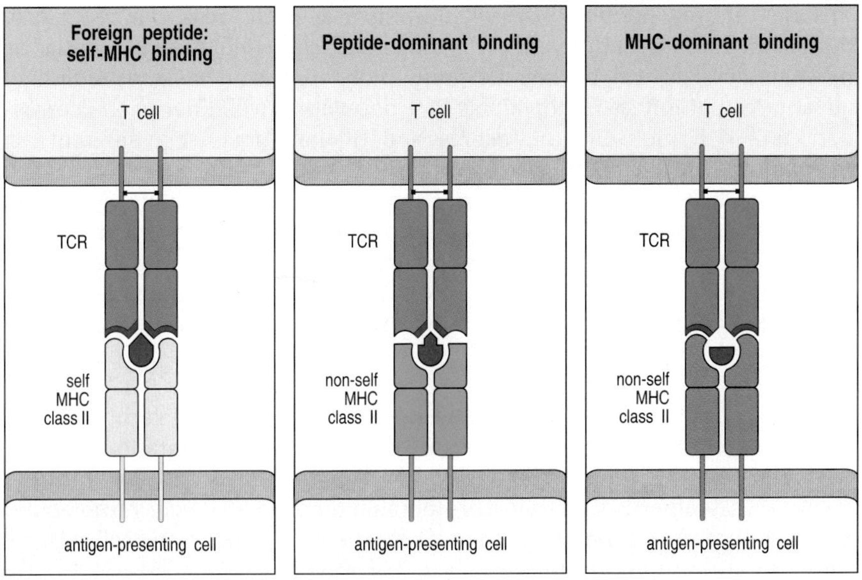

Fig. 4.26 Two modes of cross-reactive recognition that may explain alloreactivity. A T cell that is specific for one peptide:MHC combination (left panel) may cross-react with peptides presented by other, non-self (allogeneic), MHC molecules. This may come about in either of two ways. Most commonly, the peptides bound to the allogeneic MHC molecule fit well to the T-cell receptor (TCR), allowing binding even though there is not a good fit with the MHC molecule (center panel). Alternatively, but less often, the allogeneic MHC molecule may provide a better fit to the T-cell receptor, giving a tight binding that is thus less dependent on the peptide that is bound to the MHC molecule (right panel).

(see Fig. 4.26, right panel). In these cases, the recognition is less dependent on the particular peptide bound; T-cell receptor binding to unique features of the non-self MHC molecule generates a strong signal because of the high concentration of the non-self MHC molecule on the surface of the presenting cell. Both these mechanisms may contribute to the high frequency of T cells responding to the non-self MHC molecules on the transplanted tissue.

4-21 MHC polymorphism extends the range of antigens to which the immune system can respond.

Most polymorphic genes encode proteins that vary by only one or a few amino acids. As we have seen, the different allelic variants of MHC proteins differ by up to 20 amino acids. The extensive polymorphism of the MHC proteins has almost certainly evolved to outflank the evasive strategies of pathogens.

Pathogens have several possible strategies for avoiding an immune response, either by evading detection or by suppressing the ensuing response. The requirement that pathogen antigens must be presented by an MHC molecule provides two possible means of evading detection. A pathogen could escape detection by mutations that eliminated from its proteins all peptides able to bind MHC molecules. An example of this type of strategy can be seen in regions of South East China and in Papua New Guinea, where about 60% of individuals in these small isolated populations carry the HLA-A11 allele. Many isolates of the Epstein–Barr virus obtained in these populations have mutated a dominant epitope presented by HLA-A11, so that the mutant peptides no longer bind to HLA-A11 and cannot be recognized by HLA-A11-restricted T cells. This strategy is plainly much more difficult to follow if there are many different MHC molecules, and the presence of different loci encoding functionally related proteins may have been an evolutionary adaptation by hosts to this strategy by pathogens.

In large outbred populations, polymorphism at each locus can potentially double the number of different MHC molecules expressed by an individual, as most individuals will be heterozygotes. Polymorphism has the additional advantage that different individuals in a population will differ in the combinations of MHC molecules they express and will therefore present different sets of peptides from each pathogen. This makes it unlikely that all individuals in a population will be equally susceptible to a given pathogen and its spread will therefore be limited. That exposure to pathogens over an evolutionary timescale can select for expression of particular MHC alleles is indicated by the strong association of the HLA-B53 allele with recovery from a potentially lethal form of malaria; this allele is very common in people from West Africa, where malaria is endemic, and rare elsewhere, where lethal malaria is uncommon.

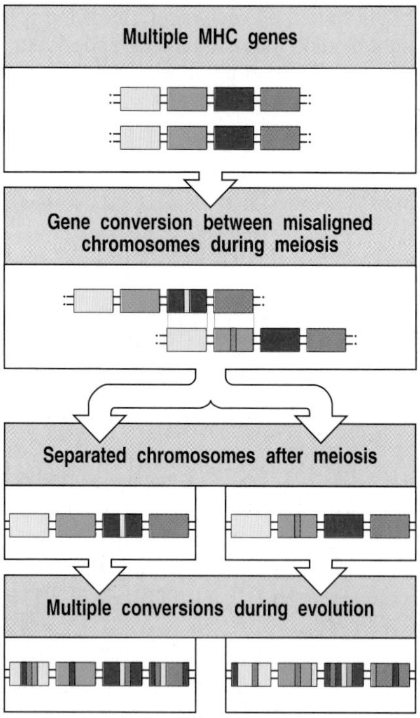

Fig. 4.27 Gene conversion can create new allelic variants by transferring sequences from one MHC gene to another. Sequences can be transferred from one gene to a similar gene by a process known as gene conversion. In this process, which can occur between two closely related genes generated by gene duplication during evolution, the two genes are apposed during meiosis. This can occur as a consequence of the misalignment of the two chromosomes when there are many copies of similar genes arrayed in tandem—somewhat like buttoning in the wrong buttonhole. The DNA sequence from one chromosome can then be copied to the other, giving rise to a new gene sequence. In this way several nucleotide changes can be inserted all at once into a gene and can cause several amino acid changes between the new gene and the original gene. The process of gene conversion has occurred many times in the evolution of MHC alleles.

Similar arguments apply to a second possibility for evading recognition. If pathogens could develop mechanisms to block the presentation of their peptides by MHC molecules, they could avoid the adaptive immune response. Adenoviruses encode a protein that binds to MHC class I molecules in the endoplasmic reticulum and prevents their transport to the cell surface, thus preventing the recognition of viral peptides by CD8 cytotoxic T cells. This MHC-binding protein must interact with a polymorphic region of the MHC class I molecule, as some alleles are retained in the endoplasmic reticulum whereas others are not. Increasing the variety of MHC molecules expressed therefore reduces the likelihood that a pathogen will be able to block presentation by all of them, and so completely evade an immune response.

These arguments raise a question: if having three MHC class I loci offers an advantage that is amplified by allelic variation, why are there not far more MHC class I loci? A full answer to this question must await a discussion of the mechanisms by which the repertoire of T-cell receptors is selected in the thymus, which will be the topic of Chapter 7. Briefly, the probable explanation is that each time a distinct MHC molecule is added to the MHC repertoire, all T cells that can recognize self peptides bound to that molecule must be removed in order to maintain self tolerance. It seems that the number of MHC loci present in humans and mice is about optimal to balance out the advantages of presenting an increased range of foreign peptides and the disadvantages of increased presentation of self peptides and the loss of T cells that accompanies it.

4-22 Multiple genetic processes generate MHC polymorphism.

MHC polymorphism appears to have been selected strongly by evolutionary pressures. However, for selection to work efficiently in organisms that reproduce slowly, such as humans, there must also be powerful mechanisms to generate the variability in MHC alleles on which selection can act. The generation of polymorphism in MHC molecules is an evolutionary problem not readily analyzed in the laboratory; however, it is clear that several genetic mechanisms contribute to the generation of new alleles. Some new alleles are the result of point mutations but many arise from combining sequences from different alleles either by genetic recombination or by **gene conversion**, in which one sequence is replaced, in part, by another from a homologous gene (Fig. 4.27).

Evidence for gene conversion comes from studies of the sequences of different MHC alleles, which reveal that some changes involve clusters of several amino acids in the MHC molecule and require multiple nucleotide changes in a contiguous stretch of the gene. Even more significantly, the same

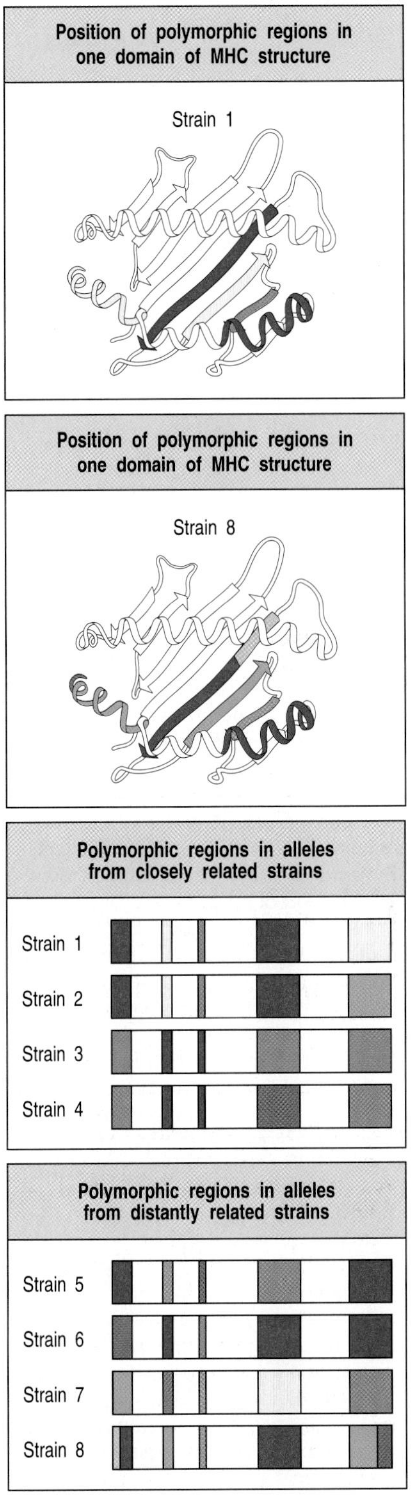

Fig. 4.28 Recombination can create new alleles by reassorting discrete polymorphic regions. Recombination differs from gene conversion in that the DNA segments are exchanged between different chromosomes rather than, as in gene conversion, being copied so that one sequence replaces sequences in another gene on the same chromosome. Analysis of many MHC allele sequences has shown that the swapping of segments of DNA has occurred many times in the evolution of MHC alleles. The variable parts of MHC domains correspond to segments of the structure, such as β strands or parts of the α helix, as shown in the upper two panels. Closely related strains of mice have MHC genes where only one or two segments have been swapped between alleles (third panel), whereas more distantly related strains show a patchwork effect that results from the accumulation of many such recombination events (bottom panel).

sequences are found within other MHC genes on the same chromosome, a prerequisite for gene conversion. Recombination between alleles at the same locus may, however, have been more important than gene conversion in generating MHC polymorphism. A comparison of sequences of MHC alleles shows that many different alleles could represent recombination events between a relatively small set of hypothetical ancestral alleles (Fig. 4.28).

The effects of selective pressure in favor of polymorphism can be seen clearly in the pattern of point mutations in the MHC genes. Point mutations can be classified as replacement substitutions, which change an amino acid, or silent substitutions, which simply change the codon but leave the amino acid the same. Replacement substitutions occur within the MHC at a higher frequency relative to silent substitutions than would be expected, providing evidence that polymorphism has been actively selected for in the evolution of the MHC.

Summary.

The major histocompatibility complex (MHC) of genes consists of a linked set of genetic loci encoding many of the proteins involved in antigen presentation to T cells, most notably the MHC glycoproteins that present peptides to the T-cell receptor. The outstanding feature of MHC genes is their extensive polymorphism. This polymorphism is of critical importance in antigen recognition by T cells. A T cell recognizes antigen as a peptide bound by a particular allelic variant of an MHC molecule, and will not recognize the same peptide bound to other MHC molecules. This behavior of T cells is called MHC restriction. Most MHC alleles differ from one another by multiple amino acid substitutions, and these differences are focused on the peptide-binding site and adjacent regions that make direct contact with the T-cell receptor. At least three properties of MHC molecules are affected by MHC polymorphism: the range of peptides bound; the conformation of the bound peptide; and the interaction of the MHC molecule directly with the T-cell receptor. Thus the highly polymorphic nature of the MHC has functional consequences, and the evolutionary selection for this polymorphism suggests that it is critical to the role of the MHC molecules in the immune response. Powerful genetic mechanisms generate the variation that is seen among MHC alleles, and a compelling argument can be made that selective pressure to maintain a wide variety of MHC molecules in the population comes from infectious agents.

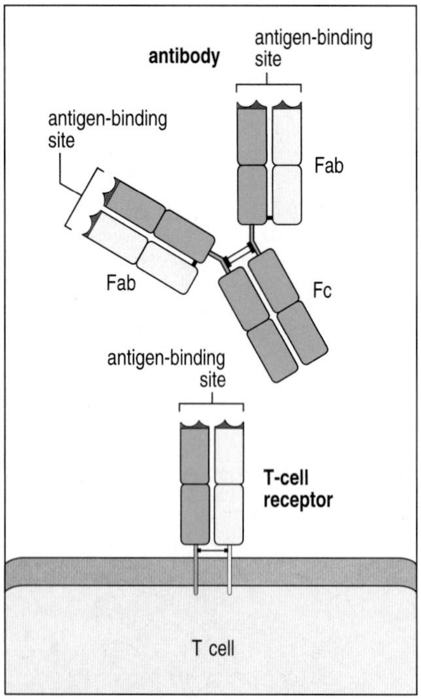

Fig. 4.29 The T-cell receptor resembles a membrane-bound Fab fragment. The Fab fragment of antibody molecules is a disulfide-linked heterodimer, each chain of which contains one immunoglobulin constant domain and one variable domain; the juxtaposition of the variable domains forms the antigen-binding site (see Section 3-6). The T-cell receptor is also a disulfide-linked heterodimer, with each chain containing an immunoglobulin constant-like domain and an immunoglobulin variable-like domain. Finally, as in the Fab fragment, the juxtaposition of the variable regions forms the site for antigen recognition.

The T-cell receptor.

The mechanisms by which a diverse set of B-cell antigen receptors is generated through recombination of a limited set of gene segments are so powerful that it is not surprising that the antigen receptor of T cells bears certain structural resemblances to immunoglobulins and is generated by the same mechanism. In this section, we shall describe the structure of the T-cell receptor proteins and the organization of the T-cell receptor genes. We shall also see how, despite their overall similarity, T-cell receptors differ from immunoglobulins in important ways, and how the MHC-binding co-receptor molecules CD4 and CD8 contribute to antigen recognition.

4-23 The T-cell receptor resembles a membrane-associated Fab fragment of immunoglobulin.

T-cell receptors were first identified by using monoclonal antibodies specific for individual cloned T-cell lines (see Section 2-18) and which could specifically inhibit antigen recognition. These **clonotypic** antibodies were then used to show that each T cell bears about 30,000 antigen receptor molecules on its surface, each receptor consisting of two different polypeptide chains, termed the **T-cell receptor α and β chains**, bound to one another by a disulfide bond. These α:β heterodimers are very similar in structure to the Fab fragment of an immunoglobulin molecule (Fig. 4.29), and they account for antigen recognition by all the functional classes of T cells we have described so far. There is an alternative type of T-cell receptor made up of different polypeptides designated γ and δ, but its discovery was unexpected and its functional significance is not yet clear, as we shall see later (see Section 4-30). T-cell receptors differ from B-cell receptors in that the T-cell receptor is monovalent, whereas immunoglobulin is bivalent, and in that the T-cell receptor is never secreted, whereas immunoglobulin is secreted upon B-cell activation (see Fig. 3.28).

Although the use of clonotypic antibodies enabled some characteristic features of the **α:β T-cell receptor** to be determined, most of what we now know about its structure and function came from studies of cloned DNA encoding the receptor chains. It was the predicted amino-acid sequence from T-cell receptor cDNAs that demonstrated clearly that both chains of the T-cell receptor have an amino-terminal variable region with homology to an immunoglobulin V domain, a constant region with homology to an immunoglobulin C domain, and a short hinge region with a cysteine residue that forms the interchain disulfide bond (Fig. 4.30). Each chain spans the lipid bilayer by a hydrophobic transmembrane domain, and ends in a short cytoplasmic domain.

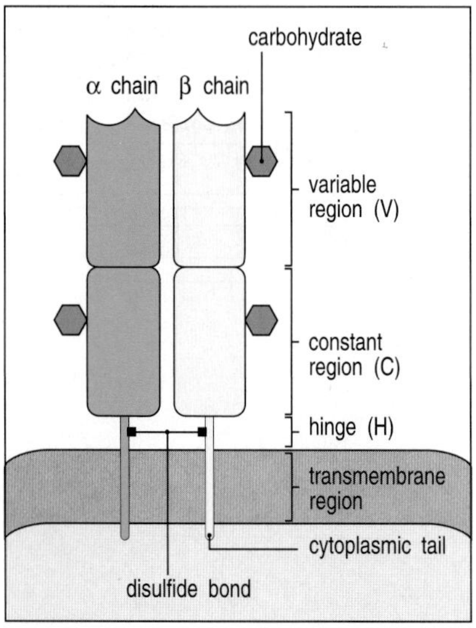

Fig. 4.30 Structure of the T-cell receptor. The T-cell receptor heterodimer is composed of two transmembrane glycoprotein chains, α and β. The external portion of each chain consists of two domains, resembling immunoglobulin variable and constant domains, respectively. Both chains have carbohydrate side chains attached to each domain. A short segment, analogous to an immunoglobulin hinge region, connects the immunoglobulin-like domains to the membrane and contains the cysteine residue that forms the interchain disulfide bond. The transmembrane helices of both chains are unusual in containing positively charged (basic) residues within the hydrophobic transmembrane segment. The α chains carry two such residues; the β chains have one.

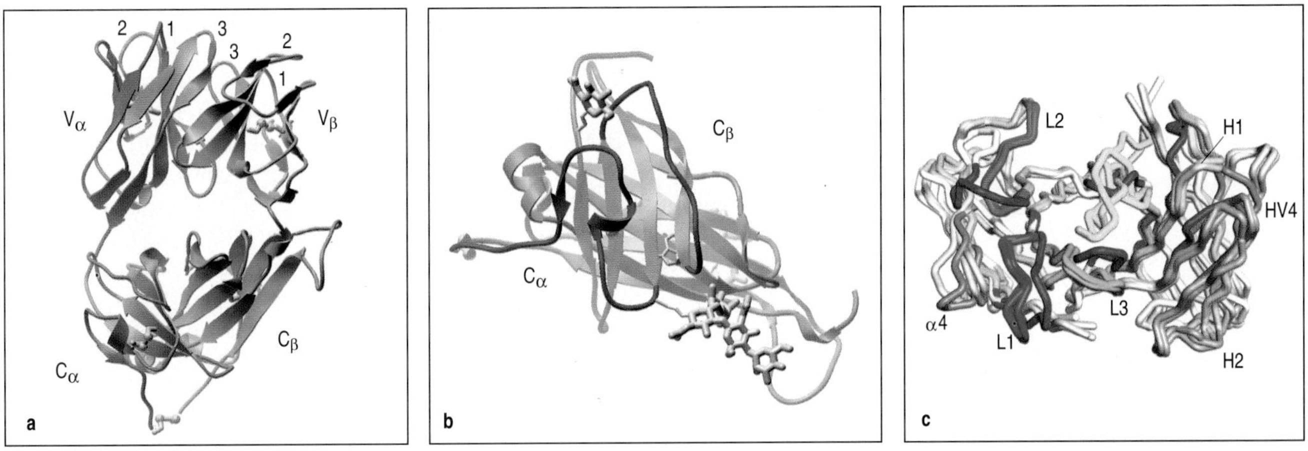

Fig. 4.31 The crystal structure of an α:β T-cell receptor resolved at 2.5 Å. In panels a and b the α chain is shown in purple and the β chain in blue. In panel a, the T-cell receptor is viewed as it would sit on a cell surface with MHC: peptide-binding CDR loops (labeled 1, 2, and 3) arrayed across its relatively flat top. In panel b, the C_α and C_β domains are shown. The C_α domain does not fold into a typical immunoglobulin-like domain, one face of the domain being mainly composed of irregular strands of polypeptide rather than β sheet. In panel c, the T-cell receptor is shown aligned with three antibody-combining sites, with the V_α domain of the T-cell receptor aligned onto the V_L domains of the antibody-combining sites and the V_β domain aligned with the V_H domains. The CDRs of the T-cell receptor and immunoglobulin molecules are colored, with the CDRs 1, 2, and 3 of the TCR shown in red and the HV4 loop shown in orange. For the immunoglobulin V domains, the CDR1 loops of the heavy and light chains are shown in light and dark blue respectively, and the CDR2 loops are shown in light and dark purple respectively. The heavy-chain CDR3 loops are shown in yellow; the light-chain CDR3s are shown in green. The HV4 loops of the TCR (orange) have no hypervariable counterparts in immunoglobulins. Photographs courtesy of I A Wilson.

The V-like domains comprise V-, D- and J-like elements in the β chain and V- and J-like elements in the α chain. It is these close similarities of T-cell receptor chains to the heavy and light immunoglobulin chains that predicted a close resemblance between the T-cell receptor heterodimer and a Fab fragment of immunoglobulin. The first few structures of the T-cell receptor heterodimer that have been determined bear this out. As expected, the T-cell receptor folds in much the same way as an antibody Fab fragment (Fig. 4.31, left panel), though appearing a little 'squashed' (that is, shorter and wider) when compared with a Fab fragment. There are, however, some distinct differences between T-cell receptors and Fab fragments. The most striking difference is in the C_α domain, where the fold is unlike that of any other immunoglobulin-like domain. One half of the domain, that which is juxtaposed to the C_β domain, forms a β sheet similar to that found in other immunoglobulin-like domains, whereas the other half of the domain contains loosely packed strands and a segment of α helix (see Fig. 4.31, center panel). The intramolecular disulphide bond, which in immunoglobulin-like domains joins two β strands, in a C_α domain joins a β strand to this segment of α helix.

There are also differences in the way in which the domains interact. The interface between the V and C domains of both T-cell receptor chains is more extensive than is seen in antibodies, which may make the elbow joint between the domains less flexible. And the interaction between the C_α and C_β domains is distinctive in being assisted by carbohydrate, with a sugar group from the C_α domain making a number of hydrogen bonds to the C_β domain (see Fig. 4.31, center panel). Finally, a comparison of the variable binding sites shows that, although the complementarity-determining region (CDR) loops align fairly closely with those of antibody molecules, there is some displacement relative to those of the antibody molecule (see Fig. 4.31, right panel). This displacement is particularly marked in the V_α CDR2 loop, which is oriented at roughly right angles to the equivalent loop in antibody V domains, as a result of a shift in the β strand that anchors one end of the loop from one face of the domain to the other. A strand displacement also causes a change in the orientation of the V_β CDR2 loop in two of the five V_β domains whose structures are known. There are as yet only four T-cell receptors whose heterodimeric structures have been solved to this level of resolution, so it remains to be seen to what degree all T-cell receptors share these features or whether, in fact, there is more variability still to be discovered.

4-24 The T-cell receptor genes resemble immunoglobulin genes.

The organization of the gene segments encoding T-cell receptor α and β chains (Fig. 4.32) is generally homologous to that of the immunoglobulin gene segments (see Sections 3-11 and 3-12 for comparison). The α chains, like immunoglobulin light chains, are assembled from V and J gene segments, although the T-cell receptor α-chain genes have many more J gene segments: 61 J_{α} gene segments are distributed over about 80 kb of DNA, whereas immunoglobulin light-chain genes have only five J gene segments. We shall see later that this has important consequences for the recognition of antigen by the T-cell receptor. The β-chain genes, like those of immunoglobulin heavy chains, have D gene segments in addition to V and J gene segments.

Like immunoglobulin genes in B cells, the T-cell receptor gene segments rearrange during development to form complete V-domain exons (Fig. 4.33). The process of T-cell receptor gene rearrangement takes place in the thymus and will be dealt with in detail in Chapter 7, where selection of the T-cell repertoire is also discussed. Essentially, however, the mechanics of gene rearrangement are similar for B and T cells. The T-cell receptor gene segments are flanked by heptamer and nonamer recombination signal sequences homologous to those found in immunoglobulin genes (see Section 3-13) and are recognized by the same enzymes: defects in three distinct genes that control gene segment rearrangement affect T- and B-cell receptor genes equally, and animals with these genetic defects lack functional lymphocytes altogether (see Section 3-17). A further shared feature of immunoglobulin and T-cell receptor gene rearrangement is the presence of P- and N-nucleotides in the junctions between the V, D, and J gene segments of the β chain. In T cells, P- and N-nucleotides are also added between the V and J gene segments of all α chains, whereas V to J gene segment joints in immunoglobulin light-chain genes are modified in only some instances.

The main differences between the immunoglobulin genes and those encoding the T-cell receptor reflect the fact that all the effector functions of B cells depend upon secreted antibodies whose distinct constant-region isotypes trigger distinct effector mechanisms. The effector functions of T cells, in contrast, depend upon cell–cell contact and are not mediated directly by the T-cell receptor, which serves only for antigen recognition. Thus, the

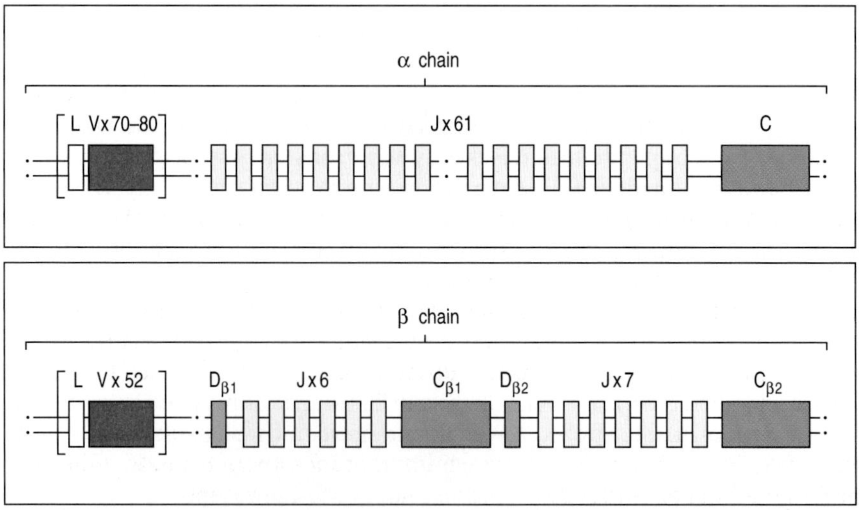

Fig. 4.32 The organization of the mouse T-cell receptor α- and β-chain genes. The arrangement of the gene segments resembles that of the immunoglobulins with separate variable (V), diversity (D), joining (J), and constant (C) gene segments. The α-chain gene consists of 70–80 variable segments, each containing an exon encoding a variable region (V) preceded by an exon encoding the leader sequence (L) that targets the protein to the endoplasmic reticulum for transport to the cell surface. How many of these are functional is not known exactly. A cluster of about 60 J gene segments is located a considerable distance from the V gene segments. The J gene segments are followed by a single constant-domain gene, which contains separate exons for the constant and hinge domains and a single exon encoding the transmembrane and cytoplasmic regions. The β-chain gene has a different organization, with a cluster of about 50 functional V gene segments located distantly from two separate clusters each containing a single D gene segment, together with six or seven J gene segments and a single constant-region gene. Each β-chain constant gene has separate exons encoding the constant, hinge, transmembrane, and cytoplasmic regions. The α-chain locus is interrupted between the J and V segments by another TCR locus—the δ-chain locus (not shown here; see Fig. 4.41).

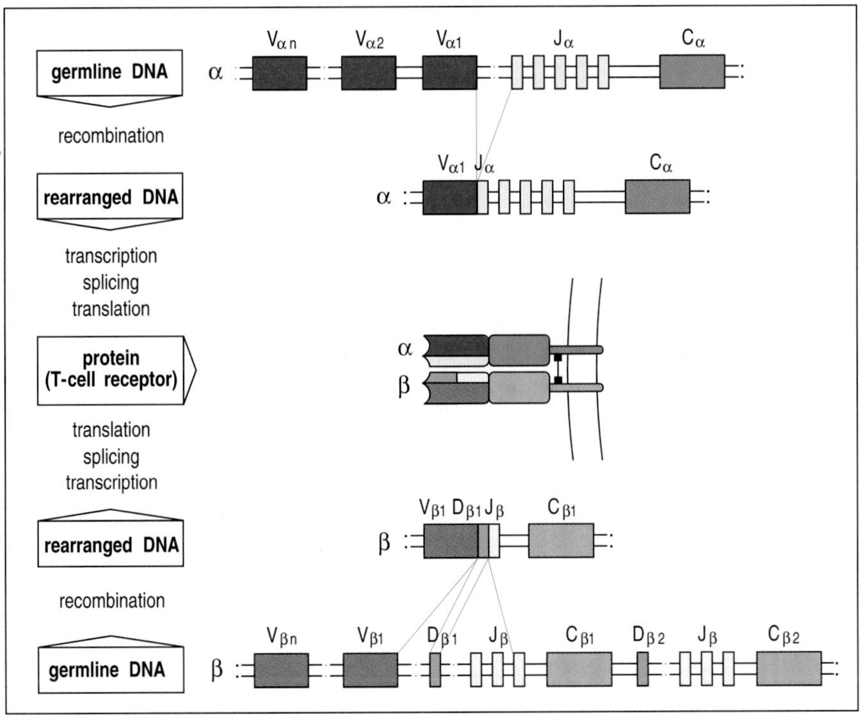

Fig. 4.33 T-cell receptor α- and β-chain gene rearrangement and expression. The T-cell receptor α- and β-chain genes are composed of discrete segments that are joined by somatic recombination during development of the T cell. Functional α- and β-chain genes are generated in the same way that complete immunoglobulin genes are created. For the α chain, a V_α gene segment rearranges to a J_α gene segment to create a functional exon. Transcription and splicing of the VJ_α exon to C_α generates the mRNA that is translated to yield the T-cell receptor α-chain protein. For the β chain, like the immunoglobulin heavy chain, the variable domain is encoded in three gene segments, V_β, D_β, and J_β. Rearrangement of these gene segments generates a functional VDJ_β exon that is transcribed and spliced to join to C_β; the resulting mRNA is translated to yield the T-cell receptor β chain. The α and β chains pair soon after their biosynthesis to yield the α:β T-cell receptor heterodimer. Not all J gene segments are shown.

constant-region genes of the T-cell receptor are much simpler than those of antibodies: there is only one C_α gene and, although there are two C_β genes, there is no known functional distinction between their products. The T-cell receptor constant-region genes encode only transmembrane polypeptides: there are no exons encoding an alternative secreted form.

4-25 T-cell receptor diversity is focused in CDR3.

The extent and pattern of diversity of T-cell receptors and immunoglobulins reflect the distinct nature of their ligands. Whereas antibodies must conform to the surfaces of an almost infinite variety of different antigens, the ligand for the T-cell receptor is always an MHC molecule; the T-cell receptors would therefore be predicted to have a relatively invariant shape, with most of the variability focused on the bound antigenic peptide occupying the center of the surface in contact with the receptor.

The three-dimensional structure of the antigen recognition site of a T-cell receptor looks much like that of an antibody molecule. In an antibody, the center of the antigen-binding site is formed by the third complementarity-determining regions (CDR3) of the heavy and light chains. The structurally equivalent loops of the T-cell receptor α and β chains, to which the D and J gene segments contribute, also form the center of the antigen-binding site on T-cell receptors, whereas the periphery of the site consists of the equivalent of the CDR1 and CDR2 loops, which are encoded within the germline V gene segments for the α and β chains. T-cell receptor genes have roughly the same number of V gene segments as do immunoglobulin genes, but only immunoglobulins show extensive diversification of expressed rearranged variable-region genes by somatic hypermutation. Thus, diversity in the CDR1 and CDR2 loops that comprise the periphery of the antigen-binding site will be far greater in an antibody molecule than in a T-cell receptor.

Fig. 4.34 The numbers of human T-cell receptor gene segments and the sources of T-cell receptor diversity compared with those of immunoglobulin. The number of κ-chain joints with N- and P-regions is shown in parentheses, as only about half of human κ chains contain N- and P-regions. Somatic hypermutation as a source of diversity in immunoglobulins is not included in this figure.

Element	Immunoglobulin		αβ receptors	
	H	κ+λ	β	α
Variable segments (V)	65	70	52	~70
Diversity segments (D)	27	0	2	0
D segments read in 3 frames	rarely	–	often	–
Joining segments (J)	6	5(κ) 4(λ)	13	61
Joints with N and P nucleotides	2	(1)	2	1
Number of V gene pairs	3.4×10^6		5.8×10^6	
Junctional diversity	$\sim 3 \times 10^7$		$\sim 2 \times 10^{11}$	
Total diversity	$\sim 10^{14}$		$\sim 10^{18}$	

The structural diversity of T-cell receptors is due entirely to combinatorial and junctional diversity generated during the process of gene rearrangement, and it can be seen from Fig. 4.34 that the variability in T-cell receptors is thereby focused on the junctional region encoded by V-, D-, J-, P-, and N-nucleotides. Because the α-chain gene of the T-cell receptor has so many J gene segments compared with the light-chain genes of immunoglobulins, and also because more than one D segment can be joined together in the β-chain gene, the variability generated in this region is even greater for T-cell receptors than for immunoglobulins. This region encodes the CDR3 loops in immunoglobulins and T-cell receptors that form the center of the antigen-binding site. Thus, the center of the T-cell receptor will be highly variable, whereas the periphery will be subject to relatively little variation.

4-26 T-cell receptors bind diagonally across the peptide-binding groove, with their two CDR3 loops making the predominant contacts with the bound peptide.

When T-cell receptors bind to MHC class I:peptide complexes, as shown in Fig. 4.35, they do so with the long axis of the complementarity-determining region oriented diagonally across the peptide-binding groove of the MHC class I molecule. The T-cell receptor lies in a saddle between two 'peaks' created by the amino-terminal ends of the α helices of the MHC α_1 and α_2 domains as shown in panel b of Fig. 4.35. It seems that there is little freedom for the T-cell receptor to be rotated with respect to the MHC molecule and yet still be able to make contact with the peptide; it is therefore likely that all T-cell receptors bind in this diagonal manner. In this orientation, the V_α domain makes contact primarily with the amino terminus of the bound peptide and the surrounding helices of the MHC class I molecule, whereas the V_β domain contacts primarily the carboxy terminus of the bound

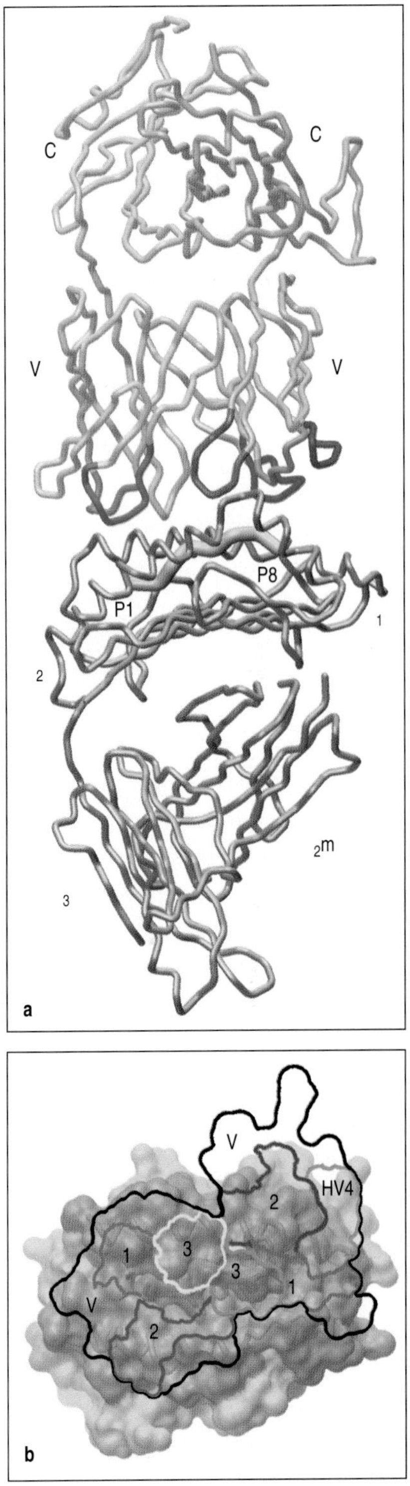

Fig. 4.35 The T-cell receptor binds to the MHC:peptide complex. Panel a: the T-cell receptor binds to the top of the MHC:peptide complex, straddling, in the case of the class I molecule shown here, both the α_1 and α_2 domain helices. The CDRs of the T-cell receptor are indicated in color: the CDR1 and CDR2 loops of the β chain in light and dark blue respectively, and the CDR1 and CDR2 loops of the α chain in light and dark purple respectively. The α chain CDR3 loop is in yellow while the β chain CDR3 loop is in green. The β chain HV4 loop is orange. Panel b: the outline of the T-cell receptor antigen-binding site (thick black line) is superimposed upon the top surface of the MHC:peptide complex (the peptide is shown in darker gray). The T-cell receptor lies diagonally across the MHC:peptide complex, with the α and β CDR3 loops of the T-cell receptor (3α, 3β, yellow and green respectively) contacting the center of the peptide. The α chains of the CDR1 and CDR2 loops (1α, 2α, light and dark purple, respectively) contact the MHC helices at the amino terminus, whereas the β chains of the CDR1 and CDR2 loops (1β, 2β, light and dark blue respectively) make contact with the helices at the carboxy end. Courtesy of I A Wilson.

peptide and the surrounding helices of the MHC class I molecule. The same orientation has been proposed for the interaction of the T-cell receptor and the MHC class II:peptide complex, but so far no X-ray crystallographic structure has been determined for this complex. The opposite orientation of the T-cell receptor, where the V_α domain makes contact with the carboxy terminus of the MHC-bound peptide, has not been observed, but it would appear possible unless other constraints, such as the interaction of the MHC molecule and T-cell receptor with the co-receptors CD4 and CD8, rule out that mode of binding.

As expected, the CDR3 loops of the T-cell receptor make the main contacts with the central residues of the peptide. In particular, the interface between the V_α and V_β CDR3 loops creates a deep pocket in which a side chain from the peptide binds. In this respect the binding of T-cell receptors to MHC:peptide complexes resembles the binding of antibodies to protein antigens (see, for example, Fig. 3.9). The CDR3 loops also make contacts with the MHC molecule as well as with the peptide, and may account for up to 50% of the interface between the T-cell receptor and the MHC:peptide complex. The T-cell receptor is not placed symmetrically over the MHC molecule, so whereas the V_α CDR1 and CDR2 loops are in close contact with the MHC:peptide complex at the amino terminus of the bound peptide, the β-chain CDR1 and CDR2 loops, which interact with the complex at the carboxy end of the bound peptide, appear to contribute much less to the binding. Whether this will remain true when the structures of more T-cell receptor:MHC:peptide complexes have been examined remains to be seen. At this stage, however, it appears that the CDR2 loops of the T-cell receptor make contact only with the MHC molecule and not with the peptide. For some T-cell receptors, the β-chain CDR2 loop makes no contact with the MHC:peptide complex, although it can contact a superantigen molecule (see below).

Although only a few residues of the peptide may be exposed within the peptide-binding cleft and make contact with the T-cell receptor, T cells are clearly able to discriminate efficiently between many different peptide:MHC complexes. There are two ways in which differences in peptide sequence might affect the binding of the T-cell receptor, in addition to a direct interaction with the exposed amino acid side chains. First, from a comparison of the structures of different peptides bound by the same MHC class I molecule, it is clear that even for peptides of the same length, the conformation of the peptide backbone can vary, depending on its sequence. Second, when different peptides are bound by the same MHC class I molecule, there can be small changes in the conformation of the MHC class I molecule itself, which could also influence the binding of the T-cell receptor.

4-27 Somatic hypermutation does not generate diversity in T-cell receptors.

When we discussed the generation of antibody diversity in Section 3-14, we saw that somatic hypermutation increases the diversity of all three complementarity-determining regions of both immunoglobulin chains. Somatic hypermutation does not occur in T-cell receptor genes, so that variability of the CDR1 and CDR2 regions is limited to that of the germline V gene segments. All the diversity in T-cell receptors is generated during rearrangement and is consequently focused on the CDR3 regions.

Why T-cell and B-cell receptors differ in their abilities to undergo somatic hypermutation is not clear, but several explanations can be suggested on the basis of the functional differences between T and B cells. Because the central role of T cells is to stimulate both humoral and cellular immune responses, it is crucially important that T cells do not react with self proteins. T cells that recognize self antigens are rigorously purged during development (see Chapter 7) and the absence of somatic hypermutation helps to ensure that somatic mutants recognizing self proteins do not arise later in the course of immune responses. This constraint does not apply with the same force to B-cell receptors, as B cells usually require T-cell help to secrete antibodies. A B cell whose receptor mutates to become self reactive would, under normal circumstances, fail to make antibody for lack of self-reactive T cells to provide this help (see Chapter 9).

An additional argument might be that T cells already interact with a self component, namely the MHC molecule that makes up the major part of the ligand for the receptor, and thus might be unusually prone to developing self-recognition capability through somatic hypermutation. In this case, the converse argument can also be made: because T-cell receptors must be able to recognize self MHC molecules as part of their ligand, it is important to avoid somatic mutation that might result in the loss of recognition and the consequent loss of any ability to respond. However, the strongest argument for this difference between immunoglobulins and T-cell receptors is the simple one that somatic hypermutation is an adaptive specialization for B cells alone, because they must make very high-affinity antibodies to capture toxin mol-ecules in the extracellular fluids. We shall see in Chapter 10 that they do this through somatic hypermutation followed by selection for antigen binding.

4-28 Many T cells respond to superantigens.

Not all antigens that bind to MHC class II molecules are presented as peptides in the peptide-binding groove. A few fall into a distinct class known as **superantigens**, which have a distinctive mode of binding that enables them to stimulate very large numbers of T cells, often with disastrous consequences. Superantigens are produced by many different pathogens, including bacteria, mycoplasmas, and viruses, and bind directly to MHC molecules without being processed previously; indeed, fragmentation of a superantigen destroys its biological activity. Instead of binding in the groove of the MHC molecule, superantigens bind to the outer surface of both the MHC class II molecule and the V_β region of the T-cell receptor. Bacterial superantigens bind mainly to the V_β CDR2 loop, and to a smaller extent to the V_β CDR1 loop and an additional loop called the hypervariable 4, or HV4, loop (Fig. 4.36). The HV4 loop is the predominant binding site for viral superantigens, at least for those encoded by the endogenous mouse

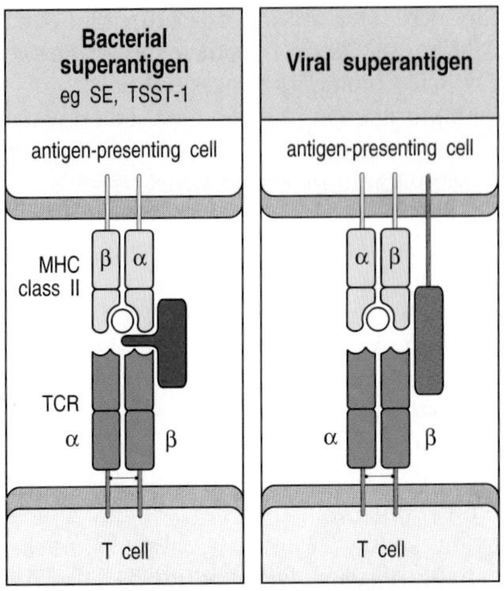

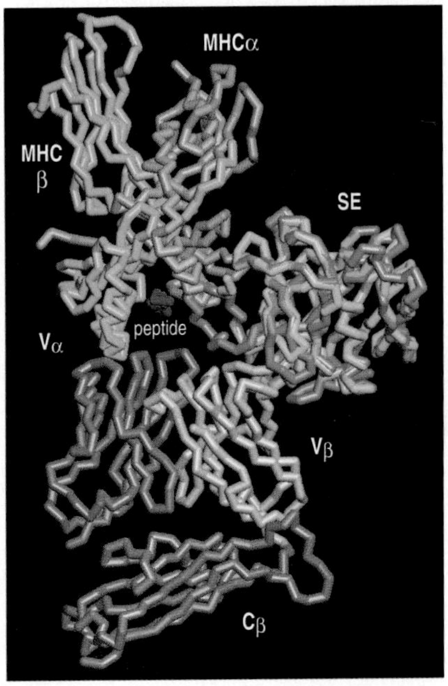

Fig. 4.36 Superantigens bind directly to T-cell receptors and to MHC molecules. Superantigens can bind independently to MHC class II molecules and to T-cell receptors, binding to the V_β domain of the T-cell receptor (TCR), away from the complementarity-determining regions, and to the outer faces of the MHC class II molecule, outside the peptide-binding site (top panels). The lower panel shows a reconstruction of the interaction between a T-cell receptor, an MHC class II molecule and a staphylococcal enterotoxin (SE) superantigen, produced by superimposing separate structures of an enterotoxin:MHC class II complex onto an enterotoxin:T-cell receptor complex. The two enterotoxin molecules (actually SEC3 and SEB) are shown in turquoise and blue, binding to the α chain of the class II molecule (yellow) and to the β chain of the T-cell receptor (colored gray for the V_β domain and pink for the C_β domain). Molecular model courtesy of H-M Li, B A Fields and R A Mariuzza.

mammary tumor viruses. Thus, the α-chain V region and the CDR3 of the β chain of the T-cell receptor have little effect on superantigen recognition, which is determined largely by the V region of the expressed β chain. Each superantigen can bind one or a few of the different products of V_β gene segments, of which there are 20–50 in mice and humans; a superantigen can thus stimulate 2–20% of all T cells.

This mode of stimulation does not prime an adaptive immune response specific for the pathogen. Instead, it causes a massive production of cytokines by CD4 T cells, the predominant responding population of T cells. These cytokines have two effects on the host: systemic toxicity and suppression of the adaptive immune response. Both these effects contribute to microbial pathogenicity. Among the bacterial superantigens are the **staphylococcal enterotoxins** (SEs), which cause common food poisoning, and the **toxic shock syndrome toxin-1** (**TSST-1**), the etiologic principle in toxic shock syndrome.

The role of viral superantigens is less clear. Endogenous viral superantigens are very common in mice, and we shall see in Chapter 7 that the study of these superantigens has played a critical role in elucidating one of the major mechanisms of self tolerance. In humans, the T-cell responses to rabies virus and the Epstein–Barr virus indicate the existence of superantigens encoded by these viruses; however, the genes encoding these superantigens have not yet been identified.

4-29 The co-receptor molecules CD4 and CD8 cooperate with the T-cell receptor in antigen recognition.

We have seen that T cells fall into two major classes that differ in the class of MHC molecule they recognize, have different effector functions, and are distinguished by the expression of the cell-surface proteins CD4 and CD8 (Fig. 4.37). CD4 and CD8 were known for some time as markers for different functional sets of T cells before it became clear that they play an important part in the differential recognition of MHC class II and MHC class I molecules. It is now known that CD4 binds to invariant parts of the MHC class II molecule and CD8 to invariant parts of the MHC class I molecule. During antigen recognition, CD4 and CD8 molecules associate on the T-cell surface with components of the T-cell receptor. For this reason, they are called **co-receptors**.

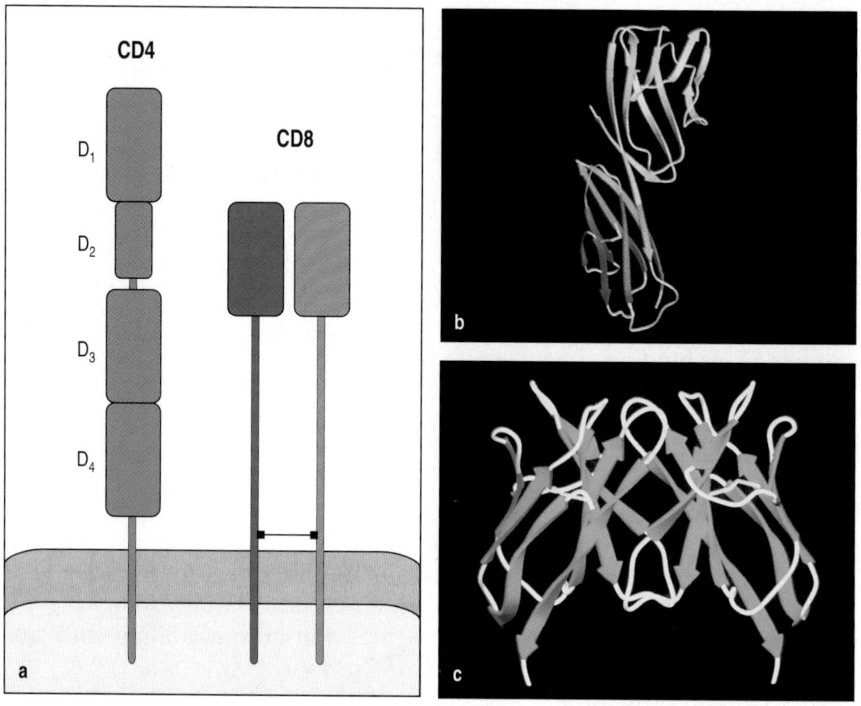

Fig. 4.37 The outline structures of the CD4 and CD8 co-receptor molecules. The CD4 molecule contains four immunoglobulin-like domains, as shown in diagrammatic form in panel a and as a ribbon diagram of the structure in panel b. The amino-terminal domain, D1, is similar in structure to an immunoglobulin variable domain. The second domain, D2, although it is clearly related to immunoglobulin domains, is different from both V and C domains and has been termed a C2 domain. The first two domains of CD4 form a rigid rod-like structure that is linked to the two carboxy-terminal domains by a flexible link. The binding site for MHC class II molecules is thought to involve both the D1 and D2 domains of CD4. The CD8 molecule (panels a and c) is a heterodimer of an α and a β chain that are covalently associated by a disulfide bond. The two chains of the dimer have very similar structures, each having a single domain resembling an immunoglobulin variable domain and a stretch of peptide believed to be in a relatively extended conformation that links the variable region-like domain to the cell membrane.

CD4 is a single-chain molecule composed of four immunoglobulin-like domains. The first two domains (D1 and D2) of the CD4 molecule are packed tightly together to form a rigid rod some 60 Å long, which is joined by a flexible hinge to a similar rod formed by the third and fourth domains (D3 and D4). The cytoplasmic domain interacts strongly with a cytoplasmic tyrosine kinase called Lck, which enables the CD4 molecule to participate in signal transduction (see Chapter 5). CD4 binds MHC class II molecules through a region that lies mainly on a lateral face of the first domain (D1), although it is thought that residues in the second domain may also be involved. CD4 binding

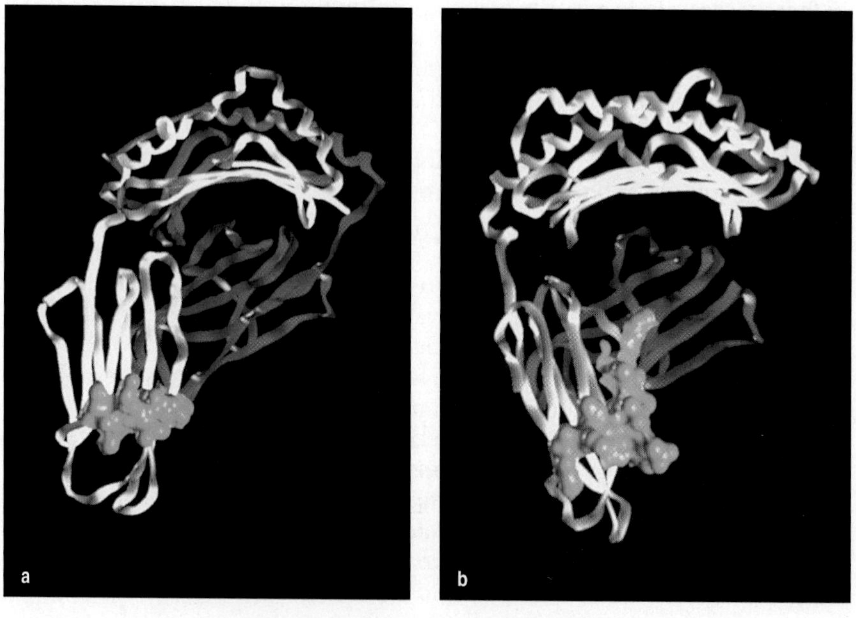

Fig. 4.38 The binding sites for CD4 and CD8 on MHC class II and class I molecules lie in the immunoglobulin-like domains. The binding sites for CD4 and CD8 on the MHC class I and class II molecules, respectively, lie in the immunoglobulin-like domains nearest to the membrane, distant from the peptide-binding cleft. In panel a, CD4 binds to a site, shown surfaced in green, at the base of the β_2 domain of an MHC class II molecule and distant from the peptide-binding site. The α chain of the class II molecule is purple, and the β chain is white. In panel b, CD8 binds to a site, also shown surfaced in green, at the base of the α_3 domain of the MHC class I molecule. The α chain of the class I molecule is white, and the β_2-microglobulin is purple. Photographs courtesy of C Thorpe.

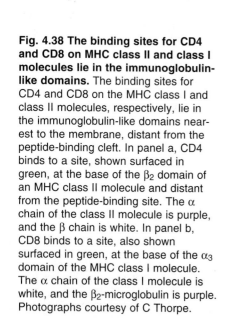

to MHC class II is weak on its own and it is not clear whether such binding is able to transmit a signal to the interior of the T cell. As CD4 binds to a site on the β_2 domain of the MHC class II molecule that is well away from the site where the T-cell receptor binds (Fig. 4.38a), the CD4 molecule and the T-cell receptor can bind the same peptide:MHC class II complex. As shown in Fig. 4.39, CD4 is capable of forming homodimers through a site in the D4 domain of the molecule, while leaving the MHC class II binding site free to interact. Thus, the CD4 molecule is capable of crosslinking two MHC class II molecules and hence also the two T-cell receptors bound to those MHC molecules. Whether this dimerization of CD4 is important in the activation of T cells is not known at present. Nevertheless, the presence of CD4, and the simultaneous binding of both CD4 and the T-cell receptor to the same MHC class II:peptide complex, result in a marked increase in the sensitivity of a T cell to antigen presented by MHC class II molecules, lowering by 100-fold the dose of antigen required for activation.

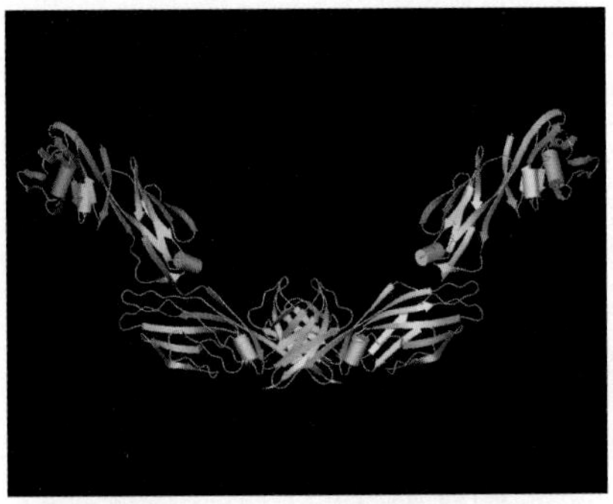

Fig. 4.39 CD4 is capable of forming dimers. The structure of the extracellular domains of the CD4 molecule has been determined by X-ray crystallography. Two molecules of CD4 can interact with each other through their D4 domains, forming homodimers. The site that binds MHC class II molecules remains available in such dimers.

Although CD4 and CD8 both function as co-receptors, their structures are quite distinct. The CD8 molecule is a disulfide-linked heterodimer consisting of an α and a β chain, each containing a single immunoglobulin-like domain linked to the membrane by a segment of extended polypeptide chain. This segment is extensively glycosylated, a feature believed to be important in maintaining this stretch of polypeptide in an extended conformation and protecting it from cleavage by proteases. CD8α chains can also form homodimers, although these are not seen when the β chains are present.

CD8 binds weakly to a site in the α_3 domain of MHC class I molecules equivalent to the site in MHC class II molecules to which CD4 binds (see Fig. 4.38). Although only the interaction of the CD8α homodimer with MHC class I is known in detail, from this it is clearly inferred that the binding site of the CD8α:β heterodimer is formed by the interaction of the CD8α and β chains. In addition, the CD8 molecule (most probably the α chain) interacts with residues in the base of the α_2 domain of the class I molecule. Binding in this way, CD8 leaves the upper surface of the class I molecule exposed and free to interact simultaneously with a T-cell receptor, as shown in Fig. 4.40. CD8 also binds Lck through the cytoplasmic tail of the α chain; thus the simultaneous binding of CD8 and the T-cell receptor to the same MHC:peptide complex brings the Lck kinase into close proximity with the receptor (see Chapter 5). This increases the sensitivity of T cells bearing the CD8 co-receptor to antigen presented by MHC class I molecules by about 100-fold. Thus, the two co-receptor proteins have similar functions and bind to the same approximate location in MHC class I and MHC class II molecules even though the structures of CD4 and CD8 are only distantly related.

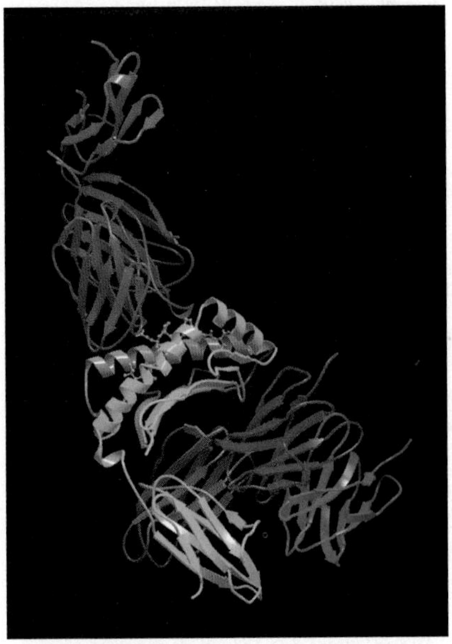

Fig. 4.40 CD8 binds to a site on MHC class I molecules distant from that to which the TCR binds. The relative positions of the T-cell receptor and CD8 molecules bound to the same MHC class I molecule can be seen in this hypothetical reconstruction of the interaction of an MHC class I molecule (the α chain is shown in green; β_2 microglobulin can be seen faintly in the background) with a T-cell receptor and CD8. The α and β chains of the T-cell receptor are shown in pink and purple, respectively. The CD8 structure is that of a CD9α homodimer, but is colored to represent the likely orientation of the subuits in the heterodimer, with the CD8α subunit in red and the CD8β subunit in blue. Photograph courtesy of G Gao.

4-30 Some T cells bear an alternative form of T-cell receptor with γ and δ chains.

At the time of the discovery of the T-cell receptor α:β heterodimer, all known specific immune responses could be accounted for by the action of T cells bearing these polypeptides. Thus, it was a real surprise when, during the search for genes encoding T-cell receptors, a third T cell-specific cDNA was discovered that also encoded a protein with homology to immunoglobulin and whose gene rearranged in T lymphocytes. As this gene clearly did not encode either the α or the β chain of the T-cell receptor, it was termed T-cell receptor γ. The γ polypeptide is found on the cell surface associated with a second polypeptide, the T-cell receptor δ chain, to form a **γ:δ heterodimer**.

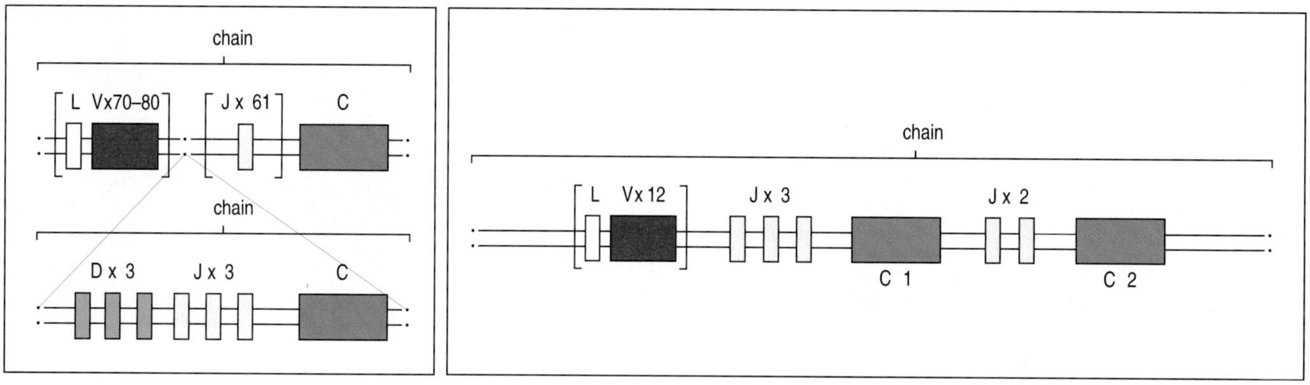

Fig. 4.41 The organization of the T-cell receptor γ- and δ-chain genes in humans. The γ and δ genes, like the α and β T-cell receptor genes, have discrete V, D, J, and C gene segments. Uniquely, the locus encoding the δ chain is located entirely within the α-chain locus. The three D_δ gene segments, three J_δ gene segments, and the single δ constant-region gene lie between the cluster of V_α gene segments and the cluster of J_α gene segments, whereas the V_δ gene segments are interspersed among the V_α gene segments; it is not known exactly how many V_δ gene segments there are but there are at least four. The human γ chain resembles the β chain, with two constant-region gene segments each with its own J gene segments. The mouse γ genes (not shown) have a more complex organization and there are three functional clusters of γ genes, each containing V and J gene segments and a constant-region gene. Rearrangement of the γ and δ genes proceeds as for the other T-cell receptor genes, with the exception that during δ-chain gene rearrangement both D segments can be used in the same gene. The use of two D segments greatly increases the variability, mainly because extra N-region nucleotides can be added at the junction between the two D gene segments as well as at the VD and DJ junctions.

The organization of the γ and δ genes (Fig. 4.41) resembles that of the α and β genes, although there are some important differences. The gene complex encoding the δ chain is found entirely within the T-cell receptor α-chain gene complex, between the V_α and the J_α gene segments. As a result, any rearrangement of the α-chain genes inactivates the genes encoding δ chains. There are many fewer V gene segments at the γ and δ loci than at either the T-cell receptor α or β loci or at any of the immunoglobulin loci. Increased junctional variability in the δ chains may compensate for the small number of possible variable regions and has the effect of focusing almost all of the variability in the γ:δ receptor in the junctional region. As we have seen, the amino acids encoded by the junctional regions lie at the center of the T-cell receptor binding site.

T cells bearing γ:δ receptors are a distinct lineage of cells whose functions are at present unknown. The ligands for these receptors are also unknown, although some γ:δ T cells can recognize products of certain class IB genes (see Section 4-16), whereas others appear to be able to recognize antigen directly, much as antibodies do, without the requirement for presentation by an MHC molecule or processing of the antigen. Detailed analysis of the rearranged variable regions of γ:δ T-cell receptors shows that they resemble variable regions of antibody molecules more than they resemble the variable regions of α:β T-cell receptors. In peripheral lymphoid tissues, only a very small percentage (generally 1–5%) of T cells express γ:δ receptors. However, in epithelial tissues, especially in the epidermis and small intestine of the mouse, most T cells express γ:δ receptors. The receptors of these epithelial γ:δ T cells show extremely restricted variability. We shall return to the possible functional significance of these findings in Chapter 10.

Summary.

T-cell receptors are structurally similar to immunoglobulins and are encoded by homologous genes. Diversity is distributed differently in T-cell receptors, which have roughly the same number of V gene segments but more J gene segments and greater diversification of the junctions between gene segments. Moreover, functional T-cell receptors are not known to diversify their V genes after rearrangement through somatic hypermutation. This leads to a T-cell receptor structure in which the highest diversity is in the central part of the receptor, which contacts the bound peptide fragment of the ligand. The co-receptor molecules CD4 and CD8 bind MHC class II and MHC class I molecules respectively and act synergistically with the T-cell receptor in signaling, resulting in about a 100-fold increase in the sensitivity of T cells to antigen.

Summary to Chapter 4.

The T-cell receptor is very similar to immunoglobulin, both in structure and in the way in which variability is introduced into the antigen-binding site. However, there are also important differences between T-cell receptors and immunoglobulins that are related to the nature of the antigens recognized by T cells. T-cell receptors recognize foreign peptide antigen bound to an MHC molecule on the target cell surface. There are two classes of MHC molecule: MHC class I molecules, which bind stably to peptides derived from proteins synthesized and degraded in the cytosol; and MHC class II molecules, which bind stably to peptides derived from proteins degraded in endocytic vesicles. In addition to being bound by the T-cell receptor, the two classes of MHC molecule are differentially recognized by two molecules, CD8 and CD4, that characterize the two major subsets of T cells and cooperate with the T-cell receptor in activating the T cell. CD8 T cells recognize MHC class I:peptide complexes and are activated to kill cells displaying foreign peptides derived from cytosolic pathogens. CD4 T cells recognize MHC class II:peptide complexes and are specialized to activate other immune effector cells, for example B cells or macrophages, to act against the foreign antigens or pathogens that they have taken up. There are several genes for each class of MHC molecule, arranged in a cluster in the MHC, together with genes involved in the degradation of protein antigens into peptides and their transport to the cell surface bound to MHC molecules. Each different MHC molecule can bind stably to a range of different peptides, and because the genes for the MHC class I and class II molecules are highly polymorphic as well as being polygenic, each cell expresses several different MHC class I and class II proteins and is thus able to bind many different peptide antigens. The binding site for the peptide on an MHC molecule lies in a cleft between two α helices; thus, the T-cell receptor recognizes a ligand that has a region of high variability (the peptide antigen) lying between regions of lesser variability (the α helices of the MHC). The generation of diversity in T-cell receptors reflects this, as the process of gene rearrangement focuses the diversity into the CDR3 loops lying at the center of the antigen-binding site and, unlike immunoglobulins, T-cell receptors are known not to use somatic hypermutation as a means of increasing receptor variability. The variability of the CDR1 and CDR2 loops that form the periphery of the binding site is therefore limited to that encoded in the germline by the different V genes. As a consequence of the T-cell receptor's binding a combined peptide:MHC ligand, T cells manifest MHC-restricted antigen recognition, such that a given T cell is specific for a specific peptide bound to a specific MHC molecule. The genes encoding the T-cell receptors appear to have evolved specifically to recognize MHC molecules, thus accounting for the high frequency of T cells responding to allogeneic MHC molecules.

General references.

Bodmer, J.G., Marsh, S.G.E., Albert, E.D., Bodmer, W.F., DuPont, B., Erlich, H.A., Mach, B., Mayr, W.R., Parham, P., Saszuki, T., et al.: **Nomenclature for factors of the HLA system, 1991.** *Tissue Antigens* 1992, **39**:161-173.

Germain, R.N.: **MHC-dependent antigen processing and peptide presentation: Providing ligands for T lymphocyte activation.** *Cell* 1994, **76**:287-299.

Klein, J.: *Natural History of the Major Histocompatibility Complex.* New York: J. Wiley & Sons; 1986.

Moller, G. (ed): Origin of major histocompatibility complex diversity. *Immunol. Rev.* 1995, **143**:5-292.

General references.

4-1 T cells recognize a complex of a peptide fragment bound to an MHC molecule.

Fremont, D.H., Rees, W.A., Kozono, H.: **Biophysical studies of T cell receptors and their ligands.** *Curr. Opin. Immunol.* 1996, **8**:93-100.

4-2 T cells with different functions recognize peptides produced in two distinct intracellular compartments.

Morrison, L.A., Lukacher, A.E., Braciale, V.L., Fan, D.P., and Braciale, T.J.: **Differences in antigen presentation to MHC class I- and class II-restricted influenza virus-specific cytolytic T-lymphocyte clones.** *J. Exp. Med.* 1986, **163**:903.

Song, R. and Harding C.V.: **Roles of proteasomes, transporter for antigen presentation (TAP), and β_2-microglobulin in the processing of bacterial or particulate antigens via an alternate class I MHC processing pathway.** *J. Immunol.* 1996, **156**:4182-4190.

4-3 The two classes of MHC molecule have distinct subunit structures but similar three-dimensional structures.

Dessen, A., Lawrence, C.M., Cupo, S., Zaller, D.M., and Wiley, D.C.: **X-ray crystal structure of HLA-DR4 (DRA*0101, DRB1*0401) complexed with a peptide from human collagen II.** *Immunity* 1997, **7**:473-481.

Fremont, D.H., Hendrickson, W.A., Marrack, P., and Kappler, J.: **Structures of an MHC class II molecule with covalently bound single peptides.** *Science* 1996, **272**:1001-1004.

Fremont, D.H., Monnaie, D., Nelson, C.A., Hendrickson, W.A., and Unanue, E.R.: **Crystal structure of I-Ak in complex with a dominant epitope of lysozyme.** *Immunity* 1998, **8**:305-317.

Murthy, V.L., and Stern, L.J.: **The class II MHC protein HLA-DR1 in complex with an endogenous peptide: implications for the structural basis of the specificity of peptide binding.** *Structure* 1997, **5**:1385-1396.

Reid, S.W., McAdam, S., Smith, K.J., Klenerman, P., O'Callaghan, C.A., Harlos, K., Jakobsen, B.K., McMichael, A.J., Bell, J.I., Stuart, D.I., and Jones, E.Y.: **Antagonist HIV-1 Gag peptides induce structural changes in HLA B8.** *J. Exp. Med.* 1996, **184**:2279-2286.

Smith, K.J., Reid, S.W., Harlos, K., McMichael, A.J., Stuart, D.I., Bell, J.I., and Jones, E.Y.: **Bound water structure and polymorphic amino acids act together to allow the binding of different peptides to MHC class I HLA-B53.** *Immunity* 1996, **4**:215-228.

4-4 Peptides are stably bound to MHC molecules.

Fremont, D.H., Matsumura, M., Stura, E.A., Peterson, P.A. and Wilson, I.: **Crystal structures of two viral peptides in complex with murine MHC class 1 H-2K^b.** *Science* 1992, **257**:919-927.

Madden, D.R., Gorga, J.C., Strominger, J.L. and Wiley, D.C.: **The three-dimensional structure of HLA-B27 at 2.1Å resolution suggests a general mechanism for tight peptide binding to MHC.** *Cell* 1992, **70**:1035-1048.

4-5 MHC class I molecules bind short peptides of 8–10 amino acids by both ends.

Bouvier, M., and Wiley, D.C.: **Importance of peptide amino and carboxyl termini to the stability of MHC class I molecules.** *Science* 1994, **265**:398-402.

Weiss, G.A., Collins, E.J., Garboczi, D.N., Wiley, D.C., and Schreiber, S.L.: **A tricyclic ring system replaces the variable regions of peptides presented by three alleles of human MHC class I molecules.** *Chem. Biol.* 1995, **2**:401-407.

4-6 The length of the peptides bound by MHC class II molecules is not constrained.

Rammensee, H.G.: **Chemistry of peptides associated with MHC class I and class II molecules.** *Curr. Opin. Immunol.* 1995, **7**:85-96.

Rudensky, A,Y., Preston-Hurlburt, P., Hong, S-C., Barlow, A., Janeway Jr, C.A.: **Sequence analysis of peptides bound to MHC class II molecules.** *Nature* 1991, **353**:622.

4-7 The two classes of MHC molecule are expressed differentially on cells.

Steimle, V., Siegrist, C.A., Mottet, A., Lisowska-Grospierre, B. and Mach, B.: **Regulation of MHC class II expression by interferon-γ mediated by the transactivator gene CIITA.** *Science* 1994, **265**:106-109.

4-8 Peptides that bind to MHC class I molecules are actively transported from the cytosol to the endoplasmic reticulum.

Shepherd, J.C., Schumacher, T.N.M., Ashton-Rickardt, P.G., Imaeda, S., Ploegh, H.L., Janeway, C.A. Jr., and Tonegawa, S.: **TAP1-dependent peptide translocation *in vitro* is ATP-dependent and peptide-selective.** *Cell* 1993, **74**:577-584.

Townsend, A., Ohlen, C., Foster, L., Bastin, J., Lunggren, H.-G., and Karre, K.: **A mutant cell in which association of class I heavy and light chains is induced by viral peptides.** *Cold Spring Harbor Symp. Quant. Biol.* 1989, **54**:299-308.

4-9 Peptides of cytosolic proteins are generated in the cytosol before transport into the endoplasmic reticulum.

Bai, A., and Foreman, J.: **The effect of proteasome inhibitor lactacystin on the presentation of transporter associated with antigen processing (TAP)-dependent and TAP-independent peptide epitopes by MHC class I molecules.** *J. Immunol.* 1997, **159**: 2139-2146.

Craiu, A.C., et al.: **Lactacystin and clasto-lactacystin β-lactone modify multiple proteasome β subunits and inhibit intracellular protein degradation and MHC class I antigen presentation.** *J. Biol. Chem.* 1997, **272**:13437-13445.

4-10 Newly synthesized MHC class I molecules are retained in the endoplasmic reticulum by binding a TAP-1 associated protein until they bind peptide.

Hengel, H., Koopman, J.O., Flohr, T., Moranyi, W., Goulmy, E., Hammerling,

G.J., Koszinowski, U.H., and Monburg, F.: **A viral ER resident glycoprotein inactivates the MHC encoded peptide transporter.** *Immunity* 1997, **6**:623-632.

Hengel, H., and Koszinowski, U.H.: **Interference with antigen processing by viruses.** *Curr. Opin. Immunol.* 1997, **9**:470-476.

Machold, R.P., Wiertz, E.J.H.J., Jones, T.R., and Ploegh, H.L.: **The HCMV gene products US11 and US2 differ in their ability to attack allelic forms of murine major histocompatibility complex (MHC) class I heavy chains.** *J. Exp. Med.* 1997, **185**: 363-366.

Ortmann, B., Copeman, J., Lehner, P.J., Sadasivan, B., Herberg, J.A., Grandea, A.G., Riddell, S.R., Tampe, R., Spies, T., Trowsdale, J., and Cresswell, P.: **A critical role for tapasin in the assembly and function of multimeric MHC class I-TAP complexes.** *Science* 1997, **277**:1306-1309.

Pamer, E., and Cresswell, P.: **Mechanisms of MHC class I-restricted antigen processing.** *Annu. Rev. Immunol.* 1998, **16**:323-358.

4-11 Peptides presented by MHC class II molecules are generated in acidified endocytic vesicles.

Chapman, H.A.: **Endosomal proteolysis and MHC class II function.** *Curr. Opin. Immunol.* 1998, **10**:93-102.

Deussing, J., Roth, W., Saftig, P., Peters, C., Ploegh, H.L., and Villadangos, J.A.: **Cathepsins B and D are dispensable for major histocompatibility complex class II-mediated antigen presentation.** *Proc. Natl. Acad. Sci. USA* 1998, **95**:4516-4521.

4-12 The invariant chain directs newly synthesized MHC class II molecules to acidified intracellular vesicles.

Brachet, V., Raposo, G., Amigorena, S., and Mellman, I.: **Ii controls the transport of major histocompatibility class II molecules to and from lysosomes.** *J. Cell Biol.* 1997, **137**:51-55.

Kleijmeer, M.J., Morkowski, S., Griffith, J.M., Rudensky, A.Y., and Geuze, H.J.: **Major histocompatibility complex class II compartments in human and mouse B lymphoblasts represent conventional endocytic compartments.** *J. Cell Biol.* 1997, **139**:639-649.

4-13 A specialized MHC class II-like molecule catalyzes loading of MHC class II molecules with endogenously processed peptides.

Denzin, L.K., Sant'Angelo, D.B., Hammond, C., Surman, M.J., and Cresswell, P.: **Negative regulation by HLA-DO of MHC class II restricted antigen processing.** *Science* 1997, **278**:106-109.

Kropshofer, H., Arndt, S.O., Moldenhauer, G., Hammerling, G.J., and Vogt, A.B.: **HLA-DM acts as a molecular chaperone and rescues empty HLA-DR molecules at lysosomal pH.** *Immunity* 1997, **6**:293-302.

Kropshofer, H., Vogt, A.B., Moldenhauer, G.J.H., Blum, J.S., and Hammerling, G.J.: **Editing of the HLA-DR peptide repertoire by HLA-DM.** *EMBO J.* 1996, **15**:6144-6154.

Van Ham, S.M., Tjin, E.P.M., Lillemeier, B.F., Gruneberg, U., Van Meijgaarden, K.E., Pastoors, L., Verwoerd, D., Tulp, A., Canas, B., Rahman, D., Ottenhoff, T.H.M., Pappin, D.J.C., Trowsdale, J., and Neefjes, J.: **HLA-DO is a negative regulator of HLA-DM mediated MHC class II peptide loading.** *Curr. Biol.* 1997, **7**:950-957.

4-14 Stable peptide binding by MHC molecules allows effective antigen presentation at the cell surface.

Lanzavecchia, A., Reid, P.A., and Watts, C.: **Irreversible association of peptides with class II MHC molecules in living cells.** *Nature* 1992, **357**:249-252.

4-15 Many proteins involved in antigen processing and presentation are encoded by genes within the major histocompatibility complex.

Herberg, J.A., Beck S., and Trowsdale, J.: **TAPASIN, DAXX, RGL2, HKE2 and four new genes (BING 1, 3 to 5) form a dense cluster at the centromeric end of the MHC.** *J. Mol. Biol.* 1998, **277**:839-857.

4-16 A variety of genes with specialized functions in immunity are also encoded in the MHC.

Groh, V., Bahram, S., Bauer, S., Herman, A., Beauchamp, M., and Spies, T.: **Cell stress-regulated human major histocompatibility complex class I gene expressed in gastrointestinal epithelium.** *Proc. Natl. Acad. Sci. USA* 1996, **93**:12445-12450.

Groh, V., Steinle, A., Bauer, S., and Spies, T.: **Recognition of stress-induced MHC molecules by intestinal epithelial gammadelta T cells.** *Science* 1998, **279**:1737-1740.

Wilson, I.A., and Bjorkman, PJ.: **Unusual MHC-like molecules: CD1, Fc receptor, the hemochromatosis gene product, and viral homologs.** *Curr. Opin. Immunol.* 1998, **10**:67-73.

Zeng, Z.H., Castano, A.R., Segelke, B., Stura, E.A., Peterson, P.A., and Wilson, I.A.: **The crystal structure of murine CD1: and MHC-like fold with a large hydrophobic binding groove.** *Science* 1997, **277**:339-345.

4-17 The protein products of MHC class I and class II genes are highly polymorphic.

Marsh, S.G.: **HLA class II region sequences.** 1998, *Tissue Antigens* **51**:467-507.

Marsh, S.G.: **Nomenclature for factors of the HLA system. WHO nomenclature committee for factors of the HLA system.** *Hum. Immunol.* 1998, **59**:256-257.

4-18 MHC polymorphism determines which peptides can be bound and presented to T cells.

Babbitt, B., Allen, P.M., Matsueda, G., Haber, P., Unuanue, E.R.: **Binding of immunogenic peptides to Ia histocompatibility molecules.** *Nature* 1985, **317**:359.

Hammer, J.: **New methods to predict MHC-binding sequences within protein antigens.** *Curr. Opin. Immunol.* 1995, **7**:263-269.

4-19 MHC polymorphism affects antigen recognition by T cells, both indirectly, by controlling peptide binding, and directly, through contacts between the T-cell receptor and the MHC molecule itself.

Katz, D.H., Hamaoka, T., Dorf, M.E., Maurer, P.H., and Benacerraf, B.: **Cell interactions between histoincompatible T and B lymphocytes. IV. Involvement of immune response (Ir) gene control of lymphocyte interaction controlled by the gene.** *J. Exp. Med.* 1973, **138**:734.

Rosenthal, A.S., and Shevach, E.M.: **Function of macrophages in antigen recognition by guinea pig T lymphocytes. I. Requirement for histocompatible macrophages and lymphocytes.** *J. Exp. Med.* 1973, **138**:1194.

Zinkernagel, R.M., and Doherty, P.C.: **Restriction of *in vivo* T-cell mediated cytotoxicity in lymphocytic choriomeningitis within a syngeneic or semiallogeneic system.** *Nature* 1974, **248**:701-702.

4-20 Non-self MHC molecules are recognized by 1-10% of T cells.

Kaye, J., and Janeway, C.A. Jr.: **The Fab fragment of a directly activating**

monoclonal antibody that precipitates a disulfide-linked heterodimer from a helper T-cell clone blocks activation by either allogeneic Ia or antigen and self-Ia. *J. Exp. Med.* 1984, **159**:1397-1412.

4-21 MHC polymorphism extends the range of antigens to which the immune system can respond.

Franco, A., Ferrari, C., Sette, A., Chisari, F.V.: **Viral mutations, TCR antagonism and escape from the immune response.** *Curr. Opin. Immunol.* 1995, **7**:524-531

Hill, A.V., Elvin, J., Willis, A.C., Aidoo, M., Allsopp, C.E.M., Gotch, F.M., Gao, X.M., Takiguchi, M., Greenwood, B.M., Townsend, A.R.M., McMichael, A.J., Whittle, H.C.: **Molecular analysis of the association of B53 and resistance to severe malaria.** *Nature* 1992, **360**:434-440.

Potts, W.K., Slev, P.R.: **Pathogen-based models favouring MHC genetic diversity.** *Immunol. Rev.* 1995, **143**:181-197.

4-22 Multiple genetic processes generate MHC polymorphism.

Gaur, L.K., and Nepom, G.T.: **Ancestral major histocompatibility complex DRB genes beget conserved patterns of localized polymorphisms.** *Proc. Natl. Acad. Sci. USA* 1996, **93**:5380-5383.

4-23 The T-cell receptor resembles a membrane-associated Fab fragment of immunoglobulin.

Garboczi, D.N., Ghosh, P., Utz, U., Fan, Q.R., Biddison, W.E., and Wiley, D.C.: **Structure of the complex between human T-cell receptor, viral peptide and HLA-A2.** *Nature* 1996, **384**:134-141.

Housset, D., Mazza, G., Gregoire, C., Piras, C., Malissen, B., and Fontecilla-Camps, J.C.: **The three-dimensional structure of a T cell antigen receptor $V_\alpha V_\beta$ domain reveals a novel arrangement of the V_β domain.** *EMBO J.* 1997, **16**:4205-4216.

4-24 The T-cell receptor genes resemble immunoglobulin genes.

Rowen, L., Koop, B.F. and Hood, L.: **The complete 685-kilobase DNA sequence of the human β T cell receptor locus.** *Science* 1996, **272**:1755-1762.

4-25 T-cell receptor diversity is focused in CDR3.

Garcia, K.C., Degano, M., Pease, L.R., Huang, M., Peterson, P.A., Leyton, L., and Wilson, I.A.: **Structural basis of plasticity in T cell receptor recognition of a self peptide-MHC antigen.** *Science* 1998, **279**:1166-1172.

4-26 T-cell receptors bind diagonally across the peptide-binding groove, with their two CDR3 loops making the predominant contacts with the bound peptide.

Ding, Y.H., Smith, K.J., Garboczi, D.N., Utz, U., Biddison, W.E., and Wiley, D.C.: **Two human T cell receptors bind in a similar diagonal mode to the HLA-A2/tax peptide complex using different TCR amino acids.** *Immunity* 1998, **8**:403-411.

Teng., M-K., Smolyar, A., Tse, A.G.D., Liu, J-H., Liu, J., Hussey, R.E., Nathenson, S.G., Chang, H-C., Reinherz, E.L., and Wang, J-H.: **Identification of a common docking topology with substantial variation among different TCR-MHC-peptide complexes.** *Curr. Biol.* 1998, **8**:409-412.

4-27 Somatic hypermutation does not generate diversity in T-cell receptors.

Zheng, B., Xue, W., and Kelsoe, G.: **Locus-specific somatic hypermutation in germinal centre T cells.** *Nature* 1994, **372**:556-559.

4-28 Many T cells respond to superantigens.

Fields, B.A., Malchiodi, E.L., Ysern, X., Stauffacher, C.V., Schlievert, P.M., Karjalainen, K., and Mariuzza, R.A.: **Crystal structure of a T cell receptor β chain complexed with a superantigen.** *Nature* 1996, **384**:188-192.

4-29 The co-receptor molecules CD4 and CD8 cooperate with the T-cell receptor in antigen recognition.

Gao, G.F., Tormo, J., Gerth, U.C., Wyer, J.R., McMichael, A.J., Stuart, D.I., Bell, J.I., Jones, E.Y., and Jakobsen, B.Y.: **Crystal structure of the complex between human CD8αα and HLA-A2.** *Nature* 1997, **387**:630-634.

Wu, H., Kwong, P.D., and Hendrickson, W.A.: **Dimeric association and segmental variability in the structure of human CD4.** *Nature* 1997, **387**:427-530.

Zamoyska, R.: **CD4 and CD8: modulators of T cell receptor recognition of antigen and of immune responses?** *Curr. Opin. Immunol.* 1998, **10**:82-86.

4-30 Some T cells bear an alternative form of T-cell receptor with γ and δ chains.

Li, H., Lebedeva, M.I., Llera, A.S., Fields, B.A., Brenner, M.B., and Mariuzza, M.B.: **Structure of the V_δ domain of a human γ:δ T-cell antigen receptor.** *Nature* 1998, **391**:502-506.

Signaling through Lymphocyte Receptors

5

Cells communicate with their environment through a variety of cell-surface receptors that recognize and bind molecules present in the extracellular environment. The main function of T and B lymphocytes is to respond to antigen and so, in their case, the receptors for antigen are the most important and the best studied. Binding of antigen to these receptors generates intracellular signals that alter the cells' behavior, and the mechanisms that bring this about will be the main topic of this chapter. Because of the diversity of antigen receptors in the normal lymphocyte population, most of our information on intracellular signaling in lymphocytes comes from tumor-derived lymphoid cell lines whose antigen receptors have been stimulated by anti-receptor antibodies. We shall use this information to infer the signaling pathways generated when a mature naive lymphocyte binds its specific antigen and is activated to undergo clonal expansion followed by differentiation to a functional effector cell. We shall, however, also consider how signaling via the antigen receptor and other receptors can lead to other responses such as inactivation or cell death, depending on the stage of development of the cell and the nature of the ligand.

The antigen receptors of B and T lymphocytes are expressed at the cell surface as distinct multiprotein complexes with different recognition properties, which have been described in Chapters 3 and 4. However, after ligand is bound, the intracellular signaling pathways that lead from the receptor are remarkably similar in B and T cells. In both cases they lead to the nucleus and to changes in gene expression that dictate the lymphocyte's response.

We shall first discuss some general principles of cell signaling and introduce some of the common mechanisms used in intracellular signaling pathways, with particular reference to the antigen receptor signaling pathways. In the second part of the chapter we shall outline the signaling pathways from the antigen receptors to the nucleus. Other signals also impinge on the response to antigen and influence the development and survival of lymphocytes. These will be the focus of the third and final part of the chapter.

General principles of transmembrane signaling.

The challenge that faces all cells that respond to external stimuli is how the recognition of a stimulus, usually by receptors on the outer surface of the cell, is able to effect changes within the cell. Extracellular signals are transmitted across the plasma membrane by receptor proteins, which are instrumental in converting extracellular ligand binding into an intracellular biochemical event. Conversion of a signal from one form into another is known as **signal transduction**, and in this part of the chapter we consider several different mechanisms of signal transduction in cell signaling. Cell-surface receptors activate intracellular signaling pathways and so convert an extracellular signal into an intracellular one that then transmits the signal onward. The signal is converted into different biochemical forms, distributed to different sites in the cell, and sustained and amplified as it proceeds towards its final destination. In antigen-receptor signaling the final destination is the nucleus, where the activation of transcription factors turns on new gene expression.

5-1 Binding of antigen leads to clustering of antigen receptors on lymphocytes.

All cell-surface receptors are transmembrane proteins or protein complexes that provide a link between the exterior and the interior of the cell. Many receptors undergo a distinct change in protein conformation on binding their ligand. In some types of receptor this conformational change opens an ion channel into the cell and the resulting change in the concentration of important ions within the cell acts as the intracellular signal, which is then converted into an intracellular response. In other receptors the conformational change affects the cytoplasmic portion of the receptor, enabling it to associate with and activate intracellular signaling proteins and enzymes.

The crucial effect of ligand binding to antigen receptors is to cause them to cluster together on the cell surface, although the mechanism of clustering is still poorly defined. The requirement for receptor clustering was first shown experimentally in somewhat artificial systems by using antibodies against the extracellular portion of the receptor to mimic antigen binding. Antibodies

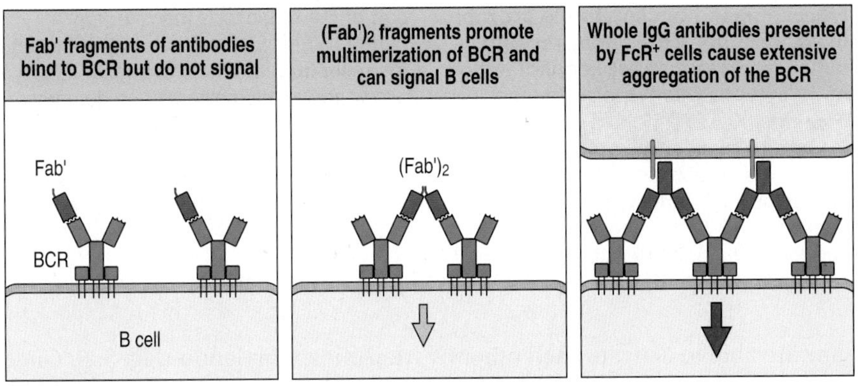
Fab' fragments of antibodies bind to BCR but do not signal
(Fab')2 fragments promote multimerization of BCR and can signal B cells
Whole IgG antibodies presented by FcR+ cells cause extensive aggregation of the BCR
Fab'
BCR
B cell
(Fab')2

Fig. 5.1 Crosslinking of antigen receptors is the first step in lymphocyte activation. The requirement for receptor crosslinking is illustrated by the use of anti-immunoglobulin antibodies to activate the B-cell antigen receptor (BCR). As shown in the left panel, Fab′ fragments of the anti-immunoglobulin can bind to the receptors but cannot crosslink them; they also fail to activate B cells. $(Fab')_2$ fragments of the same anti-immunoglobulin, which have two binding sites, can bridge two receptors (center panel), and thus signal, albeit weakly, to the B cell. The most effective activation occurs when receptors are extensively crosslinked by many identical antibody molecules presented by other cells, such as macrophages, that have Fc receptors (FcR) for the constant regions of intact antibodies, as shown in the right panel. The same situation applies to T-cell antigen receptors.

against the B-cell antigen receptor or the T-cell antigen receptor activate signaling by inducing clustering of the receptor complexes. This is a very convenient system for the analysis of early events after activation, as all the cells in a sample will be stimulated at the same time, making the course of the response easier to follow.

Antigen receptor clustering occurs when the receptors are crosslinked. The importance of crosslinking was shown by comparing the response after stimulation with $(Fab')_2$ antibody fragments, which have two binding sites, to that with Fab′ fragments, which have only one binding site (see Fig. 3.4). On lymphocytes treated with Fab′ fragments the antigen receptors do not cluster and the cells make no response, whereas on lymphocytes treated with $(Fab')_2$ fragments the receptors become clustered and the cells respond. The response is strongest when lymphocytes are stimulated with intact antibody bound to cells carrying receptors for the Fc portion of immunoglobulin (Fig. 5.1).

How antigen receptors are clustered *in vivo* is not yet completely understood. Pathogens such as intact bacteria and viruses have repetitive epitopes on their surfaces that can crosslink the receptors on B cells and cause them to cluster. Complex molecules that contain regularly repeated identical epitopes will also have the same effect. However, it is still uncertain how B-cell receptors can be clustered by soluble monomeric antigens, such as most of the experimental antigens that immunologists use to study immune responses. T cells presumably undergo receptor clustering in response to contact with another cell surface bearing multiple copies of a specific MHC:peptide complex (see Fig. 8.29), but the details of this clustering remain obscure. How ligand-binding leads to receptor clustering and generates a signal is more clearly understood for some other, simpler receptors, as we shall see in the next section.

5-2 Clustering of antigen receptors leads to activation of intracellular signals.

In most of the receptors discussed in this chapter, the intracellular signaling activity is initiated by the activation of **protein tyrosine kinases**, enzymes that affect the activity of other proteins by adding a phosphate group to certain tyrosine residues. The receptors for certain growth factors provide the simplest example of this type of receptor. They have cytoplasmic domains that contain an intrinsic tyrosine kinase activity. These enzyme domains are normally inactive, but when they are brought together by receptor clustering,

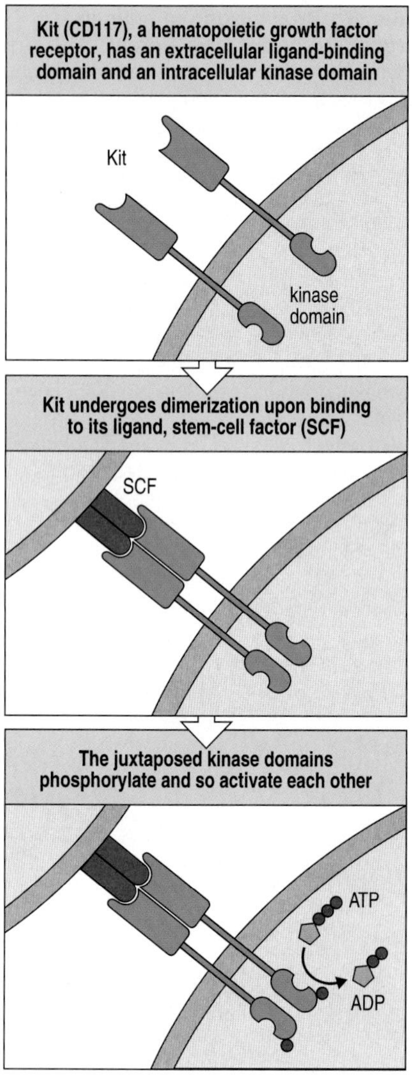
Kit (CD117), a hematopoietic growth factor receptor, has an extracellular ligand-binding domain and an intracellular kinase domain
Kit
kinase domain
Kit undergoes dimerization upon binding to its ligand, stem-cell factor (SCF)
SCF
The juxtaposed kinase domains phosphorylate and so activate each other
ATP
ADP

Fig. 5.2 Ligand binding to the growth factor receptor Kit induces receptor dimerization and transphosphorylation of its cytoplasmic tyrosine kinase domains. Kit (CD117) is a transmembrane protein with an external ligand-binding domain specific for stem-cell factor (SCF) and a cytoplasmic domain with intrinsic tyrosine kinase activity. In the unbound state, the kinase part of the receptor is inactive (top panel). When SCF binds to Kit, it causes the receptor proteins to dimerize; this allows the two tyrosine kinase domains to phosphorylate one another and so become activated. Transactivation of protein kinases by transphosphorylation is an important step in signaling from many cell-surface kinases.

they are able to activate each other by transphosphorylation (Fig. 5.2). Once activated, these tyrosine kinases can phosphorylate and activate other cytoplasmic signaling molecules.

The situation in the antigen receptors is somewhat more complex. As we shall see later, they do not themselves have intrinsic tyrosine kinase activity. Instead, the cytoplasmic portions of some of the receptor components bind to intracellular protein tyrosine kinases, which are therefore known as **receptor-associated tyrosine kinases**. When the receptors cluster, these enzymes are brought together and act on each other and on the receptor cytoplasmic tails to initiate the signaling process as in the example above.

In the case of the antigen receptors, the first tyrosine kinases associated with the receptor are members of the Src (pronounced 'Sark') family of tyrosine kinases. The Src-family kinases are common components of signaling pathways concerned with the control of cell division and differentiation in vertebrates and other animals. The prototypic family member Src was initially discovered as the **oncogene** v-*src* that is responsible for the ability of the Rous sarcoma virus to produce tumors in chickens. This viral gene was subsequently shown to be a modified form of a normal cellular gene called c-*src* that the virus had picked up from its host cell at some time in the past. Several other common components of signaling pathways that regulate cell growth and division were also first discovered through their oncogenic action when mutated or removed from their normal controls.

The receptor-associated Src-family kinases have a key role in transducing signals across the lymphocyte membrane; their activation informs the cell interior that the receptor has encountered its antigen. But this is just the first step in a multistep process. When a cell is signaled by the binding of ligand to a kinase-coupled receptor, kinase activation initiates a cascade of intracellular signaling that transfers the signal to other molecules and eventually carries it to the nucleus.

5-3 Phosphorylation of receptor cytoplasmic tails by tyrosine kinases concentrates intracellular signaling molecules around the receptors.

Phosphorylation of enzymes and other proteins by protein kinases is a common general mechanism by which cells regulate their biochemical activity and has many advantages as a control mechanism. It is rapid, not requiring new protein synthesis or protein degradation to change the biochemical activity of a cell. It can also be easily reversed by the action of **protein phosphatases**, which remove the phosphate group. Many enzymes become active when phosphorylated and inactive when dephosphorylated, or vice versa; the activity of many of the protein kinases involved in signaling is regulated in this way.

Another and equally important outcome of protein phosphorylation is the creation of a binding site for other proteins. This does not alter the intrinsic activity of the molecules concerned. Instead, phosphorylation is used as a tag, allowing the recruitment of other proteins that bind to the phosphorylated site. For example, many kinases involved in signaling are associated with the inner surface of the cell membrane and can act only inefficiently upon their target proteins when these are free in the cytosol. Receptor activation and the phosphorylation of membrane-associated proteins can, however, create binding sites for these target proteins. Cytosolic proteins that bind to phosphorylated sites at the membrane are thus concentrated with respect to the kinase and can in their turn be phosphorylated and activated (Fig. 5.3).

Proteins can be phosphorylated on three classes of amino acid: on tyrosine, on serine or threonine, or on histidine. Each of these requires a separate class of kinases to add phosphate groups; only the first two are relevant to signaling within the immune system. As we have seen, the early events of signaling, associated with the clustering of the antigen receptors, predominantly involve protein tyrosine kinases; the later events also involve protein serine/threonine kinases.

In antigen receptor signaling, the phosphotyrosines generated by tyrosine kinase action form binding sites for a protein domain known as an **SH2 domain** (Src homology 2 domain), which is found in many intracellular signaling proteins including the Src-family kinases, in which SH2 domains were first discovered. Binding of SH2 domains to phosphotyrosines is a crucial mechanism for recruiting intracellular signaling molecules to an activated receptor. As well as the SH2 domain, Src-family kinases possess another binding domain known as SH3 or Src homology 3. This domain, which is also found in other proteins, binds to proline-rich regions in diverse proteins and can thus recruit these proteins into the signaling pathway, as we shall see later. Src-family kinases are usually anchored to the cell membrane by a lipid moiety attached to their amino-terminal region. They are distributed over the inner surface of the cell membrane and during cell activation become localized to sites of receptor signaling by binding to phosphotyrosine via their SH2 domains.

As a signaling mechanism, phosphorylation also has the advantage that it is easily and rapidly reversible by protein phosphatases that can specifically remove the phosphate groups added by protein kinases. It is crucial that components of signaling pathways can be readily returned to their unstimulated state; not only does this make the signaling pathway ready to receive another signal but it sets a limit on the time that any individual signal is active, preventing cellular responses from running out of control. Therefore, it is not

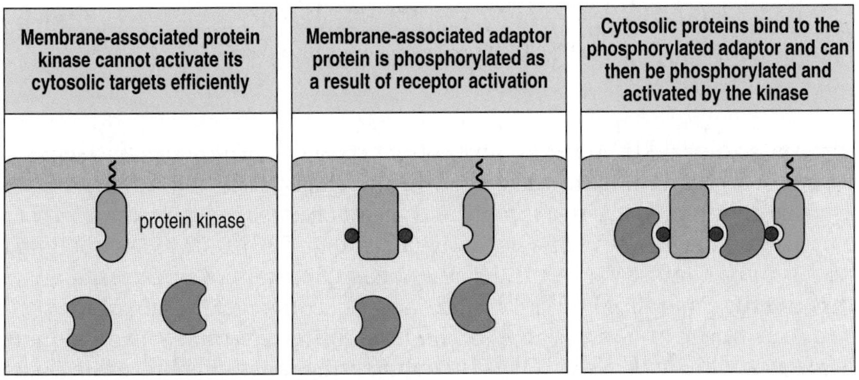

Membrane-associated protein kinase cannot activate its cytosolic targets efficiently
protein kinase
Membrane-associated adaptor protein is phosphorylated as a result of receptor activation
Cytosolic proteins bind to the phosphorylated adaptor and can then be phosphorylated and activated by the kinase

Fig. 5.3 Protein phosphorylation creates binding sites that recruit proteins to the signaling pathway. Src-family kinases are localized at the inner surface of the cell membrane and cannot activate their cytosolic targets efficiently unless these are brought to the membrane. However, the kinases can phosphorylate another membrane-associated protein, which in turn can bind cytosolic proteins containing SH2 domains. These can then be activated by the membrane-associated kinase.

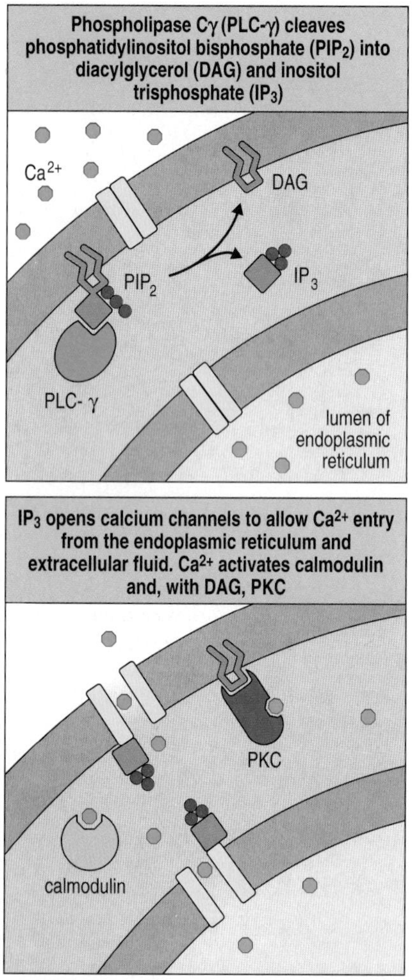

Fig. 5.4 The enzyme phospholipase C-γ cleaves inositol phospholipids to generate two important signaling molecules. The membrane phospholipid phosphatidylinositol-4,5-bisphosphate (PIP_2) is a component of the inner leaflet of the plasma membrane. When phospholipase C-γ (PLC-γ) is activated, it cleaves PIP_2 to its two component parts, inositol trisphosphate (IP_3) and diacylglycerol (DAG). Both these molecules are important in signaling. IP_3 binds to calcium channels in the cell membrane, opening the channels and allowing Ca^{2+} to enter the cytosol both from the endoplasmic reticulum, where Ca^{2+} is stored within the cell, and from the external fluids surrounding the cell. DAG attracts the protein kinase C to the cell membrane and activates it in conjunction with the increased Ca^{2+}. The active form of protein kinase C is a serine/threonine kinase with several roles in cell activation. Raised Ca^{2+} levels also activate a protein known as calmodulin, a ubiquitous Ca^{2+} -binding protein that is responsible for activating other Ca^{2+}-dependent enzymes within the cell.

surprising that the signaling pathways that link the cell surface to changes in gene expression employ protein phosphorylation and dephosphorylation to regulate the activity of many of their components.

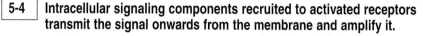

5-4 Intracellular signaling components recruited to activated receptors transmit the signal onwards from the membrane and amplify it.

Several classes of protein are typically recruited to the activated receptors and participate in signal propagation. The enzyme **phospholipase C-γ (PLC-γ)** contains an SH2 domain through which it can bind to phosphotyrosine; it is thus recruited to the site of receptor-associated tyrosine kinase activity at the cell membrane. PLC-γ has a crucial role in propagating the signal onward from the membrane and in amplifying it. Phosphorylation of a tyrosine residue in PLC-γ activates the enzyme, which then cleaves molecules of the membrane phospholipid **phosphatidylinositol bisphosphate** (**PIP_2**) into its two component parts, **inositol trisphosphate** (**IP_3**) and **diacylglycerol** (**DAG**) (Fig. 5.4). As one molecule of PLC-γ can generate many molecules of DAG and IP_3, this and similar enzymatic steps serve to amplify and sustain the signal. Production of DAG and IP_3 by activated PLC-γ is a common step in pathways from many types of receptor.

Diffusion of IP_3 away from the membrane causes the release of Ca^{2+} into the cytosol from intracellular storage sites in the endoplasmic reticulum, immediately raising intracellular free Ca^{2+} levels several-fold. This triggers the opening of calcium channels in the plasma membrane that let more Ca^{2+} into the cell, thus sustaining the signal. Increased intracellular free Ca^{2+} leads to the activation of the Ca^{2+}-binding protein calmodulin, which in turn binds to and regulates the activity of several other proteins and enzymes in the cell, transmitting the signal onwards along pathways that also eventually converge on the nucleus.

The other product of PIP_2 cleavage is DAG, which remains associated with the inner surface of the plasma membrane. DAG helps to activate **protein kinase C** (**PKC**), which is a serine/threonine protein kinase that is thought to initiate one of the signaling pathways leading to the nucleus. PKC is further activated by the Ca^{2+} released by IP_3 action. Thus, the two products of the cleavage of PIP_2 reinforce each other in activating PKC; they also have other, independent, effects.

Many of the cellular processes that are activated in lymphocytes when antigen binds to its receptor are common to many cell types. For example, resting lymphocytes proliferate when exposed to antigen, whereas other cell types proliferate in response to specific growth factors; what differs in each case is the receptor that initiates the common response pathway in the different cell types. To link these different receptors to common intracellular signaling components, specialized **adaptor proteins** are needed.

In lymphocytes, adaptor proteins that bind to the antigen receptors often contain a central SH2 domain flanked by two SH3 domains. These proteins do not have kinase activity themselves and their function in general is to recruit other molecules to the activated receptor. Such a protein can bind to a phosphotyrosine residue via its SH2 domain and to other proteins containing proline-rich motifs via its SH3 domains (Fig. 5.5). Binding to an adaptor protein positions these other proteins at or near the cell membrane, where they can in turn be phosphorylated and activated by the tyrosine kinases associated with the receptor. One important family of proteins whose members

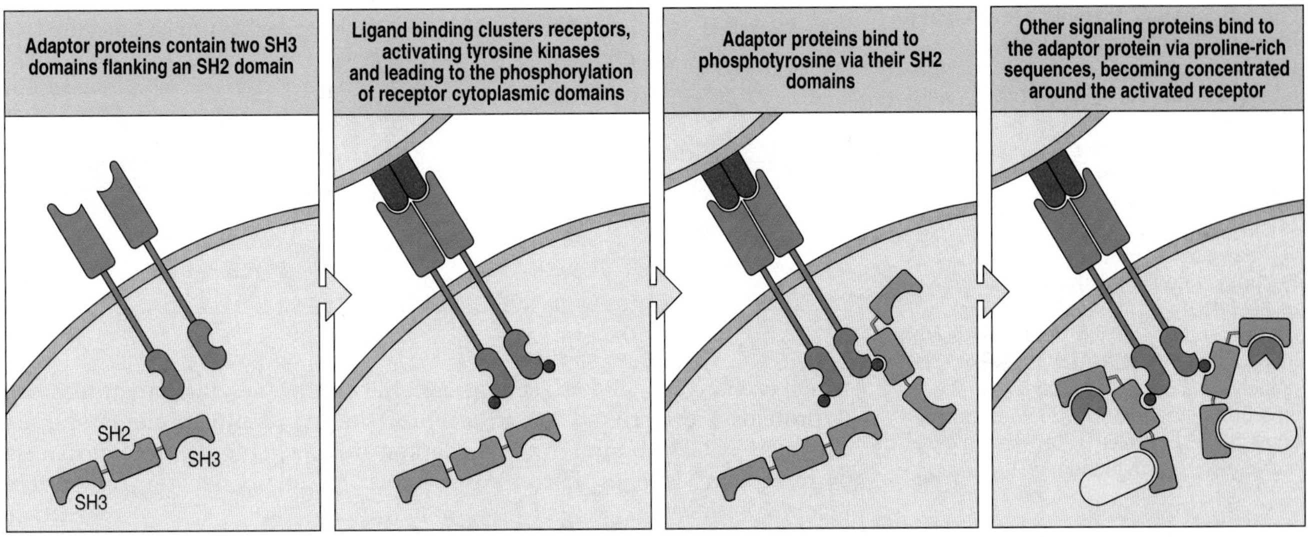

Adaptor proteins contain two SH3 domains flanking an SH2 domain
Ligand binding clusters receptors, activating tyrosine kinases and leading to the phosphorylation of receptor cytoplasmic domains
Adaptor proteins bind to phosphotyrosine via their SH2 domains
Other signaling proteins bind to the adaptor protein via proline-rich sequences, becoming concentrated around the activated receptor
SH2
SH3
SH3

Fig. 5.5 Signals are propagated from the receptor through adaptor proteins, which recruit other signaling proteins to the receptor. Adaptor proteins are specialized signaling molecules that usually have no enzymatic activity themselves. Instead, they allow other molecules to become associated with the activated receptors. Adaptors, as shown in the first panel, often contain SH2 domains flanked by SH3 domains. Once a receptor, in this case Kit, has been activated and transphosphorylated (second panel), adaptors can bind to the phosphotyrosines through their SH2 domains (third panel). Other molecules that contain proline-rich regions can now bind to the adaptors and be activated by the receptor-associated kinases (fourth panel).

are bound by adaptors and activated during signaling are the **guanine-nucleotide exchange factors** (**GEFs**) which, as we shall see in the next section, pass the signal on to another common central component of many signaling pathways, the small G proteins.

5-5 Small G proteins activate a protein kinase cascade that transmits the signal to the nucleus.

Small GTP-binding proteins or **small G proteins** are another class of protein that serves to propagate signals from tyrosine kinase-associated receptors. The family of small single-chain GTP-binding proteins is distinct from the heterotrimeric G proteins that associate with seven-span transmembrane receptors such as rhodopsin. The best-known small G protein is **Ras.** Like the Src-family kinases, Ras was discovered through its effects on cell growth. A gene encoding a mutated form of Ras was found in various animal retroviruses that cause tumors, and the corresponding cellular *RAS* gene was subsequently found to be mutated in many different human tumors. The frequent discovery of mutant *RAS* in tumors indicated that the normal gene had a critical role in the control of cell growth and focused attention on the physiological role of Ras.

Small G proteins such as Ras exist in two states, depending on whether they are binding GTP or GDP. The GTP-bound form of Ras is active, whereas its inactivation is mediated by the intrinsic GTPase activity of Ras, which removes a phosphate group from GTP to leave Ras in its inactive GDP-bound form. Because of their intrinsic GTPase activity, small G proteins do not stay permanently activated, eventually turning themselves off; mutation at a single residue can lock them in the active state, causing them to become oncogenic. The small G proteins are normally found in the inactive, GDP-containing state; their activation involves a guanine-nucleotide exchange factor (GEF),

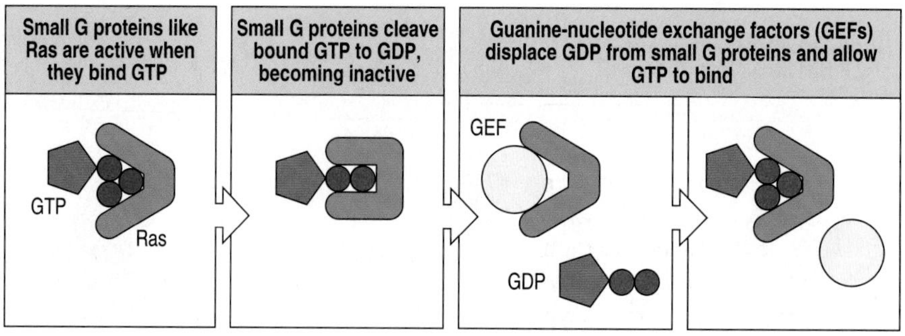
Small G proteins like Ras are active when they bind GTP
Small G proteins cleave bound GTP to GDP, becoming inactive
Guanine-nucleotide exchange factors (GEFs) displace GDP from small G proteins and allow GTP to bind
GTP
Ras
GEF
GDP

Fig. 5.6 Small G proteins are switched from inactive to active states by guanine-nucleotide exchange factors. Ras is a small GTP-binding protein with intrinsic GTPase activity. In its GTP-bound state, it is active (left panel), whereas in the GDP-bound state it is inactive. Most of the time, it is in the inactive state owing to its intrinsic GTPase activity (second panel). Receptor signaling activates guanine-nucleotide exchange factors (GEFs), which can bind to small G proteins such as Ras and displace GDP, allowing GTP to bind in its place (right panel). In the time before the intrinsic GTPase activity converts GTP to GDP, the Ras protein is active and transmits the signal onward.

which exchanges GDP for GTP (Fig. 5.6). In lymphocytes, Ras and other small G proteins are recruited to the receptor site by adaptor proteins and are activated by GEFs bound to these adaptor proteins. Thus, G proteins can act as molecular switches, becoming switched on only when the cell-surface receptor is activated.

Once activated, small G proteins activate a cascade of protein kinases known as the **mitogen-activated protein kinase** (**MAP kinase**) cascade. This kinase cascade is found in all multicellular animals and is responsible for many of the effects of activating ligands. The MAP kinase cascade leads directly to the phosphorylation and activation of transcription factors in the nucleus. In particular, the AP-1 family of transcription factors, which are heterodimers of the oncogenes *fos* and *jun*, are activated through the MAP kinase cascade. We shall discuss these activation pathways in more detail in the next part of the chapter, where the structures of the lymphocyte antigen receptor complexes are described and we look at the particular signals generated by the antigen receptors and the co-receptors that cluster with them.

Summary.

Lymphocyte antigen receptors signal for cell activation by using signal transduction mechanisms common to many intracellular signaling pathways. On ligand binding, antigen receptor clustering leads to the activation of receptor-associated protein tyrosine kinases at the cytoplasmic face of the plasma membrane. These initiate intracellular signaling by phosphorylating tyrosine residues in the clustered receptor tails. The phosphorylated tyrosines act as binding sites for additional kinases and other signaling molecules that amplify the signal and transmit it onwards. The enzyme phospholipase C-γ is recruited in this way and intiates two major pathways of intracellular signaling that are common to many other receptors. Cleavage of the membrane phospholipid PIP_2 by this enzyme produces the diffusible messenger inositol trisphosphate (IP_3) and membrane-bound diacylglycerol (DAG). IP_3 action leads to a sharp increase in the level of intracellular free Ca^{2+}, which activates various calcium-dependent enzymes. Together with Ca^{2+}, DAG initiates a second signaling pathway by activating protein kinase C. A third pathway involves small G proteins, proteins with GTPase activity that are activated by binding GTP, but then hydrolyze the GTP to GDP and become inactive. Small G proteins are recruited to the signaling pathway and activated by guanine-nucleotide exchange factors (GEFs), which catalyze the exchange of GDP for GTP. GEFs and other signaling molecules are linked to the activated receptors by adaptor proteins that bind to phosphorylated tyrosines through one protein domain, the SH2 domain, and to other signaling molecules through other domains including SH3. All these signaling pathways eventually converge on the nucleus to alter patterns of gene transcription.

Antigen receptor structure and signaling pathways.

The antigen receptors on B cells (the **B-cell receptor** or **BCR**) and T cells (the **T-cell receptor** or **TCR**) are multiprotein complexes made up of clonally variable antigen-binding chains—the heavy and light immunoglobulin chains in the B-cell receptor, and the α and β chains in the T-cell receptor—that are associated with invariant accessory proteins. The invariant chains are required both for transport of the receptors to the cell surface and, most importantly, for initiating signaling when the receptors bind to an extracellular ligand. Antigen binding to the receptor leads ultimately to the activation of nuclear transcription factors that turn on new gene expression and turn off genes typically expressed only in resting cells. In this part of the chapter we also see how clustering of the antigen receptors with co-receptors helps to generate these signals.

5-6 The variable chains of lymphocyte antigen receptors are associated with invariant accessory chains that perform the signaling function of the receptor.

The antigen-binding portion of the B-cell receptor complex is a cell-surface immunoglobulin that is formed from heavy and light chains of the same antigen specificity as the secreted antibodies that the B-cell will eventually produce. The mRNA for the cell-surface heavy chain is spliced in such a way that the carboxy terminus of the protein is made up of a transmembrane domain and a very short cytoplasmic tail (see Fig. 3.28). The heavy and light chains do not by themselves make up a complete receptor, however. When cells were transfected with heavy- and light-chain cDNA derived from a cell expressing surface immunoglobulin, the immunoglobulin produced remained inside the transfected cell rather than appearing on the surface. This implies that other molecules are necessary to enable the cell-surface expression of the immunoglobulin receptor. Two proteins associated with heavy chains on the B-cell surface were subsequently identified and called **Igα** and **Igβ**; transfection of Igα and Igβ cDNA along with that for the immunoglobulin chains results in surface expression of the B-cell receptor.

The complete B-cell receptor is generally considered as a complex of eight chains—two identical light chains, two identical heavy chains, and two each of the Igα and Igβ chains (Fig. 5.7). The Igα and Igβ genes are closely linked in the genome and encode proteins composed of a single amino-terminal immunoglobulin-like domain connected via a transmembrane domain to a cytoplasmic tail. Igα and Igβ are crucial for signaling from the B-cell receptor.

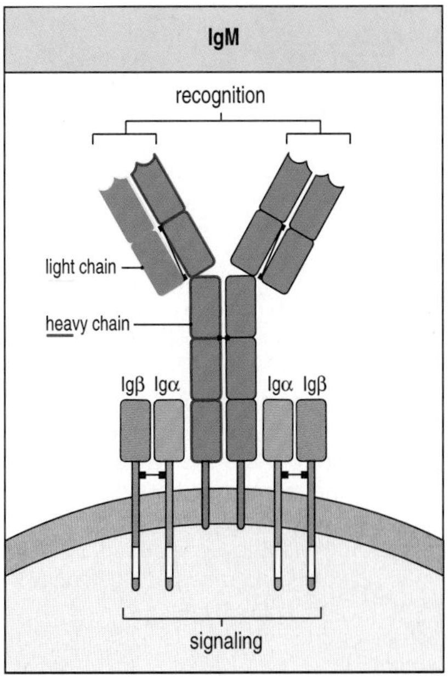

Fig. 5.7 The B-cell receptor is made up of cell-surface immunoglobulin in a complex with the invariant proteins Igα and Igβ. The immunoglobulin recognizes and binds antigen but cannot itself generate a signal. It is associated with non-specific signaling molecules, Igα and Igβ. Igα and Igβ each have a single immunoreceptor tyrosine-based activation motif (ITAM), shown as a yellow segment, in their cytosolic tails that enables them to signal when the B-cell receptor is ligated with antigen. Igα and Igβ are disulfide-linked but their exact stoichiometry in the receptor complex is not yet known, nor is it known which binds to the heavy chain. The ratio of two Igα and two Igβ chains to each immunoglobulin molecule as shown here is a guess.

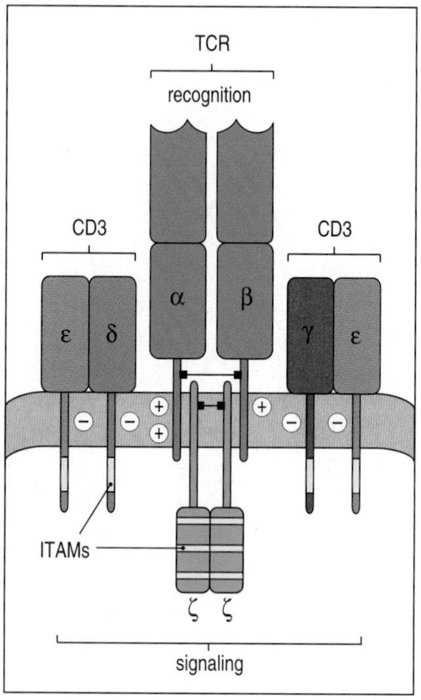

Fig. 5.8 The T-cell receptor is made up of antigen-recognition proteins and invariant signaling proteins. The α:β heterodimer (TCR) recognizes and binds antigen but cannot signal to the cell that antigen has bound. It is associated with four other signaling chains, collectively called CD3, which associate with the TCR heterodimer and transport it to the cell surface. There it is associated with a homodimer of ζ chains, which also signal to the interior of the cell that antigen binding has occurred. Each CD3 chain has one ITAM (yellow segment), whereas each ζ chain has three. Note also the matching of positive charges in the trans-membrane domains of the α and β chains with negative charges in the CD3 chains.

The transmembrane form of the immunoglobulin heavy chain has only a very short cytoplasmic tail and it was hard to understand how it could signal. The discovery of Igα and Igβ solved this intellectual problem.

Signaling from the B-cell receptor complex depends on the presence in Igα and Igβ of amino acid sequences called **immunoreceptor tyrosine-based activation motifs (ITAMs)**. These motifs were originally identified in the cytoplasmic tails of Igα and Igβ, but are now known to be present in the accessory chains involved in signaling from the T-cell receptor, and in the Fc receptors on mast cells and natural killer (NK) cells that bind antibody constant regions. ITAMs are composed of two tyrosine residues separated by around 13 amino acids; when antigen binds, the tyrosines become phosphorylated by the receptorassociated tyrosine kinases. The canonical ITAM sequence is ...**YXX[L/V]X$_{7-11}$YXX[L/V]**..., where Y is tyrosine, L is leucine, V is valine, and X represents any amino acid. Igα and Igβ each have a single ITAM in their cytosolic tails. This gives the B-cell receptor a total of four ITAMs.

The complete T-cell antigen receptor is also a complex of eight chains, two of which are the highly variable α and β chains, which provide the single antigen-binding site (see Chapter 4). The invariant accessory chains are CD3γ, CD3δ, and CD3ε, which make up the **CD3 complex** along with a largely intracyto-plasmic homodimer of ζ chains. Although the exact stoichiometry of the T-cell receptor complex is not definitively established, it seems that each α:β heterodimer is associated at the cell surface with one CD3γ, one CD3δ, two CD3εs, and one ζ homodimer (Fig. 5.8). The three CD3 proteins are encoded in adjacent genes that are regulated as a unit and are required for surface expression of the α:β heterodimer and for signaling via the receptor. Optimal expression and maximum signaling, however, also require the ζ chain, which is encoded elsewhere in the genome.

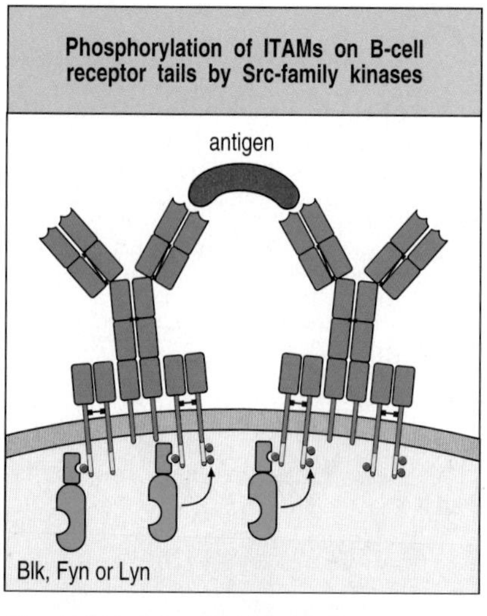

Fig. 5.9 Src-family kinases are associated with the antigen receptors and phosphorylate the tyrosines in ITAMs. The membrane-bound Src-family kinases Fyn, Blk, and Lyn are associated with the B-cell receptor. After ligand binding and receptor clustering they phosphorylate tyrosines in the ITAMs on the cytoplasmic tails of Igα and Igβ.

The CD3 proteins resemble Igα and Igβ in having an extracellular immunoglobulin-like domain and a single ITAM in their cytoplasmic tails. The ζ chain is distinct in having only a short extracellular domain, but has three ITAMs in its cytoplasmic domain. The CD3 chains have negatively charged acidic residues in their transmembrane domains, which are able to interact with the positive charges of the T-cell receptor α and β chains, as shown in Fig. 5.8. In total the T-cell receptor complex is equipped with 10 ITAMs, which might give it greater flexibility in signaling compared with the B-cell receptor, as will be discussed later in this chapter.

The antigen receptors of B and T lymphocytes are constructed similarly. Both are molecular complexes made up two types of functional component: variable chains that recognize the individual antigens and invariant chains that have a role both in the surface expression of the receptors and in transmitting signals to the cell's interior, enabling antigen recognition to be translated into action.

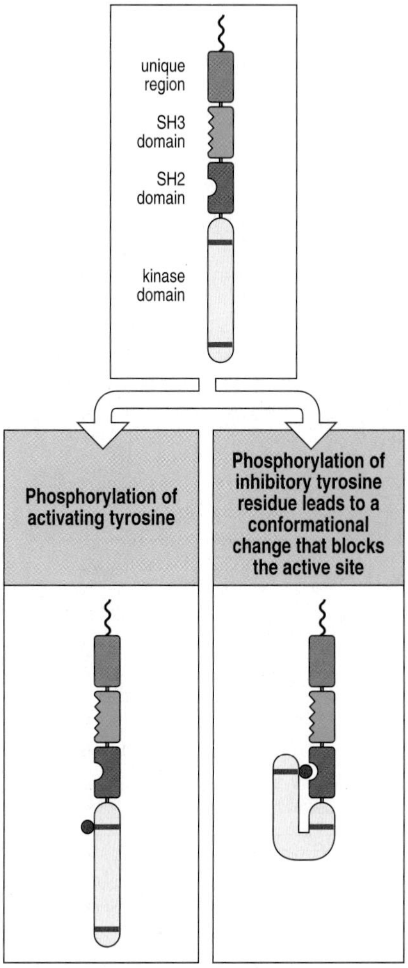
unique region
SH3 domain
SH2 domain
kinase domain
Phosphorylation of activating tyrosine
Phosphorylation of inhibitory tyrosine residue leads to a conformational change that blocks the active site

Fig. 5.10 Regulation of Src-family kinase activity. The kinase domain of a Src-family kinase contains two tyrosine residues (red bars) which are targets for phosphorylation. Phosphorylation of one of these tyrosines (bottom left panel) stimulates kinase activity, and this tyrosine is a target for phosphorylation by receptor-associated tyrosine kinases. The second tyrosine lies near the carboxy terminus and has a regulatory function. When it has been phosphorylated, the kinase is inactive as a result of an interaction between the kinase domain and the SH2 domain, as shown in the lower right panel.

5-7 The ITAMs associated with the B-cell and T-cell receptors are phosphorylated by protein tyrosine kinases of the Src family.

Phosphorylation of the tyrosines in ITAMs serves as the first intracellular signal indicating that the lymphocyte has detected its specific antigen. The B-cell receptor associates with three protein tyrosine kinases of the Src family—**Fyn**, **Blk**, and **Lyn**. When the receptors cluster after antigen binding, these kinases are thought to phosphorylate the two tyrosine residues in the ITAMs of the cytoplasmic tails of Igβ and Igα (Fig. 5.9).

The initial events in T-cell receptor signaling are also implemented by two Src-family kinases—**Lck**, which is constitutively associated with the cytoplasmic domain of the co-receptor molecules CD4 and CD8 (see Chapter 4), and Fyn, which associates with the cytoplasmic domains of the ζ and CD3ε chains upon receptor clustering. CD4 or CD8 is clustered together with the antigen receptor when the receptor binds to its peptide:MHC ligand. Both Fyn and Lck phosphorylate specific ITAMs on the accessory chains of the T-cell receptor complex.

The enzyme activity of the Src-family kinases is itself regulated by the phosphorylation status of the kinase domain, which has two regulatory tyrosine residues, one activating and the other inhibitory. Even after being phosphorylated at the activating tyrosine, the Src-family kinase can be kept inactive by a protein tyrosine kinase called **Csk** (C-terminal Src kinase), which phosphorylates the inhibitory tyrosine near the carboxy terminus (Fig. 5.10). As Csk activity is constitutive in resting cells, the Src proteins are generally inactive. An agent that counteracts the effects of Csk is the transmembrane protein tyrosine phosphatase **CD45** or **leukocyte common antigen**, which is required for receptor signaling in lymphocytes and other cells. CD45 can remove the phosphate from phosphotyrosines, especially from the inhibitory tyrosine residue in the carboxy terminus of Src-family kinases. Thus, the balance between Csk, which inactivates Src-family kinases, and CD45, which can activate them, is one methods by which the activity of Src-family kinases is regulated.

Before we consider how the signals generated by these kinases are transmitted onward, we will look at how antigen binding by both B cells and T cells directly or indirectly activates co-receptor molecules that are essential for producing a strong and effective intracellular signal.

5-8 Antigen receptor signaling is enhanced by co-receptors that bind the same ligand.

Antigen binding by surface immunoglobulin is insufficient on its own to activate B cells, and whereas bacteria and viruses express multiple repeats of particular epitopes, only rarely do soluble antigens have the repeating epitopes that are capable of crosslinking surface immunoglobulin (see Section 5-1). The

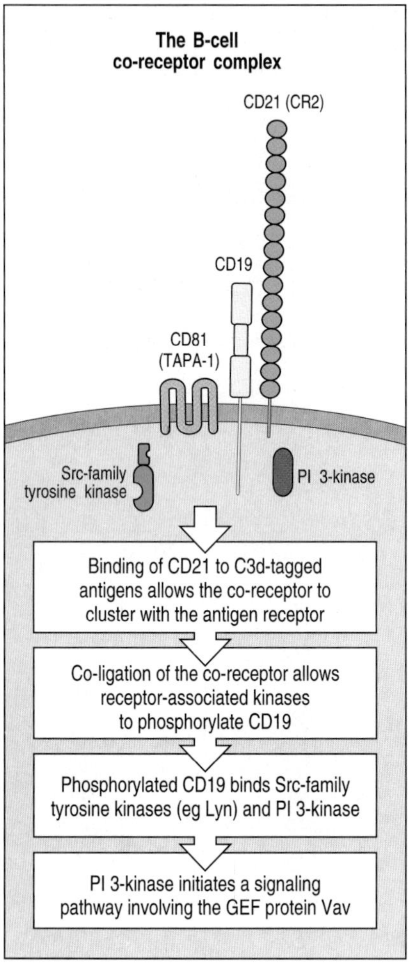
The B-cell co-receptor complex
CD21 (CR2)
CD19
CD81 (TAPA-1)
Src-family tyrosine kinase
PI 3-kinase
Binding of CD21 to C3d-tagged antigens allows the co-receptor to cluster with the antigen receptor
Co-ligation of the co-receptor allows receptor-associated kinases to phosphorylate CD19
Phosphorylated CD19 binds Src-family tyrosine kinases (eg Lyn) and PI 3-kinase
PI 3-kinase initiates a signaling pathway involving the GEF protein Vav

Fig. 5.11 B-cell antigen receptor signaling is modulated by a co-receptor complex of at least three cell-surface molecules, CD19, CD21, and CD81. Binding of the cleaved complement fragment C3d to antigen allows the tagged antigen to bind to both the B-cell receptor and the complement receptor CD21 (complement receptor 2, CR2), a component of the B-cell co-receptor complex. Crosslinking and clustering of the co-receptor with the antigen receptor results in the phosphorylation of tyrosine residues in the cytoplasmic domain of CD19 by protein kinases associated with the B-cell receptor; other Src-family kinases can bind to phosphorylated CD19 and so augment signaling through the B-cell receptor. Phosphorylated CD19 can also bind the enzyme phosphatidylinositol 3-kinase (PI 3-kinase), which is instrumental in recruiting the guanine-nucleotide exchange factor Vav to the receptor complex.

initiation of most B-cell responses (see Chapter 9) therefore requires other signals, and this is an important safeguard against self-reactive responses. For protein antigens, additional activating signals are delivered to B cells by the specialized subset of T cells known as helper T cells, which recognize the antigen bound to the surface immunoglobulin of B cells after it has been internalized and processed into peptide fragments that are presented by MHC class II molecules on the surface of antigen-specific B cells. We shall discuss these interactions in detail in Chapter 9. However, B-cell receptor signaling is also enhanced by the **B-cell co-receptor**, a complex of the cell-surface molecules CD19, CD21, and CD81 (Fig. 5.11), which can be co-ligated with the B-cell receptor.

CD19 is expressed on all B cells from an early stage in their development. It is implicated in B-cell activation by three lines of evidence. First, anti-CD19 antibodies enhance B-cell responses to antigen *in vitro*; second, genetically engineered mice that lack CD19 make deficient B-cell responses to most antigens; and third, co-ligation of CD19 and the B-cell antigen receptor complex amplifies the intracellular effects of signaling through the antigen receptor.

CD21 (complement receptor 2 or CR2), with which CD19 is physically associated in the cell membrane, is a receptor for complement fragments that are known to be important in inducing strong B-cell responses. The complement system, which we shall consider later in Chapter 9, can tag antigens by covalent binding of the C3d complement fragment to antigen:antibody complexes. These complexes can thus crosslink CD21 and its associated proteins with the B-cell receptor, which induces phosphorylation of the cytoplasmic tail of CD19 by the B-cell receptor-associated tyrosine kinases. The role of the third component of the B-cell co-receptor complex, CD81 (TAPA-1), is as yet unknown. Co-ligation of the B-cell receptor with its CD19, CD21, CD81 co-receptor increases signaling 1000 to 10,000 fold.

Optimal signaling through the T-cell receptor complex also occurs only when it clusters with the co-receptors CD4 or CD8. This is why most T cells with MHC class II-restricted receptors express CD4, which binds MHC class II molecules, whereas most T cells with MHC class I-restricted receptors express CD8, which binds MHC class I molecules. About 100 identical specific peptide:MHC complexes are required on a target cell to trigger a T cell expressing the appropriate co-receptor. In the absence of the appropriate co-receptor, 10,000 identical complexes (about 10% of all the MHC molecules on a cell) are required for optimal T-cell activation. This density is rarely, if ever, achieved *in vivo*.

Aggregation of the T-cell receptor with the appropriate co-receptor helps to activate the T cell by bringing the Lck tyrosine kinase associated with the cytoplasmic domain of the co-receptor together with its targets associated

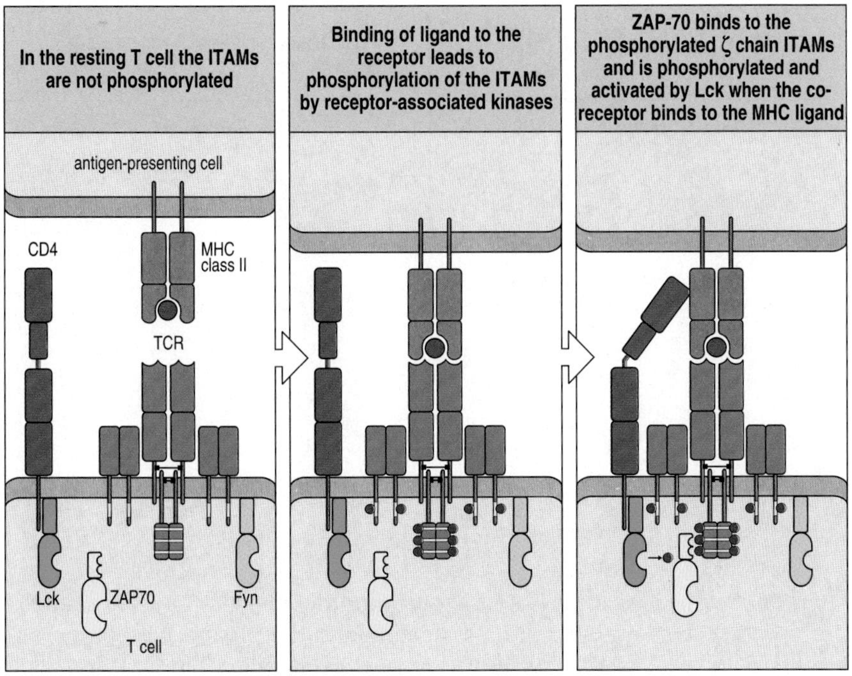
In the resting T cell the ITAMs are not phosphorylated
Binding of ligand to the receptor leads to phosphorylation of the ITAMs by receptor-associated kinases
ZAP-70 binds to the phosphorylated ζ chain ITAMs and is phosphorylated and activated by Lck when the co-receptor binds to the MHC ligand
antigen-presenting cell
CD4
MHC class II
TCR
Lck
ZAP70
Fyn
T cell

Fig. 5.12 Clustering of the T-cell receptor and a co-receptor initiates signaling within the T cell. When T-cell receptors become clustered on binding MHC:peptide complexes on the surface of an antigen-presenting cell, activation of receptor-associated kinases such as Fyn leads to phosphorylation of the CD3γ, δ, and ε ITAMs as well as those on the ζ chain. The tyrosine kinase ZAP-70 binds to the phosphorylated ITAMs of the ζ chain but is not activated until binding of the co-receptor to the MHC molecule on the antigen-presenting cell (here shown as CD4 binding to an MHC class II molecule) brings the kinase Lck into the complex. Lck then phosphorylates and activates ZAP-70.

with the cytoplasmic domains of the T-cell receptor complex (Fig. 5.12). One of the principal targets of Lck in T cells is another kinase, the ζ-chain-associated protein or ZAP-70, to which we now turn.

5-9 Fully phosphorylated ITAMs bind the protein tyrosine kinases Syk and ZAP-70 and enable them to be activated.

Once the ITAMs in the receptor cytoplasmic tails have been phosphorylated, they can recruit the next players in the signaling cascade, the protein tyrosine kinases **Syk** in B cells and **ZAP-70** in T cells. These two proteins define a second family of protein tyrosine kinases expressed mainly in lymphocytes; they have two SH2 domains in their amino-terminal halves and a carboxy-terminal kinase domain. As each SH2 domain binds to one phosphotyrosine, these proteins potentially bind to motifs with two phosphotyrosines spaced a precise distance apart; tyrosines spaced correctly are found in the ITAM motif. Thus, ZAP-70 or Syk is recruited to the receptor complex upon full phosphorylation of the ITAMs.

Until Syk has bound to the doubly phosphorylated ITAM in Igβ it is inactive enzymatically. To become active it must itself be phosphorylated, and this is thought to occur by transphosphorylation. Each B-cell receptor complex contains at least two Igβ molecules, each able to bind one molecule of Syk; once bound, these are brought into contact with each other and are thus able to phosphorylate, and hence activate, each other (Fig. 5.13). Once activated, Syk phosphorylates target proteins to initiate a cascade of intracellular signaling molecules, which will be described in the next section.

ZAP-70 is not activated by transphosphorylation after binding to the ITAMs in the ζ chain; instead, it is activated by either of the receptor-associated kinases Lck or Fyn; Lck seems to be the predominant activator of ZAP-70. Once activated, the ZAP-70 kinase phosphorylates the substrate **LAT** (**linker of activation in T cells**) and the protein called **SLP-76**, a second linker or

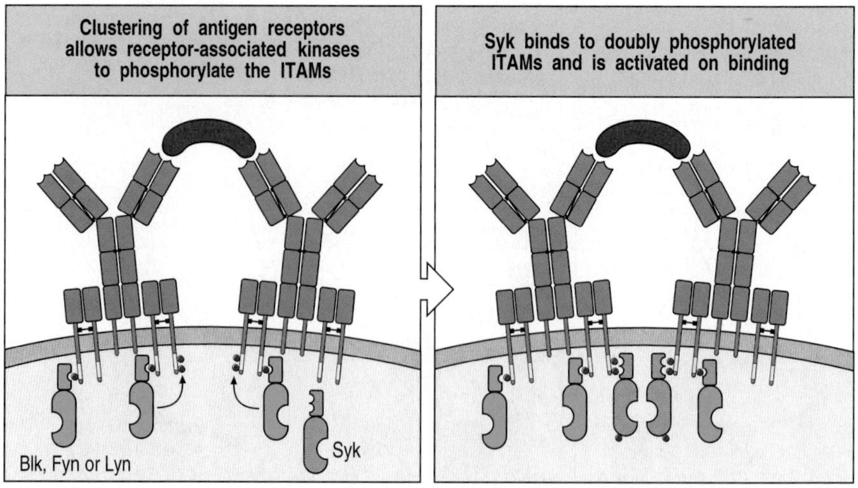

Fig. 5.13 Full phosphorylation of the ITAMs on Igβ of the B-cell receptor leads to the creation of a binding site for Syk and Syk activation via transphosphorylation. On clustering of the receptors, the receptor-associated tyrosine kinases Blk, Fyn, and Lyn phosphorylate the ITAMs on the Igα and Igβ cytoplasmic tails. Subsequently, Syk binds to the phosphorylated ITAMs of the Igβ chain. Because there are two Igβ chains in each B-cell receptor complex, two Syk molecules become bound in close proximity and can activate each other by transphosphorylation, thus initiating further signaling.

adaptor protein in T cells. LAT is a recently defined protein that contains multiple tyrosines and is found associated with the plasma membrane in T cells. This protein has a crucial role as an SH2-domain-binding protein in transmitting the signal from the T-cell membrane to downstream targets, which are recruited to the T-cell receptor by binding through their SH2 domains to phosphorylated LAT. A protein serving the same function in B cells has recently been identified as **BLNK** for **B** cell **link**er protein. This, again, has multiple sites for tyrosine phosphorylation, and interacts with many of the same proteins as LAT and SLP-76, to which it has strong homology in sequence.

5-10 Downstream events are mediated by proteins that associate with the phosphorylated tyrosines and bind to and activate other proteins.

Once the Syk or ZAP-70 tyrosine kinases have been activated and, in T cells, LAT and SLP-76 or in B cells BLNK have been phosphorylated, the next steps in the signaling pathway serve to propagate the signal at the cell membrane, and eventually to communicate it to the nucleus. Signal propagation involves various proteins that bind to phosphotyrosine via SH2 domains, and proteins that are targets for protein tyrosine kinases. Several classes of protein participate in signal propagation. These include the enzyme PLC-γ (see Section 5-4), which initiates two of the main signaling pathways that lead to the nucleus. The third main pathway is generated by activation of the small G protein Ras. This is achieved by adaptor proteins and guanine-nucleotide exchange factors (see Section 5-4) recruited to the phosphorylated receptors.

In B lymphocytes, the adaptor protein Shc binds to tyrosine residues that have been phosphorylated by the receptor-associated tyrosine kinases. Another adaptor protein, Grb2, whose SH2 domain is flanked on both sides by SH3 domains, forms a complex with Shc and this complex binds the guanine-nucleotide exchange factor SOS. SOS, in turn, is involved in activating Ras by the mechanism shown in Fig. 5.6. In T lymphocytes, the adaptor protein Grb2 is recruited by phosphorylated LAT and, via SOS, recruits Ras to the pathway. Adaptor proteins thus link ligand binding by the antigen receptor at the cell surface to the activation of Ras, which then triggers further signaling events downstream.

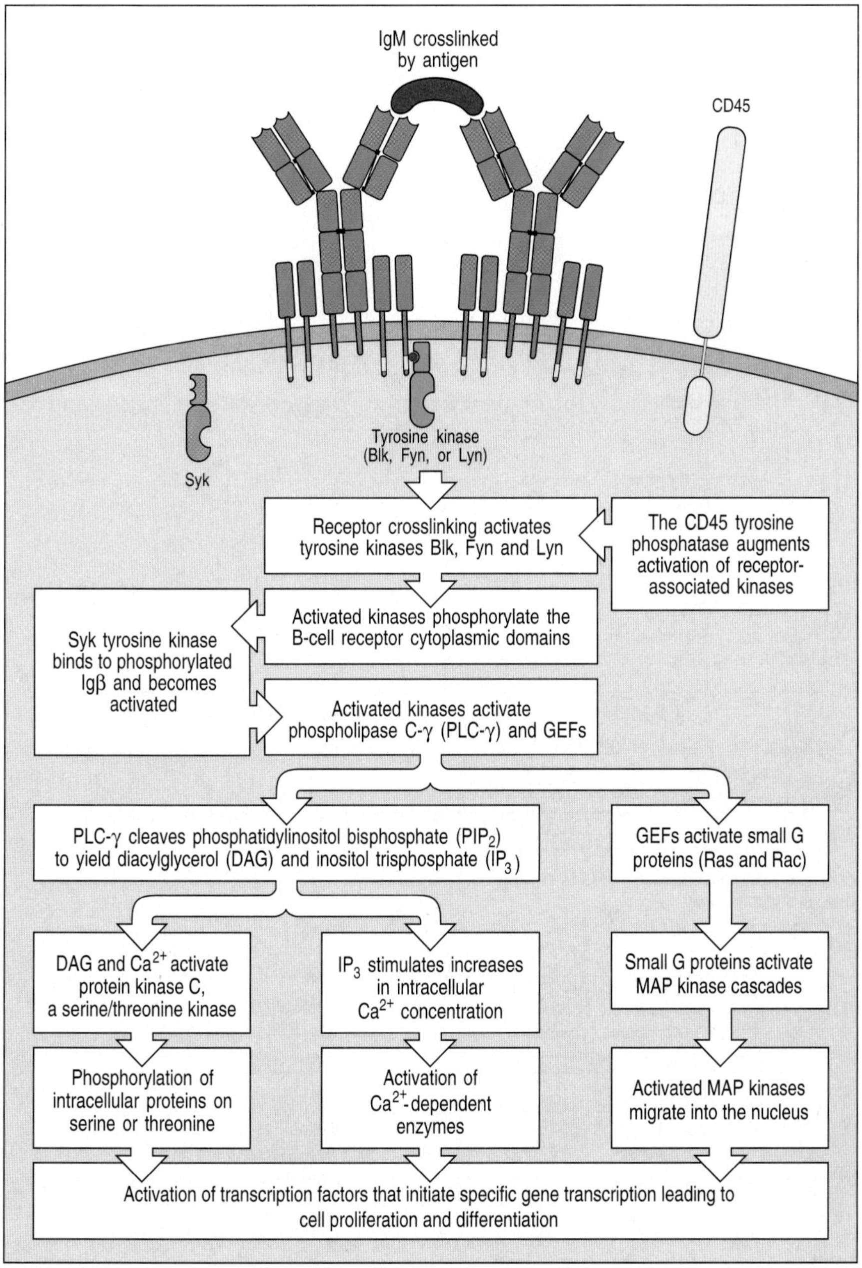

Fig. 5.14 Simplified outline of the intracellular signaling pathways initiated by crosslinking of B-cell receptors by antigen. Crosslinking-surface immunoglobulin molecules activates the receptor-associated Src-family protein tyrosine kinases Blk, Fyn, and Lyn. The CD45 phosphatase, which can remove an inhibitory phosphate from these kinases, is probably also involved in their activation. The receptor-associated kinases phosphorylate the ITAMs in the receptor complex, which bind and activate the cytosolic protein kinase Syk. Syk then initiates the phosphorylation and consequent activation of the enzyme phospholipase C-γ, which cleaves the membrane phospholipid PIP_2 into IP_3 and DAG, thus initiating two of the three main signaling pathways to the nucleus. IP_3 releases Ca^{2+} from intracellular and extracellular sources, and Ca^{2+}-dependent enzymes are activated, whereas DAG activates protein kinase C with the help of Ca^{2+}. The third main signaling pathway is initiated by guanine-nucleotide exchange factors (GEFs) that become associated with the receptor and activate small GTP-binding proteins such as Ras. These in turn trigger protein kinase cascades (MAP kinase cascades) that lead to the activation of MAP kinases that move into the nucleus and phosphorylate gene transcription regulatory proteins. This scheme is a simplification of the events that actually occur during signaling, showing only the main events and pathways.

Another small G protein is activated via the B-cell co-receptor complex (see Section 5-8). Phosphorylated CD19 binds a multifunctional intracellular signaling molecule called **Vav**; this is an adaptor protein that also contains guanine-nucleotide exchange factor activity. When Vav is activated, it can activate the small G protein called Rac, which also initiates a signaling pathway that ends in the nucleus.

Small G proteins such as Ras and Rac activate a cascade of protein kinases that leads directly to the phosphorylation and activation of transcription factors; we shall discuss this in the next section. As we have seen so far, a signal originating from activated tyrosine kinases at the cell membrane can be propagated through several different pathways involving many intracellular proteins. The main outlines of these pathways are summarized in Fig. 5.14 for B cells and Fig. 5.15 for T cells. We now turn to the question of how signals are transmitted to the nucleus to activate transcription factors that can regulate specific genes.

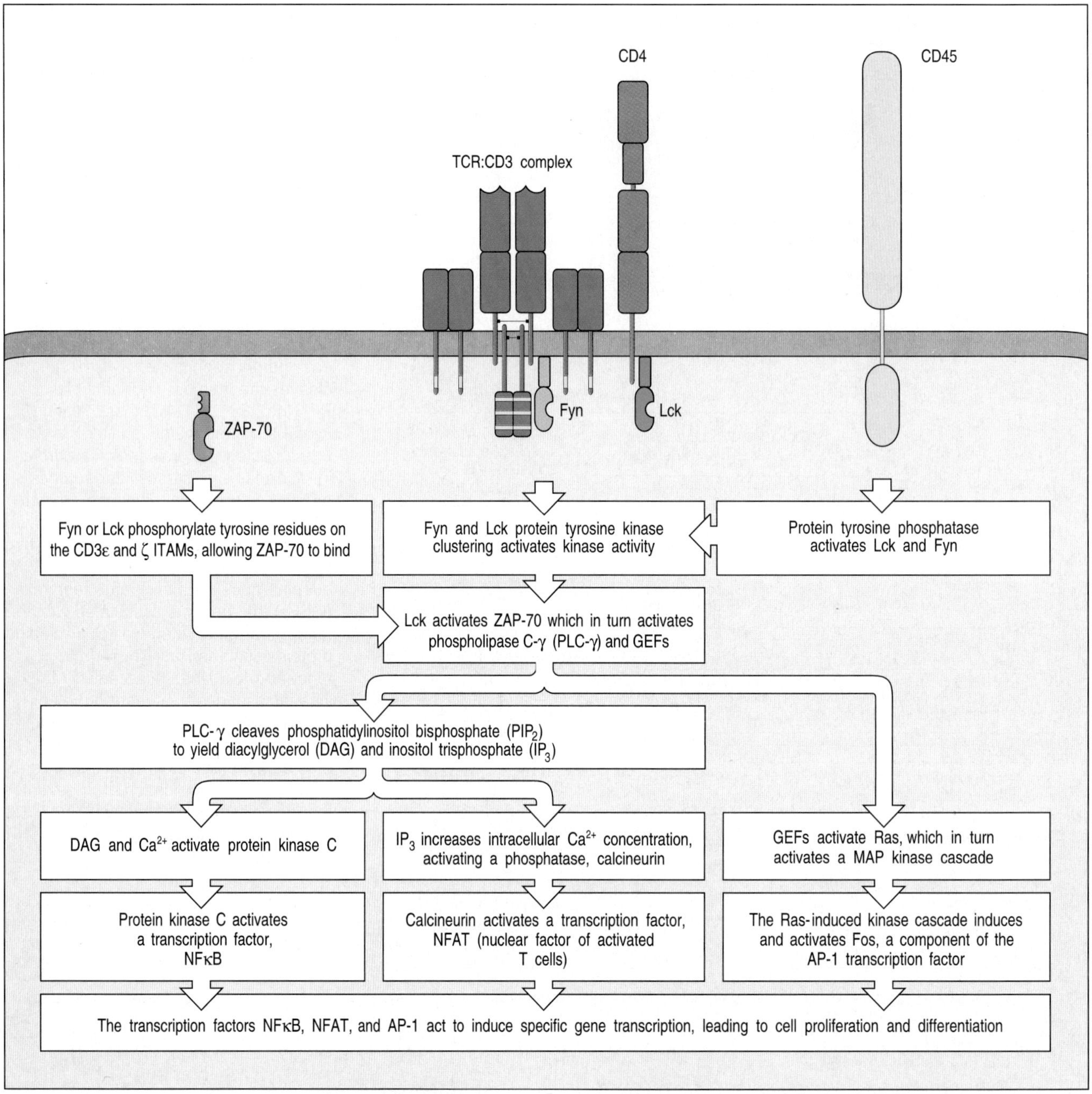

Fig. 5.15 Simplified outline of the intracellular signaling pathways initiated by the T-cell receptor complex and its co-receptor. The T-cell receptor complex and co-receptor (in this example the CD4 molecule) are associated with Src-family protein kinases, Fyn and Lck respectively. It is thought that binding of a peptide:MHC ligand to the T-cell receptor and co-receptor brings together CD4, the T-cell receptor complex, and CD45. This allows the CD45 tyrosine phosphatase to remove inhibitory phosphate groups and thereby activate Lck and Fyn. Events occurring after the activation of these tyrosine kinases are not known in detail. Phosphorylation of the ζ chains enables them to bind the cytosolic tyrosine kinase ZAP-70. In freshly isolated T cells, inactive ZAP-70 is already bound to the ζ chain, so this step is thought to occur before stimulation with specific antigen. The subsequent activation of bound ZAP-70 by phosphorylation leads to three important signaling pathways, two of which are initiated by the phosphorylation and activation of PLC-γ, which then cleaves PIP_2 into DAG and IP_3. Activation of protein kinase C by DAG leads to activation of the transcription factor NFκB. The sudden increase in intracellular free Ca^{2+} as a result of IP_3 action activates a cytoplasmic phosphatase, calcineurin, which enables the transcription factor NFAT to translocate from the cytoplasm to the nucleus. Full transcriptional activity of NFAT also requires a member of the AP-1 family of transcription factors; these are dimers of members of the Fos and Jun families of transcription regulators. A third signaling pathway initiated by activated ZAP-70 is the activation of Ras and the subsequent activation of a MAP kinase cascade. This culminates in the activation of Fos and hence of the AP-1 transcription factors. Together, NFκB, NFAT, and AP-1 act on the T-cell chromosomes, initiating new gene transcription that results in the differentiation, proliferation and effector actions of T cells. As with Fig. 5.14, this model is a simplified version of the main events only.

5-11 Antigen receptor ligation leads ultimately to the induction of new gene synthesis by activating transcription factors.

The ultimate response of lymphocytes to extracellular signals is the induction of new gene expression, which is achieved through the activation of transcription factors—proteins that control the initiation of transcription by binding to regulatory sites in the DNA. The transcription factors involved in lymphocyte responses to antigen are activated as a consequence of phosphorylation by serine/threonine kinases called **MAP kinases**. These kinases are themselves activated by phosphorylation; in the inactive, non-phosphorylated state they are resident in the cytoplasm but when activated by phosphorylation they translocate into the nucleus.

Transcription factors are activated by MAP kinases by the general pathway shown in Fig. 5.16. In lymphocytes, the MAP kinases that are thought to activate transcription factors in response to antigen receptor ligation are called Erk1 (for extracellular-regulated kinase-1) and Erk2. MAP kinases are unusual in that their full activation requires phosphorylation on both a tyrosine and a threonine residue, which are separated in the protein by a single amino acid. This can done only by MAP kinase kinases, enzymes with dual specificity for tyrosine and serine/threonine residues. In the context of antigen receptor signaling, the MAP kinase kinases that activate Erk1 and Erk2 are called Mek1 and Mek2. MAP kinase kinases are themselves activated by phosphorylation by a MAP kinase kinase kinase, which is the first kinase in the cascade. It is thought that in lymphocytes this MAP kinase kinase kinase is the serine/threonine kinase **Raf** (Fig. 5.17, left panel) and that the antigen receptor activates this cascade through the activation of Raf by the GTP-bound form of Ras.

The activation of transcription factors through activation of a MAP kinase cascade is a critical part of many cell signaling pathways. How the signal is individualized for different stimuli is not yet known, although the existence of variant components at each level of the pathway may allow this to occur. For

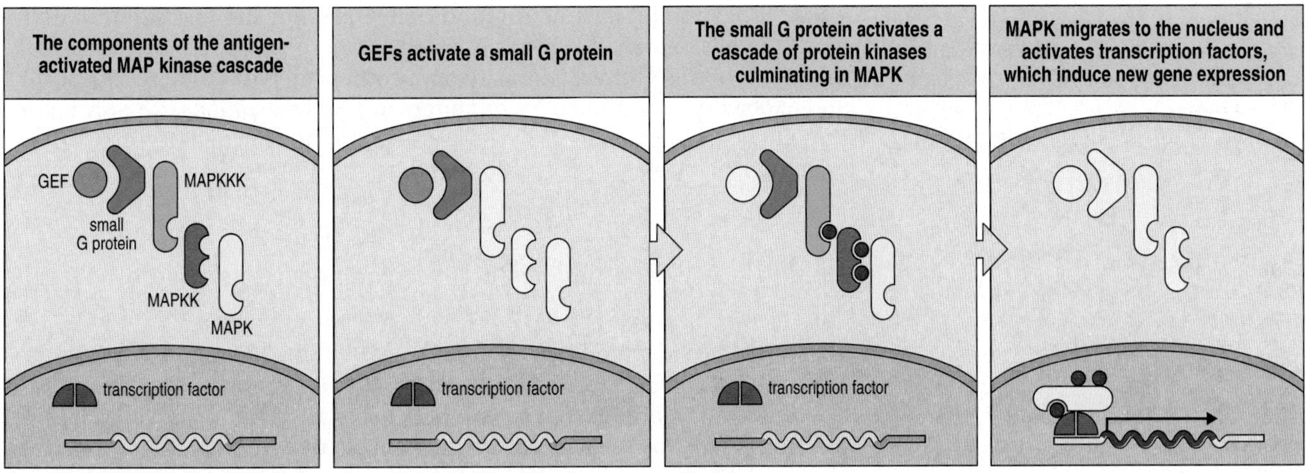

Fig. 5.16 MAP kinase cascades activate transcription factors. All MAP kinase cascades share the same general features. They are initiated by a small G protein, which is switched from an inactive state to an active state by a guanine-nucleotide exchange factor (GEF). The small G protein activates the first enzyme of the cascade, a protein kinase called a MAP kinase kinase kinase (MAPKKK). As expected from its name, the MAP kinase kinase kinase phosphorylates a second enzyme, MAP kinase kinase (MAPKK), which in its turn phosphorylates and activates the mitogen-activated protein kinase (MAP kinase, MAPK) on two sites, a tyrosine and a threonine separated by a single amino acid. This MAP kinase, when doubly phosphorylated, is both activated as a kinase and enabled to translocate from the cytosol into the nucleus, where it can phosphorylate and activate nuclear transcription factors.

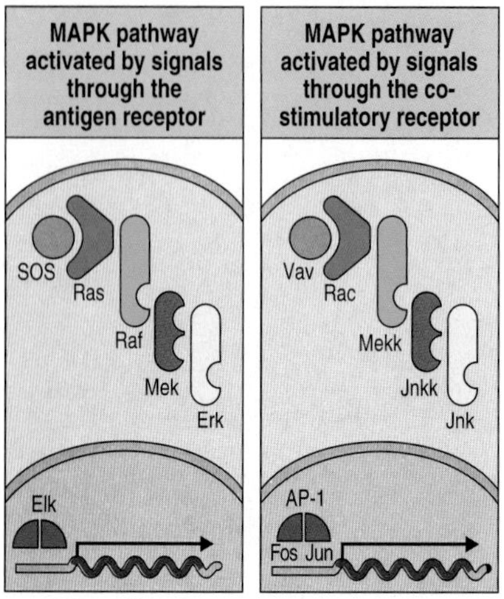

Fig. 5.17 Initiation of MAP kinase cascades by guanine-nucleotide exchange factors is involved in both antigen receptor and co-stimulatory receptor signaling. Signaling through the antigen receptors of both B and T cells (left panel) leads to the activation of the guanine-nucleotide exchange factor SOS, which activates the small G protein Ras. Ras activates the MAP kinase cascade, where Raf, Mek, and Erk activate each other in turn. Erk finally phosphorylates and activates the transcription factor Elk, which enters the nucleus to initiate new gene transcription, particularly of the Fos gene. Signaling through the B-cell co-receptor or through the co-stimulatory receptor CD28 on T cells leads to the recruitment and activation of a separate guanine-nucleotide exchange factor, Vav. Activation of Vav initiates a second MAP kinase cascade as Vav activates the small G protein Rac, which initiates the sequential activation of Mekk, Jnkk, and Jnk. Finally, Jnk (which stands for Jun N-terminal kinase) phosphorylates and activates Jun, which, together with Fos, forms the AP-1 transcription factor.

example, the MAP kinase cascade triggered by signals from the B-cell co-receptor through Vav and Rac uses different kinases from those in the antigen receptor pathway and activates different transcription factors (see Fig. 5.17, right panel). The MAP kinase cascade activated by signals through the antigen receptor activates the transcription factor Elk, which in turn upregulates the synthesis of the transcription factor Fos, whereas the MAP kinase pathway activated through the B-cell co-receptor, or the T-cell co-stimulatory molecule CD28, activates the transcription factor Jun. These pathways can combine in their effects because heterodimers of Fos and Jun form AP-1 transcription factors, which regulate the expression of many genes involved in cell growth.

A transcription factor that is activated by a different pathway is **NFAT** (**nuclear factor of activated T cells**). This is something of a misnomer as NFAT transcription factors are now known to be found not only in T cells but also in B cells, NK cells, mast cells, and monocytes, as well as in some non-hematopoietic cells. NFAT contains an amino acid sequence called a nuclear localization signal, which enables it to be translocated into the nucleus. In unstimulated cells, this sequence is rendered inoperative by phosphorylation at serine/threonine residues; hence NFAT is retained in the cytosol after its synthesis. NFAT is released from the cytosol by the action of the enzyme **calcineurin**, a serine/threonine protein phosphatase. Calcineurin is itself activated by the increase in intracellular free Ca^{2+} that accompanies lymphocyte activation (see Fig. 5.15). Once NFAT has been dephosphorylated by calcineurin, it enters the nucleus, where it acts as a transcriptional regulatory protein in combination with AP-1 transcription factors, which, as we have seen above, are dimers of Fos and Jun. In lymphocytes, NFAT interacts with Fos:Jun heterodimers that have been activated by the MAP kinase known as Jnk (or 'Junk', as it is commonly called), which is activated in T cells by a similar pathway to that shown in Fig. 5.17 (right panel). This illustrates how complex the process of signal transduction can be, and how proteins that regulate transcription can integrate signals that come from different pathways.

The importance of NFAT in T-cell activation is illustrated by the effects of the selective inhibitors of NFAT called cyclosporin A and FK506 (tacrolimus). These drugs inhibit calcineurin and hence prevent the formation of active NFAT. T cells express low levels of calcineurin, so they are more sensitive to inhibition of this pathway than are many other cell types. Both cyclosporin A and FK506 thus act as effective inhibitors of T-cell activation with only limited side effects (see Chapter 14). These drugs are used widely to prevent graft rejection, which they do by inhibiting the activation of alloreactive T cells.

5-12 Not all ligands for the T-cell receptor produce a similar response.

So far, we have assumed that all peptide:MHC complexes that are recognized by a given T-cell receptor will activate the T cell equally—that is, that the T-cell receptor is a binary switch with only two settings, 'on' and 'off.' Peptides that trigger the 'on' setting are called **agonist peptides** by analogy with the agonist drugs that activate other receptors. However, experiments originally designed to explore the structural basis for antigen recognition by T cells unexpectedly showed that recognition does not necessarily lead to activation. Indeed, some peptide:MHC complexes that do not themselves activate a given T cell can actually inhibit its response to the agonist peptide:MHC complex. These peptides are usually called **antagonist peptides**, as they antagonize the action of the agonist peptide. Other peptides that can trigger only a part of the program activated by the agonist peptide are called **altered** peptide ligands or partial agonists. Such ligands can, for instance, induce the lymphocyte to secrete cytokines but not to proliferate.

The extent to which antagonist peptides and altered peptide ligands influence physiological immune responses is not known, although they might contribute to the persistence of some viral infections. For example, mutant peptides of epitopes on cells infected with human immunodeficiency virus (HIV) can inhibit the activation of CD8 T cells specific for the original agonist epitope at a 1:100 ratio of mutant to agonist; this sort of effect could allow cells infected with a mutant virus, which arise commonly during the course of HIV infection, to survive in the presence of virus-specific cytotoxic cells. Differential signaling by variants of agonist peptides might also be important for the development of T cells in the thymus, where the immature cells are selected for their potential to recognize foreign peptide antigens in the context of self MHC molecules. This process, which we shall discuss in Chapter 7, would seem to require an immature T cell to be signaled by a self-peptide:self-MHC complex that is related to, but not identical with, the foreign peptide:self-MHC complex that the mature T cell will recognize.

The basis of the incomplete activating signal delivered by altered peptide ligands is also unknown, but it has been shown that recognition of these ligands leads to altered phosphorylation of CD3ε and ζ chains, and to the recruitment of inactive ZAP-70 tyrosine kinase to the T-cell receptor. Crosslinking the T-cell receptor alone, without any co-receptor engagement, can also generate the same partial phosphorylation events within the cell. Thus, the incomplete signal generated by altered peptide ligands might reflect a failure to recruit the co-receptors or a failure of the T-cell receptor to interact with the co-receptor productively. Also, because the affinity of the T-cell receptor for the altered peptide ligand:MHC complex is lower than that for the activating complex, the T-cell receptor might dissociate too quickly from its ligand for a full activating signal to be delivered. Another possibility, for which some evidence exists, is that conformational changes in the T-cell receptor can contribute to signaling and that altered peptide ligands fail to trigger these conformational changes.

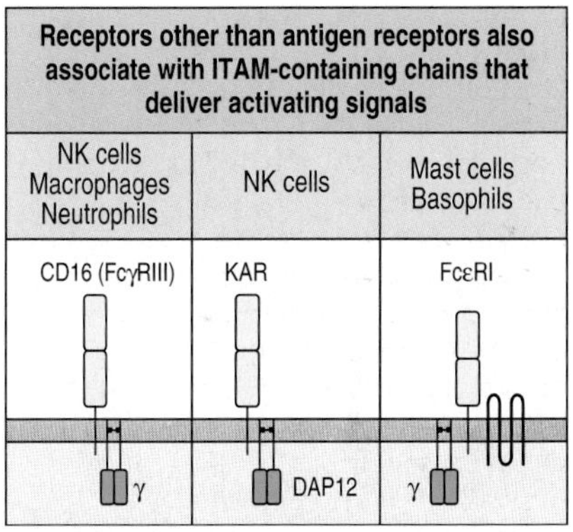

Fig. 5.18 Receptors that pair with ITAM-containing chains can deliver activating signals. Cells other than B and T cells have receptors that pair with ITAM-containing chains, which are phosphorylated when the receptor is crosslinked. These receptors deliver activating signals. The Fcγ receptor III (CD16) is found on NK cells, macrophages, and neutrophils. Binding of IgG to this receptor activates the killing function of the NK cell, leading to the process known as antibody-dependent cell-mediated cytotoxicity (ADCC). The killer activatory receptor (KAR) is also found on NK cells. The Fcε receptor is found on mast cells and basophils. It binds to IgE antibodies with very high affinity. When antigen subsequently binds to the IgE, the mast cell is triggered to release granules containing inflammatory mediators. The γ chain associated with the Fc receptors and the DAP12 associated with the KAR contain one or more ITAMs per chain and are present as homodimers.

5-13 Other receptors on leukocytes also use ITAMs to signal activation.

Although this chapter is focused on lymphocyte antigen receptors, which are the signaling machines that regulate adaptive immunity, other receptors on immune system cells also use the ITAM motif to transduce activating signals (Fig. 5.18). One example is FcγRIII (CD16) on NK cells; this is a receptor for IgG that is involved in antibody-dependent cell-mediated cytotoxicity (ADCC), which we shall learn about in Chapter 9. FcγRIII is associated with either an ITAM-containing ζ chain like those found in the T-cell receptor complex or with a second member of the same protein family known as the γ chain. The γ chain is also associated with another type of receptor—the Fcε receptor I (FcεRI) on mast cells—that is involved in allergic responses (to be discussed in Chapter 12).

A third member of the ITAM-containing ζ-chain family has recently been discovered. This protein, called DAP12, is present as a ζ-like homodimer in the membranes of NK cells, but has only a single ITAM on each chain of the homodimer. DAP12 associates with receptors called **killer activatory receptors** (**KARs**). When these receptors bind to MHC molecules on target cells they associate with DAP12, which provides a signal for the release of the cytotoxic granules by which NK cells kill their targets.

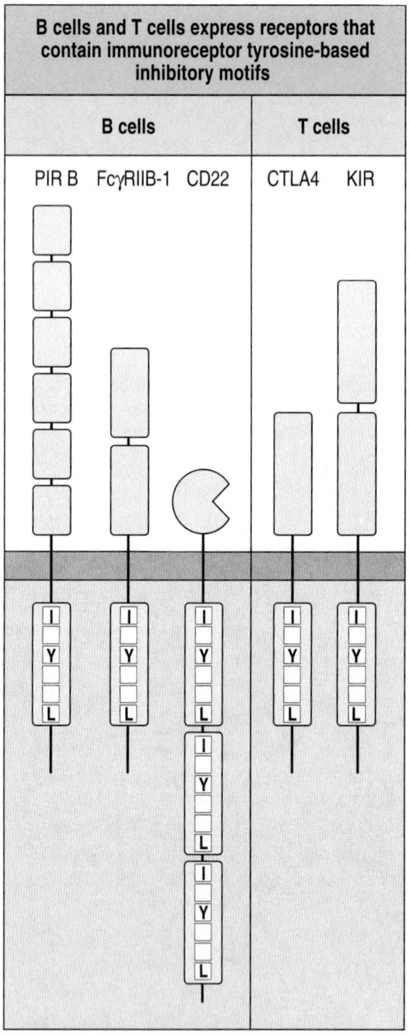

Fig. 5.19 Some lymphocyte cell-surface receptors contain motifs involved in downregulating activation. Several receptors that transduce signals that inhibit lymphocyte activation contain ITIMs (immunoreceptor tyrosine-based inhibitory motifs) in their cytoplasmic tails. These bind to various phosphatases that, when activated, inhibit signals derived from ITAM-containing receptors. Analysis of the mode of action of these receptors is in its early stages and little is known about how they inhibit immune responses activated by ITAM-derived signals.

5-14 Antigen receptor signaling can be inhibited by receptors associated with ITIMs.

Both B and T cells receive signals that can counteract and modify the activation signals delivered through antigen receptors and co-receptors. These inhibitory signals usually block the response by raising the threshold for signal transduction to occur. Most of these modifying signals are received through receptors that bear a distinct motif called an **immunoreceptor tyrosine-based inhibitory motif** (**ITIM**) in their cytoplasmic tails. In this motif, a large hydrophobic residue such as isoleucine or valine occurs two residues upstream of a tyrosine that is followed by two amino acids and a leucine to give the amino-acid sequence ...[I/V]XYXXL... (Fig. 5.19).

The ITIM motif is found in several receptors that modulate activation signals in lymphocytes. It functions by recruiting one or other of the inhibitory phosphatases SHP-1 and SHIP. These phosphatases carry an SH2 domain that preferentially binds the phosphorylated tyrosines in the ITIM. SHP-1 is a protein tyrosine phosphatase and removes the phosphate groups added by tyrosine kinases. SHIP is an inositol phosphatase and removes the 5′ phosphate from phosphatidylinositol trisphosphate (PI-3,4,5-P); the exact mechanism by which SHIP regulates signaling is not known, but it is thought to inhibit the activation of PLC-γ and thus the production of DAG and IP_3.

Receptors that bind the phosphatases SHP-1 and SHIP are involved in B-cell and T-cell inactivation. For instance, it has long been known that the activation of naive B cells can be inhibited by co-ligation of the B-cell receptor with the B-cell Fc receptor for IgG known as FcγRIIB-1. This method of inhibiting unwanted antibody responses by using IgG antibodies has been employed for many years: the activation of Rh-antibody-producing B cells in Rhesus-negative (Rh^-) mothers carrying a Rh^+ fetus can be prevented by giving the mother injections of anti-Rh IgG, as will be discussed in Chapter 10. However, only recently was the FcγRIIB-1 ITIM motif defined and shown to draw SHIP into a complex with the B-cell receptor that inhibited B-cell receptor signaling under these conditions.

Several other receptors on B and T cells contain the ITIM motif and inhibit cell activation when they are ligated along with the antigen receptors. Examples are CD22, a B-cell transmembrane protein that inhibits B-cell signaling, and CTLA4, a T-cell transmembrane protein that is induced by activation and then has a critical role in regulating T-cell signaling; it binds to the tyrosine phosphatase SHP-2. Thus, the ITIM motif clearly has the ability to modify signals originating from B-cell or T-cell receptors, even to the extent of completely negating them. In NK cells, killer inhibitory receptors (KIRs) containing ITIMs are involved in inhibiting the release of cytotoxic granules when the NK cell recognizes a healthy uninfected cell, as will be discussed in Chapter 10.

Summary.

Lymphocyte antigen receptors are multiprotein complexes made up of variable antigen-binding chains and of invariant chains that transmit the signal that antigen has bound. The cytoplasmic tails of the invariant chains contain amino-acid motifs called ITAMs, each possessing two tyrosine residues, that are targeted by receptor-associated protein tyrosine kinases of the Src family on receptor aggregation. The B-cell receptor is associated with four such ITAMs, whereas the T-cell receptor is associated with ten, the larger number allowing the T-cell receptor a greater flexibility in signaling. Once the ITAMs have been phosphorylated by a Src-family kinase, the Syk-family kinases bind and become activated. Receptor phosphorylation initiates several signaling pathways, including those through phospholipase Cγ and small G proteins, which converge on the nucleus and result in new patterns of gene expression. The small G proteins activate a cascade of serine/threonine protein kinases known as the MAP kinase cascade, which leads to the phosphorylation and activation of transcription factors. Signaling through the antigen receptors can be enhanced by signaling through the B-cell co-receptor on B cells and through the co-stimulatory molecule CD28 on T cells. Activating signals can be modulated or inhibited by signals from other, inhibitory, receptors that are associated with chains containing a different motif, ITIM, in their cytoplasmic tails. This provides a mechanism for turning cells on or off to different degrees, allowing the modulation of the adaptive immune response.

Other signals received by lymphocytes use various signal transduction pathways.

Lymphocytes are normally studied in terms of their responsiveness to antigen. However, they bear numerous other receptors that make them aware of events occurring both in their immediate vicinity and at distant sites. Among these are the presence of infection and of cytokines produced by the cell itself or reaching it from elsewhere. In the absence of infection, lymphocyte populations are also kept remarkably constant in numbers. This **homeostasis** is achieved by a host of extracellular factors that interact with receptors on lymphocytes, the most important of which is the antigen receptor. Other receptors and ligands that come into play include Fas and its ligand, various cytokine receptors and their ligands, and cellular factors such as Bcl-2 that modulate survival.

5-15 Microbes and their products release NFκB from its site in the cytosol through an ancient pathway of host defense against infection.

Lipopolysaccharide, a component of the cell wall of Gram-negative bacteria, and many other microbes and their products can activate the transcription factor known as NFκB in lymphocytes. These microbial substances are believed to trigger a protease cascade that generates a ligand for a receptor called Toll. Toll signals for NFκB activation through a pathway that seems to operate in most multicellular organisms. The cytoplasmic domain of the Toll receptor is known as a TIR domain because it is also found in the cytoplasmic tail of the receptor for the cytokine interleukin-1 (IL-1R). Ligand binding to the extracellular portion of the Toll receptor induces the TIR domain to bind and activate an adaptor protein known as MyD88. The IL-1 receptor also binds and activates MyD88 on binding IL-1 and, because both the ligand and the signaling pathway for the IL-1 receptor are more fully worked out, we illustrate the pathway in Fig. 5.20 and describe it below; the pathway from mammalian Toll will probably be very similar.

At one end, the MyD88 protein also has a TIR domain through which it interacts with the TIR domain of Toll or IL-1R. At the other end of the MyD88 protein is a domain that mediates protein–protein interactions.

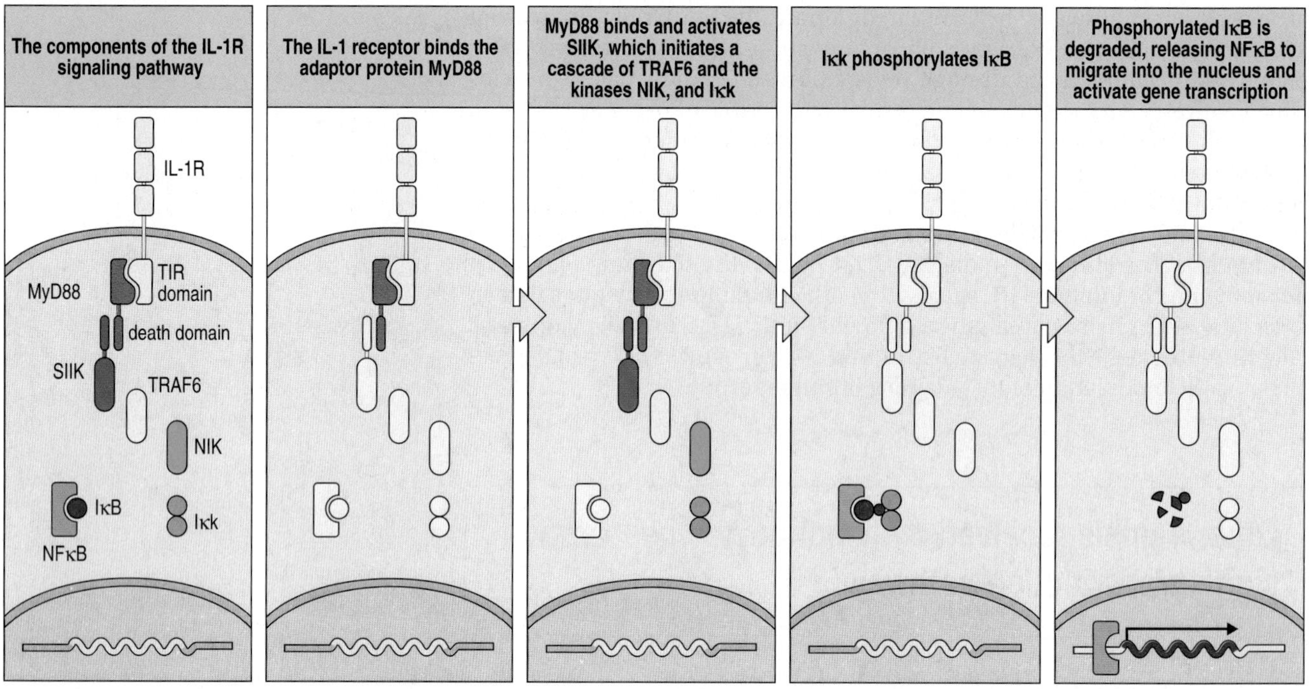
The components of the IL-1R signaling pathway
IL-1R
MyD88
TIR domain
death domain
SIIK
TRAF6
NIK
IκB
Iκk
NFκB
The IL-1 receptor binds the adaptor protein MyD88
MyD88 binds and activates SIIK, which initiates a cascade of TRAF6 and the kinases NIK, and Iκk
Iκk phosphorylates IκB
Phosphorylated IκB is degraded, releasing NFκB to migrate into the nucleus and activate gene transcription

Fig. 5.20 The transcription factor NFκB is activated by signals from receptors with the TIR cytoplasmic domain. The cytoplasmic domain of the IL-1 receptor (IL-1R) is homologous to that of the receptor protein Toll and is called a TIR (Toll/IL-1R) domain. This domain can interact with other TIR domains, including that found in the adaptor protein MyD88. MyD88 also contains a death domain through which it interacts with the death domain in a serine/threonine innate immunity kinase (SIIK), which activates TRAF6. This activates the NFκB kinase (NIK), which in turn activates Iκ kinase-α and Iκ kinase-β, which form the dimer Iκk. Iκk phosphorylates IκB, a cytosolic protein that is retaining the transcription factor NFκB in the cytoplasm. Once phosphorylated, IκB dissociates from its complex with NFκB, allowing NFκB to enter the nucleus and activate genes involved in host defense against infection.

This is known as a **death domain** because it was first found in proteins involved in programmed cell death. Through this death domain, the bound adaptor protein interacts with another death domain on a serine/threonine innate immunity kinase (SIIK) known as IRAK, or the IL-1R-associated kinase. This inititates a kinase activation cascade through which two kinases known as Iκkα and Iκkβ are activated to form a dimer (Iκk) that phosphorylates an inhibitory protein known as IκB. This protein is bound in a complex in the cytosol with the transcription factor NFκB and inhibits its action by retaining it there. When IκB is phosphorylated, it dissociates from the complex and is rapidly degraded by proteasomes. After removal of IκB, NFκB enters the nucleus and binds to various promoters, activating genes that contribute to adaptive immunity and the secretion of pro-inflammatory cytokines. Also activated is the gene for IκB itself, which is rapidly synthesized and inactivates the NFκB signal.

TIR domains and the signaling pathway that they generate were originally discovered in the fruit fly *Drosophila melanogaster* through the receptor protein Toll, which is involved in determining dorso-ventral body pattern during fly embryogenesis. Recently, however, the Toll protein was also shown to participate in defense against infection in adult flies. A close homolog of Toll has been identified in mammals, and similar proteins are also used by plants in their defense against viruses, indicating that the Toll pathway is an ancient signaling pathway that is used in innate defences in most multicellular organisms.

5-16 Cytokines signal lymphocytes by binding to cytokine receptors and triggering Janus kinases to phosphorylate and activate STAT proteins.

T cells secrete and receive a multitude of signals in the form of cytokines, small proteins that affect cellular behavior in a variety of ways. Cytokines can act on the cells that produce them (autocrine action), on other cells in the immediate vicinity (paracrine action), or on cells at a distance (endocrine action) after being carried in blood or tissue fluids. Cytokines is a general name for a collection of small proteins (of ~20,000 daltons) which each act on a specific receptor; many of these cytokine receptors use a particularly rapid and direct signaling pathway.

Cytokine binding to such receptors activates receptor-associated tyrosine kinases of the **Janus kinase** family (**JAKs**), so-called because they have two symmetrical kinase-like domains, and thus resemble the two-headed mythical Roman god Janus. These kinases then phosphorylate cytosolic proteins called **signal transducers and activators of transcription** (**STATs**). Phosphorylation of STAT proteins leads to their homo- and heterodimerization; STAT dimers can then translocate to the nucleus, where they activate various genes (Fig. 5.21). The proteins encoded by these genes contribute to the growth and differentiation of particular subsets of lymphocytes.

In this pathway, gene transcription is activated very soon after the cytokine binds to its receptor, and specificity of signaling in response to different cytokines is achieved by using different combinations of JAKs and STATs. This

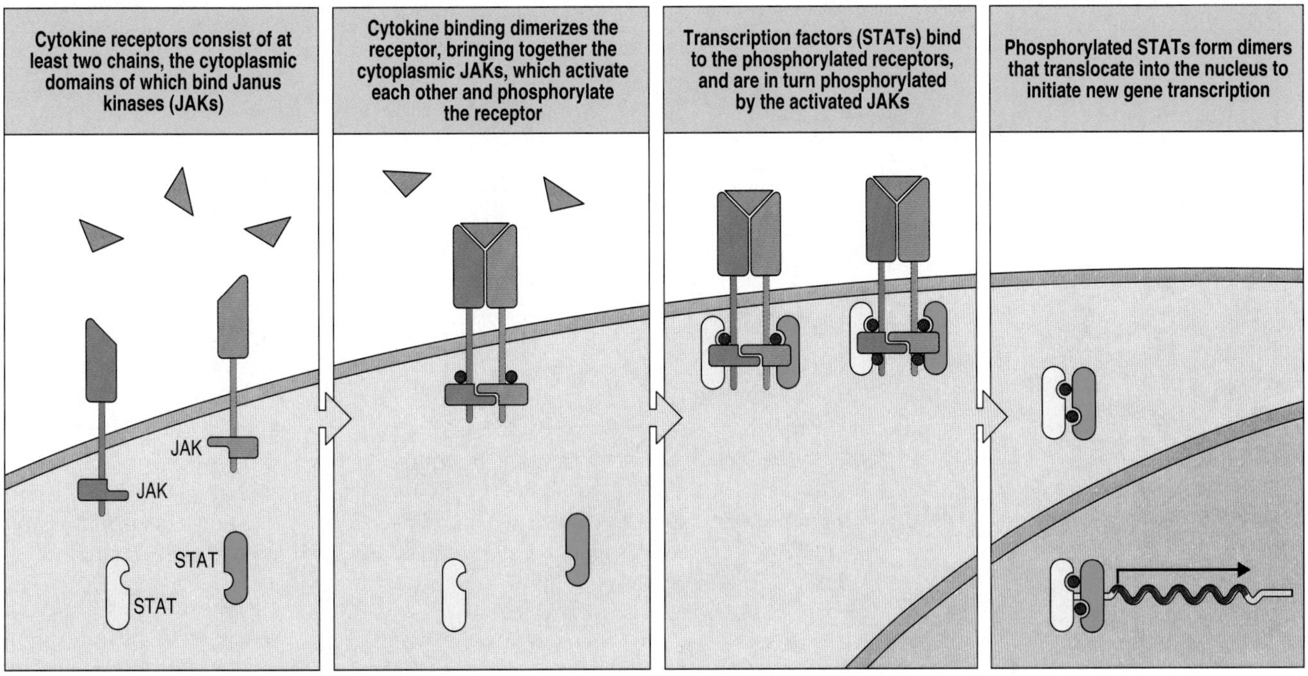

Cytokine receptors consist of at least two chains, the cytoplasmic domains of which bind Janus kinases (JAKs)
JAK
JAK
STAT
STAT
Cytokine binding dimerizes the receptor, bringing together the cytoplasmic JAKs, which activate each other and phosphorylate the receptor
Transcription factors (STATs) bind to the phosphorylated receptors, and are in turn phosphorylated by the activated JAKs
Phosphorylated STATs form dimers that translocate into the nucleus to initiate new gene transcription

Fig. 5.21 Many cytokine receptors signal by a rapid pathway using receptor-associated kinases to activate specific transcription factors. Many cytokines act via receptors that are associated with cytoplasmic Janus kinases (JAKs). The receptor consists of at least two chains, each associated with a specific JAK (left panel). Ligand binding and dimerization of the receptor chains brings together the JAKs, which can transactivate each other, subsequently phosphorylating tyrosines in the receptor tails (second panel). Members of the STAT (signal transducer and activator of transcription) family of proteins bind to the phosphorylated receptors and are themselves phosphorylated by the JAKs (third panel). On phosphorylation, STAT proteins dimerize by binding phosphotyrosine residues in SH2 pockets and go rapidly to the nucleus, where they bind to and activate transcription of a variety of genes important for adaptive immunity.

signaling pathway is used by most of the cytokines that are released by T cells in response to antigen. Although cytokines are not in themselves antigen-specific, their effects can be targeted in an antigen-specific manner by their directed release in antigen-specific cell–cell interactions and their selective action on the cell that triggers their production, as we shall see in Chapter 8.

5-17 Programmed cell death of activated lymphocytes is triggered mainly through the receptor Fas.

When antigen-specific lymphocytes are activated through their antigen receptors, they first undergo blast transformation and begin to increase their numbers exponentially by cell division. This clonal expansion can continue for up to seven or eight days so that lymphocytes specific for the infecting antigen increase vastly and can come to predominate in the population. In the response to certain viruses, nearly 50% of the CD8 T cells at the peak of the response are specific for a single virus-derived peptide:MHC class I complex. After clonal expansion, the activated T cells undergo their final differentiation into effector cells; these remove the pathogen from the body, which terminates the antigenic stimulus.

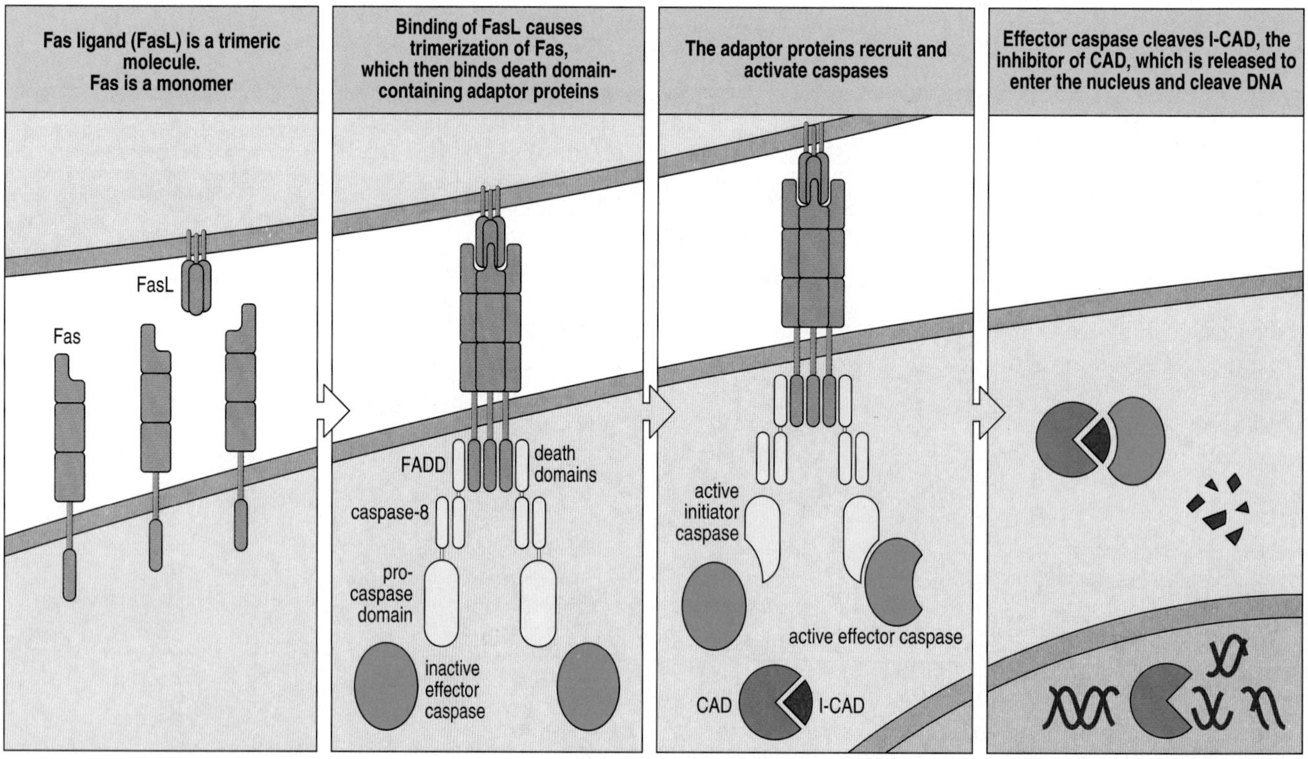

Fig. 5.22 Binding of Fas ligand to Fas initiates the process of apoptosis. The ligand, Fas ligand (FasL), recognized by Fas is a homotrimer, and when its binds it induces the trimerization of Fas, bringing death domains in the Fas cytoplasmic tails together. A number of adaptor proteins containing death domains bind to the death domains of Fas, in particular the protein FADD, which in turn interacts through a second death domain with the protease caspase-8. Clustered caspase-8 can transactivate, cleaving caspase-8 itself to release an active caspase domain that in turn can activate other caspases. The ensuing caspase cascade culminates in the activation of the caspase-activatable DNase (CAD), which is present in all cells in an inactive cytoplasmic form bound to an inhibitory protein called I-CAD. When I-CAD is broken down by caspases, CAD can enter the nucleus where it cleaves DNA into the 200 base pair fragments that are characteristic of apoptosis.

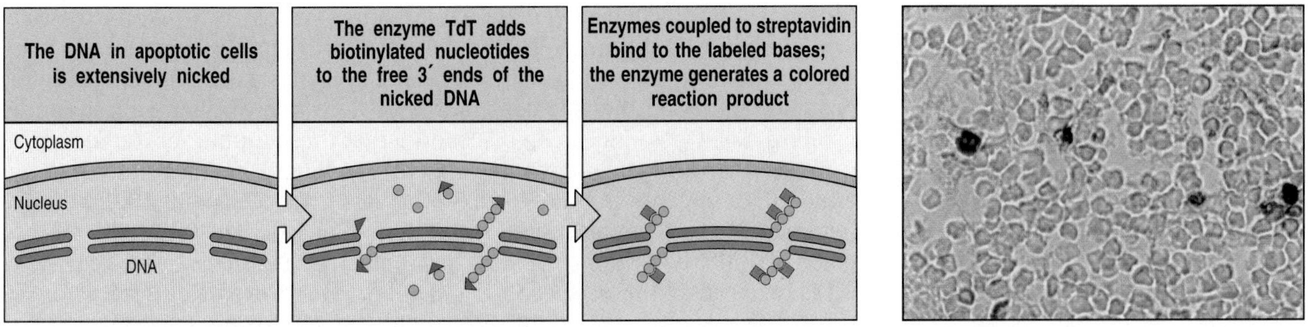

Fig. 5.23 Fragmented DNA can be labeled by terminal deoxynucleotidyl transferase (TdT) to reveal apoptotic cells. When cells undergo programmed cell death, or apoptosis, their DNA becomes fragmented (left panel). The enzyme TdT is able to add nucleotides to the ends of DNA fragments; most commonly in this assay, biotin-labeled nucleotides (usually dUTP) are added (second panel). The biotinylated DNA can be detected by using streptavidin, which binds to biotin, coupled to enzymes that convert a colorless substrate into a colored insoluble product (third panel). Cells stained in this way can be detected by light microscopy, as shown in the photograph of apoptotic cells (stained red) in the thymic cortex. Photograph courtesy of R Budd and J Russell.

When the infection has terminated, the activated effector T cells are no longer needed and cessation of the antigenic stimulus prompts them to undergo programmed cell death or apoptosis. Apoptosis can probably be induced by several mechanisms, but one that has been particularly well defined is the interaction of the receptor molecule **Fas** on T cells with its ligand **Fas ligand**. Fas ligand is a member of the tumor necrosis factor (TNF) family of membrane-associated cytokines, whereas Fas is a member of the TNF receptor family. Both Fas and its ligand are normally induced during the course of an adaptive immune response. TNF and its receptor TNFR-1 can act in a similar way to Fas ligand and Fas but their actions are far less significant.

All pathways inducing apoptosis lead to the activation of a series of cysteine proteases that cleave protein chains after aspartic acid residues and have therefore been called **caspases**. In the case of activated lymphocytes, apoptosis is initiated by stimulation of the receptors Fas or TNFR-1. The ligands for these receptors are in the form of trimers and when they bind, they induce trimerization of the receptors themselves (Fig. 5.22). The cytoplasmic tails of these receptors share a motif known as a death domain which, as we saw in Section 5-15, is a protein–protein interaction domain. The adaptor proteins that interact with the death domains in the cytosolic tails of Fas and TNFR-1 are called FADD and TRADD respectively. These in turn interact through a different region with the protein caspase-8 (also known as FLICE), whose carboxy-terminal domain is a pro-caspase (the inactive form of a caspase). Binding activates the enzymatic activity of caspase-8, leading to a protease cascade in which activated caspases cleave and activate a succession of downstream caspases. At the end of this pathway a caspase-activated DNase (CAD) enters the nucleus and cleaves DNA to produce the DNA fragments characteristic of an apoptotic cell.

Apoptotic cells can be detected by a procedure known as TUNEL staining. In this technique, the 3′ ends of the DNA fragments generated in apoptotic cells are labeled with biotin-coupled uridine by using the enzyme terminal deoxynucleotidyl transferase (TdT). The biotin label is then detected with enzyme tagged streptavidin, which binds to biotin. When the colorless substrate of the enzyme is added to a tissue section or cell culture, it is reacted upon to produce a colored precipitate only in cells that have undergone apoptosis (Fig. 5.23). This technique has revolutionized the detection of apoptotic cells.

Mutations in mice have been known for some time that cause an excessive accumulation of abnormal T cells that lack both co-receptor proteins and express the CD45 isoform usually expressed by B cells. Loss of the co-receptor proteins suggested that the cells had been activated but had subsequently failed to die. It has since been discovered that these mutations are recessive mutations in the genes encoding Fas or Fas ligand. Similar mutations in either of these genes in humans do not always cause a mutant phenotype in heterozygous individuals, but in some cases they lead to a massive accumulation of similarly abnormal T cells. The production of an effect in heterozygotes is likely to reflect the need for trimerization of Fas for efficient operation of the Fas–Fas ligand interaction, and the fact that if one of the members of the trimer is mutant, the trimer cannot transduce a signal.

5-18 Lymphocyte survival is maintained by a balance between death-promoting and death-inhibiting members of the Bcl-2 family of proteins.

Lymphocytes are being continually exposed to situations that could induce apoptosis, and to protect against inadvertent cell death they possess a separate set of proteins that inhibit programmed cell death. The first member of this family of proteins was again discovered as an oncogene. When tumors form in the B-cell lineage, they are frequently associated with chromosomal translocations in which the chromosomal DNA is broken and an active immunoglobulin locus is joined to a gene that affects cell growth, usually activating that gene in the process (see Chapter 6). By cloning the DNA breakpoints, one can isolate the gene that has been activated by the translocation. One such gene is *bcl-2*, which was isolated from the second B-cell lymphoma to have its breakpoint identified. *bcl-2* is homologous with the *Caenorhabditis elegans* gene *ced-9*, which is a cell death inhibitory gene. In cultured B cells and in transgenic mice, the expression of *bcl-2* protects against cell death. *bcl-2* is a member of a small family of closely related genes, some of which inhibit cell death whereas others promote it. These genes can be divided into

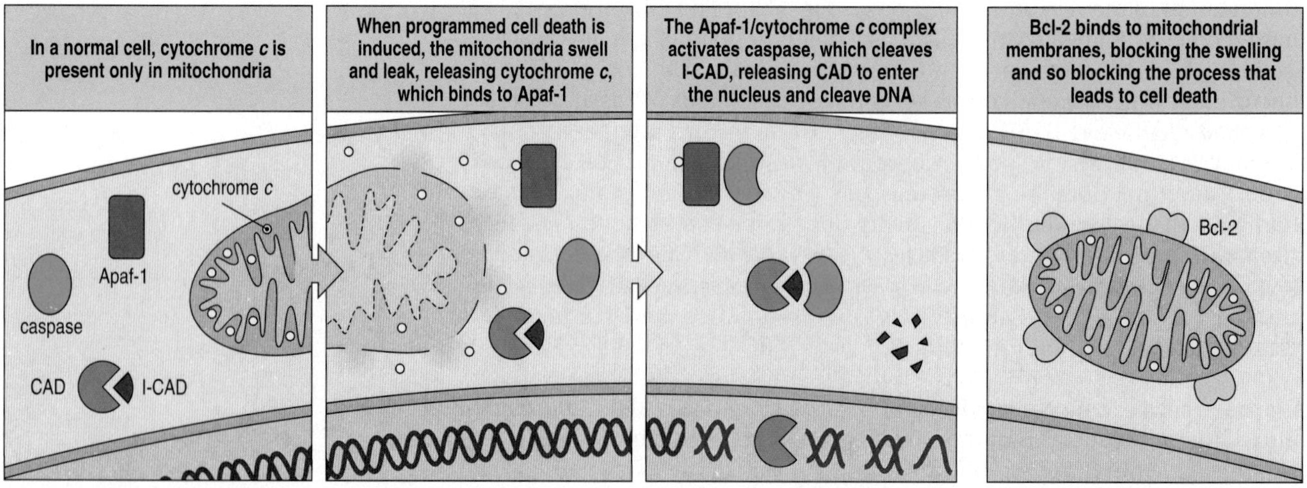

In a normal cell, cytochrome c is present only in mitochondria
When programmed cell death is induced, the mitochondria swell and leak, releasing cytochrome c, which binds to Apaf-1
The Apaf-1/cytochrome c complex activates caspase, which cleaves I-CAD, releasing CAD to enter the nucleus and cleave DNA
Bcl-2 binds to mitochondrial membranes, blocking the swelling and so blocking the process that leads to cell death
cytochrome c
Apaf-1
caspase
CAD
I-CAD
Bcl-2

Fig. 5.24 Bcl-2 inhibits the processes that lead to programmed cell death. In normal cells, cytochrome *c* is confined to the mitochondria (first panel). However, during apoptosis the mitochondria swell, allowing the cytochrome *c* to leak out into the cytosol (second panel). There it interacts with the protein Apaf-1, forming a cytochrome *c*:Apaf-1 complex that now can activate caspases. An activated caspase cleaves I-CAD (see Fig. 5.22), which leads to DNA fragmentation (third panel). Bcl-2 interacts with the mitochondrial outer membrane and blocks the mitochondrial swelling that leads to cytochrome *c* release (last panel).

death-inhibiting genes, such as *bcl-2* and *bcl-X_L*, and death-promoting genes such as *Bax* and *Bad*. The proteins encoded by these genes act as dimers, and as Bcl-2 and Bax protein can dimerize with each other to form heterodimers, the more abundant protein determines whether the cell lives or dies. One way in which Bcl-2 acts to prevent cell death is shown in Fig. 5.24.

The balance between death-promoting and death-inhibiting gene expression is critically important in lymphocytes, because lymphocyte populations are regulated so that a person will, in the absence of infection, maintain a constant level of T and B cells despite the production and death of many lymphocytes each day. The fate of individual lymphocytes is set by signals delivered mainly or entirely through the antigen-specific receptors, as we shall learn in Part III of this book, which deals with the production of the mature repertoire of receptors on B and T lymphocytes. In the last section of this chapter we shall look at the evidence for a continued role of antigen-receptor signaling in maintaining the survival of mature T and B cells.

5-19 Homeostasis of lymphocyte populations is maintained by signals that lymphocytes are continually receiving through their antigen receptors.

Most of what we know about signaling through the B-cell receptor and the T-cell receptor derives from observations in cultured B-cell and T-cell lines. In these cells, as we have seen in Section 5-9, the tyrosine kinases Syk and ZAP-70 are recruited as a consequence of receptor clustering and the subsequent phosphorylation of ITAMs in the receptor subunit tails. However, in T cells isolated directly from lymph nodes or the thymus, ZAP-70 is already bound to partly phosphorylated ζ chains but is not yet activated. It therefore seems likely that most circulating T cells have already received signals through their antigen receptors and have gone some way down the path to activation. Subsequent encounter with an antigen, by bringing the antigen receptor together with CD4 or CD8, then initiates the second step, in which Lck both fully phosphorylates the ζ chains and phosphorylates and activates ZAP-70.

How T cells are held poised along their activation pathway is not clear. However, the same state can be produced in both normal naive T cells and cultured T cells by exposure to altered peptide ligands (see Section 5-12). It therefore seems likely that the phosphorylated ITAMs and bound ZAP-70 seen in mature naive T cells *in vivo* reflect the receipt of signals from self MHC molecules bound to self peptides that behave like altered peptide ligands.

As we shall see in Chapter 7, the ability to interact with self-peptide:self-MHC ligands is a criterion for survival during T-cell development in the thymus. Developing T cells are subject to stringent testing once the α:β T-cell receptor is expressed. This selection process retains those cells whose receptors interact effectively with various self-peptide:self-MHC ligands (positive selection), and removes those T cells that either cannot participate in such interactions (death by neglect) or recognize a self-peptide:self-MHC complex so well that they could damage host cells if allowed to mature; such cells are removed by clonal deletion (negative selection). Those T cells that mature and emerge into the periphery have therefore been selected for their ability to recognize self-MHC:self-peptide complexes without being fully activated by them.

It is now known, from a number of studies, that T cells in the peripheral lymphoid tissues continue to receive signals through interactions with self-MHC:self-peptide complexes. These signals enable the cells to survive and are delivered most effectively by the cells that are most capable of T-cell activation, namely the dendritic cells. Thus, the T cells in mice transgenic for a T-cell receptor that is positively selected on a particular self-MHC: self-peptide complex can survive only if they receive similar signals from self-MHC:self-peptide complexes in the periphery. It seems as though a continuous dialog between dendritic cells presenting self peptides on self MHC molecules, and recirculating naive T cells that were positively selected by signals from the same self-peptide:self-MHC molecules during development, maintains the peripheral T-cell repertoire. This might account for the state of partial phosphorylation of the ζ chains discussed above. A requirement for repeated interaction with the dendritic cells resident in peripheral lymphoid tissue could also account for the tight regulation of the numbers of CD4 and CD8 cells.

In B cells, it is also clear that signaling through the antigen receptor determines cell survival from the time that it is first expressed on the cell surface. As we shall see in Chapter 6, autoreactive B cells are induced to die on binding antigen. However, the expression of a functional B-cell receptor at the cell surface is also essential for maturation and survival. The role of the B-cell receptor in signaling for survival in the periphery was recently demonstrated quite dramatically by using a conditional gene knock-out strategy (see Section 2-27). Mice were made transgenic for a rearranged immunoglobulin V_H gene flanked by *loxP* sites. This construct was 'knocked-in' to the site at which the rearranged V_H gene normally occurs in such a way that only the transgene could produce the heavy-chain V region. The animals were also made transgenic for the enzyme Cre recombinase, which can excise *loxP*-flanked genes; the transgene encoding the Cre recombinase was made inducible by interferon. After treatment with interferon, the induced Cre recombinase removed the transgenic rearranged V_H gene, preventing the production of a heavy chain. Most of the B cells in the treated animals lost their receptors; these receptor-negative B cells rapidly disappeared. Thus, the B-cell receptor is clearly required to keep re-circulating B cells alive, and must have a role in perceiving or transmitting survival signals to each B cell. Recent evidence suggests that receptor specificity can affect this process. However, the ligand or ligands responsible for signaling for B-cell survival are not yet known.

The identity of the ligands responsible for delivering survival signals to T and B cells through their antigen receptors remains an important question in immunology. There is also much to learn about the signaling process itself. Survival signals are likely to regulate the level of proteins in the Bcl-2 family because, as we discussed in Section 5-18, increased levels of Bcl-2 and Bcl-X_L promote the survival of lymphocytes, whereas increased levels of Bax and Bad have an opposing effect. As yet, however, a direct link between antigen-receptor signaling for survival and the regulation of the Bcl-2 family has not been shown.

Summary.

Many different signals govern lymphocyte behavior, only some of which are delivered via the antigen receptor. Lymphocyte development, activation, and longevity are clearly influenced by the antigen receptor, but these processes are also regulated by other extracellular signals. Other signals are delivered in a variety of ways. An ancient signaling pathway with a role in host defense leads rapidly from the IL-1 receptor or a similar receptor called Toll to initiate the detachment and degradation of the inhibitory protein IκB from the transcription factor NFκB, which can then enter the nucleus and activate the transcription of many genes. Many cytokines signal through an express pathway that links receptor-associated JAK kinases to preformed STAT proteins, which after phosphorylation dimerize through their SH2 domains and head for the nucleus. Activated lymphocytes are programmed to die when the Fas receptor that they express binds the Fas ligand. This transmits a death signal, which activates a protease cascade that triggers apoptosis. Lymphocyte apoptosis is inhibited by some members of the intracellular Bcl-2 family and promoted by others. Working out the complete picture of the signals processed by lymphocytes as they develop, circulate, respond to antigen, and die is an immense and daunting prospect.

Summary to Chapter 5.

In adaptive immune responses, signaling through the antigen receptors of naive mature lymphocytes induces the clonal expansion and differentiation of antigen-specific lymphocytes after engagement with foreign antigen. Lymphocyte antigen receptors belong to the general class of receptors that are associated with cytoplasmic protein tyrosine kinases. The antigen-binding chains of the receptors are associated on the cell surface with invariant chains that are responsible for generating an intracellular signal indicating that antigen has bound. These chains contain sequence motifs called ITAMs, which after phosphorylation by receptor-associated tyrosine kinases recruit intracellular signaling molecules to the activated receptor. The intracellular signaling pathway leading from the antigen receptors results in gene activation, new protein synthesis, and the stimulation of cell division. This pathway is subject to regulation at most of its steps; these control points form important checkpoints in the pathways leading to lymphocyte activation and the clonal expansion and differentiation of antigen-specific lymphocytes that occurs during an adaptive immune response. As well as enabling lymphocytes to respond to foreign antigens in an adaptive immune response, signals delivered through the antigen receptors are important in selecting lymphocytes for removal or survival during lymphocyte development in the primary lymphoid organs as well as survival later on in the periphery. What ligands are responsible for these effects remains a central question in immunology. Lymphocytes also carry receptors for many other extracellular signals, such as cytokines and Fas ligand. The latter, by interacting with the cell-surface receptor Fas on activated lymphocytes, induces apoptosis and is involved in controlling lymphocyte numbers and in removing activated lymphocytes once an infection has been cleared.

General references.

DeFranco, A.L., and Weiss, A.: **Lymphocyte activation and effector functions**. *Curr. Opin. Immunol.* 1998, **10**:243-367.

Healy, J.L., and Goodnow, C.C.: **Positive versus negative signaling by lymphocyte antigen receptors**. *Annu. Rev. Immunol.* 1998, **16**:645-670.

Weiss, A., and Littman, D.R.: **Signal transduction by lymphocyte antigen receptors.** *Cell* 1994, **76**:263-274.

Section references.

5-1 Binding of antigen leads to clustering of antigen receptors on lymphocytes.

Monks, C.R.F., Kupfer, H., Tamir, I., Barlow, A., and Kupfer, A.: **Selective modulation of protein kinase C-τ during T cell activation**. *Nature* 1997, **385**:83-86.

5-2 Clustering of antigen receptors leads to activation of intracellular signals.

Klemm, J.D., Schreiber, S.L., and Crabtree, G.R.: **Dimerization as a regulatory mechanism in signal transduction**. *Annu. Rev. Immunol.* 1998, **16**:569-592.

5-3 Phosphorylation of receptor cytoplasmic tails by tyrosine kinases concentrates intracellular signaling molecules around the receptors and activates them.

Reth, M.: **Antigen receptor tail clue**. *Nature* 1989, **338**:383-384.

5-4 Intracellular signaling components recruited to activated receptors transmit the signal onwards from the membrane and amplify it.

Pawson, T., and Scott, J.D.: **Signaling through scaffold, anchoring and adaptor proteins**. *Science* 1997, **278**:2075-2080.

Peterson, E.J., Clements, J.L., Fang, N., and Koretzky, G.A.: **Adaptor proteins in lymphocyte antigen receptor signaling**. *Curr. Opin. Immunol.* 1998, **10**:337-344.

5-5 Small G proteins activate a protein kinase cascade that transmits the signal to the nucleus.

Cantrell, D.: **T cell antigen receptor signal transduction pathways.** *Annu. Rev. Immunol.* 1996, **14**:259-274.

Henning, S.W., and Cantrell, D.A.: **GTPases in antigen receptor signaling.** *Curr. Opin. Immunol.* 1998, **10**:322-329.

5-6 The variable chains of lymphocyte antigen receptors are associated with invariant accessory chains that carry out the signaling function of the receptor.

Gold, M.R., and DeFranco, A.L.: **Biochemistry of B lymphocyte activation.** *Adv. Immunol.* 1994, **55**:221-295.

Malisson, B., and Malisson, M.: **Functions of TCR and preTCR subunits: lessons from gene ablation.** *Curr. Opin. Immunol.* 1996, **8**:394-401.

5-7 The ITAMs associated with the B-cell and T-cell receptors are phosphorylated by protein tyrosine kinases of the Src family.

Bolen, J.B., and Brugge, J.S.: **Lymphocyte protein tyrosine kinases: potential targets for drug discovery.** *Annu. Rev. Immunol.* 1997, **15**:371-404.

Trowbridge, I., and Thomas, M.L.: **CD45: and emerging role as a protein tyrosine phosphatase required for lymphocyte activation and development.** *Annu. Rev. Immunol.* 1994, **12**:85-116.

5-8 Antigen-receptor signaling requires co-receptors that bind the same ligand.

Fearon, D.G., and Carter, R.H.: **The CD19/CR2/TAPA-1 complex of B lymphocytes: linking natural to acquired immunity.** *Annu. Rev. Immunol.* 1995, **13**:127-149.

Janeway, C.A., Jr.: **The T cell receptor as a multicomponent signaling machine: CD4/CD8 co-receptors and CD45 in T cell activation.** *Annu. Rev. Immunol.* 1992, **10**:645-674.

5-9 The fully phosphorylated ITAMs bind the kinases Syk and ZAP-70 and enable them to be activated.

Kersh, E.N., Shaw, A.S., and Allen, P.M.: **Fidelity of T cell activation through multistep T cell receptor ζ phosphoylation.** *Science* 1998, **281**:572-575.

5-10 Downstream events are mediated by proteins that associate with the phosphorylated tyrosines and bind to and activate other proteins.

Agarwal, S., and Rao, A.: **Long-range transcriptional regulation of cytokine gene expression.** *Curr. Opin. Immunol.* 1998, **10**:345-352.

Zhang, W., Sloan-Lancaster, J., Kitchen, J., Trible, R.P., and Samelson, L.E.: **LAT: the ZAP-70 tyrosine kinase substrate that links T cell receptor to cellular activation**. *Cell* 1997, **92**:83-92.

5-11 Antigen-receptor ligation leads ultimately to induction of new gene synthesis by activating transcription factors.

Jacinto, E., Werlin, G., and Karin, M.: **Cooperation between Syk and Rac1 leads to synergistic JNK activation in T lymphocytes.** *Immunity* 1998, **8**:31-41.

Yoshida, H., Nisina, H., Takimoto, H., Marengere, L.E.M., Wakeham, A.C., Bouchard, D., Kong, Y.-Y., Ohteki, T., Shahinian, A., Bachmann, M., Ohashi, P.S., Penninger, J., Crabtree, G.R., and Mak, T.W.: **The transcription factor NF-ATc1 regulates lymphocyte proliferation and Th2 cytokine production**. *Immunity* 1998, **8**:115-124.

5-12 Not all ligands for the TCR produce a similar response.

Jameson, S.C., and Bevan, M.J.: **T cell receptor antagonists and partial agonists**. *Immunity* 1995, **2**:1-11.

Sloan-Lancaster, J., and Allen, P.M.: **Altered peptide ligand-induced partial T cell activation: molecular mechanisms and roles in T-cell biology.** *Annu. Rev. Immunol.* 1996, **14**:1-27.

5-13 Other receptors on leukocytes also use ITAMs to signal activation.

Lanier, L.L., Cortiss, B.C., Wu, J., Leong, C., and Phillips, J.H: **Immunoreceptor DAP12 bearing a tyrosine-based activation motif is involved in activating NK cells.** *Nature* 1998, **391**:703-707.

Torigoe, C., Inman, J.K., and Metzger, H.: **An unusual mechanism for ligand antagonism**. *Science* 1998, **281**:568-572.

5-14 Antigen-receptor signaling can be inhibited by receptors associated with ITIMs.

Coggeshall, K.M.: **Inhibitory signaling by B cell FcRγIIb**. *Curr. Opin. Immunol.* 1998, **10**:306-312.

Saito, T.: **Negative regulation of T cell activation**. *Curr. Opin. Immunol.* 1998, **10**:313-321.

Siminovitch, K.A., and Neel, B.G.: **Regulation of B cell signal transduction by SH2-containing phosphatases**. *Sem. Immunol.* 1998, **10**:329-347.

Thompson, C.B., and Allison, J.P.: **The emerging role of CTLA-4 as an immune attenuator.** *Immunity* 1997, **7**:445-450.

5-15 Microbes and their products release NFκB from its site in the cytosol through an ancient pathway of host defence against infection.

Gerondakis, S., Grumont, R., Rourke, I., and Grossman, M.: **The regulation and roles of Rel/NFκB transcription factors during lymphocyte activation**. *Curr. Opin. Immunol.* 1998, **10**:353-359.

Ghosh, S., May, M.J., and Kopp, E.B.: **NF-κB and Rel proteins: evolutionarily conserved mediators of immune responses**. *Annu. Rev. Immunol.* 1998, **16**:225-260.

Medzhitov, R., and Janeway, C.A., Jr.: **An ancient system of host defence**. *Curr. Opin. Immunol.* 1998, **10**:12-15.

5-16 Cytokines signal lymphocytes by binding to cytokine receptors and triggering JAK kinases to phosphorylate and activate STAT proteins.

Leonard, W.J., and O'Shea, J.J.: **Jaks and STATs: biological implications**. *Annu. Rev. Immunol.* 1998, **16**:293-322.

Liu, K.D., Gaffen, S.L., and Goldsmith, M.A.: **JAK/STAT signaling by cytokine receptors.** *Curr. Opin. Immunol.* 1998, **10**:271-278.

O'Shea, J.J.: **Jaks, STATs, cytokine signal transduction, and immunoregulation: are we there yet?** *Immunity* 1997, **7**:1-11.

5-17 Most programmed cell death of activated lymphocytes is mediated by the binding of Fas to its ligand.

Nagata, S.: **Apoptosis by death factor.** *Cell* 1997, **88**:355-365.

Wallach, D., Kovalenko, A.W., Varfolomeev, E.E., and Boldin, M.P.: **Death-inducing functions of ligands of the tumor necrosis factor family: a *Sanhedrin* verdict.** 1998, **10**:279-288.

5-18 Lymphocyte survival is maintained by a balance between death-promoting and death-inhibiting members of the Bcl-2 family of proteins.

Chao, D.T., and Korsmeyeer, S.J.: **BCL-2 family: regulators of cell death**. *Annu. Rev. Immunol.* 1998, **16**:395-419.

Vander Heiden, M.G., Chandel, N.S., Williamson, E.K., Schumacker, P.T., and Thompson, C.B.: **Bcl-XL regulates the membrane potential and volume homeostasis of mitochondria**. *Cell* 1997, **91**:627-637.

5-19 Homeostasis of lymphocyte populations is maintained by signals that lymphocytes are continually receiving through their antigen receptors.

Fu, C., Turck, C.W., Kurosaki, T, and Chan, A.C.: **BLNK: a central linker protein in B cell activation.** *Immunity* 1998, **9**:93-103.

Lam, K.P., Kühn, R., and Rajewsky, K.: ***In vivo* ablation of surface immunoglobulin on mature B cells by inducible gene targeting results in rapid cell death.** *Cell* 1997, **90**:1073-1083.

Tanchot, C., Lemonnier, F.A., Perarnau, B, Freitas, A.A., and Rocha, B.: **Differential requirements for suvival and proliferation of CD8 naive T cells**. *Science* 1997, **276**:2057-2062.

PART III

THE DEVELOPMENT OF LYMPHOCYTE REPERTOIRES

The Development of B Lymphocytes

6

In order that an individual may make antibodies against the wide range of pathogens encountered during a lifetime, B lymphocytes expressing a diverse repertoire of immunoglobulins must be generated continually. Each B cell expresses immunoglobulin of a single antigen specificity, which is determined early in its differentiation, when the genes encoding the immunoglobulin heavy and light chains are assembled from gene segments. In humans and mice, immunoglobulin diversity is generated largely by this process of gene rearrangement. As we saw in Chapter 3, many thousands of rearrangements are possible for both heavy- and light-chain genes; ongoing gene rearrangement in developing B cells is therefore continually providing a new population of immature B cells bearing a highly diverse repertoire of surface immunoglobulin molecules that act as specific receptors for antigen.

The expression of antigen receptors on the surface of a B lymphocyte marks a watershed in its development. The B cell can now detect ligands in its environment, and potentially self-reactive cells must therefore be eliminated before further maturation is allowed. The stimulation of B cells at this stage of their development by molecules binding to surface immunoglobulin leads to the loss of the B cell, and in this way tolerance is established to ubiquitous self antigens. The cells that survive to form part of the long-lived pool of mature peripheral B cells are only a small fraction of those generated in the bone marrow. Nonetheless, these cells express a large repertoire of receptors capable of responding to a virtually unlimited variety of non-self structures. This repertoire provides the raw material on which clonal selection acts in an adaptive immune response.

The development of B cells can be divided into four broad phases (Fig. 6.1). In this chapter, we shall define the different stages of B-cell development in the bone marrow of mice and humans, and see how the unselected B-cell receptor repertoire is generated, before discussing what is known of the mechanisms by which tolerance to self can be ensured once a B cell expresses a complete immunoglobulin molecule at the cell surface. We shall follow the fate of newly generated B cells as they leave the bone marrow and circulate through the lymphoid tissues, although the final stages in the life history of a surviving B cell, in which an encounter with foreign antigen activates it to become an antibody-producing plasma cell or a memory B cell, will be discussed in more detail in Chapter 9. In this chapter we also consider a second population of B cells that develops by a different pathway, and look at how B-cell tumors can capture features of the different stages in normal B-cell development.

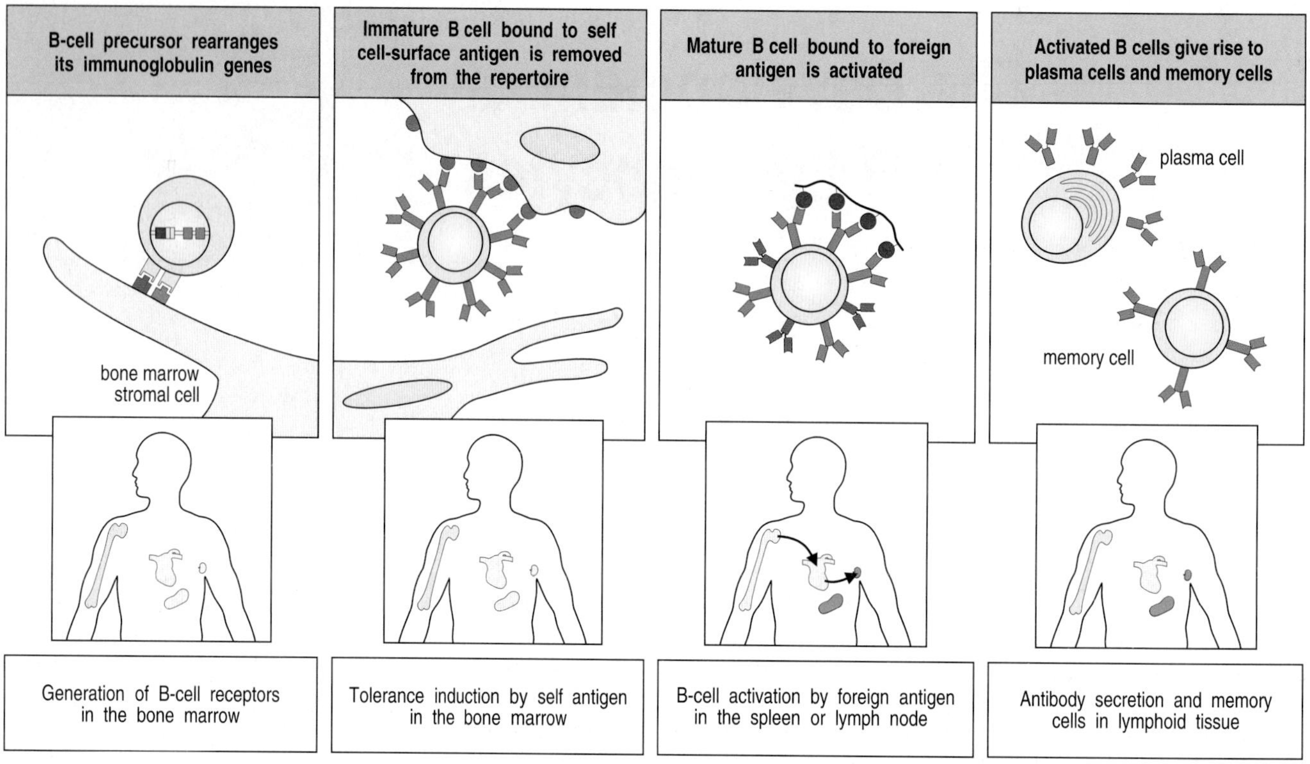
B-cell precursor rearranges its immunoglobulin genes
bone marrow stromal cell
Generation of B-cell receptors in the bone marrow
Immature B cell bound to self cell-surface antigen is removed from the repertoire
Tolerance induction by self antigen in the bone marrow
Mature B cell bound to foreign antigen is activated
B-cell activation by foreign antigen in the spleen or lymph node
Activated B cells give rise to plasma cells and memory cells
plasma cell
memory cell
Antibody secretion and memory cells in lymphoid tissue

Fig. 6.1 The development of B cells can be divided into four broad phases. In the first phase of development, progenitor B cells in the bone marrow rearrange their immunoglobulin genes. This phase is independent of antigen but dependent on interactions with bone marrow stromal cells (first panel). It ends in an immature B cell that carries an antigen receptor in the form of cell-surface IgM and can now interact with antigens in its environment (second panel). Immature B cells that are stimulated by antigen at this stage either die or are inactivated, thus removing many self-reactive B cells from the repertoire. In the third phase of development, B cells not reactive to self antigens mature to express IgD as well as IgM and emerge into the periphery, where they may be activated by encounter with foreign antigen in a secondary lymphoid organ (third panel). Finally, activated mature B cells proliferate and can differentiate into two types of cell (fourth panel). One is the antibody-secreting plasma cell, which may remain in the lymphoid organ or migrate to the bone marrow; the other is the memory B cell, which is crucial for secondary responses.

Generation of B cells.

The successive stages of B-cell differentiation in mice and humans are marked by successive steps in the rearrangement and expression of the immunoglobulin genes, as well as by changes in the expression of cell-surface and intracellular molecules. This developmental program generates B cells in the fetal liver and continues in the bone marrow. In this section, we look at the steps leading to the production of mature B cells expressing surface immunoglobulin. B-cell development must also be regulated so that each mature B cell produces only one heavy chain and one light chain, and thus bears receptors of a single specificity. We shall see that this entails two series of gene rearrangements, each of which is terminated when a protein product is made successfully, and whose success determines whether further development can occur.

6-1 B-cell development proceeds through several stages.

The stages in primary B-cell development are defined by the sequential rearrangement and expression of heavy- and light-chain immunoglobulin genes (Fig. 6.2). The earliest B-lineage cells are known as **pro-B cells**, as they are progenitor cells with limited self-renewal capacity. They are derived from pluripotent hematopoietic stem cells and are identified by the appearance of cell-surface proteins characteristic of early B-lineage cells. Rearrangement of heavy-chain immunoglobulin gene segments takes place in these pro-B cells; D_H to J_H joining at the **early pro-B cell** stage is followed by V_H to DJ_H joining at the **late pro-B cell** stage.

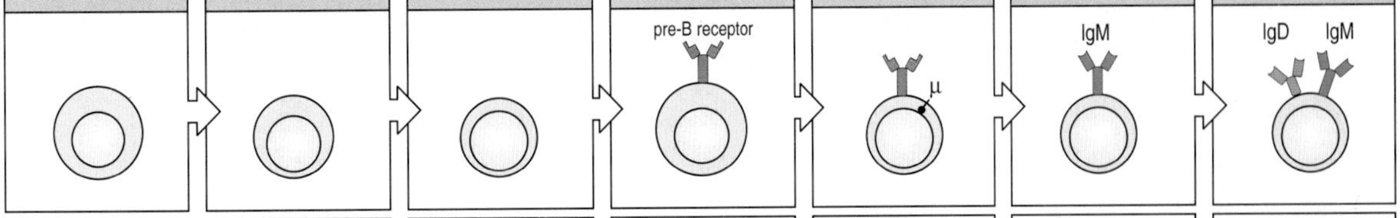

	Stem cell	Early pro-B cell	Late pro-B cell	Large pre-B cell	Small pre-B cell	Immature B cell	Mature B cell
H-chain genes	Germline	D–J rearranged	V–DJ rearranged	VDJ rearranged	VDJ rearranged	VDJ rearranged	VDJ rearranged
L-chain genes	Germline	Germline	Germline	Germline	V–J rearrangement	VJ rearranged	VJ rearranged
Surface Ig	Absent	Absent	Absent	μ chain at surface as part of pre-B receptor	μ chain in cytoplasm and at surface	IgM expressed on cell surface	IgD and IgM made from alternatively spliced H-chain transcripts

Productive VDJ_H joining leads to the expression of an intact μ chain, which is the hallmark of the next main stage of development, the **pre-B cell** stage. The μ chain in **large pre-B cells** is expressed in small amounts at the cell surface in combination with a surrogate light chain as part of a **pre-B cell receptor**; this permits the large pre-B cells to divide further before giving rise to **small pre-B cells**, in which light-chain rearrangements proceed. Once a light-chain gene is assembled and a complete IgM molecule is expressed on the cell surface, the cell is defined as an **immature B cell**. All development up to this point has taken place in the bone marrow and is independent of antigen. At this stage, immature B cells are subject to selection for self-tolerance and ability to survive in the periphery, as described in the next part of this chapter. The B cells that survive in the periphery undergo further differentiation to become **mature B cells** expressing IgD in addition to IgM. These cells, also called **naive B cells** until they encounter their specific antigen, recirculate through secondary lymphoid tissues where they may encounter and be activated by foreign antigen.

The expression of the immunoglobulin heavy and light chains are key milestones in this differentiation pathway. These events do more than simply delineate stages of the pathway; as we shall see in Section 6-6, the surface expression first of an intact heavy chain, and later of a complete immunoglobulin molecule, actively regulates progression from one stage to the next.

Fig. 6.2 The development of a B-lineage cell proceeds through several stages marked by the rearrangement and expression of the immunoglobulin genes. The stem cell has not yet begun to rearrange its immunoglobulin (Ig) genes; they are in the germline configuration as found in all non-lymphoid cells. The heavy-chain (H-chain) genes rearrange first. Rearrangements of D gene segments to J_H gene segments occur in early pro-B cells, generating late pro-B cells in which a V_H gene segment becomes joined to the rearranged DJ_H. A successful VDJ_H rearrangement leads to the expression of an intact immunoglobulin heavy chain at the cell surface as part of the pre-B cell receptor. Once this occurs the cell is defined as a large pre-B cell, which, like the stem cell, is large and actively dividing. Large pre-B cells then cease dividing to become small resting pre-B cells in which the μ heavy chain is found mainly inside the cell, and in which the light-chain (L-chain) genes can be rearranged. Small pre-B cells express low levels of cell-surface pre-B cell receptor, which has apparently lost its function, as the cells are now small, re-express the RAG proteins and start to rearrange the light-chain (L-chain) genes. Upon successfully assembling a light-chain gene, the cell becomes an immature B cell that expresses light chains and μ heavy chains as surface IgM molecules and no longer expresses the pre-B cell receptor. Mature B cells are marked by the additional appearance of IgD on the cell surface.

6-2 The bone marrow provides an essential microenvironment for early B-cell development.

B-cell development is dependent on the non-lymphoid stromal cells found in the bone marrow; stem cells isolated from the bone marrow and grown in culture fail to differentiate into B cells unless bone marrow stromal cells are also present. The stroma, whose name derives from the Greek word for a mattress, thus provides a necessary support for B-cell development. The contribution of the stromal cells is twofold. First, they form specific adhesion contacts with the developing B-lineage cells by interactions between cell adhesion molecules (CAMs) and their ligands. Second, they provide growth factors, for instance the membrane-bound stem-cell factor (SCF), which is recognized by the surface receptor Kit on early B-lineage cells, and the secreted

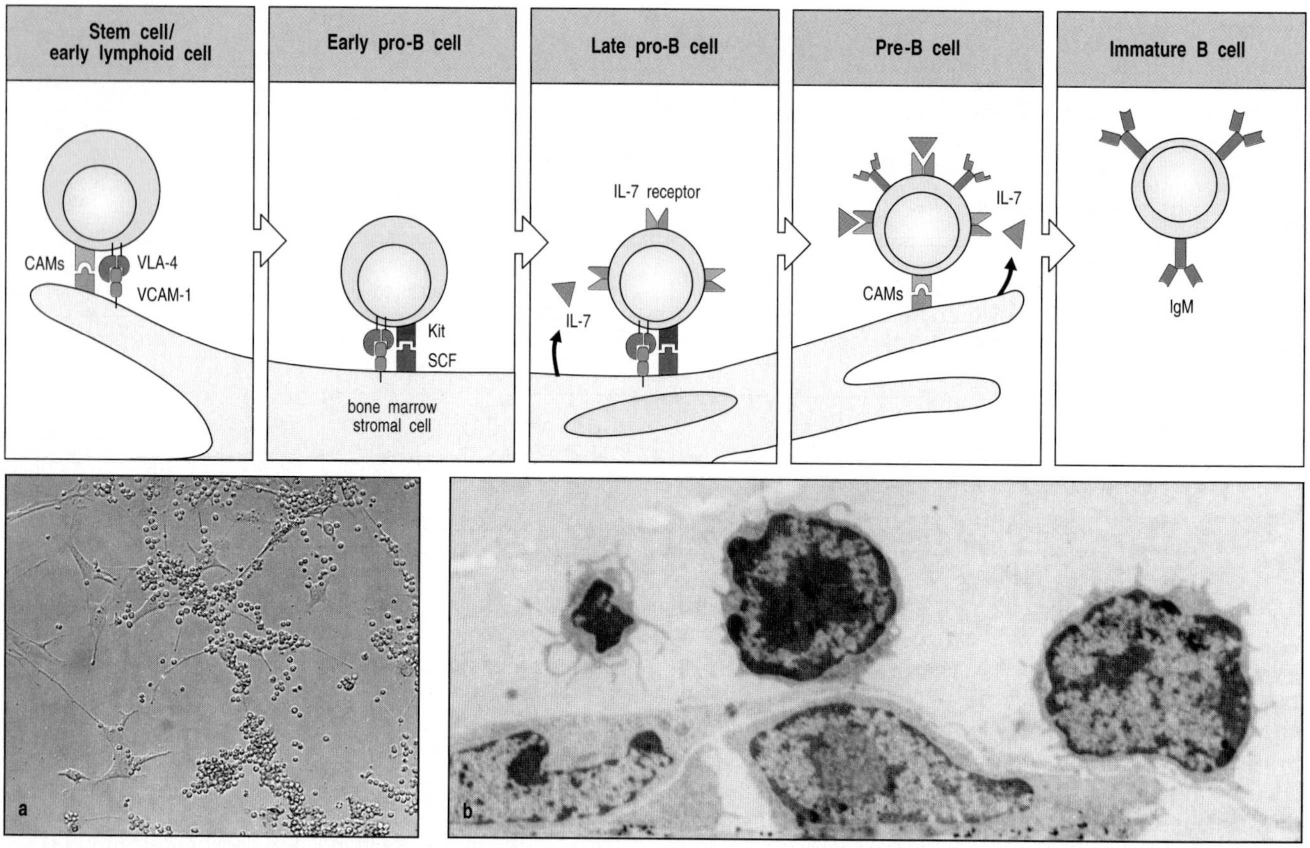

Fig. 6.3 The early stages of B-cell development are dependent on bone marrow stromal cells. The upper panels show the interactions between precursor B cells and stromal cells that are required for the development to immature B-cell stage. Lymphoid progenitor cells and early pro-B cells bind to the adhesion molecule VCAM-1 on stromal cells through the integrin VLA-4 and also interact through other cell adhesion molecules (CAMs). These adhesive interactions promote the binding of the receptor tyrosine kinase Kit on the surface of the pro-B cell to stem-cell factor (SCF) on the stromal cell, which activates the kinase and induces the proliferation of the B-cell progenitors. Later stages require interleukin-7 (IL-7) for proliferation and further development. Panel a: light micrograph showing small round cells, which are the B-lymphoid progenitors, in intimate contact with cultured stromal cells, which have extended processes fastening them to the plastic dish on which they are grown. Panel b: High-magnification electron micrograph of a similar cell culture in which two lymphoid cells are seen adhering to a flattened stromal cell. Photographs courtesy of A Rolink (a); P Kincade and P L Witte (b).

cytokine interleukin-7 (IL-7), which is recognized by late pro-B and pre-B cells (Fig. 6.3). In addition, a small chemokine known as PBSF/SDF-1, which is produced constitutively by bone marrow stromal cells, has an important role in early stages of B-cell development, as shown by the failure of B-cell development in mice lacking the gene for this chemokine. Other adhesion molecules and growth factors produced by stromal cells have a role in B-cell development; this is an active area of research and a full understanding of the factors that regulate B-cell differentiation has yet to be achieved.

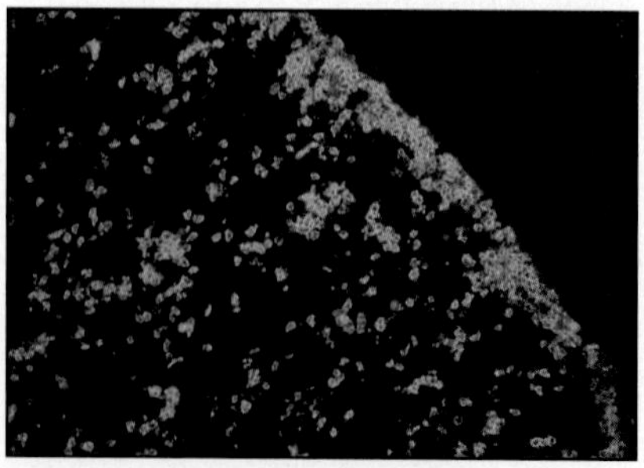

Fig. 6.4 B-lineage cells move within the bone marrow towards its central axis as they mature. Part of a transverse section of a rat femur photographed under ultraviolet illumination to identify fluorescent cells (green) stained for terminal deoxynucleotidyl transferase (TdT), which marks the pro-B stage (see Fig. 6.2). Early pro-B cells expressing TdT are concentrated near the endosteum (the inner bone surface), as seen towards the upper right of the picture. As development proceeds, cells lose TdT and pass towards the center of the marrow cavity (lower left), where they wait in sinuses ready for export. Photograph courtesy of D Opstelten and M Hermans.

As B-lineage cells mature, they migrate in contact with stromal cells within the marrow. The earliest stem cells lie in a region called the subendosteum, which is adjacent to the inner bone surface. As maturation proceeds, B-lineage cells move towards the central axis of the marrow cavity (Fig. 6.4). Later stages of maturation become less dependent on contact with stromal cells. Final development from immature B cells into mature B cells occurs mainly in the secondary lymphoid organs such as the spleen.

6-3 The survival of developing B cells depends on the productive, sequential rearrangement of a heavy- and a light-chain gene.

A large number of developing B cells are lost because they fail to rearrange their immunoglobulin gene segments productively to form one functional heavy-chain gene and one functional light-chain gene. This failure is caused by imprecise joining of the gene segments, which in turn is a consequence of the way in which diversity is created at the junctions during gene rearrangement (see Section 3-16). The imprecise joining mechanism means that it is a matter of chance whether V gene segments and J gene segments are assembled so that the J region and the C region downstream will be read in the correct reading frame; each time a V gene segment undergoes rearrangement to a J or a DJ gene segment there is a roughly two-in-three chance of generating an out-of-frame sequence downstream from the join. The sequence of immunoglobulin gene rearrangement and the points at which non-productive rearrangements may lead to cell loss are shown in Fig. 6.5. As we shall see later, far fewer cells are expected to be lost because of failure to make

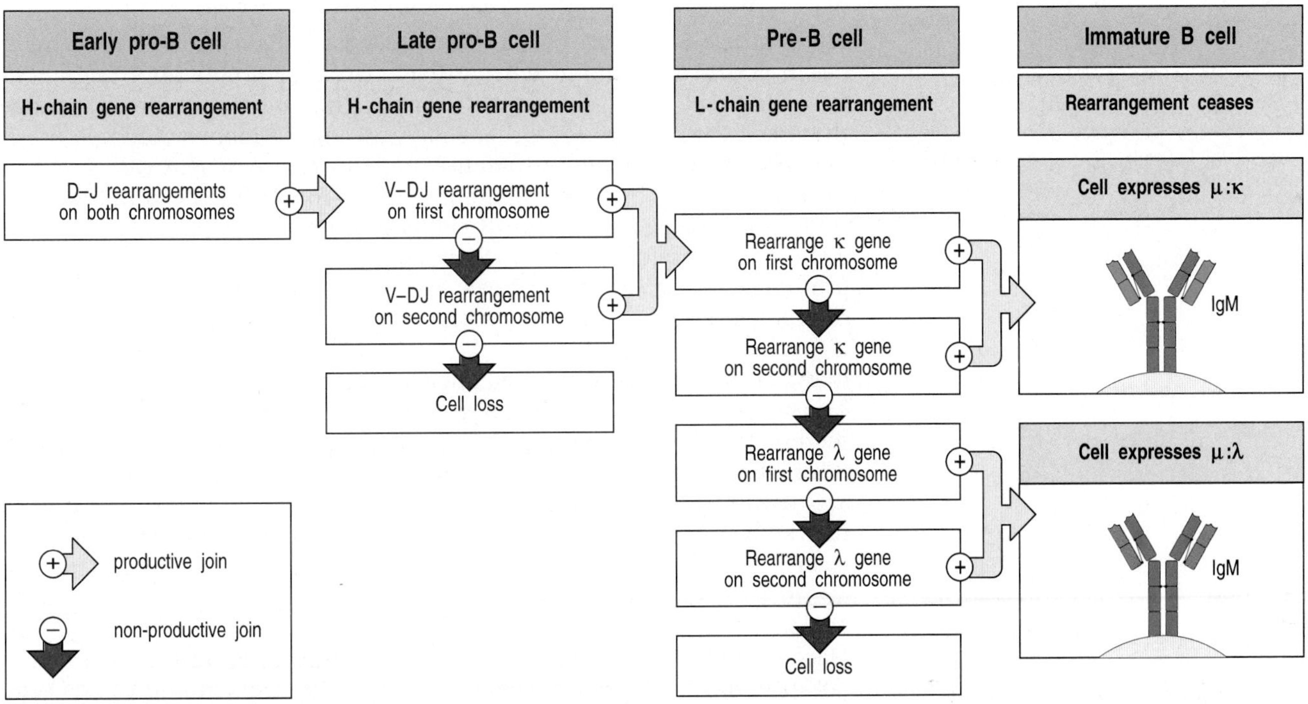

Fig. 6.5 The steps in immunoglobulin gene rearrangement at which cells can be lost. The developmental program usually rearranges heavy-chain (H-chain) genes first and then light-chain (L-chain) genes. Cells are allowed to progress to the next stage when a productive rearrangement has been achieved. Each rearrangement has about a one-in-three chance of being successful but if the first attempt is non-productive, development is suspended and there is a chance for one or more further attempts. The scope for repeated rearrangements is greater in the light-chain genes (see Fig. 6.6), so that fewer cells are lost between the pre-B and immature B cell stages than in the pro-B to pre-B transition.

productive light-chain gene rearrangements than are lost at the stage of heavy-chain gene rearrangements. This is because the opportunity for successive rearrangement attempts is much greater in light-chain genes.

Immunoglobulin heavy-chain gene rearrangement begins in early pro-B cells with a D_H gene segment's joining to a J_H gene segment. This frequently occurs on both chromosomes of the diploid set before the cell, now classified as a late pro-B cell, proceeds to rearrange a V_H gene to a DJ_H complex. Most D–J_H joins in humans are potentially useful, as almost all human D gene segments can be translated in all three reading frames.

V_H to DJ_H rearrangement occurs first on only one chromosome, and the chances of this rearrangement's generating a joint that puts the V gene in-frame with downstream sequences is one in three. Taking into account the possibility of rearranging one of the considerable number of V_H pseudogenes, the probability of generating a productive V_H–DJ_H rearrangement on the first attempt is something less than one-in-three. A successful first rearrangement means that intact μ chains are produced and the cell progresses to become a pre-B cell. In at least two out of three cases, however, the first rearrangement is non-productive, and rearrangement continues on the other chromosome, again with a chance of less than one in three of being productive. Overall this gives a rough estimate of the chance of generating a pre-B cell of something less than 55% ($^1/_3 + (^1/_3 \times {}^2/_3) = 0.55$). Pro-B cells in which both rearrangements are non-productive are unable to receive a survival signal. Some of these cells may be rescued by a secondary rearrangement, which replaces their V gene segment sequences with another V_H gene (we shall discuss this mechanism further in Section 6-10), but most seem to be lost from the lineage. A considerable proportion of pro-B cells are therefore lost at this stage.

The large pre-B cells in which a successful heavy-chain gene rearrangement has just occurred begin to actively divide. This cell proliferation is thought to occur after the transient expression of a μ heavy chain at the cell surface as part of the pre-B receptor (see Fig. 6.2). In the mouse the large pre-B cells divide several times, expanding the population of cells with successful in-frame joins by approximately 30- to 60-fold before becoming resting small pre-B cells. A large pre-B cell with a particular rearranged heavy-chain gene therefore gives rise to many progeny. Upon reaching the small pre-B cell stage, each of these can make different light-chain gene rearrangements.

In the mouse and human B-cell lineages, κ genes tend to rearrange before λ genes. This was first deduced from the observation that myeloma cells secreting λ chains generally have both their κ and λ genes rearranged, whereas in myelomas secreting κ chains it is generally only the κ genes that are rearranged. This order is occasionally reversed, however, and λ gene rearrangement does not absolutely require the prior rearrangement of the κ genes.

As with the heavy-chain genes, light-chain gene rearrangements also take place on only one chromosome at a time. Unlike the heavy-chain genes, however, there is scope for repeated rearrangements of unused V and J gene segments (Fig. 6.6). Several successive attempts at productive rearrangement of a light-chain gene can therefore be made on one chromosome before initiating any rearrangements on the second chromosome. Similar repeated rearrangements for the heavy-chain locus are much less likely because there are no spare downstream DJ_H joints to which a new V_H could join and the 12/23 rule prevents V_H from rearranging directly to J_H.

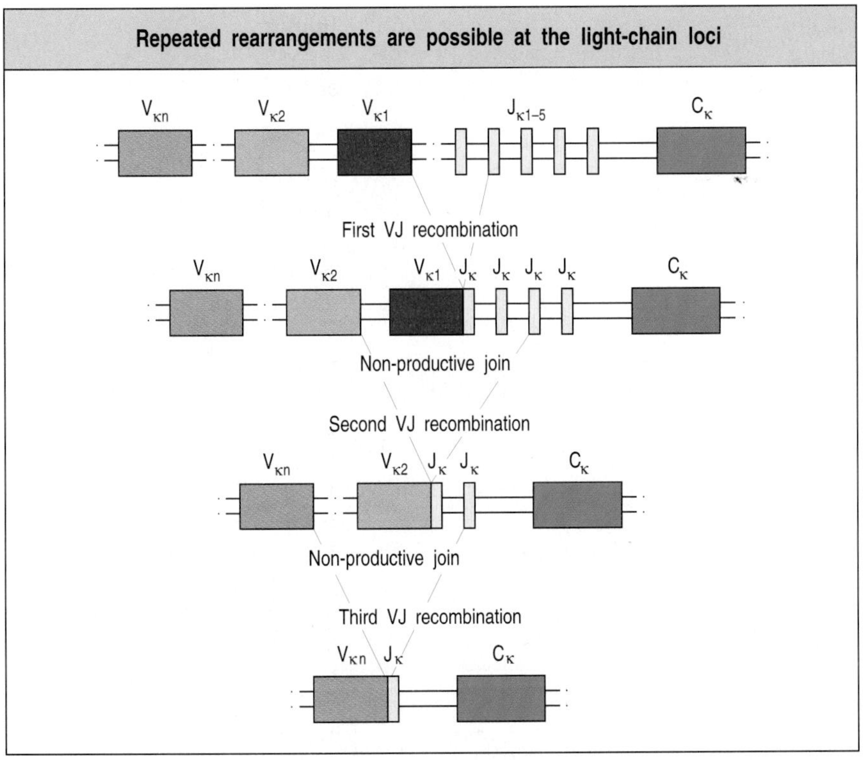

Fig. 6.6 Non-productive light-chain gene rearrangements can be rescued by further gene rearrangement. The organization of the light-chain loci in mice and humans offers many opportunities for rescue of pre-B cells that initially make an out-of-frame light-chain gene rearrangement. Light-chain rescue is illustrated for the human κ locus. If the first rearrangement is non-productive, a 5′ V_κ gene segment can recombine with a 3′ J_κ gene segment to remove the out-of-frame join and replace it. In principle, this can happen up to five times on each chromosome, because there are five functional J_κ segments in humans. If all rearrangements of κ-chain genes fail to yield a productive light-chain join, λ-chain gene rearrangement may succeed (not shown).

In the light-chain genes, each V–J join has a one-in-three chance of assembling a productive gene and the chances of eventually generating an intact light chain are greatly increased by the potential for multiple successive rearrangement events on each chromosome and at each of the two light-chain loci. As a result, most B-cell precursors that reach the pre-B cell stage succeed in generating progeny that bear intact IgM molecules and can be classified as immature B cells. Overall, the proportion of B-lineage cells that survive the process of primary receptor diversification should theoretically be less than 50%, and can be considerably less if the V_H pseudogenes are taken into account.

6-4 The expression of proteins regulating immunoglobulin gene rearrangement and function is developmentally programmed.

Although the sequential steps in immunoglobulin gene rearrangement and immunoglobulin chain production provide defining markers for the different stages in B-cell differentiation, a variety of other proteins contribute to this process. Figure 6.7 lists some of these, and shows how their expression is regulated through the different stages of B-cell development.

Immunoglobulin gene rearrangement is itself dependent on the expression of the recombination activation genes ***RAG-1*** and ***RAG-2*** (see Section 3-17). The enzymes encoded by these genes are also required for rearrangement of the T-cell receptor genes and are active at very early stages of lymphoid development, before the B- and T-cell lineages have diverged. In both these lineages there is a later temporary suppression of *RAG* gene expression after the first successful rearrangement event, which is followed by several cell divisions. During these cell divisions, the RAG proteins are inactive, but they are resynthesized later when the cells cease dividing and go on to the second rearrangement event; in B cells, this is the rearrangement of light-chain genes.

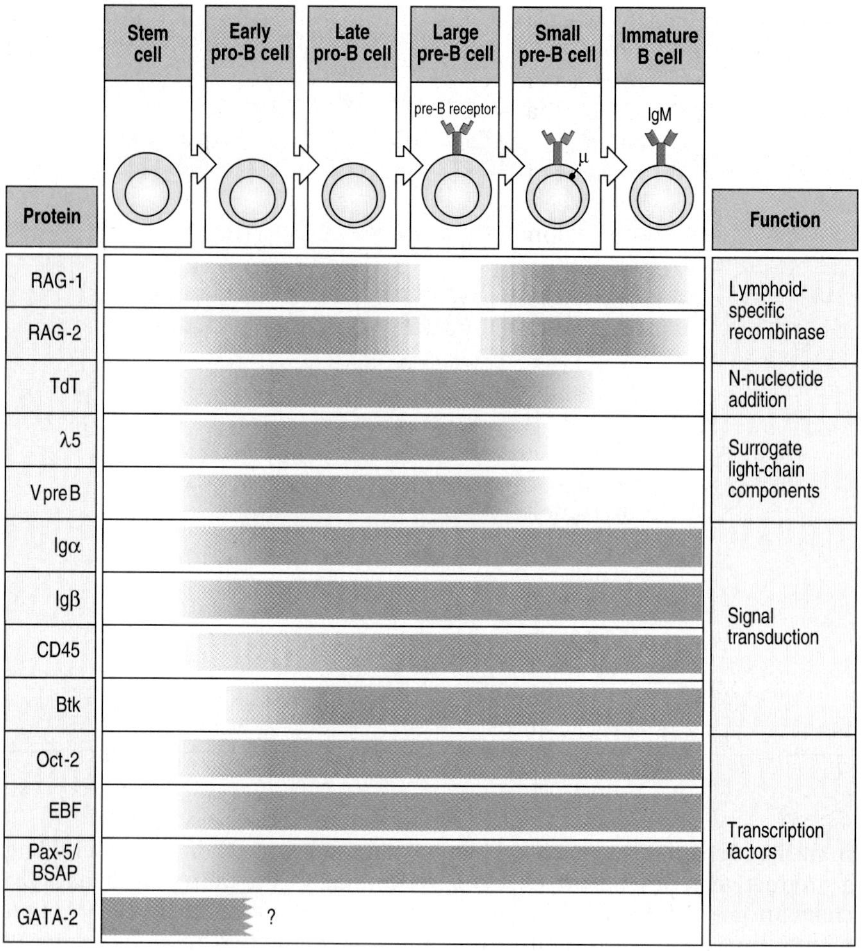

Fig. 6.7 The temporal expression of several cellular proteins known to be important for B-cell development. The proteins listed here are a selection of those known to be associated with early B-lineage development, and have been included because of their proven importance in the developmental sequence, largely on the basis of studies in mice. Their individual contributions to B-cell development are discussed in the text, with the exception of the Octamer Transcription Factor, Oct-2, which binds the octamer ATGCAAAT found in the heavy-chain promoter and elsewhere, and GATA-2, which is one example of the many transcription factors that are active in several hematopoietic lineages. The *pax-5* gene product known as B-lineage-specific activator protein (BSAP) is involved in regulating the expression of several of the other proteins listed. The tight temporal regulation of the expression of these proteins, and of the immunoglobulin genes themselves, would be expected to impose a strict sequence on the events of B-cell differentiation.

Another enzyme, **terminal deoxynucleotidyl transferase (TdT)**, contributes to the diversity of both B- and T-cell antigen receptor repertoires by adding N-nucleotides at the rearrangement joints (see Section 3-16). As with the *RAG* genes, TdT is expressed in early lymphoid progenitors but, unlike the RAG-1/RAG-2 enzyme, it is not essential to the rearrangement process; indeed, at the time that the peripheral immune system is first being supplied with T and B lymphocytes, TdT is not expressed at all. In adult humans, it is expressed in pro-B cells but it ceases to be expressed at the pre-B cell stage when heavy-chain gene rearrangement is complete and light-chain gene rearrangement has commenced. This timing explains why N-nucleotides are found in the V–D and D–J joints of heavy-chain genes and in about a quarter of human light-chain joints. N-nucleotides are rarely found in mouse light-chain V–J joints, showing that TdT is switched off earlier in mouse B cells.

Proteins required for the cell-surface expression of the immunoglobulin chains are also essential for B-cell development. These include Igα (CD79α) and Igβ (CD79β), which are components of the pre-B cell receptor and the B-cell antigen receptor complexes on the cell surface (see Fig. 6.8); Igα and Igβ transduce signals from these receptors by interacting with intracellular tyrosine kinases through their cytoplasmic tails (discussed in Chapter 5). Igα and Igβ are expressed from the pro-B cell stage until the death of the cell or its terminal differentiation into an antibody-secreting plasma cell. For reasons that are not known, mice lacking Igβ have a block in B-cell development at the pro-B cell stage before VDJ_H rearrangements are complete. Formation of the pre-B receptor complex also requires λ5 and VpreB, which together make up a surrogate light chain that combines with the product of the newly rearranged heavy-chain gene. The pre-B receptor complex appears only briefly at the cell surface, and λ5 and VpreB cease to be expressed as soon as this occurs. The cell-surface expression of the pre-B receptor is an important checkpoint in B-cell development and will be discussed further in Section 6-6.

Other signal transduction proteins with a role in B-cell development are also included in Fig. 6.7. Bruton's tyrosine kinase (Btk) has received intense scrutiny because mutations in the *Btk* gene cause a profound B-lineage-specific immune deficiency, **Bruton's X-linked agammaglobulinemia (XLA)** (see Section 11-7). In humans, the block in B-cell development caused by *XLA* locus mutations is almost total, interrupting the transition from pre-B cell to immature B cell. A similar, though less severe, defect called X-linked immunodeficiency, or **xid**, arises from mutations in the corresponding gene in mice.

Finally, several gene-regulatory proteins are essential for B-cell development, as shown by deficiencies of the B-cell lineage in genetically engineered mutants lacking these proteins. At least 10 transcription factors necessary for normal B-lineage development have been described, and there are likely to be others. One essential transcription factor is the early B-cell factor, EBF, which regulates the transcription of the gene for Igα. Another is the *pax-5* gene product, one isoform of which is the B-lineage specific activator protein (BSAP). This protein is active in late pro-B cells and permits the proper functioning of the heavy-chain enhancer (see Section 6-5); it also binds to regulatory sites in the genes for λ5, VpreB, and other B-cell specific proteins. It seems likely that these regulatory proteins and others like them together direct the developmental program of B-lineage cells. The gene-regulatory proteins involved in the tissue-specific transcriptional regulation of immunoglobulin genes are also likely to be important in regulating the order of events in gene rearrangement, as we shall discuss below.

6-5 Immunoglobulin gene rearrangements are closely coordinated with chromatin accessibility.

The **V(D)J recombinase** system (see Section 3-17) operates in both B- and T-lineage cells and employs the same core enzymes, which recognize the same conserved recombination signal sequences in both immunoglobulin and T-cell receptor genes. Yet rearrangements of T-cell receptor genes do not occur in B-lineage cells, nor do complete rearrangements of immunoglobulin genes occur in T cells. Rearrangement events are associated with low-level transcription of the gene segments about to be joined (Fig. 6.8), suggesting that lineage specificity is achieved by controlling rearrangement using gene-regulatory proteins specific to T or B cells. Such proteins, for example the B-lineage specific gene-regulatory proteins (BSAP), bind tissue-specific enhancers and influence local chromatin structure. In this way they are thought to make the chromatin accessible to transcription factors and to recombination enzymes.

As a consequence of immunoglobulin gene rearrangement, the promoter upstream of the V gene segments is brought nearer to the enhancers, and transcription of the rearranged gene segments is dramatically increased. Thus, gene rearrangement can be viewed as a powerful mechanism for

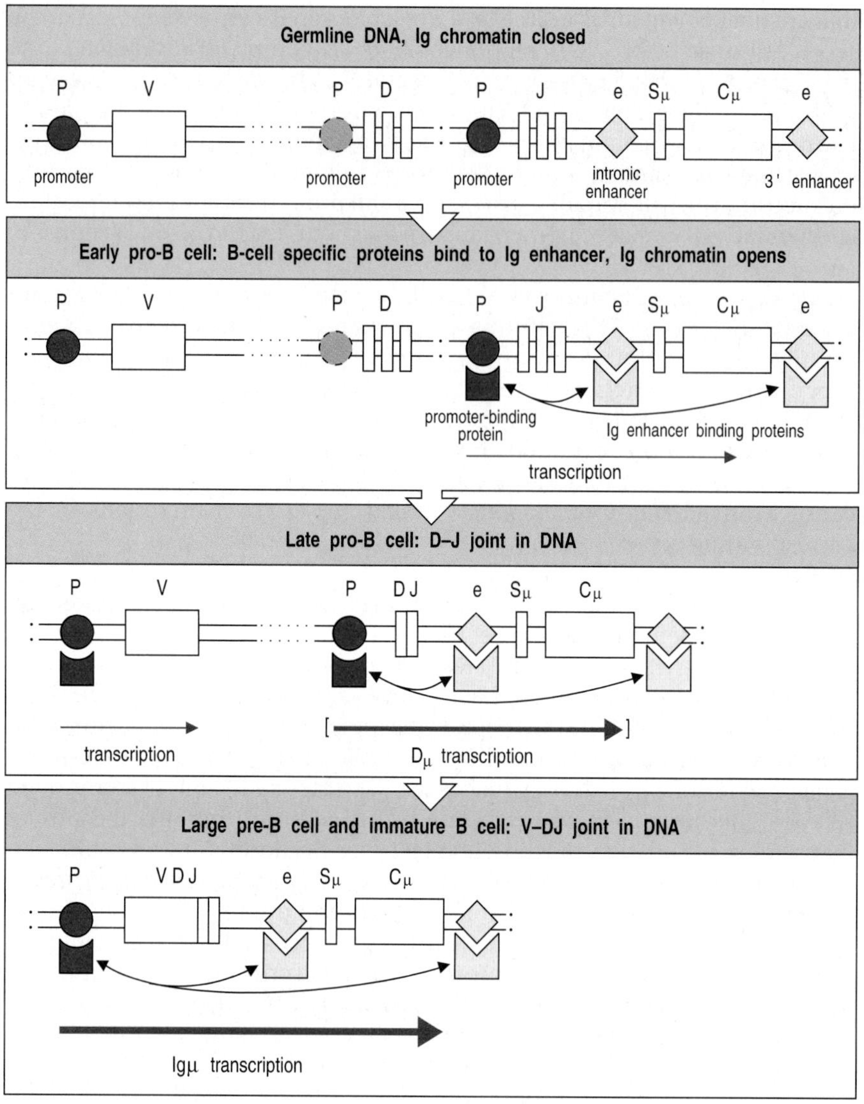

Fig. 6.8 Proteins binding to promoter and enhancer elements contribute to the sequence of gene rearrangement and regulate the level of RNA transcription. First panel: in germline DNA, stem cells, and non-lymphoid cells, the chromatin containing the immunoglobulin genes is in a closed conformation. Second panel: in the early pro-B cell, specific proteins bind to the Ig enhancer elements (e); for the heavy chain, these are in the J–C intron and 3′ to the C exons. The DNA is now in an open conformation, and low-level transcription from promoters (P) upstream of the J gene segments is seen. Third panel: the rearrangement D to J_H that follows the initiation of transcription of the J gene segments activates a low level of transcription from promoters located upstream of the D gene segments. For some D–J_H joins in the mouse this may result in the expression of a truncated heavy-chain (D_μ) at levels sufficient to abort further development but, in most cases, it is followed by initiation of low-level transcription of an upstream V_H gene segment. Fourth panel: subsequent rearrangement of a V gene segment brings its promoter under the influence of the heavy-chain enhancers, leading to enhanced production of a μ heavy-chain mRNA in large pre-B cells and their progeny. S_μ represents the switch signal sequence for isotype switching (see Section 3-22).

regulating gene expression, as well as for generating receptor diversity. Several cases of gene rearrangement's bringing genes under the control of a new promoter are known from prokaryotes and single-celled eukaryotes but, in vertebrates, only the immunoglobulin and T-cell receptor genes are known to use gene rearrangement to regulate gene expression.

6-6 Cell-surface expression of the products of rearranged immunoglobulin genes provides checkpoints in B-cell development.

Each fully differentiated B cell has only one successfully rearranged heavy-chain gene and one successfully rearranged light-chain gene. The increased transcription of the successfully rearranged genes and the rapid appearance of their protein products have an important part in ensuring this outcome. This is shown dramatically by introducing an already rearranged immunoglobulin heavy-chain gene into the germline of a mouse; virtually all B cells in this transgenic mouse will express the product of the rearranged transgene and, in these B cells, rearrangement of the endogenous heavy-chain genes is suppressed. However, the endogenous light-chain genes rearrange normally, thus generating a variety of complete immunoglobulin molecules. A similar though less efficient suppression of endogenous light-chain gene rearrangement occurs in mice transgenic for a rearranged light-chain gene. Transgenic mice carrying both a rearranged heavy-chain gene and a rearranged light-chain gene make B cells in which rearrangements of all endogenous immunoglobulin genes are overwhelmingly suppressed. These mice therefore express a B-cell repertoire with the single dominant specificity conferred by the transgenes, and have been valuable in studies of self-tolerance, as we shall see in Section 6-9.

In contrast to the effects of transgenes expressing complete immunoglobulin chains, a heavy-chain transgene lacking the transmembrane exon fails to suppress the rearrangement of endogenous heavy-chain genes. This is because membrane insertion and cell-surface expression of the heavy chain is needed for the suppression to occur. As we have seen, the heavy chain is expressed briefly at the cell surface as part of the pre-B cell receptor as soon as it is made. This appearance depends on an association between the heavy chain and two proteins made in pro-B cells, which pair non-covalently to form a **surrogate light chain** (Fig. 6.9). One of these proteins is called **λ5** because of its close similarity to C domains of λ light chains, while the other, called **VpreB**, resembles a V domain but bears an extra amino-terminal protein sequence. Together, λ5, VpreB, the μ heavy chain, and the attendant constant Igα and Igβ chains form the cell-surface pre-B receptor complex, which structurally resembles a complete cell-surface immunoglobulin receptor complex (see Fig. 6.9).

The pre-B cell receptor complex is expressed only transiently, perhaps because the production of λ5 stops as soon as it is formed. Nevertheless, it mediates an important checkpoint in B-cell development. In mice lacking λ5, or possessing mutant heavy-chain genes that cannot produce transmembrane heavy chains, the pre-B cell receptor cannot be formed and development is blocked after heavy-chain gene rearrangement.

Such mice have rearrangements of the heavy-chain genes on both chromosomes in all cells, and so about 10% of the cells have two productive VDJ_H rearrangements. In normal mice, the appearance of the pre-B cell receptor on the cell surface coincides with inactivation of the RAG-2 protein by phosphorylation, which targets it for degradation; synthesis of the mRNA for RAG-1 and RAG-2 is also suppressed, suggesting that this is the mechanism by which further rearrangement at the heavy-chain locus is blocked. Expression of the pre-B cell receptor at the cell surface is also associated with cell enlargement followed by a burst of proliferation. Such cells then undergo the transition to small resting pre-B cells, in which the RAG-1/RAG-2 enzyme is again active and the light-chain genes are rearranged.

The pre-B cell receptor therefore seems to signal to the cell that a complete heavy-chain gene has been formed, that further rearrangements at this locus should be suppressed, and that development to the next stage can proceed. The intracellular tyrosine kinase Btk is thought to play a part in transducing this signal, as in its absence B-cell development is blocked at the pre-B cell stage.

Once a light-chain gene has been rearranged successfully, its product combines with the heavy chain to form intact IgM (see Fig. 6.9), which is expressed at the cell surface in a complex with Igα and Igβ. It is not yet clear how, in the absence of specific antigen, the cell senses that a functional immunoglobulin

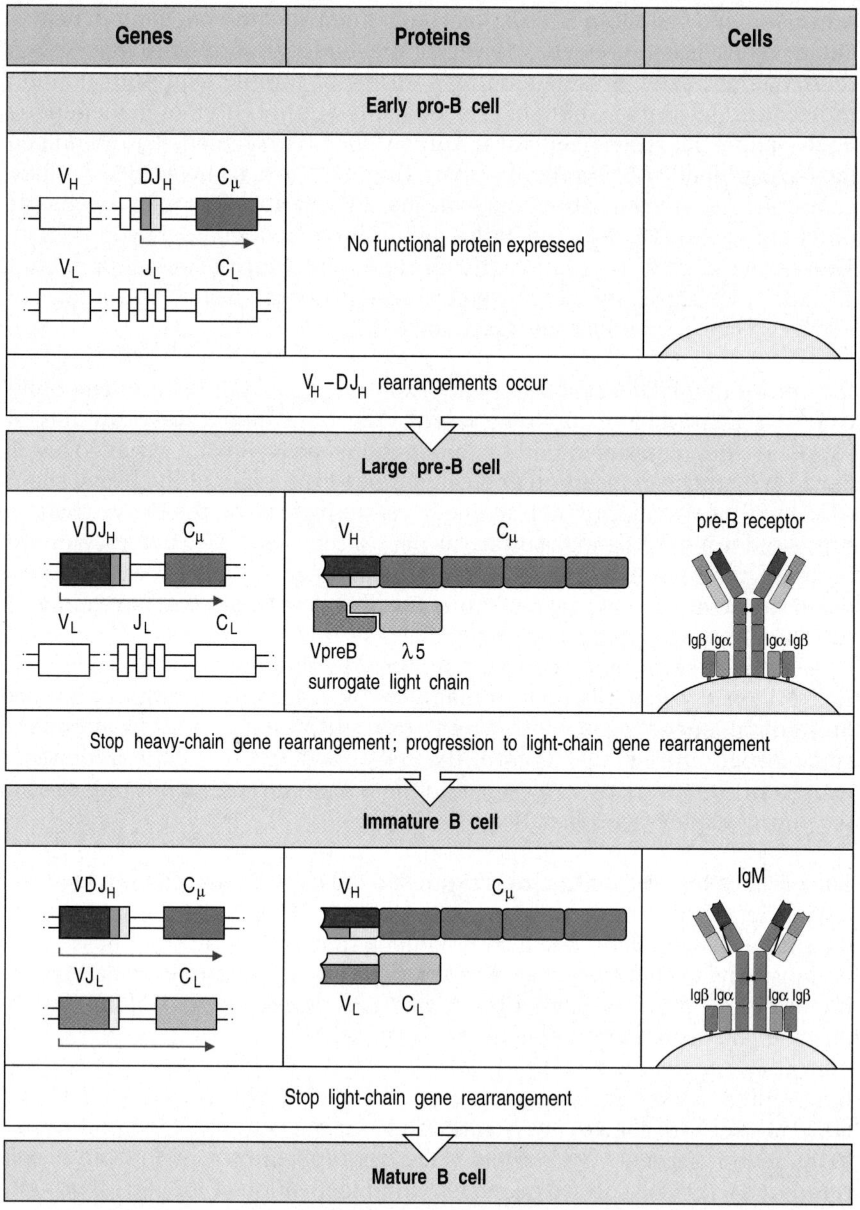

Fig. 6.9 Productively rearranged immunoglobulin genes encode proteins that are expressed immediately at the surface of the cell. Top panel: in early pro-B cells, heavy-chain gene rearrangement is not yet complete and no functional μ protein is expressed. Second panel: as soon as a productive heavy-chain gene rearrangement has taken place, μ chains are expressed at the surface of the cell in a complex with two other chains, λ5 and VpreB, which together make up a surrogate light chain. The whole immunoglobulin-like complex is known as the pre-B cell receptor. It is also associated with two other protein chains, Igα (CD79α) and Igβ (CD79β), on the cell surface. These associated chains signal the B cell to halt heavy-chain gene rearrangement, and drive the transition to the large pre-B cell stage by inducing proliferation. The progeny of large pre-B cells stop dividing and become small pre-B cells, in which light-chain gene rearrangements commence. Third panel: successful light-chain gene rearrangement results in the production of a light chain that binds the μ chain to form a complete IgM molecule which is expressed together with Igα and Igβ at the cell surface. Signaling via these surface IgM molecules is thought to trigger the cessation of light-chain gene rearrangement.

receptor has been expressed. Nevertheless, at this point light-chain gene rearrangement ceases, and it seems that this checkpoint, like that mediated by the pre-B cell receptor, requires a signal from the receptor to suppress any further light-chain gene rearrangement and to enable further maturation of the B cell.

A role for Igα in signaling at both these checkpoints is indicated by the reduction in B-lineage cells in mice expressing Igα with a truncated cytoplasmic domain. In these mice the population of immature B cells in the marrow is reduced fourfold, while the number of peripheral B cells is reduced 100-fold, showing that an ability to signal through Igα is particularly important in dictating survival once a complete immunoglobulin molecule is expressed. Other molecules associated with signaling through the B-cell receptor are also required for the B cell to complete its maturation and survive in the periphery. For example, immature B cells fail to become mature recirculating B cells if they lack the tyrosine kinase Syk, which transduces signals from the B-cell receptor to the interior of the cell (see Chapter 5). An ability to signal through surface immunoglobulin is also required for the continued survival and recirculation of mature peripheral B cells. This has been shown by using the technique of inducible gene targeting (see Section 2-27) to destroy the expression of immunoglobulin in the mature B cells of genetically engineered mice. These experiments show that if mature recirculating B cells cease to express immunoglobulin they die within a few days.

6-7 The immunoglobulin gene rearrangement program produces monospecific B cells.

We saw in Chapter 1 that individual B cells must produce only one specificity of antibody to ensure that all the antibodies secreted by an activated B cell are specific for the antigen that originally triggered its proliferation. This is a cardinal feature of clonal selection that prevents the secretion of antibodies of other specificities that could be wasteful or even harmful. We can now see how the regulated program of immunoglobulin gene rearrangement guarantees monospecificity, and how this, in turn, explains the phenomena of allelic and isotypic exclusion. **Allelic exclusion** signifies the expression of a gene from only one of the two parental chromosomes bearing that gene in each individual cell (Fig. 6.10); it occurs for both the heavy- and light-chain genes. **Isotypic exclusion** occurs only for the light chain, which is produced from only one of the four light-chain loci (two κ; two λ) in each individual cell. The phenomenon of allelic exclusion was first discovered over 30 years ago, when it provided one of the original pieces of experimental support for the clonal selection theory for lymphocytes.

Allelic exclusion seems to operate without substantial allelic preference, as the alleles at each locus are generally expressed at roughly equal frequencies; chance probably determines which allele rearranges first. However, for isotype exclusion the situation is different, and the ratios of κ-expressing versus λ-expressing mature B cells vary from extreme to extreme in different species; in mice and rats it is 95:5, in humans typically 65:35, and in cats it is the opposite of mice. These ratios correlate most strongly with the number of functional V_κ and V_λ gene segments. Additionally, they reflect the kinetics and efficiency of gene segment rearrangements. The κ:λ ratio in mature lymphocyte populations is useful in clinical diagnostics, as an aberrant κ:λ ratio indicates the dominance of one clone and the presence of a lymphoproliferative disorder, which may be malignant (see Section 6-17).

Once a complete immunoglobulin molecule is generated, the monospecific B cell, now called an immature B cell, is exposed to selection by antigens in its environment. Before immature B cells leave the bone marrow and populate the peripheral lymphoid tissues, where they can be stimulated to make antibody, those that are stimulated by self antigens must be removed or inactivated to ensure self tolerance. We shall describe some mechanisms of

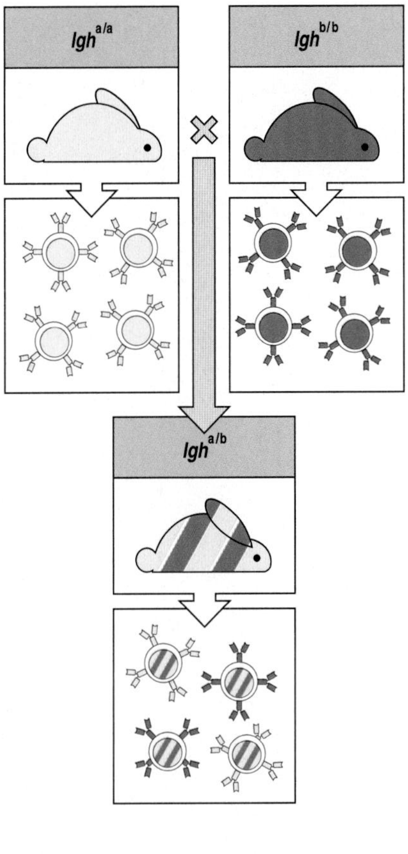

Fig. 6.10 Allelic exclusion in individual B cells. Most species have genetic polymorphisms of the constant regions of their immunoglobulin heavy- and light-chain genes known as allotypes (see Section 3-23). In rabbits, for example, all of the B cells in an individual homozygous for the *a* allele of the immunoglobulin heavy-chain locus (*Igh*) will express immunoglobulin of type a, whereas in an individual homozygous for the *b* allele, all the B cells make immunoglobulin of type b. In a heterozygous animal, which carries the *a* allele on one of the *Igh* chromosomes and the *b* allele on the other, individual B cells can be shown to carry either a-type or b-type immunoglobulin but not both. This allelic exclusion reflects productive rearrangement of only one of the two parental *Igh* loci.

ensuring B-cell tolerance in subsequent sections of this chapter, and also in Chapter 13; the activation of B cells in the periphery will be discussed in Chapter 9. First, however, we look at the fate of the new B cells generated in the bone marrow as they emerge into the periphery.

6-8 Lymphoid follicles are thought to provide a second essential environment for B cells.

After B cells leave the bone marrow, they pass from the blood via specialized vascular endothelium into the secondary lymphoid organs—the spleen, lymph nodes and mucosal-associated lymphoid tissue (MALT). Only a small proportion of these immature B cells then complete their maturation and survive to recirculate between the lymphoid organs and the blood; most die by apoptosis 3–4 days after first expressing surface IgM. This failure to survive in the periphery seems to be a consequence of competition for a place in the pool of longer-lived recirculating peripheral B cells. If recirculating cells are depleted, the proportion of immature B cells that are recruited into the recirculating pool increases until the losses are compensated. Conversely, if immature B cell production ceases, the lifespan of recirculating B cells becomes indefinite.

It is thought that the B cells are competing for survival signals derived from secondary lymphoid tissue. Within the secondary lymphoid organs, B cells are localized in discrete clusters, called **follicles**, consisting primarily of B cells and a specialized stromal cell, the **follicular dendritic cell**. Even after maturing, B lymphocytes seem to need signals from follicles, possibly from the follicular dendritic cells, to survive and continue to recirculate. This has been deduced from experiments that trace the fate of transgenic B cells expressing surface immunoglobulin of a single known specificity. When these B lymphocytes are transfused intravenously into recipients in which a space has been cleared for them, for example by irradiation, they survive. If they are introduced into a recipient with normal lymphocyte numbers, however, most are excluded from the follicles by other B cells and die. It therefore seems that the microenvironment of the lymphoid follicle provides signals essential for the survival of B cells.

The identity of these signals and their source still remain to be determined, although recent results suggest that entry into the follicles might be controlled by the follicular chemokine BLC, whose receptor, BLR1, is expressed by B cells as they mature. B cells that succeed in migrating into a follicle seem to receive a survival signal; they stay there for about a day, then move on into the efferent lymphatics and back into the blood. Their continued survival depends, however, on further recirculation through the follicles of secondary lymphoid tissue. It also depends, as we have seen in Section 6-6, on the continued expression of surface immunoglobulin, indicating that a survival signal is repeatedly delivered through the B-cell receptor for antigen.

Because the number of follicles is limited, exclusion from these follicles might explain how the size of the mature B lymphocyte pool is regulated and why the majority of newly generated B cells that emerge from the bone marrow die in the periphery after only a few days. Thus the life of a B cell can be viewed as a continuing Darwinian struggle in which B cells compete with each other for access to lymphoid follicles. As we shall see, this competition might also have an important part in the elimination of some self-reactive B cells.

Summary.

B cells are generated throughout life in the specialized environment of the bone marrow and may require a second environment, the lymphoid follicle, to maintain their existence as mature recirculating B cells. As B cells differentiate from primitive stem cells, they proceed through stages that are marked by the sequential rearrangement of immunoglobulin gene segments to generate a diverse repertoire of antigen receptors. This developmental program also involves changes in the expression of other cellular proteins and is directed by transcription factors. It is characterized by two important checkpoints; as each complete immunoglobulin chain is generated and expressed at the cell surface it signals the developing cell to cease rearrangement of the set of gene segments specifying that chain and to progress to the next developmental step. Conversely, if successive rearrangements fail to generate first a heavy chain that can form a pre-B cell receptor, and then a light chain that can be expressed as part of a complete immunoglobulin molecule at the cell surface, the developing B cell fails to progress and dies. The end product of this process is an immature B cell with surface immunoglobulin of a single specificity. At this stage in its development, the B cell is ready for selective events driven by antigen binding to this receptor. It has ended its antigen-independent phase of existence and enters the antigen-dependent phase.

Selection of B cells.

If an antigen binds specifically to the newly expressed surface IgM of an immature B lymphocyte, the cell is eliminated or inactivated. Thus, those B cells that recognize self molecules while still immature are prevented from developing further and from secreting antibodies that bind self cells or tissues. This results in a B-cell receptor repertoire tolerant of the self molecules encountered up to this stage. We shall see that the form of presentation of a self molecule, for instance whether membrane-bound or soluble, makes a difference to how it leads to tolerance. The immature B cell can die, can survive for a short time but become non-functional, or can edit its receptors to a new specificity. This screening for potential autoreactivity takes place both in the bone marrow and as the B cells first migrate to the peripheral lymphoid tissues. Here, they can encounter other self molecules that were not available to the immature B cells. A variety of supplementary mechanisms render mature self-reactive B cells unresponsive to self antigens; these are described in Section 13-28.

6-9 Immature B cells can be eliminated or inactivated by contact with self antigens.

The immature B cells generated in the bone marrow express only surface IgM. The maturation of naive B cells involves emigration from the bone marrow, migration into secondary lymphoid tissue, and the alternative splicing of heavy-chain transcripts to generate mature B cells expressing surface IgM and IgD (see Fig. 3.25). The transition into a mature B cell also involves the expression of higher levels of CR2 (CD21)—a receptor for the complement component C3d and a component of the B-cell co-receptor (see Section 5-8). A proportion of newly generated immature B cells fail to make this transition, however, and one reason for this is the elimination of potentially self-reactive cells.

Immature B cells expressing only IgM are eliminated or inactivated if they bind to abundant multivalent ligands. This can be demonstrated by using anti-μ chain antibodies to crosslink the surface IgM of immature B cells, mimicking the effect of multivalent antigens; such treatment results in the inactivation or death of the immature B cells. This response distinguishes immature B cells from mature B cells, which are normally activated by multivalent antigens. Thus, immature B cells bearing self-reactive receptors must be eliminated during the few days before the change in responsiveness occurs.

Experiments with transgenic mice have shown that two different mechanisms ensure tolerance to self antigens encountered by immature B cells. One of them operates when there are multivalent antigens, for example multiple copies of an MHC molecule on a cell surface; the other operates when the antigens are of low valence, for example small soluble proteins (Fig. 6.11). The effect of encounter with a multivalent antigen was tested in mice transgenic for both chains of an antibody specific for H-2K^b MHC class I molecules; in such mice, for the reasons discussed in Section 6-6, all B cells that develop bear the anti-MHC immunoglobulin as surface IgM. If the transgenic mouse does not express H-2K^b, normal numbers of B cells develop, all bearing transgene-encoded anti-H-2K^b receptors. However, in mice bearing both H-2K^b and the immunoglobulin transgenes, B-cell development is blocked: normal numbers of pre-B cells and immature B cells are found, but B cells expressing the anti-H-2K^b antibody as surface IgM never mature to populate

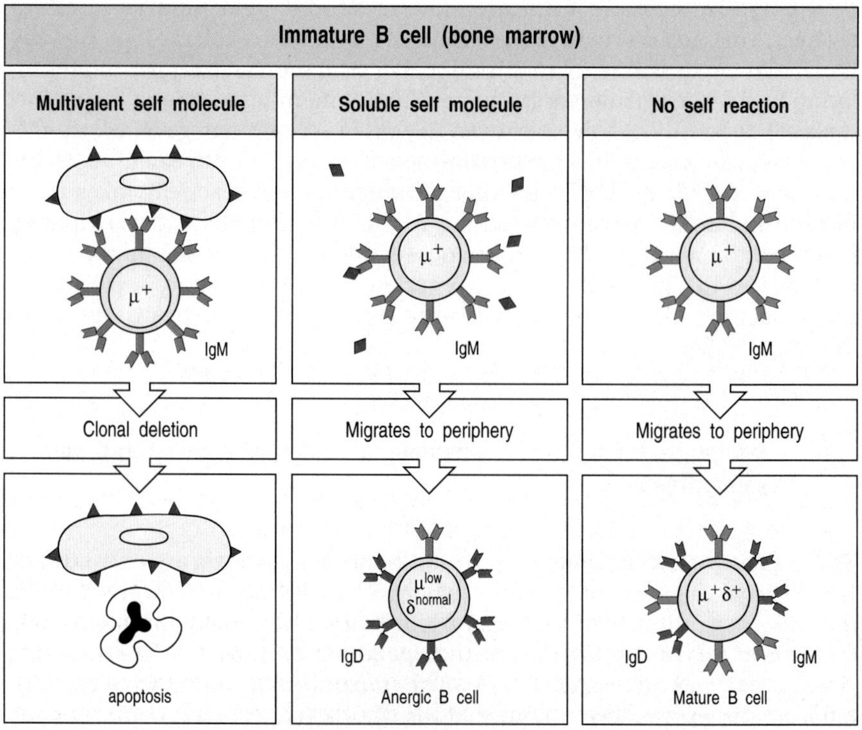

Fig. 6.11 Binding to self molecules in the bone marrow can lead to the death or inactivation of immature B cells. Left panels: when developing B cells express receptors that recognize ubiquitous self cell-surface molecules, such as those of the MHC, they are deleted from the repertoire (clonal deletion). These B cells undergo programmed cell death or apoptosis. Center panels: immature B cells that bind soluble self antigens are rendered unresponsive to the antigen (anergic) and bear little surface IgM. They migrate to the periphery where they express IgD but remain anergic; if in competition with other B cells in the periphery, they are rapidly lost. Right panels: only immature B cells that do not encounter antigen mature normally; they migrate from the bone marrow to the peripheral lymphoid tissues where they may become mature recirculating B cells bearing both IgM and IgD on their surface.

the spleen and lymph nodes; instead, most of these immature B cells die in the bone marrow by apoptosis. This antigen-induced loss from the B-cell population is known as **clonal deletion** (see Fig. 6.11, left panels).

In contrast, when soluble antigen binds an immature B cell, the cell is inactivated but not killed. Thus, when hen egg lysozyme (HEL) is expressed in soluble form (from a transgene) in mice that are also transgenic for high-affinity anti-HEL antibody, the HEL-specific B cells mature but are unable to respond to antigen. The non-responsive cells retain their IgM within the cell and transport little to the surface. In addition, they develop a partial block in signal transduction so that, despite normal levels of HEL-binding surface IgD, the cells cannot be stimulated by crosslinking this receptor. It seems that this block in signal transduction is at a step before the phosphorylation of the Igα and Igβ associated with the B-cell receptor (see Section 5-6), although its exact nature is not yet known. This state of non-reactivity is called **anergy**, and such B cells are defined as **anergic** (see Fig. 6.11, center panels).

An extension of these experiments shows that anergy is induced in a similar way in mature B cells that encounter and bind an abundant soluble antigen. In this case, the HEL transgene is placed under the control of an inducible promoter that can be regulated by changes in the diet of the mouse. It is thus possible to induce the production of lysozyme at any time and thereby study its effects on HEL-specific B cells at different stages of maturation. Such experiments have demonstrated that both mature and immature B cells are inactivated when they are chronically exposed to the soluble antigen.

In a normal animal, in which the B cells binding soluble antigen are in a minority, they are detained in the T-cell areas of the secondary lymphoid tissue and excluded from the primary lymphoid follicles. Since anergic B cells cannot be activated by T cells, and T-cell help will in any case not be available for self antigens to which the T cells themselves are tolerant, these antigen-binding B cells will not be activated to secrete antibody. Instead they are rapidly lost, thus ensuring that the long-lived pool of peripheral B cells is purged of such potentially self-reactive cells.

Peripheral antigens not present in the tissues through which naive B cells circulate will not be able to purge newly generated B cells as they emerge, and cells expressing receptors specific for such antigens might therefore survive. However, in the event of such a cell's contacting these self antigens, it will usually be held in check and then eliminated because of its dependence on T cells, as explained in the preceding paragraph and in Section 13-28.

6-10 Some potentially self-reactive B cells can be rescued by further immunoglobulin gene rearrangement.

In the discussion of success rates of rearrangement (see Section 6-3), we saw that a rearrangement attempt on a given chromosome can be repeated. Although this might normally occur after an unproductive rearrangement (see Fig. 6.6), it has also been shown to rescue immature self-reactive B cells by deleting a rearrangement that encodes a self-reactive receptor and replacing it with the product of a further rearrangement event. These **receptor editing** phenomena have been demonstrated in mice bearing transgenes for auto-antibody heavy and light chains that have been 'knocked in' to the genome.

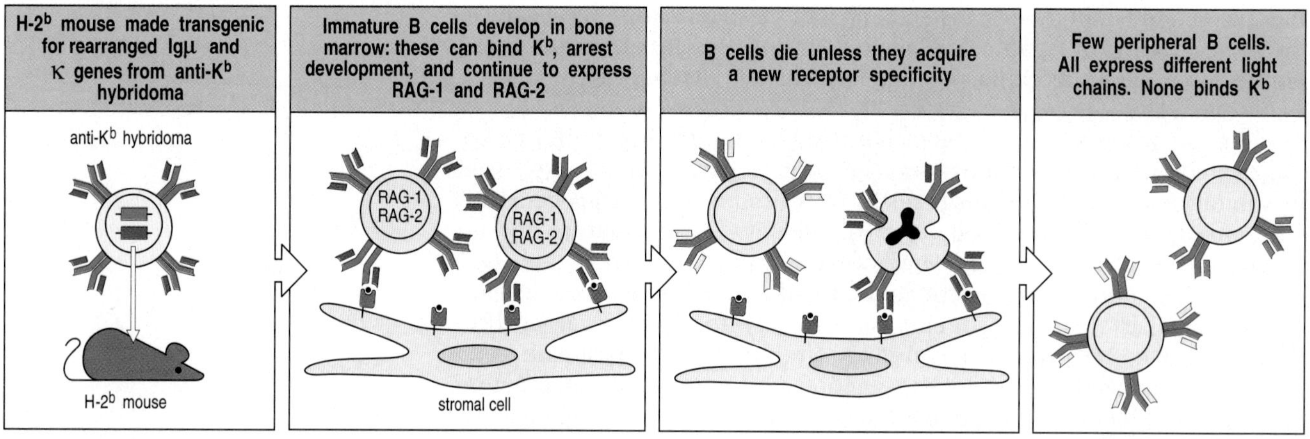
H-2b mouse made transgenic for rearranged Igμ and κ genes from anti-Kb hybridoma
Immature B cells develop in bone marrow: these can bind Kb, arrest development, and continue to express RAG-1 and RAG-2
B cells die unless they acquire a new receptor specificity
Few peripheral B cells. All express different light chains. None binds Kb
anti-Kb hybridoma
H-2b mouse
RAG-1
RAG-2
stromal cell

Fig. 6.12 Replacement of light chains by receptor editing can rescue some self-reactive B cells by changing their antigen specificity. To demonstrate receptor editing, rearranged heavy- and light-chain genes of an immunoglobulin specific for the MHC class I molecule H-2K^b are introduced into an H-2K^b mouse, which expresses the H-2K^b molecule on its tissues (first panel). The developing B cells all express the anti-H-2K^b immunoglobulin at a very early stage and thus can bind to the H-2K^b molecules in the bone marrow (second panel). Binding to H-2K^b in the bone marrow seems to signal the B cells to arrest their development and continue to express the RAG-1 and RAG-2 proteins. This enables some B cells to make further light-chain gene rearrangements that generate new receptor specificities (yellow). Most of the B cells do not make new rearrangements and are deleted (third panel). If a B cell makes a new receptor that does not bind self antigens in the bone marrow, it can mature and enter the peripheral circulation. Thus, in the H-2K^b mice that carry this transgene, there are many fewer peripheral B cells than normal and none expresses the transgenic light chain (fourth panel).

Unlike ordinary transgenes, which integrate anywhere in the genome, these transgenes are placed in their normal site within the immunoglobulin V and J gene segment loci by a method explained in Section 2-27. In this way, the transgene imitates a primary rearrangement and is surrounded by unused endogenous gene segments. In mice that express the antigen recognized by the transgene-encoded receptor, the few peripheral B cells that emerge are not self-reactive because they have used these surrounding gene segments for further rearrangements that delete the self-reactive transgene (Fig. 6.12).

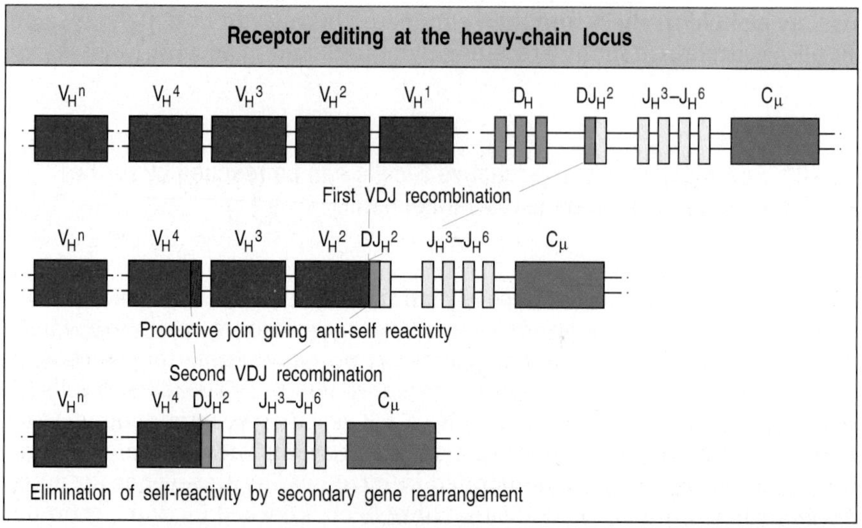
Receptor editing at the heavy-chain locus
VHn VH4 VH3 VH2 VH1 DH DJH2 JH3–JH6 Cμ
First VDJ recombination
VHn VH4 VH3 VH2 DJH2 JH3–JH6 Cμ
Productive join giving anti-self reactivity
Second VDJ recombination
VHn VH4 DJH2 JH3–JH6 Cμ
Elimination of self-reactivity by secondary gene rearrangement

Fig. 6.13 Immunoglobulin heavy-chain gene rearrangements can be rescued by further rearrangements that make use of additional recombination signal sequences. The first V to DJ recombination event that takes place at the heavy-chain locus excises all unused D_H segments so that further rearrangements with the conventional recombination signal sequences flanking unused gene segments are not possible. Instead, additional recombination signal sequences that are embedded in the 3′ ends of most V_H gene segments can be used to displace a self-reactive V_H from the first VDJ join and replace it with a different V_H gene segment.

At a light-chain locus, the conventional recombination signal sequences that flank the unused gene segments can be used for further rearrangements, as is thought to occur after a non-productive rearrangement (see Fig. 6.6). At the heavy-chain locus, however, receptor editing requires the use of embedded recombination signal sequences in a recombination event that displaces the V gene segment sequence from the self-reactive rearrangement and replaces it with a new V gene segment (Fig. 6.13). These recombination signal sequences are found in highly conserved regions at the 3′ end of most V_H gene segments and they might also be used to replace non-productive rearrangements at the heavy-chain locus, although this has yet to be demonstrated. So far, receptor editing has been demonstrated only in mice transgenic for rearranged immunoglobulin genes or in normal mouse cells treated with anti-immunoglobulin. However, active RAG-1 and RAG-2 enzymes have been found in rare immature B cells in the bone marrow and more recently in rare B cells in the peripheral lymphoid tissue (see Chapter 9) of normal mice; these cells might represent self-reactive B cells undergoing further receptor gene rearrangement.

6-11 In some species most immunoglobulin gene diversification occurs after gene rearrangement.

In some non-human species B cells develop in a different fashion. In birds, rabbits, cows, pigs, sheep, and horses there is little or no **germline diversity** in the V, D, and J gene segments that form the initial B-cell receptors, and the rearranged V-region sequences are identical or similar in most immature B cells. These B cells then migrate to a specialized micro-environment, the best known of which is the bursa of Fabricius in chickens. Here, surface immunoglobulin-positive B cells proliferate rapidly, and their rearranged receptor genes undergo further diversification. In birds and rabbits this occurs by a process of **gene conversion**, in which an unused V gene segment exchanges short sequences with the expressed rearranged V-region gene (Fig. 6.14). In sheep and cows diversification is the result of somatic hypermutation, which occurs in an organ known as the ileal Peyer's patch.

It is possible that the B-cell receptor first generated in these animals is specific for a ligand or a B-cell superantigen that is expressed by cells in the bursa of Fabricius or its equivalent, and that recognition of this ligand by the invariant B-cell receptor drives the intense B-cell proliferation found at such sites. According to this model, the cells can only stop proliferating and emerge from the bursa as mature B cells when their receptors have diversified to such an extent that they no longer recognize this self ligand (see Fig. 6.14).

It is not known why this alternative strategy of diversifying an initially homogeneous pool of receptors exists in these species. The phenomenon is in some respects analogous to the receptor editing that occurs in mice when immature B cells bind self antigens (see Section 6-10). It might turn out to be a general rule that lymphocyte repertoires are selected both positively and negatively by the signals they receive from self ligands as they mature. We shall see in Chapter 7 that the interaction of receptors on newly formed lymphocytes with self molecules in a specialized micro-environment is crucial for the development of the other major lymphocyte subset—the T cells. This positive selection in the T-cell lineage might be seen as a parallel to the proposed recognition of self ligands in the B-cell lineage, but this remains very controversial.

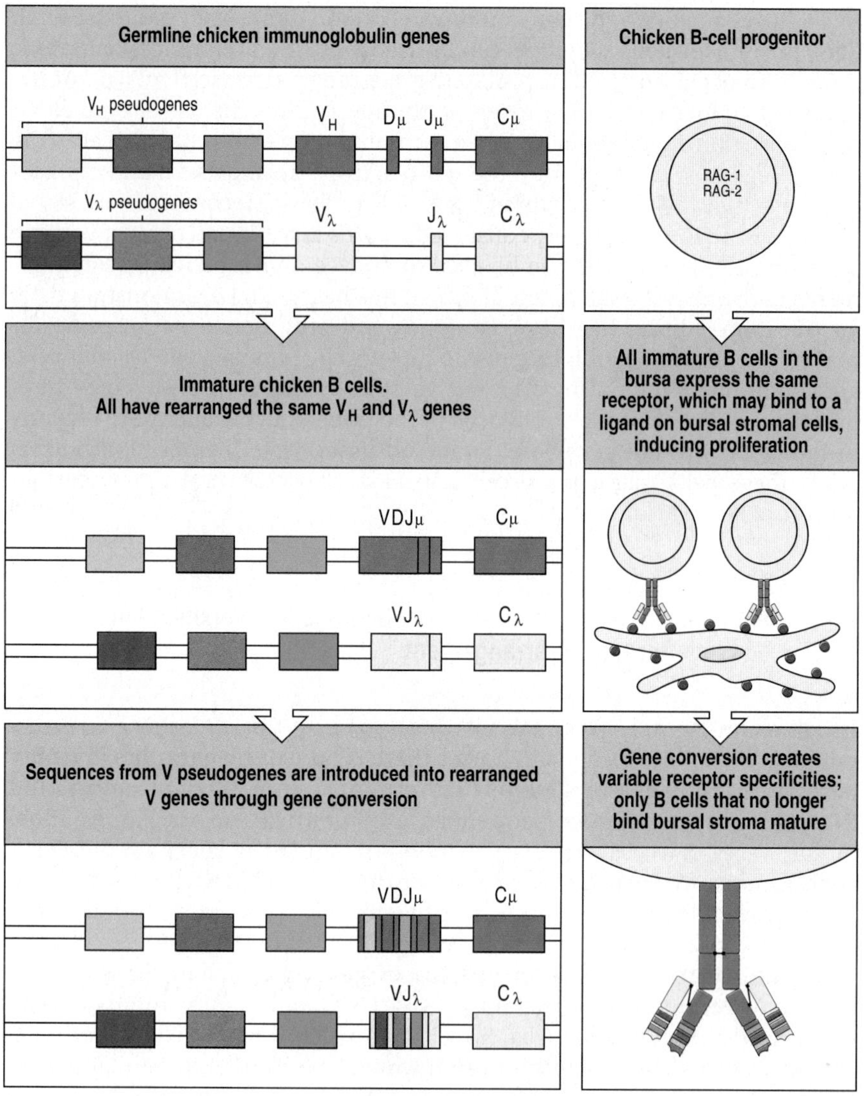

Fig. 6.14 The diversification of chicken immunoglobulins occurs through gene conversion. In chickens, all B cells express the same surface immunoglobulin initially; there is only one active V, D, and J gene segment for the chicken heavy-chain gene and one active V and J for each of the light-chain genes (top left panel). Gene rearrangement can thus produce only a single receptor specificity. Immature B cells expressing this receptor migrate to the bursa of Fabricius, where the receptor is hypothesized to engage a self cell-surface molecule that induces cell proliferation (center panels). Gene conversion events introduce sequences from adjacent V pseudogenes into the expressed gene, creating diversity in the receptors (bottom panels). In this scheme, only when a B cell loses reactivity to the bursal self molecule does it cease to proliferate and emigrate to the periphery.

6-12 B cells are produced continuously but only some contribute to a relatively stable peripheral pool.

Our knowledge of the dynamics of B-cell populations comes from labeling experiments in the bone marrow of normal young adult mice. Here, about 35×10^6 large pre-B cells enter mitosis each day. However, only $10–20 \times 10^6$ new B lymphocytes emerge as the end-products of primary development in the bone marrow. The loss of more than half of the initial pre-B cells must be caused by their failure to make a productive light-chain gene rearrangement and to the deletion of self-reactive immature B cells while still in the bone marrow. Figure 6.15 shows the possible fates of the newly made B cells that survive to leave the bone marrow and enter the periphery.

The daily output of $10–20 \times 10^6$ new B cells is roughly 5–10% of the total B-lymphocyte population in the steady-state peripheral pool. Although the size of this pool is not easy to measure, it seems to remain constant, and the stream of new B cells must be balanced by the death of an equal number of peripheral B cells. These deaths are chiefly in the short-lived peripheral B-cell population, which has a turnover time of less than a week. Most

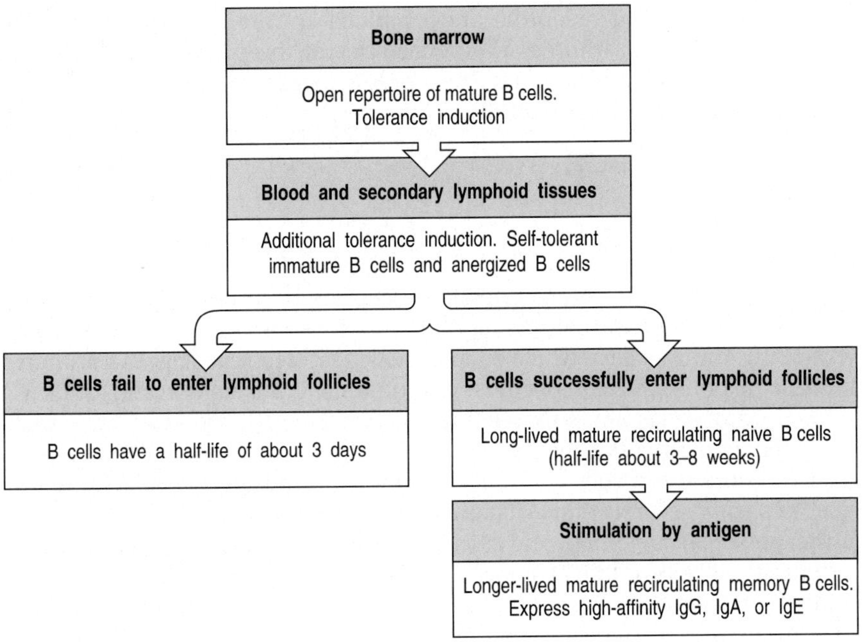

Fig. 6.15 Proposed population dynamics of conventional B cells. B cells are produced as receptor-positive immature B cells in the bone marrow. The most avidly self-reactive B cells are removed at this stage. B cells then migrate to the periphery where they enter the secondary lymphoid tissues. It is estimated that $10–20 \times 10^6$ B cells are produced by the bone marrow and exported each day in a mouse, and an equal number is lost from the periphery. There seem to be two classes of peripheral B cell: long-lived B cells and short-lived B cells. The short-lived B cells are recently formed B cells by definition. Most of the turnover of short-lived B cells might result from B cells that fail to enter lymphoid follicles; in some cases this is a consequence of being rendered anergic by binding to soluble self-antigen. About half of all peripheral B cells are relatively long-lived. These mature naive B cells recirculate through peripheral lymphoid tissues and have a half-life of 3–8 weeks. Memory B cells, which have been activated previously by antigen and T cells, have a longer life, sustained by autocrine production of nerve growth factor.

B cells have a short life span once they leave the bone marrow and enter the periphery; only a small number of newly made B cells survive to become part of the pool of relatively long-lived peripheral B cells.

As discussed in Section 6-8, the failure of most B cells to survive for more than a few days in the periphery may be due to the continual competition between peripheral B cells for access to a limited number of follicles that provide survival signals. The population dynamics indicate that this competition favors B cells that are already established in the relatively long-lived and stable peripheral B cell pool. Why this should be so is not entirely clear, although phenotypic changes that occur as the naive B cell matures, such as the expression of the BLR1 receptor for the follicular chemokine BLC, and the increased expression of the B-cell co-receptor component CR2, might have a role. Receptor specificity might also have a part in selecting a peripheral B-cell pool that continues to do well in the ongoing competition to survive. We know that some of the short-lived peripheral B cells fail to survive because they have bound soluble self antigen and are excluded from the lymphoid follicles for this reason. However, the extent to which recruitment into the long-lived pool of recirculating B cells is ligand-mediated and governed by receptor specificity is not known. Several experiments indicate a positive role for receptor signaling in the maturation and continued recirculation of peripheral B cells, but such signaling need not depend on antigen-specific interactions; the receptor could, for example, be acting as a conduit for co-receptor engagement with complement components fixed to follicular dendritic cells.

The labeling studies in mice show that long-lived peripheral B cells have a broad distribution of life spans. There are relatively long-lived naive B cells with a half-life of 1–2 months; these are thought to be sustained by signals received each time they pass through follicles. There are also non-dividing B cells that are very long-lived. These include the memory B cells that differentiate from mature B cells after their first encounter with antigen. Memory B cells persist for extended periods after antigenic stimulation, and also seem to require intermittent follicular signals; we shall return to B-cell memory in Chapter 10. The production and loss of new B cells ensures that different receptors are produced continuously to meet new antigenic challenges, whereas the persistence of the progeny of cells that have been activated ensures that those cells proven to recognize pathogens are retained to combat reinfection.

Summary.

Experiments using transgene-encoded receptors have demonstrated three different mechanisms that prevent potentially self-reactive B cells from developing to the stage at which encounter with antigen can trigger a primary immune response. B cells specific for ubiquitous multivalent ligands such as self MHC molecules are eliminated soon after their antigen receptor is first expressed, in a process known as clonal deletion. A small proportion of self-reactive immature B cells can escape this fate, however, by replacing their receptors with new receptors that are no longer autoreactive. This occurs by further gene rearrangement and is known as receptor editing. Both immature and mature B cells binding soluble self antigens can be rendered anergic on binding the antigen. Although such cells are exported from the bone marrow, they are excluded from lymphoid follicles and do not survive long in the periphery. Thus, the mature peripheral B-cell repertoire is a selected subset of the immature B-cell repertoire. Indeed, only a small proportion of newly made immature B cells survive to form a part of the long-lived peripheral B-cell pool, which indicates that this selection is extensive, although how much of it is ligand-mediated and governed by receptor specificity remains unclear. A very different strategy of generating a diverse peripheral B-cell repertoire is seen in some mammals and in birds. In these species, immature B cells expressing only limited gene rearrangements migrate to specialized organs where their receptors are diversified either by gene conversion or by somatic hypermutation. This might be a ligand-driven process whose end-products are selected for a loss of reactivity to the putative self ligand.

B-cell heterogeneity.

The B cells in peripheral lymphoid tissues are heterogeneous; different populations are found in different locations and are distinguished by the different cell-surface molecules they express. Part of this heterogeneity results from the B-cell response to antigenic stimulation: naive recirculating B cells are clearly distinct from the B lymphoblasts actively responding to antigen, whereas the very long-lived memory B lymphocytes can be distinguished from naive B cells by the expression of isotypes other than IgM and IgD, as well as by other cell-surface changes. There is also a population of B cells that can arise from distinct stem cells early in an animal's development and that renews itself by continuing division in the peripheral lymphoid tissues of adult animals. These cells are known as B-1 cells to distinguish them from the conventional B-2 cells whose development we have described in the first part of this chapter; they seem to represent a distinct developmental lineage of B cells.

In the remaining part of this chapter, we describe these populations of B cells and some malignant tumors of B-lineage cells that reflect their distinctiveness. Because tumors of B cells are a source of large numbers of cells that can be identified by their identical immunoglobulin gene rearrangements, they have proved invaluable for studying B-cell development, homing behavior, and function. Tumors of fully differentiated antibody-secreting plasma cells provided the means for understanding the genetic basis of antibody diversity and isotype switching; tumors of less differentiated B-lineage cells have illustrated the steps through which B-cell development proceeds. Some tumors representing B cells at early stages of development retain the ability to rearrange their immunoglobulin genes, and much of what we know about gene rearrangement has come from studying these B-cell tumor lines. Malignant tumors of B cells have also shed light on the normal processes by which B-cell growth is controlled, as we shall see in Section 6-16.

6-13 B cells bearing surface CD5 express a distinctive repertoire of receptors.

Not all B cells conform to the developmental pathway that we have described in previous sections. A significant subset of B cells in mice and humans, and the major population in rabbits, arises early in ontogeny and has a distinctive receptor repertoire and functional properties (Fig. 6.16). These B cells were first identified by surface expression of the protein CD5 and are also characterized by displaying surface IgM with little or no IgD even when mature. These unconventional B-lineage cells are termed **B-1 cells**, because their development precedes that of the conventional B cells, sometimes termed **B-2 cells**. They are also known as **CD5 B cells**, although CD5 itself cannot be essential for their function because cells that have similar traits develop normally in mice lacking the CD5 gene, and in rats B-1 cells do not display CD5.

Little is currently known about the function of B-1 cells. Although relatively sparse in lymph nodes or spleen, they are the predominant B-cell population in the peritoneal and pleural cavities. B-1 cells make little contribution to adaptive immune responses to protein antigens but contribute strongly to some antibody responses against carbohydrate antigens, as we shall see

Property	B-1 cells	Conventional B-2 cells
When first produced	Fetus	After birth
Mode of renewal	Self-renewing	Replaced from bone marrow
Spontaneous production of immunoglobulin	High	Low
Specificity	Polyreactive	Monospecific, especially after immunization
Isotypes secreted	IgM >> IgG	IgG > IgM
Somatic hypermutation	Low–none	High
Response to carbohydrate antigen	Yes	Maybe
Response to protein antigen	Maybe	Yes

Fig. 6.16 A comparison of the properties of B-1 cells and conventional B cells or B-2 cells. B-1 cells can develop in unusual sites in the fetus, such as the omentum, in addition to the liver, and are produced from the bone marrow for only a short time around the time of birth. An autonomous self-renewing pool of these cells then establishes itself outside the marrow. The limited diversity of the B-1 cell repertoire and the propensity of B-1 cells to produce low-affinity polyreactive antibodies suggest that they mediate a more primitive, less adaptive, immune reponse than the conventional B cells or B-2 cells.

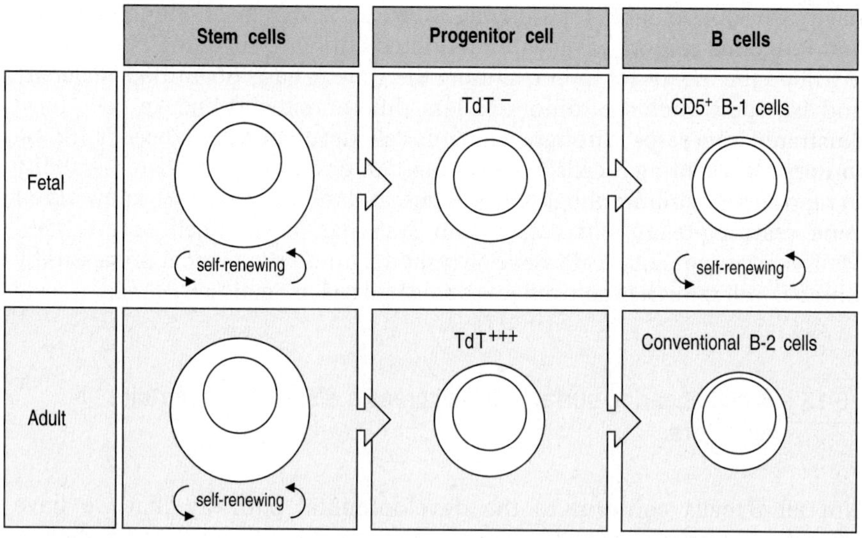

Fig. 6.17 Stem cells at different stages of development give rise to distinct B-cell populations. Stem cells in fetal mice give rise to progenitors that have little or no terminal deoxynucleotidyl transferase (TdT), which in turn give rise to B-1 cells. These cells have few N-nucleotides in their V(D)J junctions and are self-renewing. Adult bone marrow stem cells give rise only to conventional B cells that have high junctional diversity and a range of life spans; they are not self-renewing and are replaced continuously by new cells generated in the bone marrow.

in Chapter 9. Moreover, a large proportion of serum IgM in normal non-immune mice derives from B-1 cells. These distinguishing features may reflect the different developmental origin of B-1 cells and their distinct receptor repertoire.

B-1 cells arise early in ontogeny. In mice, they derive from an immature stem cell that is most active in the prenatal period (Fig. 6.17). In B-1 cells, the immunoglobulin heavy-chain gene rearrangements are dominated by those using V_H gene segments that lie closest to the D gene segments in the germline. As TdT is not active in the prenatal period, these early heavy-chain gene rearrangements are accompanied by few, if any, N-nucleotide insertions. The VDJ junctions expressed by B-1 cells are thus less diverse than those of conventional B-2 cells. The recognition properties of the receptors on B-1 cells also differ from those of conventional B-2 cells. B-1 cell receptors and the antibodies that these cells produce tend to bind numerous different ligands with relatively low affinity, a property known as **polyreactivity**. These polyreactive receptors show a preference for binding common bacterial polysaccharides. Indeed, the V gene segments that encode the receptors of B-1 cells might have evolved by natural selection to recognize common bacterial antigens, allowing them to contribute to early phases of the immune response, which we shall discuss in detail in Chapter 10.

B-1 cells that develop postnatally use a more diverse repertoire of V genes and their rearranged immunoglobulin genes have abundant N-nucleotides. As mice develop, however, the bone marrow stem cells seem to undergo a developmental change such that only B-2 cells are produced (see Fig. 6.17). In adult animals, the population of B-1 cells is maintained by continued division in peripheral sites, a process that requires IL-10. B-1 cells are of interest to clinicians, as they are the origin of the common B-cell tumor in the disease chronic lymphocytic leukemia (CLL). CLL cells often display CD5, which is a useful diagnostic clue.

There are some striking similarities between the development of B-1 cells and that of the second lineage of T cells, the $\gamma{:}\delta$ T cells. As we shall see in Chapter 7, $\gamma{:}\delta$ T cells expressing a limited repertoire of receptors with no N-nucleotide additions dominate T-cell production early in ontogeny, whereas later $\gamma{:}\delta$ T cells are generated from different V_γ gene segments and have more diverse receptors. Their production is secondary to that of the major lineage of $\alpha{:}\beta$ T cells.

6-14 B cells at different developmental stages are found in different anatomical sites.

B cells are not sessile. They change their location as they develop in the bone marrow and eventually leave the bone marrow to travel to the B-cell rich areas of peripheral lymphoid tissues, such as the follicles of spleen and lymph nodes (Fig. 6.18). Recirculating B cells migrate through the secondary lymphoid organs, entering via high endothelial venules. They pass through the outer T-cell zones and reach the follicles, where they remain for about a day before leaving the lymphoid organ in the lymph and re-entering the blood. Many B cells are found in the gut-associated lymphoid tissues, which have high B- to T-cell ratios. These tissues include the very large lymphoid follicles known as Peyer's patches, the appendix, and the tonsils, all of which provide specialized sites where B cells can become committed to synthesizing IgA.

If B cells encounter antigen before or on entering the lymphoid tissue they stay in the outer T-cell zone, where they are very efficient at finding primed T cells. If a B cell makes an antigen-specific interaction with a T cell it is induced to proliferate and its progeny migrate either to follicles, where they form germinal centers, or to extra-follicular foci, where they divide further before becoming short-lived antibody-secreting plasma cells. Such extra-follicular sites of B-cell growth and maturation are found in the medullary cords of lymph nodes or the part of the red pulp of the spleen adjacent to the T-cell zone. There are no obvious equivalent sites of extra-follicular B-cell growth in the palatine tonsils or Peyer's patches. The activated B cells that migrate to follicles form **germinal centers** (see Fig. 6.18). Here, they undergo intense proliferation accompanied by somatic hypermutation of their rearranged V-region genes and there is selection for cells bearing receptors with a higher affinity for the stimulating antigen (see Section 3-18 and Chapters 9 and 10).

As some of the progeny B cells develop into antibody-secreting plasma cells they migrate again; plasma cells are found predominantly in the medullary cords of lymph nodes, in the red pulp of the spleen, in the bone marrow and in the lamina propria of mucosal tissues. The bone marrow can be an important site of IgG antibody production. In hyperimmune animals producing IgG antibody in response to repeated immunization with antigen, up to 90% of the antibody can be derived from plasma cells in the bone marrow. By contrast, in IgA responses, which mostly develop in the gut-associated lymphoid tissue,

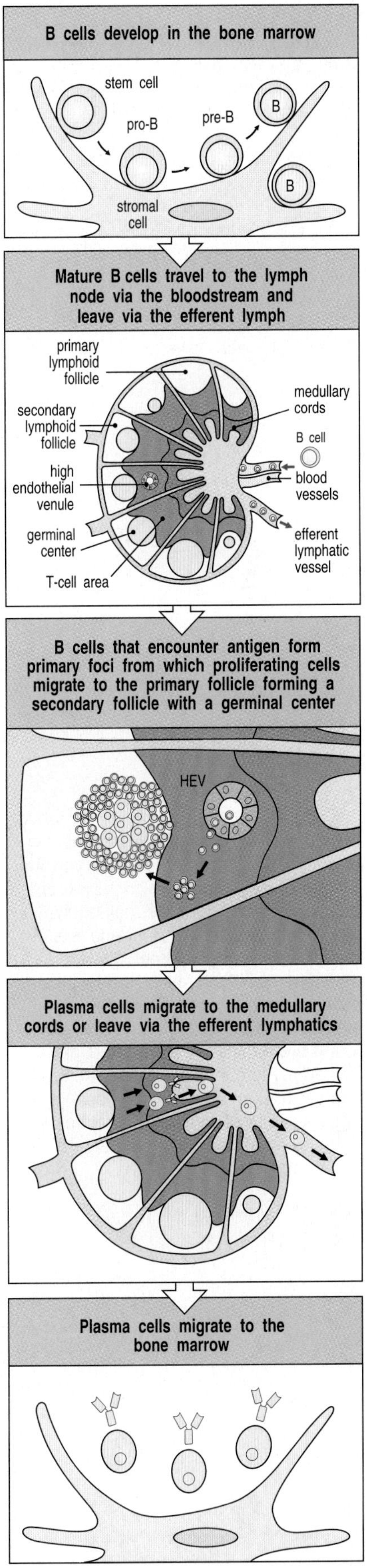
B cells develop in the bone marrow
stem cell
pro-B
pre-B
B
B
stromal cell
Mature B cells travel to the lymph node via the bloodstream and leave via the efferent lymph
primary lymphoid follicle
secondary lymphoid follicle
high endothelial venule
germinal center
T-cell area
medullary cords
B cell
blood vessels
efferent lymphatic vessel
B cells that encounter antigen form primary foci from which proliferating cells migrate to the primary follicle forming a secondary follicle with a germinal center
HEV
Plasma cells migrate to the medullary cords or leave via the efferent lymphatics
Plasma cells migrate to the bone marrow

Fig. 6.18 The distribution of conventional B cells within a secondary lymphoid organ. B-lineage cells are found in bone marrow, blood, lymphoid organs, and lymph. In adult mammals, B cells develop in the bone marrow (first panel). They then migrate via the blood to the secondary lymphoid organs, where they leave the circulation by crossing the high endothelium of specialized venules (HEV) and enter the cortex of the lymphoid organ (second panel). Their progress through a lymph node is illustrated here; the other major peripheral tissues are the spleen and Peyer's patches of the gut. If they do not encounter their specific antigen, they either die or migrate through primary follicles and leave the lymph node by the efferent lymphatic vessels, and eventually return to the blood through the thoracic duct (not shown). In the presence of appropriately presented antigen, B cells are activated by helper T cells to form primary foci of proliferating cells from which the B cells then migrate to form the germinal center within a follicle (third panel; see also Fig. 9.10). Germinal centers are sites of rapid B-cell proliferation and differentiation. Some of these cells then migrate to the medullary cords of the lymph node to complete their differentiation into antibody-secreting plasma cells (fourth panel). Other plasma cells leave the lymph node via the efferent lymphatics and migrate to the bone marrow (fifth panel). A few weeks after it forms, the germinal center reaction has died down.

the lamina propria of the mucosa is the destination of the antibody-producing cells. Other B cells leaving the germinal centers become memory B cells. These do not recirculate but are found in the marginal zones of the spleen, under the sub-capsular sinus of the lymph nodes, under the intestinal epithelium in Peyer's patches and in the crypt epithelium of tonsils. Some memory B cells also migrate through the blood. Memory B cells might have undergone switch recombination and then express IgG or IgA on their surface, but a significant proportion are not switched and express IgM without IgD.

6-15 B-cell tumors often occupy the same site as their normal counterparts.

Tumors retain many of the characteristics of the cell type from which they arose, especially when the tumor is relatively differentiated and slow-growing. This is clearly illustrated in the case of B-cell tumors. Tumors corresponding to essentially all stages of B-cell development have been found in humans, from the earliest stages to the myelomas that represent malignant outgrowths of plasma cells (Fig. 6.19). Furthermore, each type of tumor retains its characteristic homing properties. Thus, a tumor that resembles mature, naive B cells homes to follicles in lymph nodes and spleen, giving rise to a follicular center cell lymphoma, whereas a tumor of plasma cells usually disperses to many different sites in bone marrow, from which comes the clinical name of multiple myeloma (tumor of bone marrow). These similarities mean that it is possible to use tumor cells, which are available in large quantities, to study the cell-surface molecules and signaling pathways responsible for homing behavior.

Fig. 6.19 B-cell tumors represent clonal outgrowths of B cells at various stages of development. Each type of tumor cell has a normal B-cell equivalent, homes to similar sites, and has behavior similar to that cell. Thus, myeloma cells look much like the plasma cells from which they derive, they secrete immunoglobulin, and they are found predominantly in the bone marrow. Many lymphomas and myelomas can go through a preliminary less aggressive lymphoproliferative phase, and some mild lymphoproliferations seem to be benign.

Name of tumor	Normal cell equivalent	Location	Status of Ig V genes
Chronic lymphocytic leukemia	CD5 B-1 cell	Blood	Mutated
Acute lymphoblastic leukemia	Lymphoid progenitor	Bone marrow and blood	Unmutated
Pre-B cell leukemia	Pre-B cell (pre-B receptor)		Unmutated
Follicular center cell lymphoma Burkitt's lymphoma	Mature B cell	Periphery	Mutated, intraclonal variability
Waldenström's macroglobulinemia	IgM-secreting B cell		Mutated, no variability within clone
Multiple myeloma	Plasma cell. Various isotypes	Bone marrow	Mutated, no variability within clone

The status of the immunoglobulin genes in a B-cell tumor (see Fig. 6.19) provides important information on its origin; in particular it tells us whether it has been through a germinal center. Tumors arising from lymphoid precursors, pre-B cells, and B-1 cells (the vast majority of CLLs) have V genes that have not undergone somatic hypermutation. By contrast, tumors of mature B cells, such as Burkitt's lymphoma, which arises from germinal-center B cells, express mutated V genes. If the V genes from several different Burkitt's lymphoma lines from the same patient are sequenced, minor variations are seen because somatic hypermutation is an ongoing process in the tumor cells. Later-stage B-cell tumors such as multiple myelomas contain mutated genes but do not display intraclonal variation because by this stage in B cell development somatic hypermutation has ceased.

The general conclusion that a tumor represents the clonal outgrowth of a single transformed cell is very clearly illustrated by tumors of B-lineage cells. All the cells in a B-cell tumor have identical immunoglobulin gene rearrangements, decisively documenting their origin from one cell. This is useful for clinical diagnosis, as tumor cells can be detected by sensitive assays for these homogeneous rearrangements (Fig. 6.20). Only on rare occasions are biclonal tumors with two patterns of rearrangement found.

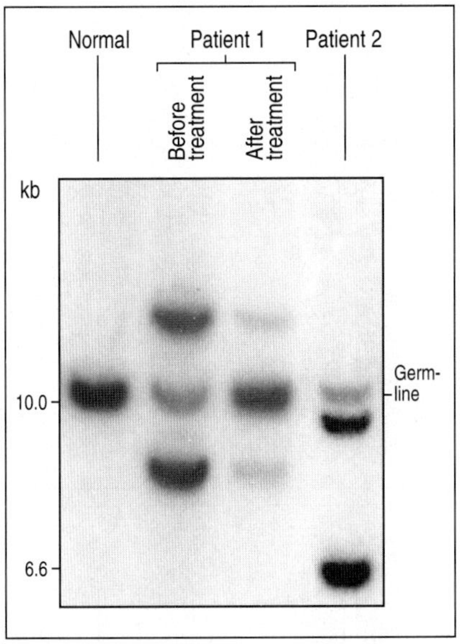

Fig. 6.20 Clonal analysis of B-cell tumors. DNA analysis of white blood cells by using Southern blotting techniques can be used to detect lymphoid malignancy. In a sample from a healthy person (left lane), immunoglobulin genes are in the germline configuration in non-B cells, so a digest of their DNA with a suitable restriction endonuclease yields a single germline DNA fragment when probed with an immunoglobulin heavy-chain J region probe (J_H). Normal B cells present in this sample make many different rearrangements to J_H, producing a spectrum of 'bands' each so faint that it is undetectable. By contrast, in samples from patients with B-cell malignancies (Patient 1 and Patient 2), where a single cell has given rise to all the tumor cells in the sample, two extra predominant bands are seen by using the J_H probe. These bands are characteristic of each patient's tumor and result from the rearrangement of both alleles of the J_H gene in the original tumor cells. The intensity of the bands compared with that of the germline band gives an indication of the abundance of the tumor cells in the sample. After anti-tumor treatment (see Patient 1), the intensity of the tumor-specific bands can be seen to diminish. Photograph courtesy of T J Vulliamy and L Luzzatto.

6-16 Malignant B cells frequently carry chromosomal translocations that join immunoglobulin loci to genes regulating cell growth.

The unregulated growth that is the most striking characteristic of tumor cells is caused by mutations that release the cell from the normal restraints on its growth. In B-cell tumors, the disruption of normal growth controls is often associated with an aberrant immunoglobulin gene rearrangement, in which one of its immunoglobulin loci is joined to a gene on another chromosome. This genetic fusion with another chromosome is known as a **translocation**, and in B-cell tumors such translocations are found to disrupt the expression and function of genes important for controlling cell growth. Cellular genes that cause cancer when their function or expression is disrupted are termed **oncogenes**.

Translocations give rise to chromosomal abnormalities that are visible microscopically in metaphase. Characteristic translocations are seen in different B-cell tumors and reflect the involvement of a particular oncogene in each tumor type. Characteristic translocations, involving the T-cell receptor loci, are also seen in T-cell tumors. Immunoglobulin and T-cell receptor loci are sites at which double-stranded DNA breaks occur during normal gene rearrangement, so it is not surprising that they are especially likely to be sites of chromosomal translocation in T and B cells.

The analysis of chromosomal abnormalities has revealed much about the regulation of B-cell growth and the disruption of growth control in tumor cells. In Burkitt's lymphoma cells, the *myc* oncogene on chromosome 8 is recombined with an immunoglobulin locus by translocations that involve either chromosome 14 (heavy chain) (Fig. 6.21), chromosome 2 (κ light chain), or chromosome 22 (λ light chain). The Myc protein is known to be involved in the control of the cell cycle in normal cells. The translocation deregulates expression of the Myc protein, which leads to increased proliferation of B cells, although other mutations elsewhere in the genome are also needed before a B-cell tumor results.

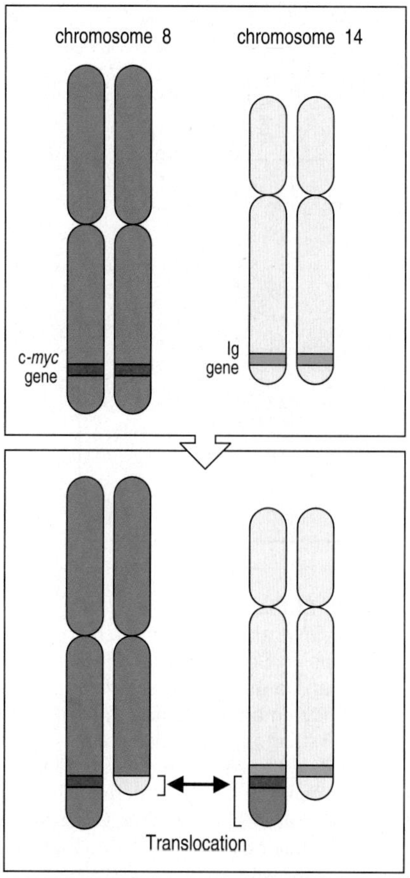

Fig. 6.21 Specific chromosomal rearrangements are found in some lymphoid tumors. If a chromosomal rearrangement joins one of the immunoglobulin genes to a cellular oncogene, it can result in the aberrant expression of the oncogene under the control of the immunoglobulin regulatory sequences. Such chromosomal rearrangements are frequently associated with B-cell tumors. In the example shown, from Burkitt's lymphoma, the translocation of the oncogene c-*myc* from chromosome 8 to the immunoglobulin heavy-chain locus on chromosome 14 results in the deregulated expression of c-*myc* and the unregulated growth of the B cell. The immunoglobulin gene located on the normal chromosome 14 is usually productively rearranged and the tumors that result from such translocations generally have a mature B-cell phenotype and express immunoglobulin.

Other B-cell lymphomas bear a chromosomal translocation of immunoglobulin genes to the oncogene *bcl-2,* increasing the production of Bcl-2 protein. The Bcl-2 protein prevents programmed cell death in B-lineage cells, so its abnormal expression allows some B cells to survive and accumulate beyond their normal life span. During this time further genetic changes can occur that lead to malignant transformation. Mice carrying an expressed *bcl-2* transgene tend to develop B-cell lymphomas late in life.

Summary.

B cells are not a single, homogeneous population. Rather, there are two major subpopulations that seem to arise from distinct stem cells and have distinctive properties. B-1 cells, many of which bear surface CD5, arise early in ontogeny and in adults form a self-renewing population, which is the dominant B-cell population in the pleural and peritoneal cavities. The repertoire of receptors expressed by B-1 cells is distinctive, and low-affinity polyreactive IgM is made by these cells. Conventional B cells, or B-2 cells, appear later in ontogeny and they continue to be generated from the bone marrow throughout the lives of both humans and mice. Conventional B cells can be divided into several subpopulations, which represent stages in their maturation. The final stage is the antibody-secreting plasma cell. Peripheral B-2 cells migrate through lymphoid tissues at various sites in the body. Cells at different stages of maturation are found in distinctive locations within the lymphoid tissues, suggesting that important microenvironmental factors act on cells at different stages of development. Our understanding of B-cell development and migratory behavior has been, and will continue to be, aided greatly by studying B-cell tumors. These tumors are readily identified by their unique immunoglobulin gene rearrangements; they often carry translocations involving genes that normally regulate B-cell growth, and their behavior often reflects the normal behavior of the cells from which they arose.

Summary to Chapter 6.

There are two major subpopulations of B cells in humans and mice. B-1 cells are generated early in ontogeny; they persist in adults as a self-renewing population and their repertoire of receptors is less diverse than that of the better-studied B-2 cells. These are also known simply as B cells; they develop in the bone marrow and are produced throughout life. The development of human B-2 cells is summarized in Fig. 6.22. The stages of B-cell development are marked by a series of irreversible changes in the immunoglobulin genes, which contribute to the diversity of antibodies, and by changes in immunoglobulin gene expression that depend on the regulation of transcription and RNA splicing (Fig. 6.23). The permanent DNA rearrangements provide molecular fingerprints that allow us to follow specific lymphocytes and their progeny throughout their life. The irreversible gene rearrangements that assemble complete immunoglobulin genes from the separate V, D, and J gene segments are regulated to ensure the monospecificity of each B cell. Heavy-chain gene segments rearrange first, and as soon as a functional heavy-chain gene is generated its product is expressed at the cell surface with a surrogate light chain, and further heavy-chain gene rearrangement ceases. Light-chain gene rearrangements are similarly halted when either an intact κ or λ chain is produced, and a complete IgM molecule is expressed at the cell surface. The generation of diversity in this primary receptor repertoire is independent of encounter with antigen and leads to some cell loss as a result

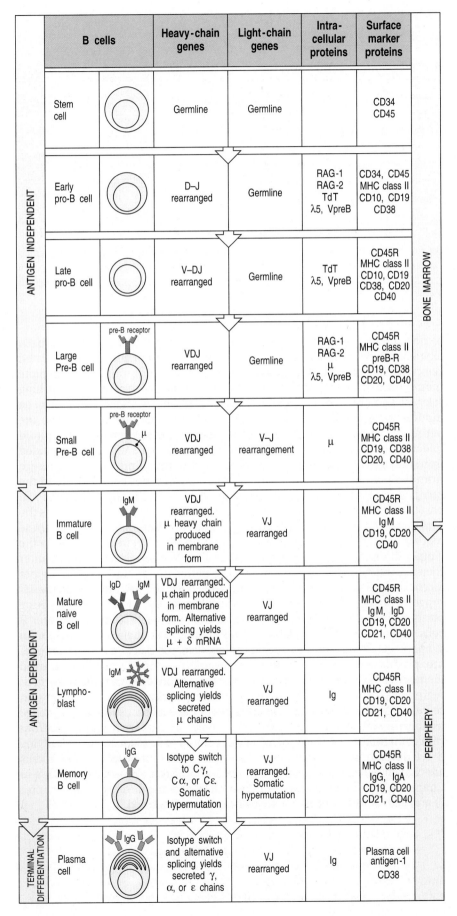

Phase	B cells		Heavy-chain genes	Light-chain genes	Intra-cellular proteins	Surface marker proteins	Location
ANTIGEN INDEPENDENT	Stem cell	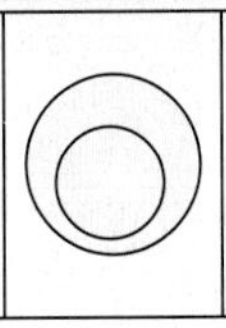	Germline	Germline		CD34 CD45	BONE MARROW
ANTIGEN INDEPENDENT	Early pro-B cell	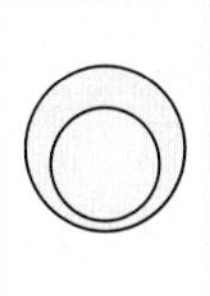	D–J rearranged	Germline	RAG-1 RAG-2 TdT λ5, VpreB	CD34, CD45 MHC class II CD10, CD19 CD38	BONE MARROW
ANTIGEN INDEPENDENT	Late pro-B cell	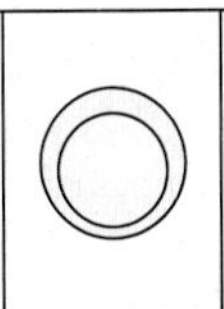	V–DJ rearranged	Germline	TdT λ5, VpreB	CD45R MHC class II CD10, CD19 CD38, CD20 CD40	BONE MARROW
ANTIGEN INDEPENDENT	Large Pre-B cell	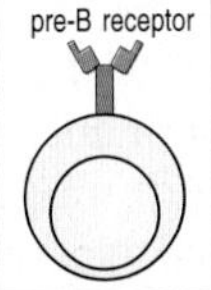 pre-B receptor	VDJ rearranged	Germline	RAG-1 RAG-2 μ λ5, VpreB	CD45R MHC class II preB-R CD19, CD38 CD20, CD40	BONE MARROW
ANTIGEN INDEPENDENT	Small Pre-B cell	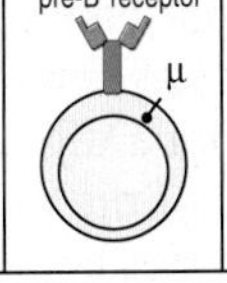 pre-B receptor μ	VDJ rearranged	V–J rearrangement	μ	CD45R MHC class II CD19, CD38 CD20, CD40	BONE MARROW
ANTIGEN DEPENDENT	Immature B cell	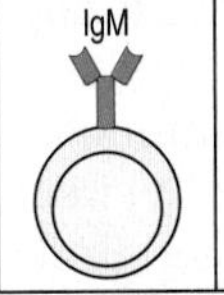 IgM	VDJ rearranged. μ heavy chain produced in membrane form	VJ rearranged		CD45R MHC class II IgM CD19, CD20 CD40	BONE MARROW
ANTIGEN DEPENDENT	Mature naive B cell	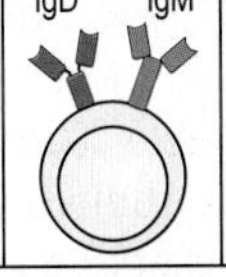 IgD IgM	VDJ rearranged. μ chain produced in membrane form. Alternative splicing yields μ + δ mRNA	VJ rearranged		CD45R MHC class II IgM, IgD CD19, CD20 CD21, CD40	PERIPHERY
ANTIGEN DEPENDENT	Lympho-blast	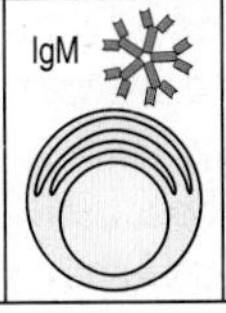 IgM	VDJ rearranged. Alternative splicing yields secreted μ chains	VJ rearranged	Ig	CD45R MHC class II CD19, CD20 CD21, CD40	PERIPHERY
ANTIGEN DEPENDENT	Memory B cell	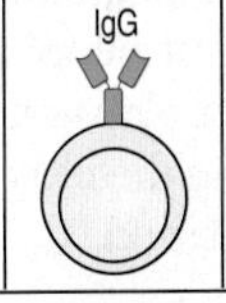 IgG	Isotype switch to Cγ, Cα, or Cε. Somatic hypermutation	VJ rearranged. Somatic hypermutation		CD45R MHC class II IgG, IgA CD19, CD20 CD21, CD40	PERIPHERY
TERMINAL DIFFERENTIATION	Plasma cell	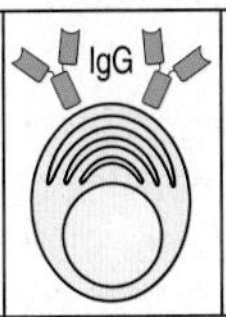 IgG	Isotype switch and alternative splicing yields secreted γ, α, or ε chains	VJ rearranged	Ig	Plasma cell antigen-1 CD38	PERIPHERY

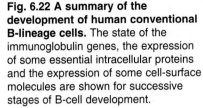

Fig. 6.22 A summary of the development of human conventional B-lineage cells. The state of the immunoglobulin genes, the expression of some essential intracellular proteins and the expression of some cell-surface molecules are shown for successive stages of B-cell development.

Event	Process	Nature of change
V-region assembly	Somatic recombination of DNA	Irreversible
Junctional diversity	Imprecise joining, N-sequence insertion in DNA	Irreversible
Transcriptional activation	Activation of promoter by proximity to the enhancer	Irreversible but regulated
Switch recombination	Somatic recombination of DNA	Irreversible
Somatic hypermutation	DNA point mutation	Irreversible
IgM, IgD expression on surface	Differential splicing of RNA	Reversible, regulated
Membrane vs secreted form	Differential splicing of RNA	Reversible, regulated

Fig. 6.23 Changes in immunoglobulin genes that accompany B-cell development and differentiation. Those changes that establish immunological diversity are all irreversible, as they involve changes in B-cell DNA. Switch recombination allows the same variable (V) region to be attached to several functionally distinct heavy-chain constant regions (as explained in Chapter 3) and thereby creates functional diversity in an irreversible manner. By contrast, the expression of IgM versus IgD, and of membrane-bound versus secreted forms of all immunoglobulin types, can in principle be reversibly regulated.

of non-productive rearrangements. Once surface immunoglobulin has been expressed, however, it can function as an antigen receptor and ligand-mediated selection can take place. Immature B cells that bind antigen in the bone marrow can die, change their receptor, or become anergic, thus establishing tolerance to these self antigens. Clonal deletion and anergy can also be induced by antigens contacted by B cells when they first emerge into the periphery. Once in the periphery, B cells seem to need access to the lymphoid follicles to survive; anergic B cells that bind soluble self antigens are excluded from the follicles and thus die. Indeed, the majority of the cells that emerge from the bone marrow survive for less than a week, perhaps because of the competition for follicular access. The repertoire of B cells that survive to form part of the relatively long-lived pool of mature peripheral B cells provides the raw material for clonal selection in the adaptive immune response. These mature naive B cells, which co-express IgM and IgD, recirculate through the lymphoid organs, including Peyer's patches in the gut, until they encounter their specific antigen. Upon interacting with antigen and specific helper T cells in the T-cell areas of lymphoid tissue, a B cell is activated to divide. Its progeny can give rise to both local extra-follicular foci of antibody-secreting plasma cells and a germinal center in a nearby follicle. Germinal-center B cells proliferate vigorously and undergo changes in their immunoglobulin V-region genes by somatic hypermutation. Selected B cells then differentiate into plasma cells, which secrete large amounts of antibody, or into long-lived memory cells, which contribute to lasting protective immunity. Many IgG-secreting plasma cells migrate to the bone marrow, whereas B cells that secrete IgA migrate to the lamina propria of the mucosal surfaces. B-lineage tumors that reflect the properties of their normal counterparts can arise from cells at many of these stages of development and have been useful in studying B-cell development and function. We shall look in more depth at how B cells are clonally selected to function in an adaptive humoral immune response in Chapter 9, but before this can be fully understood, the development (Chapter 7) and activation (Chapter 8) of T cells must be described.

General references.

Alt, F., and Marrack, P. (eds): *Current Opinion in Immunology, Vol. 10* London, Current Biology, 1998.

Casali, P., and Silberstein, L.E.S. (eds): **Immunoglobulin gene expression in development and disease.** *Ann. N.Y. Acad. Sci.* 1995, **764**.

Fischer, M.B., Goerg, S., Shen, L., Prodeus, A.P., Goodnow, C.C., Kelsoe, G., and Carroll, M.C.: **Dependence of germinal center B cells on expression of CD21/CD35 for survival.** *Science* 1998, **280**:582-585.

Goodnow, C.C., Cyster, J.G., Hartley, S.B., Bell, S.E., Cooke, M.P., Healy, J.I., Akkaraju, S., Rathmell, J.C., Pogue, S.L., and Shokat, K.P.: **Self-tolerance checkpoints in B-lymphocyte development.** *Adv. Immunol.* 1995, **59**:279-368.

Hardy, R.R., and Hayakawa, K.: **B-lineage differentiation stages resolved by multiparameter flow-cytometry.** *Ann. N.Y. Acad. Sci.* 1995, **764**:19-24.

Kelsoe, G.: ***In-situ* studies of the germinal center reaction.** *Adv. Immunol.* 1995, **60**:267-288.

Leu, T.M.J., Eastman, Q.M., and Schatz, D.G.: **Coding joint formation in a cell-free V(D)J recombination system.** *Immunity* 1997, **7**:303-314.

Linette, G.P., and Korsmeyer, S.J.: **Differentiation and cell-death—lessons from the immune-system.** *Curr. Opin. Cell Biol.* 1994, **6**:809-815.

Loffert, D., Schaal, S., Ehlich, A., Hardy, R.R., Zou, Y.R., Muller, W., and Rajewsky, K.: **Early B-cell development in the mouse—insights from mutations introduced by gene targeting.** *Immunol. Rev.* 1994, **137**:135-153.

Osmond, D.G., Rolink, A., and Melchers, F.: **Murine B lymphopoiesis: towards a unified model.** *Immunol. Today* 1998, **19**:65-68.

Rajewsky, K.: **Clonal selection and learning in the antibody system.** *Nature* 1996, **381**:751-758.

Rolink, A., and Melchers, F.: **B-Lymphopoiesis in the mouse.** *Adv. Immunol.* 1993, **53**:123-156.

Spangrude, G.J.: **Biological and clinical aspects of hematopoietic stem-cells.** *Ann. Rev. Med.* 1994, **45**:93-104.

Section references.

6-1 B cell development proceeds through several stages.

Ehrlich, A., and Kuppers, R.: **Analysis of immunoglobulin gene rearrangements in single B cells.** *Curr. Opin. Immunol.* 1995, **7**:281-284.

Kee, B.L., and Paige, C.J.: **Murine B-cell development—commitment and progression from multipotential progenitors to mature B-lymphocytes.** *Intl. Rev. Cytol.—a survey of Cell Biol.* 1995, **157**:129-179.

Lam, K.P., Khu, R., Rajewsky, K.: ***In vivo* ablation of surface immunoglobulin on mature B cells by inducible gene targeting results in rapid cell death.** *Cell* 1997, **90**:1073-1083.

Pospisil, R., Fitts, M.G., Mage, R.G.: **CD5 is a potential selecting ligand for B cell surface immunoglobulin framework region sequences.** *J. Exp. Med.* 1996, **184**:1279-1284.

Ten-Boekel, E., Melchers, F., and Rolink, A.: **The status of Ig loci rearrangements in single cells from different stages of B-cell development.** *Intl. Immunol.* 1995, **7**:1013-1019.

6-2 The bone marrow provides an essential microenvironment for early B-cell development.

Funk, P.E., Kincade, P.W., and Witte, P.L.: **Native associations of early hematopoietic stem-cells and stromal cells isolated in bone-marrow cell aggregates.** *Blood* 1994, **83**:361-369.

Jacobsen, K., Kravitz, J., Kincade, P.W., and Osmond, D.G.: **Adhesion receptors on bone-marrow stromal cells—*in vivo* expression of vascular cell adhesion molecule-1 by reticular cells and sinusoidal endothelium in normal and γ-irradiated mice.** *Blood* 1996, **87**:73-82.

Klug, C.A., Morrison, S.J., Masek, M., Hahm, K., Smale, S.T., and Weissman, I.L.: **Hematopoietic stem cells and lymphoid progenitors express different Ikaros isoforms, and Ikaros is localized to heterochromatin in immature lymphocytes.** *Proc. Natl. Acad. Sci. USA* 1998, **95**:657-662.

Nagasawa, T., Hirota, S., Tachibana, V., Takakura, N., Nishikawa, S., Kitamura, Y., Yoshida, V., Kikutani, H, Kishimoto, T.: **Defects of B-cell lymphopoiesis and bone-marrow myelopoiesis in mice lacking the CXC chemokine PBSF/SDF-1.** *Nature* 1996, **382**:635-638.

Rosenberg, N., and Kincade, P.W.: **B-lineage differentiation in normal and transformed-cells and the microenvironment that supports it.** *Curr. Opin. Immunol.* 1994, **6**:203-211.

6-3 The survival of developing B cells depends on the productive, sequential rearrangement of a heavy- and a light-chain gene.

Prak, E.L., and Weigert, M.: **Light-chain replacement—a new model for antibody gene rearrangement.** *J. Exp. Med.* 1995, **182**:541-548.

6-4 The expression of proteins regulating immunoglobulin gene rearrangement and function is developmentally programmed.

Desiderio, S.: **Lymphopoiesis—transcription factors controlling B-cell development.** *Curr. Biol.* 1995, **5**:605-608.

Khan, W.N., Alt, F.W., Gerstein, R.M., Malynn, B.A., Larsson, I., Rathbun, G., Davidson, L., Muller, S., Kantor, A.B., Herzenberg, L.A., Rosen, F.S., and Sideras, P.: **Defective B-cell development and function in *btk*-deficient mice.** *Immunity* 1995, **3**:283-299.

Knight, A.M., Lucocq, J.M., Prescott, A.R., Ponnambalam, S., and Watts, C.: **Antigen endocytosis and presentation mediated by human membrane IgG1 in the absence of the Igα/Igβ dimer.** *EMBO J.* 1997, **16**:3842-3850.

Neurath, M.F., Stuber, E.R., and Strober, W.: **BSAP—a key regulator of B-cell development and differentiation.** *Immunol. Today* 1995, **16**:564-569.

Opstelten, D.: **B lymphocyte development and transcription regulation *in vivo*.** *Adv. Immunol.* 1996, **63**:197-268.

Russell, S.M., Tayebi, N., Nakajima, H., Riedy, M.C., Roberts, J.L., Aman, M.J., Migone, T.S., Noguchi, M., Markert, M.L., Buckley, R.H., Oshea, J.J., and Leonard, W.J.: **Mutation of Jak3 in a patient with SCID—essential role of Jak3 in lymphoid development.** *Science* 1995, **270**:797-800.

Sideras, P., and Smith, C.I.E.: **Molecular and cellular aspects of X-linked agammaglobulinemia.** *Adv. Immunol.* 1995, **59**:135.

6-5 Immunoglobulin gene rearrangements are closely coordinated with chromatin accessibility.

Neuberger, M.S.: **Cells strongly expressing Ig(κ) transgenes show clonal recruitment of hypermutation: a role for both MAR and the enhancers.** *EMBO J.* 1997, **16**:3987-3994.

Sleckman, B.P., Gorman, J.R., and Alt, F.W.: **Accessibility control of antigen receptor variable region gene assembly—role of *cis*-acting elements.** *Annu. Rev. Immunol.* 1996, **14**:459-481.

Stanhope-Baker, P., Hudson, K.M., Shaffer, A.L., Constantinescu, A., and Schlissel, M.S.: **Cell-type specific chromatin structure determines the targeting of V(D)J recombinase activity *in vitro*.** *Cell* 1996, **85**:887-897.

Xu, Y., Davidson, L., Alt, F.W., and Baltimore, D.: **Deletion of the Ig-κ light-chain intronic enhancer/matrix attachment region impairs but does not abolish Vκ-Jκ rearrangement.** *Immunity* 1996, **4**:377-385.

6-6 Cell-surface expression of the products of rearranged immunoglobulin genes provides checkpoints in B-cell development.

Grawunder, U., Leu, T.M.J., Schatz, D.G., Werner, A., Rolink, A.G., Melchers, F., and Winkler, T.H.: **Down-regulation of *Rag1* and *Rag2* gene expression in pre-B** cells after functional immunoglobulin heavy-chain rearrangement. *Immunity* 1995, **3**:601-608.

Horne, M.C., Roth, P.E., and Defranco, A.L.: **Assembly of the truncated immunoglobulin heavy chain Dμ into antigen receptor-like complexes in pre-B cells but not in B cells.** *Immunity* 1996, **4**:145-158.

6-7 The immunoglobulin gene rearrangement program produces monospecific B cells.

Arakawa, H., Shimizu, T., and Takeda, S.: **Reevaluation of the probabilities for productive rearrangements on the κ-loci and λ-loci.** *Intl. Immunol.* 1996, **8**:91-99.

Gorman, J.R., van der Stoep, N., Monroe, R., Cogne, M., Davidson, L., and Alt, F.W.: **The Igk 3′ enhancer influences the ratio of Igκ versus Igl B lymphocytes.** *Immunity* 1996, **5**:241-252.

Loffert, D., Ehlich, A., Muller, W., and Rajewsky, K.: **Surrogate light-chain expression is required to establish immunoglobulin heavy-chain allelic exclusion during early B-cell development.** *Immunity* 1996, **4**:133-144.

Takeda, S., Sonoda, E., and Arakawa, H.: **The κ-λ ratio of immature B cells.** *Immunol. Today* 1996, **17**:200-200.

6-8 Lymphoid follicles are thought to provide a second essential environment for B cells.

Cyster, J.G., and Goodnow, C.C.: **Protein tyrosine phosphatase-1c negatively regulates antigen receptor signaling in B-lymphocytes and determines thresholds for negative selection.** *Immunity* 1995, **2**:13-24.

Liu, Y-J.: **Sites of B lymphocyte selection, activation, and tolerance in spleen.** *J. Exp. Med.* 1997, **186**:625-629.

6-9 Immature B cells can be eliminated or inactivated by contact with self antigens.

Cornall, R.J., Goodnow, C.C., and Cyster, J.G.: **The regulation of self-reactive B cells.** *Curr. Opin. Immunol.* 1995, **7**:804-811.

6-10 Some potentially self-reactive B cells can be rescued by further immunoglobulin gene rearrangement.

Chen, C., Nagy, Z., Prak, E.L., and Weigert, M.: **Immunoglobulin heavy chain gene replacement—a mechanism of receptor editing.** *Immunity* 1995, **3**:747-755.

Chen, C., Nagy, Z., Radic, M.Z., Hardy, R.R., Huszar, D., Camper, S.A., and Weigert, M.: **The site and stage of anti-DNA B-cell deletion.** *Nature* 1995, **373**:252-255.

Melamed, D., Benschop, R.J., Cambier, J.C., and Nemazee, D.: **Developmental regulation of B lymphocyte immune tolerance compartmentalizes clonal selecton from receptor selection.** *Cell* 1998, **92**:173-182.

Pelanda, R., Schwers, S., Sonoda, E., Torres, R.M., Nemazee, D., and Rajewsky, K.: **Receptor editing in a transgenic mouse model: site, efficiency, and role in B cell tolerance and antibody diversification.** *Immunity* 1997, **7**:765-775.

6-11 In some species most immunoglobulin gene diversification occurs after gene rearrangement.

Knight, K.L., and Crane, M.A.: **Generating the antibody repertoire in rabbit.** *Adv. Immunol.* 1994, **56**:179-218.

Reynaud, C.A., Bertocci, B., Dahan, A., and Weill, J.C.: **Formation of the chicken B-cell repertoire—ontogeny, regulation of Ig gene rearrangement, and diversification by gene conversion.** *Adv. Immunol.* 1994, **57**:353-378.

Reynaud, C.A., Garcia, C., Hein, W.R., and Weill, J.C.: **Hypermutation generating the sheep immunoglobulin repertoire is an antigen independent process.** *Cell* 1995, **80**:115-125.

Vajdy, M., Sethupathi, P., and Knight, K.L.: **Dependence of antibody somatic diversification on gut-associated lymphoid tissue in rabbits.** *J. Immunol.* 1998, **160**:2725-2729.

Weill, J.C., and Reynaud, C.A.: **Rearrangement/hypermutation/gene conversion—when, where, and why.** *Immunol. Today* 1996, **17**:92-97.

6-12 B cells are produced continuously but only some contribute to a relatively stable peripheral pool.

Crane, M.A., Kingzette, M., Knight, K.L.: **Evidence for limited B-lymphopoiesis in adult rabbits.** *J. Exp. Med.* 1996, **183**:2119-2121.

Fulcher, D.A., and Basten, A.: **Reduced lifespan of anergic self-reactive B cells in a double-transgenic model.** *J. Exp. Med.* 1994, **179**:125-134.

Osmond, D.G.: **The turnover of B-cell populations.** *Immunol. Today* 1993, **14**:34-37.

Reynaud, C.A., et al.: **Generation of diversity in mammalian gut-associated lymphoid tissues: restricted V gene usage does not preclude complex V gene organization.** *J. Immunol.* 1997, **159**:3093-3095.

6-13 B cells bearing surface CD5 express a distinctive repertoire of receptors.

Murakami, M., and Honjo, T.: **Involvement of B-1 cells in mucosal immunity and autoimmunity.** *Immunol. Today* 1995, **16**:534-539.

Murakami, M., Yoshioka, H., Shirai, T., Tsubata, T., and Honjo, T.: **Prevention of autoimmune symptoms in autoimmune-prone mice by elimination of B-1 cells.** *Intl. Immunol.* 1995, **7**:877-882.

6-14 B cells at different developmental stages are found in different anatomical sites.

Casali, P., et al.: **Human lymphocytes making rheumatoid factor and antibody to ssDNA belong to Leu-1 + B-cell subset.** *Science* 1987, **236**:77-81.

Griebel, P.J., and Hein, W.R.: **Expanding the role of Peyer's patches in B-cell ontogeny.** *Immunol. Today* 1996, **17**:30-39.

Maclennan, I.C.M.: **Germinal centers.** *Annu. Rev. Immunol.* 1994, **12**:117-139.

Tarlinton, D.: **Germinal centers: form and function.** *Curr. Opin. Immunol.* 1998, **10**:245-251.

Zheng, B., Han, S., Takahashi, Y., and Kelsoe, G.: **Immunosenescence and germinal center reaction.** *Immunol. Rev.* 1997, **160**:63-77.

6-15 B-cell tumors often occupy the same site as their normal counterparts.

Cotran, R.S., Kumar, V., Robbins, S.L.: *Diseases of white cells, lymph nodes, and spleen. Pathologic basis of disease,* 5th edn. W.B. Saunders, 1994, 629-672.

6-16 Malignant B cells frequently carry chromosomal translocations that join immunoglobulin loci to genes regulating cell growth.

Corral, J., Lavenir, I., Impey, H., Warren, A.J., Forster, A., Larson, T.A., Bell, S., McKenzie, A.N.J., King, G., and Rabbitts, T.H.: **An Mll-AF9 fusion gene made by homologous recombination causes acute leukemia in chimeric mice—a method to create fusion oncogenes.** *Cell* 1996, **85**:853-861.

Cory, S.: **Regulation of lymphocyte survival by the *Bcl-2* gene family.** *Annu. Rev. Immunol.* 1995, **13**:513-543.

Rabbitts, T.H.: **Chromosomal translocations in human cancer.** *Nature* 1994, **372**:143-149.

Yang, E., and Korsmeyer, S.J.: **Molecular thanatopsis—a discourse on the *Bcl-2* family and cell death.** *Blood* 1996, **88**:386-401.

The Thymus and the Development of T Lymphocytes

7

T-cell development has much in common with B-cell development. Like B cells, T cells derive from bone marrow stem cells and undergo gene rearrangements in a specialized microenvironment to produce a unique antigen receptor on each cell. However, unlike B cells, T cells do not differentiate in the bone marrow but migrate at a very early stage to the **thymus**. This central lymphoid organ provides the specialized microenvironment in which receptor gene rearrangement and T-cell maturation occur. There is a further crucial difference in T- and B-cell development that reflects the distinct ways in which T cells and B cells recognize antigen. As we learned in Chapter 4, T cells recognize antigen in the form of peptides bound to the highly polymorphic cell-surface molecules encoded by the MHC. T cells are said to be MHC restricted because their receptors are specific for both a particular MHC allele and the foreign or non-self peptide it binds. Because adaptive immunity depends on MHC-restricted T-cell responses, it is essential that each individual's T cells are able to respond to foreign antigenic peptides when they are bound to his or her own MHC molecules, that is, that they are self MHC restricted. It is equally important, however, that each individual's T cells should be unable to respond to self peptides bound to self MHC molecules; that is, they must also be self tolerant.

T cells are selected to fulfill these dual requirements for self MHC restriction and self tolerance during their maturation in the thymus. Here, once the immature T cells have rearranged their antigen-receptor genes and the receptor is expressed on the cell surface, they are screened by the two selective processes. They are selected for self MHC restriction by positive selection, whereas negative selection eliminates those cells that are specific for self peptides bound to self MHC molecules.

How T cells are selected for self restriction but not self reactivity during T-cell maturation is one of the most intriguing problems in immunobiology and, in consequence, one of the most active areas of research. In this chapter we shall describe what is known about the development of T cells and what mechanisms might explain positive and negative selection. As developmental studies depend largely on experimental manipulation, virtually all the information that we have about T-cell development in the thymus has been gained from experiments with mice. We shall comment on the development of human T cells where information exists.

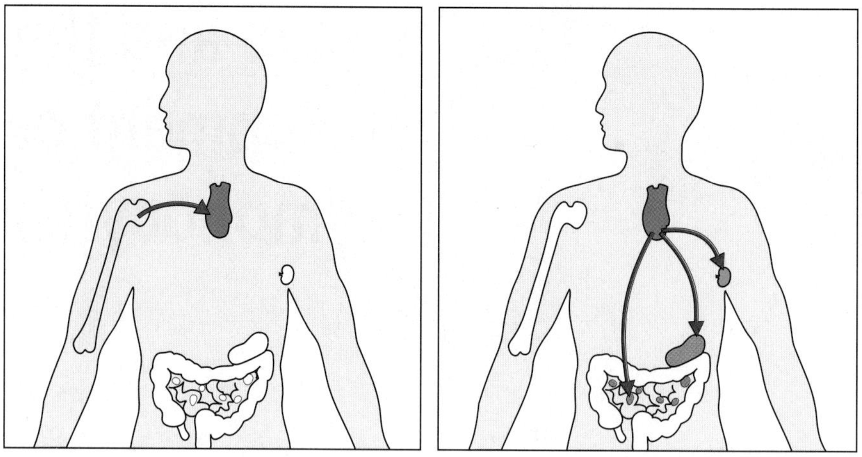

Fig. 7.1 T-cell precursors migrate to the thymus to mature. T cells derive from bone marrow stem cells, whose progeny migrate from the bone marrow to the thymus (left panel), where the development of T cells occurs. Mature T cells leave the thymus and recirculate from the bloodstream through secondary lymphoid tissues (right panel) such as lymph nodes, spleen, or Peyer's patches, where they may encounter foreign antigen. It is the developmental process within the thymus that is the subject of this chapter.

The development of T cells in the thymus.

T cells develop from bone marrow stem cells but their progenitors migrate to the thymus where they mature (Fig. 7.1): for that reason they are called **thymus dependent (T) lymphocytes** or **T cells**. In the thymus, the immature T cells, or **thymocytes**, proliferate and differentiate, passing through a series of discrete phenotypic stages that can be identified by distinctive patterns of expression of various cell-surface proteins. During their development as thymocytes the cells undergo the gene rearrangements that produce the T-cell receptor, and also undergo the positive and negative selection that shapes the mature T-cell receptor repertoire. These processes depend upon interactions of the developing cells with the cells of the thymic microenvironment. We shall therefore begin with a general overview of the stages of thymocyte development and its relationship to thymic architecture, before going on in the later sections of this chapter to consider the rearrangement of the T-cell receptor genes and the selection of the mature T-cell receptor repertoire.

7-1 T cells develop in the thymus.

T lymphocytes develop in the thymus, a lymphoid organ in the upper anterior thorax, just above the heart. In young individuals, the thymus contains many developing T-cell precursors embedded in an epithelial network known as the **thymic stroma**, which provides a unique microenvironment for T-cell development. The thymus consists of numerous lobules, each clearly differentiated into an outer cortical region—the **thymic cortex**—and an inner **medulla**. The thymic stroma arises early in embryonic development from the endodermal and ectodermal layers of the embryonic structures known as the third pharyngeal pouch and third branchial cleft. Together these epithelial tissues form a rudimentary thymus, or **thymic anlage**. The thymic anlage then attracts cells of hematopoietic origin, which colonize it; these give rise to thymocytes committed to the T-cell lineage, and to intrathymic dendritic cells; the thymus is also independently colonized by numerous macrophages, also of bone marrow origin. The thymocytes are not simply passengers within the thymus; they influence the arrangement of the thymic epithelial cells on which they depend for their survival, inducing the formation of a reticular epithelial structure that surrounds the developing thymocytes (Fig. 7.2).

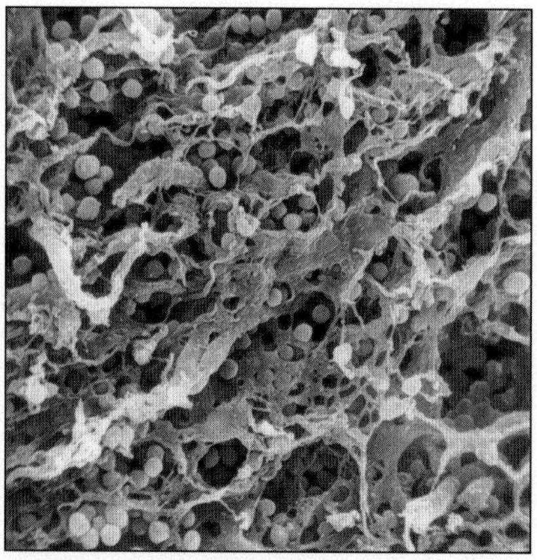

Fig. 7.2 The epithelial cells of the thymus form a network surrounding developing thymocytes. In this scanning electron micrograph of the thymus, the developing thymocytes (the spherical cells) occupy the interstices of an extensive network of epithelial cells. Photograph courtesy of W van Ewijk.

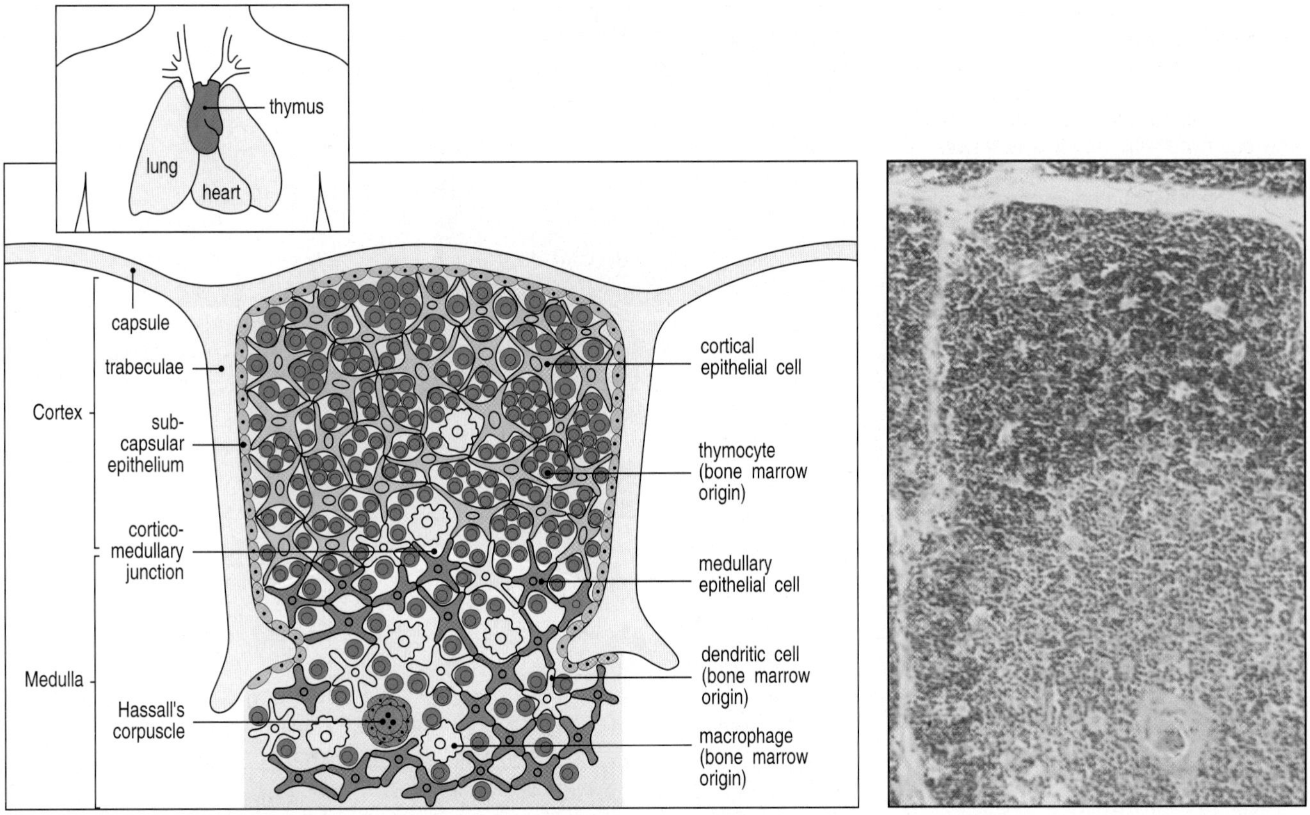

Fig. 7.3 The cellular organization of the thymus. The thymus, which lies in the midline of the body, above the heart, is made up of several lobules, each of which contains discrete cortical (outer) and medullary (central) regions. As shown in the diagram on the left, the cortex consists of immature thymocytes (dark blue), branched cortical epithelial cells (pale blue) with which the immature cortical thymocytes are closely associated, and scattered macrophages (yellow), which are involved in clearing apoptotic thymocytes. The medulla consists of mature thymocytes (dark blue), and medullary epithelial cells (orange), along with macrophages (yellow) and dendritic cells (yellow) of bone marrow origin. Hassall's corpuscles are probably also sites of cell destruction. The thymocytes in the outer cortical cell layer are proliferating immature cells, whereas the deeper cortical thymocytes are mainly cells undergoing thymic selection. The photograph shows the equivalent section of a human thymus, stained with hematoxylin and eosin. The cortex is darkly staining; the medulla is lightly stained. The large body in the medulla is a Hassall's corpuscle. Photograph courtesy of C J Howe.

The human thymus is fully developed before birth. Like the mouse thymus, it consists of a thymic stroma of epithelia and connective tissue, which is populated by very large numbers of bone marrow derived cells. These cells are differentially distributed between the thymic cortex and medulla; the cortex contains only immature thymocytes and scattered macrophages, whereas more mature thymocytes, along with dendritic cells and macrophages, are found in the medulla (Fig. 7.3). We shall see that this reflects the different developmental events that occur within these two compartments.

The rate of T-cell production by the thymus is greatest before puberty. After puberty, the thymus begins to shrink and the production of new T cells in adults is lower. However, in both mice and humans, removal of the thymus is not accompanied by any notable loss of T-cell function although adult thymectomy in mice is accompanied by some loss of T cells. Thus, it seems that once the T-cell repertoire is established, immunity can be sustained without the production of large numbers of new T cells.

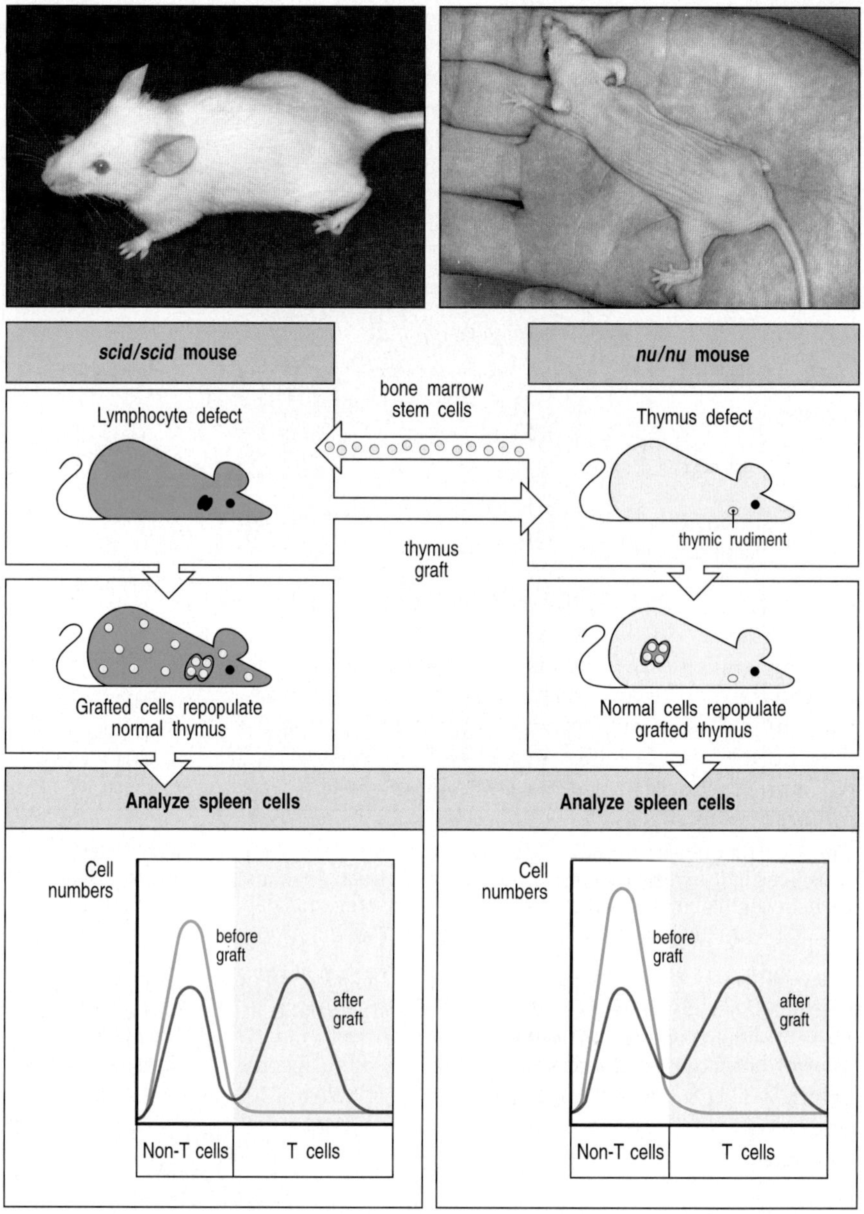

Fig. 7.4 The thymus is critical for the maturation of bone marrow derived cells into T cells. Mice with the *scid* mutation (upper left photo) have a defect that prevents lymphocyte maturation, whereas mice with the *nude* mutation (upper right photo) have a defect that affects the development of the cortical epithelium of the thymus. T cells do not develop in either strain of mouse: this can be demonstrated by staining spleen cells with antibodies specific for mature T cells and analyzing them in a flow cytometer (see Fig. 2.24), as represented by the blue line in the graphs in the bottom panels. Bone marrow cells from *nude* mice can restore T cells to *scid* mice (red line in graph on left), showing that, in the right environment, the *nude* bone marrow cells are intrinsically normal, and capable of producing T cells. Thymic epithelial cells from *scid* mice can induce the maturation of T cells in *nude* mice (red line in graph on right), demonstrating that the thymus provides the essential microenvironment for T-cell development.

7-2 The thymus is required for T-cell maturation.

The importance of the thymus in immunity was first discovered through experiments on mice. It was found that surgical removal of the thymus (thymectomy) at birth resulted in an immunodeficient mouse, focusing interest on this organ at a time when the difference between T and B cells in mammals had not yet been defined. Much evidence has accumulated since to establish the importance of the thymus in T-cell development, including observations of immunodeficient children. Thus, for example, in the **DiGeorge syndrome** in humans, and in mice with the ***nude*** mutation (which also causes hairlessness), the thymus fails to form and the affected individual produces B lymphocytes but few T lymphocytes.

The crucial role of the thymic stroma in inducing the differentiation of bone marrow derived precursor cells can be demonstrated by using two mutant mice, each lacking mature T cells for a different reason. In *nude* mice, the thymic epithelium fails to differentiate; whereas in *scid* mice, which we encountered in Section 3-17, B and T lymphocytes fail to develop because of a defect in T-cell receptor gene rearrangement. Reciprocal grafts of thymus and bone marrow between these immunodeficient strains show that *nude* bone marrow precursors develop normally in a *scid* thymus (Fig. 7.4). We shall see later (see Section 7-10) that thymic grafts between different strains of mice have been instrumental in defining the role of the thymus in selecting the mature T-cell receptor repertoire.

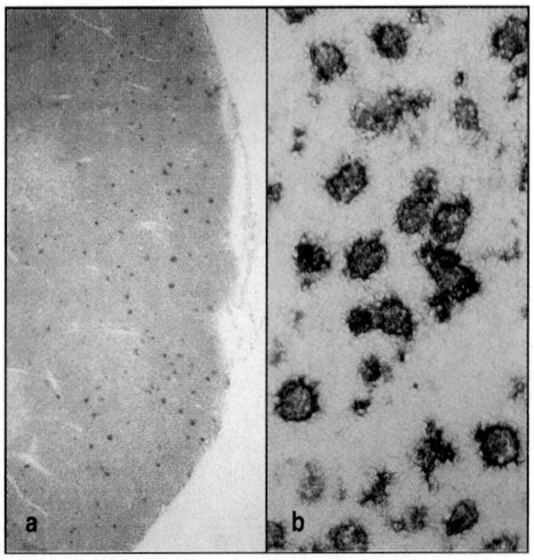

Fig. 7.5 Developing T cells that undergo apoptosis are ingested by macrophages in the thymic cortex. Panel a shows a section through the thymic cortex (to the right) and part of the medulla in which cells have been stained for apoptosis with a red dye. Apoptotic cells are scattered throughout the cortex but are rare in the medulla. Panel b shows a section of thymic cortex at higher magnification that has been stained red for apoptotic cells and blue for macrophages. The apoptotic cells can be seen situated within macrophages. Magnifications: panel a, × 45; panel b, × 164. Photographs courtesy of J Sprent and C Suhr.

7-3 Developing T cells proliferate in the thymus but most die there.

T-cell precursors arriving in the thymus from the bone marrow spend up to a week differentiating there before they enter a phase of intense proliferation. In a young adult mouse, where the thymus contains around 10^8 to 2×10^8 thymocytes, about 5×10^7 new cells are generated each day. However, only about 10^6 to 2×10^6 (roughly 3%) of these will leave the thymus each day as mature T cells. Despite the disparity between the numbers of T cells generated daily in the thymus and the number leaving, the thymus does not continue to grow in size or in cell number. This is because approximately 98% of the thymocytes that develop in the thymus also die within the thymus. No widespread damage is seen, indicating that death is occurring by apoptosis rather than necrosis.

Apoptosis is a common feature of many developmental pathways and one of its features is that changes in the plasma membrane of cells undergoing apoptosis lead to their rapid phagocytosis. Indeed, apoptotic bodies, which are residual condensed chromatin, are seen inside macrophages throughout the thymic cortex (Fig. 7.5), and as a result of the ready staining of the apoptotic bodies these macrophages are sometimes referred to as 'tingible body macrophages.' The apparently profligate wastefulness of this massive cell death is a crucial part of T-cell development, as it reflects the intensive screening that each new T cell undergoes for self MHC restriction and self tolerance.

7-4 Successive stages in the development of thymocytes are marked by changes in cell-surface molecules.

As thymocytes proliferate and mature into T cells, they pass through a series of distinct phases marked by changes in the status of T-cell receptor genes, and in the expression of the T-cell receptor, the co-receptors CD4 and CD8, and other cell-surface molecules that reflect the state of functional maturation of the cell. Specific combinations of these cell-surface molecules can thus be used as markers for T cells at different stages of differentiation: the principal stages are summarized in Fig. 7.6.

When progenitor cells first enter the thymus from the bone marrow, they lack most of the surface molecules characteristic of mature T cells and their receptor genes are unrearranged. At this stage these cells can give rise to intrathymic dendritic cells as well as the $\alpha{:}\beta$ and $\gamma{:}\delta$ thymocytes. A little later, interactions with the thymic stroma trigger differentiation, proliferation, and the expression of the first T-cell-specific surface molecules, for example CD2

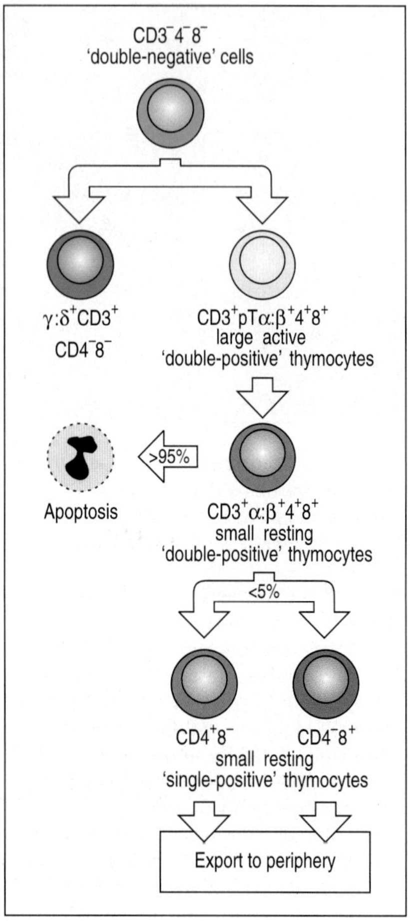

Fig. 7.6 Changes in cell-surface molecules allow thymocyte populations at different stages of maturation to be distinguished. The most important cell-surface molecules for identifying thymocyte subpopulations have been CD4, CD8, and T-cell receptor complex molecules (CD3, and α and β chains). The earliest cell population in the thymus does not express any of these. As these cells do not express CD4 or CD8, they are called 'double-negative' thymocytes. (The γ:δ T cells found in the thymus also lack CD4 or CD8 but these are a minor population.) Maturation of α:β T cells occurs through stages where both CD4 and CD8 are expressed by the same cell, along with the pre-T-cell receptor (pTα:β) and later low levels of the T-cell receptor (α:β) itself. These cells are known as 'double-positive' thymocytes. Most thymocytes (~97%) die within the thymus after becoming small double-positive cells. Those whose receptors bind self MHC molecules lose expression of either CD4 or CD8 and increase the level of expression of the T-cell receptor. The outcome of this process is the 'single-positive' thymocytes, which, after maturation, are exported from the thymus as mature single-positive T cells.

and (in mice) Thy-1. At the end of this phase, which can last about a week, the immature thymocytes bear distinctive markers of the T-cell developmental lineage but they do not express any of the three cell-surface markers that define mature T cells, namely the CD3:T-cell receptor complex, and the co-receptors CD4 or CD8. Owing to the absence of CD4 and CD8, such cells are called **'double-negative' thymocytes** (see Fig. 7.6).

In the fully developed thymus, these immature double-negative cells form part of a small, highly heterogeneous, pool of cells (about 5% of total thymocytes) that includes two populations of more mature cells that belong to minority lineages. One of these, representing about 20% of all the double-negative cells in the thymus, comprises cells that have rearranged and are expressing the genes encoding the γ:δ T-cell receptor; we shall return to these cells in Section 7-7. The second, also representing about 20% of all double negatives, includes cells bearing α:β T-cell receptors of a very limited diversity, which are activated as part of the early response to many infections (not shown in Fig. 7.6). We shall reserve the term double-negative thymocytes to describe the population of immature thymocytes that do not yet express a complete T-cell receptor molecule. These cells, which comprise 60% of all double negatives, give rise to both γ:δ and α:β T cells (see Fig. 7.6). Most of them develop along the α:β pathway, which is shown in more detail in Fig. 7.7.

The immature double-negative thymocyte stage can be subdivided on the basis of expression of the adhesion molecule CD44 and of CD25, the α chain of the interleukin (IL)-2 receptor. At first, double-negative thymocytes express CD44 but not CD25; in these cells, the genes encoding both chains of the T-cell receptor are in the germline configuration. As the thymocytes mature further, they begin to express CD25 on their surface, and still later, expression of CD44 is reduced. In these latter cells, which are known as $CD44^{low}CD25^+$ cells, rearrangement of the T-cell receptor β-chain gene occurs. Cells that fail to make a successful rearrangement of the β-chain gene remain in the $CD44^{low}CD25^+$ stage, whereas cells that make productive β-chain gene rearrangements and express the β chain lose expression of CD25 once again (see Fig. 7.7). The functional significance of the transient expression of CD25 is unclear: T cells develop normally in mice in which the IL-2 gene has been deleted by gene knock-out (see Section 2-27). By contrast, the IL-7 receptor does seem essential for early T-cell development, as T cells do not develop in either humans or mice when this receptor is defective.

The β chains expressed by $CD44^{low}CD25^+$ thymocytes pair with a surrogate α chain called pTα (pre-T-cell α), which allows them to assemble a pre-TCR that exits the endoplasmic reticulum via the *cis*-Golgi (with the CD3 molecules).

Expression of this complex on the cell surface leads to cell proliferation and the arrest of further β-chain gene rearrangements, and eventually results in the expression of both CD8 and CD4. Cells expressing both CD8 and CD4 on their surface comprise the vast majority of thymocytes and are called **'double-positive' thymocytes** (see Fig. 7.6). Once the cells cease to proliferate and become small double-positive cells, the α-chain genes rearrange, producing in most developing thymocytes a complete α:β antigen receptor.

Small double-positive thymocytes initially express only low levels of the T-cell receptor. Most of these cells express receptors that cannot recognize self MHC; they are destined to fail positive selection and die. Those double-positive cells that recognize self MHC, however, mature to express high levels of the T-cell receptor and subsequently cease to express one or other of the two co-receptor molecules, becoming either CD4 or CD8 **'single-positive' thymocytes** (see Fig. 7.6). Thymocytes also undergo negative selection during and after the double-positive stage in development, eliminating those cells capable of responding to self antigens. Approximately 2% survive this dual screening and mature as single-positive T cells that are gradually exported from the thymus to join the peripheral T-cell repertoire. From the time that a T-cell progenitor enters the thymus to the time of its export from the thymus is estimated as around 3 weeks in the mouse.

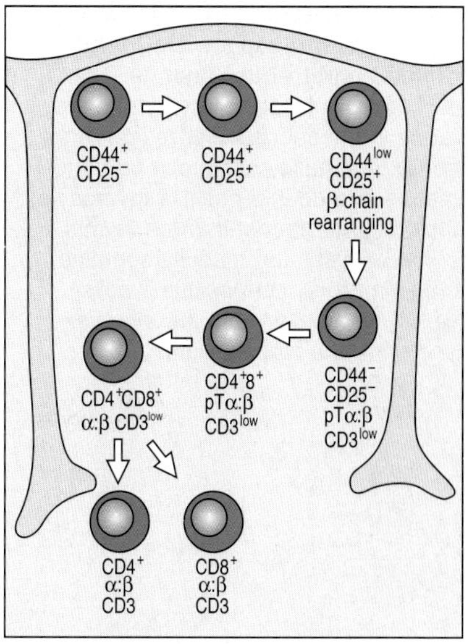

Fig. 7.7 The pathway of α:β T-cell development. Double-negative thymocytes committed to the T-cell lineage first express CD44 and, at a later stage, the α chain of the IL-2 receptor, CD25. After this, the $CD44^+$ $CD25^+$ cells begin to rearrange their β-chain genes, becoming $CD44^{low}$ as this occurs. The cells are then arrested in the $CD44^{low}$ $CD25^+$ stage until they productively rearrange their β-chain genes, allowing the in-frame β chain to pair with the surrogate pTα chain and be expressed on the cell surface, which triggers their entry into the cell cycle. Expression of pTα:β on the cell surface is associated with small amounts of CD3, and causes the loss of CD25, cessation of β-chain gene rearrangement, cell proliferation, and the expression of CD4 and CD8. After the cells cease proliferating and revert to small CD4 CD8 double-positive cells they begin rearrangement of their α-chain genes. The cells then express low levels of an α:β T-cell receptor and the associated CD3 complex and are ready for selection. Most cells die by failing to be positively selected or as a consequence of negative selection, but some are selected to mature into CD4 or CD8 single-positive cells and eventually to leave the thymus.

7-5 Thymocytes at different developmental stages are found in distinct parts of the thymus.

We have seen that the thymus is divided into two main regions, a peripheral cortex and a central medulla (see Fig. 7.3). Most T-cell development takes place in the cortex; only mature single-positive thymocytes are seen in the medulla. At the outer edge of the cortex, in the subcapsular region of the thymus (Fig. 7.8), large immature double-negative thymocytes proliferate vigorously; these cells are thought to represent the thymic progenitors and their immediate progeny. These cells will give rise to all subsequent thymocyte populations. Deeper in the cortex, most of the thymocytes are small double-positive cells. The stroma of the cortex is composed of epithelial cells with long branching processes that express MHC class II as well as MHC class I molecules on their surface. The thymic cortex is densely packed with thymocytes, and the branching processes of the thymic cortical epithelial cells make contact with most cortical thymocytes (see Fig. 7.2). Contact between the MHC molecules on thymic cortical epithelial cells and the receptors of developing T cells has a crucial role in positive selection, as we shall see later in this chapter.

The medulla of the thymus is less well characterized. It contains relatively few thymocytes, and those that are present are single-positive cells resembling mature T cells. These cells may be newly mature T cells that are leaving the thymus through the medulla, or they may represent some other population of mature T cells that remain within the medulla or return to it from the periphery to perform some specialized function, such as the elimination of infectious agents within the thymus. Before they mature, the developing thymocytes must undergo negative selection to remove self-reactive cells. We shall see that this selective process is carried out mainly by the dendritic cells, which are particularly numerous at the cortico-medullary junction, and by the macrophages that are scattered in the cortex but are also abundant in the thymic medulla. However, before discussing in more detail how thymocytes are both positively and negatively selected to mature into a population of self-MHC restricted and self-tolerant T cells, we must turn to the gene rearrangements that generate the receptor repertoire on which positive and negative selection act.

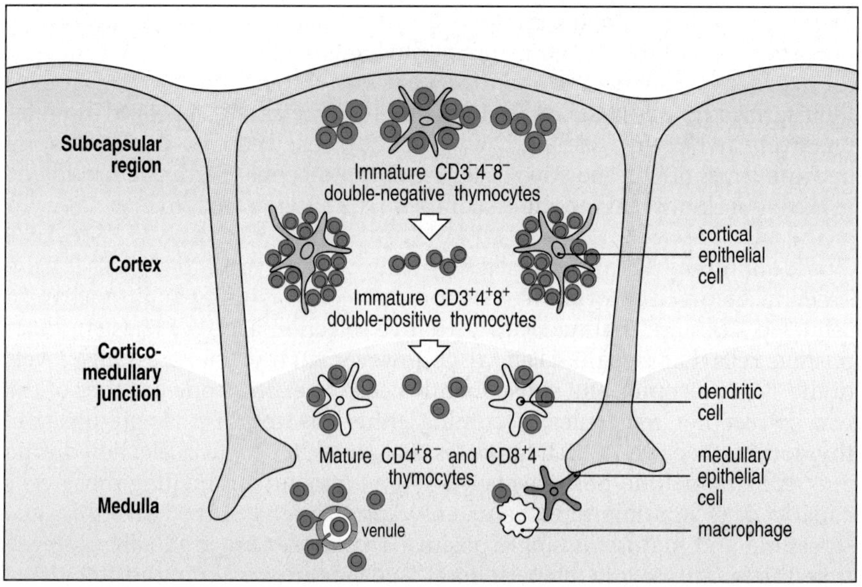

Fig. 7.8 Thymocytes of different developmental stages are found in distinct parts of the thymus. The earliest cells to enter the thymus are found in the subcapsular region of the cortex. As these cells proliferate and mature into double-positive thymocytes, they migrate deeper into the thymic cortex. Finally, the medulla contains only mature single-positive T cells, which eventually leave the thymus and enter the bloodstream.

Summary.

The thymus provides a specialized and architecturally organized microenvironment for the development of mature T cells. Precursors of T cells migrate from the bone marrow and mature in the thymus, passing through a series of stages that can be distinguished by the differential expression of CD44 and CD25, the CD3:T-cell receptor complex proteins, and the co-receptor proteins CD4 and CD8. T-cell development is accompanied by extensive cell death, reflecting the intense selection of T cells and the elimination of those with inappropriate receptor specificities. Most steps in T-cell differentiation occur in the cortex of the thymus. The thymic medulla contains mainly mature T cells.

T-cell receptor gene rearrangements and receptor expression.

In the early stages of thymocyte maturation, the T-cell receptor genes go through a programmed series of rearrangements that culminate in the large number of immature T cells each expressing a receptor of a different antigen specificity. This process is similar in many ways to that which occurs in B cells, with two important differences. First, rearrangement of two different sets of receptor genes distinguishes two T-cell lineages, most (about 95–99%) expressing α:β T-cell receptors, whereas the minority (about 1–5%) express γ:δ T-cell receptors. Second, there is much greater scope for repeated rearrangements of T-cell receptor genes, particularly at the α locus, allowing the rescue of many cells in which the initial rearrangements have been nonproductive. The end result of this process in the α:β T-cell lineage is a double-positive thymocyte expressing antigen receptors on which positive and negative selection can then act.

7-6 T cells with α:β or γ:δ receptors arise from a common progenitor.

T cells bearing γ:δ receptors differ from α:β T cells in their specificity, the pattern of expression of the CD4 and CD8 co-receptors, and in their anatomical distribution. The two types of T cell also differ in function, although relatively little is known about the function of γ:δ T cells. Nevertheless, studies of the gene rearrangements found in thymocytes and mature γ:δ and α:β T cells suggest that they diverge from a common precursor at a relatively late stage of their development, when gene rearrangements have already occurred. Mature γ:δ T cells can be found to have productively rearranged β-chain genes, whereas mature α:β T cells often contain rearranged, but mostly (about 80%) out-of-frame, γ- and δ-chain genes.

At present, the factors that regulate the commitment of a common T-cell precursor to one or other of these lineages are not known, although it might depend simply upon whether productive rearrangements of a γ and a δ gene have occurred in the same cell. The β, γ, and δ genes undergo rearrangement almost simultaneously in developing thymocytes and it seems that successful rearrangement of the γ and δ genes leads to the expression of a functional γ:δ T-cell receptor that signals the cell to differentiate along the γ:δ lineage. In most precursors, however, there is a successful rearrangement of a β-chain gene that results in the production of a functional β-chain protein. This β-chain protein can pair with the surrogate α chain, pTα, to create a pre-T-cell receptor (β:pTα), thus arresting further β-chain gene rearrangements and signaling the thymocyte to proliferate, to express its co-receptor genes, and eventually to start transcribing the α-chain genes.

The cells that express the pre-T-cell receptor molecule have taken a further maturational step towards becoming α:β cells but they are not necessarily committed to this lineage. Once the proliferative burst is over, the enzymes responsible for gene rearrangement are reactivated and the successful completion of γ:δ rearrangements might still divert such cells to become γ:δ T cells, thus accounting for the productive β-chain gene rearrangements seen in some γ:δ T cells. In most cases, however, rearrangements at the α locus will delete the intervening δ genes as an extrachromosomal circle. The further maturation of α:β T cells depends on α-gene rearrangements producing a functional α:β receptor that is positively selected for its ability to recognize self MHC molecules. The signals that drive a γ:δ T cell to mature are unknown. Certainly, γ:δ T cells seem able to develop in the absence of a functioning thymus and are present in, for example, *nude* mice.

7-7 Cells expressing particular γ- and δ-chain genes arise in an ordered sequence during embryonic development.

The first T cells to appear during embryonic development carry γ:δ T-cell receptors (Fig. 7.9). In the mouse, where the development of the immune system can be studied in detail, γ:δ T cells first appear in discrete waves or bursts, with the T cells in each wave populating distinct sites in the adult animal.

The first wave of γ:δ T cells populates the epidermis, where they are called **dendritic epidermal T cells (dETC)**, whereas the second wave homes to the epithelial layers of the reproductive tract. The receptors expressed by these early waves of γ:δ T cells are essentially homogeneous: all the cells in each wave express the same rearranged V_γ and V_δ sequences and the same

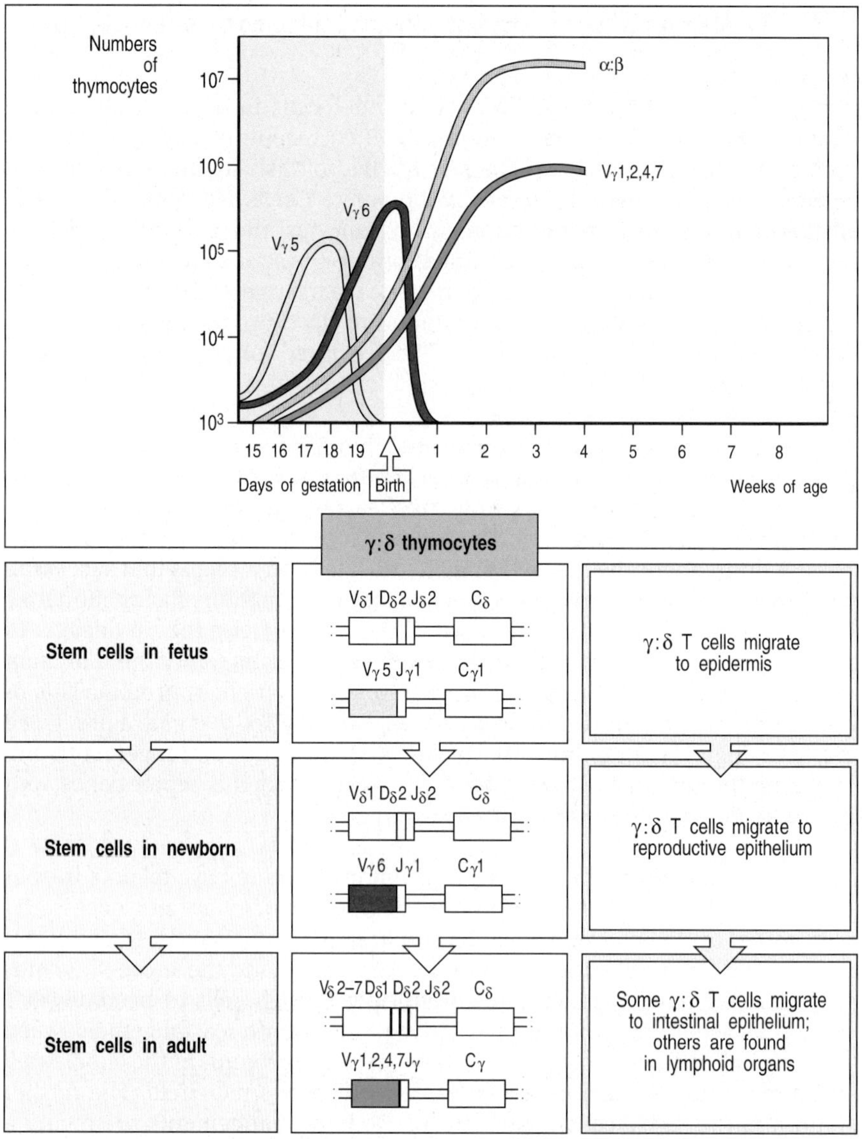

Fig. 7.9 The rearrangement of T-cell receptor γ and δ genes in the mouse proceeds in waves of cells expressing different V gene segments. At about 2 weeks of gestation, the $C_\gamma1$ locus is expressed with its closest V gene ($V_\gamma5$; also known as $V_\gamma3$*). After a few days $V_\gamma5$-bearing cells decline (upper panel) and are replaced by cells expressing the next most proximal gene, $V_\gamma6$. Both these rearranged γ chains are expressed with the same rearranged δ-chain gene, as shown in the lower panels, and there is no junctional diversity. As a consequence, all of the γ:δ T cells produced in each of these early waves have the same specificity, although the nature of the antigen recognized by the early γ:δ T cells is not known. The $V_\gamma5$-bearing cells migrate selectively to the epidermis, whereas the $V_\gamma6$-bearing cells migrate to the epithelium of the reproductive tract. After birth, the α:β T-cell lineage becomes dominant and, although γ:δ T cells are still produced, they are a much more heterogeneous population, with more junctional diversity.

*The nomenclature of V_γ gene segments is a source of confusion, as there are two different nomenclatures. One system specifies $V_\gamma1.1$, $V_\gamma1.2$, and $V_\gamma1.3$, whereas the other refers to these as $V_\gamma1$, $V_\gamma2$, and $V_\gamma3$. Thus, $V_\gamma2$, $V_\gamma3$, $V_\gamma4$, and $V_\gamma5$ in the first system become $V_\gamma4$, $V_\gamma5$, $V_\gamma6$, and $V_\gamma7$ in the second. In this book, we use the latter nomenclature.

J regions. There are no N-nucleotides to contribute additional diversity at the junctions between V, D, and J gene segments, reflecting the absence of the enzyme terminal deoxynucleotidyl transferase (TdT) from these fetal T cells.

Later in development, T cells are produced continuously rather than in waves, and α:β T cells predominate, making up more than 95% of thymocytes. The γ:δ T cells produced at this stage are different from those of the early waves, with considerably more diverse receptors containing several different V gene segments and abundant N-nucleotide additions. Most of these γ:δ T cells, like α:β T cells, are found in peripheral lymphoid tissues rather than in the epithelial sites populated by the early γ:δ T cells.

The developmental changes in V gene segment usage and N-nucleotide additions in γ:δ T cells parallel changes in B-cell populations during fetal development (see Section 6-12). Their functional significance is unclear, however, and we do not know whether similar changes in the pattern of receptors expressed by γ:δ T cells occur in humans. Certainly, the γ:δ T cells that home to the skin of mice, the dETC, do not seem to have exact human counterparts, although there are γ:δ T cells in the human reproductive and gastrointestinal tracts.

7-8 Rearrangement of the β-chain genes and production of a β chain triggers several events in developing thymocytes.

T cells expressing α:β receptors first appear a few days after the earliest γ:δ T cells and rapidly become the most abundant type of thymocyte (see Fig. 7.9). The rearrangement of β- and α-chain genes during T-cell development is summarized in Fig. 7.10 and closely parallels the rearrangement of

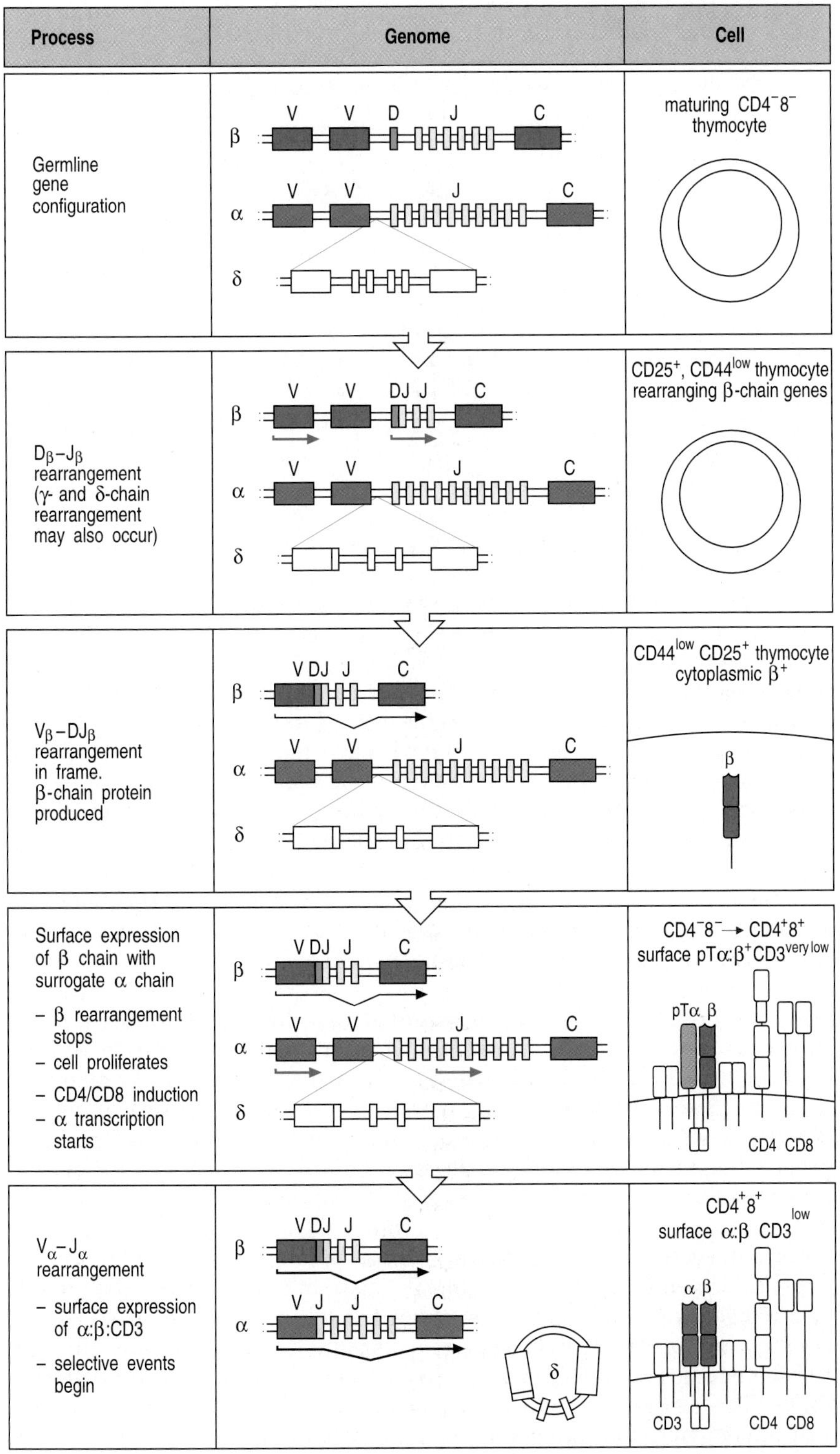

Fig. 7.10 The stages of gene rearrangement in α:β T cells. The sequence of gene rearrangements is shown, together with an indication of the stage at which the events take place and the nature of the cell-surface receptor molecules expressed at each stage. The β-chain genes rearrange first, in $CD4^-$ $CD8^-$ double-negative thymocytes expressing CD25 and low levels of CD44. As with immunoglobulin heavy-chain genes, D to J gene segments rearrange before V gene segments rearrange to DJ (second and third panels). As there are four D gene segments and two sets of J gene segments, it is possible to make up to four attempts to generate a productive rearrangement of the β-chain genes. The productively rearranged gene is expressed initially within the cell and then at low levels on the cell surface in a complex with the CD3 chains as a pTα:β heterodimer (fourth panel), where pTα is a 33 kDa surrogate α chain equivalent to λ5 in B-cell development. The expression of the pre-T-cell receptor signals the developing thymocytes via the tyrosine kinase Lck to halt β-chain gene rearrangement, and to undergo multiple cycles of division. At the end of this proliferative burst, the CD4 and CD8 molecules are expressed, the cell ceases cycling, and the α chain is now able to undergo rearrangement. The first α-chain gene rearrangement deletes all δ D, J, and C gene segments on that chromosome, although these are retained as a circular DNA, proving that these are non-dividing cells (bottom panel). This inactivates the δ-chain gene. Rearrangements at the α-chain locus can proceed through several cycles, because of the large number of V_α and J_α gene segments, so that productive rearrangements almost always occur. When a functional α chain is produced that pairs efficiently with the β chain, the $CD3^{low}$ $CD4^+$ $CD8^+$ thymocyte is ready to undergo selection for its ability to recognize self peptides in association with self MHC molecules.

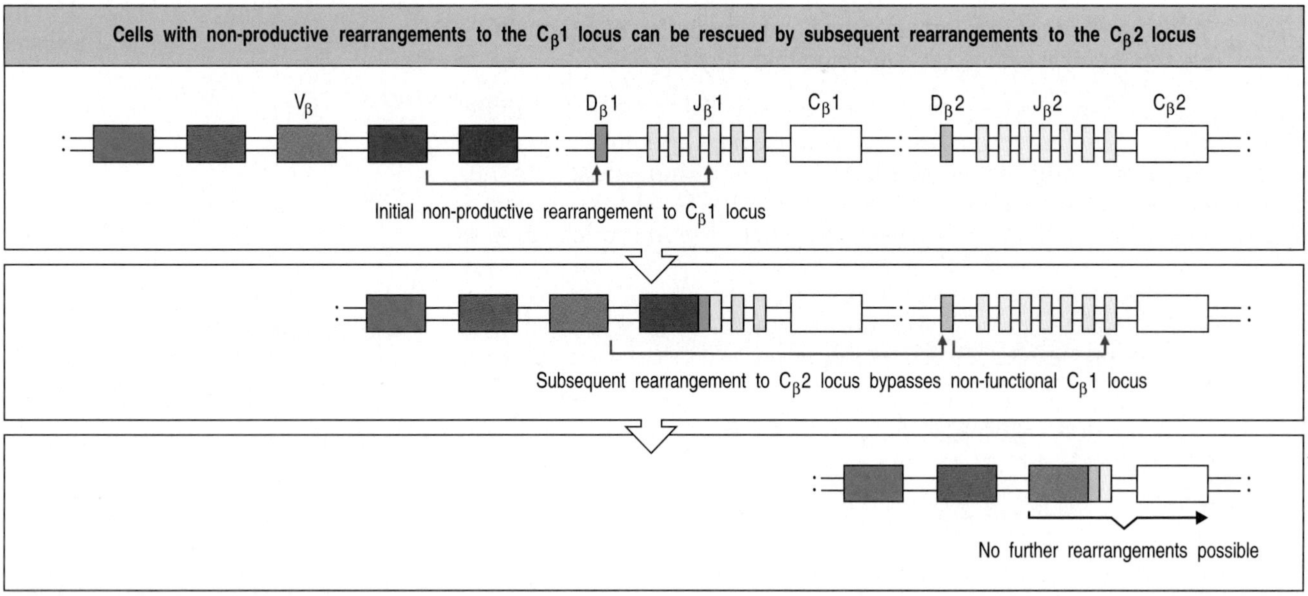

Fig. 7.11 Rescue of non-productive β-chain gene rearrangements. Successive rearrangements can rescue an initial non-productive β-chain gene rearrangement only if that rearrangement involved the D and J gene segments in the $C_\beta 1$ locus. A second rearrangement is then possible in which a second V_β gene segment rearranges to a D and a J segment in the $C_\beta 2$ locus, deleting the $C_\beta 1$ locus and the non-productively rearranged gene.

immunoglobulin heavy- and light-chain genes during B-cell development (see Section 6-2). The β-chain genes rearrange first. D_β gene segments rearrange to J_β gene segments, and this is followed by V_β to DJ_β gene rearrangement. If no functional β chain can be synthesized from these rearrangements, the cell will not be able to produce a pre-T-cell receptor and will die unless it successfully rearranges both γ- and δ-chain genes. However, unlike B cells with non-productive immunoglobulin heavy-chain gene rearrangements, thymocytes with non-productive β-chain VDJ rearrangements can be rescued by subsequent β-chain gene rearrangements, because of the organization of the D_β and J_β gene segments into two clusters upstream of two C_β constant-region genes (Fig. 7.11). This increases the likelihood of a productive VDJ join from 55% for immunoglobulin heavy-chain genes to more than 80% for T-cell receptor β-chain genes.

Once a productive β-chain gene rearrangement has occurred, the β-chain protein is expressed together with the invariant partner chain pTα and the CD3 molecules (see Fig. 7.10). The β:pTα heterodimer is a functional pre-T-cell receptor analogous to the μ:VpreB:λ5 pre-B-cell receptor complex in B-cell development (see Section 6-6). Expression of the pre-T-cell receptor triggers the phosphorylation and degradation of RAG-2, halting β-chain gene rearrangement; it also induces rapid cell proliferation, and eventually induces expression of the co-receptor proteins CD4 and CD8. All these events require the protein tyrosine kinase Lck, which later associates with the co-receptor proteins. In mice genetically deficient in Lck, T-cell development is arrested before the double-positive stage. The role of the expressed β chain in suppressing further β-chain gene rearrangement can be demonstrated in transgenic mice containing a rearranged T-cell receptor β-chain transgene: these mice express the transgenic β chain on almost all their T cells and rearrangement of the endogenous β-chain genes is strongly suppressed.

During the proliferative phase triggered by the expression of the pre-T-cell receptor, the *RAG-1* and *RAG-2* genes that mediate receptor gene segment recombination are also repressed. Hence, no rearrangement of α-chain genes occurs until the proliferative phase ends and *RAG-1* and *RAG-2* mRNA and functional RAG-1:RAG-2 protein accumulate again. This ensures that each successful rearrangement of a β-chain gene gives rise to many CD4 CD8 double-positive thymocytes. Each of these can independently rearrange its α-chain genes once the cells stop dividing, so that a single functional β chain can be associated with many different α chains in the progeny cells. During the period of α-chain gene rearrangement, α:β heterodimeric receptors are first expressed, and selection by peptide:MHC complexes in the thymus can begin.

7-9 T-cell receptor α-chain genes undergo successive rearrangements until positive selection or cell death intervenes.

The T-cell receptor α-chain genes are comparable to the immunoglobulin κ and λ light-chain genes in that they do not have D gene segments and are rearranged only after their partner receptor chain gene has been expressed. However, the presence of multiple V_α gene segments, and of some 60 J_α gene segments spread over some 80 kb of DNA, allows many successive VJ_α gene rearrangements, conferring an even greater capacity to rescue cells with non-productive rearrangements than occurs through successive rearrangements of the κ and λ light-chain genes (Fig. 7.12).

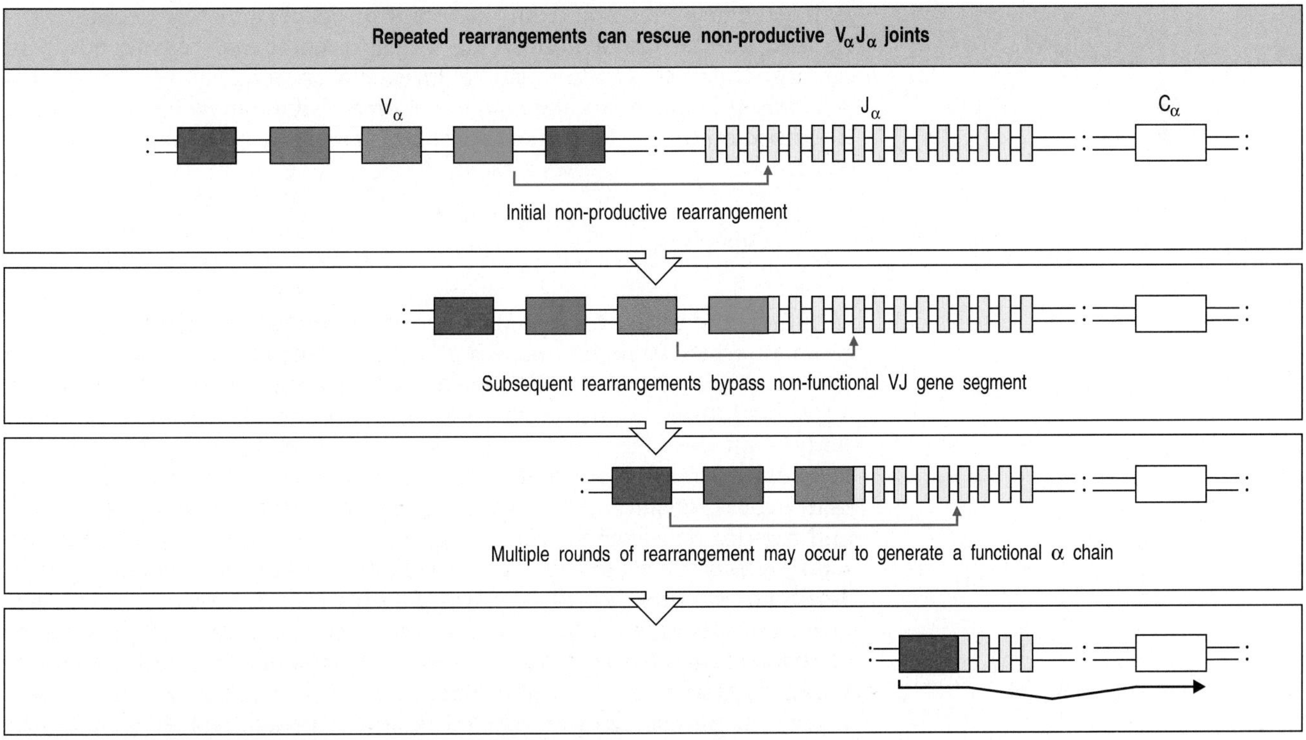

Fig. 7.12 Multiple successive rearrangement events can rescue non-productive T-cell receptor α-chain gene rearrangements. The multiplicity of V and J gene segments at the α-chain locus allows successive rearrangement events to 'leapfrog' over previously rearranged VJ segments, deleting any intervening gene segments. The α-chain rescue pathway resembles that of immunoglobulin κ light-chain genes (see Section 6-3). However, at the α-chain locus, this process can replace both non-productive and productive rearrangements as it continues until either a productive rearrangement leads to positive selection or the cell dies.

The potential for successive α-chain gene rearrangements on both chromosomes virtually guarantees that α-chain proteins will be produced in every developing T cell. Moreover, many T cells have in-frame rearrangements on both chromosomes and thus can produce two α-chain proteins. This is possible because the rearrangement of α-chain genes can continue even after production of a cell-surface receptor. We have seen that in B cells, the expression of a receptor on the cell surface shuts off further rearrangement of the receptor genes; this is accompanied by the silencing of the *RAG-1* and *RAG-2* genes. In developing T cells, by contrast, receptor expression alone is not sufficient to shut off gene rearrangement. Instead, the rearrangement machinery, including the RAG-1:RAG-2 heterodimer and TdT, can remain active, and continued rearrangements can allow several different α chains to be produced and tested for self MHC recognition in partnership with the single β chain of each developing T cell. This phase of gene rearrangement lasts for up to 3 or 4 days in the mouse and it only ceases when positive selection occurs as a consequence of receptor engagement, or when the cell dies. Thus, in the strict sense, T-cell receptor α-chain genes are not subject to allelic exclusion (see Section 6-7). However, as we shall see in the next part of this chapter, only T-cell receptors that are positively selected for self MHC recognition can function in self MHC restricted responses. The regulation of α-gene rearrangement by positive selection therefore ensures that each T cell has only a single functional specificity, even if two different alpha chains are expressed.

Clearly, the engagement of any particular T-cell receptor with a self MHC:self peptide ligand will be dependent on its specificity. Thus, the phase of α-gene rearrangement marks an important change in the forces shaping the destiny of the T cell: up to this point the development of the thymocyte has been independent of antigen; from this point on, developmental decisions will depend on the interaction of the T-cell receptor with its peptide:MHC ligands. The earliest of these antigen-dependent events, in which the T-cell receptor repertoire is shaped by self MHC:self peptide ligands, will be described more fully in the next part of this chapter.

Summary.

In differentiating T cells, receptor genes rearrange according to a defined program, which is similar to that in B cells, with the added complication that individual precursor cells can follow one of two distinct lines of development. These lead to cells bearing either γ:δ T-cell receptors or α:β T-cell receptors. Early in ontogeny, γ:δ T cells predominate, but from birth onwards more than 90% of thymocytes express T-cell receptors encoded by productively rearranged α-chain and β-chain genes. In developing thymocytes, the γ, δ, and β genes rearrange virtually simultaneously, and productive rearrangements of both the γ and δ genes can lead to the production of a functional γ:δ T-cell receptor and the development of a γ:δ T cell. In most cells this fails to occur, however, and the cells enter the α:β lineage. In these cells the successful arrangement of a β-chain gene leads to the expression of a functional β chain that forms part of a pre-T-cell receptor. The pre-T-cell receptor, like the pre-B-cell receptor, signals the developing cell to proliferate, to arrest β-chain gene rearrangement, to express CD4 and CD8, and eventually to rearrange the α-chain gene. Rearrangement of the α-chain gene continues in these CD4 CD8 double-positive thymocytes until positive selection allows the maturation of a single-positive α:β cell or the thymocyte dies. The α:β and γ:δ T-cell lineages home to different tissues and perform different functions, although the function of γ:δ T cells is not yet fully established.

Positive and negative selection of T cells.

We saw in the first part of this chapter that the T-cell precursors that first enter the thymus express neither the T-cell receptor nor either of the two co-receptor molecules CD4 and CD8, and are thus called double-negative cells. During a phase of vigorous proliferation in the subcapsular zone, these immature thymocytes differentiate into double-positive cells, expressing low levels of the T-cell receptor and both co-receptor molecules, and move on to the deeper layers of the thymic cortex. These double-positive cells have a life expectancy of only 3 or 4 days unless they are rescued by engagement of their T-cell receptor. The rescue of double-positive thymocytes from programmed cell death allows their maturation into CD4 or CD8 single-positive cells and is known as **positive selection**. By favoring the survival of thymocytes whose receptors can interact with self peptide:self MHC complexes, positive selection ensures that mature T cells can recognize foreign or non-self antigen preferentially in the context of self MHC molecules, and can thus function in the self MHC restricted responses on which adaptive immunity depends. Double-positive cells also undergo **negative selection**: those cells whose receptors recognize self peptide:self MHC complexes too well are induced to undergo apoptosis, thereby eliminating potentially self-reactive cells before they mature. This sequence of events is summarized in Fig. 7.13.

Even though the rules by which receptor engagement can signal death to some thymocytes and survival to others are not fully understood, it is clear that positive and negative selection are crucial events in generating a repertoire of self MHC restricted and self-tolerant T cells. In this section, we shall examine the crucial interactions between developing thymocytes and the different thymic components that contribute to the selection of the mature T-cell receptor repertoire, and will discuss the mechanisms by which these distinctive selective processes generate a self MHC restricted and self-tolerant repertoire of T cells.

7-10 Only T cells specific for peptides bound to self MHC molecules mature in the thymus.

Positive selection was first demonstrated in experiments in which bone marrow cells from a mouse of one MHC genotype were transferred into an irradiated mouse of a different MHC genotype. Irradiation destroys all the lymphocytes and bone marrow progenitor cells in the host animal, so that all the cells derived from bone marrow are of the donor genotype. This includes all lymphocytes and antigen-presenting cells. Mice whose bone marrow derived cells have been replaced by those of a donor mouse are known as **bone marrow chimeras** (see Section 2-23). The donor mice used in the experiments on positive selection were F1 hybrids derived from MHC^a and MHC^b parents, and thus were of the $MHC^{a \times b}$ genotype, whereas the irradiated recipients were one of the parental strains, either MHC^a or MHC^b (Fig. 7.14).

Individual T cells in $MHC^{a \times b}$ F1 hybrid mice will recognize antigen presented by either MHC^a or MHC^b, but not both, because their receptors are MHC restricted in antigen recognition. When T cells of $MHC^{a \times b}$ genotype develop in a parental MHC^a thymus and are then immunized, they will be presented with antigen bound to both MHC^a and MHC^b, as the antigen-presenting cells

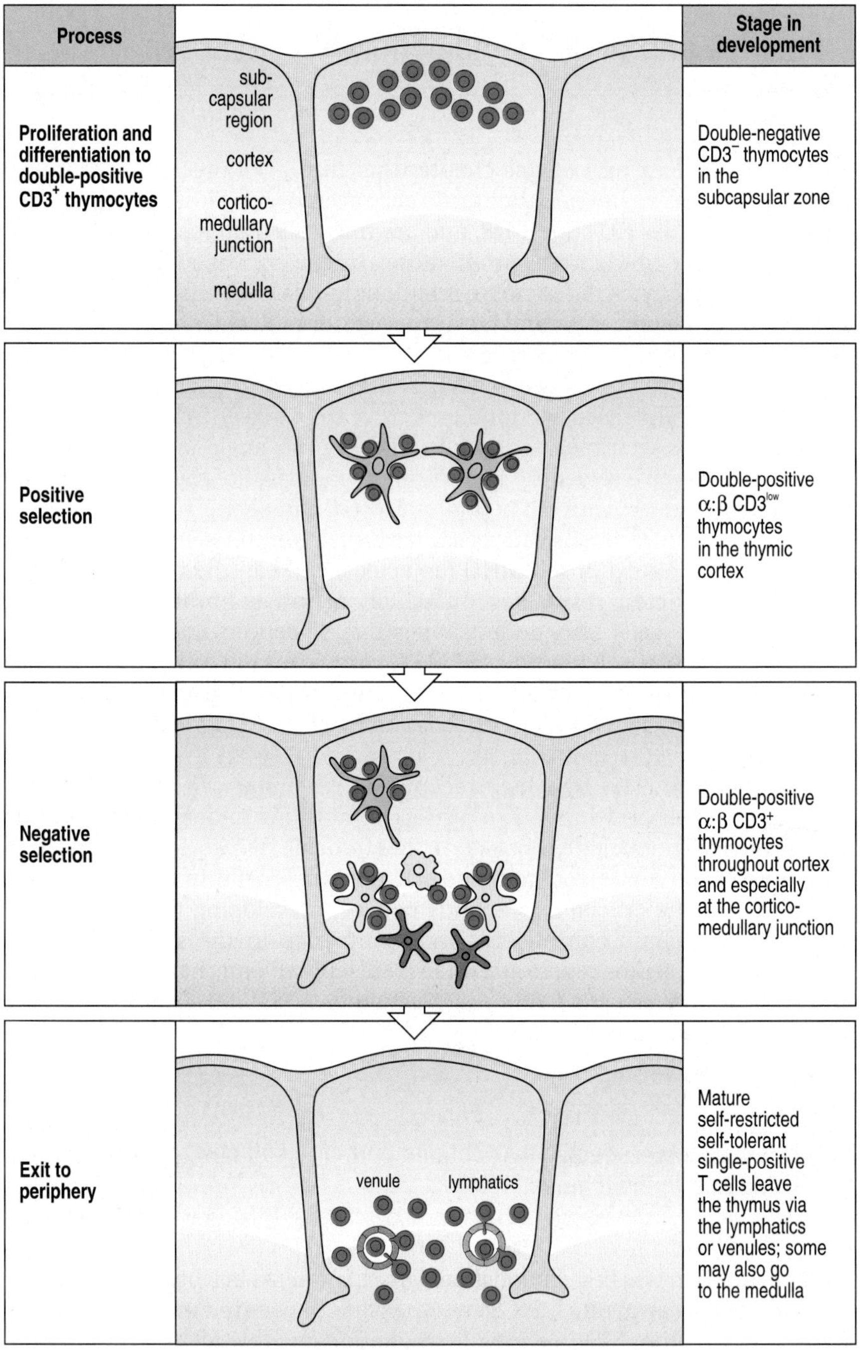

Fig. 7.13 The development of T cells can be considered as a series of discrete phases. Thymocyte progenitors enter the thymus from the venules and migrate to the subcapsular region. At this stage, they express neither the antigen receptor nor either of the two co-receptors CD4 and CD8, and are known as double-negative thymocytes. These cells proliferate in the subcapsular region of the thymus and begin the process of gene rearrangement that culminates in the expression of the pre-T-cell receptor along with the co-receptors CD4 and CD8 on the cell surface, to produce double-positive cells. As the cells mature they move deeper into the thymus. These cells then rearrange α-chain genes and become sensitive to peptide:MHC complexes. These α:β double-positive cells are found in the thymic cortex, where they undergo positive and negative selection. Negative selection is thought to be most stringent at the cortico-medullary junction, where nearly mature thymocytes encounter plentiful dendritic cells derived from the same intrathymic precursors that gave rise to the thymocytes. Finally, surviving mature single-positive T cells exit to the peripheral circulation from the medulla.

in the chimera are of bone marrow origin and thus $MHC^{a\times b}$. However, these T cells recognize antigen mainly, if not exclusively, when it is presented by MHC^{a} molecules (see Fig. 7.14). This experiment thus shows clearly that the environment in which the T cells mature determines the MHC restriction of the mature T-cell receptor repertoire.

A further experiment demonstrates that the host component responsible for the positive selection of developing T cells is the thymic stroma. For this experiment, the recipient animals were athymic *nude* or thymectomized mice of the $MHC^{a\times b}$ genotype with thymic stromal grafts of the MHC^{a} genotype, so that all of their cells carried both MHC^{a} and MHC^{b} except those of the

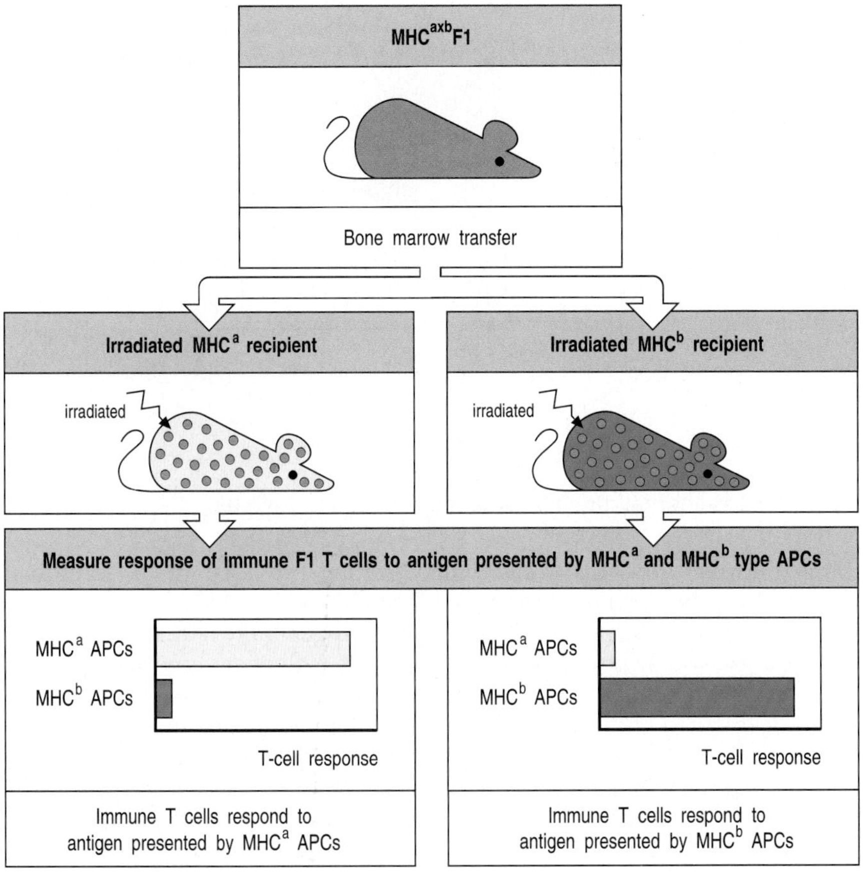

MHC axb F1
Bone marrow transfer
Irradiated MHC a recipient
Irradiated MHC b recipient
irradiated
irradiated
Measure response of immune F1 T cells to antigen presented by MHC a and MHC b type APCs
MHC a APCs
MHC b APCs
T-cell response
MHC a APCs
MHC b APCs
T-cell response
Immune T cells respond to antigen presented by MHC a APCs
Immune T cells respond to antigen presented by MHC b APCs

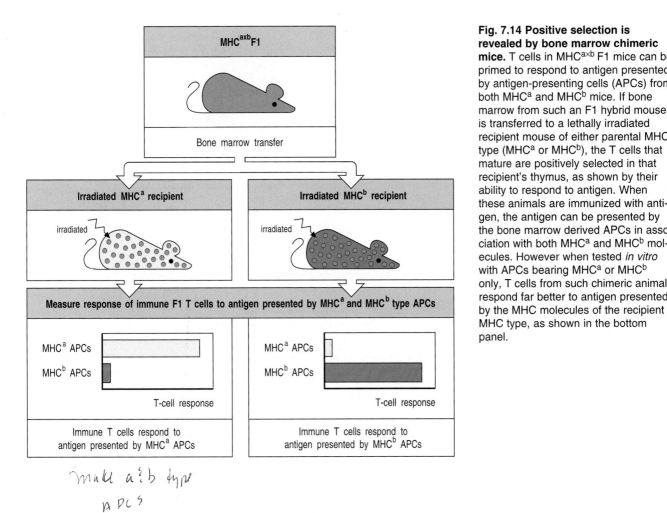

Fig. 7.14 Positive selection is revealed by bone marrow chimeric mice. T cells in MHC$^{a\times b}$ F1 mice can be primed to respond to antigen presented by antigen-presenting cells (APCs) from both MHC^{a} and MHC^{b} mice. If bone marrow from such an F1 hybrid mouse is transferred to a lethally irradiated recipient mouse of either parental MHC type (MHC^{a} or MHC^{b}), the T cells that mature are positively selected in that recipient's thymus, as shown by their ability to respond to antigen. When these animals are immunized with antigen, the antigen can be presented by the bone marrow derived APCs in association with both MHC^{a} and MHC^{b} molecules. However when tested *in vitro* with APCs bearing MHC^{a} or MHC^{b} only, T cells from such chimeric animals respond far better to antigen presented by the MHC molecules of the recipient MHC type, as shown in the bottom panel.

thymic stroma. The MHC$^{a\times b}$ bone marrow cells of these mice also mature into T cells that recognize antigens presented by MHC^{a} but not antigens presented by MHC^{b}. Thus, what mature T cells consider to be self MHC is determined by the MHC molecules expressed by the thymic stromal cells that they encounter during intrathymic development. We shall see later that the thymic cortical epithelial cell is the critical cell that governs the specificity of positive selection.

The chimeric mice used to demonstrate positive selection produce normal T-cell responses to foreign antigens. By contrast, chimeras made by injecting MHC^{a} bone marrow cells into MHC^{b} animals cannot make normal T-cell responses. This is because the T cells in these animals have been selected to recognize peptides when they are presented by MHC^{b}, whereas the antigen-presenting cells that they encounter as mature T cells in the periphery are bone marrow derived MHC^{a} cells. The T cells will therefore fail to recognize antigen presented by antigen-presenting cells of their own MHC type, and T cells can be activated in these animals only if antigen-presenting cells of the MHC^{b} type are injected together with the antigen. Thus, for a bone marrow graft to reconstitute immunity, there must be at least one MHC molecule in common between donor and recipient (Fig. 7.15). This can be an important consideration when bone marrow grafts are used in the treatment of human diseases such as leukemias.

Fig. 7.15 Summary of T-cell responses to immunization in bone marrow chimeric mice. T cells can make antigen-specific immune responses far better if the antigen-presenting cells (APCs) present in the host at the time of priming share at least one MHC molecule with the thymus in which the T cells developed.

Bone marrow donor	*In vivo* immunized recipient	Mice contain APC of type:	Secondary T-cell responses to antigen presented *in vitro* by APC of type:	
			MHC^a APC	MHC^b APC
MHC^{axb}	MHC^a	MHC^{axb}	Yes	No
MHC^{axb}	MHC^b	MHC^{axb}	No	Yes
MHC^a	MHC^b	MHC^a	No	No
MHC^a	MHC^b + MHC^b APC	MHC^a + MHC^b	No	Yes

7-11 Cells that fail positive selection die in the thymus.

Bone marrow chimeras and thymic grafting provided the first crucial evidence for the central importance of the thymus in positive selection but more detailed investigation of the process has generally required the use of mice transgenic for productively rearranged T-cell receptor genes of known specificity. When such genes are introduced into the genome of a mouse, rearrangement of the endogenous genes is inhibited, so that most developing T cells express the receptor encoded by the α- and β-chain transgenes. By introducing the transgenes into mice of known MHC genotype, it is possible to establish the effect of MHC molecules on the maturation of thymocytes with known recognition properties. Such studies have confirmed that T cells develop to maturity only if they are self MHC restricted. They have also established the fate of T cells that fail positive selection. Rearranged receptor genes from a mature T cell specific for an antigenic peptide presented by a particular MHC molecule were introduced into a recipient mouse lacking that molecule, and the fate of the thymocytes was investigated directly by staining with clonotypic antibodies specific for the transgenic receptor. Antibodies to other molecules such as CD4 and CD8 were used at the same time to mark the stages of T-cell development. In this way it was shown that cells that fail to recognize the MHC molecules present on the thymic epithelium never progress further than the double-positive stage and die in the thymus within 3 or 4 days of their last division.

In a normal thymus, the fate of each thymocyte depends on the specificity of the receptor it expresses and, as we saw in Section 7-9, the specificity can undergo several changes as the α-chain genes continue to rearrange. The ability of a single developing thymocyte to express several different rearranged α-chain genes during the time that it is susceptible to positive selection must increase the yield of useful T cells significantly; without this mechanism many more thymocytes would fail positive selection and die. However, the continued rearrangement of α-chain genes also makes it likely that a significant percentage of mature T cells will express two receptors, sharing a β chain but differing in their α chains. Indeed, one can predict that if the frequency of positive selection is sufficiently rare, roughly one in three mature T cells would have two α chains at the cell surface. This was confirmed recently for both human and mouse T cells.

T cells with dual specificity might be expected to give rise to inappropriate immune responses if the cell is activated through one receptor yet can act upon target cells recognized by the second receptor. However, although anti-bodies specific for either of the α chains present in a single cell could induce a response, only one of the two receptors will be able to recognize peptide presented by self MHC molecules. The regulation of α-chain gene rearrangement by positive selection should ensure that each cell expresses only a single receptor that can function in self MHC restricted responses. Thus, the existence of cells with two α-chain genes productively rearranged and two α chains expressed at the cell surface does not seem to challenge the importance of clonal selection, which depends on a single functional specificity being expressed by each cell.

7-12 Positive selection acts on a repertoire of receptors with inherent specificity for MHC molecules.

Positive selection acts on a repertoire of receptors whose specificity is determined by a combination of germline-encoded V regions, and junctional regions whose diversity is randomly created as the genes rearrange. This receptor repertoire must be capable of recognizing all of the hundreds of different allelic variants of MHC molecules present in the population, as the genes for the T-cell receptor α and β chains segregate in the population independently from those of the MHC. If the binding specificity of the unselected repertoire were completely random, only a relatively small proportion of thymocytes would be capable of recognizing an MHC molecule. However, it seems that the variable CDR1 and CDR2 loops of the T-cell receptor, which are encoded within the germline V gene segments (see Sections 4-24 and 4-26), confer an intrinsic specificity for MHC molecules. This has been shown by examining T cells expressing an unselected repertoire of receptors. Such T cells can be induced to develop in fetal thymic organ cultures that lack the expression of both MHC class I and MHC class II molecules. The receptor engagement that is responsible for normal positive selection is mimicked by treatment with antibodies. When the reactivity of these T cells is tested, roughly 5% are able to respond to a particular MHC class II genotype and, because they developed without selection by MHC molecules, this must reflect a specificity inherent in the germline receptor genes.

The proportion of mature T cells responding in these assays gives an estimate of the proportion of thymocytes that would be negatively selected in response to the same MHC molecules; a greater range of receptors must be susceptible to positive selection, but the percentage is not directly ascertained by these experiments. Nevertheless, the pattern of reactivity displayed by the unselected T cells suggests that most, if not all, receptors can be useful in some MHC genotypes. Moreover, the germline-encoded specificity for MHC must significantly increase the proportion of receptors that can be positively selected in any one individual.

7-13 The expression of CD4 and CD8 on mature T cells and the associated T-cell functions are determined by positive selection.

At the time of positive selection, the thymocyte expresses both CD4 and CD8 co-receptor molecules. At the end of the selection process, mature thymocytes ready for export to the periphery express only one of these two co-receptors. Moreover, almost all mature T cells that express CD4 have receptors that recognize peptides bound to self MHC class II molecules and are programmed to become cytokine-secreting cells, whereas most of those that express CD8 have receptors that recognize peptides bound to self MHC class I molecules and are programmed to become cytotoxic effector cells. Thus, positive selection also determines the cell-surface phenotype and functional potential of the mature T cell, selecting the appropriate

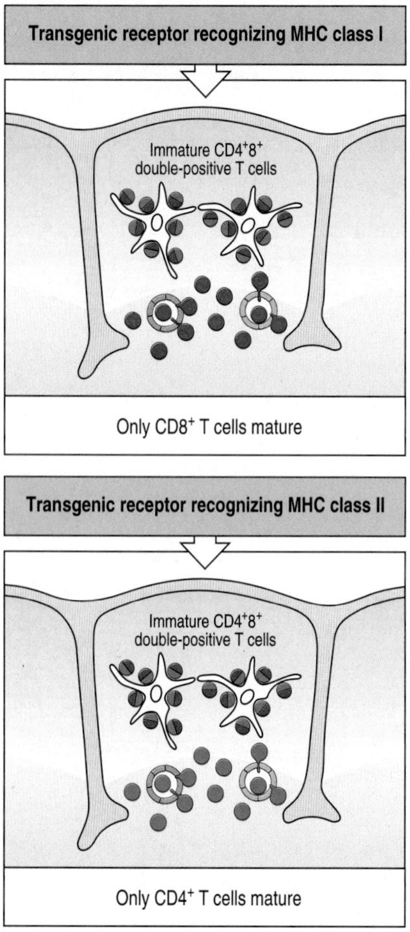

Fig. 7.16 Positive selection determines co-receptor specificity. In mice transgenic for T-cell receptors restricted by an MHC class I molecule (top panel), the only mature T cells to develop have the CD8 (red) phenotype. In mice transgenic for receptors restricted by an MHC class II molecule (bottom panel), all the mature T cells have the CD4 (blue) phenotype. In both cases, normal numbers of immature, double-positive thymocytes are found. The specificity of the T-cell receptor determines the outcome of the developmental pathway, ensuring that the only T cells that mature are those equipped with a co-receptor that is able to bind the same self MHC molecule as the T-cell receptor.

co-receptor for efficient antigen recognition and the appropriate program for the T cell's eventual functional differentiation in an immune response. Again, experiments with mice made transgenic for rearranged T-cell receptor genes show clearly that it is the specificity of the T-cell receptor for self MHC molecules that determines which co-receptor a mature T cell will express. If the T-cell receptor transgenes encode a receptor specific for antigen presented by self MHC class I molecules, mature T cells that express the transgenic receptor are CD8 T cells. Similarly, in mice made transgenic for a receptor that recognizes antigen with self MHC class II molecules, mature T cells that express the transgenic receptor also express CD4 (Fig. 7.16).

The importance of MHC molecules in such selective events can be seen in the class of human immunodeficiency diseases known as **bare lymphocyte syndromes**, which are caused by mutations that lead to an absence of MHC molecules on lymphocytes and thymic epithelial cells. People who lack MHC class II molecules have CD8 T cells but only a few, highly abnormal, CD4 T cells; a similar result has been obtained in mice in which MHC class II expression has been eliminated by targeted gene disruption (see Section 2-27). Likewise, mice and humans that lack MHC class I molecules lack CD8 T cells. Thus, MHC class II molecules are required for CD4 T-cell development, whereas MHC class I molecules are required for CD8 T-cell development.

In mature T cells, the co-receptor functions of CD8 and CD4 depend on their respective abilities to bind invariant sites on MHC class I and class II molecules (see Section 4-29). Co-receptor binding to an MHC molecule is also required for normal positive selection, as shown for CD4 in the experiment discussed below. Thus positive selection depends on engagement of both the antigen receptor and co-receptor with an MHC molecule, and determines the survival of single-positive cells that express only the appropriate co-receptor. However, the mechanism whereby the selection of antigen receptor and co-receptor are coordinated remains to be established.

There are two broad theories for how this might be accomplished: the **instructive model** and the **stochastic/selection model**. According to the instructive model, the two co-receptors deliver distinct intracellular signals: the signal delivered by CD4 shuts off the expression of the CD8 gene and induces the CD4 differentiation pathway, whereas the signal delivered by CD8 silences the CD4 gene and induces the CD8 differentiation pathway. In the stochastic/selection model, by contrast, inactivation of either the CD4 or CD8 gene is not determined by the specificity of the receptor. Instead it is random, or dictated by a mechanism of lineage commitment that does not rely on receptor specificity, and the thymocytes are then tested for the correct match of co-receptor with the specificity of the antigen receptor. Cells in which the specificity of the co-receptor is mismatched with that of the antigen receptor die, whereas those with correct matching of CD4 and MHC class II recognition, or of CD8 and MHC class I recognition, survive, mature, and leave the thymus. In practice, it has been hard to distinguish experimentally between these two models, and elements of both might be correct. The timing and mechanism of lineage commitment therefore remains uncertain. What is clear is that there must be a divergence of developmental programming, so that genes involved in the killing of target cells, for example, are activated in effector CD8 T cells while various cytokine genes are activated in effector CD4 T cells. One gene that might be involved in the developmental programming is the mammalian homolog of *Notch*, a gene that serves to regulate cell fate in many developmental systems in which a single precursor cell has the potential to differentiate along one of two distinct pathways. *Notch* protein over-expressed in thymocytes directs them into the CD8 lineage, suggesting that its normal role is to inhibit the pathway that leads to CD4 T cells.

7-14 Thymic cortical epithelial cells mediate positive selection.

The ability of thymic stromal cells to determine co-receptor specificity has been used as the basis for experiments aimed at identifying the stromal cell type that is critical in positive selection. An obvious candidate, on the basis of much circumstantial evidence, is the thymic cortical epithelial cell. These cells form a web of processes that make close contacts with the double-positive T cells undergoing positive selection (see Fig. 7.2) and at the sites of contact T-cell receptors can be seen clustering with MHC molecules.

Direct evidence that thymic cortical epithelial cells mediate positive selection comes from an ingenious manipulation of mice whose MHC class II genes have been eliminated by targeted gene disruption (Fig. 7.17). As we have already seen, mutant mice that lack MHC class II molecules do not produce CD4 T cells. To test the role of the thymic epithelium in positive selection, a MHC class II gene was placed under the control of a promoter that restricted its expression to thymic cortical epithelial cells, and introduced as a transgene into such mutant mice. CD4 T cells then developed normally. A further variant of this experiment shows that, in order to promote the normal development of CD4 T cells, the MHC class II molecule on the thymic epithelium must be able to interact with CD4. Thus, when the MHC class II transgene expressed in the thymus contains a mutation that prevents its binding to CD4, very few CD4 T cells develop. Equivalent studies of CD8 interaction with MHC class I molecules show that co-receptor binding is necessary for normal positive selection of CD8 cells as well.

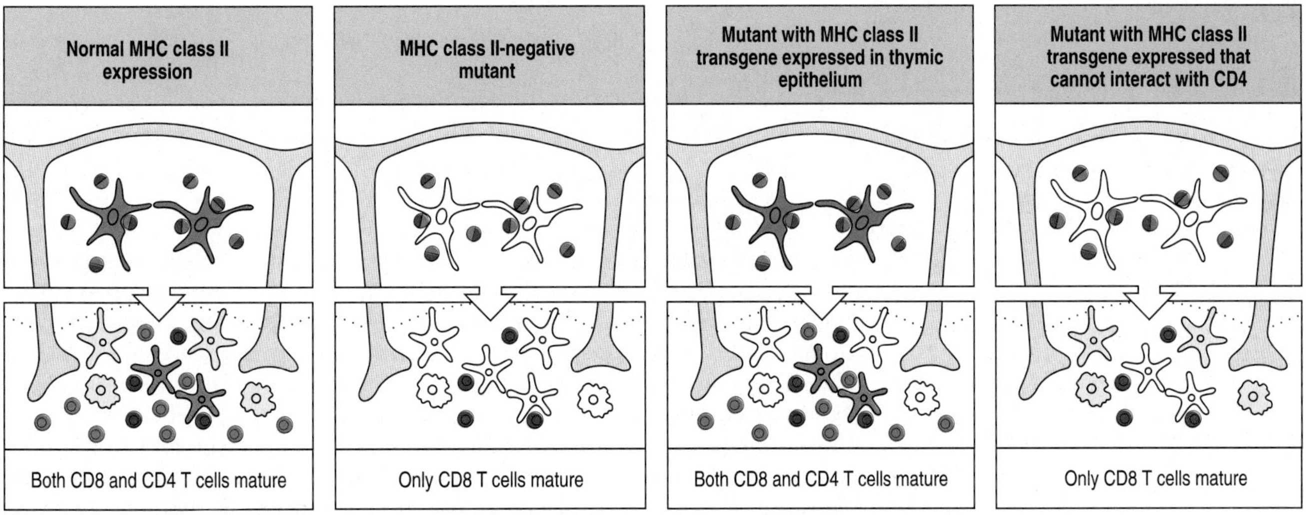

Fig. 7.17 Thymic cortical epithelial cells mediate positive selection. The expression of MHC class II molecules in the thymus of normal and mutant strains of mice is shown by coloring the stromal cells only if they are expressing MHC class II molecules. In the thymus of normal mice (first panels), which express MHC class II molecules on epithelial cells in the thymic cortex (blue) as well as on medullary epithelial cells (orange) and bone marrow derived cells (yellow), both CD4 (blue) and CD8 (red) T cells mature. Double-positive thymocytes are shown as half red/half blue. The second panels represent mutant mice in which MHC class II expression has been eliminated by targeted gene disruption; in these mice, few CD4 T cells develop, although CD8 T cells develop normally. In MHC class II-negative mice containing an MHC class II transgene engineered so that it is expressed only on the epithelial cells of the thymic cortex (third panels), normal numbers of CD4 T cells mature. The MHC class II molecule needs to be able to interact with the CD4 protein, as mutant MHC class II molecules with a defective CD4 binding site do not allow the positive selection of CD4 T cells (fourth panels). Thus, the cortical epithelial cells are the critical cell type mediating positive selection.

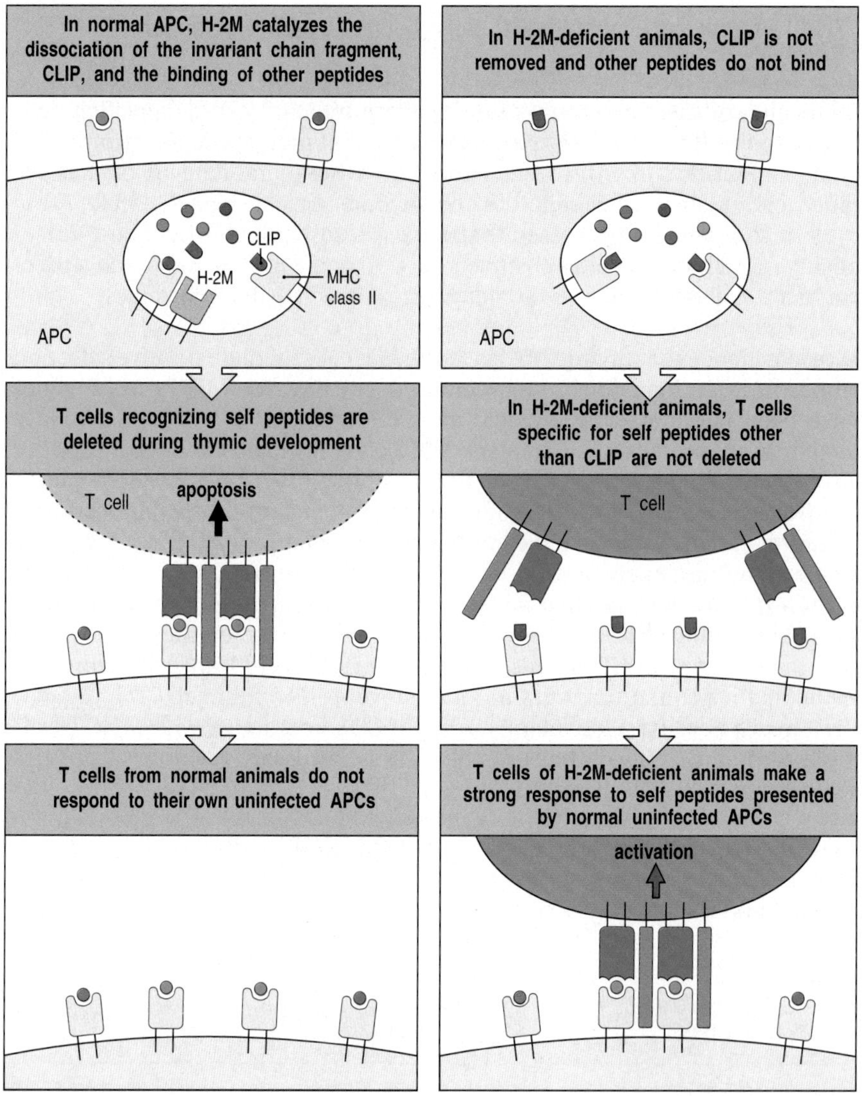

Fig. 7.18 The peptides that are bound to MHC class II molecules can affect the T-cell receptor repertoire. The left panels show the normal situation in which a range of peptides is presented by antigen-presenting cells (APCs) to immature T cells in the thymus, and the consequent deletion of self-reactive T cells. The right panels show a case in which a single peptide predominates. Mice with their H-2Mα gene disrupted by targeted mutagenesis express MHC class II molecules that predominantly carry the CLIP peptide of the invariant chain (top right panel). Their MHC class II molecules thus present only the CLIP peptide to T cells maturing in the thymus. CD4 T cells mature in the presence of this single dominant peptide:MHC complex but they are reduced in number by two- to threefold, even though negative selection of MHC class II-restricted T-cells will be limited to deleting T cells specific for the CLIP peptides (middle right panel). Mature T cells from such mice respond strongly to MHC-identical APCs, which express the normal array of self peptides (bottom right panel), showing that a majority of the T cells that are positively selected by this single dominant MHC:peptide complex are reactive with other self peptides complexed with the same MHC molecule. The T cells bearing these receptors would be deleted by negative selection in normal mice (middle and bottom left panels).

The critical role of the thymic epithelium in positive selection raises the question of whether there is anything distinctive about the antigen-presenting properties of these cells. This is not clear at present; however, as we discussed in Section 4-13, thymic epithelial cells are unusual in expressing an MHC class II-like molecule called HLA-DO (H-2O in mice). HLA-DO inhibits the action of another MHC class II-like molecule known as HLA-DM (H-2M in mice) and, because HLA-DM catalyzes the exchange of peptides bound by intravesicular MHC class II molecules, the presence of HLA-DO can affect the spectrum of peptides displayed on the surface of thymic epithelium by MHC class II molecules. Thymic epithelial cells do seem to express at the cell surface a relatively high density of MHC class II molecules that retain the invariant chain-associated peptide (CLIP) (see Fig. 4.16). However, they also present a range of other peptides, and it remains to be seen whether the MHC:peptide complexes presented by these cells have any special characteristics that are important for positive selection. Moreover, recent evidence suggests that different proteases are employed in the production of peptides in thymic cortical epithelium, where cathepsin L dominates, and in the periphery, where cathepsin S seems to be most important. T-cell development is severely impaired in cathepsin L knock-out mice.

Artificial manipulation of the peptides presented by thymic epithelium has a profound effect on positive selection. The effect of selecting in the presence of a single peptide:class II MHC complex has been demonstrated in experiments in which H-2M, the mouse homolog of human HLA-DM, is disrupted. In mice lacking a functional H-2Mα chain, the CLIP fragment of the invariant chain is not released from the newly synthesized MHC class II molecules. These CLIP-associated MHC class II molecules are unable to bind the usual array of self peptides that are normally acquired in the peptide-loading compartment of the cells and thus the main self peptide presented by MHC class II molecules on the surface of the thymic epithelium is the invariant CLIP peptide (Fig. 7.18). In these mice, the total number of CD4 T cells is reduced two- to threefold. A diversity of T-cell receptor β chains are expressed in these cells, but of six T-cell receptor transgenes examined so far, none are positively selected despite the presence of the MHC class II molecule that normally selects these receptors. Thus it seems that a significantly reduced repertoire of T cells is selected in the presence of a single predominant MHC class II:peptide complex. In addition, a high proportion of the selected T cells are reactive against stimulator cells from wild-type mice of the same MHC genotype, and would therefore be eliminated by negative selection in a non-mutant mouse. Many of these cells might have been positively selected through an interaction dominated by contacts between the T-cell receptor and MHC class II molecule, to which the bound peptide made a minimal contribution. Such cells would be more likely to react with the same MHC molecule complexed with a different self peptide than would cells that were positively selected through an interaction in which contacts with the bound peptide played a greater part. These experiments therefore suggest that a diversity of bound peptides is important for the positive selection of a sufficiently diverse repertoire of self MHC restricted T cells. We shall see in Section 7-18 how the selection of individual T-cell receptors can be mediated by specific peptides that are related, but not identical, to the antigenic peptides to which these receptors respond.

7-15 T cells specific for ubiquitous self antigens are deleted in the thymus.

When a mature T cell in the periphery encounters its corresponding antigen on a professional antigen-presenting cell, it is activated to proliferate and produce effector T cells. In contrast, when a developing thymocyte encounters its corresponding antigen on thymic stromal or bone marrow derived cells, it dies by apoptosis. This response to antigen is the basis of negative selection and has been demonstrated in mice expressing a transgenic T-cell receptor specific for a peptide of ovalbumin bound to a MHC class II molecule. When such a mouse is injected with the appropriate ovalbumin peptide, its peripheral CD4 T cells become activated, but most of the intrathymic cells die (Fig. 7.19). Similar results are obtained in thymic organ culture with T cells from normal or transgenic mice, showing that secondary effects of the induction of cytokines or corticosteroids cannot account for these results.

In unmanipulated mice, T cells developing in the thymus encounter a large selection of self peptides bound to self MHC molecules on thymic cells of various types, and cells that would be reactive to such peptides are eliminated by these encounters. The deletion of T cells that recognize self peptides in the thymus can be demonstrated experimentally in mice made transgenic for the rearranged genes encoding receptors that are specific for self peptides expressed only in male mice. Thymocytes bearing these receptors disappear from the developing T-cell population in male mice at the CD4 CD8

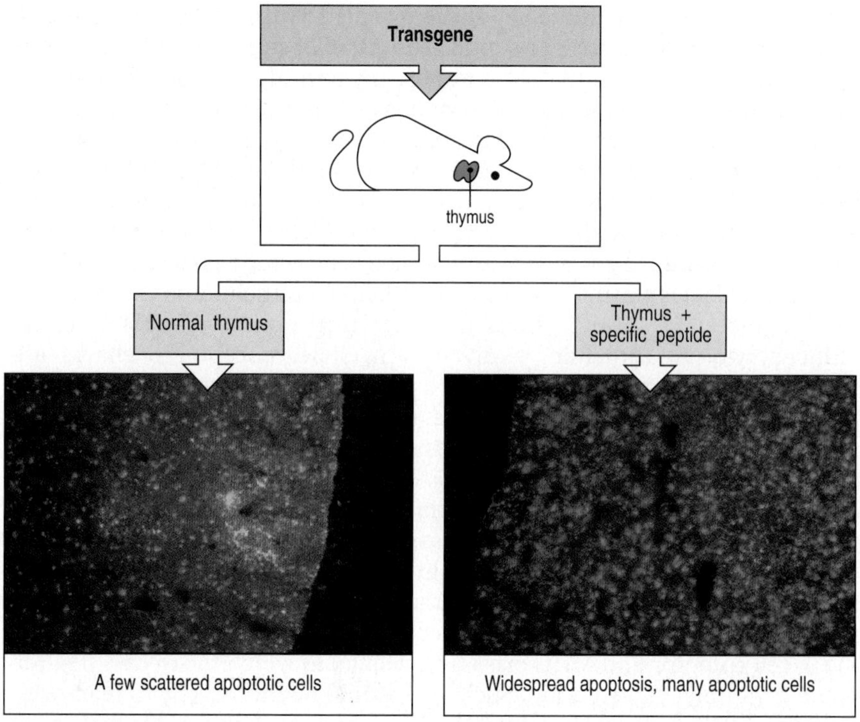

Fig. 7.19 T cells specific for self antigens are deleted in the thymus. In mice transgenic for a T-cell receptor that recognizes a known peptide antigen complexed with self MHC, all of the T cells have the same specificity; in the absence of the peptide, most thymocytes mature and migrate to the periphery. This can be seen in the bottom left panel where a normal thymus is stained with antibody to identify the medulla (in green), and by the TUNEL technique (see Fig. 5.23) in red to identify apoptotic cells. However, if the peptide antigen recognized by the transgenic T-cell receptor is added, then massive cell death occurs, as shown by the increased numbers of apoptotic cells in the right-hand bottom panel. Photographs courtesy of A Wack and D Kioussis.

double-positive stage of development, and no single-positive cells bearing the transgenic receptors mature in male mice. In female mice, by contrast, the transgenic T cells mature normally. Similar experiments have been carried out with other antigens and gave similar results.

These experiments illustrate the principle that self peptide:self MHC complexes encountered in the thymus purge the T-cell repertoire of cells bearing self-reactive receptors during development. However, not all self proteins are expressed in the thymus, and those that appear in the periphery, or are expressed at different stages in development, such as after puberty, must encounter mature T cells that have the potential to respond to them. That in most cases there is no response to such proteins suggests that some other mechanism or mechanisms must prevent T-cell responses to such antigens. The induction of self tolerance in the periphery will be discussed in Chapter 13.

7-16 Negative selection is driven most efficiently by bone marrow derived antigen-presenting cells.

Whereas thymic cortical epithelial cells mediate positive selection, negative selection in the thymus can be mediated by several different cell types. The most important of these are the bone marrow derived dendritic cells and macrophages. These are the professional antigen-presenting cells that also activate mature T cells in peripheral lymphoid tissues. The self antigens presented by these cells are therefore the most important source of potential autoimmune responses, and T cells responding to such self peptides must be eliminated in the thymus.

Bone marrow chimera experiments have shown clearly the role of thymic macrophages and dendritic cells in negative selection. In these experiments, $MHC^{a \times b}$ F1 bone marrow is grafted into one of the parental strains (MHC^a in Fig. 7.20) so that T cells developing in the grafted animals are exposed to thymic epithelium of the host strain (MHC^a) but to $MHC^{a \times b}$ F1 dendritic cells and macrophages. These bone marrow chimeras will tolerate skin grafts from animals of both MHC^a and MHC^b strains (see Fig. 7.20), provided that the skin cells do not present any tissue-specific peptides that differ between the two strains. This means that the animals must have become tolerant not only to host MHC^a:peptide antigens but also to the donor-specific MHC^b:peptide antigens carried on cells derived from the grafted marrow. As the only cells that could present MHC^b:peptide antigens are the bone marrow derived cells, these cell types are assumed to have a crucial role in negative selection.

Although bone marrow derived antigen-presenting cells are the principal mediators of negative selection, both thymocytes themselves and thymic epithelial cells also have the ability to cause the deletion of self-reactive cells. Such reactions may normally be of little significance. In patients undergoing bone marrow transplantation from an unrelated donor, however, where all the thymic macrophages and dendritic cells are of donor type, negative selection mediated by thymic epithelial cells can assume a special importance in maintaining tolerance to the recipient's own self tissue antigens.

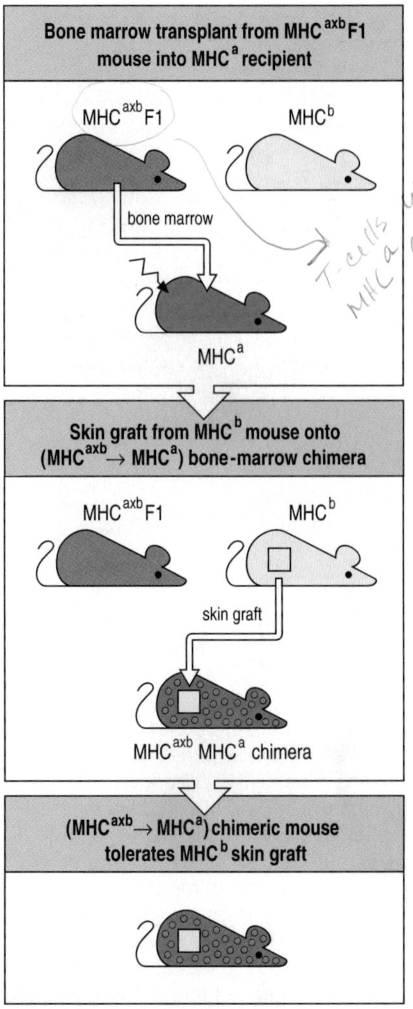

Fig. 7.20 Bone marrow derived cells mediate negative selection in the thymus. When $MHC^{a \times b}$ F1 bone marrow is injected into an irradiated MHC^a mouse, the T cells mature on thymic epithelium expressing only MHC^a molecules. Nevertheless, the chimeric mice are tolerant to skin grafts expressing MHC^b molecules (provided that these grafts do not present skin-specific peptides that differ between strains a and b). This implies that the T cells whose receptors recognize self antigens presented by MHC^b have been eliminated in the thymus. As the transplanted $MHC^{a \times b}$ F1 bone marrow cells are the only source of MHC^b molecules in the thymus, bone marrow derived cells must be able to induce negative selection.

7-17 Endogenous superantigens mediate negative selection of T-cell receptors derived from particular V_β gene segments.

It is virtually impossible to demonstrate directly the negative selection of T cells specific for any particular self antigen in the normal thymus because such T cells will be too few to detect. There is, however, one case in which negative selection can be seen on a large scale in normal mice and the point at which it occurs in T-cell development can be identified. In the most striking examples, T cells expressing receptors encoded by particular V_β gene segments are virtually eliminated in the affected mouse strains. This occurs as the consequence of the interaction of immature thymocytes with endogenous superantigens present in those strains. We learned in Chapter 4 that superantigens are viral or bacterial proteins that bind tightly to both MHC class II molecules and particular V_β domains, irrespective of the antigen specificity of the receptor and the peptide bound by the MHC molecule (see Fig. 4.36).

The endogenous superantigens of mice are encoded by mouse mammary tumor virus (MMTV) genomes that have become integrated into the mouse chromosomes, where they are inherited by successive generations of mice along with mouse genes. Like the bacterial superantigens, these viral antigens induce strong T-cell responses; indeed, they were originally designated **minor lymphocyte stimulating** (**Mls**) antigens because, although they are not major histocompatibility complex proteins (hence minor), they stimulate exceptionally strong primary T-cell responses when T cells from a strain lacking the superantigen gene are stimulated by B cells from MHC-identical mice that express it.

T cells bearing V_β regions to which the Mls proteins bind apoptose during intrathymic maturation in Mls^+ strains. For example, one variant of the Mls antigen ($Mls\text{-}1^a$) deletes all thymocytes expressing the $V_\beta6$ variable region (and also those expressing $V_\beta8.1$ and $V_\beta9$), whereas such cells are not deleted in mice that lack $Mls\text{-}1^a$. Thus, the expression of endogenous superantigens in mice has a profound impact on the T-cell receptor repertoire. This sort of deletion has not yet been seen in any other species, including humans, despite the presence of retroviral sequences in the genomes of many mammals.

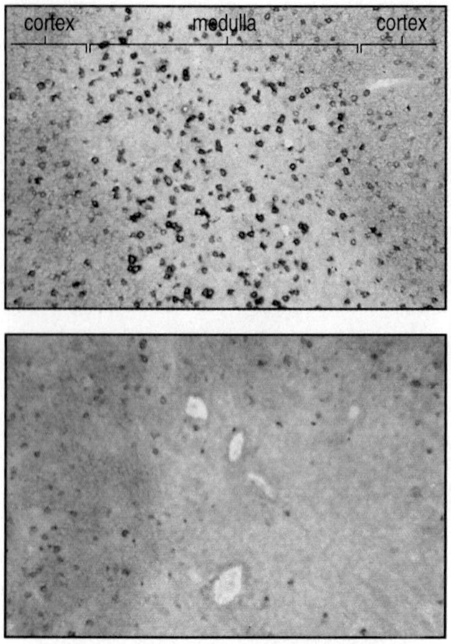

Fig. 7.21 **Clonal deletion by Mls-1a occurs late in the development of thymocytes.** T cells with Mls-1a-responsive receptors encoded by $V_{\beta}6$ are seen in both the cortex and medulla of Mls-1b mice (top panel, cells stained brown with anti-$V_{\beta}6$ antibody). Note that the mature cells in the medulla express higher levels of the receptor and thus stain more darkly than the immature cells in the cortex. In Mls-1a mice (lower panel) there is no obvious reduction in the number of immature cortical T cells expressing the $V_{\beta}6$ receptor, but the mature cells are not found. Photographs courtesy of H Hengartner.

In mice that express the superantigen and thus are tolerant to it, cells expressing receptors responsive to superantigens are found among double-positive thymocytes, and are abundant in thymic cortex but absent from the thymic medulla and the periphery. This suggests that superantigens might delete relatively mature cells as they migrate out of the cortex into the medulla, where a particularly dense network of dendritic cells marks the cortico-medullary junction (Fig. 7.21). However, although clonal deletion by superantigens is a powerful tool for examining negative selection in normal mice, it must be remembered that superantigen-driven clonal deletion might not be representative of clonal deletion by self peptide:self MHC complexes. What is clear is that clonal deletion by either superantigens or self peptide:self MHC complexes generates a repertoire of T cells that does not respond to the self antigens expressed by its own professional antigen-presenting cells.

7-18 The signals for negative and positive selection must differ.

In the preceding sections we have described some of the experiments that have contributed to the large body of evidence that T cells are selected for both self MHC restriction and self tolerance by MHC molecules expressed on stromal cells in the thymus. We now turn to the central question posed by positive and negative selection: how can engagement of the receptor with self MHC:self peptide complexes lead both to further maturation of thymocytes during positive selection, and to cell death during negative selection? The answers to these questions are still not known for certain, but two possible mechanisms have been suggested and we shall briefly describe each before going on to discuss some recent experiments bearing on the mechanism of positive selection.

There are two issues to be resolved. First, the interactions that lead to positive selection must include more receptor specificities than those that lead to negative selection. Otherwise, all the cells that are positively selected in the thymic cortex would be eliminated by negative selection and no T cells would ever leave the thymus (Fig. 7.22). Second, the consequences of the interactions leading to positive and negative selection must differ, so that cells that recognize self peptide:self MHC complexes on cortical epithelial cells are induced to mature, whereas those whose receptors might confer autoreactivity are induced to die.

Two main hypotheses have been proposed to account for these differences between positive and negative selection. The first is the **avidity hypothesis**, which states that the outcome of MHC:peptide binding by T-cell receptors of thymocytes depends upon the strength of the signal delivered by the receptor on binding, and that this will, in turn, depend upon both the affinity of the T-cell receptor for the MHC:peptide complex and the density of the complex on a thymic cortical epithelial cell. Thymocytes that are signaled weakly are rescued from programmed cell death and are therefore positively selected, whereas thymocytes that are signaled strongly are driven to programmed cell death and are therefore negatively selected. Because more complexes are likely to be weakly bound than strongly, this will also result in the positive selection of a larger repertoire of cells than are negatively selected.

Alternatively, the delivery of incomplete activating signals by self peptides could account for positive selection: we shall call this the **differential signaling hypothesis**. Under this hypothesis, it is the nature of the signal delivered by the receptor, rather than the number of receptors engaged, that distinguishes

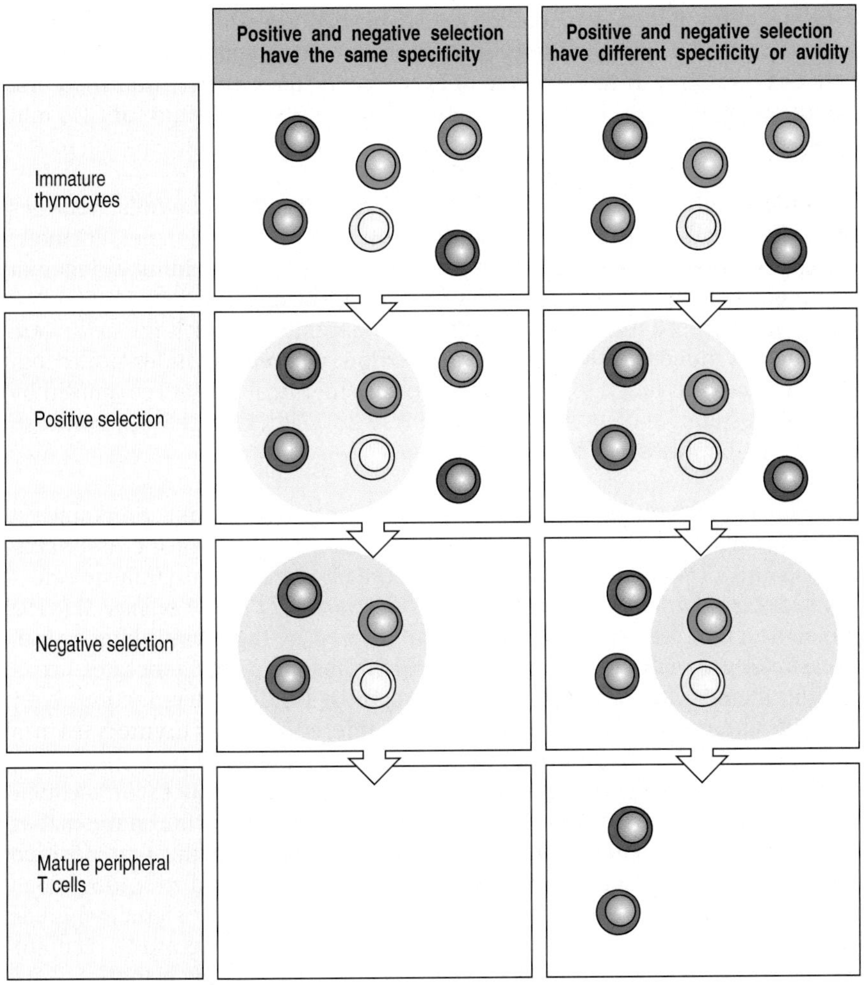

Fig. 7.22 The specificity or affinity of positive selection must differ from that of negative selection. Immature T cells are positively selected so that only those thymocytes whose receptors can engage the peptide:MHC complexes on thymic epithelium mature, giving rise to a self MHC-restricted population of thymocytes. Negative selection removes those thymocytes whose receptors recognize self peptides complexed with self MHC molecules, giving a self-tolerant population of thymocytes. If the specificity of positive and negative selection were the same (left panels), all the T cells that survive positive selection would be deleted during negative selection. Only if the specificity or affinity of negative selection is different from that of positive selection (right panels) can thymocytes mature into T cells.

positive from negative selection. Thus, the avidity hypothesis would predict that the same MHC:peptide complex could drive the positive or negative selection of the same receptor depending on its density on the cell surface, whereas the differential signaling hypothesis would predict that this could not occur, because it proposes that the signals leading to positive and negative selection are qualitatively different. A new approach to testing these hypotheses has opened up with the recent description of antagonist peptides.

7-19 Partial activating signals might account for positive selection of CD8 but not CD4 T cells.

Antagonist peptides, which we discussed in Section 5-12, are variants of the peptide component of an MHC:peptide complex that, when substituted for the original antigenic peptide, engage the T-cell receptor and block the response of the T cell to the original MHC:peptide complex. Peptides that act as antagonists deliver a partial signal to the T cell that produces a subset of the full response. Because positive selection is thought to entail the delivery of a weak or partial signal to developing thymocytes, it seemed possible that antagonist peptides might mimic the mechanism that operates in positive selection. By using thymic lobe cultures from mice transgenic for a known

T-cell receptor, it has been possible to test this hypothesis. These experiments, and more recent ones using receptor crosslinking reagents to drive thymocyte development in the absence of MHC molecules, suggest that there are important differences in the requirements for positive selection of CD8 and CD4 thymocytes.

The cultured thymic lobes used for the experiments with antagonist peptides were genetically engineered to express only the MHC molecules normally recognized by the transgenic receptor, and the effect of adding antagonist peptides could then be tested. With CD8 cells, these experiments showed that antagonist or partial agonist variants of the peptide for which the cells were specific did indeed result in positive selection, whereas peptides not recognized by the cells failed to select and peptides identical to that recognized by the cells resulted in negative selection (Fig. 7.23). With CD4 cells, by contrast, antagonist peptides also failed to select.

Experiments with reagents used to crosslink receptors and co-receptors suggest that this might reflect a differential requirement for co-receptor crosslinking in CD8 and CD4 cell development. For these experiments, thymocyte selection was tested on thymic organ cultures lacking either class I or class II MHC molecules, and the selective signal was delivered by the crosslinking ligand. Crosslinking T-cell receptor molecules with those reagents generates intracellular signals like those seen after antagonist binding (see Section 5-12) and, like antagonist binding, it induced the differentiation of thymocytes into CD8 cells. CD4 cell differentiation, by contrast, could be induced by crosslinking the receptor with either CD4 or CD8, which produces intracellular signals similar to those seen when T cells bind the MHC:peptide complex that they normally recognize. Together, these experiments imply that positive selection of CD8 T cells can occur when thymocytes bind to endogenous peptide complexes that deliver a partial or weak activating signal, whereas positive selection of CD4 T cells can occur through binding of the receptor and either of the two co-receptors to give a normal activating signal.

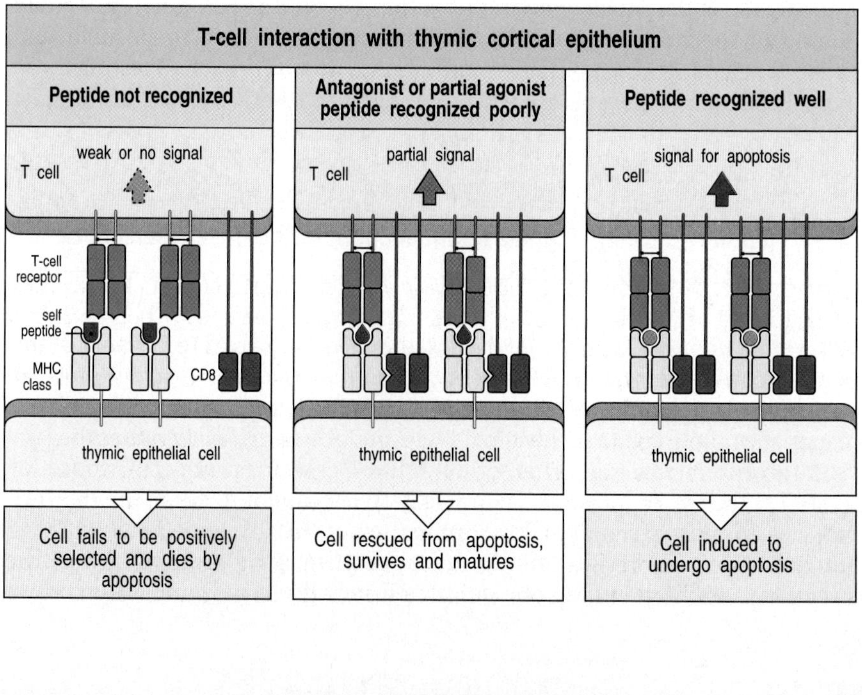

Fig. 7.23 Effects on thymic selection of peptides that deliver no signal, a partial signal, or a strong signal. Self MHC:self peptide complexes interacting with thymocytes bearing a particular T-cell receptor have different effects depending on the peptide involved. When the peptide is not recognized (left panel), the cells undergo apoptosis or death by neglect. When the peptide acts like an antagonist or partial agonist and produces a partial signal, this rescues the cells from programmed cell death, and they are thus positively selected (middle panel). When the self peptide:self MHC complex is recognized strongly and produces a strong signal, this triggers apoptosis (right panel).

This mechanism for the positive selection of CD4 T cells has important implications for tolerance to self antigens. If a normal activating signal from self MHC:self peptide complexes positively selects the mature CD4 T-cell population, instead of only negatively selecting as in CD8 T cells, then the same self peptides might be able to activate mature CD4 T cells in the periphery and induce autoimmunity. We shall see in Chapter 13 that there are cases in which this seems to occur. We now turn to the question of how the requirement for tolerance to self has been balanced against the need for recognition of diverse pathogens in the evolution of the MHC.

7-20 The requirements of T-cell activation and thymic selection might explain why the MHC is highly polymorphic and not highly polygenic.

As we saw in Chapter 4, the MHC genes are highly polymorphic, with loci having over a hundred different allelic variants within the human population. There are several different genes for each class of MHC molecule (each of which is highly polymorphic), and so the MHC is also somewhat polygenic. All these allelic and non-allelic MHC variants preferentially bind different antigenic peptides. The polymorphism and polygeny of the MHC is believed to reflect both the selective advantage to the species of individuals able to bind different sets of peptides, and the selective advantage to each individual of expressing several variants able to bind peptides from a broad range of pathogens.

From the individual's point of view, it would seem that the greater the number of different MHC genes, that is, the greater the polygeny, the better the protection from infectious organisms. However, a consideration of the constraints of thymic selection also suggests why polymorphism in a few genes might be preferable to the expression in one individual of a large number of different genes. It seems likely that around 5% of the T cells that are positively selected by one self MHC molecule will be reactive to self peptides presented by another self MHC molecule and must therefore be deleted in the thymus to avoid reactivity to self. This estimate is based on the frequency of T cells that respond to a given non-self MHC molecule (see Section 4-20). Thus, each new MHC molecule expressed will cost the animal 5% of its T-cell receptor repertoire. The counterbalancing effect of gaining new MHC molecules that can drive positive selection offers a net advantage up to a point, but this diminishes and then becomes negative as the number of MHC molecules expressed increases. A normal human expresses up to 15 different MHC molecules, and thus might delete a substantial portion (up to 75%) of the positively selected T cells that develop. It seems unlikely that the individual will improve his or her ability to respond to pathogens by adding MHC genes. Instead, MHC polymorphism seems to be the dominant mechanism.

There is an additional consideration, which derives from a limit on the number of MHC molecules that can be expressed on the cell surface. If the number of different MHC molecules is increased while the total number of MHC molecules stays constant, the concentration of any one MHC molecule decreases. As it requires around 100–200 identical peptide:MHC complexes to activate a T cell, and the maximum level of binding of any one peptide to any given MHC molecule is estimated to be 1% of those molecules, each different MHC molecule must be present at the cell surface in sufficient numbers to effectively present the peptides that bind to it. If too many different MHC molecules are expressed, then no single MHC:peptide complex will be present at a sufficient level to activate T cells. Thus the number of different

MHC molecules expressed by each individual might also be optimized to achieve a maximum diversity of specific MHC:peptide complexes that are present in numbers sufficient to activate a T cell.

7-21 A range of tumors of immune system cells throws light on different stages of T-cell development.

We saw in Chapter 6 that tumors of lymphoid cells corresponding in phenotype to intermediate stages in the development of the B cell provide invaluable tools in the analysis of B-cell differentiation. Tumors of T cells and other cells involved in T-cell development have been identified but, unlike the malignancies of B cells, few that correspond to intermediate stages in T-cell development have been identified in humans. Instead, the tumors resemble either mature T cells or, in common **acute lymphoblastic leukemia**, the earliest type of lymphoid progenitor (Fig. 7.24). One possible reason for the rarity of tumors corresponding to intermediate stages is that immature T cells are programmed to die unless rescued within a very narrow time window by positive selection (see Section 7-11). Thymocytes might simply not linger long enough at the intermediate stages of their development to provide an opportunity for malignant transformation. Thus, only cells that are already transformed at earlier stages, or that do not become transformed until the T cell has matured, are ever seen as tumors.

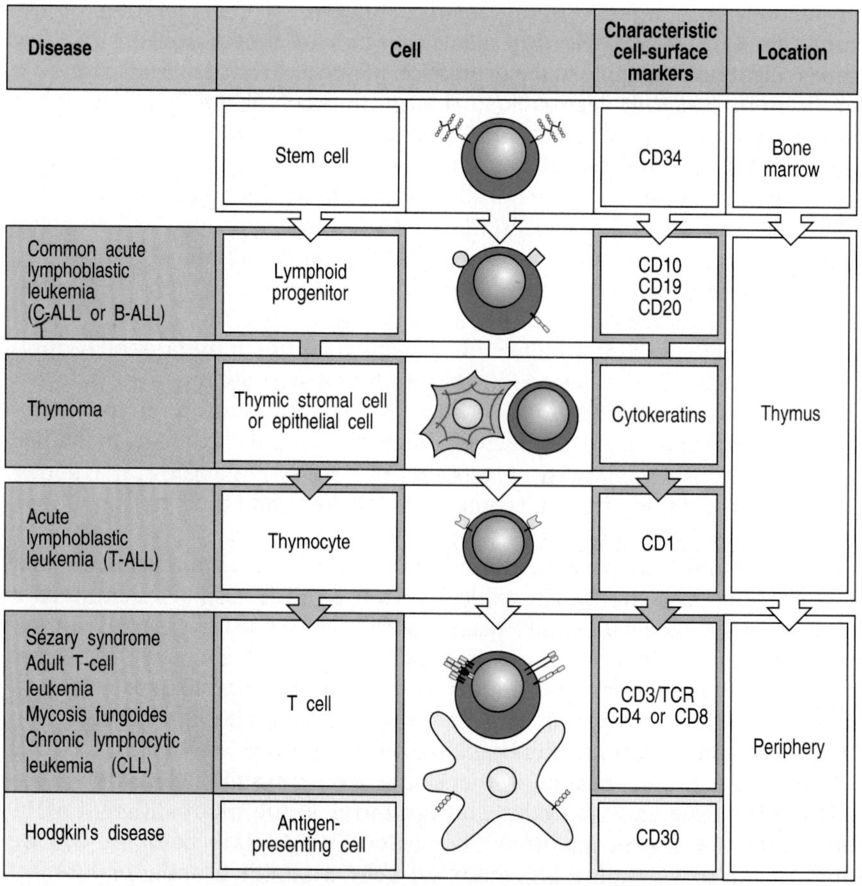

Disease
Cell
Characteristic cell-surface markers
Location
Stem cell
CD34
Bone marrow
Common acute lymphoblastic leukemia (C-ALL or B-ALL)
Lymphoid progenitor
CD10
CD19
CD20
Thymoma
Thymic stromal cell or epithelial cell
Cytokeratins
Thymus
Acute lymphoblastic leukemia (T-ALL)
Thymocyte
CD1
Sézary syndrome
Adult T-cell leukemia
Mycosis fungoides
Chronic lymphocytic leukemia (CLL)
T cell
CD3/TCR
CD4 or CD8
Periphery
Hodgkin's disease
Antigen-presenting cell
CD30

Fig. 7.24 T-cell tumors represent monoclonal outgrowths of normal cell populations. Each distinct T-cell tumor has a normal equivalent, as also seen with B-cell tumors, and retains many of the properties of the cell from which it develops. Some of these tumors represent massive outgrowth of a rare cell type; for example, common acute lymphoblastic leukemia is derived from the lymphoid progenitor cell. Two T-cell related tumors are also included. Thymomas derive from thymic stromal or epithelial cells, whereas the malignantly transformed cell in Hodgkin's disease is thought to be an antigen-presenting cell. Some characteristic cell-surface markers for each stage are also shown. For example, CD10 (common acute lymphoblastic leukemia antigen or CALLA) is a widely used marker for acute lymphoblastic leukemia. Note that T-cell chronic lymphocytic leukemia (CLL) cells express CD8, whereas the other T-cell tumors mentioned express CD4. Adult T-cell leukemia is caused by the retrovirus HTLV-1.

The behavior of T-cell and other lymphoid tumors has provided insight into different aspects of T-cell biology, and vice versa. T-cell tumors provide valuable information about the phenotype, homing properties, and receptor gene rearrangements of normal T-cell types. For example, **cutaneous T-cell lymphomas**, which home to the skin and proliferate slowly, are clonal outgrowths of a CD4 T cell that, when activated, homes to the skin. The most complex lymphoid tumor is known as **Hodgkin's disease**, and appears in several forms. The malignantly transformed cell seems to be an antigen-presenting cell, and in some patients the disease is dominated by T cells that are stimulated by the tumor cells. This form of the disease is called Hodgkin's lymphoma. Other patients have no lymphocytic abnormalities and show proliferation of a reticular cell, a condition known as nodular sclerosis. The differences between these two manifestations of Hodgkin's disease might reflect a real heterogeneity in the transformed cell or, more probably, differences in the T-cell responses of individual patients to the transformed cells. The prognosis for Hodgkin's lymphoma is far better than that for nodular sclerosis, suggesting that the responding T cells may be controlling tumor growth. Control of tumors by the immune response will be considered in Chapter 14.

As with B-cell tumors, T-cell lymphomas can be shown to be monoclonal outgrowths of a single transformed cell by examination of the rearrangements of their receptor genes (Fig. 7.25). When tissues or cells from patients are examined, the proportion of cells showing the same rearrangements is a measure of the proportion of transformed cells in the sample. The sensitivity of this approach can be increased by using the polymerase chain reaction (see Section 2-17) to identify the tumor-specific rearrangement, allowing the identification of very small numbers of tumor cells remaining in a tissue. This can be of particular importance in cases where a patient's own bone marrow has been taken for reinjection after radiotherapy. If the marrow contains any transformed cells, returning the marrow would retransplant the tumor into the patient and defeat the object of the therapy. Techniques exist to deplete the bone marrow of tumor cells; the efficiency of these techniques can be monitored by determining the persistence of the tumor-specific gene rearrangement.

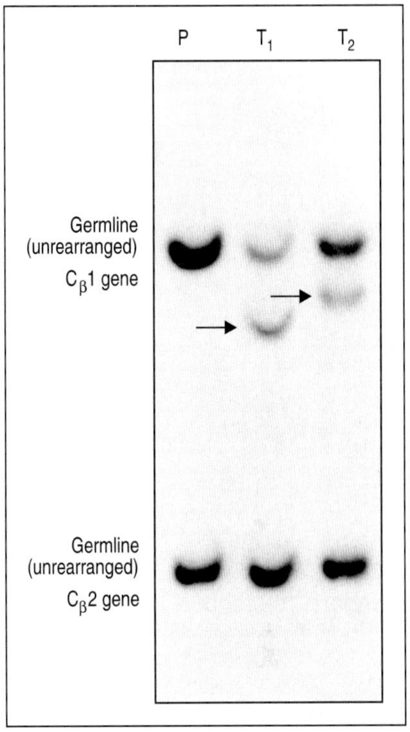

Fig. 7.25 The unique rearrangement events in each T cell can be used to identify tumors of T cells. Tumors are the outgrowth of a single transformed cell. Thus, each cell in a tumor will have an identical pattern of rearranged T-cell receptor genes. These panels show the migration in gel electrophoresis of DNA fragments containing the T-cell receptor β-chain constant regions; the DNA is either obtained from the placenta (lane P), a tissue in which the T-cell receptor genes are not rearranged, or from peripheral blood lymphocytes from two patients suffering from T-cell tumors (lanes T_1 and T_2). Bands corresponding to the unrearranged $C_{\beta}1$ and $C_{\beta}2$ genes can be seen in all lanes. Additional bands corresponding to specific rearrangements (arrowed) can be seen in each of the tumor samples, indicating that a large proportion of the cells in the sample carry an identical rearrangement. Note that these are the only discrete additional bands that can be seen in these samples; no bands deriving from rearranged genes in the normal lymphocytes also present in the patients' samples can be seen, as no one rearranged band is present at sufficient concentration to be detected in this assay. Photograph courtesy of T Diss.

Summary.

T-cell development involves two types of selection: positive selection for recognition of self MHC:self peptide complexes that provide an as yet poorly defined positive survival signal; and negative selection for cells bearing receptors specific for self peptide:self MHC complexes that would trigger the T cell in the periphery. The first process is normally mediated exclusively by thymic epithelial cells and the second largely by dendritic cells and macrophages. Receptors are selected from a diverse repertoire that has an inherent specificity for MHC molecules. Positive selection ensures that all mature T cells have functional receptors capable of responding to peptides presented by self MHC molecules on antigen-presenting cells, selects the appropriate co-receptor and determines functional commitment for the class of MHC molecule recognized. Negative selection eliminates self-reactive cells. The paradox that recognition of self MHC:self peptide ligands by the same basic receptor can lead to two opposing effects, namely positive and negative selection, is one of the central mysteries of immunology. Its solution will rest in understanding the ligands, the receptors, the signal transduction mechanisms, and the physiology of each step of the process.

Summary to Chapter 7.

T-cell development occurs in the special inductive microenvironment of the thymic cortex. In this location, T-cell receptor genes rearrange to generate the two lineages of T cells, the $\gamma{:}\delta$ T cells, whose function remains mysterious, and the $\alpha{:}\beta$ T cells, which are the primary mediators of the adaptive immune response. The specialized environment of the thymus selects for the maturation of those $\alpha{:}\beta$ T cells having useful receptors by contact of the receptor and its co-receptor with self MHC molecules on thymic cortical epithelial cells. Also within the thymus, professional antigen-presenting cells of bone marrow origin delete all T cells whose receptors recognize self antigens normally expressed by these cells, thus ensuring self tolerance. In this way, a useful and non-damaging repertoire of T-cell receptors is generated.

General references.

Moller, G (ed): **Positive T-cell selection in the thymus.** *Immunol. Rev.* 1993, **135**:5-242.

Nossal, G.J.V.: **Negative selection of lymphocytes.** *Cell* 1994, **76**:229-239.

von Boehmer, H.: **Positive selection of lymphocytes.** *Cell* 1994, **76**:219-228.

von Boehmer, H.: **The developmental biology of T lymphocytes.** *Annu. Rev. Immunol.* 1993, **6**:309-326.

Section references.

7-1 T cells develop in the thymus.

Cordier, A.C., and Haumont, S.M.: **Development of thymus, parathyroids, and ultimobranchial bodies in NMRI and nude mice.** *Am. J. Ana.* 1980, **157**:227.

Nehls, M., Kyewski, B., Messerle, M., Waldschütz, R., Schüddekopf, K., Smith, A.J.H., and Boehm, T.: **Two genetically separable steps in the differentiation of thymic epithelium.** *Science* 1996, **272**:886-889.

van Ewijk, W.: **T-cell differentiation is influenced by thymic microenvironments.** *Annu. Rev. Immunol.* 1991, **9**:591-615.

7-2 The thymus is required for T-cell maturation.

Anderson, G., Moore, N.C., Owen, J.J.T., and Jenkinson, E.J.: **Cellular interactions in thymocyte development.** *Annu. Rev. Immunol.* 1996, **14**:73-99.

Carlyle, J.R., and Zúñiga-Pflücker, J.C.: **Requirement for the thymus in alpha-beta T lymphocyte lineage commitment.** *Immunity* 1998, **9**:187-197.

DeKoning, J.L., DiMolfetto, C., Reilly, Q., Wei, W. L., Havran W.L., and Lo, D.: **Thymic cortical epithelium is sufficient for the development of mature T cells in relB-deficient mice.** *J. Immunol.* 1997, **158**:2558-2566.

Zúñiga-Pflücker, J.C., Lenardo, M.J.: **Regulation of thymocyte development from immature progenitors.** *Curr. Opin. Immunol.* 1996, **8**:215-224.

7-3 Developing T cells proliferate in the thymus but most die there.

Shortman, K., Egerton, M., Spangrude, G.J., and Scollay, R.: **The generation and fate of thymocytes.** *Semin. Immunol.* 1990, **2**:3-12.

Strasser, A.: **Life and death during lymphocyte development and function: evidence for two distinct killing mechanisms.** *Curr. Opin. Immunol.* 1995, **7**:228-234.

7-4 Successive stages in the development of thymocytes are marked by changes in cell-surface molecules.

Petrie, H.T., Hugo, P., Scollay, R., and Shortman, K.: **Lineage relationships and developmental kinetics of immature thymocytes: CD3, CD4, and CD8 acquisition *in vivo* and *in vitro*.** *J. Exp. Med.* 1990, **172**:1583-1588.

Shortman, K., and Wu, L.: **Early T lymphocyte progenitors.** *Annu. Rev. Immunol.* 1996, **14**:29-47.

7-5 Thymocyes at different developmental stages are found in distinct parts of the thymus.

Picker, L.J., and Siegelman, M.H.: Lymphoid tissues and organs, in Paul, W.E. (ed): *Fundamental Immunology*, 3rd edn. New York, Raven Press, 1993.

7-6 T cells with $\alpha{:}\beta$ or $\gamma{:}\delta$ receptors arise from a common progenitor.

Kang, J., and Raulet, D.H.: **Events that regulate differentiation of $\alpha\ \beta$ TCR+ and $\gamma\ \delta$ TCR+ T cells from a common precursor.** *Semin. Immunol.* 1997, **9**:171-179.

Kang, J., Coles, M., Cado, D., and Raulet, D.H.: **The developmental fate of T cells is critically influenced by TCR$\gamma\delta$ expression.** *Immunity* 1998, **8**:427-438.

Lauzurica, P., and Krangel, M.S.: **Temporal and lineage-specific control of T- cell receptor α/δ gene rearrangement by T-cell receptor α and δ enhancers.** *J. Exp. Med.* 1994, **179**:1913-1921.

Livak, F., Petrie, H.T., Crispe, I.N., and Schatz, D.G.: **In-frame TCR δ gene rearrangements play a critical role in the $\alpha\beta/\gamma\delta$ T cell lineage decision.** *Immunity* 1995, **2**:617-627.

7-7 Cells expressing particular γ- and β-chain genes arise in an ordered sequence during embryonic development.

Dunon, D., Courtois, D., Vainio, O., Six, A., Chen, C.H., Cooper, M.D., Dangy J.P., and Imhof, B.A.: **Ontogeny of the immune system: $\gamma\delta$ and $\alpha\beta$ T cells migrate from thymus to the periphery in alternating waves.** *J. Exp. Med.* 1997, **186**:977-988.

Havran, W.L., Boismenu, R.: **Activation and function of γδ T cells.** *Curr. Opin. Immunol.* 1994, **6**:442-446.

Itohara, S., Nakanishi, N., Kanagawa, O., Kubo, R., and Tonegawa, S.: **Monoclonal antibodies specific to native murine T cell receptor γδ analysis of γδ T cells in thymic ontogeny and peripheral lymphoid organs.** *Proc. Natl. Acad. Sci. USA* 1989, **86**:5094-5098.

7-8 Rearrangement of the β-chain genes and production of a β chain triggers several events in developing thymocytes.

Boismenu, R., Rhein, M., Fischer, W.H., and Havran, W.L.: **A role for CD81 in early T cell development.** *Science* 1996, **271**:198-200.

Borst, J., Jacobs, H., and Brouns, G.: **Composition and function of T-cell receptor and B-cell receptor complexes on precursor lymphocytes.** *Curr. Opin. Immunol.* 1996, **8**:181-190.

Dudley, E.C., Petrie, H.T., Shah, L.M., Owen, M.J., and Hayday, A.C.: **T-cell receptor β chain gene rearrangement and selection during thymocyte development in adult mice.** *Immunity* 1994, **1**:83-93.

Philpott, K.I., Viney, J.L., Kay, G., Rastan, S., Gardiner, E.M., Chae, S., Hayday, A.C., and Owen, M.J.: **Lymphoid development in mice congenitally lacking T cell receptor α β-expressing cells.** *Science* 1992, **256**:1448-1453.

Saint-Ruf, C., Ungewiss, K., Groetrrup, M., Bruno, L., Fehling, H.J., and von Boehmer, H.: **Analysis and expression of a cloned pre-T-cell receptor gene.** *Science* 1994, **266**:1208.

7-9 T-cell receptor α-chain genes undergo successive rearrangements until positive selection or cell death intervenes.

Hardardottir, F., Baron, J.L., and Janeway, C.A. Jr.: **T cells with two functional antigen-specific receptors.** *Proc. Natl. Acad. Sci. USA* 1995, **92**:354-358.

Marrack, P., and Kappler, J.: **Positive selection of thymocytes bearing alpha beta T cell receptors.** *Curr. Opin. Immunol.* 1997, **9**:250-255.

Padovan, E., Casorati, G., Dellabona, P., Meyer, S., Brockhaus, M., and Lanzavecchia, A.: **Expression of two T-cell receptor α chains: dual receptor T cells.** *Science* 1993 **262**:422-424.

Petrie, H.T., Livak, F., Schatz, D.G., Strasser, A., Crispe, I.N., and Shortman, K.: **Multiple rearrangements in T-cell receptor α-chain genes maximize the production of useful thymocytes.** *J. Exp. Med.* 1993, **178**:615-622.

7-10 Only T cells specific for peptides bound to self MHC molecules mature in the thymus.

Fink, P.J., and Bevan, M.J.: **H-2 antigens of the thymus determine lymphocyte specificity.** *J. Exp. Med.* 1978, **148**:766-775.

Hogquist, K.A., Tomlinson, A.J., Kieper, W.C., McGargill, M.A., Hart, M.C., Naylor, S., and Jameson, S.C.: **Identification of a naturally occurring ligand for thymic positive selection.** *Immunity* 1997, **6**:389-399.

Ignatowicz, L., Kappler, J., and Marrack, P.: **The repertoire of T cells shaped by a single MHC/peptide ligand.** *Cell* 1996, **84**:521-529.

Zinkernagel, R.M., Callahan, G.N., Klein, J., and Dennert, G.: **Cytotoxic T cells learn specificity for self H-2 during differentiation in the thymus.** *Nature* 1978, **271**:251-253.

7-11 Cells that fail positive selection die in the thymus.

Huessman, M., Scott, B., Kisielow, P., and von Boehmer, H.: **Kinetics and efficacy of positive selection in the thymus of normal and T-cell receptor transgenic mice.** *Cell* 1991, **66**:533-562.

Surh, C.D., and Sprent, J.: **T-cell apoptosis detected *in situ* during positive and negative selection in the thymus.** *Nature* 1994, **372**:100-103.

7-12 Positive selection acts on a repertoire of receptors with inherent specificity for MHC molecules.

Borgulya, P., Kishi, H., Uematsu, Y., and von Boehmer, H.: **Exclusion and inclusion of α and β T-cell receptor alleles.** *Cell* 1992, **65**:529-537.

Malissen, M., Trucy, J., Jouvin-Marche, E., Cazenave, P.A., Scollay, R., and Malissen, B.: **Regulation of TCR α and β chain gene allelic exclusion during T-cell development.** *Immunol. Today* 1992, **13**:315-322.

Merkenschlager, M., Graf, D., Lovatt, M., Bommhardt, U., Zamoyska, R., and. Fisher, A.G.: **How many thymocytes audition for selection?** *J. Exp. Med.* 1997, **186**:1149-1158.

Petrie, H.T., Livak, F., Burtrum. D., and Mazel, S.: **T cell receptor gene recombination patterns and mechanisms—cell death, rescue and T cell production.** *J. Exp. Med.* 1995, **182**:121-127.

Zerrahn, J., Held, W., and Raulet, D.H.: **The MHC reactivity of the T cell repertoire prior to positive and negative selection.** *Cell* 1997, **88**:627-636.

7-13 The expression of CD4 and CD8 on mature T cells and the associated T-cell functions are determined by positive selection.

Kaye, J., Hsu, M.L., Sauvon, M.E., Jameson, S.C., Gascoigne, N.R.J., and Hedrick, S.M.: **Selective development of $CD4^+$ T cells in transgenic mice expressing a class II MHC-restricted antigen receptor.** *Nature* 1989, **341**:746-748.

Lundberg, K., Heath, W., Kontgen, F., Carbone, F.R., and Shortman, K.: **Intermediate steps in positive selection: differentiation of $CD4^+8^{int}TCR^{int}$ thymocytes into $CD4^-8^+TCR^{hi}$ thymocytes.** *J. Exp. Med.* 1995, **181**:1643-1651.

von Boehmer, H., Kisielow, P., Lishi, H., Scott, B., Borgulya, P., and Teh, H.S.: **The expression of CD4 and CD8 accessory molecules on mature T cells is not random but correlates with the specificity of the αβ receptor for antigen.** *Immunol. Rev.* 1989, **109**:143-151.

7-14 Thymic cortical epithelial cells mediate positive selection.

Cosgrove, D., Chan, S.H., Waltzinger, C., Benoist, C., and Mathis, D.: **The thymic compartment responsible for positive selection of $CD4^+$ T cells.** *Intl. Immunol.* 1992, **4**:707-710.

Fowlkes, B.J., and Schweighoffer, E.: **Positive selection of T cells.** *Curr. Opin. Immunol.* 1995, **7**:188-195.

7-15 T cells specific for ubiquitous self antigens are deleted in the thymus.

Kishimoto, H., and Sprent, J.: **Negative selection in the thymus includes semimature T cells.** *J. Exp. Med.* 1997, 185:263-271.

Kruisbeek, A.M., and Amsen, D.: **Mechanisms underlying T-cell tolerance.** *Curr. Opin. Immunol.* 1996, **8**:233-244.

Zal, T., Volkmann, A., and Stockinger, B.: **Mechanisms of tolerance induction in major histocompatibility complex class II-restricted T cell specific for a blood-borne self antigen.** *J. Exp. Med.* 1994, **180**:2089-2099.

7-16 Negative selection is driven most efficiently by bone marrow derived antigen-presenting cells.

Matzinger, P., and Guerder, S.: **Does T cell tolerance require a dedicated antigen-presenting cell?** *Nature* 1989, **338**:74-76.

Sprent, J., and Webb, S.R.: **Intrathymic and extrathymic clonal deletion of T cells.** *Curr. Opin. Immunol.* 1995, **7**:196-205.

7-17 **Endogenous superantigens mediate negative selection of T-cell receptors derived from particular V_β gene segments.**

Kappler, J.W., Roehm, N., and Marrack, P.: **T-cell tolerance by clonal elimination in the thymus.** *Cell* 1987, **49**:273-280.

MacDonald, H.R., Schneider, R., Lees, R.K., Howe, R.C., Acha-Orbea, H., Festenstein, H., Zinkernagel, R.M., and Hengartner, H.: **T-cell receptor V_β use predicts reactivity and tolerance to Mls^a-encoded antigens.** *Nature* 1988, **332**:40-45.

7-18 **The signals for negative and positive selection must differ.**

Alberola-Ila, J., Hogquist, K.A., Swan, K.A., Bevan, M.J., and Perlmutter, R.M.: **Positive and negative selection invoke distinct signaling pathways**. *J. Exp. Med.* 1996, **184**:9-18.

Ashton-Rickardt, P.G., Bandeira, A., Delaney, J.R., Van Kaer, L., Pircher, H.P., Zinkernagel, R.M., and Tonegawa, S.: **Evidence for a differential avidity model of T-cell selection in the thymus.** *Cell* 1994, **74**:577.

Hogquist, K.A., Jameson, S.C., Heath, W.R., Howard, J.L., Bevan, M.J., and Carbane, F.R.: **T-cell receptor antagonist peptides induce positive selection.** *Cell* 1994, **76**:17-27.

Jameson, S.C., Hogquist, K.A., and Bevan, M.J: **Specificity and flexibility in thymic selection.** *Annu. Rev. Immunol.* 1994, **369**:750-753.

7-19 **Partial activating signals might account for positive selection of CD8 but not CD4 T cells.**

Bommhardt, U., Cole, M.S., Tso, J.Y., and Zamoyska, R.: **Signals through CD8 or CD4 can induce commitment to the CD4 lineage in the thymus**. *Eur. J. Immunol.* 1997, **27**:1152-1163.

Albert Basson, M., Bommhardt, U., Cole, M.S., Tso, J.Y., and Zamoyska, R.: **CD3 ligation on immature thymocytes generates antagonist-like signals appropriate for CD8 lineage commitment, independently of T cell receptor specificity.** *J. Exp. Med.* 1998, **187**:1249-1260.

7-20 **The requirements of T-cell activation and thymic selection might explain why the MHC is highly polymorphic and not highly polygenic.**

Alberola-Ila, J., Forbush, K.A., Seger, R., Krebs, E.G., and Perlmutter, R.M.: **Selective requirement for MAP kinase activation in thymocyte differentiation.** *Nature* 1995, **373**:620-623.

Demotz, S., Grey, H.M., and Sette, A.: **The minimal number of class II MHC-antigen complexes needed for T cell activation.** *Science* 1990, **249**:1028-1030.

Harding, C.V., and Unanue, E.R.: **Quantitation of antigen-presenting cell MHC class II/peptide complexes necessary for T cell stimulation.** *Nature* 1990, **346**:574-576.

Liao, X.C., and Littman, D.R.: **Altered T cell receptor signaling and disrupted T cell development in mice lacking Itk**. *Immunity* 1995, **3**:757-769.

Valitutti, S., Muller, S., Cella, M., Padovan, E., and Lanzavecchia, A.: **Serial triggering of many T-cell receptors by a few peptide-MHC complexes.** *Nature* 1995, **375**:148-151.

Wang, C-R., Hashimoto, K., Kubo, S., Yokochi, T., Kubo, M., Suzuki, M., Suzuki, K., Tada, T., and Nakayama, T.: **T cell receptor-mediated signaling events in CD4$^+$CD8$^+$ thymocytes undergoing thymic selection: requirement of calcineurin activated for thymic positive selection but not negative selection.** *J. Exp. Med.* 1995, **181**:927-941.

7-21 **A range of tumors of immune system cells throws light on different stages of T-cell development.**

Hwang, L-Y., and Baer, R.J.: **The role of chromosome translocations in T cell acute leukemia.** *Curr. Opin. Immunol.* 1995, **7**:659-664.

Rabbitts, T.H.: **Chromosomal translocations in human cancer.** *Nature* 1994, **372**:143-149.

PART IV THE ADAPTIVE IMMUNE RESPONSE

8 T-Cell Mediated Immunity

Once they have completed their development in the thymus, T cells enter the bloodstream, from which they migrate through the peripheral lymphoid organs, returning to the bloodstream to recirculate between blood and peripheral lymphoid tissue until they encounter antigen. To participate in an adaptive immune response, these **naive T cells** must be induced to proliferate and differentiate into cells capable of contributing to the removal of pathogens; we shall term these **armed effector T cells** because they can act very rapidly upon encountering specific antigen on other cells. The cells on which armed effector T cells act will be referred to as **target cells**.

In this chapter, we shall see how naive T cells are activated to proliferate and differentiate into armed effector T cells the first time they encounter their specific antigen in the form of a peptide:MHC complex on the surface of a **professional antigen-presenting cell** (**APC**). These specialized antigen-presenting cells are distinguished by surface molecules that synergize with specific antigen in the activation of naive T cells. Professional antigen-presenting cells are concentrated in the T-cell zones of peripheral lymphoid organs, to which they migrate after trapping antigen in the periphery. They present peptide fragments of protein antigens to recirculating naive T cells. The most important professional antigen-presenting cells are the highly specialized **dendritic cells**, whose only known function is to present antigen, and **macrophages**, which are also important as phagocytic cells in providing a first line of defense against infection and as targets for activation by armed effector T cells. B cells can also serve as professional antigen-presenting cells in some circumstances.

Effector T cells, as we learned in Chapter 4, fall into three functional classes that detect peptide antigens derived from different types of pathogen. Peptides derived from pathogens that multiply within the cytoplasm of the cell are carried to the cell surface by MHC class I molecules and presented to CD8 T cells, which differentiate into **cytotoxic T cells** that kill infected target cells. Peptide antigens derived from pathogens multiplying in intracellular vesicles, and those derived from ingested extracellular bacteria and toxins, are carried to the cell surface by MHC class II molecules and presented to CD4 T cells that can differentiate into two types of effector T cell. Pathogens that accumulate in large numbers inside macrophage vesicles tend to stimulate the differentiation of T_H1 cells, whereas extracellular antigens tend to stimulate the production of T_H2 cells. T_H1 cells activate the microbicidal properties of macrophages and induce B cells to make IgG antibodies that are very effective at opsonizing extracellular pathogens for uptake by phagocytic cells. T_H2 cells initiate the humoral immune response by activating naive antigen-specific B cells to produce IgM antibodies, and may subsequently stimulate the production of different isotypes, including IgA and IgE, as well as neutralizing and/or weakly opsonizing subtypes of IgG (Fig. 8.1).

	Cell-mediated immunity		Humoral immunity
Typical pathogens	Vaccinia virus Influenza virus Rabies virus *Listeria*	*Mycobacterium tuberculosis* *Mycobacterium leprae* *Leishmania donovani* *Pneumocystis carinii*	*Clostridium tetani* *Staphylococcus aureus* *Streptococcus pneumoniae* Polio virus *Pneumocystis carinii*
Location	Cytosol	Macrophage vesicles	Extracellular fluid
Effector T cell	Cytotoxic CD8 T cell	T_H1 cell	T_H1/T_H2 cell
Antigen recognition	Peptide:MHC class I on infected cell	Peptide:MHC class II on infected macrophage	Peptide:MHC class II on antigen-specific B cell
Effector action	Killing of infected cell	Activation of infected macrophages	Activation of specific B cell to make antibody

Fig. 8.1 The role of effector T cells in cell-mediated and humoral immunity to representative pathogens. Cell-mediated immunity involves the destruction of infected cells by cytotoxic T cells, or the destruction of intracellular pathogens by macrophages activated by T_H1 cells, and is directed principally at intracellular pathogens. However, T_H1 cells can also contribute to humoral immunity by inducing the production of strongly opsonizing antibodies, whereas T_H2 cells initiate the humoral response by activating naive B cells to secrete IgM, and induce the production of other antibody isotypes including weakly opsonizing antibodies such as IgG1 and IgG3 (mouse) and IgG2 and IgG4 (human) as well as IgA and IgE (mouse and human). All types of antibody contribute to humoral immunity, which is directed principally at extracellular pathogens. Note, however, that both cell-mediated and humoral immunity are involved in many infections, such as the response to *Pneumocystis carinii*, which requires antibody for ingestion by phagocytes and macrophage activation for effective destruction of the ingested pathogen.

The first encounter of naive T cells with antigen on a professional antigen-presenting cell results in a **primary immune response**, and at the same time generates immunological memory, which provides protection from subsequent challenge by the same pathogen. The generation of memory T cells, long-lived cells that give an accelerated response to antigen, is much less well understood than the generation of effector T cells and will be dealt with in Chapter 10. Memory T cells differ in several ways from naive T cells, but like naive T cells they are quiescent and require activation by professional antigen-presenting cells to regenerate effector T cells.

Armed effector T cells differ in many ways from their naive precursors, and these changes equip them to respond quickly and efficiently when they encounter specific antigen on their target cells. In the final sections of this chapter we shall describe the specialized mechanisms of T-cell mediated cytotoxicity and of macrophage activation by armed effector T cells, the major components of **cell-mediated immunity**. We shall leave the activation of B cells by helper T cells until Chapter 9, where the humoral or antibody-mediated immune response is discussed.

The production of armed effector T cells.

Activation of naive T cells requires recognition of a foreign peptide fragment bound to a self MHC molecule, but this is not on its own sufficient for activation, which also requires the simultaneous delivery of a **co-stimulatory signal** by a specialized antigen-presenting cell. Only professional antigen-presenting cells are able to express both classes of MHC molecule as well as the co-stimulatory surface molecules that drive the clonal expansion of naive T cells and their differentiation into armed effector T cells. The activation of naive T cells on initial encounter with antigen on the surface of a professional antigen-presenting cell is often called **priming**, to distinguish it from the responses of armed effector T cells to antigen on their target cells, and the responses of primed memory T cells.

8-1 The initial interaction of naive T cells with antigen occurs in peripheral lymphoid organs.

Adaptive immune responses are not initiated at the site where a pathogen first establishes a focus of infection. They occur in the organized peripheral lymphoid tissues, such as the lymph nodes, to which the pathogen or its products are transported in the lymph. Pathogens infecting peripheral sites will be trapped in the lymph nodes directly downstream of the site of infection; those that enter the blood will be trapped in the spleen; and pathogens infecting mucosal surfaces will accumulate in Peyer's patches or tonsils (see Fig. 1.10). All these lymphoid organs contain antigen-presenting cells specialized for capturing antigen and activating T cells; the most important of these are the dendritic cells, which capture antigen at the site of infection and then migrate to the downstream lymph node.

Naive T cells enter lymphoid tissue by crossing the walls of specialized venules known as **high endothelial venules (HEV)**. They circulate continuously from the bloodstream to the lymphoid organs and back to the blood, making contact with many thousands of antigen-presenting cells in the lymphoid tissues every day. These contacts allow the sampling of MHC:peptide complexes on the surface of the antigen-presenting cells, which is important for two reasons. One appears to reinforce the process of positive selection for self MHC recognition that occurs during T-cell development. As we discussed in Chapter 7, T-cell receptors are selected for their ability to interact with self-MHC: self peptide complexes during development, thereby selecting a repertoire of mature T cells capable of being activated by non-self peptides bound to these same MHC molecules. Recent experiments show that T-cell survival in the periphery also depends on contact with self MHC:self peptide ligands, and that the signals required for survival are delivered effectively through interactions with MHC:peptide complexes on dendritic cells. Thus, as the naive T cells migrate through the lymphoid tissue, they receive specific survival signals through their interactions with dendritic cells. At the same time, the sampling of MHC:peptide ligands ensures that each T cell has a high probability of encountering antigens derived from pathogens at any site of infection. This is crucial for the initiation of an adaptive immune response, since only one naive T cell in 10^4–10^6 is likely to be specific for a particular antigen, and adaptive immunity depends on the activation and expansion of such rare antigen-specific T cells. The T cells that do not encounter their specific antigen eventually reach the medulla of the lymph node from where they are carried by the efferent lymphatics back to the blood to continue recirculating through other lymphoid organs. Naive T cells that recognize their specific antigen on the surface of a professional antigen-presenting cell cease to migrate and embark on the steps that will lead to the generation of armed effector cells (Fig. 8.2).

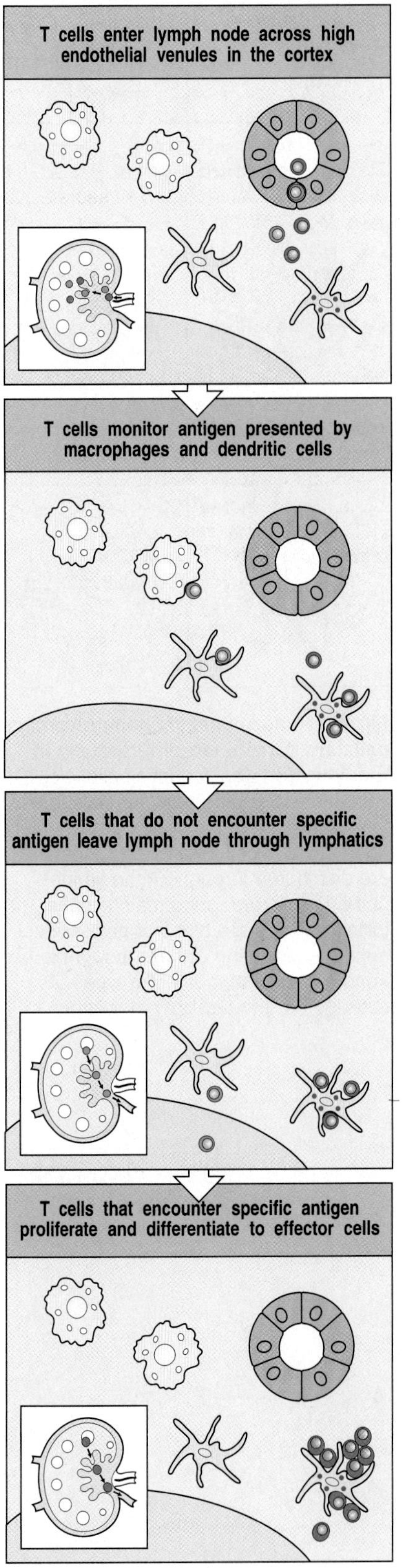
T cells enter lymph node across high endothelial venules in the cortex
T cells monitor antigen presented by macrophages and dendritic cells
T cells that do not encounter specific antigen leave lymph node through lymphatics
T cells that encounter specific antigen proliferate and differentiate to effector cells

Fig. 8.2 Naive T cells encounter antigen during their recirculation through peripheral lymphoid organs. Naive T cells recirculate through peripheral lymphoid organs, such as the lymph node shown here, entering through specialized regions of vascular endothelium called high endothelial venules. On leaving the blood vessel, the T cells enter the deep cortex of the lymph node, where they encounter many antigen-presenting cells (mainly dendritic cells and macrophages). Those T cells shown in green do not encounter their specific antigen. They receive a survival signal through their interaction with self-MHC:self-peptide complexes and leave the lymph node through the lymphatics to return to the circulation. T cells shown in blue encounter their specific antigen on the surface of an antigen-presenting cell and are activated to proliferate and to differentiate into armed effector T cells. Once this process is completed, these armed effector T cells also leave the lymph node via the efferent lymphatics and enter the circulation.

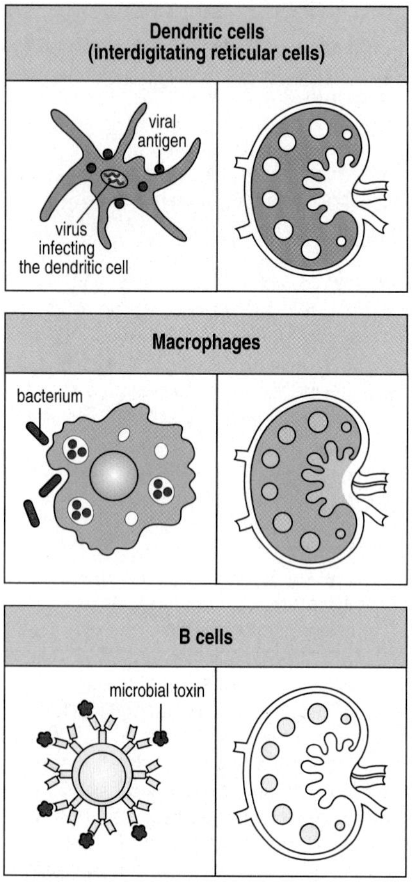

Fig. 8.3 Professional antigen-presenting cells are distributed differentially in the lymph node. Dendritic cells, also called interdigitating reticular cells, are found throughout the cortex of the lymph node in the T-cell areas. Macrophages are distributed throughout the lymph node. B cells are found mainly in the follicles. The three types of professional antigen-presenting cell are thought to be adapted to present different types of pathogen or products of pathogens.

The three main types of specialized antigen-presenting cell present in the peripheral lymphoid organs are dendritic cells, macrophages, and B cells. Each of these cell types is specialized to process and present antigens from different sources to T cells. Dendritic cells appear to function exclusively as antigen-presenting cells, whereas macrophages and B cells are also the targets of subsequent actions of armed effector CD4 T cells. Only these three cell types express the specialized co-stimulatory molecules required to activate naive T cells; furthermore, macrophages and B cells express these molecules only when suitably activated by infection.

The three types of antigen-presenting cell are distributed differently in the lymphoid organs (Fig. 8.3). Dendritic cells, which in lymphoid tissues are also known as **interdigitating reticular cells**, are present mainly in the T-cell areas of the lymph node. These cells, which were mentioned in Chapter 7 because of their role in negative selection of thymocytes, are the most potent antigen-presenting cells for naive T cells. Macrophages are found in all areas of the lymph node and actively ingest microbes and particulate antigens. As most pathogens are particulate, macrophages stimulate immune responses to many sources of infection. Finally, the B cells in the lymphoid follicles are particularly efficient at taking up soluble antigens, such as bacterial toxins, by the specific binding of the antigens to the B-cell surface immunoglobulin molecules. Degraded fragments of these antigens can return to the B-cell surface complexed with MHC class II molecules, thus enabling antigen-specific B cells to play a part in the activation of naive CD4 T cells. The generation of effector cells from a naive T cell takes several days. At the end of this period, the armed effector T cells leave the lymphoid organ and re-enter the bloodstream so that they can migrate to sites of infection.

8-2 Lymphocyte migration, activation, and effector function depend on cell–cell interactions mediated by cell adhesion molecules.

The migration of naive T cells through the lymph nodes, and their initial interactions with antigen-presenting cells, involves antigen non-specific binding to other cells. Similar interactions eventually guide the effector T cells into the peripheral tissues and play an important part in interactions with their target cells. Binding of T cells to other cells is controlled by an array of adhesion molecules on the surface of the T lymphocyte. These cell-surface proteins recognize a complementary array of adhesion molecules on the surfaces of the cells with which the T cell interacts. The main classes of adhesion molecule involved in lymphocyte interactions are the selectins, the integrins, members of the immunoglobulin superfamily, and some mucin-like molecules. Some of these molecules are concerned mainly with lymphocyte homing and migration, which we shall describe in more detail in Chapter 10, where we present an integrated view of the immune response; others have broader roles in the generation of immune responses and the interactions of armed effector T cells with their target cells, which will be discussed here.

The nomenclature of the adhesion molecules is confusing, because most were first defined either as cell-surface molecules recognized by monoclonal antibodies or in functional assays of cell–cell interactions and were only subsequently characterized biochemically. For this reason, the names of many of the adhesion molecules bear no relationship to the structural families to which they belong. A brief explanation of the terminology can be found in the legend to Fig. 8.4, where the main classes of leukocyte adhesion molecules are summarized. We begin our discussion, however, with the **selectins**, which all belong to the same small protein family.

Family	Example	Name	Tissue distribution	Ligand
Selectins Bind carbohydrates. Initiate leukocyte: endothelial interaction	L-selectin 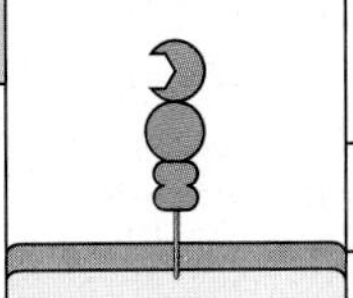	L-selectin (MEL-14, CD62L)	Naive and some memory lymphocytes, neutrophils, monocytes, macrophages, eosinophils	Sulfated sialyl Lewis^X, GlyCAM-1, CD34, MAdCAM-1
		P-selectin (PADGEM, CD62P)	Activated endothelium and platelets	Sialyl Lewis^X, PSGL-1
		E-selectin (ELAM-1, CD62E)	Activated endothelium	Sialyl Lewis^X
Mucin-like vascular addressins Bind to L-selectin. Initiate leukocyte: endothelial interaction	CD34 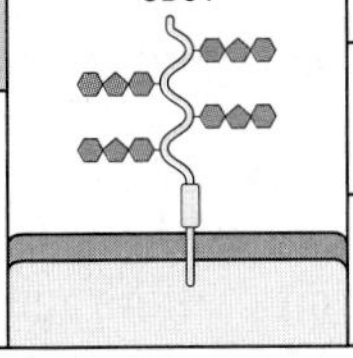	CD34	Endothelium	L-selectin
		GlyCAM-1	High endothelial venules	L-selectin
		MAdCAM-1	Mucosal lymphoid tissue venules	L-selectin, integrin $\alpha_4\beta_7$
Integrins Bind to cell-adhesion molecules and extracellular matrix. Strong adhesion	LFA-1 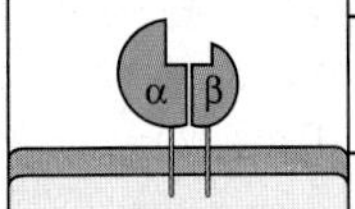	$\alpha_L\beta_2$ (LFA-1, CD11a/CD18)	Monocytes, T cells, macrophages, neutrophils, dendritic cells	ICAMs
		$\alpha_M\beta_2$ (Mac-1, CR3, CD11b/CD18)	Neutrophils, monocytes, macrophages	ICAM-1, iC3b, fibrinogen
		$\alpha_X\beta_2$ (CR4, p150.95, CD11c/CD18)	Dendritic cells, macrophages, neutrophils	iC3b
		$\alpha_4\beta_1$ (VLA-4, LPAM-2, CD49d/CD29)	Lymphocytes, monocytes, macrophages	VCAM-1 Fibronectin
		$\alpha_5\beta_1$ (VLA-5, CD49d/CD29)	Monocytes, macrophages	Fibronectin
		$\alpha_4\beta_7$ (LPAM-1)	Lymphocytes	MAdCAM-1
		$\alpha_E\beta_7$	Intraepithelial lymphocytes	E-cadherin
Immunoglobulin superfamily Various roles in cell adhesion. Ligand for integrins	CD2 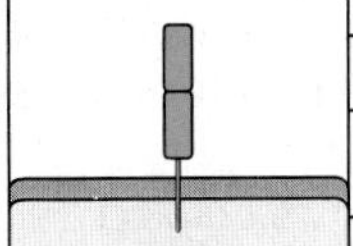	CD2 (LFA-2)	T cells	LFA-3
		ICAM-1 (CD54)	Activated vessels, lymphocytes, dendritic cells	LFA-1, Mac1
		ICAM-2 (CD102)	Resting vessels, dendritic cells	LFA-1
		ICAM-3 (CD50)	Lymphocytes	LFA-1
		LFA-3 (CD58)	Lymphocytes, antigen-presenting cells	CD2
		VCAM-1 (CD106)	Activated endothelium	VLA-4

Fig. 8.4 Adhesion molecules in leukocyte interactions. Several structural families of adhesion molecules play a part in leukocyte migration, homing and cell–cell interactions: the selectins; mucin-like vascular addressins; the integrins; and proteins of the immunoglobulin superfamily. The figure shows schematic representations of an example from each family, a list of other family members that participate in leukocyte interactions, their cellular distribution, and their partner (ligand) in adhesive interactions. The family members shown here are limited to those we consider in this text but include some that will not be encountered until Chapter 10. The nomenclature of the different molecules in these families is confusing because it often reflects the way in which the molecules were first identified rather than their related structural characteristics. Thus, whereas all the ICAMs are immunoglobulin-related, and all the VLA molecules are β_1 integrins, the CD nomenclature reflects the characterization of leukocyte cell-surface molecules by raising monoclonal antibodies against them (Appendix I contains details of all the CD molecules mentioned in this book). Thus CD molecules comprise a large and diverse collection of cell-surface molecules, which includes adhesion molecules in all the structural families. The LFA molecules were defined through experiments in which cytotoxic T-cell killing could be blocked by monoclonal antibodies against cell-surface molecules on the interacting cells, and there are LFA molecules in both the integrin and the immunoglobulin families. Alternative names for each of the adhesion molecules are given in parentheses. Sialyl Lewis^x, which is recognized by P- and E-selectin, is an oligosaccharide present on the cell-surface glycoproteins of circulating leukocytes.

The selectins (CD62) are particularly important for leukocyte homing to specific tissues, and can be expressed either on leukocytes (**L-selectin**, CD62L) or on vascular endothelium (**P-selectin**, CD62P, and **E-selectin**, CD62E, which will be discussed in Chapter 10). Selectins are cell-surface molecules with a common core structure, distinguished from each other by the presence of different lectin-like domains in their extracellular portion (Fig. 8.5). Lectins bind to specific sugar groups, and each selectin binds to a cell-surface carbohydrate. L-selectin is expressed on naive T cells. It binds to the carbohydrate moiety, sulfated sialyl Lewis^x, of mucin-like molecules called **vascular addressins**, which are expressed on the surface of vascular endothelial cells. Two of these addressins, **CD34** and **GlyCAM-1**, are

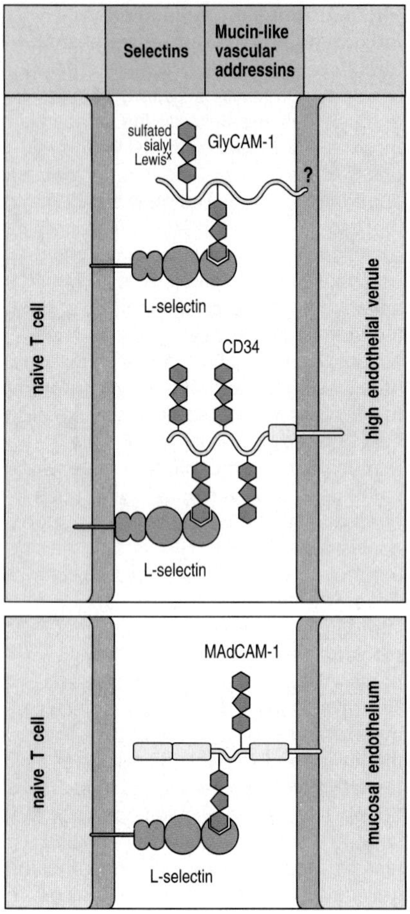

Fig. 8.5 L-selectin and the mucin-like vascular addressins direct naive lymphocyte homing to lymphoid tissues. L-selectin is expressed on naive T cells, which bind to sulfated sialyl Lewis^x moieties on the vascular addressins CD34 and GlyCAM-1 on high endothelial venules in order to enter lymph nodes. The relative importance of CD34 and GlyCAM-1 in this interaction is unclear. GlyCAM-1 is expressed exclusively on high endothelial venules but has no transmembrane region and it is unclear how it is attached to the membrane; CD34 has a transmembrane anchor and is expressed in appropriately glycosylated form only on high endothelial venule cells, although it is found in other forms on other endothelial cells. The addressin MAdCAM-1 is expressed on mucosal endothelium and guides entry into mucosal lymphoid tissue. L-selectin recognizes the carbohydrate moieties on the vascular addressins.

expressed as sulfated sialyl Lewis^x molecules on high endothelial venules in lymph nodes. A third, **MAdCAM-1**, is expressed on endothelium in mucosa, and guides lymphocyte entry into mucosal lymphoid tissue such as that of the gut (see Fig. 8.5).

The interaction between L-selectin and the vascular addressins is responsible for the specific homing of naive T cells to lymphoid organs but does not, on its own, enable the cell to cross the endothelial barrier into the lymphoid tissue; for this, molecules of two other families, the integrins and the immunoglobulin superfamily, are required. Proteins of these two families also play a critical part in the subsequent interactions of lymphocytes with antigen-presenting cells and later with their target cells.

The integrins comprise a large family of cell-surface proteins that mediate adhesion between cells, and between cells and the extracellular matrix, in immune and inflammatory responses. They are also important in many aspects of tissue organization and cell migration during development. An integrin molecule consists of a large α chain that pairs non-covalently with a smaller β chain. There are several subfamilies of integrins, broadly defined by their common β chains. We shall be concerned chiefly with the **leukocyte integrins**, which have a common β_2 chain with distinct α chains (Fig. 8.6). All T cells express a leukocyte integrin known as **lymphocyte function-associated antigen-1** (**LFA-1**). This is thought to be the most important adhesion molecule for lymphocyte activation because antibodies to LFA-1 effectively inhibit the activation of both naive and armed effector T cells.

LFA-1 and two other members of the leukocyte integrin family are also expressed on neutrophils and macrophages. In **leukocyte adhesion deficiency**, an inherited immunodeficiency disease resulting from a defect in the synthesis

Fig. 8.6 Integrins are important in leukocyte adhesion. Integrins are heterodimeric proteins containing a β chain, which defines the class of integrin, and an α chain, which defines the different integrins within a class. The α chain is larger than the β chain and contains binding sites for divalent cations that may be important in signaling. Most integrins expressed on leukocytes have a common β chain, β_2, but different α chains. VLA-4, which is a β_1 integrin, and LFA-1 are upregulated on armed effector T cells and are important in the migration and activation of these cells. Macrophages and neutrophils express all three members of the β_2 integrin family. Like LFA-1, Mac-1 binds the immunoglobulin superfamily ICAM molecules, but in addition Mac-1 is a complement receptor whose function will be discussed in Chapter 10; other functions of p150.95, which also binds complement, are unknown.

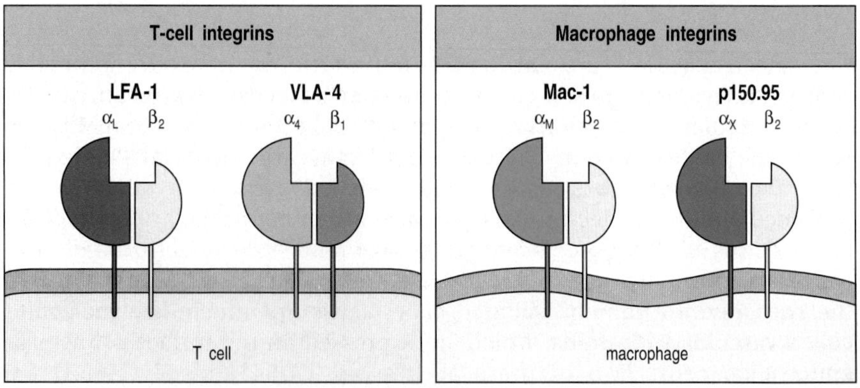

of the common β_2 chain, immunity to infection with extracellular bacteria is severely impaired because of defective neutrophil and macrophage function. Surprisingly, T-cell responses can be normal in such patients. This is probably because T cells also express other adhesion molecules, including CD2 and members of the β_1 integrin family, which may be able to compensate for the absence of LFA-1. Expression of the β_1 integrins increases significantly at a late stage in T-cell activation, and they are thus often called **VLAs** for **very late activation antigens**; we shall see in Chapter 10 that they play an important part in directing armed effector T cells to their target tissues.

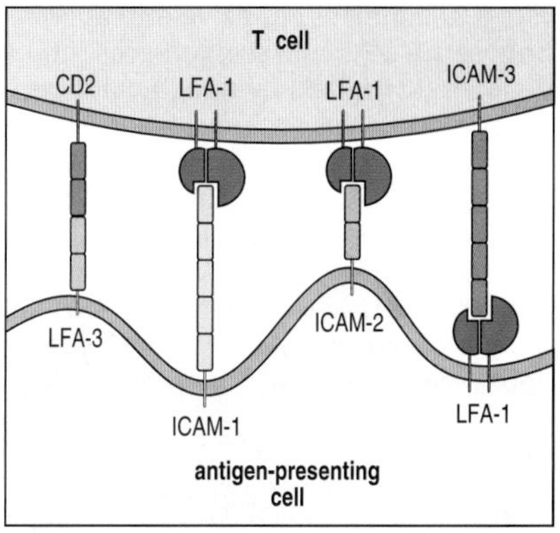

Fig. 8.7 Cell-surface molecules of the immunoglobulin superfamily are important in the interactions of lymphocytes with antigen-presenting cells. In the initial encounter of T cells with antigen-presenting cells, CD2 binding to LFA-3 on the antigen-presenting cell synergizes with LFA-1 binding to ICAM-1 and ICAM-2. ICAM-3 expressed on the T cell binds to LFA-1 on the antigen-presenting cell. Similar adhesive interactions occur between effector T cells and their targets (not shown). LFA-1 is the integrin heterodimer CD11a/CD18. LFA-3 is also known as CD58, while ICAM-1, -2, and -3 are CD54, CD102, and CD50, respectively.

Many cell-surface adhesion molecules are members of the immunoglobulin superfamily, which also includes the antigen receptors of T and B cells, the co-receptors CD4, CD8, and CD19, and the invariant domains of MHC molecules. At least five adhesion molecules of the immunoglobulin superfamily are especially important in T-cell activation (Fig. 8.7). Three very similar **intercellular adhesion molecules** (**ICAMs**)—**ICAM-1**, **ICAM-2**, and **ICAM-3**—all bind to the T-cell integrin LFA-1. ICAM-1 and ICAM-2 are expressed on endothelium as well as on antigen-presenting cells; binding to these molecules enables lymphocytes to migrate through blood vessel walls. ICAM-3 is expressed only on leukocytes and is thought to play an important part in adhesion between T cells and antigen-presenting cells. The interaction of LFA-1 with ICAM-1 and ICAM-2 synergizes with a second adhesive interaction involving the immunoglobulin superfamily members CD2 and LFA-3; CD2 is expressed on the T-cell surface, and LFA-3 is expressed on the antigen-presenting cell (see Fig. 8.7).

8-3 The initial interaction of T cells with antigen-presenting cells is also mediated by cell-adhesion molecules.

As they migrate through the cortical region of the lymph node, naive T cells bind transiently to each antigen-presenting cell they encounter. Professional antigen-presenting cells, and dendritic cells in particular, bind naive T cells very efficiently through interactions between LFA-1, CD2, and ICAM-3 on the T cell, and ICAM-1, ICAM-2, LFA-1, and LFA-3 on the antigen-presenting cell (see Fig. 8.7). These molecules synergize in the binding of lymphocytes to antigen-presenting cells, and the exact role of each has been difficult to distinguish. People lacking LFA-1 can have normal T-cell responses, and this also seems to be the case for genetically engineered mice lacking CD2. It would not be surprising if there were sufficient redundancy in the molecules mediating T-cell adhesive interactions to enable immune responses to occur in the absence of any one of them; such molecular redundancy has been observed in other complex biological processes.

The transient binding of naive T cells to professional antigen-presenting cells is crucial in providing time for T cells to sample large numbers of MHC molecules on the surface of each antigen-presenting cell for the presence of specific peptide. In those rare cases in which a naive T cell recognizes its specific peptide:MHC ligand, signaling through the T-cell receptor induces a conformational change in LFA-1, which greatly increases its affinity for ICAM-1 and ICAM-2. The mechanism of this conformational change in LFA-1 is not known. These changes stabilize the association between the antigen-specific T cell and the antigen-presenting cell (Fig. 8.8). The association can persist for several days during which the naive T cell proliferates and its progeny, which also adhere to the antigen-presenting cell, differentiate into armed effector T cells.

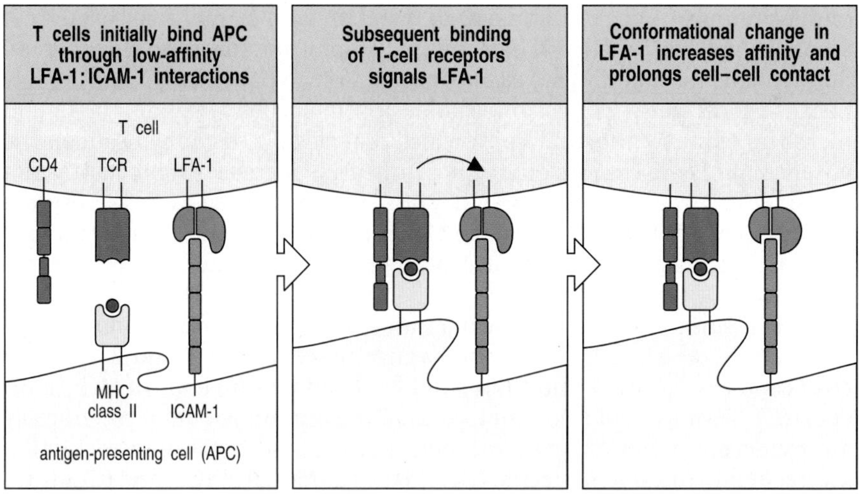

Fig. 8.8 Transient adhesive interactions between T cells and antigen-presenting cells are stabilized by specific antigen recognition. When a T cell binds to its specific ligand on an antigen-presenting cell (APC), intracellular signaling through the T-cell receptor (TCR) induces a conformational change in LFA-1 that causes it to bind with higher affinity to ICAMs on the antigen-presenting cell. The T cell shown here is a CD4 T cell.

Most T-cell encounters with antigen-presenting cells, however, do not result in the recognition of a specific antigen. In these encounters, the T cells must be able to separate efficiently from the antigen-presenting cells so that they can continue to migrate through the lymph node, eventually leaving via the efferent lymphatic vessels to re-enter the blood and continue recirculating. Dissociation, like stable binding, may also involve signaling between the T cell and the antigen-presenting cells, but little is known of its mechanism.

8-4 Both specific ligand and co-stimulatory signals provided by a professional antigen-presenting cell are required for the clonal expansion of naive T cells.

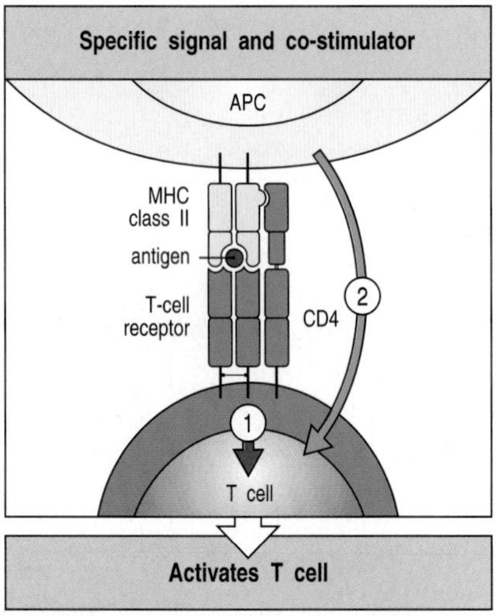

Fig. 8.9 Activation of naive T cells requires two independent signals. Binding of the peptide:MHC complex by the T-cell receptor and, in this example, the CD4 co-receptor, transmits a signal (arrow 1) to the T cell that antigen has been recognized. Activation of naive T cells requires a second signal (arrow 2), the co-stimulatory signal, to be delivered by the same antigen-presenting cell.

We saw in Chapter 4 that effector T cells are triggered when their antigen-specific receptors and either the CD4 or the CD8 co-receptors bind to peptide:MHC complexes. Nevertheless, ligation of the T-cell receptor and co-receptor does not, on its own, stimulate naive T cells to proliferate and differentiate into armed effector T cells. The antigen-specific clonal expansion of naive T cells requires a second or co-stimulatory signal (Fig. 8.9), which must be delivered by the same antigen-presenting cell on which the T cell recognizes its specific antigen. CD8 T cells appear to require a stronger co-stimulatory signal than CD4 cells and, as we shall see later, their clonal expansion may be aided by CD4 cells interacting with the same antigen-presenting cell.

The best characterized co-stimulatory molecules on antigen-presenting cells are the structurally related glycoproteins **B7.1** (CD80) and **B7.2** (CD86), which we shall call **B7 molecules** in the subsequent text, as functional differences between them have yet to be defined. The B7 molecules are homodimeric members of the immunoglobulin superfamily found exclusively on the surface of cells capable of stimulating T-cell growth. Their role in co-stimulation has been demonstrated by transfecting fibroblasts that express a T-cell ligand with genes encoding B7 molecules and showing that the fibroblasts could then stimulate the growth of naive T cells. The receptor for B7 molecules on the T cell is **CD28**, yet another member of the immunoglobulin superfamily (Fig. 8.10). Ligation of CD28 by B7 molecules or by anti-CD28 antibodies will co-stimulate the growth of naive T cells, whereas anti-B7 antibodies that inhibit the binding of B7 molecules to CD28 will inhibit T-cell responses.

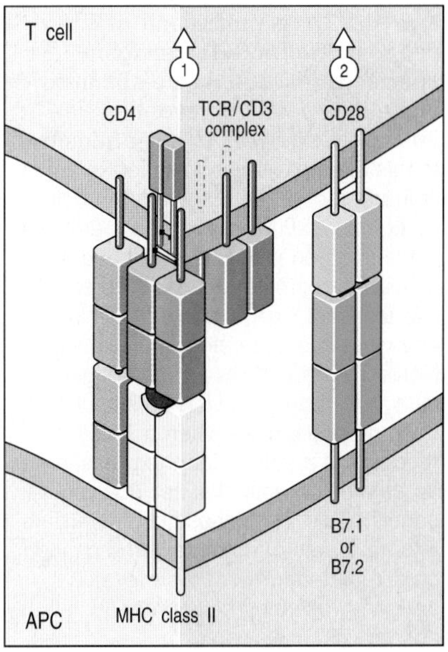

Fig. 8.10 The principal co-stimulatory molecules expressed on professional antigen-presenting cells are B7 molecules, which bind the T-cell protein CD28. Binding of the T-cell receptor (TCR) and its co-receptor CD4 to the peptide:MHC class II complex on the antigen-presenting cell (APC) delivers a signal (→ 1) that can induce the clonal expansion of T cells only when the co-stimulatory signal (→ 2) is given by binding of CD28 to B7 molecules. Both CD28 and B7 molecules are members of the immunoglobulin superfamily. B7.1 and B7.2 are homodimers, each of whose chains has one immunoglobulin V-like domain and one C-like domain. CD28 is a disulfide-linked homodimer in which each chain has one V-like domain.

On naive T cells, CD28 is the only receptor for B7 molecules. Once T cells are activated, however, they express an additional receptor called **CTLA-4** (CD152). CTLA-4 closely resembles CD28 in sequence, and the two molecules are encoded by closely linked genes. CTLA-4 binds B7 molecules about 20 times more avidly than CD28 and delivers an inhibitory signal to the activated T cell (Fig. 8.11). This makes the activated progeny of a naive T cell less sensitive to stimulation by the antigen-presenting cell and limits the amount of the T-cell growth factor interleukin-2 (IL-2) produced. Thus, binding of CTLA-4 to B7 molecules is essential for limiting the proliferative response of activated T cells to antigen and B7 on the surface of antigen-presenting cells. This was confirmed by producing mice with a disrupted CTLA-4 gene; such mice develop a fatal disorder characterized by massive proliferation of lymphocytes. Although other molecules have been reported to co-stimulate naive T cells, so far only B7.1 (CD80) and B7.2 (CD86) binding to CD28 has been shown definitively to provide co-stimulatory signals in normal immune responses.

The requirement for the simultaneous delivery of antigen-specific and co-stimulatory signals by one cell in the activation of naive T cells means that only professional antigen-presenting cells can initiate T-cell responses. This is important because not all potentially self-reactive T cells are deleted in the thymus; peptides derived from proteins made only in specialized cells in the

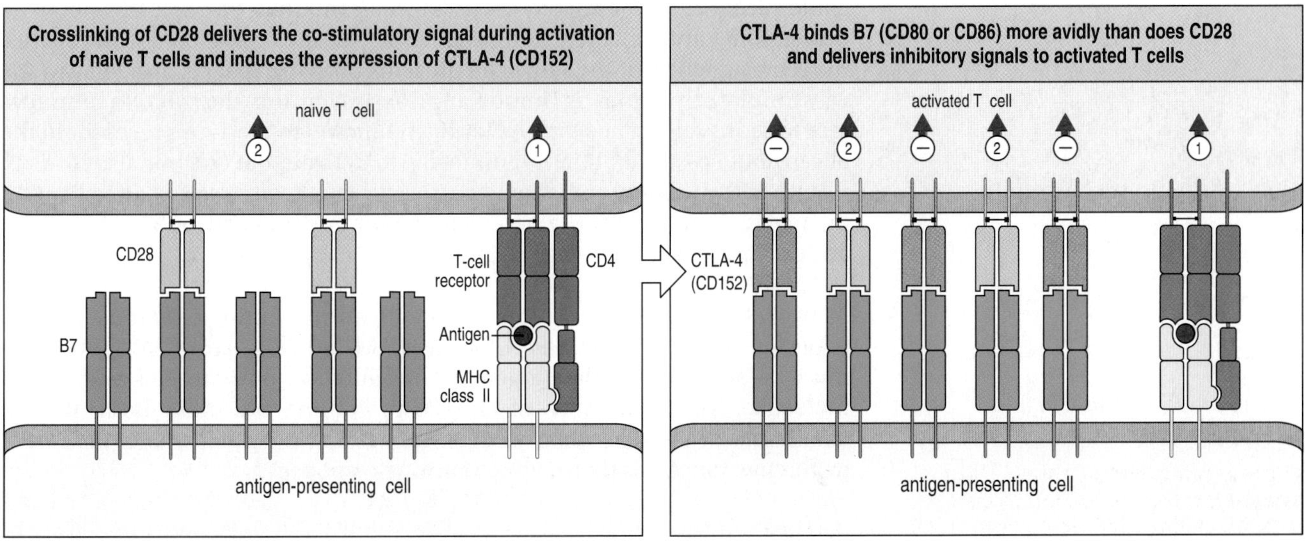

Fig. 8.11 T-cell activation through the T-cell receptor and CD28 leads to the increased expression of CTLA-4 (CD152), an inhibitory receptor for B7 molecules. Naive T cells express CD28, which delivers a co-stimulatory signal on binding B7 molecules (left panel), thereby driving the activation and expansion of T cells that encounter specific antigen presented by a professional antigen-presenting cell. Once activated, T cells express increased levels of CTLA-4 (right panel). CTLA-4 has a higher affinity for B7 molecules than does CD28 and thus binds most or all the B7 molecules, effectively shutting down the proliferative phase of the response.

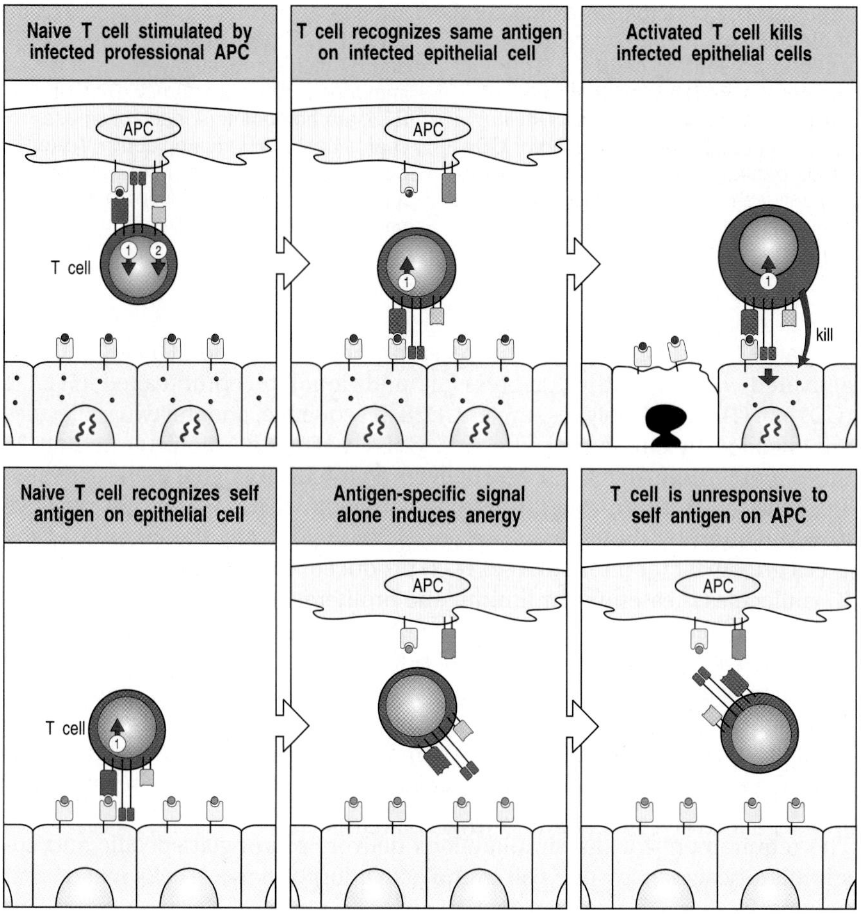

Fig. 8.12 The requirement for one cell to deliver both the antigen-specific signal and the co-stimulatory signal is crucial in preventing immune responses to self antigens. In this example of the initiation of an immune response to a virus, a T cell recognizes a viral peptide on a professional antigen-presenting cell and is activated to proliferate and differentiate into an effector cell capable of eliminating any virus-infected cell (upper panels). Naive T cells that recognize antigen on cells that cannot provide co-stimulation become anergic, as when a T-cell recognizes a self antigen expressed by an uninfected epithelial cell (lower panels). This T cell does not differentiate into an armed effector cell, and cannot be stimulated further by professional antigen-presenting cells presenting that antigen.

peripheral tissues might not be encountered during the negative selection of thymocytes. Self tolerance could be broken if naive autoreactive T cells could recognize self antigens on tissue cells and then be co-stimulated by a professional antigen-presenting cell, either locally or at a distant site. Thus, the requirement that the same cell presents both the specific antigen and the co-stimulatory signal is important in preventing destructive immune responses to self tissues. Indeed, antigen binding to the T-cell receptor in the absence of co-stimulation not only fails to activate the cell but also leads to a state called **anergy**, in which the T cell becomes refractory to activation by specific antigen even when the antigen is presented by a professional antigen-presenting cell (Fig. 8.12).

As well as B7.1 and B7.2, professional antigen-presenting cells must express adhesion molecules such as ICAM-1, ICAM-2, LFA-1, and LFA-3, and they must be able to process antigen for presentation on both classes of MHC molecule. The three types of antigen-presenting cell differ both in their co-stimulatory and in their antigen-processing properties, and thus have distinctive functions in initiating immune responses.

8-5 Mature dendritic cells are highly efficient inducers of naive T-cell activation.

The only known function of dendritic cells is to present antigen to T cells, and the mature dendritic cells found in lymphoid tissues are by far the most potent stimulators of naive T cells. This ability is not shared, however, by the immature dendritic cells found under most surface epithelia and in most solid organs such as the heart and kidneys. Dendritic cells arise from myeloid progenitors within the bone marrow, and emerge from the bone marrow to migrate via the blood to their peripheral sites. In these sites, they have an immature phenotype that is associated with low levels of MHC proteins, and they lack expression of co-stimulatory B7 molecules (Fig. 8.13, top panel). Thus, they are not yet equipped to stimulate naive T cells. However, they are very active in taking up antigens by phagocytosis using receptors such as DEC 205, whereas other extracellular antigens are taken up non-specifically by **macropinocytosis**, a process in which large volumes of surrounding fluid are engulfed.

These immature dendritic cells persist in the peripheral tissues for variable lengths of time. They are stimulated by infection to migrate via the lymphatics to the local lymphoid tissues, where they have a completely different phenotype. The lymphoid dendritic cells are no longer able to engulf antigens by phagocytosis or macropinocytosis. However, they now express very high levels of MHC class I and MHC class II proteins that are long-lived, allowing them to present peptides stably from proteins that they acquired from the infection. They also express very high levels of adhesion molecules as well as high levels of B7 molecules, and they secrete a chemotactic cytokine or chemokine that specifically attracts naive T cells (see Fig. 8.13, center panel); this chemokine, called DC-CK, is expressed only in dendritic cells in lymphoid tissues. These properties help to explain their ability to stimulate strong naive T-cell responses. Although dendritic cells will also present some self peptides, the T-cell receptor repertoire has been purged of receptors that recognize these same peptides in the thymus (see Chapter 7) and thus autoaggressive T-cell responses are avoided.

Among the antigens that are believed to be presented by dendritic cells are protein antigens from environmental sources that serve to trigger allergic reactions upon inhalation (see Chapter 12), viral antigens, bacterial antigens, and alloantigens deriving from a transplanted organ, which form the basis for graft rejection (see Chapter 13). Viruses can also infect dendritic cells by binding to any of several molecules on the cell surface, or by being engulfed but not destroyed by the phagocytic form of immature dendritic cells. Such viruses will synthesize their proteins by using the cell's own biosynthetic machinery, leading to surface expression of viral peptides by MHC class I and MHC class II molecules. In principle, any non-self antigen will be immunogenic if it is taken up and presented by a dendritic cell that is activated to migrate to nearby lymphoid tissues and mature. However, the activation of dendritic cell migration and maturation must be triggered by a signal that is generally only present in the context of infection (see Chapter 10). Dendritic cells are likely to be particularly important in stimulating T-cell responses to antigens such as viral antigens, which fail to induce co-stimulatory activity in other types of antigen-presenting cell.

Typical of the immature dendritic cells are the **Langerhans' cells** of the skin. These cells are actively phagocytic and contain large granules, known as Birbeck granules, which may be a type of phagosome. An infection triggers the migration of these cells to the regional lymph nodes (Fig. 8.14), where they rapidly lose the ability to take up and process antigen, but synthesize new MHC molecules that present the pathogen peptides at a high level. On arriving in the regional lymph node, they express the B7 molecules that can co-stimulate naive T cells and also express large numbers of adhesion molecules that allow them to interact with antigen-specific T cells as described in Section 8-3 (see Fig. 8.14). In this way the Langerhans' cells capture antigens from invading pathogens and differentiate into mature dendritic cells that are uniquely fitted for presenting these antigens and activating naive T cells.

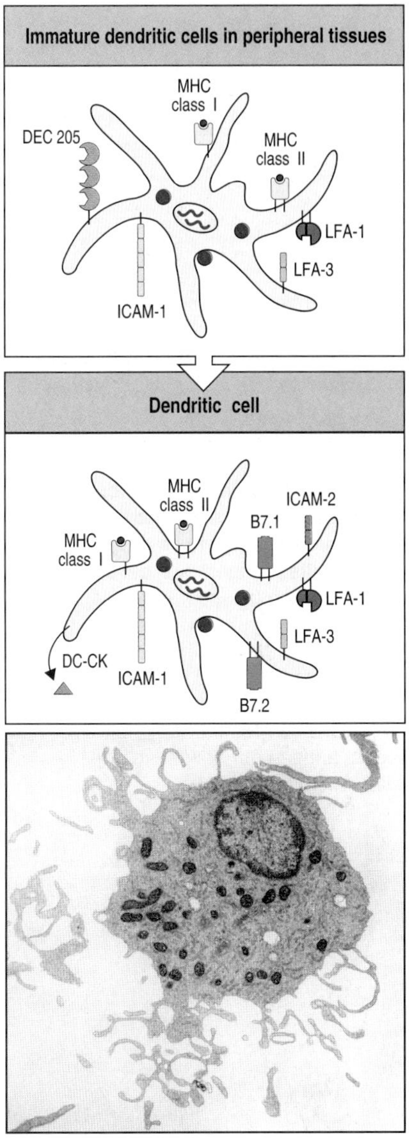

Fig. 8.13 Dendritic cells mature through at least two definable stages to become potent antigen-presenting cells in lymphoid tissue. Dendritic cells arise from bone marrow progenitors and migrate via the blood to peripheral tissues and organs, where they are highly phagocytic via receptors such as DEC 205 but do not express co-stimulatory molecules (top panel). When they pick up antigen in the peripheral tissues, they are induced to migrate via the afferent lymphatic vessels to the regional lymph node (see Fig. 8.14). Here they express high levels of T-cell activating potential but are no longer phagocytic. Dendritic cells in lymphoid tissue express B7.1, B7.2, and high levels of MHC class I and class II molecules, as well as high levels of the adhesion molecules ICAM-1, ICAM-2, LFA-1, and LFA-3 (center panel). The photograph shows a mature dendritic cell. Photograph courtesy of J Barker.

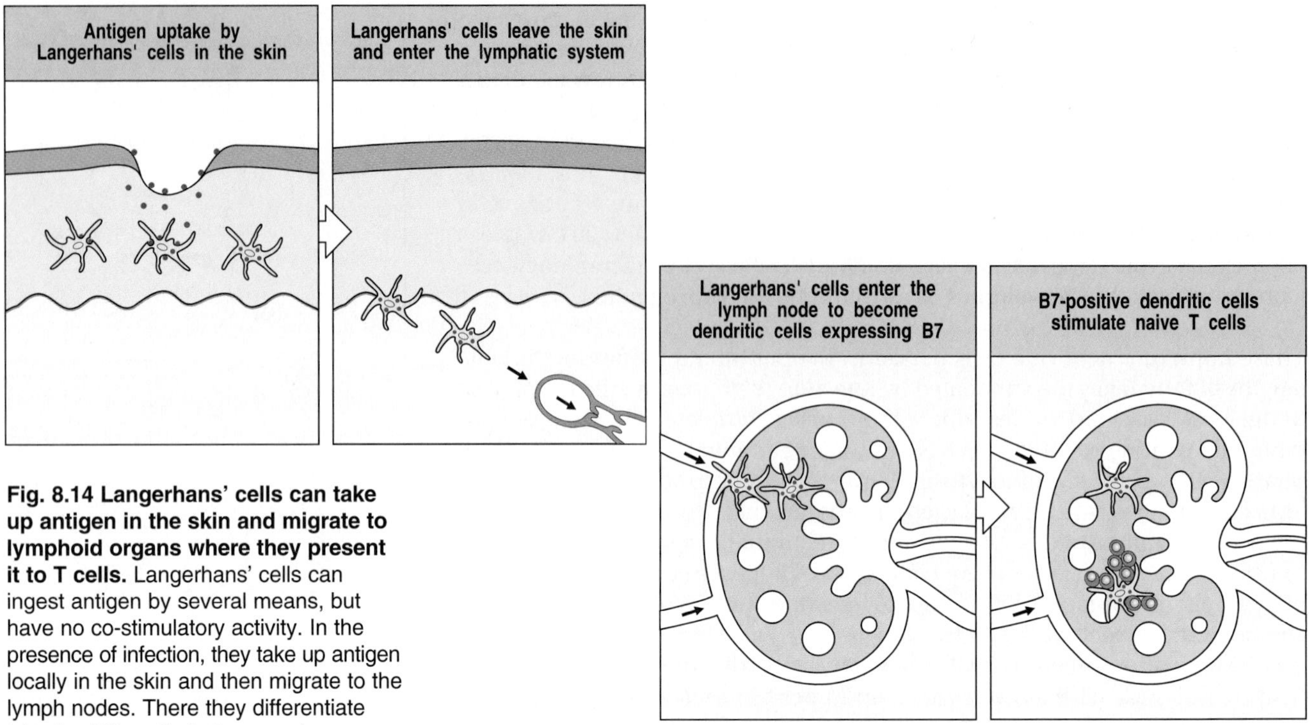

Fig. 8.14 Langerhans' cells can take up antigen in the skin and migrate to lymphoid organs where they present it to T cells. Langerhans' cells can ingest antigen by several means, but have no co-stimulatory activity. In the presence of infection, they take up antigen locally in the skin and then migrate to the lymph nodes. There they differentiate into dendritic cells that can no longer ingest antigen but now have co-stimulatory activity.

8-6 Macrophages are scavenger cells that can be induced by pathogens to present foreign antigens to naive T cells.

Many of the microorganisms that enter the body are readily engulfed and destroyed by phagocytes, which provide an innate, antigen non-specific first line of defense against infection, as will be described in Chapter 10. Microorganisms that are destroyed by phagocytes without additional help from T cells do not cause disease and do not require an adaptive immune response. Pathogens, by definition, have developed mechanisms to avoid elimination by innate immune mechanisms, and the recognition and removal of such pathogens is the function of the adaptive immune response. Mononuclear phagocytes or macrophages, in which ingested microorganisms persist, contribute to the adaptive immune response by acting as professional antigen-presenting cells. As we shall see later in this chapter and in Chapter 10, the adaptive immune response is in turn able to stimulate the microbicidal and phagocytic capacities of these cells.

Professional antigen-presenting cells must be able to present peptide fragments of the antigen on both classes of MHC molecule, and to deliver a co-stimulatory signal through the expression of B7 molecules. Resting macrophages, however, have few or no MHC class II molecules on their surface, and do not express B7. The expression of both MHC class II and B7 molecules is induced in these cells by the ingestion of microorganisms.

Macrophages have a variety of receptors for microbial constituents, including the macrophage mannose receptor and the scavenger receptor (see Chapter 10). Once bound, microorganisms are engulfed and degraded in the endosomes and lysosomes, generating peptides that can be presented by MHC class II molecules on the cell surface. At the same time, MHC class II and B7 molecules are induced on the surface of the macrophage. The receptors that recognize microbial constituents probably also mediate the induction of co-stimulatory activity, because exposure to a single microbial constituent can induce B7 molecules on most macrophages. There is evidence that these receptors evolved originally to allow the phagocytic cells in primitive eukaryotic organisms to recognize microorganisms by binding to structures such as bacterial carbohydrates or lipopolysaccharide that are not found in eukaryotes. Macrophage receptors still serve this function in innate immunity as well as playing an important part in the initiation of adaptive immune responses.

The induction of co-stimulatory activity by common microbial constituents is believed to allow the immune system to distinguish antigens borne by infectious agents from antigens associated with innocuous proteins, including self proteins. Indeed, many foreign proteins do not induce an immune response when injected on their own, presumably because they fail to induce co-stimulatory activity in antigen-presenting cells. When such protein antigens are mixed with bacteria, however, they become immunogenic, because the bacteria induce the essential co-stimulatory activity in cells that ingest the protein (Fig. 8.15). Bacteria used in this way are known as adjuvants (see Section 2-4). We shall see in Chapter 13 how self tissue proteins mixed with bacterial adjuvants can induce autoimmune diseases, illustrating the crucial importance of the regulation of co-stimulatory activity in self:non-self discrimination.

As macrophages continuously scavenge dead or senescent cells, it is particularly important that they should not normally be capable of activating T cells in the absence of microbial infection. The Kupffer cells of the liver sinusoids and the macrophages of the splenic red pulp, in particular, remove large numbers of dying cells from the blood daily. Kupffer cells express little MHC class II and its expression is not increased by the ingestion of dead cells in the absence of infection or inflammation. Moreover, Kupffer cells are not located at sites through which large numbers of naive T cells pass. Thus, although they generate large amounts of self peptides in their endosomes and lysosomes, these macrophages are not likely to elicit an autoimmune response.

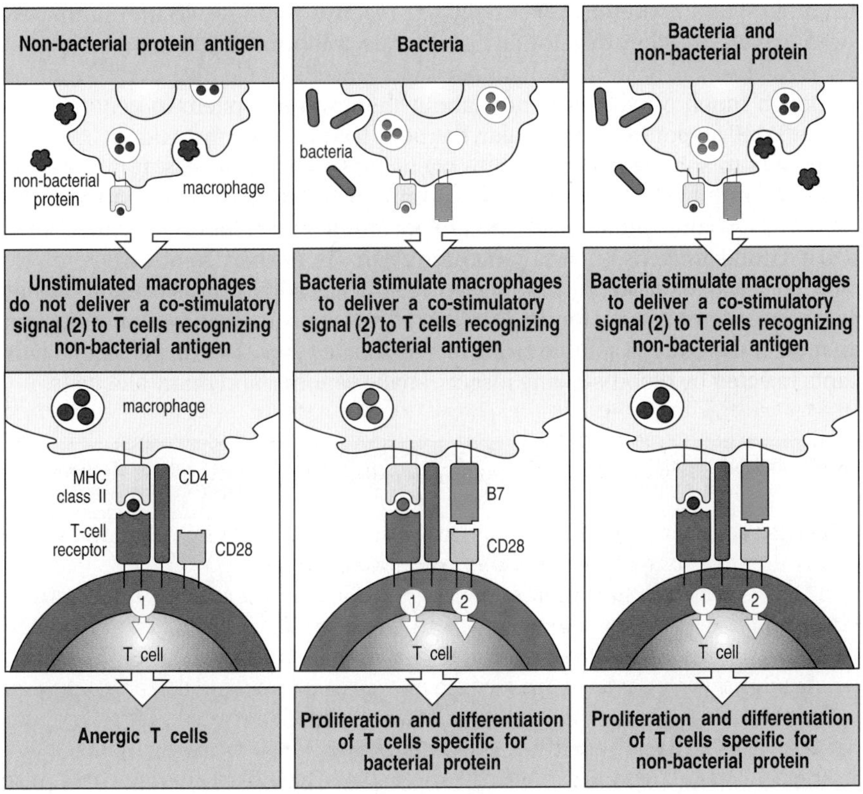

Non-bacterial protein antigen
Bacteria
Bacteria and non-bacterial protein
non-bacterial protein
macrophage
bacteria
Unstimulated macrophages do not deliver a co-stimulatory signal (2) to T cells recognizing non-bacterial antigen
Bacteria stimulate macrophages to deliver a co-stimulatory signal (2) to T cells recognizing bacterial antigen
Bacteria stimulate macrophages to deliver a co-stimulatory signal (2) to T cells recognizing non-bacterial antigen
macrophage
MHC class II
CD4
T-cell receptor
CD28
B7
CD28
1
2
T cell
T cell
T cell
Anergic T cells
Proliferation and differentiation of T cells specific for bacterial protein
Proliferation and differentiation of T cells specific for non-bacterial protein

Fig. 8.15 Microbial substances can induce co-stimulatory activity in macrophages. If protein antigens are taken up and presented by macrophages in the absence of bacterial components that induce co-stimulatory activity in the macrophage, T cells specific for the antigen will become anergic (refractory to activation). Many bacteria induce the expression of co-stimulators by antigen-presenting cells, and macrophages presenting peptide antigens derived by degradation of such bacteria can activate naive T cells. When bacteria are mixed with protein antigens, the protein antigens are rendered immunogenic because the bacteria induce B7 co-stimulatory molecules in the antigen-presenting cells. Such added bacteria act as adjuvants.

8-7 B cells are highly efficient at presenting antigens that bind to their surface immunoglobulin.

Macrophages cannot take up soluble antigens efficiently, but immature dendritic cells can take up large amounts of antigen from extracellular fluid by macropinocytosis. B cells, by contrast, are uniquely adapted to bind specific soluble molecules through their cell-surface immunoglobulin. B cells internalize the soluble antigens bound by their surface immunoglobulin receptors and then display peptide fragments of these antigens as peptide: MHC class II complexes. Because this mechanism of antigen uptake is highly efficient and B cells constitutively express high levels of MHC class II molecules, high levels of specific peptide:MHC class II complexes are generated at the B-cell surface (Fig. 8.16). This pathway of antigen presentation allows B cells to be targeted by antigen-specific CD4 T cells, which drive their differentiation, as we shall see in Chapter 9. In circumstances in which the presenting B cell is induced to express co-stimulatory activity, it also allows B cells to activate naive T cells.

B cells do not constitutively express co-stimulatory activity but they can be induced by various microbial constituents to express B7.1 and especially B7.2. Indeed, B7.1 was first identified as a molecule expressed on B cells activated by microbial lipopolysaccharide. These observations help explain why it is essential to co-inject bacterial adjuvants in order to produce an immune response to soluble proteins such as ovalbumin, hen egg-white lysozyme, and cytochrome *c*, which seem to require B cells as antigen-presenting cells.

The requirement for induced co-stimulatory activity also helps to explain why, although B cells efficiently present soluble proteins, they are unlikely to initiate responses to soluble self proteins in the absence of infection: in the absence of co-stimulatory activity, antigen not only fails to activate naive T cells but causes them to become anergic, or non-responsive (see Fig. 8.12). This provides an additional safeguard to the mechanisms discussed in Chapters 6 and 7 whereby potentially self-reactive T and B cells are eliminated or inactivated as they develop in the thymus and bone marrow.

Although much of what we know about the immune system in general, and about T-cell responses in particular, has been learned from the study of immune responses to soluble protein immunogens presented by B cells, it is not clear how important a role B cells play in priming naive T cells in natural immune responses. Soluble protein antigens are not abundant during natural infections; most natural antigens, such as bacteria and viruses, are particulate, whereas soluble bacterial toxins act by binding to cell surfaces and are thus present only at low concentrations in solution. However, there are some natural immunogens that enter the body as soluble molecules; examples are insect toxins, anticoagulants injected by blood-sucking insects, snake venoms, and many allergens.

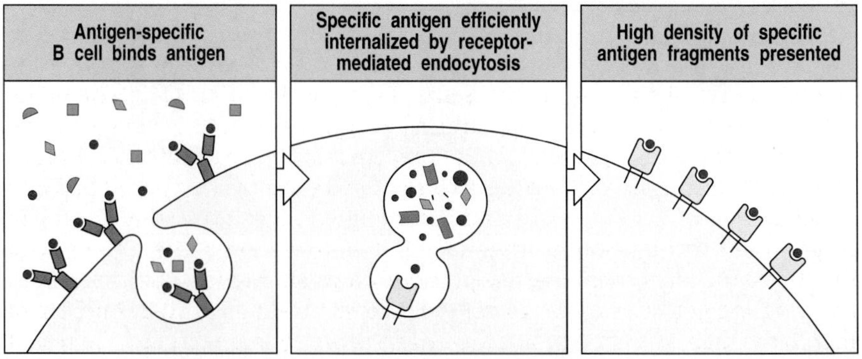
Antigen-specific B cell binds antigen
Specific antigen efficiently internalized by receptor-mediated endocytosis
High density of specific antigen fragments presented

Fig. 8.16 B cells can use their immunoglobulin receptor to present specific antigen very efficiently to T cells. Surface immunoglobulin allows B cells to bind and internalize specific antigen very efficiently. The internalized antigen is processed in intracellular vesicles where it binds to MHC class II molecules, which transport the antigenic fragments to the cell surface where they can be recognized by T cells. When the protein antigen is not recognized specifically by the B cell, its internalization is inefficient and only a low density of fragments of such proteins are subsequently presented at the B-cell surface (not shown).

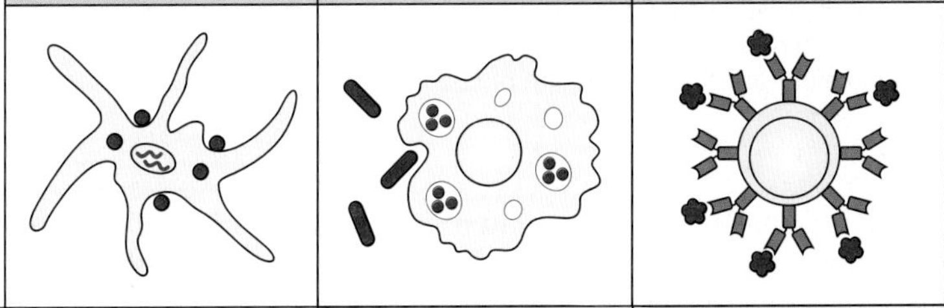

	Dendritic cells	Macrophages	B cells
Antigen uptake	+++ Macropinocytosis and phagocytosis by tissue dendritic cells. Viral infection	Phagocytosis +++	Antigen-specific receptor (Ig) ++++
MHC expression	Low on tissue dendritic cells. High on lymphoid dendritic cells	Inducible by bacteria and cytokines – to +++	Constitutive. Increases on activation +++ to ++++
Co-stimulator delivery	Constitutive by mature non-phagocytic lymphoid dendritic cells ++++	Inducible – to +++	Inducible – to +++
Antigen presented	Peptides. Viral antigens. Allergens	Particulate antigens. Intracellular and extracellular pathogens	Soluble antigens. Toxins. Viruses
Location	Lymphoid tissue. Connective tissue. Epithelia	Lymphoid tissue. Connective tissue. Body cavities	Lymphoid tissue. Peripheral blood

Fig. 8.17 The properties of professional antigen-presenting cells. Dendritic cells, macrophages, and B cells are the main cell types involved in the initial presentation of foreign antigens to naive T cells. These cells vary in their means of antigen uptake, MHC class II expression, co-stimulator expression, the type of antigen they present effectively, and their locations in the body.

Thus T-cell responses are primed by three distinct classes of professional antigen-presenting cell. Each is optimally equipped to present a particular class of antigen to naive T cells and, in each, the expression of co-stimulatory activity is controlled so as to provoke responses against pathogens while avoiding immunization against self (Fig. 8.17).

8-8 Activated T cells synthesize the T-cell growth factor interleukin-2 and its receptor.

Naive T cells can live for many years without dividing. These small resting cells have condensed chromatin and a scanty cytoplasm and synthesize little RNA or protein. On activation, they must re-enter the cell cycle and divide rapidly to produce large numbers of progeny that will differentiate into armed effector T cells. Their proliferation and differentiation are driven by a protein growth factor or cytokine called **interleukin-2** (**IL-2**), which is produced by the activated T cell itself.

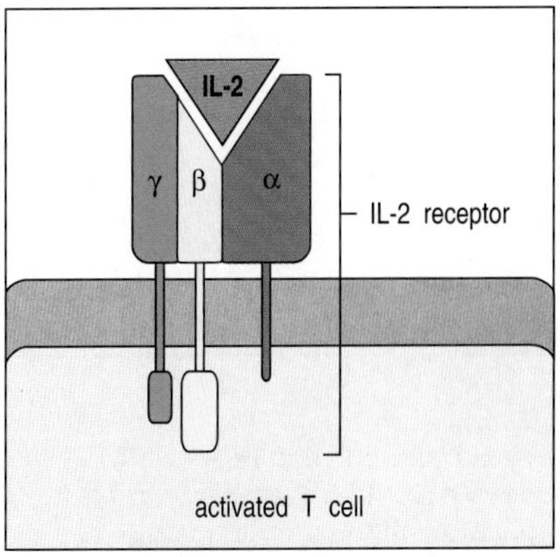

Fig. 8.18 High-affinity IL-2 receptors are three-chain structures that are produced only on activated T cells. On resting T cells, the β and γ chains are expressed constitutively. They bind IL-2 with moderate affinity. Activation of T cells induces the synthesis of the α chain and the formation of the high-affinity heterotrimeric receptor. The β and γ chains show amino acid similarities to cell-surface receptors for growth hormone and prolactin, both of which also regulate cell growth and differentiation.

The initial encounter with specific antigen in the presence of the required co-stimulatory signal triggers the entry of the T cell into the G1 phase of the cell cycle and, at the same time, induces the synthesis of IL-2 along with the α chain of the IL-2 receptor. The IL-2 receptor has three chains: α, β and γ (Fig. 8.18). Resting T cells express a form of this receptor composed of β and γ chains that binds IL-2 with moderate affinity, allowing resting T cells to respond to very high concentrations of IL-2. Association of the α chain with the β and γ chains creates a receptor with a much higher affinity for IL-2, allowing the cell to respond to very low concentrations of IL-2. Binding of IL-2 to the high-affinity receptor then triggers progression through the rest of

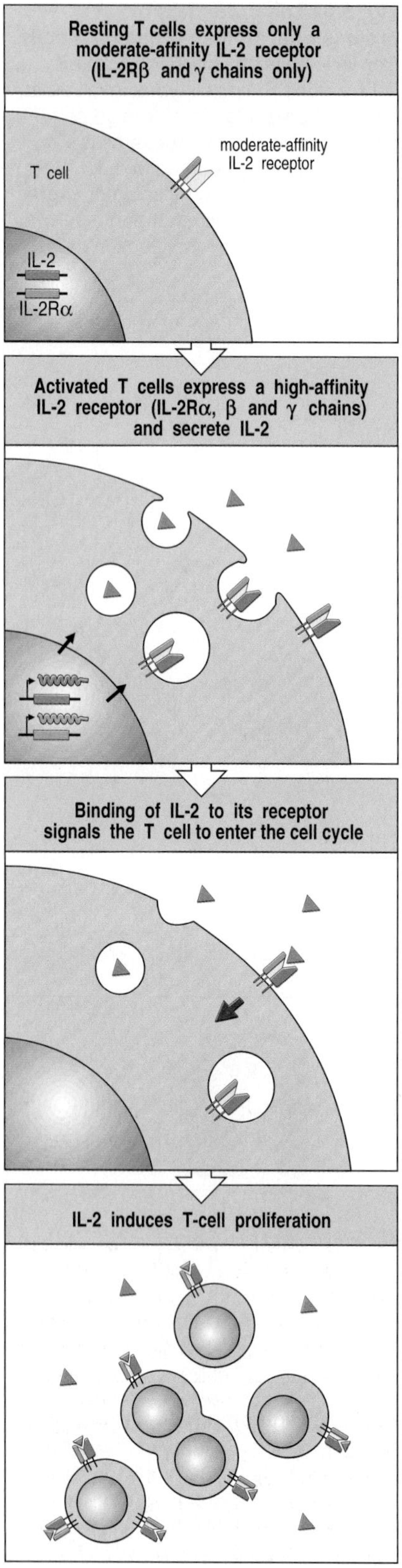
Resting T cells express only a moderate-affinity IL-2 receptor (IL-2Rβ and γ chains only)
moderate-affinity IL-2 receptor
T cell
IL-2
IL-2Rα
Activated T cells express a high-affinity IL-2 receptor (IL-2Rα, β and γ chains) and secrete IL-2
Binding of IL-2 to its receptor signals the T cell to enter the cell cycle
IL-2 induces T-cell proliferation

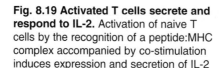

Fig. 8.19 Activated T cells secrete and respond to IL-2. Activation of naive T cells by the recognition of a peptide:MHC complex accompanied by co-stimulation induces expression and secretion of IL-2 and the expression of high-affinity IL-2 receptors. IL-2 binds to the high-affinity IL-2 receptors to promote T-cell growth in an autocrine fashion.

the cell cycle. T cells activated in this way can divide two to three times a day for several days, allowing one cell to give rise to thousands of progeny that all bear the same receptor for antigen. IL-2 also promotes the differentiation of these cells into armed effector T cells (Fig. 8.19).

Activation also causes the expression on the T-cell surface of the molecule that is called **CD40 ligand**, because it binds to the B-cell surface molecule **CD40**. The binding of CD40 ligand to CD40 activates CD40-positive cells and induces the surface expression of B7.1 and B7.2 on the B cell, further driving the T-cell response.

8-9 The co-stimulatory signal is necessary for the synthesis and secretion of IL-2.

The production of IL-2 determines whether a T cell will proliferate and become an armed effector cell, and the most important function of the co-stimulatory signal is to promote the synthesis of IL-2. Antigen recognition by the T-cell receptor ultimately induces several transcription factors (see Chapter 5). One of these factors, **NF-AT** (**nuclear factor of activation in T cells**), binds to the promoter region of the IL-2 gene and is necessary to activate its transcription. IL-2 gene transcription on its own, however, does not lead to the production of IL-2, which additionally requires CD28 ligation by B7. One effect of signaling through CD28 is thought to be the stabilization of IL-2 mRNA. Cytokine mRNAs are very short-lived because of an 'instability' sequence in their 3′ untranslated region, as we shall see later in this chapter. This instability prevents sustained cytokine production and release, and enables cytokine activity to be tightly regulated. The stabilization of IL-2 mRNA increases IL-2 synthesis by 20- to 30-fold. A second effect of CD28 ligation is to activate transcription factors (AP-1 and NF-κB) that increase transcription of IL-2 mRNA by about 3-fold. These two effects together increase IL-2 protein production by 100-fold. When a T cell recognizes specific antigen in the absence of co-stimulation through its CD28 molecule, little IL-2 is produced and the T cell does not proliferate.

The requirements for expression of the IL-2 receptor are less stringent than those for IL-2 synthesis. For example, T-cell receptor ligation alone is frequently sufficient to induce the expression of high-affinity IL-2 receptors on T cells. This can allow IL-2 made by one cell to act on IL-2 receptors expressed on neighboring antigen-specific cells. Later in this chapter we shall see how this may be important in the priming of CD8 T cells.

The central importance of IL-2 in initiating adaptive immune responses is well illustrated by the drugs that are most commonly used to suppress undesirable immune responses such as the rejection of tissue grafts. The immunosuppressive drugs cyclosporin A and FK506 (tacrolimus) inhibit IL-2 production by disrupting signaling through the T-cell receptor, whereas rapamycin inhibits signaling through the IL-2 receptor. Cyclosporin A and rapamycin act synergistically to inhibit immune responses by preventing the IL-2-driven clonal expansion of T cells. The mode of action of these drugs will be considered in detail in Chapter 14.

8-10 Antigen recognition in the absence of co-stimulation leads to T-cell tolerance.

Antigen recognition in the absence of co-stimulation inactivates naive T cells, inducing a state known as anergy (Fig. 8.20). The most important change in anergic T cells is their inability to produce IL-2. This prevents them from proliferating and differentiating into effector cells when they encounter antigen, even if the antigen is subsequently presented by professional antigen-presenting cells. This helps to ensure the tolerance of T cells to self tissue antigens. Although anergy has only been demonstrated formally *in vitro*, there is sufficiently compelling evidence from studies *in vivo* showing peripheral tolerance to various antigens to assume that it happens in this setting as well.

As we saw in Section 7-16, any protein synthesized by all cells will be presented by professional antigen-presenting cells in the thymus and will cause clonal deletion of T cells reactive to these ubiquitous self proteins. However, many proteins have specialized functions and are made only by the cells of certain tissues. Because MHC class I molecules present only those peptides derived from proteins synthesized within the cell, such tissue-specific peptides will not be displayed on the MHC molecules of thymic cells, and T cells recognizing them are unlikely to be deleted in the thymus. An important factor in avoiding autoimmune responses to such tissue-specific proteins is the absence of co-stimulatory activity on tissue cells. Naive T cells recognizing self peptides on tissue cells are not activated; instead they may be induced to enter a state of anergy (see Fig. 8.20).

Although the deletion of potentially autoreactive T cells is readily understood as a simple way to maintain self tolerance, the retention of anergic T cells specific for tissue antigens is less easy to understand. It would seem more economical and efficient to eliminate such cells; indeed, binding of the T-cell receptor on peripheral T cells in the absence of co-stimulators can lead to programmed cell death as well as to anergy. Nevertheless, some T cells persist in an anergic state *in vivo*. One possible explanation for this is that such anergic T cells have a role in preventing responses by naive, non-anergic T cells to foreign antigens that mimic self-peptide:self-MHC complexes. The persisting anergic T cells could recognize and bind to such

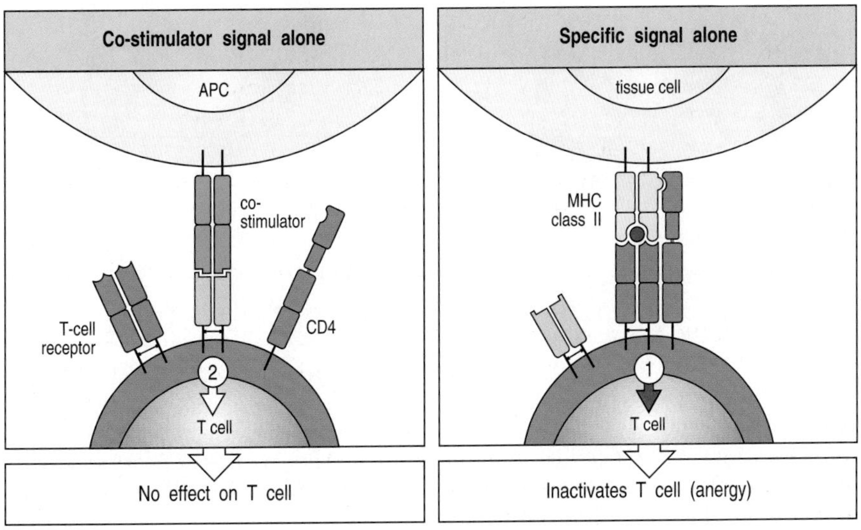

Fig. 8.20 T-cell tolerance to antigens expressed on tissue cells results from antigen recognition in the absence of co-stimulation. An antigen-presenting cell (APC) will neither activate nor inactivate a T cell if the appropriate antigen is not present on the APC surface, even if it expresses a co-stimulatory molecule and can deliver signal 2 (left panel). However, when a T cell recognizes antigen in the absence of co-stimulatory molecules, it receives signal 1 alone and is inactivated (right panel). This allows self antigens expressed on tissue cells to induce tolerance in the T cell population.

peptide:MHC complexes on professional antigen-presenting cells without responding, and thus could compete with naive, potentially autoreactive cells of the same specificity. In this way, anergic T cells could serve to prevent the accidental activation of autoreactive T cells by infectious agents, thus actively contributing to tolerance.

8-11 Proliferating T cells differentiate into armed effector T cells that do not require co-stimulation to act.

The combination of antigen and co-stimulator induces naive T cells to express IL-2 and its receptor; IL-2 then induces clonal expansion of the naive T cell and the differentiation of its progeny. Late in the proliferative phase of the response, after 4–5 days of rapid growth, these T cells differentiate into armed effector T cells that are able to synthesize all the proteins required for their specialized functions as helper or cytotoxic T cells. As well as acquiring the capacity to synthesize the appropriate arsenal of specialized effector molecules when they encounter antigen on target cells, all classes of armed effector T cells have undergone several changes that distinguish them from naive T cells. One of the most critical is in their activation requirements: once a T cell has differentiated into an armed effector cell, further encounter with its specific antigen results in immune attack without the need for co-stimulation (Fig. 8.21).

This applies to all classes of armed effector T cells. Its importance is particularly easy to understand in the case of cytotoxic CD8 T cells, which must be able to act on any cell infected with a virus, whether or not the infected cell can express co-stimulatory molecules. However, it is also important for the effector function of CD4 cells, as armed effector CD4 T cells must be able to activate B cells and macrophages that have taken up antigen, even if, as is often the case, they have too little co-stimulatory activity to activate a naive CD4 T cell.

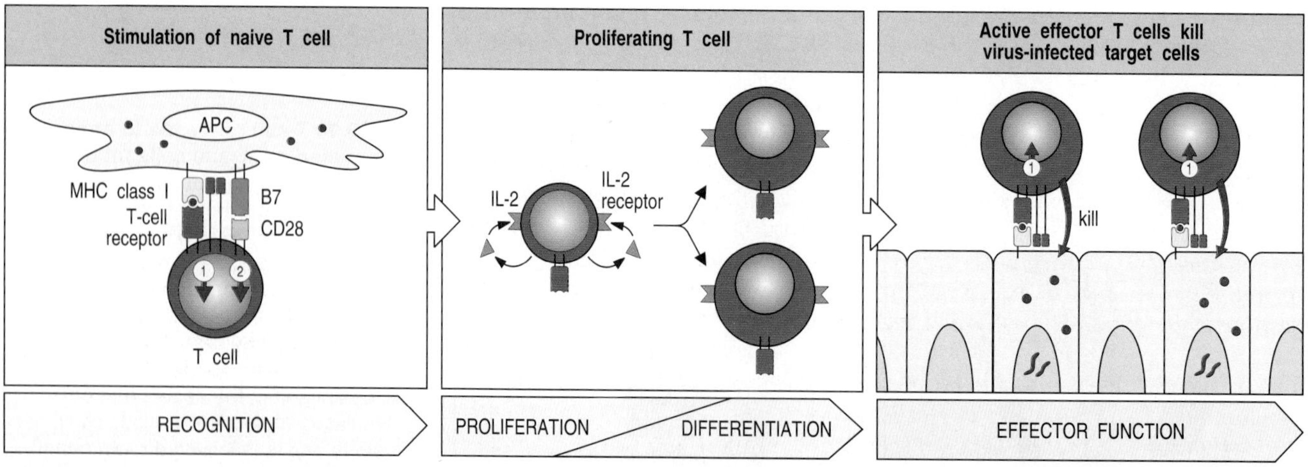

Fig. 8.21 Clonal expansion precedes differentiation to effector function. A naive T cell that recognizes antigen on the surface of a professional antigen-presenting cell and receives the required two signals (arrows 1 and 2, left panel) becomes activated and both secretes and responds to IL-2. IL-2-driven clonal expansion (center panel) is followed by the differentiation of the T cells to armed effector cell status. Once the cells have differentiated into effector T cells, any encounter with specific antigen triggers their effector actions without the need for co-stimulation. Thus, as illustrated here, a cytotoxic T cell can kill targets that express only the peptide:MHC ligand and not co-stimulatory signals (right panel).

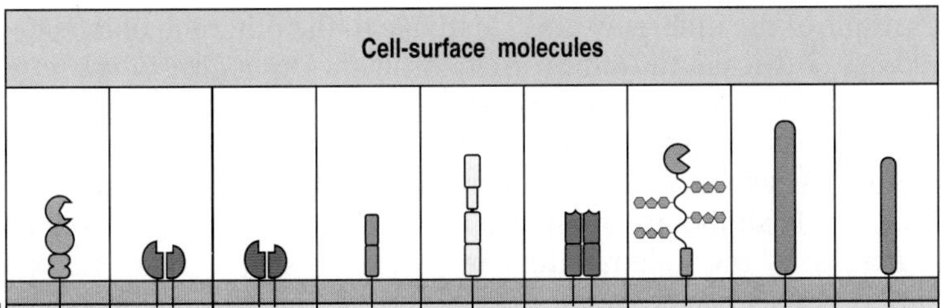

CD4 T cell	L-selectin	VLA-4	LFA-1	CD2	CD4	T-cell receptor	CD44	CD45RA	CD45RO
Resting	+	–	+	+	+	+	+	+	–
Activated	–	+	++	++	+	+	++	–	+

Fig. 8.22 Activation of T cells changes the expression of several cell-surface molecules. The example here is a CD4 T cell. Resting naive T cells express L-selectin, through which they home to lymph nodes, with relatively low levels of other adhesion molecules such as CD2 and LFA-1. Upon activation of the T cell, the expression of these molecules changes. Expression of L-selectin is lost and, instead, increased amounts of the integrin VLA-4 are expressed. VLA-4 acts as a homing receptor for vascular endothelium in sites of inflammation and ensures that activated T cells recirculate through peripheral tissues where they may encounter sites of infection. Activated T cells express higher densities of the adhesion molecules CD2 and LFA-1, increasing the avidity of the interaction of the activated T cell with potential target cells. The adhesion molecule CD44 is also increased on activated T cells. Finally, the isoform of the CD45 molecule expressed by activated cells changes, by alternative splicing of the RNA transcript of the CD45 gene, so that activated T cells now express the CD45RO isoform that associates with the T-cell receptor and CD4. The consequences of this change in CD45 make the T cell more sensitive to stimulation by low concentrations of peptide:MHC complexes.

Other changes are seen in the adhesion molecules expressed by armed effector T cells. They express higher levels of the cell adhesion molecules LFA-1 and CD2, but lose their cell-surface L-selectin and thus cease to recirculate through lymph nodes. Instead, they express the integrin VLA-4, which allows them to bind to vascular endothelium at sites of inflammation. This allows the armed effector T cells to enter sites of infection and put their armory of effector proteins to good use. These changes in the T-cell surface are summarized in Fig. 8.22.

8-12 The differentiation of CD4 T cells into T_H1 or T_H2 cells determines whether humoral or cell-mediated immunity will predominate.

Naive CD8 T cells emerging from the thymus are already predestined to become cytotoxic cells, even though they are not yet expressing any of the differentiated functions of armed effector cells. The case of CD4 T cells, however, is more complex. Naive CD4 T cells can differentiate upon activation into either T_H1 or T_H2 cells, which differ in the cytokines they produce upon activation and thus differ in their function (see Sections 8-16 and 8-17). The decision on which fate a naive CD4 T cell will follow is made during its first encounter with antigen (Fig. 8.23).

The factors that determine whether a proliferating CD4 T cell will differentiate into a T_H1 or a T_H2 cell are not fully understood. The cytokines elicited by infectious agents (principally the interleukins **IL-12** and **IL-4**), the co-stimulators used to drive the response, and the nature of the peptide:MHC ligand can all have an effect. In particular, because the decision to differentiate into T_H1 versus T_H2 cells occurs early in the immune response, the cytokines produced in response to pathogens by cells of the innate non-adaptive immune system play an important part in shaping the subsequent adaptive response; we shall learn more about this in Chapter 10.

The consequences of inducing T_H1 versus T_H2 cells are profound: the selective production of T_H1 cells leads to cell-mediated immunity, whereas the production of predominantly T_H2 cells provides humoral immunity. A striking

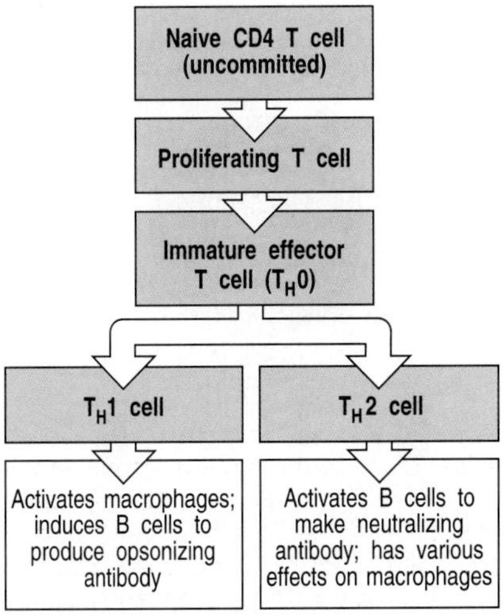

Fig. 8.23 The stages of activation of CD4 T cells. Naive CD4 T cells first respond to their specific peptide:MHC class II complexes by making IL-2 and proliferating. These cells then differentiate into a cell type known as T_H0, which has some of the effector functions characteristic of T_H1 and T_H2 cells. The T_H0 cell has the potential to become either a T_H1 cell or a T_H2 cell.

example of the difference this can make to the outcome of infection is seen in leprosy, a disease caused by infection with *Mycobacterium leprae. M. leprae,* like *M. tuberculosis,* grows in macrophage vesicles, and effective host defense requires macrophage activation by T_H1 cells. In patients with tuberculoid leprosy, in which T_H1 cells are preferentially induced, few live bacteria are found, little antibody is produced and, although skin and peripheral nerves are damaged by the inflammatory responses associated with macrophage activation, the disease progresses slowly and the patient usually survives. However, when T_H2 cells are preferentially induced, the main response is humoral, the antibodies produced cannot reach the intracellular bacteria, and the patients develop lepromatous leprosy, in which *M. leprae* grows abundantly in macrophages, causing gross tissue destruction, which is eventually fatal.

8-13 Naive CD8 T cells can be activated in different ways to become armed cytotoxic effector cells.

Naive CD8 T cells differentiate into cytotoxic cells, and perhaps because the effector actions of these cells are so destructive, naive CD8 T cells require more co-stimulatory activity to drive them to become armed effector cells than do naive CD4 T cells. This requirement can be met in two ways. The simplest is activation by antigen-presenting cells, such as dendritic cells, that have high intrinsic co-stimulatory activity. These cells can directly stimulate CD8 T cells to synthesize the IL-2 that drives their own proliferation and differentiation (Fig. 8.24). This has been exploited to generate cytotoxic T-cell responses against tumors, as we shall see in Chapter 14.

Cytotoxic T cell responses to some viruses and tissue grafts, however, seem to require the presence of CD4 T cells during the priming of the naive CD8 T cell. In these responses, both the naive CD8 T cell and the CD4 T cell must recognize antigen on the surface of the same antigen-presenting cell. In this case, it is thought that the actions of the CD4 T cell may be necessary to compensate for inadequate co-stimulation of naive CD8 T cells by the antigen-presenting cell. This compensatory effect could occur in either of two ways. If the CD4 T cell is an armed effector cell, it may activate the antigen-presenting cell to express higher levels of co-stimulatory activity. We shall see that this is one of the actions of the specialized molecules produced by effector CD4 T cells. This would enable the antigen-presenting cell to co-stimulate the CD8 T cell directly (Fig. 8.25, left panels).

Alternatively, the CD4 T cell may be a naive or memory T cell, which secretes IL-2 in response to antigen and lower levels of co-stimulatory molecules. As IL-2 receptors can be induced by receptor ligation alone, the CD8 T cell may express IL-2 receptors even though it cannot produce the IL-2 needed to drive its own proliferation. The IL-2 in this case comes instead from the adjacent responding CD4 T cell (see Fig. 8.25, right panels). Adding IL-2 can eliminate the need for a co-stimulatory signal for CD8 T-cell activation in experimental situations. This is in keeping with the general finding that the crucial role of co-stimulation for T cells is the production of sufficient IL-2 to drive their clonal expansion, allowing them to differentiate into armed effector T cells.

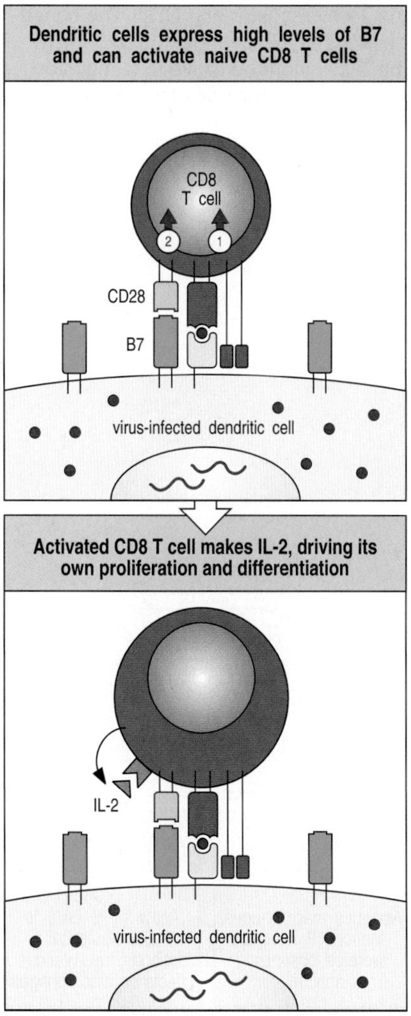

Dendritic cells express high levels of B7 and can activate naive CD8 T cells
CD8 T cell
2
1
CD28
B7
virus-infected dendritic cell
Activated CD8 T cell makes IL-2, driving its own proliferation and differentiation
IL-2
virus-infected dendritic cell

Fig. 8.24 Naive CD8 T cells can be activated directly by potent antigen-presenting cells. Naive CD8 T cells that encounter peptide:MHC class I complexes on the surface of dendritic cells, which express high levels of co-stimulatory molecules (top panel), are activated to produce IL-2 (bottom panel) and proliferate in response to it, eventually differentiating into armed cytotoxic CD8 T cells (not shown).

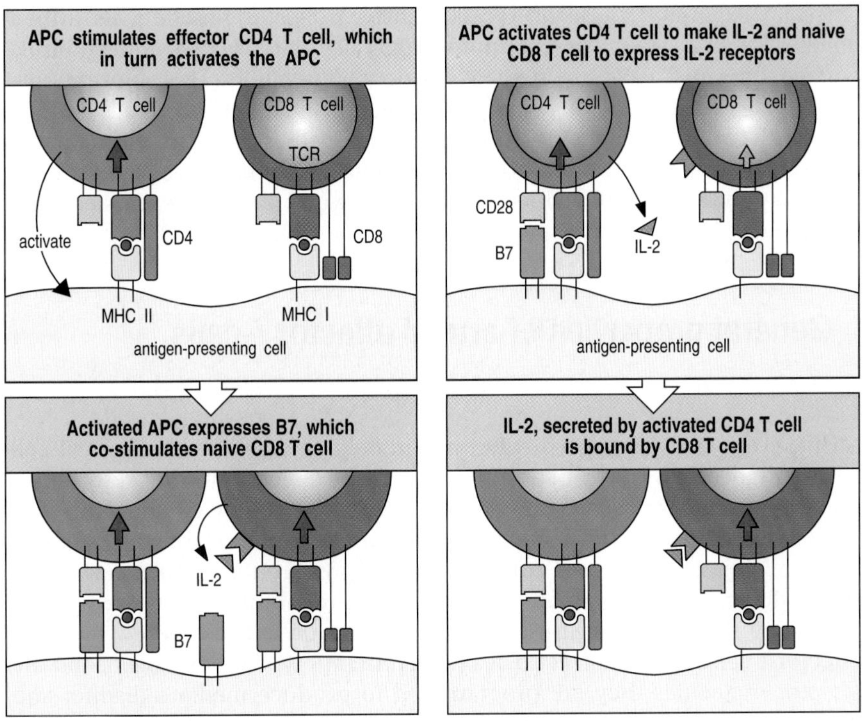

Fig. 8.25 Some CD8 T-cell responses require CD4 T cells. CD8 T cells recognizing antigen on weakly co-stimulating cells may become activated only in the presence of CD4 T cells bound to the same antigen-presenting cell. There are two ways in which CD4 T cells may contribute to the activation of CD8 T cells. Left panels: an effector CD4 T cell may recognize antigen on the antigen-presenting cell and be triggered to induce increased levels of co-stimulatory activity on the antigen-presenting cell, which in turn activates the CD8 T cell to make its own IL-2. Right panels: alternatively, a naive CD4 T cell activated by the antigen-presenting cell may provide the IL-2 required for the proliferation and differentiation of the CD8 T cell. Which of these two mechanisms operates *in vivo* is not known.

Summary.

The crucial first step in adaptive immunity is the activation of naive antigen-specific T cells by professional antigen-presenting cells. This occurs in the lymphoid tissues and organs through which naive T cells are constantly passing. The most distinctive feature of professional antigen-presenting cells is the expression of co-stimulatory activities, of which the B7.1 and B7.2 molecules are the best characterized. Naive T cells will respond to antigen only when one cell presents both specific antigen to the T-cell receptor and a B7 molecule to CD28, the receptor for B7 on the T cell. The three cell types that can serve as professional antigen-presenting cells are dendritic cells, macrophages, and B cells. Each of these cells has a distinct function in eliciting immune responses. Tissue dendritic cells take up antigens by phagocytosis and macropinocytosis and are stimulated by infection to migrate to the local lymphoid tissue, where they differentiate into mature dendritic cells expressing co-stimulatory activity constitutively. They serve as the most potent activators of naive T-cell responses. Macrophages efficiently ingest particulate antigens such as bacteria and are induced by infectious agents to express MHC class II molecules and co-stimulatory activity. The unique ability of B cells to bind and internalize soluble protein antigens via their receptors may be important in activating T cells to this class of antigen, provided that co-stimulatory molecules are also induced on the B cell.

The activation of T cells by professional antigen-presenting cells leads to their proliferation and the differentiation of their progeny into armed effector T cells. The proliferation and differentiation of T cells depends on the production of cytokines, in particular the T-cell growth factor IL-2, and their binding to a high-affinity receptor on the activated T cell. T cells whose antigen receptors are ligated in the absence of co-stimulatory signals fail to make IL-2 and instead become anergic or die. This dual requirement for both receptor ligation and co-stimulation helps to prevent naive T cells from responding to antigens on self-tissue cells, which lack co-stimulatory activity. Proliferating T cells develop into armed effector T cells, the critical event in most adaptive immune responses.

Once an expanded clone of T cells achieves effector function, its armed effector T-cell progeny can act on any target cell that displays antigen on its surface. Effector T cells can mediate a variety of functions. Their most important functions are the killing of infected cells by CD8 cytotoxic T cells and the activation of macrophages by T_H1 cells, which together make up cell-mediated immunity, and the activation of B cells by both T_H2 and T_H1 cells to produce different types of antibody, thus driving the humoral immune response.

General properties of armed effector T cells.

All T-cell effector functions involve the interaction of an armed effector T cell with a target cell displaying specific antigen. The effector proteins released by these T cells are focused on the appropriate target cell by mechanisms that are activated by the specific recognition of antigen on the target cell surface. The focusing mechanism is common to all types of effector T cells, whereas their effector actions depend on the array of membrane and secreted proteins they express or release upon receptor ligation. The different types of effector T cell are specialized to deal with different types of pathogen, and the effector molecules they are programmed to produce mediate distinct and appropriate effects on the target cell (Fig. 8.26).

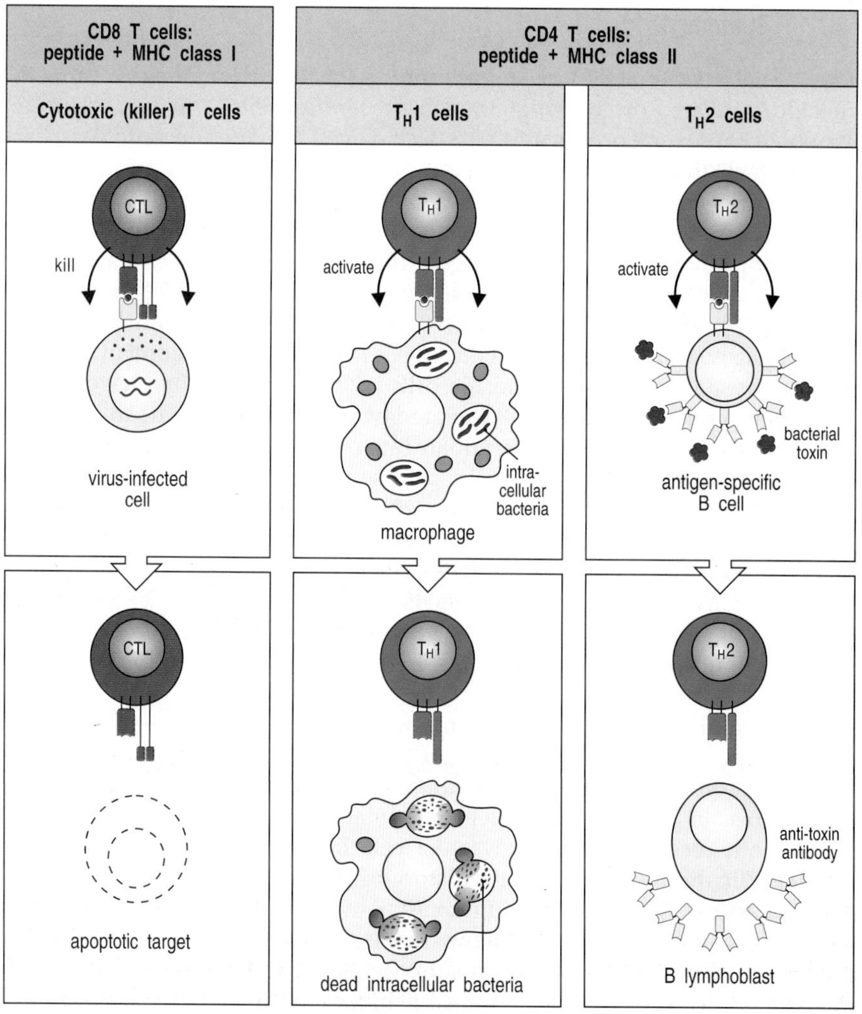

Fig. 8.26 There are three classes of effector T cell, specialized to deal with three classes of pathogen. CD8 cytotoxic cells (left panels) kill target cells that display antigenic fragments of cytosolic pathogens, most notably viruses, bound to MHC class I molecules at the cell surface. T_H1 cells (middle panels) and T_H2 cells (right panels) both express the CD4 co-receptor and recognize fragments of antigens degraded within intracellular vesicles, displayed at the cell surface by MHC class II molecules. The T_H1 cells, upon activation, activate macrophages, allowing them to destroy intracellular microorganisms more efficiently; they can also activate B cells to produce strongly opsonizing antibodies belonging to certain IgG subclasses (IgG1 and IgG3 in humans, and their homologs IgG2a and IgG2b in the mouse). T_H2 cells, on the other hand, drive B cells to differentiate and produce immunoglobulins of all other types, and are responsible for initiating B-cell responses by activating naive B cells to proliferate and secrete IgM. The various types of immunoglobulin together make up the effector molecules of the humoral immune response.

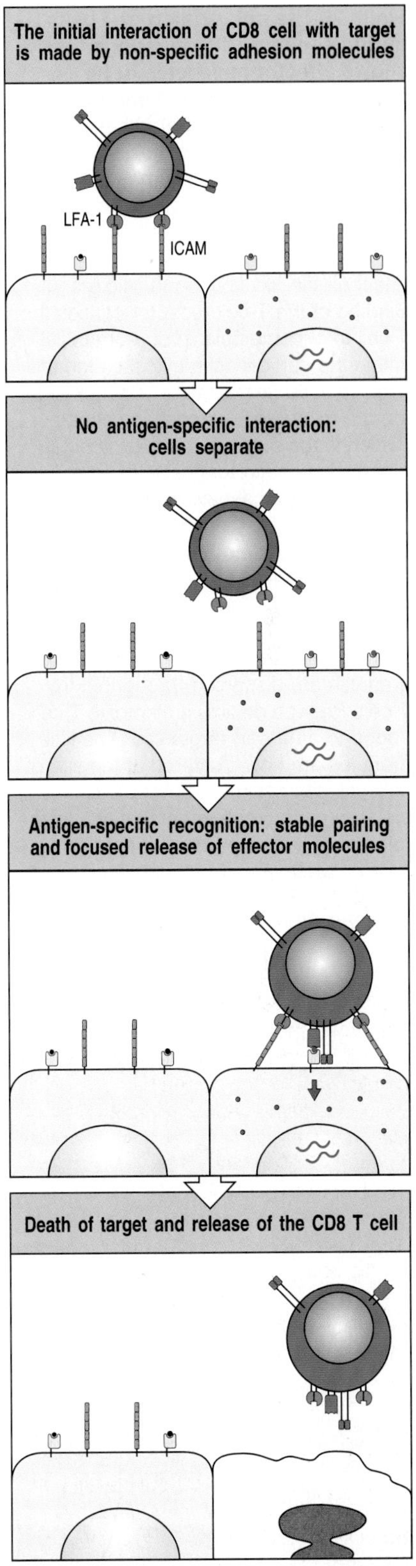

Fig. 8.27 Interactions of T cells with their targets are mediated initially by non-specific adhesion molecules. The major initial interaction is between LFA-1 expressed by the T cell, illustrated here as a CD8 cytotoxic T cell, and ICAM-1 or ICAM-2 expressed by the target cell (top panel). This binding allows the T cell to remain in contact with the target cell and to scan its surface for the presence of specific peptide:MHC complexes. If the target cell does not carry the specific antigen, the T cell disengages (second panel) and can scan other potential targets. If the target cell carries the specific antigen (third panel), signaling through the T-cell receptor increases the strength of the adhesive interactions, prolonging the contact between the two cells, and stimulating the T cell to deliver its effector molecules. The T cell then disengages (bottom panel).

8-14 Effector T cell interactions with target cells are initiated by antigen non-specific cell-adhesion molecules.

Once an effector T cell has completed its differentiation in the lymphoid tissue it must find target cells that are displaying the specific MHC:peptide complex that it recognizes. This occurs in two steps. First, the armed effector T cells emigrate from their site of activation in the lymphoid tissues and enter the blood via the thoracic duct. Second, because of the cell-surface changes that have occurred during differentiation, they then migrate into the peripheral tissues, particularly at sites of infection. They are guided to these sites by changes in the adhesion molecules expressed on the endothelium of the local blood vessels as a result of infection, and by local chemotactic factors, as we shall see in Chapter 10.

The initial binding of an effector T cell to its target, like that of a naive T cell to an antigen-presenting cell, is an antigen non-specific interaction mediated by the LFA-1 and CD2 adhesion molecules. The level of LFA-1 and of CD2 is two- to fourfold higher on armed effector T cells than on naive T cells, and so armed effector T cells can bind efficiently to target cells that have lower levels of ICAMs and LFA-3 on their surface than do professional antigen-presenting cells. This interaction is normally transient unless recognition of antigen on the target cell triggers a change in the affinity of the T-cell LFA-1 for its ligands on the target cell surface. This change causes the T cell to bind more tightly to its target, and to remain bound for long enough to release its specific effector molecules. Armed CD4 effector T cells, which activate macrophages or induce B cells to secrete antibody, must maintain contact with their targets for relatively long periods. Cytotoxic T cells, by contrast, can be observed under the microscope attaching to and dissociating from successive specific targets relatively rapidly as they kill them (Fig. 8.27). Killing of the target, or some local change in the T cell, then allows the effector T cell to detach and address new targets. How armed CD4 effector T cells disengage from their targets is not known, although current evidence suggests that CD4 binding directly to MHC class II molecules on target cells that are not displaying antigen signals the cell to detach.

8-15 Binding of the T-cell receptor complex directs the release of effector molecules and focuses them on the target cell.

The binding of the T-cell receptor to an antigen:MHC complex on a target cell not only increases the strength with which the T cell binds its target but also triggers a polarization of the T cell that focuses the secretion of effector molecules on the target cell. When binding to peptide:MHC complexes, the T-cell receptor molecules and their crosslinked co-receptors cluster at the

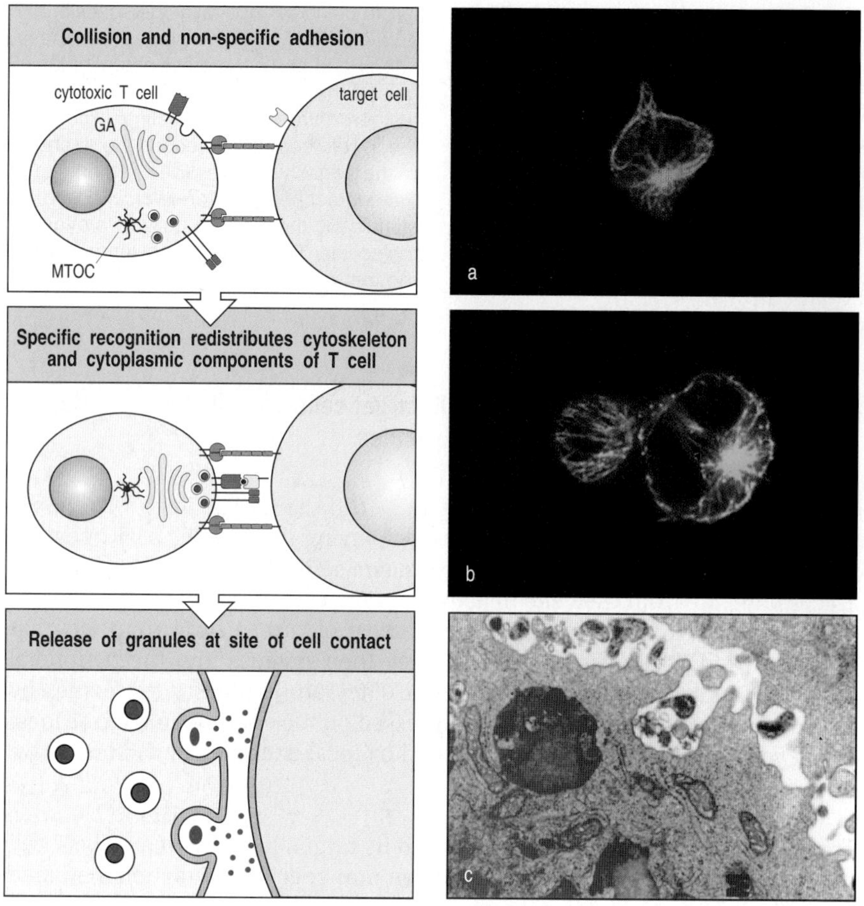

Fig. 8.28 The polarization of T cells during specific antigen recognition allows effector molecules to be focused on the antigen-bearing target cell. The example illustrated here is a CD8 cytotoxic T cell. Cytotoxic CD8 cells contain specialized lysosomes called lytic granules, which contain cytotoxic proteins. Initial binding to a target cell through adhesion molecules does not have any effect on the location of the lytic granules. Binding of the T-cell receptor causes the T cell to become polarized: reorganization within the cortical actin cytoskeleton at the site of contact has the effect of aligning the microtubule-organizing center (which is the origin of the microtubules along which secretory vesicles travel) towards the target cell. Proteins stored in lytic granules are then directed specifically onto the target cell. The photomicrograph in panel a shows an unbound, isolated cytotoxic T cell. The microtubule cytoskeleton is stained in green and the lytic granules in red. Note how the lytic granules are dispersed throughout the T cell. Panel b depicts a cytotoxic T cell bound to a (larger) target cell. The lytic granules are now clustered at the site of cell–cell contact in the bound T cell. The electron micrograph in panel c shows the release of granules from a cytotoxic T cell. Panels a and b courtesy of G Griffiths. Panel c courtesy of E R Podack.

site of cell–cell contact. The clustering of the T-cell receptors then signals a reorientation of the cytoskeleton that polarizes the effector cell so as to focus the release of effector molecules at the site of contact with the target cell, as illustrated for a cytotoxic T cell in Fig. 8.28. Polarization of the cell starts with the local reorganization of the cortical actin cytoskeleton at the site of contact; this in turn leads to the reorientation of the microtubule-organizing center, the center from which the microtubule cytoskeleton is produced, and of the Golgi apparatus, through which most proteins destined for secretion travel. In the cytotoxic T cell, the cytoskeletal reorientation focuses exocytosis of the preformed lytic granules at the site of contact with its target cell.

The polarization of a T cell also focuses the secretion of soluble effector molecules whose synthesis is induced *de novo* by ligation of the T-cell receptor. For example, the soluble cytokine IL-4, which is the principal effector molecule of T_H2 cells, is concentrated and confined to the site of contact with the target cell (see Fig. 9.6). It has recently been shown that the enhanced binding of LFA-1 to ICAM-1 creates a molecular seal surrounding the clustered T-cell receptor molecules, CD4 co-receptor, and CD28 (Fig. 8.29).

Outer ring (red)	Inner circle (green)
LFA-1: ICAM-1	TCR, CD4, CD28

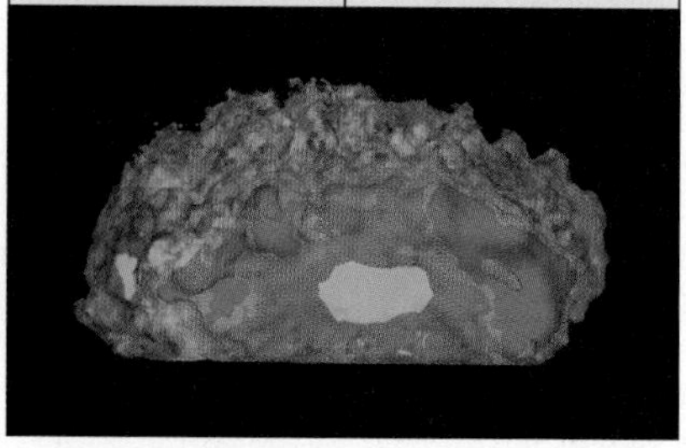

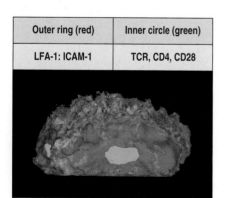

Fig. 8.29 Tight junctions are formed between armed effector T cells and their targets. Confocal fluorescence micrograph of the area of contact between a T cell and a B cell (as viewed through one of the cells). The outer red ring is made up of LFA-1 on the T cell and its counter-receptors on the target cell, whereas molecules that cluster in the center of the ring (green) include the T-cell receptor complex, the co-receptor CD4, and CD28. Photograph courtesy of A Kupfer.

Thus, the antigen-specific T-cell receptor controls the delivery of effector signals in three ways: it induces the stable binding of effector cells to their specific target cells to create a tightly held narrow space in which effector molecules can be concentrated; it focuses their delivery at the site of contact by inducing a reorientation of the secretory apparatus of the effector cell; and it triggers their synthesis and/or release. All these receptor-coordinated mechanisms contribute to the selective action of effector molecules on the cells bearing specific antigen. In this way, effector T cell activity is highly selective for those target cells that display antigen, although the effector molecules themselves are not antigen-specific.

8-16 The effector functions of T cells are determined by the array of effector molecules they produce.

The effector molecules produced by armed effector T cells fall into two broad classes: **cytotoxins**, which are stored in specialized lytic granules and released by cytotoxic CD8 T cells, and **cytokines** and related membrane-associated proteins, which are synthesized *de novo* by all effector T cells. The cytotoxins are the principal effector molecules of cytotoxic T cells and will be discussed further in Section 8-22. Their release in particular must be tightly regulated as they are not specific: they can penetrate the lipid bilayer and trigger an intrinsic death program in any target cell. By contrast, the cytokines and membrane-associated proteins act by binding to specific receptors on the target cell. Cytokines and membrane-associated proteins are the principal mediators of CD4 T-cell effector actions, and the main effector actions of CD4 cells are therefore directed at specialized cells that express receptors for these effector molecules.

The effector actions and main effector molecules of all three functional classes of effector T cells are summarized in Fig. 8.30. The cytokines are a diverse

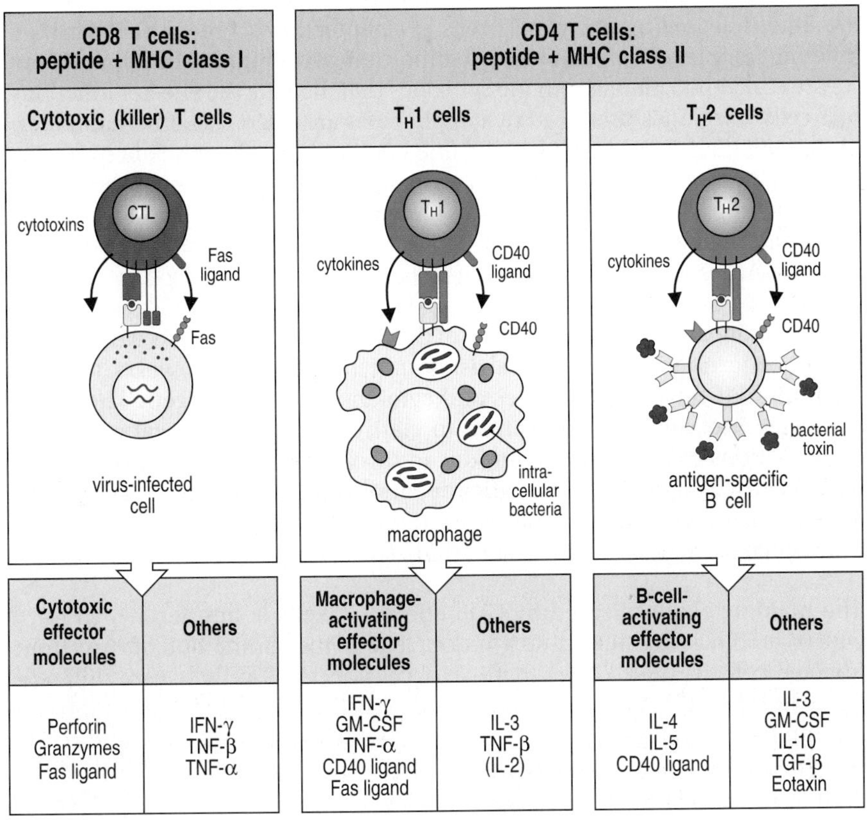

Fig. 8.30 The three main types of armed effector T cell produce distinct sets of effector molecules. CD8 T cells are predominantly killer T cells that recognize pathogen-derived peptides bound to MHC class I molecules. They release perforin (which creates holes in the target cell membranes), granzymes (which are proteases), and often the cytokine IFN-γ. A membrane-bound effector molecule expressed on CD8 T cells is the ligand for Fas, a receptor whose activation induces apoptosis. CD4 T cells recognize peptides bound to MHC class II molecules and are of two functional types: T_H1 cells and T_H2 cells. T_H1 cells are specialized for the activation of macrophages that are infected by or have ingested pathogens; they secrete IFN-γ as well as other effector molecules, and express membrane-bound CD40 ligand and/or Fas ligand. These are both members of the TNF family but CD40 ligand triggers the activation of the target cell, whereas Fas ligand triggers the death of Fas-expressing cells, so their pattern of expression has a strong influence on function. T_H2 cells are specialized for B-cell activation; they secrete the B-cell growth factors IL-4 and IL-5. The principal membrane-bound effector molecule expressed by T_H2 cells is CD40 ligand, which binds to CD40 on the B cell and induces B-cell proliferation.

group of molecules and we shall give a brief overview of them below before discussing specifically the T-cell cytokines and their contributions to the effector actions of CD8 cytotoxic T cells, T_H1 cells, and T_H2 cells. As we shall see, the soluble cytokines and membrane-associated molecules often act in combination to mediate the effects of T cells on their specific target cells.

The membrane-associated effector molecules, which we shall discuss further in Section 8-20, are all structurally related to **tumor necrosis factor** (**TNF**), and their receptors on target cells are members of the TNF receptor (TNFR) family. All three classes of effector T cell express one or more members of the TNF family upon recognizing specific antigen on the target cell. The membrane-bound TNF family member CD40 ligand is of particular importance for CD4 effector function; it is induced on T_H1 and T_H2 cells and delivers activating signals to B cells and macrophages through the TNFR protein CD40. TNF-α is made by T_H1 cells, some T_H2 cells, and cytotoxic T cells in soluble and membrane-associated forms, and can also deliver activating signals to macrophages. Some members of the family of TNF receptors can stimulate death by apoptosis. Thus **Fas ligand** (CD95L), the principal membrane-associated TNF-related molecule expressed by cytotoxic T cells, can trigger death by apoptosis in target cells bearing the receptor protein **Fas** (CD95); some T_H1 cells also express Fas ligand and can kill Fas-bearing cells with which they interact. Death by this mechanism appears to be important for removing activated Fas-bearing lymphocytes, and if it fails, a lymphoproliferative disease associated with severe autoimmunity results (see Section 8-20).

8-17 Cytokines can act locally or at a distance.

Cytokines are small soluble proteins secreted by one cell that can alter the behavior or properties of the cell itself or of another cell. They are released by many cells in addition to those of the immune system. We shall discuss the cytokines released by phagocytic cells in Chapter 10, where we deal with the inflammatory reactions that play an important part in innate immunity; here we are concerned mainly with the cytokines that mediate the effector functions of T cells. Cytokines produced by lymphocytes are often called **lymphokines**, but this nomenclature can be confusing because some lymphokines are also secreted by non-lymphoid cells; we shall therefore use the generic term 'cytokine' for all of them. Most cytokines produced by T cells are given the name **interleukin** (**IL**) followed by a number: we have already encountered IL-2. Cytokines of immunological interest are listed in Appendix II.

Most cytokines can have a multitude of different biological effects when tested at high concentration in biological assays *in vitro* but, in recent years, targeted disruption of genes for cytokines and cytokine receptors in gene knock-out mice (see Section 2-27) has helped to clarify their physiological roles. The major actions of the cytokines produced by effector T cells are given in Fig. 8.31. As the effect of a cytokine varies depending on the target cell, the actions are listed according to the major target cell types: B cells, T cells, macrophages, hematopoietic cells, and tissue cells.

The main cytokine released by CD8 effector T cells is interferon-γ (IFN-γ), which can block viral replication or even lead to the elimination of virus from infected cells without killing them. T_H1 cells and T_H2 cells release different but overlapping sets of cytokines, which define their distinct actions in immunity. T_H2 cells secrete IL-4 and IL-5, which activate B cells, and IL-10, which inhibits macrophage activation. T_H1 cells secrete IFN-γ, which is the main macrophage-activating cytokine, and lymphotoxin (LT-α or TNF-β),

Cytokine	T-cell source	Effects on					Effect of gene knock-out
		B cells	T cells	Macrophages	Hematopoietic cells	Other somatic cells	
Interleukin-2 (IL-2)	T_H0, T_H1, some CTL	Stimulates growth and J-chain synthesis	Growth	–	Stimulates NK cell growth	–	↓ T-cell responses IBD
Interferon-γ (IFN-γ)	T_H1, CTL	Differentiation IgG2a synthesis	Inhibits T_H2 cell growth	Activation, ↑ MHC class I and class II	Activates NK cells	Antiviral ↑ MHC class I and class II	Susceptible to mycobacteria
Lymphotoxin (LT, TNF-β)	T_H1, some CTL	Inhibits	Kills	Activates, induces NO production	Activates neutrophils	Kills fibroblasts and tumor cells	Absence of lymph nodes. Disorganized spleen
Interleukin-4 (IL-4)	T_H2	Activation, growth IgG1, IgE ↑ MHC class II induction	Growth, survival	Inhibits macrophage activation	↑ Growth of mast cells	–	No T_H2
Interleukin-5 (IL-5)	T_H2	Differentiation IgA synthesis	–	–	↑ Eosinophil growth and differentiation	–	–
Interleukin-10 (IL-10)	T_H2	↑ MHC class II	Inhibits T_H1	Inhibits cytokine release	Co-stimulates mast cell growth	–	IBD
Interleukin-3 (IL-3)	T_H1, T_H2, some CTL	–	–	–	Growth factor for progenitor hematopoietic cells (multi-CSF)	–	–
Tumor necrosis factor-α (TNF-α)	T_H1, some T_H2, some CTL	–	–	Activates, induces NO production	–	Activates microvascular endothelium	Resistance to Gram –ve sepsis
Granulocyte-macrophage colony-stimulating factor (GM-CSF)	T_H1, some T_H2, some CTL	Differentiation	Inhibits growth	Activation. Differentiation to dendritic cells	↑ Production of granulocytes and macrophages (myelopoiesis) and dendritic cells	–	–
Transforming growth factor-β (TGF-β)	CD4 T cells	Inhibits growth IgA switch factor	–	Inhibits activation	Activates neutrophils	Inhibits/ stimulates cell growth	Death at ~10 weeks

Fig. 8.31 The nomenclature and functions of well-defined T-cell cytokines. The major actions are noted in boxes. Each cytokine has multiple activities on different cell types. The mixture of cytokines secreted by a given cell type produces many effects through what is called a 'cytokine network'. Major activities of effector cytokines are highlighted in red. ↑ = increase; ↓ = decrease; CTL = cytotoxic lymphocyte; NK = natural killer cell; CSF = colony-stimulating factor; IBD = inflammatory bowel disease.

which activates macrophages, inhibits B cells and is directly cytotoxic for some cells. The T_H0 cells from which both these functional classes derive (see Fig. 8.23) also secrete cytokines, including IL-2, IL-4, and IFN-γ, and may therefore have a distinctive effector function.

We have already discussed in Section 8-15 how the T-cell receptor can orchestrate the polarized release of these cytokines so that they are concentrated at the site of contact with the target cell. Furthermore, most of the soluble cytokines have local actions that synergize with those of the membrane-bound effector molecules; the effect of all these molecules is therefore combinatorial, and since the membrane-bound effectors can only bind to receptors on an interacting cell, this is another mechanism by which selective effects of cytokines are focused on the target cell. The effects of some cytokines are further confined to target cells by tight regulation of their synthesis: as we shall see later, the synthesis of cytokines such as IL-2, IL-4, and IFN-γ is controlled so that secretion from T cells does not continue after the interaction with a target cell ends.

Some cytokines, however, have more distant effects. IL-3 and GM-CSF (see Fig. 8.31), for example, which are released by both types of CD4 effector T cell, help to recruit effector cells in infection by acting on bone marrow cells to stimulate the production of macrophages and granulocytes, both of which are important non-specific effector cells in both humoral and cell-mediated immunity; IL-3 and GM-CSF also stimulate the production of dendritic cells from bone marrow precursors. IL-5, produced by T_H2 cells, can increase the production of eosinophils, which contribute to the late phase of allergic reactions in which there is a predominant activation of T_H2 cells (see Chapter 12). Whether a cytokine effect is local or more distant is likely to reflect the amounts released, the degree to which this release is focused on the target cell, and the stability of the cytokine *in vivo* but, for most of the cytokines, in particular those with more distant effects, these factors are not yet known.

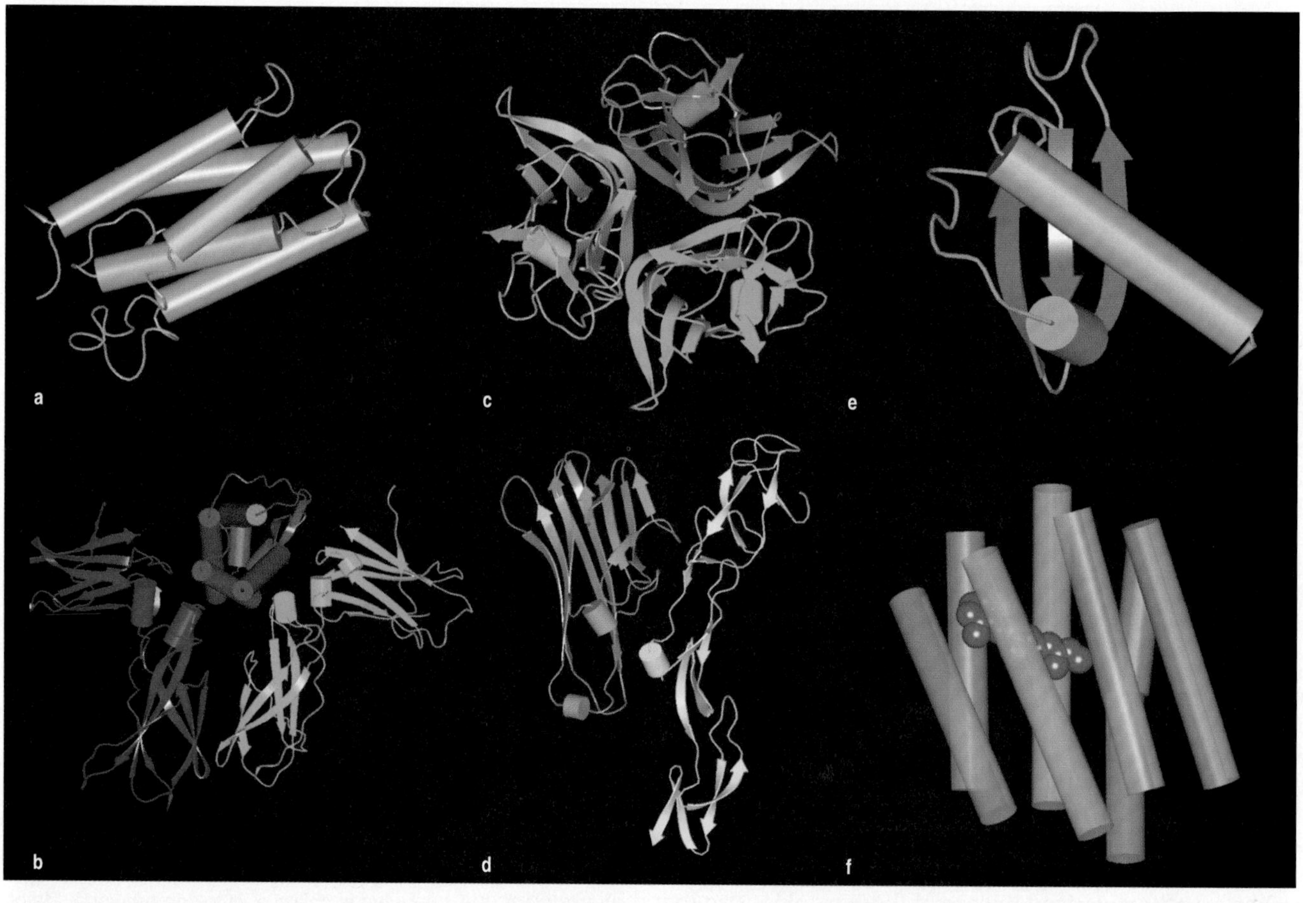

Fig. 8.32 Cytokines and their receptors can be grouped into a small number of structural families. Representatives of the main families of immunological interest except the interferons and their receptors are shown here. Cytokines are in the top row with their receptors below. The hematopoietins are represented by IL-4 (a). They are small single-chain proteins. A hypothetical model of the dimeric IL-4 receptor structure (based on the known structure of the related human growth hormone receptor) is shown in b, with bound IL-4 in red. Tumor necrosis factor (TNF) and its related molecules occur as trimers, as shown in c. The structure of one subunit of a TNF receptor binding a monomeric TNF is shown in d. The chemokines form a large family of small proteins represented here by IL-8 (e). The receptors for the chemokines are members of the large family of seven-span receptors, which also includes the photoreceptor protein rhodopsin and many other receptors. They have seven transmembrane helices, and all members of this receptor family interact with G-proteins. The only solved structure of a seven-span membrane protein is of the bacterial protein bacteriorhodopsin; it is depicted in f, showing the orientation of the seven transmembrane helices (blue) with the bound ligand (in this case retinal) in red. Cylinders represent α helices and arrows represent β strands.

Hematopoietin-receptor family	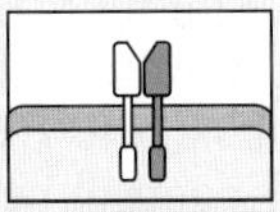	IL-2 receptor β and γ chains, IL-3, IL-4, IL-5, IL-6, IL-7, IL-9, IL-13, IL-15 and GM-CSF receptors, erythropoietin and growth hormone receptors
TNF-receptor family	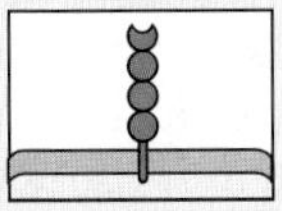	Tumor necrosis factor (TNF) receptors I and II CD40, Fas (Apo 1), CD30, CD27, nerve growth factor receptor
Chemokine-receptor family	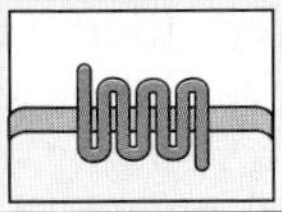	CCR1–5, CXCR1–4

Fig. 8.33 Cytokine receptors belong to families of receptor proteins, each with a distinctive structure. Some cytokine receptors are members of the hematopoietin-receptor family, some are members of the tumor necrosis factor-receptor (TNFR) family, and some are members of the chemokine-receptor family. Each family member is a variant with a distinct specificity, performing a particular function on the cell that expresses it. In the hematopoietin-receptor family, the α chain often defines the ligand specificity of the receptor, whereas the β or γ chain confers the intracellular signaling function. For the TNFR family, the ligands may be associated with the cell membrane rather than being secreted. Of the receptors listed here, some have been mentioned already in this book, some will occur in later chapters, and some are important examples from other biological systems. The diagrams indicate the representations of these receptors that you will encounter throughout this book. CCR1–5 = CC chemokine receptors; CXCR1–4 = CXC chemokine receptors.

8-18 Cytokines and their receptors fall into distinct families of structurally related proteins.

Cytokines can be grouped by structure into families: the hematopoietins, the interferons, the chemokines, and the TNF family (Fig. 8.32). Members of the TNF family act as trimers, most of which are membrane-bound and so are quite distinct in their properties from the other cytokines. Nevertheless, they share some important properties with the soluble T-cell cytokines, as they are also synthesized *de novo* upon antigen recognition by T cells, and affect the behavior of the target cell.

Cytokines act on receptors that can be grouped into equivalent families on the basis of their structure (Fig. 8.33). These families of cytokines and their receptors are also characterized by functional similarities and genetic linkage. For instance, among the hematopoietins, IL-3, IL-4, IL-5, IL-13 and GM-CSF are related structurally, their genes are closely linked in the genome, and all are major cytokines produced by T_H2 cells. In addition, they bind to closely related receptors: the IL-3, IL-4, IL-5, IL-13 and GM-CSF receptors share a common β chain. Another subgroup of hematopoietin receptors is defined by their use of the γ chain of the IL-2 receptor; this is shared by receptors for the cytokines IL-4, IL-7, IL-9, and IL-15 and is now called the γ common chain (γ_c). Overall, the structural, functional and genetic relations between the cytokines and their receptors suggest that they may have diversified together in the evolution of increasingly specialized effector functions.

These specific functional effects depend on intracellular signaling events that are triggered by the binding of cytokines to their specific receptors. The hematopoietin and interferon receptors all signal through a similar pathway, which is described in Chapter 5. The key signaling molecules of this pathway are members of the **Janus** family of cytoplasmic tyrosine kinases (JAKs) and their targets the signal transducing activators of transcription (STATs), which enter the nucleus to activate specific genes. Since the JAKs and STATs are present as families of related molecules, different members may be activated to achieve different effects.

8-19 The TNF family of cytokines are trimeric proteins that are often cell-surface associated.

TNF-α is made by T cells in soluble and membrane-associated forms, both of which are made up of three identical protein chains (a homotrimer). TNF-β (LT-α) can be produced as a secreted homotrimer, but is usually linked to the cell surface by forming heterotrimers with a third, membrane-associated, member of this family called LT-β. The receptors for these molecules, TNFR I and TNFR II, form homotrimers when bound to either TNF-α or LT. The trimeric structure is characteristic of all members of the TNF family, and the ligand-induced trimerization of their receptors seems to be the critical event in initiating signaling.

Most effector T cells express members of the TNF protein family as cell-surface molecules. The most important TNF family proteins in T-cell effector function are TNF-α and TNF-β (which can also be produced as secreted molecules), Fas ligand, and CD40 ligand, the latter two always being cell-surface associated. These molecules all bind receptors that are members of the TNFR family; TNFR I and II can each interact with either TNF-α or TNF-β, whereas Fas ligand and CD40 ligand bind respectively to the transmembrane protein Fas and to the transmembrane protein CD40, on target cells.

Fas is expressed on many cells, especially on activated lymphocytes. Activation of Fas by the Fas ligand has profound consequences for the cell as Fas contains a 'death domain' in its cytoplasmic tail; which can initiate an activation cascade of cellular proteases called **caspases**; leading to apoptotic cell death (see Fig. 5.22). Fas is important in maintaining lymphocyte homeostasis, as can be seen from the effects of mutations in the Fas or Fas ligand genes. Mice and humans with a mutant form of Fas develop a lymphoproliferative disease associated with severe autoimmunity. A mutation in the gene encoding the Fas ligand in another mouse strain creates a nearly identical phenotype. These mutant phenotypes represent the best characterized examples of generalized autoimmunity caused by single-gene defects. Other TNFR family members, including TNFR I, are also associated with death domains and can also induce programmed cell death. Thus, TNF-α and TNF-β can induce programmed cell death by binding to TNFR I.

The cytoplasmic tail of CD40 lacks a death domain; instead, it appears to be linked to proteins called **TRAFs** (TNF-receptor associated factors) about which little is known. CD40 is involved in macrophage and B-cell activation; the ligation of CD40 on B cells promotes growth and isotype switching, whereas CD40 ligation on macrophages induces them to secrete TNF-α and to become receptive to much lower concentrations of IFN-γ. Deficiency in CD40 ligand expression is associated with immunodeficiency, as we shall learn in Chapters 9 and 11.

Summary.

Interactions between armed effector T cells and their targets are initiated by transient non-specific adhesion between the cells. T-cell effector functions are elicited only when peptide:MHC complexes on the surface of the target cell are recognized by the receptor on an armed effector T cell. This recognition event triggers the armed effector T cell to adhere more strongly to the antigen-bearing target cell and to release its effector molecules directly at the target cell, leading to the activation or death of the target. The consequences of antigen recognition by an armed effector T cell are determined largely by the set of effector molecules it produces on binding a specific target cell. CD8 cytotoxic T cells store preformed cytotoxins in specialized lytic granules whose release can be tightly focused at the site of contact with the infected target cell. Cytokines, and one or more members of the TNF family of membrane-associated effector proteins, are synthesized *de novo* by all three types of effector T cell. T_H2 cells express B-cell-activating effector molecules, whereas T_H1 cells express effector molecules that activate macrophages. CD8 T cells express membrane-associated Fas ligand that induces programmed cell death; they also release IFN-γ. Membrane-associated effector molecules can deliver signals only to an interacting cell, whereas soluble cytokines can act on cytokine receptors expressed locally on the target cell, or on hematopoietic cells at a distance. The actions of cytokines and membrane-associated effector molecules through their specific receptors, together with the effects of cytotoxins released by CD8 cells, account for most of the effector functions of T cells.

T-cell mediated cytotoxicity.

All viruses, and some bacteria, multiply in the cytoplasm of infected cells; indeed, the virus is a highly sophisticated parasite that has no biosynthetic or metabolic apparatus of its own and, in consequence, can replicate only inside cells. Once inside cells, these pathogens are not accessible to antibodies and can be eliminated only by the destruction or modification of the infected cells on which they depend. This role in host defense is fulfilled by cytotoxic CD8 T cells. The critical role of cytotoxic T cells in limiting such infections is seen in the increased susceptibility of animals artificially depleted of these T cells, or of mice or humans that lack the MHC class I molecules that present antigen to CD8 T cells. As well as controlling infection by viruses and cytoplasmic bacteria, CD8 T cells are important in controlling some protozoan infections and are crucial, for example, in host defense against the protozoan *Toxoplasma gondii*, a vesicular parasite that exports peptides from the infected vesicles to the cytosol, from which they enter the MHC class I processing pathway. The elimination of infected cells without the destruction of healthy tissue requires the cytotoxic mechanisms of CD8 T cells to be both powerful and accurately targeted.

8-20 Cytotoxic T cells can induce target cells to undergo programmed cell death.

Cells can die in either of two ways. Physical or chemical injury, such as the deprivation of oxygen that occurs in heart muscle during a heart attack, or membrane damage with antibody and complement, leads to cell disintegration or **necrosis**. The dead or necrotic tissue is taken up and degraded by phagocytic cells, which eventually clear the damaged tissue and heal the wound. The other form of cell death is known as **programmed cell death** or **apoptosis**. Apoptosis is a normal cellular response that is crucial in the tissue remodeling that occurs during development and metamorphosis in all multicellular animals. As we saw in Chapter 7, most thymocytes die an apoptotic death when they fail positive selection or are negatively selected as a result of recognizing self antigens. Early changes seen in apoptotic cell death are nuclear blebbing, alteration in cell morphology and, eventually, fragmentation of the DNA. The cell then destroys itself from within, shrinking by shedding membrane-bound vesicles, and degrading itself until little is left. A hallmark of this type of cell death is the fragmentation of nuclear DNA into 200 base-pair (bp) pieces through the activation of endogenous nucleases that cleave the DNA between nucleosomes, each of which contains about 200 bp of DNA.

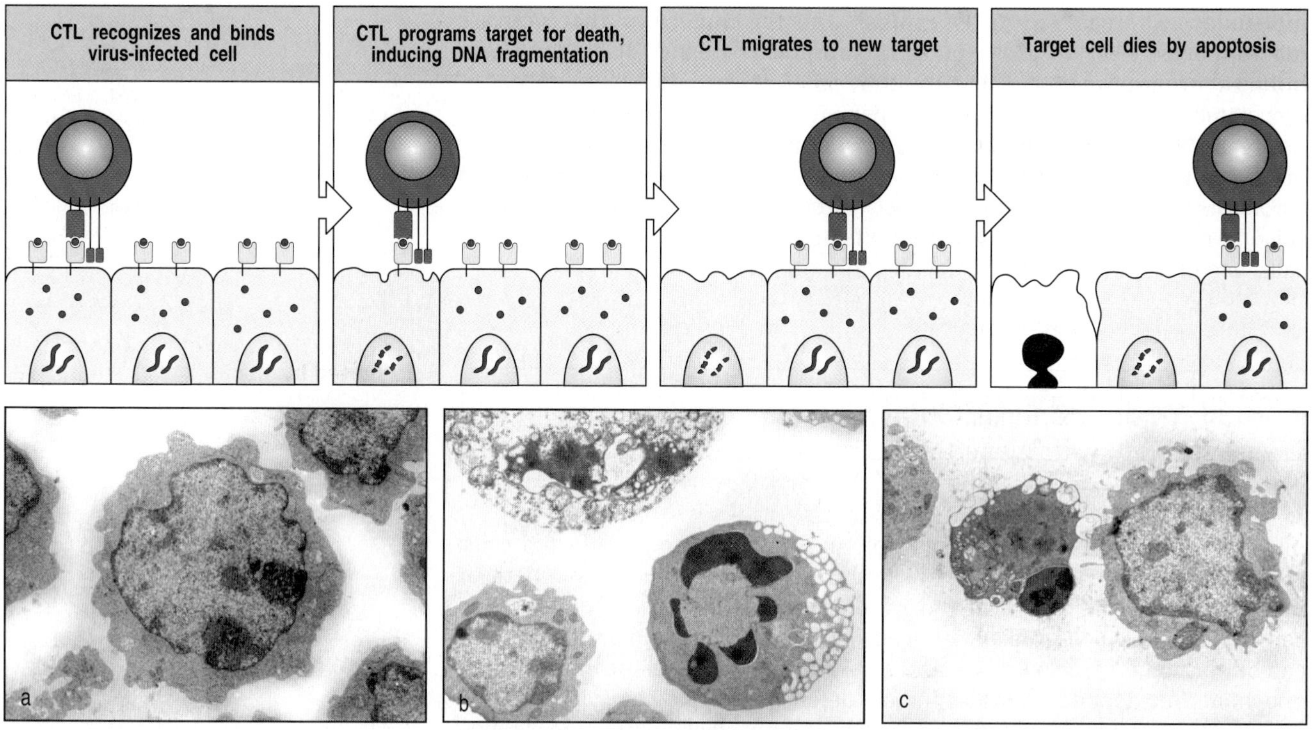
CTL recognizes and binds virus-infected cell
CTL programs target for death, inducing DNA fragmentation
CTL migrates to new target
Target cell dies by apoptosis
a
b
c

Fig. 8.34 Cytotoxic CD8 T cells can induce apoptosis in target cells. Specific recognition of peptide:MHC complexes on a target cell (top panels) by a cytotoxic CD8 T cell (CTL) leads to the death of the target cell by apoptosis. Cytotoxic T cells can recycle to kill multiple targets. Each killing requires the same series of steps, including receptor binding and directed release of cytotoxic proteins stored in lytic granules. The process of apoptosis is shown in the micrographs (bottom panels), where panel a shows a healthy cell with a normal nucleus. Early in apoptosis (panel b) the chromatin becomes condensed (red) and, although the cell sheds membrane vesicles, the integrity of the cell membrane is retained, in contrast to the necrotic cell in the upper part of the same field. In late stages of apoptosis (panel c), the cell nucleus (middle cell) is very condensed, no mitochondria are visible and the cell has lost much of its cytoplasm and membrane through the shedding of vesicles. Photographs (× 3500) courtesy of R Windsor and E Hirst.

Cytotoxic T cells kill their targets by programming them to undergo apoptosis. When cytotoxic T cells are mixed with target cells and rapidly brought into contact by centrifugation, they can program antigen-specific target cells to die within 5 minutes, although death may take hours to become fully evident. The short period required by cytotoxic T cells to program their targets to die reflects the release of preformed effector molecules, which activate an endogenous apoptotic pathway within the target cell (Fig. 8.34).

As well as killing the host cell, the apoptotic mechanism may also act directly on cytosolic pathogens. For example, the nucleases that are activated in apoptosis to destroy cellular DNA can also degrade viral DNA. This prevents the assembly of virions and thus the release of infectious virus, which could otherwise infect nearby cells. Other enzymes activated in the course of apoptosis may destroy non-viral cytosolic pathogens. Apoptosis is therefore preferable to necrosis as a means of killing infected cells; in necrosis, intact pathogens are released from the dead cell and these can continue to infect healthy cells, or can parasitize the macrophages that ingest them.

8-21 Cytotoxic effector proteins that trigger apoptosis are contained in the granules of CD8 cytotoxic T cells.

The principal mechanism through which cytotoxic T cells act is by the calcium-dependent release of specialized **lytic granules** upon recognition of antigen on the surface of a target cell. These granules are modified lysosomes that contain at least two distinct classes of cytotoxic effector protein that are expressed selectively in cytotoxic T cells (Fig. 8.35). Such proteins are stored in the lytic granules in an active form, but conditions within the granules prevent them from functioning until after their release. One of these cytotoxic proteins, known as **perforin**, polymerizes to generate transmembrane pores in target cell membranes. The other class of cytotoxic proteins comprises at least three proteases called **granzymes**, which belong to the same family of enzymes (the serine proteases) as the digestive enzymes trypsin and chymotrypsin. Granules that store perforin and granzymes can be seen in armed CD8 cytotoxic effector cells in tissue lesions.

When purified granules from cytotoxic T cells are added to target cells *in vitro*, they lyse the cells by creating pores in the lipid bilayer. The pores consist of polymers of perforin, which is a major constituent of these granules. On release from the granule, perforin forms a cylindrical structure that is lipophilic on the outside and hydrophilic down a hollow center with an inner diameter of 16 nm. It is not known whether this structure is first formed and then inserted into the lipid bilayer of the target cell membrane, or whether it is formed in the bilayer itself. The pore that is formed allows water and salts to pass rapidly into the cell (Fig. 8.36). With the integrity of the cell membrane destroyed, the cells die rapidly. Large numbers of purified granules can kill target cells *in vitro* without inducing fragmentation of cellular DNA, but this lytic mechanism of cell killing probably occurs only at artificially high levels of perforin that do not reflect the physiological activity of cytotoxic T cells.

Both perforin and granzymes are required for effective cell killing. The separate roles of perforin and granzymes have been investigated in a cell system that relies upon similarities between the lytic granules of T cells and the granules of mast cells. Release of mast cell granules occurs on crosslinking of the Fcε receptor (see Chapter 9), just as release of lytic granules from CD8 T cells occurs on crosslinking of the T-cell receptor, and by a similar mechanism. Both the Fcε receptor and the T-cell receptor have ITAM motifs in their cytoplasmic domains, and crosslinking leads to tyrosine phosphorylation of the ITAMs (see Chapter 5).

Protein in lytic granules of cytotoxic T cells	Actions on target cells
Perforin	Polymerizes to form a pore in target membrane
Granzymes	Serine proteases, which activate apoptosis once in the cytoplasm of the target cell

Fig. 8.35 Cytotoxic effector proteins released by cytotoxic T cells.

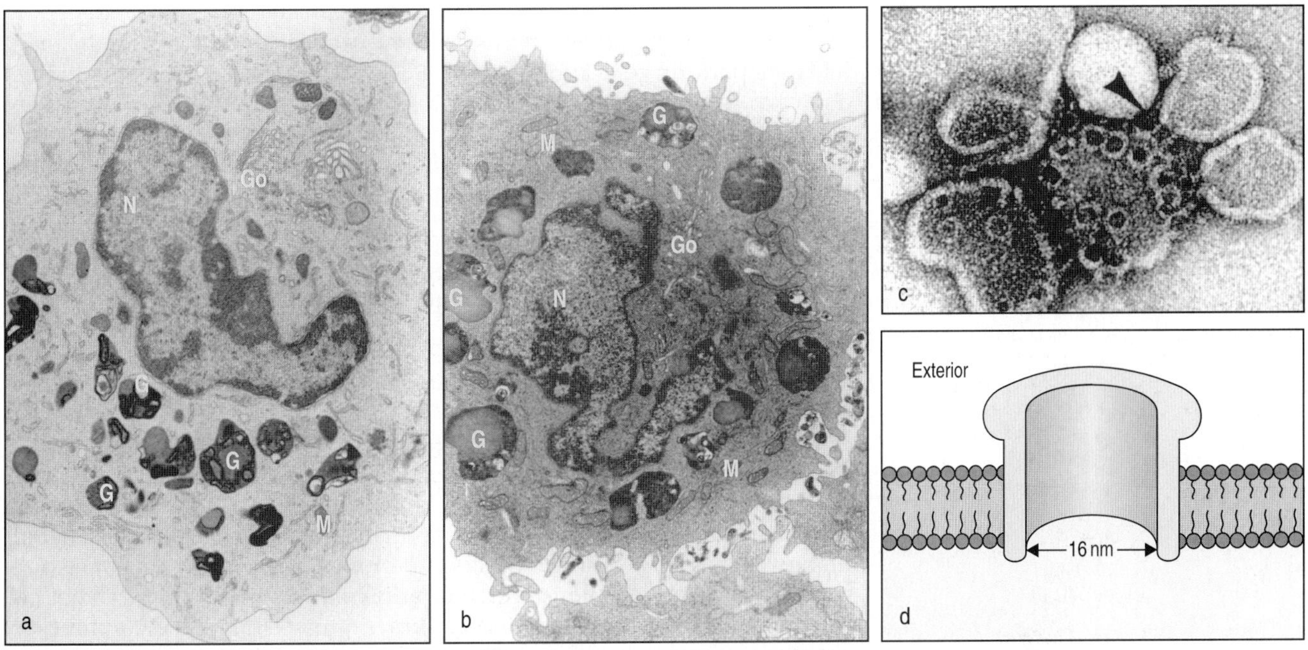

Fig. 8.36 Perforin released from the lytic granules of cytotoxic T cells can insert into the target cell membrane to form pores. Perforin molecules, as well as several other effector molecules, are contained in the granules of cytotoxic T cells (panel a). When a CD8 cytotoxic T cell recognizes its target, the granule contents are released onto the target cell (panel b, bottom right quadrant). The perforin molecules released from the granules polymerize in the membrane of the target cell to form pores. The structure of these pores is best seen when purified perforin is added to synthetic lipid vesicles (panel c: pores are seen both end on, as circles, and sideways on, arrow). The pores span the target cell membrane (panel d). G = granule; N = nucleus; M = mitochondrion; Go = Golgi apparatus. Photographs courtesy of E Podack.

When a mast cell line is transfected with the gene for perforin or for granzyme, the gene products are stored in mast cell granules. When the cell is activated through its Fcε receptor, these granules are released. When transfected with the gene for perforin alone, mast cells can kill other cells, but not very efficiently, because large numbers of the transfected cells are needed. Mast cells transfected with the gene for granzyme B alone are unable to kill other cells. However, when perforin-transfected mast cells are also transfected with the gene encoding granzyme B, the cells or their purified granules become as effective at killing targets as granules from cytotoxic cells, and granules from both types of cell induce DNA fragmentation. This suggests that perforin makes pores through which the granzymes can move into the target cell.

The granzymes are proteases, so although they have a role in triggering apoptosis in the target cell, they cannot act directly to fragment the DNA. Rather, they must activate an enzyme, or more probably an enzyme cascade, in the target cell. Granzyme B can cleave the ubiquitous cellular enzyme CPP-32, which is believed to have a key role in the programmed cell death of all cells. CPP-32 is a member of the caspase family of proteases, and activates a nuclease, called caspase-activated deoxyribonuclese or CAD, by cleaving an inhibitory protein (ICAD) that binds to and inactivates CAD. This enzyme is believed to be the final effector of DNA degradation in apoptosis.

Cells undergoing programmed cell death are rapidly ingested by phagocytic cells in their vicinity. The phagocytes recognize some change in the cell membrane, most probably the exposure of phosphatidylserine, which is normally found only in the inner leaflet of the membrane. The ingested cell is then completely broken down and digested by the phagocyte without the induction of co-stimulatory proteins. Thus, apoptosis is normally an immunologically 'quiet' process; that is, apoptotic cells do not normally contribute to or stimulate immune responses.

The importance of perforin in this process is well illustrated in mice that have had their perforin gene knocked out. Such mice are severely defective in their ability to mount a cytotoxic T-cell response to many but not all viruses, whereas mice that are defective in the granzyme B gene have a less profound defect, probably because there are several genes coding for granzymes.

8-22 Activated CD8 T cells and some CD4 effector T cells express Fas ligand, which can also activate apoptosis.

The release of granule contents accounts for most of the cytotoxic activity of CD8 effector T cells, as shown by the loss of most killing activity in perforin gene knock-out mice. This granule-mediated killing is strictly calcium-dependent, yet some cytotoxic actions of CD8 T cells survive calcium depletion. Moreover, some CD4 T cells are also capable of killing other cells, yet do not contain granules and make neither perforin nor granzymes. These observations implied that there must be a second perforin-independent mechanism of cytotoxicity. This mechanism involves the binding of Fas in the target cell membrane by the Fas ligand, which is expressed in the membranes of activated cytotoxic T cells and T_H1 cells. Activation of Fas leads to apoptosis in the target cell. As discussed in Section 8-19, the lymphoproliferative and autoimmune disorders seen in mice and humans with mutations in genes for either Fas or the Fas ligand imply that this pathway of killing is very important in regulating peripheral immune responses, and such Fas–Fas ligand interactions are important in terminating lymphocyte growth after the removal of the initiating pathogen.

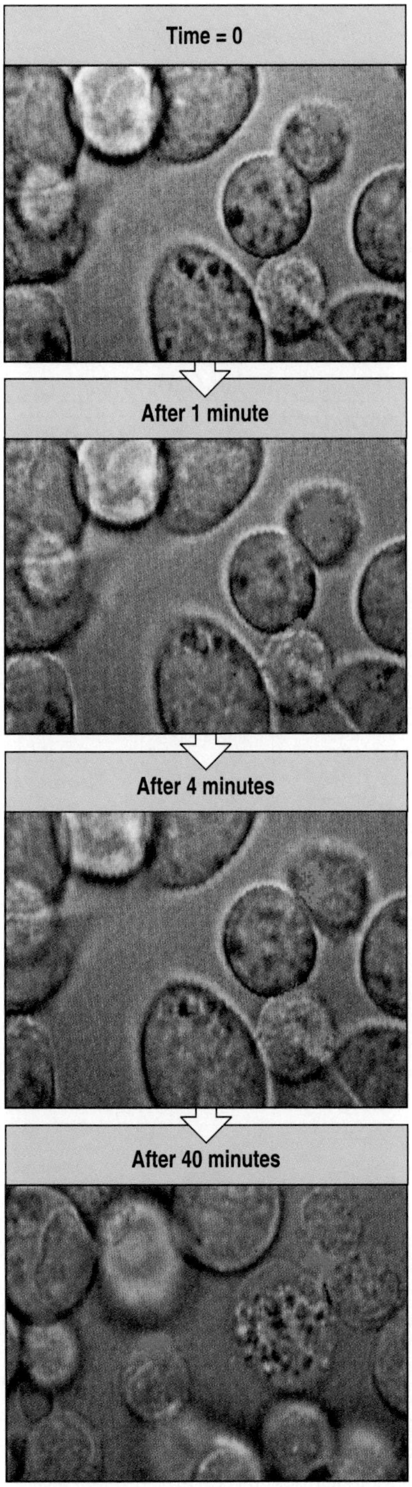

Fig. 8.37 Effector molecules are released from T-cell granules in a highly polar fashion. The granules of cytotoxic T cells can be labeled with fluorescent dyes, allowing them to be seen under the microscope, and their movements followed by time-lapse photography. Here we show a series of pictures taken during the interaction of a cytotoxic T cell with a target cell, which is eventually killed. In the top panel, at time 0, the T cell (upper right) has just made contact with a target cell (diagonally below). At this time, the granules of the T cell, labeled with a red fluorescent dye, are distant from the point of contact. In the second panel, after 1 minute has elapsed, the granules have begun to move towards the target cell, a move that has essentially been completed in the third panel, after 4 minutes. After 40 minutes, in the last panel, the granule contents have been released into the space between the T cell and the target, which has begun to undergo apoptosis (note the fragmented nucleus). The T cell will now disengage from the target cell and can recognize and kill other targets. Photographs courtesy of G Griffiths.

8-23 Cytotoxic T cells are selective and serial killers of targets expressing specific antigen.

When cytotoxic T cells are offered a mixture of equal amounts of two target cells, one bearing specific antigen and the other not, they kill only the target cell bearing the specific antigen. The 'innocent bystander' cells and the cytotoxic T cells themselves are not killed, despite the fact that cloned cytotoxic T cells can be recognized and killed by other cytotoxic T cells just like any tissue cell. At first sight, this may seem surprising, because the effector molecules released by cytotoxic T cells lack any specificity for antigen. The explanation probably lies in the highly polar release of the effector molecules. As we have already seen in Fig. 8.28, cytotoxic T cells orient their Golgi apparatus and microtubule-organizing center to focus secretion on the point of contact with a target cell. Granule movement towards the point of contact is shown in Fig. 8.37. Cytotoxic T cells attached to several different target cells reorient their secretory apparatus towards each cell in turn and kill them one by one, strongly suggesting that the mechanism whereby cytotoxic mediators are released allows attack at only one point of contact at any one time. The narrowly focused action of cytotoxic CD8 T cells allows them to kill single infected cells in a tissue without creating widespread tissue damage (Fig. 8.38) and is of critical importance in tissues where cell regeneration does not occur, as in neurons of the central nervous system, or is very limited, as in the pancreatic islets.

Cytotoxic T cells can kill their targets rapidly because they store preformed cytotoxic proteins in forms that are inactive in the environment of the lytic granule. Cytotoxic proteins are synthesized and loaded into the lytic granules during the first encounter of a naive cytotoxic precursor T cell with its specific antigen. Ligation of the T-cell receptor similarly induces *de novo* synthesis of perforin and granzymes in armed effector CD8 T cells, so that the supply of lytic granules is replenished. This makes it possible for a single CD8 T cell to kill many targets in succession.

8-24 Cytotoxic T cells also act by releasing cytokines.

Although the secretion of perforin and granzymes is the main way by which cytotoxic CD8 T cells eliminate infection, with the expression of Fas ligand playing a lesser role, most cytotoxic CD8 T cells also release the cytokines IFN-γ, TNF-α, and TNF-β, which contribute to host defense in several other ways. IFN-γ directly inhibits viral replication, and also induces the increased

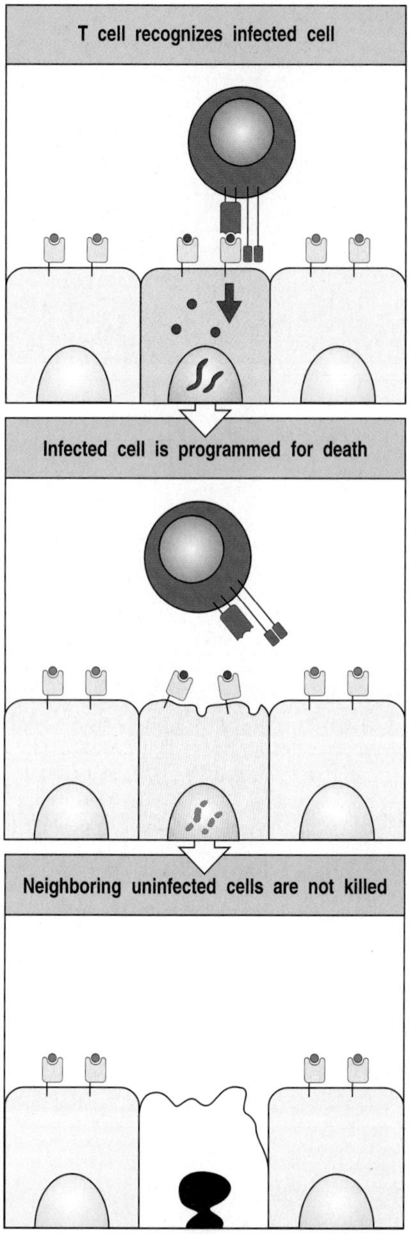
T cell recognizes infected cell
Infected cell is programmed for death
Neighboring uninfected cells are not killed

Fig. 8.38 Cytotoxic T cells kill target cells bearing specific antigen while sparing neighboring uninfected cells. All the cells in a tissue are susceptible to lysis by the cytotoxic proteins of armed effector CD8 T cells, but only infected cells are killed. Specific recognition by the T-cell receptor identifies which target cell to kill, and the polarized release of granules (not shown) ensures that neighboring cells are spared.

expression of MHC class I and other molecules involved in peptide loading of the newly synthesized MHC class I proteins in infected cells. This increases the chance that infected cells will be recognized as target cells for cytotoxic attack. IFN-γ also activates macrophages, recruiting them to sites of infection both as effector cells and as antigen-presenting cells. The activation of macrophages by IFN-γ is a critical component of the host immune response to intracellular protozoan pathogens such as *Toxoplasma gondii.* IFN-γ also has a secondary role in decreasing the tryptophan concentration within responsive cells and thus can kill intracellular parasites, effectively by starvation. TNF-α or TNF-β can synergize with IFN-γ in macrophage activation, and in killing some target cells through their interaction with TNFR I. Thus, armed cytotoxic CD8 effector T cells act in a variety of ways to limit the spread of cytosolic pathogens. The relative importance of each of these mechanisms remains to be determined.

Summary.

Armed CD8 cytotoxic effector T cells are essential in host defense against pathogens that live in the cytosol, the commonest of which are viruses. These cytotoxic T cells can kill any cell harboring such pathogens by recognizing foreign peptides that are transported to the cell surface bound to MHC class I molecules. CD8 cytotoxic T cells carry out their killing function by releasing two types of preformed cytotoxic protein: the granzymes, which seem able to induce apoptosis in any type of target cell, and the pore-forming protein perforin, which punches holes in the target-cell membrane through which the granzymes can enter. These properties allow the cytotoxic T cell to attack and destroy virtually any cell that is infected with a cytosolic pathogen. A membrane-bound molecule, the Fas ligand, expressed by CD8 and some CD4 T cells, is also capable of inducing apoptosis by binding to Fas expressed by some target cells. Cytotoxic CD8 T cells also produce IFN-γ, which is an inhibitor of viral replication and is an important inducer of MHC class I expression and macrophage activation. Cytotoxic T cells kill infected targets with great precision, sparing adjacent normal cells. This precision is critical in minimizing tissue damage while allowing the eradication of infected cells.

Macrophage activation by armed CD4 T_H1 cells.

Some microorganisms such as mycobacteria, the causative agents of tuberculosis and leprosy, are intracellular pathogens that grow primarily in phagolysosomes of macrophages. There they are shielded from the effects of both antibodies and cytotoxic T cells. These microbes maintain themselves in the usually hostile environment of the phagocyte by inhibiting the fusion of lysosomes to the phagolysosomes in which they grow, or by preventing the acidification of these vesicles that is required to activate lysosomal proteases. Such microorganisms can be eliminated when the macrophage is activated by a T_H1 cell. Armed T_H1 cells act by synthesizing membrane-associated proteins and a range of soluble cytokines whose local and distant actions coordinate the immune response to these intracellular pathogens. Armed T_H1 effector cells can also activate macrophages to kill recently ingested pathogens.

8-25 Armed T_H1 cells have a central role in macrophage activation.

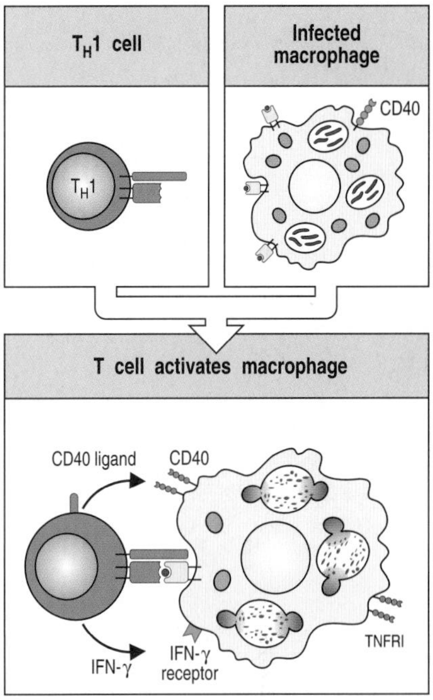

Fig. 8.39 T_H1 cells activate macrophages to become highly microbicidal. When a T_H1 cell specific for a bacterial peptide contacts an infected macrophage, the T cell is induced to secrete the macrophage-activating factor IFN-γ and to express CD40 ligand. Together, these newly synthesized T_H1 proteins activate the macrophage.

A number of important pathogens live within macrophages, whereas many others are ingested by macrophages from the extracellular fluid. In many cases the macrophage is able to destroy such pathogens without the need for T-cell activation, as we shall see in Chapter 10, but in several clinically important infections CD4 T cells are needed to provide activating signals for macrophages. The induction of antimicrobial mechanisms in macrophages is known as **macrophage activation** and is the principal effector action of T_H1 cells. Among the extracellular pathogens that are killed when macrophages are activated is *Pneumocystis carinii,* which, because of a deficiency of CD4 T cells, is a common cause of death in people with AIDS. Macrophage activation can be measured by the ability of activated macrophages to damage a broad spectrum of microbes as well as certain tumor cells. This ability to act on extracellular targets extends to healthy self cells, which means that macrophages must normally be maintained in a non-activated state.

Macrophages require two signals for activation. One of these is provided by IFN-γ; the other can be provided by a variety of means, and is required to sensitize the macrophage to respond to IFN-γ. Armed T_H1 cells can deliver both signals. IFN-γ is the most characteristic cytokine produced by armed T_H1 cells on interacting with their specific target cells, whereas the CD40 ligand expressed by the T_H1 cell delivers the sensitizing signal by contacting CD40 on the macrophage (Fig. 8.39). CD8 T cells are also an important source of IFN-γ and can activate macrophages presenting antigens derived from cytosolic proteins; mice lacking MHC class I molecules (and that therefore have no CD8 T cells) show increased susceptibility to some parasitic infections. Macrophages can be made more sensitive to IFN-γ by very small amounts of bacterial lipopolysaccharide, and this latter pathway may be particularly important when CD8 T cells are the primary source of the IFN-γ. It is also possible that membrane-associated TNF-α or TNF-β can substitute for CD40 ligand in macrophage activation. These cell-associated molecules apparently stimulate the macrophage to secrete TNF-α, and antibody to TNF-α can inhibit macrophage activation. T_H2 cells are inefficient macrophage activators because they produce IL-10, a cytokine that can deactivate macrophages, and they do not produce IFN-γ. However, they do express CD40 ligand and can deliver the contact-dependent signal required to activate macrophages to respond to IFN-γ.

8-26 The expression of cytokines and membrane-associated molecules by armed CD4 T_H1 cells requires new RNA and protein synthesis.

Within minutes of the recognition of specific antigen by armed cytotoxic effector CD8 T cells, directed exocytosis of preformed perforins and granzymes programs the target cell to die via apoptosis. In contrast, when armed T_H1 cells encounter their specific ligand, they must synthesize *de novo* the cytokines and cell-surface molecules that mediate their effects. This process requires hours rather than minutes, so T_H1 cells must adhere to their target cells for far longer than cytotoxic T cells.

Recognition of its target by a T_H1 cell rapidly induces transcription of cytokine genes and new protein synthesis begins within 1 hour of receptor triggering. The newly synthesized cytokines are then delivered directly through microvesicles of the constitutive secretory pathway to the site of contact between the

T-cell membrane and the macrophage. It is thought that the newly synthesized cell-surface CD40 ligand is also expressed in this polarized fashion. This means that, although all macrophages have receptors for IFN-γ, the macrophage actually displaying antigen to the armed T_H1 cell is far more likely to become activated by it than are neighboring uninfected macrophages.

8-27 Activation of macrophages by armed T_H1 cells promotes microbial killing and must be tightly regulated to avoid damage to host tissues.

T_H1 cells activate infected macrophages through cell contact and the focal secretion of IFN-γ. This generates a series of biochemical responses that converts the macrophage into a potent antimicrobial effector cell (Fig. 8.40). Activated macrophages fuse their lysosomes more efficiently to phagosomes, exposing intracellular or recently ingested extracellular microbes to a variety of microbicidal lysosomal enzymes. Activated macrophages also make oxygen radicals and nitric oxide (NO), both of which have potent antimicrobial activity, as well as synthesizing antimicrobial peptides and proteases that can be released to attack extracellular parasites.

Additional changes in the activated macrophage help to amplify the immune response. The number of MHC class II molecules, B7 molecules, and CD40 and TNF receptors on the macrophage surface increases, making the cell both more effective at presenting antigen to fresh T cells, which may thereby be recruited as effector cells, and more responsive to CD40 ligand and to TNF-α. TNF-α synergizes with IFN-γ in macrophage activation, particularly in the induction of the reactive nitrogen metabolite NO, which has broad antimicrobial activity. The NO is produced by the enzyme **inducible NO synthase** (**iNOS**), and mice that have had the gene for iNOS knocked out are highly susceptible to infection with several intracellular pathogens. Activated macrophages secrete IL-12, which directs the differentiation of activated naive CD4 T cells into T_H1 effector cells, as we shall learn in Chapter 10. These and many other surface and secreted molecules of activated macrophages are instrumental in the effector actions of macrophages in cell-mediated responses, and they are also important effectors in humoral immune responses, which we shall discuss in Chapter 9, and in recruiting other immune cells to sites of infection, a function to which we shall return in Chapter 10.

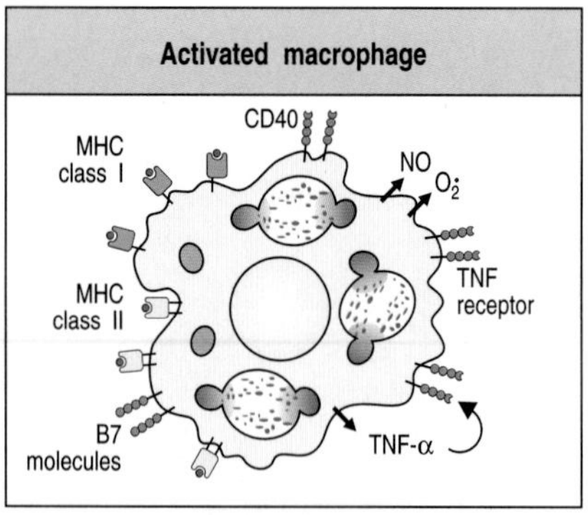

Fig. 8.40 Activated macrophages undergo changes that greatly increase their antimicrobial effectiveness and amplify the immune response. Activated macrophages increase their expression of CD40 and of TNF receptors, and secrete TNF-α. This autocrine stimulus synergizes with IFN-γ secreted by T_H1 cells to increase the antimicrobial action of the macrophage, in particular by inducing the production of nitric oxide (NO) and oxygen radicals ($O_2\cdot$). The macrophage also upregulates its B7 proteins in response to binding the CD40 ligand on the T cell, and increases its expression of class II MHC molecules, thus allowing further activation of resting CD4 T cells.

Because activated macrophages are extremely effective in destroying pathogens, one may ask why macrophages are not simply maintained in a state of constant activation. Besides the fact that macrophages consume large quantities of energy to maintain the activated state, macrophage activation *in vivo* is usually associated with localized tissue destruction that apparently results from the release of antimicrobial mediators such as oxygen radicals, NO, and proteases, which are also toxic to host cells. The ability of activated macrophages to release toxic mediators is important in host defense because it enables them to attack large extracellular pathogens that they cannot ingest, such as parasitic worms. This can only be achieved, however, at the expense of tissue damage. Tight regulation of the activity of macrophages by T_H1 cells thus allows the specific and effective deployment of this potent means of host defense while minimizing local tissue damage and energy consumption.

Controlling macrophage activation so as to avoid undesirable effects on normal tissues is contained by mechanisms that control IFN-γ synthesis. This seems to be achieved in two ways. First, the mRNA encoding IFN-γ, like that encoding a variety of other cytokines, contains a sequence $(AUUUA)_n$ in its 3′ untranslated region that greatly reduces its half-life, and this serves to limit the period of cytokine production. Second, activation of the T cell appears to induce the production of a new protein that promotes cytokine mRNA degradation: treatment of activated effector T cells with the protein synthesis inhibitor cycloheximide greatly increases the level of cytokine mRNA. The rapid destruction of cytokine mRNA, together with the focal delivery of IFN-γ at the point of contact between the activated T_H1 cell and its macrophage target, thus limits the action of the effector T cell to the infected macrophage. We shall see in Chapter 9, when we consider the activation of B cells by T_H2 cells, that the same mechanisms direct and limit T-cell help to the specific antigen-binding B cell. In addition, macrophage activation itself is markedly inhibited by cytokines such as transforming growth factor-β (TGF-β), IL-4, IL-10, and IL-13. Because several of these inhibitory cytokines are produced by T_H2 cells, the induction of CD4 T cells belonging to the T_H2 subset represents an important pathway for controlling the effector functions of activated macrophages.

8-28 T_H1 cells coordinate the host response to intracellular pathogens.

The activation of macrophages by IFN-γ secreted by armed T_H1 cells expressing CD40 ligand is central to the host response to pathogens that proliferate in macrophage vesicles. In mice in which the IFN-γ gene or the CD40 ligand gene has been destroyed by targeted gene disruption, production of antimicrobial agents by macrophages is impaired, and the animals succumb to sublethal doses of *Mycobacterium* spp., *Leishmania* spp., and vaccinia virus. Mice lacking a TNF receptor also show increased susceptibility to these pathogens. However, although IFN-γ and CD40 ligand are probably the most important effector molecules synthesized by T_H1 cells, the immune response to pathogens that proliferate in macrophage vesicles is complex, and other cytokines secreted by T_H1 cells have a crucial role in coordinating these responses (Fig. 8.41). For example, macrophages that are chronically infected with intracellular bacteria may lose the ability to become activated. Such cells could provide a reservoir of infection that is shielded from immune attack. Activated T_H1 cells can express Fas ligand and thus kill a limited range of target cells that express Fas, including macrophages, thereby destroying these infected cells.

Whereas some intravesicular bacteria pose a hazard by incapacitating chronically infected macrophages, others, including some mycobacteria and *Listeria monocytogenes,* can escape from cell vesicles and enter the cytoplasm, where they are not susceptible to macrophage activation. Their presence can, however, be detected by CD8 cytotoxic T cells, which can release them by killing the cell. The pathogens released when macrophages are killed either by T_H1 cells or by CD8 cytotoxic T cells can be taken up by freshly recruited macrophages still capable of activation to antimicrobial activity.

Another very important function of T_H1 cells is the recruitment of phagocytic cells to sites of infection. T_H1 cells recruit macrophages by two mechanisms. First, they make the hematopoietic growth factors IL-3 and GM-CSF, which stimulate the production of new phagocytic cells in the bone marrow. Second, TNF-α and TNF-β, which are secreted by T_H1 cells at sites of infection, change the surface properties of endothelial cells so that phagocytes adhere to them, while chemokines such as macrophage chemotactic factor (MCF), produced by T_H1 cells in the inflammatory response, serve to direct the migration of these phagocytic cells through the vascular endothelium to the site of the infection (see Chapter 10).

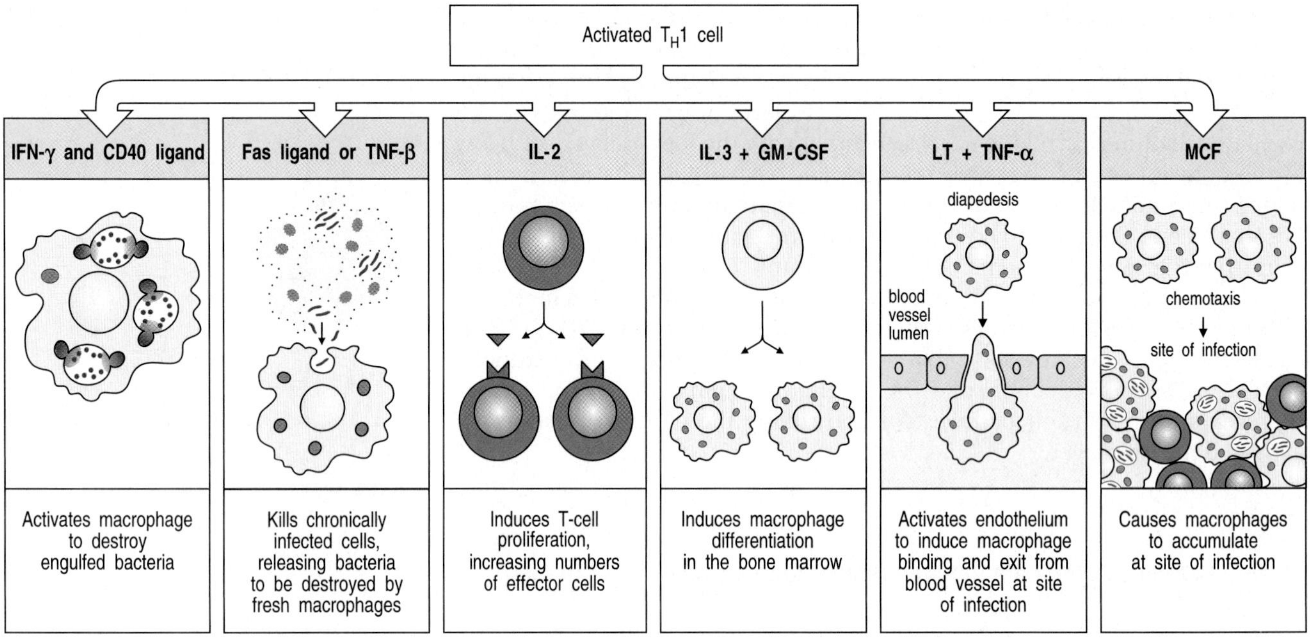

Fig. 8.41 The immune response to intracellular bacteria is coordinated by activated T_H1 cells. The activation of T_H1 cells by infected macrophages results in the synthesis of cytokines that both activate the macrophage and coordinate the immune response to intracellular pathogens. IFN-γ and CD40 ligand synergize in activating the macrophage, which allows it to kill engulfed pathogens. Chronically infected macrophages lose the ability to kill intracellular bacteria, and Fas ligand or TNF-β produced by the T_H1 cell can kill these macrophages, releasing the engulfed bacteria, which are taken up and killed by fresh macrophages. In this way, IFN-γ and TNF-β synergize in the removal of intracellular bacteria. IL-2 produced by T_H1 cells induces T-cell proliferation and potentiates the release of other cytokines. IL-3 and GM-CSF stimulate the production of new macrophages by acting on hematopoietic stem cells in the bone marrow. New macrophages are recruited to the site of infection by the action of TNF-β and TNF-α (and other cytokines) on vascular endothelium, which signal macrophages to leave the bloodstream and enter the tissues. A chemokine with macrophage chemotactic activity (MCF) signals macrophages to migrate into sites of infection and accumulate there. Thus, the T_H1 cell coordinates a macrophage response that is highly effective in destroying intracellular infectious agents.

When microbes effectively resist the microbicidal effects of activated macrophages, chronic infection with inflammation can develop. Often, this has a characteristic pattern, consisting of a central area of macrophages surrounded by activated lymphocytes. This pathological pattern is called a **granuloma** (Fig. 8.42). Giant cells consisting of fused macrophages usually form the center of these granulomas. This serves to 'wall-off' pathogens that resist destruction. T_H2 cells seem to participate in granulomas along with T_H1 cells, perhaps by regulating their activity and preventing widespread tissue damage. In tuberculosis, the center of the large granulomas can become isolated and the cells there die, probably from a combination of lack of oxygen and the cytotoxic effects of activated macrophages. As the dead tissue in the center resembles cheese, this process is called **caseation necrosis**. Thus, the activation of T_H1 cells can cause significant pathology. Their non-activation, however, leads to the more serious consequence of death from disseminated infection, which is now seen frequently in patients with AIDS and concomitant mycobacterial infection.

Summary.

CD4 T cells that can activate macrophages have a critical role in host defense against those intracellular and engulfed extracellular pathogens that resist killing in non-activated macrophages. Macrophages are activated by membrane-bound signals delivered by activated T_H1 cells as well as by the potent macrophage-activating cytokine IFN-γ secreted by activated T cells. Once activated, the macrophage can kill intracellular and ingested bacteria. Activated macrophages can also cause local tissue damage, which explains why this activity must be strictly regulated by T cells. T_H1 cells produce a range of cytokines and surface molecules that not only activate infected macrophages but can also kill chronically infected senescent macrophages, stimulate the production of new macrophages in bone marrow, and recruit fresh macrophages to sites of infection. Thus, T_H1 cells have a central role in controlling and coordinating host defense against certain intracellular pathogens. It is likely that the absence of this function explains the preponderance of infections with intracellular pathogens in adult AIDS patients.

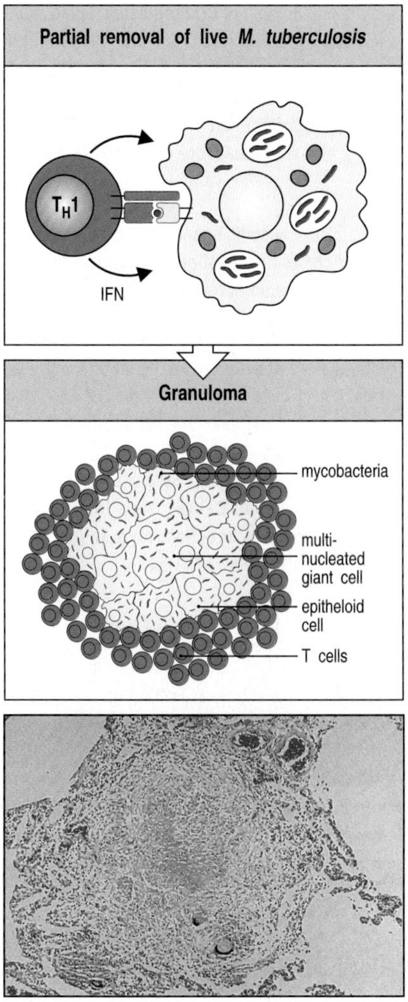

Fig. 8.42 Granulomas form when an intracellular pathogen or its constituents cannot be totally eliminated. When mycobacteria (red) resist the effects of macrophage activation, a characteristic localized inflammatory response called a granuloma develops. This consists of a central core of infected macrophages. The core may include multinucleated giant cells, which are fused macrophages, surrounded by large macrophages often called epithelioid cells. Mycobacteria can persist in the cells of the granuloma. The central core is surrounded by T cells, many of which are CD4-positive. The exact mechanisms by which this balance is achieved, and how it breaks down, are unknown. Granulomas, as seen in the bottom panel, also form in the lungs and elsewhere in a disease known as sarcoidosis, which may be caused by occult mycobacterial infection. Photograph courtesy of J Orrell.

Summary to Chapter 8.

Armed effector T cells are crucial to almost all adaptive immune responses. Adaptive immune responses are initiated when naive T cells encounter specific antigen on the surface of a professional antigen-presenting cell that also expresses the co-stimulatory molecules B7.1 and B7.2. The activated T cells produce IL-2, which drives them to proliferate and differentiate into armed effector T cells. All T-cell effector functions involve cell–cell interactions. When armed effector T cells recognize specific antigen on target cells, they release mediators that act directly on the target cell, altering its behavior. The triggering of armed effector T cells by peptide:MHC complexes is independent of co-stimulation, so that any infected target cell can be activated or destroyed by an armed effector T cell. CD8 cytotoxic T cells kill target cells infected with cytosolic pathogens, removing sites of pathogen replication. CD4 T_H1 cells activate macrophages to kill intracellular parasites. CD4 T_H2 cells are essential in the activation of B cells to secrete the antibodies that mediate humoral immune responses directed against extracellular pathogens, as will be seen in Chapter 9. Thus, effector T cells control virtually all known effector mechanisms of the adaptive immune response.

General references.

Ihle, J.N.: **Cytokine receptor signaling**. *Nature* 1995, **377**:591-594.

Janeway, C.A., and Bottomly, K.: **Signals and signs for lymphocyte responses**. *Cell* 1994, **76**:275-285.

Lenschow, D.J., Walunas, T.L., and Bluestone, J.A.: **CD28/B7 system of T cell costimulation**. *Annu. Rev. Immunol.* 1996, **14**:233-258.

Mosmann, T.R., and Coffman, R.L.: **T_H1 and T_H2 cells: different patterns of lymphokine secretion lead to different functional properties**. *Annu. Rev. Immunol.* 1989, **7**:145-173.

Pigott, R., and Power, C.: *The Adhesion Molecule Facts Book.* Academic Press, London, 1993.

Springer, T.A.: **Traffic signals for lymphocyte recirculation and leukocyte emigration: the multistep paradigm**. *Cell* 1994, **76**:301-314.

Section references.

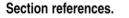

8-1 The initial interaction of naive T cells with antigen occurs in peripheral lymphoid organs.

Picker, L.J., and Butcher, E.C.: **Physiological and molecular mechanisms of lymphocyte homing**. *Annu. Rev. Immunol.* 1993, **10**:561-591.

Pierre, P., Turkey, S.J., Gatti, E., Hull, M., Meltzer, J., Mirza, A., Inaba, K., Steinman, R.M., and Mellman, I.: **Developmental regulation of MHC class II transport in mouse dendritic cells**. *Nature* 1997, **388**:787-792.

Tilney, N.L., and Gowans, J.L.: **The sensitization of rats by allografts transplanted to alymphatic pedicles of skin**. *J. Exp. Med.* 1971, **113**:951.

8-2 Lymphocyte migration, activation, and effector function depend on cell–cell interactions mediated by cell adhesion molecules.

Hogg, N., and Landis, R.C.: **Adhesion molecules in cell interactions**. *Curr. Opin. Immunol.* 1993, **5**:383-390.

Picker, L.J.: **Control of lymphocyte homing**. *Curr. Opin. Immunol.* 1994, **6**:394-406.

8-3 The initial interaction of T cells with antigen-presenting cells is also mediated by cell-adhesion molecules.

Dustin, M.L., and Springer, T.A.: **T-cell receptor crosslinking transiently stimulates adhesiveness through LFA-1**. *Nature* 1989, **341**:619-624.

Hahn, W.C., Rosenstein, Y., Clavo, V., Burakoff, S.J., and Bierer, B.: **A distinct**

cytoplasmic domain of CD2 regulates ligand avidity and T-cell responsiveness to antigen. *Proc. Natl. Acad. Sci. USA* 1992, **89**:7179-7183.

Shimizu, Y., van Seventer, G., Horgan, K.J., and Shaw, S.: **Roles of adhesion molecules in T-cell recognition: fundamental similarities between four integrins on resting human T cells (LFA-1, VLA-4, VLA-5, VLA-6) in expression, binding, and co-stimulation**. *Immunol. Rev.* 1990, **114**:109-143.

8-4 Both specific ligand and co-stimulatory signals provided by a professional antigen-presenting cell are required for the clonal expansion of naive T cells.

Liu, Y., and Janeway, C.A. Jr.: **Cells that present both specific ligand and co-stimulatory activity are the most efficient inducers of clonal expansion of normal CD4 T cells**. *Proc. Natl. Acad. Sci. USA* 1992, **89**:3845-3949.

Rudd, C.E.: **Upstream-Downstream: CD28 cosignaling pathways and T cell function**. *Immunity* 1996, **4**:527-534.

Tivol, E.A., Borriello, F., Schweitzer, A.N., Lynch, W.P., Bluestone, J.A., and Sharpe, A.H.: **Loss of CTLA-4 leads to massive lymphoproliferation and fatal multiorgan tissue destruction, revealing a critical negative regulatory role of CTLA-4**. *Immunity* 1995, **3**:541-547.

8-5 Mature dendritic cells are highly efficient inducers of naive T-cell activation.

Banchereau, J., and Steinman, R.M.: **Dendritic cells and the control of immunity**. *Nature* 1998, **392**:245-252.

Sallusto, F., Lanzavecchia, A.: **Efficient presentation of soluble antigen by cultured human dendritic cells is maintained by granulocyte/macrophage colony-stimulating factor plus interleukin-4 and downregulated by tumor necrosis factor**. *J. Exp. Med.* 1994, **179**:1109-1118.

8-6 Macrophages are scavenger cells that can be induced by pathogens to present foreign antigens to naive T cells.

Razi-Wolf, Z., Freeman, G.J., Galvin, F., Benacerraf, B., Nadler, L., and Reiser, H.: **Expression and function of the murine B7 antigen, the major co-stimulatory molecule expressed by peritoneal exudate cells**. *Proc. Natl. Acad. Sci. USA* 1992, **89**:4210-4214.

8-7 B cells are highly efficient at presenting antigens that bind to their surface immunoglobulin.

Lanzavecchia, A.: **Receptor-mediated antigen uptake and its effect on antigen presentation to class II-restricted T lymphocytes**. *Annu. Rev. Immunol.* 1993, **8**:773-793.

8-8 Activated T cells synthesize the T-cell growth factor interleukin-2 and its receptor.

Jain, J., Loh, C., and Rao, A.: **Transcriptional regulation of the IL-2 gene**. *Curr. Opin. Immunol.* 1995, **7**:333-342.

Minami, Y., Kono, T., Miyazaki, T., and Taniguchi, T.: **The IL-2 receptor complex: its structure, function, and target genes**. *Annu. Rev. Immunol.* 1993, **11**:245-267.

8-9 The co-stimulatory signal is necessary for the synthesis and secretion of IL-2.

Fraser, J.D., Irving, B.A., Crabtree, G.R., and Weiss, A.: **Regulation of interleukin-2 gene enhancer activity by the T-cell accessory molecule CD28**. *Science* 1991, **251**:313-316.

Lindsten, T., June, C.H., Ledbetter, J.A., Stella, G., and Thompson, C.B.: **Regulation of lymphokine messenger RNA stability by a surface-mediated T-cell activation pathway**. *Science* 1989, **244**:339-342.

8-10 Antigen recognition in the absence of co-stimulation leads to T-cell tolerance.

Fields, P.E., Gajewski, T.F., and Frank, W.F.: **Blocked *ras* activation in anergic CD4+ T cells**. *Science* 1996, **273**:1276-1278.

Guerder, S., Meyerhoff, J., and Flavell, R.A.: **The role of the T cell costimulator B7.1 in autoimmunity and the induction and maintenance of tolerance to peripheral antigen**. *Immunity* 1994, **1**:155-166.

Li, W., Whaley, C.D., Mondino, A., and Mueller, D.L.: **Blocked signal transduction to the ERK and JNK protein kinases in anergic CD4+ T cells**. *Science* 1996, **273**:1272-1276.

8-11 Proliferating T cells differentiate into armed effector T cells that do not require co-stimulation to act.

Wong, S.F., Visintin, I., Wen, L., Flavell, R.A., and Janeway, C.A. Jr.: **CD8 T cell clones from young NOD islets can transfer rapid onset of diabetes in NOD mice in the absence of CD4 cells**. *J. Exp. Med.* 1996, **183**:67-76.

8-12 The differentiation of CD4 T cells into T_H1 or T_H2 cells determines whether humoral or cell-mediated immunity will predominate.

Kamogawa, Y., Minasi, L.A., Carding, S.R., Bottomly, K., and Flavell, R.A.: **The relationship of IL-4 and IFN-γ producing T cells studied by lineage ablation of IL-4-producing cells**. *Cell* 1993, **75**:985-995.

8-13 Naive CD8 T cells can be activated in different ways to become armed cytotoxic effector cells.

Azuma, M., Cayabyab, M., Buck, D., Phillips, J.H., and Lanier, L.L.: **CD28 interaction with B7 co-stimulates primary allogeneic proliferative responses and cytotoxicity mediated by small, resting T lymphocytes**. *J. Exp. Med.* 1992, **175**:353-360.

8-14 Effector T-cell interactions with target cells are initiated by antigen non-specific cell-adhesion molecules.

O'Rourke, A.M., and Mescher, M.F.: **Cytotoxic T lymphocyte activation involves a cascade of signaling and adhesion events**. *Nature* 1992, **358**:253-255.

Rodrigues, M., Nussezweig, R.S., Romero, P., and Zavala, F.: **The *in vivo* cytotoxic activity of CD8+ T-cell clones correlates with their levels of expression of adhesion molecules**. *J. Exp. Med.* 1992, **175**:895-905.

van Seventer, G.A., Simuzi, Y., and Shaw, S.: **Roles of multiple accessory molecules in T-cell activation**. *Curr. Opin. Immunol.* 1991, **3**:294-303.

8-15 Binding of the T-cell receptor complex directs the release of effector molecules and focuses them on the target cell.

Griffiths, G.M.: **The cell biology of CTL killing**. *Curr. Opin. Immunol.* 1995, **7**:343-348.

8-16 The effector functions of T cells are determined by the array of effector molecules they produce.

Armitage, R.J., Fanslow, W.C., Strockbine, L., Sato, T.A., Cliffors, K.N., MacDuff,

B.M., Anderson, D.M., Gimpel, S.D., Davis Smith, T., Maliszewski, C.R.: **Molecular and biological characterization of a murine ligand for CD40**. *Nature* 1992, **357**:80-82.

8-17 **Cytokines can act locally or at a distance.**

Arai, K., Lee, F., Miyajima, A., Miyatake, S., Arai, N., and Yokota, T.: **Cytokines: co-ordinators of immune and inflammatory responses**. *Annu. Rev. Biochem.* 1990, **59**:783.

8-18 **Cytokines and their receptors fall into distinct families of structurally related proteins.**

Taga, T., Kishimoto, T.: **Signaling mechanisms through cytokine receptors that share signal tranducing receptors components**. *Curr. Opin. Immunol.* 1995, **7**:17-23.

Thompson, A.: *The Cytokine Handbook.* 2nd edn. Academic Press, San Diego, 1994.

8-19 **The TNF family of cytokines are trimeric proteins that are often cell-surface associated.**

Armitage, R.J.: **Tumor necrosis factor receptor superfamily members and their ligands**. *Curr. Opin. Immunol.* 1994, **6**:407-413.

8-20 **Cytotoxic T cells can induce target cells to undergo programmed cell death.**

Henkart, P.A.: **Lymphocyte-mediated cytotoxicology: two pathways and multiple effector molecules**. *Immunity* 1994, **1**:343-346.

Squier, M.K.T., and Cohen, J.J.: **Cell-mediated cytotoxic mechanisms**. *Curr. Opin. Immunol.* 1994, **6**:447-452.

8-21 **Cytotoxic effector proteins that trigger apoptosis are contained in the granules of CD8 cytotoxic T cells.**

Kāgi, B., Ledermann, K., and Bürki, R.: **Molecular mechanisms of lymphocyte-mediated cytotoxicity and their role in immunological protection and pathogenesis *in vivo***. *Annu. Rev. Immunol.* 1994, **12**:207-232.

Shiver, J.W., Su, L., and Henkart, P.A.: **Cytotoxicity with target DNA breakdown by rat basophilic leukemia cells expressing both cytolysin and granzyme A**. *Cell* 1992, **71**:315-322.

8-22 **Activated CD8 T cells and some CD4 effector T cells express Fas ligand, which can also activate apoptosis.**

Fisher, G.H., Rosenberg, E.J., Straus, S.E., Dale, J.K., Middleton, L.A., Lin, A.Y., Strober, W., Leonardo, M.J., and Puck, J.M.: **Dominant interfering Fas gene mutations impair apoptosis in a human autoimmune lymphoproliferative syndrome**. *Cell* 1995, **81**:935-946.

Suda, T., Takahashi, T., Goldstein P., and Nagata, S.: **Molecular cloning and expression of the Fas ligand, a novel member of the tumor necrosis factor family**. *Cell* 1993, **75**:1169-1178.

Watanbe, F.R., Branna, C.I., Copeland, N.G., Jenkins, N.A., and Nagata, S.: **Lymphoproliferation disorder in mice explained by defects in Fas antigen that mediates apoptosis**. *Nature* 1992, **356**:314-317.

8-23 **Cytotoxic T cells are selective and serial killers of targets expressing specific antigen.**

Kuppers, R.C., and Henney, C.S.: **Studies on the mechanism of lymphocyte-mediated cytolysis. IX. Relationships between antigen recognition and lytic expression in killer T cells**. *J. Immunol.* 1977, **118**:71-76.

8-24 **Cytotoxic T cells also act by releasing cytokines.**

Ramshaw, I., Ruby, J., Ramsay, A., Ada, G., and Karupiah, G.: **Expression of cytokines by recombiant vaccinia viruses: a model for studying cytokines in virus infections *in vivo***. *Immunol. Rev.* 1992, **127**:157-182.

8-25 **Armed T_H1 cells have a central role in macrophage activation.**

Munoz Fernandez, M.A., Fernandez, M.A., and Fresno, M.: **Synergism between tumor necrosis factor-α and interferon-γ on macrophage activation for the killing of intracellular *Trypanosoma crusi* through a nitric oxide-dependent mechanism**. *Eur. J. Immunol.* 1992, **22**:301-307.

Stout, R., and Bottomly, K.: **Antigen-specific activation of effector macrophages by interferon-γ producing (T_H1) T-cell clones: failure of IL-4 producing (T_H2) T-cell clones to activate effector functions in macrophages**. *J. Immunol.* 1989, **142**:760.

8-26 **The expression of cytokines and membrane-associated molecules by armed CD4 T_H1 cells requires new RNA and protein synthesis.**

Shaw, G., and Karmen, R.: **A conserved UAU sequence from the 3′ untranslated region of GM-CSF mRNA mediates selective mRNA degradation**. *Cell* 1986, **46**:659.

8-27 **Activation of macrophages by armed T_H1 cells promotes microbial killing and must be tightly regulated to avoid damage to host tissues.**

Paulnock, D.M.: **Macrophage activation by T cells**. *Curr. Opin. Immunol.* 1992, **4**:344-349.

8-28 **T_H1 cells coordinate the host response to intracellular pathogens.**

Kindler, V., Sappino, A.-P., Grau, G.E., Piquet, P.-F., and Vassali, P.: **The inducing role of tumor necrosis factor in the development of bactericidal granulomas during BCG development**. *Cell* 1989, **56**:731-740.

McInnes, A., and Rennick, D.M.: **Interleukin-4 induces cultured monocytes/macrophages to form giant multinucleated cells**. *J. Exp. Med.* 1988, **167**:598-611.

Yamamura, M., Uyemura, K., Deans, R,J., Weinberg, K., Rea, T.H., Bloom, B.R., and Modlin, R.L.: **Defining protective responses to pathogens: cytokine profiles in leprosy lesions**. *Science* 1991, **254**:277-279.

The Humoral Immune Response

9

Many of the bacteria that are most important in human infectious diseases multiply in the extracellular spaces of the body, and most intracellular pathogens must spread by moving from cell to cell through the extracellular fluids. The humoral immune response leads to the destruction of extracellular microorganisms and prevents the spread of intracellular infections. This is achieved by antibodies secreted by B lymphocytes.

There are three main ways in which antibodies contribute to immunity (Fig. 9.1). Viruses and intracellular bacteria, which need to enter cells in order to grow, spread from cell to cell by binding to specific molecules on their target cell surface. Antibodies that bind to the pathogen can prevent this and are said to **neutralize** the pathogen. Neutralization by antibodies is also important in protection from bacterial toxins. Other types of bacteria multiply outside cells, and antibodies protect against these pathogens mainly by facilitating pathogen uptake into phagocytic cells that are specialized to destroy ingested bacteria. There are two ways in which this can occur. In the first case, bound antibodies coating the pathogen are recognized by specific Fc receptors on the surface of phagocytic cells. Coating the surface of a pathogen to enhance phagocytosis in this way is called **opsonization**. Alternatively, antibodies binding to the surface of a pathogen can activate the proteins of the **complement** system. Complement proteins bound to the pathogen also opsonize it by binding complement receptors on phagocytes. Other complement components recruit phagocytic cells to the site of

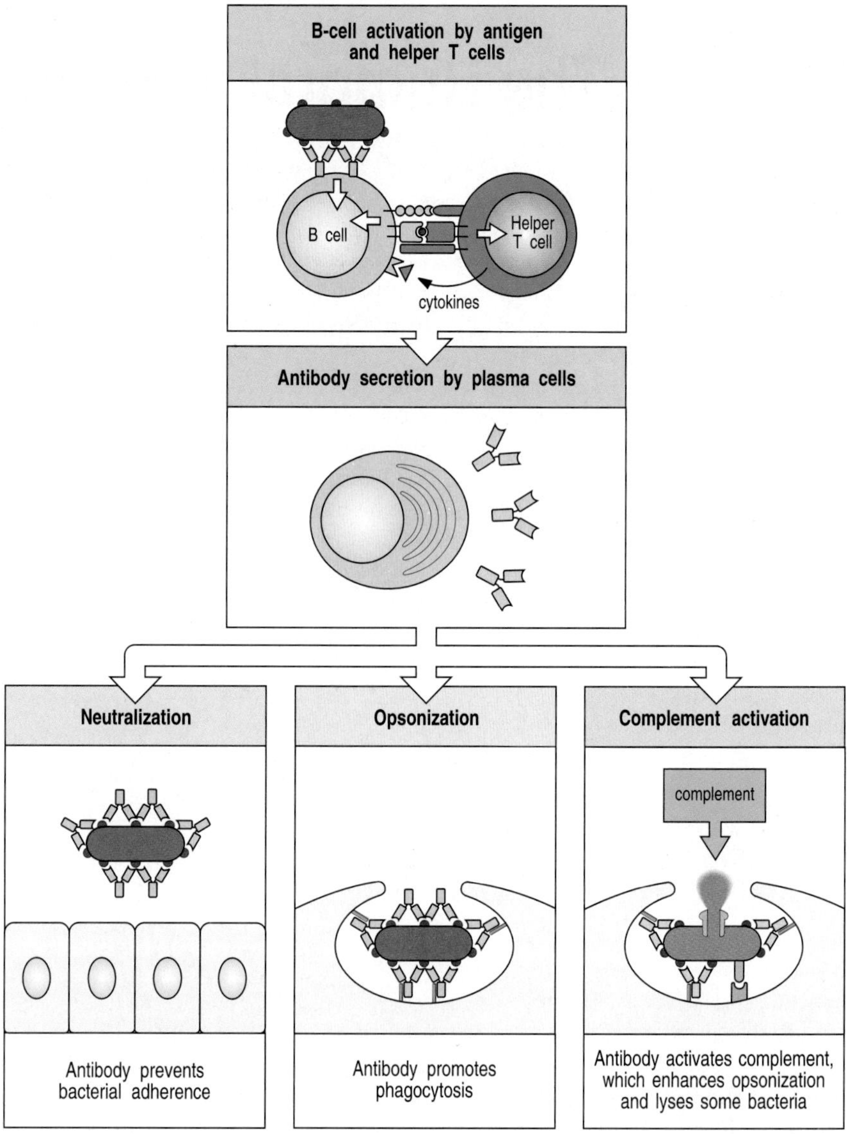

Fig. 9.1 The humoral immune response is mediated by antibody molecules that are secreted by plasma cells. Antigen that binds to the B-cell antigen receptor signals B cells and is, at the same time, internalized and processed into peptides that activate armed helper T cells. Signals from the bound antigen and from the helper T cell induce the B cell to proliferate and differentiate into a plasma cell secreting specific antibody (top two panels). There are three main ways in which these antibodies protect the host from infection (bottom panels). They can inhibit the toxic effects or infectivity of pathogens by binding to them: this is termed neutralization (left panel). By coating the pathogens, they can enable accessory cells that recognize their Fc portions of arrays of antibodies to ingest and kill the pathogen, a process called opsonization (middle panel). Antibodies can also trigger the complement cascade of proteins, which strongly enhance opsonization, and can directly kill some bacterial cells (right panel).

infection, and the terminal components of complement can lyse certain microorganisms directly by forming pores in their membranes. Which effector mechanisms are recruited in a particular response is determined by the **isotypes** of the antibodies produced.

The activation of B cells and their differentiation into antibody-secreting cells is triggered by antigen and usually requires **helper T cells**. The term 'helper T cell' is often used to mean a cell from the T_H2 class of CD4 T cells, but a subset of T_H1 cells can also help in B-cell activation. In this chapter we shall therefore use the term 'helper T cell' to mean any armed effector CD4 T cell that can activate a B cell. Helper T cells also control **isotype switching** and have a role in initiating **somatic hypermutation** of antibody variable (V)-region genes and directing the affinity maturation of antibodies that occurs during the course of a humoral immune response. In the first part of this chapter we shall describe the interactions of B cells with helper T cells and the mechanism of affinity maturation in the specialized microenvironment of peripheral lymphoid tissues. In the rest of the chapter we shall discuss in detail the mechanisms whereby antibodies contain and eliminate infections.

Antibody production by B lymphocytes.

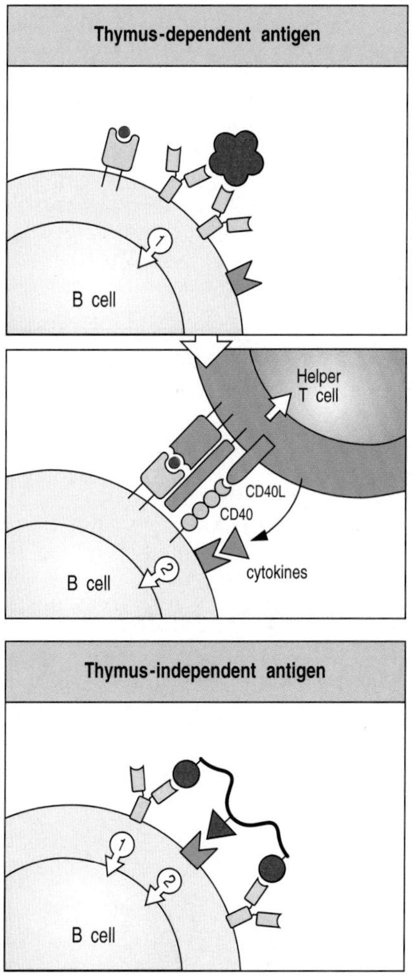

Fig. 9.2 A second signal is required for B-cell activation by either thymus-dependent or thymus-independent antigens. The first signal required for B-cell activation is delivered through the antigen receptor (top panel). For thymus-dependent antigens, the second signal is delivered by a helper T cell that recognizes degraded fragments of the antigen as peptides bound to MHC class II molecules on the B-cell surface (second panel); the interaction between CD40 ligand (CD40L) on the T cell and CD40 on the B cell contributes an essential part of this second signal. For thymus-independent antigens, the second signal can be delivered by the antigen itself (third panel), or by non-thymus-derived accessory cells (not shown).

The surface immunoglobulin that serves as the antigen receptor on B lymphocytes has two roles in their activation. First, like the antigen receptor on T cells, it transmits signals directly to the cell's interior when it binds antigen (see Section 5-1). Second, via receptor-mediated endocytosis it delivers the antigen to intracellular sites where it is degraded and from where it is returned to the B-cell surface as peptides bound to MHC class II molecules. The peptide:MHC class II complex can then be recognized by antigen-specific armed helper T cells, triggering them to make molecules that, in turn, cause the B cell to proliferate and its progeny to differentiate into antibody-secreting cells. Some microbial antigens can activate B cells directly in the absence of T-cell help and provide a means whereby antibodies can be produced rapidly against many important bacterial pathogens. However, the changes in the functional properties of antibody molecules that result from isotype switching, and the changes in the V region that occur during affinity maturation, depend upon the interaction of antigen-stimulated B cells with helper T cells and other cells in the peripheral lymphoid organs. Antibodies induced by microbial antigens alone are therefore less variable and less functionally versatile than those induced with T-cell help.

9-1 The antibody response is initiated when B cells bind antigen and are signaled by helper T cells or by certain microbial antigens.

It is a general rule in adaptive immunity that naive antigen-specific lymphocytes cannot be activated by antigen alone. Naive T cells require a co-stimulatory signal from professional antigen-presenting cells; naive B cells require accessory signals that can come either from an armed helper T cell or, in some cases, directly from microbial constituents.

Antibody responses to protein antigens require antigen-specific T-cell help. B cells become effective targets for armed helper T cells when antigen bound by surface immunoglobulin is internalized and returned to the cell surface as peptides bound to MHC class II molecules. Helper T cells that recognize the peptide:MHC complex then deliver activating signals to the B cell. Thus, protein antigens binding to B cells provide both a specific signal to the B cell and a focus for antigen-specific T-cell help. These antigens are unable to induce antibody responses in animals or humans in which the thymus fails to develop and generate peptide-specific T cells, and they are therefore known as **thymus-dependent** or **TD antigens** (Fig. 9.2, top two panels).

The B-cell co-receptor complex of CD19:CD21:CD81 (see Section 5-8) can greatly enhance B-cell responsiveness to antigen. When hen egg lysozyme is coupled with three linked molecules of the complement fragment C3dg, which is a ligand for CD21, also known as complement receptor CR2 (see Section 9-27), the antigen induces antibody without added adjuvant when used to immunize mice, and at doses as much as 10,000 times smaller than that of unmodified hen egg lysozyme. Whether this works by increasing B-cell signaling, by inducing co-stimulatory molecules on antigen-binding B cells, or by increasing the uptake of antigen is not yet known.

Although armed peptide-specific helper T cells are required for B-cell responses to protein antigens, many constituents of microbes, such as bacterial polysaccharides, can induce B cells to produce antibody in the absence of such helper T cells. These microbial antigens are known as **thymus-independent** or **TI antigens** because they induce antibody responses in individuals who lack a thymus and hence have no T lymphocytes, except those minor subsets that are able to develop extra-thymically. The second signal required to activate antibody production to TI antigens is either provided directly by recognition of a common microbial constituent (see Fig. 9.2, bottom panel) or by a non-thymus-derived accessory cell. Thymus-independent antibody responses provide some protection against extracellular bacteria and we shall return to them at the end of this part of the chapter.

9-2 Armed helper T cells activate B cells that recognize the same antigen.

Thymus-dependent antibody responses require the activation of B cells by helper T cells that respond to the same antigen; this is called **linked recognition**. This means that before B cells can be induced to make antibody to a given pathogen in an infection, a CD4 T cell specific for peptides of the pathogen must first be activated to produce appropriate armed helper T cells. Although the epitope recognized by the armed helper T cell must therefore be linked to that recognized by the B cell, the two cells need not recognize identical epitopes. Indeed, we saw in Chapter 4 that T cells can recognize internal peptides in proteins that are quite distinct from the surface epitopes on the same molecule recognized by B cells. For more complex natural antigens, such as viruses, the T cell and the B cell might not even recognize the same protein. It is, however, crucial that the peptide recognized by the T cell be a physical part of the antigen recognized by the B cell, which can thereby produce the appropriate peptide upon internalization of antigen bound to its surface immunoglobulin.

For example, by recognizing an epitope on a viral protein coat, a B cell can internalize a complete virus particle. After internalization, the virus particle is degraded and peptides from internal viral proteins as well as coat proteins can be displayed by MHC class II molecules on the B-cell surface. Helper T cells that have been primed earlier in an infection by macrophages or dendritic cells presenting these same internal peptides can then activate the B cell to make antibodies that recognize the coat protein (Fig. 9.3).

The specific activation of the B cell by T cells sensitized to the same antigen or pathogen depends on the ability of the antigen-specific B cell to concentrate the appropriate peptides on its surface MHC class II molecules. B cells binding a specific antigen are 10,000-fold more efficient at displaying peptide fragments of the antigen on their surface MHC class II molecules than are B cells that do not bind the antigen. Armed helper T cells will thus help only B cells whose receptors bind to the antigen containing the peptide that they recognize. As T-cell activation requires the recognition of peptide:MHC class

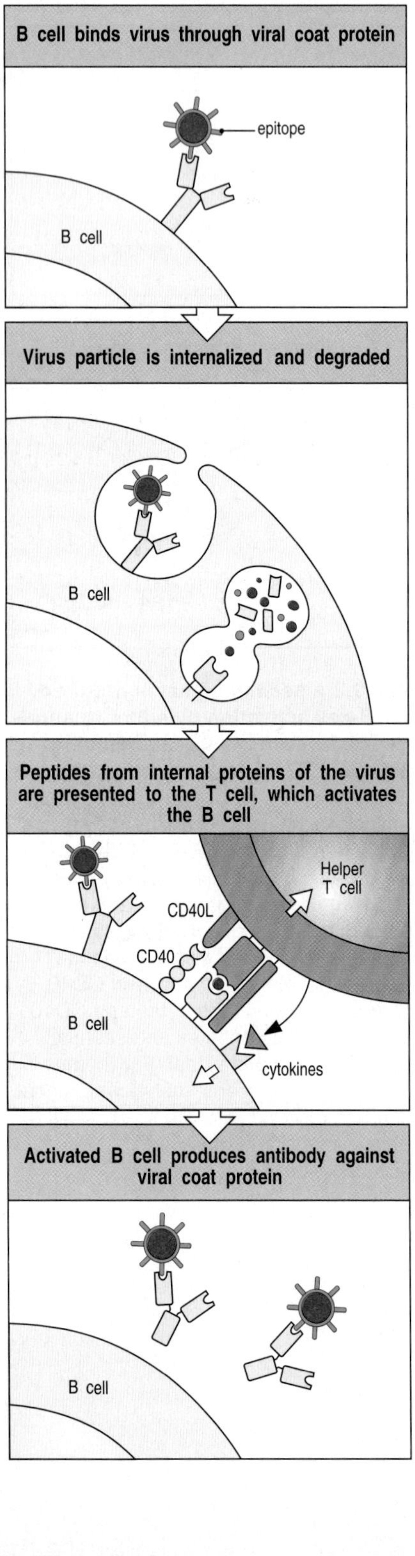

Fig. 9.3 B cells and helper T cells must recognize epitopes of the same molecular complex in order to interact. An epitope on a viral coat protein is recognized by the surface immunoglobulin on a B cell and the virus is internalized and degraded. Peptides derived from viral proteins, including internal proteins, are returned to the B-cell surface bound to MHC class II molecules. Here, these complexes are recognized by helper T cells, which help to activate the B cells to produce antibody against the coat protein.

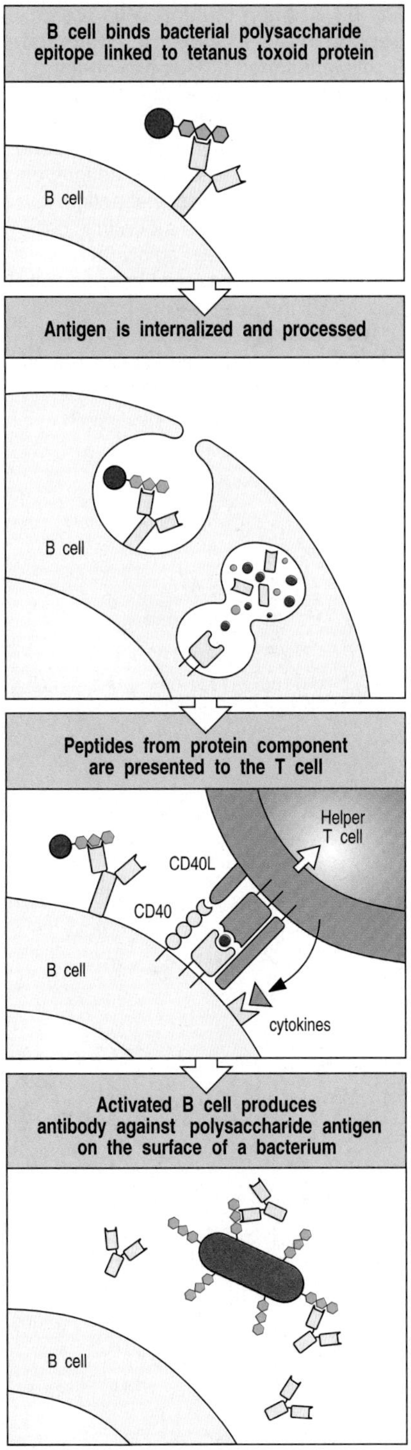

Fig. 9.4 Protein antigens attached to polysaccharide antigens allow T cells to help polysaccharide-specific B cells. *Haemophilus influenzae* B vaccine is a conjugate of bacterial polysaccharide and the tetanus toxoid protein. The B cell recognizes and binds the polysaccharide, internalizes and degrades the whole conjugate and then displays toxoid-derived peptides on surface MHC class II molecules. Helper T cells generated in response to earlier vaccination against the toxoid recognize the complex on the B-cell surface and activate the B cell to produce anti-polysaccharide antibody. This antibody can then protect against infection with *H. influenzae* B.

II complexes on the surface of the specific B cell, only T cells in direct contact with the antigen-binding B cell can participate. How these two cells find each other will be discussed in the next section.

The requirement for linked recognition has important consequences for the regulation and manipulation of the humoral immune response. One of these is to help ensure self tolerance, as will be described in Chapter 13. Another important application of linked recognition is in the design of vaccines, such as that used to immunize infants against *Haemophilus influenzae* B. This bacterial pathogen can infect the lining of the brain, called the meninges, causing meningitis and, in severe cases, neurological damage or death. Protective immunity to this pathogen is mediated by antibodies against its capsular polysaccharide. Although adults make very effective thymus-independent responses to these polysaccharide antigens, such responses are weak in the immature immune system of the infant. To make an effective vaccine for use in infants, therefore, the polysaccharide is linked chemically to tetanus toxoid, a foreign protein against which infants are routinely and successfully vaccinated (see Fig. 1.32 and Chapter 14). B cells that bind the polysaccharide component of the vaccine can be activated by helper T cells specific for peptides of the linked toxoid (Fig. 9.4).

Linked recognition was originally discovered through studies on the production of antibodies to haptens, as described in Chapter 2. Haptens are small chemical groups that cannot elicit antibody responses because they cannot recruit T-cell help. When coupled to a carrier protein, however, they become immunogenic, because T cells can be primed to peptides derived from the protein. This effect is responsible for allergic responses shown by many people to the antibiotic penicillin, which reacts with host proteins to form a coupled hapten that can stimulate an antibody response, as we shall learn in Chapter 12.

9-3 Antigen-binding B cells are trapped in the T-cell zone of lymphoid tissues and are activated by encounter with armed helper T cells.

One of the most puzzling features of the antibody response is the question of how an antigen-specific B cell manages to encounter a helper T cell with an appropriate antigen specificity. This question arises because the frequency of naive lymphocytes specific for any one antigen is estimated to be between 1 in 10,000 and 1 in 1,000,000, so the chance of an encounter between a T and B lymphocyte that recognize the same antigen should be between 1 in 10^8 and 1 in 10^{12}. Achieving such a productive encounter presents a far more difficult challenge than that of getting effector T cells activated because, in that case, only one of the two cells involved has specific receptors. Moreover, T and B cells mostly occupy quite distinct zones in peripheral lymphoid tissue (see, for example, Fig. 1.7). As in naive T-cell activation (discussed in Chapter 8), the answer seems to lie in the antigen-specific trapping of migrating lymphocytes.

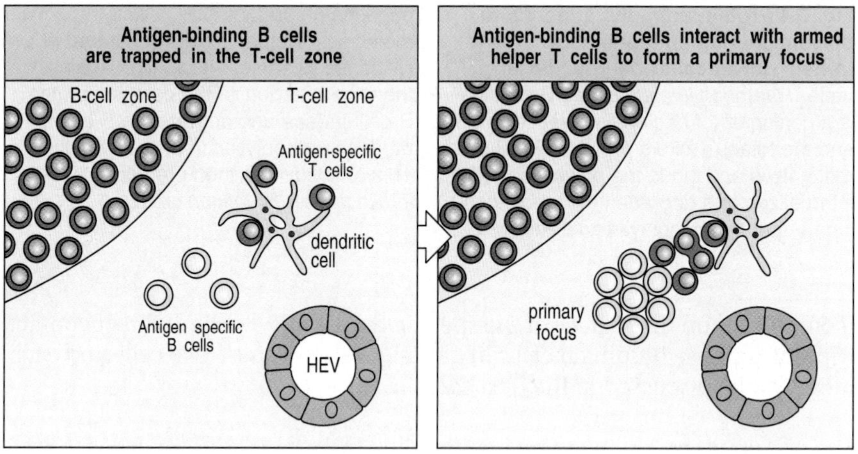

Fig. 9.5 Antigen-binding cells are trapped in the T-cell zone. T cells and B cells are sorted at their site of entry into lymphoid tissues through a high endothelial venule (HEV). Antigen-specific T cells remain in the T-cell zone provided that they encounter antigen on the surface of a professional antigen-presenting cell such as a dendritic cell. B cells normally move rapidly through this area, unless they bind specific antigen, in which case they are trapped before leaving the T-cell zone and thus can interact with antigen-specific armed helper T cells. This gives rise to a primary focus of B cells and T cells.

When an antigen is introduced into mice, it is captured and processed by professional antigen-presenting cells, especially the dendritic cells that migrate from the tissues into the T-cell zones of local lymph nodes. Recirculating naive T cells pass by such cells continuously and those rare T cells whose receptors bind peptides derived from the antigen are trapped very efficiently. This trapping clearly involves the specific antigen receptor on the T cell, although it is stabilized by the activation of adhesion molecules as we learned in Section 8-3. Ingenious experiments with mice transgenic for specific rearranged immunoglobulin genes show that, in the presence of antigen, B cells with antigen-specific receptors are also trapped in the T-cell zones of lymphoid tissue. It is not known whether this unusual arrest of migrating antigen-binding B cells occurs by a similar mechanism to that of T cells—the activation of adhesion molecules by antigen encounter—but this seems likely.

Trapping of B cells in the T-cell zones, which are also the sites of helper T-cell activation, provides an elegant solution to the problem posed at the beginning of this section. T cells are trapped and activated in the T-cell zones; as B cells migrate through high endothelial venules they first enter these same T-cell zones. Those B cells that have bound antigen are trapped, whereas most B cells move quickly through the T-cell zone into the B-cell zone. Thus, antigen-binding B cells are selectively trapped in precisely the correct location to maximize the chances of encountering a helper T cell that can activate them. Interaction with armed helper T cells activates the B cell to establish a primary focus of clonal expansion (Fig. 9.5).

9-4 Peptide:MHC class II complexes on a B cell trigger armed helper T cells to make membrane-bound and secreted molecules that activate the B cell.

Armed helper T cells activate B cells when they recognize the appropriate peptide:MHC class II complex on the B-cell surface. As with armed T_H1 cells acting on macrophages, specific recognition of peptide:MHC class II complexes on B cells triggers armed helper T cells to synthesize both cell-bound and secreted effector molecules that synergize in B-cell activation. One particularly important effector molecule, which has a role in directing all phases of the B-cell response, is a T-cell surface molecule of the tumor necrosis factor (TNF) family, known as the **CD40 ligand** (CD40L) because it binds to the B-cell surface molecule **CD40**. CD40 is a member of the TNF-receptor

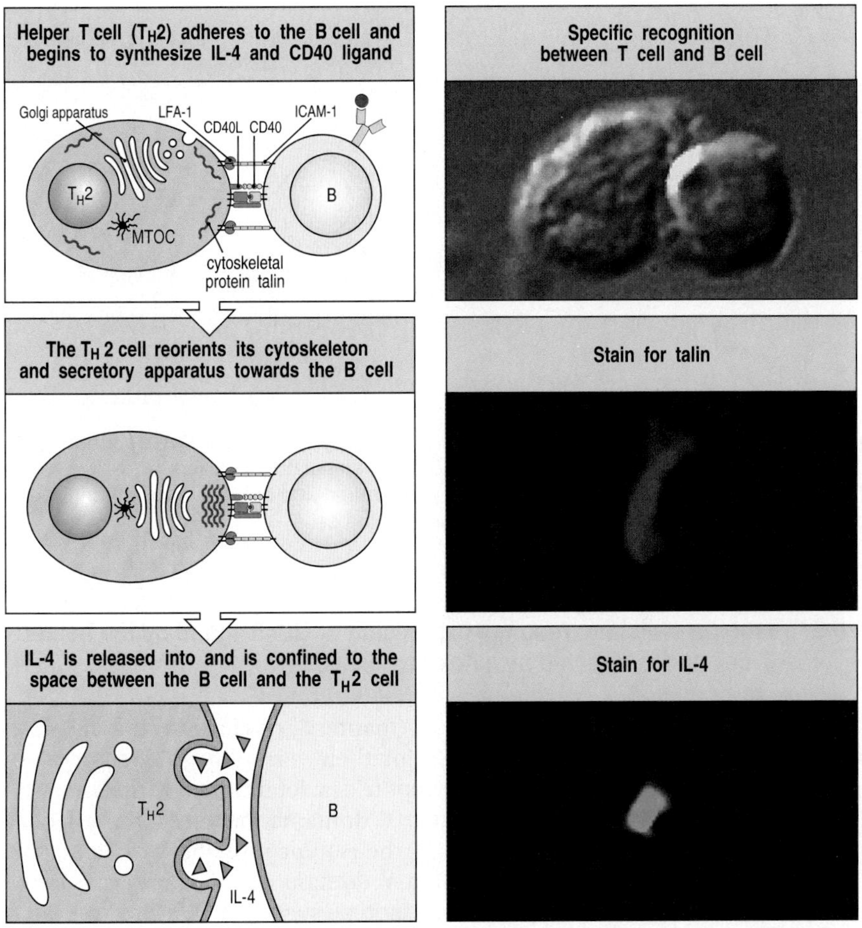
Helper T cell (T_H2) adheres to the B cell and begins to synthesize IL-4 and CD40 ligand
Golgi apparatus
LFA-1
CD40L
CD40
ICAM-1
T_H2
B
MTOC
cytoskeletal protein talin
Specific recognition between T cell and B cell
The T_H 2 cell reorients its cytoskeleton and secretory apparatus towards the B cell
Stain for talin
IL-4 is released into and is confined to the space between the B cell and the T_H2 cell
T_H2
B
IL-4
Stain for IL-4

Fig. 9.6 When an armed helper T cell encounters an antigen-binding B cell, it becomes polarized and secretes IL-4 and other cytokines at the point of cell–cell contact. On binding antigen on the B cell through its T-cell receptor, the helper T cell is induced to express CD40 ligand (CD40L), which binds to CD40 on the B cell. As shown in the top left panel, the tight junction formed between the cells on antigen-specific binding seems to be sealed by a ring of adhesion molecules, with LFA-1 on the T cell interacting with ICAM-1 on the B cell (see also Fig. 8.29). The cytoskeleton becomes polarized, as revealed by the relocation of the cytoskeletal protein talin (stained red in right center panel), to the point of cell–cell contact, and the secretory apparatus (the Golgi apparatus) is re-oriented by the cytoskeleton towards the point of contact with the B cell. As shown in the bottom panels, cytokines are released at the point of contact. The bottom right panel shows IL-4 (stained green) confined to the space between the B cell and the helper T cell. MTOC, microtubule organizing center. Photographs courtesy of A Kupfer.

family of cytokine receptors (see Section 8-19) and is analogous to the TNF receptor on macrophages and Fas on cytotoxic T-cell targets; however, it does not contain a 'death domain'. Binding of CD40 by CD40 ligand helps to drive the resting B cell into the cell cycle and is essential for B-cell responses to thymus-dependent antigens; people and mice with mutations that affect CD40 ligand make very weak and ineffective antibody responses and suffer from severe humoral immunodeficiency, as we shall see in Chapter 11.

B cells are stimulated to proliferate *in vitro* when they are exposed to a mixture of artificially synthesized CD40 ligand and the cytokine IL-4. IL-4 is also made by armed T_H2 cells when they recognize their specific ligand on the B-cell surface, and IL-4 and CD40 ligand are thought to synergize in driving the clonal expansion that precedes antibody production *in vivo*. IL-4 is secreted in a polar fashion by the T_H2 cell and is directed at the site of contact with the B cell (Fig. 9.6) so that it acts selectively on the antigen-specific target B cell.

The initial steps in the activation of B cells by helper T cells are strikingly analogous to those of the activation of macrophages by T_H1 cells. However, whereas the activation of infected macrophages leads directly to the destruction of the pathogen, naive B cells, like naive T cells, must undergo clonal expansion before they can differentiate into effector cells. The immediate effect of activation by helper T cells is therefore to trigger **primary foci** of B-cell proliferation. Some of these B cells then differentiate into antibody-secreting **plasma cells** (Fig. 9.7). Two additional cytokines, IL-5 and IL-6, both secreted by helper T cells, contribute to these later stages in B-cell activation.

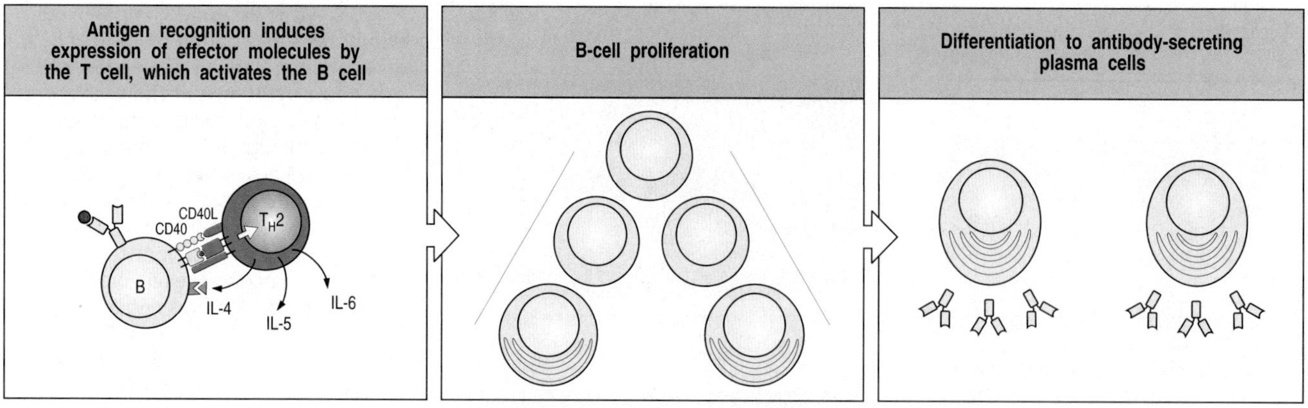
Antigen recognition induces expression of effector molecules by the T cell, which activates the B cell
B-cell proliferation
Differentiation to antibody-secreting plasma cells
CD40L
CD40
T_H2
B
IL-4
IL-5
IL-6

Fig. 9.7 Armed helper T cells stimulate the proliferation and then the differentiation of antigen-binding B cells. The specific interaction of an antigen-binding B cell with an armed helper T cell leads to the expression of the B-cell stimulatory molecule CD40 ligand (CD40L) on the helper T-cell surface and to the secretion of the B-cell stimulatory cytokines IL-4, IL-5, and IL-6, which drive the proliferation and differentiation of the B cell into antibody-secreting plasma cells.

9-5 Isotype switching requires expression of CD40 ligand by the helper T cell and is directed by cytokines.

Antibodies are remarkable, as we saw in Chapter 3, not only for the diversity of their antigen-binding sites but also for their versatility as effector molecules. The specificity of an antibody response is determined by the antigen-binding site, which consists of the two V domains; however, the effector action of the antibody is determined by the isotype of its heavy-chain constant (C) domains. A given heavy-chain V domain can become associated with the C region of any isotype through isotype switching. We shall see later in this chapter how antibodies of each isotype contribute to the elimination of pathogens. The DNA rearrangements that underlie isotype switching and confer this functional diversity on the humoral immune response are directed by cytokines, especially those released by armed effector CD4 T cells.

All naive B cells express cell-surface IgM and IgD, yet IgM makes up less than 10% of the immunoglobulin found in plasma, where the most abundant isotype is IgG. Much of the antibody in plasma has therefore been produced by B cells that have undergone isotype switching. Little IgD antibody is produced at any time, so the early stages of the antibody response are dominated by IgM antibodies. Later, IgG and IgA are the predominant isotypes, with IgE contributing a small but biologically important part of the response. The overall predominance of IgG results, in part, from its longer lifetime in the plasma (see Fig. 3.20).

Isotype switching does not occur in individuals who lack a functional CD40 ligand, which is necessary for interactions between B cells and helper T cells; such individuals make only small amounts of IgM antibodies in response to thymus-dependent antigens and have abnormally high levels of IgM in their plasma, perhaps induced by thymus-independent antigens expressed by pathogens that chronically infect these patients (see Sections 9-9 and 9-10).

Most of what is known about the regulation of isotype switching by helper T cells has come from experiments in which mouse B cells are stimulated with bacterial lipopolysaccharide (LPS) and purified cytokines *in vitro*. These experiments show that different cytokines preferentially induce switching to different isotypes. Some of these cytokines are the same as those that drive B-cell proliferation in the initiation of a B-cell response. In the mouse, IL-4

Role of cytokines in regulating Ig isotype expression

Cytokines	IgM	IgG3	IgG1	IgG2b	IgG2a	IgE	IgA
IL-4	Inhibits	Inhibits	Induces		Inhibits	Induces	
IL-5							Augments production
IFN-γ	Inhibits	Induces	Inhibits		Induces	Inhibits	
TGF-β	Inhibits	Inhibits		Induces			Induces

Fig. 9.8 Different cytokines induce switching to different isotypes. The individual cytokines induce (violet) or inhibit (red) production of certain isotypes. Much of the inhibitory effect is probably the result of directed switching to a different isotype. These data are drawn from experiments with mouse cells.

preferentially induces switching to IgG1 and IgE, whereas tumor growth factor (TGF)-β induces switching to IgG2b and IgA. T_H2 cells make both of these cytokines as well as IL-5, which induces IgA secretion by cells that have already undergone switching. Although T_H1 cells are poor initiators of antibody responses, they participate in isotype switching by releasing interferon (IFN)-γ, which preferentially induces switching to IgG2a and IgG3. The role of cytokines in directing B cells to make the different isotypes of antibody is summarized in Fig. 9.8.

Cytokines induce isotype switching by stimulating the formation and splicing of mRNA transcribed from the switch recombination sites that lie 5′ to each heavy-chain C gene (see Fig. 3.26). When activated B cells are exposed to IL-4, for example, transcription from a site upstream of the switch regions of $C_{\gamma1}$ and C_{ε} can be detected a day or two before switching occurs (Fig. 9.9). Recent data suggest that the production of a spliced switch transcript has a role in directing switching but the mechanism is not yet clear. Each of the cytokines that induces switching seems to induce transcription from the

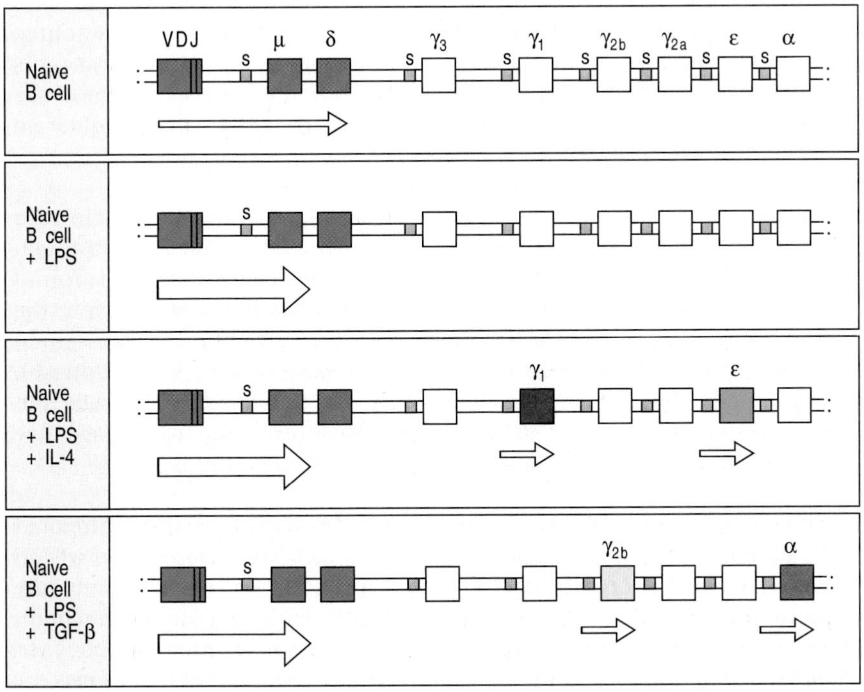

Fig. 9.9 Isotype switching is preceded by transcriptional activation of heavy-chain C-region genes. Resting naive B cells transcribe the μ and δ genes at a low rate, giving rise to surface IgM and IgD. Bacterial lipopolysaccharide (LPS), which can activate B cells independently of antigen (see Section 9-9), induces IgM secretion. In the presence of IL-4, however, $C_{\gamma1}$ and C_{ε} are transcribed at a low rate, presaging switches to IgG1 and IgE production. The transcripts originate before the 5′ end of the region to which switching occurs, and do not code for protein. Similarly, TGF-β gives rise to $C_{\gamma2b}$ and C_{α} transcripts and drives switching to IgG2b and IgA. It is not known what determines which of the two transcriptionally activated heavy-chain C genes undergoes switching. Arrows indicate transcription. The figure shows isotype switching in the mouse.

switch regions of two different heavy-chain C genes, promoting specific recombination to one or other of these genes only. Such a directed mechanism is supported by the observation that individual B cells frequently undergo switching to the same C gene on both chromosomes, even though only one of the chromosomes is producing the expressed antibody. Thus, helper T cells regulate both the production of antibody by B cells and the isotype that determines the effector function of the antibody that is ultimately produced. How the balance between different isotypes is regulated in the humoral immune response to a given pathogen is not understood.

9-6 Activated B cells proliferate extensively in the specialized microenvironment of the germinal center.

In suspension cultures of lymphocytes, the interaction of naive antigen-binding B cells and specific armed helper T cells can lead to the production of antibody of all isotypes. However, although B-cell proliferation and differentiation (including isotype switching) can all be induced in this way *in vitro*, interactions with T cells in suspension culture cannot reproduce, either in magnitude or in complexity, the responses obtained by using the same cells cultured with fragments of spleen, or the antibody response achieved *in vivo*. In particular, the gradual increase in the affinity of antibodies for the inducing antigen that is seen in the course of an antibody response requires specialized features of lymphoid tissue. This phenomenon, which is known as **affinity maturation**, is the consequence of somatic hypermutation of the immunoglobulin genes coupled with selection of B cells with high-affinity surface immunoglobulin. Affinity maturation depends upon the interaction of activated B cells with cells in the specialized microenvironment of the **germinal center**, which forms after stimulation with antigen and has already been mentioned in Chapter 6 as a site of intense B-cell proliferation in the lymph nodes and spleen.

Germinal centers are formed in the B-cell areas of lymphoid tissue a week or so after stimulation with antigen. The progeny of B cells that have been activated by helper T cells in the T-cell zones of lymphoid tissues can follow one of two fates. Some migrate to the medullary cords and differentiate into short-lived plasma cells secreting IgM or IgG, thus providing an early source of circulating antibodies (to be discussed in Chapter 10). However, others migrate along with the T cells that activated them into the B-cell areas of the lymphoid tissue and enter the **primary follicles,** where they proliferate further to form germinal centers (Fig. 9.10).

The primary follicles, which are also known simply as lymphoid follicles, are also the site to which naive recirculating B cells migrate as they pass through the lymphoid tissue. Primary follicles contain resting B cells clustered around a dense network of processes extending from a specialized cell type, the **follicular dendritic cell** (FDC), which is thought to provide signals essential to the survival and continued recirculation of naive B cells (see Chapter 6). Follicular dendritic cells seem to attract both naive and activated B cells into the follicles and are thought to make a central contribution to the selective events that underlie affinity maturation of the antibody response.

The cellular origins of follicular dendritic cells are obscure; they are unrelated to the dendritic cells that activate T cells. They lack MHC class II expression and are not derived from hematopoietic stem cells; indeed their only similarity to dendritic cells is their branched morphology ('dendritic' means branched). Their role in driving the maturation of the humoral immune response depends chiefly on their ability to hold intact antigens on their surfaces for

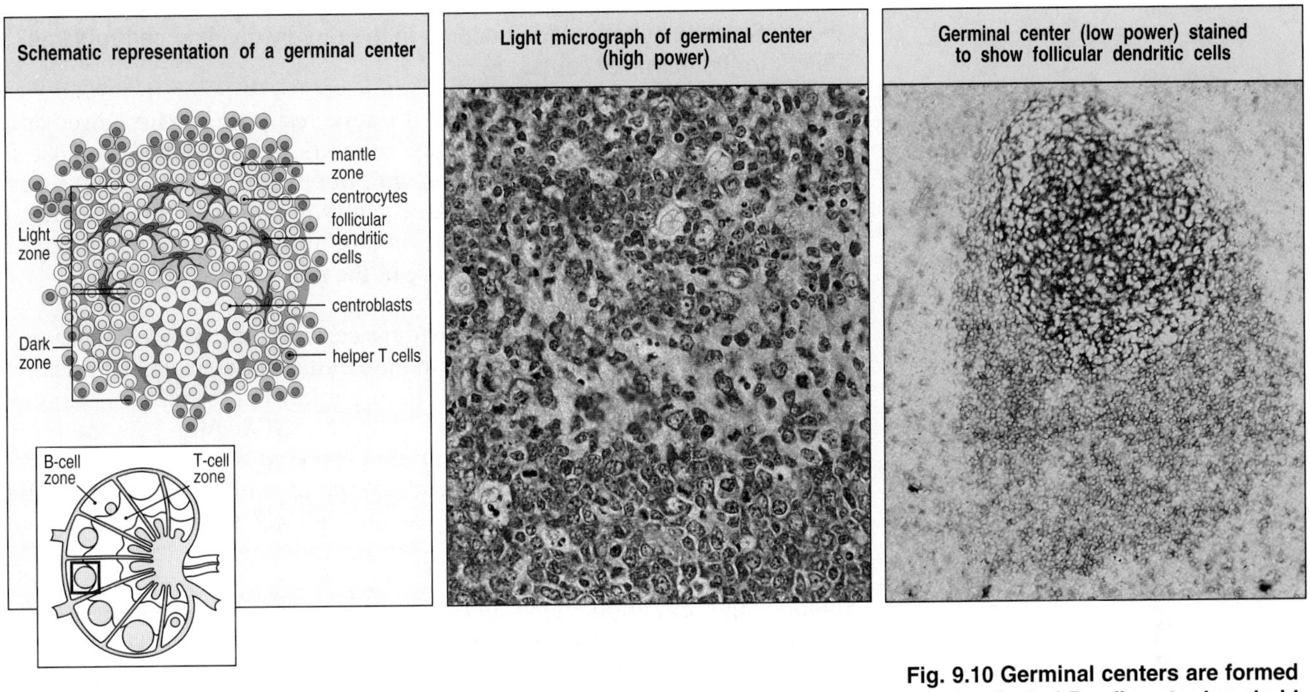

Fig. 9.10 Germinal centers are formed when activated B cells enter lymphoid follicles. The germinal center is a specialized microenvironment in which B-cell proliferation, somatic hypermutation, and selection for antigen binding all occur. Rapidly proliferating B cells in germinal centers are called centroblasts. Closely packed centroblasts form the so-called 'dark zone' of the germinal center, as can be seen in the lower part of the center panel, which shows a section through a germinal center. As these cells mature, they stop dividing and become small centrocytes, moving out into an area of the germinal center called the 'light zone' (the upper part of the center panel), where the centrocytes make contact with a dense network of follicular dendritic cell (FDC) processes. The FDCs are not stained in the center panel but can be seen clearly in the right panel where both FDCs (stained blue with an antibody against Bu10, an FDC-specific marker) in the germinal center and also the mature B cells in the mantle zone (stained brown with an antibody against IgD) can be seen. The plane of this section chiefly reveals the dense network of FDCs in the light zone, although the less dense network in the dark zone can just be seen at the bottom of the figure. Photographs courtesy of I MacLennan.

long periods, ranging from months to years in some cases. Follicular dendritic cells express the complement receptors CR1, CR2, and CR3 (see Section 9-27) and receptors for the Fc portion of immunoglobulin (see Section 9-16); these receptors might be involved in holding antibody and complement-associated antigen in this site. Other specialized properties of follicular dendritic cells, for example the ability to attract B cells to the follicles, are beginning to be characterized. However, this cell type is difficult to study outside an intact lymphoid organ and relatively little is known about the molecular basis of its function.

When activated B cells enter the primary lymphoid follicle, they start dividing to form germinal centers. The proliferating B cells in germinal centers divide about once every 6 hours and can be distinguished by their morphological characteristics, which are typical of blast cells (see Fig. 1.18): thus they are large cells with an expanded cytoplasm, which stains intensely for RNA, and with diffuse chromatin in the nucleus. These B-cell blasts are called **centroblasts**. The visible focus of centroblasts that forms in a few days within a primary lymphoid follicle is called the dark zone of the germinal center. The centroblasts give rise to **centrocytes** that enter the follicular dendritic cell network in the light zone (see Fig. 9.10). The helper T cells that migrate to the primary follicle along with the activated B cells also undergo some clonal expansion and can be seen intermingled with the centrocytes in the light zone. The remaining B cells that are not specific for antigen are pushed to the outside to form the **mantle zone**.

The rapid proliferation of cells in the germinal center greatly increases the number of B cells specific for the pathogen that initiated the antibody response. By dissecting out individual germinal centers and even individual B cells, and using the polymerase chain reaction (see Section 2-17) to analyze the DNA encoding expressed immunoglobulin chains, it has been possible to demonstrate that the B cells in each germinal center proliferate rapidly, so that after a few days most germinal center B cells are derived from only one or a few founder cells. This technique has also revealed that the germinal centers are the site of somatic hypermutation of immunoglobulin V-domain genes.

9-7 Somatic hypermutation occurs in the rapidly dividing centroblasts in the germinal center.

Affinity maturation in the course of an immune response can be viewed as a Darwinian process, requiring first the generation of variability in B-cell receptors and then selection for those with the highest affinity for antigen. The variability is generated by somatic hypermutation of the rearranged immunoglobulin V-domain genes; selection by antigen of cells bearing these mutated receptors occurs on the surface of the follicular dendritic cell.

Somatic hypermutation occurs in dividing centroblasts, whose rearranged immunoglobulin V-region genes accumulate mutations at a rate of about one base pair per 10^3 per cell division. (The mutation rate of all other known somatic cell DNA is one base pair per 10^{10} per cell division.) As there are about 360 base pairs encoding each of the expressed heavy and light-chain V-region genes in a B cell, and about three out of every four base changes results in an altered amino acid, every second cell will acquire a mutation in its receptor at each division.

Somatic hypermutation affects all the rearranged V-region genes in a B cell, whether they are expressed in immunoglobulin chains or not. These mutations also affect some DNA flanking the rearranged V gene but they generally do not extend into the C-region exons. Thus, the rearranged V genes in a B cell are somehow targeted for the introduction of random somatic point mutations. The resulting mutant receptors are expressed on the progeny of the rapidly dividing centroblasts, which are small cells called centrocytes, all derived from the few antigen-specific progenitors that founded the germinal center. As the number of centrocytes increases in the germinal center, two distinct regions begin to be distinguished: the dark zone, where proliferating centroblasts are packed closely together and where there are few follicular dendritic cells, and the light zone, where less densely packed centrocytes make contact with the numerous cells of the follicular dendritic cell network (see Fig. 9.10). The centrocytes eventually give rise to memory B cells and antibody-secreting plasma cells.

9-8 Centrocytes with the best antigen-binding receptors are selected for survival.

Centrocytes are programmed to die within a fixed period unless their surface immunoglobulin is bound to antigen and they are subsequently contacted by a helper T cell bearing CD40 ligand. After somatic hypermutation, the surface immunoglobulin on the centrocytes derived from a single progenitor B cell might bind antigen either better or worse than the immunoglobulin expressed on its precursor. Some will inevitably lose the ability to bind antigen at all and centrocytes bearing these mutations die; a characteristic feature of germinal centers is the presence of **tingible body macrophages**, which are phagocytes engulfing apoptotic cells. If, in contrast, the mutant surface immunoglobulin of the centrocyte binds antigen well, the cell is induced to express the *bcl-x_L* gene, whose product inhibits apoptotic cell death, and the cell is rescued (Fig. 9.11). Affinity maturation occurs in the primary response as well as in secondary and subsequent responses, as we shall see in Chapter 10.

Centrocyte selection involves two stages. First, the centrocytes enter the dense follicular dendritic cell network of the light zone, where they have the opportunity to bind and take up antigen from follicular dendritic cells. Centrocytes with the highest-affinity receptors for antigen are the most likely to succeed. If a centrocyte binds and internalizes the antigen, it then moves to the outer edge of the light zone where helper T cells expressing CD40 ligand are concentrated. Centrocytes that fail to take up antigen from follicular dendritic cells die and are phagocytosed by local macrophages; cell death in the germinal center is seen to occur in the area of the light zone rich in follicular dendritic cells.

In the second stage of centrocyte selection, centrocytes that have successfully taken up and processed antigen from follicular dendritic cells engage in an antigen-specific interaction with the helper T cells at the edge of the light zone. They exchange signals that induce further proliferation of the participating T and B cells, and differentiation of the latter, either to memory B cells

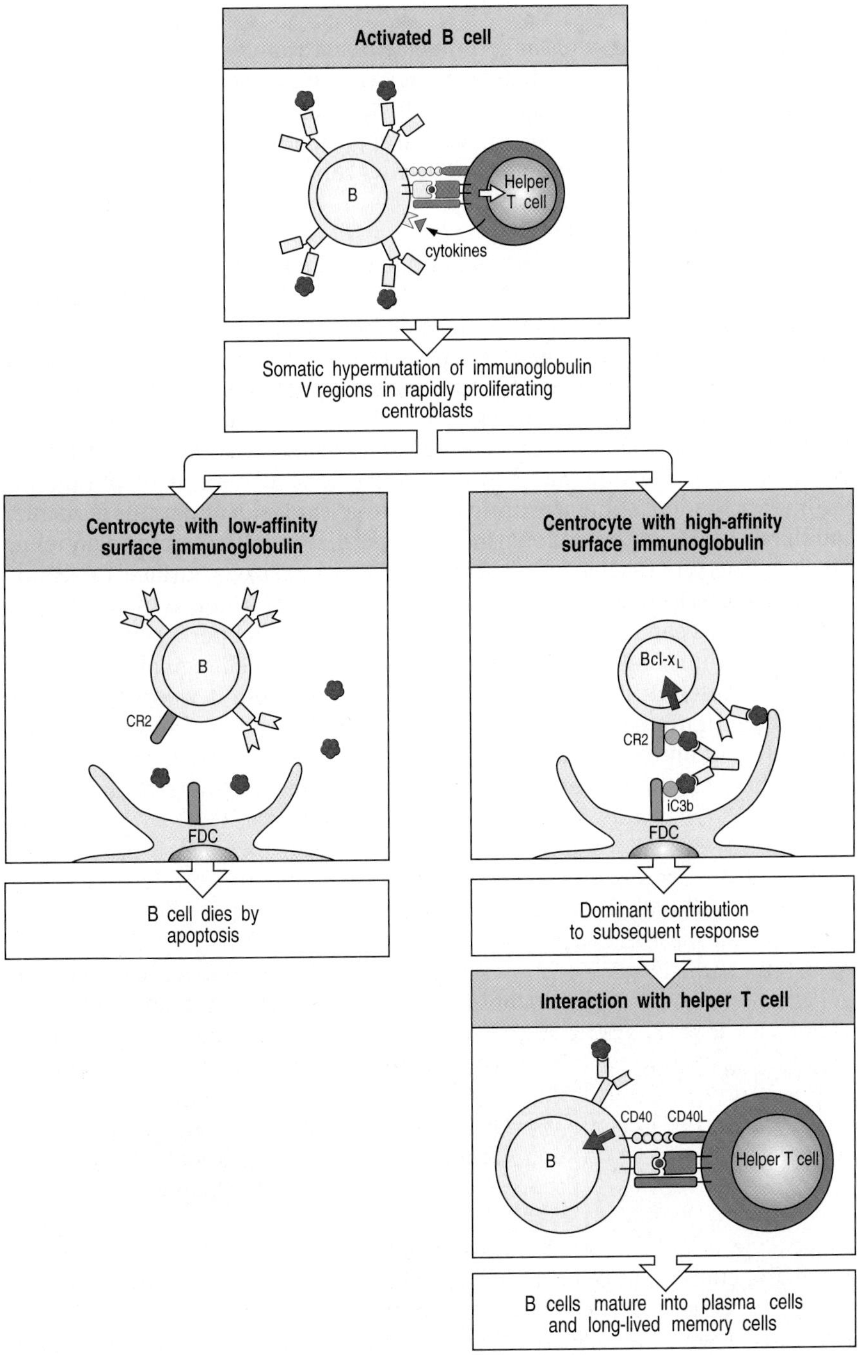

Fig. 9.11 After somatic hypermutation, B cells with high-affinity receptors for antigen are rescued from apoptosis by binding antigen on the surface of follicular dendritic cells and receiving signals from specific helper cells. Somatic hypermutation occurs during the proliferation of centroblasts in germinal centers. The centroblasts give rise to small, non-dividing centrocytes. These interact with follicular dendritic cells (FDC) that display antigen, such as bacteria and bacterial fragments, in complexes with antibody and complement on their surface. Centrocytes whose receptors no longer bind antigen die by apoptosis, whereas centrocytes with receptors that bind well are induced to express Bcl-x_L and survive. It is not known whether Bcl-x_L is induced by the B-cell receptor binding to antigen, as shown here, or by making contact with helper T cells, as shown in the bottom panel. The higher the affinity of the receptor is for antigen, the better will the centrocyte compete with other centrocytes. Binding of the B-cell co-receptor complex (CR2:CD19) to the complement component iC3b (see Section 9-27) bound to antigen on the FDC surface amplifies the signal from the antigen receptor. This process allows the selection of B cells of progressively higher affinity to contribute to the response. A B cell that binds antigen can then present fragments of the antigen to antigen-specific helper T cells that surround the germinal center (see Fig. 9.10). These further stimulate the B cells to enter the memory B-cell pool or to become plasma cells.

or to plasma cells. The involvement of T cells in the selective process serves to prevent centrocytes that have acquired specificity for self antigens from being selected. Evidence both *in vivo* and *in vitro* indicates that CD40 ligation during centrocyte interaction with helper T cells is necessary but insufficient for memory B-cell formation.

As a consequence of this selection for antigen binding, the mutations in the expressed immunoglobulin genes of the surviving cells tend to encode altered amino acids in the complementarity-determining regions (CDRs), whereas mutations in sequences encoding framework residues, which might affect the stability of the variable domain, tend to be silent, as we saw in Chapter 3. In addition, it seems that CDRs have been selected to be particularly susceptible to somatic hypermutation by using particular codons that are more readily altered by this process.

The selection of centrocytes in germinal centers resembles, in some respects, the positive selection of developing thymocytes. However, B cells are selected during the response to foreign antigen instead of during ontogeny, and the selecting antigen is the foreign antigen itself, providing a direct check on the ability of each cell to produce antibodies that can bind the invading pathogen at the time when they are needed.

B cells that have successfully bound antigen and survived selection leave the germinal center to become either memory B cells or antibody-secreting plasma cells. The differentiation of a B cell into a plasma cell is accompanied by many morphological changes that reflect its commitment to the production of large amounts of secreted antibody. The properties of resting B cells and plasma cells are compared in Fig. 9.12. Plasma cells have abundant cytoplasm that is dominated by multiple layers of rough endoplasmic reticulum (see Fig. 1.18). The nucleus shows a characteristic pattern of peripheral chromatin condensation, a prominent perinuclear Golgi apparatus is visible, and the cisternae of the endoplasmic reticulum are rich in immunoglobulin, which makes up 10–20% of all the protein synthesized. Surface immunoglobulin and MHC class II molecules are in low concentrations or are absent, so plasma cells can no longer interact with antigen or helper T cells and antibody secretion is independent of both antigen and T-cell regulation. Plasma cells have different lifespans. Some survive only about 4 weeks after their final differentiation, whereas others are very long-lived and account for the persistence of antibody responses. In the bone marrow, plasma cells obtain signals that are essential for their survival from bone marrow stromal cells.

Fig. 9.12 Plasma cells secrete antibody at a high rate but can no longer respond to antigen or helper T cells. Resting B cells display specific immunoglobulin and MHC class II molecules on their surface. They can take up antigen and present it to helper T cells, which then induce the B cells to proliferate, switch isotype and undergo somatic hypermutation; however, B cells do not secrete significant amounts of antibody. Plasma cells are terminally differentiated B cells and they secrete antibodies. They can no longer interact with helper T cells because they lack surface immunoglobulin and MHC class II molecules. They have also lost the ability to change isotype or undergo somatic hypermutation.

	Property					
	Intrinsic			Inducible		
B-lineage cell	Surface Ig	Surface MHC class II	High-rate Ig secretion	Growth	Somatic hyper-mutation	Isotype switch
 Resting B cell	Yes	Yes	No	Yes	Yes	Yes
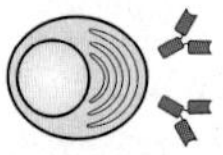 Plasma cell	No	No	Yes	No	No	No

The alternative fate of B cells leaving the germinal center is to become memory B cells that do not secrete antibody in the primary response but can be rapidly activated upon subsequent challenge with the same antigen. It is not known exactly what signals determine whether a given B cell will become a memory B cell or a plasma cell. We discuss B-cell memory further in Chapter 10.

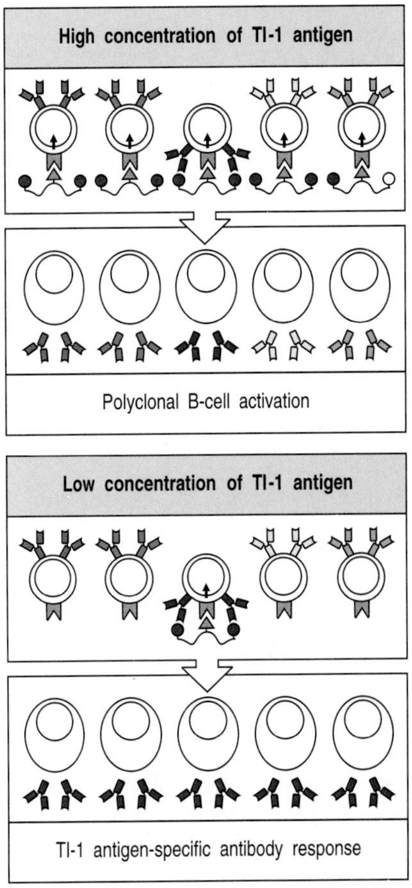

Fig. 9.13 T-cell independent type 1 antigens (TI-1 antigens) are polyclonal B-cell activators at high concentrations, whereas at low concentrations they induce an antigen-specific antibody response. At high concentrations, the signal delivered by the B-cell activating moiety of TI-1 antigens is sufficient to induce proliferation and antibody secretion by B cells in the absence of specific antigen binding to surface immunoglobulin, so all B cells respond (top two panels). At low concentrations, only specific antigen-binding B cells bind enough of the TI-1 antigen to focus its B-cell activating properties onto the B cell; this gives a specific antibody response to epitopes on the TI-1 antigen (bottom two panels).

9-9 B-cell responses to bacterial antigens with intrinsic B-cell activating ability do not require T-cell help.

Although antibody responses to protein antigens are dependent on helper T cells, humans and mice with T-cell deficiencies nevertheless make antibodies to many bacteria. This is because the special properties of some bacterial polysaccharides, polymeric proteins, and lipopolysaccharides enables them to stimulate naive B cells in the absence of peptide-specific T-cell help. These antigens are known as thymus-independent antigens (TI antigens) because they stimulate strong antibody responses in athymic animals or individuals. In normal individuals, these bacterial products induce antibody responses in the absence of classical T-cell responses, which cannot be induced by non-protein antigens. However, B-cell responses to these TI antigens can receive help from T cells that recognize non-protein antigens and can develop outside the thymus, as they are greatly diminished in animals that have no T cells at all.

Thymus-independent antigens fall into two classes, which activate B cells by different mechanisms. Antigens in the first class, the **TI-1 antigens**, contain an intrinsic activity that can directly induce the proliferation of B cells. At high concentration, these molecules cause the proliferation and differentiation of most B cells, regardless of their antigen specificity; this is known as **polyclonal activation** (Fig. 9.13, top two panels). As a result of their ability to stimulate most B cells to divide, TI-1 antigens are often called **B-cell mitogens**, a mitogen being a substance that induces cells to undergo mitosis. When B cells are exposed to concentrations of TI-1 antigens that are 10^3–10^5 times lower than those used for polyclonal activation, only those B cells whose immunoglobulin receptors bind the TI-1 molecules become activated, because only by binding to antigenic determinants on the molecule are they able to concentrate sufficient TI-1 molecules on the surface to be activated (see Fig. 9.13, bottom two panels). In the presence of large amounts of the TI-1 antigen, this concentrating effect is not required and all B cells can be stimulated.

It is likely that during normal infections *in vivo*, concentrations of TI-1 antigens are low; thus, only antigen-specific B cells are likely to be activated and these will produce antibodies specific for TI-1 antigens. Such responses have an important role in specific defense against several extracellular pathogens, as they arise earlier than thymus-dependent responses because they do not require prior priming and clonal expansion of helper T cells. However, TI-1 antigens are inefficient inducers of isotype switching, affinity maturation, or memory B cells, all of which require specific T-cell help.

9-10 B-cell responses to bacterial polysaccharides do not require peptide-specific T-cell help.

The second class of thymus-independent antigens consist of molecules such as bacterial cell-wall and capsular polysaccharides that have highly repetitive structures. These thymus-independent antigens, called **TI-2 antigens**, contain no intrinsic B-cell stimulating activity. Whereas TI-1 antigens can activate both immature and mature B cells, TI-2 antigens can activate only mature B cells; immature B cells, as we saw in Chapter 6, are inactivated by repetitive epitopes. This might be why infants do not make antibodies to polysaccharide antigens efficiently; most of their B cells are immature. Responses to several TI-2 antigens are prominent among B-1 cells (also known as CD5 B cells), which comprise an autonomously replicating subpopulation of B cells (see Chapter 6). Although these cells arise early in development, in young children they do not make a fully effective response to carbohydrate antigens until about 5 years of age.

TI-2 antigens most probably act by extensively crosslinking the cell-surface immunoglobulin of specific mature B cells (Fig. 9.14, left panels). Excessive crosslinking of receptors, however, renders mature B cells unresponsive, just as it does immature B cells. Thus, epitope density seems to be critical in the activation of B cells by TI-2 antigens: at too low a density the level of receptor crosslinking is insufficient to activate the cell; at too high a density the cell becomes anergic.

Although responses to TI-2 antigens can occur in nude mice (which lack a thymus), depletion of all T cells by knocking out the T-cell receptor β and δ loci eliminates responses to TI-2 antigens. Moreover, responses to TI-2 antigens can be augmented *in vivo* by transferring small numbers of T cells to such T-cell deficient mice. How T cells contribute to TI-2 responses is not

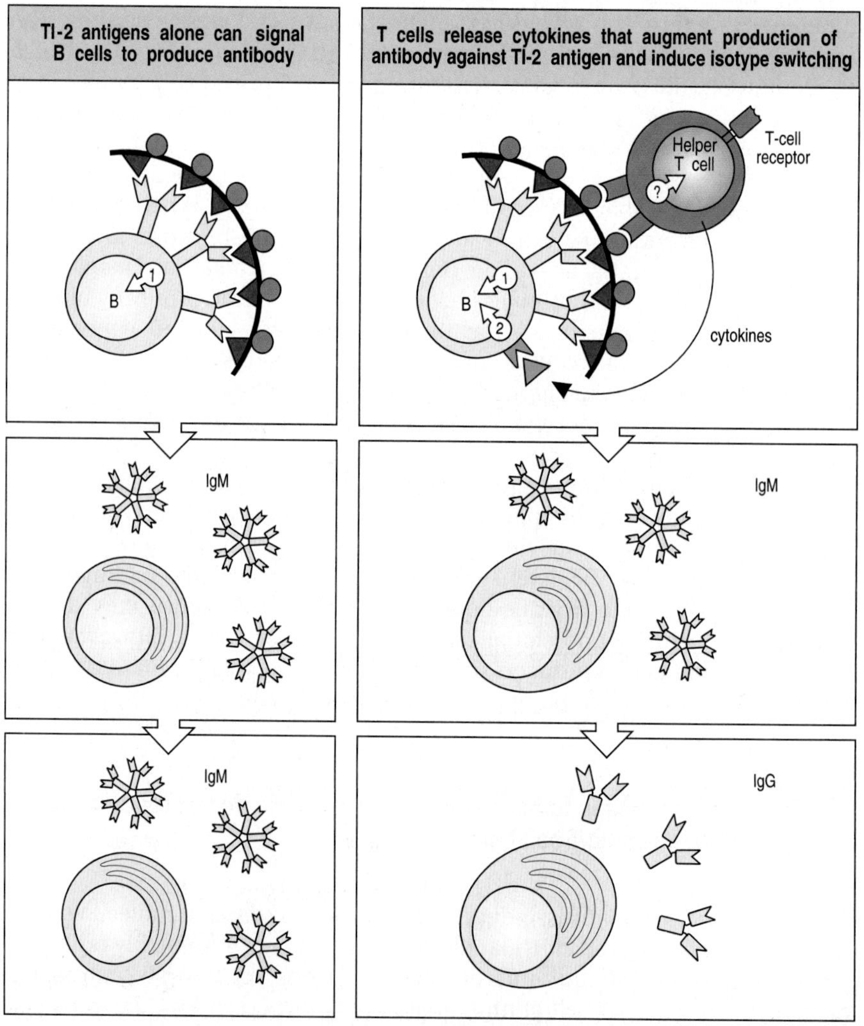

Fig. 9.14 B-cell activation by TI-2 antigens requires, or is greatly enhanced by, cytokines. Multiple crosslinking of the B-cell receptor by TI-2 antigens can lead to IgM antibody production (left panels), but there is evidence that helper T cells greatly augment these responses and lead to isotype switching as well (right panels). It is not clear how T cells are activated in this case, because polysaccharide antigens cannot produce peptide fragments that might be recognized by T cells on the B-cell surface. One possibility is that a component of the antigen binds to a cell-surface molecule common to T cells of all specificities, as shown in the figure.

clear. One possibility is that T cells can recognize and become activated by TI-2 antigens through cell-surface triggering molecules shared by all T cells (see Fig. 9.14, right panels). Alternatively, the help might come from γ:δ T cells or from CD4 CD8 double-negative α:β T cells that can recognize, through their receptors, certain polysaccharides bound to unconventional MHC class I or class I-like molecules such as CD1. Such T cells can develop outside the thymus, principally in the gut.

B-cell responses to TI-2 antigens provide a prompt and specific response to an important class of pathogens. Most extracellular bacterial pathogens have cell-wall polysaccharides that enable them to resist ingestion by phagocytes. This allows them not only to escape direct destruction by phagocytes but also to avoid stimulating T-cell responses through the presentation of bacterial peptides by macrophages. Antibody that is produced rapidly in response to this polysaccharide capsule without the help of peptide-specific T cells can coat such encapsulated **pyogenic bacteria**, promoting their ingestion and hence destruction. Both IgM and IgG antibodies are induced by TI-2 antigens and are likely to be an important part of the humoral immune response in many bacterial infections. We mentioned earlier the importance of antibodies to the capsular polysaccharide of *Haemophilus influenzae* B, a TI-2 antigen, in protective immunity to this bacterium. A further example of the importance of TI-2 responses can be seen in patients with an immunodeficiency disease known as the **Wiskott–Aldrich syndrome**. These patients can respond, although poorly, to protein antigens but fail to make antibody against polysaccharide antigens and are highly susceptible to infection with extracellular bacteria that have polysaccharide capsules. Thus, the TI responses are important components of the humoral immune response to non-protein antigens that are unable to recruit peptide-specific T-cell help; the distinguishing features of thymus-dependent, TI-1, and TI-2 antibody responses are summarized in Fig. 9.15.

	TD antigen	TI-1 antigen	TI-2 antigen
Antibody response in infants	Yes	Yes	No
Antibody production in congenitally athymic individual	No	Yes	Yes
Antibody response in absence of all T cells	No	Yes	No
Primes T cells	Yes	No	No
Polyclonal B-cell activation	No	Yes	No
Requires repeating epitopes	No	No	Yes
Examples of antigen	Diphtheria toxin Viral hemagglutinin Purified protein derivative (PPD) of *Mycobacterium tuberculosis*	Bacterial lipopoly-saccharide *Brucella abortus*	Pneumococcal poly-saccharide Salmonella polymerized flagellin Dextran Hapten-conjugated ficoll (polysucrose)

Fig. 9.15 Properties of different classes of antigen that elicit antibody responses.

Summary.

B-cell activation by many antigens, especially monomeric proteins, requires binding of the antigen by the B-cell surface immunoglobulin and interaction of the B cell with antigen-specific helper T cells. These helper T cells recognize peptide fragments derived from the antigen internalized by the B cell and act through the binding of CD40L to CD40 on the B cell and by the directed release of cytokines. The initial interaction occurs in the T-cell area of the lymphoid tissue, where both T and B cells are trapped as a consequence of binding antigen; further interactions between T and B cells occur after migration into the B-cell zone. Helper T cells induce a phase of vigorous B-cell proliferation, and direct the differentiation of the clonally expanded progeny of the naive B cells into either antibody-secreting plasma cells or memory B cells. During the differentiation of activated B cells, the antibody isotype can change in response to cytokines released by helper T cells and the antigen-binding properties of the antibody can change by somatic hypermutation of V-region genes. Somatic hypermutation and selection for high-affinity binding occur in germinal centers formed by B cells proliferating in lymphoid follicles, where antigen is displayed on the surface of follicular dendritic cells. Helper T cells control these processes by selectively activating cells that have retained their antigen specificity and by inducing proliferation and differentiation into plasma cells and memory B cells. Some non-protein antigens stimulate B cells in the absence of linked recognition by peptide-specific helper T cells. These thymus-independent antigens induce only limited isotype switching and do not induce memory B cells. However, responses to these antigens might have a critical role in host defense against pathogens whose surface antigens cannot elicit peptide-specific T-cell responses.

The distribution and functions of immunoglobulin isotypes.

Extracellular pathogens can find their way to most sites in the body and antibodies must be equally widely distributed to combat them. Most are distributed by diffusion from their site of synthesis but specialized transport mechanisms are required to deliver them to internal epithelial surfaces, such as those of the lung and intestine. The location of antibodies is determined by their isotype, which can limit their diffusion or enable them to engage specific transporters that deliver them across various epithelia. In this part of the chapter we shall describe the mechanisms whereby antibodies of different isotypes are directed to the compartments of the body in which their distinct effector functions are appropriate, and discuss the protective functions of antibodies that result solely from their binding to pathogens. In the last two parts of the chapter we shall discuss the effector cells and molecules that are specifically engaged by antibodies of the different isotypes.

9-11 Antibodies of different isotypes operate in distinct places and have distinct effector functions.

Pathogens most commonly enter the body across epithelial barriers presented by the mucosa of the respiratory, digestive, and urogenital tracts, or through damaged skin. Pathogens entering in this way can then establish infections in the tissues. Less often, insects, wounds, or hypodermic needles introduce microbes directly into the blood. The body's mucosal surfaces, tissues, and blood all need to be protected by antibodies from such infections, and antibodies of different isotypes are adapted to function in different compartments. Because a given variable region can become associated with any constant region through isotype switching, B cells can produce antibodies, all specific for the same eliciting antigen, that provide all of the protective functions appropriate for each body compartment.

The first antibodies to be produced in a humoral immune response are always IgM, because VDJ joining occurs just 5′ to the C_μ gene exons (see Figs 9.9 and 3.15). These early IgM antibodies are produced before B cells have undergone somatic hypermutation and therefore tend to be of low affinity. IgM molecules, however, form pentamers whose 10 antigen-binding sites can bind simultaneously to multivalent antigens, such as bacterial cell-wall polysaccharides, compensating for the relatively low affinity of the monomers by multipoint binding that confers high avidity. As a result of the large size of the pentamers, IgM is usually confined to the blood. Their pentameric structure also makes IgM antibodies especially potent in activating the complement system, as we shall see later. Infection of the bloodstream has serious consequences unless it is controlled quickly, and the rapid production of IgM and its efficient activation of the complement system are important in controlling such infections. IgM is also produced in secondary and subsequent responses, and after somatic hypermutation, although other isotypes dominate the later phases of a response.

Antibodies of the other isotypes—IgG, IgA, and IgE—are smaller, and diffuse easily out of the blood into the tissues. Although IgA, as we saw in Chapter 3, can form dimers, IgG and IgE are always monomeric. The affinity of the individual antigen-binding sites for antigen is therefore critical for the effectiveness of antibodies of these three isotypes, and B cells are selected for increased affinity of antigen-binding in the germinal centers, mainly after they have undergone switching to these isotypes. IgG is the principal isotype in the blood and extracellular fluid, whereas IgA is the principal isotype in secretions, the most important being those of the mucous epithelium of the intestinal and respiratory tracts. Whereas IgG efficiently opsonizes pathogens for engulfment by phagocytes and activates the complement system, IgA is a less potent opsonin and a weak activator of complement. This distinction is not surprising, as IgG operates mainly in the body tissues where accessory cells and molecules are available, whereas IgA operates mainly on body surfaces where complement and phagocytes are not normally present, and therefore functions chiefly as a neutralizing antibody.

Finally, IgE antibody is present only at very low levels in blood or extracellular fluid, but is bound avidly by receptors on mast cells that are found just beneath the skin and mucosa, and along blood vessels in connective tissue. Antigen binding to this IgE triggers mast cells to release powerful chemical mediators that induce reactions, such as coughing, sneezing, and vomiting, that can expel infectious agents. The distribution and main functions of antibodies of the different isotypes are summarized in Fig. 9.16.

Functional activity	IgM	IgD	IgG1	IgG2	IgG3	IgG4	IgA	IgE
Neutralization	+	−	++	++	++	++	++	−
Opsonization	−	−	+++	*	++	+	+	−
Sensitization for killing by NK cells	−	−	++	−	++	−	−	−
Sensitization of mast cells	−	−	+	−	+	−	−	+++
Activates complement system	+++	−	++	+	+++	−	+	−

Distribution	IgM	IgD	IgG1	IgG2	IgG3	IgG4	IgA	IgE
Transport across epithelium	+	−	−	−	−	−	+++ (dimer)	−
Transport across placenta	−	−	+++	+	++	+/−	−	−
Diffusion into extravascular sites	+/−	−	+++	+++	+++	+++	++ (monomer)	+
Mean serum level (mg ml^{-1})	1.5	0.04	9	3	1	0.5	2.1	3x10^{-5}

Fig. 9.16 Each human immunoglobulin isotype has specialized functions and a unique distribution. The major effector functions of each isotype (+++) are shaded in dark red, whereas lesser functions (++) are shown in dark pink, and very minor functions (+) in pale pink. The distributions are marked similarly, with actual average levels in serum being shown in the bottom row. *IgG2 can act as an opsonin in the presence of Fc receptors of a particular allotype, found in about 50% of Caucasians.

9-12 Transport proteins that bind to the Fc domain of antibodies carry specific isotypes across epithelial barriers.

The primary sites of synthesis of IgA antibodies, and their main loci of action, are at the epithelial surfaces of the body. IgA-secreting plasma cells are found predominantly in the connective tissue called the lamina propria, which lies immediately below the basement membrane of many surface epithelia. From there, the IgA antibodies must be transported across the epithelium to its external surface, for example to the lumen of the gut or the bronchi. IgA antibody synthesized in the lamina propria is secreted as an IgA dimeric molecule associated with a single J chain (see Fig. 3.22). This polymeric form of IgA binds specifically to a molecule called the poly-Ig receptor, which is present on the basolateral surfaces of the overlying epithelial cells (Fig. 9.17). When the poly-Ig receptor has bound a molecule of dimeric IgA, the complex is internalized and carried through the cytoplasm of the epithelial cell in a transport vesicle to its apical surface. This process is called **transcytosis**. At the apical surface of the epithelial cell, the poly-Ig receptor is cleaved enzymatically, releasing the extracellular portion of the receptor still attached to the Fc region of the dimeric IgA. This fragment of the receptor, called the **secretory component**, might help to protect the dimeric IgA from proteolytic cleavage. Some molecules of dimeric IgA diffuse from the lamina propria into the extracellular spaces of the tissues, draining into the bloodstream before being excreted into the gut via the bile. Therefore, it is not surprising that patients with obstructive jaundice, a condition in which bile is not excreted, show a marked increase in dimeric IgA in the plasma.

The principal sites of IgA synthesis and secretion are the gut, the respiratory epithelium, the lactating breast, and various other exocrine glands such as the salivary and tear glands. It is believed that the primary functional role of IgA antibodies is to protect epithelial surfaces from infectious agents, just as IgG antibodies protect the extracellular spaces of the internal milieu. IgA

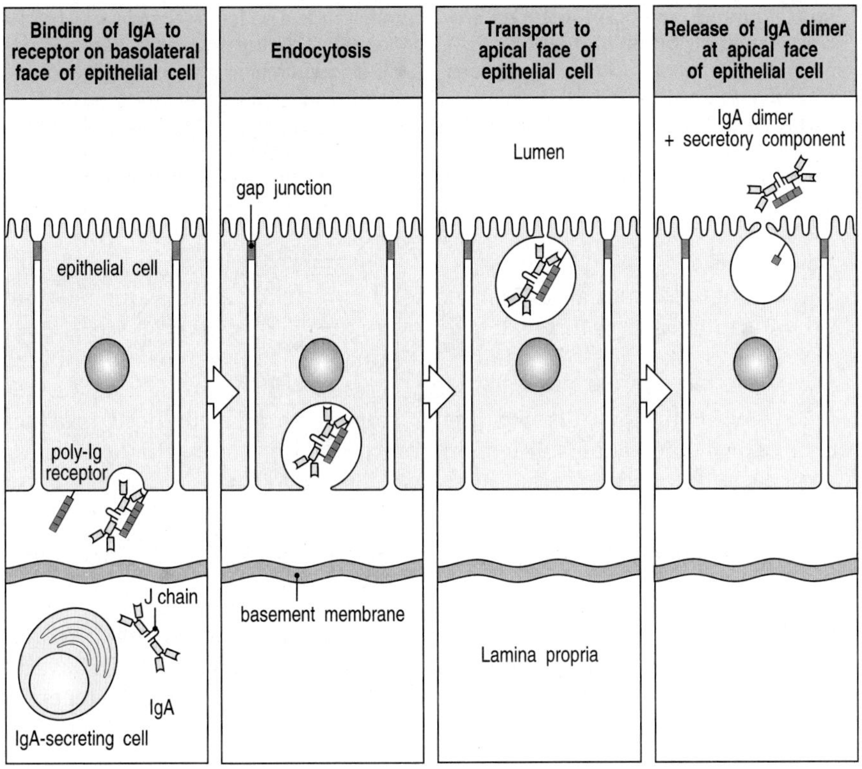

Fig. 9.17 Transcytosis of IgA antibody across epithelia is mediated by the poly-Ig receptor, a specialized transport protein. Most IgA antibody is synthesized in plasma cells lying just beneath epithelial basement membranes of the gut, respiratory epithelia, tear and salivary glands, and the lactating mammary gland. The IgA dimer bound to a J chain diffuses across the basement membrane and is bound by the poly-Ig receptor on the basolateral surface of the epithelial cell. The bound complex undergoes transcytosis in which it is transported in a vesicle across the cell to the apical surface, where the poly-Ig receptor is cleaved to leave the extracellular IgA-binding component bound to the IgA molecule as the so-called secretory component. The residual piece of the poly-Ig receptor is non-functional and is degraded. In this way, IgA is transported across epithelia into the lumen of several organs that are in contact with the external environment.

antibodies prevent the attachment of bacteria or toxins to epithelial cells or the absorption of foreign substances, and provide the first line of defense against a wide variety of pathogens. Newborn infants are especially vulnerable to infection, having had no prior exposure to the microbes in the environment they enter at birth. IgA antibodies are secreted in breast milk and are thereby transferred to the gut of the newborn infant, where they provide protection from newly encountered bacteria until the infant can synthesize its own protective antibody.

IgA is not the only protective antibody conferred on the infant by its mother. Maternal IgG is transported across the placenta directly into the bloodstream of the fetus during intrauterine life; human babies at birth have as high a level of plasma IgG as their mothers, and with the same range of specificities. The selective transport of IgG from mother to fetus results from a specific IgG transport protein in the placenta, FcRn, which is closely related in structure to MHC class I molecules. Despite this similarity, FcRn binds IgG quite differently from the binding of peptide in MHC class I, as its peptide-binding groove is occluded. It binds to the Fc portion of IgG molecules (Fig. 9.18). Two molecules of FcRn bind one molecule of IgG, bearing it across the placenta. In some rodents, FcRn also delivers IgG to the circulation of the neonate from the gut lumen. Maternal IgG is ingested by the newborn animals in colostrum, the protein-rich fluid in the postnatal mammary gland. In this case, transport is from the lumen of the gut into the blood and tissues. This receptor is found only in fetal and early postnatal life and seems to have as its counterpart in humans the transplacental FcRn.

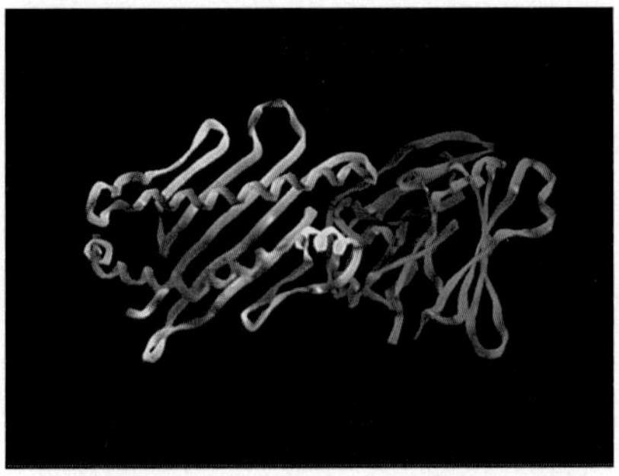

Fig. 9.18 FcRn binds to the Fc portion of IgG. The structure of a molecule of FcRn (white) bound to the Fc piece of IgG (blue) is shown. FcRn transports IgG molecules across the placenta in humans and across the gut in rats and mice. Photograph courtesy of P Björkman.

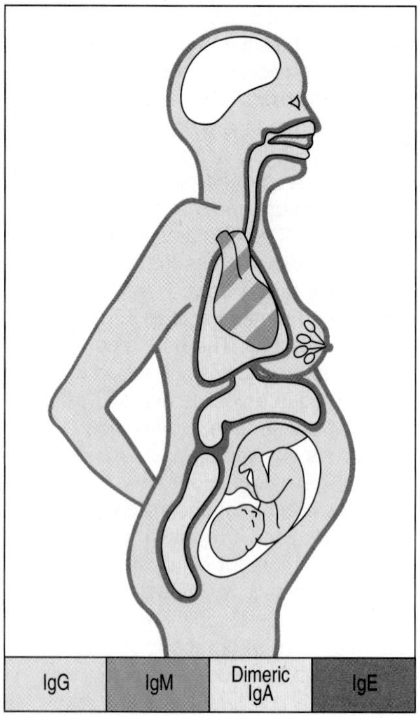

Fig. 9.19 Immunoglobulin isotypes are selectively distributed in the body. IgG and IgM predominate in plasma, whereas IgG and monomeric IgA are the major isotypes in extracellular fluid within the body. Polymeric IgA predominates in secretions across epithelia, including breast milk. The fetus receives IgG from the mother by transplacental transport. IgE is found mainly as mast-cell associated antibody just beneath epithelial surfaces (especially the respiratory tract, gastrointestinal tract, and skin). The brain is normally devoid of immunoglobulin.

By means of these specialized transport systems, mammals are supplied from birth with antibodies against pathogens common in their environments. As they mature and make their own antibodies of all isotypes, these are distributed selectively to different sites in the body (Fig. 9.19). Thus, throughout life, isotype switching and the distribution of isotypes through the body provides effective protection against infection in extracellular spaces.

9-13 High-affinity IgG and IgA antibodies can neutralize bacterial toxins.

Many bacteria cause disease by secreting proteins called bacterial toxins, which damage or disrupt the function of the host's cells (Fig. 9.20). To have an effect, a toxin must interact with a specific molecule that serves as a receptor

Fig. 9.20 Many common diseases are caused by bacterial toxins. Several examples are shown here. These toxins are all exotoxins—proteins secreted by the bacteria. Bacteria also have endotoxins, or non-secreted toxins, which are released when the bacterium dies. The endotoxins are also important in the pathogenesis of disease but there the host response is more complex (see Chapter 10).

Disease	Organism	Toxin	Effects *in vivo*
Tetanus	*Clostridium tetani*	Tetanus toxin	Blocks inhibitory neuron action leading to chronic muscle contraction
Diphtheria	*Corynebacterium diphtheriae*	Diphtheria toxin	Inhibits protein synthesis leading to epithelial cell damage and myocarditis
Gas gangrene	*Clostridium perfringens*	Clostridial toxin	Phospholipase activation leading to cell death
Cholera	*Vibrio cholerae*	Cholera toxin	Activates adenylate cyclase, elevates cAMP in cells, leading to changes in intestinal epithelial cells that cause loss of water and electrolytes
Anthrax	*Bacillus anthracis*	Anthrax toxic complex	Increases vascular permeability leading to edema, hemorrhage, and circulatory collapse
Botulism	*Clostridium botulinum*	Botulinum toxin	Blocks release of acetylcholine leading to paralysis
Whooping cough	*Bordetella pertussis*	Pertussis toxin	ADP-ribosylation of G proteins leading to lymphoproliferation
		Tracheal cytotoxin	Inhibits cilia and causes epithelial cell loss
Scarlet fever	*Streptococcus pyogenes*	Erythrogenic toxin	Vasodilation leading to scarlet fever rash
		Leukocidin Streptolysins	Kill phagocytes, allowing bacterial survival
Food poisoning	*Staphylococcus aureus*	Staphylococcal enterotoxin	Acts on intestinal neurons to induce vomiting. Also a potent T-cell mitogen (SE superantigen)
Toxic-shock syndrome	*Staphylococcus aureus*	Toxic-shock syndrome toxin	Causes hypotension and skin loss. Also a potent T-cell mitogen (TSST-1 superantigen)

on the surface of the target cell. In many toxins, the receptor-binding domain is on one polypeptide chain, whereas the toxic function is carried by a second chain. Antibodies that bind to the receptor-binding site on the toxin molecule can prevent the toxin from binding to the cell and thus protect the cell from toxic attack (Fig. 9.21). This protective effect of antibodies, as already mentioned, is called **neutralization**, and antibodies acting in this way are referred to as **neutralizing antibodies**.

Most toxins are active at nanomolar concentrations: a single molecule of diphtheria toxin can kill a cell. To neutralize toxins, therefore, antibodies must be able to diffuse into the tissues and bind the toxin rapidly and with high affinity. The diffusibility of IgG antibodies in the extracellular fluids and their high affinity make these the principal neutralizing antibodies for toxins found at this site. IgA antibodies similarly neutralize toxins at the mucosal surfaces of the body.

Diphtheria and tetanus toxins are among the bacterial toxins in which the toxic and the receptor-binding functions of the molecule are on two separate chains. It is therefore possible to immunize individuals, usually infants, with modified toxin molecules in which the toxic chain has been denatured. These modified toxin molecules, which are called toxoids, lack toxic activity but retain the receptor-binding site, so that immunization with the toxoid induces neutralizing antibodies effective in protecting against the native toxin.

With some insect or animal venoms, in which toxicity is such that a single exposure is capable of causing severe tissue damage or death, the adaptive immune response is too slow to be protective. Exposure to these venoms is a rare event and protective vaccines have not been developed for use in humans. Instead, for these toxins, neutralizing antibodies are generated by immunizing other species, such as horses, with insect and snake venoms to produce anti-venom antibodies (anti-venins) for use in protecting humans. Transfer of antibodies in this way is known as passive immunization (see Section 2-20).

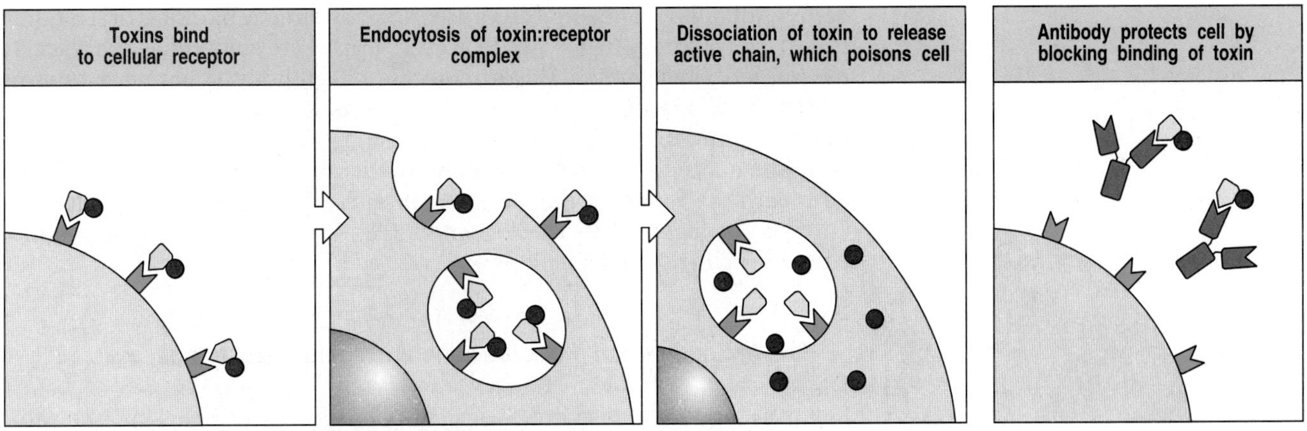
Toxins bind to cellular receptor
Endocytosis of toxin:receptor complex
Dissociation of toxin to release active chain, which poisons cell
Antibody protects cell by blocking binding of toxin

Fig. 9.21 Neutralization by IgG antibodies protects cells from toxin action. Many bacteria (as well as venomous insects and snakes) cause their damaging effects by elaborating toxic proteins (see Fig. 9.20). These toxins usually contain several distinct moieties. One part of the toxin molecule must bind a cellular receptor, which enables the molecule to be internalized. Another part of the toxin molecule then enters the cytoplasm and poisons the cell. In some cases, a single molecule of toxin can kill a cell. Antibodies that inhibit toxin binding can prevent, or neutralize, these effects.

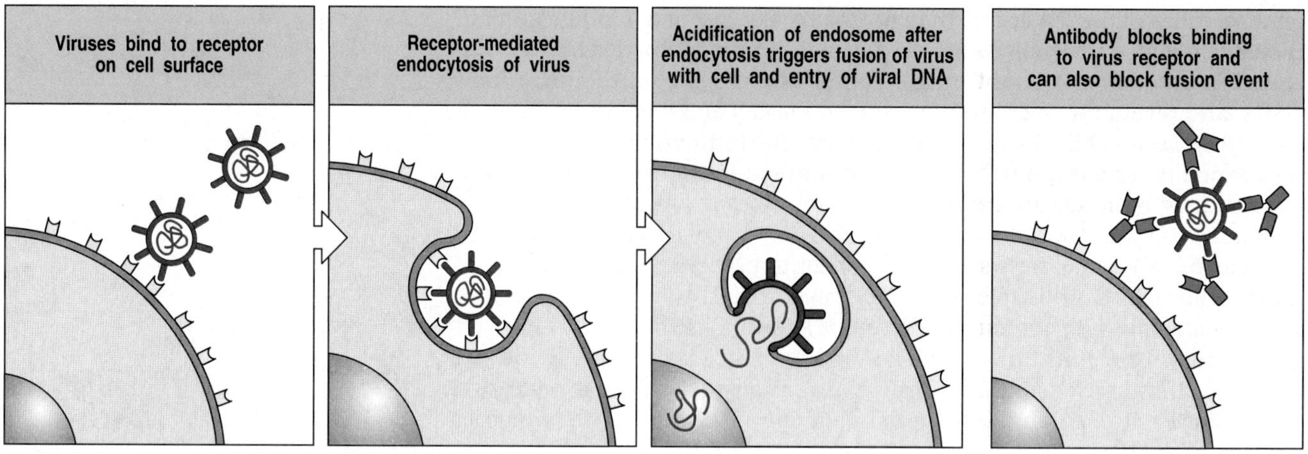

Fig. 9.22 Viral infection of cells can be blocked by neutralizing antibodies. For a virus to infect a cell, it must insert its genes into the cytoplasm. The first step in entry into a cell is usually the binding of the virus to a receptor on the cell surface. For enveloped viruses, as shown in the figure, entry into the cytoplasm requires fusion of the viral envelope and the cell membrane. For some viruses, this fusion event takes place on the cell surface (not shown); for others it can occur only within the more acidic environment of endosomes, as shown here. Non-enveloped viruses must also bind to receptors on cell surfaces but they enter the cytoplasm by disrupting endosomes. Antibodies bound to viral surface proteins neutralize the virus, inhibiting either its initial binding to the cell or its subsequent entry.

9-14 High-affinity IgG and IgA antibodies can inhibit the infectivity of viruses.

When animal viruses infect cells, they must first bind to a specific cell-surface protein, often a cell-type-specific protein that determines which cells they can infect. For example, the influenza virus carries a surface protein called **influenza hemagglutinin**, which binds to terminal sialic acid residues of the carbohydrate moieties found on certain glycoproteins expressed by epithelial cells of the respiratory tract. It is known as hemagglutinin because it recognizes similar sialic acid residues on chicken red blood cells and can agglutinate such cells by binding to these sites. Antibodies to the hemagglutinin can inhibit infection by the influenza virus. Such antibodies are called virus-neutralizing antibodies and, as with the neutralization of toxins, and for the same reasons, high-affinity IgA and IgG antibodies are particularly important in virus neutralization.

Many antibodies that neutralize viruses do so by directly blocking viral binding to surface receptors (Fig. 9.22). However, viruses are sometimes successfully neutralized when only a single molecule of antibody is bound to a virus particle that has many receptor-binding proteins on its surface. In these cases, the antibody must cause some change in the virus that disrupts its structure and either prevents it from interacting with its receptors or interferes with the fusion of the virus membrane with the cell surface after the virus has engaged its surface receptor. The viral nucleic acids thus cannot enter the cell and replicate there.

9-15 Antibodies can block the adherence of bacteria to host cells.

Many bacteria have specific cell-surface molecules, called adhesins, that allow them to bind to the surface of host cells. This adherence reaction is critical to the infectivity of these bacteria, whether they enter the cell, as occurs with some pathogens such as *Salmonella* species, or remain attached to the cell surface as extracellular pathogens (Fig. 9.23). For example, the bacterium *Neisseria gonorrhoeae*, the causative agent of the sexually transmitted disease gonorrhea, has a cell-surface protein known as pilin. Pilin allows the bacterium to adhere to the epithelial cells of the urinary and reproductive tracts and is essential to its infectivity. Antibodies against pilin can inhibit this adhesive reaction and prevent infection.

IgA antibodies secreted onto the mucosal surfaces of the intestinal, respiratory, and reproductive tracts are particularly important in preventing infection by preventing the adhesion of bacteria, viruses, or other pathogens to the epithelial cells that line these surfaces. The adhesion of bacteria to cells within the body can also contribute to pathogenesis, and IgG antibodies against adhesins can protect from damage in this way.

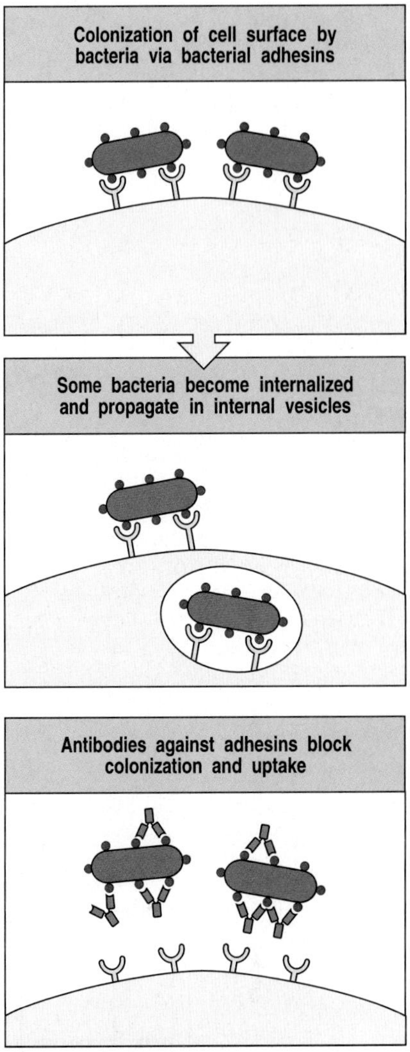

Fig. 9.23 Antibodies can prevent attachment of bacteria to cell surfaces. Many bacterial infections require an interaction between the bacterium and a cell surface. This is particularly true for infections of mucosal surfaces. The attachment process involves very specific molecular interactions between bacterial adhesins and their ligands on host cells; antibodies against bacterial adhesins can block such infections.

Summary.

The antibody response begins with antigen binding to IgM-expressing B cells and can then lead to the production of antibody of the same specificity of all different isotypes. Each isotype is specialized both in its localization in the body and in the functions it can perform. IgM antibodies are synthesized early in a response and are found mainly in blood. They are pentameric in structure and specialized to activate complement efficiently upon binding antigen. IgG antibodies are synthesized later in the response, are usually of higher affinity, and are found in blood and in extracellular fluid, where they can neutralize toxins, viruses, and bacteria, opsonize them for phagocytosis, and activate the complement system. IgA antibodies are synthesized as monomers, which enter blood and extracellular fluids, or as dimeric molecules in the lamina propria. The IgA dimers are then selectively transported across epithelia into sites such as the lumen of the gut, where they neutralize toxins and viruses, and block the entry of bacteria across the intestinal epithelium. Most IgE antibody is bound to the surface of mast cells that reside mainly just below body surfaces; antigen binding to this IgE triggers local defense reactions. Thus, each of these isotypes occupies a particular site in the body and has a particular role in defending the body against extracellular pathogens and their toxic products.

Fc receptor-bearing accessory cells in humoral immunity.

The ability of high-affinity antibodies to neutralize toxins, viruses, or bacteria can protect against infection but does not, on its own, solve the problem of how to remove the pathogens and their products from the body. Moreover, many pathogens are not neutralized by antibody and must be destroyed by other means. To dispose of neutralized microorganisms and to attack resistant extracellular pathogens, antibodies can activate a variety of **accessory effector cells** bearing receptors, the **Fc receptors**, specific for the Fc portion of antibodies of a particular isotype. These accessory cells include the phagocytic cells (macrophages and neutrophils), which ingest antibody-coated bacteria and kill them, and other cells—natural killer (NK) cells, eosinophils, basophils, and mast cells (see Fig. 1.4)—which are triggered to secrete stored mediators when their Fc receptors are engaged. Accessory cells are activated when their Fc receptors are aggregated by binding to multiple Fc pieces of antibody molecules bound to a pathogen.

9-16 The Fc receptors of accessory cells are signaling receptors specific for immunoglobulins of different isotypes.

The Fc receptors are a family of molecules that can bind to the Fc portion of immunoglobulin molecules. Each member of the family recognizes immunoglobulin of one or a few closely related isotypes through a recognition domain on the α chain of the Fc receptor. Fc receptors are themselves members of the immunoglobulin gene superfamily of proteins. Different accessory cells bear Fc receptors for antibodies of different isotypes, and the isotype of the antibody thus determines which accessory cell will be engaged in a given response. The different Fc receptors, the cells that express them, and their isotype specificity, are shown in Fig. 9.24.

Fc receptors, like the T-cell receptor, function as part of a multisubunit complex. Only the α chain is required for specific recognition; the other chains are required for transport to the cell surface and for signal transduction when Fc is bound. Indeed, signal transduction by most Fc receptors is mediated by a chain called the γ chain that is closely related to the T-cell receptor ζ chain. This is true of some Fcγ receptors, for the Fcα receptor, and for the high-affinity receptor for IgE; an exception is human FcγRII-A, in which the cytoplasmic domain of the α chain replaces the function of the γ chain. FcγRII-B1 and FcγRII-B2 function as inhibitory receptors by binding to the inositol 5′-phosphatase SHIP and are part of a regulatory mechanism that inhibits the activation of naive B cells, mast cells, macrophages, and neutrophils.

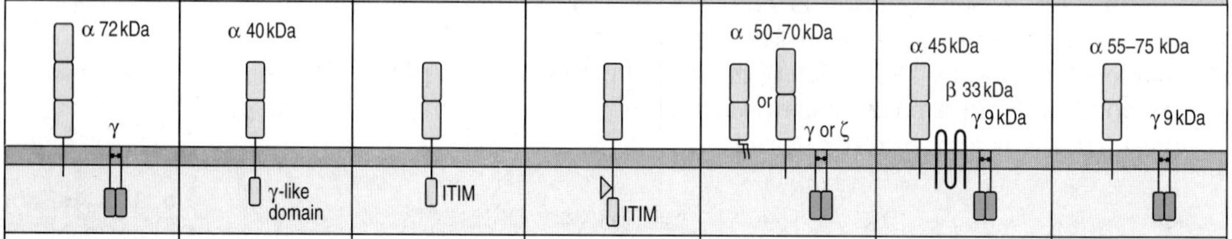

Receptor	FcγRI (CD64)	FcγRII-A (CD32)	FcγRII-B2 (CD32)	FcγRII-B1 (CD32)	FcγRIII (CD16)	FcεRI	FcαRI (CD89)
Structure	α 72kDa; γ	α 40kDa; γ-like domain	ITIM	ITIM	α 50–70kDa; or; γ or ζ	α 45kDa; β 33kDa; γ 9kDa	α 55–75 kDa; γ 9kDa
Binding	IgG1 10^8 M^{-1}	IgG1 2×10^6 M^{-1}	IgG1 2×10^6 M^{-1}	IgG1 2×10^6 M^{-1}	IgG1 5×10^5 M^{-1}	IgE 10^{10} M^{-1}	IgA1, IgA2 10^7 M^{-1}
Order of affinity	1) IgG1=IgG3 2) IgG4 3) IgG2	1) IgG1 2) IgG3=IgG2† 3) IgG4	1) IgG1=IgG3 2) IgG4 3) IgG2	1) IgG1=IgG3 2) IgG4 3) IgG2	IgG1=IgG3		IgA1=IgA2
Cell type	Macrophages Neutrophils* Eosinophils* Dendritic cells	Macrophages Neutrophils Eosinophils Platelets Langerhans' cells	Macrophages Neutrophils Eosinophils	B cells Mast cells	NK cells Eosinophils Macrophages Neutrophils Mast cells	Mast cells Eosinophils* Basophils	Macrophages Neutrophils Eosinophils††
Effect of ligation	Uptake Stimulation Activation of respiratory burst Induction of killing	Uptake Granule release (eosinophils)	Uptake Inhibition of stimulation	No uptake Inhibition of stimulation	Induction of killing (NK cells)	Secretion of granules	Uptake Induction of killing

Fig. 9.24 Distinct receptors for the Fc region of the different immunoglobulin isotypes are expressed on different accessory cells. The subunit structure and binding properties of these receptors and the cell types expressing them are shown. The complete multimolecular structure of most receptors is not yet known but they might all be multichain molecular complexes similar to the Fcε receptor I (FcεRI). The exact chain composition of any receptor can vary from one cell type to another. For example, FcγRIII in neutrophils is expressed as a molecule with a glycophosphoinositol membrane anchor, without γ chains, whereas in NK cells it is a transmembrane molecule associated with γ chains as shown. The binding affinities are taken from data on human receptors. *In these cases Fc receptor expression is inducible rather than constitutive. † Only some allotypes of FcγRII-A bind IgG2. †† In eosinophils, the molecular weight of CD89α is 70–100 kDa.

Although the most prominent function of Fc receptors is the activation of accessory cells against pathogens, they can also contribute in other ways to immune responses. For example, the FcγRII-B receptor negatively regulates some B-cell responses, whereas the same receptor on mast cells, macrophages, and neutrophils sets the threshold for activation of these cells by immune complexes. Fc receptors expressed by the Langerhans' cells of the skin enable them to ingest antigen:antibody complexes and present antigenic peptides to T cells. The stable binding of such complexes to the Fc receptors on the follicular dendritic cell surface enables them to drive the maturation of humoral immune responses.

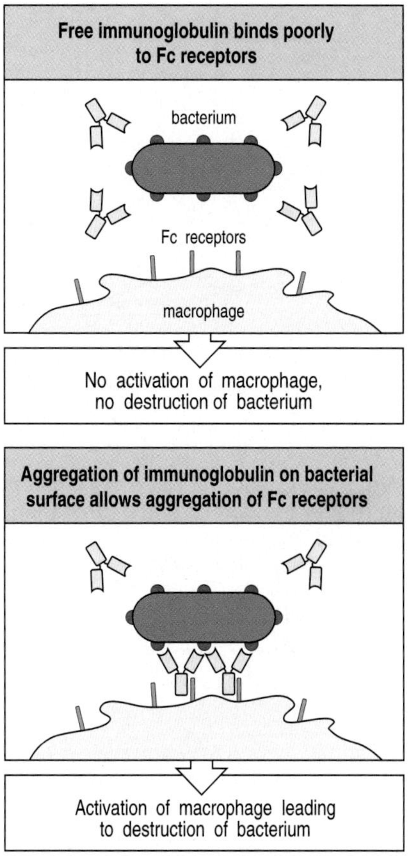

Fig. 9.25 Bound antibody is distinguishable from free immunoglobulin by its state of aggregation. Free immunoglobulin molecules bind Fc receptors with very low affinity. Antigen-bound immunoglobulin, however, can bind effectively to Fc receptors in a high-avidity interaction because several antibody molecules that are bound to the same surface bind to multiple Fc receptors on the surface of the accessory cell.

9-17 Fc receptors on phagocytes are activated by antibodies bound to the surface of pathogens.

Phagocytes are activated by IgG antibodies, especially IgG1 and IgG3, that bind to specific Fcγ receptors on the phagocyte surface (see Fig. 9.24). As phagocyte activation can initiate an inflammatory response and cause tissue damage, it is essential that the Fc receptors on phagocytes are able to distinguish antibody molecules bound to a pathogen from the majority of free antibody molecules that are not bound to anything. This condition is met by the aggregation or multimerization of antibodies that occurs when antibodies bind to multimeric antigens or to antigenic particles such as viruses and bacteria. If Fc receptors on the surface of an accessory cell bind an immunoglobulin monomer with low affinity, they will bind such antibody-coated particles with high avidity, and this is probably the principal mechanism by which bound antibodies are distinguished from free immunoglobulin (Fig. 9.25). The result is that Fc receptors enable accessory cells to detect pathogens through bound antibody molecules. Thus, specific antibody combined with Fc receptors provide the means by which accessory cells that lack intrinsic specificity can identify and remove pathogens and their products from the extracellular spaces of the body.

9-18 Fc receptors on phagocytes allow them to ingest and destroy opsonized extracellular pathogens.

The most important accessory cells in humoral immune responses are the phagocytic cells of the monocytic and myelocytic lineages, particularly the macrophages and the polymorphonuclear neutrophilic leukocytes or **neutrophils**. Phagocytosis is the ingestion of particles by cells and involves binding of the particle to the surface of the phagocyte, followed by its internalization and destruction.

Many bacteria are directly recognized, ingested, and destroyed by phagocytes, and these bacteria are not pathogenic in normal individuals (see Chapter 10). Bacterial pathogens, however, often have polysaccharide capsules that allow them to resist direct engulfment by phagocytes. These bacteria become susceptible to phagocytosis, however, when they are coated with antibody that engages the Fcγ or Fcα receptors on phagocytic cells, triggering the uptake and destruction of the bacteria (Fig. 9.26). Coating a microorganism with molecules that allow its destruction by phagocytes is known as opsonization. Bacterial polysaccharides, as we have seen, belong to the TI-2 class of thymus-independent antigens, and opsonization by thymus-independent antibodies produced in response to bacterial polysaccharides early in an immune response is important in ensuring the prompt destruction of many encapsulated bacteria.

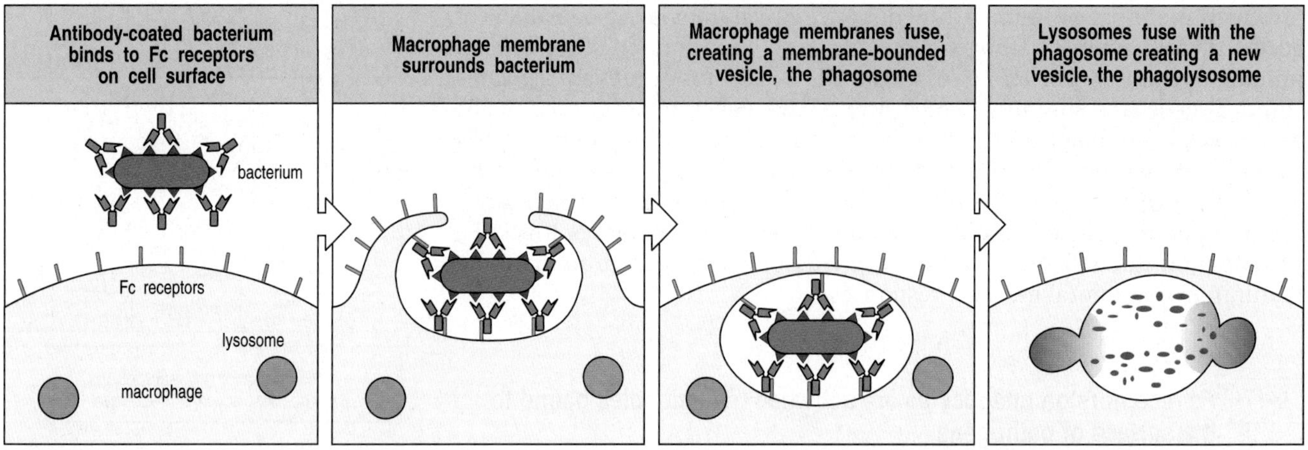
Antibody-coated bacterium binds to Fc receptors on cell surface
Macrophage membrane surrounds bacterium
Macrophage membranes fuse, creating a membrane-bounded vesicle, the phagosome
Lysosomes fuse with the phagosome creating a new vesicle, the phagolysosome
bacterium
Fc receptors
lysosome
macrophage

Fig. 9.26 A major function of Fc receptors on phagocytes is to trigger the uptake and degradation of antibody-coated bacteria. Many bacteria resist phagocytosis by macrophages and polymorphonuclear leukocytes. Antibodies binding to these bacteria, however, enable them to be ingested and degraded through interaction of the multiple Fc domains arrayed on the bacterial surface with Fc receptors on the phagocyte surface. Fc-receptor binding also signals the phagocyte to increase the rate of phagocytosis, fuse lysosomes with phagosomes, and increase its bactericidal activity.

Both the internalization and the destruction of microorganisms are greatly enhanced by interactions between the molecules coating an opsonized microorganism and their specific receptors on the phagocyte surface. When an antibody-coated pathogen binds to Fcγ receptors on the surface of a phagocytic cell, for example, the cell surface extends around the surface of the particle through successive binding of cellular Fcγ receptors to the antibody Fc domains bound to the pathogen surface. This is an active process that is triggered by the binding of Fcγ receptors. Endocytosis of the particle leads to its enclosure in an acidified cytoplasmic vesicle called a phagosome. The phagosome then fuses with one or more lysosomes to generate a phagolysosome, releasing the lysosomal enzymes into the phagosome interior where they destroy the bacterium (see Fig. 9.26).

Phagocytes can also damage bacteria through the generation of a variety of toxic products. The most important of these are hydrogen peroxide (H_2O_2), the superoxide anion (O_2^-), and nitric oxide (NO), which are directly toxic

Class of mechanism	Specific products
Acidification	pH=~3.5–4.0, bacteriostatic or bactericidal
Toxic oxygen-derived products	Superoxide O_2^-, hydrogen peroxide H_2O_2, singlet oxygen $^1O_2^{\bullet}$, hydroxyl radical $OH^{\cdot}$, hypohalite OCl
Toxic nitrogen oxides	Nitric oxide NO
Antimicrobial peptides	Defensins and cationic proteins
Enzymes	Lysozyme—dissolves cell walls of some Gram-positive bacteria. Acid hydrolases—further digest bacteria
Competitors	Lactoferrin (binds Fe) and vitamin B12 binding protein

Fig. 9.27 Ingestion of antibody-coated bacteria triggers production or release of many bactericidal agents in phagocytic cells. Most of these agents are made by both macrophages and neutrophils. Some of them are toxic; others, such as lactoferrin, work by binding essential nutrients and preventing their uptake by the bacteria. The same agents can be released by phagocytes interacting with large antibody-coated surfaces such as parasitic worms or host tissues. As these agents are also toxic to host cells, phagocyte activation can cause extensive tissue damage during an infection.

to bacteria. They are generated in a process known as the **respiratory burst**. Production of these metabolites is induced by the binding of aggregated antibodies to Fcγ receptors. The microbicidal products of activated phagocytes can also damage host cells; a series of enzymes, including catalase, which degrades hydrogen peroxide, and superoxide dismutase, which converts the superoxide anion into hydrogen peroxide, are also produced during phagocytosis. These control the action of these products so that they act primarily on pathogens within phagolysosomes. The agents whereby phagocytic cells damage and destroy ingested bacteria are summarized in Fig. 9.27.

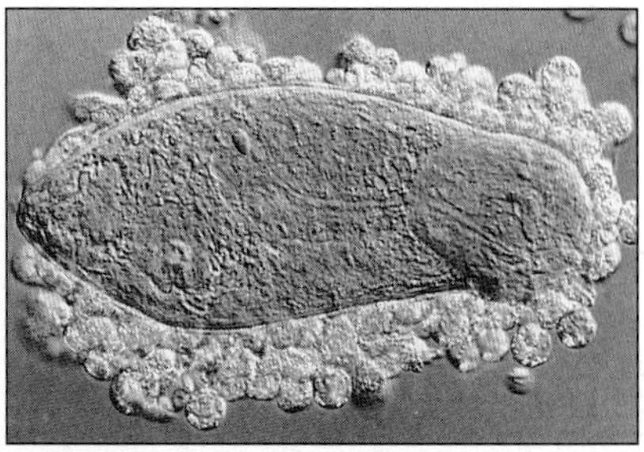

Fig. 9.28 Eosinophils attacking a schistosome larva in the presence of serum from an infected patient. Large parasites, such as worms, cannot be ingested by phagocytes; however, when the worm is coated with antibody, especially IgE, eosinophils can attack it via the high-affinity FcεRI. Similar attacks can be mounted by other Fc receptor-bearing cells on various large targets. Photograph courtesy of A Butterworth.

Some particles are too large for a phagocyte to ingest—parasitic worms are one example. In this case, the phagocyte attaches to the surface of the parasite (Fig. 9.28) via its Fcγ, Fcα or Fcε receptors, and the lysosomes fuse with the attached surface membrane. This reaction discharges the contents of the lysosome onto the surface of the antibody-coated parasite, damaging it directly in the extracellular space. Whereas the principal phagocytes in the destruction of bacteria are macrophages and neutrophils, large parasites such as helminths are more usually attacked by eosinophils. Thus, Fcγ and Fcα receptors can trigger the internalization of external particles by phagocytosis, or the externalization of internal vesicles by exocytosis. The latter process is usually mediated by antigen crosslinking of IgE bound to the high-affinity FcεRI. We shall see in the next three sections that NK cells and mast cells also release mediators stored in their vesicles when their Fc receptors are aggregated.

9-19 Fc receptors activate natural killer cells to destroy antibody-coated targets.

Infected cells are usually destroyed by T cells alerted by foreign peptides bound to cell-surface MHC molecules. However, virus-infected cells can also signal the presence of intracellular infection by expressing on their surfaces viral proteins that can be recognized by antibodies. Cells bound by such antibodies can then be killed by a specialized non-T, non-B lymphoid cell called a **natural killer cell** (**NK cell**).

Natural killer cells are large lymphoid cells with prominent intracellular granules; they make up a small fraction of peripheral blood lymphoid cells. These cells bear no known antigen-specific receptors but are able to recognize and kill a limited range of abnormal cells. They were first discovered because of their ability to kill some tumor cells but are now known to have an important role in innate immunity, as will be discussed in Chapter 10.

The destruction of antibody-coated target cells by NK cells is called **antibody-dependent cell-mediated cytotoxicity** (**ADCC**) and is triggered when antibody bound to the surface of a cell interacts with Fc receptors on the NK cell (Fig. 9.29). NK cells express the receptor FcγRIII (CD16). FcγRIII recognizes the IgG1 and IgG3 subclasses and triggers cytotoxic attack by the NK cell on antibody-coated target cells. The mechanism of attack is exactly analogous to that of cytotoxic T cells, involving the release of cytoplasmic granules containing perforin and granzymes (Section 8-21). The importance of ADCC in defense against infection with bacteria or viruses has not yet been fully established. However, ADCC represents yet another mechanism by which, through engaging an Fc receptor, antibodies can direct an antigen-specific attack by an effector cell lacking specificity for antigen.

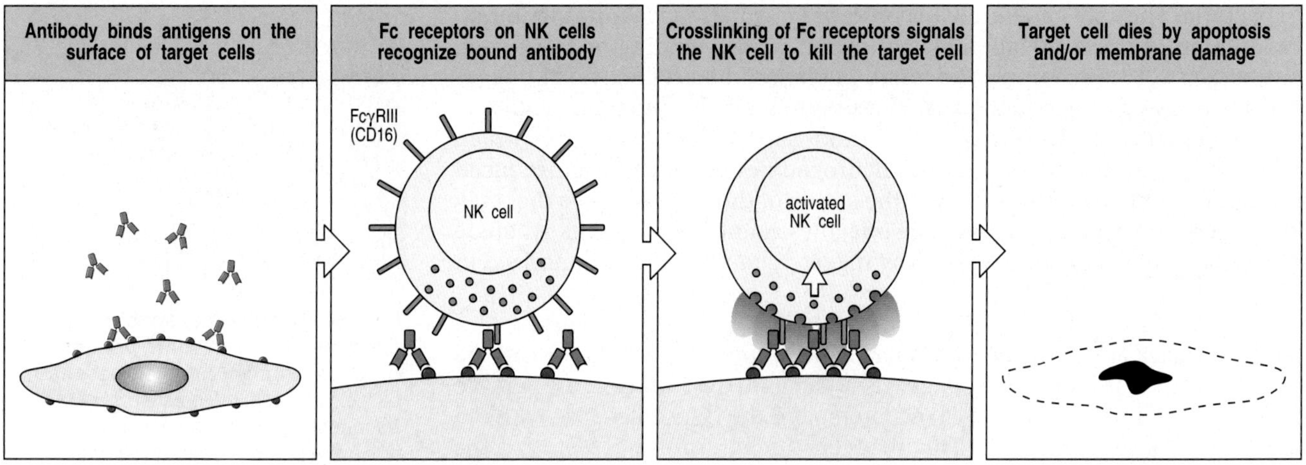
Antibody binds antigens on the surface of target cells
Fc receptors on NK cells recognize bound antibody
FcγRIII (CD16)
NK cell
Crosslinking of Fc receptors signals the NK cell to kill the target cell
activated NK cell
Target cell dies by apoptosis and/or membrane damage

Fig. 9.29 Antibody-coated target cells can be killed by natural killer cells (NK cells) in antibody-dependent cell-mediated cytotoxicity (ADCC). NK cells are large granular non-T, non-B lymphoid cells that have FcγRIII (CD16) on their surface. When these cells encounter cells coated with IgG antibody, they rapidly kill the target cell. The importance of ADCC in host defense or tissue damage is still controversial.

9-20 Mast cells, basophils, and activated eosinophils bind IgE antibody with high affinity.

When pathogens cross epithelial barriers and establish a local focus of infection, the host must mobilize its defenses and direct them to the site of pathogen growth. One mechanism by which this is achieved is to activate a specialized cell type known as a **mast cell.** Mast cells are large cells containing distinctive cytoplasmic granules that contain a mixture of chemical mediators, including **histamine**, that act rapidly to make local blood vessels more permeable. Mast cells have a distinctive appearance after staining with the dye toluidine blue that makes them readily identifiable in tissues (see Fig. 1.4). They are found in particularly high concentrations in vascularized connective tissues just beneath body epithelial surfaces, including the submucosal tissues of the gastrointestinal and respiratory tracts and the dermis that lies just below the epidermal layers of the skin.

Mast cells can be activated to release their granules, and to secrete lipid mediators and cytokines via antibody bound to Fc receptors specific for IgE (FcεRI) and IgG (FcγRIII). We have seen earlier that most Fc receptors bind stably to the Fc region of antibodies only when these are bound to antigen. By contrast, FcεRI binds monomeric IgE antibodies with a very high affinity, measured at approximately 10^{10} M^{-1}. Thus, even at the low levels of IgE found circulating in normal individuals, a substantial portion of the total IgE is bound to the FcεRI found on mast cells and their circulating counterparts, the basophilic granulocytes or **basophils. Eosinophils** are granulocytes that can express Fc receptors, but these cells can express FcεRI only when activated and recruited to an inflammatory site.

Although mast cells are usually stably associated with bound IgE, they are not activated simply by the binding of monomeric antigens to cell-surface IgE. Mast-cell activation occurs when the bound IgE is crosslinked by multivalent antigen. This signal activates the mast cell to release the contents of its prominent granules, to synthesize and release lipid mediators such as prostaglandin D_2 and leukotriene C_4, and to secrete cytokines, thereby initiating a local inflammatory response. The immediate consequence of antigen crosslinking of the IgE displayed on the mast-cell surface is degranulation, which occurs within seconds (Fig. 9.30). This releases the stored histamine,

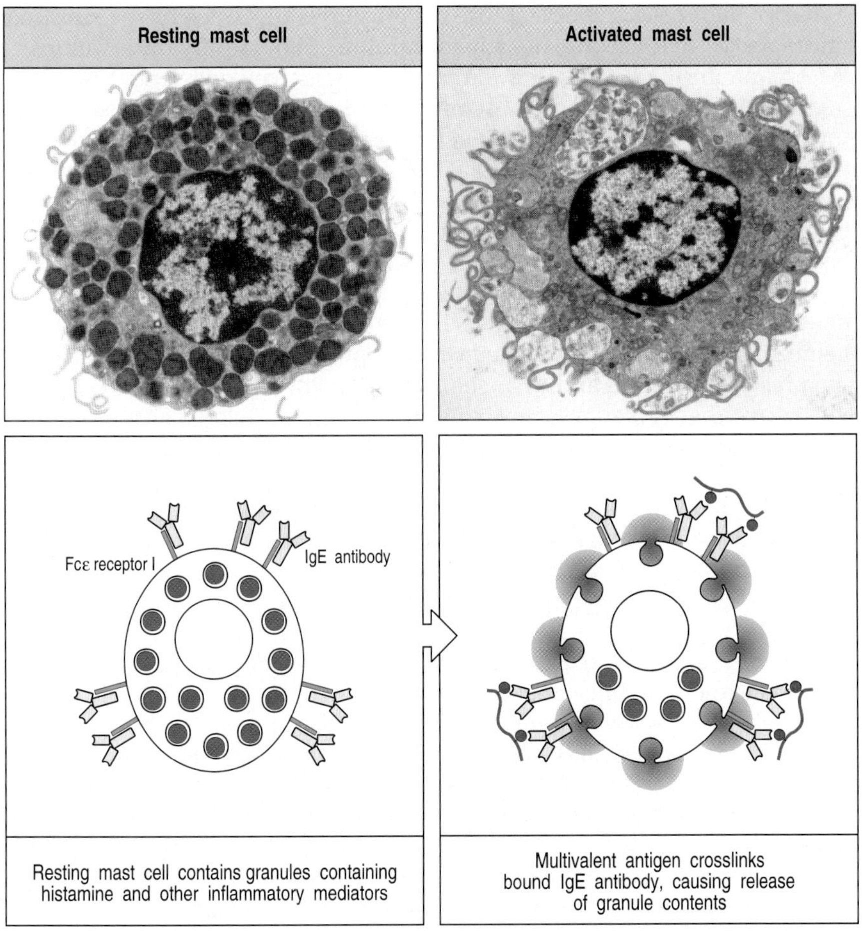
Resting mast cell
Activated mast cell
Fcε receptor I
IgE antibody
Resting mast cell contains granules containing histamine and other inflammatory mediators
Multivalent antigen crosslinks bound IgE antibody, causing release of granule contents

Fig. 9.30 IgE antibody crosslinking on mast-cell surfaces leads to a rapid release of inflammatory mediators. Mast cells are large cells found in connective tissue that can be distinguished by secretory granules containing many inflammatory mediators. They bind stably to monomeric IgE antibodies through the very high-affinity Fcε receptor I. Antigen crosslinking of the bound IgE antibody molecules triggers rapid degranulation, releasing inflammatory mediators into the surrounding tissue. These mediators trigger local inflammation, which recruits cells and proteins required for host defense to sites of infection. Photographs courtesy of A M Dvorak.

causing a local increase in blood flow and vascular permeability that quickly leads to fluid accumulation in the surrounding tissue, including antibody molecules. Shortly afterwards, there is an influx of blood-borne cells such as polymorphonuclear leukocytes and later macrophages, eosinophils, and effector lymphocytes. This influx can last from a period of a few minutes to a few hours and produces a specific inflammatory response at a site of infection. Thus, mast cells form a part of the front line of host defenses against pathogens that enter the body across epithelial barriers.

9-21 IgE-mediated activation of accessory cells has an important role in resistance to parasite infection.

Mast cells are thought to serve at least three important functions in host defense. First, their location near body surfaces allows them to recruit both specific and non-specific effector elements to sites where infectious agents are most likely to enter the internal milieu. Second, they also increase the flow of lymph from sites of antigen deposition to the regional lymph nodes, where naive lymphocytes are first activated. Third, their ability to trigger muscular contraction can contribute to the physical expulsion of pathogens from the lungs or the gut. Mast cells respond rapidly to the binding of antigen to surface-bound IgE antibodies, and their activation leads to the recruitment and activation of basophils and eosinophils, which contribute further to the IgE-mediated response. There is increasing evidence that such IgE-mediated responses are crucial to defense against parasite infestation.

A role for mast cells in the clearance of parasites is suggested by the intestinal mastocytosis that accompanies helminth infection, and by observations of W/W^V mutant mice, which have a profound mast-cell deficiency caused by mutation of the gene c-*kit*. These mutant mice show impaired clearance of the intestinal nematodes *Trichinella spiralis* and *Strongyloides* species. Clearance of *Strongyloides* species is even more impaired in W/W^V mice that lack IL-3 and therefore fail to produce basophils in response to infection with worms, as well as lacking functional mast cells. Thus both mast cells and basophils seem to contribute to defense against these helminth parasites. Other evidence points to the importance of IgE antibodies and eosinophils in defense against parasites. Infections by certain classes of parasite, particularly helminths, are strongly associated with the production of IgE antibodies and the presence of blood and tissue eosinophilia. Furthermore, experiments in mice show that depletion of eosinophils by using polyclonal anti-eosinophil antisera increases the severity of infection on exposure to the parasitic helminth *Schistosoma mansoni*. Eosinophils seem to be directly responsible for helminth destruction; examination of infected tissues shows degranulated eosinophils adherent to helminths, and experiments *in vitro* have shown that eosinophils can kill *Schistosoma mansoni* in the presence of specific IgE (see Fig. 9.28), IgG, or IgA anti-schistosome antibodies.

The role of IgE, mast cells, basophils, and eosinophils can also be seen in resistance to the feeding of blood-sucking ixodid ticks. Normal skin at the site of a tick bite shows degranulated mast cells, and an accumulation of basophils and eosinophils, which are degranulated, an indicator of recent activation. Resistance to subsequent feeding by these ticks develops after the first exposure, suggesting a specific immunological mechanism. Mast cell-deficient mice show no such acquired resistance to tick species, and in guinea pigs the depletion of either basophils or eosinophils by using specific polyclonal antibodies also reduces resistance to tick feeding. Finally, recent experiments have shown that resistance to ticks in mice is mediated by specific IgE antibody.

Thus, many data from both clinical studies and experiments support a role for this system of IgE binding to the high-affinity FcεRI in host resistance to pathogens that enter across epithelia. We shall see later, in Chapter 12, that this same system accounts for many of the symptoms in allergic diseases such as asthma, hay-fever, and the life-threatening response known as systemic anaphylaxis.

Summary.

Antibody-coated pathogens are recognized by accessory effector cells through Fc receptors that bind to the Fc domains of the bound antibodies. This binding activates the accessory cell and triggers the destruction of the pathogen. Fc receptors comprise a family of molecules, each of which recognizes immunoglobulins of specific isotypes. Fc receptors on macrophages and neutrophils recognize the constant regions of IgG or IgA antibodies bound to the surface of a pathogen and trigger the engulfment and destruction of IgG- or IgA-coated bacteria by these phagocytic cells. Binding to the Fc receptor also induces the production of microbicidal agents in the intracellular vesicles of the phagocyte. Eosinophils are important in the elimination of parasites too large to be engulfed; they bear Fc receptors specific for the constant region of IgG, as well as high-affinity receptors for IgE; aggregation of these receptors triggers the release of toxic substances onto the surface of the parasite. NK cells, tissue mast cells, and blood basophils also release their granule contents when their Fc receptors are engaged. The high-affinity receptor for IgE, which is expressed constitutively by mast cells and basophils and by activated eosinophils, differs from other Fc receptors in that it can bind free monomeric antibody, thereby allowing an immediate response to pathogens at the site of entry into the tissues. When the IgE bound to the surface of a mast cell is aggregated by binding to antigen, it triggers the release of histamine and many other mediators that increase the blood flow to sites of infection; it thereby recruits antibodies and effector cells to these sites. Mast cells are found principally below epithelial surfaces of the skin and the digestive and respiratory tracts, and their activation by innocuous substances is responsible for many of the symptoms of acute allergic reactions.

The complement system in humoral immunity.

Complement was discovered many years ago as a heat-labile component of normal plasma that augments the opsonization of bacteria by antibodies and allows some antibodies to kill bacteria. This activity was said to 'complement' the antibacterial activity of antibody, hence the name complement. The complement system is made up of a large number of distinct plasma proteins; one is activated directly by antigen-bound antibody to trigger a cascade of reactions, each of which results in the activation of another complement component. Some activated complement proteins bind covalently to bacteria, opsonizing them for engulfment by phagocytes bearing complement receptors. Small fragments of some complement proteins act as chemoattractants to recruit phagocytes to the site of complement activation and to activate them. The **terminal complement components** damage certain bacteria by creating pores in the bacterial membrane.

The effector functions of complement can be activated through three pathways (Fig. 9.31). The **classical pathway** is activated by antibody's binding to antigen. The **mannan-binding lectin pathway** (MBLectin pathway) is initiated by binding of a serum lectin, the mannan-binding lectin, to mannose-containing carbohydrates on bacteria or viruses. Finally, the **alternative pathway** can be initiated when a spontaneously activated complement component binds to the surface of a pathogen. It provides an amplification loop for the classical pathway of complement activation because one of the activated components

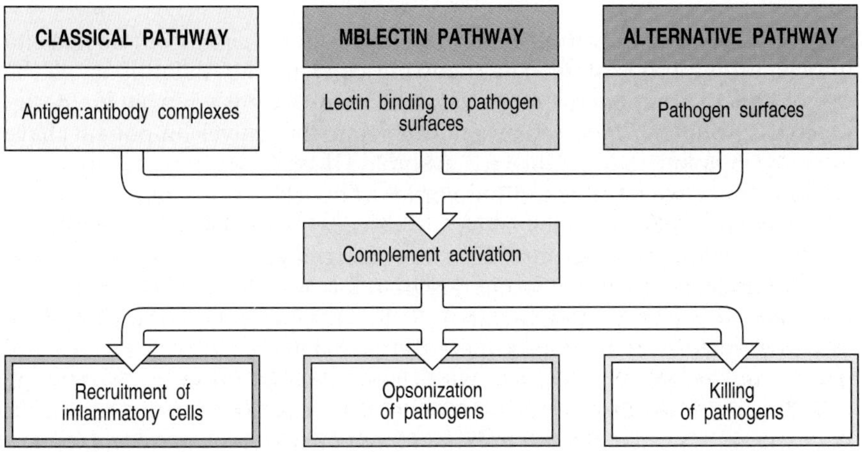

Fig. 9.31 Schematic overview of the complement cascade. There are three pathways of complement activation: the classical pathway, which is triggered by antibody; the MBLectin pathway, which is triggered by mannan-binding lectin (MBL), a normal serum constituent that binds some encapsulated bacteria; and the alternative pathway, which is triggered directly on pathogen surfaces. They all generate a crucial enzymatic activity that, in turn, generates the effector molecules of complement. The three main consequences of complement activation are opsonization of pathogens, the recruitment of inflammatory cells, and direct killing of pathogens.

of the classical pathway can also initiate the alternative pathway. Complement thus seems to be an important part of innate humoral immunity that has been harnessed by the adaptive humoral response. We shall focus here on the classical pathway, touching on the alternative pathway in its role in amplifying the classical pathway but leaving the details of how the lectin and alternative pathways can be initiated in the absence of antibody for Chapter 10, where we discuss innate mechanisms of immunity.

9-22 Complement is a system of plasma proteins that interact with bound antibodies and surface receptors to aid in the elimination of pathogens.

In the humoral immune response, the binding of either IgM or most classes of IgG to a pathogen activates the complement cascade. The events of the complement cascade can be divided into two sequences of reactions, which we shall call 'early' and 'late' events. The early events consist of a series of proteolytic steps in which an inactive precursor protein is cleaved to yield a large active fragment, which binds to the surface of a pathogen and contributes to the next cleavage, and a small peptide fragment that is released from the cell and often mediates inflammatory responses. The early events end with the production of a protease called a **C3 convertase**, which binds covalently to the pathogen surface. Here it generates large amounts of the main effector molecule of the complement system—an opsonin that binds covalently to pathogen surfaces and thereby targets them for destruction. The C3 convertase also generates a C5 convertase that produces the most important small peptide mediator of inflammation, as well as a large active fragment that initiates the late events of complement activation. These comprise a sequence of polymerization reactions in which the terminal complement components interact to form a membrane-attack complex, which creates a pore in membranes of certain pathogens that can lead to their death.

The C3 convertase thus occupies a central position in the complement cascade (Fig. 9.32). The reactions triggered by bound antibody molecules are called the classical pathway of complement activation because this pathway was discovered first. However, the alternative pathway of complement activation, in which the early events are triggered in the absence of antibody, probably arose first in evolution. The MBLectin-dependent pathway might be an evolutionary intermediate (see Chapter 10). Each pathway generates a C3 convertase by a different route but two of the three convertases are identical and all are homologous and have the same activity; so the principal effector molecules and the late events are the same for all three pathways.

The nomenclature of complement proteins is often a significant obstacle to understanding this system. The following conventional definitions will be used here. All components of the classical complement pathway are designated by the letter C followed by a number, and the native components have a simple number designation, for example C1 and C2. Unfortunately, the components were numbered in the order of their discovery rather than the sequence of reactions, which is C1, C4, C2, C3, C5, C6, C7, C8, and C9. The products of the cleavage reactions are designated by added lower-case letters, the larger fragment being designated b and the smaller a; thus, for example, C4 is cleaved to C4b and C4a. The components of the alternative pathway, instead of being numbered, are designated by different capital letters, for example B and D. As with the classical pathway, their cleavage products are designated by the addition of lowercase a and b: thus, the large fragment of B is called Bb and the small fragment Ba. Activated complement

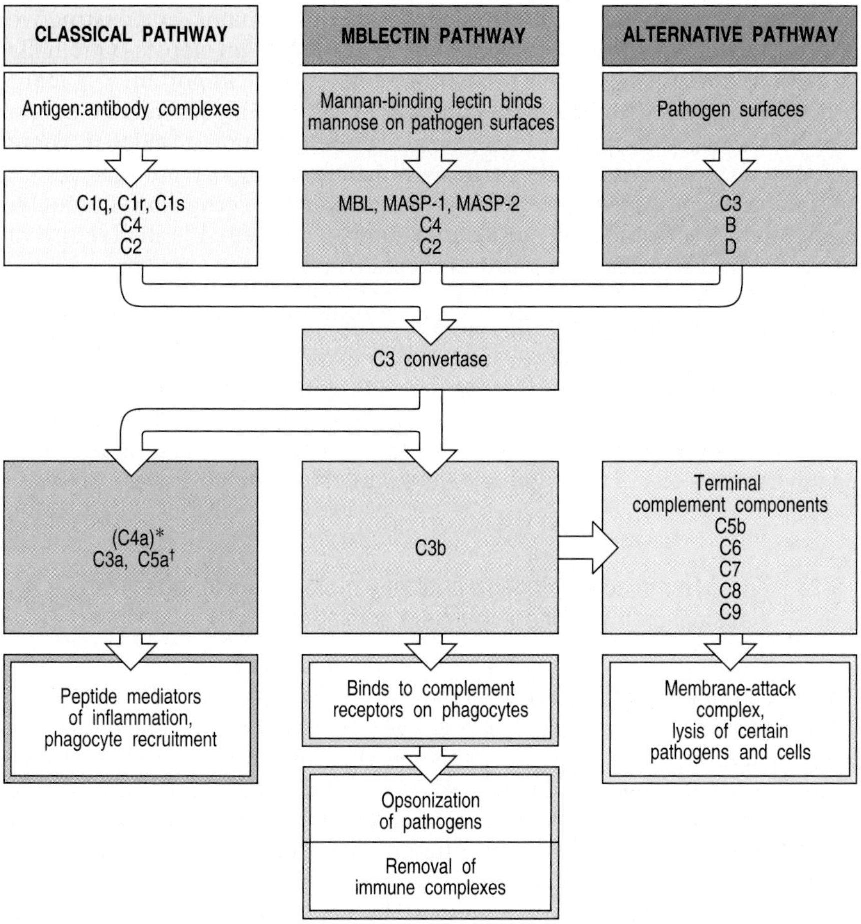

Fig. 9.32 Overview of the main components and effector actions of complement. The early events of all three pathways of complement activation involve a series of cleavage reactions culminating in the formation of an enzymatic activity called a C3 convertase, which cleaves complement component C3. In the MBLectin pathway MASP stands for *m*annan-binding lectin-*a*ssociated *s*erum *p*rotease. The production of C3 convertase activity is the point at which the three pathways converge and the effector functions of complement are generated. The larger cleavage fragment of C3 (C3b) binds covalently to the membrane and opsonizes bacteria, enabling phagocytes to internalize them. The small fragments of C5 and C3, called C5a and C3a, are peptide mediators of local inflammation. *C4a is generated by the cleavage of C4 during the early events of the classical pathway (and not by the action of C3 or C5 convertase); it is also a peptide mediator of inflammation but its effects are relatively weak. Similarly, the large cleavage fragment of C4, C4b, is a weak opsonin (not shown). Finally, the C3b bound to the C3 convertase forms a C5 convertase which binds C5 and generates C5a (marked with a † to indicate its generation by C5 convertase), and C5b, which associates with the bacterial membrane and triggers the late events, in which the terminal components of complement assemble into a membrane-attack complex that can damage the membrane of certain pathogens.

components are often designated by a horizontal line, for example $\overline{\text{C2b}}$; however, we shall not use this convention. It is also useful to be aware that the large active fragment of C2 was originally designated C2a, and is still called that in some texts and research papers. Here, for consistency, we shall call all large fragments of complement b, so the large active fragment of C2 will be designated C2b. In the classical and MBLectin pathways the C3 convertase enzyme is formed from membrane-bound C4b complexed with C2b. In the alternative pathway, a homologous C3 convertase is formed from membrane-bound C3b complexed with Bb.

An overview of the complement system is shown in Fig. 9.32. The generation of the C3 convertase, so called because it is specific for the cleavage of complement component C3, results in the rapid cleavage of many molecules of C3 to produce C3b, which can bind covalently to the pathogen surface. The cleavage of C3 and the binding of large numbers of C3b molecules to the surface of the pathogen is a pivotal event in complement activation. At this point the pathways of complement activation converge, the main effector activities of complement are generated, and the late events begin, with C3b having a central role. Bound C3b and its derivative fragments are the major opsonins of the complement system, binding to complement receptors on phagocytes and facilitating the engulfment of the pathogen. C3b also binds C5, allowing it to be cleaved by the C2b component of the C4b,2b,3b C5 convertase to initiate the assembly of the membrane-attack complex. C5a and C3a mediate local inflammatory responses, recruiting fluid, cells, and proteins to the site of infection, and activating phagocytic cells. Finally, the binding of C3b initiates the alternative pathway, thereby amplifying complement activation.

Functional protein classes in the complement system	
Binding to antigen: antibody complexes	C1q
Binding to mannose on bacteria	MBL
Activating enzymes	C1r C1s C2b Bb D MASP-1 MASP-2
Membrane-binding proteins and opsonins	C4b C3b
Peptide mediators of inflammation	C5a C3a C4a
Membrane-attack proteins	C5b C6 C7 C8 C9
Complement receptors	CR1 CR2 CR3 CR4 C1qR
Complement-regulatory proteins	C1INH C4bp CR1 MCP DAF H I P CD59

Fig. 9.33 Functional protein classes in the complement system.

It is clear that a pathway leading to such potent inflammatory and destructive effects, and which, moreover, has a built-in amplification step, is potentially dangerous and must be subject to tight regulation. One important safeguard is that key activated complement components are rapidly inactivated unless they bind to the pathogen surface on which their activation is initiated. There are also several points in the pathway at which regulatory proteins act on complement components to prevent the inadvertent activation of complement on host cells and hence accidental damage to them. We shall return to these regulatory mechanisms at the end of this part of the chapter.

We have now introduced all the relevant components of complement, albeit in a superficial manner, and we are ready for a more detailed account of their functions. To help in distinguishing the different components according to their functions, we shall use a color code in figures in this section that list the various components and their activities: this is introduced in Fig. 9.33, where all the components of complement are grouped by function.

9-23 The C1q molecule binds to antibody molecules to trigger the classical pathway of complement activation.

The first component of the classical pathway of complement activation is C1, which is a complex of three proteins called C1q, C1r, and C1s, two molecules each of C1r and C1s being bound to each molecule of C1q (Fig. 9.34). Complement activation is initiated when antibodies attached to the surface of a pathogen bind C1q. C1q can be bound by either IgM or IgG antibodies (see Fig. 9.16) but, because of the structural requirements of binding to C1q, neither of these antibody isotypes can activate complement in solution; the cascade is initiated only when they are bound to multiple sites on a cell surface, normally that of a pathogen.

The C1q molecule has six globular heads joined to a common stem by long, filamentous domains that resemble collagen molecules; the whole C1q complex has been likened to a bunch of six tulips held together by the stems. Each globular head can bind to one Fc domain, and binding of two or more globular heads activates the C1q molecule. In plasma, the pentameric IgM molecule has a planar conformation that does not bind C1q (Fig. 9.35, left panel); however, binding to the surface of a pathogen deforms the IgM pentamer so that it looks like a staple (see Fig. 9.35, right panel), and this distortion exposes binding sites for the C1q heads. Although C1q binds with low affinity to some subclasses of IgG in solution, the binding energy required

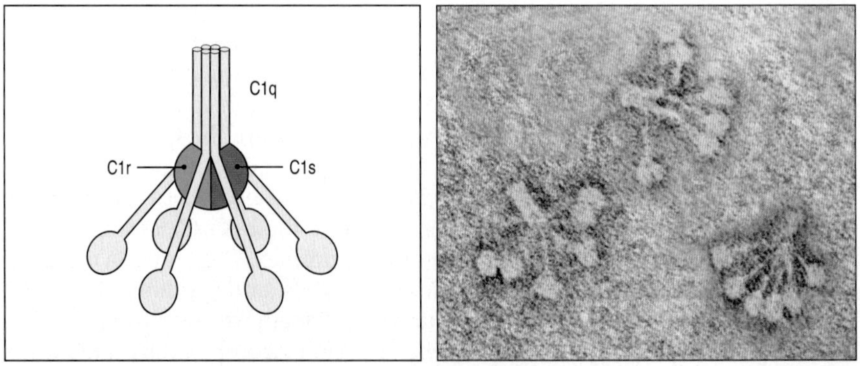

Fig. 9.34 The first protein in the classical pathway of complement activation is C1, which is a complex of C1q, C1r, and C1s. C1q is composed of six identical subunits with globular heads and long, collagen-like tails. The tails bind to two molecules each of C1r and C1s to form the C1 complex $C1q{:}C1r_2{:}C1s_2$; the heads bind to the Fc domains of immunoglobulin molecules. Photograph (× 500,000) courtesy of K B M Reid.

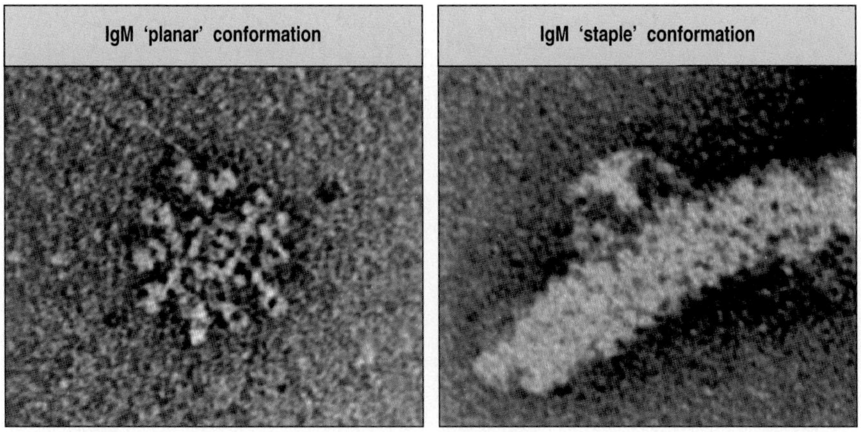

Fig. 9.35 The two conformations of IgM. The left panel shows the planar conformation of soluble IgM; the right panel shows the staple conformation of IgM bound to a bacterial flagellum. Photographs (× 760,000) courtesy of K H Roux.

for C1q activation is achieved only when a single molecule of C1q can bind two or more IgG molecules that are held through binding antigen within 30–40 nm of each other. This requires the binding of many IgG molecules to a single pathogen. For this reason, IgM is much more efficient in activating complement than is IgG.

The binding of C1q to a single bound IgM molecule, or to two or more bound IgG molecules, leads to the activation of an enzymatic activity in C1r; the active form of C1r then cleaves its associated C1s to generate an active serine protease (Fig. 9.36). The activation of C1s completes the first step in the classical pathway of complement activation.

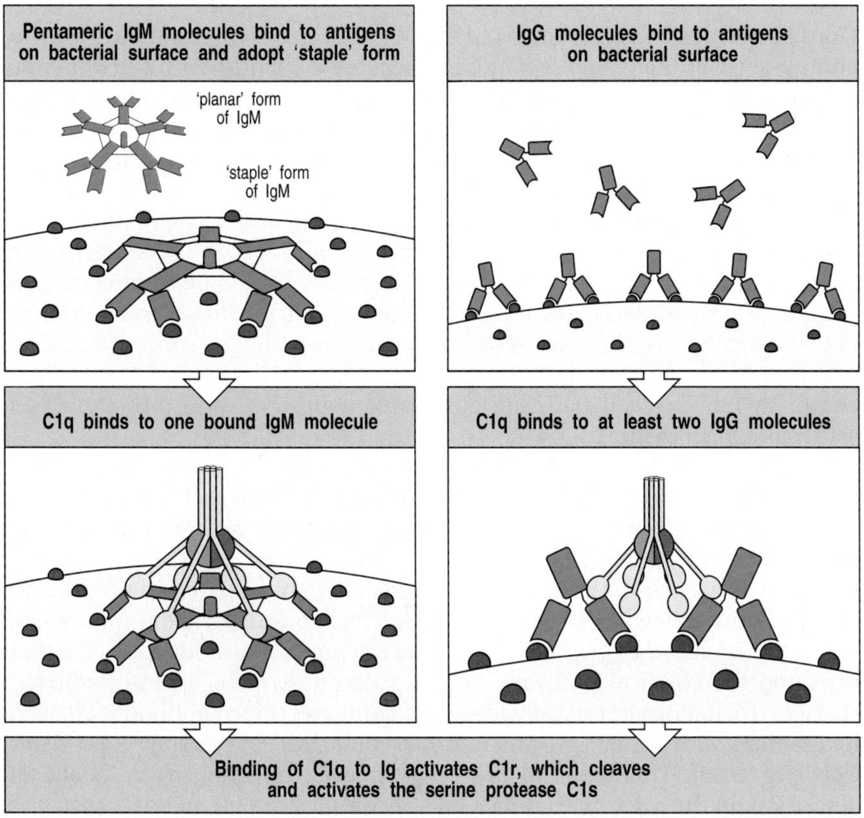

Fig. 9.36 The classical pathway of complement activation is initiated by binding of C1q to antibody on a bacterial surface. In the left panels, one molecule of IgM, bent into the 'staple' conformation by binding several identical epitopes on a pathogen surface, allows the globular heads of C1q to bind to its Fc pieces on the surface of the pathogen. In the right panels, multiple molecules of IgG bound on the surface of a pathogen allow the binding of a single molecule of C1q to two or more Fc pieces. In both cases, the binding of C1q activates the associated C1r, which becomes an active enzyme that cleaves the proenzyme C1s, generating a serine protease that initiates the classical complement cascade.

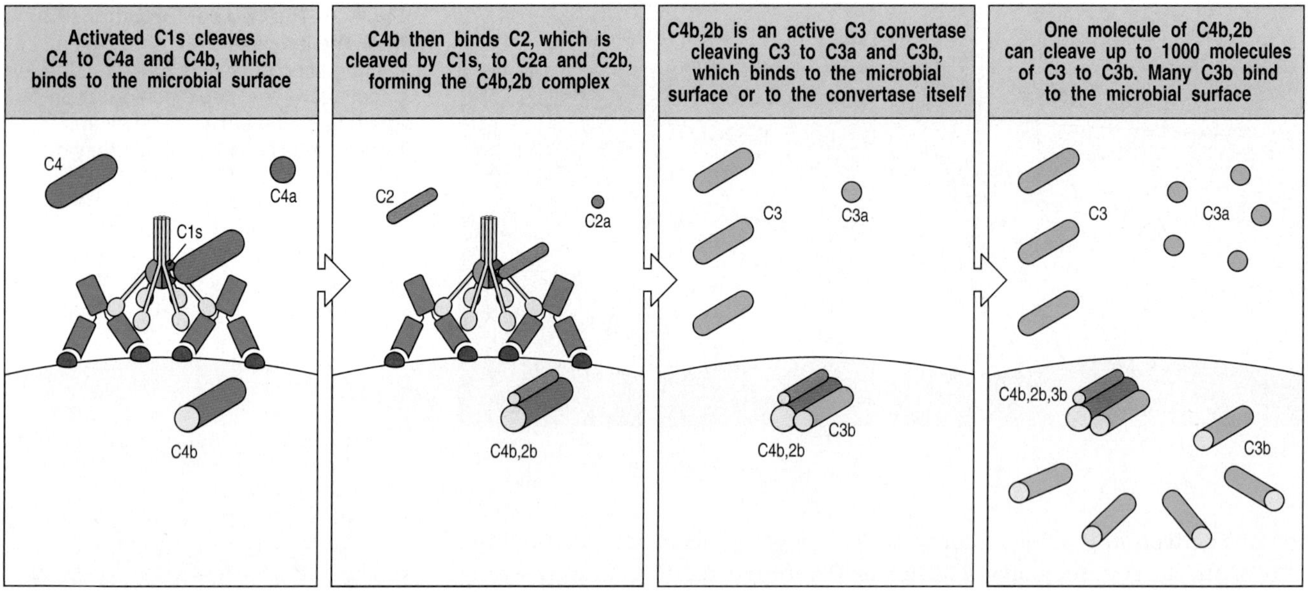

Fig. 9.37 The classical pathway of complement activation generates a C3 convertase that deposits large numbers of C3b molecules on the pathogen. The steps in the reaction are outlined here and detailed in the text. The cleavage of C4 by C1s exposes a reactive group on C4b that allows it to bind covalently to the pathogen surface. C4b then binds C2, making it susceptible to cleavage by C1s. The larger C2b fragment is the active protease component of the C3 convertase, which cleaves many molecules of C3 to produce C3b, which binds to the pathogen surface, and C3a, an inflammatory mediator.

9-24 The classical pathway of complement activation generates a C3 convertase bound to the pathogen surface.

Once bound antibody has activated C1s, the C1s enzyme acts on the next two components of the classical pathway, cleaving C4 and then C2 to generate two large fragments, C4b and C2b, which together form the C3 convertase of the classical pathway. In the first step, C1s cleaves the plasma protein C4 to produce C4b, which binds covalently to the surface of the pathogen. The covalently attached C4b then binds one molecule of C2, making it susceptible, in turn, to cleavage by C1s. C1s cleaves C2 to produce the large fragment C2b, which is itself a serine protease. The complex of C4b with the active serine protease C2b remains on the surface of the pathogen as the C3 convertase of the classical pathway. Its most important activity is to cleave large numbers of C3 molecules to C3b, some of which bind to the pathogen surface, and C3a, which initiates a local inflammatory response. These reactions, which comprise the classical pathway of complement activation, are shown in schematic form in Fig. 9.37.

It is important that the C3 convertase is attached firmly to the pathogen so that C3 activation occurs there and not in plasma nor on host-cell surfaces. This is achieved principally by the covalent binding of C4b to the pathogen surface. Cleavage of C4 exposes a highly reactive thioester bond on the C4b molecule that allows it to bind covalently to molecules in the immediate vicinity of its site of activation: this can be the bound antibody molecule that activated the classical pathway, or any adjacent protein on the pathogen surface. If C4b does not rapidly form this bond, the thioester bond is cleaved by reaction with water and this hydrolysis reaction irreversibly inactivates C4b (Fig. 9.38). This helps to prevent C4b from diffusing from its site of activation on the microbial surface and becoming coupled to host cells.

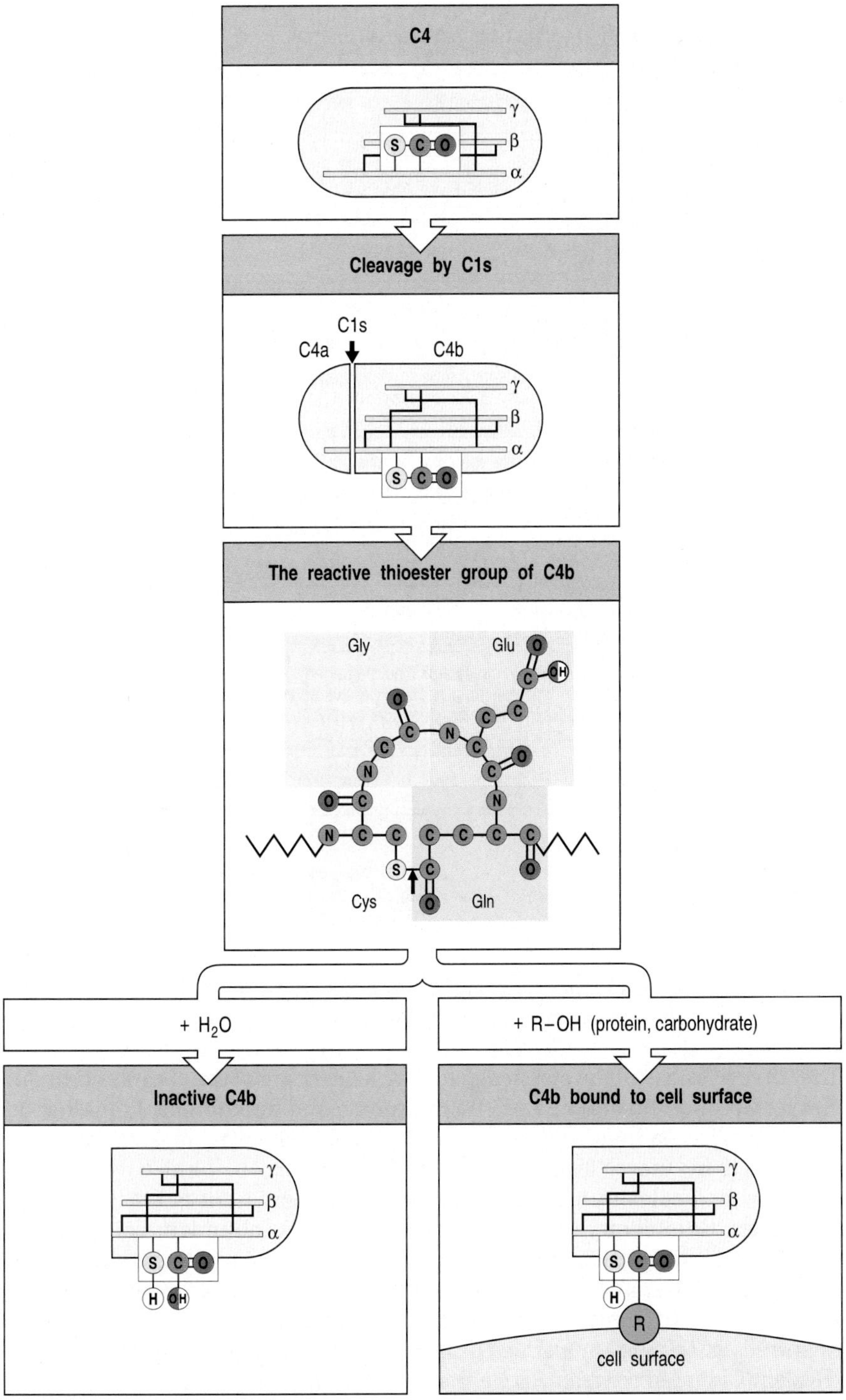

Fig. 9.38 Cleavage of C4 exposes an active thioester bond that causes the large fragment, C4b, to bind covalently to nearby molecules on the bacterial cell surface. Intact C4 consists of an α, a β and a γ chain with a shielded thioester bond on the α chain that is exposed when the α chain is cleaved by C1s to produce C4b. The thioester bond (arrowed in the third panel) is rapidly hydrolyzed (that is, cleaved by water), inactivating C4b unless it reacts with hydroxyl or amino groups to form a covalent linkage with molecules on the pathogen surface. The homologous protein C3 has an identical reactive thioester bond that is also exposed on the C3b fragment when C3 is cleaved by C2b. The covalent attachment of C3b and C4b enables these molecules to act as opsonins and is important in confining complement activation to the pathogen surface.

C2 becomes susceptible to cleavage by C1s only when it is bound by C4b, and the C2b serine protease is thereby also confined to the pathogen surface, where it remains associated with C4b, providing the enzymatic activity of the C3 convertase of the classical pathway. The activation of C3 molecules thus also occurs at the surface of the pathogen, and the C3b cleavage product of C3 binds covalently by the same mechanism as C4b, as we shall see below. The proteins of the classical pathway of complement activation and their active forms are listed in Fig. 9.39.

Fig. 9.39 The proteins of the classical pathway of complement activation.

Proteins of the classical pathway of complement activation		
Native component	**Active form**	**Function of the active form**
C1 (C1q: $C1r_2$:$C1s_2$)	C1q	Binds to antibody that has bound antigen, thus allowing autoactivation of C1r
	C1r	Cleaves C1s to active protease
	C1s	Cleaves C4 and C2
C4	C4b	Covalently binds to pathogen and opsonizes it. Binds C2 for cleavage by C1s
	C4a	Peptide mediator of inflammation (weak activity)
C2	C2b	Active enzyme of classical pathway C3/C5 convertase: cleaves C3 and C5
	C2a	Precursor of vasoactive C2 kinin
C3	C3b	Many molecules bind pathogen surface and act as opsonins. Initiates amplification via the alternative pathway. Binds C5 for cleavage by C2b
	C3a	Peptide mediator of inflammation (intermediate activity)

9-25 The cell-bound C3 convertase deposits large numbers of C3b molecules on the pathogen surface.

The C3 convertase of the classical pathway, consisting of the complex C4b,2b, cleaves C3 into C3b and C3a. C3 is structurally and functionally homologous to C4; C3b, like its molecular homolog C4b, has a reactive thioester bond that is exposed by the cleavage of C3. This allows C3b to bind covalently to adjacent molecules on the pathogen surface; otherwise it is inactivated by hydrolysis. Complement component C3 is the most abundant complement protein in plasma, existing at a concentration of 1.2 mg μl^{-1}, and up to 1000 molecules of C3b can bind in the vicinity of a single active C3 convertase (see Fig. 9.37). Thus, the main effect of complement activation is to deposit large quantities of C3b on the surface of the initiating pathogen, where it forms a covalently bonded coat that, as we shall see, can signal the ultimate destruction of the pathogen by phagocytes.

The next step in the cascade is the generation of the C5 convertase by the binding of C3b to C4b,2b to yield C4b,2b,3b. This complex binds C5 and makes C5 susceptible to cleavage by the serine protease activity of C2b, initiating the generation of the membrane-attack complex. This reaction is much more limited than cleavage of C3, as C5 can be cleaved only if it binds C3b that is part of a C4b,2b,3b C5 convertase complex. Thus, the end result of the early events of complement activation by the classical pathway is the binding of large numbers of C3b molecules on the surface of the pathogen, with the generation of a more limited number of C5b molecules, and the release of C3a and C5a.

The many C3b molecules deposited on the pathogen surface can be recognized by complement receptors on phagocytic cells, stimulating them to engulf the pathogen. The small peptides C4a, C3a, and C5a, which are generated by the cleavage of C4, C3, and C5, are local inflammatory mediators of increasing potency. Finally, the generation of C5b leads to the formation of the membrane-attack complex. Before discussing these effector functions of complement in greater detail, we shall see how bound C3b can amplify the effects of the classical pathway by initiating the activation of the alternative pathway.

9-26 Bound C3b initiates the alternative pathway of complement activation to amplify the effects of the classical pathway.

Apart from the initiating step, the events of the alternative pathway of complement activation are exactly analogous to those of the classical pathway and involve homologous activated components. Thus, in each case, a large active fragment is deposited on the surface of the pathogen where it binds a second component and renders it susceptible to cleavage by an activating protease to generate the active protease component of the resulting C3 convertase (Fig. 9.40).

In the classical pathway, the first covalently bound fragment is C4b, generated by the cleavage of C4 by activated C1s. In the alternative pathway, the first covalently bound fragment is C3b, and the alternative pathway is activated by the covalent binding of C3b to the pathogen surface. We have already seen that C3b is structurally and functionally homologous to C4b, the first active fragment to bind to the pathogen surface in the classical pathway. In the second step of the alternative pathway, C3b binds to factor B, which is structurally and functionally homologous to C2.

Binding of factor B to C3b makes it susceptible to cleavage by the plasma protease factor D. This cleavage yields a small fragment Ba and an active protease Bb, which remains bound to C3b to make the complex C3b,Bb, which is the C3 convertase of the alternative pathway of complement activation. Note that C3b,Bb is the exact structural and functional homolog of C4b,2b, the C3 convertase of the classical pathway, and that the homologous

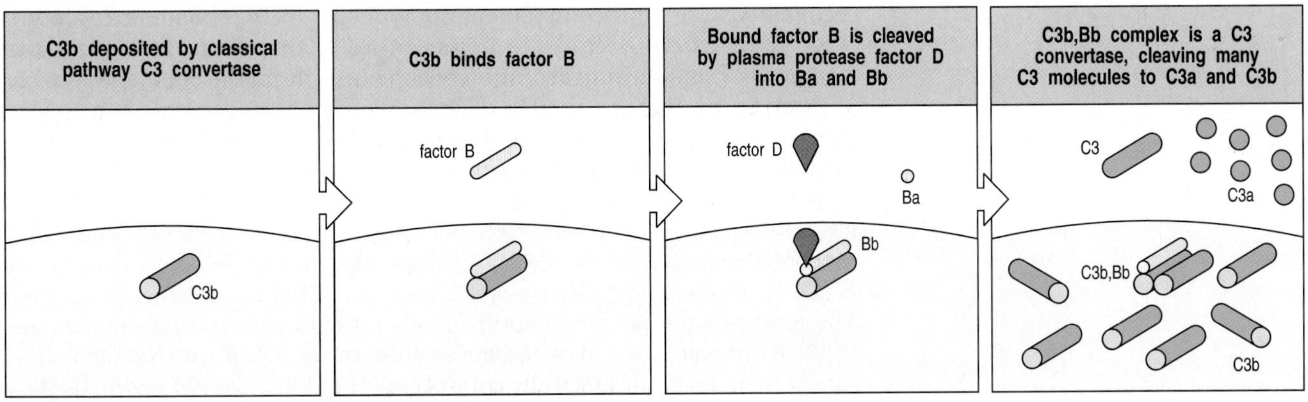

Fig. 9.40 The alternative pathway of complement activation can amplify the classical pathway by forming an alternative C3 convertase (and depositing more C3b molecules on the pathogen). C3b deposited by the classical pathway can bind factor B, making it susceptible to cleavage by factor D. The C3b,Bb complex is the C3 convertase of the alternative pathway of complement activation and its action, like that of C4b,2b, results in the deposition of many molecules of C3b on the pathogen surface.

Fig. 9.41 The proteins of the alternative pathway of complement activation.

Proteins of the alternative pathway of complement activation		
Native component	**Active fragments**	**Function**
C3	C3b	Binds to pathogen surface, binds B for cleavage by D, C3b,Bb is C3 convertase and $C3b_2Bb$ is C5 convertase
Factor B (B)	Ba	Small fragment of B, unknown function
	Bb	Bb is active enzyme of the C3 convertase C3b,Bb and C5 convertase $C3b_2Bb$
Factor D (D)	D	Plasma serine protease, cleaves B when it is bound to C3b to Ba and Bb

components C2 of the classical pathway and factor B of the alternative pathway are encoded by adjacent genes in the class III region of the MHC (see Fig. 4.20). The components of the alternative pathway of complement activation are summarized in Fig. 9.41.

The C3 convertase of the alternative pathway, like that of the classical pathway, can cleave many molecules of C3 to generate yet more active C3b on the surface of the pathogen (see Fig. 9.40). The net result of activation of the classical pathway and its amplification via the alternative pathway is the rapid saturation of the surface of a pathogen with C3b, with the release of the small inflammatory mediator C3a. Some of the bound C3b binds to pre-existing C3 convertase, yielding $C3b_2Bb$, the alternative-pathway C5 convertase. $C3b_2Bb$ can cleave C5 into C5b, which initiates the generation of the membrane-attack complex, and C5a, a potent inflammatory mediator. We now return to the effector actions initiated by C3b.

9-27 Some complement components bind to specific receptors on phagocytes and help to stimulate their activation.

The most important action of complement is to facilitate the uptake and destruction of pathogens by phagocytic cells. This occurs by the specific recognition of bound complement components by **complement receptors** (**CRs**) on phagocytes. Similar receptors on red blood cells are involved in the clearance of soluble antigen:antibody complexes from the circulation, as we shall see in the next section. The complement receptors expressed on phagocytic cells bind pathogens opsonized with bound complement components: opsonization of pathogens is a major function of C3b and its proteolytic derivatives. C4b, the functional homolog of C3b, also acts as an opsonin but has a relatively minor role, largely because so much more C3b is generated than C4b.

The five known types of receptor for bound complement components are listed, with their functions and distributions, in Fig. 9.42. Complement receptors and Fc receptors generally act in concert to facilitate phagocytosis (Fig. 9.43). The best characterized complement receptor is the C3b receptor **CR1**, which is expressed on both macrophages and polymorphonuclear leukocytes. Binding of C3b to CR1 cannot by itself stimulate phagocytosis, but it can enhance phagocytosis and microbicidal activity induced either by the binding of IgG to the Fcγ receptor (Fig. 9.44, top row) or by other immune

Receptor	Specificity	Functions	Cell types
CR1	C3b, C4b	Promotes C3b and C4b decay Stimulates phagocytosis Erythrocyte transport of immune complexes	Erythrocytes, macrophages, monocytes, polymorphonuclear leukocytes, B cells, FDC
CR2 (CD21)	C3d, iC3b, C3dg Epstein–Barr virus	Part of B-cell co-receptor Epstein–Barr virus receptor	B cells, FDC
CR3 (CD11b/ CD18)	iC3b	Stimulates phagocytosis	Macrophages, monocytes, polymorphonuclear leukocytes, FDC
CR4 (gp150, 95) (CD11c/ CD18)	iC3b	Stimulates phagocytosis	Macrophages, monocytes, polymorphonuclear leukocytes
C1q receptor	C1q (collagen region)	Binding of immune complexes to phagocytes	B cells, macrophages, monocytes, platelets, endothelial cells

Fig. 9.42 Distribution and function of receptors for complement proteins on the surfaces of cells. There are several different receptors specific for different bound complement components. CR1 and CR3 are especially important in inducing phagocytosis of bacteria with complement components on their surface. CR1 on erythrocytes also has an important role in clearing immune complexes from the circulation. CR2 is found mainly on B cells, where it is also part of the B-cell co-receptor complex and the receptor by which the Epstein–Barr virus selectively infects B cells, causing infectious mononucleosis.

mediators, such as the T-cell derived cytokine IFN-γ. The small complement fragment C5a can also activate macrophages to ingest bacteria coated with complement alone by binding to a specific receptor, the **C5a receptor**, which has seven membrane-spanning domains. Receptors of this type typically couple with guanine nucleotide-binding proteins called G-proteins, and the C5a receptor signals cells in this way. It is particularly important in the destruction of pathogens coated with complement and IgM, as phagocytes do not have Fc receptors for IgM (see Fig. 9.44, bottom row). Extracellular matrix-associated proteins such as fibronectin can also contribute to phagocyte activation; these are encountered when phagocytes are recruited to connective tissue and activated there. C3a, which has inflammatory activities

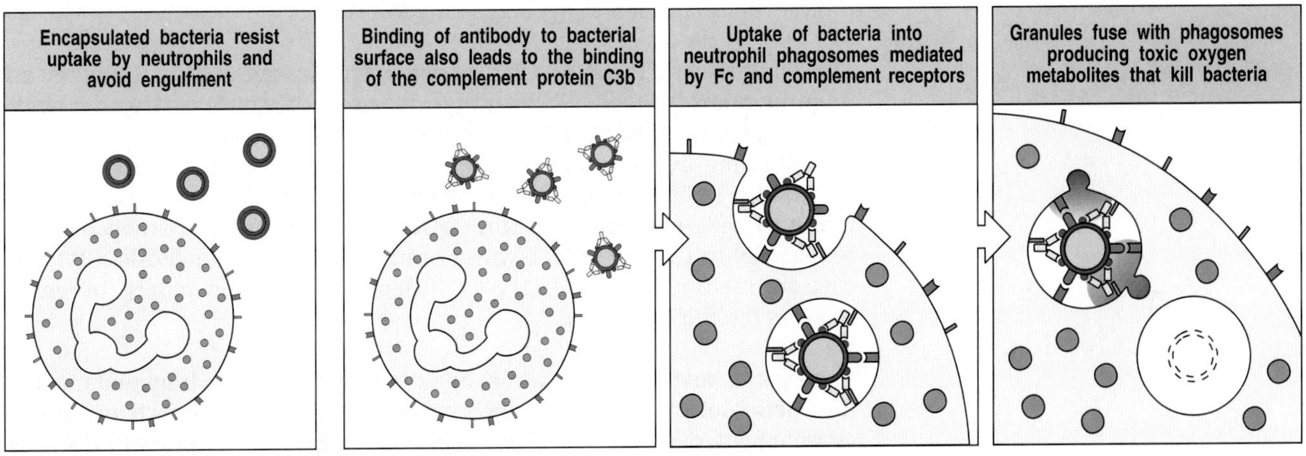

Encapsulated bacteria resist uptake by neutrophils and avoid engulfment
Binding of antibody to bacterial surface also leads to the binding of the complement protein C3b
Uptake of bacteria into neutrophil phagosomes mediated by Fc and complement receptors
Granules fuse with phagosomes producing toxic oxygen metabolites that kill bacteria

Fig. 9.43 Encapsulated bacteria are more efficiently engulfed by phagocytes when they are also coated with complement. The phagocytes shown here are neutrophils; macrophages also bear complement receptors that act in the same way. Here, the neutrophil binds the bacterium by both Fc receptors and complement receptors, which synergize in inducing pathogen uptake and neutrophil activation.

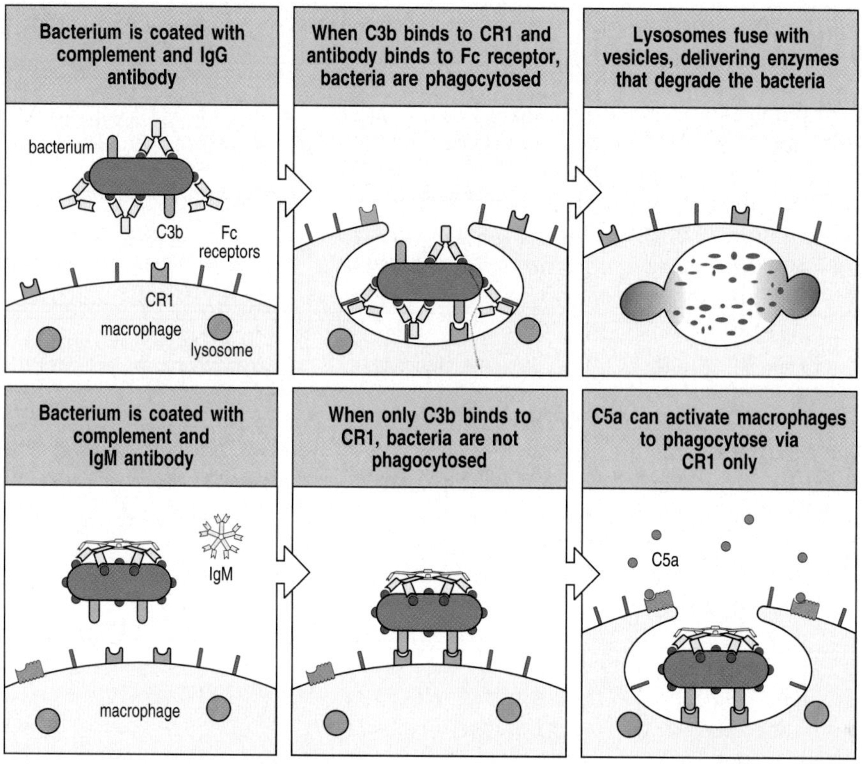

Fig. 9.44 Complement CR1 receptors require ancillary activating signals to participate in phagocytosis. Fc receptors and complement receptors synergize in inducing phagocytosis, and bacteria coated with IgG antibody and complement are therefore more readily ingested than those coated with IgG alone (upper panels). When bacteria are coated with IgM antibody and complement, however, they cannot be ingested unless the phagocyte is pre-activated, for example by T cells or by C5a, as phagocytes do not have Fc receptors for IgM (lower panels).

similar to those of C5a, although it is a less potent chemoattractant, binds to its own specific receptor, the C3a receptor, which is homologous in structure to the C5a receptor.

Three other complement receptors—CR2 (also known as CD21), CR3, and CR4—bind to inactivated forms of C3b that remain attached to the pathogen surface. Like several other key components of complement, C3b is subject to the action of regulatory mechanisms and it can be cleaved further into derivatives that cannot form an active convertase (see Section 9-31). One of the inactive derivatives of C3b, known as iC3b, remains attached to the pathogen and acts as an opsonin in its own right when bound by the complement receptors CR2 or CR3. Unlike the binding of C3b to CR1, the binding of iC3b to CR3 is sufficient on its own to stimulate phagocytosis. A second breakdown product of C3b, called C3dg, binds only to CR2.

The complement receptor CR2, which recognizes iC3b and C3dg, is an important part of the B-cell co-receptor complex. It is believed that the binding of iC3b and/or C3dg to CR2 is crucial for B-cell responses by providing a link between the B-cell antigen receptor and its co-receptor, making the B cell 100- to 10,000-fold more sensitive to antigen (see Section 9-1). CR2 also makes B cells susceptible to the **Epstein–Barr virus** (**EBV**), which binds specifically to CR2 and is the cause of **infectious mononucleosis**. CR3 and CR4 are members of the CD11/CD18 leukocyte integrin family, of which LFA-1 is the third member.

The central role of opsonization by C3b and its inactive fragments in the destruction of extracellular pathogens can be seen in the effects of various complement deficiency diseases. Whereas individuals deficient in any of the late components of complement are relatively unaffected, individuals deficient in C3 or in molecules that catalyze C3b deposition show increased susceptibility to infection by a wide range of extracellular bacteria, as we shall see in Chapter 11.

9-28 Complement receptors are important in the removal of immune complexes from the circulation.

Many small soluble antigens form antibody:antigen complexes that contain too few molecules of IgG to be readily bound to Fcγ receptors. These include toxins bound by neutralizing antibodies and debris from dead microorganisms. Such **immune complexes** are found after most infections, and are removed from the circulation through the action of complement. The soluble immune complexes trigger their own removal by directly activating complement, so that the activated components C4b and C3b bind covalently to the complex, which is then cleared from the circulation by the binding of C4b and C3b to CR1 on the surface of erythrocytes. The erythrocytes transport the bound complexes of antigen, antibody, and complement to the liver and spleen. Here, macrophages remove the complexes from the erythrocyte surface without destroying the erythrocyte, and then degrade the immune complexes (Fig. 9.45). Even larger aggregates of particulate antigen and antibody can be made soluble by activation of the classical complement pathway, and then removed by binding to complement receptors.

Immune complexes that are not removed tend to deposit in the basement membranes of small blood vessels, most notably those of the renal glomerulus where the blood is filtered to form urine. Immune complexes that pass through the basement membrane of the glomerulus bind to CR1 on the renal podocytes that lie beneath the basement membrane. The functional significance of these receptors is unknown; however, they have an important role in the pathology that can arise in some autoimmune diseases.

In the autoimmune disease systemic lupus erythematosus, which we shall describe in Chapter 13, excessive levels of circulating immune complexes cause huge deposits of antigen, antibody, and complement on the podocytes, damaging the glomerulus; kidney failure is the principal danger in this disease. Immune complexes can also be a cause of pathology in patients with deficiencies in the early components of complement. Such patients do not clear immune complexes effectively and they also suffer tissue damage, especially kidney damage, in a similar way.

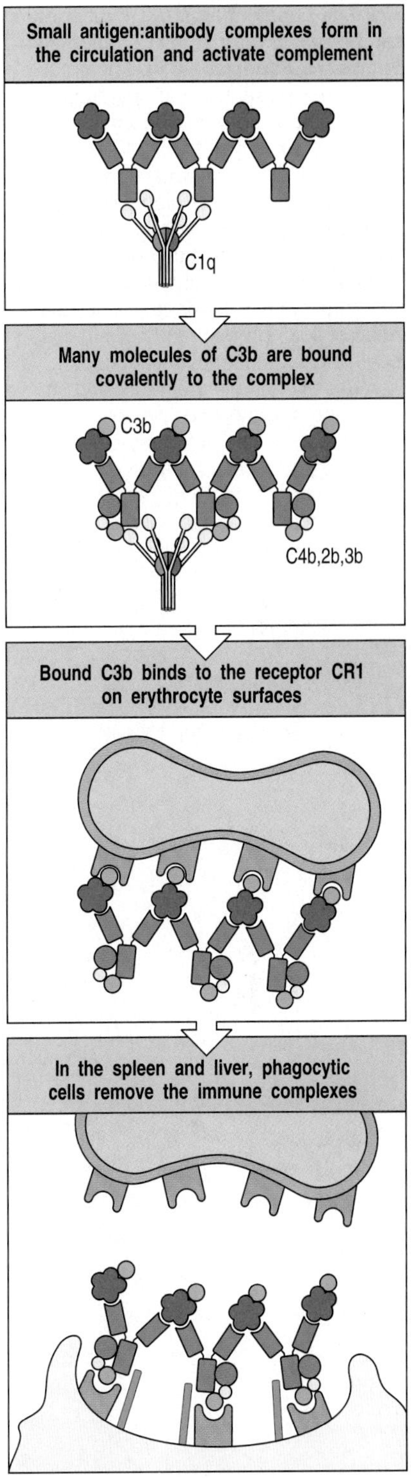
Small antigen:antibody complexes form in the circulation and activate complement
C1q
Many molecules of C3b are bound covalently to the complex
C3b
C4b,2b,3b
Bound C3b binds to the receptor CR1 on erythrocyte surfaces
In the spleen and liver, phagocytic cells remove the immune complexes

Fig. 9.45 Erythrocyte CR1 helps to clear immune complexes from the circulation. CR1 on the erythrocyte surface has an important role in the clearance of immune complexes from the circulation. Immune complexes bind to CR1 on erythrocytes, which transport them to the liver and spleen, where they are removed by macrophages expressing receptors for both Fc and bound complement components.

9-29 Small peptide fragments released during complement activation trigger a local inflammatory response to infection.

Many molecules released during immune responses induce a local inflammatory response. We have seen earlier in this chapter how mast cells can be triggered to release local inflammatory mediators, and we shall see in Chapter 10 that similar reactions can be produced by activated phagocytes.

The small complement fragments C3a, C4a, and C5a act on specific receptors to produce similar local inflammatory responses and are therefore often referred to as **anaphylatoxins** (anaphylaxis is an acute systemic inflammatory response; see Section 12-11). Of the three, C5a is the most stable and has the highest specific biological activity. All three induce smooth muscle contraction and increase vascular permeability, and C3a and C5a can activate mast cells to release mediators that cause similar effects (see Section 9-20). These changes recruit antibody, complement, and phagocytic cells to the site of an infection (Fig. 9.46), and the increased fluid in the tissues hastens the movement of pathogen-containing antigen-presenting cells to the local lymph nodes, contributing to the prompt initiation of the adaptive immune response.

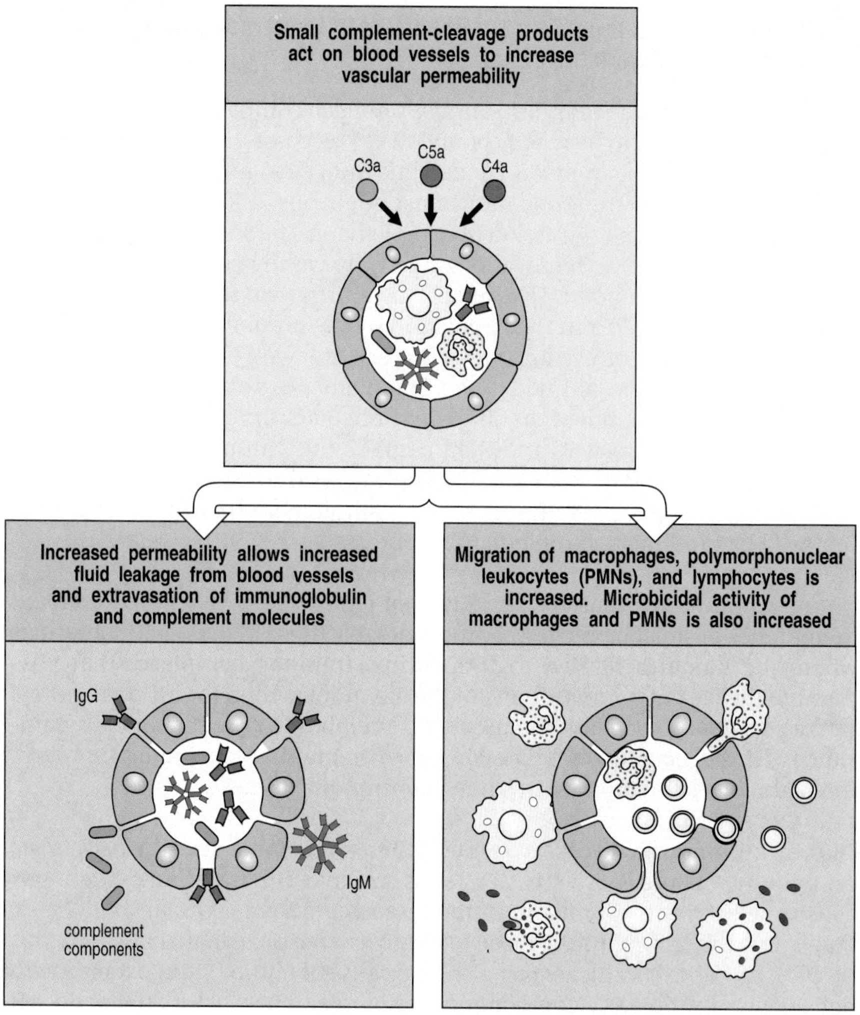

Fig. 9.46 Local inflammatory responses can be induced by small complement fragments, especially C5a. The small complement fragments are differentially active: C5a is more active than C3a, which is more active than C4a. They cause local inflammatory responses by acting directly on local blood vessels, stimulating an increase in blood flow, increased binding of phagocytes to endothelial cells, and increased vascular permeability. This leads to the accumulation of fluid, protein, and cells in the local tissues. Fluid accumulation increases lymphatic drainage, bringing antigen to nearby lymph nodes. The antibodies, complement, and cells thus recruited participate in pathogen clearance by enhancing phagocytosis. The small complement fragments also directly increase the activity of the phagocytes. C5a also activates mast cells (not shown), whose inflammatory mediators contribute to the inflammatory response.

C5a also acts directly on neutrophils and monocytes to increase their adherence to vessel walls, their migration toward sites of antigen deposition, and their ability to ingest particles, as well as increasing the expression of CR1 and CR3 on the surfaces of these cells. In this way C5a and, to a smaller extent, C3a and C4a, act in concert with other complement components to hasten the destruction of pathogens by phagocytes.

9-30 The terminal complement proteins polymerize to form pores in membranes that can kill pathogens.

The most dramatic effect of complement activation is the assembly of the terminal components of complement (Fig. 9.47) to form a membrane-attack complex. The reactions leading to the formation of this complex are shown schematically in Figs 9.48 and 9.49. The end result is a pore in the lipid bilayer membrane that destroys membrane integrity. This is thought to kill the pathogen by destroying the proton gradient across the pathogen cell membrane.

The terminal complement components that form the membrane-attack complex		
Native protein	**Active component**	**Function**
C5	C5a	Small peptide mediator of inflammation (high activity)
	C5b	Initiates assembly of the membrane-attack system
C6	C6	Binds C5b, forms acceptor for C7
C7	C7	Binds C5b,6, amphiphilic complex inserts in lipid bilayer
C8	C8	Binds C5b,6,7, initiates C9 polymerization
C9	$C9_n$	Polymerizes to C5b,6,7,8 to form a membrane-spanning channel, lysing cell

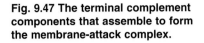

Fig. 9.47 The terminal complement components that assemble to form the membrane-attack complex.

The first step in the formation of the membrane-attack complex is the cleavage of C5 by a C5 convertase (see Fig. 9.48). The next stages are shown in Fig. 9.49. One molecule of C5b binds one molecule of C6, and the C5b,6 complex then binds one molecule of C7. This reaction leads to a conformational change in the constituent molecules, with the exposure of a hydrophobic site on C7. This hydrophobic domain of C7 inserts into the lipid bilayer; similar hydrophobic sites are exposed on the later components C8 and C9 when they are bound to the complex, allowing these proteins also to insert into the lipid bilayer. C8 is a complex of two proteins, called C8β, which binds to C5b, and C8α-γ, which inserts into the lipid bilayer. The binding of C8β to the membrane-associated C5b,6,7 complex allows the binding of the α-γ component. Finally, C8α-γ induces the polymerization of 10 to 16 molecules of C9 into the annular or ring structure called the **membrane-attack complex**. The membrane-attack complex, shown schematically and by electron microscopy in Fig. 9.49, has a hydrophobic external face, allowing it to associate with the lipid bilayer, but a hydrophilic internal channel. The diameter of this channel is about 100 Å, allowing the free passage of solutes and water across the lipid bilayer. The disruption of the lipid bilayer leads to the loss of cellular homeostasis, the disruption of the proton gradient across the membrane, the penetration of enzymes such as lysozyme into the cell, and the eventual destruction of the pathogen.

The membrane-attack complex is strikingly similar to the perforin pores generated by cytotoxic T cells and NK cells, and the main components of these two structures, C9 and perforin 1, are products of closely related genes. The diameter of the membrane-attack complex inner channel is smaller than that of the perforin ring, which has an inner diameter of about 160 Å. The larger perforin pore might be required to allow ready access of granzymes to the interior of the target cell to initiate apoptosis.

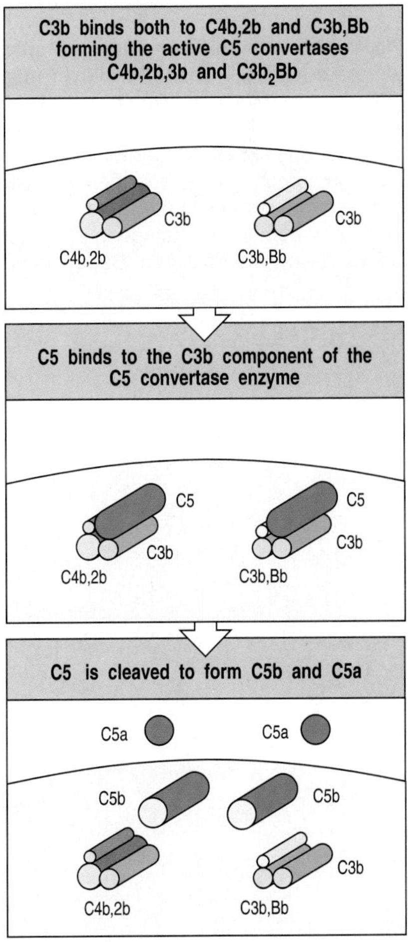

Fig. 9.48 Complement component C5 is activated by a C5 convertase when C5 is complexed to C3b. Top panel: C5 convertases are formed when C3b binds either the classical pathway C3 convertase C4b,2b to form C4b,2b,C3b, or the alternative pathway C3 convertase C3b,Bb to form $C3b_2$,Bb. Middle panel: C5 binds to the C3b in these complexes. Bottom panel: C5 is cleaved by the active enzyme C2b or Bb to form C5b and the inflammatory mediator C5a. Unlike C3b and C4b, C5b is not covalently bound to the cell surface. The production of C5b initiates the assembly of the terminal complement components.

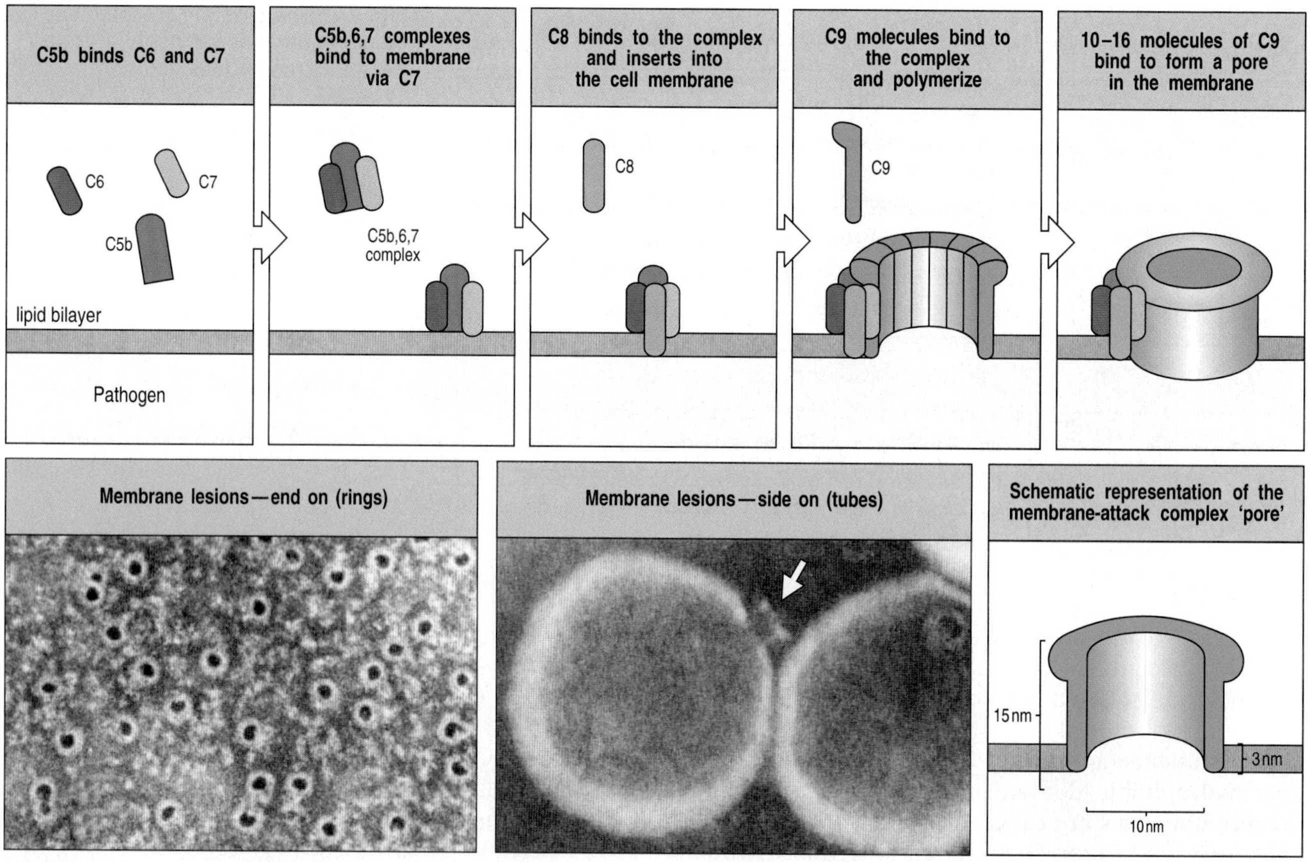

Fig. 9.49 The membrane-attack complex assembles to generate a pore in the lipid bilayer membrane. The sequence of steps and their approximate appearance are shown here in schematic form. C5b triggers the assembly of a complex of one molecule each of C6, C7, and C8, in that order. C7 and C8 undergo conformational changes that expose hydrophobic domains that insert into the membrane. This complex causes moderate membrane damage in its own right, and also serves to induce the polymerization of C9, again with the exposure of a hydrophobic site. Up to 16 molecules of C9 are then added to the assembly to generate a channel of 100 Å diameter in the membrane. This channel disrupts the bacterial cell membrane, killing the bacterium. The electron micrographs show erythrocyte membranes with membrane-attack complexes in two orientations, end on and side on. Note the resemblance of these complexes to pores caused by perforin. C9 and perforin 1, the major components of these two membrane lesions, are products of related genes. Photographs courtesy of S Bhakdi and J Tranum-Jensen.

Although the effect of the membrane-attack complex is very dramatic, particularly in experimental demonstrations in which antibodies against red blood cell membranes are used to trigger the complement cascade, the significance of these components in host defense seems to be quite limited. To date, deficiencies in complement components C5–C9 have been associated with susceptibility only to *Neisseria* species, the bacteria that cause the sexually transmitted disease gonorrhea and a common form of bacterial meningitis. The opsonizing and inflammatory actions of the earlier components of the complement cascade thus are clearly most important for host defense against infection.

9-31 Complement-regulatory proteins serve to protect host cells from the effects of complement activation.

When the components of complement are activated, as we have seen, they usually bind immediately to molecules on the pathogen surface and are thereby confined to the microbe on which their activation was initiated.

Fig. 9.50 The proteins that regulate the activity of complement.

Control proteins of the classical and alternative pathways	
Name (symbol)	Role in the regulation of complement activation
C1 inhibitor (C1INH)	Binds to activated C1r, C1s, removing it from C1q
C4-binding protein (C4BP)	Binds C4b, displacing C2b; co-factor for C4b cleavage by I
Complement-receptor 1 (CR1)	Binds C4b, displacing C2b, or C3b displacing Bb; co-factor for I
Factor H (H)	Binds C3b, displacing Bb; co-factor for I
Factor I (I)	Serine protease that cleaves C3b and C4b; aided by H, MCP, C4BP or CR1
Decay-accelerating factor (DAF)	Membrane protein that displaces Bb from C3b and C2b from C4b
Membrane co-factor protein (MCP)	Membrane protein that promotes C3b and C4b inactivation by I
CD59 (protectin)	Prevents formation of membrane-attack complex on homologous cells. Widely expressed on membranes

However, activated complement components can sometimes escape to bind proteins on host cells; and all components of complement are activated spontaneously at a low rate in plasma. These activated complement components have the potential to destroy any cells to which they bind. Host cells are protected from such inadvertent damage by a series of complement-regulatory proteins, summarized in Fig. 9.50. Some of these proteins are associated with the host cell surface, whereas others are plasma proteins. These proteins protect host cells from accidental complement activation, thereby confining these reactions to the surfaces of pathogens.

The regulatory reactions are shown in Fig. 9.51. The activation of C1 is controlled by a plasma protein, the **C1 inhibitor** (**C1INH**), which binds the active enzyme moiety, C1r:C1s, and causes it to dissociate from C1q, which remains bound to antibody on the pathogen (see Fig. 9.51, top row). In this way, C1INH limits the time during which active C1s is able to cleave C4 and C2. In the same way, C1INH serves to limit the spontaneous activation of C1 in the plasma. Its importance can be seen in the C1INH deficiency disease **hereditary angioneurotic edema**, in which chronic spontaneous complement activation leads to the production of excess cleaved fragments of C4 and C2. The small fragment of C2, C2a, is further cleaved into a peptide, the C2 kinin, that causes extensive swelling—the most dangerous is local swelling in the trachea, which can lead to suffocation. Bradykinin, which has similar actions to C2 kinin, is also produced in an uncontrolled fashion in this disease, as a result of the lack of inhibition of another plasma protease, kallikrein, that is also regulated by C1INH. This disease is fully corrected by replacing C1INH. The large activated fragments of C4 and C2, which normally combine to form the C3 convertase, do not damage host cells in such patients because C4b is rapidly inactivated in plasma (see Fig. 9.38) and the convertase does not form. Any convertase that accidentally forms on a host cell, however, is inactivated by further control mechanisms.

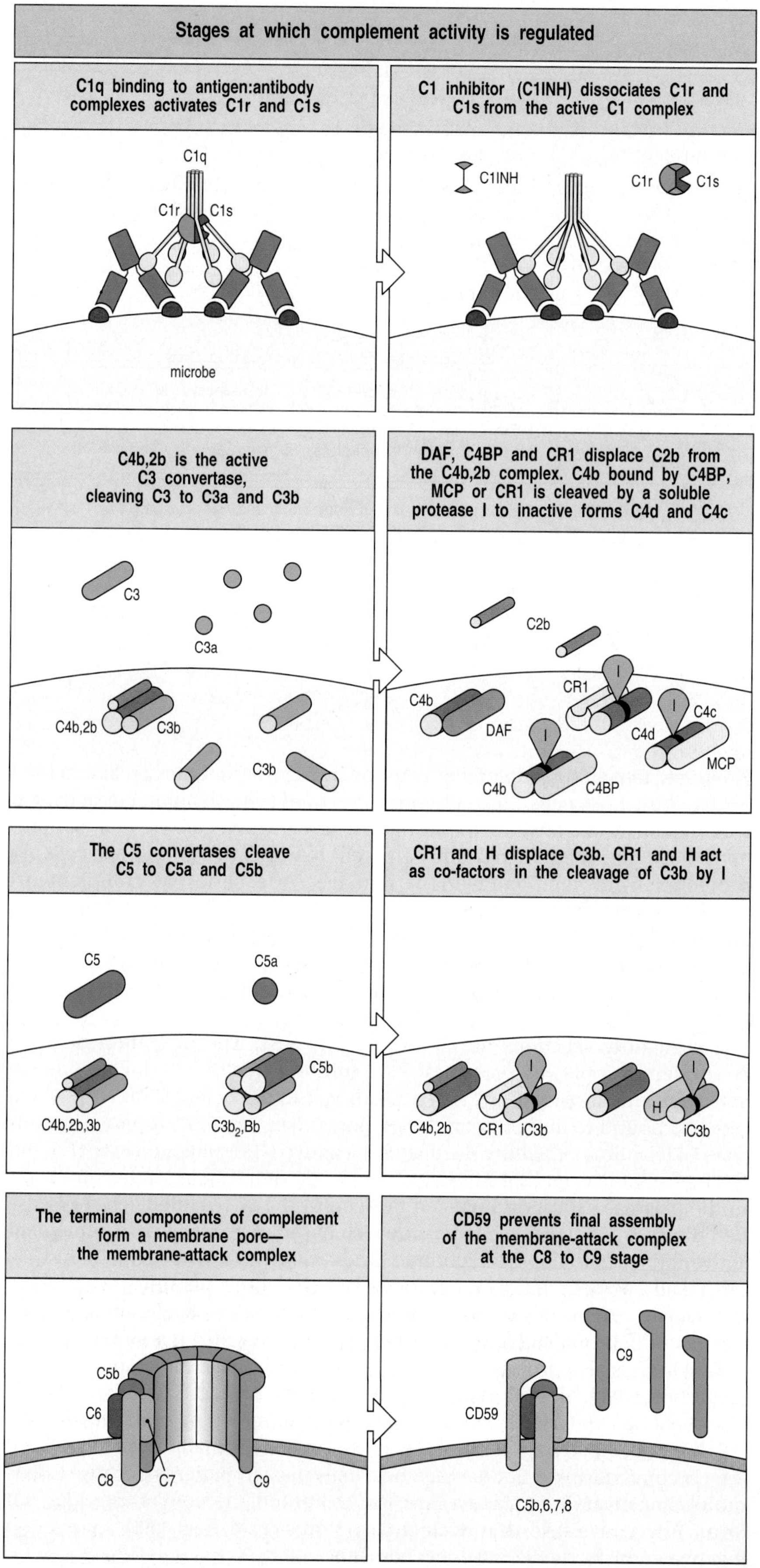

Fig. 9.51 Complement activation is regulated by a series of proteins that serve to protect host cells from accidental damage. These act on different stages of the complement cascade, dissociating complexes or catalyzing the enzymatic degradation of covalently bound complement proteins. Stages in the complement cascade are shown schematically down the left side of the figure, with the control reactions on the right.

First, C2b can be displaced from the complex by either of two proteins—a serum protein called C4-binding protein (C4BP) or a cell-surface protein called decay-accelerating factor (DAF) (see Fig. 9.51, second row). These compete with C2b for binding to C4b. When C4BP binds to C4b, C4b becomes highly susceptible to cleavage by a plasma protein called factor I. Factor I inactivates C4b by cleaving it into the subfragments C4c and C4d. An essentially analogous mechanism operates to inactivate C3b. In this case, either the complement receptor CR1 or a plasma protein called factor H bind to C3b, displacing C2b or Bb and making C3b susceptible to cleavage by factor I (see Fig. 9.51, third row). Factor H also has a binding site for sialic acid, which is abundant on mammalian cells but is absent from most bacteria. A second membrane-associated protein, called membrane co-factor protein (MCP) can bind to membrane-associated C3b and catalyze its destruction by factor I. All of the proteins that bind the homologous C4b and C3b molecules share one or more copies of a structural element called the short consensus repeat (SCR), complement control protein (CCP) repeat, or (especially in Japan) the Sushi domain.

The activity of the terminal complement components is also regulated by cell-surface proteins; the best known is protectin or CD59 (see Fig. 9.51, bottom row). CD59 and DAF are both linked to the cell surface by a phosphoinositol glycolipid tail, like many other membrane proteins. If the synthesis of these glycolipid linkages is defective, as occurs in individuals who have a somatic mutation affecting an X-encoded enzyme in a clone of hematopoietic cells, both CD59 and DAF fail to function. This causes the disease **paroxysmal nocturnal hemoglobinuria**, characterized by episodes of intravascular red blood cell lysis by complement. Cells that lack CD59 only are also susceptible to destruction from spontaneous activation of the complement cascade.

Summary.

The complement system is one of the major mechanisms by which antigen recognition is converted into an effective defense against infection and is particularly important in defense against extracellular bacteria. Complement is a system of plasma proteins that can be activated by antibody, leading to a cascade of reactions that occurs on the surface of pathogens and generates active components with various effector functions. There are three pathways of complement activation: the classical pathway, which is triggered by antibody; the MBLectin-activated pathway; and the alternative pathway, which provides an amplification loop for the classical pathway. Both the MBLectin-activated pathway and the alternative pathway are initiated independently of antibody as part of innate immunity. The early events in all pathways consist of a sequence of cleavage reactions in which the larger cleavage product binds covalently to the pathogen surface or antigen:antibody complex and contributes to the activation of the next component. The pathways converge with the formation of a C3 convertase enzyme, which cleaves C3 to produce the active complement component C3b. The binding of large numbers of C3b molecules to the pathogen is the central event in complement activation. Bound complement components, especially bound C3b and its inactive fragments, are recognized by specific complement receptors on phagocytic cells, which engulf pathogens opsonized by C3b and its inactive fragments. Erythrocytes also express a complement receptor specific for C3b, which allows them to bind and transport soluble immune complexes, leading to their clearance by Fc-receptor-expressing phagocytes in the spleen and liver. The complement system also has a major role in enhancing the humoral immune response: C3b, iC3b, and C3dg bound to antigen lower the threshold for activation of B cells, as discussed in Chapter 5. The small cleavage fragments of C3, C4, and especially C5, recruit phagocytes to sites of infection and activate them by binding to specific G-protein coupled receptors. Together, these activities promote the uptake and destruction of pathogens by phagocytes. The molecules of C3b that bind the C3 convertase itself initiate the late events, binding C5 to make it susceptible to cleavage by C2b or Bb. The larger C5b fragment triggers the assembly of a membrane-attack complex, which can result in the lysis of certain pathogens. These effects seem to be important only for the killing of a few pathogens but might have a major role in immunopathology. The activity of complement components is modulated by a system of regulatory proteins that prevent tissue damage as a result of inadvertent binding of activated complement components to host cells or spontaneous activation of complement components in plasma.

Summary to Chapter 9.

The humoral immune response to infection involves the production of antibody by plasma cells derived from B lymphocytes, the binding of this antibody to the pathogen, and the elimination of the pathogen by accessory cells and molecules of the humoral immune system. The production of antibody usually requires the action of helper T cells specific for a peptide fragment of the antigen recognized by the B cell. The B cell then proliferates and differentiates in the specialized microenvironment of lymphoid tissues, where somatic hypermutation generates diversity in the surface immunoglobulin of the B cell. The B cells that bind antigen most avidly are selected for further differentiation by contact with antigen on the surface of follicular dendritic cells and interactions with antigen-specific germinal center helper T cells. These events allow the affinity of antibodies to increase over the course of an antibody response, especially in repeated responses to the same antigen. Helper T cells also direct isotype switching, leading to the production of antibody of various isotypes that can be distributed to various body compartments. IgM is produced early in the response and has a major role in protecting against infection in the bloodstream, whereas more mature isotypes such as IgG diffuse into the tissues. Multimeric IgA is produced in the lamina propria and transported across epithelial surfaces, whereas IgE is made in small amounts and binds avidly to the surface of mast cells. Antibodies that bind with high affinity to critical sites on toxins, viruses, and bacteria can neutralize them. However, pathogens and their products are destroyed and removed from the body largely through uptake into phagocytes and degradation inside these cells. Antibodies that coat pathogens bind to Fc receptors on phagocytes, which are thereby triggered to engulf and destroy them. Fc receptors on other cells lead to exocytosis of stored mediators, and this is particularly important in allergic reactions, where mast cells are triggered by antigen binding to IgE antibody to release inflammatory mediators, as we shall learn in Chapter 12. Antibodies can also initiate pathogen destruction by activating the complement system of plasma proteins. Components of complement can opsonize pathogens for uptake by phagocytes, can recruit phagocytes to sites of infection, and can directly destroy pathogens by creating membrane pores in their surfaces. Thus, the humoral immune response is targeted to specific pathogens through the production of specific antibody; however, the effector actions of that antibody are the determined by the isotype of the antibody and are the same for all pathogens bound by antibody of a particular isotype.

General references.

Cambier, J.C., Pleissen, C.M., and Clack M.E.: **Signal transduction by B-cell antigen receptors and its conceptions**. *Annu. Rev. Immunol.* 1994, **12**:458-486.

Law, S.K.A., and Reid, K.B.M.: *Complement*, 1st edn. Oxford, IRL Press, 1988.

Liszewski, M.K., Farries, T.C., Lublin, D.M., Rooney, I.A., and Atkinson, J.P.: **Control of the complement system**. *Ad. Immunol.* 1996, **61**:201-284.

Metzger, H. (ed): *Fc Receptors and the Action of Antibodies*, 1st edn. Washington, DC, American Society for Microbiology, 1990.

Rajewsky, K.: **Clonal selection and learning in the antibody system**. *Nature* 1996, **381**:751-758.

Ross, G.D. (ed): *Immunobiology of the Complement System,* 1st edn. Orlando, Academic Press, 1986.

Section references.

9-1 The antibody response is initiated when B cells bind antigen and are signaled by helper T cells or by certain microbial antigens.

DeFranco, A.L.: **Molecular aspects of B-lymphocyte activation**. *Annu. Rev. Cell Biol.* 1987, **3**:143-178.

9-2 Armed helper T cells activate B cells that recognize the same antigen.

Parker, D.C.: **T cell-dependent B-cell activation.** *Annu. Rev. Immunol.* 1993, **11**:331-340.

9-3 Antigen-binding B cells are trapped in the T-cell zone of lymphoid tissues and are activated by encounter with armed helper T cells.

Cyster, J.G., Hartley, S.B., Goodnow, C.C.: **Competition for follicular niches excludes self-reactive cells from the recirculating B-cell repertoire**. *Nature* 1994, **371**:389-395.

9-4 Peptide:MHC class II complexes on a B cell trigger armed helper T cells to make membrane-bound and secreted molecules that activate the B cell.

Banchereau, J., Bazan, F., Blanchard, D., Briere, F., Galizzi, J.P., Vankooten, C., Liu, Y.J., Rousset, F., and Saeland, S.: **The CD40 antigen and its ligand.** *Annu. Rev. Immunol.* 1994, **12**:881-922.

Foy, T.M., Aruffo, A., Bajorath, J., Buhlmann, J.E., Noelle, R.J.: **Immune regulation by CD40 and its ligand GP39**. *Annu. Rev. Immunol.* 1996, **14**:591-617.

9-5 Isotype switching requires expression of CD40 ligand by the helper T cell and is directed by cytokines.

Aruffo, A., Farrington, M., Hollenbaugh, D., Li, X., Milatovich, A., Nonoyama, S., Bajorath, J., Grosmaire, L.S., Stenkamp, R., Neubauer, M., Roberts, R.L., Noelle, R.J., Ledbetter, J.A., Francke, U., and Ochs, H.D.: **The CD40 ligand, gp39, is defective in activated T cells from patients with X-linked hyper-IgM syndrome.** *Cell* 1993, **72**:291-300.

Lorenz, M., Jung, S., Radbruch, A.: **Switch transcripts in immunoglobulin class switching.** *Science* 1995, **267**:1825-1828.

Stavnezer, J.: **Immunoglobulin class switching.** *Curr. Opin. Immunol.* 1996, **8**:199-205.

9-6 Activated B cells proliferate extensively in the specialized microenvironment of the germinal center.

Fischer, M.B., Goerg, S., Shen, L., Prodeus, A.P., Goodnow, C.C., Kelsoe, G., and Carroll, M.C.: **Dependence of germinal center B cells on expression of CD21/CD35 for survival**. *Science* 1998, **280**:582-585.

Kelsoe, G.: **Life and death in germinal centers (Redux).** *Immunity* 1996, **4**:107-111.

MacLennan, I.C.M.: **Germinal centers**. *Annu. Rev. Immunol.* 1994, **12**:117-139.

9-7 Somatic hypermutation occurs in the rapidly dividing centroblasts in the germinal center.

Küppers, R., Zhao, M., Hansmann, M.L., Rajewsky, K.: **Tracing B cell development in human germinal centers by molecular analysis of single cells picked from histological sections.** *EMBO J.* 1993, **12**:4955-4967.

Wagner, S.D., Neuberger, M.S.: **Somatic hypermutation of immunoglobulin genes**. *Annu. Rev. Immunol.* 1996, **14**:441-457.

9-8 Centrocytes with the best antigen-binding receptors are selected for survival.

Casamayor-Palleja, M., Feuillard, J., Ball, J., Drew, M., MacLennan, I.C.M.: **Centrocytes rapidly adopt a memory B cell phenotype on co-culture with autologous germinal center T cell-enriched preparations**. *Intl. Immunol.* 1995, **8**:737-744.

Humphrey, J.H., Grennan, D., and Sundaram, V.: **The origin of follicular dendritic cells in the mouse and the mechanism of trapping of immune complexes on them.** *Eur. J. Immunol.* 1984, **14**:1859.

Kosco, M.H., Szakal, A.K., Tew, J.G.: ***In vivo* obtained antigen presented by germinal center B cells to T cells *in vitro*.** *J. Immunol.* 1988, **140**:354-360.

Tew, J.G., DiLosa, R.M., Burton, G.F., Kosco, M.H., Kupp, L.I., Masuda, A., Szakal, A.K.: **Germinal centers and antibody production in bone marrow**. *Immunol. Rev.* 1992, **126**:99-112.

9-9 B-cell responses to bacterial antigens with intrinsic B-cell activating ability do not require T-cell help.

Anderson, J., Coutinho, A., Lernhardt, W., and Melchers, F.: **Clonal growth and maturation to immunoglobulin secretion *in vitro* of every growth-inducible B lymphocyte.** *Cell* 1977, **10**:27-34.

Coutinho, A.: **The theory of the one non-specific model for B-cell activation.** *Transplant. Rev.* 1975, **23**:49-65.

9-10 B-cell responses to bacterial polysaccharides do not require peptide-specific T-cell help.

Mond, J.J., Lees, A., Snapper, C.M.: **T cell-independent antigens type 2.** *Annu. Rev. Immunol.* 1995, **13**:655-692.

9-11 Antibodies of different isotypes operate in distinct places and have distinct effector functions.

Janeway, C.A., Rosen, F.S., Merler, E., and Alper, C.A.: *The Gamma Globulins*, 2nd edn. Boston, Little Brown and Co., 1967.

9-12 Transport proteins that bind to the Fc domain of antibodies carry specific isotypes across epithelial barriers.

Simister, N.E. and Mostov, K.E.: **An Fc receptor structurally related to MHC class I antigens.** *Nature* 1989, **337**:184-187.

Mostov, K.E.: **Transepithelial transport of immunoglobulins.** *Annu. Rev. Immunol.* 1994, **12**:63-84.

Burmeister, W.P., Gastinel, L.N., Simister, N.E., Blum, M.L., Bjorkman, P.J.: **Crystal structure at 2.2 Å resolution of the MHC-related neonatal Fc receptor.** *Nature* 1994, **372**:336-343.

9-13 **High-affinity IgG and IgA antibodies can neutralize bacterial toxins.**

Robbins, F.C., Robbins, J.B.: **Current status and prospects for some improved and new bacterial vaccines.** *Am. J. Pub. Health* 1986, **7**:105-125.

9-14 **High-affinity IgG and IgA antibodies can inhibit the infectivity of viruses.**

Mandel, B.: **Neutralization of polio virus: a hypothesis to explain the mechanism and the one hit character of the neutralization reaction.** *Virology* 1976, **69**:500-510.

Possee, R.D., Schild, G.C., and Dimmock, N.J.: **Studies on the mechanism of neutralization of influenza virus by antibody: evidence that neutralizing antibody (anti-hemaglutanin) inactivates influenza virus *in vivo* by inhibiting virion transcriptase activity.** *J. Gen. Virol.* 1982, **58**:373-386.

9-15 **Antibodies can block the adherence of bacteria to host cells.**

Fischetti, V.A., and Bessen, D.: Effect of mucosal antibodies to M protein in colonization by group A streptococci, in Switalski, L., Hook, M., and Beachery, E. (eds): *Molecular Mechanisms of Microbial Adhesion.* New York, Springer, 1989.

9-16 **The Fc receptors of accessory cells are signaling receptors specific for immunoglobulins of different isotypes.**

Ravetch, J.V., and Kinet, J.: **Fc receptors.** *Annu. Rev. Immunol.* 1993, **9**:457-492.

Takai, T., Li, M., Sylvestre, D., Clynes, R., and Ravetch, J.V.: **FcRγ chain deletion results in pleiotrophic effector-cell defects.** *Cell* 1994, **76**:519-529.

9-17 **Fc receptors on phagocytes are activated by antibodies bound to the surface of pathogens.**

Burton, D.R.: The conformation of antibodies, in Metzger, H. (ed): *Fc Receptors and the Action of Antibodies*, 1st edn. Washington, DC, Raven Press, 1990.

9-18 **Fc receptors on phagocytes allow them to ingest and destroy opsonized extracellular pathogens.**

Gounni, A.S., Lamkhioued, B., Ochiai, K., Tanaka, Y., Delaporte, E., Capron, A., Kinet, J-P., and Capron, M.: **High-affinity IgE receptor on eosinophils is involved in defence against parasites.** *Nature* 1994, **367**:183-186.

Karakawa, W.W., Sutton, A., Schneerson, R., Karpas, A., and Vann, W.F.: **Capsular antibodies induce type-specific phagocytosis of capsulated *Staphylococcus aureus* by human polymorphonuclear leukocytes.** *Infect. Immun.* 1986, **56**:1090-1095.

9-19 **Fc receptors activate natural killer cells to destroy antibody-coated targets.**

Lanier, L.L., and Phillips, J.H.: **Evidence for three types of human cytotoxic lymphocyte.** *Immunol. Today* 1986, **7**:132.

Lanier, L.L., Ruitenberg, J.J., and Phillips, J.H.: **Functional and biochemical analysis of CD16 antigen on natural killer cells and granulocytes.** *J. Immunol.* 1988, **141**:3487-3485.

9-20 **Mast cells, basophils, and activated eosinophils bind IgE antibody with high affinity.**

Beaven, M.A., and Metzger, H.: **Signal transduction by Fc receptors: the FcεRI case.** *Immunol. Today* 1993, **14**:222-226.

Sutton, B.J., and Gould, H.J.: **The human IgE network.** *Nature* 1993, **366**:421-428.

9-21 **IgE-mediated activation of accessory cells has an important role in resistance to parasite infection.**

Capron, A., and Dessaint, J.P.: **Immunologic aspects of schistosomiasis.** *Annu. Rev. Med.* 1992, **43**:209-218.

Grencis, R.K., Else, K.J., Huntley, J.F., and Nishikawa, S.I.: **The *in vivo* role of stem cell factor (c-kit ligand) on mastocytosis and host protective immunity to the intestinal nematode *Trichinella spiralis* in mice.** *Parasite Immunol.* 1993, **15**:55-59.

Kasugai, T., Tei, H., Okada, M., Hirota, S., Morimoto, M., Yamada, M., Nakama, A., Arizono, N., and Kitamura, Y.: **Infection with *Nippostrongylus brasiliensis* induces invasion of mast cell precursors from peripheral blood to small intestine.** *Blood* 1995, **85**:1334-1340.

Ushio, H., Watanabe, N., Kiso, Y., Higuchi, S., and Matsuda, H.: **Protective immunity and mast cell and eosinophil responses in mice infested with larval *Haemaphysalis longicornis* ticks.** *Parasite Immunol.* 1993, **15**:209-214.

9-22 **Complement is a system of plasma proteins that interact with bound antibodies and surface receptors to aid in the elimination of pathogens.**

Tomlinson, S.: **Complement defense mechanisms.** *Curr. Opin. Immunol.* 1993, **5**:83-89.

9-23 **The C1q molecule binds to antibody molecules to trigger the classical pathway of complement activation.**

Cooper, N.R.: **The classical complement pathway. Activation and regulation of the first complement component.** *Adv. Immunol.* 1985, **37**:151-216.

Perkins, S.J., and Nealis, A.S.: **The quaternary structure in solution of human complement subcomponent C1r2Cls2.** *Biochem. J.* 1989, **263**:463-469.

Prodeus, A.P., Zhou, X., Maurer, M., Galli, S.J., and Carroll, M.C.: **Impaired mast cell-dependent natural immunity in complement C3-deficient mice.** *Nature* 1997, **390**:172-175.

9-24 **The classical pathway of complement activation generates a C3 convertase bound to the pathogen surface.**

Chan, A.R., Karp, D.R., Shreffler, D.C., and Atkinson, J.P.: **The 20 faces of the fourth component of complement.** *Immunol. Today* 1984, **5**:200-203.

Dodds, A.W., Xiang-Dong, R., Willis, A.C., Law, S.K.A.: **The reaction mechanism of the internal thioester in the human complement component C4.** *Nature* 1996, **379**:177-179.

Nagar, B., Jones, R.G., Diefenbach, R.J., Isenman, D.E., and Rini, J.M.: **X-ray crystal structure of C3d: a C3 fragment and ligend for complement receptor 2.** *Science* 1998, **280**:1277-1281.

Oglesby, T.J., Accavitti, M.A., and Volanakis, J.E.: **Evidence for a C4b binding site on the C2b domain of C2.** *J. Immunol.* 1988, **141**:926-931.

9-25 **The cell-bound C3 convertase deposits large numbers of C3b molecules on the pathogen surface.**

deBruijn, M.H.L., and Fey, G.M.: **Human complement component C3: cDNA coding sequence and derived primary structure.** *Proc. Natl. Acad. Sci. USA*

1985, **82**:708-712.

Volanakis, J.E.: **Participation of C3 and its ligand in complement activation.** *Curr. Top. Microbiol. Immunol.* 1989, **153**:1-21.

9-26 Bound C3b initiates the alternative pathway of complement activation to amplify the effects of the classical pathway.

Kolb, W.P., Morrow, P.R., and Tamerius, J.D.: **Ba and Bb fragments of Factor B activation: fragment production, biological activities, neoepitope expression and quantitation in clinical samples.** *Complement Inflamm.* 1989, **6**:175-204.

9-27 Some complement components bind to specific receptors on phagocytes and help to stimulate their activation.

Ahearn, J.M., and Fearon, D.T.: **Structure and function of the complement receptors of CR1 (CD35) and CR2 (CD21).** *Adv. Immunol.* 1989, **46**:183-219.

Croix, D.A., Ahearn, J.M., Rosengard, A.M., Han, S., Kelsoe, G., Ma, M., Carroll, M.C.: **Antibody response to a T-dependent antigen requires B cell expression of complement receptors.** *J. Exp. Med.* 1996, **183**:1857-1864.

Dempsey, P.W., Allison, M.E., Akkaraju, S., Goodnow, C.C., Fearon, D.T.: **C3d of complement as a molecular adjuvant: bridging innate and acquired immunity.** *Science* 1996, **271**:348-350.

9-28 Complement receptors are important in the removal of immune complexes from the circulation.

Schifferli, J.A., and Taylor, J.P.: **Physiologic and pathologic aspects of circulating immune complexes.** *Kidney Intl.* 1989, **35**:993-1003.

Schifferli, J.A., Ng, Y.C., and Peters, D.K.: **The role of complement and its receptor in the elimination of immune complexes.** *N. Engl. J. Med.* 1986, **315**:488-495.

9-29 Small peptide fragments released during complement activation trigger a local inflammatory response to infection.

Frank, M.M., and Fries, L.F.: **The role of complement in inflammation and phagocytosis.** *Immunol. Today* 1991, **12**:322-326.

Gerard, C., Gerard, N.P.: **C5a anaphylatoxin and its seven transmembrane-segment receptor.** *Annu. Rev. Immunol.* 1994, **12**:775-808.

Hopken, U.E., et al.: **The C5a chemoattractant receptor mediates mucosal defence to infection.** *Nature* 1996, **383**:86-89.

9-30 The terminal complement proteins polymerize to form pores in membranes that can kill pathogens.

Bhakdi, S., and Tranum-Jensen, J.: **Complement lysis: a hole is a hole.** *Immunol. Today* 1991, **12**:318-320.

Esser, A.F.: **Big MAC attack: complement proteins cause leaky patches.** *Immunol. Today* 1991, **12**:316-318.

9-31 Complement-regulatory proteins serve to protect host cells from the effects of complement activation.

Davies, A., Simmons, D.I., Hale, G., Harrison, R.A., Tighe, H., Lachmann, P.J., and Waldmann, H.: **CD59, an Ly-6-like protein expressed in human lymphoid cells, regulates the action of the complement membrane attack complex on homologous cells.** *J. Exp. Med.* 1989, **170**:637-654.

Meri, S., Pangburn, M.K.: **Discrimination between activators and nonactivators of the alternative pathway of complement: regulation via a sialic acid/polyanion binding site on factor H.** *Proc. Natl. Acad. Sci. USA* 1990, **87**:3982-3986.

PART V

THE IMMUNE SYSTEM IN HEALTH AND DISEASE

Host Defense Against Infection

10

Throughout this book we have examined the individual mechanisms by which the adaptive immune response acts to protect the host from pathogenic infectious agents. In the remaining five chapters, we consider how the cells and molecules of the immune system work as an integrated host defense system to eliminate or control the infectious agent and to provide long-lasting protective immunity, how failures of immune defense and unwanted immune responses can occur, and how the immune response can be manipulated to benefit the host. In this chapter, we shall examine the role of the immune system as a whole in host defense, including those innate, non-adaptive defenses that comprise early barriers to infectious disease.

The microorganisms that are encountered daily in the life of a normal healthy individual only occasionally cause perceptible disease. Most are detected and destroyed within hours by defense mechanisms that do not require a prolonged period of induction because they do not rely on the clonal expansion of antigen-specific lymphocytes: these are the mechanisms of **innate immunity**. Only if an infectious organism can breach these early lines of defense will an **adaptive immune response** ensue, with the generation of antigen-specific effector cells that specifically target the pathogen, and memory cells that prevent subsequent infection with the same microorganism.

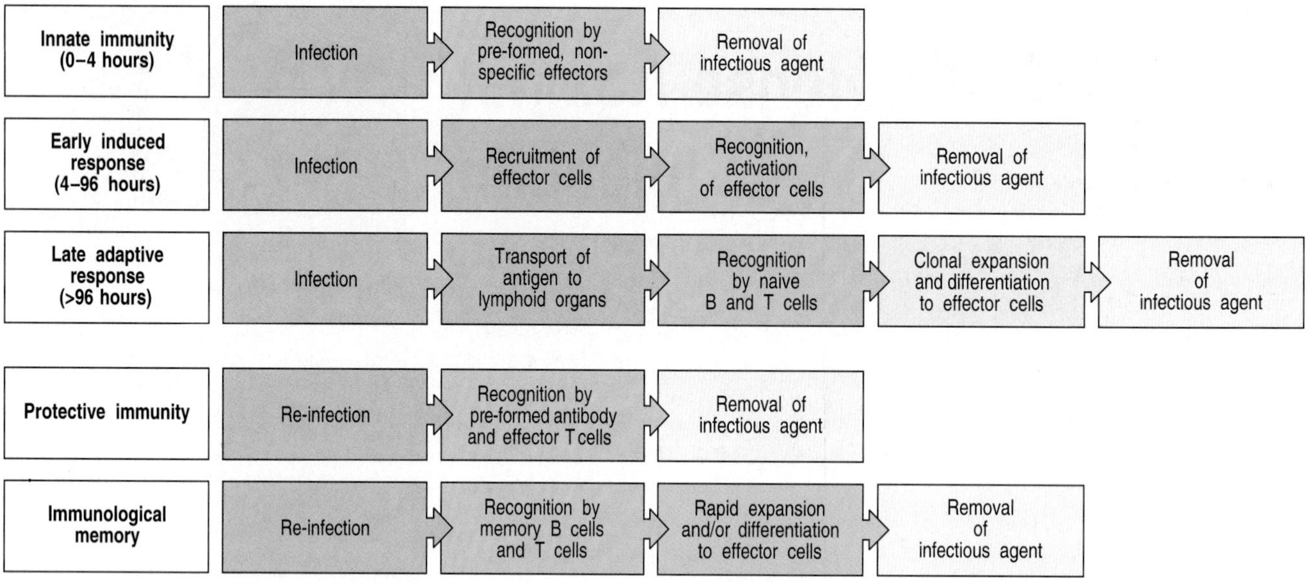

Fig. 10.1 The response to an initial infection occurs in three phases. The effector mechanisms that remove the infectious agent (eg phagocytes, NK cells, complement) are similar or identical in each phase but the recognition mechanisms differ. Adaptive immunity occurs late, because rare antigen-specific cells must undergo clonal expansion before they can differentiate into effector cells. After an adaptive immune response to a pathogen, the response to re-infection is much more rapid; pre-formed antibodies and effector cells act immediately on the pathogen, and immunological memory speeds a renewed adaptive response.

In the preceding two chapters, we have discussed how an adaptive immune response is induced, and how pathogens are eliminated or controlled by the effector cells generated in such a response. Here these mechanisms will be set in the broader context of the entire array of mammalian host defenses against infection, beginning with the innate immune mechanisms that successfully prevent most infections from becoming established. This type of immunity also has an essential role in inducing the subsequent adaptive response to those infections that do overcome the first lines of defense.

The time course of the different phases of an immune response is summarized in Fig. 10.1. The innate immune mechanisms act immediately, and are followed some hours later by **early induced responses**, which can be activated by infection but do not generate lasting protective immunity. These early phases help to keep infection under control while the antigen-specific lymphocytes of the adaptive immune response are activated. Moreover, cytokines produced during these early phases have an important role in shaping the subsequent development of the adaptive immune response and can determine whether the response is predominantly T-cell mediated or humoral. Several days are required for the clonal expansion and differentiation of naive lymphocytes into effector T cells and antibody-secreting B cells that can target the pathogen for elimination. During this period, specific immunological memory is also established; this ensures a rapid re-induction of antibody and antigen-specific effector T cells on subsequent encounters with the same pathogen, thus providing long-lasting protection against re-infection. In this chapter we shall learn how the different phases of host defense are orchestrated in space and time, and how changes in specialized cell-surface molecules guide lymphocytes to the appropriate site of action at different stages of the immune response.

Infection and innate immunity.

Microorganisms that cause pathology in humans and animals enter the body at different sites and produce disease by a variety of mechanisms. Such invasions are initially countered, in all vertebrates, by innate defense mechanisms that pre-exist in all individuals and act within minutes of infection. Only when the innate host defenses are by-passed, evaded, or overwhelmed is an induced or adaptive immune response required, and even then the same effector mechanisms that operate in innate immunity are ultimately harnessed to eliminate the pathogen. In this section we shall describe briefly the infectious strategies of microorganisms, before examining the innate host defenses that, in most cases, prevent infection from becoming established.

10-1 The infectious process can be divided into several distinct phases.

The process of infection can be broken down into stages, each of which can be blocked by different defense mechanisms. Before these are deployed, an infection must be established by infectious particles shed by an infected individual. The number, route, mode of transmission, and stability of an infectious agent outside the host determine its infectivity. Some pathogens, such as anthrax, are spread by spores that are highly resistant to heat and drying, whereas others, such as the human immunodeficiency virus, are spread only by the exchange of body fluids such as blood because they are unable to survive as isolated infectious agents.

Although the body is constantly exposed to infectious agents, infectious disease is fortunately quite rare. The epithelial surfaces of the body serve as an efficient barrier to most microorganisms and those that enter are efficiently removed by innate immune mechanisms. Only when a microorganism has crossed an epithelial barrier and established a site of infection does infectious disease occur, and little pathology will be caused unless the agent is able to spread. Extracellular pathogens spread by direct extension of the infectious center, or through the lymphatics or the bloodstream. Usually, spread through the bloodstream occurs only after the lymphatic system has been overwhelmed by the burden of infectious agent. Obligate intracellular pathogens must spread from cell to cell; they do so either by direct transmission from one cell to the next or by release into the extracellular fluid and reinfection of both adjacent and distant cells.

Most infectious agents show a significant degree of host specificity, causing disease only in one or a few related species. What determines host specificity for each agent is not known but the requirement for attachment to a particular cell-surface molecule is one factor, and other interactions with host cells are commonly needed to support replication. The molecular mechanism of host specificity is an area of intense research interest known as molecular pathogenesis.

Although most microorganisms are repelled by innate host defenses, an initial infection once established generally leads to perceptible disease followed by an effective host adaptive immune response. A cure involves the clearance of both extracellular infectious particles and intracellular residues of infection. After many infections there is little or no residual pathology after an effective primary response. In some cases, however, the infection or the response to it

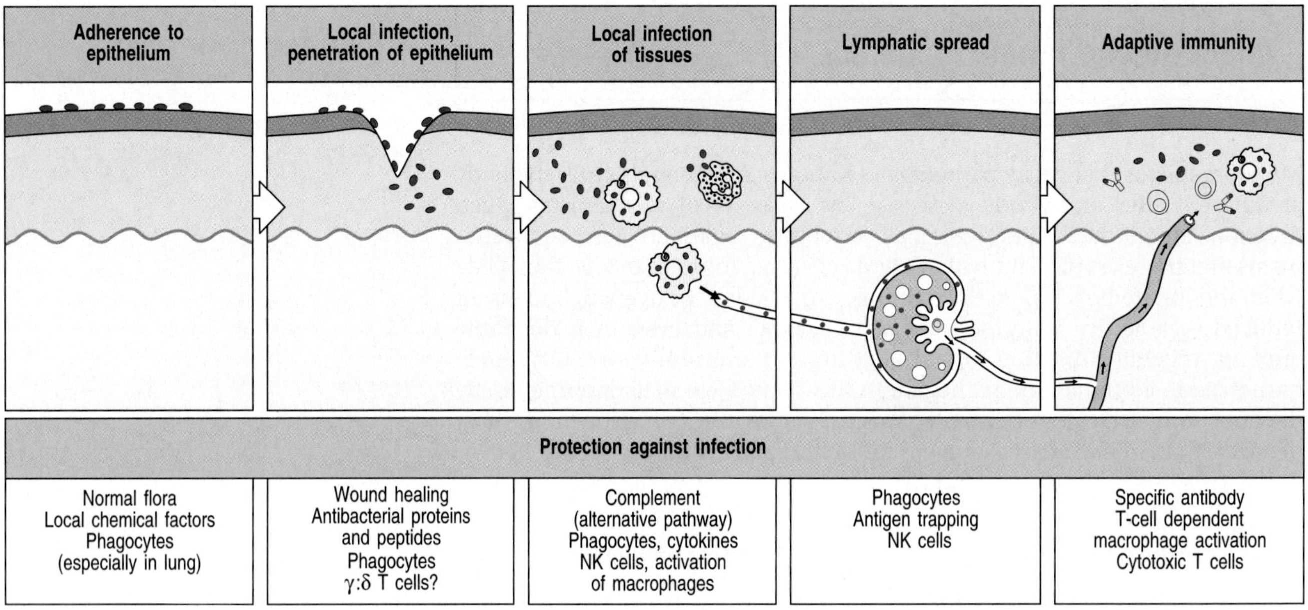
Adherence to epithelium
Local infection, penetration of epithelium
Local infection of tissues
Lymphatic spread
Adaptive immunity
Protection against infection
Normal flora
Local chemical factors
Phagocytes
(especially in lung)
Wound healing
Antibacterial proteins and peptides
Phagocytes
γ:δ T cells?
Complement
(alternative pathway)
Phagocytes, cytokines
NK cells, activation of macrophages
Phagocytes
Antigen trapping
NK cells
Specific antibody
T-cell dependent macrophage activation
Cytotoxic T cells

Fig. 10.2 Infections and the responses to them can be divided into a series of stages. These are illustrated here for an infectious microorganism entering across an epithelium, the commonest route of entry. The infectious organism must first adhere to the epithelial cells and then cross the epithelium. A local non-adaptive response helps contain the infection and delivers antigen to local lymph nodes, leading to adaptive immunity and clearance of the infection. The role of γ:δ T cells is uncertain, as indicated by the question mark.

cause significant tissue damage. Also, some pathogens such as cytomegalovirus or *Mycobacterium tuberculosis* are contained but not completely cleared. Thus, when the adaptive immune response is later weakened, as it is in acquired immune deficiency syndrome (AIDS), these diseases reappear.

In addition to clearance of the infectious agent, an effective adaptive immune response prevents re-infection. For some infectious agents, this protection is essentially absolute, whereas for others infection is reduced or attenuated upon re-exposure. The progress of an infection is illustrated in Fig. 10.2, which summarizes the defense mechanisms activated at each stage, each of which will be described in detail in the course of this chapter.

10-2 Infectious diseases are caused by diverse living agents that replicate in their hosts.

The agents that cause disease fall into five groups: viruses, bacteria, fungi, protozoa, and helminths (worms). Protozoa and worms are usually grouped together as parasites, and are the subject of the discipline of parasitology, whereas viruses, bacteria, and fungi are the subject of microbiology. In Fig. 10.3, the common classes of microorganisms and parasites are listed with typical examples of each. The remarkable variety of these pathogens has required potential hosts to develop two crucial features of adaptive immunity. First, the need to recognize a wide range of different pathogens has driven the development of receptors on B and T cells of equal or greater diversity. Second, the distinct habitats and life cycles of pathogens have to be countered by a range of distinct effector mechanisms. The characteristic features of each pathogen are its mode of transmission, its mechanism of replication, its pathogenesis or the means by which it causes disease, and the response it elicits. We will focus here on the immune responses to these pathogens.

Some common causes of disease in humans			
Viruses	DNA viruses	Adenoviruses	Human adenoviruses (eg types 3, 4, and 7)
		Herpesviruses	Herpes simplex, varicella zoster, Epstein–Barr virus, cytomegalovirus, Kaposi's sarcoma
		Poxviruses	Vaccinia virus
		Parvoviruses	Human parvovirus
		Papovaviruses	Papilloma virus
		Hepadnaviruses	Hepatitis B virus
	RNA viruses	Orthomyxoviruses	Influenza virus
		Paramyxoviruses	Mumps, measles, respiratory syncytial virus
		Coronaviruses	Common cold viruses
		Picornaviruses	Polio, coxsackie, hepatitis A, rhinovirus
		Reoviruses	Rotavirus, reovirus
		Togaviruses	Rubella, arthropod-borne encephalitis
		Flaviviruses	Arthropod-borne viruses, (yellow fever, dengue fever)
		Arenaviruses	Lymphocytic choriomeningitis, Lassa fever
		Rhabdoviruses	Rabies
		Retroviruses	Human T-cell leukemia virus, HIV
Bacteria	Gram +ve cocci	Staphylococci	*Staphylococcus aureus*
		Streptococci	*Streptococcus pneumoniae, S. pyogenes*
	Gram –ve cocci	Neisseriae	*Neisseria gonorrhoeae, N. meningitidis*
	Gram +ve bacilli		*Corynebacteria, Bacillus anthracis, Listeria monocytogenes*
	Gram –ve bacilli		*Salmonella, Shigella, Campylobacter, Vibrio, Yersinia, Pasteurella, Pseudomonas, Brucella, Haemophilus, Legionella, Bordetella*
	Anaerobic bacteria	Clostridia	*Clostridium tetani, C. botulinum, C. perfringens*
	Spirochetes		*Treponema pallidum, Borrelia burgdorferi, Leptospira interrogans*
	Mycobacteria		*Mycobacterium tuberculosis, M. leprae, M. avium*
	Rickettsias		*Rickettsia prowazeki*
	Chlamydias		*Chlamydia trachomatis*
	Mycoplasmas		*Mycoplasma pneumoniae*
Fungi			*Candida albicans, Cryptococcus neoformans, Aspergillus, Histoplasma capsulatum, Coccidioides immitis, Pneumocystis carinii*
Protozoa			*Entamoeba histolytica, Giardia, Leishmania, Plasmodium, Trypanosoma, Toxoplasma gondii, Cryptosporidium*
Worms	Intestinal		*Trichuris trichura, Trichinella spiralis, Enterobius vermicularis, Ascaris lumbricoides, Ancylostoma, Strongyloides*
	Tissues		*Filaria, Onchocerca volvulus, Loa loa, Dracuncula medinensis*
	Blood, liver		*Schistosoma, Clonorchis sinensis*

Fig. 10.3 A variety of microorganisms can cause disease. Pathogenic organisms are of five main types: viruses, bacteria, fungi, protozoa, and worms. Some common pathogens in each group are listed in the column on the right.

Infectious agents can grow in various body compartments, as shown schematically in Fig. 10.4. We have already seen that two major compartments can be defined—intracellular and extracellular. Intracellular pathogens must invade host cells in order to replicate, and must either be prevented from entering cells or detected and eliminated once they have done so. Such pathogens can be subdivided further into those that replicate freely in the cell, such as viruses and certain bacteria (species of *Chlamydia* and *Rickettsia* as well as *Listeria*), and those such as the mycobacteria, that replicate in intracellular vesicles. Many microorganisms replicate in extracellular spaces, either within the body or on the surface of epithelia. Extracellular bacteria are usually susceptible to killing by phagocytes and thus have developed means to resist engulfment. The encapsulated Gram-positive cocci, for instance, grow in extracellular spaces and resist phagocytosis by means of their polysaccharide capsule; if this mechanism of resistance is overcome by opsonization, they are readily killed after ingestion by phagocytic cells.

Different infectious agents cause markedly different diseases, reflecting the diverse processes by which they damage tissues (Fig. 10.5). Many extracellular pathogens cause disease by releasing toxic products or toxins (see Fig. 9.20). Intracellular infectious agents frequently cause disease by damaging the cells that house them. The immune response to the infectious agent can itself be a major cause of pathology in several diseases (see Fig. 10.5). The pathology caused by a particular infectious agent also depends on the site in which it grows, so that *Streptococcus pneumoniae* in the lung causes pneumonia, whereas in the blood it causes a rapidly fatal systemic illness.

Fig. 10.4 Pathogens can be found in various compartments in the body, where they must be combated by different host defense mechanisms. Virtually all pathogens have an extracellular phase where they are vulnerable to antibody-mediated effector mechanisms. However, intracellular phases are not accessible to antibody, and these are attacked by T cells.

	Extracellular		Intracellular	
	Interstitial spaces, blood, lymph	Epithelial surfaces	Cytoplasmic	Vesicular
Site of infection		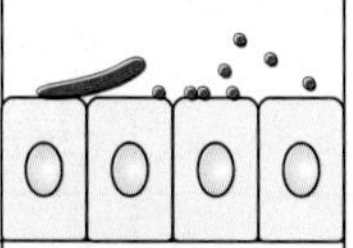		
Organisms	Viruses Bacteria Protozoa Fungi Worms	*Neisseria gonorrhoeae* Worms Mycoplasma *Streptococcus pneumoniae* *Vibrio cholerae* *Escherichia coli* *Candida albicans* *Helicobacter pylori*	Viruses *Chlamydia* spp. *Rickettsia* spp. *Listeria monocytogenes* Protozoa	Mycobacteria *Salmonella typhimurium* *Leishmania* spp. *Listeria* spp. *Trypanosoma* spp. *Legionella pneumophila* *Cryptococcus neoformans* *Histoplasma* *Yersinia pestis*
Protective immunity	Antibodies Complement Phagocytosis Neutralization	Antibodies, especially IgA Anti-microbial peptides	Cytotoxic T cells NK cells	T-cell and NK-cell dependent macrophage activation

	Direct mechanisms of tissue damage by pathogens			Indirect mechanisms of tissue damage by pathogens		
	Exotoxin production	Endotoxin	Direct cytopathic effect	Immune complexes	Anti-host antibody	Cell-mediated immunity
Pathogenic mechanism	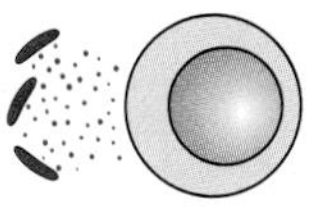		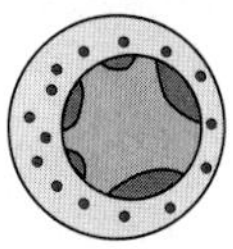	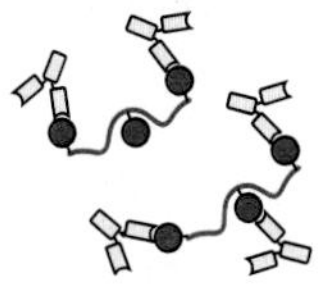	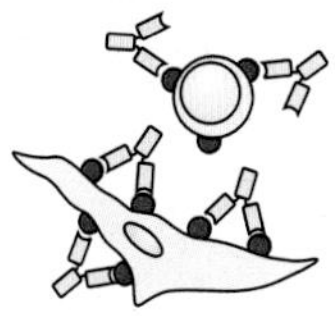	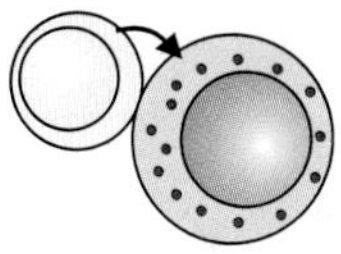
Infectious agent	*Streptococcus pyogenes* *Staphylococcus aureus* *Corynebacterium diphtheriae* *Clostridium tetani* *Vibrio cholerae*	*Escherichia coli* *Haemophilus influenzae* *Salmonella typhi* *Shigella* *Pseudomonas aeruginosa* *Yersinia pestis*	Variola Varicella-zoster Hepatitis B virus Polio virus Measles virus Influenza virus Herpes simplex virus	Hepatitis B virus Malaria *Streptococcus pyogenes* *Treponema pallidum* Most acute infections	*Streptococcus pyogenes* *Mycoplasma pneumoniae*	*Mycobacterium tuberculosis* *Mycobacterium leprae* Lymphocytic choriomeningitis virus *Borrelia burgdorferi* *Schistosoma mansoni* Herpes simplex virus
Disease	Tonsilitis, scarlet fever Boils, toxic shock syndrome, food poisoning Diphtheria Tetanus Cholera	Gram-negative sepsis Meningitis, pneumonia Typhoid Bacillary dysentery Wound infection Plague	Smallpox Chickenpox, shingles Hepatitis Poliomyelitis Measles, subacute sclerosing panencephalitis Influenza Cold sores	Kidney disease Vascular deposits Glomerulonephritis Kidney damage in secondary syphilis Transient renal deposits	Rheumatic fever Hemolytic anemia	Tuberculosis Tuberculoid leprosy Aseptic meningitis Lyme arthritis Schistosomiasis Herpes stromal keratitis

Fig. 10.5 Pathogens can damage tissues in a variety of different ways. The mechanisms of damage, representative infectious agents, and the common name of the disease associated with each are shown. Exotoxins are released by microorganisms and act at the surface of host cells, for example by binding receptors. Endotoxins, which are intrinsic components of microbial structure, trigger phagocytes to release cytokines that produce local or systemic symptoms. Many pathogens are cytopathic, directly damaging the cells they infect. Finally, adaptive immune responses to the pathogen can generate antigen:antibody complexes that can, in turn, activate neutrophils and macrophages, antibodies that cross-react with host tissues, or T cells that kill infected cells, all with some potential to damage the host's tissues. In addition, neutrophils, the most abundant cells early in infection, release many proteins and small-molecule inflammatory mediators that both control infection and cause tissue damage (not shown).

10-3 Surface epithelia make up a natural barrier to infection.

Our body surfaces are defended by epithelia, which provide a physical barrier between the internal milieu and the external world containing pathogens. These epithelia comprise the skin and the linings of the body's tubular structures, such as the gastrointestinal, respiratory, and genito-urinary tracts. Infections occur only when the pathogen can colonize or cross over these barriers. The importance of epithelia in protection against infection is obvious when the barrier is breached, as in wounds and burns, where infection is a major cause of mortality and morbidity. People with defective secretion of mucus or inhibition of ciliary movement, where bacteria can colonize the epithelial surface, frequently develop lung infections. In the absence of wounding or disruption, pathogens normally cross epithelial barriers by adhering to molecules on mucosal epithelial cells. This specific attachment allows the pathogen to infect the epithelial cell, or to damage it so that the epithelium can be crossed.

Fig. 10.6 Surface epithelia provide mechanical, chemical, and microbiological barriers to infection.

Epithelial barriers to infection	
Mechanical	Epithelial cells joined by tight junctions Longitudinal flow of air or fluid across epithelium Movement of mucus by cilia
Chemical	Fatty acids (skin) Enzymes: lysozyme (saliva, sweat, tears), pepsin (gut) Low pH (stomach) Antibacterial peptides; cryptidins (intestine)
Microbiological	Normal flora compete for nutrients and attachment to epithelium and can produce antibacterial substances

Our surface epithelia are more than mere physical barriers to infection; they also produce chemical substances that are microbicidal or inhibit microbial growth (Fig. 10.6). For instance, the acid pH of the stomach and digestive enzymes of the upper gastrointestinal tract make a substantial chemical barrier to infection. Antibacterial peptides called cryptidins are made by Paneth cells, which are resident in the base of the crypts in the small intestine beneath the epithelial stem cells. Furthermore, most epithelia are associated with a normal flora of non-pathogenic bacteria that compete with pathogenic microorganisms for nutrients and for attachment sites on cells. The normal flora can also produce antimicrobial substances, such as the colicins (antibacterial proteins made by *Escherichia coli*) that prevent colonization by other bacteria. When the non-pathogenic bacteria are killed by antibiotic treatment, pathogenic microorganisms frequently replace them and cause disease.

When a pathogen crosses an epithelial barrier and begins to replicate in the tissues of the host, the host's defense mechanisms are required to remove the pathogen. The first phase of host defense depends upon the cells and molecules that mediate innate immunity.

10-4 The alternative pathway of complement activation provides a non-adaptive first line of defense against many microorganisms.

The alternative pathway of complement activation can proceed on many microbial surfaces in the absence of specific antibody (see Fig. 9.40). In this way, it triggers the same antimicrobial actions as the classical pathway without the delay of 5–7 days required for antibody production, and can be regarded as an innate humoral response.

The reaction cascade of the alternative pathway of complement activation is shown schematically in Fig. 10.7. C3 is abundant in plasma, and C3b is produced at a significant rate by spontaneous cleavage (also known as 'tickover') due to the hydrolysis of the thioester bond in C3 to form C3(H_2O) which has an altered conformation, allowing binding of the plasma protein factor B. The binding of B by C3(H_2O) allows a plasma protease called factor D to cleave factor B to Ba and Bb. The C3(H_2O)Bb complex is a C3 convertase that can cleave many molecules of C3 to C3a and C3b. Although much of this C3b is inactivated by hydrolysis, some attaches covalently, through its reactive thioester group, to the surfaces of host cells or to pathogens. C3b bound in this way is able to bind factor B, inducing its cleavage by factor D to yield the small fragment Ba and the active protease Bb. When this occurs on the surface of a host cell, as we learned in Chapter 9, the C3b,Bb complex is

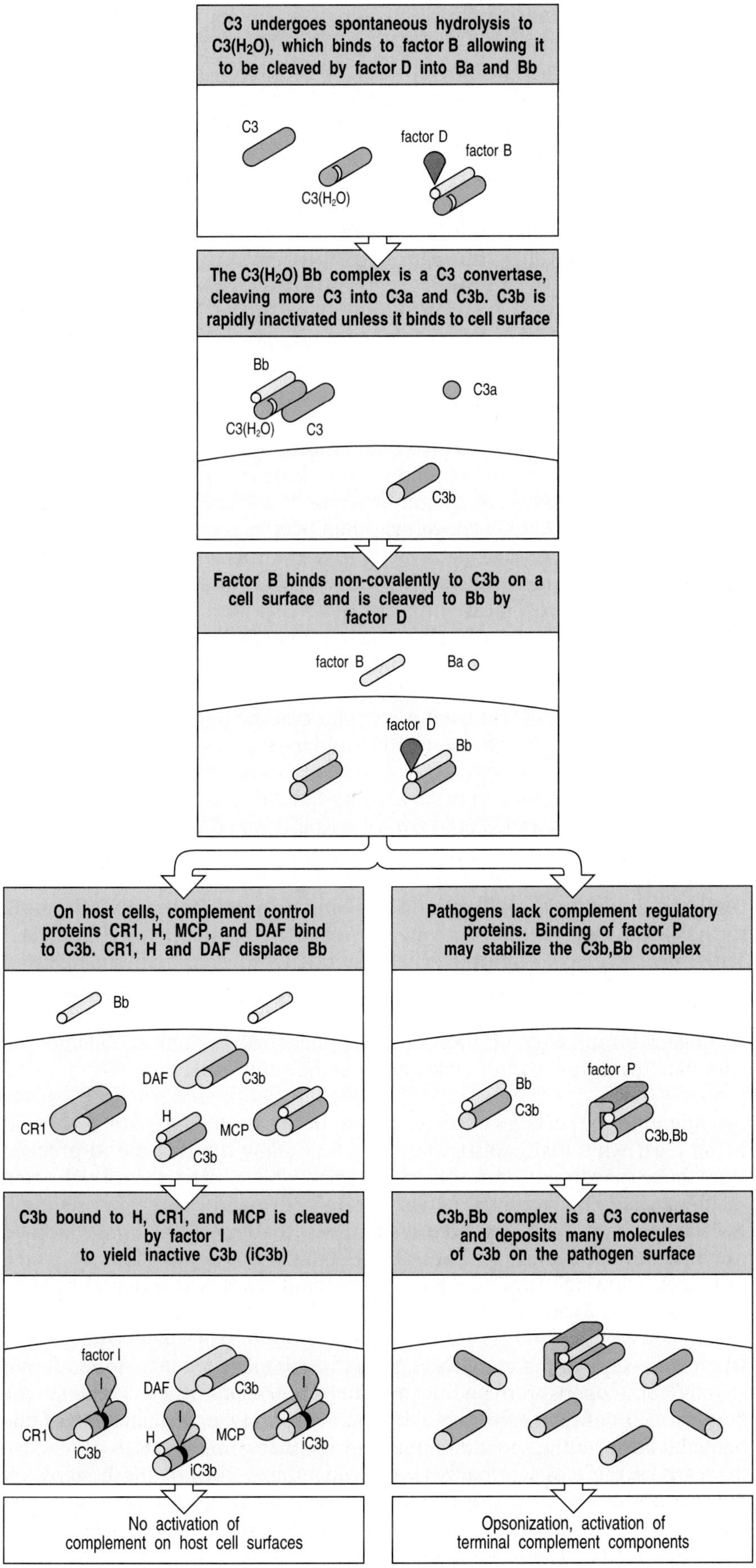

Fig. 10.7 Complement activated by the alternative pathway attacks pathogens while sparing host cells, which are protected by complement regulatory proteins. The complement component C3 is cleaved spontaneously in plasma to give C3(H_2O), which binds factor B and enables this to be cleaved by factor D (top panel). The resulting soluble C3 convertase cleaves C3 to give C3a and C3b, which can attach to host or pathogen (second panel). It binds factor B, which in turn is rapidly cleaved by factor D to Bb, which remains bound to C3b to form a C3 convertase, and Ba, which is released (third panel). If C3b,Bb forms on the surface of host cells (bottom left panels) it is rapidly inactivated by complement regulatory proteins expressed by the host cell: complement receptor 1 (CR1), decay accelerating factor (DAF), and membrane co-factor of proteolysis (MCP). Host cell surfaces also favor binding of factor H from plasma. CR1, DAF, and factor H displace Bb from C3b, and CR1, MCP, and factor H catalyze the cleavage of bound C3b by factor I to produce inactive C3b (known as iC3b). Bacterial surfaces (bottom right panels) do not express complement regulatory proteins and favor binding of factor P (properdin), which stabilizes the C3b,Bb convertase activity. This convertase is the equivalent of C4b,C2b of the classical pathway (see Fig. 10.8) and initiates the cleavage of further molecules of C3 leading to opsonization by C3b and the generation of C3b_2Bb, the alternative pathway C5 convertase, leading to activation of the terminal complement components.

prevented from initiating further activation steps by the cell-surface proteins CR1 (complement receptor 1), DAF (decay-accelerating factor), and MCP (membrane co-factor of proteolysis), and by the plasma protein factor H. CR1 and DAF are membrane-associated molecules, whereas factor H has affinity for the terminal sialic acids of host cell membrane glycoproteins and thus also binds to the surfaces of host cells. All these bind to C3b, displacing Bb and thus preventing the next step in the activation pathway. In addition, factor H, CR1, and MCP render C3b susceptible to cleavage by factor I, a serine protease that circulates in active form and cleaves C3b first into iC3b and then further to C3dg, thus permanently inactivating it (see Fig. 10.7).

Microbial cells lack the protective proteins CR1, MCP, and DAF, and also the sialic acids that allow factor H to bind preferentially to C3b on the surface of host cells. Consequently, the C3b,Bb complexes formed on the surface of a microorganism are not dissociated and function as active C3 convertases. It seems that microbial cells also favor the binding of a positive regulatory component of the alternative pathway known as **properdin**, or **factor P**, which augments activation by binding to C3b,Bb complexes and stabilizing them, preventing their dissociation by factor H and subsequent cleavage by factor I. The stabilized C3 convertase then acts in the same way as the C3 convertase of the classical pathway (see Section 9-25) and converts large numbers of free C3 molecules to C3b, which coats the adjacent surface, and C3a, which mediates local inflammation. C3b and its derivative iC3b opsonize the pathogen for uptake by complement receptors expressed on phagocytic cells. The derivative iC3b can be further cleaved to C3dg, which also remains bound to the pathogen. C3dg is the ligand for CD21, which forms part of the B-cell co-receptor complex and can engage this co-receptor to increase B-cell signaling 100- to 1000-fold in response to specific antigen.

Some molecules of C3b bind to the existing C3b,Bb complex to form $C3b_2$,Bb, the alternative pathway C5 convertase. This binds and cleaves C5, initiating the lytic pathway and releasing the potent inflammatory peptide C5a (see Section 9-29). Once initiated, the alternative pathway can promote its own feedback amplification, with bound C3b binding more molecules of factor B, further increasing C3 and C5 convertase activity on the pathogen surface. Thus the principal effector molecules generated by the alternative pathway of complement activation are the same as those generated by the classical pathway in an adaptive immune response; C3b and its derivatives opsonize pathogens for uptake and destruction by phagocytes, whereas C5a and C3a stimulate the influx of more phagocytes to the site of infection.

Not all microbial surfaces allow activation of the alternative pathway, and it is not clear what distinguishes surfaces that allow the cascade to proceed from those that do not. Some bacterial surfaces have high levels of sialic acid residues, like the surfaces of vertebrate cells, and are therefore more resistant to attack by the alternative pathway than most bacteria, which do not have surface sialic acid. These bacteria favor binding of factor H, which displaces factor B from the bound C3b and makes C3b susceptible to inactivation by factor I.

Only two events in the classical pathway of complement activation are not exactly homologous to the equivalent steps in the alternative pathway: the first is the initial cleavage that, in the classical pathway, deposits C4b on the bacterial surface; the second is the cleavage that generates C2b, the active protease of the classical pathway C3 convertase. Both of these steps are mediated in the classical pathway by activation of C1s by bound antibody; in the alternative pathway, C3 is activated spontaneously, whereas factor B is activated by the plasma protein factor D.

Step in pathway	Protein serving function in pathway			Relationship
	Alternative (innate)	Lectin	Classical	
Initiating serine protease	D	MASP	C1s	Homologous (C1s and MASP)
Covalent binding to cell surface	C3b	C4b		Homologous
C3/C5 convertase	Bb	C2b		Homologous
Control of activation	CR1 H	CR1 C4bp		Identical Homologous
Opsonization	C3b			Identical
Initiation of effector pathway	C5b			Identical
Local inflammation	C5a, C3a			Identical

Fig. 10.8 There is a close relationship between the factors of the alternative, lectin-mediated, and classical pathways of complement activation. Most of the factors are either identical or the products of genes that have duplicated and then diverged in sequence. The proteins C4 and C3 are homologous and contain the unstable thioester bond by which their large fragments, C4b and C3b, bind covalently to membranes. The genes encoding proteins C2 and B are adjacent in the class III region of the MHC and arose by gene duplication. Factor H, CR1, and C4bp regulatory proteins share a repeat sequence common to many complement regulatory proteins. The greatest divergence between the pathways is in their initiation: in the classical pathway the C1 complex serves to convert antibody binding into enzyme activity on a specific surface; in the lectin-mediated pathway, mannan-binding-lectin (MBL) associates with a serine protease, forming MBL-associated serine protease (MASP), to serve the same function; whereas in the alternative pathway this enzyme activity is provided by factor D.

The alternative pathway of complement activation thus illustrates the general principle that most of the immune effector mechanisms that can be activated by the adaptive immune response can also be induced in a non-clonal fashion as part of the early, non-adaptive host response against infection. It is almost certain that the adaptive response evolved by adding specific recognition to the original non-adaptive system. This is illustrated particularly clearly in the complement system, because here the components are defined, and the functional homologs can be seen to be evolutionarily related (Fig. 10.8).

10-5 Phagocytes provide innate cellular immunity in tissues and initiate host-defense responses.

Macrophages mature continuously from circulating monocytes (see Fig. 1.3) and leave the circulation to migrate into tissues throughout the body. They are found in especially large numbers in connective tissue, in association with the gastrointestinal tract, in the lung (where they are found in the interstitium and the alveoli), along certain blood vessels in the liver (where they are known as Kuppfer cells), and throughout the spleen where they serve to remove senescent red blood cells. The second major family of phagocytes—the neutrophils, or polymorphonuclear neutrophilic leukocytes (PMNs, or polys)—are produced and lost in large numbers each day. Both these phagocytic cells have a key role in all phases of host defense. In addition to engulfing opsonized particles coated with antibodies and/or complement, they can recognize and ingest many pathogens directly. Indeed, the same complement receptors by which they engulf opsonized particles recognize various microbial constituents. For example, the leukocyte integrins CD11b/CD18 (also known as CR3 or Mac-1) and CD11c/CD18 (CR4) are able to recognize several microbial substances, including bacterial lipopolysaccharide (LPS), the lipophosphoglycan of *Leishmania*, the filamentous hemagglutinin of *Bordetella*, and structures on yeasts such as *Candida* and *Histoplasma*. Tissue macrophages and neutrophils also have on their surface other receptors able to recognize components common to many pathogens. These receptors include the macrophage mannose receptor, which is found on macrophages

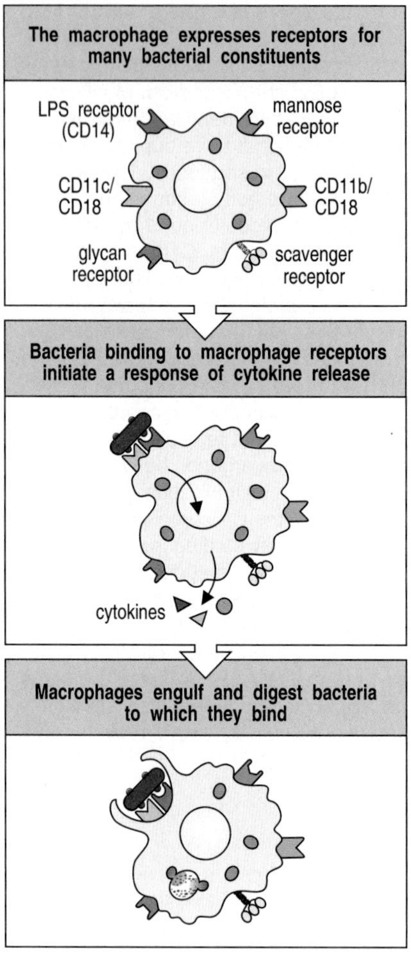
The macrophage expresses receptors for many bacterial constituents
LPS receptor (CD14)
mannose receptor
CD11c/ CD18
CD11b/ CD18
glycan receptor
scavenger receptor
Bacteria binding to macrophage receptors initiate a response of cytokine release
cytokines
Macrophages engulf and digest bacteria to which they bind

Fig. 10.9 Phagocytes bear several different receptors that recognize microbial components and induce phagocytosis and the release of cytokines. The figure illustrates this for two such receptors, CD14 and CD11c/CD18 (CR4), both of which are specific for bacterial lipopolysaccharide (LPS).

but not on monocytes or neutrophils, the scavenger receptor, which binds many sialidated ligands, and CD14, an LPS-binding molecule found predominantly on monocytes and macrophages (Fig. 10.9).

When pathogens cross an epithelial barrier, they are immediately recognized by phagocytes in the subepithelial connective tissues, with three important consequences. The first is the trapping, engulfment, and destruction of the pathogen by tissue macrophages and migrating neutrophils. In addition to being phagocytic, macrophages and neutrophils have granules that contain enzymes, proteins, and peptides that can mediate an intracellular anti-bacterial response. This innate cellular immune response is immediate and can be sufficient to prevent an infection from becoming established, even after a microbe has crossed an epithelial barrier. Indeed, the great cellular immunologist Elie Metchnikoff believed that the innate response of macrophages and neutrophils encompassed all host defense. For a microbe to become pathogenic, it must devise strategies of avoiding engulfment by phagocytes or, like the mycobacteria, devise ways of growing inside the phagosome; many extracellular bacteria coat themselves with a thick poly-saccharide capsule that is not recognized by any phagocyte receptor. Even without such devices, if sufficient bacteria enter the body and simply over-whelm the innate host defenses, they can establish a focus of infection.

The second important effect of the interaction of phagocytes with pathogens is the secretion of cytokines by the phagocyte. It is thought that the pathogen induces cytokine secretion by binding to the same receptors used for engulf-ment. Cytokines are an important component of the next phase of host defense, which comprises a series of induced but non-adaptive responses, as discussed in the next part of this chapter. Cytokine release is also induced by the small peptides released from the complement cascade.

Finally, as we learned in Chapter 8, receptors on macrophages (but not neutrophils) have an important role in antigen uptake and processing, and signals transmitted by these receptors are likely to be responsible for inducing the expression of co-stimulatory molecules that allow the macrophage to function as a professional antigen-presenting cell. Thus, macrophages are important in the induction of the adaptive immune response, and their released cytokines have an additional role in determining the form of the adaptive immune response, as we shall see in the third part of this chapter.

Summary.

The mammalian body is susceptible to infection by many pathogens, which must first make contact with the host and then establish a focus of infection in order to cause disease. These pathogens differ greatly in their lifestyles and means of pathogenesis; this requires an equally diverse set of defensive responses from the host immune system. The first phase of host defense is called innate immunity, and consists of those mechanisms that are present and ready to attack an invader at any time. The epithelial surfaces of the body keep pathogens out as a first line of defense, and many viruses and bacteria can enter tissues only through specialized cell-surface interactions. Bacteria that overcome this barrier are faced with two immediate lines of defense. First, they are subject to humoral attack by the alternative pathway of comple-ment activation, which is spontaneously active in plasma and can opsonize or destroy bacteria while sparing host cells, which are protected by comple-ment regulatory proteins. Second, they can be directly recognized and engulfed by phagocytic macrophages and neutrophils with receptors for common bacterial components.

Innate immunity involves the direct engagement of an effector mechanism by the pathogen, acts immediately on contact with it, and is unaltered in its ability to resist a subsequent challenge. This distinguishes innate immunity from the induced responses that we shall consider next and from the adaptive immune response that provides long-lasting protection against re-infection.

Non-adaptive host responses to infection.

The activation of complement by the alternative pathway and the engulfment of microorganisms by phagocytes occur in the early hours of local infection. If the microorganism evades or overwhelms these innate defenses, the infection can still be contained by a second wave of responses involving the activation of a variety of humoral and cell-mediated effector mechanisms that are strikingly similar to those discussed in Chapters 8 and 9. These are the early induced responses. Unlike the adaptive response, these responses to pathogens involve recognition mechanisms that are based on relatively invariant receptors, and they do not lead to the lasting protective immunity against the inducing pathogen that is the hallmark of adaptive immunity. Instead, as we shall see, the same response is usually made to all pathogens of a given class.

The early induced but non-adaptive responses are important for two main reasons. First, they can repel a pathogen or, more often, hold it in check until an adaptive immune response can be mounted. The early responses occur rapidly, because they do not require clonal expansion, whereas adaptive responses have a latent period of clonal expansion before the proliferating lymphocytes mature into effector cells capable of eliminating an infection. Second, these early responses influence the adaptive response in several ways, as we shall see when we consider this later phase of host defense in the next part of this chapter.

10-6 The innate immune response produces inflammatory mediators that recruit new phagocytic cells to local sites of infection.

One important function of the innate immune response is to recruit more phagocytic cells and effector molecules to the site of the infection through the release of a battery of cytokines and other inflammatory mediators that have profound effects on subsequent events. The cytokines secreted by phagocytes in response to infection are a structurally diverse group of molecules and include interleukin-1 (IL-1), interleukin-6 (IL-6), interleukin-8 (IL-8), interleukin-12 (IL-12), and tumor necrosis factor-α (TNF-α). All have important local and systemic effects, which are summarized in Fig. 10.10. Phagocytes also release other proteins with potent local effects, such as the enzymes plasminogen activator and phospholipase.

Phagocytes release a variety of other molecules in response to infectious agents, including toxic oxygen radicals, peroxides, nitric oxide (NO), and lipid mediators of inflammation such as prostaglandins, leukotrienes—particularly leukotriene B4 (LTB4)—and platelet-activating factor (PAF). In addition, the activation of complement by infectious agents contributes the inflammatory mediators C5a and the less potent C3a. As well as being an inflammatory mediator in its own right, C5a is also able to activate mast cells, causing them

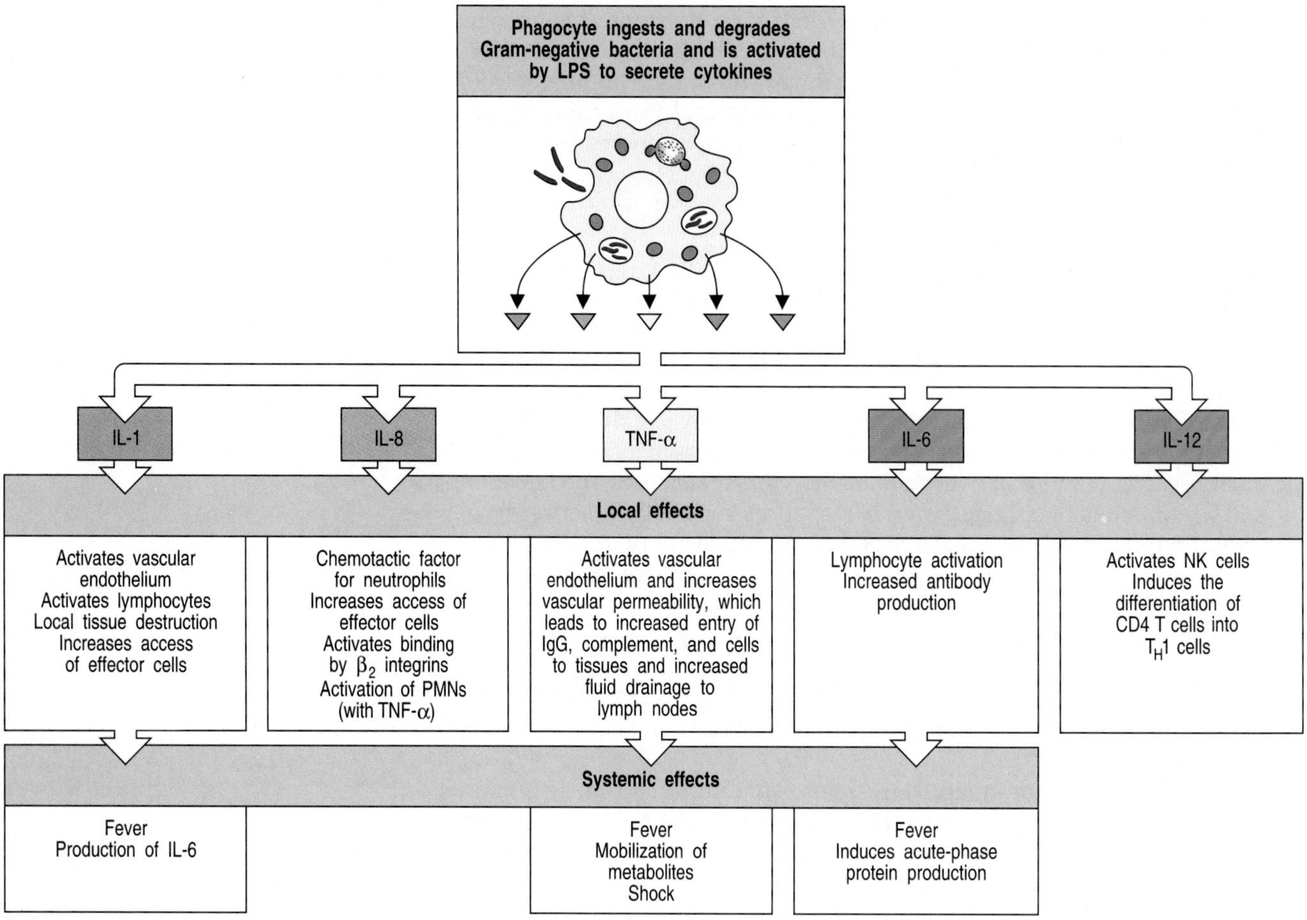

Fig. 10.10 Important cytokines secreted by macrophages in response to bacterial products include IL-1, IL-6, IL-8, IL-12 and TNF-α. TNF-α is an inducer of a local inflammatory response that helps to contain infections (see Section 10-8). It also has systemic effects, many of which are harmful. IL-1, IL-6, and TNF-α have a critical role in inducing the acute-phase response in the liver (see Section 10-11) and induce fever, which favors effective host defense in several ways. IL-8 is particularly important in directing neutrophil migration to sites of infection. IL-12 activates NK cells and favors the differentiation of T_H1 cells.

to release their granule contents, which include histamine, serotonin (in mice), and LTB4. These contribute to the changes in endothelial cells at sites of infection.

We have already discussed, in Section 9-20, the activation of mast cells and the actions of the inflammatory mediators that they release; however, when an individual first encounters a new pathogen there is unlikely to be any IgE of an appropriate specificity bound to the mast cells, so this route of activation is only likely to occur on re-infection. We shall return to the role of mast cells in inflammatory responses in Chapter 12, when we discuss allergic responses mediated by IgE.

The combined local effects of these mediators results in an **inflammatory response**, which is usually one of the immediate local reactions to infection. Inflammatory responses, which are operationally characterized by pain, redness, heat, and swelling at the site of an infection, reflect two types of change in the local blood vessels. The first of these is an increase in vascular diameter, leading to increased local blood flow—hence the heat and redness—and a reduction in the velocity of blood flow, especially along the surfaces of local blood vessels.

Under normal conditions, leukocytes are restricted to the center of blood vessels, where the flow is fastest. In inflammatory sites, where the vessels are dilated, the slower blood flow allows the leukocytes to move out of the center of the blood vessel and interact with the vascular endothelium. In addition to these changes, there is an increase in vascular permeability, leading to the local accumulation of fluid—hence the swelling and pain—as well as the accumulation of immunoglobulins, complement, and other blood proteins in the tissue.

The second effect of these mediators is to induce the expression of adhesion molecules on the endothelial cells of the local blood vessels, which bind to the surface of circulating monocytes and neutrophils and greatly increase the rate of migration of these phagocytic cells out of the blood and into the tissues. Even in the absence of infection, monocytes are migrating continuously into the tissues, where they differentiate into macrophages; during an inflammatory response, the induction of adhesion molecules on the endothelial cells, as well as induced changes in the adhesion molecules expressed on leukocytes, recruit large numbers of circulating leukocytes, initially neutrophils and later monocytes, into the site of an infection.

10-7 The migration of leukocytes out of blood vessels depends on adhesive interactions activated by the local release of inflammatory mediators.

The migration of leukocytes out of blood vessels, a process known as **extravasation**, is thought to occur in four steps. We shall describe this process as it is known to occur for monocytes and neutrophils (Fig. 10.11). Similar processes are thought to account for the homing of naive T lymphocytes to peripheral lymphoid organs and the delivery of effector T cells to sites of infection, as we shall see later.

The first step in this process involves selectins (see Fig. 8.5). The adhesive molecule P-selectin, which is carried inside endothelial cells in granules known as **Weibel–Palade bodies**, appears on endothelial cell surfaces within a few minutes of exposure to leukotriene B4, C5a, or histamine. A second selectin, E-selectin, appears a few hours after exposure to lipopolysaccharide or TNF-α. These selectins recognize carbohydrate epitopes, in this case the sialyl-Lewis^{x} moiety of certain leukocyte glycoproteins. The interaction of P-selectin and E-selectin with these surface glycoproteins allows monocytes and neutrophils to adhere reversibly to the vessel wall, so that circulating leukocytes can be seen to 'roll' along endothelium that has been treated with inflammatory cytokines (see Fig. 10.11, top panel). This adhesive interaction permits the stronger interactions of the second step in leukocyte migration.

The second step depends upon interactions between the leukocyte integrins known as LFA-1 (CD11a:CD18) and CR3 (CD11b:CD18—also called Mac-1) with molecules on endothelium such as the immunoglobulin-related adhesion molecule ICAM-1, which is also induced on endothelial cells by TNF-α (see Fig. 10.11, bottom panel). LFA-1 and CR3 normally adhere only weakly, but IL-8 or other chemoattractant cytokines (chemokines) trigger a conformational change in LFA-1 and CR3 on the rolling leukocyte, which greatly increases its adhesive capacity. In consequence, the leukocyte attaches firmly to the endothelium and the rolling is arrested.

In the third step, the leukocyte extravasates, or crosses the endothelial wall. This step also involves the leukocyte integrins LFA-1 and Mac-1, as well as a further adhesive interaction. This involves an immunoglobulin-related molecule called PECAM or CD31, which is expressed both on the leukocyte and

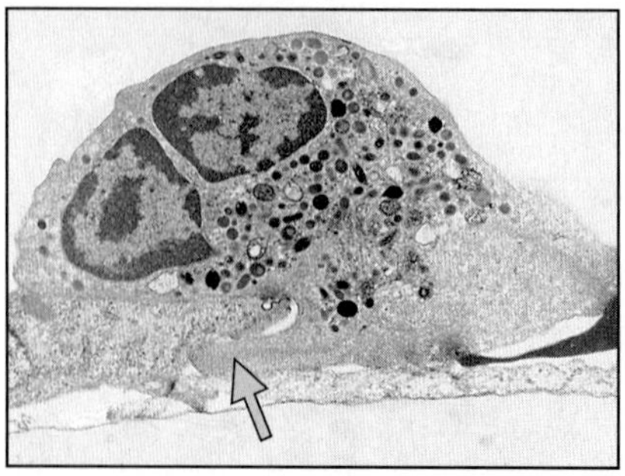

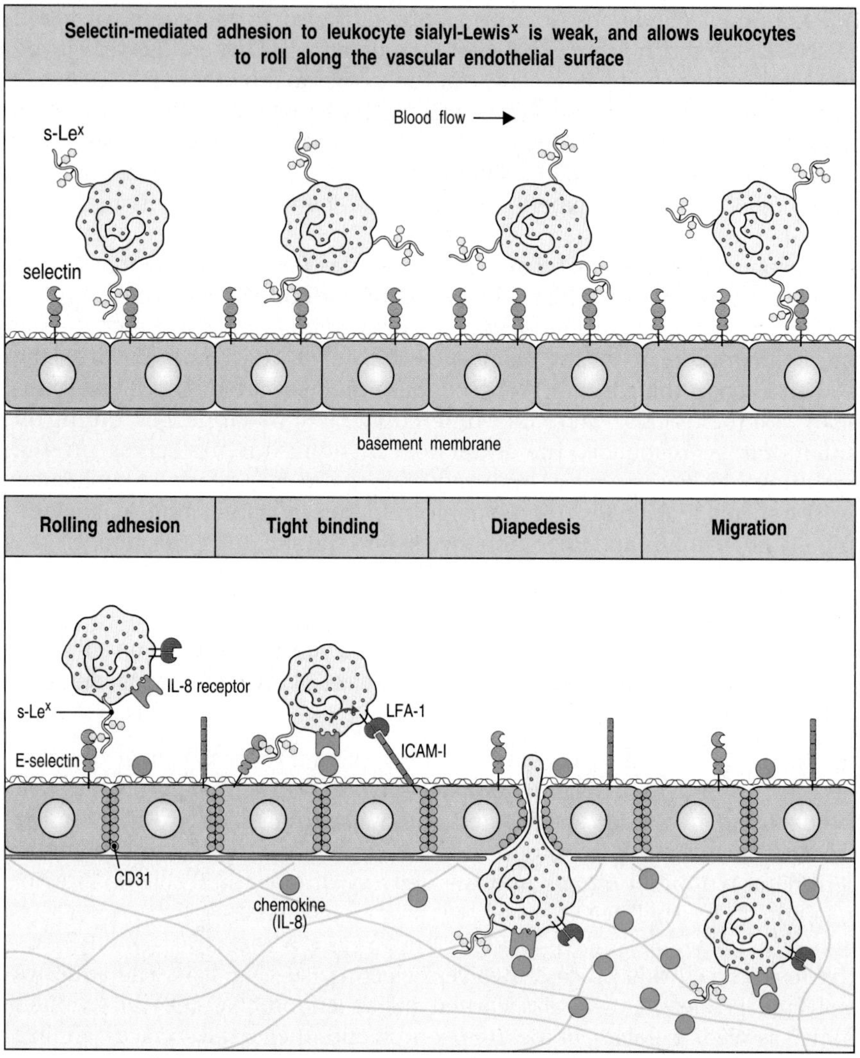

Fig. 10.11 Phagocytic leukocytes are directed to sites of infection through interactions between adhesion molecules induced by cytokines. The first step (top panel) involves the reversible binding of leukocytes to vascular endothelium through interactions between selectins induced on the endothelium and their carbohydrate ligands on the leukocyte, shown here for E-selectin and its ligand the sialyl-Lewis^x moiety (s-Le^x). This interaction cannot anchor the cells against the shearing force of the flow of blood and instead they roll along the endothelium, continually making and breaking contact. The binding does, however, allow stronger interactions, which occur as a result of the induction of ICAM-1 on the endothelium and the activation of its receptors LFA-1 and Mac-1 (not shown) on the leukocyte. Tight binding between these molecules arrests the rolling and allows the leukocyte to squeeze between the endothelial cells forming the wall of the blood vessel (ie to extravasate). The leukocyte integrins LFA-1 and Mac-1 are required for extravasation, and for migration toward chemoattractants. Adhesion between molecules of CD31, expressed on both the leukocyte and the junction of the endothelial cells, is also thought to contribute to diapedesis. The leukocyte also needs to traverse the basement membrane. Finally, the leukocyte migrates along a concentration gradient of chemokines (here shown as IL-8) secreted by cells at the site of infection. The electron micrograph shows a neutrophil extravasating between endothelial cells. Photograph (x 5500) courtesy of I Bird and J Spragg.

at the intercellular junctions of endothelial cells. These interactions enable the phagocyte to squeeze between the endothelial cells. It then penetrates the basement membrane (an extracellular matrix structure) with the aid of proteolytic enzymes that break down the proteins of the basement membrane. The movement through the vessel wall is known as **diapedesis**, and allows phagocytes to enter the site of infection. The fourth and final step in extravasation is the migration of the leukocytes through the tissues under the influence of chemokines. We shall discuss the activities of the small polypeptide cytokines known as chemokines in more detail later in this chapter.

10-8 TNF-α induces blood vessel occlusion and has an important role in containing local infection but can be fatal when released systemically.

The molecular changes induced at the endothelial cell surface by inflammatory mediators also induce the expression of molecules on endothelial cells that trigger blood clotting in the local small vessels, occluding them and cutting off blood flow. This can be important in preventing the pathogen from entering the bloodstream and spreading through the blood to organs all over the body. Instead, the fluid that has leaked into the tissue in the early phases of inflammation, carries the pathogen enclosed in phagocytic cells via the lymph to the regional lymph nodes, where an adaptive immune response can be initiated. The importance of TNF-α in the containment of local infection is illustrated by experiments in which rabbits are infected locally with a bacterium. Normally, the infection will be contained at the site of the inoculation; if, however, an injection of anti-TNF-α antibody is also given, the infection spreads via the blood to other organs.

Once an infection spreads to the bloodstream, however, the same mechanisms whereby TNF-α so effectively contains local infection instead become catastrophic (Fig. 10.12). The presence of infection in the bloodstream,

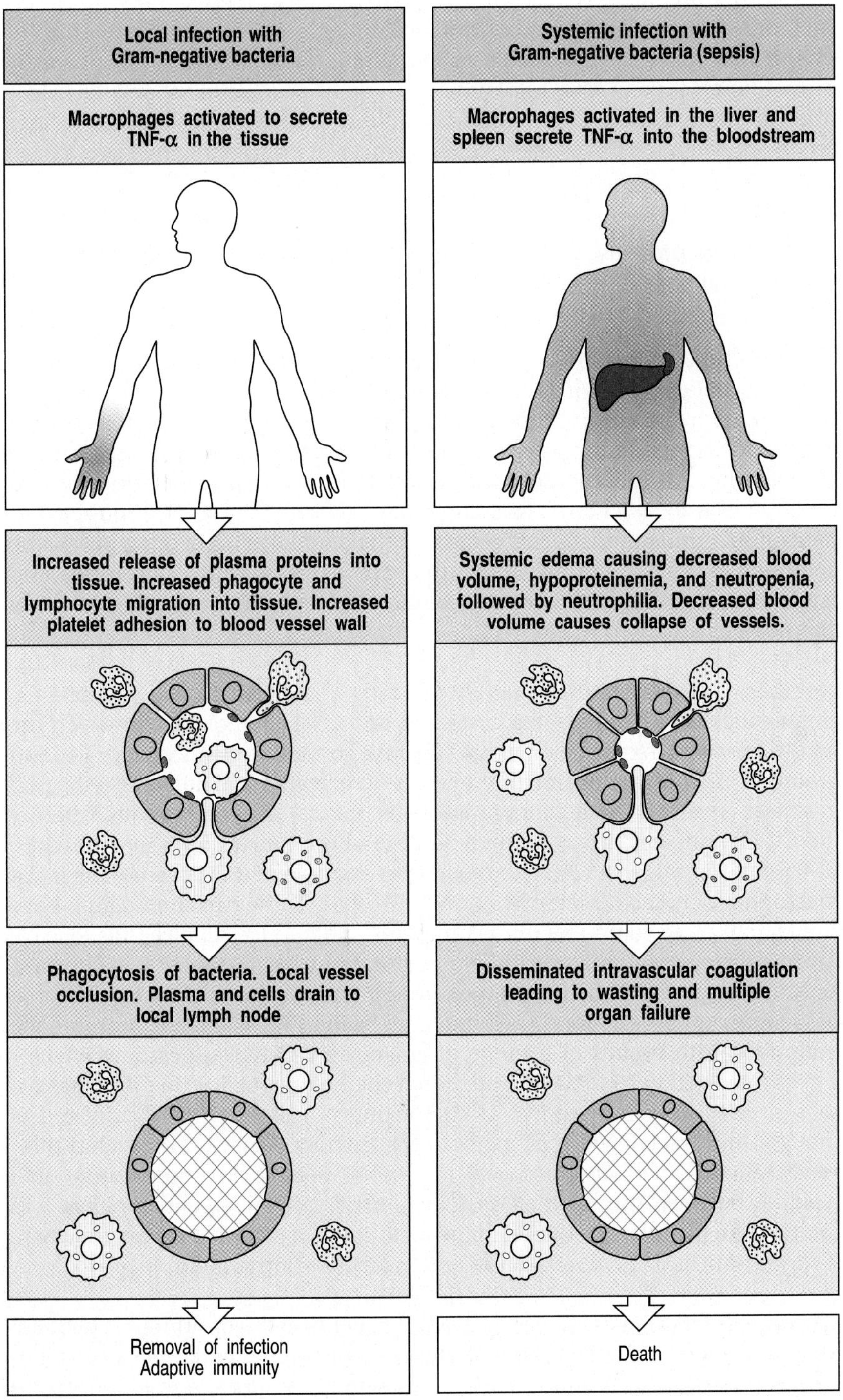
Local infection with Gram-negative bacteria
Systemic infection with Gram-negative bacteria (sepsis)
Macrophages activated to secrete TNF-α in the tissue
Macrophages activated in the liver and spleen secrete TNF-α into the bloodstream
Increased release of plasma proteins into tissue. Increased phagocyte and lymphocyte migration into tissue. Increased platelet adhesion to blood vessel wall
Systemic edema causing decreased blood volume, hypoproteinemia, and neutropenia, followed by neutrophilia. Decreased blood volume causes collapse of vessels.
Phagocytosis of bacteria. Local vessel occlusion. Plasma and cells drain to local lymph node
Disseminated intravascular coagulation leading to wasting and multiple organ failure
Removal of infection Adaptive immunity
Death

Fig. 10.12 The release of TNF-α by macrophages induces local protective effects, but TNF-α can have damaging effects when released systemically. The panels on the left show the causes and consequences of local release of TNF-α, while the panels on the right show the causes and consequences of systemic release. The central panels illustrate the common effects of TNF-α, which acts on blood vessels, especially venules, to increase blood flow, increase vascular permeability to fluid, proteins, and cells, and to increase endothelial adhesiveness for white blood cells and platelets. Local release thus allows an influx into the infected tissue of fluid, cells, and proteins that participate in host defense. The small vessels later clot, preventing spread of the infection to the blood, and the accumulated fluid and cells drain to regional lymph nodes where the adaptive immune response is initiated. When there is a systemic infection, or sepsis, with bacteria that elicit TNF-α production, then TNF-α is released into the blood and acts in a similar way on all small blood vessels. The result is shock, disseminated intravascular coagulation with depletion of clotting factors and consequent bleeding, multiple organ failure, and death.

known as **sepsis**, is accompanied by the release of TNF-α by macrophages in the liver, spleen, and other sites. The systemic release of TNF-α causes vasodilation and loss of plasma volume owing to increased vascular permeability, leading to shock. In **septic shock**, disseminated intravascular coagulation (blood clotting) is also triggered by TNF-α, leading to the generation of clots in the small vessels and the massive consumption of clotting proteins. The patient's ability to clot blood appropriately is lost. This condition frequently leads to the failure of vital organs such as the kidneys, liver, heart, and lungs, which are quickly compromised by the failure of normal perfusion; consequently, septic shock has a high mortality rate.

Mice with a mutant TNF-α receptor gene are resistant to septic shock; however, such mice are also unable to control local infection. Although the features of TNF-α that make it so valuable in containing local infection are precisely those that give it a central role in the pathogenesis of septic shock, it is clear from the evolutionary conservation of TNF-α that its benefits in the former arena outweigh the devastating consequences of its systemic release.

10-9 Small proteins called chemokines recruit new phagocytic cells to local sites of infection.

Some of the cytokines released in response to infection belong to a family of closely related proteins called **chemokines**, small polypeptides that are synthesized by phagocytes and by many other cell types. IL-8, whose contribution to extravasation we have just discussed, belongs to this subset of cytokines. All the chemokines are related in amino acid sequence and function mainly as chemoattractants for leukocytes, recruiting monocytes, neutrophils, and other effector cells from the blood to sites of infection. Some chemokines also function in lymphocyte development, migration, and angiogenesis (the growth of new blood vessels); the properties of some chemokines are listed in Fig. 10.13.

Members of the chemokine family fall mostly into two broad groups—CC chemokines with two adjacent cysteines, and CXC chemokines, in which the equivalent two cysteine residues are separated by another amino acid. The two groups of chemokines act on different sets of receptors and different cell types; in general, the CXC chemokines promote the migration of neutrophils, whereas the CC chemokines promote the migration of monocytes or other cell types. IL-8 is an example of a CXC chemokine; an example of a CC chemokine is the macrophage chemoattractant protein-1 (MCP-1). These two chemokines have similar, although complementary, functions: IL-8 induces neutrophils to leave the bloodstream and migrate into the surrounding tissues; MCP-1, in contrast, acts on monocytes, inducing their migration from the bloodstream to become tissue macrophages. Other CC chemokines such as RANTES may promote the infiltration into tissues of a range of leukocyte cell types including effector T cells (see Section 10-20), with individual chemokines acting on different subsets of cells. The only known C chemokine is called lymphotactin and is thought to attract T-cell precursors to the thymus. A newly discovered molecule called fractalkine is unusual in several ways: it has three amino acid residues between the two half-cysteines, making it a CX_3C chemokine; it is multimodular; and it is tethered to the membrane of cells that express it, where it serves both as a chemoattractant and as an adhesion protein.

The role of chemokines such as IL-8 and MCP-1 in cell recruitment is twofold: first, to convert the initial rolling of the leukocyte on the endothelial cells into stable binding; and second to direct its migration along a gradient of the

Class	Chemokine	Produced by	Receptors	Chemoattracted cells	Major effects
CXC	IL-8	Monocytes Macrophages Fibroblasts Keratinocytes Endothelial cells	CXCR1 CXCR2	Neutrophils Naive T cells	Mobilizes, activates and degranulates neutrophils Angiogenesis
	PBP β-TG NAP-2	Platelets	CXCR2	Neutrophils	Activates neutrophils Clot resorption Angiogenesis
	GROα, β, γ	Monocytes Fibroblasts Endothelium	CXCR2	Neutrophils Naive T cells Fibroblasts	Activates neutrophils Fibroplasia Angiogenesis
	IP-10	Keratinocytes Monocytes T cells Fibroblasts Endothelium	CXCR3	Resting T cells NK cells Monocytes	Immunostimulant Anti-angiogenic Promotes T_H1 immunity
	SDF-1	Stromal cells	CXCR4	Naive T cells Progenitor ($CD34^+$) B cells	B-cell development Lymphocyte homing Competes with HIV-1
CC	MIP-1α	Monocytes T cells Mast cells Fibroblasts	CCR1, 3, 5	Monocytes NK and T cells Basophils Dendritic cells	Competes with HIV-1 Anti-viral defense Promotes T_H1 immunity
	MIP-1β	Monocytes Macrophages Neutrophils Endothelium	CCR1, 3, 5	Monocytes NK and T cells Dendritic cells	Competes with HIV-1
	MCP-1	Monocytes Macrophages Fibroblasts Keratinocytes	CCR2B	Monocytes NK and T cells Basophils Dendritic cells	Activates macrophages Basophil histamine release Promotes T_H2 immunity
	RANTES	T cells Endothelium Platelets	CCR1, 3, 5	Monocytes NK and T cells Basophils Eosinophils Dendritic cells	Degranulates basophils Activates T cells Chronic inflammation
	Eotaxin	Endothelium Monocytes Epithelium T cells	CCR3	Eosinophils Monocytes T cells	Role in allergy
C	Lymphotactin	CD8>CD4 T cells	?	Thymocytes Dendritic cells NK cells	Lymphocyte trafficking and development
CXXXC (CX_3C)	Fractalkine	Monocytes Endothelium Microglial cells	CX_3CR1	Monocytes T cells	Leukocyte-endothelial adhesion Brain inflammation

Fig. 10.13 Properties of selected chemokines. Chemokines fall mainly into two related but distinct groups: the CC chemokines, which in humans are all encoded in one region of chromosome 4, have two adjacent cysteine residues; CXC chemokines, which are found in a cluster on chromosome 17, have an amino acid residue between the equivalent two cysteines. A C chemokine with only one cysteine at this location, and fractalkine, a CX_3C chemokine, are encoded elsewhere in the genome. Each chemokine interacts with one or more receptors, and affects one or more types of cell.

chemokine that increases in concentration towards the site of infection. This is achieved by the binding of the small, soluble chemokines to proteoglycan molecules in the extracellular matrix and on endothelial cell surfaces, thus displaying the chemokines on a solid substrate along which the leukocytes can migrate. Once the leukocytes have crossed the endothelium and the basement membrane to enter the tissues, their migration to the focus of infection is directed by the gradient of matrix-associated chemokine molecules.

Chemokines can be produced by a wide variety of cell types in response to bacterial products, viruses, and agents that cause physical damage, such as silica or the urate crystals that occur in gout. Thus, infection or physical damage to tissues sets in motion the recruitment of phagocytic cells to the site of damage. Both IL-8 and MCP-1 also activate their respective target cells, so that not only are neutrophils and macrophages brought to potential sites of infection but, in the process, they are armed to deal with any pathogens they may encounter. In particular, neutrophils exposed to IL-8 and TNF-α are activated to mediate a respiratory burst that generates oxygen radicals and nitric oxide, and to release their stored granule contents, thus contributing both to host defense and to local tissue destruction seen in local sites of infection with pyogenic (pus-forming) bacteria. Just as all the chemokines have similar structures, all their receptors are similar in structure; all are integral membrane proteins containing seven membrane-spanning helices. This structure is characteristic of receptors such as rhodopsin (see Fig. 8.32) and the muscarinic acetylcholine receptor, which are coupled to guanine nucleotide binding proteins (G-proteins); the chemokine receptors are also activated through coupled G-proteins.

Finally, there are several known examples of viruses that interfere with chemokine action and utilize chemokines or their receptors for their own ends. The most notorious of these is the human immunodeficiency virus-1 (HIV-1), the cause of AIDS. HIV-1 enters cells that express both CD4 and the chemokine receptor CCR5. People who are homozygous for a mutation in CCR5 are resistant to infection with HIV-1. We shall return to this in Chapter 11, where we discuss means adopted by pathogens to frustrate the immune response.

Thus, tissue phagocytes initiate host responses in tissues, and their numbers are soon augmented through the action of chemokines, which recruit large numbers of circulating phagocytic and immunocompetent cells to sites of infection and tissue damage. Why there are so many chemokines is not yet known; neither is the exact role of each one in host defense and in pathological responses.

10-10 Neutrophils predominate in the early cellular infiltrate into inflammatory sites.

Neutrophils are abundant in the blood but are absent from normal tissues. They are short-lived, surviving only a few hours after leaving the bone marrow. The innate immune response produces a variety of factors that are chemotactic for neutrophils and they rapidly emigrate from the blood to enter sites of infection, where they are the earliest phagocytic cells to be recruited. Later, they are followed by monocytes, the precursors of macrophages. Once in an inflammatory site, the neutrophils are able to eliminate many pathogens by phagocytosis.

The role of neutrophils in the phagocytosis of antibody-coated pathogens was discussed in Chapter 9, and where the individual has had a previous encounter with the pathogen this is likely to be the dominant mechanism by which microorganisms are removed. However, neutrophils are able to phagocytose bacteria even in the absence of specific antibodies and can thus provide a protective response in the first encounter with a pathogen. Bacterial cell wall components can be bound directly by several receptors on neutrophils. Neutrophils can also phagocytose microorganisms coated with the complement component C3b and its inactive derivative iC3b, which are deposited on the surface of the pathogen by the alternative pathway of complement activation (see Sections 9-26 and 10-4).

Neutrophils produce several bacteriostatic and toxic products, and phagocytosed pathogens are killed rapidly (see Section 9-18). Neutrophil granules containing enzymes, proteins, and peptides fuse with pathogen-containing phagosomes upon neutrophil activation, releasing these antibacterial agents on to the pathogen. The combination of toxic oxygen metabolites, nitric oxide, proteases, phospholipases, and antibacterial proteins and peptides is able to eliminate Gram-positive and Gram-negative bacteria, fungi, and even some enveloped viruses. The importance of neutrophils in host defense is best illustrated by considering inherited defects in neutrophil maturation or antibacterial functions; patients with such deficiencies suffer recurrent infections, often of bacteria and fungi that form part of the normal flora. In patients with no neutrophils such infections frequently escape from the local site to produce a life-threatening septicemia (infection of the blood). Neutrophils themselves are short-lived and the pus formed at sites of inflammation contains many dead and dying neutrophils. Any microorganisms that have been phagocytosed but not killed are released at this point and can be re-phagocytosed by other neutrophils or by macrophages that accumulate later in the inflammatory response. Even this sequestration of microorganisms can be important in host defense; in individuals whose neutrophils are unable to kill phagocytosed organisms, in contrast to those with no neutrophils at all, infections only rarely spread beyond the local inflammatory site.

10-11 Cytokines released by phagocytes also activate the acute-phase response.

As well as their important local effects, the cytokines produced by macrophages and neutrophils have long-range effects that contribute to host defense. One of these is the elevation of body temperature, which is caused by TNF-α, IL-1, IL-6, and other cytokines. These are termed 'endogenous pyrogens' because they cause fever and derive from an endogenous source rather than from bacterial components. Fever is generally beneficial to host defense; most pathogens grow better at lower temperatures and adaptive immune responses are more intense at raised temperatures. Host cells are also protected from deleterious effects of TNF-α at raised temperatures.

The effects of TNF-α, IL-1, and IL-6 are summarized in Fig. 10.14. One of the most important of these is the initiation of a response known as the **acute-phase response** (Fig. 10.15). This involves a shift in the proteins secreted by the liver into the blood plasma and results from the action of IL-1, IL-6,

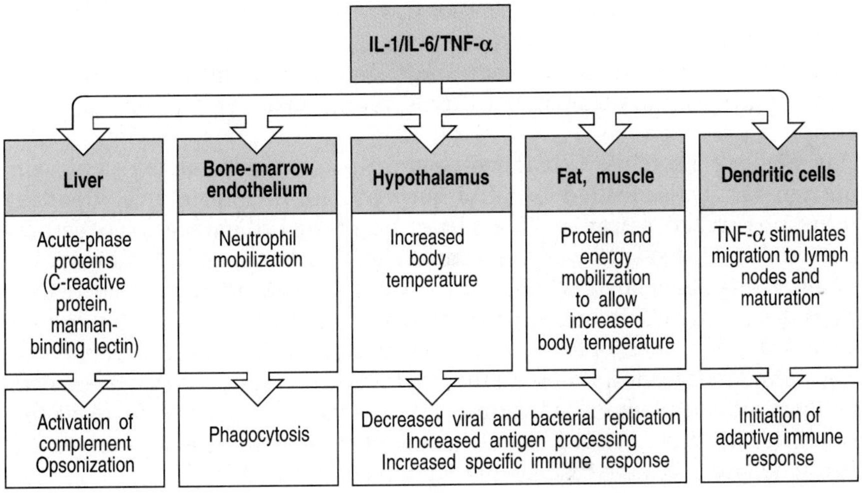

Fig. 10.14 The cytokines TNF-α, IL-1, and IL-6 have a wide spectrum of biological activities that help to co-ordinate the body's responses to infection. IL-1, IL-6, and TNF-α activate hepatocytes to synthesize acute-phase proteins, and bone-marrow endothelium to release neutrophils. The acute-phase proteins act as opsonins, whereas the disposal of opsonized pathogens is augmented by enhanced recruitment of neutrophils from the bone marrow. IL-1, IL-6, and TNF-α are also endogenous pyrogens, raising body temperature, which is believed to help to eliminate infections. A major effect of these cytokines is to act on the hypothalmus, altering the body's temperature regulation, and on muscle and fat cells, altering energy mobilization to increase the body temperature. At elevated temperatures, bacterial and viral replication are decreased, whereas processing of antigen is enhanced. Finally, they help activate B and T cells by inducing migration to lymph nodes and maturation of dendritic cells to induce an adaptive immune response.

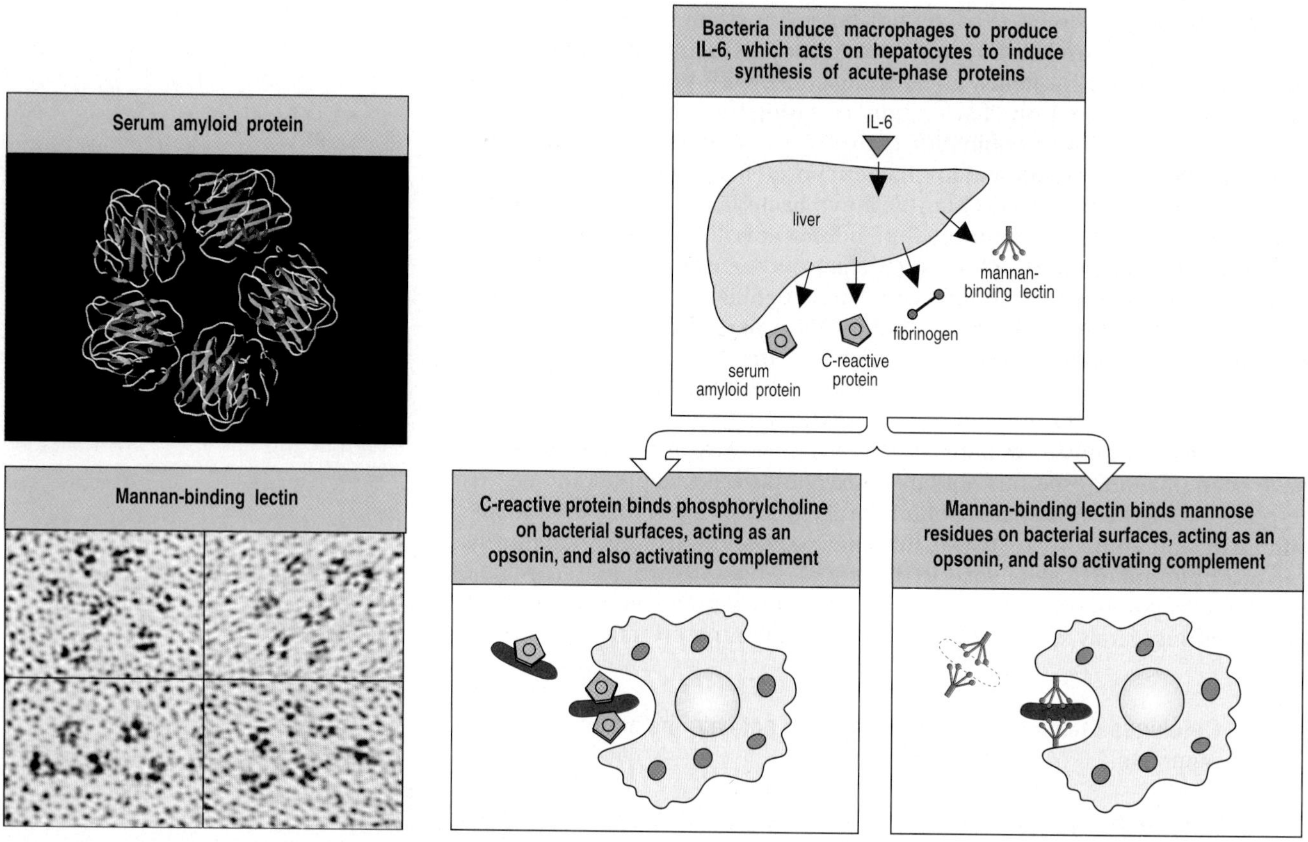

Fig. 10.15 The acute-phase response produces molecules that bind bacteria but not host cells. Acute-phase proteins are produced by liver cells in response to cytokines released by phagocytes in the presence of bacteria. They include serum amyloid protein (SAP) (in mice but not humans), C-reactive protein (CRP), fibrinogen, and mannan-binding lectin (MBL). SAP and CRP are homologous in structure; both are pentraxins, forming five-membered disks, as shown for SAP (upper photograph). CRP binds phosphorylcholine on bacterial surfaces but does not recognize it in the form in which it is found in host-cell membranes, and can both act as an opsonin in its own right and activate the classical complement pathway by binding C1q to augment opsonization. MBL is a member of the collectin family, which includes C1q, which it resembles in its structure (see lower photograph and Fig. 9.34). MBL binds mannose residues on bacterial cell surfaces and, like CRP, can both act as an opsonin in its own right and activate complement. MBL activates the lectin complement pathway by binding and activating two serine esterases, resembling C1rs, that in turn activates C4 and C2. Thus, CRP and MBL can lead to bacterial clearance in the same way as an IgM antibody. Photographs courtesy of J Emsley (SAP) and K Reid (MBL).

and TNF-α on hepatocytes. In the acute-phase response, levels of some plasma proteins go down, while levels of others increase markedly. The proteins whose synthesis is induced by TNF-α, IL-1, and IL-6 are called acute-phase proteins. Of the acute-phase proteins, two are of particular interest because they mimic the action of antibodies but, unlike antibodies, these proteins have broad specificity for pathogen molecules.

One of these proteins, **C-reactive protein**, is a member of the **pentraxin** protein family, so called because they are formed from five identical subunits. C-reactive protein binds to the phosphorylcholine portion of certain bacterial and fungal cell wall lipopolysaccharides. Phosphorylcholine is also found in mammalian cell membrane phospholipids but in a form that cannot react with C-reactive protein. When C-reactive protein binds to a bacterium, it is not only able to opsonize it but can also activate the complement cascade by binding to C1q, the first component of the classical pathway of complement activation. The interaction with C1q involves the collagen-like parts of C1q, rather than the globular heads contacted by antibody, but the same cascade of reactions is initiated.

The second acute-phase protein of interest is **mannan-binding lectin (MBL)**. This is found in normal serum at low levels but is also produced in increased amounts during the acute-phase response. It is a calcium-dependent sugar-binding protein, or lectin, a member of a structurally related family of proteins known as the **collectins**. It binds to mannose residues, which are accessible on many bacteria but are covered by other sugar groups in the carbohydrates on vertebrate cells. MBL also acts as an opsonin for monocytes, which, unlike tissue macrophages, do not express the macrophage mannose receptor. The structure of mannan-binding lectin resembles that of the C1q component of complement, although the two proteins do not share sequence homology. When it binds to bacteria, mannan-binding lectin can, like C1q, activate a proteolytic enzyme complex that cleaves C4 and C2 to initiate complement activation by the lectin pathway (see Fig. 10.8). The collectins also include the pulmonary surfactant proteins A and D (SP-A and SP-D), which are probably important in binding and opsonizing pulmonary pathogens such as *Pneumocystis carinii*.

Thus, within a day or two, the acute-phase response provides the host with two proteins with the functional properties of antibodies and which can bind a broad range of bacteria. However, unlike antibodies, they have no structural diversity, and are made in response to any stimulus that triggers the release of TNF-α, IL-1, and IL-6, so their synthesis is not specifically induced and targeted.

A final distant effect of the cytokines produced by phagocytes is to induce a leukocytosis, an increase in circulating neutrophils. The leukocytes come from two sources: the bone marrow, from which mature leukocytes are released in increased numbers; and sites in blood vessels where the leukocytes are attached loosely to endothelial cells. Finally, the migration of dendritic cells from their sites in peripheral tissues to the lymph node, and their maturation into non-phagocytic but highly co-stimulatory antigen-presenting cells is stimulated by TNF-α. Such cells are crucial to the initiation of adaptive immunity, as we shall see in Section 10-16. All these effects of cytokines produced in response to infection contribute to the control of infection while the adaptive immune response is being developed.

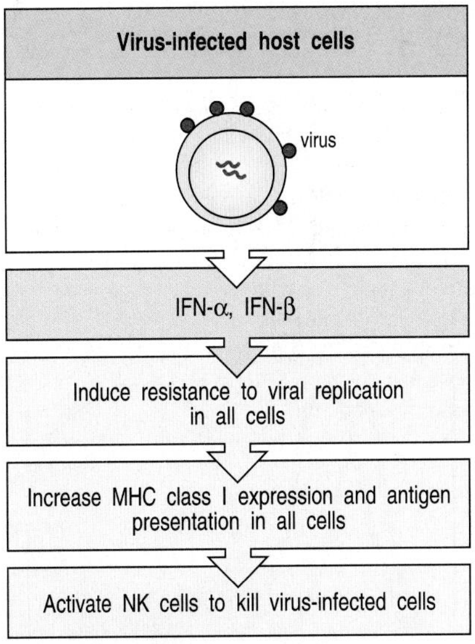

Fig. 10.16 Interferons are antiviral proteins produced by cells in response to viral infection. The α- and β-interferons have three major functions. First, they induce resistance to viral replication by activating cellular genes that destroy mRNA and inhibit the translation of viral and some host proteins. Second, they induce MHC class I expression in most uninfected cells in the body, thus enhancing the resistance to natural killer (NK) cells, and make cells newly infected by virus more susceptible to killing by CD8 cytotoxic T cells. Third, they activate NK cells, which then kill virus-infected cells selectively.

10-12 Interferons inhibit viral replication and activate certain host-defense responses.

Infection of cells with viruses induces the production of proteins known as **interferons** because they were found to interfere with viral replication in previously uninfected tissue culture cells. They are believed to have a similar role *in vivo*, blocking the spread of viruses to uninfected cells. These antiviral interferons, called **interferon-α (IFN-α)** and **interferon-β (IFN-β)**, are quite distinct from **interferon-γ (IFN-γ)**, which is produced by activated NK cells and in larger amounts by effector T cells and thus appears mainly after the induction of the adaptive immune response. IFN-α, actually a family of several closely related proteins, and IFN-β, the product of a single gene, are synthesized by many cell types after viral infection. Double-stranded RNA is a potent inducer of interferon synthesis. It is not found in mammalian cells but forms the genome of some viruses and might be made as part of the infectious cycle of all viruses; it might be the common element in interferon induction. Interferons make several contributions to host defense against viral infection (Fig. 10.16).

An obvious and important effect of interferons is the induction of a state of resistance to viral replication in all cells. IFN-α and IFN-β bind to a common cellular receptor on cells. The interferon receptor, like other cytokine receptors (see Chapter 8), is coupled to a Janus-family tyrosine kinase, which in turn phosphorylates signal-transducing activators of transcription known as STATs. The binding of phosphorylated STAT proteins to the promoters of several genes induces the synthesis of host-cell proteins that contribute to the inhibition of viral replication. One of these is the enzyme oligo-adenylate synthetase, which polymerizes ATP into a series of 2′–5′ linked oligomers (nucleotides in nucleic acids are normally linked 3′–5′). These activate an endoribonuclease that then degrades viral RNA. A second protein activated by IFN-α and IFN-β is a serine/threonine kinase called P1 kinase. This enzyme phosphorylates the eukaryotic protein synthesis initiation factor eIF-2, thereby inhibiting translation and thus contributing to the inhibition of viral replication. Another interferon-inducible protein called Mx is known to be required for cellular resistance to influenza virus replication. Mice that lack the gene for Mx are highly susceptible to infection with the influenza virus, whereas genetically normal mice are not.

The second effect of interferons in host defense is to increase the expression of MHC class I molecules, TAP transporter proteins, and the Lmp2 and Lmp7 components of the proteasome. This enhances the ability of host cells to present viral peptides to CD8 T cells should infection occur (see Section 4-10). At the same time, this increase in MHC class I expression protects uninfected host cells against attack by natural killer cells (NK cells). Natural killer cells are strongly activated by IFN-α and IFN-β, and make several important contributions to early host responses to viral infections, as we shall see next.

10-13 Natural killer cells serve as an early defense against certain intracellular infections.

Natural killer, or NK cells, which we introduced in Chapter 9 as the effectors in antibody-dependent cell-mediated cytotoxicity, are identified by their ability to kill certain lymphoid tumor cell lines *in vitro* without the need for prior immunization or activation. However, their known function in host defense is in the early phases of infection with several intracellular pathogens, particularly herpes group viruses, *Leishmania*, and *Listeria monocytogenes*, and we shall consider them here from that point of view.

Although NK cells that can kill sensitive targets can be isolated from uninfected individuals, this activity is increased by between 20- and 100-fold when NK cells are exposed to IFN-α and IFN-β or to the NK-cell activating factor IL-12, which is one of the cytokines produced early in many infections (Fig. 10.17). IL-12, in synergy with TNF-α, can also elicit the production of large amounts of IFN-γ by NK cells, and this secreted IFN-γ is crucial in controlling some infections before T cells have been activated to produce this cytokine. One example is the response to the intracellular bacterium *Listeria monocytogenes.* Mice that lack T and B lymphocytes are initially quite resistant to this pathogen; however, antibody-mediated depletion of NK cells or neutralization of TNF-α or IFN-γ or their receptors renders these mice highly susceptible, so that they die a few days after infection.

If NK cells are to mediate host defense against infection with viruses and other pathogens, they must have some mechanism for distinguishing infected from uninfected cells. Exactly how this is achieved has not yet been worked out, but recognition of 'altered self' is thought to be involved. NK cells have

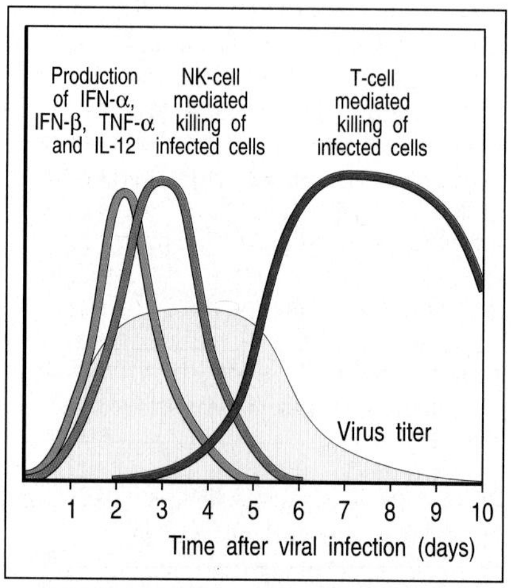

Fig. 10.17 Natural killer cells (NK cells) are an early component of the host response to virus infection. Interferons-α and -β and the cytokines TNF-α and IL-12 appear first, followed by a wave of NK cells, which together control virus replication but do not eliminate the virus. Virus elimination is accomplished when specific CD8 T cells are produced. Without NK cells, the levels of certain viruses are much higher in the early days of the infection.

two types of surface receptor that control their cytotoxic activity. One type triggers killing by NK cells; several receptors can provide this activation signal, including calcium-binding C-type lectins that recognize a wide variety of carbohydrate ligands found on many cells. A second set of receptors inhibit activation, and prevent NK cells from killing normal cells. These inhibitory receptors are specific for MHC class I alleles, which explains why NK cells selectively kill target cells bearing low levels of MHC class I molecules. Thus, one possible mechanism by which NK cells distinguish infected from uninfected cells is by recognizing alterations in MHC class I expression (Fig. 10.18). Another is that they recognize changes in cell-surface glycoproteins induced by viral or bacterial infection.

In mice, receptors that inhibit the activation of NK cells are encoded by a multigene family of C-type lectins called Ly49. Different Ly49 receptors recognize different MHC class I alleles and are differentially expressed on different subsets of NK cells. Some NK cells express Ly49 receptors specific for non-self MHC alleles, but each cell expresses at least one receptor that can recognize an MHC class I allele expressed by the host. In humans, there are inhibitory receptors that recognize distinct HLA-B and HLA-C alleles. These receptors are structurally different from those of the mouse, being members of the immunoglobulin gene superfamily; they are usually called p58 and p70, or killer inhibitory receptors (KIRs). In addition, human NK cells express a heterodimer of two C-type lectin molecules, called CD94 and NKG2. Other

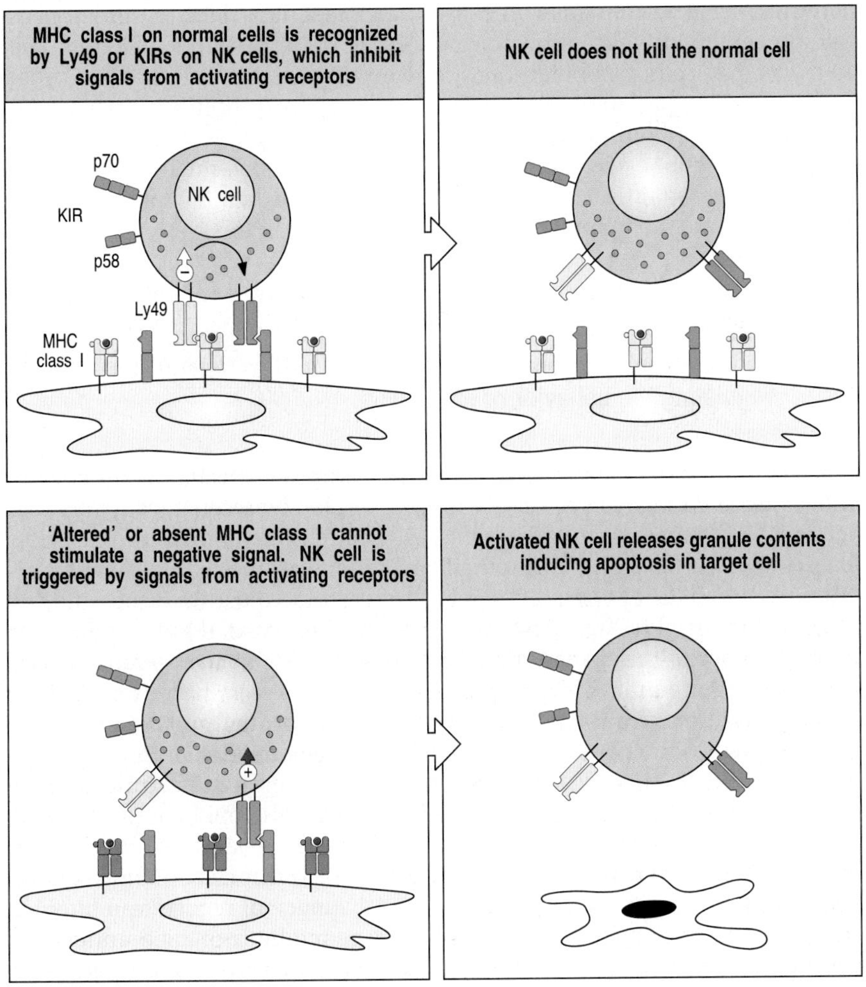

MHC class I on normal cells is recognized by Ly49 or KIRs on NK cells, which inhibit signals from activating receptors
p70
KIR
p58
NK cell
Ly49
MHC class I
NK cell does not kill the normal cell
'Altered' or absent MHC class I cannot stimulate a negative signal. NK cell is triggered by signals from activating receptors
Activated NK cell releases granule contents inducing apoptosis in target cell

Fig. 10.18 Possible mechanisms whereby NK cells distinguish infected from non-infected cells. A proposed mechanism of NK cell recognition is shown. NK cells can use several different receptors that signal them to kill, including lectin-like receptors that recognize carbohydrate on self cells. However, another set of receptors, called Ly49 in the mouse and killer inhibitory receptors (KIR) in the human, recognize MHC class I molecules and inhibit killing by NK cells by overruling the actions of the killer receptors. This inhibitory signal is lost when host cells do not express MHC class I and perhaps also in cells infected with virus, which might inhibit MHC class I expression or alter its conformation. Normal cells respond to IFN-α and -β by increasing levels of MHC class I expression, making them resistant to activated NK killing. Infected cells can fail to increase MHC class I expression, making them targets for activated NK cells. Ly49 and KIR are encoded by different families of genes—the C-type lectins for Ly49 and the immunoglobulin gene superfamily for the KIRs. The KIRs are made in two forms, p58 and p70, which differ by the presence of one immunoglobulin domain.

inhibitory NK receptors specific for the products of the MHC class I loci are being defined at a rapid rate, and all are members of either the immunoglobulin-like KIR family or the Ly49-like C-type lectin family. A common feature of all inhibitory NK receptors is the presence of an immunoreceptor tyrosine-based inhibitory motif (ITIM) in their cytoplasmic domains (see Fig. 5.19).

Because the binding of the inhibitory receptors to MHC class I molecules inhibits NK activity, normal syngeneic cells are protected from attack by NK cells. Virus-infected cells, however, can become susceptible to killing by NK cells by a variety of mechanisms. First, some viruses inhibit all protein synthesis in their host cells, so the augmented synthesis of MHC class I proteins induced by interferon would be blocked selectively in infected cells, and NK cells would no longer be inhibited through their MHC-specific receptors. Second, some viruses can selectively prevent the export of MHC class I molecules, which might allow the infected cell to evade recognition by CD8 T cells but would make it sensitive to NK cell killing. There is also evidence that introduction of new peptides into self MHC class I molecules is detectable by NK cells. It is not known whether these peptides are recognized directly or whether they alter MHC conformation. Finally, virus infection alters the glycosylation of cellular proteins, perhaps allowing dominant recognition by activation receptors or removing the normal ligand for the inhibitory receptors. Either of these last two mechanisms could allow infected cells to be detected even when the level of MHC class I expression has not been altered.

Clearly, much remains to be learned about this innate mechanism of cytotoxic attack and its physiological significance. At present, the only clue to the function of NK cells in humans comes from a rare patient deficient in NK cells who proved highly susceptible to early phases of herpes virus infection. The ability of NK cells to operate early in host defense by mechanisms that involve the recognition of self MHC molecules suggests they might represent the modern remnants of the evolutionary forebears of T cells. Two other 'primitive' lymphocyte types—γ:δ T cells and B-1 cells—might also participate in the pre-adaptive immune response. Of these, the γ:δ T cells, which we discuss first, are the more enigmatic.

10-14 T cells bearing γ:δ T-cell receptors are found in lymphoid organs and most epithelia and might contribute to host defense by regulating the behavior of other cells.

The discovery of γ:δ T cells was accidental; they were detected as a consequence of having immunoglobulin-like receptors encoded by rearranged genes (see Section 4-30), and their function remains obscure. One of their most striking features is their division into two highly distinct sets of cells. One set of γ:δ T cells is found in the lymphoid tissue of all vertebrates and, like B cells and α:β T cells, they display highly diverse receptors. By contrast, intra-epithelial γ:δ T cells occur variably in different vertebrates, and commonly display receptors of very limited diversity, particularly in the skin and the female reproductive tract of mice, where the γ:δ cells are essentially homogeneous in any one site. On the basis of this limited diversity of epithelial γ:δ T cells and their limited recirculatory behavior, it has been proposed that intra-epithelial γ:δ T cells recognize ligands that are derived from the epithelium in which they reside and signify that it has become infected. Candidate ligands are heat-shock proteins, MHC class IB molecules, and unorthodox nucleotides and phospholipids, for all of which there is evidence of recognition by γ:δ T cells. Unlike α:β T cells, γ:δ T cells do not generally recognize antigen as peptides presented by MHC molecules; instead they seem to recognize their target antigens directly, and could potentially recognize and respond rapidly to molecules expressed by many different cell types. Recognition of molecules expressed as a consequence of infection, rather than of pathogen-specific antigens themselves, would distinguish γ:δ T cells from other lymphocytes and arguably place them at the intersection between innate and adaptive immunity.

Several recent studies of mice deficient in γ:δ T cells have revealed exaggerated responses to various pathogens and even to self tissues, rather than deficiencies in pathogen control and rejection. This has led to the suggestion that at least some γ:δ T cells have a regulatory role in modulating immune responses, a function that would be consistent with their demonstrated ability to secrete regulatory cytokines when activated. Which aspects of the phenotype of γ:δ-deficient mice are attributable to which subset of γ:δ T cells remains to be clarified.

10-15 B-1 cells form a separate population of B cells, producing antibodies against common bacterial polysaccharides.

The production of antibody by conventional B cells has a major role in the adaptive immune response. However, there is a separate lineage of B cells, marked by the cell-surface protein CD5, that have properties quite distinct from those of conventional B cells (see Section 6-13). These so-called CD5 B cells, or B-1 cells, are in many ways analogous to epithelial γ:δ T cells: they arise early in ontogeny, they use a distinctive and limited set of V genes to make their receptors, they are self-renewing in the periphery, and they are the predominant lymphocyte in a distinctive microenvironment, the peritoneal cavity.

B-1 cells seem to make antibody responses mainly to polysaccharide antigens of the TI-2 type. These T-cell independent responses do not induce significant class switching or somatic hypermutation of immunoglobulin variable regions; as a consequence, the predominant antibody isotype produced is IgM (Fig. 10.19). Although these responses can be augmented by T cells, with IL-5 having an important role (see Section 9-10), they appear within 48 hours of the exposure to antigen, and the T cells involved are therefore not part of an antigen-specific adaptive immune response. The lack of an antigen-specific interaction with helper T cells might explain why immunological memory is not generated: repeated exposures to the same TI-2 antigen elicit similar or decreased responses with each exposure to antigen. Thus, these responses, although generated by lymphocytes with rearranging receptors, resemble innate rather than adaptive immune responses.

As with γ:δ T cells, the precise role of B-1 cells in host defense is uncertain. Mice that are deficient in B-1 cells are more susceptible to infection with *Streptococcus pneumoniae*; this is because they fail to produce an antibody against the phospholipid headgroup phosphorylcholine that effectively protects against this organism. The phosphorylcholine is bound to carbohydrate in the bacterial cell wall and is recognized as a TI-2 antigen because it can very effectively crosslink the B-1 cell antigen receptor. A significant fraction of the B-1 cells can make antibodies of this specificity, and because no antigen-specific T-cell help is required, a potent response can be produced early in infection with this pathogen. Whether human B-1 cells have the same role is uncertain. In terms of evolution, it is interesting to note that γ:δ T cells seem to defend the body surfaces, whereas B-1 cells defend the body cavity. Both cell types are relatively limited in their range of specificities and in the efficiency of

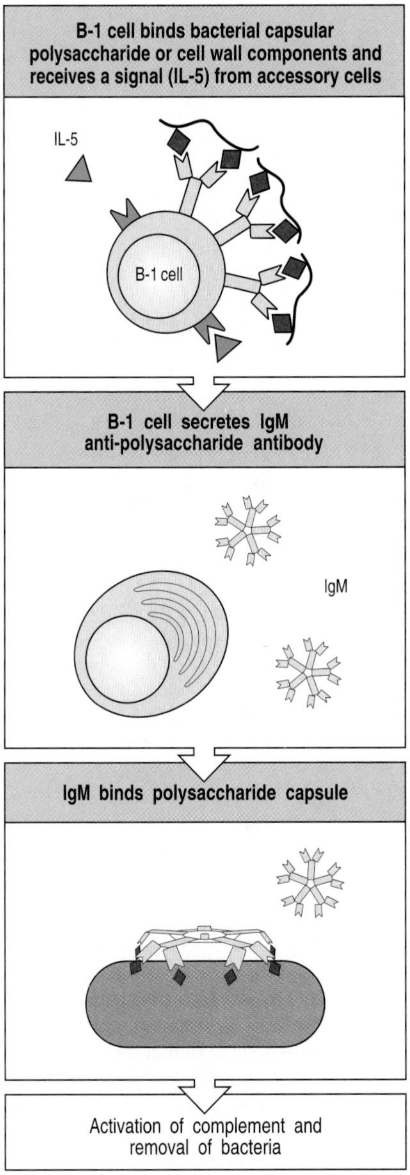

Fig. 10.19 CD5 B cells might be important in the response to carbohydrate antigens such as bacterial polysaccharides. These TI-2 responses might require IL-5 provided by T cells but this help is not antigen specific and its mechanism is not clear. These responses are rapid, with antibody appearing in 48 hours, presumably because there is a high frequency of precursors of the responding lymphocytes so that little clonal expansion is required. In the absence of antigen-specific T-cell help, only IgM is produced and, in mice, these responses therefore work mainly through the activation of complement.

their responses. It is possible that these two cell types represent a transitional phase in the evolution of the adaptive immune response, guarding the two main compartments of primitive organisms—the epithelial surfaces and the body cavity. It is not yet clear whether they are still critical to host defense or whether they represent an evolutionary relic. Nevertheless, as each cell type is prominent in certain sites in the body and contributes to certain responses, they must be incorporated into our thinking about host defense.

Summary.

The early induced but non-adaptive responses to infection involve a wide variety of effector mechanisms directed at distinct classes of pathogen. These responses are triggered by receptors that are either non-clonal or of very limited diversity, and are distinguished from adaptive immunity by their failure to provide lasting immunity or immunological memory. Some are induced by cytokines released by phagocytes in response to microbial infection. These cytokines have three major effects. First, they induce the production of acute-phase proteins by the liver, which can bind to bacterial surface molecules and activate complement or phagocytes. Second, they can elevate body temperature, which is thought to be deleterious to the microorganism but to enhance the immune response. Third, they induce inflammation, in which the surface properties and permeability of blood vessels are changed, recruiting phagocytes, immune cells, and molecules to the site of infection. Interferons are produced by cells infected with viruses, and these slow viral replication and enhance the presentation of viral peptides to cytotoxic T cells, as well as activating natural killer cells (NK cells), which can distinguish infected from uninfected host cells. NK cells, B-1 cells, and γ:δ T cells are lymphocytes with receptors of limited diversity that seem to provide early protection from a limited range of pathogens but do not generate lasting immunity or immunological memory. All these mechanisms have an important role, both on their own in holding infection in check during its early phases while the adaptive immune response is being developed, and also in their impact on the adaptive immune response that develops subsequently.

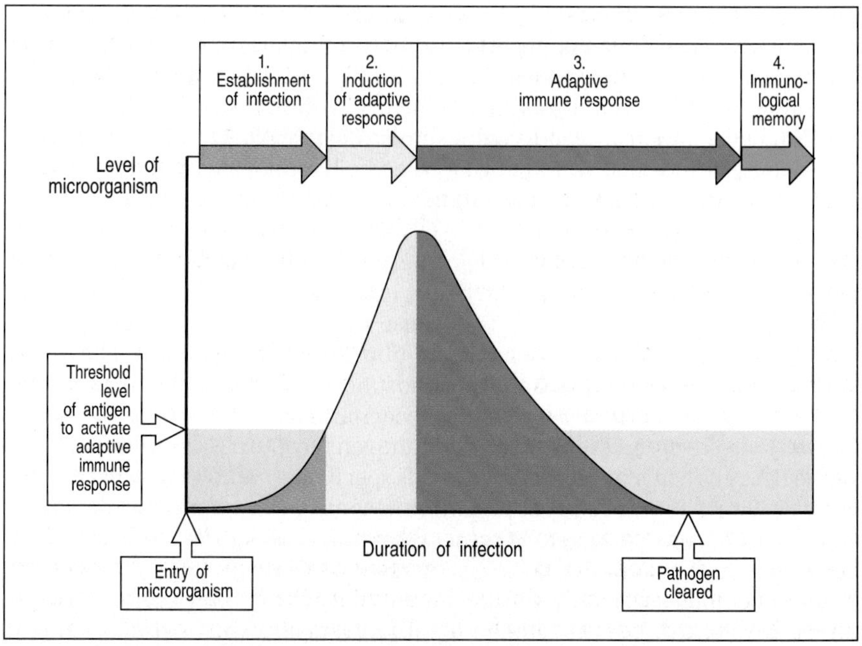

Fig. 10.20 The course of a typical acute infection. 1. The level of infectious agent increases with pathogen replication. 2. When the pathogen level exceeds the threshold dose of antigen required for an adaptive response, the response is initiated; the pathogen continues to grow, retarded only by the innate and early, non-adaptive responses. 3. After 4–5 days, effector cells and molecules of the adaptive response start to clear the infection. 4. When the infection is cleared, and the dose of antigen falls below the response threshold, the response ceases but antibody, residual effector cells, and also immunological memory provide lasting protection against re-infection.

Adaptive immunity to infection.

It is not known how many infections are dealt with solely by non-adaptive mechanisms of host defense; this is because they are eliminated early and such infections produce little in the way of symptoms or pathology. Moreover, deficiencies in non-adaptive defenses are rare, so it has seldom been possible to study their consequences. Adaptive immunity is triggered when an infection eludes the innate defense mechanisms and generates a threshold dose of antigen (Fig. 10.20). This antigen then initiates an adaptive immune response, which becomes effective only after several days, the time required for antigen-specific T and B cells to proliferate and differentiate into effector cells. Meanwhile, the pathogen continues to grow in the host, held in check mainly by innate and non-adaptive mechanisms (Fig. 10.21). In the earlier chapters of this book we discussed the cells and molecules that mediate the adaptive immune response, and the interactions between cells that stimulate individual steps in its development. We are now ready to see how each cell type is recruited in turn in the course of a primary immune response to a pathogen, and how the effector cells and molecules that are generated in response to antigen are dispersed to their sites of action, leading to clearance of the infection and the establishment of a state of protective immunity.

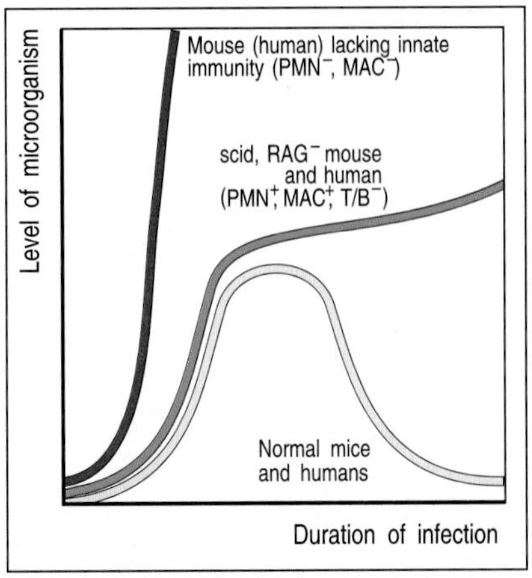

Fig. 10.21 The time course of infections in various types of immunodeficient mice and humans. The red curve shows the growth of microorganisms in the absence of innate immunity. The green curve indicates mice and humans that have innate immunity but lack adaptive immunity. The yellow curve shows the normal course of an infection in immunocompetent mice or humans. PMN, polymorphonuclear leukocytes; MAC, macrophages; T, T cells; B, B cells.

10-16 T-cell activation is initiated when recirculating T cells encounter specific antigen in draining lymphoid tissues.

The first step in any adaptive immune response leading to protective immunity is the activation of naive T cells in the draining lymphoid organs. The importance of the peripheral lymphoid organs was first shown by ingenious experiments in which a skin flap was isolated from the body wall so that it had a blood circulation but no lymphatic drainage. Antigen placed in this site did not elicit a T-cell response, showing that T cells do not become sensitized in peripheral tissues. We now know that naive T lymphocytes become sensitized in lymphoid organs by antigens taken up by dendritic cells. As noted in Chapter 8, immature dendritic cells in tissues are stimulated by infection to migrate to draining lymph nodes through which naive T cells circulate. Antigens introduced directly into the bloodstream are picked up by antigen-presenting cells in the spleen, and lymphoid cell sensitization then occurs in the splenic white pulp (see Fig. 1.9). The trapping of antigen by antigen-presenting cells that migrate to these lymphoid tissues, and the continuous recirculation of naive T cells through these tissues, ensure that rare antigen-specific T cells will encounter their specific antigen on a professional antigen-presenting cell surface.

The recirculation of naive T cells through the lymphoid organs is orchestrated by adhesive interactions between lymphocytes and endothelial cells. Naive T cells enter the lymphoid organs in essentially the same way as described earlier for the entry of phagocytes into sites of infection (see Fig. 10.11), except that in this case the selectin is expressed on the T cell rather than the endothelium. L-selectin on naive T cells binds to sulfated carbohydrates on various proteins such as the vascular addressins GlyCAM-1 and CD34. CD34 is expressed on endothelial cells in many tissues but is properly glycosylated for L-selectin binding only on the high endothelial venule cells of lymph nodes. L-selectin binding promotes a rolling interaction like that mediated by

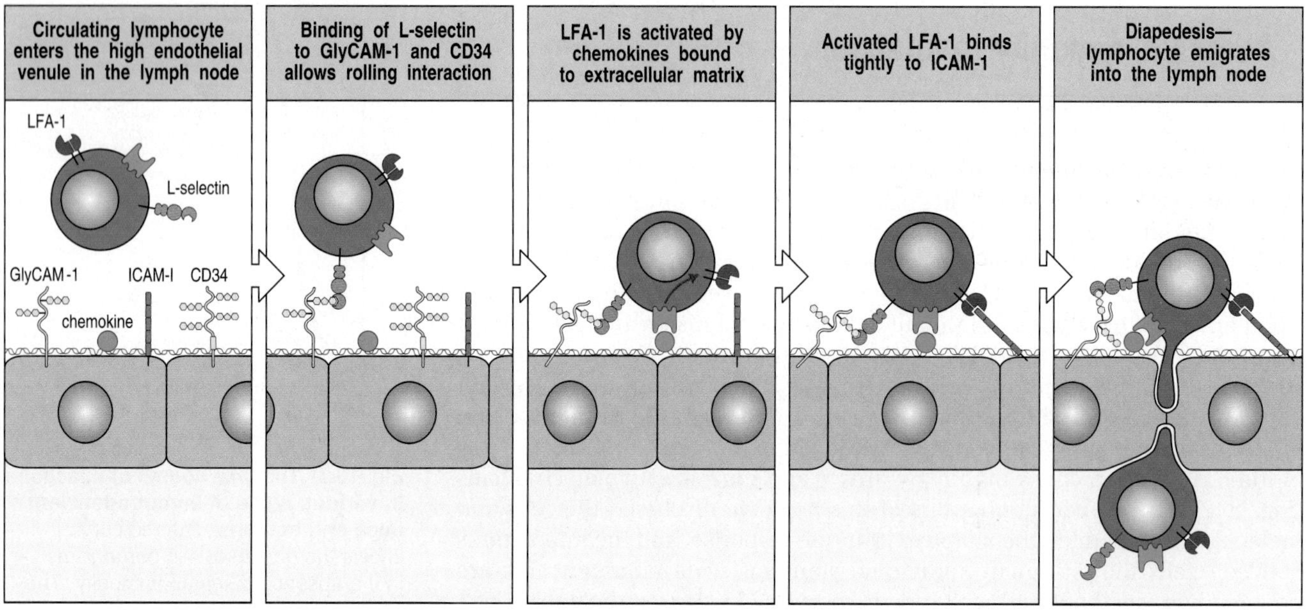
Circulating lymphocyte enters the high endothelial venule in the lymph node
Binding of L-selectin to GlyCAM-1 and CD34 allows rolling interaction
LFA-1 is activated by chemokines bound to extracellular matrix
Activated LFA-1 binds tightly to ICAM-1
Diapedesis—lymphocyte emigrates into the lymph node
LFA-1
L-selectin
GlyCAM-1
ICAM-I
CD34
chemokine

Fig. 10.22 Lymphocytes in the blood enter lymphoid tissue by crossing high endothelial venules. The first step in lymphocyte entry is the binding of L-selectin on the lymphocyte to sulfated carbohydrates of many proteins including GlyCAM-1 and CD34 on the high endothelial venule cell. Local chemokines activate LFA-1 on the lymphocyte and cause it to bind tightly to ICAM-1 on the endothelial cell, allowing transendothelial migration.

P- and E-selectin when they bind to the surface of phagocytes. This interaction is critical to the selectivity of naive lymphocyte homing. Although this interaction is too weak to promote extravasation, it is essential for the initiation of the stronger interactions that then follow between the T cell and the high endothelium, which are mediated by molecules with a relatively broad tissue distribution.

Stimulation by locally bound chemokines activates the adhesion molecule LFA-1 on the T cell, increasing its affinity for ICAM-2, which is expressed constitutively on all endothelial cells, and ICAM-1, which, in the absence of inflammation, is expressed only on the high endothelial venule cells of peripheral lymphoid tissues. The binding of LFA-1 to its ligands ICAM-1 and ICAM-2 has a major role in T-cell adhesion to and migration through the wall of the blood vessel into the lymph node (Fig. 10.22).

The high endothelial venules are located in the T-cell rich zone of the lymph nodes. This area is also inhabited by dendritic cells that have recently migrated from nearby tissues and developed potent co-stimulatory capacity. The migrating T cells scan the surface of these dendritic cells for specific peptide:MHC complexes. If they do not recognize antigen presented by these

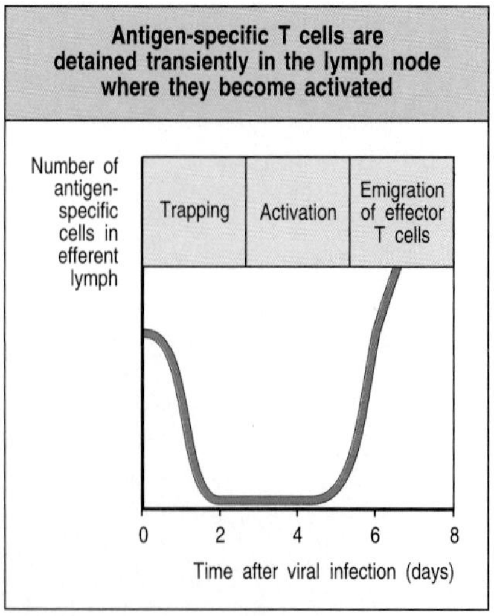
Antigen-specific T cells are detained transiently in the lymph node where they become activated
Number of antigen-specific cells in efferent lymph
Trapping
Activation
Emigration of effector T cells
0
2
4
6
8
Time after viral infection (days)

Fig. 10.23 Trapping and activation of antigen-specific naive T cells in lymphoid tissue. Naive T cells enter the lymph node from the blood and encounter many antigen-presenting dendritic cells in the lymph node cortex. T cells that do not recognize their specific antigen in the cortex leave via the efferent lymphatics and re-enter the blood. T cells that do recognize their specific antigen bind stably to the dendritic cell and are activated through their T-cell receptors, resulting in the production of armed effector T cells. Lymphocyte recirculation and recognition is so effective that all the specific naive T cells can be trapped by antigen in one node within 2 days. By 5 days after the arrival of antigen, activated effector T cells are leaving the lymph node in large numbers via the efferent lymphatics.

cells, they eventually leave the node via an efferent lymphatic vessel, which returns them to the blood so that they can recirculate through other lymph nodes. Rarely, a naive T cell recognizes its specific peptide:MHC complex on the surface of a dendritic cell, which signals the activation of LFA-1, causing the T cell to adhere strongly to the dendritic cell and cease migrating. Binding to the peptide:MHC complex and to co-stimulatory molecules on the surface of the dendritic cell also activates the naive T cell to proliferate and differentiate, resulting in the production of armed, antigen-specific effector T cells (see Fig. 8.2). The efficiency with which T cells screen each antigen-presenting cell in lymph nodes is very high, as can be seen by the rapid trapping of antigen-specific T cells in a single lymph node containing antigen: all antigen-specific T cells can be trapped in a lymph node within 48 hours of antigen deposition (Fig. 10.23).

10-17 Cytokines made in the early phases of an infection influence the functional differentiation of CD4 T cells.

It is during the initial response of naive CD4 T cells to antigen in the peripheral lymphoid tissues that the differentiation of these cells into the two major classes of CD4 effector T cell occurs. This step, at which a naive CD4 T cell becomes either an armed T_H1 cell or an armed T_H2 cell, has a critical impact on the outcome of an adaptive immune response, determining whether it will be dominated by macrophage activation or by antibody production. This is one of the most important events in the induction of an adaptive immune response.

The mechanism controlling this step in CD4 T-cell differentiation is not yet fully defined; however, it is clear that it can be profoundly influenced by cytokines present during the initial proliferative phase of T-cell activation. Experiments *in vitro* have shown that CD4 T cells initially stimulated in the presence of IL-12 and IFN-γ tend to develop into T_H1 cells (Fig. 10.24, left panels), in part because IFN-γ inhibits the proliferation of T_H2 cells. As IL-12 and IFN-γ are produced by dendritic cells, macrophages, and NK cells in the early phases of responses to viruses and some intracellular bacteria such as *Listeria* species, T-cell responses in these infections tend to be dominated by T_H1 cells. By contrast, CD4 T cells activated in the presence of IL-4 and especially IL-6 tend to differentiate into T_H2 cells, as IL-4 and IL-6 promote the differentiation of T_H2 cells, whereas IL-4 and IL-10 inhibit the generation of T_H1 cells.

One possible source of the IL-4 needed to generate T_H2 cells is a specialized subset of CD4 T cells that express the NK1.1 marker normally associated with NK cells and are called **NK1.1 CD4 T cells**. These T cells have a nearly invariant α:β T-cell receptor; in fact, essentially the same receptor seems to be used in the NK 1.1 CD4 T cells of mice and their counterparts in humans. Unlike other CD4 T cells, the development of the NK 1.1 CD4 T cells does not depend on the expression of MHC class II molecules. Instead, they recognize an MHC class IB molecule, CD1, which is not encoded within the MHC. In mice there are two CD1 genes (CD1.1 and CD1.2), whereas in humans there are five (CD1a–e) of which only CD1d is homologous to the murine CD1.1 and CD1.2. CD1 molecules are expressed by thymocytes, antigen-presenting cells, and intestinal epithelium.

Although the exact function of CD1 molecules is not well defined, CD1b is known to present a bacterial lipid, mycolic acid, to α:β T cells, whereas other CD1 molecules are recognized by γ:δ T cells. The activation of NK1.1 CD4 T cells is thought to depend on the expression of CD1 molecules induced in

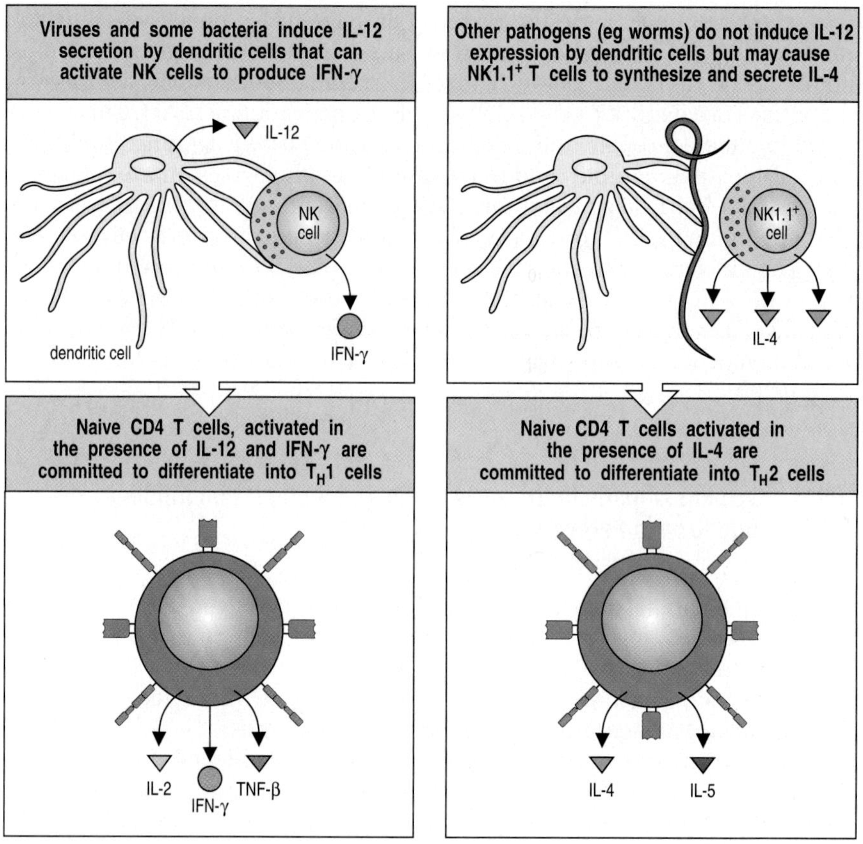

Fig. 10.24 The differentiation of naive CD4 T cells into armed effector cell types is influenced by cytokines elicited by the pathogen. Many pathogens, especially intracellular bacteria and viruses, activate dendritic cells and NK cells to produce IL-12 and IFN-γ, which act on proliferating CD4 T cells, causing them to differentiate into T_H1 cells. IL-4 can inhibit these responses. IL-4, produced by an NK1.1$^+$ CD4 T cell in response to parasitic worms or other pathogens, acts on proliferating CD4 T cells to cause them to become T_H2 cells. The mechanism by which these cytokines induce the selective differentiation of CD4 T cells is not known. They could act either when the CD4 T cell is first activated by an antigen-presenting cell or during the proliferative phase that ensues.

response to infection; whether the NK1.1 CD4 T cells recognize a specific antigen presented by these CD1 molecules is not known. Upon activation, these NK1.1 CD4 T cells secrete very large amounts of IL-4 and can therefore enhance the development of T_H2 cells (see Fig. 10.24, right panels), which promotes the production of IgG1 (mouse) and IgE (mouse, human) in subsequent humoral immune responses.

The differential capacity of pathogens to interact with dendritic cells, macrophages, NK cells, and NK1.1 CD4 T cells can therefore influence the overall balance of cytokines present early in the immune response and thus determine whether T_H1 or T_H2 cells develop preferentially to bias the adaptive immune response towards a cellular or a humoral response. This can in turn determine whether the pathogen is eliminated or survives within the host, and some pathogens might have evolved to interact with the innate immune system so as to generate responses that are beneficial to them rather than to the host.

10-18 Distinct subsets of T cells can regulate the growth and effector functions of other T-cell subsets.

The two subsets of CD4 T cells—T_H1 cells and T_H2 cells—have very different functions: T_H2 cells are the most effective activators of B cells, especially in primary responses, whereas T_H1 cells are crucial for activating macrophages. It is also clear that the two CD4 T-cell subsets can regulate each other; once one subset becomes dominant, it is often hard to shift the response to the other subset. One reason for this is that cytokines from one type of CD4 T cell

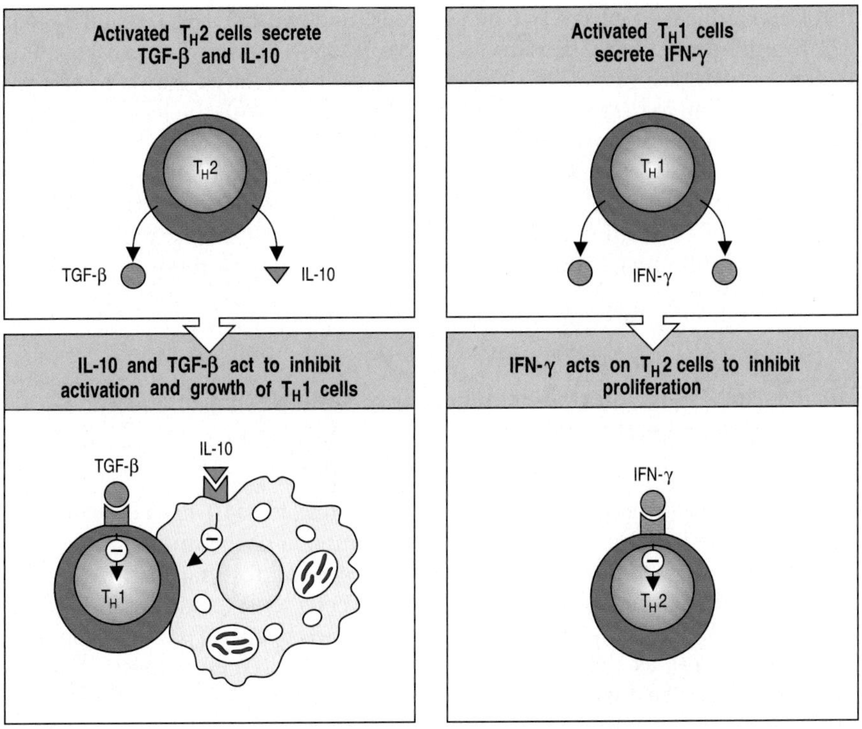

Fig. 10.25 The two subsets of CD4 T cells each produce cytokines that can negatively regulate the other subset. T_H2 cells make IL-10, which acts on macrophages to inhibit T_H1 activation, perhaps by blocking macrophage IL-12 synthesis, and TGF-β, which acts directly on the T_H1 cells to inhibit their growth (left panels). T_H1 cells make IFN-γ, which blocks the growth of T_H2 cells (right panels). These effects allow either subset to dominate a response by suppressing outgrowth of cells of the other subset.

inhibit the activation of the other. Thus, IL-10, a product of T_H2 cells, can inhibit the development of T_H1 cells by acting on the antigen-presenting cell, whereas IFN-γ, a product of T_H1 cells, can prevent the activation of T_H2 cells (Fig. 10.25). If a particular CD4 T-cell subset is activated first or preferentially in a response, it can suppress the development of the other subset. The overall effect is that certain responses are dominated by either humoral (T_H2) or cell-mediated (T_H1) immunity.

This interplay of cytokines is important in human disease, but it has been explored at present mainly in certain mouse models, where such polarized responses are easier to study. For example, when CD4 T cells in BALB/c mice are stimulated with the protozoan parasite *Leishmania*, their CD4 T cells fail to differentiate into T_H1 effector cells; instead, they preferentially make T_H2 cells in response to this pathogen. These T_H2 cells are unable to activate macrophages to inhibit leishmanial growth, resulting in susceptibility to disease. By contrast, C57BL/6 mice respond by producing T_H1 cells that protect the host by activating infected macrophages to kill the *Leishmania*. The activation of T_H2 cells in BALB/c mice can be reversed if IL-4 is blocked in the first days of infection by injecting anti-IL-4 antibody. This treatment is ineffective after a week or so of infection.

Because cytokines seem to regulate the balance between T_H1 and T_H2 cells, one might expect that it would be possible to shift this balance by administering appropriate cytokines. IL-2 and IFN-γ have been used to stimulate cell-mediated immunity in diseases such as lepromatous leprosy and can cause both a local resolution of the lesion and a systemic change in T-cell responses, as we shall see in Chapter 11. IL-12, which is a potent inducer of T_H1 cells, might be an even more attractive potential therapy.

CD8 T cells are also able to regulate the immune response by producing cytokines. It has become clear recently that CD8 T cells can, in addition to their familiar cytolytic function, also respond to antigen by secreting cytokines typical of either T_H1 or T_H2 cells. Such CD8 T cells, called T_C1 or T_C2 by analogy to the T_H subsets, seem to be responsible for the development of leprosy in its lepromatous rather than its tuberculoid form, which we discuss in detail in Chapter 11 (see Fig. 11.6). Patients with the less destructive tuberculoid leprosy make only T_H1 cells, which can activate macrophages to rid the body of leprosy bacilli. Patients with lepromatous leprosy have CD8 T cells that suppress the T_H1 response by making IL-10 and the cytokine tumor growth factor (TGF)-β. Thus, the suppression by CD8 T cells that has been observed in various situations can be explained by their expression of different sets of cytokines.

10-19 The nature and amount of antigenic peptide can also affect the differentiation of CD4 T cells.

Another factor that influences the differentiation of CD4 T cells into distinct effector subsets is the amount and exact sequence of the antigenic peptide that initiates the response. Large amounts of peptides that achieve a high density on the surface of antigen-presenting cells tend to stimulate T_H1 cell responses, whereas low-density presentation tends to elicit T_H2 cell responses. Moreover, peptides that interact strongly with the T-cell receptor tend to stimulate T_H1-like responses, whereas peptides that bind weakly tend to stimulate T_H2-like responses (Fig. 10.26).

This difference could be very important in several circumstances. For instance, allergy is caused by the production of IgE antibody, which, as we learned in Chapter 9, requires high levels of IL-4 but does not occur in the presence of IFN-γ, a potent inhibitor of IL-4-driven class switching to IgE. We shall see in Chapter 12 that antigens that elicit IgE-mediated allergy are generally delivered in minute doses, and that they elicit T_H2 cells that make IL-4 and no IFN-γ. It is also relevant that allergens do not elicit any of the known innate immune responses, which produce cytokines that tend to bias CD4 T-cell differentiation toward T_H1 cells.

Most protein antigens that elicit CD4 T-cell responses stimulate the production of both T_H1 and T_H2 cells. This reflects the presence in most proteins of several different peptide sequences that can bind to MHC class II molecules and be presented to T cells. Some of these peptides are likely to bind to MHC class II molecules with high affinity, and consequently might be present at high density on the antigen-presenting cell, whereas others might bind with low affinity and be present only at low density. Naive T cells specific for peptide antigens that have high affinity for MHC molecules are therefore likely to encounter a high density of their ligand, whereas others might only encounter a low density, and these differences in ligand density might affect the subsequent response of the T cell. Indeed, it can be demonstrated experimentally that some peptides in a protein tend to elicit T_H2 cells, whereas others tend to elicit T_H1 cells.

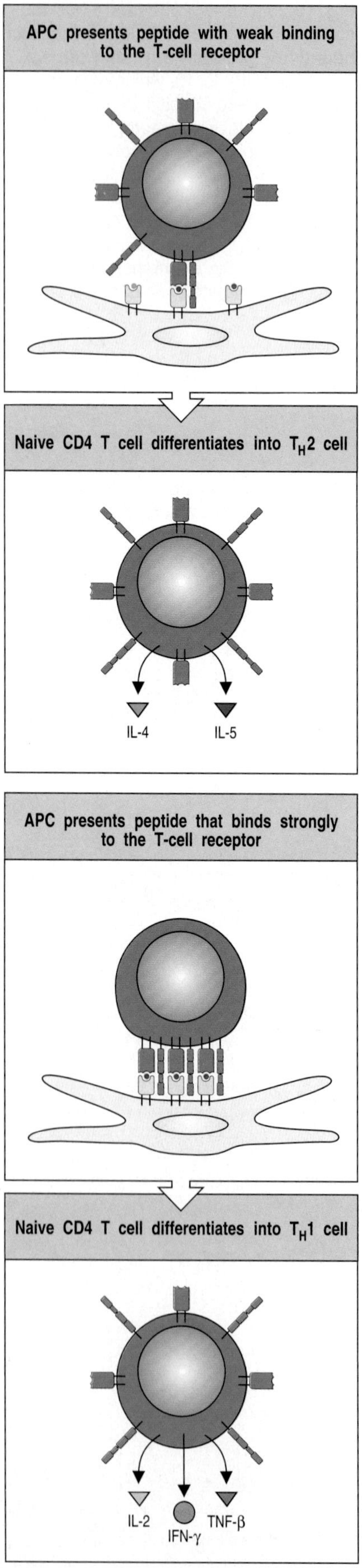
APC presents peptide with weak binding to the T-cell receptor
Naive CD4 T cell differentiates into T_H2 cell
IL-4
IL-5
APC presents peptide that binds strongly to the T-cell receptor
Naive CD4 T cell differentiates into T_H1 cell
IL-2
IFN-γ
TNF-β

Fig. 10.26 The nature and amount of ligand presented to a CD4 T cell during primary stimulation can determine its functional phenotype. CD4 T cells presented with low levels of a ligand that binds the T-cell receptor poorly differentiate preferentially into T_H2 cells making IL-4 and IL-5. Such T cells are most active in stimulating naive B cells to differentiate into plasma cells and make antibody. T cells presented with a high density of a ligand that binds the T-cell receptor strongly differentiate into T_H1 cells that secrete IL-2, TNF-β, and IFN-γ, and are most effective in activating macrophages.

10-20 Armed effector T cells are guided to sites of infection by newly expressed surface molecules.

The full activation of naive T cells takes 4–5 days and is accompanied by marked changes in the homing behavior of these cells. These occur because of changes in the expression of several cell-surface adhesion molecules that direct the migration of T cells. Thus, many armed effector T cells lose expression of the L-selectin molecule that mediates homing to the lymph nodes, whereas the expression of other adhesion molecules is increased (Fig. 10.27). One important change is a marked increase in the expression of the α_4 integrin VLA-4 (see Section 8-2), which binds to vascular cell adhesion molecule-1 (VCAM-1); cytokines induce the expression of VCAM-1 on endothelial cells at sites of infection in peripheral tissues. In this way, newly differentiated armed effector T cells are directed to the site of infection in the peripheral tissues.

Differential expression of adhesion molecules can also direct different subsets of armed effector T cells to specific sites. Some, for example, migrate to the lamina propria of the gut, which involves the binding of both L-selectin and the $\alpha_4\beta_7$ integrin expressed on the T cell to separate sites on MAdCAM-1. T cells that home to the epithelium of the gut express a novel integrin called $\alpha_e\beta_7$ and bind to E-cadherin expressed on epithelial cells. Cells that home to the skin, by contrast, express the **cutaneous lymphocyte antigen** (CLA) and bind to E-selectin.

Not all infections trigger innate immune responses that activate local endothelial cells, and it is not so clear how T cells are guided to the sites of infection in these cases. Armed effector T cells seem to enter all tissues in very small numbers, perhaps via adhesive interactions such as the binding of LFA-1 to ICAM-2, which is constitutively expressed on all endothelial cells. If these T cells recognize specific antigen in the tissue they enter, they produce cytokines such as TNF-α, which activates endothelial cells to express E-selectin, VCAM-1, and ICAM-1, and chemokines such as RANTES (see Fig. 10.13), which can then act on effector T cells to activate their adhesion molecules. The increased levels of VCAM-1 and ICAM-1 on endothelial cells bind VLA-4 and LFA-1, respectively, on armed effector T cells, recruiting more of these cells into tissues that contain antigen. At the same time, monocytes and polymorphonuclear leukocytes are recruited to these sites by adhesion to E-selectin. The TNF-α and IFN-γ released by the activated T cells also act synergistically to change the shape of endothelial cells, allowing increased blood flow, increased vascular permeability, and increased emigration of leukocytes, fluid, and protein into a site of infection.

Thus, one or a few specific effector T cells encountering antigen in a tissue can initiate a potent local inflammatory response that recruits both more specific effector cells and many accessory cells to that site. Most of the armed effector T cells that migrate at random into tissues will of course not

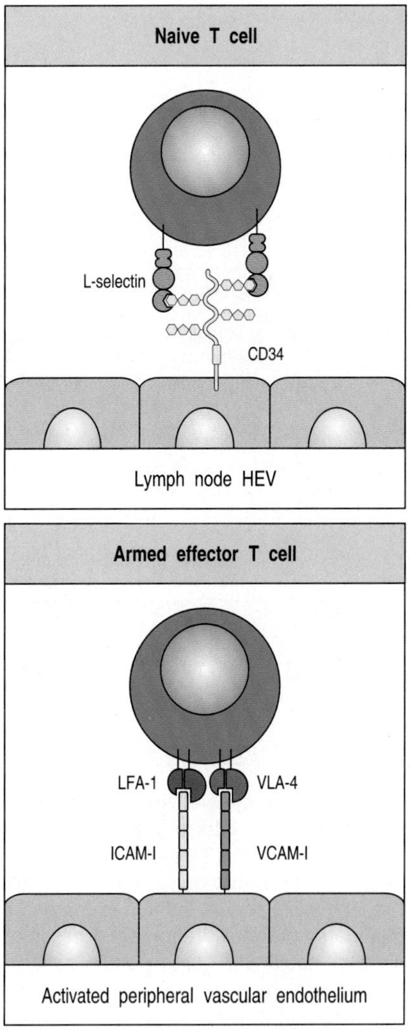

Fig. 10.27 Armed effector T cells change their surface molecules so that they can home to sites of infection via the blood. Naive T cells home to lymph nodes through the binding of L-selectin to sulfated carbohydrates displayed by various proteins, such as CD34 and GlyCAM-1 (upper panel). If they encounter antigen and differentiate into effector cells, many lose expression of L-selectin, leave the lymph node about 4–5 days later, and now express VLA-4 and increased levels of LFA-1. These bind to VCAM-1 and ICAM-1 respectively on peripheral vascular endothelium at sites of inflammation (lower panel). HEV, high endothelial venule.

encounter specific antigen, and these cells either enter afferent lymph and return to the bloodstream, or undergo apoptotic death in the tissues. Most of the T cells in afferent lymph draining peripheral tissues are memory or effector T cells that express CD45R0 and lack L-selectin. They seem to be committed to migration through potential sites of infection.

10-21 Antibody responses develop in lymphoid tissues under the direction of armed helper T cells.

Migration into the periphery is clearly important for the effector actions of CD8 cytotoxic T cells, and for T_H1 cells, which need to activate macrophages at the site of an infection. However, the most important functions of helper T cells, and of T_H2 cells in particular, depend on their interactions with B cells, and these interactions occur in the lymphoid tissues themselves. B cells specific for protein antigens cannot be activated to proliferate, form germinal centers, or differentiate into plasma cells until they encounter a helper T cell that is specific for one of the peptides derived from that antigen or antigenic complex. It follows, therefore, that humoral immune responses to protein antigens cannot occur until after antigen-specific helper T cells have been generated.

One of the most interesting questions in immunology is how two antigen-specific lymphocytes, the naive antigen-binding B cell and the armed helper T cell, find one another to initiate a T-cell dependent antibody response. As we learned in Chapter 9, the likely answer lies in the migratory path of B cells through the lymphoid tissues and the presence of armed helper T cells on that path (Fig. 10.28).

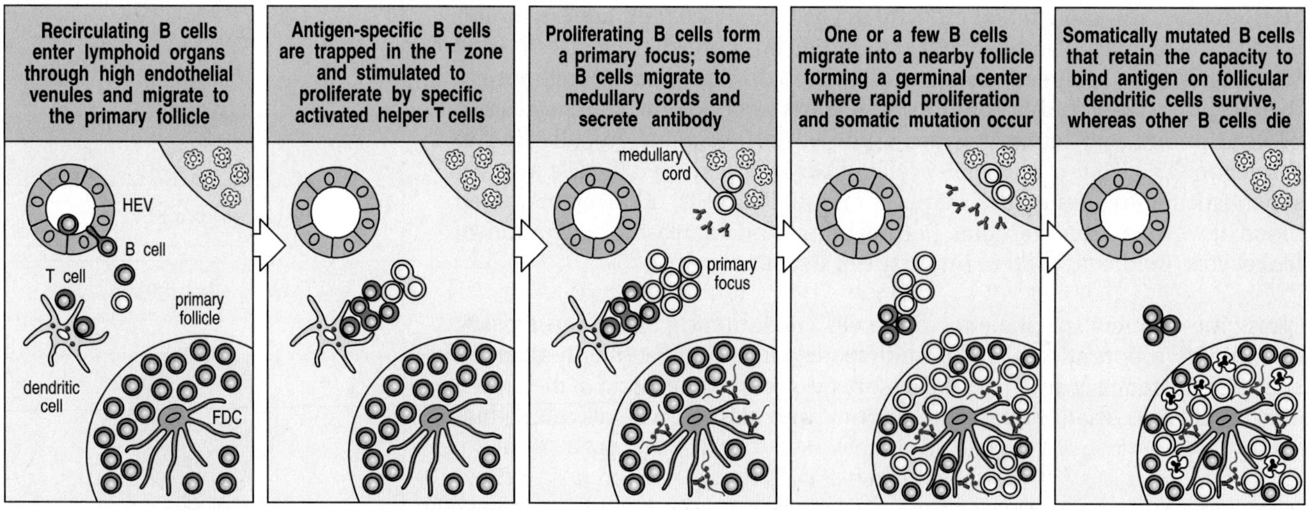

Fig. 10.28 The specialized regions of lymphoid tissue provide an environment where antigen-specific B cells can interact with armed helper T cells specific for the same antigen. The initial encounter of antigen-specific B cells with the appropriate helper T cells occurs in the T-cell areas in lymphoid tissue and stimulates the proliferation of B cells in contact with the helper T cells, resulting in some isotype switching. Some activated B-cell blasts then migrate to medullary cords, where they divide, differentiate into plasma cells, and secrete antibody for a few days. Other cells migrate into primary lymphoid follicles where they proliferate rapidly to form a germinal center under the influence of antigen trapped by follicular dendritic cells (FDC) and of helper T cells. The germinal center is the site of somatic hypermutation and selection of high-affinity B cells on the FDC network. In the primary response, antigen trapping is thought to occur only after initial production of antibody by B cells in the primary focus and the medullary cords. HEV, high endothelial venule.

B cells migrate through peripheral lymphoid organs in much the same way as T cells (see Fig. 10.28, first panel), and it is thought that the trapping and activation of naive CD4 T cells in the T-cell areas of lymphoid tissues provides a concentration of antigen-specific helper T cells capable of activating those rare B cells that are specific for the same antigen. Antigen-specific B cells are also enriched in these same areas by binding their cognate antigen; such cells are observed to accumulate in T-cell areas of the spleen and lymph nodes when exposed to their specific antigen. If the B cells receive specific signals from armed helper T cells, they proliferate in the T-cell areas of lymphoid tissues (see Fig. 10.28, second panel). In the absence of T-cell signals, these antigen-specific B cells die in less than 24 hours after arriving in the T cell zone.

About 5 days after primary immunization, primary foci of proliferating B cells appear in the T-cell areas, which correlates with the time needed for helper T cells to differentiate. Some of the B cells activated in the primary focus may migrate to the medullary cords of the lymph node or to those parts of the red pulp that are next to the T-cell zones of the spleen and secrete specific antibody for a few days (see Fig. 10.28, third panel). Others migrate to the follicle (see Fig. 10.28, fourth panel), where they proliferate further, forming a germinal center in which they undergo somatic hypermutation (see Chapter 9). The antibodies secreted by B cells differentiating early in the response not only provide early protection; they may also be important in trapping antigen, in the form of antigen:antibody complexes, on the surface of the local follicular dendritic cells. This contributes to the selection of B cells by antigen that underlies the affinity maturation observed during an antibody response. The antigen is held by a non-phagocytic Fc receptor on the follicular dendritic cells in the form of antigen:antibody complexes. Antigen can be retained in lymphoid follicles in this form for very long periods.

The proliferation, somatic hypermutation, and selection that occur in the germinal centers during a primary antibody response have been described in Chapter 9. The adhesion molecules that govern the migratory behavior of B cells are likely to be very important to this process but, as yet, little is known of their nature or of the ligands to which they bind.

10-22 Antibody responses are sustained in medullary cords and bone marrow.

Antibody-secreting cells are generated either as the result of B cells' proliferating in primary foci or after the migration of B cells to follicles. The proliferative response is started by B cells that have taken up antigen interacting in the T-cell zone with helper T cells specific for the same antigen. The B cells activated in primary foci then migrate either to adjacent follicles or to local extrafollicular sites of proliferation. The extrafollicular sites in lymph nodes are the medullary cords and in the spleen are those parts of the red pulp directly adjoining the T-cell zone. B cells grow exponentially in these sites for 2–3 days and undergo six or seven cell divisions before the progeny come out of the cell cycle and form antibody-producing plasma cells *in situ*. Most of these plasma cells have a life-span of 2–3 days, after which they undergo apoptosis. About 10% of plasma cells in these extrafollicular sites live longer; their origin and ultimate fate are unknown (see Fig. 10.29, lower panel). B cells that migrate to the primary follicles to form germinal centers have been discussed in Chapter 9 (see Sections 9-6 to 9-8). Some B cells leave germinal centers as **plasmablasts** (pre-plasma cells). Plasmablasts originating in the follicles of Peyer's patches and mesenteric lymph nodes migrate via lymph and

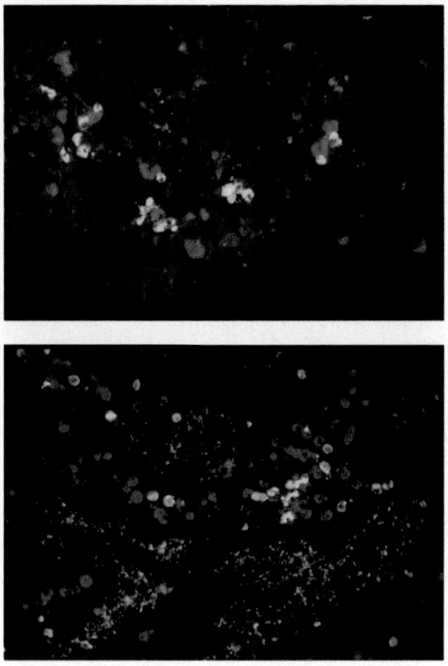

Fig. 10.29 Plasma cells are dispersed in medullary cords and bone marrow. In these sites they secrete antibody at high rates directly into the blood for distribution to the rest of the body. In the upper micrograph, longer-lived plasma cells (3 weeks to 3 months or more) in the bone marrow are revealed with antibodies specific for light chains (fluorescein anti-λ and rhodamine anti-κ stain). Plasma cells secreting immunoglobulins containing λ light chains stain green, whereas those secreting immunoglobulins containing κ light chains stain red. In the lower micrograph, plasma cells in lymph node medullary cords are stained green (with fluorescein anti-IgA) if they are secreting IgA, and red (with rhodamine anti-IgG) if they are secreting IgG. These plasma cells are short lived (2–4 days). The lymphatic sinuses are outlined by granular staining selective for IgA. Photographs courtesy of P Bramdtzaeg.

blood to the lamina propria of the gut and other epithelial surfaces. Those originating in peripheral lymph node or splenic follicles migrate to the bone marrow (see Fig. 10.29, upper panel). In these distant sites of antibody production, the plasmablasts differentiate into plasma cells that mostly have a life span of about 1 month, though a fraction of these cells can persist for much longer.

Studies of the responses to non-replicating antigens show that germinal centers are present for only 3–4 weeks after the supply of extrafollicular antigen has been exhausted. Small numbers of B cells, however, continue to proliferate in the follicles for months, and are likely to be the precursors of antigen-specific plasma cells in the mucosa and bone marrow throughout the subsequent months and years.

10-23 The effector mechanisms used to clear an infection depend on the infectious agent.

A primary adaptive immune response to an infection serves to clear the primary infection from the body and to provide protection against re-infection with the same pathogen in most cases. However, some pathogens evade complete clearance and persist for the life of the host, for example, *Leishmania*, toxoplasma, and herpes viruses. Fig. 10.30 summarizes the different types of infection and the ways in which they can be eliminated effectively by an initial adaptive immune response.

Immunity to re-infection is called **protective immunity**, and inducing protective immunity is the goal of vaccine development. Protective immunity consists of two components, immune reactants generated in the initial infection or by vaccination, and long-lived immunological memory (Fig. 10.31), which we shall consider in the last part of this chapter. Protective immunity might require the presence of preformed reactants, such as antibody molecules or armed effector T cells. For instance, effective protection against polio virus requires pre-existing antibody, because the virus will rapidly infect motor neurons and lead to their destruction unless it is neutralized by antibody and prevented from spreading within the body. Specific IgA on epithelial surfaces can also neutralize the virus before it enters the body. Thus, protective immunity can involve effector mechanisms (IgA in this case) that do not operate in the elimination of the primary infection. Pre-formed reactants can also allow the immune system to respond more rapidly and efficiently to a second exposure to a pathogen. Thus, when antibody is present, opsonization and phagocytosis of pathogens will be more efficient. If specific IgE is present, then pathogens will also be able to activate mast cells, rapidly initiating an inflammatory response through the release of histamine and leukotrienes.

	Infectious agent	Disease	Humoral immunity				Cell-mediated Immunity	
			IgM	IgG	IgE	IgA	CD4 T cells (macrophages)	CD8 killer T cells
Viruses	Variola	Smallpox					▨	■
	Varicella zoster	Chickenpox	▨	▨				▨
	Epstein–Barr virus	Mononucleosis		▨				■
	Influenza virus	Influenza		■		■		▨
	Mumps virus	Mumps		■				▨
	Measles virus	Measles		■				■
	Polio virus	Poliomyelitis		▨				▨
	Human immunodeficiency virus	AIDS		▨				▨
Bacteria	*Staphylococcus aureus*	Boils	■	■				
	Streptococcus pyogenes	Tonsilitis	■	■				
	Streptococcus pneumoniae	Pneumonia	■	■				
	Neisseria gonorrhoeae	Gonorrhea		▨		▨		
	Neisseria meningitidis	Meningitis		■				
	Corynebacterium diphtheriae	Diphtheria		▨				
	Clostridium tetani	Tetanus		▨				
	Treponema pallidum	Syphilis		▨	Transient			
	Borrelia burgdorferi	Lyme disease		▨	Transient			
	Salmonella typhi	Typhoid		■				
	Vibrio cholerae	Cholera				▨		
	Legionella pneumophila	Legionnaire's disease		▨			■	
	Rickettsia prowazeki	Typhus					■	▨
	Chlamydia trachomatis	Trachoma					▨	▨
	Mycobacteria	Tuberculosis, leprosy					■	
Fungi	*Candida albicans*	Candidiasis		▨			▨	
Protozoa	*Plasmodium* spp.	Malaria		▨			▨	
	Toxoplasma gondii	Toxoplasmosis		▨			▨	
	Trypanosoma spp.	Trypanosomiasis		■				
	Leishmania spp.	Leishmaniasis					■	
Worms	Schistosome	Schistosomiasis					■	

Fig. 10.30 Different effector mechanisms are used to clear primary infections with different pathogens and to protect against subsequent re-infection. The pathogens are listed in order of increasing complexity, and the defense mechanisms used to clear a primary infection are identified by the red shading of the boxes where these are known. Yellow shading indicates a role in protective immunity. Paler shades indicate less well-established mechanisms. Much has to be learned about such host–pathogen interactions. It is clear that classes of pathogens elicit similar protective immune responses, reflecting similarities in their lifestyles.

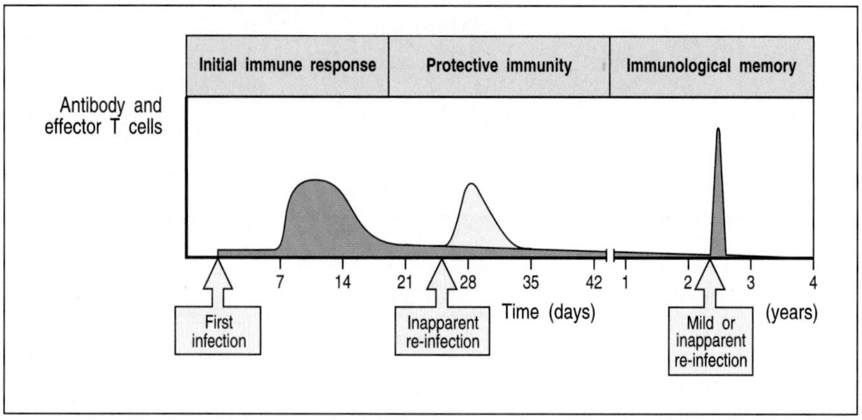

Fig. 10.31 Protective immunity consists of preformed immune reactants and immunological memory. Antibody levels and effector T-cell activity gradually decline after an infection is cleared. An early re-infection is rapidly cleared by these immune reactants. There are few symptoms but levels of immune reactants increase. Re-infection at later times leads to rapid increases in antibody and effector T cells owing to immunological memory, and infection can be mild or even inapparent.

Summary.

The adaptive immune response is required for effective protection of the host against pathogenic microorganisms. Adaptive immune responses occur when pathogens have overwhelmed or evaded non-adaptive mechanisms of host defense and established a focus of infection. The antigens of the pathogen are transported to local lymphoid organs by migrating antigen-presenting cells. This antigen is processed and presented to antigen-specific naive T cells that continuously recirculate through the lymphoid organs. T-cell priming and the differentiation of armed effector T cells occurs here, and the armed effector T cells either leave the lymphoid organ to effect cell-mediated immunity in sites of infection in the tissues, or remain in the lymphoid organ to participate in humoral immunity by activating antigen-binding B cells. Which response occurs is determined by the differentiation of CD4 T cells into T_H1 or T_H2 cells, which is in turn determined by the cytokines produced in the early non-adaptive phase. CD4 T-cell differentiation is also affected by ill-defined characteristics of the activating antigen and by its overall abundance. Ideally, the adaptive immune response eliminates the infectious agent and provides the host with a state of protective immunity against re-infection with the same pathogen.

Immunological memory.

Perhaps the most important consequence of an adaptive immune response is the establishment of a state of immunological memory. Immunological memory is the ability of the immune system to respond more rapidly and effectively to pathogens that have been encountered previously, and reflects the pre-existence of a clonally expanded population of antigen-specific lymphocytes. Memory responses, which are called secondary, tertiary, and so on, depending on the number of exposures to antigen, also differ qualitatively from primary responses.

This is particularly clear in the case of the antibody response, where the characteristics of antibodies produced in secondary and subsequent responses are distinct from those produced in the primary response to the same antigen. How immunological memory is maintained, however, is still poorly understood. The principal focus of this section will therefore be the altered character of memory responses, although we shall also outline the mechanisms that have been suggested to explain the persistence of immunological memory after exposure to antigen.

10-24 Immunological memory is long-lived after infection or vaccination.

Most children in the United States are now vaccinated against measles virus; before vaccination was widespread, most were naturally exposed to this virus and suffered from an acute, unpleasant, and potentially dangerous viral illness. Whether through vaccination or through infection, children exposed to the virus acquire long-term protection from measles. The same is true of many other acute infectious diseases: this state of protection is a consequence of immunological memory.

The basis of immunological memory has been hard to explore experimentally; although the phenomenon was first recorded by the ancient Greeks and has been exploited routinely in vaccination programs for over 200 years, it is still not clearly established whether memory reflects a long-lived population of specialized memory cells or depends on the persistence of undetectable levels of antigen that continuously re-stimulate antigen-specific lymphocytes. It can be demonstrated, however, that only individuals who were themselves previously exposed to a given infectious agent are immune, and that memory is not dependent on repeated exposure to infection as a result of contacts with other infected individuals. This was established by observations on remote island populations, where a virus such as measles can cause an epidemic, infecting all people living on the island at that time, after which the virus disappears for many years. On reintroduction from outside the island, the virus does not affect the original population but causes disease in those people born since the initial epidemic. This means that immunological memory cannot be caused by repeated exposure to infectious virus and leaves two alternative explanations.

The first is that memory is sustained by long-lived lymphocytes, induced by the original exposure, that persist in a resting state until a second encounter with the pathogen. The second is that the lymphocytes activated by the original exposure to antigen are repeatedly restimulated, even in the absence of re-infection with the pathogen. This could occur in several ways. One possibility is that the pathogen persists in small amounts that are sufficient to restimulate the activated cells but not to spread the infection to others; for example restimulation might occur through the persistence of pathogen antigens in immune complexes bound to follicular dendritic cells. Another possibility is that restimulation occurs through exposure to other, cross-reactive, antigens; these might be able to stimulate previously activated lymphocytes specific for the pathogen even though they would not have activated their naive precursor cells. Finally, restimulation could also be mediated by cytokines produced during the course of antigen-specific immune responses directed at non cross-reactive antigens; bystander memory cells, but not naive cells, might be stimulated in this way.

The experimental measurement of immunological memory has been carried out in various ways. Adoptive transfer assays of lymphocytes from animals immunized with simple, non-living antigens have been favored for such studies, as the antigen cannot proliferate. When an animal is immunized with a protein antigen, helper T cell memory appears abruptly and at its maximal level after 5 days or so. Antigen-specific memory B cells appear some days later, because B-cell activation cannot begin until armed helper T cells are available, and B cells must then enter a phase of proliferation and selection in lymphoid tissue. By 1 month after immunization, memory B cells will be present at their maximal levels. These levels are then maintained with little alteration for the lifetime of the animal. In these experiments, the existence of memory cells is measured purely in terms of the transfer of specific responsiveness from an immunized or 'primed' animal to an irradiated, immuno-incompetent host (see Sections 2-20 and 2-23). In succeeding sections, we shall look in more detail at the changes that occur in lymphocytes after antigen priming and discuss the mechanisms that might account for these changes.

10-25 Both clonal expansion and clonal differentiation contribute to immunological memory in B cells.

Immunological memory in B cells can be examined by isolating B cells from immunized mice and restimulating them with antigen in the presence of armed helper T cells specific for the same antigen. In this way, it is possible to show that antigen-specific memory B cells differ both quantitatively and qualitatively from naive B cells. B cells that can respond to antigen increase in frequency after priming by about 10- to 100-fold (Fig. 10.32) and produce antibody of higher average affinity than unprimed B lymphocytes; the affinity of that antibody continues to increase during the ongoing secondary and subsequent antibody responses (Fig. 10.33). The secondary antibody response is characterized in its first few days by the production of small amounts of IgM antibody and larger amounts of IgG antibody, with some IgA and IgE. These antibodies are produced by memory B cells that have already switched from IgM to these more mature isotypes and express IgG, IgA, or IgE on their surface, as well as a somewhat higher level of MHC class II molecules than is characteristic of naive B cells. Increased affinity for antigen and increased levels of MHC class II expression facilitate antigen uptake and presentation, and allow memory B cells to initiate their critical interactions with armed helper T cells at lower doses of antigen. Recent evidence from mice injected with antibody against nerve growth factor shows a profound loss of the capacity to make IgG antibody in a secondary response. These results suggest a role for this hormone in sustaining B-cell memory.

The distinction between primary and secondary antibody responses is most clearly seen in those cases where the primary response is dominated by antibodies that are closely related and show few if any somatic hypermutations.

	Source of B cells	
	Unimmunized donor Primary response	Immunized donor Secondary response
Frequency of specific B cells	$1{:}10^4 - 1{:}10^5$	$1{:}10^3$
Isotype of antibody produced	IgM > IgG	IgG, IgA
Affinity of antibody	Low	High
Somatic hypermutation	Low	High

Fig. 10.32 The generation of secondary antibody responses from memory B cells is distinct from the generation of the primary antibody response. The primary response usually consists of antibody molecules made by plasma cells derived from a relatively large number of different precursor B cells. The antibodies are of relatively low affinity with few somatic mutations. The secondary response derives from far fewer, high-affinity precursor B cells, which have undergone significant clonal expansion. Their receptors and antibodies are of high affinity for the antigen and show extensive somatic mutation. Thus, there is usually only a 10- to 100-fold increase in the frequency of activatable B cells after priming; however, the quality of the antibody response is altered radically, such that these precursors induce a far more intense and effective response.

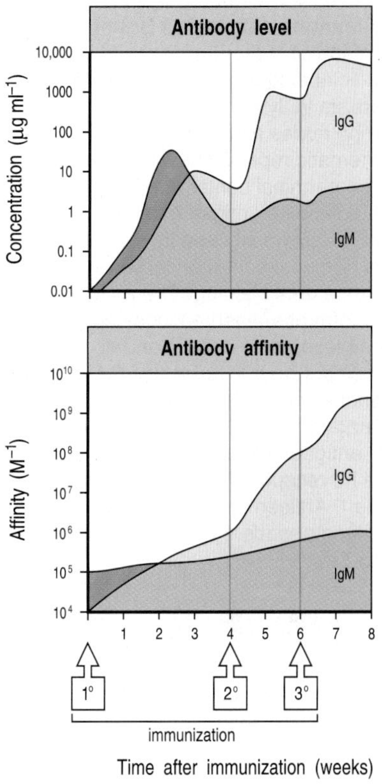

Fig. 10.33 The affinity as well as the amount of antibody increase with repeated immunization. The upper panel shows the increase in the level of antibody with time after primary, followed by secondary and tertiary, immunization; the lower panel shows the increase in the affinity of the antibodies. The increase in affinity (affinity maturation) is seen largely in IgG antibody (as well as in IgA and IgE, which are not shown) coming from mature B cells that have undergone isotype switching and somatic hypermutation to yield higher-affinity antibodies. Although some affinity maturation occurs in the primary antibody response, most arises in later responses to repeated antigen injections. Note that these graphs are on a logarithmic scale.

This occurs in inbred mouse strains in response to certain haptens that might by chance activate a pre-existing set of naive B cells poised to respond to such antigens. Such antibodies are encoded by the same V_H and V_L genes in all animals of the strain, suggesting that these variable regions might have been selected during evolution for recognition of determinants on pathogens that happen to cross-react with some haptens. As a result of the uniformity of these primary responses, changes in the antibody molecules produced in secondary responses to the same antigens are easy to observe. These differences include not only numerous somatic mutations in antibodies containing the dominant variable regions but also the addition of antibodies containing V_H and V_L gene segments not detected in the primary response. These are thought to derive from B cells that were activated at low frequency during the primary response (and thus were not detected) and differentiated into memory B cells.

10-26 Repeated immunizations lead to increasing affinity of antibody owing to somatic hypermutation and selection by antigen in germinal centers.

As we saw in Section 10-21 and Fig. 10.28, in a primary antibody response naive B cells stimulated by armed helper T cells form a primary extrafollicular focus in lymphoid tissues, where some differentiate and secrete antibody that helps to localize antigen on the surface of follicular dendritic cells (Fig. 10.34). Some B cells that have not yet undergone terminal differentiation migrate into the follicle and become germinal center B cells. Stimulated by the antigen-bearing follicular dendritic cells, these B cells enter a second proliferative phase, during which the DNA encoding their immunoglobulin variable domains undergoes somatic hypermutation before the B cells differentiate into antibody-secreting plasma cells (see Section 9-7).

The antibodies produced by plasma cells in the primary response have an important role in driving affinity maturation in the secondary response. In secondary and subsequent immune responses, any persisting antibodies produced by the B cells that differentiated in the primary response are immediately available to bind to the newly introduced antigen. Some of these antibodies divert antigen to phagocytes for degradation and disposal; however, some seem to be trapped by special antigen-transporting cells in the marginal zones of the spleen and the marginal sinus of lymph nodes. These cells bind antigen:antibody complexes and, instead of ingesting them, transport them to the lymphoid follicles, where the complexes are subsequently found on the surface of follicular dendritic cells. It is possible that the antigen-transporting cells in the spleen are B cells. In the lymph nodes, the transporter cells are resistant to ionizing radiation and their nature is obscure.

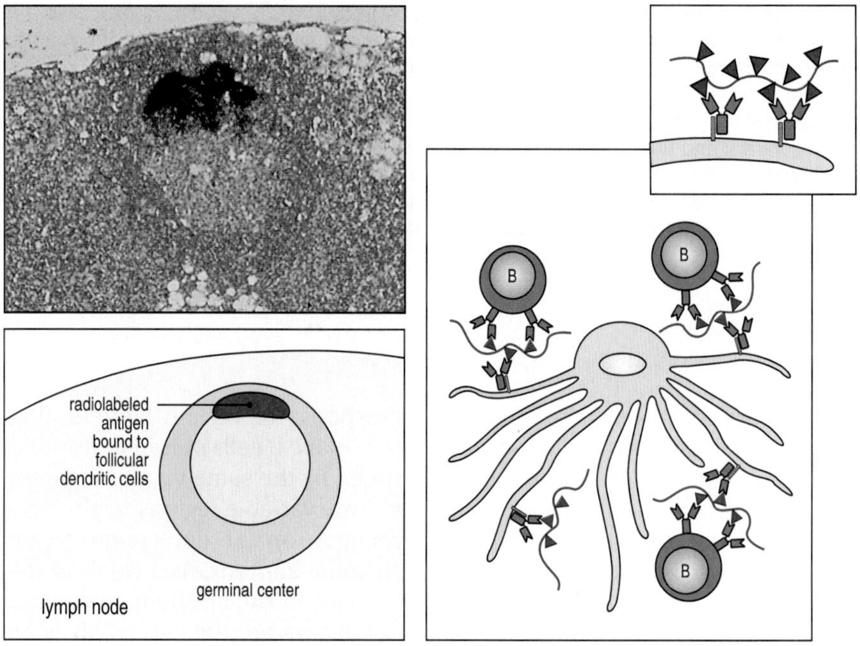

Fig. 10.34 B cells recognize antigen as immune complexes bound to the surface of follicular dendritic cells. Radiolabeled antigen localizes to, and persists in, lymphoid follicles of draining lymph nodes (see light micrograph and schematic representation below, showing a germinal center in a lymph node). Radiolabeled antigen has been injected 3 days previously and its localization in the germinal center is shown by the intense dark staining. The antigen is in the form of antigen:antibody:complement complexes bound to Fc and complement receptors on the surface of the follicular dendritic cell. These complexes are not internalized, as depicted schematically for antigen:antibody complexes bound to the Fc receptor in the right panel and insert. Antigen can persist in this form for long periods. Photograph courtesy of J Tew.

The follicular dendritic cells package the antigen into bundles of membrane coated with antigen:antibody complexes that bud off the follicular dendritic cell surface; these structures are called **iccosomes** (Fig. 10.35). It is believed that B cells whose receptors bind the antigen with sufficient avidity to compete with the existing antibody take up these iccosomes, process the antigen into peptide fragments, and present these peptides bound to MHC class II molecules to armed helper T cells surrounding and infiltrating the germinal centers (see Section 9-8). Contact between B cells presenting antigen fragments and armed helper T cells specific for the same peptides leads to an exchange of activating signals and the rapid proliferation of both

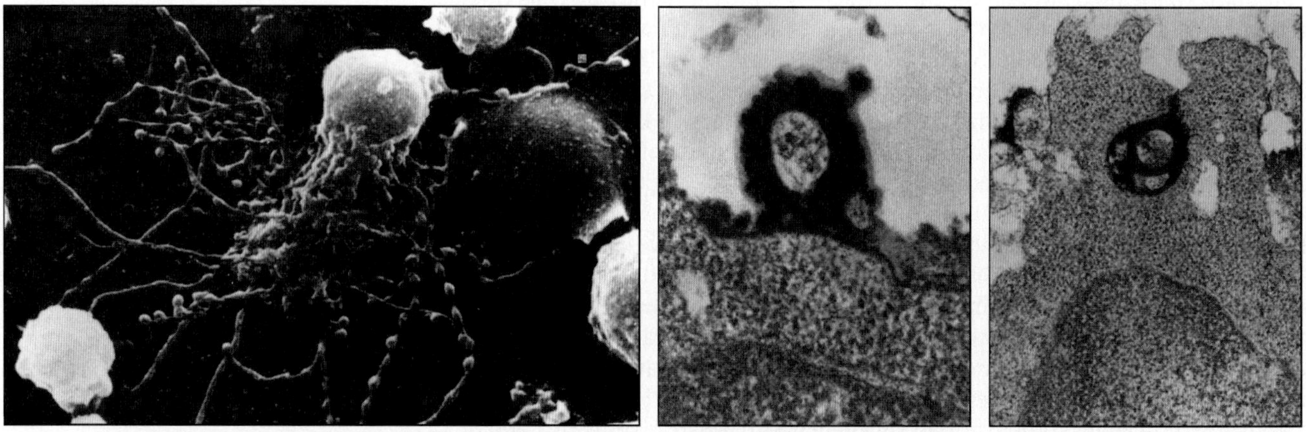

Fig. 10.35 Immune complexes bound to follicular dendritic cells form iccosomes, which are released and can be taken up by B cells in the germinal center. Follicular dendritic cells have a prominent cell body and many dendritic processes. Immune complexes, bound to Fc receptors on the follicular dendritic cell surface, become clustered, forming prominent 'beads' along the dendrites. An intermediate form of follicular dendritic cell is shown (left panel) with both straight filiform dendrites and those that are becoming beaded. These beads are shed from the cell as iccosomes (immune complex coated bodies), which can bind (center panel) and be taken up by B cells in the germinal center (right panel). In the center and right panels, the iccosome has been formed with immune complexes containing horseradish peroxidase, which is electron-dense and thus appears dark in the transmission electron micrographs. Photographs courtesy of A K Szakal.

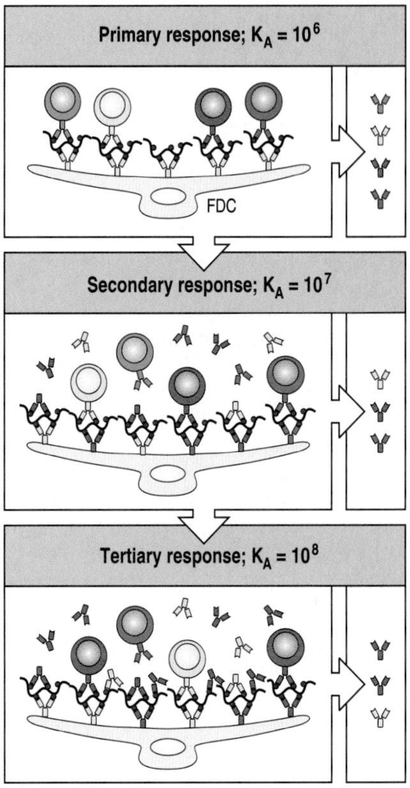

Fig. 10.36 The mechanism of affinity maturation in an antibody response. At the beginning of a primary response, high concentrations of antigen in the presence of small amounts of antibody lead to the formation of antigen:antibody complexes on follicular dendritic cells (FDCs), with little free antibody. B cells with receptors of a wide variety of affinities (K_A), most of which will bind antigen with low affinity, can thus interact with the FDCs, producing antibody of varying and relatively low affinity (top panel). Those B cells with receptors of the highest affinity are most efficient at extracting antigen from FDCs in germinal centers, and these are then selected to survive by interaction with helper T cells, even if the high-affinity B cells are actually quite infrequent. On the re-introduction of antigen, antibody produced in the primary response competes with B-cell receptors for binding to antigen:antibody complexes on FDCs, and in the secondary response only B cells with receptors of high enough affinity to compete with existing antibodies can bind antigen and contribute to the response (middle panel). In the tertiary response, the same mechanism selects for B-cell receptors with still higher affinity (bottom panel).

activated antigen-specific B cells and helper T cells. This process depends on bidirectional signaling through CD40L on the activated T cell and CD40 on the B cell, and on the induction of other co-stimulatory molecules. In this way, the affinity of the antibody produced rises progressively, as only B cells with high-affinity antigen receptors can bind antigen efficiently and be driven to proliferate by antigen-specific helper T cells (Fig. 10.36).

10-27 Memory T cells are increased in frequency and have distinct activation requirements and cell-surface proteins that distinguish them from armed effector T cells.

Because the T-cell receptor does not undergo isotype switching or affinity maturation, memory T cells have been more difficult to characterize than memory B cells. The number of T cells reactive to an antigen increases markedly after immunization, persisting at a level significantly (10- to 100-fold) above the initial frequency for the rest of the animal's or person's life. These cells carry cell-surface proteins more characteristic of armed effector cells than of naive T cells. However, it is not easy to establish whether these cells really are long-lived armed effector T cells, or whether they are cells with distinct properties that should be specifically designated memory T cells. This issue does not arise with B cells because effector B cells, as we saw in Chapter 9, are terminally differentiated plasma cells, many of which die between 3 days and 6 weeks after antigen exposure.

A major problem in experiments aimed at establishing the existence of memory T cells is that most assays for T-cell effector function take several days, during which the putative memory T cells are re-induced to armed effector cell status, so that the assays do not distinguish pre-existing effector cells from memory T cells. This problem does not apply to cytotoxic T cells, however, as cytotoxic effector T cells can program a target cell for lysis in 5 minutes. Memory CD8 T cells need to be re-activated to become cytotoxic, but they can be so without undergoing DNA synthesis, as shown by studies carried out in the presence of mitotic inhibitors. Recently it has become possible to enumerate CD8 T cells by staining them with tetrameric MHC:peptide complexes. It has been found that the number of antigen-specific CD8 T cells increases dramatically during an infection, and then drops by up to 100-fold; nevertheless, this level is distinctly higher than before priming. These experiments should soon elucidate the state of long-term memory cells.

The issue is more difficult to address for CD4 T-cell responses, and the identification of memory CD4 T cells rests largely on the existence of a population of cells with the surface characteristics of activated armed effector T cells (Fig. 10.37) but distinct from them in that they require additional restimulation before acting on target cells. Changes in three cell-surface proteins—L-selectin, CD44, and CD45—are particularly significant after exposure to antigen. L-selectin is lost on most memory T cells, whereas CD44 levels increase on all memory T cells after priming, and the isoform of CD45 changes because of alternative splicing of exons that encode the extracellular domain of CD45 (Fig. 10.38), leading to isoforms that bind to the T-cell receptor and facilitate antigen recognition. These changes are characteristic of cells that have been activated to become armed effector T cells, yet some of the cells on which these changes have occurred have many characteristics of resting CD4 T cells, suggesting that they represent memory CD4 T cells. Only after re-exposure to antigen on a professional antigen-presenting cell do they achieve armed effector T-cell status, and acquire all the characteristics of armed T_H2 or T_H1 cells, secreting IL-4 and IL-5, or IFN-γ and TNF-β, respectively. As with memory CD8 T cells, the field will soon be revolutionized by direct staining of CD4 T cells with peptide:MHC class II dimers or tetramers.

It thus seems reasonable to designate these cells as memory CD4 T cells. Together these observations suggest that naive CD4 T cells can differentiate into armed effector T cells or into memory T cells; whether armed effector T cells can persist *in vivo*, and whether they can differentiate into memory T cells, is not yet clear.

Fig. 10.37 Many cell-surface molecules alter their expression on memory T cells. This is seen most clearly with CD45, where there is a change in the isoforms expressed (see Fig. 10.38). Many of these changes are also seen on cells that have been activated to become armed effector T cells. The changes increase the adhesion of the T cell to antigen-presenting cells and to endothelial cells. They also increase the sensitivity of the memory T cell to antigen stimulation.

Molecule	Other names	Relative expression on cells of indicated subset		Comments
		Naive	Memory	
LFA-3	CD58	1	>8	Ligand for CD2, involved in adhesion and signaling
CD2	T11	1	3	Mediates T-cell adhesion and activation
LFA-1	CD11a/ CD18	1	3	Mediates leukocyte adhesion and signaling
α_4 integrin	VLA4	1	4	Involved in T-cell homing to tissues
CD44	Ly24 Pgp-1	1	2	Lymphocyte homing to tissues
CD45RO		1	30	Lowest molecular weight isoform of CD45
CD45RA		10	1	High molecular weight isoform of CD45
L-selectin		High	Most low, some high	Lymph node homing receptor
CD3		1.0	1.0	Part of antigen-specific receptor complex

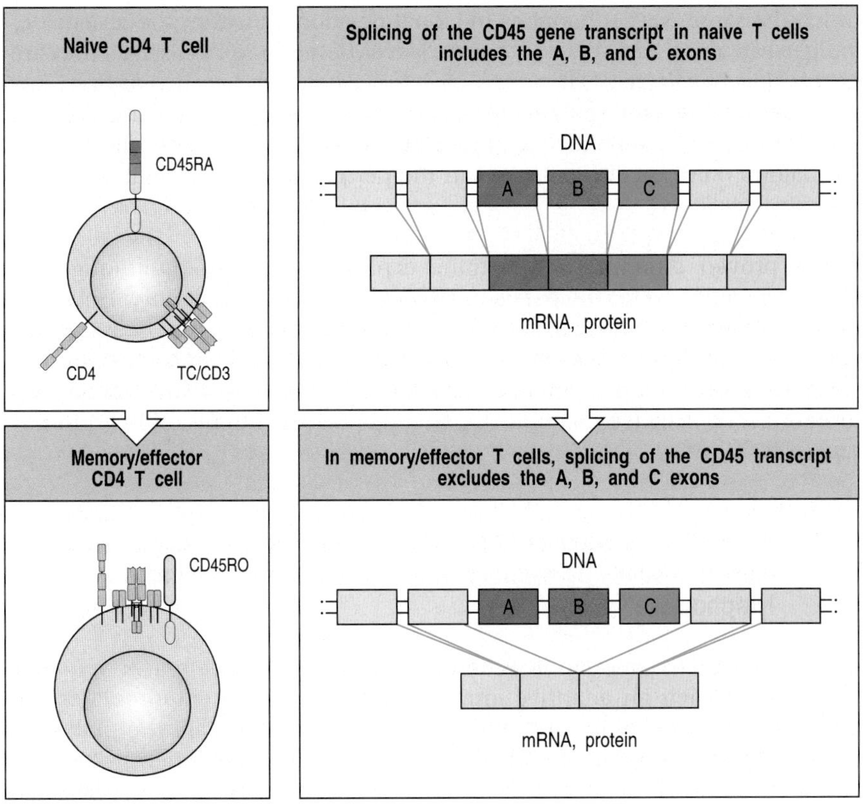

Fig. 10.38 Memory CD4 T cells express altered CD45 isoforms that regulate the interaction of the T-cell receptor with its co-receptors. CD45 is a transmembrane tyrosine phosphatase with three variable exons (A, B, and C) that encode part of its external domain. In naive T cells, high molecular weight isoforms (CD45RA) are found that do not associate with either the T-cell receptor (TC/CD3) or co-receptors (CD4). In memory T cells, the variable exons are removed by alternative splicing of CD45 RNA, and this isoform, known as CD45RO, associates with both the T-cell receptor and the co-receptor. This receptor complex seems to transduce signals more effectively than the receptor on naive T cells.

10-28 Retained antigen might have a role in immunological memory.

A successful adaptive immune response clears antigen from the body, halting further activation of naive lymphocytes. Antibody levels gradually decline, and effector T cells can no longer be detected. Residual antigen is difficult to detect, either in the form of surviving infectious agents or as antigens derived from them. Nevertheless, some antigen is probably retained for long periods as immune complexes bound to follicular dendritic cells in lymphoid follicles (see Fig. 10.34). Some intact virions might persist in this site as well, and a few infected cells might escape immune elimination. It has been proposed that this residual antigen is crucial for sustaining the cells that mediate immunological memory.

The long-lived cells that mediate immunological memory might be derived from activated naive T cells that differentiate directly into memory T cells, or they might first differentiate into effector T cells, which then either become long-lived memory T cells or are short-lived and undergo apoptosis (Fig. 10.39). Antigen has a critical role in determining the fate of the activated T cells, in a fashion reminiscent of positive selection in the thymus (see Chapter 7). Thus, high doses of antigen can trigger apoptosis of the effector T cells, in

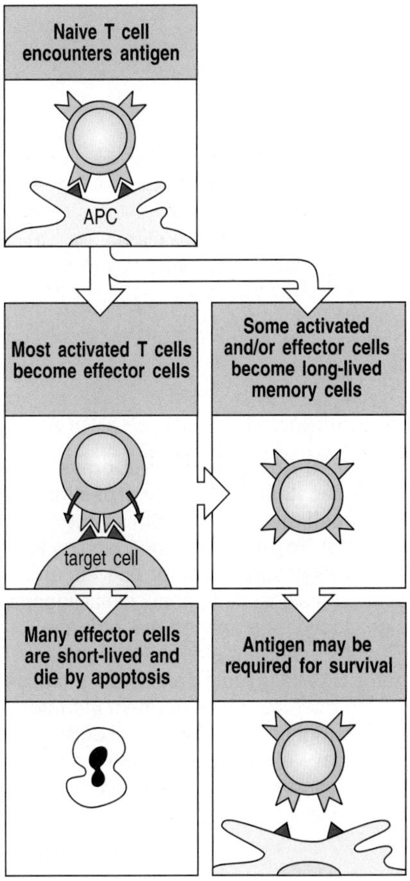

Fig. 10.39 Encounter with antigen generates effector T cells and long-lived memory T cells. Most of the effector T cells that are derived from antigen-stimulated naive T cells are relatively short lived, dying either from antigen overload or the absence of antigenic stimulus or sustaining cytokines. Some become long-lived memory T cells, which can also differentiate directly from armed effector T cells; antigenic stimulation might be required for these cells to persist. APC, antigen presenting cell.

much the same way as happens in clonal deletion; however, the absence of antigen can also lead to their apoptosis, just as developing T cells die if they are not positively selected. Memory T cells persist either because antigen has programmed them for a longer lifespan, because a low level of residual antigen preserves them by repetitive subthreshold signaling, or because the process that allows naive T cells to survive in the periphery acts more effectively on memory T cells.

It has proved difficult to determine experimentally whether antigen is absolutely required for the persistence of immunological memory. However, persistent antigen can clearly help to maintain a population of lymphocytes able to respond rapidly to the priming antigen. Thus, antigen retention in specialized sites, such as on follicular dendritic cells, where antigen persists for months or years (see Section 10-21), might be very important in immunological memory.

10-29 In immune individuals, secondary and subsequent responses are mediated solely by memory lymphocytes and not by naive lymphocytes.

In the normal course of an infection, a pathogen first proliferates to a level sufficient to elicit an adaptive immune response and then stimulates the production of antibodies and effector T cells that eliminate the pathogen from the body. Most of the armed effector T cells subsequently die and antibody levels gradually decline after the pathogen is eliminated, because the antigens that elicited the response are no longer present at the level needed to sustain it; we can think of this as feedback inhibition of the response. However, memory T and B cells remain, and maintain a heightened ability to mount a response to a recurrence of the infection.

The antibody and effector T cells remaining in an immunized individual also prevent the activation of naive B and T cells by the same antigen. Such a response would be wasteful, given the presence of memory cells that can respond much more quickly. The suppression of naive lymphocyte activation can be shown by passively transferring antibody or effector T cells to naive recipients; when the recipient is then immunized, naive lymphocytes do not respond to the original antigen but responses to other antigens are unaffected. This has been put to practical use to prevent the response of Rh^- mothers to their Rh^+ children (see Section 2-9); if anti-Rh antibody is given to the mother before she reacts to her child's red blood cells, her response will be inhibited. The mechanism of this suppression is known to involve the

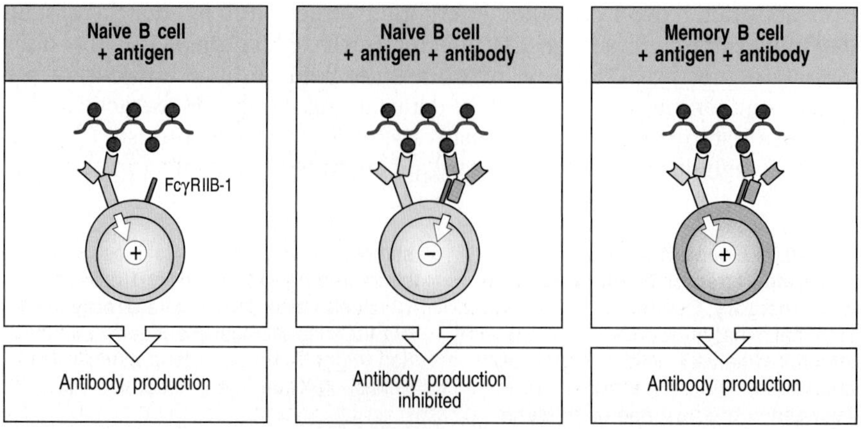

Fig. 10.40 Antibody can suppress naive B-cell activation by crosslinking the B-cell antigen receptor to the receptor FcγRIIB-1. Antigen binding to the B-cell antigen receptor delivers an activating signal (left panel); simultaneous signaling via the antigen receptor and FcγRIIB-1 delivers a negative signal to naive B cells (middle panel). Such cross-linking does not seem to affect memory B cells (right panel). This mechanism might have a role in suppressing naive B-cell responses in already primed individuals.

crosslinking of the B-cell antigen receptor to the isoform of FcγRII on the B-cell surface (FcγRIIB-1). Soluble antibody bound to antigen is able to link these two structures, which inhibits the activation of naive B cells (Fig. 10.40). FcγRIIB-1 has, in its intracellular domain, an immunoreceptor tyrosine-based inhibitory motif (ITIM) that inactivates signaling via the B-cell antigen receptor through a mechanism described in Section 5-14. For some reason, memory B-cell responses are not inhibited by antibody against the antigen, so the Rh^- mothers at risk must be identified and treated before a response has occurred. The ability of memory B cells to be activated to produce antibody even when exposed to pre-existing antibody allows secondary antibody responses to occur in individuals who are already immune.

Adoptive transfer of immune T cells to naive syngeneic mice also prevents the activation of naive T cells by antigen. This has been shown most clearly for cytotoxic T cells. It is possible that these memory CD8 T cells are activated to regain cytotoxic activity sufficiently rapidly that they can kill the antigen-presenting cells that are required to activate naive CD8 T cells, thereby inhibiting their activation.

These mechanisms might also explain the phenomenon known as **original antigenic sin**. This term was coined to describe the tendency of people to make antibodies only to epitopes expressed on the first influenza virus variant to which they were exposed, even in subsequent infections with variants that bear additional, highly immunogenic, epitopes (Fig. 10.41). Antibodies against the original virus will tend to suppress responses of naive B cells specific for the new epitopes by crosslinking their antigen receptors to FcγRIIB-1. This might benefit the host by using only those B cells that can respond most rapidly and effectively to the virus. This pattern is broken only if the person is exposed to an influenza virus that lacks all epitopes seen in the original infection, as now no pre-existing antibodies bind the virus and naive B cells are able to respond.

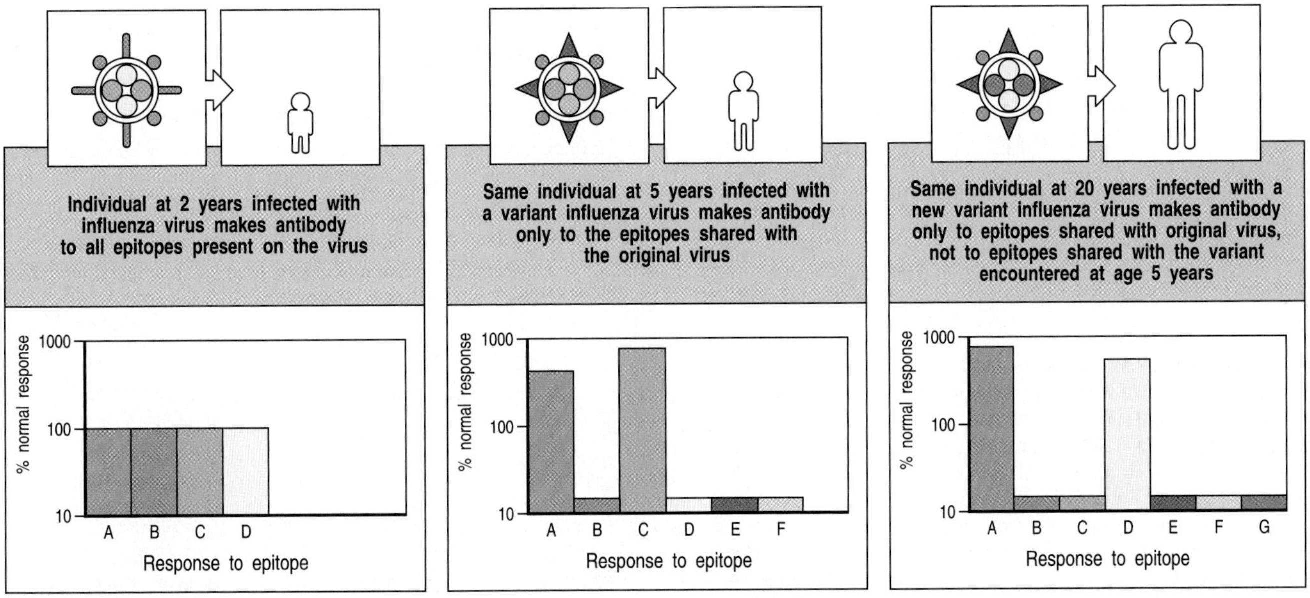

Fig. 10.41 Individuals who have already been infected with one variant of influenza virus make antibodies only to epitopes that were present on the initial virus variant when infected with a second variant. A child infected for the first time with an influenza virus makes a response to all epitopes (left panel). At age 5 years, the same child exposed to a variant virus responds preferentially to those epitopes shared with the original virus, and makes a smaller than normal response to new epitopes on the virus (middle panel). Even at age 20 years, this commitment to respond to epitopes shared with the original virus, and the subnormal response to new epitopes, is retained (right panel). This phenomenon is called 'original antigenic sin'.

Summary.

Protective immunity against re-infection is one of the most important consequences of adaptive immunity operating through clonal selection of lymphocytes. Protective immunity depends not only on pre-formed antibody and armed effector T cells, but mainly on immunological memory—an increased responsiveness to previously encountered pathogens that is long lived—and upon the establishment of a new population of memory lymphocytes. The capacity of these cells to respond rapidly to antigen can be transferred to naive recipients with primed B and T cells. The precise changes that distinguish naive, effector, and memory lymphocytes are not well characterized and, in T cells, the relative contributions of clonal expansion and differentiation to the memory phenotype are not yet clear. Memory B cells, however, can be distinguished by changes in their immunoglobulin genes because of isotype switching and somatic hypermutation, and secondary and subsequent immune responses are characterized by antibodies with increasing affinity for the antigen. It seems probable that residual antigen or infection is important in sustaining memory lymphocytes in some infections, although it is clearly not essential.

Summary to Chapter 10.

Vertebrates resist infection by pathogenic microorganisms in several ways. First, innate defenses against infection exclude infectious agents or kill them on first contact. For those pathogens that establish an infection, several early, non-adaptive responses are crucial to control infections and hold them in check until an adaptive immune response can be generated. Adaptive immunity takes several days to develop, as T and B lymphocytes must encounter their

	Phases of the immune response		
	Immediate (0–4 hours)	Early (4–96 hours)	Late (after 96 hours)
	Non-specific Innate No memory No specific T cells	Non-specific + specific Inducible No memory No specific T cells	Specific Inducible Memory Specific T cells
Barrier functions	Skin, epithelia	Local inflammation (C5a) Local TNF-α	IgA antibody in luminal spaces IgE antibody on mast cells
Response to extracellular pathogens	Phagocytes Alternative complement pathway	Mannan-binding lectin C-reactive protein T-cell independent B-cell antibody plus complement	IgG antibody and Fc receptor-bearing cells IgG, IgM antibody + classical complement pathway
Response to intracellular bacteria	Macrophages	Activated NK-dependent macrophage activation IL-1, IL-6, TNF-α, IL-12	T-cell activation of macrophages by IFN-γ
Response to virus-infected cells	Natural killer (NK) cells	Interferon-α and -β IL-12-activated NK cells	Cytotoxic T cells IFN-γ

Fig. 10.42 The components of the three phases of the immune response involved in defense against different classes of microorganisms. There are striking similarities in the effector mechanisms at each phase of the response; the main change is in the recognition structures used.

specific antigen, proliferate, and differentiate into effector cells. T-cell dependent B-cell responses cannot be initiated until antigen-specific T cells have had a chance to proliferate and differentiate. The same final effector mechanisms are used in all three phases of immunity; only the recognition mechanism changes (Fig. 10.42). Once an adaptive immune response has occurred, the infection is usually controlled, the pathogen contained or eliminated, and a state of protective immunity ensues. This state consists of the presence of effector cells and molecules produced in the initial response, and immunological memory. Immunological memory is manifest as a heightened ability to respond to pathogens that have been encountered previously and successfully eliminated. It is a property of memory T and B lymphocytes, which can transfer immune memory to naive recipients. However, the precise mechanism of immunological memory, which is arguably the most crucial feature of adaptive immunity, remains obscure. The artificial induction of protective immunity, including immunological memory, by vaccines is the most outstanding accomplishment of immunology in the field of medicine. Understanding how this is accomplished still lags behind its practical success.

General references.

Ezekowitz, R.A.B., and Hoffman, J.: **Innate immunity**. *Curr. Opin. Immunol.* 1998, **10**:9-53.

Fearon, D.T., and Locksley, R.M.: **The instructive role of innate immunity in the acquired immune response**. *Science* 1996, **272**:50-53.

Gallin, J.I., Goldstein, I.M., and Snyderman, R. (eds): *Inflammation—Basic Principles and Clinical Correlates,* 3rd edn. New York, Raven Press, 1999.

Mandell, G.L., Bennett, J.E., and Dolin, R. (eds): *Principles and Practice of Infectious Diseases,* 4th edn. New York, Churchill Livingstone, 1995.

Picker, L.J., and Butcher, E.C.: **Physiological and molecular mechanisms of lymphocyte homing.** *Annu. Rev. Immunol.* 1993, **10**:561-591.

Salyers, A.A., and Whitt, D.D.: *Bacterial Pathogenesis, A Molecular Approach.* Washington, DC, ASM Press, 1994.

Section references.

10-1 The infectious process can be divided into several distinct phases.

&

10-2 Infectious diseases are caused by diverse living agents that replicate in their hosts.

Gibbons, R.J.: How microorganisms cause disease, in Gorbach, S.L., Bartlett, J.G., and Blacklow, N.R. (eds): *Infectious Diseases,* 1st edn. 1992.

10-3 Surface epithelia make up a natural barrier to infection.

Boman, H.G.: **Peptide antibiotics: holy or heretic grails of innate immunity?** *Scand. J. Immunol.* 1996, **43**:475-482.

Lehrer, R.I., Lichtenstein, A.K., Ganz, T.: **Defensins: antimicrobial and cytotoxic peptides of mammalian cells.** *Annu. Rev. Immunol.* 1993, **11**:105-128.

10-4 The alternative pathway of complement activation provides a non-adaptive first line of defense against many microorganisms.

Liszewski, M.K., Post, T.W., and Atkinson, J.P.: **Membrane co-factor protein (MCP or CD46): newest member of the regulators of complement activation gene cluster.** *Annu. Rev. Immunol.* 1993, **9**:431-455.

Pangburn, M.K.: The alternative pathway, in Ross, G.D. (ed): *Immunobiology of the Complement System.* Orlando, Academic Press, 1986.

10-5 Phagocytes provide innate cellular immunity in tissues and initiate host-defense responses.

Ezekowitz, R.A.B., Williams, D.J., Koziel, H., Armstrong, M.Y.K., Warner, A., Richards, F.F., and Rose, R.M.: **Uptake of *Pneumocystis carinii* mediated by the macrophage mannose receptor.** *Nature* 1991, **351**:155-158.

Fenton, M.J., and Golenbeck, D.T.: **LPS-binding proteins and receptors.** *J. Leukoc. Biol.* 1998, **64**:25-32.

Ulevitch, R.J., and Tobias, P.S.: **Receptor-dependent mechanism of cell stimulation by bacterial endotoxin**. *Annu. Rev. Immunol.* 1995, **13**:437-457.

10-6 The innate immune response produces inflammatory mediators that recruit new phagocytic cells to local sites of infection.

Bevilacqua, M.P.: **Endothelial leukocyte adhesion molecules.** *Annu. Rev. Immunol.* 1993, **11**:767-804.

Downey, G.P.: **Mechanisms of leukocyte motility and chemotaxis.** *Curr. Opin. Immunol.* 1994, **6**:113-124.

Springer, T.A.: **Traffic signals for lymphocyte recirculation and leukocyte emigration: the multi-step paradigm.** *Cell* 1994, **76**:301-304.

10-7 The migration of leukocytes out of blood vessels depends on adhesive interactions activated by the local release of inflammatory mediators.

Campbell, J.J., Hedrick, J., Zlotnik, A., Siani, M.A., Thompson, D.A., and Butcher, E.C.: **Chemokines and the arrest of lymphocytes rolling under flow conditions.** *Science* 1998, **279**:381-384.

Ebnet, K., Kaldjian, E.P., Anderson, A.O., Shaw, S.: **Orchestrated information transfer underlying leukocyte endothelial interactions.** *Annu. Rev. Immunol.* 1996, **14**:155-177.

10-8 TNF-α induces blood vessel occlusion and has an important role in containing local infection but can be fatal when released systemically.

Lamping, N., Dettmer, R., Schroder, N.W., Pfiel, D., Hallatschek, W., Burger, R., and Schumann, R.R.: **LPS-binding protein protects mice from septic shock**

caused by LPS or gram-negative bacteria. *J. Clin. Invest.* 1998, **101**:2065-2071.

Pfieffer, K., Matsuyama, T., Kundig, T.M., Wakeham, A., Kishihara, K., Shahinian, A., Wiegmann, K., Ohashi, P.S., Kromke, M., and Mak, T.W.: **Mice deficient for the 55kd tumor necrosis factor receptor are resistant to endotoxic shock, yet succumb to *L. monocytogenes* infection.** *Cell* 1993, **73**:457-467.

10-9 Small proteins called chemokines recruit new phagocytic cells to local sites of infection.

Nelson, P.J., and Krensky, A.M.: **Chemokines, lymphocytes, and viruses: what goes around comes around.** *Curr. Opin. Immunol.* 1998, **10**:265-270.

Ward, S.G., Bacon, K., and Westwick, J.: **Chemokines and T lymphocytes: more that an attraction.** *Immunity* 1998, **9**:1-11.

10-10 Neutrophils predominate in the early cellular infiltrate into inflammatory sites.

Rosales, C., and Brown, E.J.: Neutrophil receptors and modulation of the immune response, in Abramson, J.S., and Wheeler, J.G. (eds): *The Natural Immune System*, New York, IRL Press, 1993.

10-11 Cytokines released by phagocytes also activate the acute-phase response.

Emsley, J., White, H.E., O'Hara, B.P., Oliva, G., Srinivasan, N., Tickle, I.J., Blundell, T.L., Pepys, M.B., and Wood, S.P.: **Structure of pentameric human serum amyloid P component.** *Nature* 1994, **367**:338.

Fraser, I.P., Koziel, H., and Ezekowitz, R.A.B.: **The serum mannose-binding protein and the macrophage mannose receptor are pattern recognition molecules that link innate and adaptive immunity.** *Semin. Immunol.* 1998, **10**: 363-372.

Weiss, W.I., Drickamer, K., Hendrickson, W.A.: **Structure of a C-type mannose-binding protein complexed with an oligosaccharide.** *Nature* 1992, **360**:127-134.

10-12 Interferons inhibit viral replication and activate certain host-defense responses.

Biron, C.A.: **Role of early cytokines, including α and β interferons in innate and adaptive immune responses to viral infections.** *Semin. Immunol.* 1998, **10**:383-390.

Sen, G.C., and Lengyel, P.: **The interferon system. A bird's eye view of its biochemistry.** *J. Biol. Chem.* 1992, **267**:5017-5020.

10-13 Natural killer cells serve as an early defense against certain intracellular infections.

Borrego, F., Ulbrecht, M, Weiss, E.H., Coligan, J.E., and Brooks, A.G.: **Recognition of human histocompatibilty leukocyte antigen (HLA)-E complexed with HLA class I signal sequence-derived peptides by CD94/NKG2 confers protection from natural killer cell-mediated lysis.** *J. Exp. Med.* 1998, **187**:813-818.

Lanier, L.L.: **NK cell receptors.** *Ann. Rev. Immunol.* 1998, **16**:359-393.

Moretta, A., Bottino, C., Vitale, M., Pende, D., Biassoni, R., Mingari, M.C., Moretta, L.: **Receptors for HLA class-1 molecules in human natural killer cells.** *Annu. Rev. Immunol.* 1996, **14**:619-648.

10-14 T cells bearing γ:δ T-cell receptors are found in lymphoid organs and most epithelia and might contribute to host defense by regulating the behavior of other cells.

Chien, Y-H., Jores, R., and Crowley, M.P.: **Recognition by γ:δ T cells.** *Annu. Rev. Immunol.* 1996, **14**:316-318.

Haas, W., Pereira, P., and Tonegawa, S.: **γ:δ cells.** *Annu. Rev. Immunol.* 1993, **11**:637-685.

10-15 B-1 cells form a separate population of B cells, producing antibodies to common bacterial polysaccharides.

Kantor, A.B., and Herzenberg, L.A.: **Origin of murine B-cell lineages.** *Annu. Rev. Immunol.* 1993, **11**:501-538.

10-16 T-cell activation is initiated when recirculating T cells encounter specific antigen in draining lymphoid tissues.

Finger, E.B., Purl, K.D., Alon, R., Lawrence, M.B., von Andrian, U.H., Springer, T.A.: **Adhesion through L-selectin requires a threshold hydrodynamic shear.** *Nature* 1996, **379**:266-269.

Roake, J.A., Rao, A.S., Morris, P.J., Larson, C.P., Hankins, D.F., Austyn, J.M.: **Dendritic cell loss from nonlymphoid tissues after systemic administration of lipopolysaccharide, tumor necrosis factor, and interleukin-1.** *J. Exp. Med.* 1995, **181**:2237-2247.

Shaw, S., Ebnet, K., Kaldjian, E.P., and Anderson, A.O.: **Orchestrated information transfer underlying leukocyte:endothelial interactions.** *Annu. Rev. Immunol.* 1996, 155-177.

10-17 Cytokines made in the early phases of an infection influence the functional differentiation of CD4 T cells.

Bendelac, A., Rivera, M.N., Parck, S.H., and Roark, J.H.: **Mouse CD1-specific NK1 T cells: development, specificity, and function.** *Annu. Rev. Immun.* 1997, **15**:535-562.

Finkelman, F.D., Shea-Donohue, T., Goldhill, J., Sullivan, C.A., Morris, S.C., Madden, K.B., Gauser, W.C., and Urban, J.F., Jr.: **Cytokine regulation of host defense against parasitic intestinal nemotodes.** *Annu. Rev. Immunol.* 1997, **15**:505-533.

Hsieh, C.-S., Macatonia, S.E., Tripp, C.S., Wolf, S.F., O'Garra, A., and Murphy, K.M.: **Development of T_H1 $CD4^+$ T cells through IL-12 produced by Listeria-induced macrophages.** *Science* 1993, **260**:547-549.

10-18 Distinct subsets of T cells can regulate the growth and effector functions of other T-cell subsets.

Croft, M., Carter, L., Swain, S.L., Dutton, R.W.: **Generation of polarized antigen-specific CD8 effector populations: reciprocal action of interleukin-4 and IL-12 in promoting type 2 versus type 1 cytokine profiles.** *J. Exp. Med.* 1994, **180**:1715-1728.

Seder, R.A., and Paul, W.E.: **Acquistion of lymphokine producing phenotype by $CD4^+$ T cells.** *Annu. Rev. Immunol.* 1994, **12**:635-673.

10-19 The nature and amount of antigenic peptide can also affect the differentiation of CD4 T cells.

Constant, S.L., and Bottomly, K.: **Induction of Th1 and Th2 CD4+ T cell responses: the alternative approaches.** *Annu. Rev. Immunol.* 1997, **15**:297-322.

Wang, L-F., Lin J-Y., Hsieh, K-H., Lin, R-H.: **Epicutaneous exposure of protein antigen induces a predominant T_H2-like response with high IgE production in mice.** *J. Immunol.* 1996, **156**:4079-4082.

10-20 Armed effector T cells are guided to sites of infection by newly expressed surface molecules.

MacKay, C.R., Marston, W., and Dudler, L.: **Altered patterns of T-cell migration through lymph nodes and skin following antigen challenge.** *Eur. J. Immunol.* 1992, **22**:2205-2210.

Romanic, A.M., Graesser, D., Baron, J.L., Visintin, I., Janeway, C.A., Jr., and Madri, J.A.: **T cell adhesion to endothelial cells and extracellular matrix is modulated upon transendothelial cell migration.** *Lab. Invest.* 1997, **76**:11-23.

10-21 **Antibody responses develop in lymphoid tissues under the direction of armed helper T cells.**

Kelsoe, G.: **Life and death in germinal centres (Redux).** *Immunity* 1996, **4**:107-111.

MacLennan, I.C.M.: **Germinal centres.** *Annu. Rev. Immunol.* 1994, **12**:117-139.

Wilson, P.C., deBouteiller, O., Liu, Y.J., Potter, K., Banchereau, J., Capra, J.D., and Pasqual, V.: **Somatic hypermutation introduces insertions and deletions into immunoglobulin V genes.** *J. Exp. Med.* 1998, **187**:59-70.

Zheng, B., Han, S., Spanopoulou, E., and Kelsoe, G.: **Immunoglobulin gene hypermutation in germinal centers is independent of the RAG-1 V(D)J recombinase.** *Immunol. Rev.* 1998, **162**:133-141.

10-22 **Antibody responses are sustained in medullary cords and bone marrow.**

Benner, R., Hijmans, W., and Haaijman, J.J.: **The bone marrow: the major source of serum immunoglobulins, but still a neglected site of antibody formation.** *Clin. Exp. Immunol.* 1981, **46**:1-8.

MacLennan, I.C.M., and Gray, D.: **Antigen-driven selection of virgin and memory B cells.** *Immunol. Rev.* 1986, **91**:61-83.

Manz, R.A., Thiel, A., and Radbruch, A.: **Lifetime of plasma cells in the bone marrow.** *Nature* 1997, **388**:133-134.

10-23 **The effector mechanisms used to clear an infection depend on the infectious agent.**

Mims, C.A.: *The Pathogenesis of Infectious Disease,* 3rd edn. London, Academic Press, 1987.

10-24 **Immunological memory is long-lived after infection or vaccination.**

Black, F.L. and Rosen, L.: **Patterns of measles antibodies in residents of Tahiti and their stability in the absence of re-exposure.** *J. Immunol.* 1962, **88**:725-731.

Sprent, J.: **T and B memory cells.** *Cell* 1994, **76**:315-322.

Sprent, J., Tough, D.F., and Sun, S.: **Factors controlling the turnover of T memory cells.** *Immunol. Rev.* 1997, **156**:79-85.

10-25 **Both clonal expansion and clonal differentiation contribute to immunological memory in B cells.**

Linton, P.J., Lai, L., Lo, D., Thorbecke, G.R., and Klinman, N.R.: **Among naive precursor cell subpopulations only progenitors of memory B cells originate germinal centers.** *Eur. J. Immunol.* 1992, **22**:1293-1297.

Liu, Y.J., and Arpin, C.: **Germinal center development.** *Immunol. Rev.* 1997, **156**:111-126.

Tarlinton, D.: **Germinal centers: form and function.** *Curr. Opin. Immunol.* 1998, **10**:245-251.

Torcia, M., Bracci-Laudiero, L., Lucibello, M., Nencioni, L., Labardi, D., Rubartelli, A., Cozzolino, F., Aloe, L., and Garaci, E.: **Nerve growth factor is an autocrine survival factor for memory B lymphocytes.** *Cell* 1996, **85**:345-356.

10-26 **Repeated immunizations lead to increasing affinity of antibody owing to somatic hypermutation and selection by antigen in germinal centers.**

Szakal, A.K., Gieringer, R.L., Kosco, M.H., Tew, J.G.: **Isolated follicular dendritic cells: cytochemical antigen localization, Nomarski, SEM and TEM morphology.** *J. Immunol.* 1985, **134**:1349-1359.

Szakal, A.K., Kosco, M.H., Tew, J.G.: **Microanatomy of lymphoid tissue during humoral immune responses: structure function relationships.** *Annu. Rev. Immunol.* 1989, **7**:91-109.

10-27 **Memory T cells are increased in frequency and have distinct activation requirements and cell-surface proteins that distinguish them from armed effector T cells.**

MacKay, C.R.: **Immunological memory.** *Adv. Immunol.* 1993, **53**:217-265.

Michie, C.A., McLean, A., Alcock, C., Beverly, P.C.L.: **Lifespan of human lymphocyte subsets defined by CD45 isoforms.** *Nature* 1992, **360**:264-265.

Novak. T.J., Farber, D., Leitenberg, D., Hong, S., Johnson, P., and Bottomly, K.: **Isoforms of the transmembrane tyrosine phosphatase CD45 differentially affect T-cell recognition.** *Immunity* 1994, **1**:81-92.

Young, J.L., Ramage, J.M., Gaston, J.S., and Beverley, P.C.: ***In vitro* responses of human CD45R0 bright, RA_ and CD45RO-RAbright T cell subsets and their relationship to memory and naive T cells.** *Eur. J. Immunol.* 1997, **27**:2383-2390.

10-28 **Retained antigen might have a role in immunological memory.**

Gray, D.: **The dynamics of immunological memory.** *Semin. Immunol.* 1992, **4**:29-34.

Sprent, J.: **Immunological Memory.** *Curr. Opin. Immunol.* 1993, **9**:371-379.

10-29 **In immune individuals, secondary and subsequent responses are mediated solely by memory lymphocytes and not by naive lymphocytes.**

Fazekas de St Groth, B., and Webster, R.G.: **Disquisitions on original antigenic sin. I. Evidence in man.** *J. Exp. Med.* 1966, **140**:2893-2898.

Fridman, W.H.: **Regulation of B cell activation and antigen presentation by Fc receptors.** *Curr. Opin. Immunol.* 1993, **5**:355-360.

Pollack, W., et al.: **Results of clinical trials of RhoGAm in women.** *Transfusion* 1968, **8**:151.

Failures of Host Defense Mechanisms

11

In the normal course of an infection, disease is followed by an adaptive immune response that clears the infection and establishes a state of protective immunity. This does not always happen, however, and in this chapter we shall examine three circumstances in which there are failures of host defense against infection: avoidance or subversion of a normal immune response by the pathogen; inherited failures of defense because of gene defects; and the acquired immune deficiency syndrome (AIDS), a generalized susceptibility to infection that is itself due to the failure of the host to control and eliminate the human immunodeficiency virus (HIV).

The propagation of a pathogen depends on its ability to replicate in a host and to spread to new hosts. Common pathogens must therefore grow without activating too vigorous an immune response and, conversely, must not kill the host too quickly. The most successful pathogens persist either because they do not elicit an immune response, or because they evade the response once it has occurred. Over millions of years of co-evolution with their hosts, pathogens have developed various strategies for avoiding destruction by the immune system, and we have encountered some of them in earlier chapters. In the first part of this chapter we shall examine these in more detail, and discuss some that have not yet been mentioned.

In the second part of the chapter we shall turn to the **immunodeficiency diseases**, in which host defense fails. In most of these diseases, a defective gene results in the elimination of one or more components of the immune system, leading to heightened susceptibility to infection with specific classes of pathogen. Immunodeficiency diseases caused by defects in T- or B-lymphocyte development, phagocyte function, and components of the complement system have all been discovered. Finally, we shall consider how the persistent infection of immune system cells by the human immunodeficiency virus, HIV, leads to the acquired immune deficiency syndrome, AIDS. The analysis of all these diseases has already made an important contribution to our understanding of host defense mechanisms and, in the longer term, might help to provide new methods of controlling or preventing infectious diseases, including AIDS.

Pathogens have evolved various means of evading or subverting normal host defenses.

Just as vertebrates have developed many different defenses against pathogens, so pathogens have evolved elaborate strategies to evade these defenses. Many pathogens use one or more of these strategies to evade the immune system. At the end of this chapter we shall see how HIV can succeed in defeating the immune response by using several of them in combination.

11-1 Antigenic variation can allow pathogens to escape from immunity.

One way in which an infectious agent can evade immune surveillance is by altering its antigens; this is particularly important for extracellular pathogens, against which the principal defense is the production of antibody against their surface structures. There are three ways in which **antigenic variation** can occur. First, many infectious agents exist in a wide variety of antigenic types. There are, for example, 84 known types of *Streptococcus pneumoniae*, an important cause of bacterial pneumonia. Each type differs from the others in the structure of its polysaccharide capsule. The different types are distinguished by serological tests and so are often known as serotypes. Infection with one serotype of such an organism can lead to type-specific immunity, which protects against re-infection with that type but not with a different serotype. Thus, from the point of view of the immune system, each serotype of *S. pneumoniae* represents a distinct organism. The result is that essentially the same pathogen can cause disease many times in the same individual (Fig. 11.1).

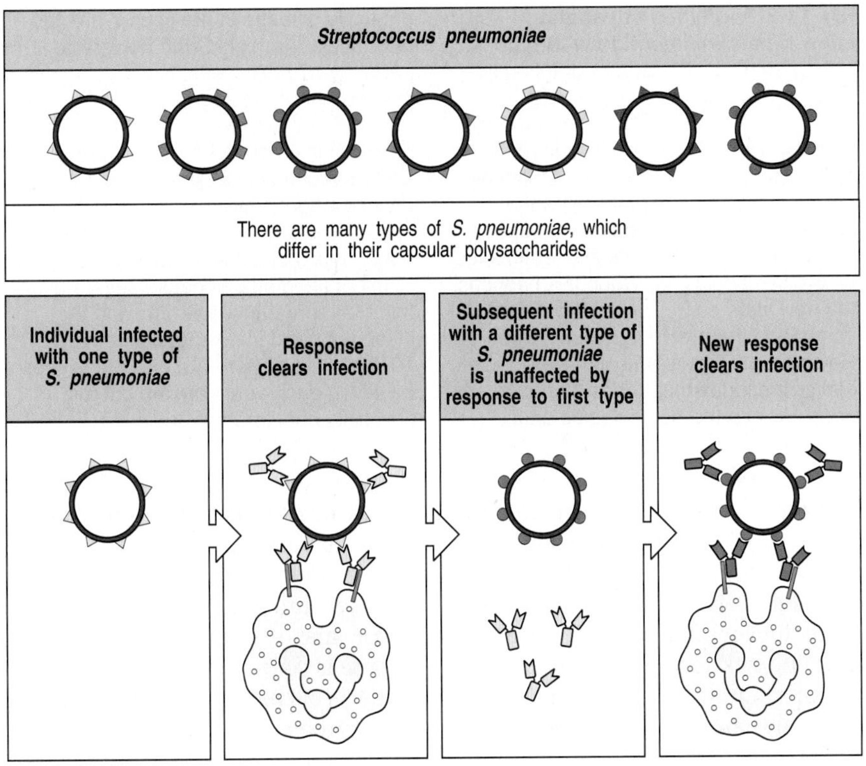
Streptococcus pneumoniae
There are many types of S. pneumoniae, which differ in their capsular polysaccharides
Individual infected with one type of S. pneumoniae
Response clears infection
Subsequent infection with a different type of S. pneumoniae is unaffected by response to first type
New response clears infection

Fig. 11.1 Host defense against *Streptococcus pneumoniae* is type specific. The different strains of *S. pneumoniae* have antigenically distinct capsular polysaccharides. The capsule prevents effective phagocytosis until the bacterium is opsonized by specific antibody and complement, allowing phagocytes to destroy it. Antibody to one type of *S. pneumoniae* does not cross-react with the other types, so an individual immune to one type has no protective immunity to a subsequent infection with a different type. An individual must generate a new immune response each time he or she is infected with a different type of *S. pneumoniae*.

A second, more dynamic mechanism of antigenic variation is seen in the influenza virus. At any one time, a single virus type is responsible for most infections throughout the world. The human population gradually develops protective immunity to this virus type, chiefly by directing neutralizing antibody against the major surface protein of the influenza virus, its hemagglutinin. Because the virus is rapidly cleared from individual hosts, its survival depends on having a large pool of unprotected individuals among whom it spreads very readily. The virus might therefore be in danger of running out of potential hosts if it had not evolved two distinct ways of changing its antigenic type (Fig. 11.2).

The first of these, **antigenic drift**, is caused by point mutations in the genes encoding hemagglutinin and a second surface protein, neuraminidase. Every 2–3 years, a variant arises with mutations that allow the virus to evade neutralization by antibodies in the population; other mutations affect epitopes that are recognized by T cells and, in particular, CD8 T cells, so that cells infected with the mutant virus also escape destruction. Individuals who were previously infected with, and hence are immune to, the old variant are thus susceptible to the new variant. This causes an epidemic that is relatively mild because there is still some cross-reaction with antibodies and T cells produced against the previous variant of the virus, and therefore most of the population have some level of immunity (see Section 10-29).

Major influenza pandemics resulting in widespread and often fatal disease occur as the result of the second process, which is termed **antigenic shift**. This happens when there is reassortment of the segmented RNA genome of the influenza virus and related animal influenza viruses in an animal host, leading to major changes in the hemagglutinin protein on the viral surface. The resulting virus is recognized poorly, if at all, by antibodies and T cells directed against the previous variant, so that most people are highly susceptible to the new virus, and severe infection results.

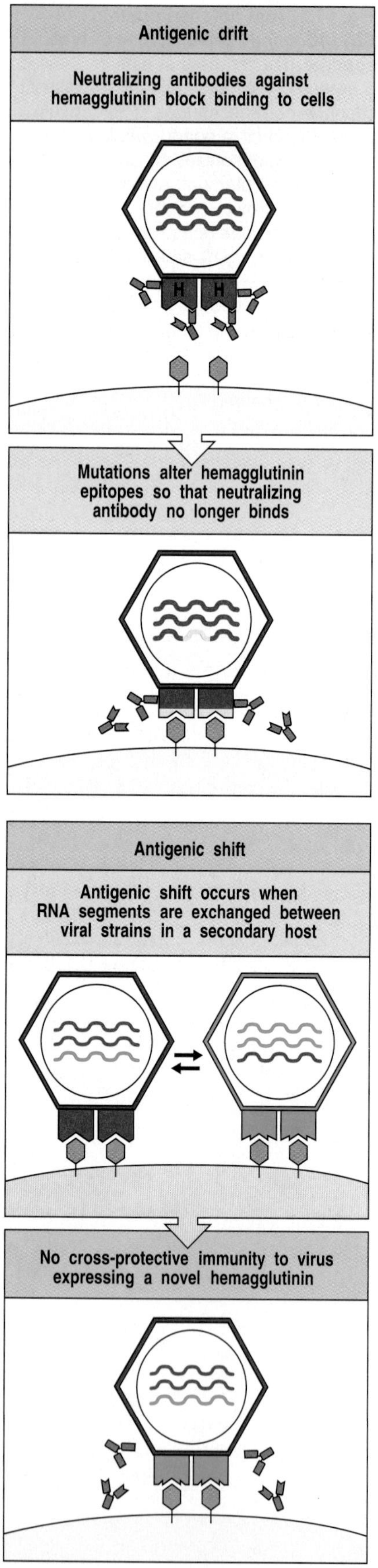

Antigenic drift
Neutralizing antibodies against hemagglutinin block binding to cells
H
H
Mutations alter hemagglutinin epitopes so that neutralizing antibody no longer binds
Antigenic shift
Antigenic shift occurs when RNA segments are exchanged between viral strains in a secondary host
No cross-protective immunity to virus expressing a novel hemagglutinin

Fig. 11.2 Two types of variation allow repetitive infection with type A influenza virus. Neutralizing antibody that mediates protective immunity is directed at the surface protein hemagglutinin (H), which is responsible for viral binding to and entry into cells. Antigenic drift (upper two panels) involves the emergence of point mutants that alter the binding sites for protective antibodies on the hemagglutinin. When this happens, the new virus can grow in a host that is immune to the previous strain of virus. However, as T cells and some antibodies can still recognize epitopes that have not been altered, the new variants cause only mild disease in previously infected individuals. Antigenic shift (lower two panels) is a rare event involving re-assortment of the segmented RNA viral genome between two influenza viruses, probably in avian hosts. These antigen-shifted viruses have large changes in their hemagglutinin molecule and therefore T cells and antibodies produced in earlier infections are not protective. These shifted strains cause severe infection that spreads widely, causing the influenza pandemics that occur every 10–50 years. (There are eight RNA molecules in each viral genome but for simplicity only three are shown.)

The third mechanism of antigenic variation involves programmed rearrangements in the DNA of the pathogen. The most striking example occurs in African trypanosomes, where changes in the major surface antigen occur repeatedly within a single infected host. Trypanosomes are insect-borne protozoa that replicate in the extracellular tissue spaces of the body and cause sleeping sickness in humans. The trypanosome is coated with a single type of glycoprotein, the variant-specific glycoprotein (VSG), which elicits a potent protective antibody response that rapidly clears most of the parasites. The trypanosome genome, however, contains about 1000 VSG genes, each encoding a protein with distinct antigenic properties. Only one of these is expressed at any one time by being placed into an active 'expression site' in the genome. The VSG gene expressed can be changed by gene rearrangement that places a new VSG gene into the expression site (Fig. 11.3). So, by having their own system of gene rearrangement that can change the VSG protein produced, trypanosomes keep one step ahead of an immune system capable of generating many distinct antibodies by gene rearrangement. A few trypanosomes with such changed surface glycoproteins escape the antibodies made by the host, and these soon grow and cause a recurrence of disease (see Fig. 11.3, bottom panel). Antibodies are then made against the new VSG, and the whole cycle repeats. This chronic cycle of antigen clearance leads to immune-complex damage and inflammation, and eventually to neurological damage, finally resulting in coma. This gives African trypanosomiasis its common name of sleeping sickness. These cycles of evasive action make trypanosome infections very difficult for the immune system to defeat and they are a major health problem in Africa. Malaria is another major disease caused by a protozoan parasite that varies its antigens to evade elimination by the immune system.

Antigenic variation also occurs in bacteria: DNA rearrangements help to account for the success of two important bacterial pathogens—*Salmonella typhimurium*, a common cause of salmonella food poisoning, and *Neisseria gonorrhoeae*, which causes gonorrhea, a major sexually transmitted disease and an increasing public health problem in the USA. *S. typhimurium* regularly alternates its surface flagellin protein by inverting a segment of its DNA containing the promoter for one flagellin gene. This turns off expression of the gene and allows the expression of a second flagellin gene, which encodes an antigenically distinct protein. *N. gonorrhoeae* has several variable antigens, the most striking of which is the pilin protein, which, like the variable surface glycoproteins of the African trypanosome, is encoded by several variant genes, only one of which is active at any given time. Silent versions of the gene from time to time replace the active version downstream of the pilin promoter. All of these mechanisms help the pathogen to evade an otherwise specific and effective immune response.

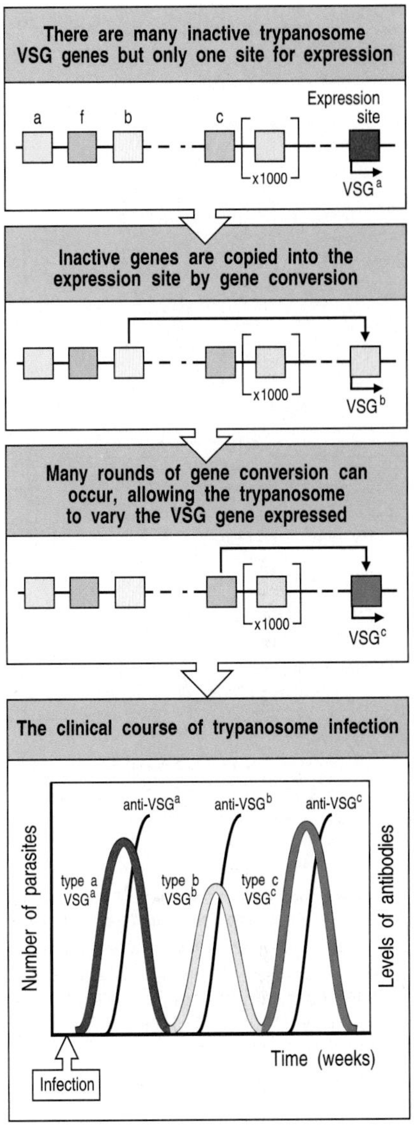

Fig. 11.3 Antigenic variation in trypanosomes allows them to escape immune surveillance. The surface of a trypanosome is covered with a variant-specific glycoprotein (VSG). Each trypanosome has about 1000 genes encoding different VSGs, but only the gene in a specific expression site within the telomere at one end of the chromosome is active. Although several genetic mechanisms have been observed for changing the VSG gene expressed, the usual mechanism is gene duplication. An inactive gene, which is not at the telomere, is copied and transposed into the telomeric expression site, where it becomes active. When an individual is first infected, antibodies are raised against the VSG initially expressed by the trypanosome population. A small number of trypanosomes spontaneously switch their VSG gene to a new type, and while the host antibody eliminates the initial variant, the new variant is unaffected. As the new variant grows, the whole sequence of events is repeated.

11-2 Some viruses persist *in vivo* by ceasing to replicate until immunity wanes.

Viruses usually betray their presence to the immune system once they have entered cells by directing the synthesis of viral proteins, fragments of which are displayed on the surface MHC molecules of the infected cell, where they are detected by T lymphocytes. To replicate, a virus must make viral proteins, and rapidly replicating viruses that produce acute viral illnesses are therefore readily detected by T cells, which normally control them. Some viruses, however, can enter a state known as **latency** in which the virus is not being replicated. In the latent state, the virus does not cause disease but, because there are no viral peptides to flag its presence, the virus cannot be eliminated. Such latent infections can be reactivated and this results in recurrent illness.

Herpes viruses often enter latency. Herpes simplex virus, the cause of cold sores, infects epithelia and spreads to sensory neurons serving the area of infection. After an effective immune response controls the epithelial infection, the virus persists in a latent state in the sensory neurons. Factors such as sunlight, bacterial infection, or hormonal changes reactivate the virus, which then travels down the axons of the sensory neuron and re-infects the epithelial tissues (Fig. 11.4). At this point, the immune response again becomes active and controls the local infection by killing the epithelial cells, producing a new sore. This cycle can be repeated many times. There are two reasons why the sensory neuron remains infected: first, the virus is quiescent in the nerve and therefore few viral proteins are produced, generating few virus-derived peptides to present on MHC class I; and second, neurons carry very low levels of MHC class I molecules, which makes it harder for CD8 T cells to recognize infected neurons and attack them. This low level of MHC class I expression might be beneficial, as it reduces the risk that neurons, which cannot regenerate, will be attacked inappropriately by CD8 T cells. It also makes neurons unusually vulnerable to persistent infections. Another example of this is provided by herpes zoster (or varicella zoster), the virus that causes chickenpox. This virus remains latent in one or a few dorsal root ganglia after the acute illness is over and can be reactivated by stress or immunosuppression to spread down the nerve and re-infect the skin. The re-infection causes the reappearance of the classic rash of varicella in the area of skin served by the infected dorsal root, a disease commonly called shingles. Herpes simplex reactivation is frequent but herpes zoster usually reactivates only once in a lifetime in an immunocompetent host.

The Epstein–Barr virus (EBV), yet another herpes virus, causes an acute infection of B lymphocytes known as infectious mononucleosis or glandular fever. EBV infects B cells by binding to CR2 (CD21), a component of the B-cell CD19 co-receptor complex. The infection causes most of the infected cells to

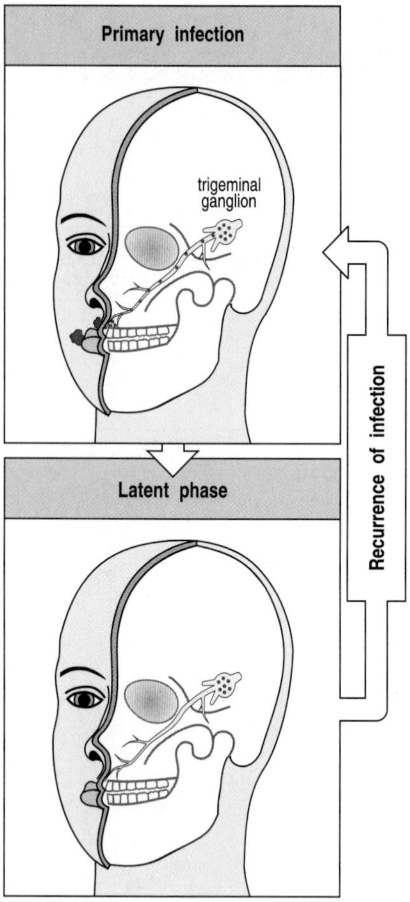

Fig. 11.4 Persistence and reactivation of herpes simplex virus infection. The initial infection in the skin is cleared by an effective immune response but residual infection persists in sensory neurons such as those of the trigeminal ganglion, whose axons innervate the lips. When the virus is reactivated, usually by some environmental stress and/or alteration in immune status, the skin in the area served by the nerve is re-infected from virus in the ganglion and a new cold sore results. This process can be repeated many times.

proliferate and produce virus, leading in turn to the proliferation of antigen-specific T cells and the excess of mononuclear white cells in the blood that gives the disease its name. The infection is controlled eventually by specific CD8 T cells, which kill the infected proliferating B cells. A fraction of B lymphocytes become latently infected, however, and EBV remains quiescent in these cells. Latently infected cells express a viral protein, EBNA-1, which is needed to maintain the viral genome, but EBNA-1 interacts with the proteasome (see Section 4-9) to prevent its own degradation into peptides that would elicit a T-cell response.

Latently infected B cells can be isolated by taking B cells from individuals who have apparently cleared their EBV infection and placing them in tissue culture: in the absence of T cells, the latently infected cells that have retained the EBV genome transform into continuously growing cell lines. *In vivo*, EBV-infected B cells sometimes undergo malignant transformation, giving rise to a B-cell lymphoma called Burkitt's lymphoma (see Section 6-16). This is a rare event and it seems likely that a crucial part of this process is a failure of T-cell surveillance. Further evidence in support of this hypothesis comes from the observations of the increased risk of development of EBV-associated B-cell lymphomas in patients with acquired and inherited immunodeficiencies of T-cell function (see Sections 11-6 and 11-23). Recent evidence suggests that EBV might also infect dendritic cells and transform them into the malignant cell in Hodgkin's disease (see Section 7-21).

11-3 Some pathogens resist destruction by host defense mechanisms or exploit them for their own purposes.

Some pathogens induce a normal immune response but have evolved specialized mechanisms for resisting its effects. For instance, some bacteria that are engulfed in the normal way by macrophages have evolved means of avoiding destruction by these phagocytes; indeed, they use macrophages as their primary host. *Mycobacterium tuberculosis*, for example, is taken up by macrophages but prevents the fusion of the phagosome with the lysosome, protecting itself from the bactericidal actions of the lysosomal contents.

Other microorganisms, such as *Listeria monocytogenes*, escape from the phagosome into the cytoplasm of the macrophage, where they can multiply readily. They then spread to adjacent cells in tissues without emerging from the cell into the extracellular environment. They do this by hijacking the host cytoskeletal protein actin, which assembles into filaments at the rear of the bacteria. The actin filaments drive the bacteria forward into vacuolar projections into adjacent cells; these vacuoles are then lysed by the *Listeria*, releasing the bacteria directly into the cytoplasm of the adjacent cell. In this way they avoid attack by antibodies. Cells infected with *L. monocytogenes* are, however, susceptible to killing by cytotoxic T cells.

The protozoan parasite *Toxoplasma gondii* can apparently generate its own vesicle, which isolates it from the rest of the cell because it does not fuse with any cellular vesicle. This might actually enable *T. gondii* to avoid making peptides derived from its proteins accessible for loading onto MHC molecules, and thereby to remain invisible to the immune system.

Two prominent spirochetal infections, **Lyme disease** and syphilis, avoid elimination by antibodies through less well understood mechanisms, and establish a persistent and extremely damaging infection in tissues. Lyme disease is caused by the spirochete bacterium *Borrelia burgdorferi*, whereas

Viral strategy	Specific mechanism	Result	Virus examples
Inhibition of humoral immunity	Virally encoded Fc receptor	Blocks effector functions of antibodies bound to infected cells	Herpes simplex Cytomegalovirus
	Virally encoded complement receptor	Blocks complement-mediated effector pathways	Herpes simplex
	Virally encoded complement control protein	Inhibits complement activation of infected cell	Vaccinia
Inhibition of inflammatory response	Virally encoded cytokine homolog, eg β-chemokine receptor	Sensitizes infected cells to effects of β-chemokine; advantage to virus unknown	Cytomegalovirus
	Virally encoded soluble cytokine receptor, eg IL-1 receptor homolog, TNF receptor homolog, γ-interferon receptor homolog	Blocks effects of cytokines by inhibiting their interaction with host receptors	Vaccinia Rabbit myxoma virus
	Viral inhibition of adhesion molecule expression, eg LFA-3 ICAM-1	Blocks adhesion of lymphocytes to infected cells	Epstein–Barr virus
Blocking of antigen processing and presentation	Inhibition of MHC class I expression	Impairs recognition of infected cells by cytotoxic T cells	Herpes simplex Cytomegalovirus
	Inhibition of peptide transport by TAP	Blocks peptide association with MHC class I	Herpes simplex
Immunosuppression of host	Virally encoded cytokine homolog of IL-10	Inhibits T_H1 lymphocytes Reduces γ-interferon production	Epstein–Barr virus

Fig. 11.5 Mechanisms of subversion of the host immune system by viruses of the herpes and pox families.

syphilis, the more widespread and much the better understood of the two diseases, is caused by *Treponema pallidum. T. pallidum* is believed to avoid recognition by antibodies by coating its surface with host molecules until it has invaded tissues such as the central nervous system, where it is less easily reached by antibodies.

Finally, many viruses have evolved mechanisms to subvert various arms of the immune system. These range from capturing cellular genes for cytokines or cytokine receptors, to synthesizing complement-regulatory molecules or inhibiting MHC class I synthesis or assembly. This area is one of the most rapidly expanding areas in the field of host–pathogen relationships. Examples of how members of the herpes and poxvirus families subvert host responses are shown in Fig. 11.5.

11-4 Immunosuppression or inappropriate immune responses can contribute to persistent disease.

Many pathogens suppress immune responses in general. For example, staphylococcal bacteria produce toxins, such as the **staphylococcal enterotoxins** and **toxic shock syndrome toxin-1**, that act as superantigens. Superantigens are proteins that bind the antigen receptor of very large numbers of T cells (see Section 4-28), stimulating them to produce cytokines that cause significant suppression of all immune responses. The details of this suppression are not understood. The stimulated T cells proliferate and then rapidly undergo apoptosis, leaving a generalized immunosuppression together with the deletion of many T cells in the periphery.

Many other pathogens cause mild or transient immunosuppression during acute infection. These forms of suppressed immunity are poorly understood but important, as they often make the host susceptible to secondary infections by common environmental microorganisms. A crucially important example of immune suppression follows trauma, burns, or even major surgery. The burned patient has a clearly diminished capability to respond to infection, and generalized infection is a common cause of death in these patients. The reasons for this are not fully understood.

Measles virus infection, in spite of the widespread availability of an effective vaccine, still accounts for 10% of the global mortality of children under 5 and is the eighth leading cause of death worldwide. Malnourished children are the main victims and the cause of death is usually secondary bacterial infection, particularly pneumonia caused by measles-induced immunosuppression. The immunosuppression that follows measles infection can last for several months and is associated with reduced T- and B-cell function. There is reduced or absent delayed-type hypersensitivity and, during this period of acquired immunodeficiency, children have markedly increased susceptibility to mycobacterial infection, reflecting the important role of macrophage activation by T_H1 cells in host defense against mycobacteria. An important mechanism for measles-induced immunosuppression is the infection of dendritic cells by measles virus. Infected dendritic cells cause unresponsiveness of T lymphocytes by mechanisms that are not yet understood and it seems likely that this is the proximate cause of the immunosuppression induced by measles virus.

The most extreme case of immune suppression caused by a pathogen is the acquired immune deficiency syndrome caused by infection with HIV. The ultimate cause of death in AIDS is usually infection with an **opportunistic pathogen**, a term used to describe a microorganism that is present in the environment but does not usually cause disease because it is well controlled by normal host defenses. HIV infection leads to a gradual loss of immune competence, allowing infection with organisms that are not normally pathogenic.

Leprosy, which we discussed in Section 8-12, is a more complex case, in which the causal bacterium, *Mycobacterium leprae*, is associated either with the suppression of cell-mediated immunity or with a strong cell-mediated antibacterial response. This leads to two major forms of the disease—lepromatous and tuberculoid leprosy. In **lepromatous leprosy**, cell-mediated immunity is profoundly depressed, *M. leprae* are present in great profusion, and cellular immune responses to many antigens are suppressed. This leads to a phenotypic state in such patients called **anergy**, here meaning the absence of delayed-type hypersensitivity to a wide range of antigens unrelated to *M. leprae*. In **tuberculoid leprosy**, by contrast, there is potent cell-mediated immunity with macrophage activation, which controls but does not eradicate infection. Few viable microorganisms are found in tissues, the patients usually survive, and most of the symptoms and pathology are caused by the inflammatory response to these persistent microorganisms (Fig. 11.6). The difference between the two forms of disease might lie in a difference in the ratio of T_H1 to T_H2 cells, and this is thought to be caused by cytokines produced by CD8 T cells, as we learned in Chapter 10 (see Section 10-18).

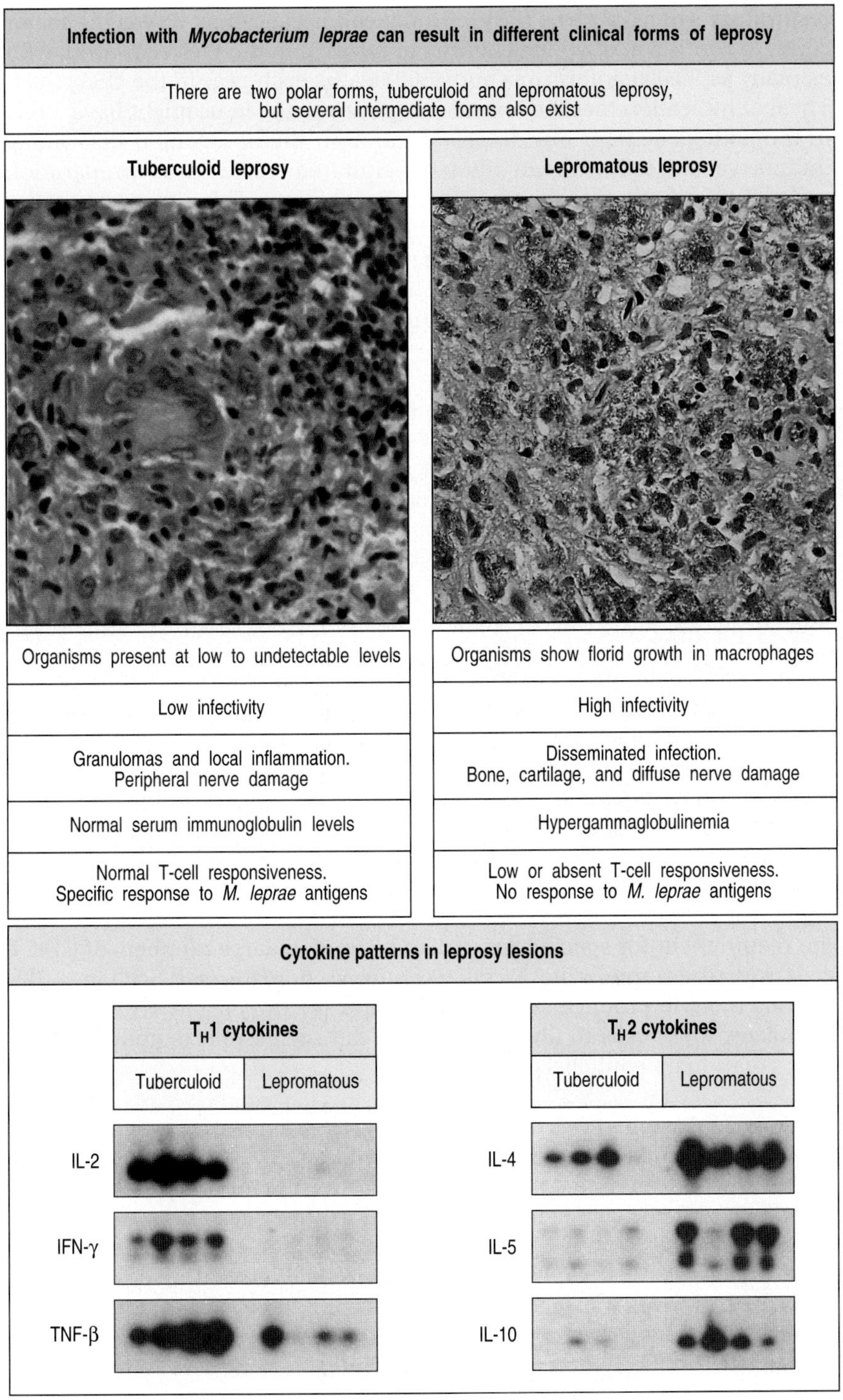

Infection with Mycobacterium leprae can result in different clinical forms of leprosy
There are two polar forms, tuberculoid and lepromatous leprosy, but several intermediate forms also exist
Tuberculoid leprosy
Lepromatous leprosy
Organisms present at low to undetectable levels
Organisms show florid growth in macrophages
Low infectivity
High infectivity
Granulomas and local inflammation. Peripheral nerve damage
Disseminated infection. Bone, cartilage, and diffuse nerve damage
Normal serum immunoglobulin levels
Hypergammaglobulinemia
Normal T-cell responsiveness. Specific response to M. leprae antigens
Low or absent T-cell responsiveness. No response to M. leprae antigens
Cytokine patterns in leprosy lesions
TH1 cytokines
Tuberculoid
Lepromatous
TH2 cytokines
Tuberculoid
Lepromatous
IL-2
IFN-γ
TNF-β
IL-4
IL-5
IL-10

Fig. 11.6 T-cell and macrophage responses to *Mycobacterium leprae* are sharply different in the two polar forms of leprosy. Infection with *M. leprae*, which stain as small dark red dots in the photographs, can lead to two very different forms of disease. In tuberculoid leprosy (left), growth of the organism is well controlled by T_H1-like cells that activate infected macrophages. The tuberculoid lesion contains granulomas and is inflamed but the inflammation is local and causes only local effects, such as peripheral nerve damage. In lepromatous leprosy (right), infection is widely disseminated and the bacilli grow uncontrolled in macrophages; in the late stages of disease there is major damage to connective tissues and to the peripheral nervous system. There are several intermediate stages between these two polar forms. The cytokine patterns in the two polar forms of the disease are sharply different, as shown by the analysis of RNA isolated from lesions of three patients with lepromatous leprosy and three patients with tuberculoid leprosy (northern blot, lower panel). Cytokines typically produced by T_H2 cells (IL-4, IL-5, and IL-10) dominate in the lepromatous form, whereas cytokines produced by T_H1 cells (IL-2, IFN-γ, and TNF-β) dominate in the tuberculoid form. It therefore seems that T_H1-like cells predominate in tuberculoid leprosy, and T_H2-like cells in lepromatous leprosy. IFN-γ would be expected to activate macrophages, enhancing killing of *M. leprae*, whereas IL-4 can actually inhibit the induction of bactericidal activity in macrophages. High levels of IL-4 would also explain the hypergammaglobulinemia observed in lepromatous leprosy. The determining factors in the initial induction of T_H1- or T_H2-like cells are not known, and the mechanism for the anergy or generalized loss of effective cell-mediated immunity in lepromatous leprosy is also not understood. Photographs courtesy of G Kaplan; cytokine patterns courtesy of R L Modlin.

11-5 Immune responses can contribute directly to pathogenesis.

Tuberculoid leprosy is just one example of an infection in which the pathology is caused largely by the immune response. This is true to some degree in most infections; for example, the fever that accompanies a bacterial infection is caused by the release of cytokines by macrophages. One medically important example of immunopathology is the wheezy broncheolitis caused by

respiratory syncytial virus (**RSV**). Broncheolitis caused by RSV is the major cause of admission of young children to hospital in the Western world, with as many as 90,000 admissions and 4500 deaths each year in the USA alone. The first indication that the immune response to the virus might have a role in the pathogenesis of this disease came from the observation that young infants vaccinated with an alum-precipitated killed virus preparation suffered a worse disease than unvaccinated children. This occurred because the vaccine failed to induce neutralizing antibodies but succeeded in producing T_H2 cells. On infection, the T_H2 cells released interleukin (IL)-3, IL-4, and IL-5, which induced bronchospasm, increased mucus secretion, and tissue eosinophilia. Mice can be infected with RSV and develop a disease similar to that seen in humans.

Another example of a pathogenic immune response is the response to the eggs of the schistosome. Schistosomes are parasitic worms that lay eggs in the hepatic portal vein. Some of the eggs reach the intestine and are shed in the feces, spreading the infection; others lodge in the portal circulation of the liver, where they elicit a potent immune response leading to chronic inflammation, hepatic fibrosis, and eventually to liver failure. This process reflects the excessive activation of T_H2 cells, and can be modulated by T_H1 cells, interferon-γ (IFN-γ), or CD8 T cells, which can also act by producing IFN-γ.

In the case of the **mouse mammary tumor virus** (**MMTV**), which causes mammary tumors in mice, the immune response is required for the infective cycle of the pathogen (Fig. 11.7). MMTV is transferred from the mother's mammary gland to her pups in milk. The virus then enters the B lymphocytes of the new host, where it must replicate to be transported to the mammary epithelium to continue its life cycle. As it is a retrovirus, however, MMTV can replicate only in dividing cells. The virus ensures that infected B cells will proliferate by causing them to express on their surface a superantigen encoded within the MMTV genome. This superantigen enables the B cells to bypass the requirement for specific antigen and stimulate large numbers of CD4 T cells with the appropriate T-cell receptor V_β domain (see Section 4-28), causing them to produce cytokines and express CD40 ligand, which in turn stimulates the B cells to divide. The virus can then replicate and infect the host's mammary epithelial cells.

One way to block this cycle of transmission is by deleting the particular subset of T cells carrying the V_β domain recognized by the viral superantigen. This has been done experimentally by taking mice that are normally susceptible to a particular MMTV virus, and using the superantigen gene from this virus to construct transgenic mice. As we learned in Section 7-17, superantigens that are expressed in the thymus induce the clonal deletion of developing T cells. Thus the expressed transgene induced the loss of T cells bearing the appropriate V_β domains. An infected B cell cannot activate any of the remaining T cells and so no B cells are stimulated to divide, blocking MMTV replication.

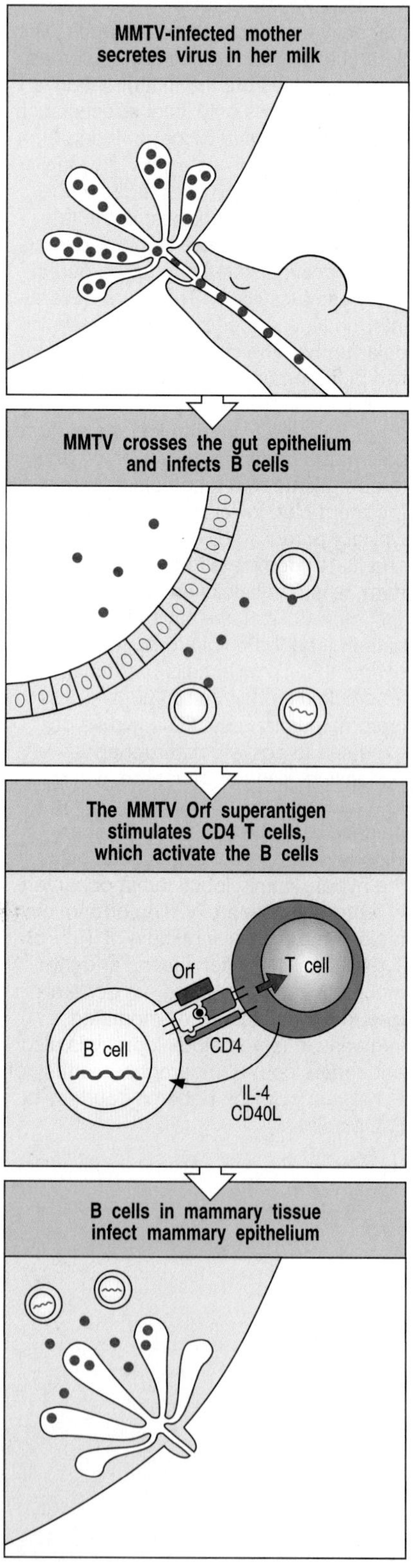
MMTV-infected mother secretes virus in her milk
MMTV crosses the gut epithelium and infects B cells
The MMTV Orf superantigen stimulates CD4 T cells, which activate the B cells
Orf
T cell
B cell
CD4
IL-4
CD40L
B cells in mammary tissue infect mammary epithelium

Fig. 11.7 Activation of T cells by the MMTV superantigen in mice is crucial for the virus life cycle. MMTV is transferred from mother to pup in milk, and crosses the gut epithelium to reach the lymphoid tissue of its new host and thus infect B lymphocytes. The superantigen encoded by MMTV, called Orf, is expressed on the surface of the B cell and binds to appropriate T-cell receptor V_β domains on CD4 T cells. The superantigen also has binding sites for MHC class II molecules, so that a complex between superantigen, MHC molecule, T-cell receptor, and CD4 is formed, activating the T cell. The T cell in turn activates the B cell to divide, allowing the virus to replicate within the B cell and subsequently to infect the mammary epithelium. CD40L, CD40 ligand.

Consequently, the transgenic mice, unlike their non-transgenic littermates, were unable to transmit the relevant strain of MMTV.

This mode of protection against MMTV might explain the finding that most mouse strains have MMTV genomes stably integrated into their DNA. These defective endogenous retroviruses have lost certain essential genes and are unable to produce virions, but they have retained the genes encoding their superantigens, which are expressed on the cells of the host. Although a section of the T-cell repertoire is lost as a result of carrying these endogenous retroviruses, the mice are protected against infection with non-defective MMTV encoding the same superantigen. There are several different strains of MMTV whose superantigens bind to different V_β domains, and these are matched by different endogenous MMTV strains. Mice containing different endogenous MMTV genomes delete different parts of their T-cell receptor repertoire, reducing the risk that whole mouse populations will be susceptible to a given MMTV strain. No human diseases dependent on such mechanisms have yet been described.

Summary.

Infectious agents can cause recurrent or persistent disease by avoiding normal host defense mechanisms or by subverting them to promote their own replication. There are many different ways of evading or subverting the immune response. Antigenic variation, latency, resistance to immune effector mechanisms, and suppression of the immune response all contribute to persistent and medically important infections. In some cases, the immune response is part of the problem; some pathogens use immune activation to spread infection, others would not cause disease if it were not for the immune response. Each of these mechanisms teaches us something about the nature of the immune response and its weaknesses, and each requires a different medical approach to prevent or to treat infection.

Inherited immunodeficiency diseases.

Immunodeficiencies occur when one or more components of the immune system is defective. The commonest cause of immune deficiency worldwide is malnutrition; however, in developed countries, most immunodeficiency diseases are inherited, and these are usually seen in the clinic as recurrent or overwhelming infections in very young children. Less commonly, acquired immunodeficiencies with causes other than malnutrition can manifest later in life. Although the pathogenesis of many of these acquired disorders has remained obscure, some are caused by known agents, such as drugs or irradiation that damage lymphocytes, or infection with measles or HIV. By examining which infections accompany a particular inherited or acquired immunodeficiency, we can see which components of the immune system are important in the response to particular infectious agents. The inherited immunodeficiency diseases also reveal how interactions between different cell types contribute to the immune response and to the development of T and B lymphocytes. Finally, these inherited diseases can lead us to the defective gene, often revealing new information about the molecular basis of immune processes and providing the necessary information for diagnosis, genetic counseling, and eventually gene therapy.

11-6 Inherited immunodeficiency diseases are caused by recessive gene defects.

Before the advent of antibiotic therapy, it is likely that most individuals with inherited immune defects died in infancy or early childhood because of their susceptibility to particular classes of pathogen (Fig. 11.8). Such cases would not

Fig. 11.8 Human immunodeficiency syndromes. The specific gene defect, the consequence for the immune system, and the resulting disease susceptibilities are listed for some common and some rare human immunodeficiency syndromes. ADA, adenosine deaminase; PNP, purine nucleotide phosphorylase; TAP, transporters associated with antigen processing; WASP, Wiskott–Aldrich syndrome protein; EBV, Epstein–Barr virus; NK, natural killer.

Name of deficiency syndrome	Specific abnormality	Immune defect	Susceptibility
Severe combined immune deficiency	ADA deficiency	No T or B cells	General
	PNP deficiency	No T or B cells	General
	X-linked *scid*, γ_c chain deficiency	No T cells	General
	Autosomal *scid* DNA repair defect	No T or B cells	General
DiGeorge syndrome	Thymic aplasia	Variable numbers of T and B cells	General
MHC class I deficiency	TAP mutations	No CD8 T cells	Viruses
MHC class II deficiency	Lack of expression of MHC class II	No CD4 T cells	General
Wiskott–Aldrich syndrome	X-linked; defective WASP gene	Defective polysaccharide antibody responses	Encapsulated extracellular bacteria
Common variable immunodeficiency	Unknown; MHC-linked	Defective antibody production	Extracellular bacteria
X-linked agamma-globulinemia	Loss of Btk tyrosine kinase	No B cells	Extracellular bacteria, viruses
X-linked hyper-IgM syndrome	Defective CD40 ligand	No isotype switching	Extracellular bacteria
Selective IgA and/or IgG deficiency	Unknown; MHC-linked	No IgA synthesis	Respiratory infections
Phagocyte deficiencies	Many different	Loss of phagocyte function	Extracellular bacteria and fungi
Complement deficiencies	Many different	Loss of specific complement components	Extracellular bacteria especially *Neisseria* spp.
Natural killer (NK) cell defect	Unknown	Loss of NK function	Herpes viruses
X-linked lympho-proliferative syndrome	SAP mutant	Inability to control B cell growth	EBV-driven B cell tumors
Ataxia telangiectasia	Gene with PI-3 kinase homology	T cells reduced	Respiratory infections
Bloom's syndrome	Defective DNA helicase	T cells reduced Reduced antibody levels	Respiratory infections

have been easy to identify, as many normal infants also died of infection. Thus, although many inherited immunodeficiency diseases have now been identified, the first immunodeficiency disease was not described until 1952. Most of the gene defects that cause these inherited immunodeficiencies are recessive and, for this reason, many of the known immunodeficiencies are caused by mutations in genes on the X chromosome. Recessive defects cause disease only when both chromosomes are defective. However, as males have only one X chromosome, all males who inherit an X chromosome carrying a defective gene will manifest disease, whereas female carriers, having two X chromosomes, are perfectly healthy. Immunodeficiency diseases that affect various steps in B- and T-lymphocyte development have been described, as have defects in surface molecules that are important for T- or B-cell function. Defects in phagocytic cells, in complement, in cytokines, in cytokine receptors, and in molecules that mediate effector responses also occur (see Fig. 11.8). Thus, immunodeficiency can be caused by defects in either the adaptive or the innate immune system. Individual examples of these diseases will be described in later sections.

More recently, the use of gene knock-out techniques in mice has allowed the creation of many immunodeficient states that are adding rapidly to our knowledge of the contribution of individual molecules to normal immune function. Nevertheless, human immunodeficiency disease is still the best source of insight into normal pathways of host defense against infectious diseases in humans. For example, a deficiency of antibody, of complement, or of phagocytic function each increases the risk of infection by certain pyogenic bacteria. This shows that the normal pathway of host defense against such bacteria is the binding of antibody followed by fixation of complement, which allows the uptake of opsonized bacteria by phagocytic cells. Breaking any one of the links in this chain of events leading to bacterial killing causes a similar immunodeficient state.

The study of immunodeficiency also teaches us about the redundancy of mechanisms of host defense against infectious disease. The first two humans to be discovered with a hereditary deficiency of complement were healthy immunologists. This teaches us two lessons. The first is that there are multiple protective immune mechanisms against infection; for example, although there is abundant evidence that complement deficiency increases susceptibility to pyogenic infection, not every human with complement deficiency suffers from recurrent infections. The second lesson concerns the phenomenon of **ascertainment artifact**. When an unusual observation is made in a patient with disease, there is a temptation to seek a causal link. However, no one would suggest that complement deficiency causes a genetic predisposition to becoming an immunologist. Complement deficiency was discovered in immunologists because they used their own blood in their experiments. If a particular measurement is made only in a group of patients with a particular disease, it is inevitable that the only abnormal results will be discovered in patients with that disease. This is an ascertainment artifact and emphasizes the importance of studying appropriate controls.

11-7 The main effect of low levels of antibody is an inability to clear extracellular bacteria.

Pyogenic or pus-forming bacteria have polysaccharide capsules that make them resistant to phagocytosis. Normal individuals can clear infections by such bacteria because antibody and complement opsonize the bacteria, making it possible for phagocytes to ingest and destroy them. The principal effect of deficiencies in antibody production is therefore a failure to control this class of bacterial infection. In addition, susceptibility to some viral infections, most notably those caused by enteroviruses, is also increased because of the importance of antibodies in neutralizing infectious viruses that enter the body through the gut (see Chapter 9).

The first description of an immunodeficiency disease was Ogden C. Bruton's account, in 1952, of the failure of a male child to produce antibody. As this defect is inherited in an X-linked fashion and is characterized by the absence of immunoglobulin in the serum, it was called **Bruton's X-linked agammaglobulinemia** (XLA). The absence of antibody can be detected by using immunoelectrophoresis (Fig. 11.9). Since then, many more diseases of antibody production have been described, most of them the consequence of failures in the development or activation of B lymphocytes.

The defective gene in XLA is now known to encode a protein tyrosine kinase called Btk (Bruton's tyrosine kinase). This protein is expressed in neutrophils as well as in B cells, although only B cells are defective in these patients, in whom B-cell maturation halts at the pre-B-cell stage. Thus it is likely that Btk is required to couple the pre-B-cell receptor (which consists of heavy chains, surrogate light chains, and Igα and Igβ) to nuclear events that lead to pre-B-cell growth and differentiation. A homologous kinase called Itk has been found in T cells and is required for normal T-cell development. Defects in Btk are analogous to defects in Lck in T-cell development (see Section 7-8). In the mouse, Lck deficiency leads to the arrest of thymocyte development at the double-negative stage, after T-cell receptor β-chain gene rearrangement and cell-surface expression but before the rearrangement

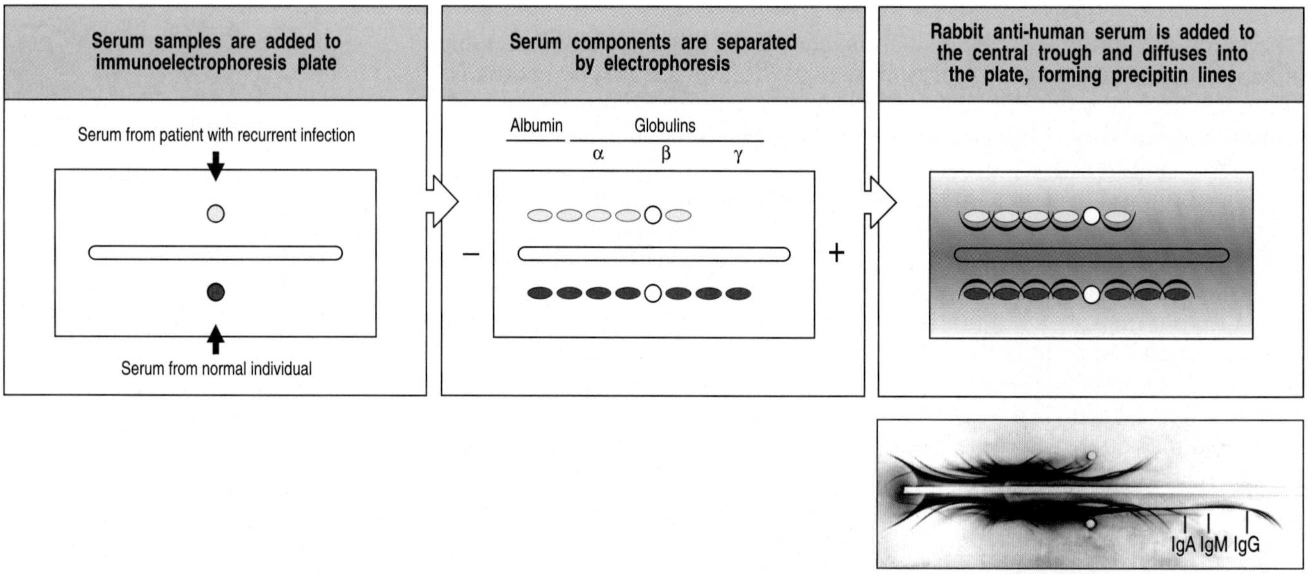

Fig. 11.9 Immunoelectrophoresis reveals the absence of several distinct immunoglobulin isotypes in serum from a patient with X-linked agammaglobulinemia (XLA). Serum samples from a normal control and from a patient with recurrent bacterial infection caused by an absence of antibody production, as reflected in an absence of gamma globulins, are separated by electrophoresis on an agar-coated slide. Antiserum raised against whole normal human serum and containing antibodies against many of its different proteins is put in a trough down the middle; each antibody forms an arc of precipitation with the protein it recognizes. The position of each arc is determined by the electrophoretic mobility of the serum protein; immunoglobulins migrate to the gamma globulin region of the gel. The absence of immunoglobulins in a patient who has X-linked agammaglobulinemia is shown in the photograph at the bottom, where several arcs are missing from the patient's serum (upper set). These are IgM, IgA, and several subclasses of IgG, each recognized in normal serum (lower set) by antibodies in the antiserum against human serum proteins. Photograph from the collection of the late C A Janeway Snr.

of the α-chain genes. Thus, there might be a cascade of signaling tyrosine kinases, involving Lck and Itk in double-negative thymocytes, and Blk and Btk in pre-B cells, which is important for lymphocyte development. In both Lck and Btk deficiencies, some B or T cells mature despite the defect in the signaling kinase, suggesting that signals transmitted by these kinases promote the rearrangement of light-chain or α-chain genes, respectively, but are not absolutely required.

As the gene responsible for XLA is found on the X chromosome, it is possible to identify female carriers by analyzing X-chromosome inactivation in their B cells. During development, female cells randomly inactivate one of their two X chromosomes. Because the product of a normal *btk* gene is required for normal B-lymphocyte development, only cells in which the normal allele of *btk* is active can develop into mature B cells (with a very few exceptions; see earlier). Thus, in female carriers of mutant *btk* genes, almost all B cells have the normal X chromosome as the active X. By contrast, the active X chromosomes in the T cells and macrophages of carriers are an equal mixture of the normal and *btk* mutant X chromosomes. This fact allowed female carriers of XLA to be identified even before the nature of *btk* was known. Non-random X inactivation only in B cells also demonstrates conclusively that the *btk* gene is required for normal B-cell development but not for the development of other cell types, and that Btk must act within B cells rather than on stromal cells or other cells required for B-cell development (Fig. 11.10).

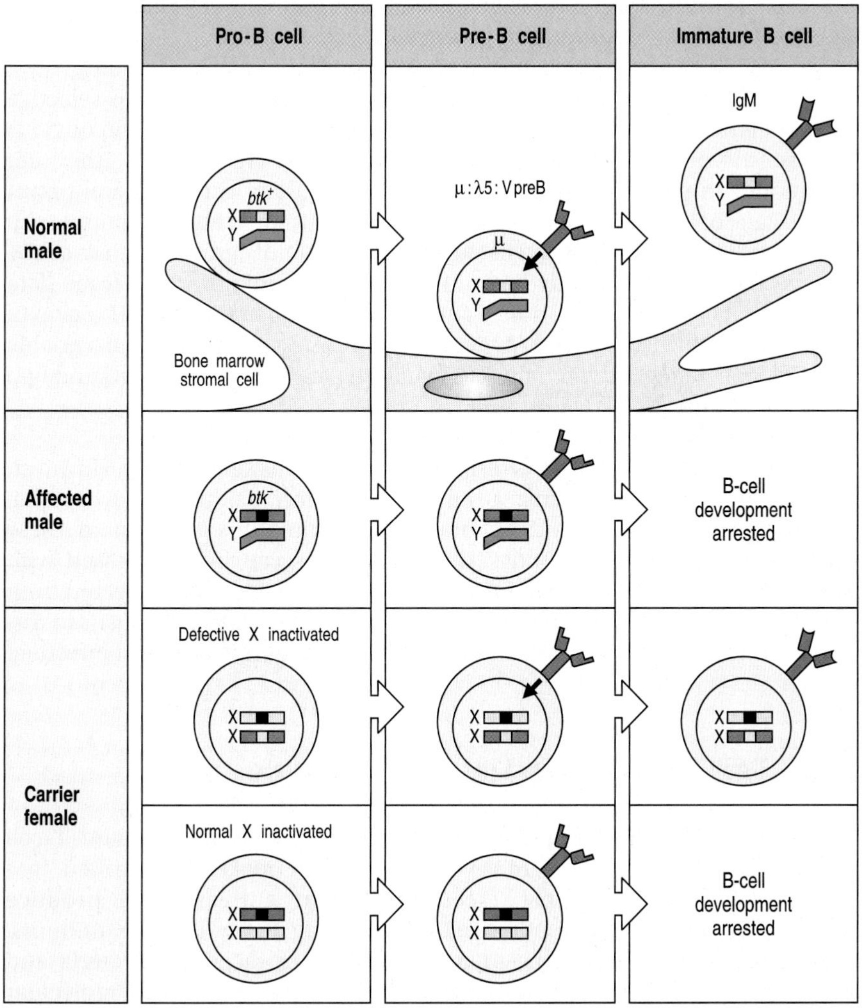

Fig. 11.10 The product of the *btk* gene is important for B-cell development. In X-linked agammaglobulinemia (XLA), a protein tyrosine kinase called Btk, encoded on the X chromosome, is defective. In normal individuals, B-cell development proceeds through a stage in which the pre-B-cell receptor consisting of μ:λ5:Vpre-B transduces a signal via Btk, triggering further B-cell development. In males with XLA, no signal can be transduced and although the pre-B-cell receptor is expressed, the B cells develop no further. In female mammals, including humans, one of the two X chromosomes in each cell is permanently inactivated early in development. Because the choice of which chromosome to inactivate is random, half of the pre-B cells in a carrier female express a wild-type *btk*, and half express the defective gene. None of the B cells that express *btk* from the defective chromosome can develop into mature B cells. Therefore, in the carrier, mature B cells always have the non-defective X chromosome active. This is in sharp contrast to all other cell types, which express the non-defective chromosome in only half of the population. Non-random X chromosome inactivation in a particular cell lineage is a clear indication that the product of the X-linked gene is required for the development of cells of that lineage. It is also sometimes possible to identify the stage at which the gene product is required, by detecting the point in development at which X-chromosome inactivation develops bias. Using this kind of analysis, one can identify carriers of X-linked traits such as XLA without needing to know the nature of the mutant gene.

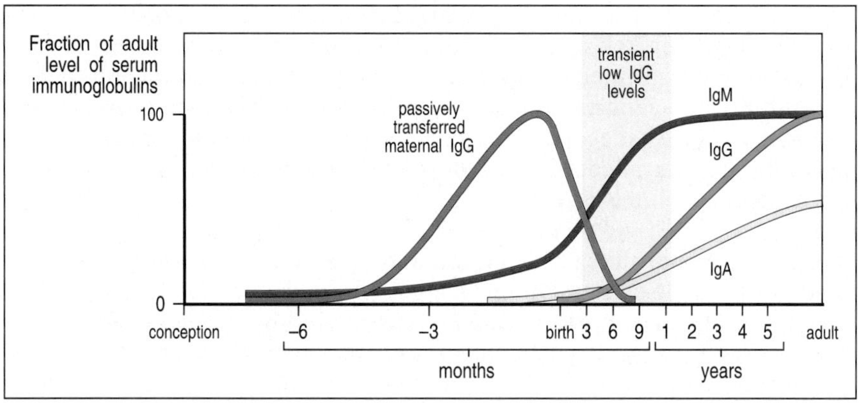

Fig. 11.11 Immunoglobulin levels in newborn infants fall to low levels around 6 months of age. Newborn babies have high levels of IgG, transported across the placenta from the mother during gestation. After birth, the production of IgM starts almost immediately; the production of IgG, however, does not begin for about 6 months, during which time the total level of IgG falls as the maternally acquired IgG is catabolized. Thus, IgG levels are low from about the age of 3 months to 1 year, which can lead to susceptibility to disease.

The commonest humoral immune defect is the transient deficiency in immunoglobulin production that occurs in the first 6–12 months of life. The newborn infant has initial antibody levels comparable to those of the mother, because of the transplacental transport of maternal IgG (see Chapter 9). As the transferred IgG is catabolized, antibody levels gradually decrease until the infant begins to produce useful amounts of its own IgG at about 6 months of age (Fig. 11.11). Thus, IgG levels are quite low between the ages of 3 months and 1 year and active IgG antibody responses are poor. In some infants this can lead to a period of heightened susceptibility to infection. This is especially true for premature babies, who begin with lower levels of maternal IgG and also reach immune competence later after birth.

The most common inherited form of immunoglobulin deficiency is selective IgA deficiency, which is seen in about 1 person in 800. Although no obvious disease susceptibility is associated with selective IgA defects, they are commoner in people with chronic lung disease than in the general population. Lack of IgA might thus result in a predisposition to lung infections with various pathogens and is consistent with the role of IgA in defense at the body's surfaces. The genetic basis of this defect is unknown but some data suggest that a gene of unidentified function mapping in the class III region of the MHC could be involved. A related syndrome called common variable immunodeficiency, in which there is usually a deficiency in both IgG and IgA, also maps to the MHC region.

People with pure B-cell defects resist many pathogens successfully. However, effective host defense against a subset of extracellular pyogenic bacteria, including staphylococci and streptococci, requires opsonization of these bacteria with specific antibody. These infections can be suppressed with antibiotics and periodic infusions of human immunoglobulin collected from a large pool of donors. As there are antibodies against many pathogens in this pooled immunoglobulin, it serves as a fairly successful shield against infection.

11-8 T-cell defects can result in low antibody levels.

Patients with **X-linked hyper-IgM syndrome** have normal B- and T-cell development and high serum levels of IgM but make very limited IgM antibody responses against T-cell dependent antigens and produce immuno-globulin isotypes other than IgM and IgD only in trace amounts. This makes them susceptible to infection with extracellular bacteria and certain opportunistic organisms such as *Pneumocystis carinii*. The molecular

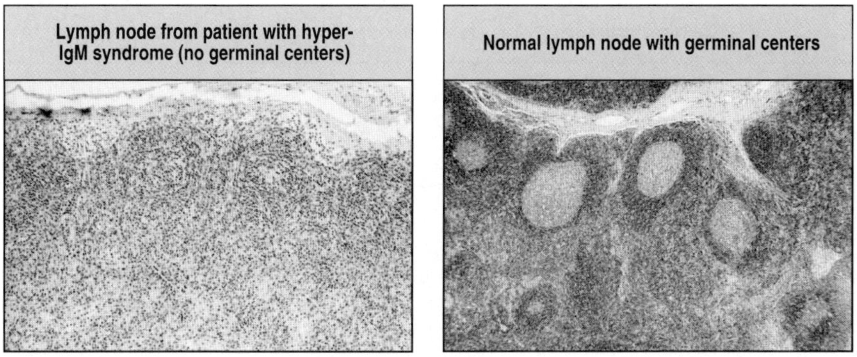

Fig. 11.12 Patients with X-linked hyper-IgM syndrome are unable to activate their B cells fully. Lymphoid tissues in patients with hyper-IgM syndrome are devoid of germinal centers (left panel), unlike a normal lymph node (right panel). B-cell activation by T cells is required both for isotype switching and for the formation of germinal centers, where extensive B-cell proliferation takes place. Photographs courtesy of R Geha and A Perez-Atayde.

defect in this disease is in the CD40 ligand on activated T cells, which therefore cannot engage CD40; the B cells themselves are normal. We learned in Chapter 9 that CD40 ligand is critical in the T-cell dependent activation of B-cell proliferation and these patients show that CD40 ligand is also essential for the induction of the isotype switch and formation of germinal centers (Fig. 11.12). The defects in cell-mediated immunity in these individuals might be due, at least in part, to the inability of their T cells to deliver an activating signal to antigen-presenting cells by engaging the CD40 expressed on these cells (see Section 8-28). A defect in T-cell activation could also contribute to the profound immunodeficiency suffered by these patients, as studies on mice that lack CD40 ligand have revealed a failure of antigen-specific T cells to expand in response to primary immunization with antigen.

In XLA, the hunt for the cause of the disease led to the discovery of a previously unidentified gene product. In X-linked hyper-IgM syndrome, the gene for CD40 ligand was cloned independently and only then identified as the defective gene in this disorder. Thus, inherited immunodeficiencies can either lead us to new genes or help us to determine the roles of known genes in normal immune system function.

11-9 Defects in complement components cause defective humoral immune function and persistence of immune complexes.

Not surprisingly, the spectrum of infections associated with complement deficiencies overlaps substantially with that seen in patients with deficiencies in antibody production. Defects in the activation of C3, and in C3 itself, are associated with a wide range of pyogenic infections, emphasizing the important role of C3 as an opsonin, promoting the phagocytosis of bacteria (Fig. 11.13). In contrast, defects in the membrane-attack components of complement (C5–C9) have more limited effects and result exclusively in susceptibility to *Neisseria* species. This indicates that host defense against these bacteria, which are capable of intracellular survival, is mediated by extracellular lysis by the membrane-attack complex of complement.

The early components of the classical complement pathway are particularly important for the elimination of immune complexes, which can cause significant pathology in autoimmune diseases such as systemic lupus erythematosus (see Chapter 13), and occasionally in persistent infections. As we learned in Chapter 9, complement components attached to soluble immune complexes allow them to be transported, ingested, and degraded by cells bearing complement receptors. They are transported by erythrocytes that capture

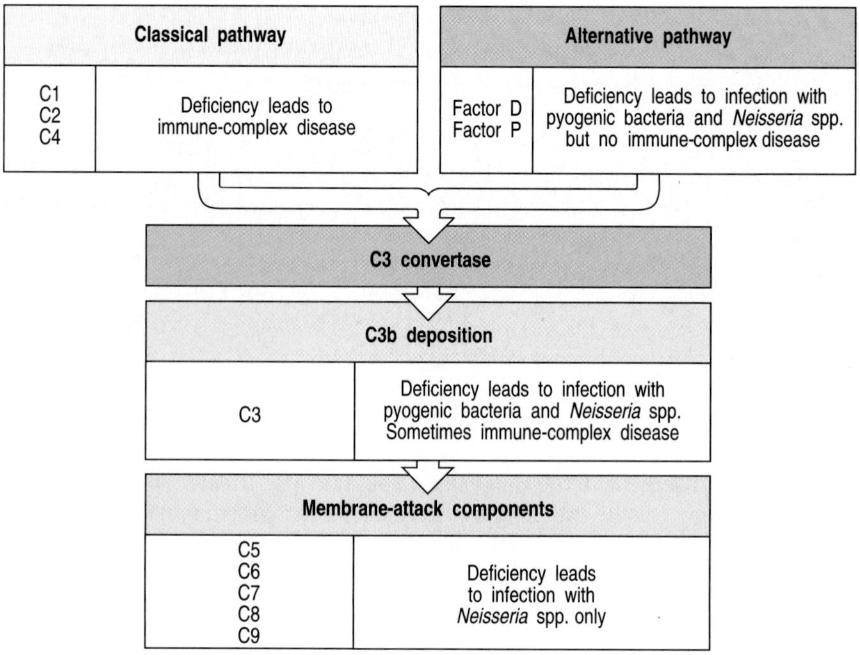

Fig. 11.13 Defects in complement components are associated with susceptibility to certain infections and accumulation of immune complexes. Defects in the early components of the alternative pathway lead to susceptibility to extracellular pathogens. Defects in the early components of the classical pathway predominantly affect the processing of immune complexes, leading to immune-complex disease. Finally, defects in the membrane-attack components are associated only with susceptibility to strains of *Neisseria* species, the causative agents of meningitis and gonorrhea, implying that the effector pathway is important chiefly in defense against these organisms.

immune complexes via the complement receptor CR1, which binds specifically to C4b and C3b. When this mechanism is inoperative, immune complexes are deposited in the tissues. Accumulating immune complexes activate phagocytes, causing inflammation and local tissue damage.

Deficiencies in control proteins that regulate complement activation can cause either immunodeficiency or autoimmune-like disease. People with defects in properdin (factor P), which enhances the activity of the alternative pathway, have a heightened susceptibility to *Neisseria* species. In contrast, patients lacking decay-accelerating factor (DAF) and CD59, which protect host cell surfaces from alternative pathway activation, destroy their own red blood cells. This results in the disease paroxysmal nocturnal hemoglobinuria, as we learned in Chapter 9. A more striking consequence arising from the loss of a regulatory protein is seen in patients with C1-inhibitor defects. These individuals fail to control the inappropriate activation of the classical pathway of complement activation, and the uncontrolled cleavage of C2 allows the generation of a vasoactive fragment of C2a, causing fluid accumulation in the tissues and epiglottal swelling that can lead to suffocation. This syndrome is called **hereditary angioneurotic edema**.

11-10 Defects in phagocytic cells permit widespread bacterial infections.

Defects in the recruitment of phagocytic cells to extravascular sites of infection can cause serious immunodeficiency. Leukocytes reach such sites by emigrating from blood vessels in a tightly regulated process consisting of three stages. The first is the rolling adherence of leukocytes to endothelial cells, through the binding of a fucosylated tetrasaccharide ligand known as sialyl-Lewis^x to E-selectin and P-selectin. Sialyl-Lewis^x is expressed on monocytes and neutrophils, whereas E-selectin and P-selectin are expressed on endothelium activated by mediators from the site of inflammation. The second stage is the tight adherence of the leukocytes to the endothelium through the binding of leukocyte β_2 integrins such as CD11b/CD18 (MAC-1/CR3) to

Type of defect / name of syndrome	Associated infectious or other diseases
Leukocyte adhesion deficiency	Widespread pyogenic bacterial infections
Chronic granulomatous disease	Intra and extracellular infection, granulomas
G6PD deficiency	Defective respiratory burst, chronic infection
Myeloperoxidase deficiency	Defective intracellular killing, chronic infection
Chediak–Higashi syndrome	Intra and extracellular infection, granulomas

Fig. 11.14 Defects in phagocytic cells are associated with persistence of bacterial infection. Defects in the leukocyte integrins with a common β subunit (CD18) or defects in the selectin ligand, sialyl-Lewis^x, prevent phagocytic cell adhesion and migration to sites of infection. The respiratory burst is defective in chronic granulomatous disease, glucose-6-phosphate dehydrogenase (G6PD) deficiency and myeloperoxidase deficiency. In chronic granulomatous disease, infections persist because macrophage activation is defective, leading to chronic stimulation of CD4 T cells and hence to granulomas. Vesicle fusion in phagocytes is defective in Chediak–Higashi syndrome. These diseases illustrate the critical role of phagocytes in removing and killing pathogenic bacteria.

counter-receptors on endothelial cells. The third and final stage is the transmigration of leukocytes through the endothelium along gradients of chemotactic molecules originating from the site of tissue injury. Deficiencies in the molecules involved in each of these stages can prevent neutrophils and macrophages from reaching sites of infection to ingest and destroy bacteria. Reduced rolling adhesion has been described in patients with a lack of sialyl-Lewis^x caused by a deficiency in the fucosylation pathway responsible for its biosynthesis. Similarly, deficiencies have been identified that prevent tight leukocyte adhesion in the leukocyte integrin common β_2 subunit CD18. Each of these deficiencies leads to infections that are resistant to antibiotic treatment and persist despite an apparently effective cellular and humoral adaptive immune response. Neutropenia associated with chemotherapy, malignancy, or aplastic anemia is associated with a similar spectrum of severe pyogenic bacterial infections.

Most of the other known defects in phagocytic cells affect their ability to kill intracellular and/or ingested extracellular bacteria (Fig. 11.14). In **chronic granulomatous disease**, phagocytes cannot produce the superoxide radical and their antibacterial activity is thereby seriously impaired. Several different genetic defects, affecting any one of the four constituent proteins of the NADPH oxidase system, can cause this. Patients with this disease have chronic bacterial infections, which in some cases lead to the formation of granulomas. Deficiencies in the enzymes glucose-6-phosphate dehydrogenase and myeloperoxidase also impair intracellular killing and lead to a similar, although less severe, phenotype. Finally, in **Chediak–Higashi syndrome**, a complex syndrome characterized by partial albinism, abnormal platelet function, and severe immunodeficiency, a defect in a gene encoding a protein involved in intracellular vesicle formation causes a failure to fuse lysosomes properly with phagosomes; the cells in these patients have enlarged granules and impaired intracellular killing.

11-11 Defects in T-cell function result in severe combined immunodeficiencies.

Although patients with B-cell defects can deal with many pathogens adequately, patients with defects in T-cell development are highly susceptible to a broad range of infectious agents. This demonstrates the central role of T cells in adaptive immune responses to virtually all antigens. As such patients make neither specific T-cell dependent antibody responses nor cell-mediated immune responses, and thus cannot develop protective immunity, they are said to suffer from **severe combined immune deficiency** (**SCID**).

Several different defects can lead to the SCID phenotype. In X-linked SCID, T cells fail to develop because of a mutation in the common γ chain of several cytokine receptors, including those for the interleukins IL-2, IL-4, IL-7, IL-9, and IL-15. We shall examine this defect further in Section 11-12. Two other defects that give rise to T-cell deficiencies and a SCID phenotype are **adenosine deaminase** (**ADA**) **deficiency** and **purine nucleotide phosphorylase** (**PNP**) **deficiency**. These defects are in enzymes involved in purine degradation, and both result in an accumulation of nucleotide metabolites that are particularly toxic to developing T cells. B cells are also somewhat compromised in these patients.

One class of SCID individuals lack expression of all MHC class II gene products on their cells. This condition is referred to as the **bare lymphocyte syndrome**. As the thymus in such individuals lacks MHC class II molecules, CD4 T cells cannot be positively selected and therefore few develop. The antigen-presenting cells in these individuals also lack MHC class II molecules and so the few CD4 T cells that do develop cannot be stimulated by antigen. In these individuals, MHC class I expression is normal and CD8 T cells develop normally. However, such people suffer from severe combined immunodeficiency, illustrating the central importance of CD4 T cells in adaptive immunity to most pathogens. The syndrome is caused not by mutations in the MHC genes themselves, but by mutations in one of several different genes encoding *trans*-acting regulatory proteins that are required for the transcriptional activation of MHC class II promoters. Four complementing gene defects (known as Groups A, B, C, and D) have been defined in patients who fail to express MHC class II molecules, which implies that at least four different genes are required for normal MHC class II gene expression. One of these, named the **MHC class II transactivator**, or **CIITA**, is the gene mutated in Group A. The genes mutated in Groups C and D are named *RFX5* and *RFXAP* and encode two proteins that are associated in a multimeric transcriptional complex that binds a sequence present in the promoter of all MHC class II genes. The genetic basis of the Group B complementing group has not yet been characterized.

In contrast, a more limited immunodeficiency, associated with chronic respiratory viral infections, has been observed in a family showing almost complete absence of cell-surface MHC class I molecules. These patients have normal levels of mRNA encoding MHC class I molecules and normal production of MHC class I proteins, but these proteins fail to reach the cell surface. The defect was shown to be similar to that in the TAP mutant cells mentioned in Section 4-8 and, indeed, affected members of this family had mutations in a *TAP* gene. These people are immunodeficient owing to a lack of CD8 T cells.

Defects in either the *RAG-1* or *RAG-2* genes result in the arrest of lymphocyte development because of a failure to rearrange the antigen receptor genes. Thus there is a complete lack of T and B cells in mice with genetically engineered defects in the *RAG* genes, and in patients with autosomally inherited forms of SCID who have mutations of *RAG-1* or *RAG-2* genes. Another group of patients with autosomal SCID have a phenotype very similar to that of a mutant mouse strain called *scid*; *scid* mice suffer from an abnormal sensitivity to ionizing radiation as well as from severe combined immunodeficiency. They produce very few mature B and T cells, as there is a failure of DNA rearrangement in their developing lymphocytes; only rare VJ or VDJ joints are seen and most of these have abnormal features. The underlying defect has now been shown to be in the enzyme DNA-dependent kinase, which binds to the end of the double-stranded breaks that occur during the process of antigen receptor gene rearrangement. These ends are found as DNA hairpin structures in the immature thymocytes of *scid* mice. Thus, it seems likely that DNA-dependent kinase is involved in resolving the hairpin structure (see Fig. 3.18).

Other defects in DNA repair and metabolizing enzymes are associated with a combination of immunodeficiency, increased sensitivity to the damaging effects of ionizing radiation, and cancer development. One example is Bloom's syndrome, a disease caused by mutations in a DNA helicase enzyme, which unwinds DNA. Another is ataxia telangiectasia, in which the underlying defect is in a protein called ATM, which contains a kinase domain thought to be involved in intracellular signaling in response to DNA damage. Because repair of double-stranded DNA breaks and lymphocyte division are central to the function of the adaptive immune system, it is not surprising that defects such as these are associated with the development of immunodeficiency.

Finally, in patients with **DiGeorge syndrome** the thymic epithelium fails to develop normally. Without the proper inductive environment T cells cannot mature, and both T-cell dependent antibody production and cell-mediated immunity are absent. Such patients have some serum immunoglobulin and variable numbers of B and T cells. As with all the severe combined immunodeficiency diseases it is the defect in T cells that is crucial. These diseases abundantly illustrate the central role of T cells in virtually all adaptive immune responses. In many cases B-cell development is normal, yet the response to nearly all pathogens is profoundly impaired.

11-12 Defective T-cell signaling, cytokine production, or cytokine action can cause immunodeficiency.

As we learned in Chapter 8, virtually all adaptive immune responses require the activation of antigen-specific T lymphocytes and their differentiation into cells producing cytokines that act on specific cytokine receptors. Several gene defects have been described that interfere with these processes. Thus, patients who lack CD3γ chains have low levels of surface T-cell receptors and defective T-cell responses. Patients making low levels of mutant CD3ε chains are also deficient in T-cell activation. Patients who make a defective form of the cytosolic protein tyrosine kinase ZAP-70 (see Section 5-9) have recently been described. Their CD4 T cells emerge from the thymus in normal numbers, whereas CD8 T cells are absent. However, the CD4 T cells that mature fail to respond to stimuli that normally activate via the T-cell receptor and the patients are thus very immunodeficient.

Another group of patients show an absence of IL-2 production upon receptor ligation, and these patients have a severe immunodeficiency; however, T-cell development is normal in these individuals, as it is in mice in which mutations have been made in their IL-2 genes by gene knock-out (see Section 2-27). These IL-2-negative patients have heterogeneous defects; some of them fail to activate the transcription factor NF-AT (see Section 5-11), which induces the transcription of several cytokine genes in addition to the IL-2 gene. This might explain why their immunodeficiency is more profound than that of mice whose IL-2 gene has been disrupted. IL-2-deficient mice can mount adaptive immune responses through an IL-2-independent pathway, possibly involving the cytokine IL-15, which shares many activities with IL-2; nevertheless, they are susceptible to a variety of infectious agents.

In contrast to the normal development of T cells in patients deficient in IL-2, there is a failure of T-cell development in patients with **X-linked severe combined immunodeficiency** (X-linked SCID), who have a defect in the γc chain of the IL-2 receptor. Thus, the γc chain must be important in T-cell development for reasons unrelated to IL-2 binding or IL-2 responses. The demonstration that the IL-2 receptor γ chain is also part of other cytokine receptors, including the IL-7 receptor, helps to explain its role in early T-cell development. The γ chain seems to function in transducing the signal from this group of receptors and interacts with a kinase, JAK3 kinase, which is known to be defective in patients with an autosomally inherited immunodeficiency similar in phenotype to X-linked SCID.

As in all serious T-cell deficiencies, X-linked SCID patients do not make effective antibody responses to most antigens, although their B cells seem normal. However, as the gene defect is on the X chromosome, one can determine whether the lack of B-cell function is solely a consequence of the lack of T-cell help by examining X-chromosome inactivation (see Section 11-7) in B cells of unaffected carriers. The majority of naive IgM-positive B cells from female carriers of X-linked SCID have inactivated the defective X chromosome rather than the normal one, showing that B-cell development is affected by, but not wholly dependent on, the common γ chain. However, mature memory B cells that have switched to isotypes other than IgM carry an inactive defective X chromosome almost without exception. This might reflect the fact that the IL-2 receptor γ chain is also part of the IL-4 receptor. Thus, B cells that lack this chain will have defective IL-4 receptors and will not proliferate in T-cell dependent antibody responses. X-linked SCID is so severe that children who inherit it can survive only in a completely pathogen-free environment, unless given antibodies and successfully treated by bone-marrow transplantation. A famous case in Houston became known as the 'bubble baby' because of the plastic bubble in which he was enclosed to protect him from infection.

The production of defects in several cytokine and cytokine receptor genes in gene knock-out mice is rapidly increasing our understanding of the role of individual cytokines in immunity. Mice lacking transforming growth factor-β (TGF-β) die of overwhelming inflammatory disease, whereas mice lacking IFN-γ or the IFN-γ receptor succumb to infection with a range of intracellular pathogens, including *M. tuberculosis.* Several human families have recently been identified with children who are susceptible to early onset mycobacterial infections and have a homozygous deficiency of the IFN-γ receptor.

Wiskott–Aldrich syndrome (**WAS**) is a disease that has shed new light on the molecular basis of T-cell signaling and its importance for immune function. The disease affects platelets and was first described as a blood-clotting disorder, but it is also associated with immunodeficiency due to impaired T-cell function, reduced T-cell numbers, and a failure of antibody responses to capsulated bacteria. WAS is caused by a defective gene on the X chromosome, encoding a protein called WAS protein (WASP). This protein has been shown to bind Cdc42, a small GTP-binding protein that is known to regulate the organization of the actin cytoskeleton and to be important for the effective collaboration of T and B cells. WASP might have a role in regulating changes in the actin cytoskeleton in response to external stimuli. It has the capability of binding to SH3 domains, which, as we saw in Chapter 5, have an affinity for amino-acid sequences rich in proline that are found on some proteins of intracellular signaling pathways. In WAS patients, and in mice whose WASP gene has been knocked-out, T cells fail to respond normally to mitogens or to the crosslinking of surface receptors. Cytotoxic T-cell responses are also impaired, and T-cell help for B-cell responses to polysaccharide antigens is lacking. WASP is expressed in all hematopoietic cell lineages, and is likely to be a key regulator of lymphocyte and platelet development and function.

11-13 Bone marrow transplantation or gene therapy can be useful to correct genetic defects.

It is frequently possible to correct the defects in lymphocyte development that lead to the SCID phenotype by replacing the defective component, generally by bone marrow transplantation. The major difficulties in these therapies result from MHC polymorphism. To be useful, the graft must share some MHC alleles with the host. As we learned in Section 7-14, the MHC alleles expressed by the thymic epithelium determine which T cells can be positively selected. When bone marrow cells are used to restore immune function to individuals with a normal thymic stroma, both the T cells and the antigen-presenting cells are derived from the graft. Therefore, unless the graft shares at least some MHC alleles with the recipient, the T cells that are selected on host thymic epithelium cannot be activated by graft-derived antigen-presenting cells (Fig. 11.15). There is also a danger that mature, post-thymic T cells in donor bone marrow might recognize the host as foreign and attack

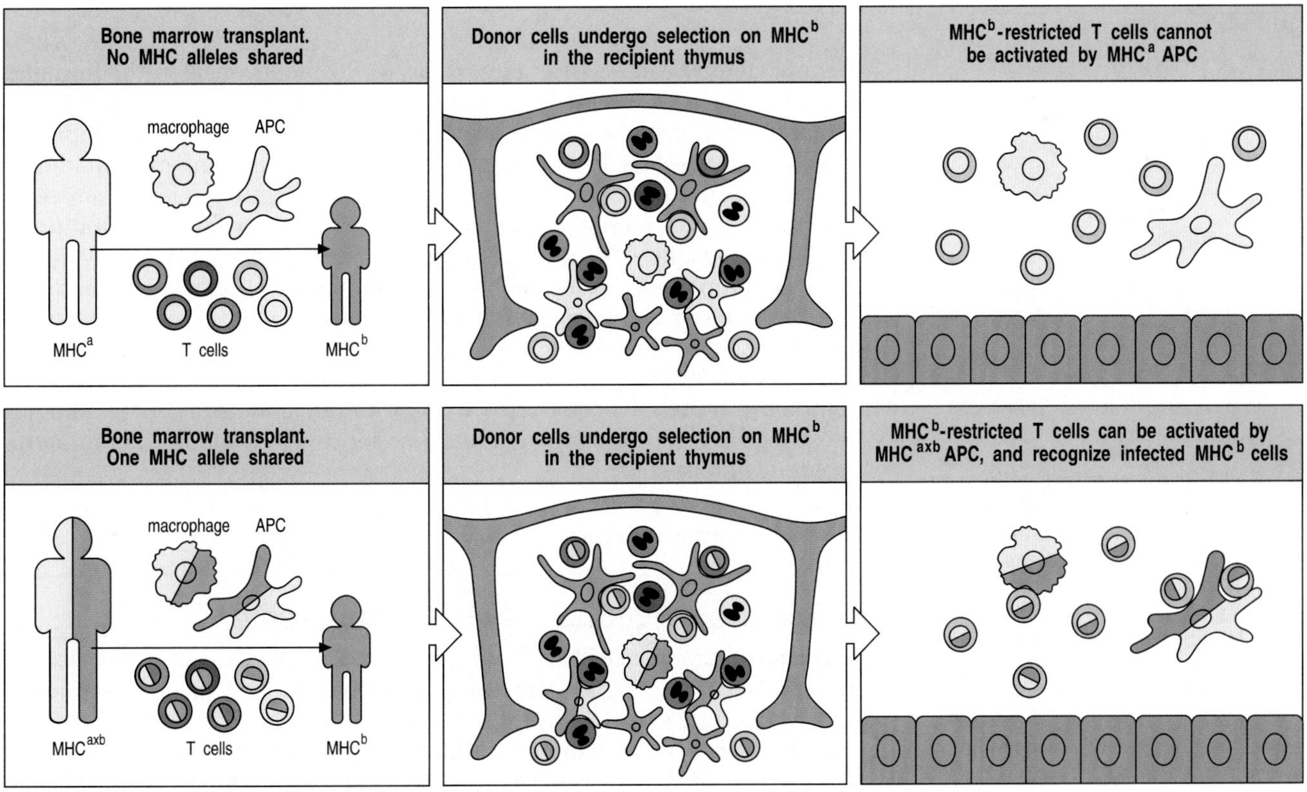

Fig. 11.15 Bone marrow donor and recipient must share at least some MHC molecules to restore immune function. If the bone marrow and the recipient thymus do not share any MHC alleles, T cells will mature in the thymus with receptors selected to recognize peptides presented by MHC molecules that are not expressed on the donor-derived antigen-presenting cells (APCs). These cells will not therefore be competent to mediate protective immunity (top panels). In the bottom panels, donor and recipient share the MHC^b allele, and T cells able to recognize MHC^b molecules are selected in the thymus. The antigen-presenting cells in the periphery can activate T cells that recognize MHC^b molecules; the activated T cells can then recognize infected MHC^b-bearing cells.

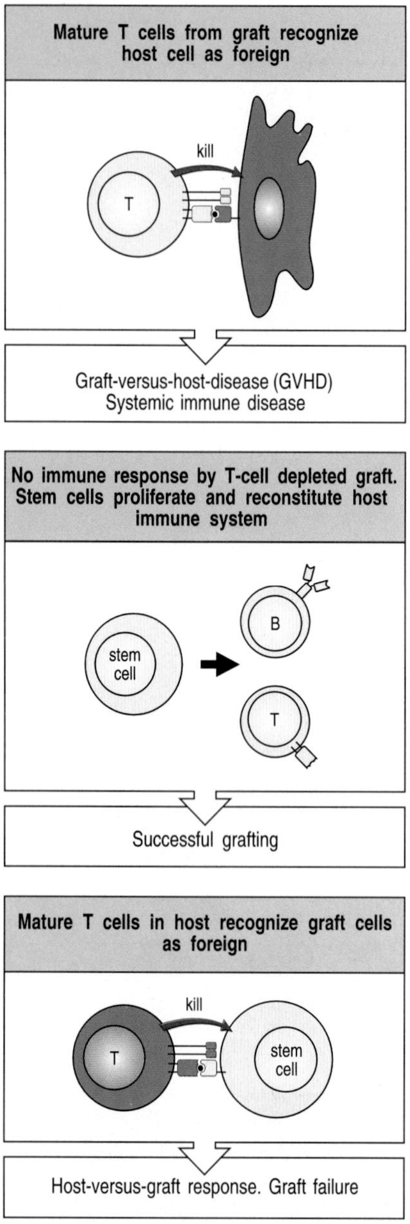

Fig. 11.16 Bone marrow grafting can be used to correct immunodeficiencies caused by defects in lymphocyte maturation but two problems can arise. First, if there are mature T cells in the bone marrow, they can attack cells of the host by recognizing their MHC antigens, causing graft-versus-host disease (top panel). This can be prevented by T-cell depletion of the donor bone marrow (middle panel). Second, if the recipient has competent T cells, these can attack the bone marrow stem cells (bottom panel). This causes failure of the graft by the usual mechanism of transplant rejection (see Chapter 13).

it, causing **graft-versus-host disease** (**GVHD**) (Fig. 11.16). This can be overcome by depleting the donor bone marrow of mature T cells. In patients with the SCID phenotype, there is little problem with the host response to the graft, as the patient is immunodeficient.

Now that specific gene defects are being identified, a different approach to correcting these inherited immune deficiencies can be attempted. The strategy involves extracting a sample of the patient's own bone marrow cells, inserting a normal copy of the defective gene into them, and returning them to the patient by transfusion. This approach, called **somatic gene therapy**, should correct the gene defect. Moreover, in immunodeficient patients, it might be possible to re-infuse the bone marrow into the patient without the usual irradiation to suppress the recipient's bone marrow function. There is no risk of graft-versus-host disease in this case, although the host might respond to the replaced gene product and reject the engineered cells. Although this kind of approach is theoretically attractive, efficient transfer of genes into bone marrow stem cells is technically difficult and has been achieved only in mouse models. The few trials of gene therapy for correcting immunodeficiency, such as the treatment of a child with ADA deficiency at the National Institute of Health (NIH) in 1990, have used the patient's lymphocytes as the vehicle for gene introduction. However, because most lymphocytes are short-lived, the treatment has to be repeated regularly.

Summary.

Genetic defects can occur in almost any molecule involved in the immune response. These defects give rise to characteristic deficiency diseases, which, although rare, provide a great deal of information about the development and functioning of the immune system in normal humans. Inherited immunodeficiencies illustrate the vital role of the adaptive immune response and T cells in particular, without which both cell-mediated and humoral immunity fail. They have provided information about the separate roles of B lymphocytes in humoral immunity and of T lymphocytes in cell-mediated immunity, the importance of phagocytes and complement in humoral and innate immunity, and the specific functions of several cell-surface or signaling molecules in the adaptive immune response. There are also some inherited immune disorders whose causes we still do not understand. The study of these diseases will undoubtedly teach us more about the normal immune response and its control.

Acquired immune deficiency syndrome.

The first cases of the **acquired immune deficiency syndrome** (**AIDS**) were reported in 1981 but it is now clear that cases of the disease had been occurring unrecognized for about 4 years before its identification. The disease is characterized by a susceptibility to infection with opportunistic pathogens or by the occurrence of an aggressive form of Kaposi's sarcoma or B-cell lymphoma, accompanied by a profound decrease in the number of CD4 T cells. As it seemed to be spread by contact with body fluids, it was early suspected to be caused by a new virus, and by 1983 the agent now known to be responsible for AIDS, called the **human immunodeficiency virus** (**HIV**),

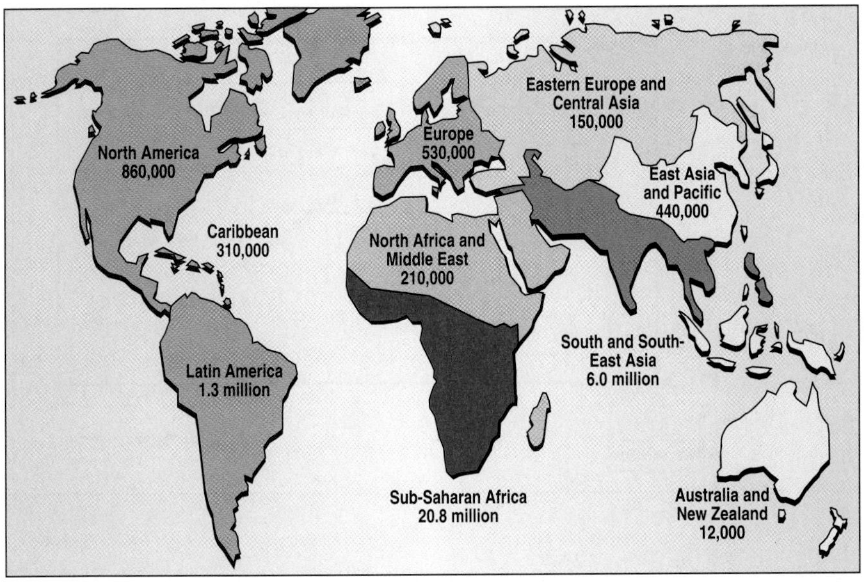

Fig. 11.17 HIV infection is spreading on all continents. The number of HIV-infected individuals is large (data are numbers of adults and children living with HIV/AIDS at the end of 1997, as estimated by the World Health Organisation) and is increasing rapidly, especially in developing countries.

was isolated and identified. There are now known to be at least two types of HIV—HIV-1 and HIV-2—which are closely related to each other. HIV-2 is endemic in West Africa and is now spreading in India. Most AIDS worldwide is, however, caused by the more virulent HIV-1.

HIV infection does not immediately cause AIDS, and the issues of how it does, and whether all HIV-infected patients will progress to overt disease, remain controversial. Nevertheless, accumulating evidence clearly implicates the growth of the virus in CD4 T cells, and the immune response to it, as the central keys to the puzzle of AIDS. HIV is a world pandemic and, although great strides are being made in understanding the pathogenesis and epidemiology of the disease, the number of infected people worldwide continues to grow at an alarming rate, presaging the death of many people from AIDS for many years to come. Estimates from the World Health Organisation are that 11.7 million people have died from AIDS since the beginning of the epidemic and that there are currently 30.6 million people alive with HIV infection (Fig. 11.17), of whom the majority are living in sub-Saharan Africa, where prevalence rates for infection are approximately 7% among young adults.

11-14 Most individuals infected with HIV progress over time to AIDS.

Many viruses cause an acute but limited infection inducing lasting protective immunity. Others, such as herpes viruses, set up a latent infection that is not eliminated but is controlled adequately by an adaptive immune response. However, infection with HIV seems rarely, if ever, to lead to an immune response that can eliminate the virus. Although the initial acute infection does seem to be controlled by the immune system, HIV continues to replicate rapidly and infect new cells.

The initial infection with HIV generally occurs after transfer of body fluids from an infected person. The virus is carried in infected CD4 T cells, dendritic cells, and macrophages, and as a free virus in blood, semen, vaginal fluid, or milk. It is most commonly spread by sexual intercourse, contaminated needles used for intravenous drug delivery, and the therapeutic use of infected blood

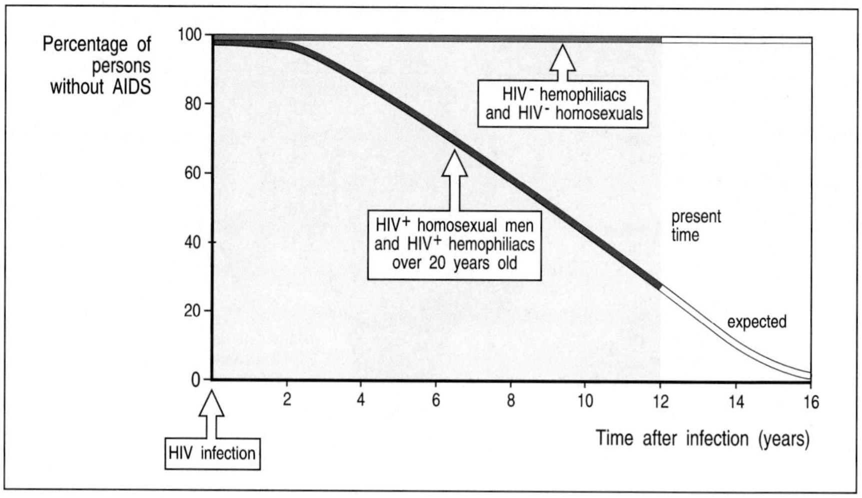

Fig. 11.18 Most HIV-infected individuals progress to AIDS over a period of years. The incidence of AIDS increases progressively with time after infection, and is predicted to reach ~95% within 15 years, although data are only available for 12 years. Homosexuals and hemophiliacs are two of the groups at highest risk in the West—homosexuals from sexually transmitted virus and hemophiliacs from infected human blood used to replace clotting factor VIII. In Africa, spread is mainly by heterosexual intercourse. Hemophiliacs are now protected by the screening of blood products and the use of recombinant factor VIII. Neither homosexuals nor hemophiliacs who have not been infected with HIV show any evidence of AIDS. There are a few individuals who, while infected with HIV, seem not to have progressed to develop AIDS. One protective mechanism is an inherited defect in the major HIV co-receptor, CCR5.

or blood products, although this last route of transmission has largely been eliminated in the developed world where blood products are screened routinely for the presence of HIV. An important route of virus transmission is from an infected mother to her baby at birth or through breast milk. In Africa, the perinatal transmission rate is approximately 25%, but this can largely be prevented by treating infected pregnant women with the drug zidovudine (AZT) (see Section 11-19). Mothers who are newly infected and breastfeed their infants transmit HIV 40% of the time, showing that HIV can also be transmitted in breast milk.

Primary infection with HIV is probably asymptomatic in most cases but sometimes causes an influenza-like illness with an abundance of virus in the peripheral blood and a marked drop in the numbers of circulating CD4 T cells. This acute viremia is associated in virtually all patients with the activation of CD8 T cells, which kill HIV-infected cells, and subsequently with antibody production, or **seroconversion**. The cytotoxic T-cell response is thought to be important in controlling virus levels, which peak and then decline, as the CD4 T-cell counts rebound to around 800 cells μl^{-1} (the normal value is 1200 cells μl^{-1}). At present, the best indicator of future disease is the level of virus that persists in the blood plasma once the symptoms of acute viremia have passed.

Most patients who are infected with HIV will eventually develop AIDS, after a period of apparent quiescence of the disease known as clinical latency or the asymptomatic period (Fig. 11.18). This period is not silent, however, for there is persistent replication of the virus, and a gradual decline in function and numbers of CD4 T cells until eventually patients have few CD4 T cells left. At this point, which can occur anywhere between 2 and 15 years or more after the primary infection, the period of clinical latency ends and opportunistic infections begin to appear.

The typical course of an infection with HIV is illustrated in Fig. 11.19. However, it has become increasingly clear that the course of the disease can vary widely. Thus, although most people infected with HIV go on to develop AIDS and ultimately to die of opportunistic infection or cancer, this is not true of all individuals. A small percentage of people seroconvert, making antibodies against many HIV proteins, but do not seem to have progressive disease, in that their CD4 T-cell counts and other measures of immune competence are maintained. These long-term non-progressors have unusually low levels of circulating virus and are being studied intensively to determine

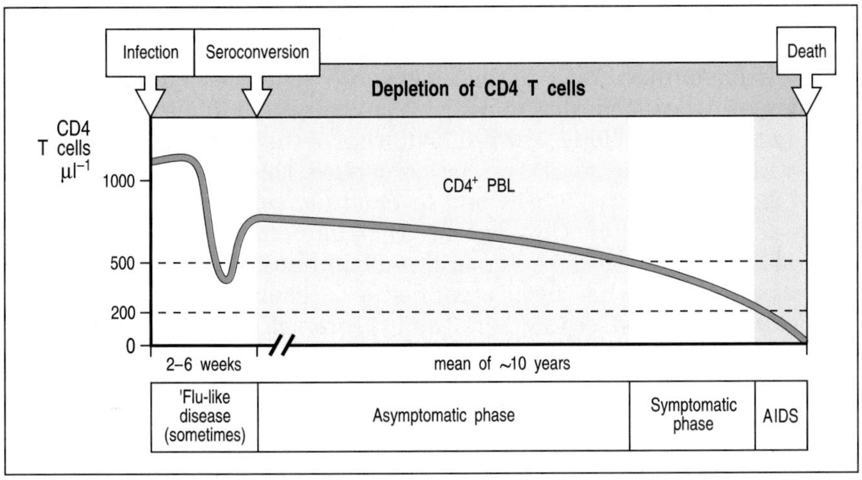

Fig. 11.19 The typical course of infection with HIV. The first few weeks are typified by an acute influenza-like viral illness, sometimes called seroconversion disease, with high titers of virus in the blood. An adaptive immune response follows, which controls the acute illness and largely restores levels of CD4 T cells (CD4⁺ PBL) but does not eradicate the virus. Opportunistic infections and other symptoms become more frequent as the CD4 T-cell count falls, starting at around 500 cells per μl^{-1}. The disease then enters the symptomatic phase. When CD4 T-cell counts fall below 200 cells per μl^{-1} the patient is said to have AIDS. Note that CD4 T-cell counts are measured for clinical purposes in cells per microliter (μl^{-1}), rather than cells per milliliter (cells ml^{-1}) used elsewhere in this book.

how they are able to control the infection with HIV. A second group consists of seronegative people who have been highly exposed to HIV yet remain disease-free and virus-negative. Some of these people have specific cytotoxic lymphocytes and T_H1 lymphocytes directed against infected cells, which confirms that they have been exposed to HIV or possibly to non-infectious HIV antigens. It is not clear whether this immune response accounts for clearing the infection, but it is a focus of considerable interest for the development and design of vaccines, which we shall discuss later.

Before going on to discuss in more detail the interactions of HIV with the immune system and the prospects for manipulating them, we must first describe the viral life cycle and the proteins on which it depends. Some of these proteins are the targets of the most successful drugs in use at present for the treatment of AIDS.

11-15 HIV is a retrovirus that infects CD4 T cells, dendritic cells, and macrophages.

HIV is an enveloped retrovirus. Each virus particle contains two copies of an RNA genome, which are transcribed into DNA in the infected cell and integrated into the host cell chromosome. The RNA transcripts produced from the integrated viral DNA serve both as mRNA to direct the synthesis of the viral proteins and later as the RNA genomes of new viral particles, which escape from the cell by budding from the plasma membrane, each in a membrane envelope. The structure of the viral particle, or virion, is shown in Fig. 11.20. HIV belongs to a group of retroviruses called the **lentiviruses**, from the Latin *lentus*, meaning slow, because of the gradual course of the diseases that they cause. These viruses persist and continue to replicate for many years before causing overt signs of disease.

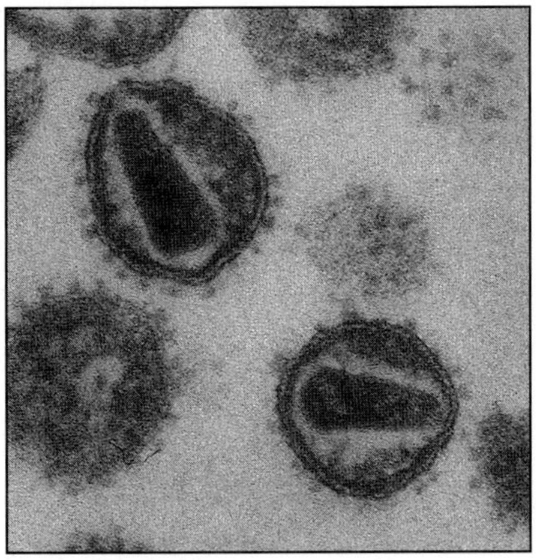

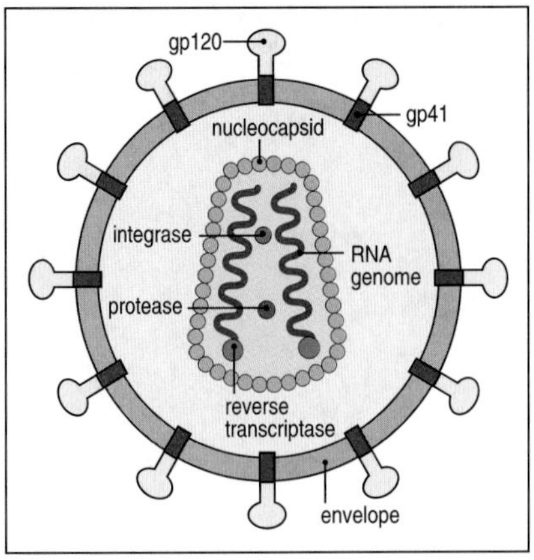

Fig. 11.20 The virion of human immunodeficiency virus (HIV). The virus illustrated is HIV-1, the leading cause of AIDS. The reverse transcriptase, integrase, and viral protease enzymes are packaged in the virion and are shown schematically in the viral capsid. In reality, many molecules of these enzymes are contained in each virion. Some structural proteins of the virus have been omitted for simplicity. Photograph courtesy of H Gelderblom.

The ability of HIV to enter particular types of cell, known as the cellular tropism of the virus, is determined by the expression of specific receptors for the virus on the surface of those cells. HIV enters cells by means of a complex of two noncovalently associated viral glycoproteins, gp120 and gp41, in the viral envelope. The gp120 portion of the glycoprotein complex binds with high affinity to CD4 on the cell surface. This glycoprotein thereby draws the virus to CD4 T cells and to dendritic cells and macrophages, which also express some CD4. Before fusion and entry of the virus, gp120 must also bind to a co-receptor in the membrane of the host cell. This co-receptor for HIV has been identified as a chemokine receptor. The chemokine receptors (see Chapters 8 and 10) are a close family of G-protein coupled receptors with seven transmembrane-spanning domains. Two chemokine receptors, known as CCR5, which is predominantly expressed on dendritic cells, macrophages, and CD4 T cells, and CXCR4, expressed on T cells, are the major co-receptors for HIV. After binding of gp120 to the receptor and co-receptor, the gp41 then causes fusion of the viral envelope and the plasma membrane of the cell, allowing the viral genome and associated viral proteins to enter the cytoplasm.

There are different variants of HIV; the cell types that they infect (their cell tropisms) are determined to a large degree by which chemokine receptor they bind as co-receptor. The tropisms of the different variants of HIV were originally defined by their ability to infect cells *in vitro*. The variants of HIV that are associated with primary infections are 'macrophage-tropic' and infect dendritic cells, macrophages, and T cells *in vivo*. These variants use CCR5, which binds the CC chemokines RANTES, MIP-1α, and MIP-1β (see Chapter 10), as a co-receptor, and require only a low level of CD4 on the cells they infect. These variants of HIV do not infect T-cell lines *in vitro*, but can infect freshly isolated T cells *in vitro*.

In contrast, 'lymphocyte-tropic' variants of HIV infect only CD4 T cells *in vivo* and use CXCR4, which binds the CXC-chemokine stromal-derived factor-1 (SDF-1), as a co-receptor. The lymphocyte-tropic variants of HIV can grow *in vitro* in T-cell lines, and require high levels of CD4 on the cells that they infect.

Evidence is accumulating that macrophage-tropic isolates of HIV are preferentially transmitted by sexual contact as they are the dominant viral phenotype found in newly infected individuals. Virus is disseminated from an initial reservoir of infected dendritic cells and macrophages and there is evidence for an important role for mucosal lymphoid tissue in this process. Mucosal epithelia, which are constantly exposed to foreign antigens, provide a milieu of immune system activity in which HIV replication occurs readily. Late in infection, in approximately 50% of cases, the viral phenotype switches to a T-lymphocyte tropic type, which is then followed by a rapid decline in CD4 T-cell count and progression to AIDS.

11-16 Genetic deficiency of the macrophage chemokine co-receptor for HIV confers resistance to HIV infection *in vivo*.

Further evidence for the importance of chemokine receptors in HIV infection has come from studies in a small group of individuals with high-risk exposure to HIV-1 but who remain seronegative. Cultures of lymphocytes and macrophages from these people were relatively resistant to macrophage-tropic HIV infection and were found to secrete high levels of RANTES, MIP-1α and MIP-1β in response to inoculation with HIV. In other experiments, the addition of these same chemokines to lymphocytes sensitive to HIV blocked their infection because of competition between these CC-chemokines and the virus for the cell-surface receptor CCR5.

The resistance of these rare individuals to HIV infection has now been explained by the discovery that they are homozygous for an allelic, non-functional variant of CCR5 caused by a 32 base-pair deletion from the coding region that leads to a frame shift and truncation of the translated protein. The gene frequency of this mutant allele in Caucasoid populations is quite high at 0.09 (meaning that about 10% of the Caucasoid population are heterozygous carriers of the allele and about 1% are homozygous). The mutant allele has not been found in Japanese or black Africans from Western or Central Africa. Heterozygous deficiency of CCR5 might provide some protection against sexual transmission of HIV infection and a modest reduction in the rate of progression of the disease. These results provide a dramatic confirmation of experimental work suggesting that CCR5 is the major macrophage and T-lymphocyte co-receptor used by HIV to establish primary infection *in vivo*, and offers the possibility that primary infection might be blocked by therapeutic antagonists of the CCR5 receptor. Indeed, there is preliminary evidence that low molecular weight inhibitors of this receptor can block infection of macrophages by HIV *in vitro*. Such low molecular weight inhibitors might be the precursors of useful drugs that could be taken by mouth.

Such drugs are very unlikely to provide complete protection against infection. A very small number of individuals who are homozygous for the non-functional variant of CCR5 are infected with HIV. These individuals seem to have suffered from primary infection by lymphocyte-tropic strains of the virus.

11-17 HIV RNA is transcribed by viral reverse transcriptase into DNA that integrates into the host cell genome.

One of the proteins that enters the cell with the viral genome is the viral reverse transcriptase, which transcribes the viral RNA into a complementary DNA (cDNA) copy. The viral cDNA is then integrated into the host cell genome by the viral integrase, which also enters the cell with the viral RNA. The integrated cDNA copy is known as the **provirus**. The infectious cycle up to the integration of the provirus is shown in Fig. 11.21. In activated CD4 T cells, virus replication is initiated by transcription of the provirus, as we shall see in the next section. However, HIV can, like other retroviruses, establish a latent infection in which the provirus remains quiescent. This seems to occur in memory CD4 T cells and in dormant macrophages, and these cells are thought to be an important reservoir of infection.

The entire HIV genome consists of nine genes flanked by long terminal repeat sequences (LTRs), which are required for the integration of the provirus into the host cell DNA and contain binding sites for gene regulatory proteins that control the expression of the viral genes. HIV shares with all other retroviruses three major genes—*gag*, *pol*, and *env*. The *gag* gene encodes the structural proteins of the viral core, *pol* encodes the enzymes involved in viral replication and integration, and *env* encodes the viral envelope glycoproteins. The *gag* and *pol* mRNAs are translated to give polyproteins—long polypeptide chains that are then cleaved by the viral protease (also encoded by *pol*) into individual functional proteins. The product of the *env* gene, gp160, has to be cleaved by a host cell protease into gp120 and gp41, which are then assembled as trimers into the **viral envelope**. The other six genes encode proteins

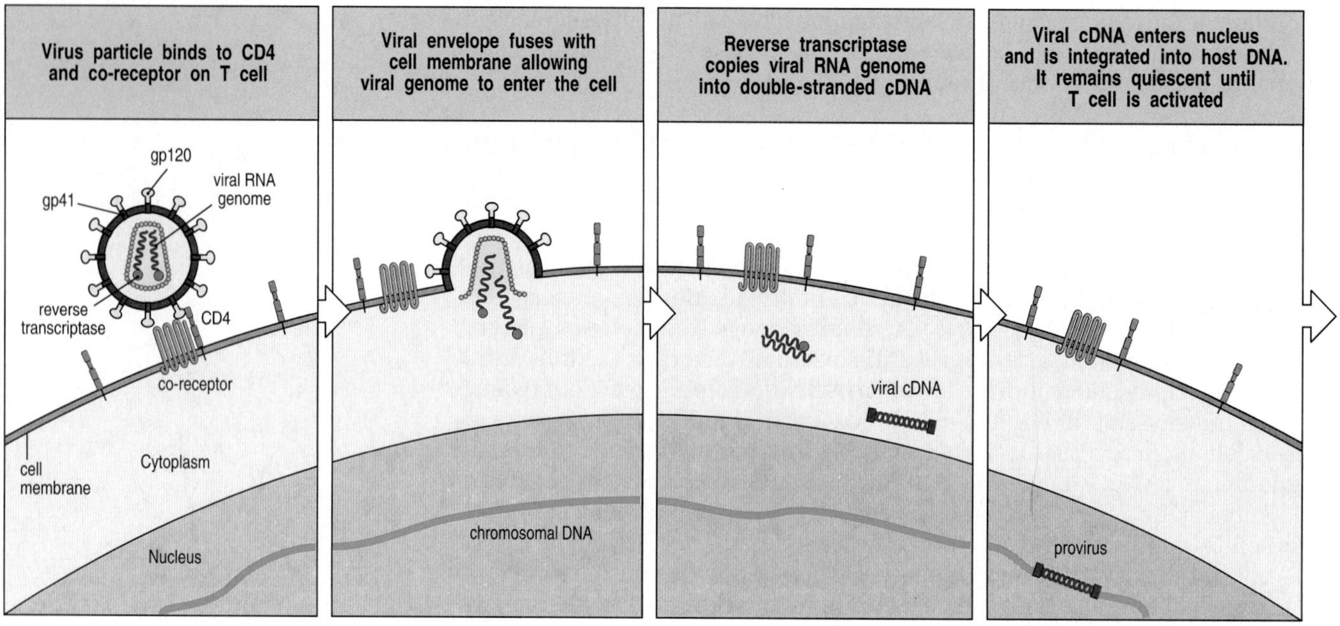
Virus particle binds to CD4 and co-receptor on T cell
Viral envelope fuses with cell membrane allowing viral genome to enter the cell
Reverse transcriptase copies viral RNA genome into double-stranded cDNA
Viral cDNA enters nucleus and is integrated into host DNA. It remains quiescent until T cell is activated
gp120
viral RNA genome
gp41
reverse transcriptase
CD4
co-receptor
cell membrane
Cytoplasm
Nucleus
chromosomal DNA
viral cDNA
provirus

Fig. 11.21 The infection of CD4 T cells by HIV. The virus binds to CD4 by using gp120, which is altered by CD4 binding so that it now also binds a specific seven-span chemokine co-receptor at the surface of the cell. This binding releases gp41, which then causes fusion of the viral envelope with the cell membrane, and the release of the viral core into the cytoplasm. Once in the cytoplasm, the viral core releases the RNA genome, which is then reverse transcribed into double-stranded cDNA. The double-stranded cDNA migrates to the nucleus in association with the viral integrase and the Vpr protein, where it is integrated into the cell genome, becoming a provirus.

that affect viral replication and infectivity in various ways that we shall discuss in the next sections. The viral genes and the functions of their products are summarized in Fig. 11.22.

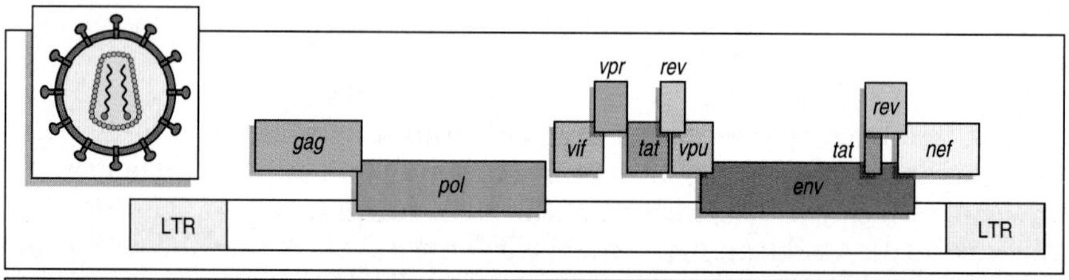
vpr
rev
gag
vif
tat
vpu
rev
tat
nef
pol
env
LTR
LTR

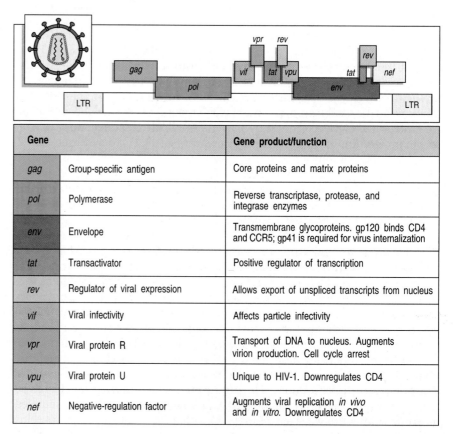

Gene		Gene product/function
gag	Group-specific antigen	Core proteins and matrix proteins
pol	Polymerase	Reverse transcriptase, protease, and integrase enzymes
env	Envelope	Transmembrane glycoproteins. gp120 binds CD4 and CCR5; gp41 is required for virus internalization
tat	Transactivator	Positive regulator of transcription
rev	Regulator of viral expression	Allows export of unspliced transcripts from nucleus
vif	Viral infectivity	Affects particle infectivity
vpr	Viral protein R	Transport of DNA to nucleus. Augments virion production. Cell cycle arrest
vpu	Viral protein U	Unique to HIV-1. Downregulates CD4
nef	Negative-regulation factor	Augments viral replication *in vivo* and *in vitro*. Downregulates CD4

Fig. 11.22 The genes and proteins of HIV-1. Like all retroviruses, HIV-1 has an RNA genome flanked by long terminal repeats (LTR) involved in viral integration and in regulation of the viral genome. The genome can be read in three frames and several of the viral genes overlap in different reading frames. This allows the virus to encode many proteins in a small genome. The three main protein products—Gag, Pol, and Env—are synthesized by all infectious retroviruses. The known functions of the different genes and their products are listed. The products of *gag*, *pol* and *env* are known to be present in the mature viral particle, together with the viral RNA. The mRNAs for Tat and Rev proteins are produced by splicing of viral transcripts, so their genes are split in the viral genome. The other gene products affect the infectivity of the virus in various ways that are not fully understood.

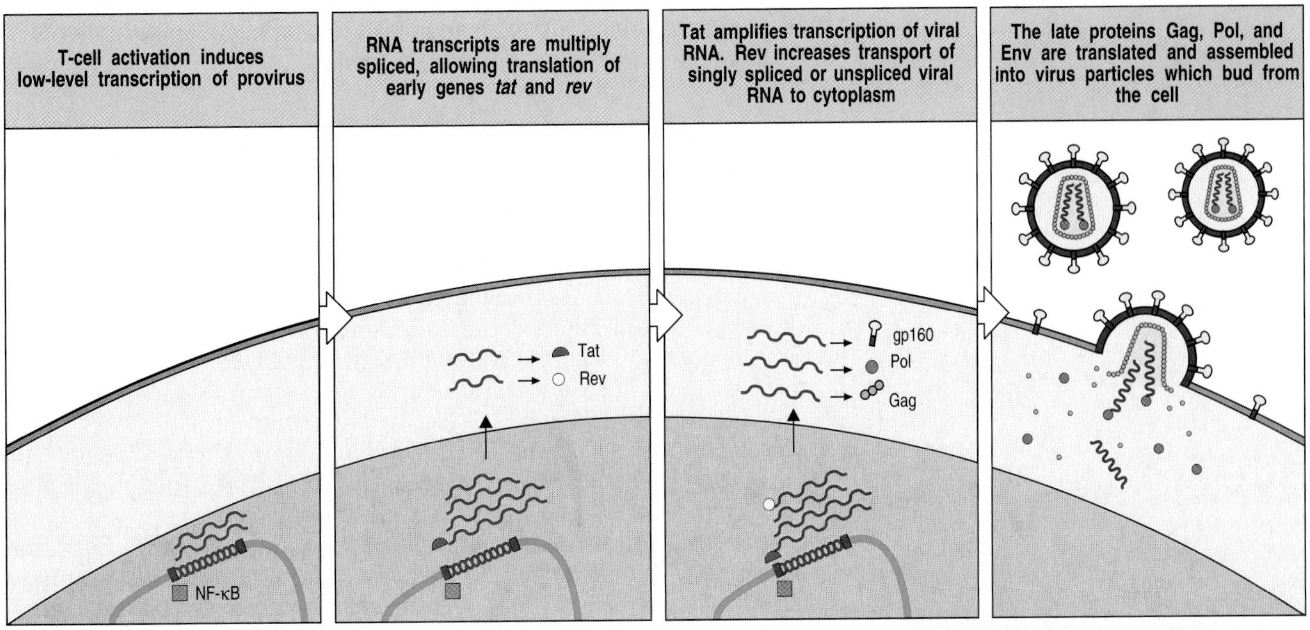

T-cell activation induces low-level transcription of provirus
RNA transcripts are multiply spliced, allowing translation of early genes *tat* and *rev*
Tat amplifies transcription of viral RNA. Rev increases transport of singly spliced or unspliced viral RNA to cytoplasm
The late proteins Gag, Pol, and Env are translated and assembled into virus particles which bud from the cell
NF-κB
Tat
Rev
gp160
Pol
Gag

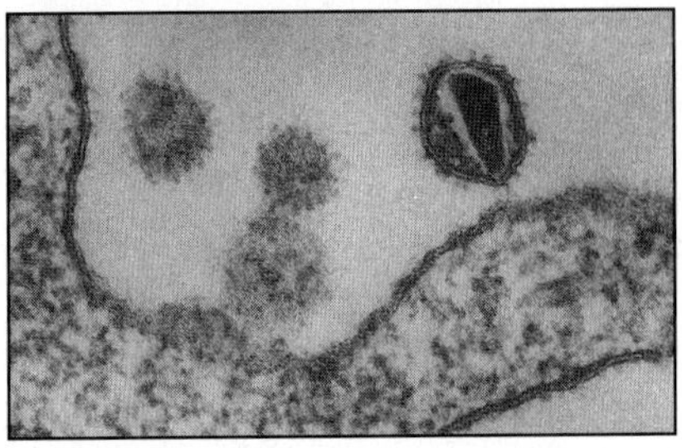

Fig. 11.23 Cells infected with HIV must be activated for the virus to replicate. Activation of CD4 T cells induces the expression of the transcription factor NF-κB, which binds to the proviral LTR and initiates the transcription of the HIV genome into RNA. The viral RNA encodes several regulatory proteins. Tat both enhances transcription from the provirus and binds to the RNA transcripts, stabilizing them in a form that can be translated. The protein Rev binds the RNA transcripts and transports them to the cytosol. The viral RNA also encodes the structural proteins that are required for the production of new viral particles. Early in the infectious cycle, levels of Rev are low and the transcript is retained in the nucleus and processed extensively, producing mRNAs encoding the regulatory proteins, such as Tat, which are required for viral replication. Later, as levels of Rev increase, Rev acts to transport less extensively spliced and unspliced transcripts out of the nucleus. The spliced and unspliced transcripts encode the structural proteins of the virus and the unspliced transcripts, which are the new viral genomes, are packaged with these to form many new virus particles. Photograph courtesy of H Gelderblom.

11-18 Transcription of the HIV provirus depends on host-cell transcription factors induced upon the activation of infected T cells.

The production of infectious virus particles from an integrated HIV provirus is stimulated by a cellular transcription factor that is present in all activated T cells. Activation of CD4 T cells induces the transcription factor NF-κB, which binds to promoters not only in the cellular DNA but also in the viral LTR, thereby initiating the transcription of viral RNA by the cellular RNA polymerase. This transcript is spliced in various ways to produce mRNAs for the viral proteins. At least two of the viral genes, *tat* and *rev*, encode proteins, **Tat** and **Rev** respectively, that promote viral replication in activated T cells. Tat is a potent transcriptional regulator that binds to an RNA sequence in the LTR known as the transcriptional activation region (TAR) and greatly enhances the rate of viral genome transcription.

The *rev* gene also has a complex function. It makes a protein that binds to a sequence called RRE (Rev responsive element) in the RNA transcript of HIV, which controls the delivery of the RNA to the cytoplasm. To express the Tat and Rev proteins, the viral transcripts must be spliced twice, whereas for other viral proteins only one splicing event is required. When the provirus is first activated, Rev levels are low, the transcripts are translocated slowly from the nucleus and thus multiple splicing events can occur. Thus more Tat and Rev are produced, and Tat in turn ensures that more viral transcripts are made. Later, when Rev levels have increased, the transcripts are translocated rapidly from the nucleus unspliced or only singly spliced. These unspliced or singly spliced transcripts are translated to produce the structural components of the viral core and envelope, together with the reverse transcriptase, the integrase, and the viral protease, all of which are needed to make new viral particles. The complete, unspliced transcripts that are exported from the nucleus late in the infectious cycle are required for the translation of *gag* and *pol* and are also destined to be packaged with the proteins as the RNA genomes of the new virus particles (Fig. 11.23).

11-19 Drugs that block HIV replication lead to a rapid decrease in titer of infectious virus and a rise in CD4 T cells.

Studies with powerful drugs that completely block the cycle of HIV replication indicate that the virus is replicating rapidly at all phases of infection, including the asymptomatic phase. Two viral proteins in particular have been the target of drugs aimed at arresting viral replication. These are the viral reverse transcriptase, which is required for synthesis of the provirus, and the viral protease, which cleaves the viral polyproteins to produce the virion proteins and viral enzymes. Inhibitors of these enzymes prevent the establishment of further infection in uninfected cells, although cells that are already infected can continue to produce virions.

Because of the great efficacy of the protease inhibitors, it is possible to learn much about the kinetics of HIV replication *in vivo* by measuring the decline in viremia after the initiation of protease inhibitor therapy. For the first 2 weeks after starting treatment there is an exponential fall in plasma virus levels with a half-life of viral decay of about 2 days (Fig. 11.24). This phase reflects the decay in virus production from cells that were actively infected at the start of drug treatment, and indicates that the half-life of productively infected cells is similarly about 2 days. The results also show that free virus is cleared from the circulation very rapidly, with a half-life of about 6 hours. After 2 weeks, levels of virus in plasma have dropped by more than 95%, representing an almost total loss of productively infected CD4 lymphocytes. After this time, the rate of decline of plasma virus levels is much slower, reflecting the very slow decay of virus production from cells that provide a longer-lived reservoir of infection, such as dendritic cells and tissue macrophages, and from activated latently infected CD4 memory T cells. Very long-term sources of infection might be CD4 memory T cells carrying integrated provirus and virus stored as immune complexes on follicular dendritic cells. These very long-lasting reservoirs of infection might prove to be resistant to drug therapy for HIV.

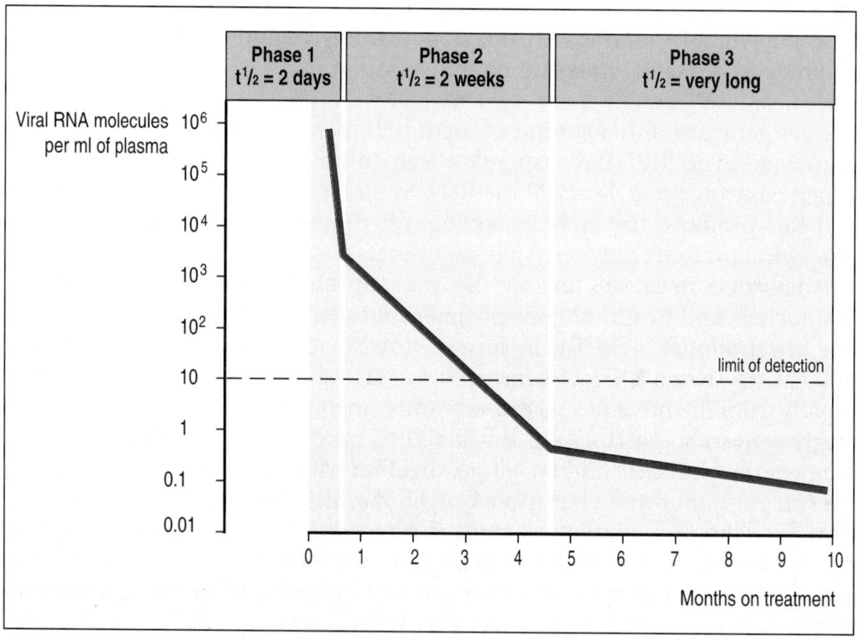

Fig. 11.24 Clearance of HIV from plasma. The production of new HIV virus particles can be arrested for prolonged periods by combinations of protease inhibitors and viral reverse transcriptase inhibitors. After the initiation of such treatment, the half-life of virus decay occurs in three phases. The first phase has a half-life of approximately 2 days and lasts for approximately 2 weeks, during which productively infected lymphocytes die and released virus is cleared from plasma. There is a decrease in virus levels in plasma of more than 95% during this first phase of viral clearance. The second phase lasts for about 6 months and has a half-life of about 2 weeks. During this phase, virus is released from infected macrophages and from resting CD4 T cells stimulated to divide and develop productive infection. It is thought that there is then a third phase of unknown length that results from the re-activation of integrated provirus in memory T cells and other long-lived reservoirs of infection. This reservoir of latently infected cells might remain present for many years. Measurement of this phase of viral decay is impossible at present as viral levels in plasma are below detectable levels.

These studies show that most of the HIV present in the circulation of an infected individual is the product of rounds of replication in newly infected cells, and that virus from these productively infected cells is released into, and rapidly cleared from, the circulation at the rate of 10^9 to 10^{10} virions every day. This raises the question of what is happening to these virus particles: how are they removed so rapidly from the circulation? It seems most likely that HIV particles are opsonized by specific antibody and complement and removed by phagocytic cells of the mononuclear phagocyte system. Opsonized HIV particles can also be trapped on the surface of follicular dendritic cells, which are known to capture antigen:antibody complexes and retain them for prolonged periods (see Chapters 9 and 10).

The other issue raised by these studies is the effect of HIV replication on the population dynamics of CD4 T cells. The decline in plasma viremia is accompanied by a steady increase in CD4 T lymphocyte counts in peripheral blood: what is the source of the new CD4 T cells that appear once treatment is started? It seems highly unlikely that they are the recent progeny of stem cells that have developed in the thymus, because CD4 T cells are not normally produced in large numbers from the thymus even at its maximum rate of production in adolescents. Some investigators believe that these cells are emerging from sites of sequestration and add little to the total numbers of CD4 T cells in the body, whereas others advocate their origin from mature CD4 T cells that replicate, and argue that the production of such cells is an ongoing process that compensates for the continual loss of productively infected CD4 T cells.

11-20 HIV accumulates many mutations in the course of infection in a single individual and drug treatment is soon followed by the outgrowth of drug-resistant variants of the virus.

The rapid replication of HIV, with the generation of 10^9 to 10^{10} virions every day, coupled with a mutation rate of approximately 3×10^{-5} per nucleotide base per cycle of replication, leads to the generation of many variants of HIV in a single infected patient in the course of an infection. Replication of a retroviral genome depends on two error-prone steps. Reverse transcriptase lacks the proofreading mechanisms associated with cellular DNA polymerases, and the RNA genomes of retroviruses are therefore copied into DNA with relatively low fidelity; the transcription of the proviral DNA into RNA copies by the cellular RNA polymerase is similarly a low-fidelity process. A rapidly replicating persistent virus that is going through these two steps repeatedly in the course of an infection can thereby accumulate many mutations, and numerous variants of HIV, sometimes called quasi-species, are found within a single infected individual. This very high variability was first recognized in HIV and has since proved to be common to the other lentiviruses.

As a consequence of its high variability, HIV rapidly develops resistance to anti-viral drugs. When anti-viral drugs are administered, variants of the virus that carry mutations conferring resistance to their effects emerge and expand until former levels of plasma virus are regained. Resistance to some of the protease inhibitors appears after only a few days (Fig. 11.25). Resistance to the reverse-transcriptase inhibitor zidovudine, the drug most widely used for treating AIDS, takes months to develop. This is because resistance to zidovudine requires three or four mutations in the viral reverse transcriptase, whereas a single mutation can confer resistance to the protease inhibitors and other reverse-transcriptase inhibitors. As a result of the relatively rapid appearance of resistance to all known anti-HIV drugs, successful drug treatment might depend on the development of a range of anti-viral drugs that can be

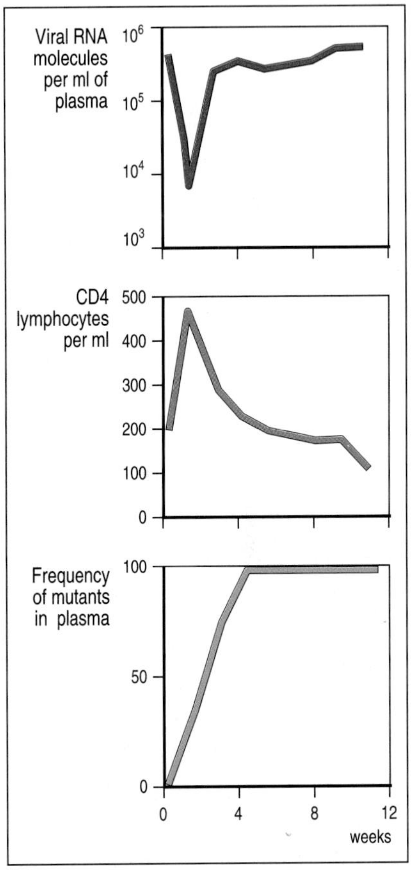

Fig. 11.25 Resistance of HIV to protease inhibitors. After the administration of a single protease inhibitor to a patient with HIV there is a precipitous fall in plasma RNA levels with a half-life of approximately 2 days (top panel). This is accompanied by an initial rise in the number of CD4 T cells in peripheral blood (middle panel). Within days of starting the drug, mutant drug-resistant variants can be detected in plasma (bottom panel) and in peripheral blood lymphocytes. After only 4 weeks of treatment, viral RNA levels and CD4 lymphocyte levels have returned to baseline levels, and 100% of plasma HIV is present as drug-resistant mutant.

used in combination. It might also be important to treat early in the course of an infection, thereby reducing the chances of a variant virus's accumulating all the necessary mutations to resist the entire cocktail.

11-21 Lymphoid tissue is the major reservoir of HIV infection.

Although viral load and turnover are usually measured by detecting the viral RNA present in viral particles in the blood, the major reservoir of HIV infection is in lymphoid tissue, in which infected CD4 T cells, monocytes, macrophages, and dendritic cells are found. In addition, HIV is trapped in the form of immune complexes on the surface of follicular dendritic cells. These cells are not themselves infected but can act as a store of infective virions.

HIV infection takes different forms within different cells. As we have seen, more than 95% of the virus that can be detected in the plasma is derived from productively infected cells, which have a very short half-life of about 2 days. Productively infected CD4 lymphocytes are found in the T-cell areas of lymphoid tissue, and these are thought to succumb to infection in the course of being activated in an immune response. Latently infected memory CD4 cells that are activated in response to antigen presentation also become productively infected. Such cells have a longer half-life of 2 to 3 weeks from the time that they are infected, and HIV can spread from these cells by rounds of replication in other activated CD4 T cells. In addition to the cells that are infected productively or latently, there is a further large population of cells infected by defective proviruses; such cells are not a source of infectious virus.

Macrophages and dendritic cells seem to be able to harbor replicating virus without necessarily being killed by it, and are therefore believed to be an important reservoir of infection, as well as a means of spreading virus to other tissues such as the brain. Although the function of macrophages as antigen-presenting cells does not seem to be compromised by HIV infection, it is thought that the virus causes abnormal patterns of cytokine secretion that could account for the wasting that commonly occurs in AIDS patients.

11-22 An immune response controls but does not eliminate HIV.

Infection with HIV generates an adaptive immune response that contains the virus but only rarely eliminates it. The time course of various elements in the adaptive immune response to HIV is shown, with the levels of infectious virus in plasma, in Fig. 11.26.

Seroconversion is the clearest evidence for an adaptive immune response to infection with HIV, but the generation of T lymphocytes responding to infected cells is thought by most workers in the field to be central in controlling the infection. Both CD8 cytotoxic T cells and T_H1 cells specifically responsive to infected cells are associated with the decline in detectable virus after the initial infection. These T-cell responses are unable to clear the infection completely and can cause some pathology. Nevertheless, there is evidence that the virus itself is cytopathic, and T-cell responses that reduce viral spread should therefore, on balance, reduce the pathology of the disease.

Strong circumstantial evidence for the destruction of infected cells by cytotoxic lymphocytes comes from studies of peripheral blood cells from infected individuals, in which cytotoxic T cells specific for viral peptides can

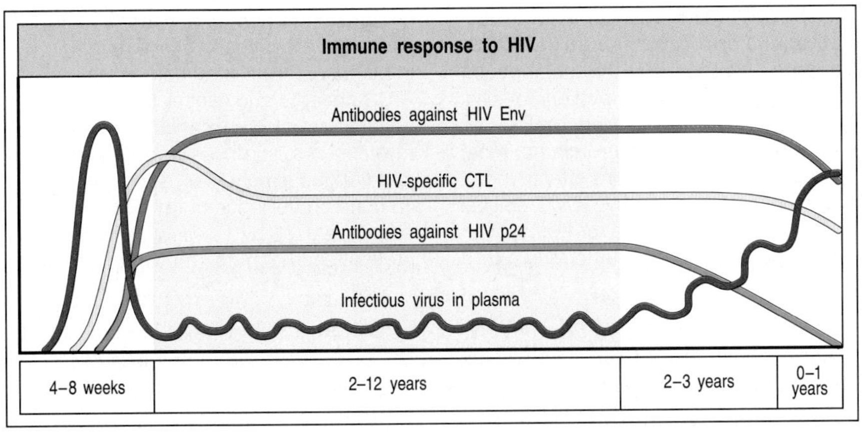

Fig. 11.26 The immune response to HIV. Infectious virus is present at relatively low levels in the peripheral blood of infected individuals during a prolonged asymptomatic phase but is replicated persistently in lymphoid tissues. During this period, CD4 T-cell counts gradually decline, although antibodies and CD8 cytotoxic T cells directed against the virus remain at high levels. Two different antibody responses are shown in the figure, one to the envelope protein (Env) of HIV, and one to the core protein p24. Eventually, the levels of antibody and HIV-specific cytotoxic T lymphocytes (CTLs) also decline, and there is a progressive increase of infectious HIV in the peripheral blood.

be shown to kill infected cells *in vitro*. Cytotoxic T cells have also been shown to invade sites of HIV replication *in vivo*. CD8 T cells specific for HIV-infected cells in a given patient often have a single dominant V_β-chain gene rearrangement that enables them to be identified. Using such a dominant V_β rearrangement as a marker for HIV-specific cytotoxic cells, these cells were shown to have infiltrated infected splenic white pulps, suggesting that the splenic white pulps are sites where the replication of HIV within infected CD4 T cells and the elimination of infected cells by cytotoxic CD8 T cells occur simultaneously. Whether such cytotoxic T-cell responses are on balance helpful or harmful is not entirely clear. In one striking case of AIDS, rapid progression of the disease was associated with the appearance of viruses expressing mutant peptides followed by the appearance of cytotoxic T cells capable of recognizing these mutant peptides and a rapid loss of CD4 T cells.

Mutations that occur as HIV replicates can allow variants of the virus to escape recognition by antibody or cytotoxic T cells and can contribute to the failure of the immune system to contain the infection in the long term. Direct escape of virus-infected cells from killing by cytotoxic T lymphocytes has been shown by the occurrence of mutations of immunodominant viral peptides presented by MHC class I molecules. In other cases, variant peptides produced by the virus have been found to act as antagonists (see Section 5-12) for T cells responsive to the wild-type epitope, thus allowing both mutant and wild-type viruses to survive. Mutant peptides acting as antagonists have also been reported in hepatitis B virus infections, and similar mutant peptides might contribute to the persistence of some viral infections, especially when, as often happens, the immune response of an individual is dominated by T cells specific for a particular epitope.

11-23 HIV infection leads to low levels of CD4 T cells, increased susceptibility to opportunistic infection, and eventually to death.

There are three dominant mechanisms for the loss of CD4 T cells in HIV infection. First, there is evidence for direct viral killing of infected cells; second, there is increased susceptibility to the induction of apoptosis in infected cells; and third, there is killing of infected CD4 T cells by CD8 cytotoxic lymphocytes that recognize viral peptides.

In addition, the binding of CD4 by gp120 can damage CD4 T cells, even if they are not actually infected with HIV. There is some experimental support for this: CD4 T cells from HIV-infected patients are more susceptible to apoptosis

Infections	
Parasites	*Toxoplasma* spp. *Cryptosporidium* spp. *Leishmania* spp. *Microsporidium* spp.
Bacteria	*Mycobacterium tuberculosis* *Mycobacterium avium intracellulare* *Salmonella* spp.
Fungi	*Pneumocystis carinii* *Cryptococcus neoformans* *Candida* spp. *Histoplasma capsulatum* *Coccidioides immitis*
Viruses	Herpes simplex Cytomegalovirus Varicella zoster
Malignancies	
Kaposi's sarcoma Non-Hodgkin's lymphoma, including EBV-positive Burkitt's lymphoma Primary lymphoma of the brain	

Fig. 11.27 A variety of opportunistic pathogens and cancers can kill AIDS patients. Infections are the major cause of death in AIDS, with respiratory infection with *Pneumocystis carinii* and mycobacteria being the most prominent. Most of these pathogens require effective macrophage activation by CD4 T cells or effective cytotoxic T cells for host defense. Opportunistic pathogens are present in the normal environment but cause severe disease primarily in immunocompromised hosts, such as AIDS patients and cancer patients. AIDS patients are also susceptible to several rare cancers, such as Kaposi's sarcoma and various lymphomas, suggesting that immune surveillance of their causative herpes viruses by T cells can normally prevent such tumors (see Chapter 14).

driven by CD4 crosslinking, and this effect can be replicated in normal CD4 T cells *in vitro* by crosslinking CD4 either with anti-CD4 or with gp120 complexed with anti-gp120. Moreover, memory CD4 T cells have been shown to be much more susceptible to inhibition through CD4 than naive T cells, perhaps accounting for the early effects of HIV infection on memory T cells.

When CD4 T-cell numbers decline below a critical level, cell-mediated immunity is lost, and infections with a variety of opportunistic microbes appear (Fig. 11.27). Typically, resistance is lost early to oral *Candida* species and to *Mycobacterium tuberculosis*, which shows as an increased prevalence of thrush (oral candidiasis) and tuberculosis. Later, patients suffer from shingles, caused by the activation of latent herpes zoster, from EBV-induced B-cell lymphomas, and from Kaposi's sarcoma, a tumor of endothelial cells that probably represents a response both to cytokines produced in the infection and to a novel herpes virus called HHV-8 that was recently identified in these lesions. *Pneumocystis carinii* pneumonia is an important opportunistic infection in patients with AIDS. In the final stages of AIDS, infection with cytomegalovirus or *Mycobacterium avium* complex is more prominent. It is important to note that not all patients with AIDS get all these infections or tumors, and there are other tumors and infections that are less prominent but still significant. Rather, this is a list of the commonest opportunistic infections and tumors, most of which are normally controlled by robust CD4 T-cell mediated immunity that wanes as the CD4 T-cell counts drop toward zero (see Fig. 11.19).

11-24 Vaccination against HIV is an attractive solution but poses many difficulties.

The development of a safe and effective vaccine for the prevention of HIV infection and AIDS is an attractive goal but its development is fraught with difficulties that have not been faced in developing vaccines against other diseases. The first problem is the nature of the infection itself, featuring a virus that proliferates extremely rapidly and causes sustained infection in the face of strong cytotoxic T-cell and antibody responses. As we discussed in Section 11-22, HIV evolves in individual patients by the selective proliferative advantage of mutant virions encoding peptide sequence changes that escape recognition by antibodies and by cytotoxic T lymphocytes. This evolution means that the development of therapeutic vaccination strategies to block the development of AIDS in HIV-infected patients will be extremely difficult. Even after the viremia has been largely cleared by drug therapy, immune responses to HIV fail to prevent drug-resistant virus from rebounding and replicating at pretreatment levels.

The second problem is our uncertainty over what form protective immunity to HIV might take. It is not known whether antibodies, cytotoxic T lymphocyte responses or both are necessary to achieve protective immunity, and which epitopes might provide the targets of protective immunity. Third, if strong cytotoxic responses are necessary to provide protection against HIV, these might be difficult to develop and sustain through vaccination. Other effective viral vaccines rely on the use of live, attenuated viruses and there are concerns over the safety of pursuing this approach for HIV. Another possible approach is the use of DNA vaccination, a technique that we discuss in Section 14-25. Both of these approaches are being tested in animal models.

The fourth problem is the ability of the virus to persist in latent form as a transcriptionally silent provirus, which is invisible to the immune system and might prevent the immune system from clearing the infection once it has been established. Thus, the ability of the immune system to clear infectious virus remains uncertain.

However, against this pessimistic background, there are grounds for hope that successful vaccines can be developed. Of particular interest are rare groups of people who have been exposed often enough to HIV to make it virtually certain that they should have become infected but who have not developed the disease. A small group of Gambian prostitutes who are estimated to have been exposed to one HIV-infected male partner each month for up to five years were found to lack antibody responses but to have cytotoxic T lymphocyte responses to a variety of peptide epitopes from HIV. These women seem to have been naturally immunized against HIV. A second 'experiment of nature' is the finding that patients infected with HIV-2 show significant protection against the more virulent HIV-1, which is the major cause of AIDS. This shows that there is some degree of cross-protection for HIV-2 and HIV-1, which are very similar structurally.

Although there is no perfect animal model for the development of HIV vaccines, one model system is based on simian immunodeficiency virus (SIV), which is closely related to HIV and infects macaques. SIV causes a similar disease to AIDS in Asian macaques such as the cynomolgus monkey, but does not cause disease in African cercopithecus monkeys such as the African green monkey, with which it has probably co-existed for up to a million years. Live attenuated SIV vaccines, lacking the *nef* gene, and hybrid HIV–SIV vaccines have been developed to test the principles of vaccination in primates, and both have proved successful in protecting primates against subsequent infection by fully virulent viruses. However, there are substantial difficulties to be overcome in the development of live attenuated HIV vaccines for use in at-risk populations, not least the worry of recombination between vaccine strains and wild-type viruses leading to reversion to a virulent phenotype. The alternative approach of DNA vaccination is being piloted in primate experiments, with some early signs of success.

Subunit vaccines, which induce immunity to only some proteins in the virus, have also been made. One such vaccine has been made from the envelope protein gp120 and has been tested on chimpanzees. This vaccine proved to be specific to the precise strain of virus used to make it, and was therefore useless in protection against natural infection. Subunit vaccines are also less efficient at inducing prolonged cytotoxic T-cell responses.

Finally, there are difficult ethical issues in the development of a vaccine. It would be unethical to conduct a vaccine trial without trying at the same time to minimize the exposure of a vaccinated population to the virus itself. However, the effectiveness of a vaccine can only be assessed in a population in which the exposure rate to the virus is high enough to assess whether vaccination is protective against infection. This means that initial vaccine trials might have to be conducted in countries where the incidence of infection is very high and public health measures have not yet succeeded in reducing the spread of HIV.

11-25 Prevention and education are one way in which the spread of HIV and AIDS can be controlled.

The one way in which we know we can protect against infection with HIV is by avoiding contact with body fluids, such as semen, blood, blood products, or milk, from people who are infected. Indeed, it has been demonstrated repeatedly that this precaution, simple enough in the developed world, is sufficient to prevent infection, as health-care workers can take care of AIDS patients for long periods without seroconversion or signs of infection.

For this strategy to work, however, two things are necessary. First, one must be able to test people at risk of infection with HIV periodically, so that they can take the steps necessary to avoid passing the virus to others. For this to work, strict confidentiality is absolutely required. One intelligent suggestion to emerge from President Reagan's AIDS Advisory Group was that before anything could be accomplished in the fight against AIDS, the rights of HIV-infected people would have to be guaranteed in all aspects of life. This step, unfortunately, was never taken.

A barrier to the control of HIV is the reluctance of individuals to find out whether they are infected, especially as one of the consequences of a positive HIV test is stigmatization by society. As a result, infected individuals can unwittingly infect many others. Balanced against this is the success of therapy with combinations of the new protease inhibitors with nucleoside inhibitors, which provides an incentive for potentially infected people to identify the presence of infection to gain the benefits of treatment. Responsibility is at the heart of AIDS prevention, and a law guaranteeing the rights of people infected with HIV might go a long way to encouraging responsible behavior. The rights of HIV-infected people are protected in The Netherlands and Sweden. The problem in the less-developed nations, where elementary health precautions are extremely difficult to establish, is more profound.

Summary.

Infection with the human immunodeficiency virus (HIV) is the cause of acquired immune deficiency syndrome (AIDS). This worldwide epidemic is now spreading at an alarming rate, especially through heterosexual contact in less-developed countries. HIV is an enveloped retrovirus that replicates in cells of the immune system. Viral entry requires the presence of CD4 and a particular chemokine receptor, and the viral cycle is dependent on transcription factors found in activated T cells. Infection with HIV causes a loss of CD4 T cells and an acute viremia that rapidly subsides as cytotoxic T-cell responses develop, but HIV infection is not eliminated by this immune response. HIV establishes a state of persistent infection in which the virus is continually replicating in newly infected cells. The current treatment consists of combinations of viral protease inhibitors together with nucleoside analogs and causes a rapid decrease in virus levels and a slower increase in CD4 T-cell counts. The main effect of HIV infection is the destruction of CD4 T cells, which occurs through the direct cytopathic effects of HIV infection and through killing by CD8 cytotoxic T cells. As the CD4 T-cell counts wane, the body becomes progressively more susceptible to opportunistic infection with intracellular microbes. Eventually, most HIV-infected individuals develop AIDS and die; some people (3–7%), however, remain healthy for many years, with no apparent ill effects of infection. We hope to be able to learn from these individuals how infection with HIV can be controlled. The study of such people gives hope that it will be possible to develop effective vaccines against HIV.

Summary to Chapter 11.

Whereas most infections elicit protective immunity, most successful pathogens have developed some means of evading a fully effective immune response, and some result in serious, persistent disease. In addition, some individuals have inherited deficiencies in different components of the immune system, making them highly susceptible to certain classes of infectious agent. Persistent infection and immunodeficiency illustrate the importance of innate and adaptive immunity in effective host defense against infection and present huge challenges for future immunological research. The human immunodeficiency virus (HIV) combines the characteristics of a persistent infectious agent with the ability to create immunodeficiency in its human host, a combination that is usually lethal to the patient. The key to fighting new pathogens like HIV is to develop our understanding of the basic properties of the immune system and its role in combating infection more fully.

General references.

Primary immunodeficiency diseases. Report of a WHO scientific group. *Clin. Exp. Immunol.* 1997, **109S1**:1-28.

Bloom. B., Zinkernagel, R.: **Immunity to infection**. *Curr. Opin. Immunol.* 1996, **8**:465-6.

Cohen, O.J., Kinter, A., Fauci, A.S.: **Host factors in the pathogenesis of HIV disease**. *Immunol. Rev.* 1997, **159**:31-48.

Fischer, A., Cavazzana-Calvo, M., De-Saint-Basile, G., DeVillartay, J.P., Di-Santo, J.P., Hivroz, C., Rieux-Laucat, F., Le-Deist, F.: **Naturally occurring primary deficiencies of the immune system**. *Annu. Rev. Immunol.* 1997, **15**:93-124.

Fischer, A., Malissen, B.: **Natural and engineered disorders of lymphocyte development**. *Science* 1998, **280**:237-43.

Kotwal, G.J.: **Microorganisms and their interaction with the immune system**. *J. Leukoc. Biol.* 1997, **62**:415-29.

Rosen, F.S., Cooper, M.D., and Wedgwood, R.J.: **The primary immunodeficiencies**. *N. Engl. J. Med.* 1995, **333**:431-440.

Royce, R.A., Sena, A., Cates, W., Jr., Cohen, M.S.: **Sexual transmission of HIV**. *N. Engl. J. Med.* 1997, **336**:1072-8.

Section references.

11-1 Antigenic variation can allow pathogens to escape from immunity.

Clegg, S., Hancox, L.S., and Yeh, K.S.:***Salmonella typhimurium* fimbrial phase variation and FimA expression.** *J. Bacteriol.* 1996, **178**:542-545.

Cossart, P.: **Host/pathogen interactions. Subversion of the mammalian cell cytoskeleton by invasive bacteria.** *J. Clin. Invest.* 1997, **99**:2307-11.

Donelson, J.E., Hill, K.L., El-Sayed, N.M.: **Multiple mechanisms of immune evasion by African trypanosomes**. *Mol. Biochem. Parasitol.* 1998, **91**:51-66.

Ito, T., Couceiro, J.N., Kelm, S., Baum, L.G., Krauss, S., Castrucci, M.R., Donatelli, I., Kida, H., Paulson, J.C., Webster, R.G., Kawaoka, Y.: **Molecular basis for the generation in pigs of influenza A viruses with pandemic potential.** *J. Virol.* 1998, **72**:7367-73.

Rudenko, G., Cross, M., Borst, P.: **Changing the end: antigenic variation orchestrated at the telomeres of African trypanosomes**. *Trends Microbiol.* 1998, **6**:113-6.

Seifert, H.S., Wright, C.J., Jerse, A.E., Cohen, M.S., and Cannon, J.G.: **Multiple gonococcal pilin antigenic variants are produced during experimental human infections**. *J. Clin. Invest.* 1994, **93**:2744-2749.

Shu, L.L., Bean, W.J., and Webster, R.G.: **Analysis of the evolution and variation of the human influenza A virus nucleoprotein gene from 1933 to 1990.** *J. Virol.* 1993, **67**:2723-2729.

Webster, R.G., Bean, W.J., Gorman, O.T., Chambers, T.M., and Kawaoka, Y.: **Evolution and ecology of influenza A viruses**. *Microbiol. Rev.* 1992, **56**:152-179.

11-2 Some viruses persist *in vivo* by ceasing to replicate until immunity wanes.

Bruggeman, C.A.: **Cytomegalovirus and latency: an overview**. *Virchows Arch. B. Cell Pathol. Incl. Mol. Pathol.* 1993, **64**:325-333.

Ehrlich, R.: **Selective mechanisms utilized by persistent and oncogenic viruses to interfere with antigen processing and presentation**. *Immunol. Res.* 1995, **14**:77-97.

Garcia Blanco, M.A. and Cullen, B.R.: **Molecular basis of latency in pathogenic human viruses**. *Science* 1991, **254**:815-820.

Ho, D.Y.: **Herpes simplex virus latency: molecular aspects**. *Prog. Med Virol.* 1992, **39**:76-115.

Kadin, M.E.: **Pathology of Hodgkin's disease**. *Curr. Opin. Oncol.* 1994, **6**:456-463.

Longnecker, R., Miller, C.L.: **Regulation of Epstein–Barr virus latency by latent membrane protein 2.** *Trends Microbiol.* 1996, **4**:38-42.

Steiner, I. and Kennedy, P.G.: **Molecular biology of herpes simplex virus type 1 latency in the nervous system**. *Mol. Neurobiol.* 1993, **7**:137-159.

11-3 Some pathogens resist destruction by host defense mechanisms or exploit them for their own purposes.

Hengel, H., Koszinowski, U.H.: **Interference with antigen processing by viruses**. *Curr. Opin. Immunol.* 1997, **9**:470-6.

Nocton, J.J., and Steere, A.C.: **Lyme disease**. *Adv. Intern. Med.* 1995, **40**:69-117.

Sinai, A.P., Joiner, K.A.: **Safe haven: the cell biology of nonfusogenic pathogen vacuoles**. *Annu. Rev. Microbiol.* 1997, **51**:415-62.

Smith, G.L.: **Virus proteins that bind cytokines, chemokines or interferons**. *Curr. Opin. Immunol.* 1996, **8**:467-71.

Smith, G.L., Symons, J.A., Khanna, A., Vanderplasschen, A., Alcami, A.: **Vaccinia virus immune evasion**. *Immunol. Rev.* 1997, **159**:137-54.

11-4 Immunosuppression or inappropriate immune responses can contribute to persistent disease.

Bhardwaj, N.: **Interactions of viruses with dendritic cells: a double-edged sword**. *J. Exp. Med.* 1997, **186**:795-9.

Bloom, B.R., Modlin, R.L., and Salgame, P.: **Stigma variations: observations on suppressor T cells and leprosy**. *Annu. Rev. Immunol.* 1992, **10**:453-488.

Fleischer, B.: **Superantigens**. *APMIS* 1994, **102**:3-12.

Salgame, P., Abrams, J.S., Clayberger, C., Goldstein, H., Convit, J., Modlin, R.L., and Bloom, B.R.: **Differing lymphokine profiles of functional subsets of human CD4 and CD8 T cell clones**. *Science* 1991, **254**:279-282.

11-5 Immune responses can contribute directly to pathogenesis.

Cheever, A.W., Yap, G.S.: **Immunologic basis of disease and disease regulation in schistosomiasis**. *Chem. Immunol.* 1997, **66**:159-176.

Doherty, P.C., Topham, D.J., Tripp, R.A., Cardin, R.D., Brooks, J.W., Stevenson, P.G.: **Effector CD4+ and CD8+ T-cell mechanisms in the control of respiratory virus infections**. *Immunol. Rev.* 1997, **159**:105-117.

Openshaw, P.J.: **Immunopathological mechanisms in respiratory syncytial virus disease**. *Springer Semin. Immunopathol.* 1995, **17**:187-201.

Ross, R.: **Mouse mammary tumor virus and its interaction with the immune system**. *Immunol. Res.* 1998; **17**:209-216.

11-6 Inherited immunodeficiency diseases are caused by recessive gene defects.

Fischer, A.: **Inherited disorders of lymphocyte development and function**. *Curr. Opin. Immunol.* 1996, **8**:445-447.

Kokron, C.M., Bonilla, F.A., Oettgen, H.C., Ramesh, N., Geha, R.S., Pandolfi, F.: **Searching for genes involved in the pathogenesis of primary immunodeficiency diseases: lessons from mouse knockouts**. *J. Clin. Immunol.* 1997, **17**:109-126.

Smart, B.A., Ochs, H.D.: **The molecular basis and treatment of primary immunodeficiency disorders**. *Curr. Opin. Pediatr.* 1997, **9**:570-576.

Smith, C.I., Notarangelo, L.D.: **Molecular basis for X-linked immunodeficiencies**. *Adv. Genet.* 1997, **35**:57-115.

11-7 The main effect of low levels of antibody is an inability to clear extracellular bacteria.

Bruton, O.C.: **Agammaglobulinemia.** *Pediatrics* 1952, **9**:722-728.

Desiderio, S.: **Role of Btk in B cell development and signaling.** *Curr. Opin. Immunol.* 1997, **9**:534-40.

Burrows, P.D., Cooper, M.D.: **IgA deficiency**. *Adv. Immunol.* 1997, **65**:245-76.

Fuleihan, R., Ramesh, N., and Geha, R.S.: **X-linked agammaglobulinemia and immunoglobulin deficiency with normal or elevated IgM: immunodeficiencies of B cell development and differentiation**. *Adv. Immunol.* 1995, **60**:37-56.

Lee, M.L., Gale, R.P., Yap, P.L.: **Use of intravenous immunoglobulin to prevent or treat infections in persons with immune deficiency**. *Annu. Rev. Med.* 1997, **48**:93-102.

Notarangelo, L.D.: **Immunodeficiencies caused by genetic defects in protein kinases**. *Curr. Opin. Immunol.* 1996, **8**:448-53.

Ochs, H.D., and Wedgwood, R.J.: **IgG subclass deficiencies**. *Annu. Rev. Med.* 1987, **38**:325-340.

Preud'homme, J.L., and Hanson, L.A.: **IgG subclass deficiency**. *Immunodefic. Rev.* 1990, **2**:129-149.

11-8 T-cell defects can result in low antibody levels.

Ramesh, N., Seki, M., Notarangelo, L.D., Geha, R.S.: **The hyper-IgM (HIM) syndrome**. *Springer Semin. Immunopathol.* 1998, **19**:383-9.

11-9 Defects in complement components cause defective humoral immune function and persistence of immune complexes.

Botto, M., Dell'Agnola, C., Bygrave, A.E., Thompson, E.M., Cook, H.T., Petry, F., Loos, M., Pandolfi, P.P., Walport, M.J.: **Homozygous C1q deficiency causes glomerulonephritis associated with multiple apoptotic bodies.** *Nat. Genet.* 1998, **19**:56-9.

Colten, H.R., and Rosen, F.S.: **Complement deficiencies**. *Annu. Rev. Immunol.* 1992, **10**:809-834.

Morgan, B.P., and Walport, M.J.: **Complement deficiency and disease**. *Immunol. Today* 1991, **12**:301-306.

11-10 Defects in phagocytic cells permit widespread bacterial infections.

Fischer, A., Lisowska Grospierre, B., Anderson, D.C., and Springer, T.A.: **Leukocyte adhesion deficiency: molecular basis and functional consequences**. *Immunodefic. Rev.* 1988, **1**:39-54.

Jackson, S.H., Gallin, J.I., and Holland, S.M.: **The p47phox mouse knock-out model of chronic granulomatous disease**. *J. Exp. Med.* 1995, **182**:751-758.

Karsan, A., Cornejo, C.J., Winn, R.K., Schwartz, B.R., Way, W., Lannir, N., Gershoni-Baruch, R., Etzioni, A., Ochs, H.D., Harlan, J.M.: **Leukocyte Adhesion Deficiency Type II is a generalized defect of de novo GDP-fucose biosynthesis. Endothelial cell fucosylation is not required for neutrophil rolling on human nonlymphoid endothelium**. *J. Clin. Invest.* 1998, **101**:2438-45.

Malech, H.L., Nauseef, W.M.: **Primary inherited defects in neutrophil function: etiology and treatment**. *Semin. Hematol.* 1997, **34**:279-90.

Rotrosen, D., and Gallin, J.I.: **Disorders of phagocyte function**. *Annu. Rev. Immunol.* 1987, **5**:127-150.

Spritz, R.A.: **Genetic defects in Chediak-Higashi syndrome and the beige mouse**. *J. Clin. Immunol.* 1998, **18**:97-105.

11-11 Defects in T-cell function result in severe combined immunodeficiencies.

Bosma, M.J., and Carroll, A.M.: **The SCID mouse mutant: definition, characterization, and potential uses**. *Annu. Rev. Immunol.* 1991, **9**:323-350.

Grusby, M.J., and Glimcher, L.H.: **Immune responses in MHC class II-deficient mice.** *Annu. Rev. Immunol.* 1995, **13**:417-435.

Hirschhorn, R.: **Adenosine deaminase deficiency: molecular basis and recent developments.** *Clin. Immunol. Immunopathol.* 1995, **76**:S219-S227.

Lavin, M.F., Shiloh, Y.: **The genetic defect in ataxia-telangiectasia.** *Annu. Rev. Immunol.* 1997, **15**:177-202.

Schwarz, K., Bartram, C.R.: **V(D)J recombination pathology.** *Adv. Immunol.* 1996, **61**:285-326.

Steimle, V., Reith, W., Mach, B.: **Major histocompatibility complex class II deficiency: a disease of gene regulation.** *Adv. Immunol.* 1996, **61**:327-40.

11-12 Defective T-cell signaling, cytokine production or cytokine action can cause immunodeficiency.

Arnaiz Villena, A., Timon, M., Corell, A., Perez Aciego, P., Martin Villa, J.M., and Regueiro, J.R.: **Brief report: primary immunodeficiency caused by mutations in the gene encoding the CD3-gamma subunit of the T-lymphocyte receptor.** *N. Engl. J. Med.* 1992, **327**:529-533.

Castigli, E., Pahwa, R., Good, R.A., Geha, R.S., and Chatila, T.A.: **Molecular basis of a multiple lymphokine deficiency in a patient with severe combined immunodeficiency.** *Proc. Natl. Acad. Sci. USA* 1993, **90**:4728-4732.

DiSanto, J.P., Keever, C.A., Small, T.N., Nicols, G.L., O'Reilly, R.J., and Flomenberg, N.: **Absence of interleukin 2 production in a severe combined immunodeficiency disease syndrome with T cells.** *J. Exp. Med.* 1990, **171**:1697-1704.

DiSanto, J.P., Rieux Laucat, F., Dautry Varsat, A., Fischer, A., and de Saint Basile, G.: **Defective human interleukin 2 receptor gamma chain in an atypical X chromosome-linked severe combined immunodeficiency with peripheral T cells.** *Proc. Natl. Acad. Sci. USA* 1994, **91**:9466-9470.

Kirchhausen, T. and Rosen, F.S.: **Disease mechanisms: unravelling Wiskott–Aldrich syndrome.** *Curr. Biol.* 1996, **6**:676-678.

Leonard, W.J.: **The molecular basis of X linked severe combined immunodeficiency.** *Annu. Rev. Med.* 1996, **47**:229-239.

Levin, M., Newport, M.: **Unravelling the genetic basis of susceptibility to mycobacterial infection.** *J. Pathol.* 1997, **181**:5-7.

Ochs, H.D.: **The Wiskott–Aldrich syndrome.** *Springer Semin. Immunopathol.* 1998, **9**:435-58.

Snapper, S.B., Rosen, F.S., Mizoguchi, E., Cohen, P., Khan, W., Liu, C.H., Hagemann, T.L., Kwan, S.P., Ferrini, R., Davidson, L., Bhan, A., Alt, F.W.: **Wiskott–Aldrich syndrome protein-deficient mice reveal a role for WASP in T but not B cell activation.** *Immunity.* 1998, **9**:81-91.

11-13 Bone marrow transplantation or gene therapy can be useful to correct genetic defects.

Anderson, W.F.: **Human gene therapy.** *Nature* 1998, **392**:25-30.

Blaese, R.M., Culver, K.W., Miller, A.D., Carter, C.S., Fleisher, T., Clerici, M., Shearer, G., Chang, L., Chiang, Y., Tolstoshev, P., Greenblatt, J.J., Rosenberg, S.A., Klein, H., Berger, M., Mullen, C.A., Ramsey, W.J., Muul, L., Morgan, R.A., and Anderson, W.F.: **T lymphocyte-directed gene therapy for ADA- SCID: initial trial results after 4 years.** *Science* 1995, **270**:475-480.

Candotti, F., Blaese, R.M.: **Gene therapy of primary immunodeficiencies.** *Springer Semin. Immunopathol.* 1998, **19**:493-508.

Cournoyer, D., and Caskey, C.T.: **Gene therapy of the immune system.** *Annu. Rev. Immunol.* 1993, **11**:297-329.

Fischer, A., Haddad, E., Jabado, N., Casanova, J.L., Blanche, S., Le Deist, F., Cavazzana-Calvo, M.: **Stem cell transplantation for immunodeficiency.** *Springer Semin. Immunopathol.* 1998, **19**:479-492.

Onodera, M., Ariga, T., Kawamura, N., Kobayashi, I., Ohtsu, M., Yamada, M., Tame, A., Furuta, H., Okano, M., Matsumoto, S., Kotani, H., McGarrity, G.J., Blaese, R.M., Sakiyama, Y.: **Successful peripheral T-lymphocyte-directed gene transfer for a patient with severe combined immune deficiency caused by adenosine deaminase deficiency.** *Blood* 1998, **91**:30-36.

11-14 Most individuals infected with HIV progress over time to AIDS.

Baltimore, D.: **Lessons from people with nonprogressive HIV infection.** *N. Engl. J. Med.* 1995, **332**:259-260.

Barre-Sinoussi, F.: **HIV as the cause of AIDS.** *Lancet* 1996, **348**:31-5.

Kirchhoff, F., Greenough, T.C., Brettler, D.B., Sullivan, J.L., and Desrosiers, R.C.: **Brief report: absence of intact nef sequences in a long-term survivor with nonprogressive HIV-1 infection.** *N. Engl. J. Med.* 1995, **332**:228-232.

Pantaleo, G., Menzo, S., Vaccarezza, M., Graziosi, C., Cohen, O.J., Demarest, J.F., Montefiori, D., Orenstein, J.M., Fox, C., Schrager, L.K., Margolick, J.B., Buchbinder, S., Giorgi, J.V., amd Fauci, A.S.: **Studies in subjects with long-term nonprogressive human immunodeficiency virus infection.** *N. Engl. J. Med.* 1995, **332**:209-216.

Peckham, C. and Gibb, D.: **Mother-to-child transmission of the human immunodeficiency virus.** *N. Engl. J. Med.* 1995, **333**:298-302.

Volberding, P.A.: **Age as a predictor of progression in HIV infection.** *Lancet* 1996, **347**:1569-70.

Wang, W.K., Essex, M., McLane, M.F., Mayer, K.H., Hsieh, C.C., Brumblay, H.G., Seage, G., and Lee, T.H.R.: **Pattern of gp120 sequence divergence linked to a lack of clinical progression in human immunodeficiency virus type 1 infection.** *Proc. Natl. Acad. Sci. USA* 1996, **93**:6693-6697.

11-15 HIV is a retrovirus that infects CD4 T cells, dendritic cells, and macrophages.

Chan, D.C., Kim, P.S.: **HIV entry and its inhibition.** *Cell* 1998, **93**:681-4.

Connor, R.I., Sheridan, K.E., Ceradini, D., Choe, S., Landau, N.R.: **Change in coreceptor use coreceptor use correlates with disease progression in HIV-1—infected individuals.** *J. Exp. Med.* 1997, **185**:621-8.

Grouard, G., Clark, E.A.: **Role of dendritic and follicular dendritic cells in HIV infection and pathogenesis.** *Curr. Opin. Immunol.* 1997, **9**:563-7.

Moore, J.P., Trkola, A., Dragic, T.: **Co-receptors for HIV-1 entry.** *Curr. Opin. Immunol.* 1997, **9**:551-62.

Unutmaz, D., Littman, D.R.: **Expression pattern of HIV-1 coreceptors on T cells: implications for viral transmission and lymphocyte homing.** *Proc. Natl. Acad. Sci. USA* 1997, **94**:1615-8.

Wyatt, R., Sodroski, J.: **The HIV-1 envelope glycoproteins: fusogens, antigens, and immunogens.** *Science* 1998, **280**:1884-8.

11-16 Genetic deficiency of the macrophage chemokine co-receptor for HIV confers resistance to HIV infection *in vivo.*

Liu, R., Paxton, W.A., Choe, S., Ceradini, D., Martin, S.R., Horuk, R., Macdonald, M.E., Stuhlmann, H., Koup, R.A., and Landau, N.R.: **Homozygous defect in HIV 1 coreceptor accounts for resistance of some multiply exposed individuals to HIV 1 infection.** *Cell* 1996, **86**:367-377.

Murakami, T., Nakajima, T., Koyanagi, Y., Tachibana, K., Fujii, N., Tamamura, H., Yoshida, N., Waki, M., Matsumoto, A., Yoshie, O., Kishimoto, T., Yamamoto, N., Nagasawa, T.: **A small molecule CXCR4 inhibitor that blocks T cell line-tropic HIV-1 infection.** *J. Exp. Med.* 1997, **186**:1389-93.

Nolan, G.P.: **Harnessing viral devices as pharmaceuticals: fighting HIV-1's fire with fire.** *Cell* 1997, **90**:821-4.

Samson, M., Libert, F., Doranz, B.J., Rucker, J., Liesnard, C., Farber, C.M., Saragosti, S., Lapoumeroulie, C., Cognaux, J., Forceille, C., Muyldermans, G., Verhofstede, C., Burtonboy, G., Georges, M., Imai, T., Rana, S., Yi, Y.J., Smyth, R.J., Collman, R.G., Doms, R.W., Vassart, G., and Parmentier, M.R.: **Resistance to HIV 1 infection in Caucasian individuals bearing mutant alleles of the CCR 5 chemokine receptor gene.** *Nature* 1996, **382**:722-725.

Yang, A.G., Bai, X., Huang, X.F., Yao, C., Chen, S.: **Phenotypic knockout of HIV type 1 chemokine coreceptor CCR-5 by intrakines as potential therapeutic approach for HIV-1 infection.** *Proc. Natl. Acad. Sci. USA* 1997, **94**:11567-72.

11-17 HIV RNA is transcribed by viral reverse transcriptase into DNA that integrates into the host cell genome.

Andrake, M.D., and Skalka, A.M.R.: **Retroviral integrase, putting the pieces together**. *J. Biol. Chem.* 1995, **271**:19633-19636.

Baltimore, D.: **The enigma of HIV infection**. *Cell* 1995, **82**:175-176.

McCune, J.M.: **Viral latency in HIV disease**. *Cell* 1995, **82**:183-188.

11-18 Transcription of the HIV provirus depends on host-cell transcription factors induced upon activation of infected T cells..

Cullen, B.R.: **HIV-1 auxiliary proteins: making connections in a dying cell**. *Cell* 1998, **93**:685-92

Emerman, M., Malim, M.H.: **HIV-1 regulatory/accessory genes: keys to unraveling viral and host cell biology**. *Science* 1998, **280**:1880-4.

Kinoshita, S., Su, L., Amano, M., Timmerman, L.A., Kaneshima, H., Nolan, G.P.: **The T cell activation factor NF-ATc positively regulates HIV-1 replication and gene expression in T cells**. *Immunity* 1997, **6**:235-44.

Subbramanian, R.A., Cohen, E.A.: **Molecular biology of the human immunodeficiency virus accessory proteins**. *J. Virol.* 1994, **68**:6831-5.

Trono, D.: **HIV accessory proteins: leading roles for the supporting cast**. *Cell* 1995, **82**:189-192.

11-19 Drugs that block HIV replication lead to a rapid decrease in titer of infectious virus and a rise in CD4 T cells.

Ho, D.D.: **Perspectives series: host/pathogen interactions. Dynamics of HIV-1 replication *in vivo*.** *J. Clin. Invest.* 1997, **99**:2565-7.

Lipsky, J.J.: **Antiretroviral drugs for AIDS**. *Lancet* 1996, **348**:800-3.

Wei, X., Ghosh, S.K., Taylor, M.E., Johnson, V.A., Emini, E.A., Deutsch, P., Lifson, J.D., Bonhoeffer, S., Nowak, M.A., Hahn, B.H., Saag, M.S., and Shaw, G.M.: **Viral dynamics in human immunodeficiency virus type 1 infection**. *Nature* 1995, **373**:117-122.

11-20 HIV accumulates many mutations in the course of infection in a single individual and drug treatment is soon followed by the outgrowth of drug-resistant variants of the virus.

Bonhoeffer, S., May, R.M., Shaw, G.M., Nowak, M.A.: **Virus dynamics and drug therapy**. *Proc. Natl. Acad. Sci. USA* 1997, **94**:6971-6.

Coffin, J.M.: **HIV population dynamics in vivo: implications for genetic variation, pathogenesis, and therapy**. *Science* 1995, **267**:483-489.

Condra, J.H., Schleif, W.A., Blahy, O.M., Gabryelski, L.J., Graham, D.J., Quintero, J.C., Rhodes, A., Robbins, H.L., Roth, E., Shivaprakash, M., Titus, D., Yang, T., Teppler, H., Squires, K.E., Deutsch, P.J., and Emini, E.A.: ***In vivo* emergence of HIV-1 variants resistant to multiple protease inhibitors**. *Nature* 1995, **374**:569-571.

Katzenstein, D.: **Combination therapies for HIV infection and genomic drug resistance**. *Lancet* 1997, **350**:970-1.

Moutouh, L., Corbeil, J., and Richman, D.D.: **Recombination leads to the rapid emergence of HIV 1 dually resistant mutants under selective drug pressure**. *Proc. Natl. Acad. Sci. USA* 1996, **93**:6106-6111.

11-21 Lymphoid tissue is the major reservoir of HIV infection.

Burton, G.F., Masuda, A., Heath, S.L., Smith, B.A., Tew, J.G., Szakal, A.K.: **Follicular dendritic cells (FDC) in retroviral infection: host/pathogen perspectives**. *Immunol. Rev.* 1997, **156**:185-97.

Cameron, P., Pope, M., Granellipiperno, A., and Steinman, R.M.: **Dendritic cells and the replication of HIV 1**. *J. Leuk. Biol.* 1996, **59**:158-171.

Chun, T.W., Carruth, L., Finzi, D., Shen, X., DiGiuseppe, J.A., Taylor, H., Hermankova, M., Chadwick, K., Margolick, J., Quinn, T.C., Kuo, YH., Brookmeyer, R., Zeiger, M.A., Barditch-Crovo, P., Siliciano, R.F.: **Quantification of latent tissue reservoirs and total body viral load in HIV-1 infection.** *Nature* 1997, **387**:183-8.

Clark, E.A.: **HIV: dendritic cells as embers for the infectious fire**. *Curr. Biol.* 1996, **6**:655-7.

Dianzani, F., Antonelli, G., Riva, E., Uccini, S., and Visco, G.: **Plasma HIV viremia and viral load in lymph nodes**. *Nat. Med.* 1996, **2**:832-833.

Finzi, D., Hermankova, M., Pierson, T., Carruth, LM., Buck, C., Chaisson, R.E., Quinn, T.C., Chadwick, K., Margolick, J., Brookmeyer, R., Gallant, J., Markowitz, M., Ho, D.D., Richman, D.D., Siliciano, R.F.: **Identification of a reservoir for HIV-1 in patients on highly active antiretroviral therapy.** *Science* 1997, **278**:1295-300.

Haase, A.T., Henry, K., Zupancic, M., Sedgewick, G., Faust, R.A., Melroe, H., Cavert, W., Gebhard, K., Staskus, K., Zhang, Z.Q., Dailey, P.J., Balfour, HH., Jr., Erice, A., Perelson, A.S.: **Quantitative image analysis of HIV-1 infection in lymphoid tissue**. *Science* 1996, **274**:985-9.

Knight, S.C., Patterson, S.: **Bone marrow-derived dendritic cells, infection with human immunodeficiency virus, and immunopathology**. *Annu. Rev. Immunol.* 1997, **15**:593-615.

Orenstein, J.M., Fox, C., Wahl, S.M.: **Macrophages as a source of HIV during opportunistic infections**. *Science* 1997, **276**:1857-61.

Wong, J.K., Hezareh, M., Gunthard, H.F., Havlir, D.V., Ignacio, C.C., Spina, C.A., Richman, D.D.: **Recovery of replication-competent HIV despite prolonged suppression of plasma viremia.** *Science* 1997, **278**:1291-5.

11-22 An immune response controls but does not eliminate HIV.

Bevan, M.J., and Braciale, T.J.: **Why can't cytotoxic T cells handle HIV?** *Proc. Natl. Acad. Sci. USA* 1995, **92**:5765-5767.

Goulder, P.J., Sewell, A.K., Lalloo, D.G., Price, D.A., Whelan, J.A., Evans, J., Taylor, G.P., Luzzi, G., Giangrande, P., Phillips, R.E., McMichael, A.J.: **Patterns of immunodominance in HIV-1-specific cytotoxic T lymphocyte responses in two human histocompatibility leukocyte antigens (HLA)-identical siblings with HLA-A*0201 are influenced by epitope mutation**. *J. Exp. Med.* 1997, **185**:1423-33.

McMichael, A.J., Phillips, R.E.: **Escape of human immunodeficiency virus from immune control**. *Annu. Rev. Immunol.* 1997, **15**:271-96.

Moss, P.A., Rowland Jones, S.L., Frodsham, P.M., McAdam, S., Giangrande, P., McMichael, A.J., and Bell, J.I.: **Persistent high frequency of human immunodeficiency virus-specific cytotoxic T cells in peripheral blood of infected donors**. *Proc. Natl. Acad. Sci. USA* 1995, **92**:5773-5777.

Oldstone, M.B.: **HIV versus cytotoxic T lymphocytesÑthe war being lost.** *N. Engl. J. Med.* 1997, **337**:1306-8.

Price, D.A., Goulder, P.J., Klenerman, P., Sewell, A.K., Easterbrook, P.J., Troop, M., Bangham, C.R., Phillips, R.E.: **Positive selection of HIV-1 cytotoxic T lymphocyte escape variants during primary infection**. *Proc. Natl. Acad. Sci. USA* 1997, **94**:1890-5.

Sattentau, Q.J.: **Neutralization of HIV 1 by antibody**. *Curr. Opin. Immunol.* 1996, **8**:540-545.

11-23 HIV infection leads to low levels of CD4 T cells, increased susceptibility to opportunistic infection, and eventually to death.

Badley, A.D., Dockrell, D., Simpson, M., Schut, R., Lynch, D.H., Leibson, P., Paya, C.V.: **Macrophage-dependent apoptosis of CD4+ T lymphocytes from HIV-infected individuals is mediated by FasL and tumor necrosis factor**. *J. Exp. Med.* 1997, **185**:55-64.

Ho, D.D., Neumann, A.U., Perelson, A.S., Chen, W., Leonard, J.M., and Markowitz, M.: **Rapid turnover of plasma virions and CD4 lymphocytes in HIV-1 infection**. *Nature* 1995, **373**:123-126.

Katlama, C., and Dickinson, G.M.: **Update on opportunistic infections**. *AIDS* 1993, **7S1**:S185-S194.

Kedes, D.H., Operskalski, E., Busch, M., Kohn, R., Flood, J., and Ganem, D.R.: **The seroepidemiology of human herpesvirus 8 (Kaposi's sarcoma associated**

herpesvirus): distribution of infection in KS risk groups and evidence for sexual transmission. *Nat. Med.* 1996, **2**:918-924.

Kolesnitchenko, V., Wahl, L.M., Tian, H., Sunila, I., Tani, Y., Hartmann, D.P., Cossman, J., Raffeld, M., Orenstein, J., Samelson, L.E., and Cohen, D.I.: **Human immunodeficiency virus 1 envelope-initiated G2-phase programmed cell death**. *Proc. Natl. Acad. Sci. USA* 1995, **92**:11889-11893.

Miller, R.: **HIV-associated respiratory diseases**. *Lancet* 1996, **348**:307-12.

Pantaleo, G. and Fauci, A.S.: **Apoptosis in HIV infection**. *Nat. Med.* 1995, **1**:118-120.

Zhong, W.D., Wang, H., Herndier, B., and Ganem, D.R.: **Restricted expression of Kaposi sarcoma associated herpesvirus (human herpesvirus 8) genes in Kaposi sarcoma**. *Proc. Natl. Acad. Sci. USA* 1996, **93**:6641-6646.

11-24 Vaccination against HIV is an attractive solution but poses many difficulties.

Bangham, C.R., Phillips, R.E.: **What is required of an HIV vaccine?** *Lancet* 1997, **350**:1617-1621.

Burton, D.R.: **A vaccine for HIV type 1: the antibody perspective**. *Proc. Natl. Acad. Sci. USA* 1997, **94**:10018-23.

Letvin, N.L.: **Progress in the development of an HIV-1 vaccine**. *Science* 1998, **280**:1875-80.

MacQueen, K.M., Buchbinder, S., Douglas, J.M., Judson, F.N., McKirnan, D.J., and Bartholow, B.: **The decision to enroll in HIV vaccine efficacy trials: concerns elicited from gay men at increased risk for HIV infection**. *AIDS Res. Hum. Retroviruses* 1994, **10 Suppl 2**:S261-S264.

Rowland-Jones, S., Tan, R., McMichael, A.: **Role of cellular immunity in protection against HIV infection**. *Adv. Immunol.* 1997; **65**:277-346.

Rowland Jones, S., Sutton, J., Ariyoshi, K., Dong, T., Gotch, F., McAdam, S., Whitby, D., Sabally, S., Gallimore, A., Corrah, T., Takiguchi, M., Schultz, T., McMichael, A., Whittle, H.: **HIV-specific cytotoxic T-cells in HIV-exposed but uninfected Gambian women**. *Nat. Med.* 1995, **1**:59-64.

Salk, J., Bretscher, P.A., Salk, P.L., Clerici, M., and Shearer, G.M.: **A strategy for prophylactic vaccination against HIV**. *Science* 1993, **260**:1270-1272.

11-25 Prevention and education are one way in which the spread of HIV and AIDS can be controlled.

Coates, T.J., Aggleton, P., Gutzwiller, F., Des-Jarlais, D., Kihara, M., Kippax, S., Schechter, M., van-den-Hoek, J.A.: **HIV prevention in developed countries**. *Lancet* 1996, **348**:1143-8.

Decosas, J., Kane, F., Anarfi, J.K., Sodji, K.D., and Wagner, H.U.: **Migration and AIDS**. *Lancet* 1995, **346**:826-828.

Dowsett, G.W.: **Sustaining safe sex: sexual practices, HIV and social context**. *AIDS* 1993, **7 Suppl 1**:S257-S262.

Kimball, A.M., Berkley, S., Ngugi, E., and Gayle, H.: **International aspects of the AIDS/HIV epidemic**. *Annu. Rev. Public. Health* 1995, **16**:253-282.

Kirby, M.: **Human rights and the HIV paradox**. *Lancet* 1996, **348**:1217-8.

Nelson, K.E., Celentano, D.D., Eiumtrakol, S., Hoover, D.R., Beyrer, C., Suprasert, S., Kuntolbutra, S., and Khamboonruang, C.: **Changes in sexual behavior and a decline in HIV infection among young men in Thailand**. *N. Engl. J. Med.* 1996, **335**:297-303.

Weniger, B.G. and Brown, T.: **The march of AIDS through Asia**. *N. Engl. J. Med.* 1996, **335**:343-345.

Allergy and Hypersensitivity

12

Allergic reactions occur when an individual who has produced IgE antibody in response to an innocuous antigen, or **allergen**, subsequently encounters the same allergen. The allergen triggers the activation of IgE-binding mast cells in the exposed tissue, leading to a series of responses that are characteristic of **allergy**. As we learned in Chapter 9, there are circumstances in which IgE is involved in protective immunity, especially in response to parasitic worms, which are prevalent in underdeveloped countries. In more advanced countries, however, IgE responses to innocuous antigens predominate and allergy is one of the most prevalent diseases (Fig. 12.1). Allergic reactions to common environmental antigens affect up to half the population in North America and Europe and, although they are rarely life-threatening, cause much distress and lost time from school and work. Because of the medical importance of allergy in industrialized societies, much more is known about the pathophysiology of IgE-mediated responses than about the normal physiological role of IgE.

Fig. 12.1 IgE-mediated reactions to extrinsic antigens. All IgE-mediated responses involve mast-cell degranulation, but the symptoms experienced by the patient can be very different depending on whether the allergen is injected, inhaled, or eaten, and depending also on the dose of the allergen.

IgE-mediated allergic reactions			
Syndrome	**Common allergens**	**Route of entry**	**Response**
Systemic anaphylaxis	Drugs Serum Venoms Peanuts	Intravenous (either directly or following rapid absorption)	Edema Increased vascular permeability Tracheal occlusion Circulatory collapse Death
Acute urticaria (wheal-and-flare)	Insect bites Allergy testing	Subcutaneous	Local increase in blood flow and vascular permeability
Allergic rhinitis (hay fever)	Pollens (ragweed, timothy, birch) Dust-mite feces	Inhaled	Edema of nasal mucosa Irritation of nasal mucosa
Asthma	Pollens Dust-mite feces	Inhaled	Bronchial constriction Increased mucus production Airway inflammation
Food allergy	Shellfish Milk Eggs Fish Wheat	Oral	Vomiting Diarrhea Pruritis (itching) Urticaria (hives) Anaphylaxis (rarely)

The term allergy was originally defined by Clemens Von Pirquet as 'an altered capacity of the body to react to a foreign substance,' which was an extremely broad definition that included all immunological reactions. Allergy is now defined in a much more restricted manner as 'disease following a response by the immune system to an otherwise innocuous antigen.' Allergy is one of a class of immune system responses that are termed **hypersensitivity reactions**; these are harmful immune responses that produce tissue injury and may cause serious disease. Hypersensitivity reactions were classified into four types by Coombs and Gell (Fig. 12.2). Allergy is often equated with type I hypersensitivity (immediate-type hypersensitivity reactions mediated by IgE), and will be used in this sense here.

In this chapter we shall first consider the mechanisms that favor the production of IgE. We shall then describe the pathophysiological consequences of the interaction between antigen and IgE that is bound by the high-affinity Fcε receptor (FcεRI) on mast cells. Finally, we shall consider the causes and consequences of other types of immunological hypersensitivity reactions.

The production of IgE.

IgE differs from other antibody isotypes in being located predominantly in tissues, where it is bound to mast cells by high-affinity surface receptors called FcεRI. Binding of antigen to IgE crosslinks these receptors and this causes the release of mediators from the mast cells, which may lead to the development of a **type I hypersensitivity reaction**. Basophils and activated eosinophils also express FcεRI; they can therefore display surface-bound IgE and also take part

	Type I	Type II		Type III	Type IV		
Immune reactant	IgE	IgG		IgG	T_H1 cells	T_H2 cells	CTL
Antigen	Soluble antigen	Cell- or matrix-associated antigen	Cell-surface receptor	Soluble antigen	Soluble antigen	Soluble antigen	Cell-associated antigen
Effector mechanism	Mast-cell activation	Complement, FcR+ cells (phagocytes, NK cells)	Antibody alters signaling	Complement Phagocytes	Macrophage activation	Eosinophil activation	Cytotoxicity
Example of hypersensitivity reaction	Allergic rhinitis, asthma, systemic anaphylaxis	Some drug allergies (eg penicillin)	Chronic urticaria (antibody to FCε R1α)	Serum sickness, Arthus reaction	Contact dermatitis, tuberculin reaction	Chronic asthma, chronic allergic rhinitis	Contact dermatitis

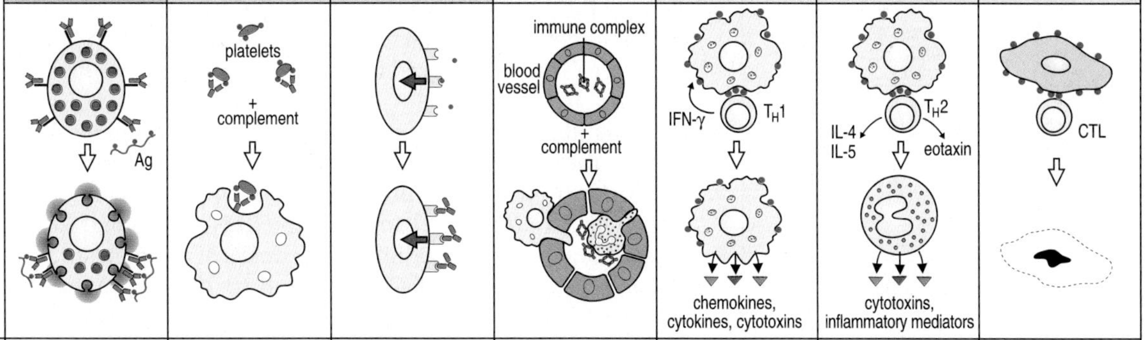

Fig. 12.2 There are four types of hypersensitivity reaction mediated by immunological mechanisms that tissue damage. Types I–III are antibody-mediated and are distinguished by the different types of antigens recognized and the different classes of antibody involved. Type I responses are mediated by IgE, which induces mast-cell activation, whereas types II and III are mediated by IgG, which can engage complement-mediated and phagocytic effector mechanisms to varying degrees, depending on the subclass of IgG and the nature of the antigen involved. Type II responses are directed against cell-surface or matrix antigens, whereas type III responses are directed against soluble antigens, and the tissue damage involved is caused by responses triggered by immune complexes. A special category of type II responses involves IgG antibodies against cell-surface receptors that disrupt the normal functions of the receptor, either by causing uncontrollable activation or by blocking receptor function. Type IV hypersensitivity reactions are T-cell mediated and can be subdivided into three groups. In the first group, tissue damage is caused by the activation of macrophages by T_H1 cells, which results in an inflammatory response. In the second, damage is caused by the activation of eosinophilic inflammatory responses by T_H2 cells; in the third, damage is caused directly by cytotoxic T cells (CTL).

in the production of type I hypersensitivity reactions. The factors that lead to an antibody response dominated by IgE are still being worked out. Here we shall describe our current understanding of these processes before turning to the question of how IgE mediates allergic reactions.

12-1 Allergens are often delivered transmucosally at low dose, a route that favors IgE production.

There are certain antigens and routes of antigen presentation to the immune system that favor the production of IgE. As we learned in Chapter 9, T_H2 cells can switch the antibody isotype from IgM to IgE, or they can cause switching to IgG2 and IgG4 (human) or IgG1 and IgG3 (mouse). Antigens that selectively evoke T_H2 cells that drive an IgE response are known as allergens.

Much human allergy is caused by a limited number of inhaled small protein allergens that reproducibly elicit IgE production in susceptible individuals. Because we inhale many different proteins that do not induce IgE production,

Features of inhaled allergens that may promote the priming of T_H2 cells that drive IgE responses	
Protein	Only proteins induce T-cell responses
Enzymatically active	Allergens are often proteases
Low dose	Favors activation of IL-4-producing CD4 T cells
Low molecular weight	Diffuses out of particle into mucus
High solubility	Readily eluted from particle
Stable	Allows survival in desiccated particle
Contains peptides that bind host MHC class II	Required for T-cell priming

Fig. 12.3 Properties of inhaled allergens. The typical characteristics of inhaled allergens are described in this table.

what is unusual about the proteins that are common allergens? Although we do not yet have a complete answer, some general principles have emerged (Fig. 12.3). Most allergens are relatively small, highly soluble, proteins that are carried on desiccated particles such as pollen grains or mite feces. On contact with the mucosa of the airways, for example, the soluble allergen elutes from the particle and diffuses into the mucosa. Allergens are typically presented to the immune system at very low doses. It has been estimated that the maximum exposure of a person to the common pollen allergens in ragweed (*Artemisia artemisiifolia*) does not exceed 1 μg per year! Yet many people develop irritating and even life-threatening T_H2-driven IgE antibody responses to these minute doses of allergen. It is important to note that only some of the people who are exposed to these substances make IgE antibodies against them. Possible factors that influence which individuals will respond to allergens are considered in Section 12-4.

It seems likely that presenting an antigen transmucosally and at very low doses is a particularly efficient way of inducing T_H2-driven IgE responses. IgE antibody production requires IL-4-producing T_H2 cells and it can be inhibited by T_H1 cells that produce interferon-γ (IFN-γ) (see Fig. 9.8). We have already learned that the presentation of low doses of antigen can favor the activation of T_H2 cells over T_H1 cells (see Section 10-19), and many common allergens are delivered to the respiratory mucosa by inhalation of a low dose. The dominant antigen-presenting cell type in the respiratory mucosa is a cell with characteristics similar to those of Langerhans' cells. These cells very efficiently take up and process protein antigens, a step that is accompanied by cellular activation. This in turn induces their migration to regional lymph nodes and differentiation into cells that are highly co-stimulatory, with characteristics that favor T_H2 differentiation.

12-2 Enzymes are frequent triggers of allergy.

In Chapter 9, we learned that several lines of evidence suggest that IgE is important in host defense against parasites. Many parasites invade their hosts by secreting proteolytic enzymes that break down connective tissue and allow the parasite access to host tissues, and it has been proposed that these enzymes are particularly active at promoting T_H2 responses. This idea receives some support from the many examples of allergens that are enzymes. The major allergen of the house dust mite (*Dermatophagoides pteronyssimus*) which is responsible for allergy in up to 20% of the North American population, is a cysteine protease homologous to papain. Papain itself, derived from the papaya fruit, is used as a meat tenderizer and causes allergy in workers preparing the enzyme; such allergies are called industrial allergies. Another industrial allergy is the asthma caused by inhalation of the bacterial enzyme subtilisin, the 'biological' component of some laundry detergents.

Injection of enzymatically active papain (but not inactivated papain) into mice stimulates an IgE response. A closely related enzyme, chymopapain, is used medically to destroy intervertebral disks in patients with sciatica; the major (although rare) complication of this procedure is anaphylaxis, an acute systemic response to allergens (see Section 12-11). Not all allergens are enzymes, however; for example, two allergens identified from filarial worms are enzyme inhibitors. Many protein allergens derived from plants have been identified and sequenced, but their functions are currently obscure. Thus, the association between enzymatic activity and allergenicity is intriguing but of unproven importance.

12-3 Class switching to IgE in B lymphocytes is favored by specific accessory signals.

IgE production requires cytokines that are released by T_H2 cells, in particular interleukin (IL)-4. T_H2 cells arise when naive CD4 T cells first encounter antigen in the presence of IL-4. The importance of IL-4 in driving IgE production is seen in mice lacking a functional IL-4 gene: the major abnormality in these mice seems to be reduced IgE synthesis. In mice, the early production of IL-4 has been shown to be the result of activation of a small subset of CD4 T cells with unusual properties. These cells, the $NK1.1^+$ subset, express T-cell receptors made up of a restricted set of β chains and an invariant α chain, and develop in response to CD1, an MHC class I-like molecule found in humans as well as mice (Fig. 12.4). Evidence for their development in response to CD1 derives from the absence of $NK1.1^+$ T cells in mice that cannot express CD1 molecules because of engineered defects in their β_2-microglobulin genes. The invariant T-cell receptor α chain expressed by $NK1.1^+$ T cells is encoded in a single V_α gene segment and a single J_α gene segment; similar cells in humans are also specific for the human homolog of mouse CD1, called CD1d, and use the homologous V_α and J_α gene segments. These T cells produce IL-4 almost immediately upon encountering CD1, which is present on cortical thymocytes and on Langerhans' cells and other antigen-presenting cells. In mice, CD1-specific T cells are the only known source of early IL-4; mice lacking β_2-microglobulin fail to make early IL-4 and are deficient in IgE production. It has not yet been established whether this pathway is important in humans.

As noted above, IL-4 drives naive CD4 T cells to differentiate into T_H2 cells and inhibits T_H1 development. Once T_H2 cells are primed, they can deliver several molecular signals that favor class switching in B lymphocytes to IgE (see Section 9-5); for example, when their T-cell receptors are ligated, they secrete cytokines such as IL-4 and IL-13. Activated T_H2 cells also express CD40 ligand (CD40L) and CD23 (the low-affinity receptor for IgE), and these co-stimulatory molecules then ligate their counter-receptors CD40 and CR2 on B lymphocytes. The combination of these signals causes class switching to IgE and B-cell proliferation.

The IgE response, once initiated, can be further amplified by basophils, mast cells, and eosinophils, which can also drive IgE production (Fig. 12.5). All three cell types express FcεRI, although eosinophils only express it when activated. When these specialized granulocytes are activated by antigen crosslinking of their FcεRI-bound IgE, they can express cell-surface CD40L

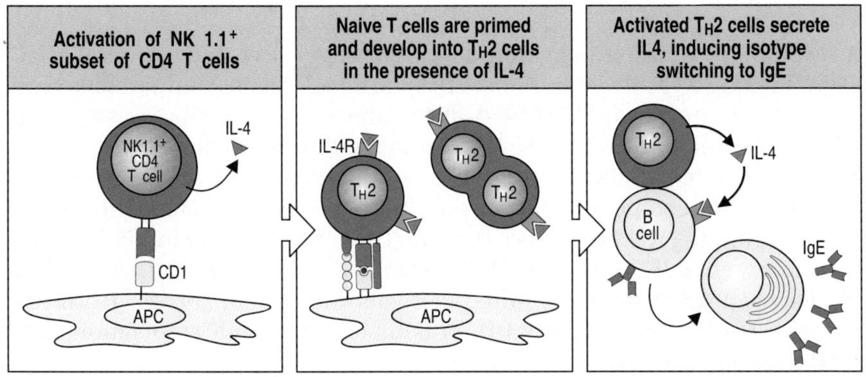

Fig. 12.4 IgE class switching in B cells is initiated by T_H2 cells, which develop in the presence of an early burst of IL-4. IL-4 is secreted early in some immune responses by a small subset of CD4 T cells ($NK1.1^+$ CD4 T cells) which interact with antigen-presenting cells bearing the non-classical MHC class I-like molecule CD1. Naive T cells being primed by their first encounter with antigen are driven to differentiate into T_H2 cells in the presence of this early burst of IL-4. These mechanisms have been characterized in mice; it is not yet known whether the same pathways operate in humans.

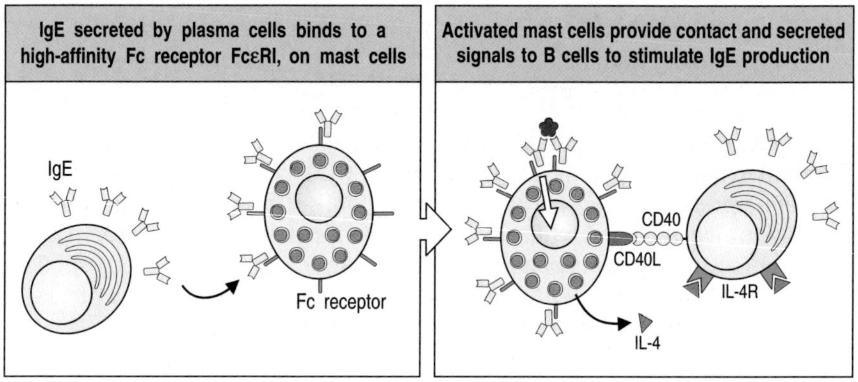

Fig. 12.5 Antigen ligation of IgE bound to mast cells leads to amplification of IgE production. IgE secreted by plasma cells binds to the high-affinity IgE receptor on mast cells, basophils, and activated eosinophils. When the surface-bound IgE is crosslinked by antigen these cells express CD40L and secrete IL-4, which in turn stimulates isotype switching by B cells and the production of more IgE. These interactions can occur *in vivo* at the site of allergen-triggered inflammation, for example in bronchial-associated lymphoid aggregates.

and secrete IL-4; like T_H2 cells, therefore, they can drive class switching and IgE production by B cells. The interaction between these specialized granulocytes and B cells can occur at the site of the allergic reaction, as B cells are observed to form germinal centers at inflammatory foci. Blocking this amplification process is a goal of therapy, as allergic reactions can otherwise become self sustaining.

12-4 Genetic factors contribute to the development of IgE-mediated allergy, but environmental factors may also be important.

Up to 40% of people in Western populations show an exaggerated tendency to mount IgE responses to a wide variety of common environmental allergens. This state is called **atopy** and seems to be influenced by several genetic loci. Atopic individuals have higher total levels of IgE in the circulation and higher levels of eosinophil than their normal counterparts. They are more susceptible to allergic diseases such as hay fever and asthma. Studies of atopic families have implicated loci on chromosomes 11q and 5q that might be important in determining the presence of atopy, and candidate genes that might affect IgE responses are found in these regions. The candidate gene on chromosome 11 encodes the β subunit of the high-affinity IgE receptor, whereas on chromosome 5 there is a cluster of tightly linked genes that includes those for IL-3, IL-4, IL-5, IL-9, IL-13, and GM-CSF. These cytokines are important in IgE isotype switching, eosinophil survival, and mast-cell proliferation. Of particular note, an inherited genetic variation in the promoter region of the IL-4 gene is associated with raised IgE levels in atopic individuals; the variant promoter will direct increased expression of a reporter gene in experimental systems. It is too early to know whether this polymorphism is important in the complex genetics of atopy.

A second type of inherited variation in IgE responses is linked to the MHC class II region and affects responses to specific allergens. Many studies have shown that specific IgE production to individual allergens is associated with particular HLA class II alleles, implying that particular MHC:peptide combinations might favor a strong T_H2 response. For example, IgE responses to several ragweed pollen allergens are particularly associated with haplotypes containing the MHC class II allele, DRB1*1501. Many individuals are therefore generally predisposed to make T_H2 responses and specifically predisposed to respond to some allergens more than others. However, allergies to common drugs such as penicillin show no association with MHC class II or the presence or absence of atopy.

The prevalence of atopic allergy and, in particular, of asthma is increasing in economically advanced regions of the world, an observation that is best explained by environmental factors. Three candidate environmental factors for which there is little supporting evidence are changes in allergen levels, environmental pollution, and dietary changes. Alterations in exposure to microbial pathogens is the most plausible explanation at present for the increase in atopic allergy. Atopy is negatively associated with a history of infection by measles or hepatitis A virus, and with positive tuberculin skin tests (suggesting prior exposure and immune response to *Mycobacterium tuberculosis*). It is possible that infection by an organism that evokes a T_H1 immune response early in life might reduce the likelihood of T_H2 responses later in life.

Summary.

Allergic reactions are the result of the production of specific IgE antibody to common, innocuous antigens. Allergens are small antigens that commonly provoke an IgE antibody response. Such antigens normally enter the body at very low doses by diffusion across mucosal surfaces and trigger a T_H2 response. Naive allergen-specific T cells are induced to develop into T_H2 cells in the presence of an early burst of IL-4, which seems to be derived from a specialized subset of T cells. The allergen-specific T_H2 cells drive allergen-specific B cells to produce IgE, which binds to the high-affinity receptor for IgE on mast cells, basophils, and activated eosinophils. IgE production can be amplified by these cells because, upon activation, they produce IL-4 and CD40L. The tendency to the production of IgE is influenced by genetic and environmental factors. Once IgE is produced in response to an allergen, re-exposure to the allergen triggers an allergic response by mechanisms to which we now turn.

Effector mechanisms in allergic reactions.

Allergic reactions are triggered when allergens crosslink preformed IgE bound to the high-affinity receptor FcεRI on mast cells. Mast cells line the body surfaces and serve to alert the immune system to local infection. In allergy, they provoke very unpleasant reactions to innocuous antigens that are not associated with invading pathogens that need to be expelled. Mast cells act by releasing stored mediators by granule exocytosis, and also by synthesizing leukotrienes and cytokines (see Fig. 12.11). The consequences of IgE-mediated mast-cell activation depend on the dose of antigen and its route of entry; symptoms range from the irritating sniffles of hay fever when pollen is inhaled, to the life-threatening circulatory collapse that occurs in systemic anaphylaxis (Fig. 12.6). The immediate allergic reaction caused by mast-cell degranulation is followed by a more sustained inflammation, known as the late-phase response. This late response involves the recruitment of other effector cells, notably T_H2 lymphocytes, eosinophils, and basophils, which contribute significantly to the immunopathology of an allergic response.

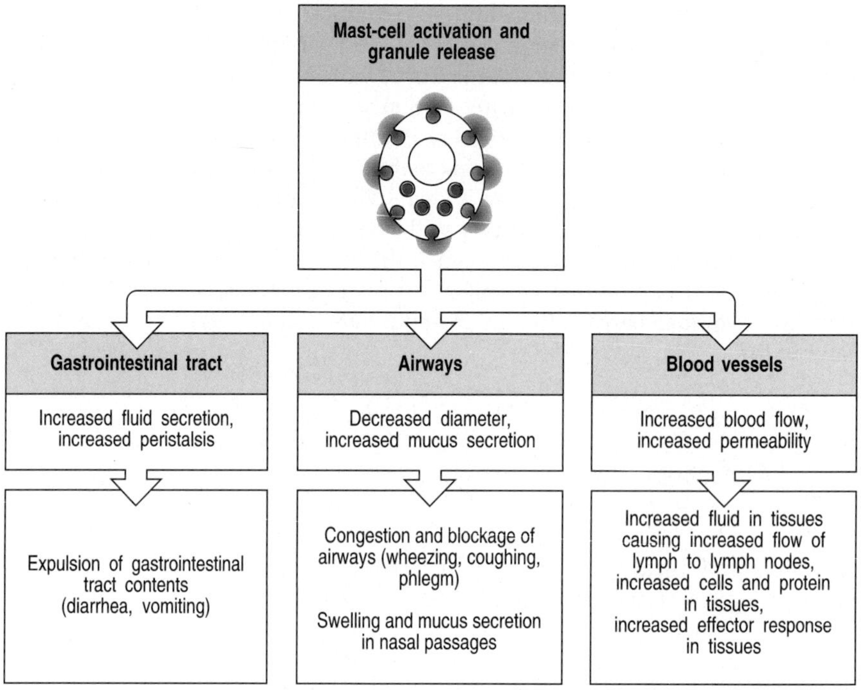

Fig. 12.6 Mast-cell products have different effects on different tissues. Mast-cell products can be divided into two categories: first, those molecules, both preformed and rapidly synthesized, that mediate acute inflammatory events after mast-cell activation; and second, cytokines and lipid mediators, which induce a late-phase chronic inflammatory response with influx and activation of T_H2 lymphocytes, monocytes, eosinophils, and neutrophils. There is some overlap between mediators that induce acute and chronic inflammatory responses, particularly among the lipid mediators, which have rapid effects causing smooth muscle contraction, increased vascular permeability, and mucus secretion, and also induce the influx and activation of leukocytes, which contribute to the late-phase response.

12-5 Most IgE is cell-bound and engages effector mechanisms of the immune system by different pathways from other antibody isotypes.

Most antibodies are found in body fluids and engage effector cells (through receptors specific for the Fc constant regions) only after binding specific antigen (through their variable regions). IgE is an exception, however, as it is captured by high-affinity receptors specific for the IgE Fc region in the absence of bound antigen. This means that IgE is mostly found fixed in the tissues on mast cells that bear this receptor, as well as on circulating basophils and activated eosinophils. The ligation of cell-bound IgE by antigen triggers activation of these cells at the site of antigen entry into the tissues. The release of inflammatory lipid mediators, cytokines, and chemokines at sites of IgE-triggered reactions results in the recruitment of eosinophils and basophils to augment the type I response.

There are two types of IgE-binding Fc receptor. The first, FcεRI, is a high-affinity receptor of the immunoglobulin superfamily that binds IgE on mast cells, basophils, and activated eosinophils (see Chapter 9). When the cell-bound IgE is crosslinked, FcεRI transduces an activating signal. High levels of IgE, such as those that exist in subjects with allergic diseases or parasite infections, can result in marked increases in surface FcεRI expression on mast cells, enhanced sensitivity of such cells to activation by low concentrations of specific antigen, and markedly increased IgE-dependent release of mediators and cytokines. The second IgE receptor, **CD23**, is a structurally unrelated molecule that binds IgE with low affinity. CD23 is found on many different cell types, including B cells, activated T cells, monocytes, eosinophils, platelets, follicular dendritic cells, and some thymic epithelial cells. This receptor was thought to be crucial for the regulation of IgE antibody levels; however, a mouse strain in which the CD23 gene was deleted by homologous recombination (see Section 2-27) shows no major abnormality in the development of polyclonal IgE responses. These mice did not show antigen-specific IgE-mediated enhancement of antibody responses, however. This demonstrates a role for CD23 on antigen-presenting cells in the capture of antigen by specific IgE.

12-6 Mast cells reside in tissues and orchestrate allergic reactions.

Mast cells were described by Ehrlich in the mesentery of rabbits and named *Mastzellen* ('fattened cells'). Like basophils, mast cells contain granules rich in acidic molecules that take up basic dyes. However, in spite of this resemblance, and the similar range of mediators stored in these basophilic granules, mast cells are derived from a different myeloid lineage from basophils and eosinophils. Mast cells are highly specialized cells, and are prominent residents of mucosal and epithelial tissues in the vicinity of small blood vessels and postcapillary venules, where they are well placed to guard against invading pathogens (see Sections 9-20 and 9-21). Mast cells are also found in subendothelial connective tissue. They home to tissues as agranular cells; their final differentiation, accompanied by granule formation, occurs after they have arrived in the tissues. The major mast-cell growth factor is stem-cell factor (SCF), whose receptor, c-Kit (CD117), is encoded by a proto-oncogene. Mice with defective c-Kit lack differentiated mast cells and studies of these mice have shown that IgE-mediated inflammatory responses are dependent almost exclusively on mast cells.

Mast cells express FcεRI constitutively on their surface and they are activated when antigens crosslink FcεRI-bound IgE. Degranulation occurs within seconds, releasing a variety of preformed mediators (see Figs 12.11 and 9.30). Among these are histamine—a short-lived vasoactive amine that causes an immediate increase in local blood flow and vessel permeability—and the enzymes mast-cell chymase, tryptase, and serine esterases. The latter might in turn activate matrix metalloproteinases, which collectively break down tissue matrix proteins. Tumor necrosis factor (TNF)-α is also stored in mast-cell granules and is released in large amounts from both preformed and newly synthesized pools on mast-cell activation. It causes endothelial activation with upregulation of the expression of adhesion molecules, which promotes the influx of inflammatory leukocytes and lymphocytes.

On mast-cell activation, chemokines, lipid mediators such as leukotrienes and platelet-activating factor (PAF), and further cytokines such as IL-4, are synthesized and act to sustain the inflammatory response. Thus, the IgE-mediated activation of mast cells orchestrates an important inflammatory cascade that is amplified by the recruitment of eosinophils, basophils, and T_H2 lymphocytes. The physiological importance of this is as a host defense mechanism, as we learned in Chapter 9. In allergy, however, the acute and chronic inflammatory reactions triggered by mast-cell activation can also have important pathophysiological consequences, as seen in the diseases associated with allergic responses to environmental antigens.

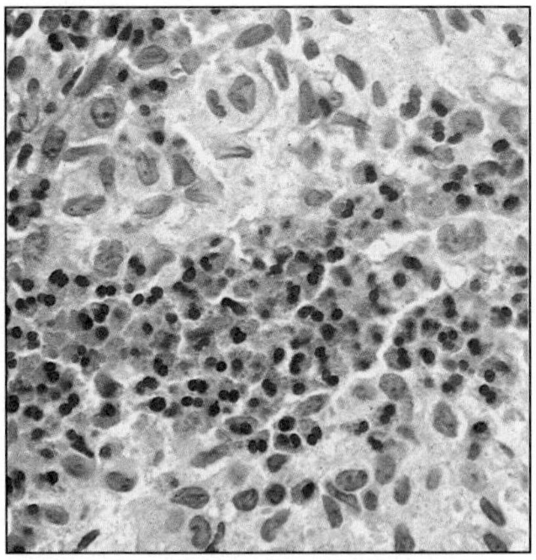

Fig. 12.7 Eosinophils can be detected easily in tissue sections by their bright orange coloration. In this light micrograph, a large number of eosinophils are seen infiltrating a Langerhans' cell histiocytosis. The tissue section is stained with hematoxylin and eosin; it is the eosin that imparts the characteristic orange color to the eosinophils. Photograph courtesy of T Krausz.

12-7 Eosinophils are normally under tight control to prevent inappropriate toxic responses.

Eosinophils are bone marrow-derived granulocytic leukocytes, so named because their granules, which contain arginine-rich basic proteins, are colored bright orange by the acidic stain eosin (Fig. 12.7). Only very small numbers of these cells are normally present in the circulation; most eosinophils are found in tissues, especially in the connective tissue immediately underneath respiratory, gut, and urogenital epithelium, implying a likely role for these cells in defense against invading organisms. Eosinophils have two kinds of effector function. First, they release highly toxic granule proteins and free radicals, which can kill microorganisms and parasites but can also cause significant tissue damage in allergic reactions. Second, they produce molecules including prostaglandins, leukotrienes, and cytokines, which amplify the inflammatory response by recruiting and activating further eosinophils, leukocytes, and epithelial cells (Fig. 12.8).

Important regulatory mechanisms inhibit the inappropriate activation and degranulation of eosinophils, which could otherwise be very harmful to the host. The first level of control regulates the production of eosinophils by the bone marrow, which is low in the absence of infection or other immune stimulation. When T_H2 cells are activated, cytokines such as IL-5 are released that increase the production of eosinophils in the bone marrow and promote their release into the circulation. However, transgenic animals overexpressing IL-5 show eosinophilia in the circulation but not in tissues. This demonstrates that a second level of control on eosinophil activity regulates the migration of eosinophils from the circulation into tissues. The key molecules in this response are CC chemokines (see Section 10-9). Most chemokines cause chemotaxis of several types of leukocyte; two of the newest members of the CC family are specific for eosinophils and have been named **eotaxin 1** and **eotaxin 2**.

The eotaxin receptor on eosinophils, CCR3, is a member of the chemokine family of receptors (see Section 10-9). As well as the eotaxins, this receptor also binds the chemokines MCP-3, MCP-4, and RANTES, providing an explanation for the finding that these chemokines can also induce eosinophil activation

Fig. 12.8 Eosinophils secrete a range of highly toxic granule proteins and other inflammatory mediators.

Class of product	Examples	Biological effects
Enzyme	Eosinophil peroxidase	Toxic to targets by catalyzing halogenation Triggers histamine release from mast cells
	Eosinophil collagenase	Remodeling of connective tissue matrix
Toxic protein	Major basic protein	Toxic to parasites and mammalian cells Triggers histamine release from mast cells
	Eosinophil cationic protein	Toxic to parasites Neurotoxin
	Eosinophil-derived neurotoxin	Neurotoxin
Cytokine	IL-3, IL-5, GM-CSF	Amplify eosinophil production by bone marrow Cause eosinophil activation
Chemokine	IL-8	Promotes influx of leukocytes
Lipid mediator	Leukotrienes C4 and D4	Smooth muscle contraction Increased vascular permeability Mucus secretion
	Platelet-activating factor	Chemotactic to leukocytes Amplifies production of lipid mediators Neutrophil, eosinophil, and platelet activation

and chemotaxis. T_H2 cells have also been found to carry CCR3, showing that, as well as cytokines, families of chemokine molecules can coordinate certain kinds of immune response.

The third level of eosinophil regulation is control of their state of activation. In their non-activated state, eosinophils do not express high-affinity IgE receptors and have a high threshold for release of their granule contents. After activation by cytokines and chemokines this threshold drops, FcεRI is expressed, and the numbers of surface complement and Fcγ receptors increase. The eosinophil is now primed to express effector activity.

The potential of eosinophils to cause tissue injury is illustrated by rare hypereosinophilic syndromes. These are sometimes seen in association with T-cell lymphomas in which unregulated IL-5 secretion drives a marked increase in the numbers of eosinophils in the blood (hypereosinophilia). The clinical manifestations of hypereosinophilia are damage to the endocardium (Fig. 12.9) and to nerves, leading to heart failure and neuropathy, both thought to be caused by the toxic effects of eosinophil granule proteins.

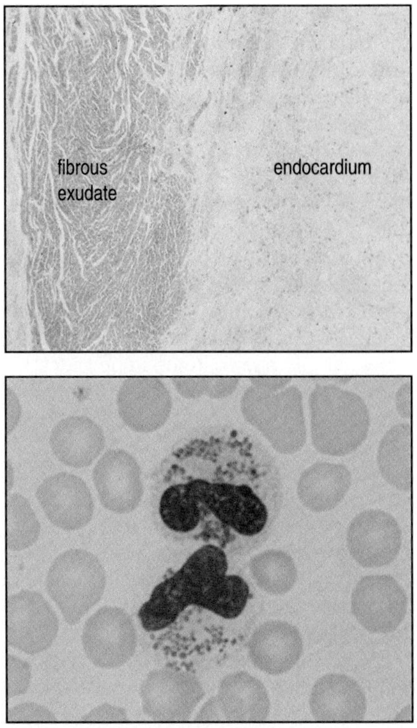

Fig. 12.9 Hypereosinophilia can cause injury to the endocardium. The top panel shows a section of the endocardium from a patient with hypereosinophilic syndrome. There is an organized fibrous exudate and the underlying endocardium is thickened by fibrous tissue. Although there are large numbers of circulating eosinophils, these cells are not seen in the injured endocardium, which is thought to be damaged by granules released from circulating eosinophils. The panel on the bottom shows two partly degranulated eosinophils (center) surrounded by erythrocytes in a peripheral blood film. Photographs courtesy of D Swirsky and T Krausz.

12-8 Eosinophils and basophils cause inflammation and tissue damage in allergic reactions.

In a local allergic reaction, mast-cell degranulation and T_H2 activation cause eosinophils to accumulate in large numbers and to become activated. Their continued presence is characteristic of chronic allergic inflammation and they are thought to be major contributors to the tissue damage that occurs.

Basophils are also present at the site of the reaction. These are bone marrow-derived granulocytes, which share a common stem-cell precursor with eosinophils. Growth factors for basophils are very similar to those for eosinophils and include IL-3, IL-5, and GM-CSF. There is evidence for reciprocal control of the maturation of the stem-cell population into basophils or eosinophils. For example, transforming growth factor (TGF)-β in the presence of IL-3 suppresses eosinophil differentiation and enhances that of basophils. Basophils are normally present in very low numbers in the circulation and seem to have a similar role to that of eosinophils in host defense against invading pathogens. Like eosinophils, they are recruited to the sites of allergic reactions. Basophils express FcεRI on the cell surface and, on activation, they release toxic mediators from the basophilic granules after which they are named.

Eosinophils, mast cells, and basophils can interact with each other. Eosinophil degranulation causes the release of **major basic protein**, which in turn causes mast cell and basophil degranulation. This effect is augmented by the presence of any of the cytokines that affect eosinophil and basophil growth, differentiation, and activation, such as IL-3, IL-5, and GM-CSF.

12-9 The allergic reaction after ligation of IgE on mast cells is divided into an immediate response and a late-phase response.

The inflammatory response after IgE-mediated mast-cell activation occurs as an immediate reaction, starting within seconds, and a late reaction, which takes up to 8–12 hours to develop. These reactions can be distinguished clinically (Fig. 12.10). The **immediate reaction** follows from the activity of histamine, prostaglandins, and other preformed or rapidly synthesized toxic mediators

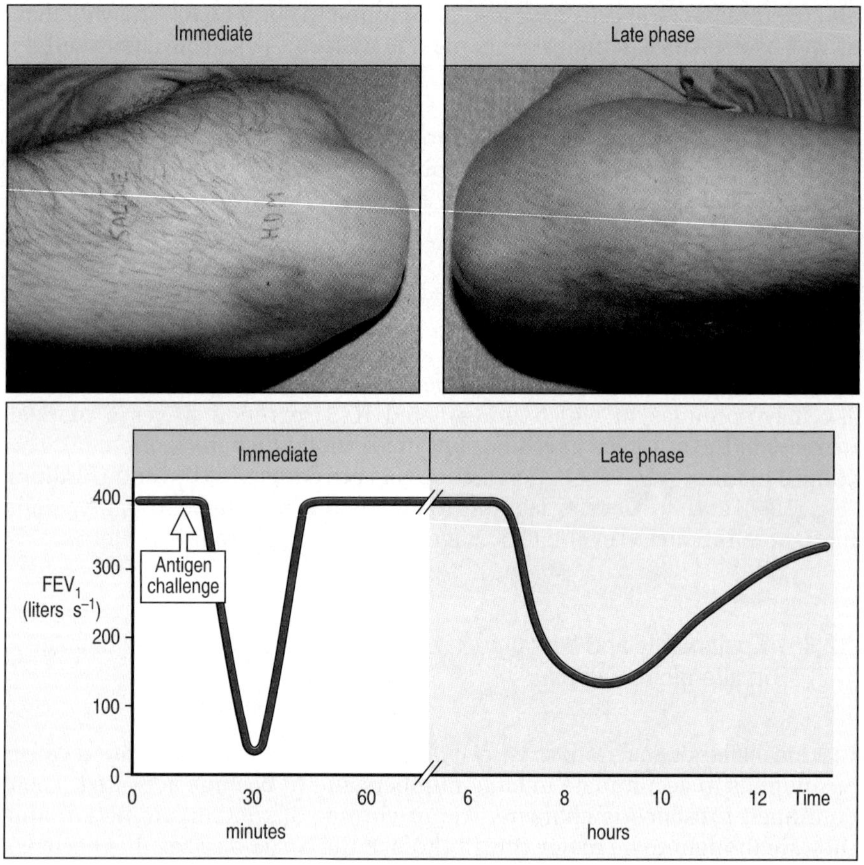

Fig. 12.10 Allergic reactions can be divided into an immediate response and a late-phase response. A wheal-and-flare allergic reaction develops within a minute or two of superficial injection of antigen into the epidermis and lasts for up to 30 minutes. The reaction to an intracutaneous injection of house dust mite antigen is shown in the upper left panel and is labeled HDM; the area labeled saline shows the absence of any response to a control injection of saline solution. A more widespread edematous response, as shown in the upper right panel, develops approximately 8 hours later and can persist for some hours. Similarly, the response to an inhaled antigen can be divided into early and late responses (bottom panel). An asthmatic response in the lungs with narrowing of the airways caused by the constriction of bronchial smooth muscle can be measured as a fall in the forced expired volume of air in one second (FEV_1). The immediate response peaks within minutes after antigen inhalation and then subsides. Approximately 8 hours after antigen challenge, there is a late-phase response that also results in a fall in the FEV_1. The immediate response is caused by the direct effects on blood vessels and smooth muscle of rapidly metabolized mediators such as histamine released by mast cells. The late-phase response is caused by the effects of an influx of inflammatory leukocytes attracted by chemokines and other mediators released by mast cells during and after the immediate response. Photographs courtesy of A B Kay.

that cause a rapid increase in vascular permeability and the contraction of smooth muscle. The **late-phase reaction** is caused by the induced synthesis and release of mediators including leukotrienes, chemokines, and cytokines from the activated mast cells. These recruit leukocytes, including eosinophils and T_H2 lymphocytes, to the site. Although the late-phase reaction is clinically less marked than the immediate response, it is associated with a second phase of smooth muscle contraction and sustained edema. The molecules synthesized and released by mast cells after activation are listed in Fig. 12.11.

The late-phase reaction is an important cause of much more serious long-term illness, as, for example, in chronic asthma. This is because the late reaction induces the recruitment of inflammatory leukocytes, especially eosinophils and T_H2 lymphocytes, to the site of the allergen-triggered mast-cell response. This late response can easily convert into a chronic inflammatory response if antigen persists and stimulates allergen-specific T_H2 cells, which in turn promote eosinophilia and further IgE production.

12-10 The clinical effects of allergic reactions vary according to the site of mast-cell activation.

When re-exposure to allergen triggers an allergic reaction, the effects are focused on the site at which mast-cell degranulation occurs. In the immediate response, the preformed mediators released are short-lived, and their potent effects on blood vessels and smooth muscles are therefore confined to the immediate vicinity of the activated mast cell. The more sustained effects of the late-phase response are also focused on the site of initial allergen-

Class of product	Examples	Biological effects
Enzyme	Tryptase, chymase, cathepsin G, carboxypeptidase	Remodeling of connective tissue matrix
Toxic mediator	Histamine, heparin	Toxic to parasites Increase vascular permeability Cause smooth muscle contraction
Cytokine	IL-4, IL-13	Stimulate and amplify T_H2 cell response
	IL-3, IL-5, GM-CSF	Promote eosinophil production and activation
	TNF-α (some stored pre-formed in granules)	Promotes inflammation, stimulates cytokine production by many cell types, activates endothelium
Chemokine	MIP-1α	Chemokinetic for monocytes, macrophages and neutrophils
Lipid mediator	Leukotrienes C4 and D4	Smooth muscle contraction Increased vascular permeability Mucus secretion
	Platelet-activating factor	Chemotactic to leukocytes Amplifies production of lipid mediators Neutrophil, eosinophil, and platelet activation

Fig. 12.11 Molecules synthesized and released by mast cells on stimulation by antigen binding to IgE. Mast cells release a wide variety of biologically active proteins and other chemical mediators. The lipid mediators derive from membrane phospholipids, which are cleaved to release the precursor molecule arachidonic acid. This molecule can be modified by two pathways to give rise to prostaglandins, thromboxanes, and leukotrienes. The leukotrienes, especially C4, D4 and E4, are important products of mast cells that sustain inflammatory responses in the tissues. Many anti-inflammatory drugs are inhibitors of arachidonic acid metabolism. Aspirin, for example, is an inhibitor of the enzyme cyclo-oxygenase and blocks the production of prostaglandins.

triggered activation, and the particular anatomy of this site may determine how readily the inflammation can be resolved. Thus, the clinical syndrome produced by an allergic reaction depends critically on three variables: the amount of allergen-specific IgE antibody present; the route by which the allergen is introduced; and the dose of allergen (Fig. 12.12).

12-11 The degranulation of mast cells in blood vessel walls after systemic absorption of allergen can cause generalized cardiovascular collapse.

If an allergen is given systemically or is rapidly absorbed from the gut, the connective tissue mast cells associated with all blood vessels can become activated. This activation causes a very dangerous syndrome called **systemic anaphylaxis**. Disseminated mast-cell activation causes a widespread increase in vascular permeability, leading to a catastrophic loss of blood pressure, constriction of the airways, and epiglottal swelling that can cause suffocation; this syndrome is called **anaphylactic shock**. This type of reaction can occur if drugs are administered to people with a specific allergy to that drug, or after an insect bite in individuals allergic to insect venom. Some foods, for example peanuts or brazil nuts, can be associated with systemic anaphylaxis. This syndrome can be rapidly fatal but can usually be controlled by the immediate injection of epinephrine (see Section 12-15).

The most frequent allergic reactions to drugs occur with penicillin and its relatives. In people with IgE antibodies against penicillin, administration of the drug by injection can cause anaphylaxis and even death. Great care should taken to avoid giving drugs to patients with a past history of allergy to the same drug or to one that is closely related structurally. Penicillin acts as a hapten (see Section 9-2); it is a small molecule with a highly reactive β-lactam

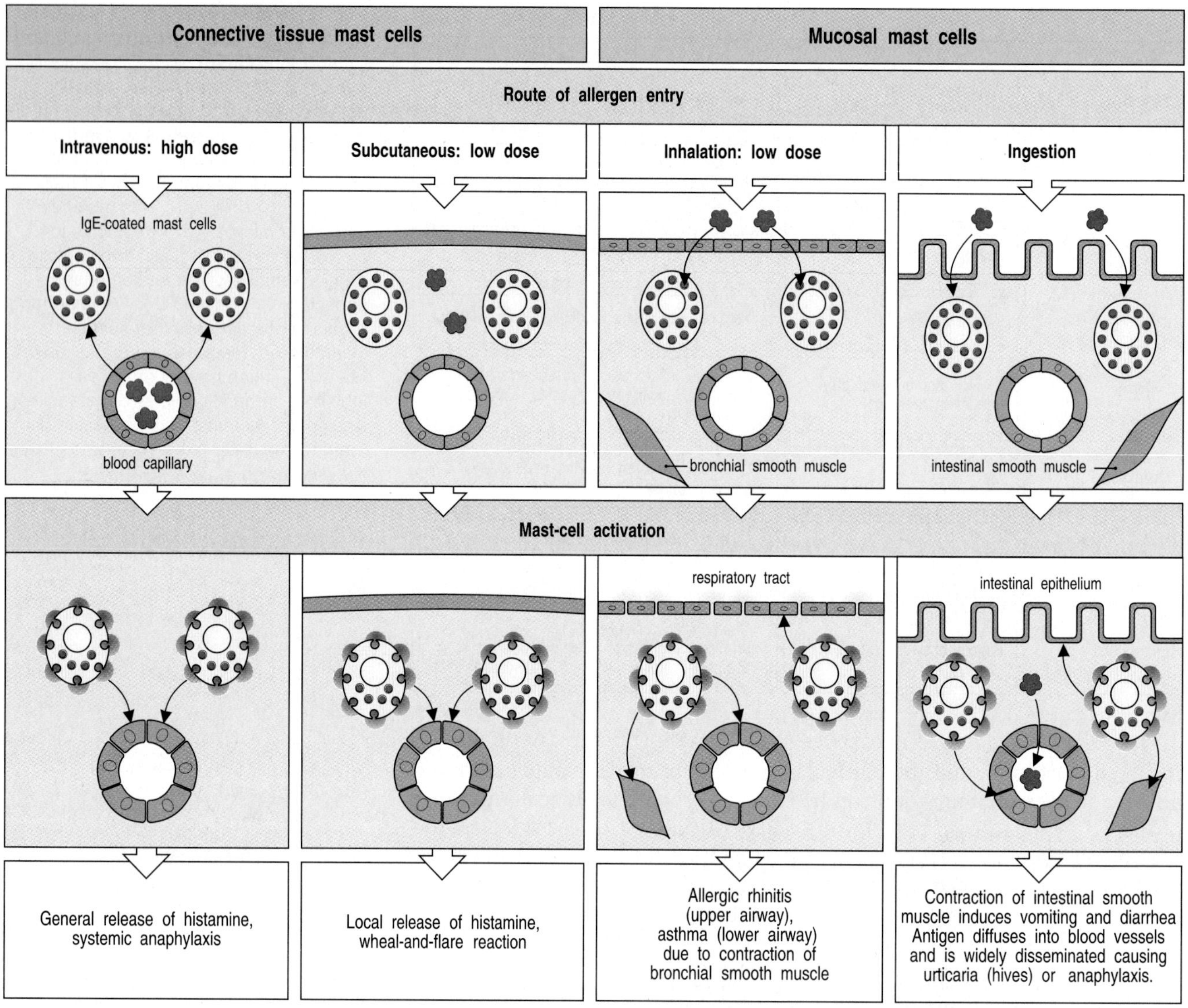

Fig. 12.12 The dose and route of allergen administration determine the type of IgE-mediated allergic reaction that results. There are two main anatomical distributions of mast cells: those associated with vascularized connective tissues, called connective tissue mast cells, and those found in submucosal layers of the gut and respiratory tract, called mucosal mast cells. In an allergic individual, all of these are loaded with IgE directed against specific allergens. The overall response to an allergen then depends on which mast cells are activated. Allergen in the bloodstream activates connective tissue mast cells throughout the body, resulting in the systemic release of histamine and other mediators. Subcutaneous administration of allergen activates only local connective tissue mast cells, leading to a local inflammatory reaction. Inhaled allergen, penetrating across epithelia, activates mainly mucosal mast cells, causing smooth muscle contraction in the lower airways; this leads to bronchoconstriction and difficulty in expelling inhaled air. Mucosal mast-cell activation also increases the local secretion of mucus by epithelial cells and causes irritation. Similarly, ingested allergen penetrates across gut epithelia, causing vomiting due to intestinal smooth muscle contraction; food allergens can also be disseminated in the bloodstream, causing urticaria (hives) when they reach the skin.

ring, crucial for its antibiotic activity. This ring reacts with amino groups on host proteins to form covalent conjugates. When penicillin is ingested or injected, it forms conjugates with self proteins, and these penicillin-modified self peptides can provoke a T_H2 response in some individuals. These T_H2 cells then activate penicillin-binding B cells to produce IgE antibody to the penicillin hapten. Thus, penicillin acts both as the B-cell antigen and, by modifying self peptides, as the T-cell antigen. When penicillin is injected intravenously into an allergic individual, the penicillin-modified proteins crosslink IgE molecules on the mast cells to cause anaphylaxis.

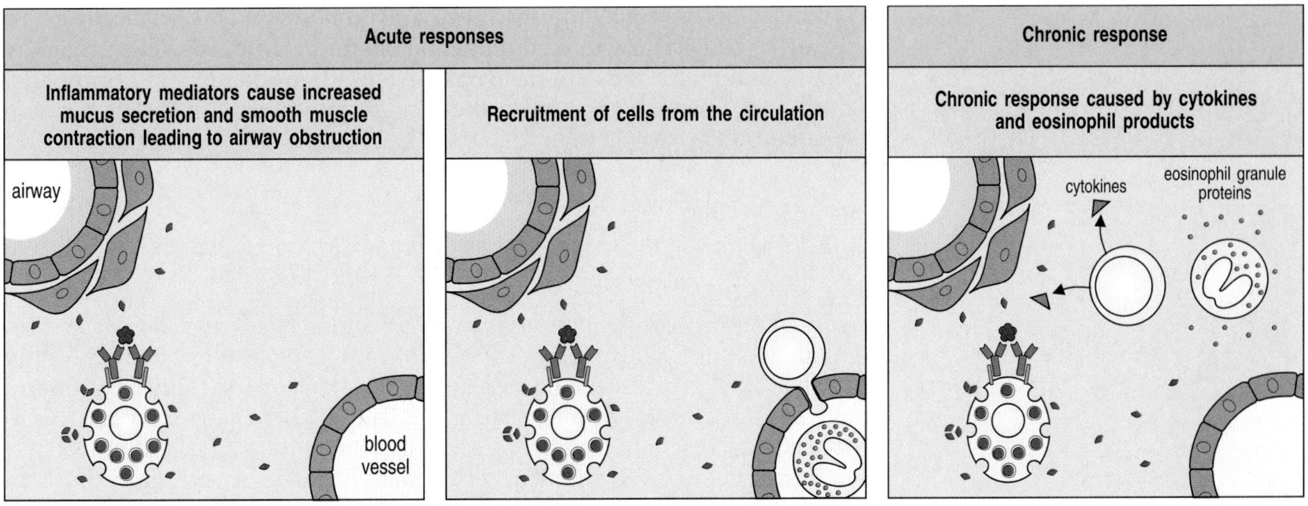

Fig. 12.13 The acute response in allergic asthma leads to T_H2-mediated chronic inflammation of the airways. In sensitized individuals, crosslinking of specific IgE on the surface of mast cells by inhaled allergen triggers them to secrete inflammatory mediators, causing bronchial smooth muscle contraction and an influx of inflammatory cells, including eosinophils, and T_H2 lymphocytes. Activated mast cells and T_H2 cells secrete cytokines that also augment eosinophil activation and degranulation, which causes further tissue injury and influx of inflammatory cells. The end result is chronic inflammation, which might then cause irreversible damage to the airways.

12-12 Allergen inhalation is associated with the development of rhinitis and asthma.

Inhalation is the most common route of allergen entry. Many people have mild allergies to inhaled antigens, manifesting as sneezing and a runny nose. This is called **allergic rhinitis**, and results from the activation of mucosal mast cells beneath the nasal epithelium by allergens that diffuse across the mucous membrane of the nasal passages. Allergic rhinitis is characterized by local edema leading to nasal obstruction, a nasal discharge, which is typically rich in eosinophils, and irritation of the nose from histamine release. A similar reaction to airborne allergens deposited on the conjunctiva of the eye is called allergic conjunctivitis. Allergic rhinitis and conjunctivitis are commonly caused by environmental allergens that are only present during certain seasons of the year. For example, hay fever is caused by a variety of allergens, including certain grass and tree pollens. Autumnal symptoms may be caused by weed pollen. These reactions are annoying but cause little lasting damage.

A more serious syndrome is **allergic asthma**, which is triggered by allergen-induced activation of submucosal mast cells in the lower airways (Fig. 12.13). This leads within seconds to bronchial constriction and increased secretion of fluid and mucus, making breathing more difficult by trapping inhaled air in the lungs. Patients with allergic asthma often need treatment, and asthmatic attacks can be life-threatening. An important feature of asthma is chronic inflammation of the airways, which is characterized by the continued presence of increased numbers of T_H2 lymphocytes, eosinophils, neutrophils, and other leukocytes (Fig. 12.14).

Although allergic asthma is initially driven by a response to a specific allergen, the subsequent chronic inflammation seems to be perpetuated even in the apparent absence of further exposure to allergen, and factors other than re-exposure to antigen can then trigger subsequent asthmatic

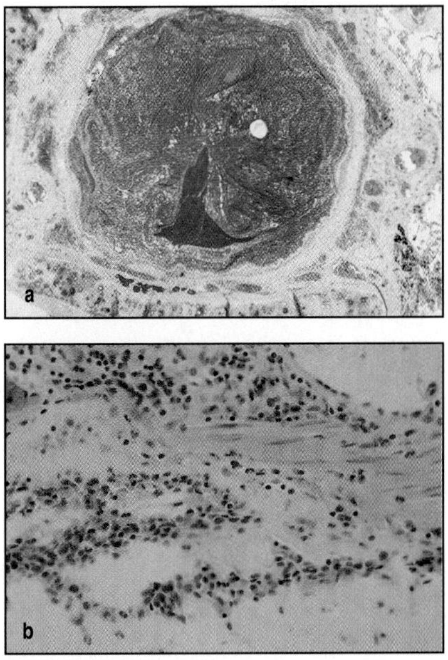

Fig. 12.14 Morphological evidence of chronic inflammation in the airways of an asthmatic patient. Panel a shows a section through a bronchus of a patient who died of asthma; there is almost total occlusion of the airway by a mucus plug. In panel b, a close-up view of the bronchial wall shows injury to the epithelium lining the bronchus, accompanied by a dense inflammatory infiltrate that includes eosinophils, neutrophils, and lymphocytes. Photographs courtesy of T Krausz.

attacks. For example, the airways of asthmatics characteristically show hyper-responsiveness to environmental chemical irritants such as cigarette smoke and sulfur dioxide. Bacterial, or more importantly viral, respiratory tract infections can exacerbate the disease by inducing a T_H2-dominated local response.

12-13 Skin allergy is manifest as urticaria or chronic eczema.

The same dichotomy between immediate and delayed responses is seen in cutaneous allergic responses. The skin forms an effective barrier to the entry of most allergens, but can be breached by local injection of small amounts of allergen, for example by a stinging insect. The entry of allergen causes a localized allergic reaction. Local mast-cell activation in the skin leads immediately to a local increase in vascular permeability, which causes extravasation of fluid. The mast-cell activation also stimulates a nerve axon reflex, causing the vasodilation of surrounding cutaneous blood vessels. The resulting skin lesion is called a **wheal-and-flare reaction**. About 8 hours later, a more widespread and sustained edematous response appears in some individuals as a consequence of the late-phase response (see Fig. 12.10). A disseminated form of the wheal-and-flare reaction, known as **urticaria** or hives, sometimes appears when ingested allergens enter the bloodstream and reach the skin. Histamine released by mast cells activated by allergen in the skin causes large, itchy red swellings beneath the skin.

Allergists take advantage of the immediate response to test for allergy by injecting minute amounts of potential allergens intracutaneously. Although the reaction after the administration of antigen by intracutaneous injection is usually very localized, there is a small risk of inducing systemic anaphylaxis. Another standard test for allergy is to measure levels of IgE antibody specific for a particular allergen in a sandwich ELISA (see Section 2-7).

Although acute urticaria is commonly caused by allergens, the causes of chronic urticaria, in which the urticarial rash can recur over long periods, are less well understood. In up to a third of cases, it seems likely that chronic urticaria is an autoimmune disease caused by autoantibodies against the α chain of FcεRI. This is an example of a type II hypersensitivity reaction (see Fig. 12.2) in which an autoantibody against a cellular receptor triggers cellular activation, in this case causing mast-cell degranulation with resulting urticaria.

A more prolonged inflammatory response is sometimes seen in the skin, most often in atopic children. They develop a persistent skin rash called **eczema**, due to a chronic inflammatory response similar to that seen in the bronchial walls of patients with asthma. The etiology of eczema is not well understood and it usually clears in adolescence, unlike rhinitis and asthma, which can persist throughout life.

12-14 Allergy to foods can cause symptoms limited to the gut but can also cause systemic reactions.

When an allergen is eaten, two types of allergic response are seen. Activation of mucosal mast cells associated with the gastrointestinal tract can lead to transepithelial fluid loss and smooth muscle contraction, generating vomiting and diarrhea. For reasons that are not understood, connective tissue mast cells in the deeper layers of the skin are also activated, presumably by IgE antibodies binding to the ingested and absorbed allergen borne by the blood, resulting in urticaria. This is a common reaction when penicillin is ingested by a patient with penicillin-specific IgE antibodies.

Ingestion of food allergens can also lead to the development of generalized anaphylaxis, accompanied by cardiovascular collapse and acute asthmatic symptoms. Certain foods are particularly associated with this type of life-threatening response, an important one being peanuts.

12-15 Allergy can be treated by inhibiting either IgE production or the effector pathways activated by crosslinking of cell-surface IgE.

The approaches to the treatment and prevention of allergy are set out in Fig. 12.15. Two treatments are commonly used in clinical practice—desensitization and blockade of the effector pathways. There are also several approaches still in the experimental stage. In **desensitization**, the aim is to shift the antibody response away from an IgE-dominated response towards one dominated by IgG, which can prevent the allergen from activating IgE-mediated effector pathways. Patients are injected with escalating doses of allergen, starting with tiny amounts. This injection schedule seems gradually to divert the IgE-dominated response, driven by T_H2 cells, to one driven by T_H1 cells, with the consequent downregulation of IgE production. Recent evidence shows that desensitization is also associated with a reduction in the numbers of mast cells at the site of the allergic reaction. This procedure carries the risk of inducing IgE-mediated allergic responses as a complication of treatment.

An alternative and still experimental approach to desensitization is vaccination with peptides derived from common allergens. This procedure induces T-cell anergy *in vivo* (see Section 8-10) associated with multiple changes in the T-cell phenotype, including downregulation of cytokine production and of expression of the T-cell receptor:CD3 complex. IgE-mediated responses are not induced because IgE can recognize only the intact antigen. A major

Target step	Mechanism of treatment	Specific approach
T_H2 activation	Reverse T_H2/T_H1 balance	Injection of specific antigen or peptides Administration of cytokines e.g. IFN-γ, IL-10, IL-12, TGF-β
Activation of B cell to produce IgE	Block co-stimulation Inhibit T_H2 cytokines	Inhibit CD40L Inhibit IL-4 or IL-13
Mast-cell activation	Inhibit effects of IgE binding to mast cell	Blockade of IgE receptor
Mediator action	Inhibit effects of mediators on specific receptors Inhibit synthesis of specific mediators	Antihistamine drugs Lipo-oxygenase inhibitors
Eosinophil-dependent inflammation	Block cytokine and chemokine receptors that mediate eosinophil recruitment and activation	Inhibit IL-5 Block CCR3

Fig. 12.15 Approaches to the treatment of allergy. Possible methods of inhibiting allergic reactions are shown. Two approaches are in regular clinical use. The first is the injection of specific antigen in desensitization regimes, which are believed to divert the immune response to the allergen from a T_H2 to a T_H1 type, so that IgG is produced in place of IgE. The second clinically useful approach is the use of specific inhibitors to block the synthesis or effects of inflammatory mediators produced by mast cells.

difficulty with this approach is that individual responses to peptides are restricted by specific MHC class II alleles, and therefore different patients carrying different MHC class II alleles can respond to different allergen-derived peptides. As the human population is outbred, expressing a wide variety of MHC class II alleles, the number of peptides required to treat all allergic individuals might be very large.

The signaling pathways that enhance the IgE response in allergic disease are also potential targets for therapy. Inhibitors of IL-4, IL-5, and IL-13 would be predicted to reduce IgE responses, although redundancy between some of the activities of these cytokines might make this approach difficult to implement in practice. A second possible approach to manipulating the response is to give cytokines that promote T_H1-type responses. IFN-γ, IFN-α, IL-10, IL-12, and TGF-β have each been shown to reduce IL-4-stimulated IgE synthesis *in vitro*, and IFN-γ and IFN-α to reduce IgE synthesis *in vivo*.

Another target for therapeutic intervention might be the high-affinity IgE receptor. An effective competitor for IgE binding at this receptor could prevent the binding of antigen-specific IgE to the surfaces of mast cells, basophils, and eosinophils. Candidate competitor molecules include modified IgE Fc constructs that lack variable regions and are thus unable to bind antigen. Yet another approach would be to block the recruitment of eosinophils to sites of allergic inflammation. The eotaxin receptor CCR-3 is a potential target for this type of therapy. The production of eosinophils in bone marrow and their exit into the circulation might also be reduced by a blockade of IL-5 action.

The mainstays of therapy at present, however, are drugs that treat the symptoms of allergic disease and limit the inflammatory response that follows the activation of cells by the crosslinking of surface IgE by antigen. Anaphylactic reactions are treated with epinephrine, which stimulates the reformation of endothelial tight junctions, promotes the relaxation of constricted bronchial smooth muscle, and also stimulates the heart. Inhaled bronchodilators that act on β-adrenergic receptors to relax constricted muscle are also used to relieve acute asthma attacks. Antihistamines that block the histamine H1 receptor reduce the urticaria that follows histamine release from mast cells and eosinophils. Relevant H1 receptors include those on blood vessels that cause increased permeability to plasma and on unmyelinated nerve fibres that are thought to mediate the sensation of itch. In chronic allergic disease it is extremely important to treat and prevent the chronic inflammatory tissue injury. Topical or systemic corticosteroids (see Chapter 14) are used to suppress the chronic inflammatory changes seen in asthma, rhinitis, and eczema.

Summary.

The allergic response to innocuous antigens reflects the pathophysiological aspect of a defensive response with the physiological role of protecting hosts against helminthic parasites. It is triggered by IgE antibodies bound to the high-affinity IgE receptor FcεRI on mast cells. Mast cells are strategically distributed beneath the mucosal surfaces of the body and in connective tissue. The resulting inflammation can be divided into early events, characterized by rapidly dispersed mediators such as histamine, and later events that involve leukotrienes, cytokines, and chemokines, which recruit and activate eosinophils in particular, but also basophils. The late phase of this response can evolve into chronic inflammation, characterized by the presence of effector T cells and eosinophils, which is most clearly seen in allergic asthma.

Hypersensitivity diseases.

In the first part of this chapter we saw how IgE is involved in allergic disease, also known as type I hypersensitivity. Immunological responses involving IgG antibodies or specific T cells can also cause adverse hypersensitivity reactions. Although these effector arms of the immune response normally participate in protective immunity to infection, they occasionally react with non-infectious antigens to produce acute or chronic hypersensitivity reactions. We shall describe common examples of such reactions in this last part of the chapter.

12-16 Innocuous antigens can cause type II hypersensitivity reactions in susceptible individuals by binding to the surfaces of circulating blood cells.

Antibody-mediated destruction of red blood cells (hemolytic anemia) or platelets (thrombocytopenia) is an uncommon side-effect associated with the intake of certain drugs such as the antibiotic penicillin, the anti-cardiac arrhythmia drug quinidine, or the anti-hypertensive agent methyldopa. These are examples of **type II hypersensitivity reactions** in which the drug binds to the cell surface and serves as a target for anti-drug IgG antibodies (see Fig. 12.2). The anti-drug antibodies are made in only a minority of individuals and it is not clear why these individuals are susceptible to developing them. The cell-bound antibody triggers clearance of the cell from the circulation, predominantly by tissue macrophages in the spleen, which bear Fcγ receptors.

12-17 Systemic disease caused by immune complex formation can follow the administration of large quantities of poorly catabolized antigens.

Type III hypersensitivity reactions can arise when the antigen is soluble. The pathology is caused by the deposition of antigen:antibody aggregates or **immune complexes** in certain tissue sites. Immune complexes are generated in all antibody responses but their pathogenic potential is determined, in part, by their size. Larger aggregates fix complement and are readily cleared from the circulation by the mononuclear phagocytic system. The small complexes that form at antigen excess, however, tend to deposit in blood vessel walls. There they can ligate Fc receptors on leukocytes, leading to leukocyte activation and tissue injury.

A local type III hypersensitivity reaction can be triggered in the skin of sensitized individuals possessing IgG antibodies against the sensitizing antigen. When antigen is injected into the skin, IgG antibody that has diffused into the tissues forms immune complexes locally. The immune complexes bind Fc receptors on mast cells and other leukocytes, which creates a local inflammatory response with increased vascular permeability. The enhanced vascular permeability allows fluid and cells, especially polymorphonuclear leukocytes, to enter the site from the local vessels. This reaction is called an **Arthus reaction** (Fig. 12.16). The immune complexes also activate complement, releasing C5a, which contributes to the inflammatory reaction. The

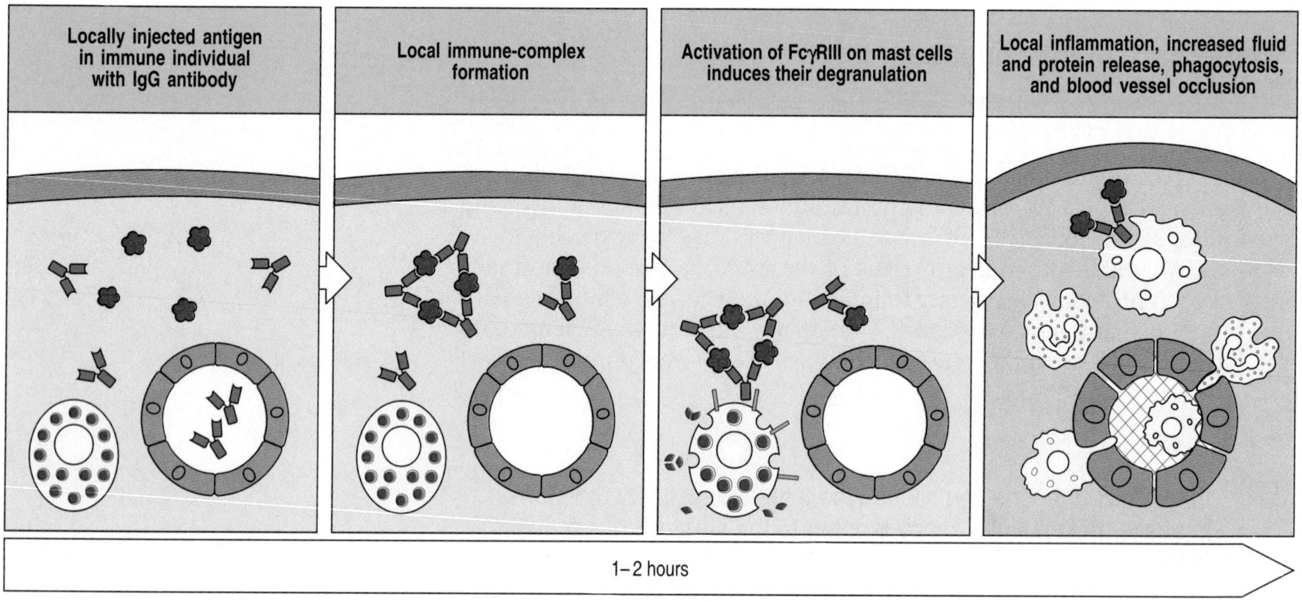
Locally injected antigen in immune individual with IgG antibody
Local immune-complex formation
Activation of FcγRIII on mast cells induces their degranulation
Local inflammation, increased fluid and protein release, phagocytosis, and blood vessel occlusion
1–2 hours

Fig. 12.16 The deposition of immune complexes in local tissues causes a local inflammatory response known as an Arthus reaction (type III hypersensitivity reaction). In individuals who have already made IgG antibody against an antigen, the same antigen injected into the skin forms immune complexes with IgG antibody that has diffused out of the capillaries. Because the dose of antigen is low, the immune complexes are only formed close to the site of injection, where they activate Fcγ receptor-bearing mast cells. As a result of mast-cell activation, inflammatory cells invade the site, and blood vessel permeability and blood flow are increased. Platelets also accumulate inside the vessel at the site, ultimately leading to vessel occlusion.

Arthus reaction is absent in mice lacking expression of the α or γ chain of the FcγRIII receptor (CD16) on mast cells, but remains largely unperturbed in complement-deficient mice, showing the primary importance of FcγRIII in triggering inflammatory responses.

A systemic type III hypersensitivity reaction, known as **serum sickness**, can result from the injection of large quantities of a poorly catabolized foreign antigen. This illness was so named because it frequently followed the administration of therapeutic horse antiserum. In the pre-antibiotic era, antiserum made by immunizing horses was often used to treat pneumococcal pneumonia; the specific anti-pneumococcal antibodies in the horse serum would help the patient to clear the infection. In much the same way, antivenin (serum from horses immunized with snake venoms) is still used today as a source of neutralizing antibodies to treat people suffering from the bites of poisonous snakes.

Serum sickness occurs 7–10 days after the injection of the horse serum, a time interval that corresponds to the time for a primary immune response to be developed against the foreign antigen. The clinical features of serum sickness are chills, fevers, rash, arthritis, and sometimes glomerulonephritis. Urticaria is a prominent feature of the rash, implying a role for histamine derived from mast-cell degranulation. In this case the mast-cell degranulation is triggered by the ligation of cell-surface FcγRIII by immune complexes.

The course of serum sickness is illustrated in Fig. 12.17. The onset of disease coincides with the development of antibodies against the abundant soluble proteins in horse serum; these antibodies form immune complexes with their antigens throughout the body. These immune complexes fix complement and can bind to and activate leukocytes bearing Fc and complement receptors; these in turn cause widespread tissue injury. The formation of immune

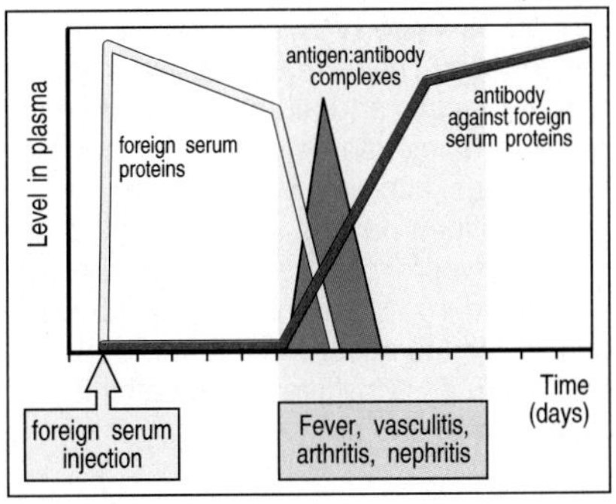

Fig. 12.17 Serum sickness is a classic example of a transient immune-complex-mediated syndrome. An injection of a foreign protein or proteins, in this case those of horse serum, leads to an antibody response. These antibodies form immune complexes with the circulating foreign proteins. The complexes are deposited in small vessels and activate complement and phagocytes, inducing fever and the symptoms of vasculitis, nephritis, and arthritis. All these effects are transient and resolve when the foreign protein is cleared.

complexes causes clearance of the foreign antigen and so serum sickness is usually a self-limiting disease. Serum sickness after a second dose of horse antiserum follows the kinetics of a secondary antibody response and the onset of disease occurs typically within a day or two. Serum sickness is nowadays seen after the use of anti-lymphocyte globulin, which is used as an immunosuppressive agent in transplant recipients (see Chapter 14), and also rarely after the administration of streptokinase, a bacterial enzyme that is used as a thrombolytic agent to treat patients with a myocardial infarction or heart attack.

A similar type of immunopathological response is seen in two other situations in which antigen persists. The first is when an adaptive antibody response fails to clear an infectious agent, for example in subacute bacterial endocarditis or in chronic viral hepatitis. In this situation, the multiplying bacteria or viruses are continuously generating new antigen in the presence of a persistent antibody response, which fails to eliminate the organism. Immune complex disease ensues, with injury to small blood vessels in many tissues and organs including the skin, kidneys, and nerves. Immune complexes also form in autoimmune diseases such as systemic lupus erythematosus where, because the antigen persists, the deposition of immune complexes continues, and serious disease can result (see Section 13-7).

Some inhaled allergens provoke IgG rather than IgE antibody responses, perhaps because they are present at relatively high levels in inhaled air. When a person is re-exposed to high doses of such inhaled antigens, immune complexes form in the alveolar wall of the lung. This leads to the accumulation of fluid, protein, and cells in the alveolar wall, slowing blood–gas interchange and compromising lung function. This type of reaction occurs in certain occupations such as farming, where there is repeated exposure to hay dust or mold spores. The disease that results is therefore called **farmer's lung**. If exposure to antigen is sustained, the alveolar membranes can become permanently damaged.

12-18 Delayed-type hypersensitivity reactions are mediated by T_H1 cells and CD8 cytotoxic T cells.

Unlike the immediate hypersensitivity reactions described so far, which are mediated by antibodies, **delayed-type hypersensitivity** or **type IV hypersensitivity reactions** are mediated by specific T cells. Such effector T cells function in essentially the same way as during a response to an infectious pathogen, as described in Chapter 8. The causes and consequences of some syndromes in which type IV hypersensitivity responses predominate are listed in Fig. 12.18. These responses are clearly caused by T cells, because they can be seen in individuals who lack immunoglobulin. Such responses can also be transferred between experimental animals by using pure T cells or cloned T-cell lines.

Fig. 12.18 Type IV hypersensitivity responses. These reactions are mediated by T cells and all take some time to develop. They can be grouped into three syndromes, according to the route by which antigen passes into the body. In delayed-type hypersensitivity the antigen is injected into the skin; in contact hypersensitivity it is absorbed into the skin; and in gluten-sensitive enteropathy it is absorbed by the gut.

Type IV hypersensitivity reactions are mediated by antigen-specific effector T cells

Syndrome	Antigen	Consequence
Delayed-type hypersensitivity	Proteins: Insect venom Mycobacterial proteins (tuberculin, lepromin)	Local skin swelling: Erythema Induration Cellular infiltrate Dermatitis
Contact hypersensitivity	Haptens: Pentadecacatechol (poison ivy) DNFB Small metal ions: Nickel Chromate	Local epidermal reaction: Erythema Cellular infiltrate Contact dermatitis
Gluten-sensitive enteropathy (celiac disease)	Gliadin	Villous atrophy in small bowel Malabsorption

The prototypic delayed-type hypersensitivity reaction is an artifact of modern medicine—the tuberculin test (see Section 2-21). This is used to determine whether an individual has previously been infected with *Mycobacterium tuberculosis*. When small amounts of a protein from *M. tuberculosis* are injected subcutaneously, a T-cell mediated local inflammatory reaction evolves over 24–72 hours in individuals who have previously responded to this pathogen. The response is mediated by T_H1 cells, which enter the site of antigen injection, recognize complexes of peptide:MHC class II on antigen-presenting cells, and release inflammatory cytokines that increase local blood vessel permeability, bringing plasma into the tissue and recruiting accessory cells to the site; this causes a visible swelling (Fig. 12.19). Each of these phases takes several hours and so the mature response appears only 24–48 hours after challenge. The cytokines produced by the activated T_H1 cells and their actions are shown in Fig. 12.20.

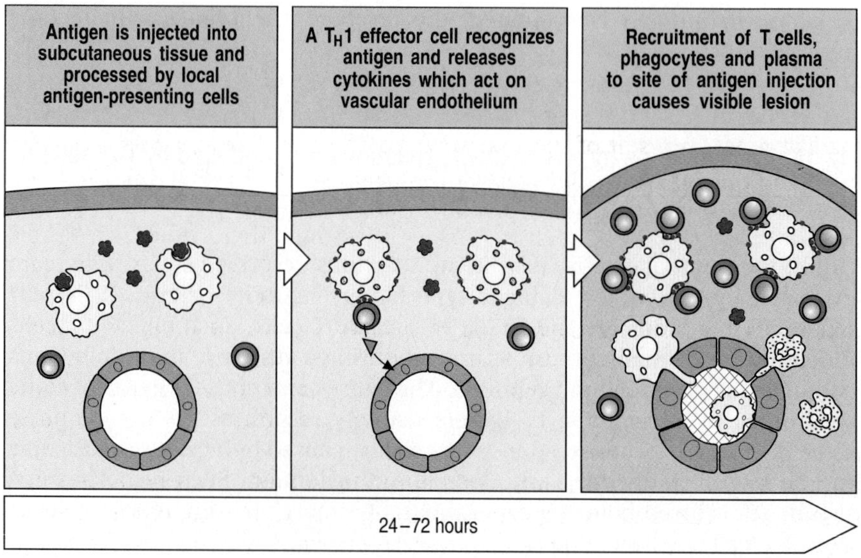
Antigen is injected into subcutaneous tissue and processed by local antigen-presenting cells
A TH1 effector cell recognizes antigen and releases cytokines which act on vascular endothelium
Recruitment of T cells, phagocytes and plasma to site of antigen injection causes visible lesion
24–72 hours

Fig. 12.19 The stages of a delayed-type hypersensitivity reaction. The first phase involves uptake, processing, and presentation of the antigen by local antigen-presenting cells. In the second phase, T_H1 cells that were primed by a previous exposure to the antigen migrate into the site of injection and become activated. Because these specific cells are rare, and because there is no inflammation to attract cells into the site, it can take several hours for a T cell of the correct specificity to arrive. These cells release mediators that activate local endothelial cells, recruiting an inflammatory cell infiltrate dominated by macrophages and causing the accumulation of fluid and protein. At this point, the lesion becomes apparent.

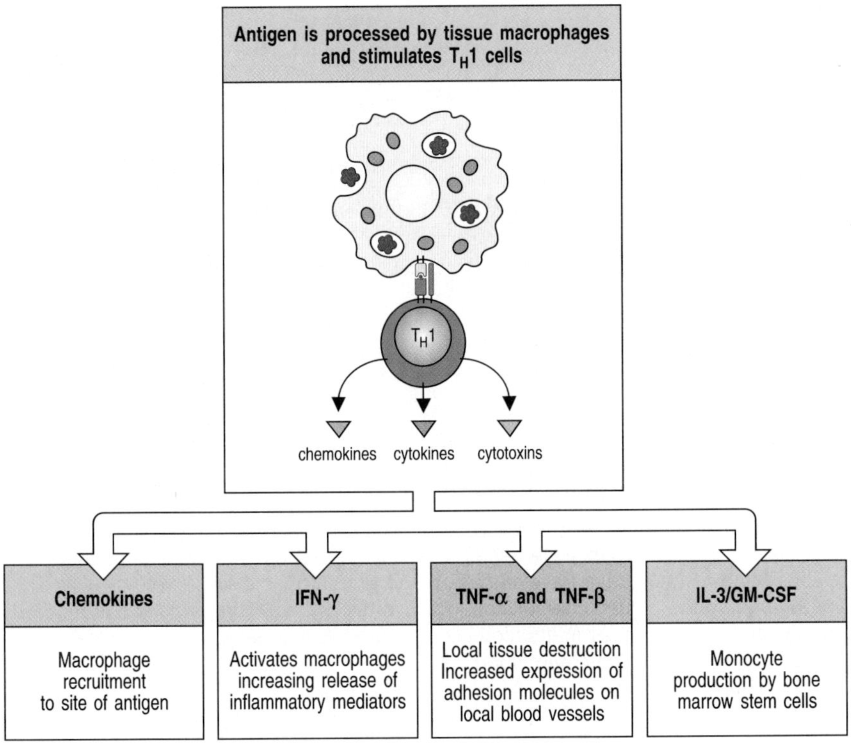

Fig. 12.20 The delayed-type (type IV) hypersensitivity response is directed by cytokines released by T_H1 cells stimulated by antigen. Antigen in the local tissues is processed by antigen-presenting cells and presented on MHC class II molecules. Antigen-specific T_H1 cells that recognize the antigen locally at the site of injection release chemokines and cytokines that recruit macrophages to the site of antigen deposition. Antigen presentation by the newly recruited macrophages then amplifies the response. T cells can also affect local blood vessels through the release of TNF-α and the cytotoxin TNF-β, and stimulate the production of macrophages through the release of IL-3 and GM-CSF. Finally, T_H1 cells activate macrophages through the release of IFN-γ and TNF-α, and kill macrophages and other sensitive cells through TNF-β or by the cell-surface expression of the Fas ligand.

Very similar reactions are observed in several cutaneous hypersensitivity responses. For instance, the rash produced by poison ivy (Fig. 12.21) is caused by a T-cell response to a chemical in the poison ivy leaf called pentadecacatechol. This compound binds covalently to host proteins. The modified self proteins are then cleaved into modified self peptides, which can bind to self MHC class II molecules and be recognized by T_H1 cells. When specifically sensitized T cells recognize these complexes, they can produce extensive inflammation (Fig. 12.22). As the chemical is delivered by contact with the skin, this is called a **contact hypersensitivity reaction**. The compounds that cause such reactions must be chemically active so that they can form stable complexes with host proteins.

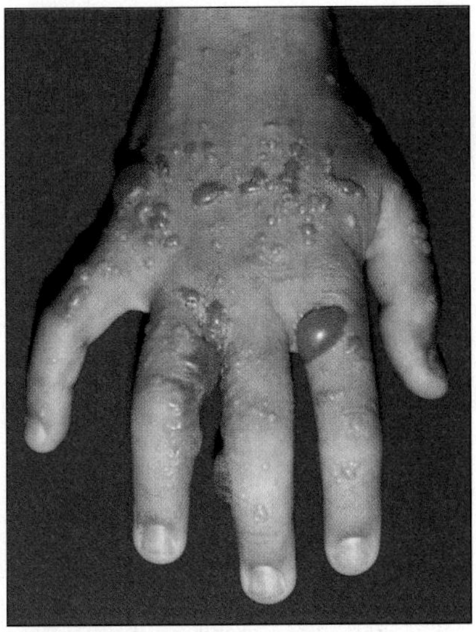

Fig. 12.21 Blistering skin lesions on hand of patient with poison ivy contact dermatitis. Photograph courtesy of R Geha.

Some insect proteins also elicit delayed-type hypersensitivity responses. However, the early phases of the host reaction to an insect bite are often IgE-mediated or the result of the direct effects of insect venoms. Finally, some unusual delayed-type hypersensitivity responses to divalent cations have been observed, for example to nickel, which can alter the conformation or peptide binding of MHC class II molecules, and thus provoke a T-cell response.

Type IV hypersensitivity reactions can also involve CD8 T cells, which damage tissues mainly by cell-mediated cytotoxicity. Some chemicals, including pentadecacatechol, are soluble in lipid and can therefore cross the cell membrane and modify intracellular proteins. These modified proteins generate modified peptides within the cytosol, which are translocated into the endoplasmic reticulum and are delivered to the cell surface by MHC class I molecules. These are recognized by CD8 T cells, which can cause damage either by killing the eliciting cell or by secreting cytokines such as IFN-γ.

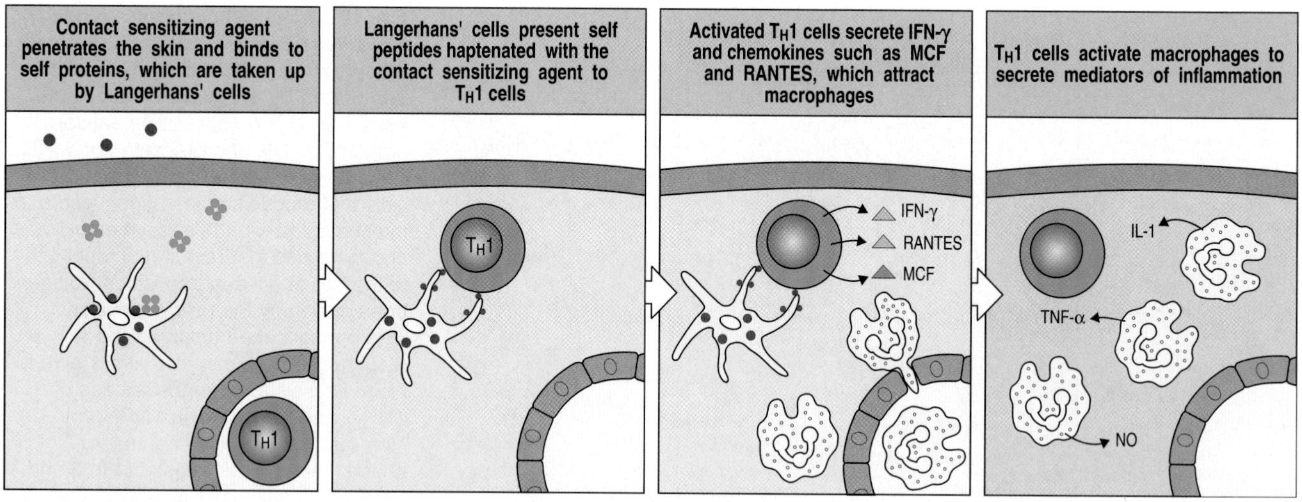
Contact sensitizing agent penetrates the skin and binds to self proteins, which are taken up by Langerhans' cells
Langerhans' cells present self peptides haptenated with the contact sensitizing agent to TH1 cells
Activated TH1 cells secrete IFN-γ and chemokines such as MCF and RANTES, which attract macrophages
TH1 cells activate macrophages to secrete mediators of inflammation
TH1
IFN-γ
RANTES
MCF
IL-1
TNF-α
NO

Fig. 12.22 Development of a delayed-type hypersensitivity response to a contact-sensitizing agent such as the pentadecacatechol in poison ivy. The contact-sensitizing agent is a small lipid-soluble molecule that can easily penetrate intact skin. It binds covalently as a hapten to a variety of endogenous proteins, which are taken up and processed by Langerhans' cells, the major antigen-presenting cells of skin. These present haptenated peptides to effector T_H1 cells (which must have been previously primed in lymph nodes and then have traveled back to the skin). These then secrete cytokines and chemokines, which in turn attract monocytes and induce their maturation into activated tissue macrophages, which contribute to the inflammatory lesions depicted in Fig. 12.21.

Summary.

Hypersensitivity diseases reflect normal immune mechanisms directed against innocuous antigens. They can be mediated by IgG antibodies bound to modified cell surfaces, or by complexes of antibodies bound to poorly catabolized antigens, as occurs in serum sickness. Hypersensitivity reactions mediated by T cells can be activated by modified self proteins, or by injected proteins such as the mycobacterial extract tuberculin. These T-cell mediated responses require the induced synthesis of effector molecules and develop more slowly, which is why they are termed delayed-type hypersensitivity.

Summary to Chapter 12.

Immune responses to otherwise innocuous antigens produce allergic or hypersensitive reactions upon re-exposure to the same antigen. Most allergies involve the production of IgE antibody to common environmental allergens. Some people are intrinsically prone to making IgE antibodies against many allergens, and such people are said to be atopic. IgE production is driven by antigen-specific T_H2 cells, which are initially primed in the presence of a burst of IL-4 released by specialized T cells early in the immune response. The IgE produced binds to the high-affinity IgE receptor FcεRI on mast cells, basophils, and activated eosinophils. The physiological role of this system is to provide front-line defense against pathogens but, in economically developed societies, it is more frequently involved in allergic reactions. Eosinophils and specific effector T cells have an extremely important role in chronic allergic inflammation, which is the major cause of the chronic morbidity of asthma. Antibodies of other isotypes and specific effector T cells contribute to hypersensitivity to other antigens.

General references.

Abbas, A.K., Murphy, K.M., and Sher, A.: **Functional diversity of helper T lymphocytes**. *Nature* 1996, **383**:787-793.

Bernstein, D.I.: **Allergic reactions to workplace allergens**. *JAMA* 1997, **278**:1907-1913.

Costa, J.J., Weller, P.F., and Galli, S.J.: **The cells of the allergic response: mast cells, basophils, and eosinophils.** *JAMA* 1997, **278**:1815-1822,

Kay, A.B.: *Allergy and Allergic Diseases.* Oxford, Blackwell Science, 1997.

Kay, A.B.: **T cells as orchestrators of the asthmatic response**. *Ciba. Found. Symp.* 1997, **206**:56-67.

Luster, A.D., and Rothenberg, M.E.: **Role of the monocyte chemoattractant protein and eotaxin subfamily of chemokines in allergic inflammation**. *J. Leukoc. Biol.* 1997, **62**:620-633.

Middleton, E. Jr., Reed, C.E., Ellis, E.F., Adkinson, N.F., Yunginger, J.W., and Busse, W.W.: *Allergy: Principles and Practice,* 4th edn. St Louis, Mosby, 1993.

Paul, W.E., Seder, R.A., and Plaut, M.: **Lymphokine and cytokine production by Fc epsilon RI+ cells**. *Adv. Immunol.* 1993, **53**:1-29.

Romagnani, S.: **Atopic allergy and other hypersensitivities interactions between genetic susceptibility, innocuous and/or microbial antigens and the immune system**. *Curr. Opin. Immunol.* 1997, **9**:773-775.

Rosen, F.S.: **Urticaria, angioedema, and anaphylaxis**. *Pediatr. Rev.* 1992, **13**:387-390.

Section references.

12-1 Allergens are often delivered transmucosally at low dose, a route that favors IgE production.

O'Hehir, R.E., Garman, R.D., Greenstein, J.L., and Lamb, J.R.: **The specificity and regulation of T-cell responsiveness to allergens**. *Annu. Rev. Immunol.* 1991, **9**:67-95.

Parronchi, P., Macchia, D., Piccinni, M.P., Biswas, P., Simonelli, C., Maggi, E., Ricci, M., Ansari, A.A., and Romagnani, S.: **Allergen and bacterial antigen-specific T-cell clones established from atopic donors show a different profile of cytokine production**. *Proc. Natl. Acad. Sci. USA* 1991, **88**:4538-4542.

Romagnani, S.: **Regulation of the development of type 2 T-helper cells in allergy**. *Curr. Opin. Immunol.* 1994, **6**:838-846.

Sertl, K., Takemura, T., Tschachler, E., Ferrans, V.J., Kaliner, M.A., and Shevach, E.M.: **Dendritic cells with antigen-presenting capability reside in airway epithelium, lung parenchyma, and visceral pleura**. *J. Exp. Med.* 1986, **163**:436-451.

12-2 Enzymes are frequent triggers of allergy.

Garraud, O., Nkenfou, C., Bradley, J.E., Perler, F.B., and Nutman, T.B.: **Identification of recombinant filarial proteins capable of inducing polyclonal and antigen-specific IgE and IgG4 antibodies**. *J. Immunol.* 1995, **155**:1316-1325.

Grammer, L.C., and Patterson, R.: **Proteins: chymopapain and insulin**. *J. Allergy. Clin. Immunol.* 1984, **74**:635-640.

Hewitt, C.R., Brown, A.P., Hart, B.J., and Pritchard, D.I.: **A major house dust mite allergen disrupts the immunoglobulin E network by selectively cleaving CD23: innate protection by antiproteases**. *J. Exp. Med.* 1995, **182**:1537-1544.

Schulz, O., Sewell, H.F., and Shakib, F.: **Proteolytic cleavage of CD25, the alpha subunit of the human T cell interleukin 2 receptor, by Der p 1, a major mite allergen with cysteine protease activity**. *J. Exp. Med.* 1998, **187**:271-275.

Thomas, W.R., Smith, W., and Hales, B.J.: **House dust mite allergen characterisation: implications for T-cell responses and immunotherapy**. *Int. Arch. Allergy. Immunol.* 1998, **115**:9-14.

Tomee, J.F., van Weissenbruch, R., de Monchy, J.G., and Kauffman, H.F.: **Interactions between inhalant allergen extracts and airway epithelial cells: effect on cytokine production and cell detachment**. *J. Allergy. Clin. Immunol.* 1998, **102**:75-85.

12-3 Class switching to IgE in B lymphocytes is favored by specific accessory signals.

Bacharier, L.B., Jabara, H., and Geha, R.S.: **Molecular mechanisms of immunoglobulin E regulation**. *Int. Arch. Allergy. Immunol.* 1998, **115**:257-269.

Bendelac, A., Rivera, M.N., Park, S.H., and Roark, J.H.: **Mouse CD1-specific NK1 T cells: development, specificity, and function**. *Annu. Rev. Immunol.* 1997, **15**:535-562.

Burd, P.R., Thompson, W.C., Max, E.E., and Mills, F.C.: **Activated mast cells produce interleukin 13**. *J. Exp. Med.* 1995, **181**:1373-1380.

Chen, H., and Paul, W.E.: **Cultured NK1.1+ CD4+ T cells produce large amounts of IL-4 and IFN-gamma upon activation by anti-CD3 or CD1**. *J. Immunol.* 1997, **159**:2240-2249.

Gauchat, J.F., Henchoz, S., Fattah, D., Mazzei, G., Aubry, J.P., Jomotte, T., Dash, L., Page, K., Solari, R., Aldebert, D., et al.: **CD40 ligand is functionally expressed on human eosinophils**. *Eur. J. Immunol.* 1995, **25**:863-865.

Gauchat, J.F., Henchoz, S., Mazzei, G., Aubry, J.P., Brunner, T., Blasey, H., Life, P., Talabot, D., Flores Romo, L., Thompson, J., et al.: **Induction of human IgE synthesis in B cells by mast cells and basophils**. *Nature* 1993, **365**:340-343.

Paul, W.E.: **Interleukin 4: signaling mechanisms and control of T cell differentiation.** *Ciba. Found. Symp.* 1997, **204**:208-216.

Romagnani, S., Parronchi, P., D'Elios, M.M., Romagnani, P., Annunziato, F., Piccinni, M.P., Manetti, R., Sampognaro, S., Mavilia, C., De-Carli, M., Maggi, E., and Del-Prete, G.F.: **An update on human Th1 and Th2 cells**. *Int. Arch. Allergy. Immunol.* 1997, **113**:153-156.

Sallusto, F., Lenig, D., Mackay, C.R., and Lanzavecchia, A.: **Flexible programs of chemokine receptor expression on human polarized T helper 1 and 2 lymphocytes**. *J. Exp. Med.* 1998, **187**:875-883.

Sallusto, F., Mackay, C.R., and Lanzavecchia, A.: **Selective expression of the eotaxin receptor CCR3 by human T helper 2 cells**. *Science* 1997, **277**:2005-2007.

Szabo, S.J., Glimcher, L.H., and Ho, I.C.: **Genes that regulate interleukin-4 expression in T cells.** *Curr. Opin. Immunol.* 1997, **9**:776-781.

12-4 Genetic factors contribute to the development of IgE-mediated allergy, but environmental factors may also be important.

Barnes, K.C., and Marsh, D.G.: **The genetics and complexity of allergy and asthma**. *Immunol. Today* 1998, **19**:325-332.

Casolaro, V., Georas, S.N., Song, Z., and Ono, S.J.: **Biology and genetics of atopic disease**. *Curr. Opin. Immunol.* 1996, **8**:796-803.

Hopkin, J.M.: **Mechanisms of enhanced prevalence of asthma and atopy in developed countries**. *Curr. Opin. Immunol.* 1997, **9**:788-792.

Matricardi, P.M., Rosmini, F., Ferrigno, L., Nisini, R., Rapicetta, M., Chionne, P., Stroffolini, T., Pasquini, P., and D'Amelio, R.: **Cross sectional retrospective study of prevalence of atopy among Italian military students with antibodies against hepatitis A virus**. *BMJ* 1997, **314**:999-1003.

Shaheen, S.O., Aaby, P., Hall, A.J., Barker, D.J., Heyes, C.B., Shiell, A.W., and Goudiaby, A.: **Measles and atopy in Guinea-Bissau**. *Lancet* 1996, **347**:1792-1796.

Shirakawa, T., Enomoto, T., Shimazu, S., and Hopkin, J.M.: **The inverse association between tuberculin responses and atopic disorder**. *Science* 1997, **275**:77-79.

12-5 Most IgE is cell-bound and engages effector mechanisms of the immune system by different pathways from other antibody isotypes.

Adamczewski, M., and Kinet, J.P.: **The high-affinity receptor for immunoglobulin E**. *Chem. Immunol.* 1994, **59**:173-190.

Bonnefoy, J.Y., Aubry, J.P., Gauchat, J.F., Graber, P., Life, P., Flores Romo, L., and Mazzei, G.: **Receptors for IgE**. *Curr. Opin. Immunol.* 1993, **5**:944-949.

Delespesse, G., Sarfati, M., Wu, C.Y., Fournier, S., and Letellier, M.: **The low-affinity receptor for IgE**. *Immunol. Rev.* 1992, **125**:77-97.

Fujiwara, H., Kikutani, H., Suematsu, S., Naka, T., Yoshida, K., Tanaka, T., Suemura, M., Matsumoto, N., Kojima, S., et al.: **The absence of IgE antibody-mediated augmentation of immune responses in CD23-deficient mice**. *Proc. Natl. Acad. Sci. USA* 1994, **91**:6835-6839.

Metzger, H.: **The receptor with high affinity for IgE**. *Immunol. Rev.* 1992, **125**:37-48.

Scharenberg, A.M., and Kinet, J.P.: **Early events in mast cell signal transduction**. *Chem. Immunol.* 1995, **61**:72-87.

Scharenberg, A.M., and Kinet, J.P.: **Initial events in Fc epsilon RI signal transduction**. *J. Allergy. Clin. Immunol.* 1994, **94**:1142-1146.

Stief, A., Texido, G., Sansig, G., Eibel, H., Le Gros, G., and van der Putten, H.: **Mice deficient in CD23 reveal its modulatory role in IgE production but no role in T and B cell development**. *J. Immunol.* 1994, **152**:3378-3390.

12-6 Mast cells reside in tissues and orchestrate allergic reactions.

Austen, K.F.: **The Paul Kallos Memorial Lecture. From slow-reacting substance of anaphylaxis to leukotriene C4 synthase**. *Int. Arch. Allergy. Immunol.* 1995, **107**:19-24.

Charlesworth, E.N.: **The role of basophils and mast cells in acute and late reactions in the skin**. *Allergy* 1997, **52**:31-43.

Galli, S.J.: **The Paul Kallos Memorial Lecture. The mast cell: a versatile effector cell for a challenging world**. *Int. Arch. Allergy. Immunol.* 1997, **113**:14-22.

Metcalfe, D.D., Baram, D., and Mekori, Y.A.: **Mast cells**. *Physiol. Rev.* 1997, **77**:1033-1079.

Rodewald, H.R., Dessing, M., Dvorak, A.M., and Galli, S.J.: **Identification of a committed precursor for the mast cell lineage**. *Science* 1996, **271**:818-822.

Vliagoftis, H., Worobec, A.S., and Metcalfe, D.D.: **The protooncogene c-kit and c-kit ligand in human disease**. *J. Allergy Clin. Immunol.* 1997, **100**:435-440.

12-7 Eosinophils are normally under tight control to prevent inappropriate toxic responses.

Capron, M., and Desreumaux, P.: **Immunobiology of eosinophils in allergy and inflammation.** *Res. Immunol.* 1997, **148:**29-33.

Collins, P.D., Marleau, S., Griffiths Johnson, D.A., Jose, P.J., and Williams, T.J.: **Cooperation between interleukin-5 and the chemokine eotaxin to induce eosinophil accumulation *in vivo***. *J. Exp. Med.* 1995, **182**:1169-1174.

Gounni, A.S., Lamkhioued, B., Delaporte, E., Dubost, A., Kinet, J.P., Capron, A., and Capron, M.: **The high-affinity IgE receptor on eosinophils: from allergy to parasites or from parasites to allergy?** *J. Allergy. Clin. Immunol.* 1994, **94**:1214-1216.

Kay, A.B., Barata, L., Meng, Q., Durham, S.R., and Ying, S.: **Eosinophils and eosinophil-associated cytokines in allergic inflammation**. *Int. Arch. Allergy. Immunol.* 1997, **113**:196-199.

Kita, H., and Gleich, G.J.: **Eosinophils and IgE receptors: a continuing controversy**. *Blood* 1997, **89**:3497-3501.

Matthews, A.N., Friend, D.S., Zimmermann, N., Sarafi, M.N., Luster, A.D., Pearlman, E., Wert, S.E., and Rothenberg, M.E.: **Eotaxin is required for the baseline level of tissue eosinophils**. *Proc. Natl. Acad. Sci. USA* 1998, **95**:6273-6278.

Palframan, R.T., Collins, P.D., Williams, T.J., and Rankin, S.M.: **Eotaxin induces a rapid release of eosinophils and their progenitors from the bone marrow**. *Blood* 1998, **91**:2240-2248.

Parker, C.W.: **Lipid mediators produced through the lipoxygenase pathway**. *Annu. Rev. Immunol.* 1987, **5**:65-84.

Rothenberg, M.E.: **Eosinophilia**. *N. Engl. J. Med.* 1998, **338**:1592-1600.

Rothenberg, M.E., MacLean, J.A., Pearlman, E., Luster, A.D., and Leder, P.: **Targeted disruption of the chemokine eotaxin partially reduces antigen-induced tissue eosinophilia**. *J. Exp. Med.* 1997, **185**:785-790.

12-8 Eosinophils and basophils cause inflammation and tissue damage in allergic reactions

Schroeder, J.T., and MacGlashan, D.W., Jr.: **New concepts: the basophil.** *J. Allergy Clin. Immunol.* 1997, **99**:429-33.

Thomas, L.L.: **Basophil and eosinophil interactions in health and disease**. *Chem. Immunol.* 1995, **61**:186-207.

12-9 The allergic reaction after ligation of IgE on mast cells is divided into an immediate response and a late-phase response.

Bentley, A.M., Kay, A.B., and Durham, S.R.: **Human late asthmatic reactions**. *Clin. Exp. Allergy* 1997, **Suppl 1**:71-86.

Liu, M.C., Hubbard, W.C., Proud, D., Stealey, B.A., Galli, S.J., Kagey Sobotka, A., Bleecker, E.R., and Lichtenstein, L.M.: **Immediate and late inflammatory responses to ragweed antigen challenge of the peripheral airways in allergic asthmatics. Cellular, mediator, and permeability changes**. *Am. Rev. Respir. Dis.* 1991, **144**:51-58.

Varney, V.A., Hamid, Q.A., Gaga, M., Ying, S., Jacobson, M., Frew, A.J., Kay, A.B., and Durham, S.R.: **Influence of grass pollen immunotherapy on cellular infiltration and cytokine mRNA expression during allergen-induced late-phase cutaneous responses**. *J. Clin. Invest.* 1993, **92**:644-651.

Werfel, S., Massey, W., Lichtenstein, L.M., and Bochner, B.S.: **Preferential recruitment of activated, memory T lymphocytes into skin chamber fluids during human cutaneous late-phase allergic reactions**. *J. Allergy. Clin. Immunol.* 1995, **96**:57-65.

12-10 The clinical effects of allergic reactions vary according to the site of mast-cell activation.

deShazo, R.D., and Kemp, S.F.: **Allergic reactions to drugs and biologic agents**. *JAMA* 1997, **278**:1895-1906.

12-11 The degranulation of mast cells in blood vessel walls after systemic absorption of allergen can cause generalized cardiovascular collapse.

Bochner, B.S. and Lichtenstein, L.M.: **Anaphylaxis**. *N. Engl. J. Med.* 1991, **324**:1785-1790.

Dombrowicz, D., Flamand, V., Brigman, K.K., Koller, B.H., and Kinet, J.P.: **Abolition of anaphylaxis by targeted disruption of the high affinity immunoglobulin E receptor alpha chain gene**. *Cell* 1993, **75**:969-976.

Fernandez, M., Warbrick, E.V., Blanca, M., and Coleman, J.W.: **Activation and hapten inhibition of mast cells sensitized with monoclonal IgE anti-penicillin antibodies: evidence for two-site recognition of the penicillin derived determinant**. *Eur. J. Immunol.* 1995, **25**:2486-2491.

Kemp, S.F., Lockey, R.F., Wolf, B.L., and Lieberman, P.: **Anaphylaxis. A review of 266 cases**. *Arch. Intern. Med.* 1995, **155**:1749-1754.

Martin, T.R., Galli, S.J., Katona, I.M., and Drazen, J.M.: **Role of mast cells in anaphylaxis. Evidence for the importance of mast cells in the cardiopulmonary alterations and death induced by anti-IgE in mice**. *J. Clin. Invest.* 1989, **83**:1375-1383.

Oettgen, H.C., Martin, T.R., Wynshaw Boris, A., Deng, C., Drazen, J.M., and Leder, P.: **Active anaphylaxis in IgE-deficient mice**. *Nature* 1994, **370**:367-370.

Reisman, R.E.: **Insect stings**. *N. Engl. J. Med.* 1994, **331**:523-527.

Weltzien, H.U., and Padovan, E.: **Molecular features of penicillin allergy**. *J. Invest. Dermatol.* 1998, **110**:203-206.

12-12 Allergen inhalation is associated with the development of rhinitis and asthma.

Arm, J.P., and Leĕ, T.H.: **The pathobiology of bronchial asthma**. *Adv.*

Immunol. 1992, **51**:323-382.

Baraniuk, J.N.: **Pathogenesis of allergic rhinitis**. *J. Allergy. Clin. Immunol.* 1997 **99**:S763-72.

Bochner, B.S., Undem, B.J., and Lichtenstein, L.M.: **Immunological aspects of allergic asthma**. *Annu. Rev. Immunol.* 1994, **12**:295-335.

Busse, W.W., Gern, J.E., and Dick, E.C.: **The role of respiratory viruses in asthma**. *Ciba. Found. Symp.* 1997, **206**:208-213.

Corrigan, C.J., and Kay, A.B.: **T cells and eosinophils in the pathogenesis of asthma**. *Immunol. Today* 1992, **13**:501-507.

Drazen, J.M., Arm, J.P., and Austen, K.F.: **Sorting out the cytokines of asthma**. *J. Exp. Med.* 1996, **183**:1-5.

Galli, S.J.: **Complexity and redundancy in the pathogenesis of asthma: reassessing the roles of mast cells and T cells**. *J. Exp. Med.* 1997, **186**:343-347.

Holgate, S.T.: **Asthma: a dynamic disease of inflammation and repair**. *Ciba. Found. Symp.* 1997, **206**:5-28; discussion 28-34, 106-110.

Naclerio. R., and Solomon, W.: **Rhinitis and inhalant allergens**. *JAMA* 1997, **278**:1842-1848.

Platts-Mills, T.A.: **The role of allergens in allergic airway disease**. *J. Allergy Clin. Immunol.* 1998, **101**:S364-S366.

12-13 Skin allergy is manifest as urticaria or chronic eczema.

Fiebiger, E., Stingl, G., and Maurer, D.: **Anti-IgE and anti-Fc epsilon RI autoantibodies in clinical allergy**. *Curr. Opin. Immunol.* 1996, **8**:784-789.

Leung, D.Y.: **Immune mechanisms in atopic dermatitis and relevance to treatment**. *Allergy Proc.* 1991, **12**:339-346.

Ring, J., Bieber, T., Vieluf, D., Kunz, B., and Przybilla, B.: **Atopic eczema, Langerhans cells and allergy**. *Int. Arch. Allergy. Appl. Immunol.* 1991, **94**:194-201.

Sabroe, R.A., Greaves, M.W.: **The pathogenesis of chronic idiopathic urticaria**. *Arch. Dermatol.* 1997, **133**:1003-1008.

12-14 Allergy to foods can cause symptoms limited to the gut but can also cause systemic reactions.

Bindslev-Jensen, C.: **Food allergy**. *BMJ* 1998, **316**:1299-1302.

Ewan, P.W.: **Clinical study of peanut and nut allergy in 62 consecutive patients: new features and associations**. *BMJ* 1996, **312**:1074-1078.

Nordlee, J.A., Taylor, S.L., Townsend, J.A., Thomas, L.A., and Bush, R.K.: **Identification of a Brazil-nut allergen in transgenic soybeans**. *N. Engl. J. Med.* 1996, **334**:688-692.

Rumsaeng, V., and Metcalfe, D.D.: **Food allergy**. *Semin. Gastrointest. Dis.* 1996, **7**:134-143.

Sampson, H.A.: **Food allergy**. *JAMA* 1997, **278**:1888-1894.

12-15 Allergy can be treated by inhibiting either IgE production or the effector pathways activated by crosslinking of cell-surface IgE.

Adkinson, N.F. Jr., Eggleston, P.A., Eney, D., Goldstein, E.O., Schuberth, K.C., Bacon, J.R., Hamilton, R.G., Weiss, M.E., Arshad, H., Meinert, C.L., Tonascia, J., and Wheeler, B.: **A controlled trial of immunotherapy for asthma in allergic children**. *N. Engl. J. Med.* 1997, **336**:324-331.

Bertrand, C., and Geppetti, P.: **Tachykinin and kinin receptor antagonists: therapeutic perspectives in allergic airway disease**. *Trends Pharmacol. Sci.* 1996, **17**:255-259.

Creticos, P.S., Reed, C.E., Norman, P.S., Khoury, J., Adkinson, N.F. Jr., Buncher, C.R., Busse, W.W., Bush, R.K., Gadde, J., Li, J.T., et al.: **Ragweed immunotherapy in adult asthma**. *N. Engl. J. Med.* 1996, **334**:501-506.

Douglass, J.A., Thien, F.C., and O'Hehir, R.E.: **Immunotherapy in asthma**. *Thorax* 1997, **52 Suppl 3**:S22-29.

Drazen, J.: **Clinical pharmacology of leukotriene receptor antagonists and 5-lipoxygenase inhibitors**. *Am. J. Respir. Crit. Care. Med.* 1998, **157**:S233-237; discussion S247-248.

Durham, S.R., Ying, S., Varney, V.A., Jacobson, M.R., Sudderick, R.M., Mackay, I.S., Kay, A.B., and Hamid, Q.A.: **Grass pollen immunotherapy inhibits allergen-induced infiltration of CD4⁺ T lymphocytes and eosinophils in the nasal mucosa and increases the number of cells expressing messenger RNA for interferon-gamma**. *J. Allergy. Clin. Immunol.* 1996, **97**:1356-1365.

Heusser, C., and Jardieu, P.: **Therapeutic potential of anti-IgE antibodies**. *Curr. Opin. Immunol.* 1997, **9**:805-813.

Lord, C.J., and Lamb, J.R.: **TH2 cells in allergic inflammation: a target of immunotherapy**. *Clin. Exp. Allergy* 1996, **26**:756-765.

Platts Mills, T.A.: **Allergen-specific treatment for asthma: III**. *Am. Rev. Respir. Dis.* 1993, **148**:553-555.

van Neerven, R.J., Ebner, C., Yssel, H., Kapsenberg, M.L., and Lamb, J.R.: **T-cell responses to allergens: epitope-specificity and clinical relevance**. *Immunol. Today* 1996, **17**:526-532.

Wenzel, S.E.: **New approaches to anti-inflammatory therapy for asthma**. *Am. J. Med.* 1998, **104**:287-300.

12-16 Innocuous antigens can cause type II hypersensitivity reactions in susceptible individuals by binding to the surfaces of circulating blood cells.

Greinacher, A., Potzsch, B., Amiral, J., Dummel, V., Eichner, A., and Mueller Eckhardt, C.: **Heparin-associated thrombocytopenia: isolation of the antibody and characterization of a multimolecular PF4-heparin complex as the major antigen**. *Thromb. Haemost.* 1994, **71**:247-251.

Murphy, W.G., and Kelton, J.G.: **Immune haemolytic anaemia and thrombocytopenia: drugs and autoantibodies**. *Biochem. Soc. Trans.* 1991, **19**:183-186.

Petz, L.D.: **Drug-induced autoimmune hemolytic anemia**. *Transfus. Med. Rev.* 1993, **7**:242-254.

Salama, A., Santoso, S., and Mueller Eckhardt, C.: **Antigenic determinants responsible for the reactions of drug-dependent antibodies with blood cells**. *Br. J. Haematol.* 1991, **78**:535-539.

12-17 Systemic disease caused by immune complex formation can follow the administration of large quantities of poorly catabolized antigens.

Bielory, L., Gascon, P., Lawley, T.J., Young, N.S., and Frank, M.M.: **Human serum sickness: a prospective analysis of 35 patients treated with equine antithymocyte globulin for bone marrow failure**. *Medicine Baltimore* 1988, **67**:40-57.

Cochrane, C.G., and Koffler, D.: **Immune complex disease in experimental animals and man**. *Adv. Immunol.* 1973, **16**:185-264.

Davies, K.A., Mathieson, P., Winearls, C.G., Rees, A.J., and Walport, M.J.: **Serum sickness and acute renal failure after streptokinase therapy for myocardial infarction**. *Clin. Exp. Immunol.* 1990, **80**:83-88.

Lawley, T.J., Bielory, L., Gascon, P., Yancey, K.B., Young, N.S., and Frank, M.M.: **A prospective clinical and immunologic analysis of patients with serum sickness**. *N. Engl. J. Med.* 1984, **311**:1407-1413.

Ravetch, J.V., and Clynes, R.: **Divergent roles for Fc receptors and complement *in vivo***. *Annu. Rev. Immunol.* 1998, **16**:421-432.

Schifferli, J.A., Ng, Y.C., and Peters, D.K.: **The role of complement and its receptor in the elimination of immune complexes**. *N. Engl. J. Med.* 1986, **315**:488-495.

Theofilopoulos, A.N. and Dixon, F.J.: **Immune complexes in human diseases: a review**. *Am. J. Pathol.* 1980, **100**:529-594.

12-18 Delayed-type hypersensitivity reactions are mediated by T_H1 cells and CD8 cytotoxic T cells.

Bernhagen, J., Bacher, M., Calandra, T., Metz, C.N., Doty, S.B., Donnelly, T., and Bucala, R.: **An essential role for macrophage migration inhibitory factor in the tuberculin delayed-type hypersensitivity reaction**. *J. Exp. Med.* 1996, **183**:277-282.

Grabbe, S., and Schwarz, T.: **Immunoregulatory mechanisms involved in elicitation of allergic contact hypersensitivity**. *Immunol. Today* 1998, **19**:37-44.

Ishii, N., Sugita, Y., Nakajima, H., Tanaka, S., and Askenase, P.W.: **Elicitation of nickel sulfate (NiSO4)-specific delayed-type hypersensitivity requires early-occurring and early-acting, NiSO4-specific DTH-initiating cells with an unusual mixed phenotype for an antigen-specific cell**. *Cell Immunol.* 1995, **161**:244-255.

Kalish, R.S., Wood, J.A., and LaPorte, A.: **Processing of urushiol (poison ivy) hapten by both endogenous and exogenous pathways for presentation to T cells *in vitro***. *J. Clin. Invest.* 1994, **93**:2039-2047.

Larsen, C.G., Thomsen, M.K., Gesser, B., Thomsen, P.D., Deleuran, B.W., Nowak, J., Skodt, V., Thomsen, H.K., Deleuran, M., Thestrup Pedersen, K., et al.: **The delayed-type hypersensitivity reaction is dependent on IL-8. Inhibition of a tuberculin skin reaction by an anti-IL-8 monoclonal antibody**. *J. Immunol.* 1995, **155**:2151-2157.

Muller, G., Saloga, J., Germann, T., Schuler, G., Knop, J., and Enk, A.H.: **IL-12 as mediator and adjuvant for the induction of contact sensitivity *in vivo***. *J. Immunol.* 1995, **155**:4661-4668.

Immune Responses in the Absence of Infection

13

We have learned in preceding chapters that the adaptive immune response is a critical component of host defense against infection and therefore essential for normal health. Unfortunately, adaptive immune responses are also sometimes elicited by antigens not associated with infectious agents, and this may cause serious disease. These responses are essentially identical to adaptive immune responses to infectious agents; only the antigens differ. In Chapter 12, we saw how responses to certain environmental antigens cause allergic diseases and other hypersensitivity reactions. In this chapter we will examine responses to two particularly important categories of antigen: responses to self tissue antigens, called **autoimmunity**, which can lead to **autoimmune diseases** characterized by tissue damage; and responses to transplanted organs that lead to **graft rejection**. We will examine these disease processes and the mechanisms that lead to the undesirable adaptive immune responses that are their root cause.

Autoimmunity: responses to self antigens.

Autoimmune disease occurs when a specific adaptive immune response is mounted against self antigens. The normal consequence of an adaptive immune response against a foreign antigen is the clearance of the antigen from the body. Virus-infected cells, for example, are destroyed by cytotoxic T cells, whereas soluble antigens are cleared by formation of immune complexes, which are taken up by cells of the mononuclear phagocytic system such as macrophages. However, when an adaptive immune response develops against self antigens, it is usually impossible for immune effector mechanisms to eliminate the antigen completely, and thus the response is sustained. The consequence is that the effector pathways of immunity cause chronic inflammatory injury to tissues, which may prove lethal. The

Some common autoimmune diseases classified by immunopathogenic mechanism		
Syndrome	**Autoantigen**	**Consequence**
Type II antibody to cell-surface or matrix antigens		
Autoimmune hemolytic anemia	Rh blood group antigens, I antigen	Destruction of red blood cells by complement and phagocytes, anemia
Autoimmune thrombocytopenic purpura	Platelet integrin GpIIb:IIIa	Abnormal bleeding
Goodpasture's syndrome	Non-collagenous domain of basement membrane collagen type IV	Glomerulonephritis Pulmonary hemorrhage
Pemphigus vulgaris	Epidermal cadherin	Blistering of skin
Acute rheumatic fever	Streptococcal cell-wall antigens. Antibodies cross-react with cardiac muscle	Arthritis, myocarditis, late scarring of heart valves
Type III immune-complex disease		
Mixed essential cryoglobulinemia	Rheumatoid factor IgG complexes (with or without hepatitis C antigens)	Systemic vasculitis
Systemic lupus erythematosus	DNA, histones, ribosomes, snRNP, scRNP	Glomerulonephritis, vasculitis, arthritis
Type IV T-cell mediated disease		
Insulin-dependent diabetes mellitus	Pancreatic β-cell antigen	β-cell destruction
Rheumatoid arthritis	Unknown synovial joint antigen	Joint inflammation and destruction
Experimental autoimmune encephalomyelitis (EAE), multiple sclerosis	Myelin basic protein, proteolipid protein, myelin oligodendrocyte glycoprotein	Brain invasion by CD4 T cells, paralysis

Fig. 13.1 Autoimmune diseases classified by the mechanism of tissue damage. All the mechanisms outlined in Fig. 12.2 except type I responses also occur in autoimmune diseases. Some additional autoimmune diseases in which the antigen is a cell-surface receptor are listed later in Fig. 13.9.

mechanisms of tissue damage in autoimmune diseases are essentially the same as those that operate in protective immunity and in hypersensitivity diseases. Some common autoimmune diseases are listed in Fig. 13.1.

Adaptive immune responses are initiated by activating antigen-specific T cells, and it is believed that autoimmunity is initiated in the same way. T-cell responses to self-antigens can inflict tissue damage either directly or indirectly. Cytotoxic T-cell responses and inappropriate activation of macrophages by T_H1 cells can cause extensive tissue damage, whereas inappropriate T-cell help to self-reactive B cells can initiate harmful autoantibody responses. Autoimmune responses are a natural consequence of the open repertoires of both B-cell and T-cell receptors that allows them to recognize any pathogen. Although these repertoires are purged of receptors that respond to self antigens encountered during development, they still include receptors reactive to some self antigens. It is not known what triggers autoimmunity, but both environmental and genetic factors, especially MHC genotype, are clearly important. Transient autoimmune responses are common, but it is only when they are sustained and cause lasting tissue damage that they attract medical attention. In this section, we shall examine the nature of autoimmune responses and how autoimmunity leads to tissue damage. In the last section of this chapter, we shall examine the mechanisms by which self tolerance is lost and autoimmune responses are initiated.

13-1 Specific adaptive immune responses to self antigens can cause autoimmune disease.

It was realized early in the study of immunity that the powerful effector mechanisms used in host defense could, if turned against the host, lead to severe tissue damage; Ehrlich termed this *horror autotoxicus*. Healthy individuals do not mount sustained adaptive immune responses to their own antigens and, although transient responses to damaged self tissues occur, these rarely cause additional tissue damage. However, although self tolerance is the general rule, sustained immune responses to self tissues occur, and these autoimmune responses cause the severe tissue damage that Ehrlich predicted.

In certain genetically susceptible species of experimental animals, autoimmune disease can be induced artificially by injection of 'self' tissues from a genetically identical animal mixed with strong adjuvants containing bacteria (see Section 2-4). This shows how autoimmunity can be provoked by inducing a specific, adaptive immune response to self antigens and forms the basis for our understanding of how autoimmune disease arises. In humans, autoimmunity usually arises spontaneously; that is, we do not know what events initiated the immune response to self that leads to the autoimmune disease. However, as we shall learn in the last section of this chapter, there is a strong association between infection and the onset of autoimmunity, suggesting that infectious agents have a critical role in the process. Although anyone can, in principle, develop an autoimmune disease, it seems that some individuals are more at risk than others of developing particular diseases. We shall first consider those factors that contribute to susceptibility.

13-2 Susceptibility to autoimmune diseases is controlled by environmental and genetic factors, especially MHC genes.

The best evidence in humans that there are disease susceptibility genes for autoimmunity comes from family studies and especially from studies of twins. A semiquantitative technique for measuring what proportion of the susceptibility to a particular disease arises from genetic factors is to compare the incidence of disease in monozygotic and dizygotic twins. If a disease shows a high concordance in all twins, it could be caused by shared genetic or environmental factors. This is because both monozygotic and dizygotic twins tend to be brought up in similarly shared environmental conditions. However, if the high concordance is restricted to monozygotic rather than dizygotic twins, then genetic factors are likely to be more important than environmental factors.

Studies with twins have been undertaken for several human diseases in which autoimmunity is important, including insulin-dependent diabetes mellitus, rheumatoid arthritis, multiple sclerosis, and systemic lupus erythematosus. In each case, around 20% of pairs of monozygotic twins show disease concordance, compared with less than 5% of dizygotic twins. A similar technique is to compare the frequency of a disease such as diabetes in the siblings of patients who have diabetes with the frequency of that disease in the general population. The ratio of these two frequencies gives a measure of the heritability of the disease, although shared environmental factors within families could also be at least partly responsible for an increased frequency.

Results from both twin and family studies show an important role for both inherited and environmental factors in the induction of autoimmune disease. In addition to this evidence from humans, certain inbred mouse strains reliably develop particular spontaneous or experimentally induced autoimmune diseases whereas others do not. These findings have led to an extensive search for genes that determine susceptibility to autoimmune disease.

Fig. 13.2 Associations of HLA serotype and of sex with susceptibility to autoimmune disease. The 'relative risk' for an HLA allele in an autoimmune disease is calculated by comparing the observed number of patients carrying the HLA allele with the number that would be expected, given the prevalence of the HLA allele in the general population. For insulin-dependent diabetes mellitus, the association is in fact with the HLA-DQ gene, which is tightly linked to the DR genes but is not detectable by serotyping. Some diseases show a significant bias in the sex ratio; this is taken to imply that sex hormones are involved in pathogenesis. Consistent with this, the difference in the sex ratio in these diseases is greatest between the menarche and the menopause, when levels of such hormones are highest.

Associations of HLA serotype with susceptibility to autoimmune disease

Disease	HLA allele	Relative risk	Sex ratio (♀:♂)
Ankylosing spondylitis	B27	87.4	0.3
Acute anterior uveitis	B27	10.04	<0.5
Goodpasture's syndrome	DR2	15.9	~1
Multiple sclerosis	DR2	4.8	10
Graves' disease	DR3	3.7	4–5
Myasthenia gravis	DR3	2.5	~1
Systemic lupus erythematosus	DR3	5.8	10–20
Insulin-dependent diabetes mellitus	DR3/DR4 heterozygote	~25	~1
Rheumatoid arthritis	DR4	4.2	3
Pemphigus vulgaris	DR4	14.4	~1
Hashimoto's thyroiditis	DR5	3.2	4–5

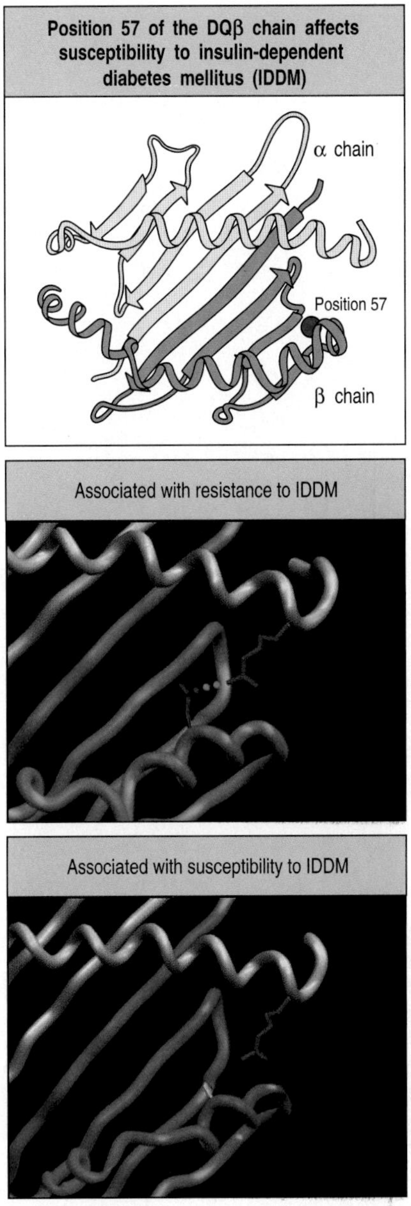

Fig. 13.3 Amino-acid changes in the sequence of an MHC class II protein correlate with susceptibility to and protection from diabetes. The HLA-DQβ_1 chain contains an aspartic acid (Asp) at position 57 in most people; in Caucasoid populations, patients with insulin-dependent diabetes mellitus (IDDM) more often have valine, serine, or alanine at this position instead, as well as other differences. Asp 57, shown in red on the backbone structure of the DQβ chain, forms a salt bridge (shown in green in the middle panel) to an arginine residue (shown in pink) in the adjacent α chain (gray). The change to an uncharged residue (for example, alanine, shown in yellow in the bottom panel) disrupts this salt bridge, altering the stability of the DQ molecule. The non-obese diabetic (NOD) strain of mice, which develops spontaneous diabetes, shows a similar replacement of serine for aspartic acid at position 57 of the homologous I-Aβ chain, and NOD mice transgenic for β chains with Asp 57 have a marked reduction in diabetes incidence. Photographs courtesy of C Thorpe.

So far the most consistent association for susceptibility to autoimmune disease has been with the MHC genotype. Many human autoimmune diseases show HLA-linked disease associations (Fig. 13.2) and these have been defined more exactly as HLA genotyping has become more precise through polymerase chain reaction analysis of HLA allele sequences. For example, the association between insulin-dependent diabetes mellitus, and the DR3 and DR4 alleles—initially discovered by HLA serotyping using antibodies—is now known to be between the disease and the DQβ genotype, which is linked closely to DR3 and DR4. The normal DQβ amino acid sequence has an aspartic acid at position 57, whereas in Caucasoid populations, patients with diabetes mostly have valine, serine, or alanine at that position (Fig. 13.3); the non-obese diabetic (NOD) strain of mice, which develops spontaneous diabetes, also has a serine at that position in the homologous MHC class II molecule. In most autoimmune diseases, susceptibility is linked most closely with MHC class II alleles but in some cases there are strong associations with particular MHC class I alleles.

The association of MHC genotype with autoimmune disease is not surprising, because autoimmune responses involve T cells, and the ability of T cells to respond to a particular antigen depends on MHC genotype. Thus the associations can be explained by a simple model in which susceptibility to an autoimmune disease is determined by differences in the ability of different allelic variants of MHC molecules to present autoantigenic peptides to autoreactive T cells. This would be consistent with what we know of T-cell involvement in specific diseases. For example, in diabetes, there are associations with both MHC class I and class II alleles; both CD8 and CD4 T cells, which respond to antigens presented by MHC class I and MHC class II molecules respectively, are known to mediate the autoimmune response.

An alternative hypothesis for the association between MHC genotype and susceptibility to autoimmune diseases emphasizes the role of MHC alleles in shaping the T-cell receptor repertoire (see Chapter 7). This hypothesis proposes that self peptides associated with certain MHC molecules may drive the positive selection of developing thymocytes that are specific for particular autoantigens. Such autoantigenic peptides might be expressed at too low a level to drive negative intrathymic selection, but be present at a sufficient level to drive positive intrathymic selection. Evidence that events in the thymus have such a role comes from the finding that insulin genes that are transcribed at a high level in the thymus tend to protect against the development of diabetes, whereas genes transcribed at a lower level are associated with disease susceptibility.

The association of MHC genotype with disease is assessed initially by comparing the frequency of different alleles in patients with their frequency in the normal population. For insulin-dependent diabetes mellitus, this approach originally demonstrated an association between disease susceptibility and the MHC class II alleles HLA-DR3 and HLA-DR4 (Fig. 13.4), which are linked tightly to HLA-DQ. In addition, such studies showed that the MHC class II allele HLA-DR2 has a dominant protective effect; individuals carrying HLA-DR2, even in association with one of the susceptibility alleles, rarely develop diabetes. Another way of determining whether MHC genes are important in autoimmune disease is to study the families of affected patients; it has been shown that two siblings affected with the same autoimmune disease are far more likely than expected to share the same MHC haplotypes (Fig. 13.5).

However, MHC genotype alone does not determine whether a person develops disease. Identical twins, sharing all of their genes, are far more likely to develop the same autoimmune disease than MHC-identical siblings, demonstrating that genetic factors other than the MHC also affect disease susceptibility. Recent studies of the genetics of autoimmune diabetes in humans and mice have shown that there are several independently segregating disease susceptibility loci in addition to the MHC.

There is evidence that several other families of genes might be important in increasing susceptibility to autoimmune disease. In humans, inherited homozygous deficiency of the early proteins of the classical pathway of complement (C1, C4, or C2) is very strongly associated with the development of systemic lupus erythematosus (SLE). The mechanism of this association is unknown but might involve the abnormal processing of immune complexes in the absence of a functional classical pathway of complement fixation. In mice and humans, abnormalities in the genes encoding proteins involved in the regulation of lymphocyte apoptosis, including Fas (CD95) and Fas ligand (CD95 ligand), are strongly associated with the development of SLE. There is preliminary evidence that inherited variation in the level of expression of certain cytokines such as tumor necrosis factor-α (TNF-α) might also increase susceptibility to autoimmune disease.

A further very important factor in disease susceptibility is the hormonal status of the patient. Many autoimmune diseases show a strong sex bias (see Fig. 13.2). Where a bias towards disease in one sex is observed in experimental animals, castration or the administration of estrogen to males usually normalizes disease incidence between the two sexes. Furthermore, many autoimmune diseases that are more common in females show peak incidence in the years of active childbearing, when production of the female sex hormones estrogen and progesterone is at its greatest. A thorough understanding of how these genetic and hormonal factors contribute to disease susceptibility might allow us to prevent the autoimmune response.

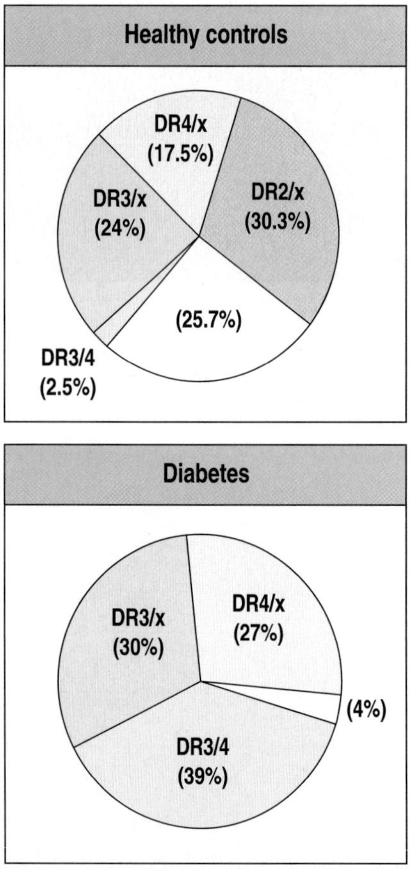

Fig. 13.4 Population studies show association of susceptibility to insulin-dependent diabetes mellitus (IDDM) with HLA genotype. The HLA genotypes (determined by serotyping) of diabetic patients (bottom panel) are not representative of those found in the population (top panel). Almost all diabetic patients express HLA-DR3 and/or HLA-DR4, and HLA-DR3/DR4 heterozygosity is greatly over-represented in diabetics compared with controls. These alleles are linked tightly to HLA-DQ alleles that confer susceptibility to IDDM. By contrast, HLA-DR2 protects against the development of IDDM and is found only extremely rarely in diabetic patients. The small letter x represents any allele other than DR2, DR3 or DR4.

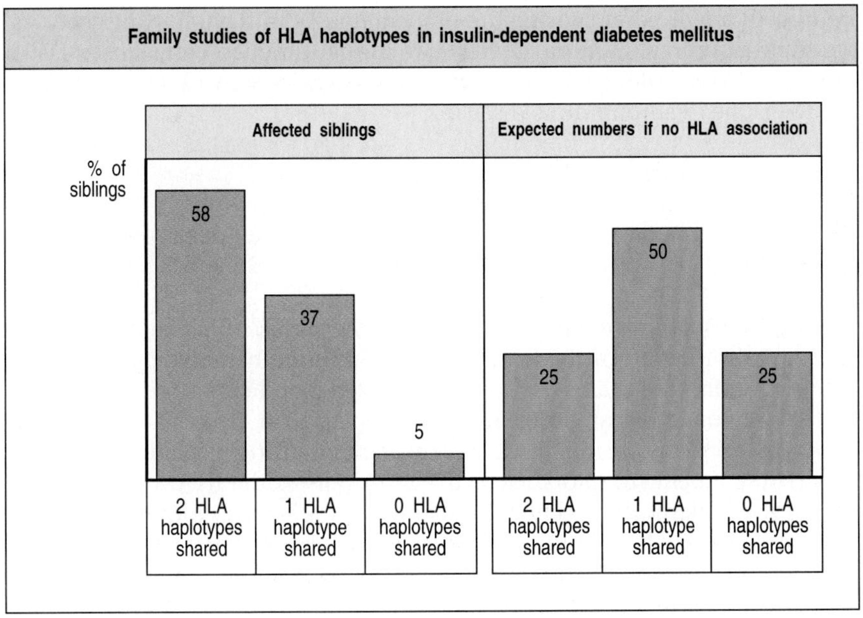

Fig. 13.5 Family studies show strong linkage of susceptibility to insulin-dependent diabetes mellitus (IDDM) with HLA genotype. In families in which two or more siblings have IDDM, it is possible to compare the HLA genotypes of affected siblings. Affected siblings share two HLA haplotypes much more frequently than would be expected if the HLA genotype did not influence disease susceptibility.

13-3 Either antibody or T cells can cause tissue damage in autoimmune disease.

Autoimmune diseases are mediated by sustained adaptive immune responses specific for self antigens. Tissue injury results because the antigen is an intrinsic component of the body and consequently the effector mechanisms of the immune system are directed at self tissues. Also, because the adaptive immune response is incapable of removing the offending autoantigen from the body, the immune response persists, and there is a constant supply of new autoantigen, which amplifies the response.

The mechanisms of tissue injury in autoimmunity can be classified according to the scheme adopted for hypersensitivity reactions (see Figs 13.1 and 12.2). As with the hypersensitivity reactions, tissue damage can be mediated by the effector actions of both T cells and antibodies. The antigen or group of antigens against which the autoimmune response is directed, and the mechanism by which the antigen-bearing tissue is damaged, together determine the pathology and clinical expression of the disease (see Fig. 13.1).

Autoimmune diseases differ from hypersensitivity responses in that Type I IgE-mediated responses do not seem to have a major role. However, IgE autoantibodies have been found in autoimmune disease, and although they have not been proved to mediate any autoimmune disease, there are diseases where this may be so. For example, asthma and eosinophilia (see Chapter 12) are found in a rare autoimmune vasculitis (an inflammatory disease of blood vessels), known as Churg–Strauss vasculitis.

By contrast, autoimmunity causing tissue injury by mechanisms analogous to type II hypersensitivity reactions is quite common. In this form of autoimmunity, IgG or IgM responses to autoantigens located on cell surfaces or extracellular matrix cause the tissue damage. In other cases of autoimmunity, tissue damage can be due to type III responses involving immune complexes containing autoantibodies to soluble autoantigens; these autoimmune diseases are systemic and are characterized by autoimmune vasculitis. Finally, in a number of organ-specific autoimmune diseases, T-cell responses are directly involved in causing the tissue damage. In the following sections we shall examine in more detail how autoantibodies can cause tissue damage, before ending with a consideration of self-reactive T-cell responses and their role in autoimmune disease.

13-4 Autoantibodies against blood cells promote their destruction.

IgG or IgM responses to antigens located on the surface of blood cells lead to the rapid destruction of these cells. In **autoimmune hemolytic anemia**, for instance, antibodies against self antigens on red blood cells trigger red blood cell destruction. This can occur in two ways (Fig. 13.6). Red cells with bound IgG or IgM antibody are rapidly cleared from the circulation by interaction with Fc or complement receptors respectively on cells of the fixed mononuclear phagocytic system; this occurs particularly in the spleen. Alternatively, the autoantibody-sensitized red cells are lysed by formation of the membrane-attack complex of complement. In **autoimmune thrombocytopenic purpura**, autoantibodies against the GpIIb:IIIa fibrinogen receptor on platelets cause thrombocytopenia (a depletion of platelets), which can in turn cause hemorrhage.

Lysis of nucleated cells by complement is less common because these cells are better defended by complement regulatory proteins. These proteins protect cells against immune attack by interfering with the activation of complement components and their assembly into a membrane-attack complex (see Section 9-31). Although the activation of complement by the bound autoantibody can proceed to a limited degree, nucleated cells are able to resist lysis by exocytosis or endocytosis of parts of the cell membrane bearing the

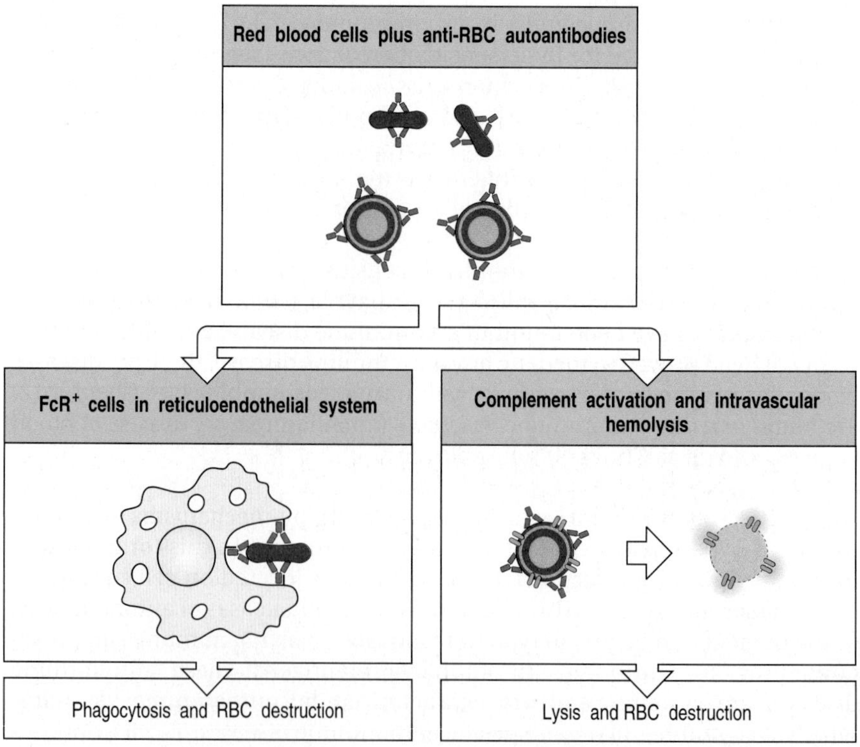

Fig. 13.6 Antibodies specific for cell-surface antigens can destroy cells. In autoimmune hemolytic anemias, red cells coated with IgG autoantibodies against a cell-surface antigen are rapidly cleared from the circulation by uptake by Fc receptor-bearing macrophages in the fixed mononuclear phagocytic system (left panel). Red cells coated with IgM autoantibodies fix C3 and are cleared by CR1- and CR3-bearing macrophages in the fixed mononuclear phagocytic system (not shown). Uptake and clearance by these mechanisms occurs mainly in the spleen. The binding of certain rare autoantibodies that fix complement extremely efficiently causes the formation of the membrane-attack complex on the red cells, leading to intravascular hemolysis (right panel).

membrane-attack complex. Nevertheless, nucleated cells targeted by autoantibodies are still destroyed by cells of the reticuloendothelial system. Autoantibodies against neutrophils cause neutropenia, which increases susceptibility to pyogenic infection. In all of these cases, accelerated clearance of autoantibody-sensitized cells is the cause of their depletion from the blood. One therapeutic approach to this type of autoimmunity is removal of the spleen, the organ in which the main clearance of red cells, platelets, and leukocytes occurs.

13-5 The fixation of sub-lytic doses of complement to cells in tissues stimulates a powerful inflammatory response.

The binding of IgG and IgM autoantibodies to cells in tissues causes inflammatory injury by a variety of mechanisms. Although nucleated cells are relatively resistant to lysis by complement, as described above, the assembly of sub-lytic amounts of the membrane-attack complex on their surface provides a powerful activating stimulus. Depending on the type of cell, the interaction of sub-lytic doses of the membrane-attack complex with the cell membrane can cause cytokine release, generation of a respiratory burst, or the mobilization of membrane phospholipids to generate arachidonic acid—the precursor to prostaglandins and leukotrienes.

Most cells in tissues are fixed in place and cells of the inflammatory system are attracted to them by the complement fragment C5a, which is released as a result of the complement activation triggered by autoantibody binding, and by other chemoattractants, such as leukotriene B4, that can be released by the autoantibody-targeted cells. Inflammatory leukocytes are further activated by binding to autoantibody Fc regions and fixed complement C3 fragments on the tissue cells. Tissue injury can then result from the products of the activated leukocytes and by antibody-dependent cellular cytotoxicity mediated by natural killer (NK) cells (see Section 9-19).

A probable example of this type of autoimmunity is **Hashimoto's thyroiditis**, in which autoantibodies against tissue-specific antigens such as thyroid peroxidase and thyroglobulin are found at extremely high levels for prolonged periods. Direct T-cell mediated cytotoxicity, which we shall discuss later, is probably also important in this disease.

13-6 Autoantibodies against receptors cause disease by stimulating or blocking receptor function.

A special class of type II hypersensitivity reaction occurs when the autoantibody binds to a cell-surface receptor. Antibody binding to a receptor can either stimulate the receptor or block its stimulation by its natural ligand. In **Graves' disease**, autoantibody against the thyroid-stimulating hormone receptor on thyroid cells stimulates the production of excessive thyroid hormone. The production of thyroid hormone is normally controlled by feedback regulation; high levels of thyroid hormone inhibit release of thyroid-stimulating hormone by the pituitary. In Graves' disease, feedback inhibition fails because the autoantibody continues to stimulate the thyroid-stimulating hormone receptor in the absence of thyroid-stimulating hormone, and the patients become hyperthyroid (Fig. 13.7).

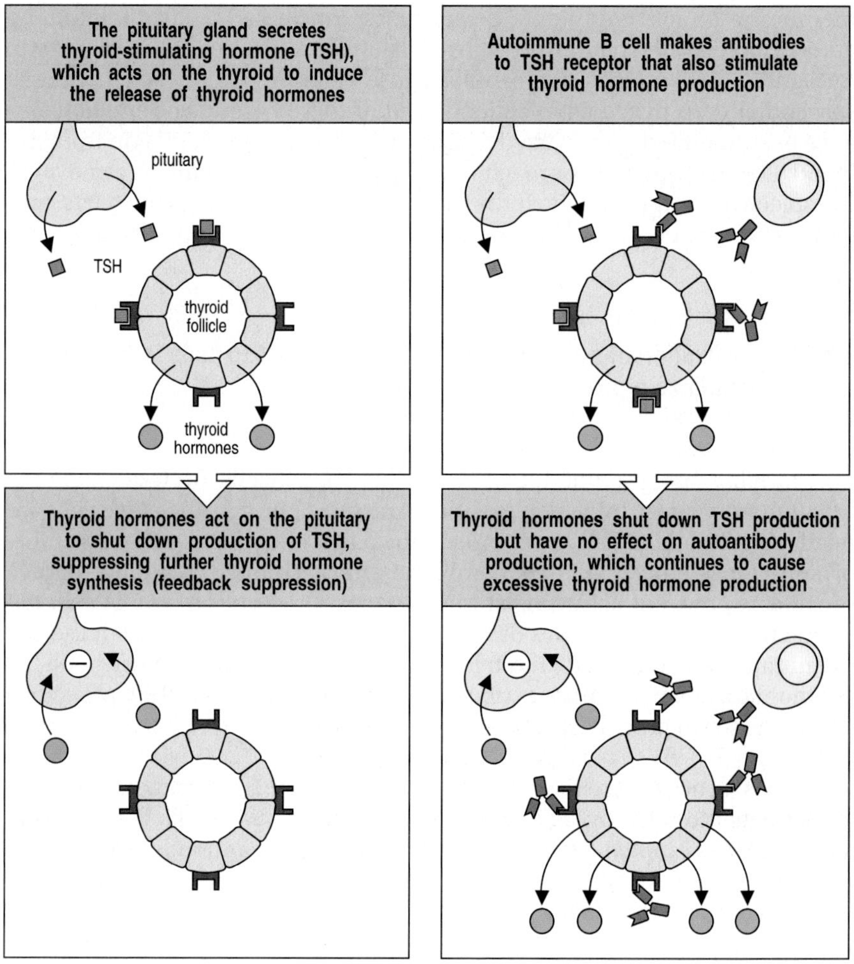

Fig. 13.7 Feedback regulation of thyroid hormone production is disrupted in Graves' disease. Graves' disease is caused by antibodies specific for the receptor for thyroid-stimulating hormone (TSH). Normally, thyroid hormones are produced in response to TSH and limit their own production by inhibiting the production of TSH by the pituitary (left panels). In Graves' disease, the autoantibodies are agonists for the TSH receptor and therefore stimulate production of thyroid hormones (right panels). The thyroid hormones inhibit TSH production in the normal way but do not affect production of the autoantibody; the excessive thyroid hormone production induced in this way causes hyperthyroidism.

In **myasthenia gravis**, autoantibodies against the α chain of the nicotinic acetylcholine receptor, which is present on skeletal muscle cells at neuromuscular junctions, can block neuromuscular transmission. The antibodies are believed to drive the internalization and intracellular degradation of acetylcholine receptors (Fig. 13.8). Patients with myasthenia gravis develop progressive weakness and may eventually die as a result of their autoimmune disease. Diseases caused by autoantibodies that act as agonists or antagonists for cell-surface receptors are listed in Fig. 13.9.

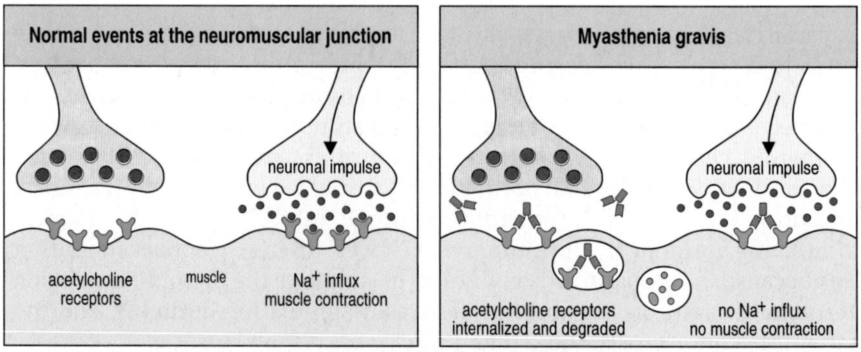

Fig. 13.8 Autoantibodies inhibit receptor function in myasthenia gravis. In normal circumstances, acetylcholine released from stimulated motor neurons at the neuromuscular junction binds to acetylcholine receptors on skeletal muscle cells, triggering muscle contraction (left panel). Myasthenia gravis is caused by antibodies against the α subunit of the receptor for acetylcholine. These antibodies bind to the receptor without activating it and also cause receptor internalization and degradation (right panel). As the number of receptors on the muscle is decreased, the muscle becomes less responsive to acetylcholine.

Diseases mediated by antibodies against cell-surface receptors		
Syndrome	**Antigen**	**Consequence**
Graves' disease	Thyroid-stimulating hormone receptor	Hyperthyroidism
Myasthenia gravis	Acetylcholine receptor	Progressive weakness
Insulin-resistant diabetes	Insulin receptor (antagonist)	Hyperglycemia, ketoacidosis
Hypoglycemia	Insulin receptor (agonist)	Hypoglycemia

Fig. 13.9 Autoimmune diseases caused by autoantibodies against cell-surface receptors. These antibodies produce different effects depending on whether they are agonists (which stimulate) or antagonists (which inhibit) for the receptor. Note that different autoantibodies against the insulin receptor can either stimulate or inhibit signaling.

13-7 Autoantibodies against extracellular antigens cause inflammatory injury by mechanisms akin to type II or type III hypersensitivity reactions.

Antibody responses to extracellular matrix molecules are infrequent but can be very damaging when they occur. In Goodpasture's syndrome, antibodies are formed against the α_3 chain of basement membrane collagen (type IV collagen). These antibodies bind to the basement membranes of renal glomeruli (Fig. 13.10a) and, in some cases, to the basement membranes of pulmonary alveoli, causing a rapidly fatal disease if untreated. The autoantibodies bound to basement membrane ligate $Fc\gamma$ receptors, leading to the activation of monocytes and neutrophils. These release chemokines that attract a further influx of neutrophils into the glomeruli, causing severe tissue injury (Fig. 13.10b). The autoantibodies also cause local activation of complement, which may amplify the tissue injury.

A much more common disease, affecting as many as one in 500 African-American or Asian women living in westernized societies, is **systemic lupus erythematosus (SLE)**. This is an immune complex-mediated disease characterized by chronic IgG antibody production directed at ubiquitous self antigens present in all nucleated cells. SLE is therefore classified as a **systemic autoimmune disease**, as opposed to the **tissue-** or **organ-specific autoimmune diseases** such as Graves' disease, which affect only one organ or tissue.

Immune complexes are produced whenever there is an antibody response to a soluble antigen (see Section 2-8). Normally, these are cleared efficiently by red blood cells bearing complement receptors and by phagocytes of the mononuclear phagocytic system that have both complement and Fc receptors, and such complexes cause little tissue damage. This system can, however, fail in

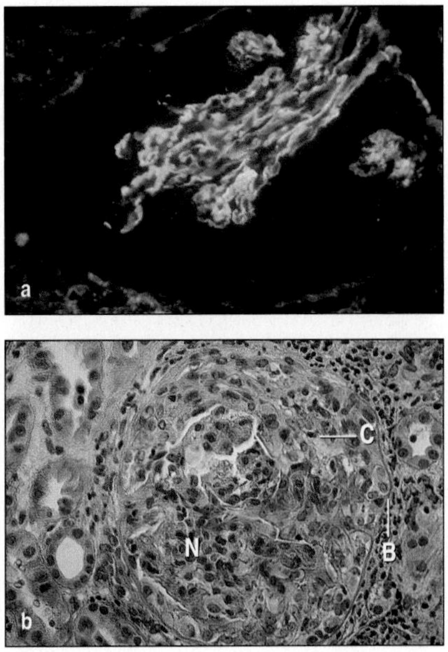

Fig. 13.10 Autoantibodies reacting with glomerular basement membrane cause the inflammatory glomerular disease of Goodpasture's syndrome. The panels show sections of renal glomeruli in serial biopsies taken from patients with Goodpasture's syndrome. Panel a, glomerulus stained for IgG deposition by immunofluoresence. Anti-glomerular basement membrane antibody (stained green) is deposited in a linear fashion along the glomerular basement membrane. The autoantibody causes local activation of cells bearing Fc receptors, complement activation, and influx of neutrophils. Panel b, hematoxylin and eosin staining of a section through a renal glomerulus shows that the glomerulus is compressed by formation of a crescent (C) of proliferating mononuclear cells within the Bowman's capsule (B) and there is influx of neutrophils (N) into the glomerular tuft. Photographs courtesy of M Thompson and D Evans.

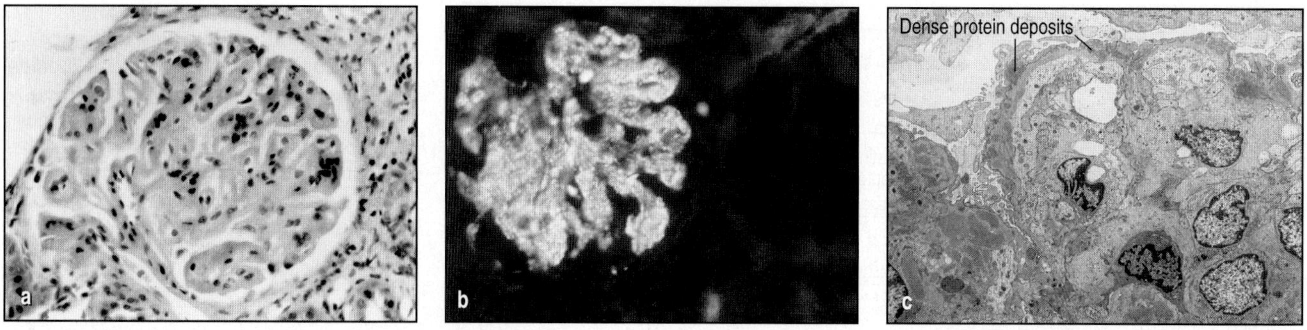

Fig. 13.11 Deposition of immune complexes in the renal glomerulus causes renal failure in systemic lupus erythematosus (SLE). Panel a, a section through a renal glomerulus from a patient with SLE shows that the deposition of immune complexes has caused thickening of the glomerular basement membrane. Panel b, a similar section stained with fluorescent anti-immunoglobulin reveals immunoglobulin deposits in the basement membrane. Panel c, by electron microscopy the immune complexes are seen as dense protein deposits between the glomerular basement membrane and the renal epithelial cells. Polymorphonuclear neutrophilic leukocytes are also present, attracted by the deposited immune complexes. Photographs courtesy of M Kashgarian.

three circumstances. The first follows the injection of large amounts of antigen, leading to the formation of large amounts of immune complexes that overwhelm the normal clearance mechanisms. An example of this is serum sickness (Section 12-17), which is caused by injection of large amounts of serum proteins. This is a transient disease, lasting only until the immune complexes have been cleared. The second circumstance is seen in chronic infections such as bacterial endocarditis, where the immune response to bacteria lodged on a cardiac valve is incapable of clearing infection. The persistent release of bacterial antigens from the valve infection in the presence of a strong anti-bacterial antibody response causes widespread immune complex injury to small blood vessels in organs such as the kidney and the skin.

The third type of failure to clear immune complexes is seen in SLE. In this disease, a wide range of autoantibodies is produced to common cellular constituents. The main antigens are three intracellular nucleoprotein particles—the nucleosome, the spliceosome, and a small cytoplasmic ribonucleoprotein complex containing two proteins known as Ro and La (named after the first two letters of the surnames of the two patients in which autoantibodies against these proteins were discovered). In SLE, large quantities of antigen are available, so large numbers of small immune complexes are produced continuously and are deposited in the walls of small blood vessels in the renal glomerulus (and in glomerular basement membrane, Fig. 13.11), joints, and other organs. This leads to the activation of phagocytic cells; the consequent tissue damage releases more nucleoprotein complexes, which in turn form more immune complexes. Eventually, the inflammation induced in small blood vessel walls, especially in the kidney and brain, can cause sufficient damage to kill the patient.

13-8 Environmental co-factors can influence the expression of autoimmune disease.

The presence of an autoantibody by itself is not sufficient to cause autoimmune disease. For disease to occur, the autoantigen must be available for binding by the autoantibody. Two examples illustrate how the availability of autoantigens and the resulting expression of disease can be modulated by environmental co-factors. As described above, in untreated Goodpasture's disease auto-antibodies against type IV collagen typically cause a fatal glomerulonephritis. Type IV collagen is distributed widely in basement membranes throughout the body, including those of the alveoli of the lung, the renal glomeruli and the cochlea of the inner ear. All patients with Goodpasture's disease develop glomerulonephritis, about 40% develop pulmonary hemorrhage, but none becomes deaf.

This pattern of disease expression was explained when it was discovered that pulmonary hemorrhage was found almost exclusively in those patients who smoked cigarettes. What differs between basement membrane in glomeruli, alveoli, and the cochlea is the availability of the antigen to antibodies. The major function of glomerular basement membrane is the filtration of plasma, and the endothelium lining glomerular capillaries is fenestrated to allow access of plasma to the basement membrane. Glomerular basement membrane is therefore immediately accessible to circulating autoantibodies. In the alveoli, in contrast, the basement membrane separates the alveolar epithelium from the capillary endothelium, whose cells are joined together by tight junctions. Injury to the endothelial lining of pulmonary capillaries is therefore necessary before antibodies can gain access to the basement membrane. Cigarette smoke stimulates an inflammatory response in the lungs, which damages alveolar capillaries and exposes the autoantigen to antibody. Finally, in the inner ear, the cochlear basement membrane seems to remain inaccessible to autoantibodies at all times.

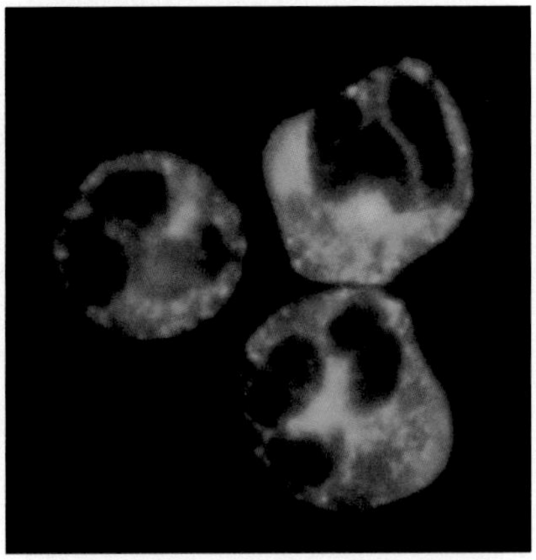

Fig. 13.12 Serum from patients with Wegener's granulomatosis contains autoantibodies reactive with neutrophil cytoplasmic granules. Normal neutrophils with permeabilized cell membranes have been incubated with serum from a patient with Wegener's granulomatosis. IgG antibodies in the serum reactive with cytoplasmic granules are detected by addition of fluorescein-conjugated antibodies against IgG. Photograph courtesy of C Pusey.

A second example of the importance of environmental influences on the expression of autoimmunity is the effect of infection on the vasculitis associated with **Wegener's granulomatosis**. This disease, which is characterized by a severe necrotizing vasculitis, is strongly associated with the presence of autoantibodies to a granule proteinase of neutrophils (Fig. 13.12); the antibodies are known as anti-neutrophil cytoplasm antibodies (commonly abbreviated as ANCA). The autoantigen is proteinase-3, an abundant serine proteinase of neutrophil granules. Although there is a general correlation between the levels of ANCA and the expression of disease, it is quite common to find patients with high levels of ANCA who remain asymptomatic. If such an individual develops an infection, however, this frequently induces a severe flare-up of the vasculitis.

It is thought that resting neutrophils do not express proteinase-3 on the cell surface, which means that in the absence of infection the antigen is inaccessible to anti-proteinase-3 autoantibodies. After infection, a variety of cytokines causes neutrophil activation, with translocation of proteinase-3 to the cell surface. Anti-proteinase-3 antibodies can now bind neutrophils and stimulate degranulation and release of free radicals. In parallel, activation of vascular endothelial cells by the infection causes the expression of vascular adhesion molecules, such as E-selectin, which promote the binding of activated neutrophils to vessel walls with resultant injury. In this way a variety of nonspecific infections can exacerbate an autoimmune disease.

13-9 The pattern of inflammatory injury in autoimmunity can be modified by anatomical constraints.

We have seen that the distribution of organ injury in Goodpasture's syndrome can be explained by the accessibility of basement membrane collagen to autoantibodies and that environmental factors can influence the availability of antigen in different organs. Another example of how the expression of autoimmune inflammation can be modified by anatomical factors is seen in

membranous glomerulonephritis. In this disease, patients develop heavy proteinuria (the excretion of protein in the urine), which can cause life-threatening depletion of plasma protein levels. Biopsy of an affected kidney reveals evidence of deposition of antibody and complement beneath the basement membrane of the glomerulus but, in contrast to Goodpasture's disease, there is no significant influx of inflammatory cells. The autoantigen in this disease has not been characterized. However, an excellent animal model of membranous glomerulonephritis is **Heymann's nephritis**, in which autoantibodies against a glycoprotein on the surface of tubular epithelial cells are induced by injection of tubular epithelial tissue. In this disease, proteinuria can be abolished by depletion of any of the proteins of the membrane-attack complex of complement but is unaltered by depletion of neutrophils. Thus, antibodies deposited beneath the glomerular basement membrane in this disease cause tissue injury by activation of complement but the glomerular basement membrane acts as a complete barrier to inflammatory leukocytes.

In other autoimmune diseases, high levels of autoantibodies against intracellular antigens can be found in the absence of any evidence of antibody-induced inflammation. One such example is a rare myositis (inflammation of muscle) associated with pulmonary fibrosis. Some patients with this disease develop high levels of autoantibodies reactive with aminoacyl-tRNA synthetases, the intracellular enzymes responsible for charging tRNAs with amino acids. Addition of these autoantibodies to cell-free extracts *in vitro* stops translation and protein synthesis completely. However, there is no evidence that these antibodies cause any injury *in vivo*, where it is unlikely that they can enter living cells. In this disease, the autoantibody is thought to be a marker of a particular pattern of tissue injury, possibly stimulated by an unknown infectious agent, and does not contribute to the immunopathology of the myositis. Other examples of autoantibodies that are useful diagnostic markers of the presence of disease, but that might play no part in causing organ injury, are antibodies against mitochondrial antigens associated with primary biliary cirrhosis and antibodies against smooth muscle antigens in chronic active hepatitis.

13-10 The mechanism of autoimmune tissue damage can often be determined by adoptive transfer.

To classify a disease as autoimmune one must show that an adaptive immune response to a self antigen causes the observed pathology. Initially, the demonstration that antibodies against the affected tissue could be detected in the serum of patients suffering from various diseases was taken as evidence that the diseases had an autoimmune basis. However, such autoantibodies are also found when tissue damage is caused by trauma or infection. This suggests that autoantibodies can result from, rather than be the cause of, tissue damage. Thus, one must show that the observed autoantibodies are pathogenic before classifying a disease as autoimmune.

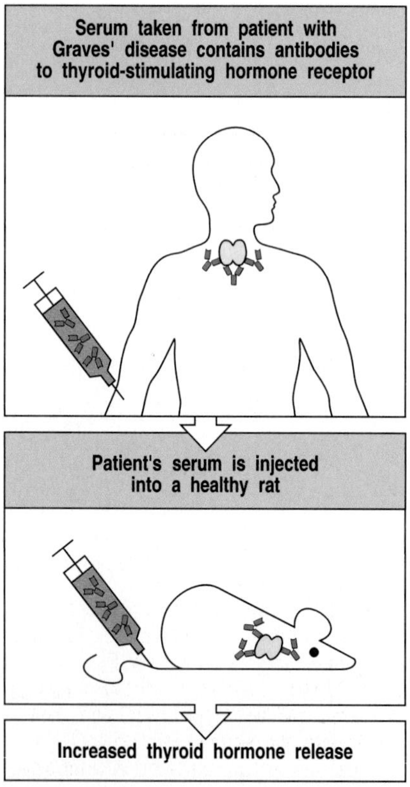

Serum taken from patient with Graves' disease contains antibodies to thyroid-stimulating hormone receptor
Patient's serum is injected into a healthy rat
Increased thyroid hormone release

Fig. 13.13 Serum from some patients with autoimmune disease can transfer the same disease to experimental animals. When the autoantigen is very similar in humans and mice or rats, the transfer of antibody from an affected human can cause the same symptoms in an experimental animal. For example, antibody from patients with Graves' disease frequently produces thyroid activation in rats.

Autoimmune diseases transferred across the placenta to the fetus and newborn infant		
Disease	**Autoantibody**	**Symptom**
Myasthenia gravis	Anti-acetylcholine receptor	Muscle weakness
Graves' disease	Anti-thyroid stimulating hormone (TSH) receptor	Hyperthyroidism
Thrombocytopenic purpura	Anti-platelet antibodies	Bruising and haemorrhage
Subacute cutaneous lupus erythematosus and/or congenital heart block	Anti-Ro antibodies	Photosensitive rash and/or bradycardia
Pemphigus vulgaris	Anti-desmoglein-3	Blistering rash

Fig. 13.14 Some autoimmune diseases that can be transferred across the placenta by pathogenic IgG autoantibodies. These diseases are caused mostly by autoantibodies to cell-surface or tissue-matrix molecules. This suggests that an important factor determining whether an autoantibody that crosses the placenta causes disease in the fetus or newborn baby is the accessibility of the antigen to the autoantibody. Autoimmune congenital heart block is caused by fibrosis of the developing cardiac conducting tissue, and there is evidence that this expresses abundant Ro antigen (see Section 13-7). Ro protein is a constituent of an intracellular small cytoplasmic RNP and it is not yet established whether it is expressed at the cell surface of cardiac conducting tissue to act as a target for autoimmune tissue injury.

It is often possible to transfer disease to experimental animals through the transfer of autoantibodies, causing pathology similar to that seen in the patient from whom the antibodies were obtained (Fig. 13.13). However, this does not always work, presumably because of species differences in autoantigen structure. Some autoimmune diseases can also be transferred from mother to fetus (Fig. 13.14) and are observed in the newborn babies of diseased mothers. When babies are exposed to IgG autoantibodies transferred across the placenta, they will often manifest pathology similar to the mother's (Fig. 13.15). This natural experiment is one of the best proofs that particular autoantibodies exert pathogenic effects. The symptoms of the disease in the newborn disappear rapidly as the maternal antibody is catabolized. The process can be speeded up by a complete exchange of the infant's blood or plasma (plasmapheresis).

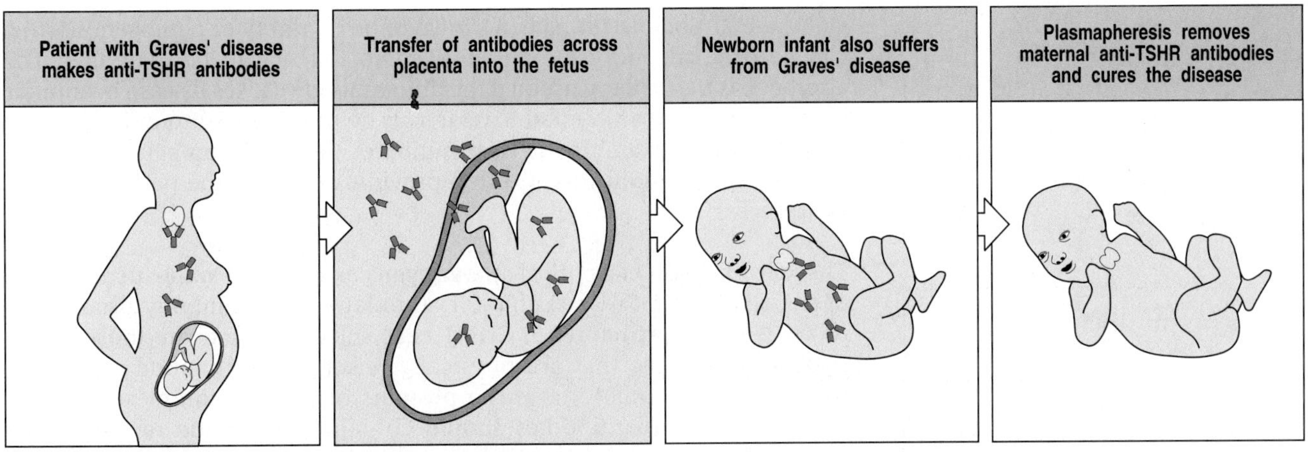

Fig. 13.15 Antibody-mediated autoimmune diseases can appear in the infants of affected mothers as a consequence of transplacental antibody transfer. In pregnant women, IgG antibodies cross the placenta and accumulate in the fetus before birth (see Fig. 9.19). Babies born to mothers with IgG-mediated autoimmune disease therefore frequently show symptoms similar to those of the mother in the first few weeks of life. Fortunately, there is little lasting damage as the symptoms disappear along with the maternal antibody. In Graves' disease, the symptoms are caused by antibodies against the thyroid-stimulating hormone receptor (TSHR). Children of mothers making thyroid-stimulating antibody are born with hyperthyroidism, but this can be corrected by replacing the plasma with normal plasma (plasmapheresis), thus removing the maternal antibody.

13-11 T cells specific for self antigens can cause direct tissue injury and have a role in sustained autoantibody responses.

Activated effector T cells specific for self peptide:self MHC complexes can cause local inflammation by activating macrophages or can damage tissue cells directly. Diseases in which these actions of T cells are likely to be important include insulin-dependent diabetes mellitus, rheumatoid arthritis and multiple sclerosis. Affected tissues in patients with these diseases are heavily infiltrated by T lymphocytes and by activated macrophages. These autoimmune diseases are mediated by specifically reactive T cells. T cells are, of course, also required to sustain all autoantibody responses.

It is much more difficult to demonstrate the existence of autoreactive T cells than it is to demonstrate the presence of autoantibodies. First, autoreactive human T cells cannot be used to transfer disease to experimental animals because T-cell recognition is MHC-restricted and animals and humans have different MHC alleles. Second, it is difficult to identify the antigen recognized by a T cell; for example, autoantibodies can be used to stain self tissues to reveal the distribution of the autoantigen, whereas T cells cannot. Nevertheless, there is strong evidence for the involvement of autoreactive T cells in several autoimmune diseases. **Insulin-dependent diabetes mellitus (IDDM)** is a disease in which the insulin-producing β cells of the pancreatic islets are selectively destroyed by specific T cells. When such diabetic patients are transplanted with half a pancreas from an identical twin donor, the β cells in the grafted tissue are rapidly and selectively destroyed by CD8 T cells. Recurrence of disease can be prevented by the immunosuppressive drug cyclosporin A (see Chapter 14), which inhibits T-cell activation. Progress towards identifying the targets of such autoreactive T cells and proving that these cells cause disease will be discussed in Section 13-13.

13-12 Autoantibodies can be used to identify the target of the autoimmune process.

Once autoantibodies have been shown to be required for pathogenesis, they can be used to purify the autoantigen so that it can be identified. This approach is particularly useful if the autoantibody causes disease in animals, from which large amounts of tissue can be obtained. Autoantibodies can also be used to examine the distribution of the target antigen in cells and tissues by immunohistology, often providing clues to the pathogenesis of the disease.

The identification of a critical autoantigen can also lead to the identification of the CD4 T cells responsible for stimulating autoantibody production. As we learned in Chapter 8, CD4 T cells selectively activate those B cells that bind epitopes that are physically linked to the peptide recognized by the T cell. It follows that the proteins or protein complexes isolated by means of autoantibodies should contain the peptide recognized by the autoreactive CD4 T cell. For example, in myasthenia gravis the autoantibodies that cause disease bind mainly to the α chain of the acetylcholine receptor and can be used to isolate the receptor from lysates of skeletal muscle cells. CD4 T cells that recognize peptide fragments of this receptor subunit can also be found in patients with myasthenia gravis (Fig. 13.16). Thus, both autoreactive B cells and autoreactive T cells are required for this disease.

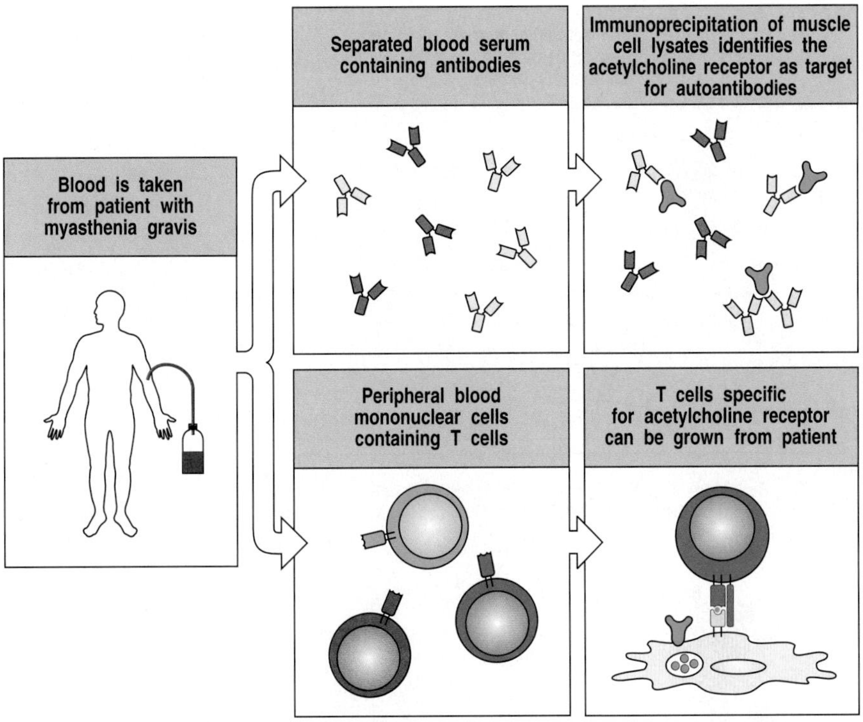

Fig. 13.16 Autoimmune disease caused by antibodies also requires autoreactive T cells. Autoantibodies from the serum of myasthenia gravis patients immunoprecipitate the acetylcholine receptor from lysates of skeletal muscle cells (top panels). To be able to produce antibodies, the same patients should also have CD4 T cells that respond to a peptide derived from the acetylcholine receptor. To detect them, T cells from myasthenia gravis patients are isolated and grown in the presence of the acetylcholine receptor plus antigen-presenting cells of the correct MHC type (bottom panels). T cells specific for epitopes of the acetylcholine receptor are stimulated to proliferate and can thus be detected.

The same phenomenon is seen in SLE. Tissue damage in this disease is caused by immune complexes of autoantibodies directed against a variety of nucleoprotein antigens (see Section 13-7). These autoantibodies show a high degree of somatic hypermutation, which has all the hallmarks of being antigen-driven (see Section 3-18), and the B cells that produce them can be shown to have undergone extensive clonal expansion. Thus, the autoantibodies have the characteristic properties of antibodies formed in response to chronic stimulation of B cells by antigen and specific CD4 T cells, strongly suggesting that they are produced in response to autoantigens containing peptides recognized by specific autoreactive CD4 T cells. Further evidence for this comes from the pattern of autoantibody specificities observed in individual patients. The autoantibodies in any one individual tend to bind all constituents of a particular nucleoprotein particle; this strongly suggests that there must be CD4 T cells present that are specific for a peptide constituent of this particle. A B cell whose receptor binds any component of this particle will internalize and process the particle, present the peptide to these autoreactive T cells, and receive help from them (Fig. 13.17). Such B-cell–T-cell interactions initiate the antibody response and promote clonal expansion and somatic hypermutation, thus accounting for the observed characteristics of the autoantibody response as well as the clustering of autoantibody specificities in individual patients.

13-13 The target of T-cell mediated autoimmunity is difficult to identify owing to the nature of T-cell ligands.

Although there is good evidence that T cells are involved in many autoimmune diseases, the T cells that cause particular diseases are hard to isolate, and their targets are difficult to identify. In part, this is because one cannot grow the specific T cells needed to identify the autoantigenic peptide without supplying them with their specific antigen in the first place. It is also difficult

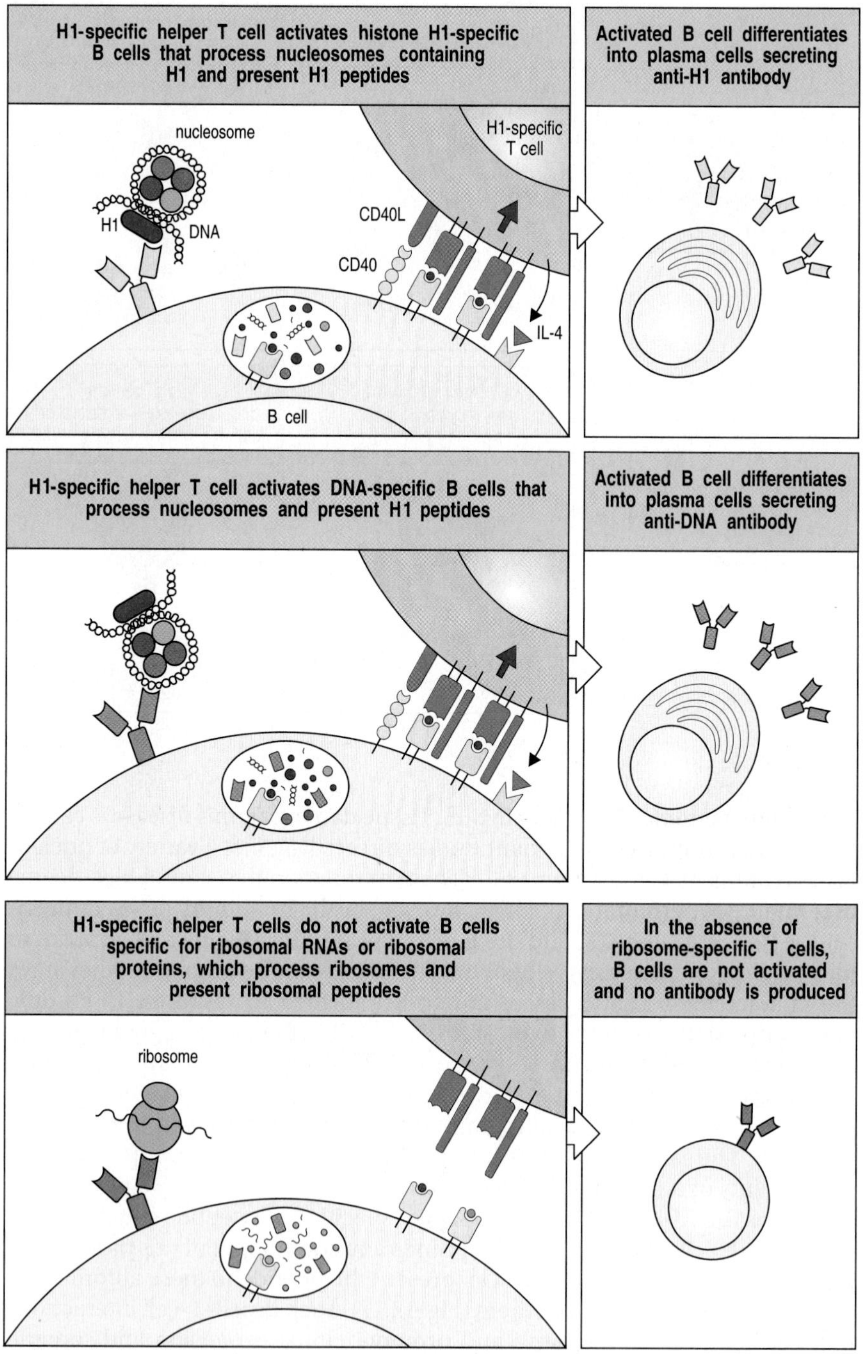

Fig. 13.17 Autoreactive helper T cells of one specificity can drive the production of autoantibodies with several different specificities. In an SLE patient, a B cell specific for the H1 histone protein in nucleosomes, for example, will bind and internalize the whole nucleosome, and present peptides derived from H1 histone as well as other peptides. This B cell can receive help from a T cell specific for one of the peptides derived from H1 (top panels). A B cell that recognizes the DNA in the nucleosome can also internalize the nucleosome, process it, and present the H1 peptide to that T cell and be activated by it (center panels). Thus, a single autoreactive helper T cell can stimulate a diverse antibody response, but the antibodies will be restricted to those specific for the constituents of a single type of particle. B cells able to bind ribosomes, for example, do not present the H1 peptide and so will not be activated to produce anti-ribosomal antibodies in this patient (bottom panels).

to assay the T cells for their ability to cause disease, because any assay requires target cells of the same MHC genotype as the patient. This problem becomes more tractable in animal models. As many autoimmune diseases in animals are induced by immunization with self tissue, the nature of the autoantigen can be determined by fractionating an extract of the tissue and testing the fractions for their ability to induce disease. It is also possible to clone T-cell lines that will transfer the disease from an affected animal to another animal with the same MHC genotype.

This has made it possible to identify the autoantigens recognized by T cells in many experimental autoimmune diseases; they are individual peptides that bind to specific MHC molecules. The peptide antigen, when made immunogenic, is able to elicit disease symptoms in animals of the appropriate MHC genotype. An example of such an experimental autoimmune disease is **experimental allergic encephalomyelitis (EAE)**, which can be induced in certain susceptible strains of mice and rats by injection of central nervous system tissue together with Freund's complete adjuvant. This disease resembles human **multiple sclerosis**, in which characteristic plaques of tissue injury are disseminated throughout the central nervous system. Plaques of active disease show infiltration of nervous tissue by lymphocytes, plasma cells and macrophages, which cause destruction of the myelin sheaths that surround nerve cell axons in the brain and spinal cord.

Further analysis of EAE showed that injection with various purified components of the myelin sheath, notably myelin basic protein (MBP), proteolipid protein (PLP) and myelin oligodendroglial protein (MOG), can induce EAE. The disease can be transferred to syngeneic animals by using cloned T-cell lines derived from animals with EAE (Fig. 13.18). Many of these cloned T-cell lines are stimulated by peptides of MBP. When animals with the appropriate MHC allele to present the peptide are immunized with the peptide recognized by these T cells, active disease results. Activated T cells specific for myelin proteins have also been identified in patients with multiple sclerosis. Although it has not yet been proved that these cells cause the demyelination in multiple sclerosis, this finding suggests that animal models might provide clues to the identity of autoantigenic proteins in human disease.

In a variety of inflammatory autoimmune diseases, it seems that self antigens are presented to T_H1 cells. EAE, for example, can be caused by T_H1 cells specific for MBP, as shown by the ability of specific clones of T_H1, but not T_H2, cells to cause disease on adoptive transfer. Although MBP is an intracellular protein, it is processed for presentation by the vesicular pathway and thus its peptides are presented by MHC class II molecules and recognized by CD4 T cells. Another inflammatory autoimmune disease, **rheumatoid arthritis**, can be caused by T_H1 cells specific for an antigen present in joints. Engagement

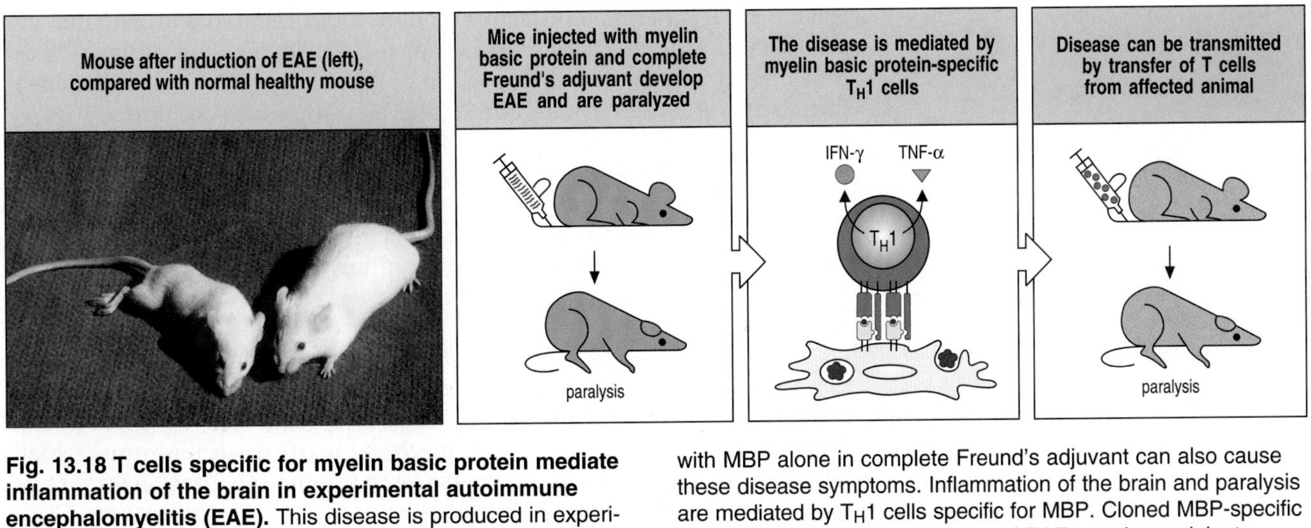

Fig. 13.18 T cells specific for myelin basic protein mediate inflammation of the brain in experimental autoimmune encephalomyelitis (EAE). This disease is produced in experimental animals by injecting them with isolated spinal cord homogenized in complete Freunds' adjuvant. EAE is due to an inflammatory reaction in the brain that causes a progressive paralysis affecting first the tail and hind limbs (as shown in the mouse on the left of the photograph, compared with a healthy mouse on the right) before progressing to forelimb paralysis and eventual death. One of the autoantigens identified in the spinal cord homogenate is myelin basic protein (MBP). Immunization with MBP alone in complete Freund's adjuvant can also cause these disease symptoms. Inflammation of the brain and paralysis are mediated by T_H1 cells specific for MBP. Cloned MBP-specific T_H1 cells can transfer symptoms of EAE to naive recipients provided that the recipients carry the correct MHC allele. In this system it has therefore proved possible to identify the peptide: MHC complex recognized by the T_H1 clones that transfer disease. Other purified components of the myelin sheath can also induce the symptoms of EAE, so there is more than one autoantigen in this disease.

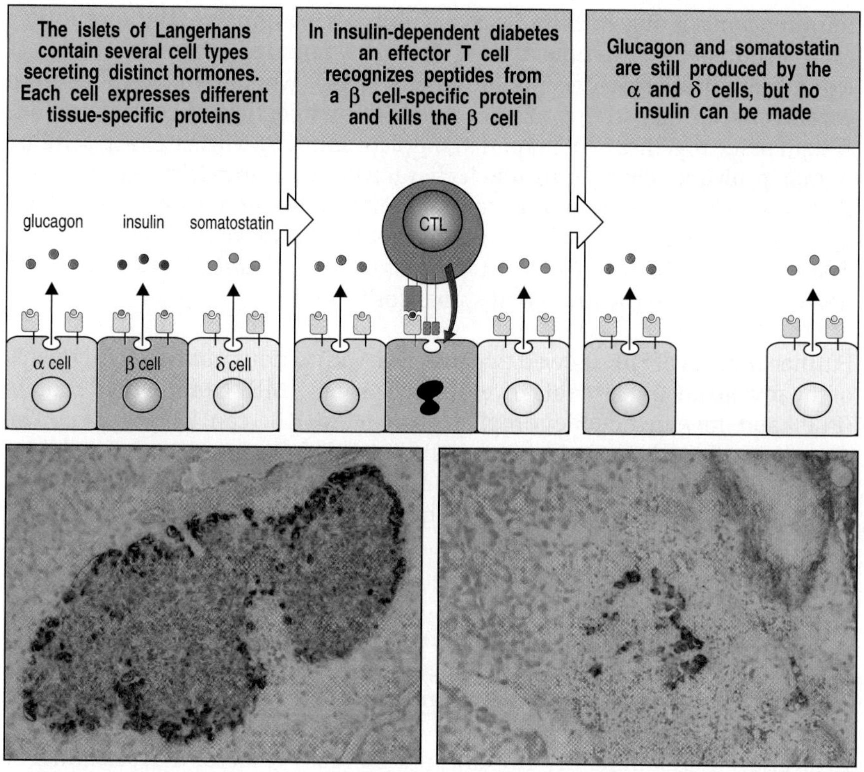

Fig. 13.19 Selective destruction of pancreatic β cells in insulin-dependent diabetes mellitus (IDDM) indicates that the autoantigen is produced in β cells and recognized on their surface. In IDDM, there is highly specific destruction of insulin-producing β cells in the pancreatic islets of Langerhans, sparing other islet cell types (α and δ). This is shown schematically in the upper panels. In the lower panels, islets from normal (left) and diabetic (right) mice are stained for insulin (brown), which shows the β cells, and glucagon (black), which shows the α cells. Note the lymphocytes infiltrating the islet in the diabetic mouse (right) and the selective loss of the β cells (brown) whereas the α cells (black) are spared. The characteristic morphology of the islet is also disrupted with the loss of the β cells. Photographs courtesy of I Visintin.

with this antigen triggers the T cells to release lymphokines that initiate local inflammation within the joint. This causes swelling, accumulation of polymorphonuclear leukocytes and macrophages, and damage to cartilage, leading to the destruction of the joint. Rheumatoid arthritis is a complex disease and also involves antibodies, often including an IgM anti-IgG autoantibody called **rheumatoid factor**. Like the SLE autoantibodies described in Section 13-12, rheumatoid factors isolated from the joints of patients with rheumatoid arthritis show evidence of a T-cell dependent antigen-driven B-cell response against the Fc portion of IgG. Some of the tissue damage in this disease is caused by the resultant IgM:IgG immune complexes.

Identification of autoantigenic peptides is particularly difficult in autoimmune diseases caused by CD8 T cells. Autoantigens recognized by CD4 T cells can be identified by adding cell extracts to cultures of blood mononuclear cells and testing for recognition by CD4 cells derived from an autoimmune patient. If the autoantigen is present in the cell extract it should be effectively presented, as phagocytes in the blood cultures can take up extracellular protein, degrade it in intracellular vesicles and present the resulting peptides bound to class II MHC molecules. By contrast, autoantigens recognized by CD8 T cells are not effectively presented in such cultures. Peptides presented by class I MHC molecules must usually be made by the target cells themselves (see Chapter 4); intact cells of target tissue from the patient must therefore be used to study autoreactive CD8 T cells that cause tissue damage. Conversely, the pathogenesis of the disease can itself give clues to the identity of the antigen in some CD8 T-cell mediated diseases. For example, in insulin-dependent diabetes mellitus, the insulin-producing β cells of the pancreatic islets of Langerhans seem to be specifically targeted and destroyed by CD8 T cells (Fig. 13.19). It is therefore likely that a protein unique to β cells is the source of the peptide that is recognized by the pathogenic CD8 T cells.

CD4 T cells also seem to be involved in insulin-dependent diabetes mellitus, consistent with the linkage of disease susceptibility to particular MHC class II alleles, as discussed earlier in this chapter (see Fig. 13.4). Identifying the autoantigen recognized by T cells in these diseases is an important goal. Not only might it help us to understand disease pathogenesis but it might also result in several innovative approaches to treatment (see Chapter 14).

Summary.

For a disease to be defined as autoimmune, the tissue damage must be shown to be caused by an adaptive immune response to self antigens. Autoimmune diseases can be mediated by autoantibodies or by autoreactive T cells, and tissue damage can result from direct attack on the cells bearing the antigen, from immune-complex formation, or from local inflammation. Autoimmune diseases caused by antibodies that bind to cellular receptors, causing either excess activity or inhibition of receptor function, fall into a special class. T cells can be involved directly in inflammation or cellular destruction, and they are also required to sustain autoantibody responses. The most convincing proof that the immune response is causal in autoimmunity is transfer of disease by transferring the active component of the immune response to an appropriate recipient. However, this is often not practicable. The current challenge is to identify the autoantigens recognized by T cells in autoimmunity, and to use this information to control the activity of these T cells, or to prevent their activation in the first place. The deeper, more important question is how the autoimmune response is induced. Much has been learned about the induction of immune responses to tissue antigens by examining the response to non-self tissues in transplantation experiments. We shall therefore examine the immune response to grafted tissues in the next section before turning in the last section of this chapter to the problem of how tolerance is maintained normally, and why responses occur in autoimmune disease.

Transplant rejection: responses to alloantigens.

The transplantation of tissues to replace diseased organs is now an important medical therapy. In most cases, adaptive immune responses to the grafted tissues are the major impediment to successful transplantation. In blood transfusion, which is the earliest and still the commonest tissue transplant, blood must be matched for ABO and Rh blood group antigens to avoid the rapid destruction of mismatched red blood cells by antibodies (see Fig. 2.7). Because there are only four major ABO types and two Rh blood types, this is relatively easy. However, when tissues containing nucleated cells are transplanted, T-cell responses to the highly polymorphic MHC molecules almost always trigger a response against the grafted organ. Matching the MHC type of donor and recipient increases the success rate of grafts, but perfect matching is possible only when donor and recipient are related and, in these cases, genetic differences at other loci still trigger rejection. In this section, we shall examine the immune response to tissue grafts, and ask why such responses do not reject the one foreign tissue graft that is tolerated routinely, the mammalian fetus.

13-14 The rejection of grafts is an immunological response mediated primarily by T cells.

The basic rules of tissue grafting were first elucidated by using skin transplanted between inbred strains of mice. Skin can be grafted with 100% success between different sites on the same animal or person (an **autograft**), or between genetically identical animals or people (a **syngeneic graft**). However, when skin is grafted between unrelated or **allogeneic** individuals (an **allograft**), the graft is initially accepted but is then rejected about 10–13 days after grafting (Fig. 13.20). This response is called a **first-set rejection** and is quite consistent. It depends on a recipient T-cell response, because skin grafted onto *nude* mice, which lack T cells, is not rejected. The ability to reject skin can be restored to *nude* mice by the adoptive transfer of normal T cells.

When a recipient that has previously rejected a graft is regrafted with skin from the same donor, the second graft is rejected more rapidly (6–8 days) in a **second-set rejection** (see Fig. 13.20). Skin from a third-party donor grafted onto the same recipient at the same time does not show this faster response but follows a first-set rejection course. The rapid course of second-set rejection can be transferred to normal or irradiated recipients by transferring T cells from the initial recipient, showing that graft rejection is caused by a specific immunological reaction.

Immune responses are a major barrier to effective tissue transplantation, destroying grafted tissue by an adaptive immune response to its foreign proteins. These responses can be mediated by cytotoxic CD8 T cells, by T_H1 cells, or by both. Antibodies can also contribute to second-set rejection of tissue grafts.

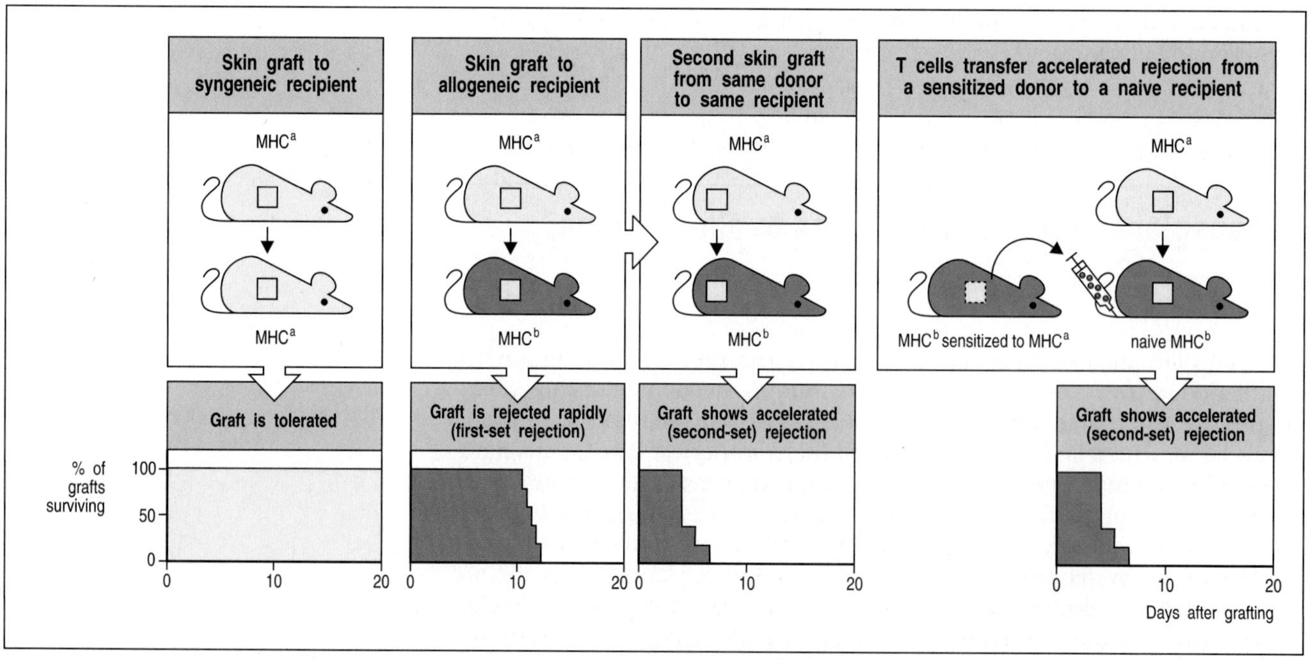

Fig. 13.20 Skin graft rejection is the result of a T-cell mediated anti-graft response. Grafts that are syngeneic are permanently accepted (first panels) but grafts differing at the MHC are rejected around 10–13 days after grafting (first-set rejection, second panels). When a mouse is grafted for a second time with skin from the same donor, it rejects the second graft faster (third panels). This is called a second-set rejection and the accelerated response is MHC-specific; skin from a second donor of the same MHC type is rejected equally fast, whereas skin from an MHC-different donor is rejected in a first-set pattern (not shown). Naive mice that are given T cells from a sensitized donor behave as if they had already been grafted (final panels).

13-15 Matching donor and recipient at the MHC improves the outcome of transplantation.

When donor and recipient differ at the MHC, the immune response is directed at the non-self MHC molecule or molecules expressed by the graft. In most tissues, these will be predominantly MHC class I antigens. Once a recipient has rejected a graft of a particular MHC type, any further graft bearing the same non-self MHC molecule will be rapidly rejected in a second-set response. As we learned in Chapter 4, the frequency of T cells specific for any non-self MHC molecule is high, making differences at MHC loci the most potent trigger of the rejection of initial grafts; indeed, the major histocompatibility complex was originally named because of its central role in graft rejection.

Once it became clear that recognition of non-self MHC molecules is a major determinant of graft rejection, a considerable amount of effort was put into MHC matching between recipient and donor. Although HLA matching significantly improves the success rate of clinical organ transplantation it does not in itself prevent rejection reactions. There are two main reasons for this. First, HLA typing is imprecise, owing to the polymorphism and complexity of the human MHC; unrelated individuals who type as HLA-identical with antibodies against MHC proteins rarely have identical MHC genotypes. However, this should not be a problem with HLA-identical siblings: because siblings inherit their MHC genes as a haplotype, one sibling in four should be truly HLA-identical. Nevertheless, grafts between HLA-identical siblings are invariably rejected, albeit more slowly, unless donor and recipient are identical twins. This rejection is the result of differences between minor histocompatibility antigens, which will be discussed in the next section.

Thus, unless donor and recipient are identical twins, all graft recipients must be given immunosuppressive drugs to prevent rejection. Indeed, the current success of clinical transplantation of solid organs is more the result of advances in immunosuppressive therapy, discussed in Chapter 14, than of improved tissue matching. The limited supply of cadaveric organs, coupled with the urgency of identifying a recipient once a donor becomes available, means that accurate matching of tissue types is achieved only rarely.

13-16 In MHC-identical grafts, rejection is caused by non-self peptides bound to graft MHC molecules.

When donor and recipient are identical at the MHC but differ at other genetic loci, graft rejection is not so rapid (Fig. 13.21). The polymorphic antigens responsible for the rejection of MHC-identical grafts are therefore termed **minor histocompatibility antigens** or **minor H antigens**. Responses to single minor H antigens are much less potent than responses to MHC differences because the frequency of responding T cells is much lower. However, most inbred mouse strains that are identical at the MHC differ at multiple minor H antigen loci, so grafts between them are still uniformly and relatively rapidly rejected. The cells that respond to minor H antigens are generally CD8 T cells, implying that most minor H antigens are peptides bound to self MHC class I molecules. However, peptides bound to self MHC class II molecules can also participate in the response to MHC-identical grafts.

Minor H antigens are now known to be peptides derived from polymorphic proteins that are presented by the MHC molecules on the graft (Fig. 13.22). MHC class I molecules bind and present a selection of peptides derived from

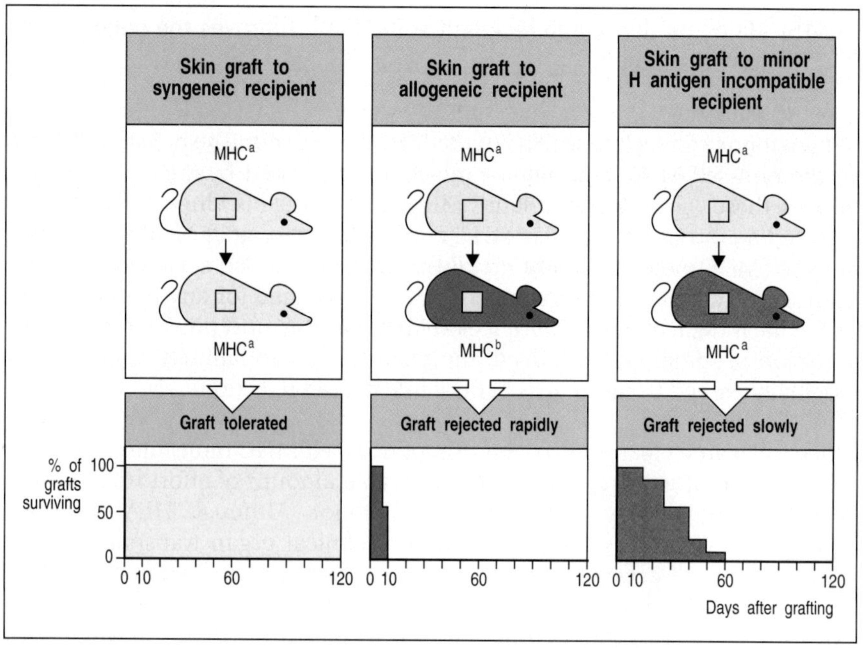

Fig. 13.21 Even complete matching at the MHC does not ensure graft survival. Although syngeneic grafts are not rejected (left panels), MHC-identical grafts from donors that differ at other loci (minor H antigen loci) are rejected (right panels), albeit more slowly than MHC-disparate grafts (center panels).

proteins made in the cell, and if polymorphisms in these proteins mean that different peptides are produced in different members of a species, these can be recognized as minor H antigens. One set of proteins that induce minor H responses is encoded on the male-specific Y chromosome. Responses induced by these proteins are known collectively as H-Y. Since these Y-chromosome specific genes are not expressed in females, female anti-male minor H responses occur; however, male anti-female responses are not seen, because both males and females express X-chromosome genes. One H-Y antigen has been identified in mice and humans as peptides from a protein encoded by the Y-chromosome gene *Smcy*. An X-chromosome homolog of *Smcy*, called *Smcx*, does not contain these peptide sequences, which are therefore expressed uniquely in males. The nature of the majority of minor H antigens, encoded by autosomal genes, is unknown, but one, HA-2, has been identified as a peptide derived from a myosin protein.

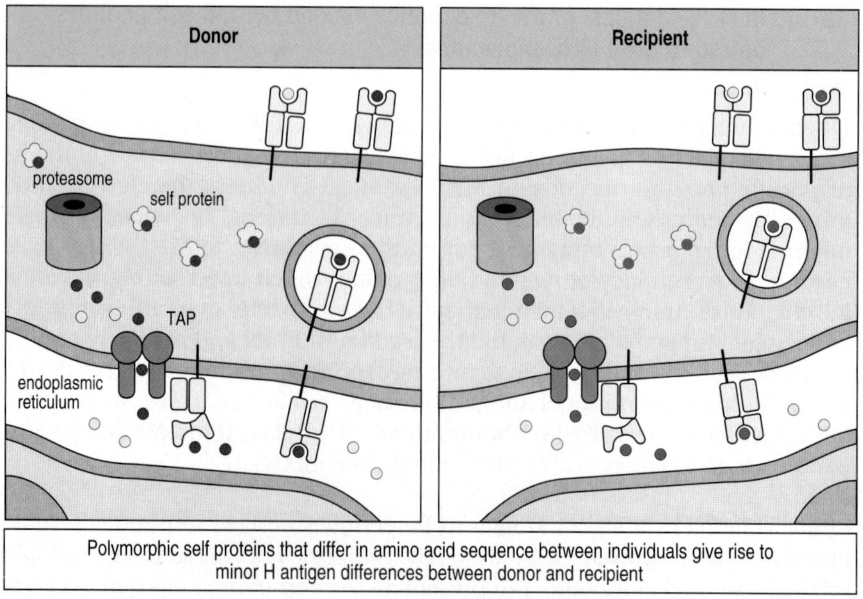

Fig. 13.22 Minor H antigens are peptides derived from polymorphic cellular proteins bound to MHC class I molecules. Self proteins are routinely digested by proteasomes within the cell's cytosol, and peptides derived from them are delivered to the endoplasmic reticulum, where they can bind to MHC class I molecules and be delivered to the cell surface. If a polymorphic protein differs between the graft donor (shown in red on the left) and the recipient (shown in blue on the right), it can give rise to an antigenic peptide (red on the donor cell) that can be recognized by the recipient's T cells as non-self and elicit an immune response. Such antigens are the minor H antigens.

The response to minor H antigens is in every way analogous to the response to viral infection. However, the antiviral response eliminates only infected cells, whereas all cells in the graft express minor H antigens, and the entire graft is destroyed in the response against these antigens, just as analogous responses to tissue-specific peptides can destroy an entire tissue in autoimmunity. Thus, even though MHC genotype might be matched exactly, polymorphism in any protein could elicit potent T-cell responses that will destroy the entire graft. It is no wonder that successful transplantation requires the use of potent immunosuppressive drugs.

13-17 There are two ways of presenting alloantigens on the transplant to the recipient's T lymphocytes.

We saw in Section 13-15 that alloreactive effector T cells that bind directly to allogeneic MHC class I molecules in mismatched organ grafts are an important cause of graft rejection; this is called direct allorecognition. Before they can cause rejection, naive alloreactive T cells must be activated by antigen-presenting cells that both bear the allogeneic MHC molecules and have co-stimulatory activity. Organ grafts carry with them antigen-presenting cells of donor origin and these are an important stimulus to alloreactivity (Fig. 13.23). This route for sensitization to a graft seems to involve donor antigen-presenting

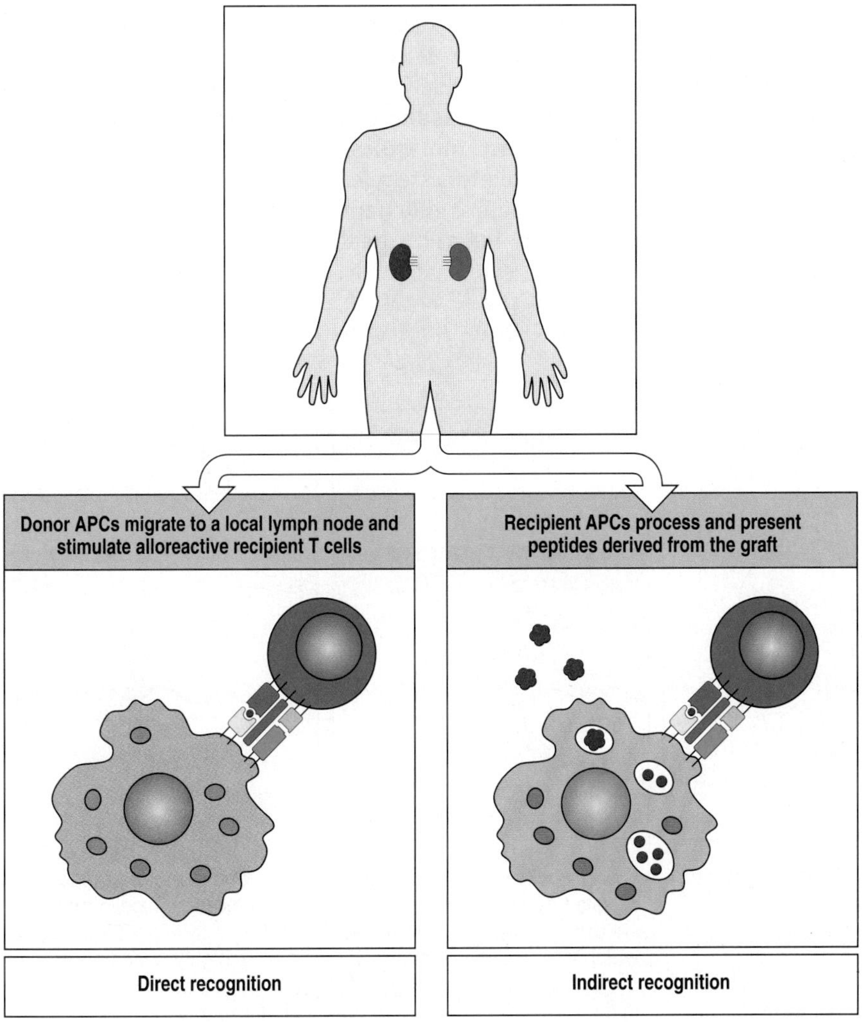
Donor APCs migrate to a local lymph node and stimulate alloreactive recipient T cells
Recipient APCs process and present peptides derived from the graft
Direct recognition
Indirect recognition

Fig. 13.23 There are two mechanisms for the recognition of alloantigens in grafted organs. Direct recognition of a grafted organ (red in upper panel) is mediated by T lymphocytes whose receptors have specificity for the allogeneic MHC class I or class II molecule in combination with peptide. These alloreactive T cells are stimulated by donor antigen-presenting cells (APC), which express both the allogeneic MHC molecule and co-stimulatory activity (bottom left panel). Indirect recognition of the graft (bottom right panel) is mediated by T cells whose receptors are specific for allogeneic peptides that are derived from the grafted organ. Proteins from the graft are processed by self antigen-presenting cells and therefore presented by self (that is, recipient) MHC class I or class II molecules.

cells' leaving the graft and migrating via the lymph to regional lymph nodes. Here, they can activate those host T cells that bear specific alloreactive receptors. The activated alloreactive effector T cells are then carried back to the graft, which they attack directly (Fig. 13.24). Indeed, if the grafted tissue is depleted of antigen-presenting cells by treatment with antibodies or by prolonged incubation, rejection occurs only after a much longer time. Also, if the site of grafting lacks lymphatic drainage, no response against the graft results.

A second mechanism of allograft recognition leading to graft rejection is the uptake of allogeneic proteins by the recipient's own antigen-presenting cells and their presentation to T cells by self MHC molecules (see Fig. 13.23). The recognition by T cells of allogeneic proteins presented in this way is known as indirect allorecognition, in contrast to the direct recognition by T cells of allogeneic MHC class I and class II molecules expressed on the graft itself. Among the graft-derived peptides presented by the recipient's antigen-presenting cells are the minor H antigens and also peptides from the foreign MHC molecules themselves, which are a major source of the polymorphic peptides recognized by the recipient's T cells.

Interestingly, when grafts that are MHC-identical to the host but mismatched at minor H antigens are depleted of antigen-presenting cells, they are rejected more rapidly than similarly depleted grafts that are MHC-different from the host but identical at minor H antigens. The slower rejection of the MHC-different grafts shows that rejection cannot be due to antigen presentation by a few residual donor antigen-presenting cells, and also that the co-stimulation needed to induce a direct response to the non-self MHC molecules on the graft cannot be delivered separately by host antigen-presenting cells. In the MHC-identical grafts, the activated cytotoxic T cells can attack the graft directly, whereas the MHC-different grafts present different peptides and lack self MHC molecules, so they may evade this type of response. These MHC-different grafts are rejected by inflammation, probably triggered when host T cells recognize graft peptides presented by host antigen-presenting macrophages, rather than by the direct attack of cytotoxic T cells on cells of the graft.

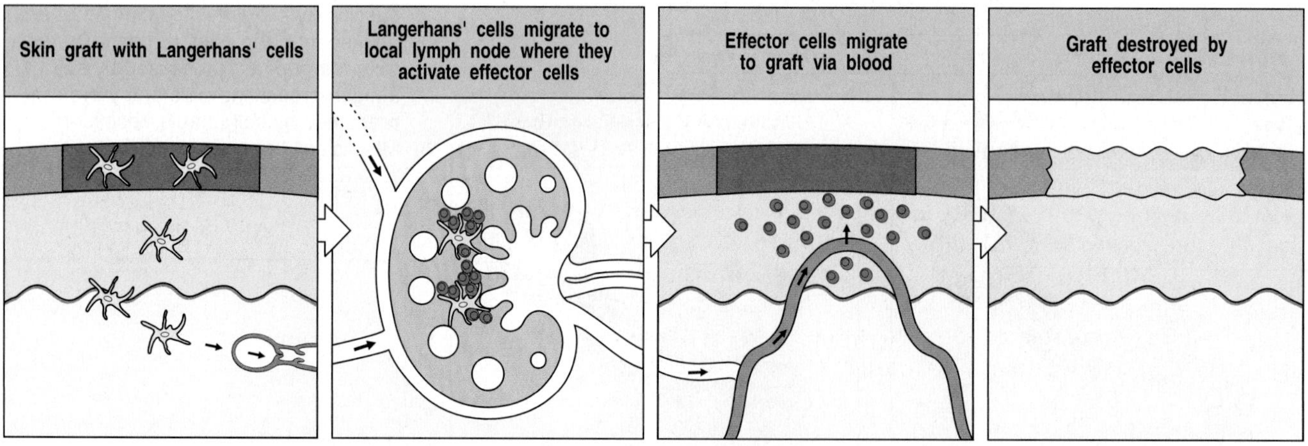

Fig. 13.24 The initiation of graft rejection normally involves migration of donor antigen-presenting cells from the graft to local lymph nodes. The example of a skin graft is illustrated here, in which Langerhans' cells are the antigen-presenting cells. They display graft peptides on their surface. In the lymph node, these antigen-presenting cells encounter recirculating naive T cells specific for graft antigens, and stimulate these T cells to divide. The resulting activated effector T cells migrate via the thoracic duct to the blood and home to the grafted tissue, which they rapidly destroy. Destruction is highly specific for donor-derived cells, suggesting that it is mediated by direct cytotoxicity and not by non-specific inflammatory processes.

The relative contributions of direct and indirect allorecognition in graft rejection are not known. Direct allorecognition is thought to be largely responsible for acute rejection, especially when MHC mismatches mean that the frequency of directly alloreactive recipient T cells is high. Furthermore, a direct cytotoxic T-cell attack on graft cells can be mediated only by T cells that recognize the graft MHC molecules directly. Nonetheless, T cells with indirect allospecificity can contribute to graft rejection by activating macrophages and are likely to be important in the development of an alloantibody response to a graft.

13-18 Antibodies reacting with endothelium cause hyperacute graft rejection.

Antibody responses are also an important potential cause of graft rejection. Pre-existing alloantibodies to blood group antigens and polymorphic MHC antigens can cause rapid rejection of transplanted organs in a complement-dependent reaction that can occur within minutes. This type of reaction is known as **hyperacute graft rejection**. Most grafts that are transplanted routinely in clinical medicine are vascularized organ grafts linked directly to the recipient's circulation. In some cases, the recipient might already have circulating antibodies to donor graft antigens, which were produced in response to a previous transplant or a blood transfusion. Such antibodies can cause very rapid rejection of vascularized grafts, because they react with antigens on the vascular endothelial cells of the graft and initiate the complement and blood clotting cascades, blocking the vessels of the graft and causing its immediate death. Such grafts become engorged and purple-colored from hemorrhaged blood, which becomes deoxygenated (Fig. 13.25). This problem can be avoided by **cross-matching** donor and recipient. Cross-matching involves determining whether the recipient has antibodies that react with the white blood cells of the donor. If antibodies of this type are found, they are a serious contraindication to transplantation, as they lead to near-certain hyperacute rejection.

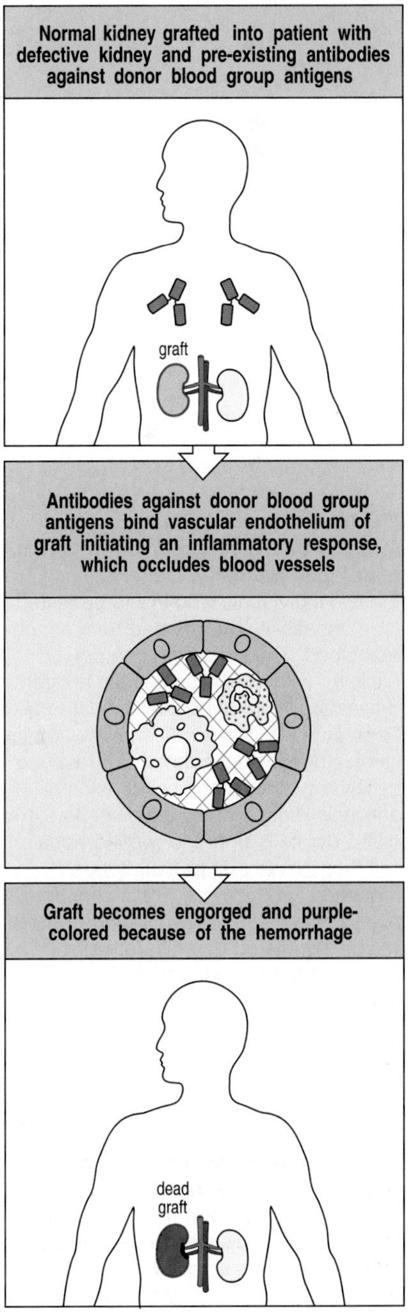

Normal kidney grafted into patient with defective kidney and pre-existing antibodies against donor blood group antigens
graft
Antibodies against donor blood group antigens bind vascular endothelium of graft initiating an inflammatory response, which occludes blood vessels
Graft becomes engorged and purple-colored because of the hemorrhage
dead graft

A very similar problem prevents the routine use of animal organs—**xenografts**—in transplantation. If xenogeneic grafts could be used, it would circumvent the major limitation in organ replacement therapy, namely the severe shortage of donor organs. Pigs have been suggested as a potential source of organs for xenografting as they are a similar size to humans and are readily farmed. Most humans and other primates have antibodies that react with endothelial cell antigens of other mammalian species, including pigs. When pig xenografts are placed in humans, these antibodies trigger hyperacute rejection by binding to the endothelial cells of the graft and initiating the complement and clotting cascades. The problem of hyperacute rejection is exacerbated in xenografts because complement-regulatory proteins such as CD59, DAF (CD55), and MCP (CD46) (see Section 9-31) work less efficiently across a species barrier; the complement-regulatory proteins of the xenogeneic endothelial cells cannot protect them from attack by human complement. A recent step towards xenotransplantation has been the development of transgenic pigs expressing human DAF.

Fig. 13.25 Pre-existing antibody against donor graft antigens can cause hyperacute graft rejection. In some cases, recipients already have antibodies to donor antigens. When the donor organ is grafted into such recipients, these antibodies bind to vascular endothelium, initiating the complement and clotting cascades. Blood vessels in the graft become obstructed by clots and leak, causing hemorrhage of blood into the graft. This becomes engorged and turns purple from the presence of deoxygenated blood.

Preliminary experiments have shown prolonged survival of organs transplanted from these pigs into recipient cynomolgus monkeys, under cover of heavy immunosuppression. However, hyperacute rejection is only the first barrier faced by a xenotransplanted organ. The T-lymphocyte mediated graft rejection mechanisms might be extremely difficult to overcome with present immunosuppressive regimes.

13-19 The corollary of graft rejection is graft-versus-host disease.

Allogeneic bone marrow transplantation is a successful therapy in some forms of leukemia and in primary immunodeficiency (see Chapter 11). In leukemia, it is first necessary to destroy the recipient's bone marrow, the source of the leukemia, by aggressive cytotoxic chemotherapy. One of the major complications of allogeneic bone marrow transplantation is graft-versus-host disease (GVHD), in which the allogeneic bone marrow recognizes the tissues of the recipient as foreign, causing a severe inflammatory disease, characterized by rashes, diarrhea, and pneumonitis. Graft-versus-host disease occurs not only when there is a mismatch of a major MHC class I or class II antigen but also in the context of disparities between minor H antigens. Graft-versus-host disease is a common complication in recipients of bone marrow transplants from HLA-identical siblings, who typically differ from each other in many polymorphic proteins encoded by genes unlinked to the MHC.

The presence of alloreactive T cells can be easily demonstrated experimentally by the mixed lymphocyte reaction (MLR), in which lymphocytes from a potential donor are mixed with irradiated lymphocytes from the potential recipient. If the donor lymphocytes contain alloreactive T cells, these will respond by cell division (Fig. 13.26). The MLR is sometimes used in the selection

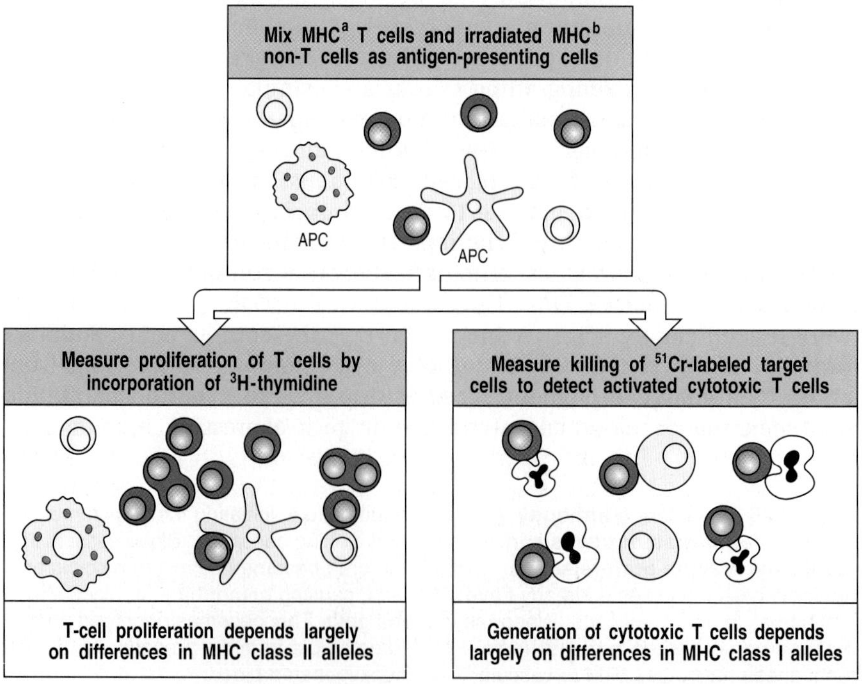

Fig. 13.26 The mixed lymphocyte reaction (MLR) can be used to detect histoincompatibility. Lymphocytes from the two individuals who are to be tested for compatibility are isolated from peripheral blood. The cells from one person (yellow), which will also contain antigen-presenting cells, are either irradiated or treated with mitomycin C; they will act as stimulator cells but cannot now respond by DNA synthesis and cell division to antigenic stimulation by the other person's cells. The cells from the two individuals are then mixed (top panel). If the unirradiated lymphocytes (the responders, blue) contain alloreactive T cells, these will be stimulated to proliferate and differentiate to effector cells. Between 3 and 7 days after mixing, the culture is assessed for T-cell proliferation (bottom left panel), which is mainly the result of CD4 T cells recognizing differences in MHC class II molecules, and for the generation of activated cytotoxic T cells (bottom right panel), which respond to differences in MHC class I molecules. When the MLR is used to select a bone marrow donor, the prospective donor's cells are used as stimulator cells.

of donors for bone marrow transplants, when the lowest possible alloreactive response is essential. However, the limitation of the MLR in selection of bone marrow donors is that the test does not accurately quantitate alloreactive T cells. A more accurate test is a version of the limiting dilution assay (see Section 2-18), which precisely counts the frequency of alloreactive T cells.

Although graft-versus-host disease is usually harmful to the recipient of a bone marrow transplant, there can be some beneficial effects. Some of the therapeutic effect of bone marrow transplantation for leukemia can be due to a graft-versus-leukemia effect, in which the allogeneic bone marrow recognizes minor H antigens or tumor-specific antigens expressed by the leukemic cells and kills the cells. One such minor H antigen, HB-1, which is a B-cell lineage marker, is expressed by acute lymphoblastic leukemia cells, which are B-lineage cells, and by Epstein–Barr virus (EBV)-transformed B lymphocytes.

One of the treatment options for suppressing the development of graft-versus-host disease is the elimination of mature T cells from the donor bone marrow *in vitro* before transplantation, thereby removing alloreactive T cells. Those T cells that subsequently mature from the donor marrow *in vivo* in the recipient are tolerant to the recipient's antigens. Although the elimination of graft-versus-host disease has benefits for the patient, there is an increase in the risk of leukemic relapse, which provides strong evidence in support of the graft-versus-leukemia effect.

13-20 Chronic organ rejection is caused by inflammatory vascular injury to the graft.

The success of modern immunosuppression means that approximately 85% of cadaveric kidney grafts are still functioning a year after transplantation. However, there has been no improvement in rates of long-term graft survival: the half-life for functional survival of renal allografts remains about 8 years. The major cause of late organ failure is chronic rejection, characterized by concentric arteriosclerosis of graft blood vessels, accompanied by glomerular and tubular fibrosis and atrophy.

Mechanisms that contribute to chronic rejection can be divided into those due to alloreactivity and those due to other pathways, and into early and late events after transplantation. Alloreactive mechanisms include the early injury associated with acute rejection and later alloreactivity, which is a largely silent process. Other important causes of chronic graft rejection include ischemia–reperfusion injury, which occurs at the time of grafting, and later-developing adverse factors such as chronic cyclosporin toxicity or cytomegalovirus infection.

Infiltration of the graft vessels and tissues by macrophages, followed by scarring, are prominent histological features of late graft rejection. A model of injury has been developed in which alloreactive T cells infiltrating the graft secrete cytokines stimulating the expression of endothelial adhesion molecules and also secrete chemokines such as RANTES, which lead to the recruitment of monocytes that mature to macrophages in the graft. A second phase of chronic inflammation then supervenes, dominated by macrophage products, including interleukin (IL)-1, TNF-α and the chemokine MCP, which leads to further macrophage recruitment. These mediators conspire to cause chronic inflammation and scarring, which eventually leads to irreversible organ failure.

Tissue transplanted	5 year graft survival*	No. of grafts in USA (1992)
Kidney	80–90%	9736
Liver	40–50%	3064
Heart	70%	2172
Lung	30–40%	535
Cornea	~70%	40,000
Bone marrow	80%	N/A

Fig. 13.27 Tissues commonly transplanted in clinical medicine. All grafts except corneal and some bone marrow grafts require long-term immunosuppression. The number of organ grafts performed in the USA in 1992 is shown (N/A = not available). *The 5 year survival values are an average; closer matching between donor and recipient generally gives better survival.

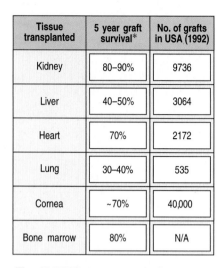

13-21 A variety of organs are transplanted routinely in clinical medicine.

Although the immune response makes organ transplantation difficult, there are few alternative therapies for organ failure. Three major advances have made it possible to use organ transplantation routinely in the clinic. First, the technical skill to carry out organ replacement surgery has been mastered by many people. Second, networks of transplantation centers have been organized to ensure that the few healthy organs that are available are HLA-typed and so matched with the most suitable recipient. Third, the use of potent immunosuppressive drugs, especially cyclosporin A and FK-506 to inhibit T-cell activation (see Chapter 14), has increased graft survival rates dramatically. Figure 13.27 lists the different organs that are transplanted in the clinic. Some of these operations are performed routinely with a very high success rate. By far the most frequently transplanted solid organ is the kidney, the organ first successfully transplanted between identical twins in the 1950s. Transplantation of the cornea is even more frequent; this tissue is a special case, as it is not vascularized, and corneal grafts between unrelated people are usually successful even without immunosuppression (see Section 13-29).

There are, however, many problems other than graft rejection associated with organ transplantation. First, donor organs are difficult to obtain; this is especially a problem when the organ involved is a vital one, such as the heart or liver. Second, the disease that destroyed the patient's organ might also destroy the graft. Third, the immunosuppression required to prevent graft rejection increases the risk of cancer and infection. Finally, the procedure is very costly. All of these problems need to be addressed before clinical transplantation can become commonplace. The problems most amenable to scientific solution are the development of more effective means of immunosuppression, the induction of graft-specific tolerance, and the development of xenografts as a practical solution to organ availability.

13-22 The fetus is an allograft that is tolerated repeatedly.

All of the transplants discussed so far are artifacts of modern medical technology. However, one tissue that is repeatedly grafted and repeatedly tolerated is the mammalian fetus. The fetus carries paternal MHC and minor H antigens that differ from those of the mother (Fig. 13.28), and yet a mother can successfully bear many children expressing the same non-self MHC proteins derived from the father. The mysterious lack of rejection of the fetus has puzzled generations of reproductive immunologists and no comprehensive explanation has yet emerged. One problem is that acceptance of the fetal allograft is so much the norm that it is difficult to study the mechanism that prevents rejection; if the mechanism for rejecting the fetus is rarely activated, how can one analyze the mechanisms that control it?

Various hypotheses have been advanced to account for the tolerance normally shown to the fetus. It has been proposed that, for some reason, the fetus is simply not recognized as foreign in the first place. There is evidence against this hypothesis, as women who have borne several children

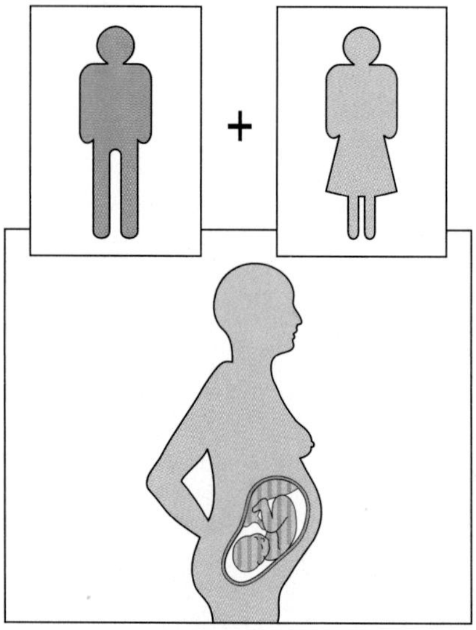

Fig. 13.28 The fetus is an allograft that is not rejected. Although the fetus carries MHC molecules derived from the father, and other foreign antigens, it is not rejected. Even when the mother bears several children to the same father, no sign of immunological rejection is seen.

usually make antibodies directed at the father's MHC proteins; indeed, this is the best source of antibodies for human MHC typing. However, the placenta, which is a fetus-derived tissue, seems to sequester the fetus away from the mother's T cells. The outer layer of the placenta, the interface between fetal and maternal tissues, is the trophoblast. This does not express classical MHC class I and class II proteins, making it resistant to recognition and attack by maternal T cells. Tissues lacking class I expression are, however, vulnerable to attack by NK cells (see Chapter 10). The trophoblast might be protected from attack by NK cells by expression of a non-classical and minimally polymorphic HLA class I molecule, HLA-G. This molecule has been shown to bind to the two major inhibitory NK receptors, KIR1 and KIR2, and to inhibit NK killing.

It is likely that fetal tolerance is a multifactorial process. The trophoblast does not act as an absolute barrier between mother and fetus, and fetal blood cells can cross the placenta and be detected in the maternal circulation, albeit in very low numbers. There is direct evidence from experiments in mice of specific T-cell tolerance against paternal MHC alloantigens. Pregnant female mice whose T cells bear a transgenic receptor specific for a paternal alloantigen showed reduced expression of this T-cell receptor during pregnancy. These same mice lost the ability to control the growth of an experimental tumor bearing the same paternal MHC alloantigen. After pregnancy, tumor growth was controlled and the level of the T-cell receptor increased. This experiment demonstrates that the maternal immune system must have been exposed to paternal MHC alloantigens.

A further factor that might contribute to maternal tolerance of the fetus is the secretion of cytokines at the feto-maternal interface. Both uterine epithelium and trophoblast secrete cytokines, including transforming growth factor (TGF)-β, IL-4, and IL-10. This cytokine pattern tends to promote T_H2 responses and suppress T_H1 responses (see Chapter 10). Induction or injection of cytokines such as interferon (IFN)-γ and IL-12, which promote T_H1 responses in experimental animals, promote fetal resorption, the equivalent of spontaneous abortion in humans.

The fetus is thus tolerated for two main reasons: it occupies a site protected by a non-immunogenic tissue barrier, and it promotes an immunosuppressive response in the mother. We shall see in Section 13-27 that several sites in the body have these characteristics and allow prolonged acceptance of foreign tissue grafts. They are usually called immunologically privileged sites.

Summary.

Clinical transplantation is now an everyday reality, its success built on MHC matching, immunosuppressive drugs, and technical skill. However, even accurate MHC matching does not prevent graft rejection; other genetic differences between host and donor can encode proteins whose peptides are presented as minor H antigens by MHC molecules on the grafted tissue. Responses to these can lead to rejection. As we lack the ability to suppress the response to the graft specifically without compromising host defense, most transplants require generalized immunosuppression of the recipient, which can cause significant toxicity and increases the risk of cancer and infection. The fetus is a natural allograft that must be accepted—it almost always is—or the species will not survive. Tolerance to the fetus might hold the key to specific graft tolerance, or it might be a special case not applicable to organ replacement therapy.

Tolerance and loss of tolerance to self tissues.

Tolerance to self is acquired by clonal deletion or inactivation of developing lymphocytes. Tolerance to antigens expressed by grafted tissues can be induced artificially, but it is very difficult to establish once a full repertoire of functional lymphocytes has been produced, which occurs during fetal life in humans and around the time of birth in mice. We have already discussed the two important mechanisms of self tolerance—clonal deletion by ubiquitous self antigens and clonal inactivation by tissue-specific antigens presented in the absence of co-stimulatory signals (see Chapters 6–8). These processes were first discovered by studying tolerance to non-self, where the absence of tolerance could be studied in the form of graft rejection. In this section, we shall consider tolerance to self and tolerance to non-self as two aspects of the same basic mechanisms. These mechanisms consist of direct induction of tolerance in the periphery either by deletion or anergy. There is also a mechanism referred to as immunological ignorance, in which T cell and antigen co-exist without affecting one or the other. Finally, there are mechanisms that involve T-cell–T-cell interactions, known variously as immune deviation or immune suppression. In an attempt to understand the related phenomena of autoimmunity and graft rejection, we also examine instances where tolerance to self or non-self is lost.

13-23 Autoantigens are not so abundant that they induce clonal deletion or anergy but are not so rare as to escape recognition entirely.

We saw in Chapter 7 that clonal deletion removes T cells that recognize ubiquitous self antigens and in Chapter 8 that antigens expressed abundantly in the periphery induce anergy or clonal deletion in lymphocytes that encounter them on tissue cells. Most self proteins are expressed at levels that are too low to serve as targets for T-cell recognition on any cell type and thus cannot serve as autoantigens. It is likely that only a very few self proteins contain peptides that are presented by a given MHC molecule at a level that is sufficient for effector T-cell recognition but too low to induce tolerance. T cells able to recognize these rare antigens will be present in the individual but will not normally be activated; they are said to be in a state of **immunological ignorance**. Most autoimmunity is likely to reflect the activation of such immunologically ignorant cells.

Autoimmunity is unlikely to reflect a general failure of the main mechanisms of tolerance—clonal deletion and clonal inactivation. For example, clonal deletion of developing lymphocytes mediates tolerance to self MHC molecules. If such tolerance were not induced, the reactions to self tissues would be similar to those seen in graft-versus-host disease (see Section 13-19). To estimate the impact of clonal deletion on the developing T-cell repertoire, we should remember that the frequency of T cells able to respond to any set of non-self MHC molecules can be as high as 10% (see Section 4-20), yet responses to self MHC antigens are not detected in naturally self-tolerant individuals. Moreover, mice given bone marrow cells from a foreign donor at birth, before significant numbers of mature T cells have appeared, can be rendered fully and permanently tolerant to the bone marrow donor's tissues, provided that the bone marrow donor's cells continue to be produced so as to induce tolerance in each new cohort of developing T cells

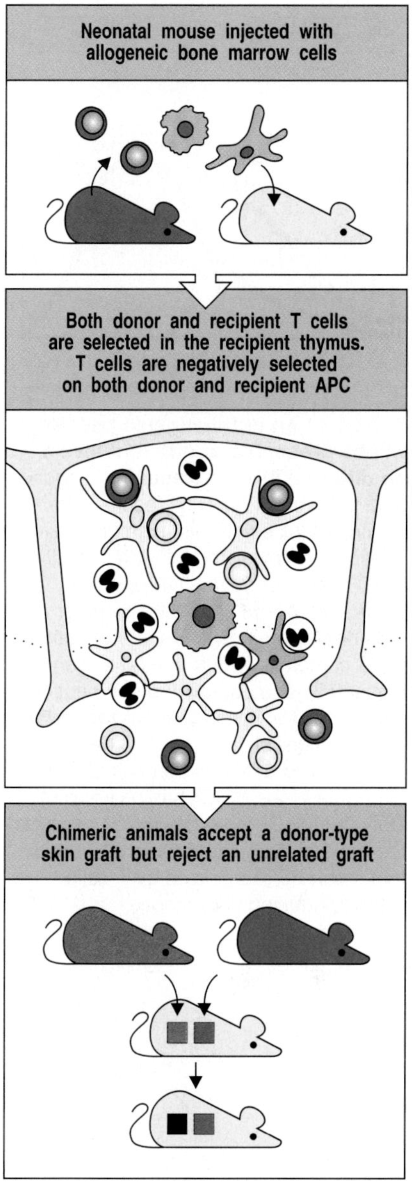

Fig. 13.29 Tolerance to allogeneic skin can be established in bone marrow chimeric mice. If mice are injected with allogeneic bone marrow at birth (top panel) before they achieve immune competence, they become chimeric, with T cells and antigen-presenting cells (APCs) deriving from both host and donor bone marrow stem cells. T cells developing in these mice are negatively selected on APCs of both host and donor (middle panel), so that mature T cells are tolerant to the MHC molecules of the bone marrow donor. This allows the chimeric mouse to accept skin derived from the bone marrow donor. Such acquired tolerance is specific, because skin from an unrelated or 'third party' donor is rejected normally (bottom panel).

(Fig. 13.29). This experiment, performed by Medawar, validated Burnet's prediction that developing lymphocytes with an open repertoire of receptors must be purged of self-reactive cells before they achieve functional maturity; it won them a Nobel Prize.

Clonal deletion reliably removes all T cells that can mount aggressive responses against self MHC molecules; autoimmune diseases, which involve rare T-cell responses to a particular self peptide bound to a self MHC molecule, are therefore unlikely to reflect a general failure in clonal deletion, nor are they likely to be caused by a random failure in the mechanisms responsible for anergy. Rather, the lymphocytes that mediate autoimmune responses seem not to be subject to clonal deletion or inactivation. Such autoreactive cells are present in all of us, but they do not normally cause autoimmunity because they are activated only under special circumstances.

A striking demonstration that autoreactive T cells can be present in healthy individuals comes from a strain of mice carrying transgenes encoding an autoreactive T-cell receptor specific for a peptide of myelin basic protein bound to self MHC class II molecules. The autoreactive receptor is present on every T cell, yet the mice are healthy unless their T cells are activated. We shall discuss these mice further in Section 13-27. As the level of the specific peptide:MHC class II complex is low except in the central nervous system, a site not visited by naive T cells, the autoreactive T cells remain in a state of immunological ignorance. When these T cells are activated, for example by deliberate immunization, as in EAE, they migrate into all tissues, including the central nervous system, where they can recognize their myelin basic protein:MHC class II ligand. Recognition triggers cytokine production by the activated T cells, causing inflammation in the brain and the destruction of myelin and neurons that ultimately causes the symptoms of paralysis in this disease.

It is likely that only a small fraction of proteins will be able to serve as autoantigens. An autoantigen must be presented by an MHC molecule at a level sufficient for the antigen to be recognized by effector T cells, but must not be presented to naive T cells at a level sufficient to induce tolerance. Many self proteins are expressed at levels too low to be detected even by effector T cells. It has been estimated that we can make approximately 10^5 proteins of average length 300 amino acids, capable of generating about 3×10^7 distinct self peptides. As MHC molecules are rarely expressed at levels above 10^5 molecules per cell, and as the MHC molecules on a cell must bind 10–100 identical peptides for T-cell recognition to occur, fewer than 10,000 self peptides (<1/3000) can be presented by any given MHC molecule at levels detectable by T cells. Thus, most peptides will be presented at levels that are insufficient to engage effector T cells, whereas many of the peptides that can be detected by T cells will be presented at a high enough level to induce clonal deletion or anergy. However, as shown in Fig. 13.30, a few peptides may fail to induce tolerance, yet are present at high enough levels

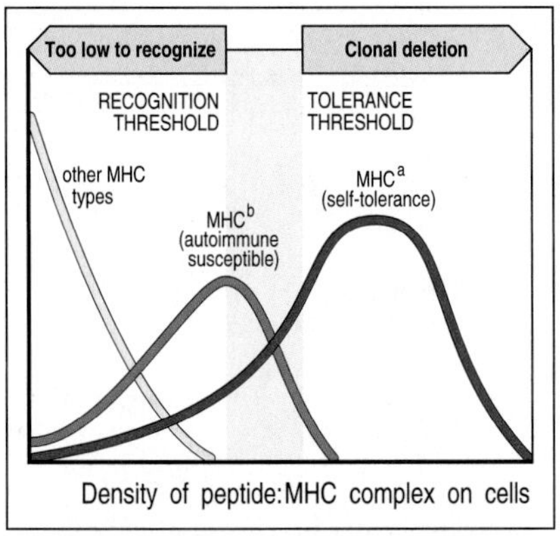

Fig. 13.30 An autoantigenic peptide will be presented at different levels on different MHC molecules. Peptides bind to different MHC molecules with varying affinities; in this example, a self peptide binds well to MHC^a, less well to MHC^b, and poorly to all other MHC types. The graph shows the number of cells displaying a given density of peptide: MHC complex for different MHC genotypes. It illustrates how the self peptide is displayed at levels that are too low for T-cell recognition in most MHC haplotypes (yellow), and at levels that would induce tolerance in the MHC^a haplotype (red). Only the MHC^b molecule (blue) presents the peptide at an intermediate level so that it can be recognized by T cells without inducing tolerance.

for recognition by effector T cells. Autoreactivity probably arises most frequently when the antigen is expressed selectively in a tissue rather than ubiquitously, because tissue-specific antigens are less likely to induce clonal deletion of developing T cells in the thymus. The nature of such peptides will vary depending on the MHC genotype of the individual, because MHC polymorphism profoundly affects peptide binding (see Section 4-18). This argument leaves aside the crucial issue of how T cells specific for such autoantigens are activated to become effector T cells, which we shall consider later.

If the idea that only a few peptides can act as autoantigens is correct, then it is not surprising that there are relatively few distinct autoimmune syndromes, and that all individuals with a particular autoimmune disease tend to recognize the same antigens. If all antigens could give rise to autoimmunity, one would expect that different individuals with the same disease might recognize different antigens on the target tissue, which does not seem to be true. Finally, because the level of autoantigenic peptide presented is determined by polymorphic residues in MHC molecules that govern the affinity of peptide binding, this idea could also explain the association of autoimmune diseases with particular MHC genotypes (see Fig. 13.2).

13-24 The induction of a tissue-specific response requires the presentation of antigen by professional antigen-presenting cells with co-stimulatory activity.

As we learned in Chapter 8, only professional antigen-presenting cells that express co-stimulatory activity can initiate clonal expansion of T cells—an essential step in all adaptive immune responses, including graft rejection and, presumably, autoimmunity. In tissue grafts, it is the donor antigen-presenting cells in the graft that initially stimulate host T cells, starting the direct allorecognition response that leads to graft rejection. Antigen-presenting cells bearing both graft antigens and co-stimulatory activity travel to regional lymph nodes. Here they are examined by large numbers of naive host T cells and can activate those that bear specific receptors (see Fig. 13.24). Grafts depleted of antigen-presenting cells are tolerated for long periods, but are eventually rejected. This rejection is due to the recipient's T cells' responding to graft antigens, both MHC and minor H antigens, after they have been processed and presented by recipient antigen-presenting cells (see Fig. 13.23).

The ability of the recipient's professional antigen-presenting cells to pick up antigens in tissues and initiate graft rejection may be relevant to the initiation of autoimmune tissue damage as well. Transplantation experiments show that host antigen-presenting cells can stiimulate both cytotoxic T-cell responses and inflammatory T_H1 responses against the transplanted tissue; thus tissue antigens can be taken up and presented in conjunction with both MHC class I and class II molecules by professional antigen-presenting cells. In autoimmunity, tissues may be similarly attacked by MHC class I-restricted cytotoxic T cells or injured by inflammatory damage mediated by T_H1 cells, as a consequence of the uptake and presentation of tissue antigens by professional antigen-presenting cells.

To induce a response to tissue antigens, the antigen-presenting cell must express co-stimulatory activity. As we saw in Chapter 8, the expression of co-stimulatory molecules in professional antigen-presenting cells is regulated to occur in response to infection. Transient autoimmune responses are seen in the context of such events, and it is thought that one trigger for autoimmunity is infection. This is discussed further in Section 13-29.

13-25 In the absence of co-stimulation, tolerance is induced.

Activation of naive T cells requires interaction with cells expressing both the appropriate peptide:MHC complex and co-stimulatory molecules; in the absence of co-stimulation, specific antigen recognition leads to T-cell anergy or deletion (see Section 8-10). Tissue cells are not known to express B7 or other co-stimulatory molecules, and can therefore induce tolerance. Experiments with transgenes show that the expression of foreign antigens in peripheral tissues can in some cases induce tolerance, whereas in other cases the foreign antigen seems not to be presented to naive T cells at a sufficient level and is ignored. Autoimmunity can be induced by co-expression of a foreign antigen and B7 in the same target tissue, but as B7 expression on peripheral tissue cells is not by itself a sufficient stimulus for autoimmunity, it is clear that the loss of tolerance to self tissues requires the co-expression of both a suitable target antigen and co-stimulatory molecules. As tissue cells are not known to express B7 or other co-stimulatory molecules, tolerance to self tissues is the norm. As discussed in Section 13-23, antigens that are unable to induce clonal anergy or deletion, but that can nonetheless act as targets for effector T cells, can serve as autoantigens; these antigens are likely to be tissue-specific and relatively few.

By analogy with graft rejection, it seems likely that autoimmunity is initiated when a professional antigen-presenting cell picks up a tissue-specific autoantigen and migrates to the regional lymph node, where it is induced to express co-stimulatory activity. Once an autoantigen is expressed on a cell with co-stimulatory potential, naive T cells specific for the autoantigen can become activated and can home to the tissues, where they interact with their target antigens. At this stage, the absence of co-stimulatory molecules on tissue cells that present the autoantigen can again limit the response. Armed effector T cells kill only a limited number of antigen-expressing tissue cells if these lack co-stimulatory activity; after killing a few targets, the effector cell dies. Thus, not only can responses not be initiated in the absence of co-stimulatory activity, they also cannot be sustained. Therefore, in addition to the question of how autoimmunity is avoided, we must ask: Why does it ever occur? How are responses to self initiated, and how they are sustained?

13-26 Dominant immune suppression can be demonstrated in models of tolerance and can affect the course of autoimmune disease.

In some models of tolerance, it can be demonstrated that specific T cells have an active role in suppressing the activity of other T cells that are capable of causing tissue damage. Tolerance in these cases is dominant in that it can be transferred by T cells, which are usually called **suppressor T cells**. Furthermore, depletion of the suppressor T cells leads to aggravated responses to self or graft antigens. Although it is clear that these phenomena of immune suppression exist, the mechanisms responsible have been the subject of much controversy. Here, we shall examine the phenomenon in three animal models.

In experiments using skin graft rejection as a model, neonatal rats can be rendered tolerant to allogeneic grafts by injection of allogeneic bone marrow. The tolerance induced is highly specific and can be transferred to normal adult recipient rats. This shows that tolerance in this model is dominant and active, as the lymphocytes of the recipient are prevented by the transferred cells from mediating graft rejection. In order to transfer this tolerance, cells of both the allogeneic graft donor and the neonatal tolerized host must be transferred. Removal of either cell type abolishes the transfer of tolerance.

This finding is reminiscent of the studies of Medawar on tolerance in neonatal bone marrow chimeric mice discussed in Section 13-23. In both cases, even injection of massive numbers of normal syngeneic lymphocytes, which would react vigorously against the foreign cells in the normal environment of the cell donor, did not break tolerance. Tolerance could be broken only with cells from an animal that had been immunized with cells from the allogeneic donor before transfer; such cells probably break tolerance by killing the allogeneic donor cells. Thus, an active host response prevents graft rejection in this model. The tolerance is specific for cells of the original donor, and so the suppression must also be specific.

In the NOD mouse model for insulin-dependent diabetes, transfer of a certain insulin-specific T-cell clone can prevent the destruction of pancreatic β cells by autoreactive T cells. This suggests that the insulin-specific T cells can suppress the activity of other autoaggressive T cells in an antigen-dependent manner. They do this by homing to the islet, where they react with insulin peptides presented on the β cells. This stimulates the secretion of cytokines, prominent among which is TGF-β, a known immunosuppressive cytokine. There are interesting hints that such cells naturally affect the course of the autoimmune response that causes human diabetes; β-cell destruction in humans occurs over a period of several years before diabetes is manifest, yet when new islets are transplanted from an identical twin into his or her diabetic sibling, they are destroyed within weeks. This suggests that, in the normal course of the disease, specific T cells protect the β cells from attack by effector T cells and the disease therefore progresses slowly. It might be that after the host islets have been destroyed, these protective mechanisms decline in activity, but that the effector cells responsible for β-cell destruction do not.

If specific suppression of autoimmune responses could be elicited at will, autoimmunity would not be a problem. Feeding with specific antigen is known to elicit a local immune response in the intestinal mucosa, and responses to the same antigen given subsequently by a systemic route are suppressed (see Section 2-3). This response has been exploited in experimental autoimmune diseases by feeding proteins from target tissues to mice; mice fed with insulin are protected from diabetes, whereas mice fed with myelin basic protein are resistant to EAE (Fig. 13.31).

Experimental allergic encephalomyelitis is normally caused by T_H1 cells that produce IFN-γ in response to myelin basic protein; in mice fed with this protein, CD4 T cells that produce cytokines such as TGF-β and IL-4 are found in the brain instead (see Fig. 13.31). TGF-β, in particular, suppresses the function of inflammatory T_H1 lymphocytes. In both these cases, the protection seems to be tissue-specific rather than antigen-specific. Thus, feeding with insulin protects against diabetes, yet insulin is not thought to be the target of autoimmune attack on the β cells. Likewise, feeding with myelin basic protein will protect against EAE elicited by immunization with other brain antigens. Feeding with antigen might induce the production of T cells producing TGF-β and IL-4 because these cytokines are also required for IgA production against antigens ingested in food. If feeding with antigen works as a clinical therapy, it would have the advantage over treatments with immunosuppressive drugs in that it does not alter the general immune competence of the host. Unfortunately, early studies of this approach to treatment in humans with multiple sclerosis or rheumatoid arthritis have shown minimal, if any, benefit.

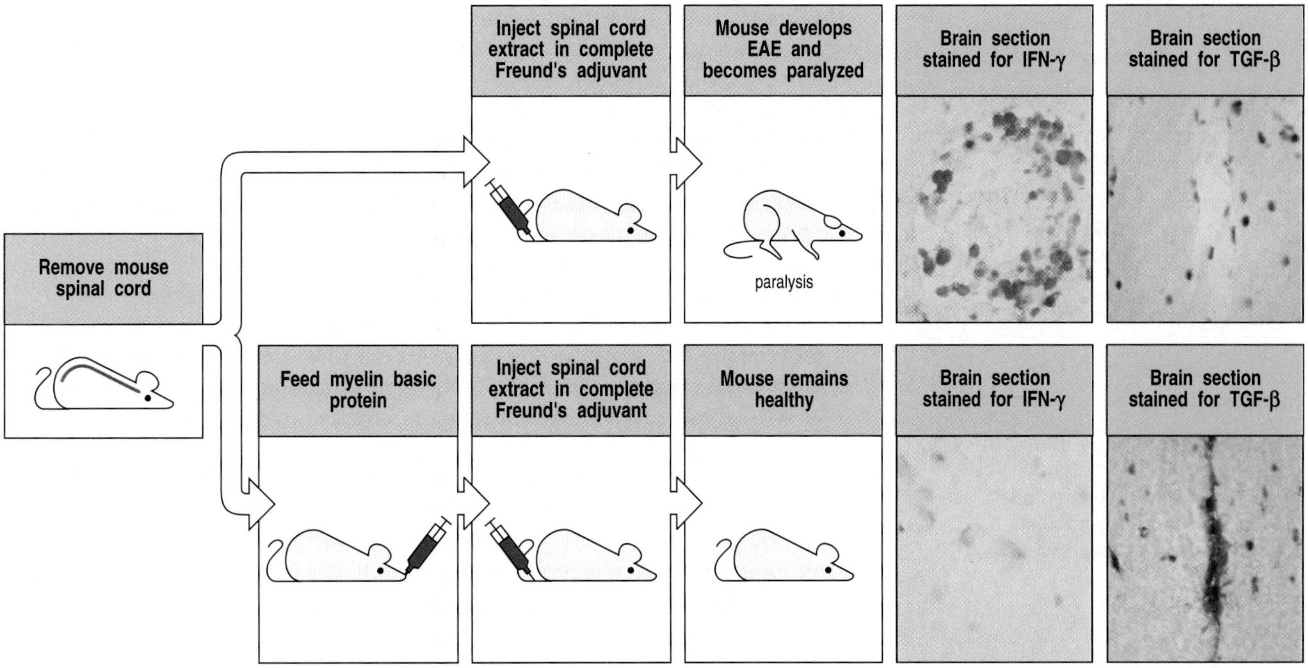

Fig. 13.31 Antigen given orally can lead to protection against autoimmune disease. Experimental allergic encephalomyelitis (EAE) is induced in mice by immunization with spinal cord homogenate in complete Freund's adjuvant (upper panels); the disease is mediated by T_H1 cells specific for myelin antigens. These cells produce the cytokine IFN-γ (top left photo, where the brown staining reveals the presence of IFN-γ), but not TGF-β. These T cells are presumably responsible for the damage that results in paralysis. When mice are first fed with myelin basic protein (MBP), later immunization with spinal cord or MBP fails to induce the disease (lower panels). In these orally tolerized mice, IFN-γ-producing cells are absent (lower left photo), whereas TGF-β-producing T cells (lower right photo, brown staining) are found in the brain in place of the auto-aggressive T_H1 cells and presumably protect the brain from autoimmune attack. Photographs courtesy of S Khoury, W Hancock, and H Weiner.

Like human diabetes, multiple sclerosis is a chronic relapsing disease with acute episodes followed by periods of quiescence. This again suggests a balance between autoimmune and protective T cells, which can alter at different stages of the disease. However, it remains to be proven whether the specific suppressive cells discussed in this section exist naturally and contribute to self tolerance, or whether they arise only upon artificial stimulation or in response to autoimmune attack. Nevertheless, as they can play an active, dominant part in self tolerance, they are particularly attractive targets for immunotherapy of autoimmune disease.

13-27 Antigens in immunologically privileged sites do not induce immune attack but can serve as targets.

Tissue grafts placed in some sites in the body do not elicit immune responses. For instance, the brain and the anterior chamber of the eye are sites in which tissues can be grafted without inducing graft rejection. Such locations in the body are termed **immunologically privileged sites** (Fig. 13.32). It was originally believed that immunological privilege arose from the failure of antigens to leave privileged sites and induce responses. However, subsequent studies have shown that antigens do leave immunologically privileged sites, and that these antigens do interact with T cells; however, instead of eliciting a destructive immune response, they induce tolerance or a response that is

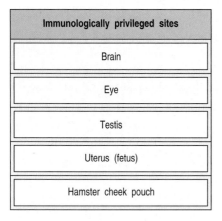

Immunologically privileged sites
Brain
Eye
Testis
Uterus (fetus)
Hamster cheek pouch

Fig. 13.32 Some body sites are immunologically privileged. Tissue grafts placed in these sites often last indefinitely, and antigens placed in these sites do not elicit destructive immune responses.

not destructive to the tissue. Immunologically privileged sites seem to be unusual in two ways. First, the communication between the privileged site and the body is atypical in that extracellular fluid in these sites does not pass through conventional lymphatics, although proteins placed in these sites do leave them and can have immunological effects. Second, humoral factors, presumably cytokines, that affect the behavior of the immune response are produced in privileged sites and leave them together with antigens. Again, the anti-inflammatory cytokine TGF-β seems to be particularly important in this regard: antigens mixed with TGF-β seem to induce T-cell responses that do not damage tissues, such as T_H2 rather than T_H1 responses.

Paradoxically, it is often the antigens sequestered in immunologically privileged sites that are the targets of autoimmune attack; for example, multiple sclerosis is directed at brain autoantigens such as myelin basic protein, and, as we have seen, EAE is induced in some strains of rats and mice upon immunization with myelin basic protein in adjuvant. It is therefore clear that this antigen does not induce deletional tolerance and anergy. As we saw in Section 13-23, mice transgenic for a T-cell receptor specific for myelin basic protein carry this autoreactive receptor on most of their T cells, yet develop normally. These T cells are readily activated by the appropriate peptide of the protein; nevertheless, the mice do not become diseased unless they are deliberately immunized with myelin basic protein, in which case they become acutely sick, show severe infiltration of the brain with specific T_H1 cells, and often die. Their non-transgenic littermates have a milder, transient illness after immunization.

Thus, at least some antigens expressed in immunologically privileged sites induce neither tolerance nor activation, but if activation is induced elsewhere they can become targets for autoimmune attack. It seems plausible that T cells specific for antigens that are sequestered in immunologically privileged sites are more likely to remain in the state of immunological ignorance described in Section 13-23. This is further shown in the eye disease **sympathetic ophthalmia** (Fig. 13.33). If one eye is ruptured by a blow or other trauma, an autoimmune response to eye proteins can occur, although it happens only rarely. Once the response is induced, it often attacks both eyes. Immunosuppression and removal of the damaged eye, the source of antigen, is frequently required to preserve vision in the undamaged eye.

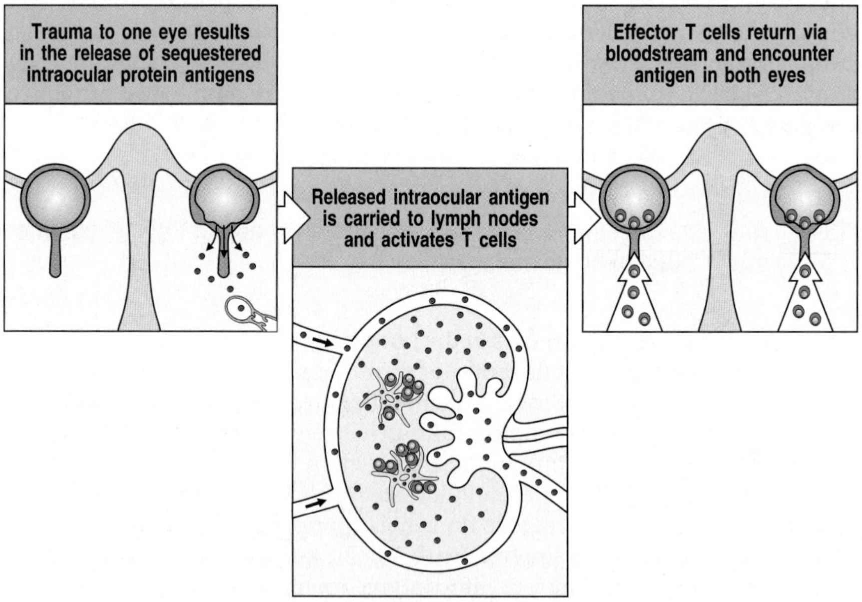
Trauma to one eye results in the release of sequestered intraocular protein antigens
Released intraocular antigen is carried to lymph nodes and activates T cells
Effector T cells return via bloodstream and encounter antigen in both eyes

Fig. 13.33 Damage to an immunologically privileged site can induce an autoimmune response. In the disease sympathetic ophthalmia, trauma to one eye releases the sequestered eye antigens into the surrounding tissues, making them accessible to T cells. The effector cells that are elicited attack the traumatized eye, and also infiltrate and attack the healthy eye. Thus, although the sequestered antigens do not induce a response by themselves, if a response is induced elsewhere they can serve as targets for attack.

It is not surprising that effector T cells can enter immunologically privileged sites: such sites can become infected and effector cells must be able to enter these sites during infection. As we learned in Chapter 10, effector T cells enter most or all tissues after activation, but accumulations of cells are seen only when antigen is recognized in the site, triggering the production of cytokines that alter tissue barriers.

13-28 B cells with receptors specific for peripheral autoantigens are held in check by a variety of mechanisms.

When B-cell antigen receptors are first expressed in the bone marrow, receptors specific for self molecules are produced as a consequence of the random generation of the repertoire. If a self molecule is expressed in the bone marrow in an appropriate form, clonal deletion and receptor editing can remove all of these self-reactive B cells while they are still immature (see Sections 6-9 and 6-10). There are, however, many self molecules available only in the periphery whose expression is restricted to particular organs. An example is thyroglobulin (the precursor of thyroxine), which is expressed only in the thyroid and at extremely low levels in plasma. Back-up mechanisms exist to ensure that B cells reactive to these self molecules do not cause autoimmune disease. When a mature B cell in the periphery encounters self molecules that bind its receptor, four proposed mechanisms could bring about the observed non-reactivity. Failure of any one of these mechanisms could lead to autoimmunity.

First, B cells that recognize a self antigen arrest their migration in the T-cell zone of peripheral lymphoid tissues (Fig. 13.34), just like B cells that bind a foreign antigen (see Chapter 9). However, unlike the response to foreign antigens, in which activated effector CD4 T cells are present, B cells binding self antigens will not be able to interact with helper CD4 T cells because no such cells exist for self antigens. This lack of interaction prevents the B cells migrating out of the T-cell zones into the follicles; instead, the trapped self-reactive B cells undergo apoptosis.

A second mechanism for inactivation of autoreactive B cells in the periphery is the induction of B-cell anergy, which is associated with downregulation of surface IgM expression and partial inhibition of the linked B-cell signaling pathways (Fig. 13.35). B-cell anergy can be induced by exposure to soluble circulating antigen; if mice are inoculated intravenously with protein solutions from which all trace of aggregates has been rigorously removed so as to eliminate multivalent complexes, the peripheral B cells that bind these proteins can be inactivated. The lifespan of anergic B cells is short as they are usually eliminated after failing to enter the primary lymphoid follicles or after interacting with antigen-specific T cells as described below. This form of B-cell tolerance can therefore be viewed as a form of delayed B-cell deletion.

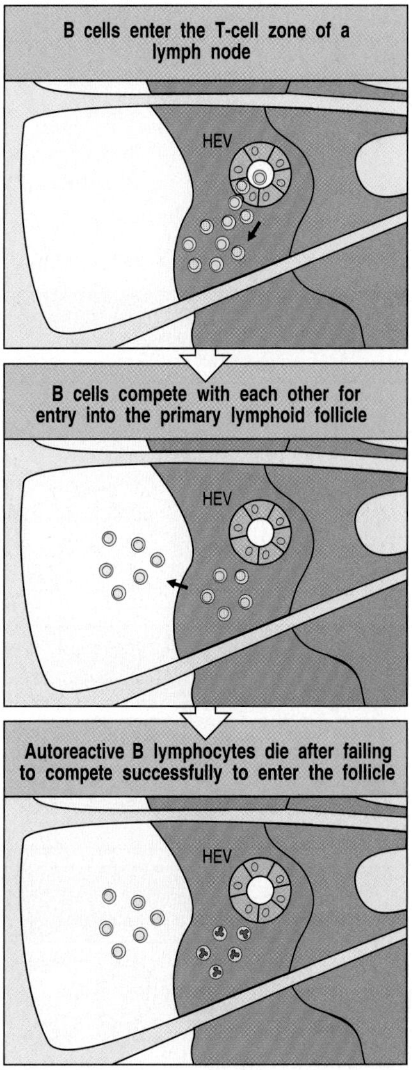

Fig. 13.34 Autoreactive B cells do not compete effectively in peripheral lymphoid tissue to enter primary lymphoid follicles. In the top panel, B cells are seen entering the T-cell zone of a lymph node through high endothelial venules (HEVs). Those with reactivity for foreign antigens are shown in yellow; autoreactive cells are gray. The auto-reactive cells fail to compete with B cells specific for foreign antigens for exit from the T-cell zone and entry into primary follicles (center panel). This is because B cells reactive to foreign antigens receive signals from antigen-specific T cells that promote their activation and survival. Thus, the autoreactive B cells fail to receive survival signals and undergo apoptosis *in situ* in the T-cell zone (bottom panel).

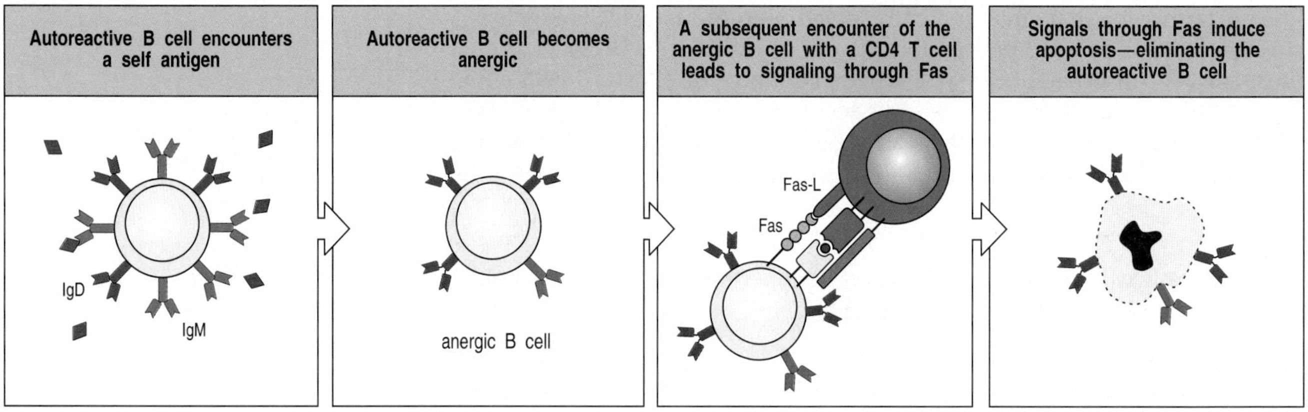

Fig. 13.35 Peripheral B-cell anergy. An autoreactive B cell encounters its soluble autoantigen in the periphery (first panel), which leads to the development of B-cell anergy (second panel). This is characterized by a reduction in both the expression of surface IgM and of the signaling pathways after ligation of surface immunoglobulin. A further mechanism to maintain peripheral B-cell tolerance is shown in the two panels on the right. If an anergized, self-reactive B cell subsequently encounters a T cell specific for a self peptide from the relevant autoantigen, Fas ligand on the T-cell surface binds to Fas on the B cell, inducing apoptosis in the B cell.

The third mechanism is dependent on the presence of T cells that are specific for the self antigen and express Fas ligand. In the rare instances when such an autoreactive T cell matures and is activated, it is able to kill autoreactive anergic B cells in a Fas-dependent manner (see Fig. 13.35). In the absence of the normal pathways of co-stimulation, anergized B lymphocytes show enhanced sensitivity to apoptosis after ligation of Fas by Fas ligand because they have been chronically exposed to self antigen. They are therefore not subject to the stimulatory signals that oppose apoptosis in B cells whose surface receptors have just been ligated by foreign antigen. The importance of this mechanism is nicely illustrated by the consequences of mutation in the genes for Fas or Fas ligand. Mice and humans deficient in Fas or Fas ligand develop severe autoimmune disease, similar to SLE, associated with overproduction of lymphocytes.

Finally, there is evidence for a distinct mechanism for dealing with B cells that develop self-reactive specificities as a result of somatic hypermutation during a response to a foreign antigen (Fig. 13.36). At a crucial phase at the height of the germinal center reaction, an encounter with a large dose of soluble antigen causes a wave of apoptosis in germinal center B cells within a few hours.

All these mechanisms re-emphasize the fact that the mere existence in the body of some B lymphocytes with receptor specificities directed against self is not in itself harmful. Before an immune response can be initiated they need to receive effective help, the B-cell antigen receptors must be ligated, and their intracellular signaling machinery must be set to respond normally.

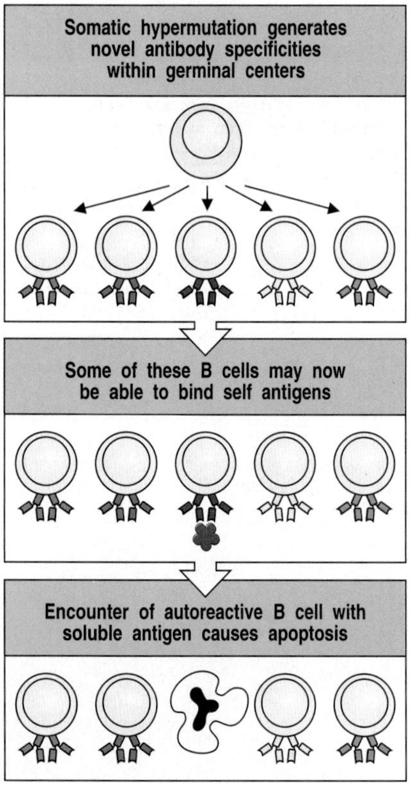

Somatic hypermutation generates novel antibody specificities within germinal centers
Some of these B cells may now be able to bind self antigens
Encounter of autoreactive B cell with soluble antigen causes apoptosis

Fig. 13.36 Elimination of autoreactive B cells in germinal centers. During the process of somatic hypermutation in germinal centers, depicted in the top panel, autoreactive antibody specificities can arise. Ligation of these antibody receptors by soluble autoantigen induces apoptosis of the autoreactive B cell by signaling through the B-cell antigen receptor.

13-29 Autoimmunity could be triggered by infection in a variety of ways.

Human autoimmune diseases often appear gradually, making it difficult to find out how the process is initiated. Nevertheless, there is a strong suspicion that infections can trigger autoimmune disease in genetically susceptible individuals. Indeed, many experimental autoimmune diseases are induced by mixing tissue cells with adjuvants that contain bacteria. For example, to induce EAE, it is necessary to emulsify the spinal cord or myelin basic protein used for immunization in complete Freund's adjuvant, which includes killed *Mycobacterium tuberculosis* (see Section 2-4); however, when the mycobacteria are omitted from the adjuvant, not only is no disease elicited but the animals become refractory to any subsequent attempt to induce the disease by antigen in complete Freund's adjuvant, and T cells can transfer this resistance to syngeneic recipients (Fig. 13.37). There are several other systems in which infection is important in the induction of disease; for example, transgenic mice that express a T-cell receptor specific for myelin basic protein (see Section 13-23) often develop spontaneous autoimmune encephalomyelitis if they become infected. One possible mechanism for this loss of tolerance is that the infectious agents induce co-stimulatory activity on antigen-presenting cells expressing low levels of peptides from myelin basic protein, thus activating the autoreactive T cells.

It has also been suggested that autoimmunity can be initiated by a mechanism known as **molecular mimicry**, in which antibodies or T cells generated in the response to an infectious agent cross-react with self antigens. To show that infectious agents can trigger responses that have the capacity to destroy tissues, mice were made transgenic for a viral nuclear protein driven by the insulin promoter, so that the protein was expressed only in pancreatic β cells. As the amount of protein expressed was low, the T cells that recognized the

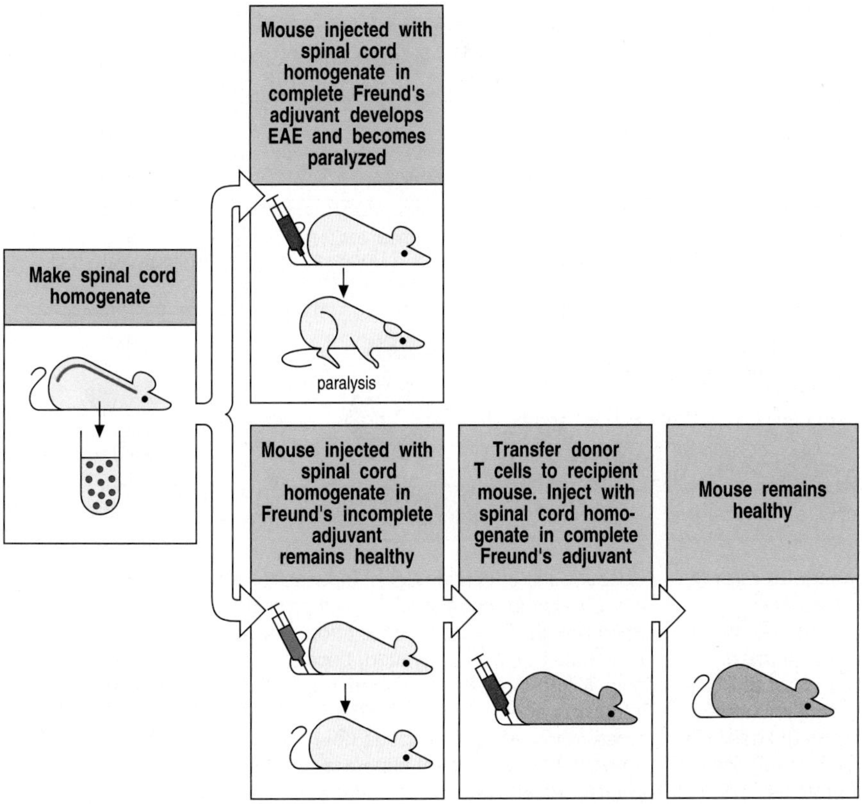

Fig. 13.37 Bacterial adjuvants are required for induction of experimental autoimmune disease. Mice immunized with spinal cord homogenate in complete Freund's adjuvant, which contains large numbers of killed *Mycobacterium tuberculosis*, get experimental allergic encephalomyelitis (EAE). Mice immunized with the same antigen in incomplete Freund's adjuvant, which lacks the *M. tuberculosis*, not only do not become diseased but are actually protected from subsequent disease induction. Moreover, T cells from these mice can transfer protection from disease to naive, syngeneic recipients.

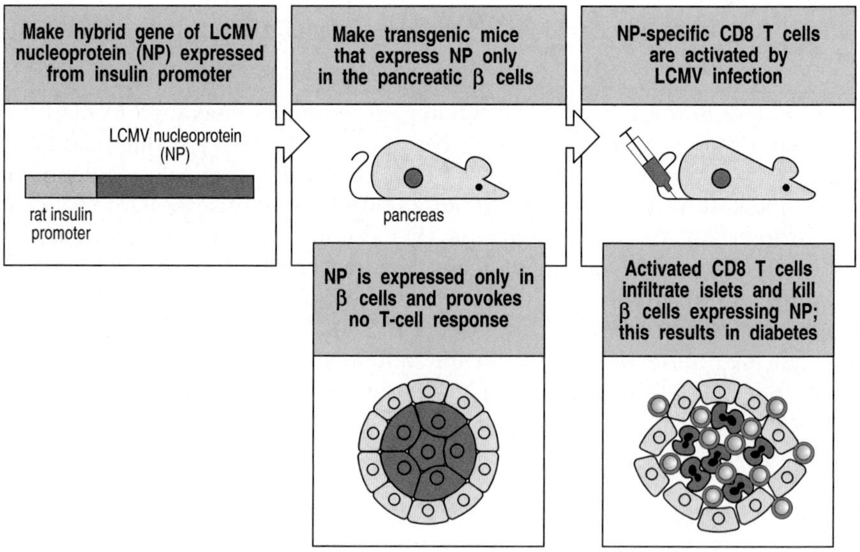

Fig. 13.38 Virus infection can break tolerance to a transgenic viral protein expressed in pancreatic β cells. Transgenic mice that express a protein from the lymphocytic choriomeningitis virus (LCMV) in their pancreatic β cells do not respond to the protein and therefore do not develop an autoimmune diabetes. However, if the transgenic mice are infected with LCMV, a potent anti-viral cytotoxic T-cell response is elicited, and this kills the β cells, leading to diabetes. It is thought that infectious agents can sometimes elicit T-cell responses that cross-react with self peptides (a process known as molecular mimicry) and that this could cause autoimmune disease in a similar way.

viral protein remained ignorant, that is, they were neither tolerant to the viral protein nor activated by it, and the animals showed no sign of disease. However, if they were infected with the live virus, they responded by making cytotoxic CD8 T cells specific for the viral protein, and these armed CD8 cytotoxic T cells could destroy the β cells, causing diabetes (Fig. 13.38). Recent work suggests that the immune response to natural infection occasionally elicits destructive T cells that cause autoimmune disease because of molecular mimicry, by cross-reacting with peptides of self antigens.

Mechanism	Disruption of cell or tissue barrier	Infection of antigen-presenting cell	Binding of pathogen to self protein	Molecular mimicry	Superantigen
Effect	Release of sequestered self antigen; activation of non-tolerized cells	Induction of co-stimulatory activity on antigen-presenting cells	Pathogen acts as carrier to allow anti-self response	Production of cross-reactive antibodies or T cells	Polyclonal activation of autoreactive T cells
Example	Sympathetic ophthalmia	Effect of adjuvants in induction of EAE	? Interstitial nephritis	Rheumatic fever ? Diabetes ? Multiple sclerosis	? Rheumatoid arthritis

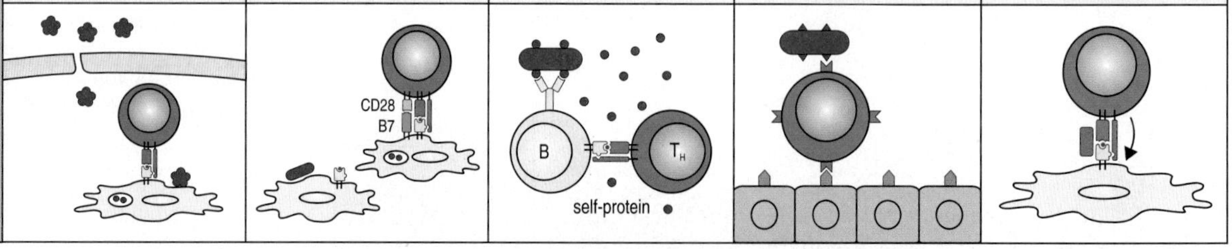

Fig. 13.39 There are several ways in which infectious agents could break self tolerance. Because some antigens are sequestered from the circulation, either behind a tissue barrier or within the cell, an infection that breaks cell and tissue barriers might expose hidden antigens (first panel). A second possibility is that infectious agents might trigger expression of co-stimulators on antigen-presenting cells that have taken up tissue antigens, thereby inducing an autoimmune response (second panel). In some cases, infectious agents might bind to self proteins. Because the infectious agent induces a helper T-cell response, any B cell that recognizes the self protein will also receive help (third panel). Such responses should be self-limiting once the infectious agent is eliminated, because at this point the T-cell help will no longer be provided. Molecular mimicry might result in infectious agents inducing either T- or B-cell responses that can cross-react with self antigens (fourth panel). Polyclonal T-cell activation by a bacterial superantigen could overcome clonal anergy, allowing an autoimmune process to begin (last panel).

Associations of infection with immune-mediated tissue damage		
Infection	**HLA association**	**Consequence**
Group A *Streptococcus*	?	Rheumatic fever (carditis, polyarthritis)
Chlamydia trachomatis	HLA-B27	Reiter's syndrome (arthritis)
Shigella flexneri, Salmonella typhimurium, Salmonella enteritidis, Yersinia enterocolitica, Campylobacter jejuni	HLA-B27	Reactive arthritis
Borrelia burgdorferi	HLA-DR2, DR4	Chronic arthritis in Lyme disease

Fig. 13.40 Association of infection with autoimmune diseases. Several autoimmune diseases occur after specific infections and are presumably triggered by the infection. The case of post-streptococcal disease is best known but is now rare because effective antibiotic therapy of Group A streptococcal infection usually prevents post-infection complications. Most of these post-infection autoimmune diseases also show susceptibility linkage to the MHC.

It has long been known that molecular mimicry can operate in antibody-mediated autoimmunity; microbial antigens can elicit antibody responses that react not only with the pathogen but also with host antigens that are similar in structure. This type of response occurs after infection with some *Streptococcus* species that elicit antibodies that cross-react with kidney, joint, and heart antigens to produce rheumatic fever. However, such responses are usually transient and do not lead to sustained autoantibody production, as the helper T cells are specific for the microbe and not for self proteins. Host proteins that form a complex with bacteria can induce a similar transient response; in this case, the antibody response is not cross-reactive but the bacterium is acting as a carrier, allowing B cells that express an autoreactive receptor to receive inappropriate T-cell help. These and some other mechanisms that could allow an infectious agent to break tolerance are summarized in Fig. 13.39. All of these mechanisms can be shown to act in experimental systems, and some evidence supports their importance in human autoimmune disease as well.

The argument that autoimmunity might be initiated by infection is strengthened by the fact that there are several human autoimmune diseases in which a prior infection with a specific agent or class of agents leads to a particular disease (Fig. 13.40). Again, disease susceptibility in these cases is determined largely by MHC genotype. However, in most autoimmune diseases there is still no firm evidence that a particular infectious agent is associated with onset of the disease; this might be partly because it often takes many years for a patient with an autoimmune disease to show noticeable symptoms.

Summary.

Tolerance to self is a normal state that is maintained chiefly by clonal deletion of developing T and B cells and clonal deletion or inactivation of mature peripheral T and B cells. In addition, some antigens are ignored by the immune system, many by being present in immunologically privileged sites. When the state of self tolerance is disrupted, perhaps by infection, autoimmunity can result. The process of clonal deletion sets limits on the kinds of autoimmune disease that can occur; indeed, only antigens that do not trigger clonal deletion in the thymus, either because they are not abundant enough or because they are tissue-specific and not expressed in the thymus, are candidate autoantigens. Tolerance to non-self can be acquired by mimicking the mechanisms responsible for tolerance to self. A third mechanism for self tolerance—dominant suppression—has been noted in several experimental systems of autoimmunity and graft rejection; if this mechanism could be understood, it is possible that it could be used to prevent both graft rejection and autoimmunity, which are closely related problems.

Summary to Chapter 13.

The response to non-infectious antigens causes three types of medical problem: allergy, the subject of Chapter 12, and autoimmunity and graft rejection, the subjects of this chapter. These responses have many features in common because all use the normal mechanisms of the adaptive immune response to produce symptoms and pathology. What is unique to these syndromes is their initiation and the nature of the antigens recognized, not the underlying nature of the response itself. For each of these undesirable categories of response, the question is how to control them without adversely affecting protective immunity to infection. The answer might lie in a more complete understanding of the regulation of the immune response, especially the suppressive mechanisms that seem to be important in tolerance. The deliberate control of the immune response is examined further in the next chapter.

General references.

Andre, I., Gonzalez, A., Wang, B., Katz, J., Benoist, C., and Mathis, D.: **Checkpoints in the progression of autoimmune disease: lessons from diabetes models**. *Proc. Natl. Acad. Sci. USA* 1996, **93**:2260-2263.

Charlton, B., Auchincloss, H., Jr., and Fathman, C.G.: **Mechanisms of transplantation tolerance**. *Annu. Rev. Immunol.* 1994, **12**:707-734.

Goodnow, C.C.: **Balancing immunity and tolerance: deleting and tuning lymphocyte repertoires**. *Proc. Natl. Acad. Sci. USA* 1996, **93**:2264-2271.

Goodnow, C.C.: **Glimpses into the balance between immunity and self-tolerance.** *Ciba Found. Symp.* 1997, **204**:190-202.

Gianani, R., and Sarvetnick, N.: **Viruses, cytokines, antigens, and autoimmunity.** *Proc. Natl. Acad. Sci. USA* 1996, **93**:2257-2259.

Moller, G. (ed): **Chronic autoimmune diseases** *Immunol. Rev.* 1995, **144**: 1-314.

Moller, G. (ed): **Models of autoimmunity.** *Immunol. Rev.* 1993, **118**:1-310.

Romagnani, S.: **Lymphokine production by human T cells in disease states.** *Annu. Rev. Immunol.* 1994, **12**:227-257.

Steinman, L.: **A few autoreactive cells in an autoimmune infiltrate control a vast population of nonspecific cells: a tale of smart bombs and the infantry.** *Proc. Natl. Acad. Sci. USA* 1996, **93**:2253-2256.

Tan, E.M.: **Autoantibodies in pathology and cell biology**. *Cell* 1991, **67**:841-842.

Section references.

13-1 Specific adaptive immune responses to self antigens can cause autoimmune disease.

Naparstek, Y., and Plotz, P.H.: **The role of autoantibodies in autoimmune disease**. *Annu. Rev. Immunol.* 1993, **11**:79-104.

Steinman, L.: **Multiple sclerosis: a coordinated immunological attack against myelin in the central nervous system.** *Cell* 1996, **85**:299-302.

13-2 Susceptibility to autoimmune diseases is controlled by environmental and genetic factors, especially MHC genes.

Bennett, S.T., and Todd, J.A.: **Human type 1 diabetes and the insulin gene: principles of mapping polygenes**. *Annu. Rev. Genet.* 1996, **30**:343-370.

Gautam, A.M., Lock, C.B., Smilek, D.E., Pearson, C.I., Steinman, L., and McDevitt, H.O.: **Minimum structural requirements for peptide presentation by major histocompatibility complex class II molecules: implications in induction of autoimmunity**. *Proc. Natl. Acad. Sci. USA* 1994, **91**:767-771.

Gonzalez, A., Katz, J.D., Mattei, M.G., Kikutani, H., Benoist, C., and Mathis D.: **Genetic control of diabetes progression**. *Immunity* 1997, **7**:873-883.

Haines, J.L., Ter Minassian, M., Bazyk, A., Gusella, J.F., Kim, D.J., Terwedow,

H., et al.: **A complete genomic screen for multiple sclerosis underscores a role for the major histocompatibility complex. The Multiple Sclerosis Genetics Group.** *Nat. Genet.* 1996, **13**:469-471.

Ridway, W.M., and Fathman, C.G.: **The association of MHC with autoimmune disease: understanding the pathogenesis of autoimmune diabetes.** *Clin. Immunol. Immunopathol.* 1998, **86**:3-10.

Todd, J.A., and Steinman, L.: **The environment strikes back**. *Curr. Opin. Immunol.* 1993, **5**:863-865.

Vyse, T.J., and Todd, J.A.: **Genetic analysis of autoimmune disease**. *Cell* 1996, **85**:311-318.

Waksman, B.H.: **Multiple sclerosis. More genes versus environment**. *Nature* 1995, **377**:105-106.

Wilder, R.L.: **Hormones and autoimmunity: animal models of arthritis**. *Baillieres. Clin. Rheumatol.* 1996, **10**:259-271.

13-3 Either antibody or T cells can cause tissue damage in autoimmune disease.

Couser, W.G.: **Pathogenesis of glomerulonephritis**. *Kidney Int. Suppl.* 1993, **42**:S19-S26.

Jennette, J.C., and Falk, R.J.: **The pathology of vasculitis involving the kidney**. *Am. J. Kidney Dis.* 1994, **24**:130-141.

McFarland, H.F.: **Significance of autoreactive T cells in diseases such as multiple sclerosis using an innovative primate model.** *J. Clin. Invest.* 1994, **94**:921-922.

Rapoport, B.: **Pathophysiology of Hashimoto's thyroiditis and hypothyroidism**. *Annu. Rev. Med.* 1991, **42**:91-96.

13-4 Autoantibodies against blood cells promote their destruction.

Domen, R.E.: **An overview of immune hemolytic anemias**. *Cleve. Clin. J. Med.* 1998, **65**:89-99.

Kiefel, V., Santoso, S., and Mueller Eckhardt, C.: **Serological, biochemical, and molecular aspects of platelet autoantigens**. *Semin. Hematol.* 1992, **29**:26-33.

Silberstein, L.E.: **Natural and pathologic human autoimmune responses to carbohydrate antigens on red blood cells**. *Springer Semin. Immunopathol.* 1993, **15**:139-153.

13-5 The fixation of sub-lytic doses of complement to cells in tissues stimulates a powerful inflammatory response.

Hansch, G.M.: **The complement attack phase: control of lysis and non-lethal effects of C5b-9**. *Immunopharmacol.* 1992, **24**:107-117.

Rother, K., Hansch, G.M., and Rauterberg, E.W.: **Complement in inflammation: induction of nephritides and progress to chronicity**. *Intl. Arch. Allergy Appl. Immunol.* 1991, **94**:23-37.

Shin, M.L., and Carney, D.F.: **Cytotoxic action and other metabolic consequences of terminal complement proteins**. *Prog. Allergy* 1988, **40**:44-81.

13-6 Autoantibodies against receptors cause disease by stimulating or blocking receptor function.

Bahn, R.S., and Heufelder, A.E.: **Pathogenesis of Graves' ophthalmopathy**. *N. Engl. J. Med.* 1993, **329**:1468-1475.

Feldmann, M., Dayan, C., Grubeck Loebenstein, B., Rapoport, B., and Londei, M.: **Mechanism of Graves thyroiditis: implications for concepts and therapy of autoimmunity**. *Int. Rev. Immunol.* 1992, **9**:91-106.

Newsom Davis, J., and Vincent, A.: **Antibody-mediated neurological disease**. *Curr. Opin. Neurobiol.* 1991, **1**:430-435.

13-7 Autoantibodies against extracellular antigens cause inflammatory injury by mechanisms akin to type II or type III hypersensitivity reactions.

Bach, J.F., and Koutouzov, S.: **Immunology. New clues to systemic lupus**. *Lancet* 1997, **350 S3**:III11.

Hardin, J.A., and Craft, J.E.: **Patterns of autoimmunity to nucleoproteins in patients with systemic lupus erythematosus**. *Rheum. Dis. Clin. North Am.* 1987, **13**:37-46.

Kotzin, B.L.: **Systemic lupus erythematosus**. *Cell* 1996, **85**:303-306.

Mamula, M.J.: **Lupus autoimmunity: from peptides to particles**. *Immunol. Rev.* 1995, **144**:301-314.

Tan, E.M.: **Antinuclear antibodies: diagnostic markers for autoimmune diseases and probes for cell biology**. *Adv. Immunol.* 1989, **44**:93-151.

13-8 Environmental co-factors can influence the expression of autoimmune disease.

Donaghy, M., and Rees, A.J.: **Cigarette smoking and lung haemorrhage in glomerulonephritis caused by autoantibodies to glomerular basement membrane**. *Lancet* 1983, **2**:1390-1393.

Kallenberg, C.G., Brouwer, E., Weening, J.J., and Tervaert, J.W.: **Anti-neutrophil cytoplasmic antibodies: current diagnostic and pathophysiological potential**. *Kidney Int.* 1994, **46**:1-15.

Pinching, A.J., Rees, A.J., Russell, B.A., Lockwood, C.M., Mitchison, R.S., and Peters, D.K.: **Relapses in Wegener's granulomatosis: the role of infection**. *BMJ* 1980, **281**:836-838.

Rees, A.J., Lockwood, C.M., and Peters, D.K.: **Enhanced allergic tissue injury in Goodpasture's syndrome by intercurrent bacterial infection**. *BMJ* 1977, **2**:723-726.

13-9 The pattern of inflammatory injury in autoimmunity can be modified by anatomical constraints.

Cavallo, T.: **Membranous nephropathy. Insights from Heymann nephritis**. *Am. J. Pathol.* 1994, **144**:651-658.

Couser, W.G., and Abrass, C.K.: **Pathogenesis of membranous nephropathy**. *Annu. Rev. Med.* 1988, **39**:517-530.

Plotz, P.H., Rider, L.G., Targoff, I.N., Raben, N., O'Hanlon, T.P., and Miller, F.W.: **NIH conference. Myositis: immunologic contributions to understanding cause, pathogenesis, and therapy**. *Ann. Intl. Med.* 1995, **122**:715-724.

Tan, E.M.: **Do autoantibodies inhibit function of their cognate antigens *in vivo?*** *Arthritis Rheum.* 1989, **32**:924-925.

Targoff, I.N.: **Immune mechanisms in myositis**. *Curr. Opin. Rheumatol.* 1990, **2**:882-888.

13-10 The mechanism of autoimmune tissue damage can often be determined by adoptive transfer.

Bottazzo, G.F., and Doniach, D.: **Autoimmune thyroid disease**. *Ann. Rev. Med.* 1986, **37**:353-359.

Vernet der Garabedian, B., Lacokova, M., Eymard, B., Morel, E., Faltin, M., Zajac, J., Sadovsky, O., Dommergues, M., Tripon, P., and Bach, J.F.: **Association of neonatal myasthenia gravis with antibodies against the fetal acetylcholine receptor**. *J. Clin. Invest.* 1994, **94**:555-559.

Vincent, A., Newsom Davis, J., Wray, D., Shillito, P., Harrison, J., Betty, M., Beeson, D., Mills, K., Palace, J., Molenaar, P., et al.: **Clinical and experimental observations in patients with congenital myasthenic syndromes.** *Ann. N. Y. Acad. Sci.* 1993, **681**:451-460.

Willcox, N.: **Myasthenia gravis**. *Curr. Opin. Immunol.* 1993, **5**:910-917.

Zamvil, S., Nelson, P., Trotter, J., Mitchell, D., Knobler, R., Fritz, R., and

Steinman, L.: **T-cell clones specific for myelin basic protein induce chronic relapsing paralysis and demyelination**. *Nature* 1985, **317**:355-358.

13-11 **T cells specific for self antigens can cause direct tissue injury and have a role in sustained autoantibody responses.**

Haskins, K., and Wegmann, D.: **Diabetogenic T-cell clones**. *Diabetes* 1996, **45**:1299-1305.

Nepom, G.T.: **Glutamic acid decarboxylase and other autoantigens in IDDM**. *Curr. Opin. Immunol.* 1995, **7**:825-830.

Tisch, R., and McDevitt, H.: **Insulin-dependent diabetes mellitus**. *Cell* 1996, **85**:291-297.

Zekzer, D., Wong, F.S., Ayalon, O., Altieri, M., Shintani, S., Solimena, M., and Sherwin, R.S.: **GAD-reactive CD4$^+$ Th1 cells induce diabetes in NOD/SCID mice**. *J. Clin. Invest.* 1998, **101**:68-73.

13-12 **Autoantibodies can be used to identify the target of the autoimmune process.**

Craft, J., and Fatenejad, S.: **Self antigens and epitope spreading in systemic autoimmunity.** *Arthritis Rheum.* 1997, **40**:1374-1382.

James, J.A., Gross, T., Scofield, R.H., and Harley, J.B.: **Immunoglobulin epitope spreading and autoimmune disease after peptide immunization: Sm B/B'-derived PPPGMRPP and PPPGIRGP induce spliceosome autoimmunity**. *J. Exp. Med.* 1995, **181**:453-461.

Protti, M.P., Manfredi, A.A., Horton, R.M., Bellone, M., and Conti Tronconi, B.M.: **Myasthenia gravis: recognition of a human autoantigen at the molecular level**. *Immunol. Today* 1993, **14**:363-368.

Roth, R., Gee, R.J., and Mamula, M.J.: **B lymphocytes as autoantigen-presenting cells in the amplification of autimmunity.** *Ann. N. Y. Acad. Sci.* 1997, **815**:88-104.

Topfer, F., Gordon, T., and McCluskey, J.: **Intra- and intermolecular spreading of autoimmunity involving the nuclear self-antigens La (SS-B) and Ro (SS-A)**. *Proc. Natl. Acad. Sci. USA* 1995, **92**:875-879.

13-13 **The target of T-cell mediated autoimmunity is difficult to identify owing to the nature of T-cell ligands.**

Bell, R.B., and Steinman, L.: **Trimolecular interactions in experimental autoimmune demyelinating disease and prospects for immunotherapy.** *Semin. Immunol.* 1991, **3**:237-245.

Feldman, M., Brennan, F.M., and Maini, R.N.: **Rheumatoid arthritis**. *Cell* 1996, **85**:307-310.

Hafler, D.A., Saadeh, M.G., Kuchroo, V.K., Kilford, E., and Steinman, L.: **TCR usage in human and experimental demyelinating disease**. *Immunol. Today* 1996, **17**:152-159.

Hafler, D.A., and Weiner, H.L.: **Immunologic mechanisms and therapy in multiple sclerosis**. *Immunol. Rev.* 1995, **144**:75-107.

Steinman, L.: **Multiple sclerosis: a coordinated immunological attack against myelin in the central nervous system**. *Cell* 1996, **85**:299-302.

Tisch, R., and McDevitt, H.O.: **Antigen-specific immunotherapy: is it a real possibility to combat T-cell-mediated autoimmunity?** *Proc. Natl. Acad. Sci. USA* 1994, **91**:437-438.

13-14 **The rejection of grafts is an immunological response mediated primarily by T cells.**

Lafferty, K.J.: **A contemporary view of transplantation tolerance: an immunologist's perspective**. *Clin. Transplant.* 1994, **8**:181-187.

Lechler, R., Gallagher, R.B., and Auchincloss, H.: **Hard graft? Future challenges in transplantation**. *Immunol. Today* 1991, **12**:214-216.

Lee, R.S. and Auchincloss, H., Jr.: **Mechanisms of tolerance to allografts**. *Chem. Immunol.* 1994, **58**:236-258.

Rosenberg, A.S. and Singer, A.: **Cellular basis of skin allograft rejection: an *in vivo* model of immune-mediated tissue destruction**. *Anuu. Rev. Immunol.* 1992, **10**:333-358.

Shi, C., Lee, W.S., He, Q., Zhang, D., Fletcher, D.L. Jr., Newell, J.B., and Haber, E.: **Immunologic basis of transplant-associated arteriosclerosis**. *Proc. Natl. Acad. Sci. USA* 1996, **93**:4051-4056.

13-15 **Matching donor and recipient at the MHC improves the outcome of transplantation.**

Benichou, G., Takizawa, P.A., Olson, C.A., McMillan, M., and Sercarz, E.E.: **Donor major histocompatibility complex (MHC) peptides are presented by recipient MHC molecules during graft rejection**. *J. Exp. Med.* 1992, **175**:305-308.

Martin, S. and Dyer, P.A.: **The case for matching MHC genes in human organ transplantation** *Nat. Genet.* 1993, **5**:210-213.

Matas, A.J.: **Is MHC matching as a primary criterion in kidney allocation justified?** *Nat. Genet.* 1993, **5**:210-213.

Opelz, G.: **Impact of HLA compatibility on survival of kidney transplants from unrelated live donors.** *Transplantation* 1997, **64**:1473-1475.

Opelz, G. and Wujciak, T.: **Cadaveric kidneys should be allocated according to the HLA match**. *Transplant. Proc.* 1995, **27**:93-99.

Opelz, G. and Wujciak, T.: **The influence of HLA compatibility on graft survival after heart transplantation. The Collaborative Transplant Study.** *N. Engl. J. Med.* 1994, **330**:816-819.

Tay, G.K., Witt, C.S., Christiansen, F.T., Charron, D., Baker, D., Herrmann, R., Smith, L.K., Diepeveen, D., Mallal, S., McCluskey, J., Lester, S., Loiseau, P., Teisserenc, H., Chapman, J., Tait, B., and Dawkins, R.L.: **Matching for MHC haplotypes results in improved survival following unrelated bone marrow transplantation**. *Bone Marrow Transplant.* 1995, **15**:381-385.

13-16 **In MHC-identical grafts, rejection is caused by non-self peptides bound to graft MHC molecules.**

Goulmy, E.: **Human minor histocompatibility antigens: new concepts for marrow transplantation and adoptive immunotherapy.** *Immunol. Rev.* 1997, **157**:125-140.

Scott, D.M., Ehrmann, I.E., Ellis, P.S., Bishop, C.E., Agulnik, A.I., Simpson, E., and Mitchell, M.J.: **Identification of a mouse male-specific transplantation antigen, H-Y**. *Nature* 1995, **376**:695-698.

Warrens, A.N., Lombardi, G., and Lechler, R.I.: **Presentation and recognition of major and minor histocompatibility antigens**. *Transpl. Immunol.* 1994, **2**:103-107.

13-17 **There are two ways of presenting alloantigens on the transplant to the recipient's T lymphocytes.**

Auchincloss, H. Jr., and Sultan, H.: **Antigen processing and presentation in transplantation.** *Curr. Opin. Immunol.* 1996, **8**:681-687.

Auchincloss, H. Jr., Lee, R., Shea, S., Markowitz, J.S., Grusby, M.J., and Glimcher, L.H.: **The role of "indirect" recognition in initiating rejection of skin grafts from major histocompatibility complex class II-deficient mice**. *Proc. Natl. Acad. Sci. USA* 1993, **90**:3373-3377.

13-18 **Antibodies reacting with endothelium cause hyperacute graft rejection.**

Dorling, A., Riesbeck, K., Warrens, A., and Lechler, R.: **Clinical xenotransplantation of solid organs**. *Lancet* 1997, **349**:867-871.

Kaufman, C.L., Gaines, B.A., and Ildstad, S.T.: **Xenotransplantation**. *Annu. Rev. Immunol.* 1995, **13**:339-367.

Kissmeyer Nielsen, F., Olsen, S., Petersen, V.P., and Fjeldborg, O.: **Hyperacute rejection of kidney allografts, associated with pre-existing humoral antibodies against donor cells**. *Lancet* 1966, **2**:662-665.

Moller, G. (ed): **Xenotransplantation**. *Immunol. Rev.* 1994, **141**:1-276.

Sharma, A., Okabe, J., Birch, P., McClellan, S.B., Martin, M.J., Platt, J.L., and Logan, J.S.: **Reduction in the level of Gal(alpha1,3)Gal in transgenic mice and pigs by the expression of an alpha(1,2)fucosyltransferase**. *Proc. Natl. Acad. Sci. USA* 1996, **93**:7190-7195.

Steele, D.J., and Auchincloss, H., Jr.: **Xenotransplantation**. *Annu. Rev. Med.* 1995, **46**:345-360.

Williams, G.M., Hume, D.M., Hudson, R.P., Jr., Morris, P.J., Kano, K., and Milgrom, F.: **"Hyperacute" renal-homograft rejection in man**. *N. Engl. J. Med.* 1968, **279**:611-618.

13-19 The corollary of graft rejection is graft-versus-host disease.

Barrett, A.J., and Malkovska, V.: **Garft-versus-leukaemia: understanding and using the alloimmune response to treat hematological malignancies**. *Br. J. Hematol.* 1996, **93**:754-761.

Barrett, A.J., and van Rhee, F.: **Graft-versus-leukaemia**. *Baillieres Clin. Haematol.* 1997, **10**:37-355.

Goulmy, E., Schipper, R., Pool, J., Blokland, E., Flakenburg, J.H., Vossen, J., Grathwohl, A., Vogelsang, G.B., van Houwelingen, H.C., and van Rood, J.J.: **Mismatches of minor histocompatibility antigens between HLA-identical donors and recipients and the development of graft-versus-host disease after bone marrow transplantation**. *N. Engl. J. Med.* 1996, **334**:281-285.

Nash, R.A., and Storb, R.: **Graft-versus-host effect after allogeneic hematopoietic stem cell transplantation: GVHD and GVL**. *Curr. Opin. Immunol.* 1996, **8**:674-680.

Rocha, M., Umansky, V., Lee, K.H., Hacker, H.J., Benner, A., and Schirrmacher, V.: **Differences between graft-versus-leukemia and graft-versus-host reactivity. I. Interaction of donor immune T cells with tumor and/or host cells**. *Blood* 1997, **89**:2189-2202.

13-20 Chronic organ rejection is caused by inflammatory vascular injury to the graft.

Hayry, P., Isoniemi, H., Yilmaz, S., Mennander, A., Lemstrom, K., Raisanen Sokolowski, A., Koskinen, P., Ustinov, J., Lautenschlager, I., Taskinen, E., Krogerus, L., Aho, P., and Paavonen, T.: **Chronic allograft rejection**. *Immunol. Rev.* 1993, **134**:33-81.

Lautenschlager, I., Soots, A., Krogeus, L., Kauppinen, H., Saarinen, O., Bruggeman, C., and Ahonen, J.: **CMV increases inflammation and accelerates chronic rejection in rat kidney allografts**. *Transplant Proc.* 1997, **29**:802-803.

Orosz, C.G., and Peletier, R.P.: **Chronic remodeling pathology in grafts**. *Curr. Opin. Immunol.* 1997, **9**:676-680.

Takada, M., Nadeau, K.C., Shaw, G.D., Marquette, K.A., and Tilney, N.L.: **The cytosine-adhesion molecule cascade in ischemia/reperfusion injury of the rat kidney. Inhibition by a soluble P-selectin ligand**. *J. Clin. Invest.* 1997, **99**:2682-2690.

13-21 A variety of organs are transplanted routinely in clinical medicine.

Murray, J.E.: **Human organ transplantation: background and consequences**. *Science* 1992, **256**:1411-1416.

13-22 The fetus is an allograft that is tolerated repeatedly.

Carosella, E.D., Dausset, J., and Kirszenbaum, M.: **HLA-G revisited**. *Immunol. Today* 1996, **17**:407-409.

Flanagan, J.R., Murata, M., Burke, P.A., Shirayoshi, Y., Appella, E., Sharp, P.A., and Ozato, K.: **Negative regulation of the major histocompatibility complex class I promoter in embryonal carcinoma cells**. *Proc. Natl. Acad. Sci. USA* 1991, **88**:3145-3149.

Hunt, J.S.: **Immunobiology of pregnancy**. *Curr. Opin. Immunol.* 1992, **4**:591-596.

Parham, P.: **Immunology: keeping mother at bay**. *Curr. Biol.* 1996, **6**:638-641.

Pazmany, L., Mandelboim, O., Vales Gomez, M., Davis, D.M., Reyburn, H.T., and Strominger, J.L.: **Protection from natural killer cell-mediated lysis by HLA-G expression on target cells**. *Science* 1996, **274**:792-795.

Rouas Freiss, N., Gonclaves, R.M., Menier, C., Dausset, J., and Carosella, E.D.: **Direct evidence to support the role of HLA-G in protecting the fetus from maternal uterine natural killer cytolysis**. *Proc. Natl. Acad. Sci. USA* 1997, **94**:11520-11525.

13-23 Autoantigens are not so abundant that they induce clonal deletion or anergy but are not so rare as to escape recognition entirely.

Billingham, R.E., Brent, L., and Medawar, P.B.: **Actively acquired tolerance of foreign cells.** *Nature* 1953, **172**:603-606.

Brent, L.: **Medawar Prize Lecture: tolerance and graft-vs-host disease: two sides of the same coin.** *Transplant. Proc.* 1995, **27**:12-14.

Goverman, J., Woods, A., Larson, L., Weiner, L.P., Hood, L., and Zaller, D.M.: **Transgenic mice that express a myelin basic protein-specific T cell receptor develop spontaneous autoimmunity**. *Cell* 1993, **72**:551-560.

Katz, J.D., Wang, B., Haskins, K., Benoist, C., and Mathis, D.: **Following a diabetogenic T cell from genesis through pathogenesis**. *Cell* 1993, **74**:1089-1100.

Margulies, D.H.: **Interactions of TCRs with MHC-peptide complexes: a quantitative basis for mechanistic models**. *Curr. Opin. Immunol.* 1997 **9**:390-395.

Wang, R., Nelson, A., Kimachi, K., Grey, H.M., and Farr, A.G.: **The role of peptides in thymic positive selection of class II major histocompatibility complex-restricted T cells**. *Proc. Natl. Acad. Sci. USA* 1998, **95**:3804-3809.

13-24 The induction of a tissue-specific response requires the presentation of antigen by professional antigen-presenting cells with co-stimulatory activity.

Bluestone, J.A.: **Costimulation and its role in organ transplantation**. *Clin. Transplant.* 1996, **10**:104-109.

Guerder, S., Picarella, D.E., Linsley, P.S., and Flavell, R.A.: **Costimulator B7-1 confers antigen-presenting-cell function to parenchymal tissue and in conjunction with tumor necrosis factor alpha leads to autoimmunity in transgenic mice**. *Proc. Natl. Acad. Sci. USA* 1994, **91**:5138-5142.

Lafferty, K.J., Prowse, S.J., Simeonovic, C.J., and Warren, H.S.: **Immunobiology of tissue transplantation: a return to the passenger leukocyte concept**. *Annu. Rev. Immunol.* 1983, **1**:143-173.

13-25 In the absence of co-stimulation, tolerance is induced.

Hammerling, G.J., Schonrich, G., Ferber, I., and Arnold, B.: **Peripheral tolerance as a multi-step mechanism**. *Immunol. Rev.* 1993, **133**:93-104.

Lenschow, D.J. and Bluestone, J.A.: **T cell co-stimulation and *in vivo* tolerance**. *Curr. Opin. Immunol.* 1993, **5**:747-752.

Miller, J.F., and Basten, A.: **Mechanisms of tolerance to self**. *Curr. Opin. Immunol.* 1996, **8**:815-821.

13-26 Dominant immune suppression can be demonstrated in models of tolerance and can affect the course of autoimmune disease.

Olsson, T.: **Critical influences of the cytokine orchestration on the outcome of myelin antigen-specific T-cell autoimmunity in experimental autoimmune encephalomyelitis and multiple sclerosis**. *Immunol. Rev.* 1995, **144**:245-268.

Qin, S., Cobbold, S.P., Pope, H., Elliott, J., Kioussis, D., Davies, J., and Waldmann, H.: **"Infectious" transplantation tolerance**. *Science* 1993, **259**:974-977.

Rocken, M., and Shevachm E.M.: **Immune deviation—the third dimension of nondeletional T cell tolerance**. *Immunol. Rev.* 1996, **149**:175-194.

Tian, J., Atkinson, M.A., Clare Salzler, M., Herschenfeld, A., Forsthuber, T., Lehmann, P.V., and Laufman, D.L.: **Nasal administration of glutamate decarboxylase (GAD65) peptides induces T_H2 responses and prevents murine insulin-dependent diabetes**. *J. Exp. Med.* 1996, **183**:1561-1567.

Weigle, W.O., and Romball, C.G.: **CD4$^+$ T-cell subsets and cytokines involved in peripheral tolerance**. *Immunol. Today* 1997, **18**:533-538.

Weiner, H.L.: **Oral tolerance for the treatment of autoimmune diseases**. *Annu. Rev. Med.* 1997, **48**:341-351.

13-27 **Antigens in immunologically privileged sites do not induce immune attack but can serve as targets.**

Alison, J., Georgiou, H.M., Strasser, A., and Vaux, D.L.: **Transgenic expression of CD95 ligand on islet beta cells induces a granulocytic infiltration but does not confer immune privilege upon islet allografts**. *Proc. Natl. Acad. Sci. USA* 1997, **94**: 3943-3947.

Ferguson, T.A., and Griffith, T.S.: **A vision of cell death: insights into immune privilege**. *Immunol. Rev.* 1997, **156**:167-184.

Streilein, J.W., Ksander, B.R., and Taylor, A.W.: **Immune deviation in relation to ocular immune privilege.** *J. Immunol.* 1997, **158**:3557-3560.

13-28 **B cells with receptors specific for peripheral autoantigens are held in check by a variety of mechanisms.**

Akkaraju, S., Canaan, K., and Goodnow, C.C.: **Self-reactive B cells are not eliminated or inactivated by autoantigen expressed on thyroid epithelial cells.** *J. Exp. Med.* 1997, **186**:2005-2012.

Fulcher, D.A., and Basten, A.: **B-cell activation versus tolerance—the central role of immunoglobulin receptor engagement and T-cell help**. *Int. Rev. Immunol.* 1997, **15**:33-52.

Goodnow, C.C., Cyster, J.G., Hartley, S.B., Bell, S.E., Cooke, M.P., Healy, J.I., Akkaraju, S., Rathmell, J.C., Pogue, S.L., and Shokat, K.P.: **Self-tolerance checkpoints in B lymphocyte development**. *Adv. Immunol.* 1995, **59**:279-368.

Rathmell, J.C., Cooke, M.P., Ho, W.Y., Grein, J., Townsend, S.E., Davis, M.M., and Goodnow, C.C.: **CD95 (Fas)-dependent elimination of self-reactive B cells upon interaction with CD4$^+$ T cells**. *Nature* 1995, **376**:181-184.

Shokat, K.M., and Goodnow, C.C.: **Antigen-induced B-cell death and elimination during germinal-centre immune responses**. *Nature* 1995, **375**:334-338.

13-29 **Autoimmunity could be triggered by infection in a variety of ways.**

Aichele, P., Bachmann, M.F., Hengarter, H., Zinkernagel, R.M.: **Immunopathology or organ-specific autoimmunity as a consequence of virus infection**. *Immunol. Rev.* **152**:21-45.

Brocke, S., Veromaa, T., Weissman, I.L., Gijbels, K., and Steinman, L.: **Infection and multiple sclerosis: a possible role for superantigens?** *Trends Microbiol.* 1994, **2**:250-254.

Gianani, R., and Sarvetnick, N.: **Viruses, cytokines, antigens, and autoimmunity**. *Proc. Natl. Acad. Sci. USA* 1996, **93**:2257-2259.

Moens, U., Seternes, O.M., Hey, A.W., Silsand, Y., Traavik, T., Johansen, B., and Rekvig, O.P.: ***In vivo* expression of a single viral DNA-binding protein generates systemic lupus erythematosus-related autoimmunity to double-stranded DNA and histones**. *Proc. Natl. Acad. Sci. USA* 1995, **92**:12393-12397.

Rocken, M., Urban, J.F., and Shevach, E.M.: **Infection breaks T-cell tolerance**. *Nature* 1992, **359**:79-82.

Steinhoff, U., Burkhart, C., Arnheiter, H., Hengartner, H., and Zinkernagel, R.: **Virus or a hapten-carrier complex can activate autoreactive B cells by providing linked T help**. *Eur. J. Immunol.* 1994, **24**:773-776.

Steinman, L., and Conlon, P.: **Viral damage and the breakdown of self-tolerance**. *Nat. Med.* 1997, **3**:1085-1087.

Von Herrath, M.G., Evans, C.F., Horwitz, M.S., and Oldstone, M.B.: **Using transgenic mouse models to dissect the pathogenesis of virus-induced autoimmune disorders of the islets of Langerhans and the central nervous system**. *Immunol. Rev.* 1996, **152**:111-143.

Von Herrath, M.G., and Oldstone, M.B.: **Virus-induced autoimmune disease**. *Curr. Opin. Immunol.* 1996, **8**:878-895.

Wucherpfennig, K.W., and Strominger, J.L.: **Molecular mimicry in T cell-mediated autoimmunity: viral peptides activate human T cell clones specific for myelin basic protein**. *Cell* 1995, **80**:695-705.

Manipulation of the Immune Response

14

Most of this book has been concerned with the mechanisms whereby the immune system successfully protects us from disease. In the preceding three chapters, however, we have seen examples of the failure of immunity to some important infections, and conversely, with allergy and autoimmunity, how inappropriate immune responses can themselves cause disease. We have also discussed the problems arising from immune responses to grafted tissues.

In this chapter we shall consider the ways in which the immune system can be manipulated or controlled, both to suppress unwanted immune responses in autoimmunity, allergy and graft rejection, and to stimulate protective immune responses to some of the diseases that, at present, largely elude the immune system. It has long been felt that it should be possible to deploy the powerful and specific mechanisms of adaptive immunity to destroy tumors, and we shall discuss the present state of progress toward that goal. In the final section of the chapter we shall discuss present vaccination strategies and how a more rational approach to the design and development of vaccines promises to increase their efficacy and widen their scope.

Extrinsic regulation of unwanted immune responses.

The unwanted immune responses that occur in autoimmune disease, transplant rejection, and allergy present slightly different problems, and the approach to developing effective treatment is correspondingly different for each. We have already discussed the treatment of allergy in Chapter 12: the problems in this case are due to the production of IgE, and the goals are, accordingly, to treat the adverse consequences of an IgE response, or to induce the production of IgG instead of IgE against the allergenic antigens. In autoimmune disease and graft rejection the problem is an immune response to tissue antigens, and the goal is to downregulate the response to avoid damage to the tissues or disruption of their function. From the point of view of management, the single most important difference between allograft rejection and autoimmunity is that allografts are a deliberate surgical intervention and the immune response to them can be foreseen, whereas autoimmune responses are not detected until they are already established. Effective treatment of an established immune response is much harder to achieve than prevention of a response before it has had a chance to develop, and autoimmune diseases are generally harder to control than a *de novo* immune response to an allograft. The relative difficulty of suppressing established immune responses is seen in animal models of autoimmunity, in which methods able to prevent the induction of autoimmune disease generally fail to halt established disease.

Current treatments for immunological disorders are nearly all empirical in origin, using immunosuppressive drugs identified by screening large numbers of natural and synthetic compounds. The drugs currently used to suppress the immune system can be divided into three categories: first, powerful anti-inflammatory drugs of the corticosteroid family such as prednisone; second, cytotoxic drugs such as azathioprine and cyclophosphamide; and third, fungal and bacterial derivatives, such as cyclosporin A, FK506 (tacrolimus), and rapamycin (sirolimus), which inhibit signaling events within T lymphocytes. These drugs are all very broad in their actions and inhibit protective functions of the immune system, as well as harmful ones. Opportunistic infection is therefore a common complication of immunosuppressive drug therapy. The ideal immunosuppressive agent would be one that targets the specific part of the adaptive immune response that causes tissue injury. Paradoxically, antibodies themselves, by virtue of their exquisite specificity, might offer the best possibility for the therapeutic inhibition of specific immune responses. We shall also consider experimental approaches to controlling specific immune responses by manipulating the local cytokine environment or by manipulating antigen so as to divert the response from a pathogenic pathway to an innocuous one. We have discussed in Chapters 12 and 13 how the pathological responses that cause allergy, autoimmunity, or graft rejection can be prevented by innocuous, non-pathological T-cell responses.

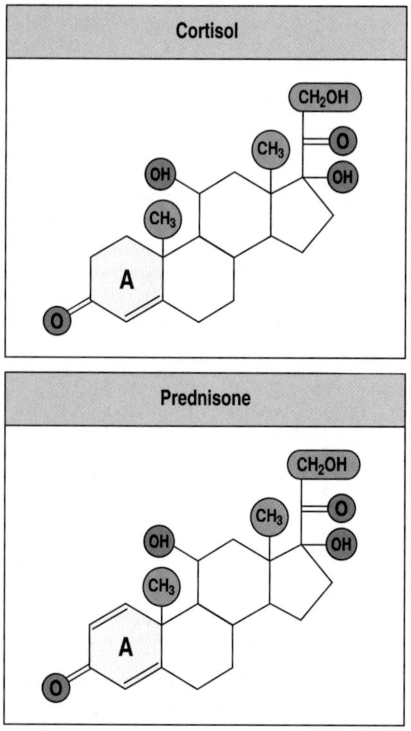

Fig. 14.1 The structure of the anti-inflammatory corticosteroid drug, prednisone. Prednisone is a synthetic analog of the natural adrenocorticosteroid cortisol. Introduction of the 1,2 double bond into the A ring increases anti-inflammatory potency approximately fourfold compared with cortisol, without modifying the sodium-retaining activity of the compound.

14-1 Corticosteroids are powerful anti-inflammatory drugs that alter the transcription of many genes.

Corticosteroid drugs are powerful anti-inflammatory agents that are used widely to suppress the harmful effects of immune responses of autoimmune or allergic origin, as well as those induced by graft rejection. **Corticosteroids** are pharmacological derivatives of members of the glucocorticoid family of steroid hormones; one of the most widely used is **prednisone**, which is a synthetic analog of cortisol (Fig. 14.1), which acts through intracellular receptors that are expressed in almost every cell of the body. On binding hormone, these receptors regulate the transcription of specific genes, as illustrated in Fig. 14.2.

The expression of as many as 1% of genes might be regulated by glucocorticoids, which can either upregulate or, less commonly, downregulate responsive genes. The pharmacological effects of corticosteroid drugs result from exposure of the glucocorticoid receptors to supraphysiological concentrations of ligand. The abnormally high level of ligation of glucocorticoid receptors causes exaggerated glucocorticoid-mediated responses, which have both beneficial and toxic effects.

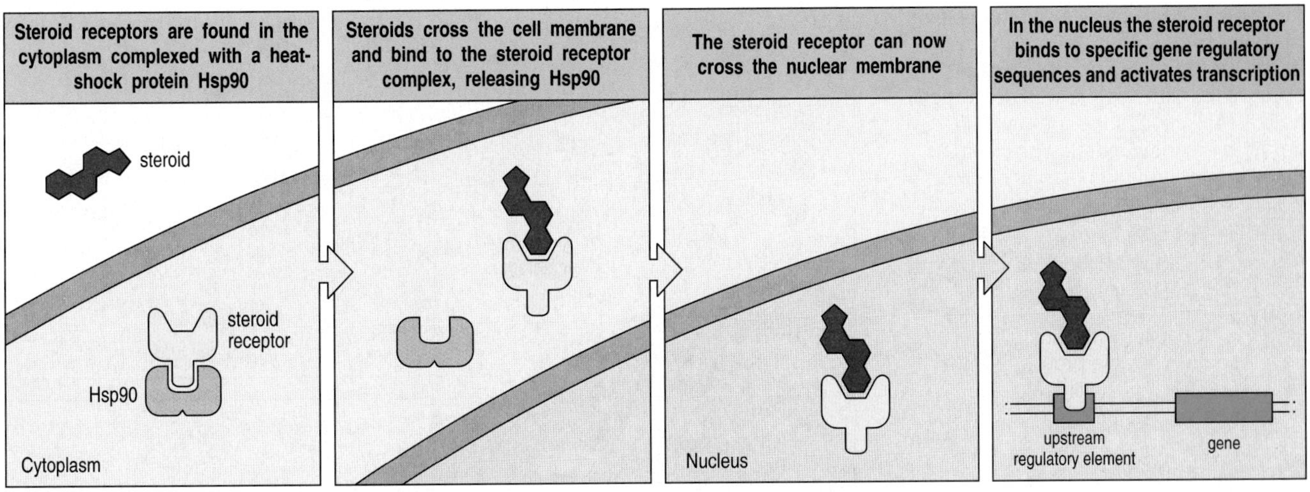

Steroid receptors are found in the cytoplasm complexed with a heat-shock protein Hsp90
Steroids cross the cell membrane and bind to the steroid receptor complex, releasing Hsp90
The steroid receptor can now cross the nuclear membrane
In the nucleus the steroid receptor binds to specific gene regulatory sequences and activates transcription
steroid
steroid receptor
Hsp90
Cytoplasm
Nucleus
upstream regulatory element
gene

Fig. 14.2 Mechanisms of steroid action. Corticosteroids are lipid-soluble compounds that enter cells by diffusing across the plasma membrane and bind to their receptors in the cytosol. The binding of corticosteroid to the receptor displaces a dimer of a heat-shock protein named Hsp90, exposing the DNA-binding region of the receptor, which then enters the nucleus and binds to specific DNA sequences in the promoter regions of steroid-responsive genes. Corticosteroids exert their effects by modulating the transcription of a wide variety of genes.

Given the large number of genes regulated by corticosteroids, it is hardly surprising that the effects of steroid therapy are very complex. The beneficial effects are anti-inflammatory and are summarized in Fig. 14.3; however, there are also many adverse effects, including fluid retention, weight gain, diabetes, bone mineral loss and thinning of the skin. The use of corticosteroids to control disease requires a careful balance between helping the patient by reducing the inflammatory manifestations of disease and avoiding harm from the toxic side-effects of the drug. For this reason, corticosteroids used in transplant recipients and to treat inflammatory autoimmune and allergic disease are often administered in combination with other drugs in an effort to keep the dose and toxic effects to a minimum. In autoimmunity and allograft rejection, corticosteroids are commonly combined with cytotoxic immunosuppressive drugs.

Corticosteroid therapy	
Effect on	**Physiological effects**
↓ IL-1, TNF-α, GM-CSF IL-3, IL-4, IL-5, IL-8	↓ Inflammation caused by cytokines
↓ NOS	↓ NO
↓ Phospholipase A_2 ↓ Cyclo-oxygenase type2 ↑ Lipocortin-1	↓ Prostaglandins ↓ Leukotrienes
↓ Adhesion molecules	Reduced emigration of leukocytes from vessels
↑ Endonucleases	Induction of apoptosis in lymphocytes and eosinophils

Fig. 14.3 Anti-inflammatory effects of corticosteroid therapy. Corticosteroids regulate the expression of many genes, with a net anti-inflammatory effect. First, they reduce the production of inflammatory mediators including cytokines, prostaglandins, and nitric oxide. Second, they inhibit inflammatory cell migration to sites of inflammation by inhibiting the expression of adhesion molecules. Third, corticosteroids promote the death by apoptosis of leukocytes and lymphocytes.

14-2 Cytotoxic drugs cause immunosuppression by killing dividing cells and have serious side-effects.

The two cytotoxic drugs most commonly used as immunosuppressants are **azathioprine** and **cyclophosphamide** (Fig. 14.4). Both interfere with DNA synthesis and have their major pharmacological action on dividing tissues. They were developed originally to treat cancer and, after observations that they were cytotoxic to dividing lymphocytes, were found to be immunosuppressive as well. The use of these compounds is limited by a range of toxic effects on tissues in the body, which have in common the property of continuous cell division. These effects include decreased immune function, as well as anemia, leukopenia, thrombocytopenia, damage to intestine epithelium, hair loss, and fetal death or injury. As a result of these toxic effects, they are used at high doses only when the aim is to eliminate all dividing lymphocytes, as in the treatment of some recipients of bone marrow transplants. They are used at lower doses, and in combination with other drugs such as corticosteroids, to treat unwanted immune responses.

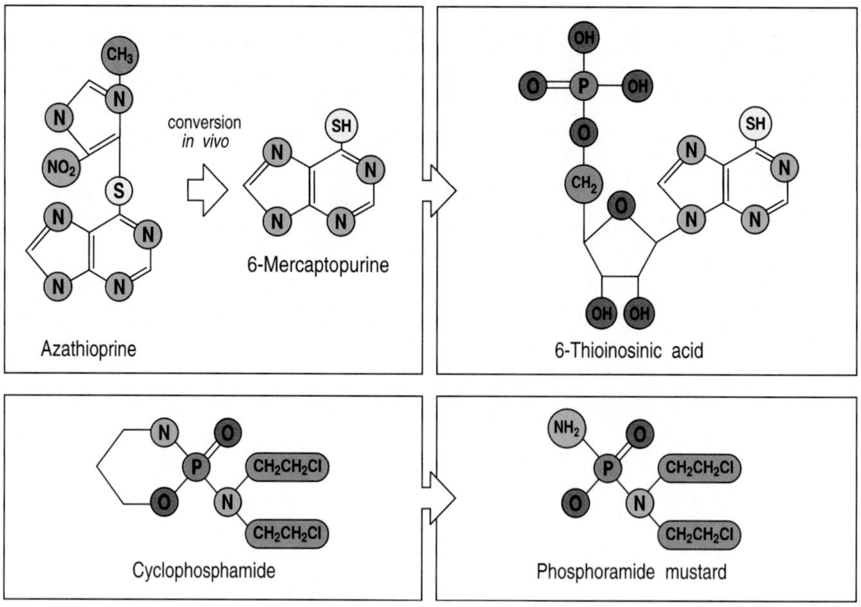

Fig. 14.4 The structure and metabolism of the cytotoxic immunosuppressive drugs azathioprine and cyclophosphamide. Azathioprine was developed as a modification of the anti-cancer drug 6-mercaptopurine; by blocking the reactive thiol group, the metabolism of this drug is slowed down. It is slowly converted *in vivo* to 6-mercaptopurine, which is then metabolized to 6-thioinosinic acid, which blocks the pathway of purine biosynthesis. Cyclophosphamide was similarly developed as a stable pro-drug, which is activated enzymatically in the body to phosphoramide mustard, a powerful and unstable DNA-alkylating agent.

Azathioprine is converted *in vivo* to a purine antagonist that interferes in the synthesis of nucleic acids and is toxic to dividing cells. It is metabolized to 6-thioinosinic acid, which competes with inosine monophosphate, thereby blocking the synthesis of adenosine monophosphate and guanosine monophosphate and thus inhibiting DNA synthesis. It is less toxic than cyclophosphamide, which is metabolized to phosphoramide mustard, which alkylates DNA. Cyclophosphamide is a member of the nitrogen mustard family of compounds, which were originally developed as chemical weapons. With this pedigree goes a range of highly toxic effects including inflammation of and hemorrhage from the bladder, known as hemorrhagic cystitis, and induction of bladder neoplasia.

14-3 Cyclosporin A, FK506 (tacrolimus), and rapamycin (sirolimus) are powerful immunosuppressive agents that interfere with T-cell signaling.

There are now relatively non-toxic alternatives to the cytotoxic drugs used for immunosuppression in transplant patients. The systematic study of products from bacteria and fungi has led to the development of a large array of important medicines including the two immunosuppressive drugs **cyclosporin A** and **FK506** or **tacrolimus**, which are now widely used to treat transplant patients. Cyclosporin A is a cyclic decapeptide derived from a soil fungus from Norway, *Tolypocladium inflatum*. FK506, now known as tacrolimus, is a macrolide compound from the filamentous bacterium *Streptomyces tsukabaensis* found in Japan (macrolides are compounds that contain a many-membered lactone ring to which is attached one or more deoxy sugars). Another *Streptomyces* macrolide, called **rapamycin** or **sirolimus**, is being tested in clinical studies and might also turn out to be an important drug in the prevention of transplant rejection; rapamycin is derived from *Streptomyces hygroscopicus*, found on Easter Island ('Rapa ui' in Polynesian—hence the name of the drug). All three compounds exert their pharmacological effects by binding to members of a family of intracellular proteins known as the immunophilins, forming complexes that interfere with signaling pathways important for the clonal expansion of lymphocytes (see Chapter 5).

Immunological effects of cyclosporin A and tacrolimus	
Cell type	**Effects**
T lymphocyte	Reduced expression of IL-2, IL-3, IL-4, GM-CSF, TNF-α. Reduced proliferation following decreased IL-2 production. Reduced Ca^{2+}-dependent exocytosis of granule-associated serine esterases. Inhibition of antigen-driven apoptosis.
B lymphocyte	Inhibition of proliferation secondary to reduced cytokine production by T lymphocytes. Inhibition of proliferation following ligation of surface immunoglobulin. Induction of apoptosis following B-cell activation.
Granulocyte	Reduced Ca^{2+}-dependent exocytosis of granule-associated serine esterases.

Fig. 14.5 Cyclosporin A and tacrolimus inhibit lymphocyte and some granulocyte responses.

Cyclosporin A and tacrolimus block T-cell proliferation by reducing the expression of several cytokine genes that are normally induced on T-cell activation (Fig. 14.5). These include interleukin (IL)-2, whose synthesis by T lymphocytes is an important growth signal for T cells. These drugs inhibit T-cell proliferation in response to either specific antigens or allogeneic cells and are used extensively in medical practice to prevent the rejection of allogeneic organ grafts. Although the major immunosuppressive effects of both drugs are probably the result of inhibition of T-cell proliferation, they have a large variety of other immunological effects (see Fig. 14.5), some of which might turn out to be important pharmacologically.

Cyclosporin A and tacrolimus are effective but they are not problem-free. First, as with the cytotoxic agents, they affect all immune responses indiscriminately. The only way of controlling their immunosuppressive action is by varying the dose; at the time of grafting, high doses are required but, once a graft is established, the dose can be decreased to allow useful protective immune responses while maintaining adequate suppression of the residual response to the grafted tissue. This is a difficult balance that is not always achieved. Furthermore, as the immunophilins are found in many cells, it is to be expected that these drugs will have effects on many other tissues. Cyclosporin A and tacrolimus are both toxic to kidneys and other organs. Finally, treatment with these drugs is expensive because they are complex natural products that must be taken for prolonged periods. Thus there is room for improvement in these compounds, and better and less expensive analogs are being sought. Nevertheless, at present, they are the drugs of choice in clinical transplantation, and they are also being tested in a variety of autoimmune diseases, especially those that, like graft rejection, are mediated by T cells.

14-4 Immunosuppressive drugs are valuable probes of intracellular signaling pathways in lymphocytes.

The mechanism of action of cyclosporin A and tacrolimus is now fairly well understood. Each binds to a different group of immunophilins: cyclosporin A to the cyclophilins, and tacrolimus to the FK-binding proteins (FKBP). These immunophilins are peptidyl-prolyl *cis-trans* isomerases but their isomerase

Fig. 14.6 Cyclosporin A and tacrolimus inhibit T-cell activation by interfering with the serine/threonine-specific phosphatase calcineurin. Signaling via T-cell receptor-associated tyrosine kinases leads to the activation and increased synthesis of the transcription factor AP-1, as well as increasing the concentration of Ca^{2+} in the cytoplasm (left panels). The Ca^{2+} binds to calcineurin and thereby activates it to dephosphorylate the cytoplasmic form of the nuclear factor of activated T cells (NF-AT). Once dephosphorylated, the active NF-AT migrates to the nucleus to form a complex with AP-1; the NF-AT:AP-1 complex can then induce the transcription of genes required for T-cell activation, including the IL-2 gene. When cyclosporin A (CsA) or tacrolimus are present, they form complexes with their immunophilin targets, cyclophilin (CyP) and FK-binding protein (FKBP), respectively (right panels). The complex of cyclophilin with cyclosporin A can bind to calcineurin, blocking its ability to activate NF-AT. The complex of tacrolimus with FKBP binds to calcineurin at the same site, also blocking its activity.

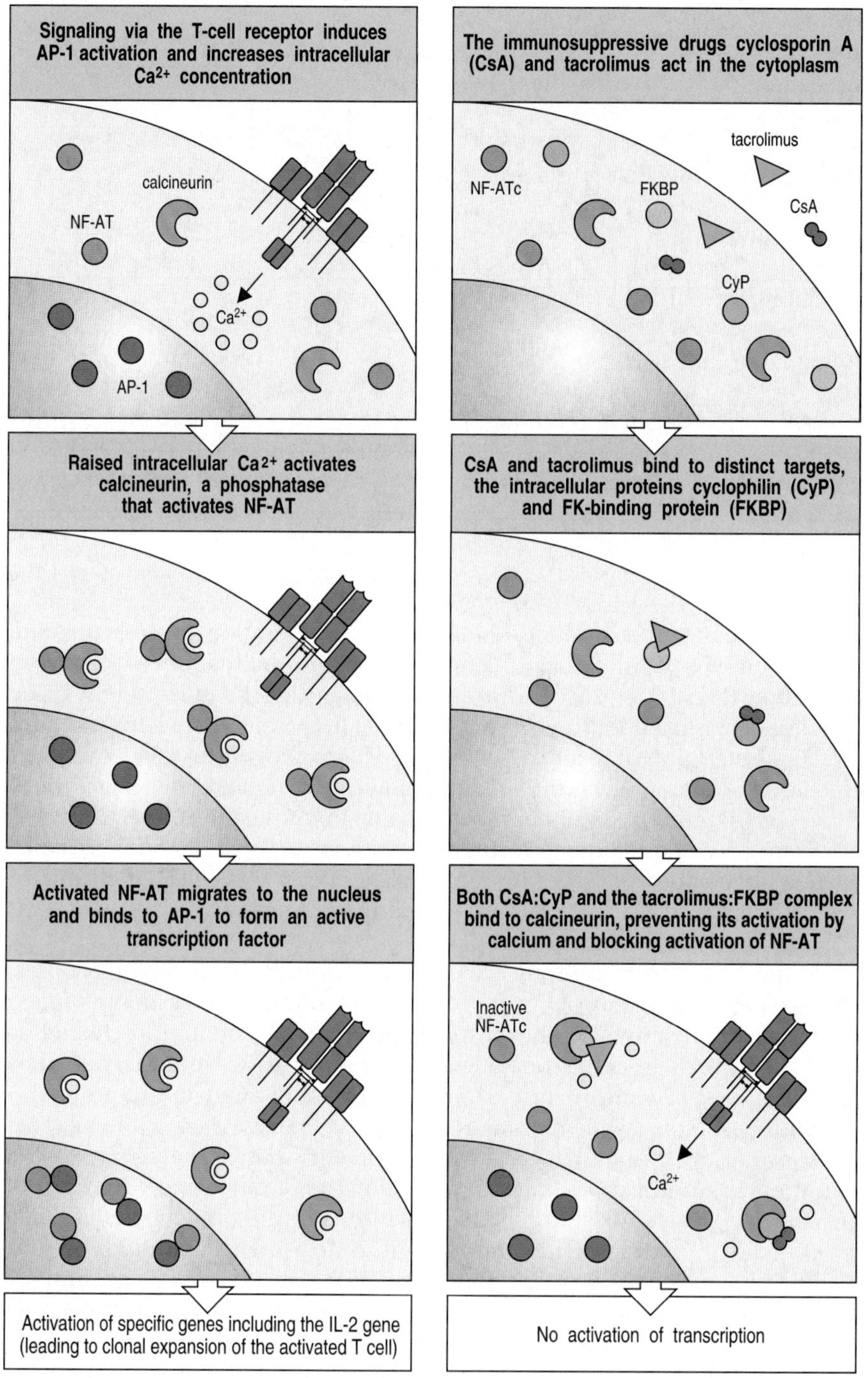

activity does not seem to be relevant to the immunosuppressive activity of the drugs that bind them. Rather, the immunophilin:drug complexes bind and inhibit the Ca^{2+}-activated serine/threonine phosphatase **calcineurin**. Calcineurin is activated in T cells when intracellular calcium ion levels rise after T-cell receptor binding; on activation it dephosphorylates the transcription factor NF-AT in the cytoplasm, allowing it to migrate to the nucleus, where it forms a complex with a second transcription factor, AP-1, and induces transcription of the IL-2 gene (Fig. 14.6, and see Sections 5-11 and 8-9). This pathway is inhibited by cyclosporin A and tacrolimus, which thus inhibit the clonal expansion of activated T cells. Calcineurin is found in other cells besides T cells but at higher levels; T cells are therefore particularly susceptible to the inhibitory effects of these drugs.

The primary physiological role of the immunophilins is to catalyze the slow steps in protein folding and to maintain the integrity of certain proteins, and so far it is unclear why calcineurin binds to these two complexes of drug and immunophilin. It is tempting to speculate that there are endogenous equivalents for cyclosporin A and tacrolimus that perform similar regulatory functions but, so far, none has been discovered.

Rapamycin has a different mode of action from either cyclosporin A or tacrolimus. Like tacrolimus, it binds to the FK BP family of immunophilins. However, the rapamycin:immunophilin complex has no effect on calcineurin activity but, instead, blocks the signal transduction pathway triggered by ligation of the IL-2 receptor. Rapamycin also inhibits lymphocyte proliferation driven by IL-4 and IL-6, implying a common post-receptor pathway of signaling by these cytokines. The rapamycin:immunophilin complex acts by binding to a protein kinase named mTOR (mammalian Target Of Rapamycin; also known as FRAP, RAFT1, and RAPT1). This kinase phosphorylates two downstream intracellular targets. The first is another kinase, p70 S6 kinase, which in turn regulates the translation of many proteins. The second is PHAS-I, a repressor of protein translation, which is inhibited by phosphorylation mediated by mTOR. It seems that the inhibition of PHAS-1 phosphorylation by the binding of the rapamycin:immunophilin complex to mTOR is responsible for the growth arrest of T lymphocytes.

14-5 Antibodies against cell-surface molecules have been used to remove specific lymphocyte subsets or to inhibit cell function.

Cytotoxic drugs kill all proliferating cells and therefore indiscriminately affect all types of activated lymphocyte and any other cell that is proliferating. Cyclosporin A, tacrolimus and rapamycin are more selective but still inhibit most adaptive immune responses. In contrast, antibodies can interfere with immune responses in a non-toxic and much more specific manner. The potential of antibodies for removal of unwanted lymphocytes is demonstrated by anti-lymphocyte globulin, a preparation of immunoglobulin from horses immunized with human lymphocytes, which has been used for many years to treat acute graft rejection episodes. Anti-lymphocyte globulin does not, however, discriminate between useful lymphocytes and those responsible for unwanted responses. Moreover, horse immunoglobulin is highly antigenic in humans and the large doses used in therapy are often followed by the development of serum sickness, caused by the formation of immune complexes of horse immunoglobulin and human anti-horse immunoglobulin antibodies (see Chapter 12). Nevertheless, anti-lymphocyte globulins are still in use to treat acute rejection and have stimulated the quest for monoclonal antibodies to achieve more specifically targeted effects.

Immunosuppressive monoclonal antibodies act by one of two general mechanisms. Some monoclonal antibodies trigger the destruction of lymphocytes *in vivo*, and are referred to as depleting, whereas others are non-depleting and act by blocking the function of their target protein without killing the cell that bears it. Many antibodies are being tested for their ability to inhibit allograft rejection and to modify the expression of autoimmune disease. We shall discuss some of these examples after looking at the measures being taken to prepare monoclonal antibodies for therapy in humans.

14-6 Antibodies can be engineered to reduce their immunogenicity in humans.

The major impediment to therapy with monoclonal antibodies in humans is that these antibodies are most readily made by using mouse cells and humans rapidly develop antibody responses to mouse antibodies. This not only blocks the actions of the mouse antibodies but leads to allergic reactions and can result in anaphylaxis with continued treatment (see Section 12-11). Once this has happened, future treatment with any mouse monoclonal antibody is ruled out. This problem can, in principle, be avoided by making antibodies that are not recognized as foreign by the human immune system, and three strategies are being explored for their construction. One approach is to clone human V regions into a phage display library and select for binding to human cells, as described in Section 2-10. In this way, monoclonal antibodies that are entirely human in origin can be obtained. Second, mice lacking endogenous immunoglobulin genes can be made transgenic (see Section 2-26) for human heavy and light immunoglobulin chain loci by using yeast artificial chromosomes. B cells in these mice have receptors encoded by human immunoglobulin genes but are not tolerant to most human proteins. It is therefore possible to induce in these mice the production of human monoclonal antibodies against determinants on human cells or proteins.

Finally, one can graft the complementarity-determining regions (CDRs) of a mouse monoclonal antibody, which form the antigen-binding loops, onto the framework of a human immunoglobulin molecule, a process known as **humanization**. Because antigen-binding specificity is determined by the structure of the CDRs (see Chapter 3), and because the overall frameworks of mouse and human antibodies are so similar, this approach produces a monoclonal antibody that is antigenically identical to human immunoglobulin but binds the same antigen as the mouse monoclonal antibody from which the CDR sequences were derived. These recombinant antibodies are far less immunogenic in humans than the parent mouse monoclonal antibodies, and thus they can be used for the treatment of humans with far less risk of anaphylaxis.

14-7 Monoclonal antibodies can be used to inhibit allograft rejection.

Antibodies specific for various physiological targets have been used in attempts to prevent the development of allograft rejection by inhibiting the development of harmful inflammatory and cytotoxic responses. One approach is to perfuse the organ before transplantation with antibodies that react with antigen-presenting cells and thus target them for destruction within the reticuloendothelial system. Depletion of antigen-presenting cells in the graft by this method is effective at preventing allograft rejection in animal models, although there is no convincing evidence that it is successful in humans. Antibodies have, however, been used to treat episodes of graft rejection in humans. Anti-CD3 antibodies are moderately effective as an adjunct to immunosuppressive drugs in the treatment of episodes of transplanted kidney rejection.

In animal studies of graft rejection, a fusion protein made from CTLA-4 and the Fc portion of human immunoglobulin has allowed the long-term survival of certain grafted tissues. Presumably this fusion protein, which binds to both B7.1 and B7.2, blocks the co-stimulation of the T cells that recognize donor antigen, and thus allows the cells to induce anergy in the alloreactive cells in the recipient (see Section 8-4).

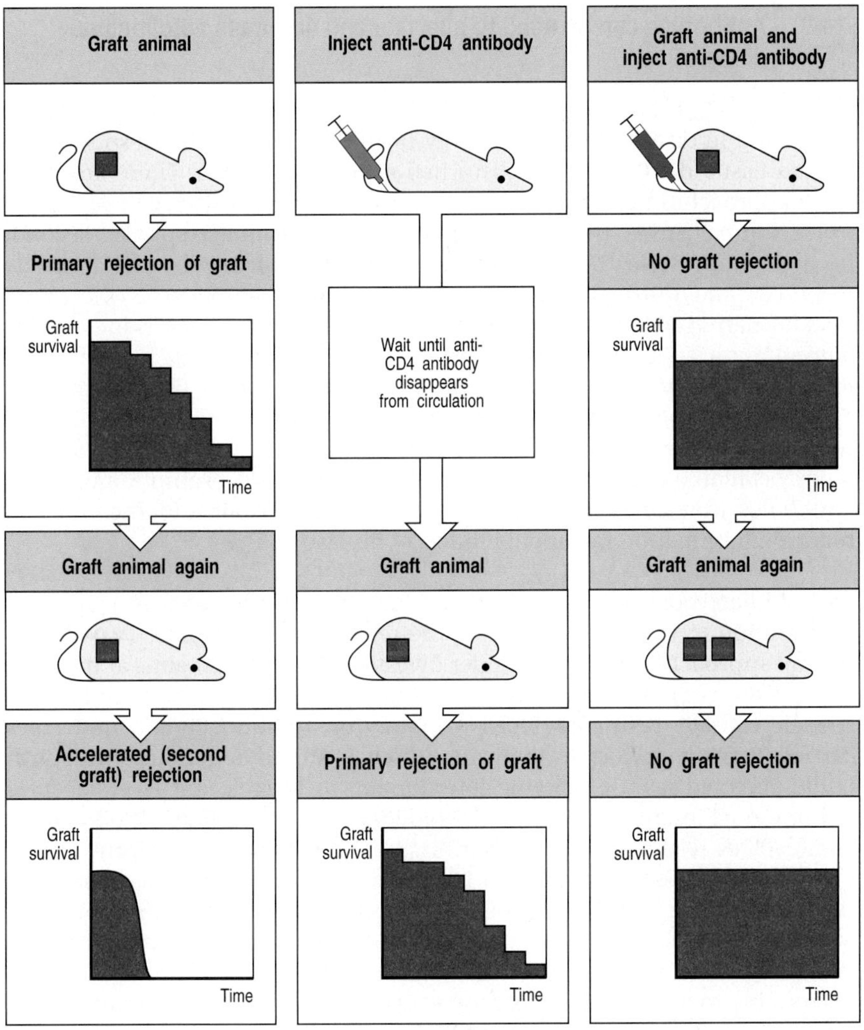

Fig. 14.7 A tissue graft given together with anti-CD4 antibody can induce specific tolerance. Mice grafted with tissue from a genetically different mouse reject that graft. Having been primed to respond to the antigens in the graft, they then reject a subsequent graft of identical tissue more rapidly (left panels). Mice injected with anti-CD4 antibody alone can recover immune competence when the antibody disappears from the circulation, as shown by a normal primary rejection of graft tissue (center panels). However, when tissue is grafted and anti-CD4 antibody is administered at the same time, the primary rejection response is markedly inhibited (right panels). An identical graft made later in the absence of anti-CD4 antibody is not rejected, showing that the animal has become tolerant to the graft antigen. This tolerance can be transferred with T cells to naive recipients (not shown).

Monoclonal antibodies against several targets have also had some success in preventing graft rejection in animals. Of particular interest are certain non-depleting anti-CD4 antibodies: when given for a short time during primary exposure to grafted tissue, these lead to a state of tolerance in the recipient (Fig. 14.7). This tolerant state is an example of the dominant immune suppression discussed in Section 13-26. It is long-lived and can be transferred to naive recipients with CD4 T cells producing cytokines typical of T_H2 cells, although T cells producing other patterns of cytokines might also be involved (see Section 14-9). The presence of anti-CD4 at the time of transplantation might favor the development of a non-damaging T_H2 response, rather than an inflammatory T_H1 response, because of a reduced avidity of interaction between the graft cells and responding T cells, as discussed in Section 10-19.

In human bone marrow transplantation, depleting antibodies directed at mature T lymphocytes have proved particularly useful. Elimination of mature T lymphocytes from donor bone marrow before infusion into a recipient is very effective at reducing the incidence of graft-versus-host disease, in which the T lymphocytes in the donor bone marrow recognize the recipient as foreign and mount a damaging alloreaction against the recipient, causing rashes, diarrhea, and pneumonia, which is often fatal (see Section 13-19).

14-8 Antibodies can be used to alleviate and suppress autoimmune disease.

Autoimmune disease is detected only once the autoimmune response has caused tissue damage or has disturbed specific functions. There are three main approaches to treatment. First, anti-inflammatory therapy can reduce tissue injury caused by an inflammatory autoimmune response; second, immunosuppressive therapy can be aimed at reducing the autoimmune response; and third, treatment can be directed specifically at the organ systems damaged by the disease, for example diabetes, which is induced by autoimmune attack on pancreatic cells and is treated with insulin. Anti-inflammatory therapy for autoimmune disease includes the use of anti-cytokine antibodies; anti-TNF-α antibodies induce striking temporary remissions in rheumatoid arthritis (Fig. 14.8). Antibodies can also be used to block cellular migration to sites of inflammation, for example anti-CD18 antibodies prevent tight leukocyte adhesion to vascular endothelium and reduce inflammation in animal models of disease.

The ultimate goal of immunotherapy for autoimmune disease is specific intervention to restore tolerance to the relevant autoantigens. Two experimental approaches to this are under investigation. The first aims at blocking the specific response to autoantigen. One way to attempt this is to identify specific clonally restricted T-cell receptors or immunoglobulin molecules carried by the lymphocytes that cause disease, and to target these with antibodies directed against idiotypic determinants on the relevant T-cell receptor or immunoglobulin. Another way is to identify particular MHC class I or class II molecules responsible for presenting peptides from autoantigens and to inhibit their antigen-presenting function selectively with antibodies or blocking peptides. This approach has been successful in some animal models of autoimmunity, for example experimental autoimmune encephalomyelitis (EAE) (Fig. 14.9), in which it seems that a limited number of clones of T cells, responding to a single peptide, might cause disease. However, autoimmune disease in humans and most animal models is driven by a polyclonal

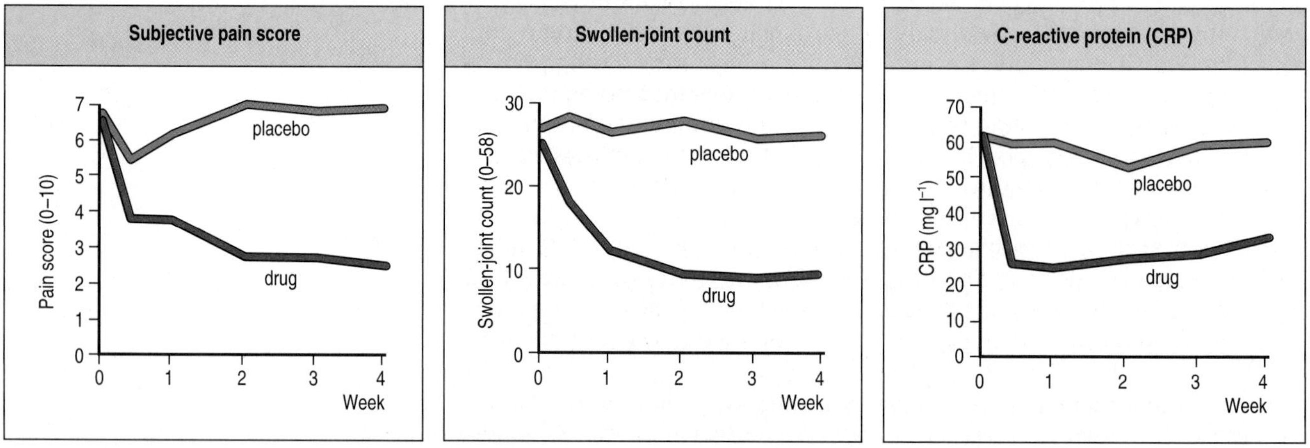

Fig. 14.8 Anti-inflammatory effects of anti-TNF-α therapy in rheumatoid arthritis. The clinical course of 24 patients was followed for 4 weeks after treatment with either a placebo or monoclonal antibody against TNF-α at a dose of 10 mg kg^{-1}. The antibody therapy was associated with a reduction in both subjective and objective parameters of disease activity (as measured by pain score and swollen joint count respectively) and in the systemic inflammatory acute-phase response, measured as a fall in the concentration of the acute-phase reactant, C-reactive protein. Data courtesy of R N Maini.

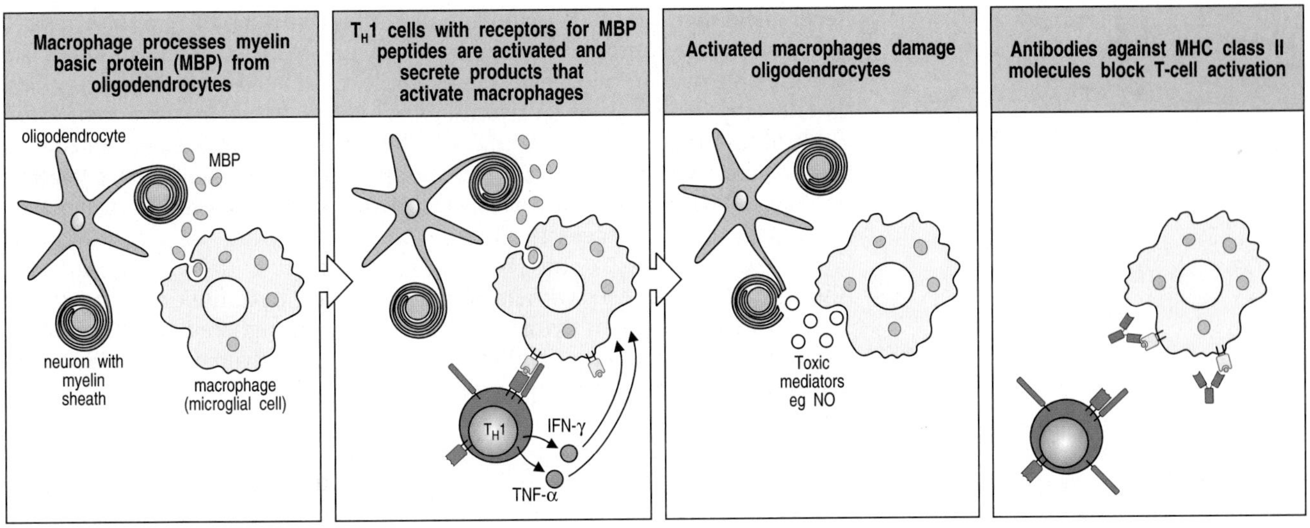
Macrophage processes myelin basic protein (MBP) from oligodendrocytes
oligodendrocyte
MBP
neuron with myelin sheath
macrophage (microglial cell)
T_H1 cells with receptors for MBP peptides are activated and secrete products that activate macrophages
T_H1
IFN-γ
TNF-α
Activated macrophages damage oligodendrocytes
Toxic mediators eg NO
Antibodies against MHC class II molecules block T-cell activation

Fig. 14.9 Anti-MHC class II antibody can inhibit the development of experimental autoimmune encephalomyelitis. In mice with experimental autoimmune encephalomyelitis (EAE), macrophages process myelin basic protein (MBP) and present MBP peptides to T_H1 lymphocytes in conjunction with co-stimulatory signals. Activated T_H1 cells secrete cytokines, which in turn activate macrophages, which cause tissue injury to oligodendrocytes. Antibodies against MHC class II molecules block this process by blocking the interaction between T_H1 cells and antigen-presenting macrophages.

response to autoantigens by T and B lymphocytes. For this reason, immunotherapy based on the identification of specific receptors carried by pathogenic lymphocytes is unlikely to succeed. Immunotherapy based on the identification of the particular MHC molecules that drive an autoimmune response is more likely to be effective, but such therapy would also inhibit some protective immune responses.

14-9 Modulation of the pattern of cytokine expression by T lymphocytes can inhibit autoimmune disease.

The second approach to immunotherapy for autoimmune disease is to try to turn a pathological autoimmune response into an innocuous one. This approach is being pursued experimentally because, as we learned in Chapter 13, tolerance to tissue antigens does not always depend on the absence of a T-cell response; instead, it can be actively maintained by T cells secreting cytokines that suppress the development of a harmful, inflammatory T-cell response. As the pattern of cytokines expressed by T lymphocytes is critical in determining the perpetuation and expression of autoimmune disease, the manipulation of cytokine expression offers a way of controlling it. There are various techniques, collectively known as **immune modulation**, that can alter cytokine expression by T lymphocytes. These involve manipulating the cytokine environment in which T-cell activation takes place, or manipulating the way antigen is presented, as these factors have been observed to influence the differentiation and cytokine-secreting function of the activated T cells (see Sections 8-12, 10-17, and 10-18).

As discussed in earlier chapters, CD4 T lymphocytes can be subdivided into two major subsets, the T_H1 cells, which secrete interferon (IFN)-γ, and the T_H2 cells, which secrete IL-4, IL-5, IL-10, and transforming growth factor (TGF)-β. In many cases, autoimmune disease is associated with the activation of T_H1 cells, which activate macrophages and drive an inflammatory immune response. In animal models of experimentally induced autoimmune disease, such as EAE, the relative activation of the T_H1 and T_H2 subsets of T lymphocytes can be manipulated to give either a T_H1 response and disease, or a T_H2 response that confers protection against disease. The preferential activation of T_H1 or T_H2 cells can be achieved by direct manipulation of the cytokine environment or by administering antigen by particular routes, for example by feeding (see Section 14-10).

Recent evidence shows that cytokine patterns secreted by T lymphocytes are very complicated and that the T_H1 and T_H2 subdivision of T lymphocytes is a considerable oversimplification. For example, CD4 T cells have been identified that develop in a cytokine environment rich in IL-10, and in turn secrete high levels of IL-10 and little IL-2 and IL-4. This pattern of cytokine secretion has bystander effects on other T cells and suppresses antigen-induced activation of other CD4 T lymphocytes. These cells have been provisionally designated T_r1 cells (T regulatory cells 1).

Another subset of T cells with immunosuppressive bystander effects secretes TGF-β as the dominant cytokine and has been designated T_H3. Such cells might be predominantly of mucosal origin and activated by the mucosal presentation of antigen (see Section 14-10).

A further subset of T cells also seems to be implicated in immunoregulation. These are the NK1.1 CD4 T cells, so named because they bear the receptor NK1.1, which is usually found on NK cells. NK1.1 T cells, which we discussed in Section 10-17, recognize antigens, including lipid antigens, presented by the class I-like molecule CD1 (see Section 4-16) and respond by secreting IL-4. Thus, when stimulated, the NK1.1 T cells can act to promote T_H2 responses. Although there is no direct evidence that NK1.1 T cells are involved in immunomodulation in humans, in mice that suffer spontaneous autoimmune disease this population of cells is either missing or decreased. Furthermore, transfer of NK1.1 T cells into such mice prevents the onset of the autoimmune disease.

Immune modulation aims at altering the balance between different subsets of responding T cells such that helpful responses are promoted and damaging responses are suppressed. As a therapy for autoimmunity it has the advantage that one might not need to know the precise nature of the autoantigen stimulating the autoimmune disease. This is because the administration of cytokines or antigen to modulate the immune response causes changes in the pattern of cytokine expression that have bystander effects on lymphocytes with the presumed autoreactive receptors. However, the drawback of this approach is the unpredictability of the results. In murine models of diseases such as diabetes and EAE, most of the results suggest that a T_H2 response can protect against T_H1-mediated disease but there is evidence that T_H2 lymphocytes can also contribute to the pathology of these diseases.

An additional problem is the difficulty of modulating established immune responses. Experiments in animals have shown that anti-cytokine antibodies (or recombinant cytokines) present at the time of immunization with an autoantigen can sometimes divert a pathogenic immune response. In contrast, the modification of an ongoing immune response is much harder to achieve with this approach, although there have been some examples of experimental success, as we shall see later.

14-10 Controlled administration of antigen can be used to manipulate the nature of an antigen-specific response.

When the target antigen of an unwanted response is identified, it is possible to manipulate the response by using antigen directly rather than by using antibodies or relying on the bystander effects discussed in the previous section. This is because the way in which antigen is presented to the immune system affects the nature of the response, and the induction of one type of response to an antigen can inhibit a pathogenic response to the same antigen.

As mentioned in Chapter 12, this principle has been applied with some success to the treatment of allergies caused by an IgE response to very low doses of antigen. Repeated treatment of allergic individuals with higher doses of allergen seems to divert this response to one dominated by T cells that favor the production of IgG and IgA antibodies. These antibodies are thought to desensitize the patient by binding the small amounts of allergen normally encountered and preventing it from binding to IgE.

With T-cell mediated autoimmune disease, there has been considerable interest in using peptide antigens to suppress pathogenic responses. The type of CD4 T-cell response induced by a peptide depends on the way in which it is presented to the immune system. For instance, peptides given orally tend to prime T_H2 T cells that make IL-4 or T cells that make predominantly TGF-β (see Section 14-9) without activating T_H1 cells or inducing a great deal of systemic antibody. These mucosal immune responses have relatively little pathogenic potential. Indeed, experiments in animal models indicate that they can protect against induced autoimmune disease. Experimental autoimmune encephalomyelitis is induced by injection of myelin basic protein in complete Freund's adjuvant and resembles multiple sclerosis, whereas collagen arthritis is similarly induced by injection of collagen type II and has features in common with rheumatoid arthritis. Oral administration of myelin basic protein or type II collagen inhibits the development of disease in animals (see Fig. 13.31), and has some beneficial effects in reducing the activity of pre-established disease. Trials using this approach in humans with multiple sclerosis or rheumatoid arthritis have found marginal therapeutic effects. Intravenous delivery of peptides can also inhibit inflammatory responses stimulated by the same peptide presented in a different context. When a soluble peptide is given intravenously, it binds preferentially to MHC class II molecules on resting B cells and tends to induce anergy in T_H1 cells. Thus, a careful choice of the dose or structure of antigen, or its route of administration, can allow us to control the type of response that results. Whether such approaches can be effective in manipulating the established immune responses driving human autoimmune diseases remains to be seen.

Summary.

Existing treatments for unwanted immune responses, such as allergic reactions, autoimmunity, and graft rejection, depend largely on three types of drug. Anti-inflammatory drugs, of which the most potent are the corticosteroids, are used for all three types of response. These have a broad spectrum of actions, however, and a correspondingly wide range of toxic side-effects; their dose must be controlled carefully. They are therefore normally used in combination with either cytotoxic or immunosuppressive drugs. The cytotoxic drugs kill all dividing cells and thereby prevent lymphocyte proliferation, but they suppress all immune responses indiscriminately and also kill other types of dividing cells. The immunosuppressive drugs act by intervening in the intracellular signaling pathways of T cells and, although they are less generally toxic than the cytotoxic drugs, they also suppress all immune responses indiscriminately. They are also much more expensive than cytotoxic drugs.

Immunosuppressive drugs are now the drugs of choice in the treatment of transplant patients, where they can be used to suppress the immune response to the graft before it has become established. Autoimmune responses are already well established at the time of diagnosis and, in consequence, are much more difficult to suppress. They are therefore less responsive to the immunosuppressive drugs and, for that reason, they are usually controlled with a combination of corticosteroids and cytotoxic drugs. In animal experiments attempts have been made to target immunosuppression more specifically, by blocking the response to autoantigen with the use of antibodies or antigenic peptides, or by diverting the immune response into a non-pathogenic pathway by manipulating the cytokine environment, or by administering antigen through the oral route where a non-pathogenic immune response is likely to be invoked. None of these treatments is yet proven in humans, and most require that the relevant antigen be known. For that reason, and because they are relatively ineffective against established immune responses, the promise of these approaches in animal models might be difficult to realize in a clinical context.

Using the immune response to attack tumors.

Cancer is one of the three leading causes of death in industrialized nations. As treatments for infectious diseases and the prevention of cardiovascular disease continue to improve, and the average life expectancy increases, cancer is likely to become the most common fatal disease in these countries. Cancers are caused by the progressive growth of the progeny of a single transformed cell. Therefore, curing cancer requires that all the malignant cells be removed or destroyed without killing the patient. An attractive way to achieve this would be to induce an immune response against the tumor that would discriminate between the cells of the tumor and their normal cell counterparts. Immunological approaches to the treatment of cancer have been attempted for over a century with tantalizing but unsustainable results. Experiments in animals have, however, provided evidence for immune responses to tumors and have shown that T cells are a critical mediator of tumor immunity. More recently, advances in our understanding of antigen presentation and the molecules involved in T-cell activation have provided new immunotherapeutic strategies based on a better molecular understanding of the immune response. These are showing some success in animal models and are now being tested in human patients.

14-11 The development of transplantable tumors in mice led to the discovery that mice could mount a protective immune response against tumors.

The finding that tumors could be induced in mice after treatment with chemical carcinogens or irradiation, coupled with the development of inbred strains of mice, made it possible to undertake the key experiments that led to the discovery of immune responses to tumors. These tumors could be transplanted between mice, and the experimental study of tumor rejection has generally been based on the use of such tumors. If these bear foreign MHC molecules they are readily recognized and destroyed by the immune system, a fact that was exploited to develop the first MHC-congenic strains of mice. Specific immunity to tumors must therefore be studied within inbred strains, so that host and tumor can be matched for MHC type.

Transplantable tumors in mice exhibit a variable pattern of growth when injected into syngeneic recipients. Most tumors grow progressively and eventually kill the host. However, if mice are injected with irradiated tumor cells that cannot grow, they are frequently protected against subsequent injection with a normally lethal dose of viable cells of the same tumor. Among transplantable tumors there seems to be a spectrum of immunogenicity: injections of irradiated tumor cells seem to induce varying degrees of protective immunity against a challenge injection of viable tumor cells at a distant site. These protective effects are not seen in T-cell deficient mice but can be conferred by adoptive transfer of T cells from immune mice, showing the need for T cells to mediate all these effects.

These observations indicate that the tumors express antigenic peptides that can become targets of a tumor-specific T-cell response. The antigens expressed by experimentally induced murine tumors, often termed **tumor-specific transplantation antigens** (**TSTAs**), or **tumor rejection antigens** (**TRAs**), are usually specific for an individual tumor. Thus immunization with irradiated tumor cells from tumor X protects a syngeneic mouse from challenge with live cells from tumor X but not from challenge with a different syngeneic tumor Y, and vice versa (Fig. 14.10).

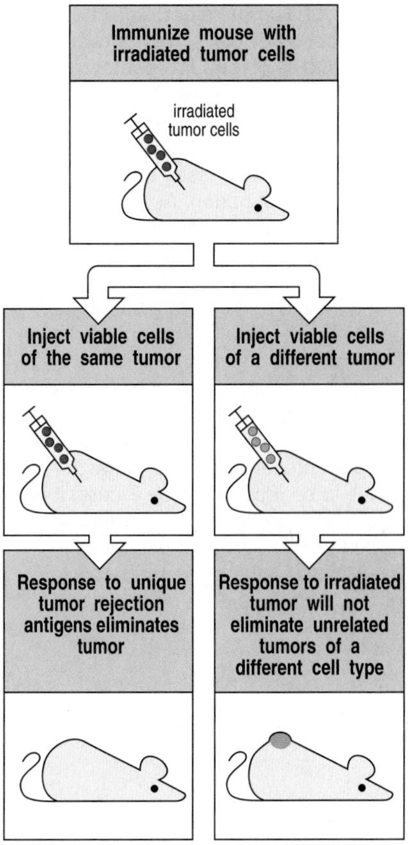

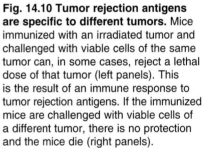

Fig. 14.10 Tumor rejection antigens are specific to different tumors. Mice immunized with an irradiated tumor and challenged with viable cells of the same tumor can, in some cases, reject a lethal dose of that tumor (left panels). This is the result of an immune response to tumor rejection antigens. If the immunized mice are challenged with viable cells of a different tumor, there is no protection and the mice die (right panels).

14-12 T lymphocytes can recognize certain types of spontaneous tumor in humans.

Tumor rejection antigens are peptides of tumor cell proteins that are presented to T cells by MHC molecules. These peptides can become the targets of a tumor-specific T-cell response because they are not displayed on the surface of normal cells, at least not at levels sufficient to be recognized by T cells. They might be proteins normally expressed only in the embryo, but which become expressed abnormally in tumor cells in adults, or they might be normal self proteins that become overexpressed (Fig. 14.11). Human tumor cells often also contain mutant proteins, which can be antigenic. Many tumors are characterized by an abnormal pattern of expression of proteins that could make them targets for immunological rejection (Fig. 14.12). Of these, only the melanoma antigens, renal cell carcinoma antigen and MUC-1 (mucin-1) on breast or pancreatic tumors have been shown to be recognized by T lymphocytes from patients with these tumors. Malignant melanoma and renal cell carcinoma are unlike other tumors in that, occasionally, even quite advanced disease shows spontaneous remission.

In melanoma, tumor-specific antigens were discovered by culturing irradiated tumor cells with autologous lymphocytes, a reaction known as the mixed lymphocyte tumor cell culture. From such cultures cytotoxic T lymphocytes could be identified that would kill, in an MHC-restricted fashion, tumor cells bearing the relevant tumor-specific antigen. Melanomas have been studied in detail by using this approach. Cytotoxic T cells reactive against melanoma peptides have been cloned and used to characterize melanomas by the array of tumor-specific antigens displayed. These studies have yielded three important findings. The first is that melanomas carry at least five different antigens that can be recognized by cytotoxic T lymphocytes. The second is that cytotoxic T lymphocytes reactive against melanoma antigens are not expanded *in vivo*, suggesting that these antigens are not immunogenic *in vivo*. The third is that the expression of these antigens can be selected against *in vitro* and possibly also *in vivo* by the presence of specific cytotoxic T cells. These discoveries offer the hope of tumor immunotherapy, an indication that

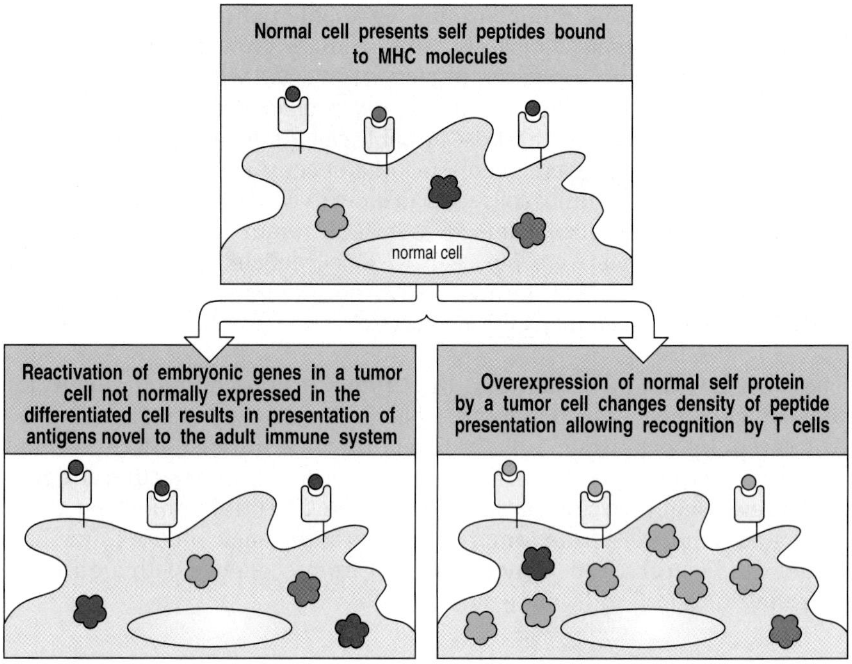

Fig. 14.11 Tumor rejection antigens are peptides of cell proteins presented by self MHC class I molecules. Tumor rejection antigens arise in two main ways. In some cases, proteins that are normally expressed only in embryonic tissues are re-expressed by the tumor cells (lower left panel). As these proteins are normally expressed at a time when the immune system is not fully developed, T cells are not tolerant of these self antigens and can respond to them as though they were foreign proteins. In other tumors, overexpression of a self protein increases the density of presentation of a normal self peptide on tumor cells (lower right panel). Such peptides are then presented at high enough levels to be recognized by T cells. It is often the case that the same embryonic or self proteins are overexpressed in many tumors of a given tissue origin, giving rise to shared tumor rejection antigens.

these antigens are not naturally strongly immunogenic, and also a caution about the possibility of selecting, *in vivo*, tumor cells that can escape recognition and killing by cytotoxic T cells.

Consistent with these findings, functional melanoma-specific T cells can be propagated from peripheral blood lymphocytes, from tumor-infiltrating lymphocytes, or by draining the lymph nodes of patients in whom the melanoma is growing. Interestingly, none of the peptides recognized by these T cells derives from the mutant proto-oncogenes or tumor suppressor genes that are likely to be responsible for the initial transformation of the cell into a cancer cell, although a few are the products of mutant genes. The rest fall into the two categories described in Fig. 14.11. Antigens of the MAGE family are not expressed in any normal adult tissues with the exception of the testis, which is an immunologically privileged site. They probably represent early developmental antigens re-expressed in the process of tumorigenesis. Only a minority of melanoma patients have T cells reactive against the MAGE antigens, indicating that these antigens are either not expressed or are not immunogenic in most cases. The most common melanoma antigens are peptides from the enzyme tyrosinase or from three other proteins—gp100, MART1, and gp75. These are differentiation antigens specific to the melanocyte lineage from which melanomas arise. It is likely that overexpression of these antigens in tumor cells leads to an abnormally high density of specific peptide:MHC complexes and this makes them immunogenic. Although in most cases tumor rejection antigens are presented as peptides complexed with MHC class I molecules, tyrosinase has been shown to stimulate CD4 T-cell responses in some melanoma patients by being ingested and presented by cells expressing MHC class II.

Tumor rejection antigens shared between most examples of a tumor, and against which tolerance can be broken, represent candidate antigens for tumor vaccines. The MAGE family of antigens are candidates because of their limited tissue distribution and their shared expression by many melanomas. It might seem dangerous to use tumor vaccines based on antigens that are not truly tumor-specific because of the risk of inducing autoimmunity. Often,

Potential tumor rejection antigens have a variety of origins			
Class of antigen	**Antigen**	**Nature of antigen**	**Tumor type**
Embryonic	MAGE-1 MAGE-3	Normal testicular proteins	Melanoma Breast Glioma
Abnormal post-translational modification	MUC-1	Underglycosylated mucin	Breast Pancreas
Differentiation	Tyrosinase	Enzyme in pathway of melanin synthesis	Melanoma
	Surface Ig	Specific antibody after gene rearrangements in B-cell clone	Lymphoma
Mutated oncogene or tumor-suppressor	Cyclin-dependent kinase 4	Cell-cycle regulator	Melanoma
	β-catenin	Relay in signal transduction pathway	Melanoma
	Caspase-8	Regulator of apoptosis	Squamous cell carcinoma
Oncoviral protein	HPV type 16, E6 and E7 proteins	Viral transforming gene products	Cervical carcinoma

Fig. 14.12 Proteins selectively expressed in human tumors are candidate tumor rejection antigens. The molecules listed here have all been found to be recognized by cytotoxic T lymphocytes raised from patients with the tumor type listed.

however, the tissues from which tumors arise are dispensable; the prostate is perhaps the best example of this. With melanoma, however, some melanocyte-specific tumor rejection antigens are also expressed in certain retinal cells, in the inner ear, in the brain, and in the skin. Despite this, melanoma patients receiving immunotherapy with whole tumor cells or tumor-cell extracts, although occasionally developing vitiligo—a destruction of pigmented cells in the skin that correlates well with a good response to the tumor—do not develop abnormalities in the visual, vestibular, and central nervous systems, perhaps because of the low level of expression of MHC class I molecules in these sites (see Section 4-7).

In addition to the tumor antigens in humans that have been shown to induce cytotoxic T-cell responses (see Fig. 14.12), there are many other candidate tumor rejection antigens that have been identified by studies of the molecular basis of cancer development. These include the products of mutated cellular oncogenes or tumor suppressors, such as Ras and p53, and also fusion proteins, such as the Bcr–Abl tyrosine kinase that results from the chromosomal translocation (t9;22) found in chronic myeloid leukemia. It is intriguing that, in each of these cases, no cytotoxic T-cell response has been identified in cultures of autologous lymphocytes with tumor cells bearing these mutated antigens. However, cytotoxic T lymphocytes specific for these antigens can be developed *in vitro* by using peptide sequences derived either from the mutated sequence or the fusion sequence of these common oncogenic proteins. In chronic myeloid leukemia, it is known that after treatment and bone marrow transplantation, mature lymphocytes from the bone marrow donor infused into the patient can help to eliminate any residual tumor, although at present it is not clear whether this is a graft-versus-host effect, where the donor lymphocytes are responding to alloantigens expressed on the leukemia cells, or whether there is a specific anti-leukemic response. The ability to prime the donor cells against leukemia-specific peptides offers the prospect of enhancing the anti-leukemic effect while minimizing the risk of graft-versus-host disease. It is a challenge for immunologists to understand why these mutated proteins do not prime cytotoxic T cells in the patients in which the tumors arise. They are excellent targets for therapy, as they are unique to the tumor and have a causal role in oncogenesis.

14-13 Tumors can escape rejection in many ways.

Burnet called the ability of the immune system to detect tumor cells and destroy them **immune surveillance**. However, it is difficult to show that tumors are subject to surveillance by the immune system; after all, cancer is a common disease, and most tumors show little evidence of immunological control. The incidence of the common tumors in mice that lack lymphocytes is little different from their incidence in mice with normal immune systems; the same is true for humans deficient in T cells. The major tumor types that occur with increased frequency in immunodeficient mice or humans are virus-associated tumors; immune surveillance thus seems to be critical for control of virus-associated tumors, but the immune system does not normally respond to the novel antigens deriving from the multiple genetic alterations in spontaneous tumors. The goal in the development of anti-cancer vaccines is to break the tolerance of the immune system for antigens expressed mainly or exclusively by the tumor.

It is not surprising that spontaneously arising tumors are rarely rejected by T cells, as in general they probably lack either distinctive antigenic peptides or the adhesion or co-stimulatory molecules needed to elicit a primary T-cell response. Moreover, there are other mechanisms whereby tumors can avoid immune attack or evade it when it occurs (Fig. 14.13). Tumors tend to be

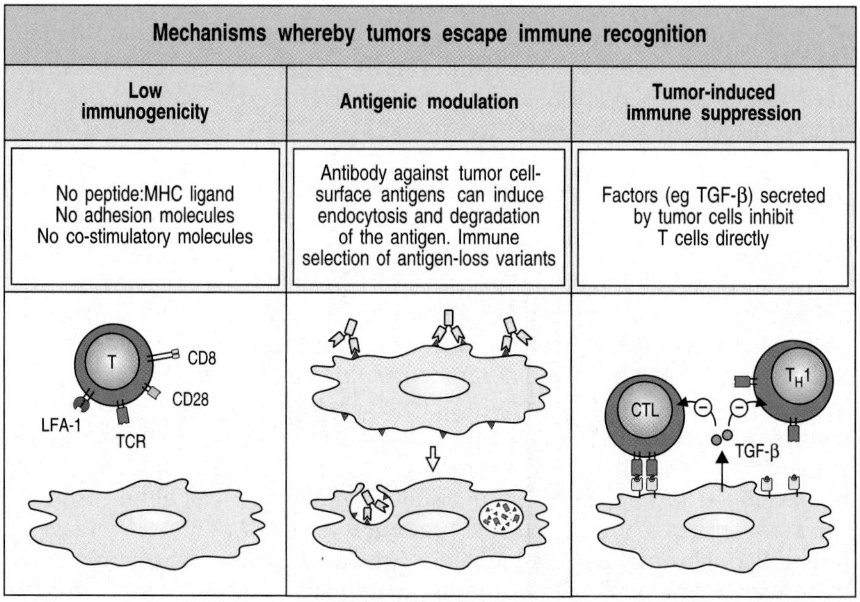

Mechanisms whereby tumors escape immune recognition
Low immunogenicity
Antigenic modulation
Tumor-induced immune suppression
No peptide:MHC ligand
No adhesion molecules
No co-stimulatory molecules
Antibody against tumor cell-surface antigens can induce endocytosis and degradation of the antigen. Immune selection of antigen-loss variants
Factors (eg TGF-β) secreted by tumor cells inhibit T cells directly
T
CD8
CD28
LFA-1
TCR
CTL
TH1
TGF-β

Fig. 14.13 Tumors can escape immune surveillance in a variety of ways. First, tumors can have low immunogenicity (left panel). Some tumors do not have peptides of novel proteins that can be presented by MHC molecules, and therefore appear normal to the immune system. Others have lost one or more MHC molecules, and most do not express co-stimulatory proteins, which are required to activate naive T cells. Second, tumors can initially express antigens to which the immune system responds but lose them by antibody-induced internalization or antigenic variation. When tumors are attacked by cells responding to a particular antigen, any tumor that does not express that antigen will have a selective advantage (center panel). Third, tumors often produce substances, such as TGF-β, that suppress immune responses directly (right panel).

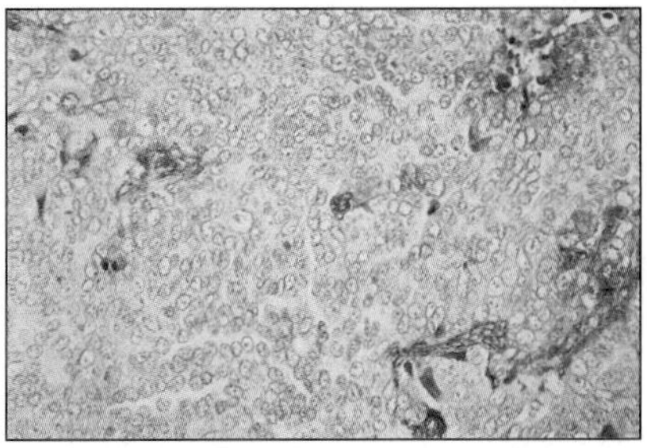

Fig. 14.14 Loss of MHC class I expression in a prostatic carcinoma. Some tumors can evade immune surveillance by loss of expression of MHC class I molecules, preventing their recognition by CD8 T cells. A section of a human prostate cancer that has been stained with a peroxidase-conjugated antibody to HLA class I, is shown. The brown stain correlating with HLA class I expression is restricted to infiltrating lymphocytes and tissue stromal cells. The tumor cells that occupy most of the section show no staining. Photograph courtesy of G Stamp.

genetically unstable and can lose their antigens by mutation; in the event of an immune response, this instability might generate escape mutants. Some tumors, such as colon cancers, lose the expression of a particular MHC class I molecule (Fig. 14.14), perhaps through immunoselection by T cells specific for a peptide presented by that MHC class I molecule. In experimental studies, when a tumor loses expression of all MHC class I molecules, it can no longer be recognized by cytotoxic T cells, although it might become susceptible to NK cells (Fig. 14.15). However, tumors that lose only one MHC

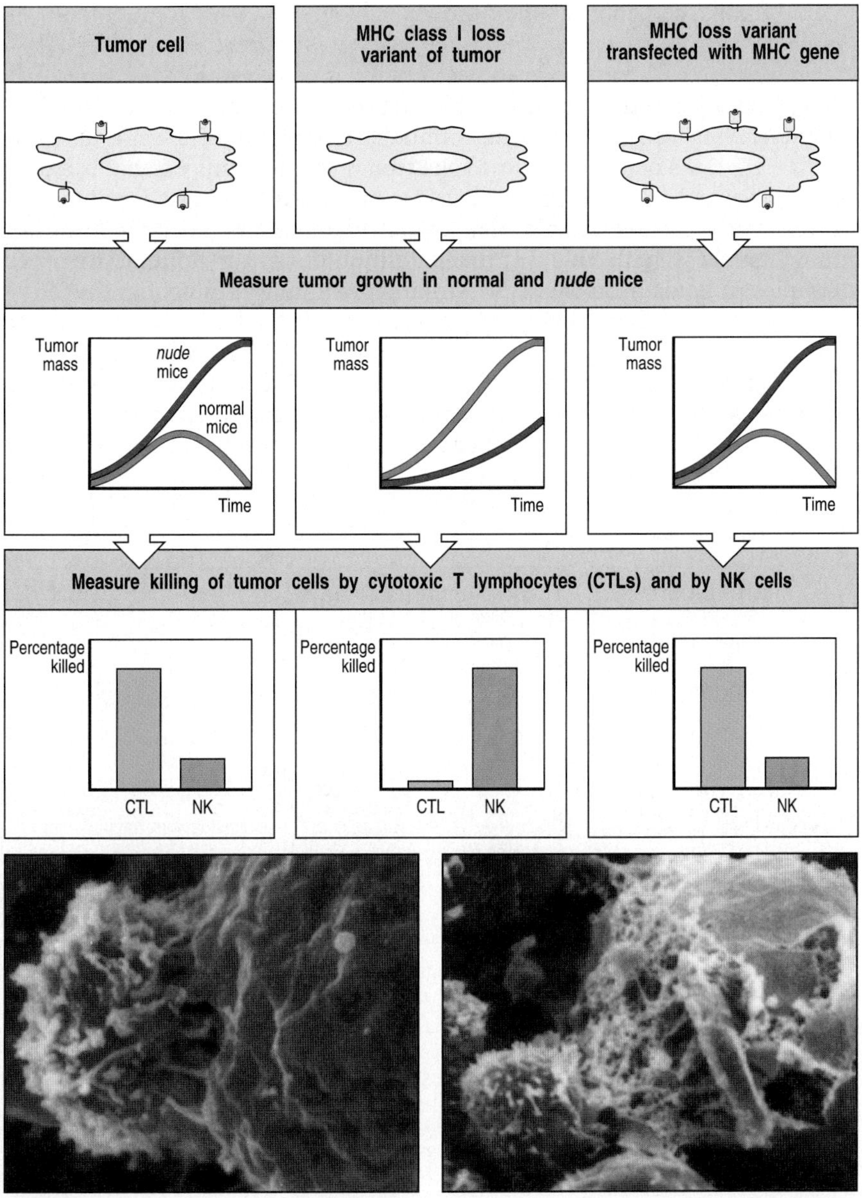

Fig. 14.15 Tumors that lose expression of all MHC class I molecules as a mechanism of escape from immune surveillance are more susceptible to natural killer (NK) cell killing. Regression of transplanted tumors is largely due to the actions of cytotoxic T cells (CTL), which recognize novel peptides bound to MHC class I antigens on the surface of the cell (left panels). NK cells have inhibitory receptors that bind MHC class I molecules, so variants of the tumor that have low levels of MHC class I, although less sensitive to CD8 cytotoxic T cells, become susceptible to NK cells (center panels). Although nude mice lack T cells, they have higher than normal levels of NK cells, and so tumors that are sensitive to NK cells grow less well in nude mice than in normal mice. Transfection with MHC class I genes can restore both resistance to NK cells and susceptibility to CD8 cytotoxic T cells (right panels). However, tumors that lose only one MHC class I molecule can escape a specific cytotoxic CD8 T-cell response while remaining NK resistant. The bottom panels show scanning electron micrographs of NK cells attacking leukemia cells. Left panel: shortly after binding to the target cell, the NK cell has put out numerous microvillous extensions and established a broad zone of contact with the leukemia cell. The NK cell is the smaller cell on the left in both photographs. Right panel: 60 minutes after mixing, long microvillous processes can be seen extending from the NK cell (bottom left) to the leukemia cell and there is extensive damage to the leukemia cell membrane; the plasma membrane of the leukemia cell has rolled up and fragmented under the NK cell attack. Photographs courtesy of J C Hiserodt.

class I molecule might be able to avoid recognition by specific CD8 cytotoxic T cells while remaining resistant to NK cells, conferring a selective advantage *in vivo*.

Yet another way in which tumors might evade rejection is by making immunosuppressive cytokines. Many tumors make these, although in most cases little is known of their precise nature. TGF-β was first identified in the culture supernatant of a tumor (hence its name) and, as we have seen, tends to suppress inflammatory T-cell responses and cell-mediated immunity, which are needed to control tumor growth. A number of tumors of different tissue origins, such as melanoma, ovarian carcinoma, and B-cell lymphoma, have been shown to produce the immunosuppressive cytokine IL-10, which can reduce dendritic cell development and activity. Thus, there are many ways in which tumors avoid recognition and destruction by the immune system.

14-14 Monoclonal antibodies against tumor antigens, alone or linked to toxins, can control tumor growth.

The advent of monoclonal antibodies suggested the possibility of targeting and destroying tumors by making antibodies against tumor-specific antigens (Fig. 14.16). This depends on finding a tumor-specific antigen that is a cell-surface molecule. Some of the cell surface molecules that have been targeted in experimental clinical trials are shown in Fig. 14.17. So far there has been limited success with this approach, although, as an adjunct to other therapies, it holds promise. Some striking preliminary results have recently been reported in the treatment of breast cancer with a monoclonal antibody against a growth factor receptor, $p185^{HER2}$, that is overexpressed in about a quarter of breast cancer patients and is associated with a poorer prognosis. The effects of this antibody can be potentiated when it is combined with

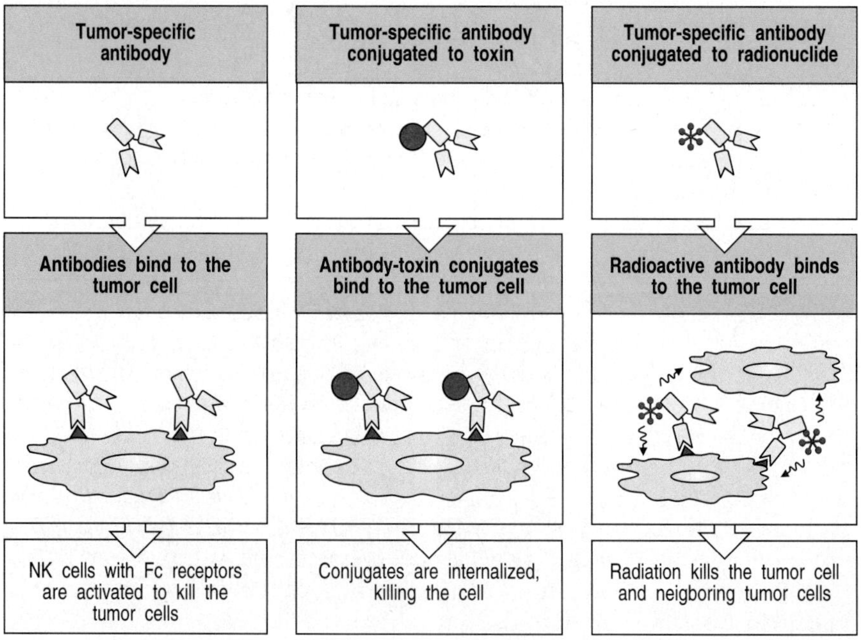
Tumor-specific antibody
Tumor-specific antibody conjugated to toxin
Tumor-specific antibody conjugated to radionuclide
Antibodies bind to the tumor cell
Antibody-toxin conjugates bind to the tumor cell
Radioactive antibody binds to the tumor cell
NK cells with Fc receptors are activated to kill the tumor cells
Conjugates are internalized, killing the cell
Radiation kills the tumor cell and neigboring tumor cells

Fig. 14.16 Ways in which a monoclonal antibody or antibody conjugate could eliminate or reduce a tumor. Monoclonal antibodies that recognize tumor-specific antigens might be used in a variety of ways to help eliminate tumors. Tumor-specific antibodies of the correct isotypes might be able to direct the lysis of the tumor cells by NK cells, activating the NK cells via their Fc receptors (left panels). A more useful strategy might be to couple the antibody to a powerful toxin (center panels). When the antibody binds to the tumor cell and is endocytosed, the toxin is released from the antibody and can kill the tumor cell. If the antibody is coupled to a radionuclide (right panels), binding of the antibody to a tumor cell will deliver a dose of radiation sufficient to kill the tumor cell. In addition, nearby tumor cells could also receive a lethal radiation dose, even though they did not bind the antibody.

Tumor tissue origin	Type of antigen	Antigen	Tumor type
Lymphoma/ leukemia	Differentiation antigen	CD5 Idiotype CAMPATH-1 (CDw52)	T-cell lymphoma B-cell lymphoma T- and B-cell lymphoma
Solid tumors	Cell-surface antigens Glycoprotein Carbohydrate	 CEA, mucin-1 Lewis^y CA-125	 Epithelial tumors (breast, colon, lung) Epithelial tumors Ovarian carcinoma
	Growth factor receptor	Epidermal growth factor receptor P185^{HER2} IL-2 receptor	Lung, breast, head, and neck tumors Breast, ovarian tumors T- and B-cell tumors
	Stromal extracellular antigen	FAP-α Tenascin Metalloproteinases	Epithelial tumors Glioblastoma multiforme Epithelial tumors

Fig. 14.17 Examples of tumor antigens that have been targeted by monoclonal antibodies in therapeutic trials. CEA, carcinoembryonic antigen.

conventional chemotherapy. Monoclonal antibodies coupled to γ-emitting radioisotopes have also been used to image tumors, for the purpose of diagnosis and monitoring tumor spread (Fig. 14.18).

The first reported successful treatment of a tumor with monoclonal antibodies used anti-idiotypic antibodies to target B-cell lymphomas whose surface immunoglobulin expressed the specific idiotype. The initial course of treatment usually leads to a remission, but the tumor always reappears in a mutant form that no longer binds to the antibody used for the initial treatment. This case represents a clear example of genetic instability enabling a tumor to evade treatment.

Other problems with tumor-specific or tumor-selective monoclonal antibodies as therapeutic agents include inefficient killing of cells after binding of the monoclonal antibody and inefficient penetration of the antibody into the tumor mass. The first problem can often be circumvented by linking the antibody to a toxin, producing a reagent called an **immunotoxin**; two favored toxins are ricin A chain and *Pseudomonas* toxin. Both approaches require the antibody to be internalized to allow the cleavage of the toxin from the antibody in the endocytic compartment (see Fig. 14.16), allowing the toxin chain to penetrate and kill the cell.

Two other approaches using monoclonal antibody conjugates involve linking the antibody molecule to chemotherapeutic drugs such as adriamycin or to radioisotopes. In the first case, the specificity of the monoclonal antibody for a cell-surface antigen on the tumor concentrates the drug to the site of the tumor. After internalization, the drug is released in the endosomes and exerts its cytostatic or cytotoxic effect. Monoclonal antibodies linked to radionuclides (see Fig. 14.16) concentrate the radioactive source in the tumor site. Both these approaches have the advantage of killing neighboring tumor cells, because the released drug or radioactive emissions can affect cells adjacent to those that actually bind the antibody. Ultimately, combinations of toxin-, drug-, or radionuclide-linked monoclonal antibodies, together with vaccination strategies aimed at inducing T-cell mediated immunity, might provide the most effective cancer immunotherapy.

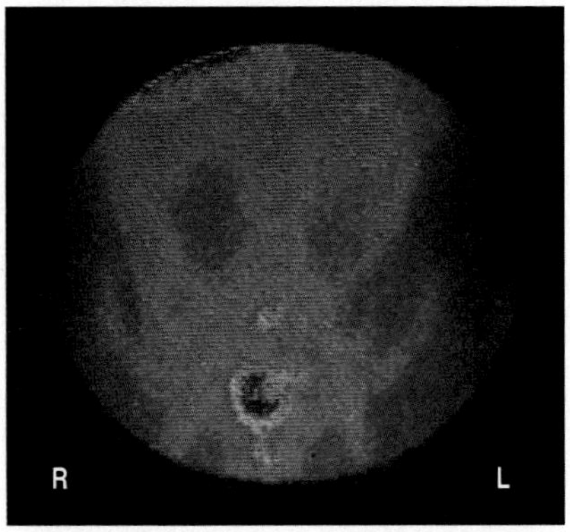

Fig. 14.18 Recurrent colorectal cancer can be detected with a radiolabeled monoclonal antibody against carcinoembryonic antigen. A patient with a possible recurrence of a colorectal cancer was injected intravenously with an indium-111 labeled monoclonal antibody to carcinoembryonic antigen. The recurrent tumor is seen as two red spots located in the pelvic region. The blood vessels are faintly outlined by circulating antibody that has not bound to the tumor. Photograph courtesy of A M Peters.

14-15 Enhancing the immunogenicity of tumors holds promise for cancer therapy.

Although antigen-specific vaccines based on dominant shared tumor antigens are, in principle, the ideal approach to T-cell mediated cancer immunotherapy, it may be many decades before the dominant tumor antigens for common cancers are identified. Even then, it is not clear how widely the relevant epitopes will be shared between tumors, and peptides of tumor rejection antigens will be presented only by particular MHC alleles. MAGE1 antigens, for example, are recognized only by T cells in melanoma patients expressing the HLA-A1 haplotype.

Thus, the individual patient's tumor removed at surgery might be, in practice, the best source of vaccine antigens. Until recently, most cell-based cancer vaccines have involved mixing either irradiated tumor cells or tumor extracts with bacterial adjuvants such as BCG or *Corynebacterium parvum* (see Section 2-4). These adjuvants have generated modest therapeutic results in melanomas but have, in general, been disappointing.

Where candidate tumor rejection antigens have been identified, for example in melanoma, experimental vaccination strategies include the use of whole proteins, peptide vaccines based on sequences recognized by cytotoxic T lymphocytes (either administered alone or presented by the patient's own dendritic cells), and recombinant viruses encoding these peptide epitopes. A novel experimental approach to tumor vaccination is the use of heat-shock proteins isolated from tumor cells. The underlying principle of this therapy is that one of the physiological activities of heat-shock proteins is to act as intracellular chaperones of antigenic peptides, and so they should be isolated with such peptides bound to them. There is evidence for receptors on the surface of professional antigen-presenting cells that take up certain heat-shock proteins together with any bound peptides. Uptake of heat-shock proteins via these receptors delivers the accompanying peptide into the antigen-processing pathways leading to peptide presentation by MHC class I proteins. This experimental technique for tumor vaccination has the advantage that it does not depend on any prior knowledge of the nature of the relevant tumor rejection antigens of the tumor, but the disadvantage that the heat-shock proteins purified from the cell carry very many peptides, so that any tumor rejection antigen might constitute only a tiny fraction of the peptides bound to the heat-shock protein.

A further experimental approach to tumor vaccination in mice is to increase the immunogenicity of tumor cells by introducing genes that encode co-stimulatory molecules or cytokines. This is intended to make the tumor itself more immunogenic. The basic scheme of such experiments is shown in Fig. 14.19. A tumor cell transfected with the gene encoding the co-stimulatory molecule B7 (see Section 8-4) is implanted in a syngeneic animal. These B7-positive cells are able to activate tumor-specific naive T cells to become armed effector T cells able to reject the tumor cells. These armed effector T cells can then target the tumor cells whether they express B7 or not; this can be shown by re-implanting non-transfected tumor cells, which are also rejected.

The second strategy, that of introducing cytokine genes into tumors so that they secrete the relevant cytokine, is aimed at attracting antigen-presenting cells to the tumor and takes advantage of the paracrine nature of cytokines. In mice, the most effective tumor vaccines so far are tumor cells that secrete granulocyte-macrophage colony-stimulating factor (GM-CSF), which

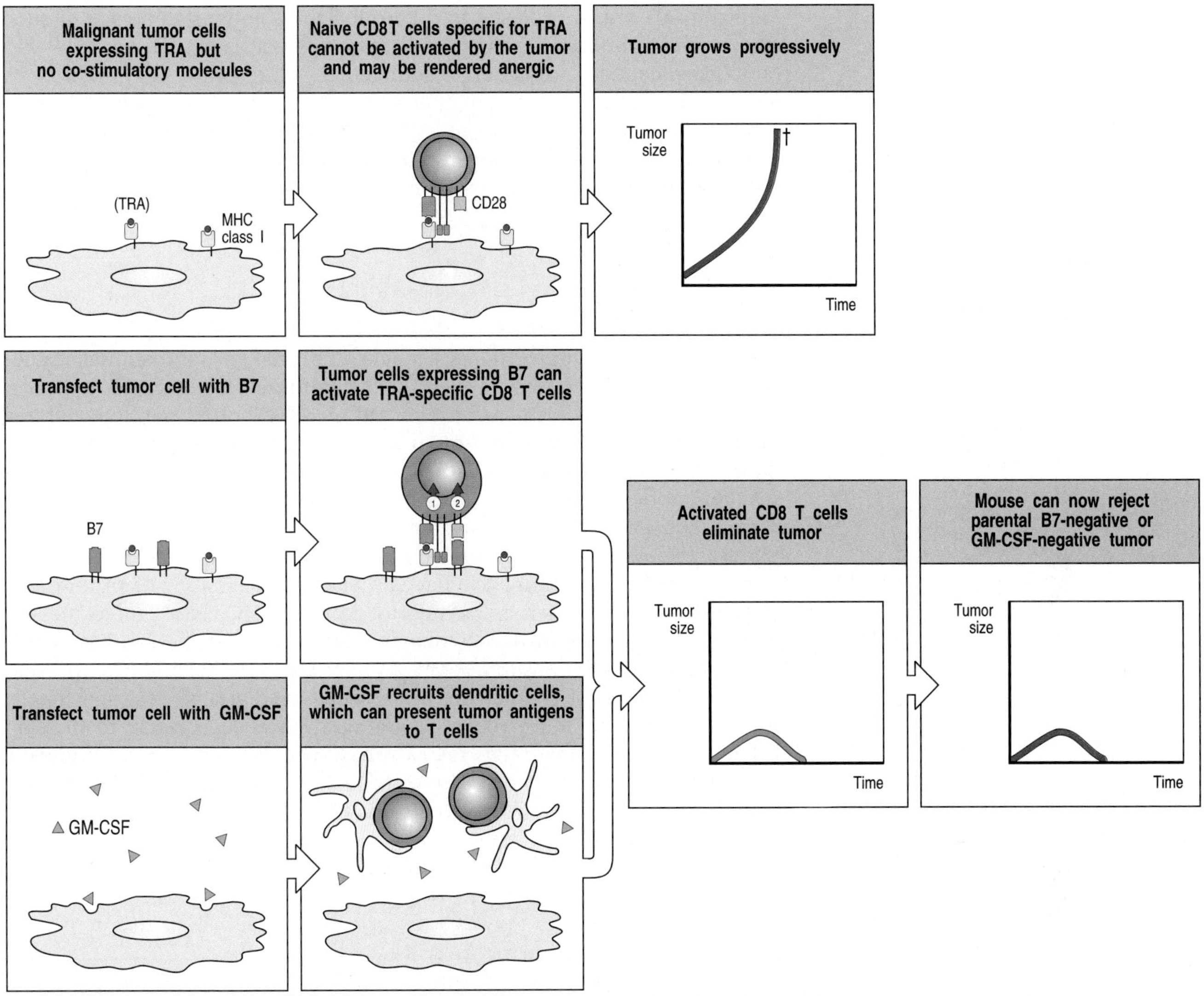

Fig. 14.19 Transfection of tumors with the gene for B7 or for GM-CSF enhances tumor immunogenicity. A tumor that does not express co-stimulatory molecules will not induce an immune response, even though it might express tumor rejection antigens (TRAs), because naive CD8 T cells specific for the TRA cannot be activated by the tumor. The tumor therefore grows progressively in normal mice (top panels). If such tumor cells are transfected with a co-stimulatory molecule, such as B7, TRA-specific CD8 T cells now receive both signal 1 and signal 2 from the same cell (see Section 8-4) and can therefore be activated (center panels). The same effect can be obtained by transfecting the tumor with the gene encoding GM-CSF, which attracts and stimulates the differentiation of dendritic cell precursors (bottom panels). Both of these strategies have been tested in mice and shown to elicit memory T cells, although results with GM-CSF are more impressive. Because TRA-specific CD8 cells have now been activated, even the original B7-negative or GM-CSF negative tumor cells can be rejected. † Death of animal.

attracts hematopoietic precursors to the site and induces their differentiation into dendritic cells. It is believed that these cells process the tumor antigens and migrate to the local lymph nodes where they induce potent anti-tumor responses. The B7-transfected cells seem less potent in inducing anti-tumor responses, perhaps because the bone marrow derived dendritic cells express more of the molecules required to activate naive T cells than do B7-transfected tumor cells (see Fig. 14.19). The use of antigen-pulsed dendritic cells provides a further strategy for developing therapeutically useful cytotoxic T-cell responses to tumors in experimental models.

Clinical trials are in progress to determine the safety and efficacy of such approaches in human patients. What is not certain is whether people with established cancers can generate sufficient T-cell responses to eliminate all their tumor cells under circumstances in which any tumor-specific naive T cells might have been rendered tolerant to the tumor. Moreover, there is always the risk that immunogenic transfectants will elicit an autoimmune response against the normal tissue from which the tumor derived.

Summary.

Tumors represent outgrowths of a single abnormal cell, and animal studies have shown that some tumors elicit specific immune responses that suppress their growth. These seem to be directed at MHC-bound peptides derived from antigens that might be mutated, inappropriately expressed or over-expressed in the tumor cells. T-cell deficient individuals, however, do not develop more tumors than normal individuals. This is probably chiefly because most tumors do not make distinctive antigenic proteins and do not express the co-stimulatory molecules necessary to initiate an adaptive immune response. Tumors also have other ways of avoiding or suppressing immune responses, such as ceasing to express MHC class I molecules, or making immunosuppressive cytokines. Monoclonal antibodies have been developed for tumor immunotherapy by conjugation to toxins or to cytotoxic drugs or radionuclides, which are thereby delivered at high dose specifically to the tumor cells. More recently, attempts have been made to develop vaccines based on tumor cells taken from patients and made immunogenic by the addition of adjuvants. This approach has been extended in animal experiments to transfection of tumor cells with genes encoding co-stimulatory molecules or cytokines.

Manipulating the immune response to fight infection.

Infection is the leading cause of death in human populations. The two most important contributions to public health in the past 100 years have been sanitation and vaccination, which together have dramatically reduced deaths from infectious disease. Modern immunology grew from the success of Jenner's and Pasteur's vaccination against smallpox and cholera, respectively, and its greatest triumph has been the global eradication of smallpox, announced by the World Health Organization in 1980. A global campaign to eradicate polio is now under way.

Induced immunity to an infectious agent can be achieved in several ways. One early strategy was to deliberately cause a mild infection with the unmodified pathogen. This was the principle of variolation, in which the inoculation of a small amount of dried material from a smallpox pustule would cause a mild infection followed by long-lasting protection against re-infection. However, infection following variolation was not always mild: fatal smallpox ensued in about 3% of cases, which would not meet modern criteria for safety. Jenner's achievement was the realization that infection with a bovine analog of smallpox, vaccinia (from *vacca*—a cow), which caused cowpox, would provide protective immunity against smallpox in

Current immunization schedule for children (USA)

Vaccine given	2 months	4 months	6 months	15 months	18 months	24 months	4–6 years	14–15 years
Diphtheria–pertussis–tetanus (DPT)	■	■	■		■		■	
Trivalent oral polio (TVOP)	■	■			■		■	
Measles				■				
Rubella				■				
Mumps				■				
Haemophilus B polysaccharide						■		
Diphtheria–tetanus toxoids								■

Fig. 14.20 Childhood vaccination schedules (in red) in the USA.

humans, without the risk of significant disease. He named the process vaccination, and Pasteur, in his honour, extended the term to the stimulation of protection to other infectious agents. Humans are not a natural host of vaccinia, which establishes only a brief and limited subcutaneous infection but contains antigens that stimulate an immune response that is cross-reactive with smallpox antigens and thereby confers protection from the human disease.

This established the general principles of safe and effective vaccination, and vaccine development in the early part of the 20th century followed two empirical pathways: first, the search for attenuated organisms with reduced pathogenicity that would stimulate protective immunity; and second, the development of vaccines based on killed organisms and subsequently purified components of organisms that would be as effective as live whole organisms because any live vaccine, including vaccinia, can cause lethal systemic infection in the immunosuppressed.

Immunization is now considered so safe and so important that most states in the USA require all children to be immunized against measles, mumps and polio viruses with live attenuated vaccines, as well as against tetanus (caused by *Clostridium tetani*), diphtheria (caused by *Corynebacterium diphtheriae*), and whooping cough (caused by *Bordetella pertussis*), with inactivated toxins or toxoids prepared from these bacteria (see Fig. 1.32). More recently, a vaccine has become available against *Haemophilus* B, one of the causative agents of meningitis. Current vaccination schedules for children in the USA are shown in Fig. 14.20. Impressive as these accomplishments are, there are still many diseases for which we lack effective vaccines, as shown in Fig. 14.21. Even where a vaccine such as measles or polio can be used effectively in developed countries, technical and economic problems can prevent its widespread use in developing countries, where mortality from these diseases is still high. The development of vaccines therefore remains an important goal of immunology and the latter half of this century has seen a shift to a more rational approach, based on a detailed molecular understanding of microbial pathogenicity, analysis of the protective host response to pathogenic organisms, and the understanding of the regulation of the immune system to generate effective T- and B-lymphocyte responses.

Fig. 14.21 Diseases for which effective vaccines are still needed. *Current measles vaccines are effective but heat-sensitive, which makes their use difficult in tropical countries. Data from C J L Murray, and A D Lopez: Global Health Statistics: a compendium of incidence, prevalence and mortality estimates for over 200 conditions. Cambridge, Harvard University Press, 1996.

Some diseases for which effective vaccines are not yet available		
Disease	**Annual mortality**	**Annual incidence**
Malaria	856,000	213,743,000
Schistosomiasis	8,000	no numbers available
Worm infestation	22,000	no numbers available
Tuberculosis	1,960,000	6,346,000
Diarrhea	2,946,000	4,073,920,000
Respiratory disease	4,299,000	362,424,000
AIDS	138,000	411,000
Measles*	1,158,000	44,334,000

14-16 There are several requirements for an effective vaccine.

The specific requirements for successful vaccination vary according to the nature of the infecting organism. For extracellular organisms, antibody provides the most important adaptive mechanism of host defense, whereas for control of intracellular organisms an effective CD8 T-lymphocyte response is also essential. The ideal vaccination provides host defense at the point of entry of the infectious agent; stimulation of mucosal immunity is therefore an important goal of vaccination against those many organisms that enter through mucosal surfaces.

Effective protective immunity against some organisms requires the presence of pre-existing antibody at the time of exposure to the infection. For example, the clinical manifestations of tetanus and diphtheria are entirely due to the effects of extremely powerful exotoxins. Pre-existing antibody against the bacterial exotoxin is necessary to provide a defense against these diseases. Pre-existing antibodies are also required to protect against some intracellular pathogens, such as the poliomyelitis virus, which infect critical host cells within a short period after entering the body and are not easily controlled by T lymphocytes once intracellular infection is established.

Immune responses to infectious agents usually involve antibodies directed at multiple epitopes and only some of these antibodies confer protection. The particular T-cell epitopes recognized can also affect the nature of the response. For example, as we saw in Chapter 11, the predominant epitope recognized by T cells after vaccination with respiratory syncytial virus induces a vigorous inflammatory response but fails to elicit neutralizing antibodies and thus causes pathology without protection. Thus, an effective vaccine must lead to the generation of antibodies and T cells directed at the correct epitopes of the infectious agent. For some of the modern vaccine techniques, in which only one or a few epitopes are used, this consideration is particularly important.

A number of very important additional constraints need to be satisified by a successful vaccine (Fig. 14.22). First, it must be safe. Vaccines must be given to huge numbers of people, relatively few of whom are likely to die of, or sometimes even catch, the disease that the vaccine is designed to prevent, and this means that even a low level of toxicity is unacceptable. Second, the vaccine must be able to produce protective immunity in a very high proportion of the people to whom it is given. Third, because it is impracticable to give large or dispersed rural populations regular 'booster' vaccinations, a successful vaccine must generate long-lived immunological memory. This means that both B and T lymphocytes must be primed by the vaccine. Fourth, vaccines must be very cheap if they are to be administered to large populations. Vaccines are one of the most cost-effective measures in health care but this benefit is eroded as the cost-per-dose rises.

An effective vaccination program provides herd immunity—by lowering the number of susceptible members of a population, the natural reservoir of infected individuals in that population falls, reducing the probability of transmission of infection. Thus, even non-vaccinated members of a population can be protected from infection if the majority are vaccinated.

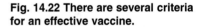

Features of effective vaccines	
Safe	Vaccine must not itself cause illness or death
Protective	Vaccine must protect against illness resulting from exposure to live pathogen
Gives sustained protection	Protection against illness must last for several years
Induces neutralizing antibody	Some pathogens (such as poliovirus) infect cells that cannot be replaced (eg neurons). Neutralizing antibody is essential to prevent infection of such cells
Induces protective T cells	Some pathogens, particularly intracellular, are more effectively dealt with by cell-mediated responses
Practical considerations	Low cost-per-dose Biological stability Ease of administration Few side-effects

Fig. 14.22 There are several criteria for an effective vaccine.

14-17 The history of vaccination against *Bordetella pertussis* illustrates the importance of developing an effective vaccine that is perceived to be safe.

The history of vaccination against the bacterium that causes whooping cough, *Bordetella pertussis,* provides a good example of the challenges of developing and disseminating an effective vaccine. At the turn of the 20th century, whooping cough killed approximately 0.5% of American children under the age of 5 years. In the early 1930s, a trial of a killed, whole bacterial cell vaccine on the Faröe Islands provided evidence of a protective effect. In the USA, systematic use of a whole-cell vaccine in combination with diphtheria and tetanus toxoids (the DPT vaccine) since the 1940s resulted in a decline in the annual infection rate from 200 to less than 2 cases per 100,000 of the population. First vaccination with DPT was typically given at the age of 3 months.

Whole-cell pertussis vaccine causes side-effects, typically redness, pain and swelling at the site of the injection; less commonly, vaccination is followed by high temperature and persistent crying. Very rarely, fits and a short-lived sleepiness or a floppy unresponsive state ensue. During the 1970s, widespread concern developed after several anecdotal observations that encephalitis leading to irreversible brain damage might very rarely follow pertussis vaccination. In Japan, in 1972, approximately 85% of children were given the pertussis vaccine, and fewer than 300 cases of whooping cough and no deaths were reported. As a result of two deaths after vaccination in Japan in 1975, DPT was temporarily suspended and then reintroduced with the first vaccination at 2 years of age rather than 3 months. In 1979 there were approximately 13,000 cases of whooping cough and 41 deaths. The possibility that pertussis vaccine very rarely causes severe brain damage has been studied extensively and expert consensus is that pertussis vaccine is not a primary cause of brain injury. There is no doubt that there is greater morbidity from whooping cough than from the vaccine.

The public and medical perception that whole-cell pertussis vaccination may be unsafe provided a powerful incentive to develop safer pertussis vaccines. Study of the natural immune response to *B. pertussis* showed that infection induced antibodies against four components of the bacterium—pertussis toxin, filamentous hemagglutinin, pertactin, and fimbrial antigens. Immunization of mice with these antigens in purified form protected them against challenge with pertussis. This has led to the development of acellular pertussis vaccines, all of which contain purified pertussis toxoid, that is, toxin inactivated by chemical treatment, for example with hydrogen peroxide or formaldehyde, or more recently by genetic engineering of the toxin. Some also contain one or more of the filamentous hemagglutinin, pertactin, and fimbrial antigens. Current evidence shows that these are probably as effective as whole-cell pertussis vaccine and are free of the common minor side-effects of the whole-cell vaccine.

The main messages of the history of pertussis vaccination are: first, that vaccines must be extremely safe and free of side-effects; second, that the public and the medical profession must perceive the vaccine to be safe; and third, that careful study of the nature of the protective immune response can lead to acellular vaccines that are safer than and as effective as whole-cell vaccines.

14-18 Conjugate vaccines have been developed as a result of understanding how T and B cells collaborate in an immune response.

Although acellular vaccines are inevitably safer than vaccines based on whole organisms, a fully effective vaccine cannot normally be made from a single isolated constituent of a microorganism, and it is now clear that this is because of the need to activate more than one cell type to initiate an immune response. One consequence of this insight has been the development of conjugate vaccines. We have already described briefly one of the most important of these in Section 9-2.

Many bacteria, including *Neisseria meningitidis* (meningococcus), *Streptococcus pneumoniae* (pneumococcus), and *Haemophilus* spp., have an outer capsule composed of polysaccharides that are species- and type-specific for particular strains of bacteria. The most effective defense against these microorganisms is opsonization of the polysaccharide coat with antibody. The aim of vaccination is therefore to elicit antibodies against the polysaccharide capsules of the bacteria.

Capsular polysaccharides can be harvested from bacterial growth medium and because they are T-independent antigens they can be used on their own as vaccines. However, young children under the age of two cannot make good T-cell independent responses and cannot be vaccinated effectively with polysaccharide vaccines. An efficient way of overcoming this problem (see Fig. 9.4) is to chemically conjugate bacterial polysaccharides to protein carriers, which provide peptides for activating T cells and convert a T-cell independent response into a T-cell dependent anti-polysaccharide response. By using this approach, various conjugate vaccines have been developed against *Haemophilus influenzae*, an important cause of serious childhood chest infections and meningitis, and these are now widely applied.

14-19 The use of adjuvants is another important approach to enhancing the immunogenicity of vaccines.

Even conjugate vaccines are not usually strongly immunogenic on their own: most require the addition of adjuvants, which we defined in Section 2-4 as substances that enhance the immunogenicity of antigens. It is thought that most, if not all, adjuvants act on antigen-presenting cells and reflect the importance of these cells in initiating immune responses. *H. influenzae* polysaccharides, for example, can conveniently be conjugated to tetanus toxoid because infants are vaccinated routinely with this protein and their T cells are already primed against it. However, tetanus toxoid is not immunogenic in the absence of adjuvants, and tetanus toxoid vaccines often contain aluminum salts, which bind polyvalently to the toxoid by ionic interactions and selectively stimulate antibody responses. Pertussis toxin, produced by *B. pertussis,* has adjuvant properties in its own right and, when given mixed with tetanus and diphtheria toxoids, not only vaccinates against whooping cough but also acts as an adjuvant for the other two toxoids. This mixture comprises the DPT triple vaccine given to infants in the first year of life.

Several small molecules, for example muramyl dipeptide extracted from the mycobacterial cell wall, also act as adjuvants. Their mechanism of action is, however, unknown. They might act by stimulating the expression of co-stimulatory activity in antigen-presenting cells or by mimicking co-stimulatory signals in T cells. Alternatively, they might enhance uptake of the antigen by dendritic cells that already express co-stimulatory molecules. Other adjuvants stimulate mucosal immune responses, which are particularly important in defense against organisms entering through the digestive or respiratory tracts. These adjuvants will be discussed later when we describe strategies for stimulating mucosal immunity.

Yet another approach to enhancing the effectiveness of vaccines is to co-administer cytokines. For example, IL-12 is a cytokine produced by macrophages and B cells that stimulates T lymphocytes and NK cells to release IFN-γ and promotes a T_H1 response. It has been used as an adjuvant to promote protective immunity against the protozoan parasite *Leishmania major.* Certain strains of mice are susceptible to severe cutaneous and systemic infection by *L. major* and mount an immune response that is predominantly T_H2 in type and is ineffective in eliminating the organism. The co-administration of IL-12 with a vaccine containing leishmania antigens generated a T_H1 response and protected these mice against challenge with *L. major.* The use of IL-12 to promote a T_H1 response has also proved valuable in reducing the pathogenic consequences of experimental parasitic infection by *Schistosoma mansoni* and will be considered in Section 14-27. These are important examples of how an understanding of the regulation of immune responses can enable rational intervention to enhance the effectiveness of vaccines.

14-20 Live-attenuated viral vaccines are usually more potent than 'killed' vaccines and can be made safer by using recombinant DNA technology.

Most anti-viral vaccines currently in use consist of inactivated or live attenuated viruses. Inactivated, or 'killed' viral vaccines consist of viruses treated so that they are unable to replicate. Live-attenuated viral vaccines are generally far more potent, perhaps because they elicit a greater number of relevant effector mechanisms, including cytotoxic CD8 T cells: inactivated viruses cannot produce proteins in the cytosol, so peptides from the viral antigens cannot be presented by MHC class I molecules and thus cytotoxic CD8 T cells are not generated by these vaccines. Attenuated viral vaccines are now in use for polio, measles, mumps, rubella, and varicella.

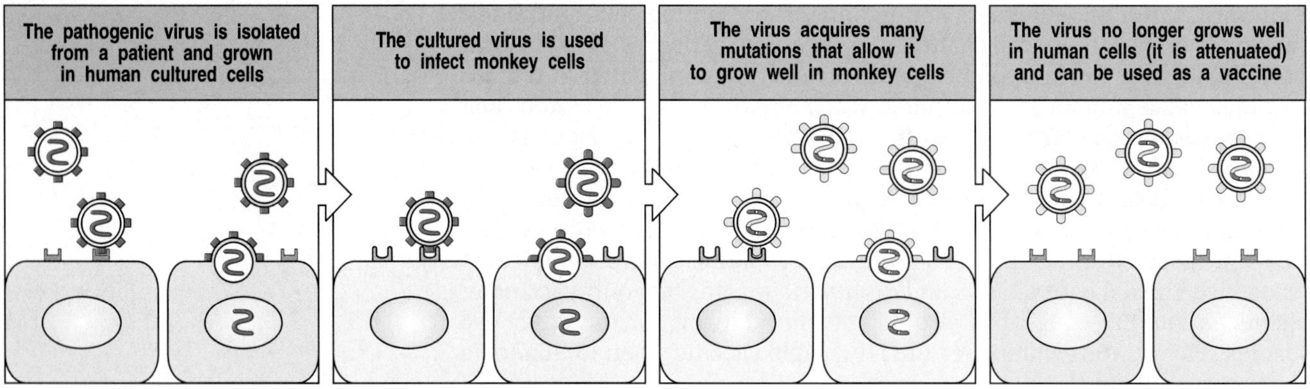
The pathogenic virus is isolated from a patient and grown in human cultured cells
The cultured virus is used to infect monkey cells
The virus acquires many mutations that allow it to grow well in monkey cells
The virus no longer grows well in human cells (it is attenuated) and can be used as a vaccine

Fig. 14.23 Viruses are traditionally attenuated by selecting for growth in non-human cells. To produce an attenuated virus, the virus must first be isolated by growing it in cultured human cells. The adaptation to growth in cultured human cells can cause some attenuation in itself; the rubella vaccine, for example, was made in this way. In general, however, the virus is then adapted to growth in cells of a different species, until it grows only poorly in human cells. The adaptation is a result of mutation, usually a combination of several point mutations. It is usually hard to tell which of the mutations in the genome of an attenuated viral stock are critical to attenuation. An attenuated virus will grow poorly in the human host, and will therefore produce immunity but not disease.

Traditionally, attenuation is achieved by growing the virus in cultured cells. Viruses are normally selected for preferential growth in non-human cells and, in the course of selection, become less able to grow in human cells (Fig. 14.23). Because these attenuated strains replicate poorly in human hosts, they induce immunity but not disease when given to people. Although attenuated virus strains contain multiple mutations in genes encoding several of their proteins, it might be possible for a pathogenic virus strain to re-emerge by a further series of mutations. For example, the type 3 Sabin polio vaccine strain differs at only 10 of 7429 nucleotides from a wild-type progenitor strain. On extremely rare occasions, reversion of the vaccine to a neurovirulent strain can occur, causing paralytic disease in the unfortunate recipient.

Attenuated organisms can also pose particular risks to immunodeficient recipients in whom they often behave as virulent opportunistic infections. Patients with immunoglobulin deficiencies show abnormal susceptibility to chronic infection by enteroviruses, and can develop chronic, and ultimately lethal, echovirus encephalitis. Infants who are vaccinated with live attenuated polio before their inherited immunoglobulin deficiences have been diagnosed are at risk because they cannot clear the virus from their gut, and there is therefore an increased chance that mutation of the virus will lead to fatal paralytic disease.

An empirical approach to attenuation is still in use but might be superseded by two new approaches that use recombinant DNA technology. One is the isolation and *in vitro* mutagenesis of specific viral genes. The mutated genes are used to replace the wild-type gene in a reconstituted virus genome, and this deliberately attenuated virus can then be used as a vaccine (Fig. 14.24). The advantage of this approach is that mutations can be engineered so that reversion to wild type is virtually impossible.

Such an approach might be useful in developing live influenza vaccines. As we learned in Chapter 11, the influenza virus can re-infect the same host several times, because it undergoes antigenic shift and thus escapes the original immune response. The current approach to vaccination against influenza is to use a killed virus vaccine that is re-formulated annually on the basis of the prevalent strains of virus. The vaccine is moderately effective, reducing mortality in elderly populations and morbidity in healthy adults. The ideal influenza vaccine would be an attenuated live organism delivered to the respiratory mucosal surface, where it would evoke a pattern of immune response typical of a natural infection. One strategy for obtaining such a vaccine is the introduction of a series of attenuating mutants into the gene encoding a viral polymerase protein, PB2. The mutated gene segment from the attenuated virus could then be substituted for the wild-type gene in a virus carrying the relevant hemagglutinin and neuraminidase antigenic variants of the current epidemic or pandemic strain, thus creating a new attenuated vaccine.

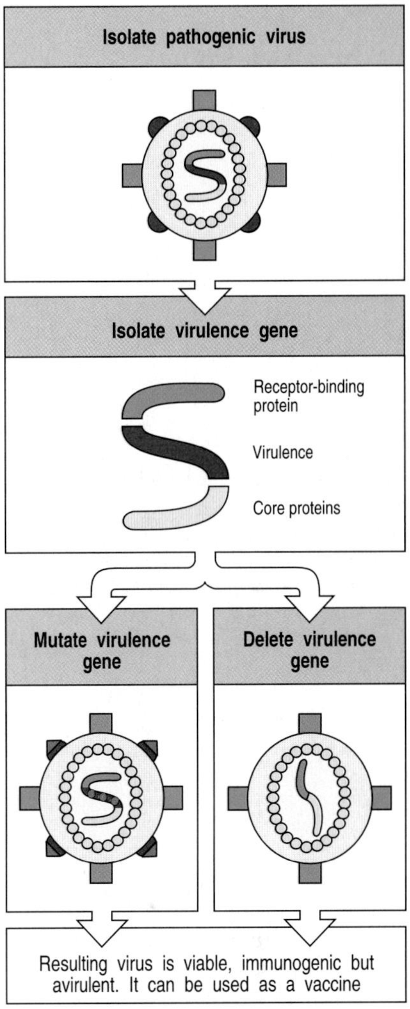

Fig. 14.24 Attenuation can be achieved more rapidly and reliably with recombinant DNA techniques. If a gene in the virus that is required for virulence but not for growth or immunogenicity can be identified, this gene can be either multiply mutated (left lower panel) or deleted from the genome (right lower panel) by using recombinant DNA techniques. This procedure creates an avirulent (non-pathogenic) virus that can be used as a vaccine. The mutations in the virulence gene are usually large, so that it is very difficult for the virus to revert to the wild type.

14-21 Live-attenuated bacterial vaccines can be developed by selecting non-pathogenic or disabled mutants.

Similar approaches are being used for bacterial vaccine development. *Salmonella typhi*, the causative agent of typhoid, has been manipulated to develop a live vaccine. A strain of wild-type bacteria was mutated by using nitrosoguanidine; a new strain was selected to be defective in the enzyme UDP-galactose epimerase blocking the pathway for synthesis of lipopolysaccharide, an important determinant of bacterial pathogenesis. Recent approaches to the rational design of attenuated *Salmonella* vaccines have involved the specific targeting of genes encoding enzymes in the biosynthetic pathways of amino acids containing aromatic rings, such as tyrosine and phenylalanine. Mutating these genes makes auxotrophic organisms, which are dependent for growth on an external supply of an essential nutrient that wild-type bacteria would be capable of biosynthesizing. These bacteria grow poorly in the gut but should survive long enough as a vaccine to induce an effective immune response.

It is not only vaccination of humans against *Salmonella* that is important. Modern methods of mass production of chickens for food has led to extensive infection of poultry with *Salmonella* strains that are pathogenic to humans and an increasingly important cause of food poisoning. Thus, in parts of the world where typhoid is prevalent, vaccinating humans has a high priority. In other parts, where food poisoning caused by *Salmonella typhimurium* and *S. enteritidis* infection is common, vaccination of chickens would contribute to public health.

14-22 Attenuated microorganisms can serve as vectors for vaccination against many pathogens.

An effective live-attenuated typhoid vaccine would not only be valuable in its own right but could also serve as a vector for presenting antigens from other organisms. Attenuated strains of *Salmonella* have been used as carriers of heterologous genes encoding tetanus toxoid and antigens from organisms as diverse as *Listeria monocytogenes, Bacillus anthracis, Leishmania major, Yersinia pestis* and *Schistosoma mansoni.* Each of these has been used as an oral vaccine to protect mice against experimental challenge with the respective pathogen.

Viral vectors can similarly be engineered to carry heterologous peptides or proteins from other microorganisms. Although vaccinia is no longer needed to protect against the development of smallpox, it remains a candidate as an avirulent carrier of heterologous antigens. It has been proposed that genes encoding protective antigens from several different organisms could be placed in a single vaccine strain. This approach makes it possible to immunize individuals against several different pathogens at once, but such a vaccine could not be used twice because the vaccinia vector itself generates long-lasting immunity that would neutralize its effectiveness on a second administration; this is an example of the phenomenon called 'original antigenic sin' (see Fig. 10.41). The development of successful heterologous vaccines requires the identification of protective antigens; it therefore depends on the analytical power of recombinant DNA methods, as well as their use to manipulate gene structure.

14-23 Synthetic peptides of protective antigens can elicit protective immunity.

One route to vaccine development is the identification of the T-cell peptide epitopes that stimulate protective immunity. This can be approached in two ways. One possibility is to synthesize systematically overlapping peptides from immunogenic proteins and to test each in turn for its ability to stimulate protective immunity. An alternative, but no less arduous approach, 'reverse' immunogenetics, has been used in developing a vaccine against malaria (Fig. 14.25).

The immunogenicity of T-cell peptide epitopes depends on their specific associations with particular polymorphic variants of MHC molecules, and the starting point for the studies on malaria was an association between the human class I molecule HLA-B53 and resistance to cerebral malaria—a relatively infrequent complication of infection that is usually fatal. The hypothesis is that these MHC molecules are protective because they present peptides that are particularly good at evoking cytotoxic T lymphocytes. A direct route to identifying the relevant peptides is is to elute them from MHC molecules of cells infected with the pathogen. In HLA-B53, a high proportion of the peptides eluted had proline in the second of nine positions; this information was used to identify candidate protective peptides from four proteins of *Plasmodium falciparum* expressed in the early phase of hepatocyte infection, an important phase of infection to target in an effective immune response. One of the candidate peptides, from liver stage antigen-1, is recognized by cytotoxic T cells when bound to HLA-B53.

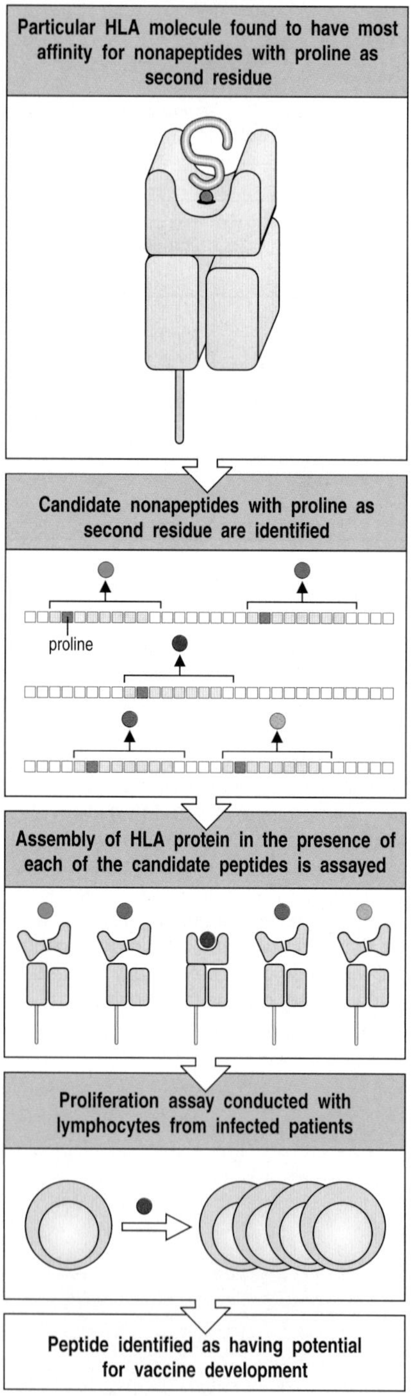

Particular HLA molecule found to have most affinity for nonapeptides with proline as second residue
Candidate nonapeptides with proline as second residue are identified
proline
Assembly of HLA protein in the presence of each of the candidate peptides is assayed
Proliferation assay conducted with lymphocytes from infected patients
Peptide identified as having potential for vaccine development

Fig. 14.25 'Reverse' immunogenetics can be used to identify protective T-cell epitopes against infectious diseases. Population studies show that the MHC class I variant HLA-B53 is associated with resistance to cerebral malaria. Self nonapeptides were eluted from HLA-B53 and found to have a strong preference for proline at the second position (top panel). Candidate nonapeptide sequences containing proline at position 2 were then identified in several malarial protein sequences (second panel) and synthesized. These synthetic nonapeptides were then tested to see whether they fitted well into the peptide groove of HLA-B53 by assaying whether HLA-B53 would assemble to form a stable cell surface heterodimer in the presence of peptide (third panel). The final panel shows the testing of peptide sequences identified by this approach to see whether they would induce the proliferation of T cells from patients infected by malaria. Such sequences are good candidates for incorporation into vaccines.

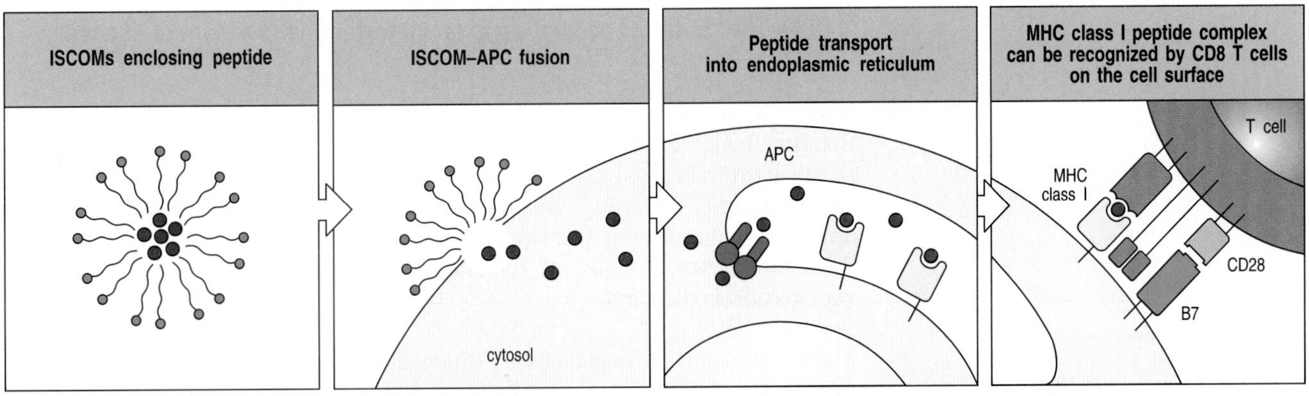

Fig. 14.26 ISCOMs can be used to deliver peptides to the MHC class processing pathway. ISCOMs (immune stimulatory complexes) are lipid micelles that will fuse with cell membranes. Peptides trapped in ISCOMs can be delivered to the cytosol of an antigen-presenting cell (APC), allowing the peptide to be transported into the endoplasmic reticulum, where it can be bound by newly synthesized MHC class I molecules and hence transported to the cell surface as peptide:MHC class I complexes. This is a possible means for delivering vaccine peptides to activate CD8 cytotoxic T cells. ISCOMs can also be used to deliver proteins to the cytosol of other types of cell, where they can be processed and presented as though they were a protein produced by the cell.

This approach is being extended to other class I and class II molecules associated with protective immune responses against infection. Recently, a protective peptide epitope was eluted from MHC class II molecules in *Leishmania*-infected macrophages and used as a guide to isolate the gene from *Leishmania.* The gene was then used to make a protein-based vaccine that primed mice from susceptible strains for responses to *Leishmania.*

These results show considerable promise but they also illustrate one of the major drawbacks to the approach. A malaria peptide that is restricted by HLA-B53 might not be immunogenic in an individual lacking HLA-B53: indeed, this presumably accounts for the higher susceptibility of these individuals to natural infections. Because of the very high polymorphism of MHC molecules in humans it will be necessary to identify panels of protective T-cell epitopes and construct vaccines containing arrays of these to develop vaccines that will protect the majority of a susceptible population.

There are other problems with peptide vaccines. Peptides are not strongly immunogenic and it is particularly difficult to generate MHC class I-specific responses by *in vivo* immunization with peptides. One approach to this problem is to integrate peptides by genetic engineering into carrier proteins such as hepatitis B core antigen, which are then processed *in vivo* via natural antigen-processing pathways. A second possible technique is the use of ISCOMs (see Fig. 2.4). **ISCOMs** (immune stimulatory complexes) are lipid carriers that act as adjuvants but have minimal toxicity. They seem to load peptides and proteins into the cell cytoplasm, allowing MHC class I-restricted T-cell responses to peptides (Fig. 14.26). These carriers are being developed for use in human immunization.

A novel source of vaccines to deliver protective peptides is the use of plant viruses, which are non-pathogenic to humans. These can be engineered to incorporate relevant peptide antigens into chimeric coat proteins. With this approach, mice have been protected against lethal challenge with rabies virus by prior feeding with spinach leaves infected by recombinant alfalfa mosaic virus incorporating a rabies virus peptide. Popeye may need rejuvenation as a role model to encourage the eating of spinach by the young.

14-24 The route of vaccination is an important determinant of success.

Most vaccines are given by injection. This route has two disadvantages, the first practical, the second immunological. Injections are painful and expensive, requiring needles, syringes and a trained injector. They are unpopular with the recipient, reducing vaccine uptake, and mass vaccination by this approach is laborious. The immunological drawback is that injection is not the usual route of entry of the majority of pathogens against which the vaccination is directed.

Many important pathogens infect mucosal surfaces or enter the body through mucosal surfaces. Examples include respiratory microorganisms such as *Bordetella pertussis*, rhinoviruses and influenza viruses, and enteric microorganisms such as *Vibrio cholerae*, *Salmonella typhi*, enteropathogenic *Escherichia coli* and *Shigella*. The enteric microorganisms are particularly important pathogens in underdeveloped countries. It is therefore important to understand how these organisms stimulate mucosal immunity and to develop vaccines that can be administered to the mucosa orally or by nasal inhalation.

The power of this approach is illustrated by the effectiveness of live-attenuated polio vaccines. The Sabin polio vaccine consists of three attenuated polio virus strains and is highly immunogenic. Moreover, just as polio itself can be transmitted by fecal contamination of public swimming pools and other failures of hygiene, the vaccine can be transmitted from one individual to another by the orofecal route. Infection with *Salmonella* likewise stimulates a powerful mucosal and systemic immune response and, as we saw in Section 14-21, has been attenuated for use as a vaccine and carrier of heterologous antigens for presentation to the mucosal immune system.

The rules of mucosal immunity are poorly understood. On the one hand, presentation of soluble protein antigens by the oral route often results in tolerance, which is important given the enormous load of food and airborne antigens presented to the gut and respiratory tract. As discussed in Sections 14-10 and 13-26, the ability to induce tolerance by oral or nasal administration of antigens is being explored as a therapeutic mechanism for reducing unwanted immune responses. On the other hand, the mucosal immune system is capable of responding to and eliminating mucosal infections such as pertussis, cholera, and polio. The proteins from these organisms that stimulate immune responses are therefore of special interest. One group of powerfully immunogenic proteins at mucosal surfaces is a series of bacterial toxins that have the property of binding to eukaryotic cells and are protease-resistant. A recent finding of potential practical importance is that certain of these molecules, such as the *E. coli* heat-labile toxin and pertussis toxin, have adjuvant properties that are retained even when the parent molecule has been engineered to eliminate its toxic properties. These molecules can be used as adjuvants for oral or nasal vaccines. In mice, nasal insufflation of either of these mutant toxins together with tetanus toxoid resulted in the development of protection against lethal challenge with tetanus toxin.

14-25 Protective immunity can be induced by injecting DNA encoding microbial antigens and human cytokines into muscle.

The latest development in vaccination has come as a surprise even to the scientists who initially developed the method. The story begins with attempts to use non-replicating bacterial plasmids encoding proteins for gene therapy:

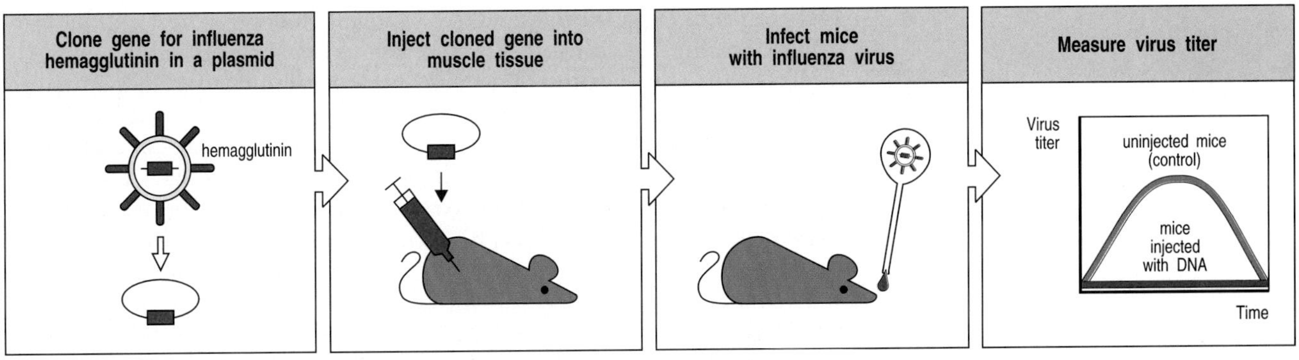

Fig. 14.27 DNA vaccination by injection of DNA encoding a protective antigen and cytokines directly into muscle. Influenza hemagglutinin contains both B- and T-cell epitopes. When a naked DNA plasmid containing the gene for this protein is injected directly into muscle, an influenza-specific immune response consisting of both antibody and cytoxic CD8 T cells results. The response can be enhanced by including a plasmid encoding GM-CSF. Presumably, the plasmid DNAs are expressed by some of the cells in the muscle tissue into which it was injected, allowing an immune response that involves both antibody and cytotoxic T cells. The details of this process are not yet understood.

proteins expressed *in vivo* from these plasmids were found to stimulate an immune response. When DNA encoding a viral immunogen is injected intramuscularly, it leads to the development of antibody responses and cytotoxic T cells that allow the mice to reject a later challenge with whole virus (Fig. 14.27). This response does not appear to damage the muscle tissue, is safe and effective and, because it uses only a single microbial gene, does not carry the risk of active infection. This procedure has been termed 'DNA vaccination.' DNA coated onto minute metal projectiles can be administered by 'biolistic' (biological ballistic) gun, so that several metal particles penetrate the skin and enter the muscle beneath. This technique has been shown to be effective in animals and might be suitable for mass immunization, although it has yet to be tested in humans. Mixing in plasmids that encode cytokines such as GM-CSF makes immunization with genes encoding protective antigens much more effective, as was seen earlier for tumor immunity.

It is not yet clear how DNA vaccination works. Why is plasmid DNA effectively expressed in muscle tissue? Do the muscle cells elicit the immune response, or do tissue dendritic cells take up the DNA and express it? How do lymphocytes encounter the antigen if it is expressed in muscle cells? How safe is the approach, and how generally applicable will it be?

14-26 The effectiveness of a vaccine can be enhanced by targeting it to sites of antigen presentation.

An important way of enhancing the effectiveness of a vaccine is to target it efficiently to antigen-presenting cells. This is an important mechanism of action of vaccine adjuvants. There are three complementary approaches. The first is to prevent proteolysis of the antigen on its way to antigen-presenting cells. Preserving antigen structure is an important reason why so many vaccines are given by injection rather than by the oral route, which exposes the vaccine to digestion in the gut. The second and third approaches are to target the vaccine selectively, once in the body, to antigen-presenting cells and to devise methods of engineering the selective uptake of the vaccine into antigen-processing pathways within the cell.

Techniques to enhance the uptake of antigens by antigen-presenting cells include coating the antigen with mannose to enhance uptake by mannose receptors on antigen-presenting cells, and presenting the antigen as an immune complex to take advantage of antibody and complement binding by Fc and complement receptors. The effects of DNA vaccination have been enhanced experimentally by injecting DNA encoding antigen coupled to CTLA-4, which enables the selective binding of the expressed protein to antigen-presenting cells carrying B7, the receptor for CTLA-4 (see Section 8-4).

A more complicated set of strategies involves targeting vaccine antigens selectively into antigen-presenting pathways within the cell. For example, human papillomavirus E7 antigen has been coupled to the signal peptide that targets a lysosomal-associated membrane protein to lysosomes and endosomes. This directs the E7 antigen directly to the intracellular compartments in which antigens are cleaved to peptides before binding to MHC class II (see Section 4-11). A vaccinia virus incorporating this chimeric antigen induced a greater response in mice to E7 antigen than did vaccinia incorporating wild-type E7 antigen alone. A second approach is the use of ISCOMs, which seem to encourage the entry of peptides into the cytoplasm, thus enhancing the loading of peptides onto MHC class I molecules (see Section 14-23).

An improved understanding of the mechanisms of mucosal immunity has led to the development of techniques to target antigens to M cells overlying Peyer's patches (see Fig. 1.10). These specialized epithelial cells lack the mucin barrier and digestive properties of other mucosal epithelial cells. Instead, they can bind and endocytose macromolecules and microorganisms, which are transcytosed intact and delivered to the underlying lymphoid tissue. In view of these properties, it is not surprising that some pathogens target M cells to gain entry to the body. The counter-attack by immunologists is to gain a detailed molecular understanding of this mechanism of bacterial pathogenesis and subvert it as a delivery system for vaccines. For example, the outer membrane fimbrial proteins of *Salmonella typhimurium* have a key role in the binding of these bacteria to M cells. It might be possible to use these fimbrial proteins or, ultimately, just their binding motifs, as targeting agents for vaccines. A related strategy to encourage the uptake of mucosal vaccines by M cells is to encapsulate antigens in particulate carriers that are taken up selectively by M cells.

14-27 An important question is whether vaccination can be used therapeutically to control existing chronic infections.

There are many chronic diseases in which infection persists because of a failure of the immune system to eliminate disease. These can be divided into two groups, those infections in which there is an obvious immune response that fails to eliminate the organism, and those in which the infection seems to be invisible to the immune system and evokes a barely detectable immune response.

In the first category, the immune response is often partly responsible for the pathogenic effects. Infection by the helminth *Schistosoma mansoni* is associated with a powerful T_H2-type response, characterized by high IgE levels, circulating and tissue eosinophilia and a harmful fibrotic response to schistosome ova, leading to hepatic fibrosis. Other common parasites, such as *Plasmodium* and *Leishmania* species, cause damage because they are not eliminated effectively by the immune response in many patients.

Mycobacteria causing tuberculosis and leprosy cause persistent intracellular infection; a T_H1 response helps to contain these infections but also causes granuloma formation and tissue necrosis. Among viruses, hepatitis B and hepatitis C infections are commonly followed by persistent viral carriage and hepatic injury, resulting in ultimate death from hepatitis or from hepatoma. HIV infection, as we have seen in Chapter 11, is only followed very rarely by viral clearance and protection against subsequent re-infection.

There is a second category of chronic infection, predominantly viral, in which the immune response fails to clear infection because of the relative invisibility of the infectious agent to the immune system. A good example is herpes simplex type 2, which is transmitted venereally, becomes latent in nerve tissue, and causes genital herpes, which is frequently recurrent. This invisibility seems to be caused by a viral protein, ICP-47, which binds to the TAP complex and inhibits peptide transport into the endoplasmic reticulum in infected cells (see Chapter 4). Thus viral peptides are not presented to the immune system by MHC class I molecules. Another example in this category of chronic infection is genital warts, caused by certain papilloma viruses, to which very little immune response is evoked.

There are two main immunological approaches to the treatment of chronic infection. One is to try to boost or change the pattern of the host immune response by using cytokine therapy. The second is to attempt therapeutic vaccination to see whether the host immune response can be supercharged by immunization with antigens from the infectious agent in combination with adjuvant. There has been substantial pharmaceutical investment in therapeutic vaccination but it is too early to know whether the approach will be successful.

Some promise for the cytokine therapy approach comes from the experimental treatment of leprosy: one can clear certain leprosy lesions by the injection of cytokines directly into the lesion and cause reversal of the type of leprosy seen. Another example in which cytokine therapy has been shown to be effective in treating an established infection depends on combining a cytokine with an anti-parasitic drug. In a proportion of mice infected with *Leishmania* and subsequently treated with a combination of drug therapy and IL-12, the immune response deviated from a T_H2 to a T_H1 pattern and the infection was cleared. In most of the animal studies, however, it seems that the anti-cytokine antibody or the cytokine needs to be present at the first encounter with the antigen to modulate the response effectively. For example, in experimental leishmaniasis in mice, susceptible BALB/c mice injected with anti-IL-4 antibody at the time of infection clear their infection (Fig. 14.28). However, if administration of anti-IL-4 is delayed by just one week, there is progressive growth of the parasite and a dominant T_H2 response.

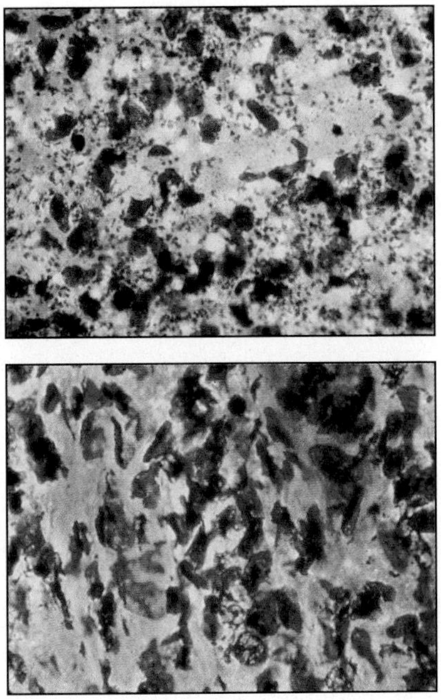

Fig. 14.28 Treatment with anti-IL-4 antibody at the time of infection with *Leishmania major* allows normally susceptible mice to clear the infection. The top panel shows a hematoxylin–eosin stained section through the footpad of a mouse of the Balb/c strain infected with *Leishmania major*. Large numbers of parasites are present in tissue macrophages. The bottom panel shows a similar preparation from a mouse infected in the same experiment but simultaneously treated with a single injection of anti-IL-4 monoclonal antibody. Very few parasites are present. Photographs courtesy of R M Locksley.

14-28 Modulation of the immune system might be used to inhibit immunopathological responses to infectious agents.

We have mentioned several times the possibility of modulating immunity by cytokine manipulation of the immune response. This approach is being explored as a means of inhibiting harmful immune responses to a number of important infections. As we have seen in the preceding section, the pathogenesis of the liver fibrosis in schistosomiasis results from the powerful T_H2-type response. The co-adminstration of *S. mansoni* ova together with IL-12 does not protect mice against subsequent infection with *S. mansoni* cercariae but has a striking effect in reducing hepatic granuloma formation and fibrosis in response to ova. IgE levels are reduced, with reduced tissue eosinophilia, and the cytokine response indicates the activation of T_H1 rather than T_H2 cells.

Although these results indicate that it might be possible to use a combination of antigen and cytokines to vaccinate against the pathology of diseases for which a fully protective vaccine is unavailable, they do not solve the difficulty of applying this approach in patients whose infection is already established.

Summary.

The greatest triumphs of modern immunology have come from vaccination, which has eradicated or virtually eliminated several human diseases. It is the single most successful manipulation of the immune system so far, because it takes advantage of the immune system's natural specificity and inducibility. Nevertheless, there are many important infectious diseases for which there is still no effective vaccine. The most effective vaccines are based on attenuated live microorganisms, but these carry some risk and are potentially lethal to immunosuppressed or immunodeficient individuals. Better techniques for developing live attenuated vaccines, or vaccines based on immunogenic components of pathogens, are therefore being sought. Most current viral vaccines are based on live attenuated virus, but many bacterial vaccines are based on components of the microorganism, including components of the toxins that it produces. The immunogenicity of these components can be enhanced by conjugation to a protein. Vaccines based on peptide epitopes are still at an experimental stage and have the problem that the peptide is likely to be specific for particular variants of the MHC molecules to which they must bind, as well as being only very weakly immunogenic. A vaccine's immunogenicity often depends on adjuvants that can help, directly or indirectly, to activate antigen-presenting cells that are necessary for the initiation of immune responses. The development of oral vaccines is particularly important for stimulating immunity to the many pathogens that enter through the mucosa. Cytokines have been used experimentally as adjuvants to boost the immunogenicity of vaccines or to bias the immune response along a specific path.

Summary to Chapter 14.

One of the great future challenges of immunology is the control of the immune response, so that unwanted immune responses can be suppressed and desirable responses elicited. Current methods of suppressing unwanted responses rely, to a great extent, on drugs that suppress adaptive immunity indiscriminately and are thus inherently flawed. We have seen in this book that the immune system can suppress its own responses in an antigen-specific manner and, by studying these endogenous regulatory events, it might be possible to devise strategies to manipulate specific responses, while sparing general immune competence. This should allow the development of new treatments that selectively suppress the responses that lead to allergy, autoimmunity, or the rejection of grafted organs. Similarly, as we understand more about tumors and infectious agents, better strategies to mobilize the immune system against cancer and infection should become possible. To achieve all this, we need to learn more about the induction of immunity and the biology of the immune system, and to apply what we have learned to human disease.

General references.

Coulie, P.G.: **Human tumour antigens recognized by T cells: new perspectives for anti-cancer vaccines?** *Mol. Med. Today* 1997, **3**:261-8.

Miller, J.F., Heath, W.R., Allison, J., Morahan, G., Hoffmann, M., Kurts, C., Kosaka, H.: **T cell tolerance and autoimmunity**. *Ciba Found. Symp.* 1997, **204**:159-171.

Nossal, G.J.: **Host immunobiology and vaccine development**. *Lancet* 1997, **350**:1316-1319.

Plotkin, S.A., and Mortimer, E.A.: *Vaccines*, 2nd edn. Philadelphia, W.B. Saunders Co., 1994.

Rappuoli, R.: **Rational design of vaccines**. *Nat. Med.* 1997, **3**:374-376.

Rocken, M., Shevach., E.M.: **Immune deviation—the third dimension of non-deletional T cell tolerance**. *Immunol. Rev.* 1996, **149**:175-194.

Suthanthiran, M., Morris, R.E., Strom, T.B.: **Immunosuppressants: cellular and molecular mechanisms of action**. *Am. J. Kidney. Dis.* 1996, **28**:159-172.

Waldmann, T.A.: **Monoclonal antibodies in diagnosis and therapy**. *Science* 1991, **252**:1657-1662.

Waldmann, T.A.: **Immune receptors: targets for therapy of leukemia/lymphoma, autoimmune diseases and for the prevention of allograft rejection**. *Annu. Rev. Immunol.* 1992, **10**:675-704.

Section references.

14-1 Corticosteroids are powerful anti-inflammatory drugs that alter the transcription of many genes.

Barnes, P.J., and Adcock, I.: **Anti-inflammatory actions of steroids: molecular mechanisms**. *Trends Pharmacol. Sci.* 1993, **14**:436-441.

Boumpas, D.T., Chrousos, G.P., Wilder, R.L., Cupps, T.R., and Balow, J.E.: **Glucocorticoid therapy for immune-mediated diseases: basic and clinical correlates**. *Ann. Intern. Med.* 1993, **119**:1198-1208.

Cronstein, B.N., Kimmel, S.C., Levin, R.I., Martiniuk, F., and Weissmann, G.: **Corticosteroids are transcriptional regulators of acute inflammation**. *Trans. Assoc. Am. Physicians.* 1992, **105**:25-35.

Cupps, T.R., and Fauci, A.S.: **Corticosteroid-mediated immunoregulation in man**. *Immunol. Rev.* 1982, **65**:133-155.

Demoly, P., Chung, K.F.: **Pharmacology of corticosteroids**. *Respir. Med.* 1998, **92**:385-94.

14-2 Cytotoxic drugs cause immunosuppression by killing dividing cells and have serious side effects.

Aarbakke, J., Janka-Schaub, G., Elion, G.B.: **Thiopurine biology and pharmacology**. *Trends Pharmacol. Sci.* 1997, **18**:3-7.

Zhu, L.P., Cupps, T.R., Whalen, G., and Fauci, A.S.: **Selective effects of cyclophosphamide therapy on activation, proliferation, and differentiation of human B cells**. *J Clin. Invest.* 1987, **79**:1082-1090.

14-3 Cyclosporin A, FK506 (tacrolimus), and rapamycin (sirolimus) are powerful immunosuppressive agents that interfere with T-cell signaling.

Brazelton, T.R., Morris, R.E.: **Molecular mechanisms of action of new xenobiotic immunosuppressive drugs: tacrolimus (FK506), sirolimus (rapamycin), mycophenolate mofetil and leflunomide**. *Curr. Opin. Immunol.* 1996, **8**:710-20.

Ho, S., Clipstone, N., Timmermann, L., Northrop, J., Graef, I., Fiorentino, D., Nourse, J., Crabtree, G.R.: **The mechanism of action of cyclosporin A and FK506**. *Clin. Immunol. Immunopathol.* 1996, **80**:S40-5.

14-4 Immunosuppressive drugs are valuable probes of intracellular signaling pathways in lymphocytes.

Abraham, R.T., Wiederrecht, G.J.: **Immunopharmacology of rapamycin**. *Annu. Rev. Immunol.* 1996, **14**:483-510.

Abraham, R.T.: **Mammalian target of rapamycin: immunosuppressive drugs uncover a novel pathway of cytokine receptor signaling**. *Curr. Opin. Immunol.* 1998, **10**:330-336.

Brown, E.J., Schreiber, S.L.: **A signaling pathway to translational control**. *Cell* 1996, **86**:517-20.

14-5 Antibodies to cell-surface molecules have been used to remove specific lymphocyte subsets or to inhibit cell function.

Cobbold, S.P., Qin, S., Leong, L.Y., Martin, G., and Waldmann, H.: **Reprogramming the immune system for peripheral tolerance with CD4 and CD8 monoclonal antibodies**. *Immunol. Rev.* 1992, **129**:165-201.

Waldmann, H. and Cobbold, S.: **The use of monoclonal antibodies to achieve immunological tolerance**. *Immunol. Today* 1993, **14**:247-251.

14-6 Antibodies can be engineered to reduce their immunogenicity in humans.

Hayden, M.S., Gilliland, L.K., Ledbetter, J.A.: **Antibody engineering**. *Curr. Opin. Immunol.* 1997, **9**:201-212.

Winter, G., and Harris, W.J.: **Humanized antibodies**. *Immunol. Today* 1993, **14**:243-246.

Winter, G., Griffiths, A.D., Hawkins, R.E., and Hoogenboom, H.R.: **Making antibodies by phage display technology**. *Annu. Rev. Immunol.* 1994, **12**:433-455.

14-7 Monoclonal antibodies can be used to inhibit allograft rejection.

Charlton, B., Auchincloss, H.Jr., and Fathman, C.G.: **Mechanisms of transplantation tolerance**. *Annu. Rev. Immunol.* 1994, **12**:707-734.

Chatenoud, L.: **Tolerogenic antibodies and fusion proteins to prevent graft rejection and treat autoimmunity**. *Mol. Med. Today* 1998, **4** 25-30.

Lechler, R., Bluestone, J.A.: **Transplantation tolerance—putting the pieces together**. *Curr. Opin. Immunol.* 1997, **9**:631-633.

Waldmann, H., Cobbold, S.: **How do monoclonal antibodies induce tolerance? A role for infectious tolerance?** *Annu. Rev. Immunol.* 1998, **16**:619-44.

14-8 Antibodies can be used to alleviate and suppress autoimmune disease.

Bach, J.F.: **Tolerance induction in transplantation and autoimmune diseases**. *Mol. Med. Today* 1995, **1**:302-3.

Cobbold, S.P., Adams, E., Marshall, S.E., Davies, J.D., Waldmann, H.: **Mechanisms of peripheral tolerance and suppression induced by monoclonal antibodies to CD4 and CD8.** *Immunol. Rev.* 1996, **149**:5-33.

Feldmann, M., Elliott, M.J., Woody, J.N., Maini, R.N.: **Anti-tumor necrosis factor-alpha therapy of rheumatoid arthritis**. *Adv. Immunol.* 1997, **64**:283-350.

Moreau, T., Coles, A., Wing, M., Thorpe, J., Miller, D., Moseley, I., Issacs, J., Hale, G., Clayton, D., Scolding, N., Waldmann, H., Compston, A.: **CAMPATH-IH in multiple sclerosis**. *Mult. Scler.* 1996, **1**:357-365.

Riethmuller, G., Rieber, E.P., Kiefersauer, S., Prinz, J., van der Lubbe, P., Meiser, B., Breedveld, F., Eisenburg, J., Kruger, K., Deusch, K. et al.: **From antilymphocyte serum to therapeutic monoclonal antibodies: first experiences with a chimeric CD4 antibody in the treatment of autoimmune disease**. *Immunol. Rev.* 1992, **129**:81-104.

14-9 **Modulation of the pattern of cytokine expression by T lymphocytes can inhibit autoimmune disease.**

Adorini, L., Sinigaglia, F.: **Pathogenesis and immunotherapy of autoimmune diseases**. *Immunol. Today* 1997, **18**:209-211.

Bach, J.F.: **Cytokine-based immunomodulation of autoimmune diseases: an overview**. *Transplant. Proc.* 1996, **28**:3023-5.

Fukaura, H., Kent, S.C., Pietrusewicz, M.J., Khoury, S.J., Weiner, H.L., Hafler, D.A.: **Induction of circulating myelin basic protein and proteolipid protein-specific transforming growth factor-beta1-secreting Th3 T cells by oral administration of myelin in multiple sclerosis patients**. *J. Clin. Invest.* 1996, **98**:70-7.

Groux, H., O'Garra, A., Bigler, M., Rouleau, M., Antonenko, S., de-Vries, J.E., Roncarolo, M.G.: **A CD4+ T-cell subset inhibits antigen-specific T-cell responses and prevents colitis**. *Nature* 1997, **389**:737-42.

Smith, J.A., Bluestone, J.A.: **T cell inactivation and cytokine deviation promoted by anti-CD3 mAbs**. *Curr. Opin. Immunol.* 1997, **9**:648-54.

14-10 **Controlled administration of antigen can be used to manipulate the nature of an antigen-specific response.**

Fairchild, P.J.: **Altered peptide ligands: prospects for immune intervention in autoimmune disease**. *Eur. J. Immunogenet.* 1997, **24**:155-67.

Liblau, R., Tisch, R., Bercovici, N., McDevitt, H.O.: **Systemic antigen in the treatment of T-cell-mediated autoimmune diseases**. *Immunol. Today* 1997, **18**:599-604.

Magee, C.C., Sayegh, M.H.: **Peptide-mediated immunosuppression**. *Curr. Opin. Immunol.* 1997, **9**:669-675.

Weiner, H.L.: **Oral tolerance for the treatment of autoimmune diseases**. *Annu. Rev. Med.* 1997, **48**:341-51.

Wraith, D.C.: **Antigen-specific immunotherapy of autoimmune disease: a commentary.** *Clin. Exp. Immunol.* 1996, **103**:349-52.

14-11 **The development of transplantable tumors in mice led to the discovery that mice could mount a protective immune response against tumors.**

Jaffee, E.M., Pardoll, D.M.: **Murine tumor antigens: is it worth the search?** *Curr. Opin. Immunol.* 1996, **8**:622-627.

14-12 **T lymphocytes can recognize certain types of spontaneous tumor in humans.**

Boon, T., Coulie, P.G., Van den Eynde, B.: **Tumor antigens recognized by T cells.** *Immunol. Today* 1997, **18**:267-8.

Disis, M.L., Cheever, M.A.: **Oncogenic proteins as tumor antigens.** *Curr. Opin. Immunol.* 1996, **8**:637-42.

Robbins, P.F., Kawakami, Y.: **Human tumor antigens recognized by T cells.** *Curr. Opin. Immunol.* 1996, **8**:628-36.

14-13 **Tumors can escape rejection in many ways.**

Bodmer, W.F., Browning, M.J., Krausa, P., Rowan, A., Bicknell, D.C., and Bodmer, J.G.: **Tumor escape from immune response by variation in HLA expression and other mechanisms**. *Ann N. Y. Acad. Sci.* 1993, **690**:42-49.

Ikeda, H., Lethe, B., Lehmann, F., van Baren, N., Baurain, J.F., de-Smet, C., Chambost, H., Vitale, M., Moretta, A., Boon, T., Coulie, P.G.: **Characterization of an antigen that is recognized on a melanoma showing partial HLA loss by CTL expressing an NK inhibitory receptor.** *Immunity* 1997, **6**:199-208.

Moller, P., and Hammerling, G.J.: **The role of surface HLA-A,B,C molecules in tumour immunity**. *Cancer Surv.* 1992, **13**:101-127.

Tada, T., Ohzeki, S., Utsumi, K., Takiuchi, H., Muramatsu, M., Li, X.F., Shimizu, J., Fujiwara, H., and Hamaoka, T.: **Transforming growth factor-beta-induced inhibition of T cell function. Susceptibility difference in T cells of various phenotypes and functions and its relevance to immunosuppression in the tumor-bearing state**. *J. Immunol.* 1991, **146**:1077-1082.

Torre Amione, G., Beauchamp, R.D., Koeppen, H., Park, B.H., Schreiber, H., Moses, H.L., and Rowley, D.A.: **A highly immunogenic tumor transfected with a murine transforming growth factor type beta 1 cDNA escapes immune surveillance**. *Proc. Natl. Acad. Sci. USA* 1990, **87**:1486-1490.

14-14 **Monoclonal antibodies to tumor antigens, alone or linked to toxins, can control tumor growth.**

Gruber, R., Holz, E., Riethmuller, G.: **Monoclonal antibodies in cancer therapy.** *Springer Semin. Immunopathol.* 1996, **18**:243-51.

von-Mehren, M., Weiner, L.M.: **Monoclonal antibody-based therapy.** *Curr. Opin. Oncol.* 1996, **8**:493-498.

Thrush, G.R., Lark, L.R., Clinchy, B.C., Vitetta, E.S.: **Immunotoxins: an update.** *Annu. Rev. Immunol.* 1996, **14**:49-71.

14-15 **Enhancing the immunogenicity of tumors holds promise for cancer therapy.**

Hellstrom, K.E., Gladstone, P., Hellstrom, I.: **Cancer vaccines: challenges and potential solutions.** *Mol. Med. Today* 1997, **3**:286-90.

Li, Y., Hellstrom, K.E., Newby, S.A., Chen, L.: **Costimulation by CD48 and B7-1 induces immunity against poorly immunogenic tumors**. *J. Exp. Med.* 1996, **183**:639-44.

Melief, C.J., Offringa, R., Toes, R.E., Kast, W.M.: **Peptide-based cancer vaccines**. *Curr. Opin. Immunol.* 1996, **8**:651-7.

Pardoll, D.M.: **Paracrine cytokine adjuvants in cancer immunotherapy**. *Annu. Rev. Immunol.* 1995, **13**:399-415.

Pardoll, D.M.: **Cancer vaccines**. *Nat. Med.* 1998, **4**:525-31.

Ragnhammar, P.: **Anti-tumoral effect of GM-CSF with or without cytokines and monoclonal antibodies in solid tumors**. *Med. Oncol.* 1996, **13**:167-76.

Schuler, G., Steinman, R.M.: **Dendritic cells as adjuvants for immune-mediated resistance to tumors**. *J. Exp. Med.* 1997, **186**:1183-7.

Tamura, Y., Peng, P., Liu, K., Daou, M., Srivastava, P.K.: **Immunotherapy of tumors with autologous tumor-derived heat shock protein preparations**. *Science* 1997, **278**:117-20.

14-16 **There are several requirements for an effective vaccine.**

Ada, G.L.: **The immunological principles of vaccination**. *Lancet* 1990, **335**:523-526.

Anderson, R.M., Donnelly, C.A., Gupta, S.: **Vaccine design, evaluation, and community-based use for antigenically variable infectious agents**. *Lancet* 1997, **350**:1466-70.

Levine, M.M., Levine, O.S.: **Influence of disease burden, public perception, and other factors on new vaccine development, implementation, and continued use**. *Lancet* 1997, **350**:1386-92.

Rabinovich, N.R., McInnes, P., Klein, D.L., and Hall, B.F.: **Vaccine technologies: view to the future**. *Science* 1994, **265**:1401-1404.

Nichol, K.L., Lind, A., Margolis, K.L., Murdoch, M., McFadden, R., Hauge, M., Magnan, S., and Drake, M.: **The effectiveness of vaccination against influenza in healthy, working adults**. *N. Engl. J. Med.* 1995, **333**:889-893.

14-17 **The history of vaccination against *Bordetella pertussis* illustrates the importance of developing an effective vaccine that is perceived to be safe.**

Mortimer, E.A.: Pertussis vaccines. In Plotkin, S.A., and Mortimer, E.A.: *Vaccines*, 2nd edn. Philadelphia, W.B. Saunders Co., 1994.

Poland, G.A.: **Acellular pertussis vaccines: new vaccines for an old disease.** *Lancet* 1996, **347**:209-210.

Rappuoli, R.: **Acellular pertussis vaccines: a turning point in infant and adolescent vaccination.** *Infect. Agents. Dis.* 1996, **5**:21-28.

14-18 Conjugate vaccines have been developed as a result of understanding how T and B cells collaborate in an immune response.

van den Dobbelsteen, G.P. and van Rees, E.P.: **Mucosal immune responses to pneumococcal polysaccharides: implications for vaccination**. *Trends. Microbiol.* 1995, **3**:155-159.

Kroll, J.S., Booy, R.: ***Haemophilus influenzae*: capsule vaccine and capsulation genetics**. *Mol. Med. Today* 1996, **2**:160-165.

Peltola, H., Kilpi, T., and Anttila, M.: **Rapid disappearance of *Haemophilus influenzae* type b meningitis after routine childhood immunisation with conjugate vaccines**. *Lancet* 1992, **340**:592-594.

14-19 The use of adjuvants is another important approach to enhancing the immunogenicity of vaccines.

Alving, C.R., Koulchin, V., Glenn, G.M., and Rao, M.: **Liposomes as carriers of peptide antigens: induction of antibodies and cytotoxic T lymphocytes to conjugated and unconjugated peptides**. *Immunol. Rev.* 1995, **145**:5-31.

Audibert, F.M., and Lise, L.D.: **Adjuvants: current status, clinical perspectives and future prospects**. *Immunol. Today* 1993, **14**:281-284.

Gupta, R.K., and Siber, G.R.: **Adjuvants for human vaccines—current status, problems and future prospects**. *Vaccine* 1995, **13**:1263-1276.

Rhodes, J., Chen, H., Hall, S.R., Beesley, J.E., Jenkins, D.C., Collins, P., and Zheng, B.: **Therapeutic potentiation of the immune system by costimulatory Schiff-base-forming drugs**. *Nature* 1995, **376**:71-75.

Takahashi, H., Takeshita, T., Morein, B., Putney, S., Germain, R.N., and Berzofsky, J.A.: **Induction of CD8+ cytotoxic T cells by immunization with purified HIV-1 envelope protein in ISCOMs**. *Nature* 1990, **344**:873-875.

Vogel, F.R.: **Immunologic adjuvants for modern vaccine formulations**. *Ann N. Y. Acad. Sci.* 1995, **754**:153-160.

14-20 Live-attenuated viral vaccines are usually more potent than 'killed' vaccines and can be made safer by using recombinant DNA technology.

Brochier, B., Kieny, M.P., Costy, F., Coppens, P., Bauduin, B., Lecocq, J.P., Languet, B., Chappuis, G., Desmettre, P., Afiademanyo, K., et al.: **Large-scale eradication of rabies using recombinant vaccinia-rabies vaccine**. *Nature* 1991, **354**:520-522.

Parkin, N.T., Chiu, P., Coelingh, K.: **Genetically engineered live attenuated influenza A virus vaccine candidates**. *J. Virol.* 1997, **71**:2772-2778.

14-21 Live-attenuated bacterial vaccines can be developed by selecting non-pathogenic or disabled mutants.

Hassan, J.O., Curtiss, R. 3rd.: **Effect of vaccination of hens with an avirulent strain of *Salmonella typhimurium* on immunity of progeny challenged with wild-type *Salmonella* strains**. *Infect. Immun.* 1996, **64**:938-944.

Guleria, I., Teitelbaum, R., McAdam, R.A., Kalpana, G., Jacobs, W.R. Jr., Bloom, B.R.: **Auxotrophic vaccines for tuberculosis**. *Nat. Med.* 1996, **2**:334-337.

Levine, M.M., Galen, J., Barry, E., Noriega, F., Tacket, C., Sztein, M., Chatfield, S., Dougan, G., Losonsky, G., Kotloff, K.: **Attenuated *Salmonella typhi* and *Shigella* as live oral vaccines and as live vectors**. *Behring Inst. Mitt.* 1997, 120-123.

14-22 Attenuated microorganisms can serve as vectors for vaccination against many pathogens.

Chatfield, S.N., Roberts, M., Dougan, G., Hormaeche, C., and Khan, C.M.: **The development of oral vaccines against parasitic diseases utilizing live attenuated Salmonella**. *Parasitology* 1995, **110 Suppl**:S17-S24.

Cirillo, J.D., Stover, C.K., Bloom, B.R., Jacobs, W.R.Jr., and Barletta, R.G.: **Bacterial vaccine vectors and bacillus Calmette-Guerin.** *Clin. Infect. Dis.* 1995, **20**:1001-1009.

Moss, B.: **Genetically engineered poxviruses for recombinant gene expression, vaccination, and safety**. *Proc. Natl. Acad. Sci. USA* 1996, **93**:11341-8.

Paoletti, E.: **Applications of pox virus vectors to vaccination: an update**. *Proc. Natl. Acad. Sci. USA* 1996, **93**:11349-11353.

14-23 Synthetic peptides of protective antigens can elicit protective immunity.

Berzofsky, J.A.: **Mechanisms of T cell recognition with application to vaccine design.** *Mol. Immunol.* 1991, **28**:217-223.

Berzofsky, J.A.: **Epitope selection and design of synthetic vaccines. Molecular approaches to enhancing immunogenicity and cross-reactivity of engineered vaccines**. *Ann N. Y. Acad. Sci.* 1993, **690**:256-264.

Davenport, M.P., Hill, A.V.: **Reverse immunogenetics: from HLA-disease associations to vaccine candidates**. *Mol. Med. Today* 1996, **2**:38-45.

Modelska, A., Dietzschold, B., Sleysh, N., Fu, Z.F., Steplewski, K., Hooper, D.C., Koprowski, H., Yusibov, V.: **Immunization against rabies with plant-derived antigen**. *Proc. Natl. Acad. Sci. USA* 1998, **95**:2481-2485.

14-24 The route of vaccination is an important determinant of success

Burnette, W.N.: **Bacterial ADP-ribosylating toxins: form, function, and recombinant vaccine development**. *Behring Inst. Mitt.* 1997, :434-41.

Douce, G., Fontana, M., Pizza, M., Rappuoli, R., Dougan, G.: **Intranasal immunogenicity and adjuvanticity of site-directed mutant derivatives of cholera toxin**. *Infect. Immun.* 1997, **65**:2821-8.

Dougan, G.: **The molecular basis for the virulence of bacterial pathogens: implications for oral vaccine development**. *Microbiology.* 1994, **140**:215-224.

Ivanoff, B., Levine, M.M., and Lambert, P.H.: **Vaccination against typhoid fever: present status**. *Bull. World Health Organ.* 1994, **72**:957-971.

Levine, M.M.: **Modern vaccines. Enteric infections**. *Lancet* 1990, **335**:958-961.

14-25 Protective immunity can be induced by injecting DNA encoding microbial antigens and human cytokines into muscle.

Donnelly, J.J., Ulmer, J.B., Shiver, J.W., Liu, M.A.: **DNA vaccines**. *Annu. Rev. Immunol.* 1997, **15**:617-48.

14-26 The effectiveness of a vaccine can be enhanced by targeting it to sites of antigen presentation.

Boyle, J.S., Brady, J.L., Lew, A.M.: **Enhanced responses to a DNA vaccine encoding a fusion antigen that is directed to sites of immune induction**. Nature 1998, **392**:408-11.

Hahn, H., Lane-Bell, P.M., Glasier, L.M., Nomellini, J.F., Bingle, W.H., Paranchych, W., Smit, J.: **Pilin-based anti-Pseudomonas vaccines: latest developments and perspectives**. *Behring Inst. Mitt.* 1997, :315-25.

Neutra, M.R.: **Current concepts in mucosal immunity. V Role of M cells in transepithelial transport of antigens and pathogens to the mucosal immune system**. *Am. J. Physiol.* 1998, **274**:G785-91.

Shen, Z., Reznikoff, G., Dranoff, G., Rock, K.L.: **Cloned dendritic cells can present exogenous antigens on both MHC class I and class II molecules**. *J. Immunol.* 1997, **158**:2723-30.

Tan, M.C., Mommaas, A.M., Drijfhout, J.W., Jordens, R., Onderwater, J.J., Verwoerd, D., Mulder, A.A., van-der-Heiden, A.N., Scheidegger, D., Oomen, L.C., Ottenhoff, T.H., Tulp, A., Neefjes, J.J., Koning, F.: **Mannose receptor-mediated uptake of antigens strongly enhances HLA class II-restricted antigen presentation by cultured dendritic cells.** *Eur. J. Immunol.* 1997, **27**:2426-35.

Thomson, S.A., Burrows, S.R., Misko, I.S., Moss, D.J., Coupar, B.E., Khanna, R.: **Targeting a polyepitope protein incorporating multiple class II-restricted viral epitopes to the secretory/endocytic pathway facilitates immune recognition by CD4+ cytotoxic T lymphocytes: a novel approach to vaccine design**. *J. Virol.* 1998, **72**:2246-52.

14-27 **An important question is whether vaccination can be used therapeutically to control existing chronic infections.**

Burke, R.L.: **Contemporary approaches to vaccination against herpes simplex virus.** *Curr. Top. Microbiol. Immunol.* 1992, **179**:137-158.

Grange, J.M., Stanford, J.L.: **Therapeutic vaccines**. *J. Med. Microbiol.* 1996, **45**:81-3.

Hill, A., Jugovic, P., York, I., Russ, G., Bennink, J., Yewdell, J., Ploegh, H., and Johnson, D.: **Herpes simplex virus turns off the TAP to evade host immunity**. *Nature* 1995, **375**:411-415.

Modlin, R.L.: **Th1-Th2 paradigm: insights from leprosy**. *J Invest. Dermatol.* 1994, **102**:828-832.

Reiner, S.L., and Locksley, R.M.: **The regulation of immunity *to Leishmania major*.** *Annu. Rev. Immunol.* 1995, **13**:151-177.

Stanford, J.L.: **The history and future of vaccination and immunotherapy for leprosy**. *Trop. Geogr. Med.* 1994, **46**:93-107.

14-28 **Modulation of the immune system might be used to inhibit immunopathological responses to infectious agents.**

Biron, C.A., and Gazzinelli, R.T.: **Effects of IL-12 on immune responses to microbial infections: a key mediator in regulating disease outcome**. *Curr. Opin. Immunol.* 1995, **7**:485-496.

Grau, G.E., and Modlin, R.L.: **Immune mechanisms in bacterial and parasitic diseases: protective immunity versus pathology**. *Curr. Opin. Immunol.* 1991, **3**:480-485.

Kaplan, G.: **Recent advances in cytokine therapy in leprosy**. *J Infect. Dis.* 1993, **167 Suppl 1**:S18-S22.

Locksley, R.M.: **Interleukin 12 in host defense against microbial pathogens**. *Proc. Natl. Acad. Sci. USA* 1993, **90**:5879-5880.

Murray, H.W.: **Interferon-gamma and host antimicrobial defense: current and future clinical applications.** *Am. J. Med.* 1994, **97**:459-467.

Scott, P., Trinchieri, G.: **IL-12 as an adjuvant for cell-mediated immunity**. *Semin. Immunol.* 1997, **9**:285-291.

Sher, A., Gazzinelli, R.T., Oswald, I.P., Clerici, M., Kullberg, M., Pearce, E.J., Berzofsky, J.A., Mosmann, T.R., James, S.L., and Morse, H.C.: **Role of T-cell derived cytokines in the downregulation of immune responses in parasitic and retroviral infection**. *Immunol. Rev.* 1992, **127**:183-204.

Sher, A., Jankovic, D., Cheever, A., Wynn, T.: **An IL-12-based vaccine approach for preventing immunopathology in schistosomiasis**. *Ann. N. Y. Acad. Sci.* 1996, **795**:202-207.

APPENDICES

Appendix I. The CD antigens.

CD antigen	Cellular expression	Molecular weight (kDa)	Functions	Other names	Family relationships
CD1a,b,c,d	Cortical thymocytes, Langerhans' cells, dendritic cells, B cells (CD1c), intestinal epithelium, smooth muscle, blood vessels (CD1d)	43–49	MHC class 1-like molecule, associated with β_2-microglobulin. May have specialized role in presentation of lipid antigens		Immunoglobulin
CD2	T cells, thymocytes, NK cells	45–58	Adhesion molecule, binding CD58 (LFA-3). Binds Lck intracellularly and activates T cells	T11, LFA-2	Immunoglobulin
CD3	Thymocytes, T cells	γ: 25–28 δ: 20 ε: 20	Associated with the T-cell antigen receptor (TCR). Required for cell-surface expression of and signal transduction by the TCR	T3	Immunoglobulin
CD4	Thymocyte subsets, T_H1 and T_H2 T cells (about two thirds of peripheral T cells), monocytes, macrophages	55	Co-receptor for MHC class II molecules. Binds Lck on cytoplasmic face of membrane. Receptor for HIV-1 and HIV-2 gp120	T4, L3T4	Immunoglobulin
CD5	Thymocytes, T cells, subset of B cells	67		T1, Ly1	Scavenger receptor
CD6	Thymocytes, T cells, B cells in chronic lymphatic leukemia	100–130	Binds CD166	T12	Scavenger receptor
CD7	Pluripotential hematopoietic cells, thymocytes, T cells	40	Unknown, cytoplasmic domain binds PI-3 kinase on crosslinking. Marker for T cell acute lymphatic leukemia and pluripotential stem cell leukemias		Immunoglobulin
CD8	Thymocytes subsets, cytotoxic T cells (about one third of peripheral T cells)	α: 32–34 β: 32–34	Co-receptor for MHC class I molecules. Binds Lck on cytoplasmic face of membrane	T8, Lyt2,3	Immunoglobulin
CD9	Pre-B cells, monocytes, eosinophils, basophils, platelets, activated T cells, brain and peripheral nerves, vascular smooth muscle	24	Mediates platelet aggregation and activation via FcγRIIa, may play a role in cell migration		Tetraspanning membrane protein, also called transmembrane 4 (TM4)
CD10	B- and T- cell precursors, bone marrow stromal cells	100	Zinc metalloproteinase, marker for pre-B acute lymphatic leukemia (ALL)	Neutral endopeptidase, common acute lymphocytic leukemia antigen (CALLA)	
CD11a	Lymphocytes, granulocytes, monocytes and macrophages	180	αL subunit of integrin LFA-1 (associated with CD18); binds to CD54 (ICAM-1), CD102 (ICAM-2), and CD50 (ICAM-3)	LFA-1	Integrin α

CD antigen	Cellular expression	Molecular weight (kDa)	Functions	Other names	Family relationships
CD11b	Myeloid and NK cells	170	αM subunit of integrin CR3 (associated with CD18): binds CD54, complement component iC3b, and extracellular matrix proteins	Mac-1	Integrin α
CD11c	Myeloid cells	150	αX subunit of integrin CR4 (associated with CD18); binds fibrinogen	CR4, p150, 95	Integrin α
CD11d	Leukocytes	125	αD subunits of integrin; associated with CD18; binds to CD50		Integrin α
CDw12	Monocytes, granulocytes, platelets	90–120	Unknown		
CD13	Myelomonocytic cells	150–170	Zinc metalloproteinase	Aminopeptidase N	
CD14	Myelomonocytic cells	53–55	Receptor for complex of lipopolysaccharide and lipopolysaccharide binding protein (LBP)		
CD15	Neutrophils, eosinophils, monocytes		Terminal trisaccharide expressed on glycolipids and many cell-surface glycoproteins	Lewis X (Le^x)	
CD15s	Leukocytes, endothelium		Ligand for CD62E, P	Sialyl-Lewis X (sLe^x)	poly-N-acetyl-lactosamine
CD16	Neutrophils, NK cells, macrophages	50–80	Component of low affinity Fc receptor, FCγRIII, mediates phagocytosis and antibody-dependent cell-mediated cytotoxicity	FcγRIII	Immunoglobulin
CDw17	Neutrophils, monocytes, platelets		Lactosyl ceramide, a cell-surface glycosphingolipid		
CD18	Leukocytes	95	Intergrin β2 subunit, associates with CD11a, b, c, and d		Integrin β
CD19	B cells	95	Forms complex with CD21 (CR2) and CD81 (TAPA-1); co-receptor for B cells—cytoplasmic domain binds cytoplasmic tyrosine kinases and PI-3 kinase		Immunoglobulin
CD20	B cells	33–37	Oligomers of CD20 may form a Ca^{2+} channel; possible role in regulating B-cell activation		Contains 4 transmembrane segments
CD21	Mature B cells, follicular dendritic cells	145	Receptor for complement component C3d, Epstein–Barr virus. With CD19 and CD81, CD21 forms co-receptor for B cells	CR2	Complement control protein (CCP)
CD22	Mature B cells	α: 130 β: 140	Binds sialoconjugates	BL-CAM	Immunoglobulin
CD23	Mature B cells, activated macrophages, eosinophils, follicular dendritic cells, platelets	45	Low-affinity receptor for IgE, regulates IgE synthesis; ligand for CD19:CD21:CD81 co-receptor	FcεRII	C-type lectin
CD24	B cells, granulocytes	35–45	Unknown	Possible human homolog of mouse heat stable antigen (HSA)	
CD25	Activated T cells, B cells, and monocytes	55	IL-2 receptor α chain	Tac	CCP
CD26	Activated B and T cells, macrophages	110	Exopeptidase, cleaves N terminal X-Pro or X-Ala dipeptides from polypeptides	Dipeptidyl peptidase IV	Type II trans-membrane glycoprotein
CD27	Medullary thymocytes, T cells, NK cells, some B cells	55	Binds CD70; can function as a co-stimulator for T and B cells		TNF receptor

CD antigen	Cellular expression	Molecular weight (kDa)	Functions	Other names	Family relationships
CD28	T-cell subsets, activated B cells	44	Activation of naive T cells, receptor for co-stimulatory signal (signal 2) binds CD80 (B7.1) and CD86 (B7.2)	Tp44	Immunoglobulin and CD86 (B7.2)
CD29	Leukocytes	130	Integrin β1 subunit, associates with CD49a integrin VLA-1 integrin		Integrin β
CD30	Activated T, B, and NK cells, monocytes	120	Binds CD30L (CD153); crosslinking CD30 enhances proliferation of B and T cells	Ki-1	TNF receptor
CD31	Monocytes, platelets, granulocytes, T-cell subsets, endothelial cells	130–140	Adhesion molecule, mediating both leukocyte/endothelial and endothelial/endothelial interactions	PECAM-1	Immunoglobulin
CD32	Monocytes, granulocytes, B cells, eosinophils	40	Low affinity Fc receptor for aggregated immunoglobulin/immune complexes	FcγRII	Immunoglobulin
CD33	Myeloid progenitor cells, monocytes	67	Binds sialoconjugates		Immunoglobulin
CD34	Hematopoietic precursors, capillary endothelium	105–120	Ligand for CD62L (L-selectin)		Mucin
CD35	Erythrocytes, B cells, monocytes, neutrophils, eosinophils, follicular dendritic cells	250	Complement receptor 1, binds C3b and C4b, mediates phagocytosis	CR1	CCP
CD36	Platelets, monocytes, endothelial cells	88	Platelet adhesion molecule; involved in recognition and phagocytosis of apoptosed cells	Platelet GPIV, GPIIIb	
CD37	Mature B cells, mature T cells, myeloid cells	40–52	Unknown, may be involved in signal transduction; forms complexes with CD53, CD81, CD82, and MHC class II		Transmembrane 4
CD38	Early B and T cells, activated T cells, germinal center B cells, plasma cells	45	NAD glycohydrolase, augments B cell proliferation	T10	
CD39	Activated B cells, activated NK cells, macrophages, dendritic cells	78	Unknown, may mediate adhesion of B cells		
CD40	B cells, macrophages, dendritic cells, basal epithelial cells	48	Binds CD154 (CD40L); receptor for co-stimulatory signal for B cells, promotes growth, differentiation and isotype switching of B cells, and cytokine production by macrophages and dendritic cells		TNF receptor
CD41	Platelets, megakaryocytes	Dimer: GPIIba: 125 GPIIbb: 22	αIIb integrin, associates with CD61 to form GPIIb, binds fibrinogen, fibronectin, von Willebrand factor, and thrombospondin	GPIIb	Integrin α
CD42a,b,c,d	Platelets, megakaryocytes	a: 23 b: 135, 23 c: 22 d: 85	Binds von Willebrand factor, thrombin; essential for platelet adhesion at sites of injury	a: GPIX b: GPIbα c: GPIbβ d: GPV	Leucine-rich repeat
CD43	Leukocytes, except resting B cells	115–135 (neutrophils) 95–115 (T cells)	Has extended structure, approx. 45 nm long and may be anti-adhesive	Leukosialin, sialophorin	Mucin
CD44	Leukocytes, erythrocytes	80–95	Binds hyaluronic acid, mediates adhesion of leukocytes	Hermes antigen, Pgp-1	Link protein
CD45	All hematopoietic cells	180–240 (multiple isoforms)	Tyrosine phosphatase, augments signaling through antigen receptor of B and T cells, multiple isoforms result from alternative splicing (see below)	Leukocyte common antigen (LCA), T200, B220	Fibronectin type III

CD antigen	Cellular expression	Molecular weight (kDa)	Functions	Other names	Family relationships
CD45RO	T-cell subsets, B-cell subsets, monocytes, macrophages	180	Isoform of CD45 containing none of the A, B, and C exons		Fibronectin type II
CD45RA	B cells, T-cell subsets (naive T cells), monocytes	205–220	Isoforms of CD45 containing the A exon		Fibronectin type II
CD45RB	T-cell subsets, B cells, monocytes, macrophages, granulocytes	190–220	Isoforms of CD45 containing the B exon	T200	Fibronectin type II
CD46	Hematopoietic and non-hematopoietic nucleated cells	56/66 (splice variants)	Membrane co-factor protein, binds to C3b and C4b to permit their degradation by Factor I	MCP	CCP
CD47	All cells	47–52	Unknown, associated with Rh blood group		
CD48	Leukocytes	40–47	Unknown	Blast-1	Immunoglobulin
CD49a	Activated T cells, monocytes, neuronal cells, smooth muscle	200	α1 integrin, associates with CD29, binds collagen, laminin-1	VLA-1	Integrin α
CD49b	B cells, monocytes, platelets, megakaryocytes, neuronal, epithelial and endothelial cells, osteoclasts	160	α2 integrin, associates with CD29, binds collagen, laminin	VLA-2, platelet GPIa	Integrin α
CD49c	B cells, many adherent cells	125, 30	α3 integrin, associates with CD29, binds laminin-5, fibronectin, collagen, entactin invasin	VLA-3	Integrin α
CD49d	Broad distribution includes B cells, thymocytes, monocytes, granulocytes, dendritic cells	150	α4 integrin, associates with CD29, binds fibronectin, MAdCAM-1, VCAM-1	VLA-4	Integrin α
CD49e	Broad distribution includes memory T cells, monocytes, platelets	135, 25	α5 integrin, associates with CD29, binds fibronectin, invasin	VLA-5	Integrin α
CD49f	T lymphocytes, monocytes, platelets, megakaryocytes, trophoblasts	125, 25	α6 integrin, associates with CD29, binds laminin, invasin, merosine	VLA-6	Integrin α
CD50	Thymocytes, T cells, B cells, monocytes, granulocytes	130	Binds integrin CD11a/CD18	ICAM-3	Immunoglobulin
CD51	Platelets, megakaryocytes	125, 24	αV integrin, associates with CD61, binds vitronectin, von Willebrand factor, fibrinogen, and thrombospondin; may be receptor for apoptotic cells	Vitronectin receptor	Integrin α
CD52	Thymocytes, T cells, B cells (not plasma cells), monocytes, granulocytes, spermatozoa	25	Unknown, target for antibodies used therapeutically to deplete T cells from bone marrow	CAMPATH-1, HE5	
CD53	Leukocytes	35–42	Unknown	MRC OX44	Transmembrane 4
CD54	Hematopoietic and non-hematopoietic cells	75–115	Intercellular adhesion molecule (ICAM)-1 binds CD11a/CD18 integrin (LFA-1) and CD11b/CD18 integrin (Mac-1), receptor for rhinovirus	ICAM-1	Immunoglobulin
CD55	Hematopoietic and non-hematopoietic cells	60–70	Decay accelerating factor (DAF), binds C3b, disassembles C3/C5 convertase	DAF	CCP
CD56	NK cells	135–220	Isoform of neural cell adhesion molecule (NCAM), adhesion molecule	NKH-1	Immunoglobulin
CD57	NK cells, subsets of T cells, B cells, and monocytes		Oligosaccharide, found on many cell-surface glycoproteins	HNK-1, Leu-7	
CD58	Hematopoietic and non-hematopoietic cells	55–70	Leukocyte function-associated antigen-3 (LFA-3), binds CD2, adhesion molecule	LFA-3	Immunoglobulin
CD59	Hematopoietic and non-hematopoietic cells	19	Binds complement components C8 and C9, blocks assembly of membrane-attack complex	Protectin, Mac inhibitor	Ly-6

CD antigen	Cellular expression	Molecular weight (kDa)	Functions	Other names	Family relationships
CDw60	T-cells subsets, platelets, monocytes		9-O-acetylayed disialyl group present on gangliosides, predominantly ganglioside D3		
CD61	Platelets, megakaryocytes, macrophages	110	Intergrin β3 subunit, associates with CD41 (GPIIb/IIIa) or CD51 (vitronectin receptor)		Integrin β
CD62E	Endothelium	140	Endothelium leukocyte adhesion molecule (ELAM), binds sialyl-Lewis X, mediates rolling interaction of neutrophils on endothelium	ELAM-1, E-selectin	C-type lectin, EGF and CCP
CD62L	B cells, T cells, monocytes, NK cells	150	Leukocyte adhesion molecule (LAM), binds CD34, GlyCAM, mediates rolling interactions with endothelium	LAM-1, L-selectin, LECAM-1	C-type lectin, EGF and CCP
CD62P	Platelets, megakaryocytes, endothelium	140	Adhesion molecule, binds CD162 (PSGL-1), mediates interaction of platelets with endothelial cells, monocytes and rolling leukocytes on endothelium	P-selectin, PADGEM	C-type lectin, EGF and CCP
CD63	Activated platelets, monocytes, macrophages	53	Unknown, is lysosomal membrane protein translocated to cell surface after activation	Platelet activation antigen	Transmembrane 4
CD64	Monocytes, macrophages	72	High-affinity receptor for IgG, binds IgG3>IgG1>IgG4>>>IgG2, mediates phagocytosis, antigen capture, ADCC	FCγRI	Immunoglobulin
CD65	Myeloid cells		Oligosaccharide component of a ceramide dodecasaccharide		
CD66a	Neutrophils	160–180	Unknown, member of carcino-embryonic antigen (CEA) family (see below)	Biliary glyco-protein-1 (BGP-1)	Immunoglobulin
CD66b	Granulocytes	95–100	Unknown, member of carcino-embryonic antigen (CEA) family	Previously CD67	Immunoglobulin
CD66c	Neutrophils, colon carcinoma	90	Unknown, member of carcino-embryonic antigen (CEA) family	Non-specific cross-reacting antigen (NCA)	Immunoglobulin
CD66d	Neutrophils	30	Unknown, member of carcino-embryonic antigen (CEA) family		Immunoglobulin
CD66e	Adult colon epithelium, colon carcinoma	180–200	Unknown, member of carcino-embryonic antigen (CEA) family	Carcino-embryonic antigen (CEA)	Immunoglobulin
CD66f	Unknown		Unknown, member of carcino-embryonic antigen (CEA) family	Pregnancy specific glycoprotein	Immunoglobulin
CD68	Monocytes, macrophages, neutrophils, basophils, large lymphocytes	110	Unknown	Macrosialin	Mucin
CD69	Activated T and B cells, activated macrophages and NK cells	28,32 homodimer	Unknown, early activation antigen	Activation inducer molecule (AIM)	C-type lectin
CD70	Activated T and B cells, and macrophages	75, 95, 170	Ligand for CD27, may function in co-stimulation of B and T cells	Ki-24	TNF
CD71	All proliferating cells, hence activated leukocytes	95 homodimer	Transferrin receptor	T9	
CD72	B cells (not plasma cells)	42 homodimer	Unknown	Lyb-2	C-type lectin
CD73	B-cell subsets, T-cell subsets	69	Ecto-5′-nucleotidase, dephosphorylates nucleotides to allow nucleoside uptake		

CD antigen	Cellular expression	Molecular weight (kDa)	Functions	Other names	Family relationships
CD74	B cells, macrophages, monocytes, MHC class II positive cells	33, 35, 41, 43 (alternative initiation and splicing)	MHC class II-associated invariant chain	Ii, Iγ	
CD75	Mature B cells, T-cells subsets		Sialoglycan moiety, ligand for CD22, mediates B-cell:B-cell adhesion		
CD76	Mature B cells, T-cell subsets		α2,6 sialylated polylactosamine expressed on glycosphingolipids and glycoproteins		
CD77	Germinal center B cells		Neutral glycosphingolipid (Galα1→4Galβ1→4Galcβ1→ ceramide), binds Shiga toxin, crosslinking induces apoptosis	Globotriaocyl-ceramide (Gb3) Pk blood group	
CD79α,β	B cells	α: 40–45 β: 37	Components of B-cell antigen receptor analogous to CD3, required for cell-surface expression and signal transduction	Igα, Igβ	Immunoglobulin
CD80	B-cell subset	60	Co-stimulator, ligand for CD28 and CTLA-4	B7 (now B7.1), BB1	Immunoglobulin
CD81	Lymphocytes	26	Associates with CD19, CD21 to form B cell co-receptor	Target of anti-proliferative anti-body (TAPA-1)	Transmembrane 4
CD82	Leukocytes	50–53	Unknown	R2	Transmembrane 4
CD83	Dendritic cells, B cells, Langerhans' cells	43	Unknown	HB15	Immunoglobulin
CDw84	Monocytes, platelets, circulating B cells	73	Unknown	GR6	Immunoglobulin
CD85	Monocytes, circulating B cells	120, 83	Unknown	GR4	
CD86	Monocytes, activated B cells, dendritic cells	80	Ligand for CD28 and CTLA4	B7.2	Immunoglobulin
CD87	Granulocytes, monocytes, macrophages, T cells, NK cells, wide variety of non-hematopoietic cell types	35–59	Receptor for urokinase plasminogen activator	uPAR	Ly-6
CD88	Polymorphonuclear leukocytes, macrophages, mast cells	43	Receptor for complement component C5a	C5aR	G protein coupled receptor
CD89	Monocytes, macrophages, granulocytes, neutrophils, B-cell subsets, T-cell subsets	50–70	IgA receptor	FcαR	Immunoglobulin
CD90	CD34$^+$ prothymocytes (human), thymocytes, T cells (mouse)	18	Unknown	Thy-1	Immunoglobulin
CD91	Monocytes, many non-hematopoietic cells	515, 85	α2-macroglobulin receptor		EGF, LDL receptor
CDw92	Neutrophils, monocytes, platelets, endothelium	70	Unknown	GR9	
CD93	Neutrophils, monocytes, endothelium	120	Unknown	GR11	
CD94	T-cell subsets, NK cells	43	Unknown	KP43	C-type lectin
CD95	Wide variety of cell lines, *in vivo* distribution uncertain	45	Binds TNF-like Fas ligand, induces apoptosis	Apo-1, Fas	TNF receptor
CD96	Activated T cells, NK cells	160	Unknown	T-cell activation increased late expression (TACTILE)	Immunoglobulin

CD antigen	Cellular expression	Molecular weight (kDa)	Functions	Other names	Family relationships
CD97	Activated B and T cells, monocytes, granulocytes	75–85	Binds CD55	GR1	EGF, G protein coupled receptor
CD98	T cells, B cells, natural killer cells, granulocytes, all human cell lines	80, 45 heterodimer	May be amino-acid transporter	4F2, FRP-1	
CD99	Peripheral blood lymphocytes, thymocytes	32	Unknown	MIC2, E2	
CD100	Hematopoietic cells	150 homodimer	Unknown	GR3	Semaphorin
CD101	Monocytes, granulocytes, dendritic cells, activated T cells	120 homodimer	Unknown	BPC#4	Immunoglobulin
CD102	Resting lymphocytes, monocytes, vascular endothelium cells (strongest)	55–65	Binds CD11a/CD18 (LFA-1) but not CD11b/CD18 (Mac-1)	ICAM-2	Immunoglobulin
CD103	Intraepithelial lymphocytes, 2–6% peripheral blood lymphocytes	150, 25	αE integrin	HML-1, α6, αE integrin	Integrin α
CD104	$CD4^-$ $CD8^-$ thymocytes, neuronal, epithelial, and some endothelial cells, Schwann cells, trophoblasts	220	Integrin β4 associates with CD49f, binds laminins	β4 integrin	Integrin β
CD105	Endothelial cells, activated monocytes and macrophages, bone-marrow cell subsets	90 homodimer	Binds TGF-β	Endoglin	
CD106	Endothelial cells	100–110	Adhesion molecule, ligand for VLA-4 (α4β1 integrin)	VCAM-1	Immunoglobulin
CD107a	Activated platelets, activated T cells, activated neutrophils, activated endothelium	110	Unknown, is lysosomal membrane protein translocated to the cell surface after activation	Lysosomal associated membrane protein-1 (LAMP-1)	
CD107b	Activated platelets, activated T cells, activated neutrophils, activated endothelium	120	Unknown, is lysosomal membrane protein translocated to the cell surface after activation	LAMP-2	
CDw108	Erythrocytes, circulating lymphocytes, lymphoblasts	80	Unknown	GR2, John Milton-Hagen blood group antigen	
CD109	Activated T cells, activated platelets, vascular endothelium	170	Unknown	Platelet activation factor, GR56	
CD110–CD113	Not yet assigned				
CD114	Granulocytes, monocytes	150	Granulocytes colony stimulating factor (G-CSF) receptor		Immunoglobulin, fibronectin type III
CD115	Monocytes, macrophages	150	Macrophage colony stimulating factor (M-CSF) receptor	M-CSFR, c-fms	Immunoglobulin, tyrosine kinase
CD116	Monocytes, neutrophils, eosinophils, endothelium	70–85	Granulocyte macrophage colony stimulating factor (GM-CSF) receptor α chain	GM-CSFRα	Cytokine receptor, fibronectin type III
CD117	Hematopoietic progenitors	145	Stem-cell factor (SCF) receptor	c-KIT	Immunoglobulin, tyrosine kinase
CD118	Broad cellular expression		Interferon-α,β receptor	IFN-α, βR	
CD119	Macrophages, monocytes, B cells, endothelium	90–100	Interferon-γ receptor	IFN-γR	Fibronectin type III
CD120a	Hematopoietic and non-hematopoietic cells, highest on epithelial cells	50–60	TNF receptor, binds both TNF-α and TNF-β	TNFR-I	TNF receptor
CD120b	Hematopoietic and non-hematopoietic cells, highest on myeloid cells	75–85	TNF receptor, binds both TNF-α and TNF-β	TNFR-II	TNF receptor

CD antigen	Cellular expression	Molecular weight (kDa)	Functions	Other names	Family relationships
CD121a	Thymocytes, T cells	80	Type I interleukin-1 receptor, binds IL-1α and IL-1β	IL-1R type I	Immunoglobulin
CDw121b	B cells, macrophages, monocytes	60–70	Type II interleukin-1 receptor, binds IL-1α and IL-1β	IL-1R type II	Immunoglobulin
CD122	NK cells, resting T-cell subsets, some B-cell lines	75	IL-2 receptor β chain	IL-2Rβ	Cytokine receptor, fibronectin type III
CD123	Bone marrow stem cells, granulocytes, monocytes, megakaryocytes	70	IL-3 receptor α chain	IL-3Rα	Cytokine receptor, fibronectin type III
CD124	Mature B and T cells, hematopoietic precursor cells	130–150	IL-4 receptor	IL-4R	Cytokine receptor, fibronectin type III
CD125	Eosinophils, basophils, activated B cells	55–60	IL-5 receptor	IL-5R	Cytokine receptor, fibronectin type III
CD126	Activated B cells and plasma cells (strong), most leukocytes (weak)	80	IL-6 receptor α subunit	IL-6Rα	Immunoglobulin, cytokine receptor, fibronectin type III
CD127	Bone marrow lymphoid precursors, pro-B cells, mature T cells, monocytes	68–79, possibly forms homodimers	IL-7 receptor	IL-7R	Fibronectin type III
CDw128	Neutrophils, basophils, T-cell subsets	58–67	IL-8 receptor	IL-8R	G protein coupled receptor
CD129	Not yet assigned				
CD130	Most cell types, strong on activated B cells and plasma cells	130	Common subunit of IL-6, IL-11, oncostatin-M (OSM) and leukemia inhibitory factor (LIF) receptors	IL-6Rβ, IL-11Rβ, OSMRβ, IFRβ	Immunoglobulin, cytokine receptor, fibronectin type III
CDw131	Myeloid progenitors, granulocytes	140	Common β subunit of IL-3, IL-5, and GM-CSF receptors	IL-3Rβ, IL-5Rβ, GM-CSFRβ	Cytokine receptor, fibronectin type III
CD132	B cells, T cells, NK cells, mast cells, neutrophils	64	IL-2 receptor γ chain, common subunit of IL-2, IL-4, IL-7, IL-9, and IL-15 receptors		Cytokine receptor
CD134	Activated T cells	50	May act as adhesion molecule costimulator	OX40	TNF receptor
CD135	Multipotential precursors, myelomonocytic and B-cell progenitors	130, 155	Growth factor receptor	FLK2, STK-1	Immunoglobulin, tyrosine kinase
CDw136	Monocytes, epithelial cells, central and peripheral nervous system	180	Chemotaxis, phagocytosis, cell growth, and differentiation	MSP-R, RON	Tyrosine kinase
CDw137	T and B lymphocytes, monocytes, some epithelial cells		Co-stimulator of T-cell proliferation	ILA (induced by lymphocyte activation), 4-1BB	TNF receptor
CD138	B cells		Heparan sulphate proteoglycan binds collagen type I	Syndecan-1	
CD139	B cells	209, 228	Unknown		
CD140a,b	Stromal cells, some endothelial cells	a: 180 b: 180	Platelet derived growth factor (PDGF) receptor α and β chains		
CD141	Vascular endothelial cells	105	Anticoagulant, binds thrombin, the complex then activates protein C	Thrombomodulin fetomodulin	C-type lectin, EGF
CD142	Epidermal keratinocytes, various epithelial cells, astrocytes, Schwann cells. Absent from cells in direct contact with plasma unless induced by inflammatory mediators	45–47	Major initiating factor of clotting. Binds Factor VIIa; this complex activates Factors VII, IX, and X	Tissue factor, thromboplastin	Fibronectin type III

CD antigen	Cellular expression	Molecular weight (kDa)	Functions	Other names	Family relationships
CD143	Endothelial cells, except large blood vessels and kidney, epithelial cells of brush borders of kidney and small intestine, neuronal cells, activated macrophages and some T cells. Soluble form in plasma	170–180	Zn^{2+} metallopeptidase dipeptidyl peptidase, cleaves angiotensin I and bradykinin from precursor forms	Angiotensin converting enzyme (ACE)	
CD144	Endothelial cells	130	Organizes adherens junction in endothelial cells	cadherin-5, VE-cadherin	Cadherin
CD145	Endothelial cells, some stromal cells	25, 90, 110	Unknown		
CD146	Endothelium	130	Potential adhesion molecule, localized at cell:cell junctions	MCAM, MUC18, S-ENDO	Immunoglobulin
CD147	Leukocytes, red blood cells, platelets, endothelial cells	55–65	Potential adhesion molecule	M6, neurothelin, EMMPRIN, basigin, OX-47	Immunoglobulin
CD148	Granulocytes, monocytes, dendritic cells, T cells, fibroblasts, nerve cells	240–260	Contact inhibition of cell growth	HPTPη	Fibronectin type III, protein tyrosine phosphatase
CD150	Thymocytes, activated lymphocytes	75–95	Unknown	SLAM	Immunoglobulin
CD151	Platelets, megakaryocytes, epithelial cells, endothelial cells	32	Associates with β1 integrins	PETA-3, SFA-1	Transmembrane 4
CD152	Activated T cells	33	Receptor for B7.1 (CD80), B7.2 (CD86); negative regulator of T-cell activation	CTLA-4	Immunoglobulin
CD153	Activated T cells, activated macrophages, neutrophils, B cells	38–40	Ligand for CD30, may co-stimulate T cells	CD30L	TNF
CD154	Activated CD4 T cells	30 (trimer)	Ligand for CD40, inducer of B cell proliferation and activation	CD40L, TRAP, T-BAM, gp39	TNF receptor
CD155	Monocytes, macrophages, thymocytes, CNS neurons	80–90	Normal function unknown; receptor for poliovirus	Poliovirus receptor	Immunoglobulin
CD156	Neutrophils, monocytes	69	Unknown, may be involved integrin leukocyte extravasation	MS2, ADAM 8 (A disintegrin and metallo-protease)	
CD157	Granulocytes, monocytes, bone marrow stromal cells, vascular endothelial cells, follicular dendritic cells	42–45 (50 on monocytes)	ADP-ribosyl cyclase, cyclic ADP-ribose hydrolase	BST-1	
CD158a	NK-cell subsets	50 or 58	Inhibits NK cell cytotoxicity on binding MHC class I molecules	p50.1, p58.1	Immunoglobulin
CD158b	NK-cell subsets	50 or 58	Inhibits NK cell cytotoxicity on binding HLA-Cw3 and related alleles	p50.2, p58.2	Immunoglobulin
CD161	NK cells, T cells	44	Regulates NK cytotoxicity	NKRP1	C-type lectin
CD162	Neutrophils, lymphocytes, monocytes	120 (homodimer)	Ligand for CD62P	PSGL-1	Mucin
CD163	Monocytes, macrophages	130	Unknown	M130	
CD164	Epithelial cells, monocytes, bone marrow stromal cells	80	Unknown	MUC-24 (multi-glycosylated protein 24)	Mucin
CD165	Thymocytes, thymic epithelial cells, CNS neurons, pancreatic islets, Bowman's capsule	37	Adhesion between thymocytes and thymic epithelium	Gp37, AD2	
CD166	Activated T cells, thymic epithelium, fibroblasts, neurons	100–105	Ligand for CD6, involved integrin neurite extension	ALCAM, BEN, DM-GRASP, SC-1	Immunoglobulin
TCRζ	T cells, NK cells	12 (homodimer)	Component of T-cell receptor, contains 3 ITAMs per chain		ζ chain

Appendix II. Cytokines and their receptors.

Family	Cytokine (alternative names)	Size (no. of amino acids) and form	Receptors (c denotes common subunit)	Producer cells	Actions	Effect of cytokine or receptor knock-out (where known)
Hematopoietins (four-helix bundles)	Epo (erythropoietin)	165, monomer*	EpoR	Kidney cells hepatocytes	Stimulates erythroid progenitors	Epo or EpoR: embryonic lethal
	IL-2 (T-cell growth factor)	133, monomer	CD25 (α), CD122 (β), CD132 (γc)	T cells	T-cell proliferation	IL-2: deregulated T-cell proliferation IL-2Rα: incomplete T-cell development IL-2Rβ: increased T-cell autoimmunity IL-2γ_c: severe combined immunodeficiency
	IL-3 (multicolony CSF)	133, monomer	CD123, βc	T cells, thymic epithelial cells	Synergistic action in early hematopoiesis	IL-3: impaired eosinophil development. Bone marrow unresponsive to IL-5, GM-CSF
	IL-4 (BCGF-1, BSF-1)	129, monomer	CD124, CD132 (γc)	T cells, mast cells	B-cell activation, IgE switch supresses T_H1 cells	IL-4: decreased IgE synthesis
	IL-5 (BCGF-2)	115, homodimer	CD125, βc	T cells, mast cells	Eosinophil growth, differentiation	IL-5: decreased IgE, IgG1 synthesis (in mice); decreased levels of IL-9, IL-10, and eosinophils
	IL-6 (IFN-β_2, BSF-2, BCDF)	184, monomer	CD126, CD130	T cells, macrophages endothelial cells	T- and B-cell growth and differentiation, acute phase protein production, fever	IL-6: decreased acute phase reaction, reduced IgA production
	IL-7	152, monomer*	CD127, CD132 (γc)	Non-T cells	Growth of pre-B cells and pre-T cells	IL-7: Early thymic and lymphocyte expansion severely impaired
	IL-9	125, monomer	IL-9R, CD132 (γc)	T cells	Mast cell enhancing activity	
	IL-11	178, monomer	IL-11R, CD130	Stromal fibroblasts	Synergistic action with IL-3 and IL-4 in hematopoiesis	
	IL-13 (P600)	132, monomer	IL-13R, CD132 (γc) (may also include CD24)	T cells	B-cell growth and differentiation, inhibits macrophage inflammatory cytokine production and T_H1 cells	IL-13: defective regulation of isotype-specific responses
	G-CSF	?, monomer*	G-CSFR	Fibroblasts and monocytes	Stimulates neutrophil development and differentiation	Defective myelopoiesis, neutropenia
	IL-15 (T-cell growth factor)	114, monomer	IL-15R, CD122 (IL-Rβ) CD132 (γc)	T cells	Il-2-like, stimulates growth of intestinal epithelium, T cells, and NK cells	
	GM-CSF (granulocyte macrophage colony stimulating factor)	127, monomer*	CD116, βc	Macrophages, T cells	Stimulates growth and differentiation of myelomonocytic lineage	GM-CSF, GM-CSFR: pulmonary alveolar proteinosis
	OSM (OM, oncostatin M)	196, monomer	OSMR or LIFR, CD130	T cells, macrophages	Stimulates Kaposis's sarcoma cells, inhibits melanoma growth	

*May function as dimers

Family	Cytokine (alternative names)	Size (no. of amino acids) and form	Receptors (c denotes common subunit)	Producer cells	Actions	Effect of cytokine or receptor knock-out (where known)
	LIF (leukemia inhibitory factor)	179, monomer	LIFR, CD130	Bone marrow stroma, fibroblasts	Maintains embryonic stem cells, like IL-6, IL-11, OSM	LIFR: die at or soon after birth; decreased hematopoietic stem cells
Interferons	IFN-γ	143, homodimer	CD119, IFNGR2	T cells, natural killer cells	Macrophage activation, increased expression of MHC molecules and antigen processing components, Ig class switching	IFN-γ, IFN-γR: Decreased resistance to bacterial infection, especially mycobacteria and certain viruses
	IFN-α	166, monomer	CD118, IFNAR2	Leukocytes	Anti-viral, increased MHC class I expression	IFN-α: Impaired antiviral defences
	IFN-β	166, monomer	CD118, IFNAR2	Fibroblasts	Anti-viral, increased MHC class I expression	
Immunoglobulin superfamily	B7.1 (CD80)	262, dimer	CD28, CTLA-4	Antigen-presenting cells	Co-stimulation of T-cell responses	CD28: decreased T-cell responses
	B7.2 (B70, CD86)		CD28, CTLA-4	Antigen-presenting cells	Co-stimulation of T-cell responses	B7.2: decreased co-stimulator response to alloantigen. CTLA-4: Massive lymphoproliferation, early death
TNF family	TNF-α (cachectin)	157, trimers	p55, p75 CD120a, CD120b	Macrophages, NK cells, T cells	Local inflammation, endothelial activation	TNF-αR: resistance to septic shock, susceptibility to *Listeria*
	TNF-β (lymphotoxin, LT, LT-α)	171, trimers	p55, p75 CD120a, CD120b	T cells, B cells	Killing, endothelial activation	TNF-β: absent lymph nodes, decreased antibody, increased IgM
	LT-β	Transmembrane, trimerizes with TNF-β, (LT-α)	LTβR or HVEM	T cells, B cells	Lymph node development	Defective development of peripheral lymph nodes, Peyer's patches, and spleen
	CD40 ligand (CD40L)	Trimers	CD40	T cells, mast cells	B-cell activation, class switching	CD40L: poor antibody response, no class switching, diminished T-cell priming (hyper IgM syndrome)
	Fas ligand (FasL)	Trimers	CD95 (Fas)	T cells, stroma?	Apoptosis, Ca^{2+}-independent cytotoxicity	Fas, FasL: mutant forms lead to lymphoproliferation, and autoimmunity
	CD27 ligand (CD27L)	Trimers (?)	CD27	T cells	Stimulates T cell proliferation	
	CD30 ligand (CD30L)	Trimers (?)	CD30	T cells	Stimulates T and B cell proliferation	CD30: Increased thymic size, alloreactivity
	4-1BBL	Trimers (?)	4-1BB	T cells	Co-stimulates T and B cells	
Unassigned	TGF-β	112, homo- and heterotrimers	TGF-βR	Chondrocytes, monocytes, T cells	Inhibits cell growth, anti-inflammatory	TGFβ: lethal inflammation

Family	Cytokine (alternative names)	Size (no. of amino acids) and form	Receptors (c denotes common subunit)	Producer cells	Actions	Effect of cytokine or receptor knock-out (where known)
	IL-1α	159, monomer	CD121a (IL-1RI) and CD121b (IL-1RII)	Macrophages, epithelial cells	Fever, T-cell activation, macrophage activation	IL-1RI: decreased IL-6 production
	IL-1β	153, monomer	CD121a (IL-1RI) and CD121b (IL-1RII)	Macrophages, epithelial cells	Fever, T-cell activation, macrophage activation	IL-1β: impaired acute phase response
	IL-1 RA	?, monomer	CD121a	Monocytes, macrophages, neutrophils, hepatocytes	Binds to but doesn't trigger IL-1 receptor, acts as a natural antagonist of IL-1 function	IL-1RA: reduced body mass, increased sensitivity to endotoxins (septic shock)
	IL-10 (cytokine synthesis inhibitor F)	160, homodimer	IL-10Rα, CRF2-4 (IL-10Rβ)	T cells, macrophages, EBV-transformed B cells	Potent suppressant of macrophage functions	Il-10 or CRF2-4: reduced growth, anemia, chronic enterocolitis
	IL-12 (NK cell stimulatory factor)	197 and 306, heterodimer	IL-12Rβ1 IL-12Rβ2	B cells, macrophages	Activates NK cells, induces CD4 T-cell differentiation to T_H1-like cells	IL-12: impaired in IFN-γ production and in T_H1 responses
	MIF	115, monomer		T cells, pituitary cells	Inhibits macrophage migration, stimulates macrophage activation	
	IL-16	130, homotetramer	CD4	T cells, mast cells, eosinophils	Chemoattractant for CD4 T cells, monocytes and eosinophils, anti-apoptotic for IL-2-stimulated T cells	
	IL-17 (mCTLA-8)	150, monomer		CD4 memory cells	Induce cytokine production by epithelia, endothelia, and fibroblasts	
	IL-18 (IGIF, interferon-γ inducing factor)	157, monomer	Il-1Rrp (IL-1R related protein)	Activated macrophages and Kupffer cells	Induces IFN-γ production by T cells and NK cells, favors T_H1 induction	Defective NK activity and T_H1 responses

Chemokine	Chromosome	Target cell	Specific receptor
†ELR⁺ CXC			
IL-8	4	Neutrophil, basophil, T cell	CXCR1, 2
GROα	4	Neutrophil	CXCR2 >> 1
GROβ	4	Neutrophil	CXCR2
GROγ	4	Neutrophil	CXCR2
ENA-78	4	Neutrophil	CXCR2
LDGF-PBP	4	Fibroblast, neutrophil	CXCR2
GCP-2	4	Neutrophil	CXCR2
†ELR⁻ CXC			
PF4	4	Fibroblast	Unknown
Mig	4	Activated T cell	CXCR3
IP-10	4	Activated T cell ($T_H1 > T_H2$)	CXCR3
SDF-1α/β	10	$CD34^+$ bone marrow cell, T cell, dendritic cell	CXCR4
CC			
MIP-1α	17	Monocyte/macrophage, T cell ($T_H1 > T_H2$), NK cell, basophil, dendritic cell, bone marrow cell, B cell	CCR1, 5, 9
MIP-1β	17	Monocyte/macrophage, T cell ($T_H1 > T_H2$), NK cell, basophil, dendritic cell, bone marrow cell	CCR1, 5, 9
MDC	16	Monocyte (?), dendritic cell, IA NK cell, T cell ($T_H2 > T_H1$)	CCR4
TECK	?	Macrophage, thymocytes, dendritic cell	Unknown
TARC	16	T-cell lines	CCR4
RANTES	17	Monocyte/macrophage, T cell (memory T cell > T cell; $T_H1 > T_H2$), NK cell, basophil, eosinophil, dendritic cell	CCR1, 3, 4, 5
HCC-1	17	Monocyte	CCR9
HCC-4	17	Monocyte	Unknown
DC-CK1	17	Naive T cell > T cell	Unknown
MIP-3α	2	T cell (memory T cell > T cell), peripheral blood mononuclear cell, bone marrow cell–dendritic cell	CCR6
MIP-3β	9	Naive T cell	CCR7
MCP-1	17	T cell, monocyte, basophil	CCR2, 9
MCP-2	17	T cell, monocyte, eosinophil, basophil	CCR2, 9
MCP-3	17	T cell, monocyte, eosinophil, basophil, dendritic cell	CCR2, 9
MCP-4	17	T cell, monocyte, eosinophil, basophil, dendritic cell	CCR2, 3, 9
None	(11)	Eosinophil, monocyte, T cell	CCR2
Eotaxin	17	Eosinophil	CCR3, 9
Eotaxin-2/MPIF-2	?	T cell (?), eosinophil, basophil	CCR3
6-Cysteine CC			
I-309	17	Neutrophil (TCA-3 only), T cell	CCR8
None	?	Monocyte, T cell	Unknown
None	(11)	T cell, neutrophil (?)	Unknown
MIP-5/HCC-2	17	T cell, monocyte, neutrophil (?), dendritic cell	CCR1, 3
MPIF-1	17 (?)	Monocyte, T cell (resting), neutrophil (?)	Unknown
6Ckine	9 (?)	Naive T cell, B cell, mesangial cells (?)	Unknown
C and CX3C			
Lymphotactin	1 (1)	T cell, NK cell	Unknown
Fractalkine	16	T cell, monocyte, neutrophil (?)	CX3CR1

† ELR refers to the three amino acids that precede the first cysteine residue of the CXC motif. If these amino acids are Glu-Leu-Arg (ie ELR^+), then the chemokine is chemotactic for neutrophils while if they are not (ELR^-) then the chemokine is chemotactic for lymphocytes.

BIOGRAPHIES

Emil von Behring (1854–1917) discovered antitoxin antibodies with Shibasaburo Kitasato.

Baruj Benacerraf (1920–) discovered immune response genes and collaborated in the first demonstration of MHC restriction.

Jules Bordet (1870–1961) discovered complement as a heat-labile component in normal serum that would enhance the anti-microbial potency of specific antibodies.

Frank MacFarlane Burnet (1899–1985) proposed the first generally accepted clonal selection hypothesis of adaptive immunity.

Jean Dausset (1916–) was an early pioneer in the study of the human major histocompatibility complex or HLA.

Peter Doherty (1940–) and **Rolf Zinkernagal** (1944–) showed that antigen recognition by T cells is MHC-restricted, thereby establishing the biological role of the proteins encoded by the major histocompatibility complex and leading to an understanding of antigen processing and its importance in the recognition of antigen by T cells.

Gerald Edelman (1929–) made crucial discoveries about the structure of immunoglobulins, including the first complete sequence of an antibody molecule.

Paul Ehrlich (1854–1915) was an early champion of humoral theories of immunity, and proposed a famous side-chain theory of antibody formation that bears a striking resemblance to current thinking about surface receptors.

James Gowans (1924–) discovered that adaptive immunity is mediated by lymphocytes, focusing the attention of immunologists on these small cells.

Michael Heidelberger (1888–1991) developed the quantitative precipitin assay, ushering in the era of quantitative immunochemistry.

Edward Jenner (1749–1823) described the successful protection of humans against smallpox infection by vaccination with cowpox or vaccinia virus. This founded the field of immunology.

Niels Jerne (1911–1994) developed the hemolytic plaque assay and several important immunological theories, including an early version of clonal selection, a prediction that lymphocyte receptors would be inherently biased to MHC recognition, and the idiotype network.

Shibasaburo Kitasato (1892–1931) discovered antibodies in collaboration with Emil von Behring.

Robert Koch (1843–1910) defined the criteria needed to characterize an infectious disease, known as Koch's postulates.

Georges Köhler (1946–1995) pioneered monoclonal antibody production from hybrid antibody-forming cells with Cesar Milstein.

Karl Landsteiner (1868–1943) discovered the ABO blood group antigens. He also carried out detailed studies of the specificity of antibody binding using haptens as model antigens.

Peter Medawar (1915–1987) used skin grafts to show that tolerance is an acquired characteristic of lymphoid cells, a key feature of clonal selection theory.

Elie Metchnikoff (1845–1916) was the first champion of cellular immunology, focusing his studies on the central role of phagocytes in host defence.

Cesar Milstein (1927–) pioneered monoclonal antibody production with Georges Köhler.

Louis Pasteur (1822–1895) was a French microbiologist and immunologist who validated the concept of immunization first studied by Jenner. He prepared vaccines against chicken cholera and rabies.

Rodney Porter (1920–1985) worked out the polypeptide structure of the antibody molecule, laying the groundwork for its analysis by protein sequencing.

George Snell (1903–1996) worked out the genetics of the murine major histocompatibility complex and generated the congenic strains needed for its biological analysis, laying the groundwork for our current understanding of the role of the MHC in T-cell biology.

Susumu Tonegawa (1939–) discovered the somatic recombination of immunological receptor genes that underlies the generation of diversity in human and murine antibodies and T-cell receptors.

GLOSSARY

The **12/23** rule states that gene segments of immunoglobulin or T-cell receptors can be joined only if one has a recognition signal sequence with a 12 base pair spacer, and the other has a 23 base pair spacer.

In the context of immunoglobulins, α is the type of heavy chain in IgA.

The **ABO blood group system** antigens are expressed on red blood cells. They are used for typing human blood for transfusion. If they do not express A or B antigens on their red blood cells, people naturally form antibodies that interact with them.

The removal of antibodies specific for one antigen from an antiserum to render it specific for another antigen or antigens is called **absorption**.

α:β T-cell receptor: see **T-cell receptor**.

Accessory cells in adaptive immunity are cells that aid in the response but do not directly mediate specific antigen recognition. They include phagocytes, mast cells, and NK cells, and are also known as **accessory effector cells**.

The **acquired immune deficiency syndrome (AIDS)** is a disease caused by infection with the human immunodeficiency virus (HIV). AIDS occurs when an infected patient has lost most of his or her CD4 T cells, so that infections with opportunistic pathogens occur.

Acquired immune response: see **adaptive immune response**.

Immunization with antigen is called **active immunization** to distinguish it from the transfer of antibody to an unimmunized individual, which is called passive immunization.

Acute lymphoblastic leukemia is a highly aggressive, undifferentiated form of lymphoid malignancy derived from a progenitor cell that is thought to be able to give rise to both lineages of lymphoid cells.

Acute-phase proteins are a series of proteins found in the blood shortly after the onset of an infection. These proteins participate in early phases of host defense against infection. An example is the mannose-binding protein.

The **acute-phase response** is a change in the blood that occurs during early phases of an infection. It includes the production of acute-phase proteins and also of cellular elements.

The **adaptive immune response** or **adaptive immunity** is the response of antigen-specific lymphocytes to antigen, including the development of immunological memory. Adaptive immune responses are generated by clonal selection of lymphocytes. Adaptive immune responses are distinct from innate and non-adaptive phases of immunity, which are not mediated by clonal selection of antigen-specific lymphocytes. Adaptive immune responses are also known as **acquired immune responses**.

The **adaptor proteins** are key linkers between receptors and downstream members of signalling pathways. These proteins are functionally heterogeneous, but share an SH2 domain as the means of interacting with the phoshotyrosine residues generated by the receptor-associated tyrosine kinases. The protein known as **Vav** is an adaptor protein of this type.

The **adenoids** are **mucosal-associated lymphoid tissues** located in the nasal cavity.

The enzyme defect **adenosine deaminase deficiency** leads to the accumulation of toxic purine nucleosides and nucleotides, resulting in the death of most developing lymphocytes within the thymus. It is a common cause of severe combined immunodeficiency.

Adhesion molecules mediate the binding of one cell to other cells or to extracellular matrix proteins. Integrins, selectins, members of the immunoglobulin gene superfamily, and CD44 and related proteins are all adhesion molecules important in the operation of the immune system.

An **adjuvant** is any substance that enhances the immune response to an antigen with which it is mixed.

Adoptive immunity is immunity conferred on a naive or irradiated recipient by transfer of lymphoid cells from an actively immunized donor. This is called **adoptive transfer** or **adoptive immunization**.

Afferent lymphatic vessels drain fluid from the tissues and carry antigens from sites of infection in most parts of the body to the lymph nodes.

Affinity is the strength of binding of one molecule to another at a single site, such as the binding of a monovalent Fab fragment of antibody to a monovalent antigen. See also **avidity**.

Affinity chromatography is the purification of a substance by means of its affinity for another substance immobilized on a solid support; an antigen can be purified by affinity chromatography on a column of specific antibody molecules covalently linked to beads.

Affinity maturation refers to the increase in the affinity of the antibodies produced during the course of a humoral immune response. It is particularly prominent in secondary and subsequent immunizations.

Agammaglobulinemia: see **X-linked agammaglobulinemia**.

Agglutination is the clumping together of particles, usually by antibody molecules binding to antigens on the surfaces of adjacent particles. Such particles are said to **agglutinate**. When the particles are red blood cells, the phenomenon is called hemagglutination.

Agonist peptides are peptide antigens that activate their specific T cells, inducing them to make cytokines and to proliferate.

AIDS: see **acquired immune deficiency syndrome**.

Alleles are variants of a single genetic locus.

Allelic exclusion refers to the fact that in a heterozygous individual, only one of the alternative C-region alleles of the heavy or light chain is expressed in an immunoglobulin molecule. The term has come to be used more generally to describe the expression of a single receptor specificity in cells with the potential to express two or more receptors.

Allergens are antigens that elicit hypersensitivity or allergic reactions.

Allergic asthma is constriction of the bronchial tree due to an allergic reaction to inhaled antigen.

An **allergic reaction** is a response to innocuous environmental antigens or allergens due to pre-existing antibody or T cells. There are various immune mechanisms of allergic reactions, but the most common is the binding of allergen to IgE antibody on mast cells that causes asthma, hay fever, and other common allergic reactions.

Allergic rhinitis is an allergic reaction in the nasal mucosa, also known as hay fever, that causes runny nose, sneezing and tears.

Allergy is the symptomatic reaction to a normally innocuous environmental antigen. It results from the interaction between the antigen and the antibody or T cells produced by earlier exposure to the same antigen.

Two individuals or two mouse strains that differ at the MHC are said to be **allogeneic**. The term can also be used for allelic differences at other loci. See also **syngeneic**; **xenogeneic**.

Rejection of grafted tissues from unrelated donors usually results from T-cell responses to **allogeneic** MHC molecules expressed by the grafted tissues.

An **allograft** is a graft of tissue from an allogeneic or non-self donor of the same species; such grafts are invariably rejected unless the recipient is immunosuppressed.

Alloreactivity describes the stimulation of T cells by MHC molecules other than self; it marks the recognition of allogeneic MHC.

Allotypes are allelic polymorphisms that can be detected by antibodies specific for the polymorphic gene products; in immunology, **allotypic** differences in immunoglobulin molecules were important in deciphering the genetics of antibodies.

An **altered peptide ligand** is a peptide, usually closely related to an agonist peptide in amino acid sequence, that induces only a partial response from T cells specific for the agonist peptide.

The **alternative pathway** of complement activation is not triggered by antibody, as is the classical pathway of complement activation, but by the binding of complement protein C3b to the surface of a pathogen; it is therefore a feature of innate immunity. The alternative pathway also amplifies the classical pathway of complement activation.

Anaphylactic shock or systemic anaphylaxis is an allergic reaction to systemically administered antigen that causes circulatory collapse and suffocation due to tracheal swelling. It results from binding of antigen to IgE antibody on connective tissue mast cells throughout the body, leading to the disseminated release of inflammatory mediators.

Anaphylatoxins are small fragments of complement proteins released by cleavage during complement activation, that recruit fluid and inflammatory cells to sites of antigen deposition. The fragments C5a, C3a, and C4a are all anaphylatoxins, listed in order of decreasing potency *in vivo*.

Peptide fragments of antigens are bound to specific MHC class I molecules by **anchor residues,** which are amino acid side chains of the peptide that bind into pockets lining the peptide-binding groove of the MHC class I molecule. Each MHC class I molecule binds different patterns of anchor residues, each called an anchor motif, giving some specificity to peptide binding. Anchor residues are less obvious for peptides that bind to MHC class II molecules.

Anergy is a state of non-responsiveness to antigen. People are said to be anergic when they cannot mount delayed-type hypersensitivity reactions to challenge antigens, whereas T and B cells are said to be anergic when they cannot respond to their specific antigen under optimal conditions of stimulation.

Antagonist peptides are peptides, usually closely related in sequence to an agonist peptide, that inhibit the response of a cloned T-cell line specific for the agonist peptide.

An **antibody** is a protein that binds specifically to a particular substance—its antigen. Each antibody molecule has a unique structure that enables it to bind specifically to its corresponding antigen, but all antibodies have the same overall structure and are known collectively as immunoglobulins. Antibodies are produced by plasma cells in response to infection or immunization, and bind to and neutralize pathogens or prepare them for uptake and destruction by phagocytes.

Antibody-dependent cell-mediated cytotoxicity (**ADCC**) is the killing of antibody-coated target cells by cells with Fc receptors that recognize the Fc region of the bound antibody. Most ADCC is mediated by NK cells that have the Fc receptor FcγRIII or CD16 on their surface.

The **antibody repertoire** describes the total variety of antibodies that an individual can make.

An **antigen** is any molecule that can bind specifically to an antibody. Their name arises from their ability to **gen**erate **anti**bodies. However, some antigens do not, by themselves, elicit antibody production; those antigens that can induce antibody production are called immunogens.

Antigen:antibody complexes are non-covalently associated groups of antigen and antibody molecules which can vary in size from small, soluble complexes to large, insoluble complexes that precipitate out of solution; they are also known as immune complexes.

The **antigen-binding site** of an antibody is the surface of the antibody molecule that makes physical contact with the antigen. Antigen-binding sites are made up of six hypervariable loops, three from the light-chain V region and three from the heavy-chain V region.

T and B lymphocytes collectively bear on their surface highly diverse **antigen receptors** capable of recognizing a wide diversity of antigens. Each individual lymphocyte bears receptors of a single antigen specificity.

An **antigenic determinant** is the portion of an antigenic molecule that is bound by a given antibody; it is also known as an epitope.

Influenza virus varies from year to year by a process of **antigenic drift** in which point mutations of viral genes cause small differences in the structure of viral surface antigens. Periodically, influenza viruses undergo an **antigenic shift** through reassortment of their segmented genome with another influenza virus, changing their surface antigens radically. Such antigenic shift variants are not recognized by individuals immune to influenza, so when antigenic shift variants arise, there is widespread and serious disease.

Many pathogens evade the adaptive immune response by **antigenic variation** in which new antigens are displayed that are not recognized by antibodies or T cells elicited in earlier infections.

Antigen presentation describes the display of antigen as peptide fragments bound to MHC molecules on the surface of a cell; all T cells recognize antigen only when it is presented in this way.

Antigen-presenting cells are highly specialized cells that can process antigens and display their peptide fragments on the cell surface together with molecules required for lymphocyte activation. The main antigen-presenting cells for T cells are dendritic cells, macrophages, and B cells, whereas the main antigen-presenting cells for B cells are follicular dendritic cells.

Antigen processing is the degradation of proteins into peptides that can bind to MHC molecules for presentation to T cells. All antigens except peptides must be processed into peptides before they can be presented by MHC molecules.

Anti-immunoglobulin antibodies are antibodies against immunoglobulin constant domains, useful for detecting bound antibody molecules in immunoassays and other applications.

An **antiserum** (plural: **antisera**) is the fluid component of clotted blood from an immune individual that contains antibodies against the molecule used for immunization. Antisera contain heterogeneous collections of antibodies, which bind the antigen used for immunization, but each has its own structure, its own epitope on the antigen, and its own set of cross-reactions. This heterogeneity makes each antiserum unique.

Aplastic anemia is a failure of bone marrow stem cells so that formation of all cellular elements of the blood ceases; it can be treated by bone marrow transplantation.

Apoptosis, or programmed cell death, is a form of cell death in which the cell activates an internal death program. It is characterized by nuclear DNA degradation, nuclear degeneration and condensation, and the phagocytosis of cell residua. Proliferating cells frequently undergo apoptosis, which is a natural process in development, and proliferating lymphocytes undergo high rates of apoptosis in development and during immune responses. It contrasts with necrosis, death from without, which occurs in situations such as poisoning and anoxia.

The **appendix** is a gut-associated lymphoid tissue located at the beginning of the colon.

In this book, we have termed activated effector T cells **armed effector T cells**, because these cells are triggered to perform their effector functions immediately on contact with cells bearing the peptide:MHC complex for which they are specific. They contrast with memory T cells, which need to be activated by antigen-presenting cells before they can mediate effector responses.

The **Arthus reaction** is a skin reaction in which antigen is injected into the dermis and reacts with IgG antibodies in the extracellular spaces, activating complement and phagocytic cells to produce a local inflammatory response.

Ascertainment artifact refers to data that seem to demonstrate some finding, but fail to do so because they are collected from a population that is selected in a biased fashion.

Atopic allergy, or **atopy**, is the increased tendency seen in some people to produce immediate hypersensitivity reactions (usually mediated by IgE antibodies) against innocuous substances.

Pathogens are said to be **attenuated** when they can grow in their host and induce immunity without producing serious clinical disease.

Antibodies specific for self antigens are called **autoantibodies**.

A graft of tissue from one site to another on the same individual is called an **autograft**.

Diseases in which the pathology is caused by immune responses to self antigens are called **autoimmune diseases**.

Autoimmune hemolytic anemia is a pathological condition with low levels of red blood cells (anemia), which is caused by autoantibodies that bind red blood cell surface antigens and target the red blood cell for destruction.

An adaptive immune response directed at self antigens is called an **autoimmune response**; likewise, adaptive immunity specific for self antigens is called **autoimmunity**.

In the disease **autoimmune thrombocytopenic purpura**, antibodies against a patient's platelets are made. Antibody binding to platelets causes them to be taken up by cells with Fc receptors and complement receptors, causing a fall in platelet counts that leads to purpura (bleeding).

Autoreactivity describes immune responses directed at self antigens.

Avidity is the sum total of the strength of binding of two molecules or cells to one another at multiple sites. It is distinct from affinity, which is the strength of binding of one site on a molecule to its ligand.

The **avidity hypothesis** (formerly called the affinity hypothesis) of T-cell selection in the thymus states that T cells must have a measurable affinity for self MHC molecules in order to mature, but not so great an affinity as to cause activation of the cell when it matures, as this would require that the cell be deleted to maintain self tolerance.

Azathioprine is a potent immunosuppressive drug that is converted to its active form *in vivo* and then kills rapidly proliferating cells, including lymphocytes responding to grafted tissues.

B cells are divided into two classes, known as **B-1 B cells**, also known as CD5 B cells, and **B-2 B cells**, also known as conventional B cells.

The major T-cell co-stimulatory molecules are **B7.1** and **B7.2**, closely related members of the immunoglobulin gene superfamily. They are expressed differentially on various antigen-presenting cell types, and they may have different consequences for the responding T cells. **B7 molecules** refers to both B7.1 and B7.2.

A **β sheet** is one of the fundamental structural building blocks of proteins, consisting of adjacent, extended strands of amino acids (**β strands**) which are bonded together by interactions between backbone amide and carbonyl groups. β sheets can be parallel, in which case the adjacent strands run in the same direction, or antiparallel, where adjacent strands run in opposite directions. All immunoglobulin domains are made up of antiparallel β-sheet structures. A **β barrel** or a **β sandwich** is another way of describing the structure of the immunoglobulin domain.

Many infectious diseases are caused by **bacteria**, which are prokaryotic microorganisms that exist as many different species and strains. Bacteria can live on body surfaces, in extracellular spaces, in cellular vesicles, or in the cytosol, and different bacterial species cause distinctive infectious diseases.

BALT: see **bronchial-associated lymphoid tissue**.

Bare lymphocyte syndrome is an immunodeficiency in which MHC class II molecules are not expressed on cells as a result of one of several different regulatory gene defects. Patients with bare lymphocyte syndrome are severely immunodeficient and have few CD4 T cells.

Basophils are white blood cells containing granules that stain with basic dyes, and which are thought to have a function similar to mast cells.

A **B cell**, or **B lymphocyte**, is one of the two major types of lymphocyte. The antigen receptor on B lymphocytes, sometimes called the B-cell receptor, is a cell-surface immunoglobulin. On activation by antigen, B cells differentiate into cells producing antibody molecules of the same antigen-specificity as this receptor.

The **B-cell antigen receptor**, or **B-cell receptor**, is the cell-surface receptor of B cells for specific antigen. It is composed of a transmembrane immunoglobulin molecule associated with the invariant Igα and Igβ chains in a non-covalent complex.

A complex of CD19, TAPA-1, and CR2 makes up the **B-cell co-receptor**; co-ligation of this complex with the B-cell antigen receptor increases responsiveness to antigen by about 100-fold.

The **B-cell corona** in the spleen is the zone of the white pulp primarily made up of B cells.

B-cell mitogens are substances that cause B cells to proliferate.

Blk: see **tyrosine kinase**.

Blood group antigens are surface molecules on red blood cells that are detectable with antibodies from other individuals. The major blood group antigens are called ABO and Rh (Rhesus), and are used in routine blood banking to type blood. There are many other blood group antigens that can be detected in cross-matching.

In transfusion medicine, **blood typing** is used to determine whether donor and recipient have the same ABO and Rh blood group antigens. A cross-match, in which serum from the donor is tested on the cells of the recipient, and vice versa, is used to rule out other incompatibilities. Transfusion of incompatible blood causes a transfusion reaction, in which red blood cells are destroyed and the released hemoglobin causes toxicity.

B lymphocytes: see **B cells**.

The **bone marrow** is the site of hematopoiesis, the generation of the cellular elements of blood, including red blood cells, monocytes, polymorphonuclear leukocytes, and platelets. The bone marrow is also the site of B-cell development in mammals and the source of stem cells that give rise to T cells upon migration to the thymus. Thus, bone marrow transplantation can restore all the cellular elements of the blood, including the cells required for adaptive immunity.

A **bone marrow chimera** is formed by transferring bone marrow from one mouse to an irradiated recipient mouse, so that all of the lymphocytes and blood cells are of donor genetic origin. Bone marrow chimeras have been crucial in elucidating the development of lymphocytes and other blood cells.

A **booster immunization** is given after a primary immunization, to increase the state of immunity.

The lymphoid cells and organized lymphoid tissues in the respiratory tract have been termed the **bronchial-associated lymphoid tissues** (**BALT**). These tissues are very important in the induction of immune responses to inhaled antigens and respiratory infection.

Bruton's X-linked agammaglobulinemia: see **X-linked agammaglobulinemia**.

The **bursa of Fabricius** is an outpouching of the cloaca found in birds. It is an aggregate of epithelial tissue and lymphoid cells and is the site of intense early B-cell proliferation. The bursa of Fabricius is required for B-cell development in birds, as its removal or **bursectomy** early in life causes an absence of B cells in adult birds. An equivalent structure has not been detected in humans, where B-cell development follows a different pathway.

C1 inhibitor (**C1INH**) is a protein that inhibits the activity of activated complement component C1 by binding to and inactivating its C1r:C1s enzymatic activity. Deficiency in C1INH is the cause of the disease hereditary angioneurotic edema, in which spontaneous complement activation causes episodes of epiglottal swelling and suffocation.

The generation of the enzyme **C3/C5 convertase** on the surface of a pathogen or cell is a crucial step in complement activation. The C3/C5 convertase then catalyzes the deposition of large numbers of C3 molecules on the pathogen surface, leading to opsonization and the activation of the effector cascade that causes membrane lesions.

The receptor for the C5a fragment of complement, the **C5a receptor**, is a seven transmembrane spanning receptor that couples to a heterotrimeric G protein.

The cytosolic serine/threonine phosphatase **calcineurin** has a crucial but undefined role in signaling via the T-cell receptor. The immunosuppressive drugs cyclosporin A and FK506 form complexes with cellular proteins called immunophilins that bind and inactivate calcineurin, suppressing T-cell responses.

The protein **calnexin** is an 88 kDa protein found in the endoplasmic reticulum. It binds to partly folded members of the immunoglobulin superfamily of proteins and retains them in the endoplasmic reticulum until folding is completed.

Antibodies or antigens can be measured in various **capture** assays. In these assays, antigens are captured by antibodies bound to plastic (or vice versa). Antibody binding to a plate-bound antigen can be measured by using labeled antigen or anti-immunoglobulin. Antigen binding to plate-bound antibody can be measured by using an antibody that binds to a different epitope on the antigen.

Carriers are foreign proteins to which small non-immunogenic antigens, or haptens, can be coupled to render the hapten immunogenic. *In vivo*, self proteins can also serve as carriers if they are correctly modified by the hapten; this is important in allergy to drugs.

Caseation necrosis is a form of necrosis seen in the center of large granulomas, such as the granulomas in tuberculosis. The term comes from the white cheesy appearance of the central necrotic area.

Caspases are a family of closely related cysteine proteases that cleave proteins at aspartic acid residues. They have important roles in apoptosis.

CD: see **clusters of differentiation** and Appendix I.

The **CD3 complex** is the complex of $\alpha{:}\beta$ or $\gamma{:}\delta$ T-cell receptor chains with the invariant subunits CD3γ, δ, and ε, and the dimeric ζ chains.

The cell-surface protein **CD4** is important for recognition by the T-cell receptor of antigenic peptides bound to MHC class II molecules. It acts as a co-receptor by binding to the lateral face of the MHC class II molecules.

CD5 B cells are a class of atypical, self-renewing B cells found mainly in the peritoneal and pleural cavities in adults and which have a far less diverse receptor repertoire than conventional B cells.

The cell-surface protein **CD8** is important for recognition by the T-cell receptor of antigenic peptides bound to MHC class I molecules. It acts as a co-receptor by binding to the lateral face of MHC class I molecules.

CD19:CR2:TAPA-1 complex: see **B-cell co-receptor**.

CD23: see Appendix I.

CD28: see Appendix I.

CD34: see Appendix I.

B-cell growth is triggered in part by the binding of **CD40 ligand**, expressed on activated helper T cells, to **CD40** on the B-cell surface.

CD45, or the leukocyte common antigen, is a transmembrane tyrosine phosphatase found on all leukocytes. It is expressed in different isoforms on different cell types, including the different subtypes of T cells. These isoforms are commonly denoted by the designation of CD45R followed by the exon whose presence gives rise to distinctive antibody-binding patterns.

The constant regions of the polypeptide chains of immunoglobulin molecules are made up of one or more constant domains or **C domains** of similar structure. A C domain is one of the two main types of immunoglobulin domain.

CDR: see **complementarity determining region**.

Cell adhesion molecules (**CAMs**) are cell-surface proteins that are involved in binding cells together in tissues and also in less permanent cell–cell interactions.

Cell-mediated immunity, or a **cell-mediated immune response**, describes any adaptive immune response in which antigen-specific T cells have the main role. It is defined operationally as all adaptive immunity that cannot be transferred to a naive recipient with serum antibody. Cf. **humoral immunity**.

Cell-surface immunoglobulin is the B-cell receptor for antigen. See **B cell**.

Cellular immunology is the study of the cellular basis of immunity.

Central lymphoid organs are sites of lymphocyte development. In humans, B lymphocytes develop in bone marrow, whereas T lymphocytes develop within the thymus from bone-marrow derived progenitors.

Central tolerance is tolerance that is established in lymphocytes developing in central lymphoid organs. Cf. **peripheral tolerance**.

Centroblasts are large, rapidly dividing cells found in germinal centers, and are the cells in which somatic hypermutation is believed to occur. Antibody-secreting and memory B cells derive from these cells.

Centrocytes are the small, non-proliferating B cells in germinal centers that derive from centroblasts. They may mature into antibody-secreting plasma cells or memory B cells, or may undergo apoptosis, depending on their receptor's interaction with antigen.

Chediak–Higashi syndrome is a defect in phagocytic cell function, of unknown causes, in which lysosomes fail to fuse properly with phagosomes and there is impaired killing of ingested bacteria.

Chemokines are small cytokines that are involved in the migration and activation of cells, especially phagocytic cells and lymphocytes. They have a central role in inflammatory responses. Chemokines and their receptors are listed in Appendix III.

Most lymphoid tumors, and many other tumors, bear **chromosomal translocations** that mark points of breakage and rejoining of different chromosomes. These chromosomal breaks are particularly frequent in lymphomas and leukemias.

Chronic granulomatous disease is an immunodeficiency disease in which multiple granulomas form as a result of defective elimination of bacteria by phagocytic cells. It is caused by a defect in the NADPH oxidase system of enzymes that generate the superoxide radical involved in bacterial killing.

The **class II-associated invariant chain peptide** (**CLIP**) is a peptide of variable length cleaved from the class II invariant chain by proteases. It remains associated with the MHC class II molecule in an unstable form until it is removed by the HLA-DM protein.

Class II transactivator (**CIITA**): see **MHC class II transactivator**.

Class switching: see **isotype switching**.

Classes: see **isotypes**.

The **classical pathway** of complement activation is the pathway activated by antibody bound to antigen, and involves complement components C1, C4, and C2 in the generation of the C3/C5 convertase. See also **alternative pathway**.

Clonal deletion is the elimination of immature lymphocytes on binding to self antigens to produce tolerance to self, as required by the clonal selection theory. Clonal deletion is the main mechanism of central tolerance and can also occur in peripheral tolerance.

Clonal expansion is the proliferation of antigen-specific lymphocytes in response to antigenic stimulation and precedes their differentiation into effector cells. It is an essential step in adaptive immunity, allowing rare antigen-specific cells to increase in number so that they can effectively combat the pathogen that elicited the response.

The **clonal selection theory** is a central paradigm of adaptive immunity. It states that adaptive immune responses derive from individual antigen-specific lymphocytes that are self-tolerant. These specific lymphocytes proliferate in response to antigen and differentiate into antigen-specific effector cells that eliminate the eliciting pathogen, and memory cells to sustain immunity. The theory was formulated by Sir Macfarlane Burnet and in earlier forms by Niels Jerne and David Talmage.

A **clone** is a population of cells all derived from a single progenitor cell.

A **cloned T-cell line** is a continuously growing line of T cells derived from a single progenitor cell. Cloned T-cell lines must be stimulated with antigen periodically to maintain growth. They are useful for studying T-cell specificity, growth, and effector functions.

A feature unique to individual cells or members of a clone is said to be **clonotypic**. Thus, a monoclonal antibody that reacts with the receptor on a cloned T-cell line is said to be a clonotypic antibody and to recognize its clonotype or the clonotypic receptor of that cell. See also **idiotype**; **idiotypic**.

Clusters of differentiation (**CD**) are groups of monoclonal antibodies that identify the same cell-surface molecule. The cell-surface molecule is designated CD followed by a number (e.g. CD1, CD2, etc.). For a current listing of CD see Appendix I.

The expression of a gene is said to be **co-dominant** when both alleles at one locus are expressed in roughly equal amounts in heterozygotes. Most genes show this property, including the highly polymorphic MHC genes.

A **coding joint** is formed by the imprecise joining of a V gene segment to a (D)J gene segment in immunoglobulin or T-cell receptor genes.

Co-isogenic: see **congenic**.

Collectins are a structurally related family of calcium-dependent sugar-binding proteins or lectins containing collagen-like sequences. An example is mannose-binding protein.

Antigen receptors manifest two distinct types of **combinatorial diversity** generated by the combination of separate units of genetic information. Receptor gene segments are joined in many different combinations to generate diverse receptor chains, and then two different receptor chains (heavy and light in immunoglobulins; α and β or γ and δ in T-cell receptors) are combined to make the antigen-recognition site.

Common lymphoid progenitors are stem cells that give rise to all lymphocytes. They are derived from pluripotent hematopoietic stem cells.

Common variable immunodeficiency is a relatively common deficiency in antibody production whose pathogenesis is not yet understood. There is a strong association with genes mapping within the MHC.

Competitive binding assays are serological assays in which unknowns are detected and quantitated by their ability to inhibit the binding of a labeled known ligand to its specific antibody. This is also referred to as a competitive inhibition assay.

When antibodies or antigens are assayed by binding of a known antibody or antigen to a known amount of labeled antibody or antigen, and then known or unknown sources of antibody or antigen are used as competitive inhibitors, it is referred to as a **competitive inhibition assay**.

The **complement** system is a set of plasma proteins that act together to attack extracellular forms of pathogens. Complement can be activated spontaneously on certain pathogens or by antibody binding to the pathogen. The pathogen becomes coated with complement proteins that facilitate pathogen removal by phagocytes and can also kill the pathogen directly.

Complement receptors (**CR**) are cell-surface proteins on various cells that recognize and bind complement proteins that have bound an antigen such as a pathogen. Complement receptors on phagocytes allow them to identify pathogens coated with complement proteins for uptake and destruction. Complement receptors include CR1, CR2, CR3, CR4, and the receptor for C1q.

The **complementarity determining regions** (**CDRs**) of immunoglobulins and T-cell receptors are the parts of these molecules that determine their specificity and make contact with specific ligand. The CDRs are the most variable part of the molecule, and contribute to the diveristy of these molecules. There are three such regions (CDR1, CDR2, and CDR3) in each V domain.

Conformational epitopes, or discontinuous epitopes, on a protein antigen are formed from several separate regions in the primary sequence of a protein brought together by protein folding. Antibodies that bind conformational epitopes bind only native folded proteins. Cf. **continuous epitopes**.

Congenic strains of mice are genetically identical at all loci except one. Each strain is generated by the repetitive backcrossing of mice carrying the desired trait onto a strain that provides the genetic background for the set of congenic strains. The most important congenic strains in immunology are the congenic resistant strains, developed by George Snell, that differ from each other at the MHC.

Conjugate vaccines are vaccines made from capsular polysaccharides bound to proteins of known immunogenicity, such as tetanus toxoid.

The **constant region** (**C region**) of an immunoglobulin or T-cell receptor is that part of the molecule that is relatively constant in amino acid sequence between different molecules. In an antibody molecule the constant regions of each chain are composed of one or more C domains. The constant region of an antibody determines its particular effector function. Cf. **variable region.**

A **contact hypersensitivity reaction** is a form of delayed-type hypersensitivity in which T cells respond to antigens that are introduced by contact with the skin. Poison ivy hypersensitivity is a contact hypersensitivity reaction due to T-cell responses to the chemical antigen pentadecacatechol in poison ivy leaves.

Continuous epitopes, or linear epitopes, are antigenic determinants on proteins that are contiguous in the amino acid sequence and therefore do not require the protein to be folded into its native conformation for antibody to bind. Cf. **conformational epitopes**.

A **convertase** is an enzymatic activity that converts a complement protein into its reactive form by cleaving it. Generation of the C3/C5 convertase is the pivotal event in complement activation.

The **Coombs test** is a test for antibody binding to red blood cells. Red blood cells that are coated with antibody are agglutinated if they are exposed to an anti-immunoglobulin antibody. The Coombs test is important in detecting the non-agglutinating antibodies against red blood cells produced in Rh incompatibility in pregnancy.

Two binding sites are said to demonstrate **cooperativity** in binding to their ligands when the effect of binding to both is greater than the sum of each binding site acting on its own.

A **co-receptor** is a cell-surface protein that increases the sensitivity of the antigen receptor to antigen by binding to associated ligands and participating in signaling for activation. CD4 and CD8 are MHC-binding co-receptors on T cells, whereas CD19 is part of a complex that makes up a co-receptor on B cells.

Corticosteroids are steroids related to those produced in the adrenal cortex, such as cortisone. Corticosteroids can kill lymphocytes, especially developing thymocytes, inducing apoptotic cell death. They are useful anti-inflammatory and immunosuppressive agents.

The proliferation of lymphocytes requires both antigen binding and the receipt of a **co-stimulatory signal**, usually delivered by a cell-surface molecule on the cell presenting antigen. For T cells, the co-stimulatory signals are B7 and B7.2, related molecules that act on the T-cell surface molecules CD28 and CTLA-4. For B cells, CD40 ligand acting on CD40 serves an analogous role.

Cowpox is the common name of the disease produced by vaccinia virus, used by Edward Jenner in the successful vaccination against smallpox, which is caused by the related variola virus.

CR: see **complement receptors**.

CR1 is one of several receptors on cells for various components of complement. It is used to remove immune complexes from the plasma.

C-reactive protein is an acute-phase protein that binds to phosphatidylcholine, which is a constituent of the C-polysaccharide of the bacterium *Streptococcus pneumoniae*, hence its name. Many other bacteria also have surface phosphatidylcholine that is accessible to C-reactive protein, so the protein can bind many different bacteria and opsonize them for uptake by phagocytes.

C region: see **constant region**.

Cross-matching is used in blood typing and histocompatibility testing to determine whether donor and recipient have antibodies against each other's cells that might interfere with successful transfusion or grafting.

A **cross-reaction** is the binding of antibody to an antigen not used to elicit that antibody. Thus, if antibody raised against antigen A also binds antigen B, it is said to cross-react with antigen B. The term is used generically to describe the reactivity of antibodies or T cells with antigens other than the eliciting antigen.

The protein known as **Csk** or **C-terminal Src kinase** is constitutively active in lymphocytes and has the function of phosphorylating the C-terminal tyrosine of Src-family tyrosine kinases, thus inactivating them.

CTLA-4 is the high-affinity receptor for B7 molecules on T cells.

Cutaneous lymphocyte antigen (**CLA**) is a cell-surface molecule that is involved in lymphocyte homing to the skin in humans.

Cutaneous T-cell lymphoma is a malignant growth of T cells that home to the skin.

Cyclophosphamide is an alkylating agent that is used as an immunosuppressive drug. It acts by killing rapidly dividing cells, including lymphocytes proliferating in response to antigen.

Cyclosporin A is a powerful immunosuppressive drug that inhibits signaling from the T-cell receptor, preventing T-cell activation and effector function. It binds to cyclophilin, and this complex binds to and inactivates the serine/threonine phosphatase calcineurin.

Cytokines are proteins made by cells that affect the behavior of other cells. Cytokines made by lymphocytes are often called lymphokines or interleukins (abbreviated IL), but the generic term cytokine is used in this book and most of the literature. Cytokines act on specific cytokine receptors on the cells that they affect. Cytokines and their receptors are listed in Appendix II. See also **chemokines**.

Cytokine receptors are cellular receptors for cytokines. Binding of the cytokine to the cytokine receptor induces new activities in the cell, such as growth, differentiation, or death. Cytokine receptors are listed in Appendix II.

Cytotoxic T cells are T cells that can kill other cells. Most cytotoxic T cells are MHC class I-restricted CD8 T cells, but CD4 T cells can also kill in some cases. Cytotoxic T cells are important in host defense against cytosolic pathogens.

Cytotoxins are proteins made by cytotoxic T cells that participate in the destruction of target cells. Perforins and granzymes or fragmentins are the major defined cytotoxins.

In the context of immunoglobulins, **δ** is the type of heavy chain in IgD.

Death domains were originally defined in proteins encoded by genes involved in programmed cell death or apoptosis, and are now known to be involved in protein–protein interactions.

Defective endogenous retroviruses are partial retroviral genomes integrated into host cell DNA and carried as host genes. There are a great many defective endogenous retroviruses in the mouse genome.

Delayed-type hypersensitivity is a form of cell-mediated immunity elicited by antigen in the skin and is mediated by CD4 T_H1 cells. It is called delayed-type hypersensitivity because the reaction appears hours to days after antigen is injected. Cf. **immediate hypersensitivity**.

Dendritic cells, also known as interdigitating reticular cells, are found in T-cell areas of lymphoid tissues. They have a branched or dendritic morphology and are the most potent stimulators of T-cell responses. Non-lymphoid tissues also contain dendritic cells, but these do not seem to stimulate T-cell responses until they are activated and migrate to lymphoid tissues. The dendritic cell derives from bone marrow precursors. It is distinct from the follicular dendritic cell that presents antigen to B cells.

Dendritic epidermal cells (**dEC**), also known as **dendritic epidermal T cells** (**dETC**) are a specialized class of γ:δT cells found in the skin of mice and some other species, but not humans. All dEC have the same γ:δ T-cell receptor; their function is unknown.

Desensitization is a procedure in which an allergic individual is exposed to increasing doses of allergen in hopes of inhibiting their allergic reactions. It probably involves shifting the balance between CD4 T_H1 and T_H2 cells and thus changing the antibody produced from IgE to IgG.

D gene segments, or **diversity gene segments**, are short DNA sequences that join the V and J gene segments in rearranged immunoglobulin heavy-chain genes and in T-cell receptor β and δ chain genes. See **gene segments**.

Diacylglycerol (**DAG**) is most commonly released from inositol phospholipids by the action of phospholipase C-γ. Diacylglycerol production is stimulated by the ligation of many receptors and it acts as an intracellular signaling molecule, activating cytosolic protein kinase C, which further propagates the signal.

Diapedesis is the movement of blood cells, particularly leukocytes, from the blood across blood vessel walls into tissues.

The **differential signaling hypothesis** proposes that qualitatively different antigens might mediate the positive and negative selection of T cells in the thymus. Cf. **avidity hypothesis**.

Differentiation antigens are proteins detected on some cells by means of specific antibodies. Many differentiation antigens have important functional roles characteristic of the differentiated phenotypes of the cell on which they are expressed, such as cell-surface immunoglobulin on B cells.

DiGeorge syndrome is a recessive genetic immunodeficiency disease in which there is a failure to develop thymic epithelium, and is associated with absent parathyroid glands and large vessel anomalies. It seems to be due to a developmental defect in neural crest cells.

The **direct Coombs test** uses anti-immunoglobulin to agglutinate red blood cells as a way of detecting whether they are coated with antibody *in vivo* due to autoimmunity or maternal anti-fetal immune responses (see Coombs test, indirect Coombs test).

Discontinuous epitopes: see **conformational epitopes**.

Diversity gene segments: see **D gene segments**.

A truncated heavy chain known as a **Dμ protein** can be produced during B-cell development as a result of heavy-chain transcription from a previously inactive promoter 5′ to the D gene segments.

The genetic defect in *scid* mice, which cannot rearrange their T- or B-cell receptor genes and have a severe combined immunodeficiency phenotype, is in the enzyme **DNA-dependent kinase**. This enzyme is part of a complex of proteins that bind to the hairpin ends of double-stranded breaks in DNA, and its catalytic subunit is critical for VDJ recombination.

In tissue grafting experiments, the grafted tissues come from a **donor** and are placed in a recipient or host.

Double-negative thymocytes are immature T cells within the thymus that lack expression of the two co-receptors, CD4 and CD8.

Double-positive thymocytes are an intermediate stage in T-cell development in the thymus and are characterized by expression of both the CD4 and the CD8 co-receptor proteins.

In the context of immunoglobulins, ε (epsilon) is the heavy chain of IgE.

The **early induced responses** or early non-adaptive responses are a series of host defense responses that are triggered by infectious agents early in infection. They are distinct from innate immunity because there is an inductive phase, and from adaptive immunity in that they do not operate by clonal selection of rare antigen-specific lymphocytes.

Early pro-B cell: see **pro-B cell**.

The common skin disease **eczema** is seen mainly in children; its etiology is poorly understood.

Effector cells are lymphocytes that can mediate the removal of pathogens from the body without the need for further differentiation, as distinct from naive lymphocytes, which must proliferate and differentiate before they can mediate effector functions, and memory cells, which must differentiate and often proliferate before they become effector cells. They are also called armed effector cells in this book, to indicate that they can be triggered to effector function by antigen binding alone.

Effector mechanisms are those processes by which pathogens are destroyed and cleared from the body. Innate and adaptive immune responses use most of the same effector mechanisms to eliminate pathogens.

Lymphocytes leave a lymph node through the **efferent lymphatic vessel**.

Electrophoresis is the movement of molecules in a charged field. In immunology, many forms of electrophoresis are used to separate molecules, especially protein molecules, to determine their charge, size, and subunit composition.

ELISA: see **enzyme-linked immunosorbent assay**.

ELISPOT assay is an adaptation of ELISA in which cells are placed over antibodies or antigens attached to a plastic surface. The antigen or antibody traps the cells' secreted products, which can then be detected by using an enzyme-coupled antibody that cleaves a colorless substrate to make a localized colored spot.

Embryonic stem (**ES**) **cells** are mouse embryonic cells that will grow continuously in culture and that retain the ability to contribute to all cell lineages. ES cells can be genetically manipulated in tissue culture and then inserted into mouse blastocysts to generate mutant lines of mice; most often, genes are deleted in ES cells by homologous recombination and the mutant ES cells are then used to generate gene knock-out mice.

Encapsulated bacteria have thick carbohydrate coats that protect them from phagocytosis. Encapsulated bacteria can cause extracellular infections and are effectively engulfed and destroyed by phagocytes only if they are first coated with antibody and complement produced in an adaptive immune response.

Some anti-carbohydrate antibodies are called **end-binders** because they bind the ends of oligosaccharide antigens, whereas others bind the sides of these molecules.

Cytokines that can induce a rise in body temperature are called **endogenous pyrogens**, as distinct from exogenous substances such as endotoxin from Gram-negative bacteria that induce fever by triggering endogenous pyrogen synthesis and release.

Endotoxins are bacterial toxins that are released only when the bacterial cell is damaged, as opposed to exotoxins, which are secreted bacterial toxins. The most important endotoxin is the lipopolysaccharide of Gram-negative bacteria, which is a potent inducer of cytokine synthesis.

The **enzyme-linked immunosorbent assay** (ELISA) is a serological assay in which bound antigen or antibody is detected by a linked enzyme that converts a colorless substrate into a colored product. The ELISA assay is widely used in biology and medicine as well as immunology.

Eosinophils are white blood cells thought to be important chiefly in defense against parasitic infections; they are activated by the lymphocytes of the adaptive immune response.

Eotaxin-1 and **eotaxin-2** are CC-chemokines that act specifically on eosinophils.

An **epitope** is a site on an antigen recognized by an antibody; epitopes are also called antigenic determinants. A T-cell epitope is a short peptide derived from a protein antigen. It binds to an MHC molecule and is recognized by a particular T cell.

Epitope spreading describes the fact that responses to autoantigens tend to become more diverse as the response persists.

The **Epstein–Barr virus** (**EBV**) is a herpesvirus that selectively infects human B cells by binding to complement receptor 2 (CR2, also known as CD21). It causes infectious mononucleosis and establishes a life-long latent infection in B cells that is controlled by T cells. Some B cells latently infected with EBV will proliferate *in vitro* to form lymphoblastoid cell lines.

The affinity of an antibody for its antigen can be determined by **equilibrium dialysis**, a technique in which antibody in a dialysis bag is exposed to varying amounts of a small antigen able to diffuse across the dialysis membrane. The amount of antigen inside and outside the bag at the equilibrium diffusion state is determined by the amount and affinity of the antibody in the bag.

E-rosettes are human T cells that will bind to treated red blood cells from sheep; the many red blood cells bound to each T cell give it the appearance of a rosette and increase its buoyant density so that the T cells can be isolated by gradient centrifugation. E-rosetting is often used for isolating human T cells.

Erythroblastosis fetalis is a severe form of Rh hemolytic disease in which maternal anti-Rh antibody enters the fetus and produces a hemolytic anemia so severe that the fetus has mainly immature erythroblasts in the peripheral blood.

E-selectin: see **selectins**.

Experimental allergic encephalomyelitis (**EAE**) is an inflammatory disease of the central nervous system that develops after mice are immunized with neural antigens in a strong adjuvant.

The movement of cells or fluid from within blood vessels to the surrounding tissues is called **extravasation**.

IgG antibody molecules can be cleaved into three fragments by the enzyme papain. Two of these are identical **Fab fragments**, so called because they are the **F**ragment with specific **a**ntigen **b**inding. The Fab fragment consists of the light chain and the N-terminal half of the heavy chain held together by an interchain disulfide bond. See also **Fc fragment**.

FACS®: see **fluorescence-activated cell sorter**.

Factor P: see **properdin**.

Farmer's lung is a hypersensitivity disease caused by the interaction of IgG antibodies with large amounts of an inhaled allergen in the alveolar wall of the lung, causing alveolar wall inflammation and compromising gas exchange.

Fas is a member of the TNF receptor family; it is expressed on certain cells and makes them susceptible to killing by cells expressing **Fas ligand**, a cell-surface member of the TNF family of proteins. Binding of Fas ligand to Fas triggers apoptosis in the Fas-bearing cell.

IgG antibody molecules can be cleaved into three fragments by the enzyme papain. One of these is the **Fc fragment**, so-called for **F**ragment **c**rystallizable. The Fc fragment consists of the C-terminal halves of the two heavy chains disulfide-bonded to each other by the residual hinge region. See also **Fab fragments**.

Fv: see **single-chain Fv**.

Fc receptors are receptors for the Fc portion of immunoglobulin isotypes. They include the Fcγ and Fcε receptors.

The high-affinity **Fcε receptor** (**FcεRI**) on the surface of mast cells and basophils binds free IgE. When antigen binds this IgE and crosslinks FcεRI it causes mast cell activation.

Fcγ receptors, including **FcγRI, RII, and RIII,** are cell-surface receptors that bind the Fc portion of IgG molecules. Most Fcγ receptors bind only aggregated IgG, allowing them to discriminate bound antibody from free IgG. They are expressed on phagocytes, B lymphocytes, NK cells, and follicular dendritic cells. They have a key role in humoral immunity, linking antibody binding to effector cell functions.

When tissue or organ grafts are placed in an unmatched recipient, they are rejected by a **first set rejection**, which is an immune response by the host against foreign antigens in the graft. Cf. **second set rejection**.

FK506, or tacrolimus, is an immunosuppressive polypeptide drug that inactivates T cells by inhibiting signal transduction from the T-cell receptor. FK506 and cyclosporin A are the most commonly used immunosuppressive drugs in organ transplantation.

Individual cells can be characterized and separated in a machine called a **fluorescence-activated cell sorter** (**FACS®**) that measures cell size, granularity, and fluorescence due to bound fluorescent antibodies as single cells pass in a stream past photodetectors. The analysis of single cells in this way is called flow cytometry and the instruments that carry out the measurements and/or sort cells are called flow cytometers or cell sorters.

The **follicular dendritic cells** of lymphoid follicles are cells of uncertain origin with long branching processes that make intimate contact with many different B cells. They have Fc receptors that are not internalized by receptor-mediated endocytosis and thus hold antigen:antibody complexes on the surface for long periods. These cells are crucial in selecting antigen-binding B cells during antibody responses.

Fragmentins, or granzymes, are serine esterases present in the granules of cytotoxic lymphocytes including T cells and NK cells. When fragmentins enter the cytosol of other cells they induce apoptosis, inducing nuclear DNA fragmentation into 200 base pair multimers, hence their name.

The V domains of immunoglobulins and T-cell receptors contain relatively invariant **framework regions** that provide a protein scaffold for the hypervariable regions that make contact with antigen.

Fungi are single-celled and multicellular eukaryotic organisms, including the yeasts and molds, that can cause a variety of diseases. Immunity to fungi is complex and involves both humoral and cell-mediated responses.

Fyn: see **tyrosine kinase**.

In the context of immunoglobulins, **γ** is the heavy chain of IgG.

GALT: see **gut-associated lymphoid tissues**.

Most T lymphocytes have α:β heterodimeric T-cell receptors, but the receptor on γ:δ T cells has distinct antigen recognition chains assembled in a **γ:δ heterodimer**. The specificity and function of these cells are uncertain.

Plasma proteins can be separated on the basis of electrophoretic mobility into albumin and the α, β, and γ globulins. Most antibodies migrate in electrophoresis as **γ globulins** (or **gamma globulins**), and patients who lack antibodies are said to have agammaglobulinemia.

In birds and rabbits, immunoglobulin receptor diversity is generated mainly by **gene conversion**, in which homologous inactive V gene segments exchange short sequences with an active, rearranged V-region gene.

Gene knock-out is jargon for gene disruption by homologous recombination.

The V domains of the polypeptide chains of antigen receptors are encoded in sets of **gene segments** that must first undergo somatic recombination to form a complete V-domain exon. There are three types of gene segment: V gene segments that encode the first 95 amino acids, D gene segments that encode about 5 amino acids, and J gene segments that form the last 10–15 amino acids of the V region. There are multiple copies of each type of gene segment in the germline DNA, but only one is expressed for each type of receptor chain in a receptor-bearing lymphocyte.

A gene can be specifically disrupted by a technique known as **gene targeting** or gene knock-out. Usually this involves homologous recombination in embryonic stem cells followed by the preparation of chimeric mice by injection of these cells into the blastocyst.

Gene therapy is the correction of a genetic defect by the introduction of a normal gene into bone marrow or other cell types. It is also known as somatic gene therapy because it does not affect the germline genes of the individual.

Genetic immunization is a novel technique for inducing adaptive immune responses. Plasmid DNA encoding a protein of interest is injected into muscle; for unknown reasons it is expressed, and elicits antibody and T-cell responses to the protein encoded by the DNA.

Germinal centers in secondary lymphoid tissues are sites of intense B-cell proliferation, selection, maturation, and death during antibody responses. Germinal centers form around follicular dendritic cell networks when activated B cells migrate into lymphoid follicles.

Immunoglobulin and T-cell receptor genes are said to be in the **germline configuration** in the DNA of germ cells and in all somatic cells in which somatic recombination has not occurred.

The **germline diversity** of antigen receptors is due to the inheritance of multiple gene segments that encode V domains; such diversity is distinguished from the diversity that is generated during gene rearrangement or after receptor gene expression, which is somatically generated.

One theory of antibody diversity, the **germline theory**, proposed that each antibody was encoded in a separate germline gene.

GlyCAM-1 is a mucin-like molecule found on the high endothelial venules of lymphoid tissues. It is an important ligand for the L-selectin molecule expressed on naive lymphocytes, directing these cells to leave the blood and enter the lymphoid tissues.

Goodpasture's syndrome is an autoimmune disease in which autoantibodies against basement membrane or type IV collagen are produced and cause extensive vasculitis. It is rapidly fatal.

G proteins are proteins that bind GTP and convert it to GDP in the process of cell signal transduction. There are two kinds of G protein, the trimeric (α, β, γ), receptor-associated G proteins, and the small G proteins, such as Ras and Raf, that act downstream of many transmembrane signaling events.

Tissue and organ grafts between genetically distinct individuals almost always elicit an immune response that causes **graft rejection**, the destruction of the grafted tissue by attacking lymphocytes.

When mature T lymphocytes are injected into a non-identical immunoincompetent recipient, they can attack the recipient, causing a **graft-versus-host** (**GVH**) reaction; in human patients, mature T cells in allogeneic bone marrow grafts can cause **graft-versus-host disease** (**GVHD**).

Granulocyte: see **polymorphonuclear leukocyte**.

Granulocyte–macrophage colony-stimulating factor (**GM–CSF**): is a cytokine involved in the growth and differentiation of myeloid and monocytic lineage cells, including dendritic cells, monocytes and tissue macrophages, and cells of the granulocyte lineage.

A **granuloma** is a site of chronic inflammation usually triggered by persistent infectious agents such as mycobacteria or by a non-degradable foreign body. Granulomas have a central area of macrophages, often fused into multinucleate giant cells, surrounded by T lymphocytes.

Granzymes: see **fragmentins**.

Graves' disease is an autoimmune disease in which antibodies against the thyroid-stimulating hormone receptor cause over-production of thyroid hormone and thus hyperthyroidism.

The **guanine nucleotide exchange factors** (**GEFs**) are proteins that share the ability to remove the bound GDP from small G proteins; this allows GTP to bind and activate the G protein.

The **gut-associated lymphoid tissues** (**GALT**) are lymphoid tissues closely associated with the gastrointestinal tract, including the palatine tonsils, Peyer's patches, and intraepithelial lymphocytes. The GALT has a distinctive biology related to its exposure to antigens from food and normal intestinal microbial flora.

The major histocompatibility complex of the mouse is called **histocompatibility-2** or more commonly **H-2**. Haplotypes are designated by a lower-case superscript, as in H-2^b.

Histocompatibility antigens (**H antigens**) are antigens that mediate the graft rejection of allogeneic tissue. They can be MHC molecules or minor H antigens. The former are studied *in vivo* by using MHC congenic, mutant, or recombinant strains.

A **haplotype** is a linked set of genes associated with one haploid genome. The term is used mainly in connection with the linked genes of the major histocompatibility complex, which are usually inherited as one haplotype from each parent. Some MHC haplotypes are over-represented in the population, a phenomenon known as linkage disequilibrium.

Haptens are molecules that can bind antibody but cannot by themselves elicit an adaptive immune response. Haptens must be chemically linked to protein carriers to elicit antibody and T-cell responses.

Hashimoto's thyroiditis is an autoimmune disease characterized by persistent high levels of antibody against thyroid-specific antigens. These antibodies recruit NK cells to the tissue, leading to damage and inflammation.

All immunoglobulin molecules have two types of chain, a **heavy (H) chain** of 50–70 kDa and a light chain of 25 kDa. The basic immunoglobulin unit consists of two identical heavy chains and two identical light chains. Heavy chains come in a variety of heavy-chain classes or isotypes, each of which confers a distinctive functional activity on the antibody molecule.

Helper CD4 T cells are CD4 T cells that can help B cells make antibody in response to antigenic challenge. The most efficient helper T cells are also known as T_H2, cells that make the cytokines IL-4 and IL-5. Some experts refer to all CD4 T cells, regardless of function, as helper T cells; we do not accept this usage because function can be determined only in cellular assays, and some CD4 T cells kill the cells they interact with.

A **hemagglutinin** is any substance that causes red blood cells to agglutinate, a process known as **hemagglutination**. The hemagglutinins in human blood are antibodies that recognize the ABO blood group antigens. Influenza and some other viruses have hemagglutinin molecules that bind to glycoproteins on host cells to initiate the infectious process.

Hematopoiesis is the generation of the cellular elements of blood, including the red blood cells, leukocytes and platelets. These cells all originate from pluripotent **hematopoietic stem cells** whose differentiated progeny divide under the influence of **hematopoietic growth factors**.

A **hematopoietic lineage** is any developmental series of cells that derives from hematopoietic stem cells and results in the production of mature blood cells.

Hemolytic disease of the newborn, or erythroblastosis fetalis, is caused by a maternal IgG antibody response to paternal antigens expressed on fetal red blood cells. The usual target of this response is the Rh blood group antigen. The maternal anti-Rh IgG antibodies cross the placenta to attack the fetal red blood cells.

The **hemolytic plaque assay** detects antibody-forming cells by the ability of their secreted antibodies to produce a **hemolytic plaque**, an area of localized destruction of a thin layer of red blood cells around each antibody-producing cell. The antibodies secreted by the B cell are trapped by antigens on the red blood cells immediately surrounding it, and then complement is added that is triggered by the bound antibody to lyse the red blood cells.

Recombination signal sequences (RSS) flanking gene segments consist of a seven-nucleotide **heptamer** and a nine-nucleotide nonamer of conserved sequence, separated by 12 or 23 nucleotides. RSSs form the target for the site-specific recombinase that joins the gene segments.

Hereditary angioneurotic edema is the clinical name for a genetic deficiency of the C1 inhibitor of the complement system. In the absence of C1 inhibitor, spontaneous activation of the complement system can cause diffuse fluid leakage from blood vessels, the most serious consequence of which is epiglottal swelling leading to suffocation.

Individuals **heterozygous** for a particular gene have two different alleles of that gene.

An excellent model for membranous glomerulonephritis is **Heymann's nephritis**, a disease induced by injecting animals with tubular epithelial tissue.

High endothelial venules (**HEV**) are specialized venules found in lymphoid tissues. Lymphocytes migrate from blood into lymphoid tissues by attaching to and migrating across the **high endothelial cells** of these vessels.

Tolerance to injected protein antigens occurs at low or high doses of antigen. Tolerance induced by the injection of high doses of antigen is called **high-zone tolerance**, whereas tolerance produced with low doses of antigen is called low-zone tolerance.

The **hinge region** of antibody molecules is a flexible domain that joins the Fab arms to the Fc piece. The flexibility of the hinge region in IgG and IgA molecules allows the Fab arms to adopt a wide range of angles, permitting binding to epitopes spaced variable distances apart.

Histamine is a vasoactive amine stored in mast cell granules. Histamine released by antigen binding to IgE molecules on mast cells causes dilation of local blood vessels and smooth muscle contraction, producing some of the symptoms of immediate hypersensitivity reactions. Anti-histamines are drugs that counter histamine action.

Histocompatibility is literally the ability of tissues (Greek: *histo*) to get along with each other. It is used in immunology to describe the genetic systems that determine the rejection of tissue and organ grafts resulting from immunological recognition of histocompatibility (H) antigens.

HIV: see **human immunodeficiency virus**.

HLA, the acronym for **H**uman **L**eukocyte **A**ntigen, is the genetic designation for the human MHC. Individual loci are designated by upper-case letters, as in HLA-A, and alleles are designated by numbers, as in HLA-A*0201.

The invariant **HLA-DM** molecule in humans is involved in loading peptides onto MHC class II molecules. It is encoded in the MHC within a set of genes resembling MHC class II genes. A homologous protein in mice is called H-2M.

Hodgkin's disease is a malignant disease in which antigen-presenting cells that resemble dendritic cells seem to be the transformed cell type. **Hodgkin's lymphoma** is a form of Hodgkin's disease in which lymphocytes predominate, and it has a much better prognosis than the nodular sclerosis form of this disease in which the predominant cell type is non-lymphoid.

Homeostasis is a generic term describing the status of physiological normality. In the case of lymphocytes, homeostasis refers to an uninfected individual who has normal numbers of lymphocytes.

Cellular genes can be disrupted by **homologous recombination** with copies of the gene into which erroneous sequences have been inserted. When these exogenous DNA fragments are introduced into cells, they recombine selectively with the cellular gene through remaining regions of sequence homology, replacing the functional gene with a non-functional copy.

The **human immunodeficiency virus** (**HIV**) is the causative agent of the acquired immune deficiency syndrome (AIDS). HIV is a retrovirus of the lentivirus family that selectively infects CD4 T cells, leading to their slow depletion, which eventually results in immunodeficiency.

Humanization is a term used to describe the genetic engineering of mouse hypervariable loops of a desired specificity into otherwise human antibodies. The DNA encoding hypervariable loops of mouse monoclonal antibodies or V regions selected in phage display libraries is inserted into the framework regions of human immunoglobulin genes. This allows the production of antibodies of a desired specificity that do not cause an immune response in humans treated with them.

Humoral immunity is the antibody-mediated specific immunity made in a **humoral immune response**. Humoral immunity can be transferred to unimmunized recipients by using immune serum containing specific antibody.

Monoclonal antibodies are most commonly produced from **hybridomas**. These are hybrid cell lines formed by fusing a specific antibody-producing B lymphocyte with a myeloma cell that is selected for its ability to grow in tissue culture and an absence of immunoglobulin chain synthesis.

Hyperacute graft rejection of an allogenic tissue graft is an immediate reaction caused by natural preformed antibodies that react against antigens on the graft. The antibodies bind to endothelium and trigger the blood clotting cascade, leading to an engorged, ischemic graft and rapid loss of the organ.

Repetitive immunization to achieve a heightened state of immunity is called **hyperimmunization**.

Immune responses to innocuous antigens that lead to symptomatic reactions upon re-exposure are called **hypersensitivity reactions**. These can cause **hypersensitivity diseases** if they occur repetitively. This state of heightened reactivity to antigen is called **hypersensitivity**. Hypersensitivity reactions are classified by mechanism: type I hypersensitivity reactions involve IgE antibody triggering of mast cells; type II hypersensitivity reactions involve IgG antibodies against cell-surface or matrix antigens; type III hypersensitivity reactions involve antigen:antibody complexes; and type IV hypersensitivity reactions are T-cell mediated.

The **hypervariable (HV) regions** of immunoglobulin and T-cell receptor V domains are small regions that make contact with the antigen and differ extensively from one receptor to the next. Cf. **framework regions**.

ICAM: see **intercellular adhesion molecule**.

Iccosomes are small fragments of membrane coated with immune complexes that fragment off the processes of follicular dendritic cells in lymphoid follicles early in a secondary or subsequent antibody response.

Each immunoglobulin molecule has the potential of binding a variety of antibodies directed at its unique features or **idiotype**. An idiotype is made up of a series of **idiotopes**.

Lymphocyte antigen receptors can recognize one another through idiotope:anti-idiotope interactions, forming an **idiotypic network** of receptors that may be important for the generation and maintenance of the receptor repertoire. The proposed components of idiotype networks exist but their functional significance is uncertain.

IFN: see **interferon**.

Ig: standard abbreviation for **immunoglobulin**. Different immunoglobulin isotypes are called IgM, IgD, IgG, IgA, and IgE.

Igα, **Ig**β: see **B-cell antigen receptor**.

IgA is the class of immunoglobulin characterized by α heavy chains. IgA antibodies are secreted mainly by mucosal lymphoid tissues.

IgD is the class of immunoglobulin characterized by δ heavy chains. It appears as surface immunoglobulin on mature naive B cells but its function is unknown.

IgE is the class of immunoglobulin characterized by ε heavy chains. It is involved in allergic reactions.

IgG is the class of immunoglobulin characterized by γ heavy chains. It is the most abundant class of immunoglobulin found in the plasma.

IgM is the class of immunoglobulin characterized by μ heavy chains. It is the first immunoglobulin to appear on the surface of B cells.

IL-12: see Appendix II.

Immature B cells are B cells that have rearranged a heavy- and a light-chain V-region gene and express surface IgM, but have not yet matured sufficiently to express surface IgD as well.

Hypersensitivity reactions that occur within minutes of exposure to antigen are called **immediate hypersensitivity reactions**; such reactions are antibody mediated. Cf. **delayed-type hypersensitivity reactions**.

When large amounts of antigen are injected into the blood, they are initially removed slowly by normal catabolic processes that also degrade plasma proteins. However, if the antigen elicits an antibody response, then antigen is removed at an accelerated rate as antigen:antibody complexes, a process known as **immune clearance**.

The binding of antibody to a soluble antigen forms an **immune complex**. Large immune complexes form when sufficient antibody is available to crosslink the antigen; these are readily cleared by the reticuloendothelial system of cells bearing Fc and complement receptors. Small soluble immune complexes that form when antigen is in excess can be deposited in and damage small blood vessels.

Immune deviation is a term used to describe the polarization of an immune response to T_H1-dominated or T_H2-dominated by the injection of antigen.

The **immune response** is the response made by the host to defend itself against a pathogen.

Immune response (Ir) genes are genetic polymorphisms that control the intensity of the immune response to a particular antigen. Virtually all Ir phenotypes are due to the differential binding of peptide fragments of antigen to MHC molecules, especially MHC class II molecules. The term is little used now. An **immune response (Ir) gene defect** is usually, but not always, due to failure to bind an immunogenic peptide, so that no T-cell response is observed.

It has been proposed that most tumors that arise are detected and eliminated by **immune surveillance** mediated by lymphocytes specific for tumor antigens. There is little evidence for the efficacy of this proposed process, but it remains an important concept in tumor immunology.

The **immune system** is the name used to describe the tissues, cells, and molecules involved in adaptive immunity, or sometimes the totality of host defense mechanisms.

Immunity is the ability to resist infection.

Immunization is the deliberate provocation of an adaptive immune response by introducing antigen into the body. See also **active immunization**; **passive immunization**.

Immunobiology is the study of the biological basis for host defense against infection.

Immunoblotting is a common technique in which proteins separated by gel electrophoresis are blotted onto a nitrocellulose membrane and revealed by the binding of specific labeled antibodies.

Immunodeficiency diseases are a group of inherited or acquired disorders in which some aspect or aspects of host defense are absent or functionally defective.

Immunodiffusion is the detection of antigen or antibody by the formation of an antigen:antibody precipitate in a clear agar gel.

Immunoelectrophoresis is a technique in which antigens are first separated by their electrophoretic mobility and are then detected and identified by immunodiffusion.

Immunofluorescence is a technique for detecting molecules by using antibodies labeled with fluorescent dyes. The bound fluorescent antibody can be detected by microscopy, by flow cytometry, or by fluorimetry, depending on the application being used. **Indirect immunofluorescence** uses anti-immunoglobulin antibodies labeled with fluorescent dyes to detect the binding of a specific unlabeled antibody.

Any molecule that can elicit an adaptive immune response on injection into a person or animal is called an **immunogen**. In practice, only proteins are fully immunogenic because only proteins can be recognized by T lymphocytes.

Immunogenetics was originally the analysis of genetic traits by means of antibodies against genetically polymorphic molecules such as blood group antigens or MHC proteins. Immunogenetics now refers to the genetic analysis, by any technique, of molecules important in immunology.

All antibody molecules belong to a family of plasma proteins called **immunoglobulins (Ig)**. Membrane-bound immunoglobulin serves as the specific antigen receptor on B lymphocytes.

Many proteins are partly or entirely composed of protein domains known as **immunoglobulin domains** or **Ig domains** because they were first described in antibody molecules. Immunoglobulin domains are characteristic of proteins of the immunoglobulin superfamily, which includes antibodies, T-cell receptors, MHC molecules, and many other proteins described in this book. The immunoglobulin domain consists of a sandwich of two β sheets held together by a disulfide bond and called the **immunoglobulin fold**. There are two main types of immunoglobulin domain: C domains and V domains. Domains less closely related to the canonical Ig domains are sometimes also called **immunoglobulin-like domains**.

Immunoglobulin fold: see **immunoglobulin domains**.

Immunoglobulin-like domain: see **immunoglobulin domains**.

Many proteins involved in antigen recognition and cell–cell interaction in the immune system and other biological systems are members of a protein family called the **immunoglobulin superfamily**, or **Ig superfamily**, because their shared structural features were first defined in immunoglobulin molecules. All members of the immunoglobulin superfamily have at least one immunoglobulin or immunoglobulin-like domain.

The detection of antigens in tissues by means of visible products produced by the degradation of a colorless substrate by antibody-linked enzymes is called **immunohistochemistry**. This technique has the advantage that it can be combined with other stains to be viewed in the light microscope, whereas immunofluorescence microscopy requires a special dark-field or UV microscope.

Immunological ignorance describes a form of self tolerance in which reactive lymphocytes and their target antigen are both detectable within an individual, yet no autoimmune attack occurs. Most autoimmune diseases probably occur when immunological ignorance is broken.

When an antigen is encountered more than once, the adaptive immune response to each subsequent encounter is speedier and more effective, a crucial feature of protective immunity known as **immunological memory**. Immunological memory is specific for a particular antigen and is long-lived.

Allogeneic tissue placed in certain sites in the body, such as the brain, does not elicit graft rejection. Such sites are called **immunologically privileged sites**. Immunological privilege results from the effects of both physical barriers to cell and antigen migration, and soluble immunosuppressive mediators such as certain cytokines.

Immunology is the study of all aspects of host defense against infection and of adverse consequences of immune responses.

Immunophilins are proteins with peptidyl-prolyl *cis–trans* isomerase activity that bind the immunosuppressive drugs cyclosporin A, FK506, and rapamycin.

Soluble proteins, or membrane proteins solubilized in detergents, can be labeled and then detected by **immunoprecipitation analysis** using specific antibodies. The immunoprecipitated labeled protein is usually detected by SDS-PAGE followed by autoradiography.

The T and B cell antigen receptors are associated with trans-membrane molecules with **immunoreceptor tyrosine-based activation motifs (ITAMs)** in their cytoplasmic domains. These tyrosine-containing motifs are sites of tyrosine phosphorylation and of association with tyrosine kinases and other phosphotyrosine-binding moieties involved in receptor signaling. Related motifs with opposing effects are **immunoreceptor tyrosine-based inhibitory motifs (ITIMs)**, which recruit phosphatases to the receptor site that remove the phosphate groups added by tyrosine kinases.

The ability of the immune system to sense and regulate its own responses is called **immunoregulation**.

Compounds that inhibit adaptive immune responses are called **immunosuppressive drugs**. They are used mainly in the treatment of graft rejection and severe autoimmune disease.

Immunotoxins are antibodies that are chemically coupled to toxic proteins usually derived from plants or microbes. The antibody targets the toxin moiety to the required cells. Immunotoxins are being tested as anti-cancer agents and as immunosuppressive drugs.

The **indirect Coombs test** is a variation of the direct Coombs test in which an unknown serum is tested for antibodies against normal red blood cells by first mixing the two and then washing the red blood cells and reacting them with anti-immunoglobulin antibody. If antibody in the unknown serum binds to the red blood cells, agglutination by anti-immunoglobulin occurs.

Indirect immunofluorescence: see **immunofluorescence**.

Macrophages and many other cells have an **inducible NO synthase**, or **iNOS**, that is induced by many different stimuli to activate NO synthesis. This is a major mechanism of host resistance to intracellular infection in mice, and probably in humans as well.

Infectious mononucleosis, or glandular fever, is the term used to describe the common form of infection with the Epstein–Barr virus. It consists of fever, malaise, and swollen lymph nodes.

Inflammation is a general term for the local accumulation of fluid, plasma proteins, and white blood cells that is initiated by physical injury, infection, or a local immune response. This is also known as an **inflammatory response**. Acute inflammation is the term used to describe early and often transient episodes, whereas chronic inflammation occurs when the infection persists or during autoimmune responses. Many different forms of inflammation are seen in different diseases. The cells that invade tissues undergoing inflammatory responses are often called **inflammatory cells** or an **inflammatory infiltrate**.

Inflammatory CD4 T cells, also known as T_H1, are armed effector T cells that make the cytokines IFN-γ and TNF on recognition of antigen. Their major function is the activation of macrophages. Some T_H1 also have cytotoxic activity.

Influenza hemagglutinin: see **hemagglutinin**.

The early phases of the host response to infection depend on **innate immunity** in which a variety of **innate resistance mechanisms** recognize and respond to the presence of a pathogen. Innate immunity is present in all individuals at all times, does not increase with repeated exposure to a given pathogen, and does not discriminate between pathogens.

When inositol phospholipid is cleaved by phospholipase C-γ, it yields **inositol trisphosphate (IP_3)** and diacylglycerol. Inositol trisphosphate releases calcium ions from intracellular stores in the endoplasmic reticulum.

The **instructive model** of T-cell development states that CD4,CD8 double-positive thymocytes with an MHC class II restricted receptor should always give rise to CD4 T cells, and those with a class I restricted receptor to CD8 T cells, because a different signal is transmitted by co-ligation of CD4 and CD8. Cf. **stochastic/selection model**.

In **insulin-dependent diabetes mellitus (IDDM)**, the β cells of the pancreatic islets of Langerhans are destroyed so that no insulin is produced. The disease is believed to result from an autoimmune attack on the β cells.

Integrins are heterodimeric cell-surface proteins involved in cell–cell and cell–matrix interactions. They are important in adhesive interactions between lymphocytes and antigen-presenting cells and in lymphocyte and leukocyte migration into tissues. The **β_1-integrins**, or very late antigens (VLA), are a family of integrins with shared β_1 chains and different α chains that mediate adhesion to other cells and to extracellular matrix proteins.

The **intercellular adhesion molecules (ICAMs)** ICAM-1, ICAM-2, and ICAM-3 are cell-surface ligands for the leukocyte integrins and are crucial in the binding of lymphocytes and other leukocytes to certain cells, including antigen-presenting cells and endothelial cells. They are members of the immunoglobulin superfamily.

Interdigitating reticular cells: see **dendritic cells**.

Interferons are cytokines that can induce cells to resist viral replication. **Interferon-α (IFN-α)** and **interferon-β (IFN-β)** are produced by leukocytes and fibroblasts respectively, as well as by other cells, whereas **interferon-γ (IFN-γ)** is a product of CD4 T_H1 cells, CD8 T cells, and NK cells. IFN-γ has as its primary action the activation of macrophages.

Interleukin, abbreviated **IL**, is a generic term for cytokines produced by leukocytes. We use the more general term cytokine in this book, but the term interleukin is used in the naming of specific cytokines such as IL-2. The interleukins are listed in Appendix II.

Interleukin-2 (IL-2) is the cytokine that is most central to the development of an adaptive immune response.

The major histocompatibility complex (MHC) class II proteins are assembled in the endoplasmic reticulum with the **invariant chain (Ii)**, which is involved in shielding the MHC class II molecules from binding peptides and in delivering them to cellular vesicles. There Ii is degraded, leaving the MHC class II molecules able to bind peptide fragments of antigen.

ISCOMs are **immune stimulatory complexes** of antigen held within a lipid matrix that acts as an adjuvant and enables the antigen to be taken up into the cytoplasm after fusion of the lipid with the plasma membrane.

Isoelectric focusing is an electrophoretic technique in which proteins migrate in a pH gradient until they reach the place in the gradient at which their net charge is neutral—their isoelectric point. Uncharged proteins no longer migrate; thus each protein is focused at its isoelectric point.

Immunoglobulins are made in several distinct **isotypes** or classes—IgM, IgG, IgD, IgA, and IgE—each of which has a distinct heavy-chain C region encoded by a distinct C-region gene. The isotype of an antibody determines the effector mechanisms that it can engage on binding antigen.

The first antibodies produced in a humoral immune response are IgM, but activated B cells subsequently undergo **isotype switching** or class switching to secrete antibodies of different isotypes: IgG, IgA, and IgE. Isotype switching does not affect antibody specificity significantly, but alters the effector functions that an antibody can engage. Isotype switching occurs by a site-specific recombination involving the deletion of the intervening DNA.

Isotypic exclusion describes the use of one or other of the light-chain isotypes, κ or λ, by a given B cell or antibody.

Cytokine receptors signal via **Janus kinases** (**JAK**)—tyrosine kinases that are activated by the aggregation of cytokine receptors. These kinases phosphorylate proteins known as STATs, for Signal Transducers and Activators of Transcription. STATs are normally found in the cytosol, but move to the nucleus on phosphorylation and activate a variety of genes.

The **J gene segments**, or **joining gene segments,** are found some distance 5′ to the C genes in immunoglobulin and T-cell receptor loci. A V and D gene segment must rearrange to a J gene segment to form a complete V-region exon.

Junctional diversity is the diversity present in antigen-specific receptors that is created during the process of joining V, D, and J gene segments.

Killer activatory receptors (**KARs**) are cell-surface receptors on NK cells or cytotoxic T cells; they are receptors that can activate killing by these cells.

Killer T cell is another commonly used term for cytotoxic T cell.

The cell-surface receptor **Kit**, present on developing B cells and other developing white blood cells, binds to the stem cell factor on bone marrow stromal cells. Kit has protein tyrosine kinase activity.

Kupffer cells are phagocytes lining the hepatic sinusoids; they remove debris and dying cells from the blood, but are not known to elicit immune responses.

λ5: see **pre-B-cell receptor**.

Langerhans' cells are phagocytic dendritic cells found in the epidermis. They can migrate from the epidermis to regional lymph nodes via the lymphatics. In the lymph node they differentiate into dendritic cells.

The **large pre-B cells** have a cell-surface **pre-B-cell receptor**, which is lost on the transition to small pre-B cells, in which light-chain gene rearrangement occurs.

In type 1 immediate hypersensitivity reactions, the **late-phase reaction** persists and is resistant to treatment with anti-histamine.

Some viruses can enter a cell but not replicate, a state known as **latency**. Latency can be established in various ways; when the virus is reactivated and replicates, it can produce disease.

Late pro-B cell: see **pro-B cell**.

LCMV: see **lymphocytic choriomeningitis virus**.

Lentiviruses are a group of retroviruses that include the human immunodeficiency virus (HIV). They cause disease after a long incubation period and can take years to become apparent.

Leprosy is caused by *Mycobacterium leprae* and occurs in a variety of forms. There are two polar forms: lepromatous leprosy, which is characterized by abundant replication of leprosy bacilli and abundant antibody production without cell-mediated immunity; and tuberculoid leprosy, in which few organisms are seen in the tissues, there is little or no antibody, but cell-mediated immunity is very active. The other forms of leprosy are intermediate between the polar forms.

Leukemia is the unrestrained proliferation of a malignant white blood cell characterized by very high numbers of the malignant cells in the blood. Leukemias can be lymphocytic, myelocytic, or monocytic.

Leukocyte is a general term for a white blood cell. Leukocytes include lymphocytes, polymorphonuclear leukocytes, and monocytes.

Leukocyte adhesion deficiency is an immunodeficiency disease in which the common β chain of the leukocyte integrins is not produced. This mainly affects the ability of leukocytes to enter sites of infection with extracellular pathogens, so that infections cannot be effectively eradicated.

Leukocyte common antigen: see **CD45**.

Leukocyte integrins: see **lymphocyte function-associated antigens**.

Leukocytosis is the presence of increased numbers of leukocytes in the blood. It is commonly seen in acute infection.

LFA-1, LFA-3: see **lymphocyte function-associated antigen**.

The immunoglobulin **light (L) chain** is the smaller of the two types of polypeptide chain that make up all immunoglobulins. It consists of one V and one C domain, and is disulfide-bonded to the heavy chain. There are two classes of light chain, known as κ and λ.

Linear epitope: see **continuous epitope**.

Alleles at linked loci within the major histocompatibility complex are said to be in **linkage disequilibrium** if they are inherited together more frequently than predicted from their individual frequencies.

Epitopes recognized by B cells and helper T cells must be physically linked for the helper T cell to activate the B cell. This is called **linked recognition**.

The membrane-associated protein known as **Linker of Activation in T cells** (**LAT**) is a protein with several tyrosines that become phosphorylated by the tyrosine kinase ZAP-70. It coordinates downstream signaling events in T-cell activation.

Low-zone tolerance: see **high-zone tolerance**.

L-selectin is an adhesion molecule of the selectin family found on lymphocytes. L-selectin binds to CD34 and GlyCAM-1 on high endothelial venules to initiate the migration of naive lymphocytes into lymphoid tissue.

Lyme disease is a chronic infection with *Borrelia burgdorferi*, a spirochete that can evade the immune response.

The **lymphatic system** is the system of lymphoid channels that drains extracellular fluid from the periphery via the thoracic duct to the blood. It includes the lymph nodes, Peyer's patches, and other organized lymphoid elements apart from the spleen, which communicates directly with the blood.

Lymphatic vessels, or **lymphatics**, are thin-walled vessels that carry lymph—the extracellular fluid that accumulates in tissues—back through the lymph nodes to the thoracic duct.

Lymph nodes are one type of secondary lymphoid organ, in which adaptive immune responses are initiated. They are found in many locations where lymphatic vessels converge, delivering antigen to antigen-presenting cells, which display it to the many recirculating lymphocytes that migrate through the lymph node. Some of these cells will recognize the antigen and respond to it, triggering an adaptive immune response.

A **lymphoblast** is a lymphocyte that has enlarged and increased its rate of RNA and protein synthesis.

Lymphocyte function-associated antigen-1 (**LFA-1**) is one of the leukocyte integrins, which are heterodimeric molecules involved in the interaction of leukocytes with other cells, such as endothelial cells and antigen-presenting cells. LFA-1 is particularly important in T-cell adhesion to these cells.

Lymphocyte function-associated antigen-3 (**LFA-3**) is a molecule found on many cells that is the ligand for CD2 (also known as LFA-2). It is a member of the immunoglobulin superfamily.

All adaptive immune responses are mediated by **lymphocytes**. Lymphocytes are a class of white blood cells that bear variable cell-surface receptors for antigen. These receptors are encoded in rearranging gene segments. There are two main classes of lymphocyte—B lymphocytes (B cells) and T lymphocytes (T cells)—which mediate humoral and cell-mediated immunity respectively. Small lymphocytes have little cytoplasm and condensed nuclear chromatin; on antigen recognition, the cell enlarges to form a lymphoblast and then proliferates and differentiates into an antigen-specific effector cell.

Lymphocytic choriomeningitis virus (**LCMV**) is a virus that causes a non-bacterial meningitis in mice and occasionally humans. It is used extensively in experimental studies.

Lymphoid organs are organized tissues characterized by very large numbers of lymphocytes interacting with a non-lymphoid stroma. The primary lymphoid organs, where lymphocytes are generated, are the thymus and bone marrow. The main secondary lymphoid organs, in which adaptive immune responses are initiated, are the lymph nodes, spleen, and mucosal-associated lymphoid tissues such as tonsils and Peyer's patches.

Lymphokines are cytokines produced by lymphocytes.

Lymphomas are tumors of lymphocytes that grow in lymphoid and other tissues but do not enter the blood in large numbers. There are many types of lymphoma, which represent the transformation of various classes of lymphoid cell.

Lymphotoxin (**LT**) is also known as tumor necrosis factor-β (TNF-β), a cytokine secreted by inflammatory CD4T cells, and is directly cytotoxic for some cells.

Lyn: see **tyrosine kinase**.

Lytic granules containing perforin and granzymes are a defining characteristic of armed effector cytotoxic cells.

In the context of immunoglobulins, μ is the heavy chain of IgM.

Macroglobulin describes plasma proteins that are globulins of high molecular weight, including immunoglobulin M (IgM).

Macrophages are large mononuclear phagocytic cells important in innate immunity, in early non-adaptive phases of host defense, as antigen-presenting cells, and as effector cells in humoral and cell-mediated immunity. They are migratory cells deriving from bone marrow precursors and are found in most tissues of the body. They have a crucial role in host defense.

Resting macrophages will not destroy certain intracellular bacteria unless the macrophage is activated by a T cell. **Macrophage activation** is important in controlling infection and also causes damage to neighboring tissues.

Dendritic cells are unique in having a process known as **macropinocytosis** in which large amounts of extracellular fluid are taken up in single vesicles. This is one means of antigen uptake.

MadCAM-1 is the mucosal cell adhesion molecule-1 or mucosal addressin that is recognized by the lymphocyte surface proteins L-selectin and VLA-4, allowing specific homing of lymphocytes to mucosal tissues.

Eosinophils can be triggered to release their **major basic protein,** which can then act on mast cells to cause their degranulation.

The **major histocompatibility complex** (**MHC**) is a cluster of genes on human chromosome 6 or mouse chromosome 17. It encodes a set of membrane glycoproteins called the MHC molecules. The MHC class I molecules present peptides generated in the cytosol to CD8 T cells, and the MHC class II molecules present peptides degraded in intracellular vesicles to CD4 T cells. The MHC also encodes proteins involved in antigen processing and other aspects of host defense. The MHC is the most polymorphic gene cluster in the human genome, having large numbers of alleles at several different loci. Because this polymorphism is usually detected by using antibodies or specific T cells, the MHC molecules are often called major histocompatibility antigens.

The **mannan-binding lectin** (**MBL**), also called **mannose-binding protein**, is an acute-phase protein that binds to mannose residues. It can opsonize pathogens bearing mannose on their surfaces and can activate the complement system via the **mannan-binding lectin pathway**. It is important in innate immunity.

The follicular **mantle zone** is a rim of B lymphocytes that surrounds lymphoid follicles. The precise nature and role of mantle zone lymphocytes have not yet been determined.

Mast cells are large cells found in connective tissues throughout the body, most abundantly in the submucosal tissues and the dermis. They contain large granules that store a variety of mediator molecules including the vasoactive amine histamine. Mast cells have high-affinity Fcε receptors (FcεRI) that allow them to bind IgE monomers. Antigen binding to this IgE triggers mast cell degranulation and mast cell activation, producing a local or systemic immediate hypersensitivity reaction. Mast cells have a crucial role in allergic reactions.

Mature B cells are B cells that have acquired surface IgM and IgD and have become able to respond to antigen.

Antigens and pathogens enter the body from the intestines through cells called **M cells**, which are specialized for this function. They are found over the gut-associated lymphoid tissue, or GALT.

The **medulla** is generally the central or collecting point of an organ. The thymic medulla is the central area of each thymic lobe, rich in bone-marrow derived antigen-presenting cells and cells of a distinctive medullary epithelium. The medulla of the lymph node is a site of macrophage and plasma cell concentration through which the lymph flows on its way to the efferent lymphatics.

The **membrane-attack complex** is made up of the terminal complement components, which assemble to generate a membrane-spanning hydrophilic pore, damaging the membrane.

Membranous glomerulonephritis is a disease of the kidneys characterized by proteinuria and heavy deposits of antibody and complement.

MHC: see **major histocompatibility complex**.

The **MHC class IB** molecules encoded within the MHC are not highly polymorphic like the class I and class II molecules, and present a restricted set of antigens.

The **MHC class II compartment** (**MIIC**) is a site in the cell where MHC class II molecules accumulate, encounter HLA-DM, and bind antigenic peptides, before migrating to the surface of the cell.

The protein that activates the transcription of MHC class II genes, the **MHC class II transactivator** (**CIITA**), is defective in the disease bare lymphocyte syndrome, a lack of expression of MHC class II gene products on all cells.

Various specialized strains of mice are used to explore the role of MHC polymorphism *in vivo*. These are called **MHC congenic**, meaning that the mice differ only at the MHC complex, **MHC recombinant**, meaning that the mice have a crossover within the MHC, or **MHC mutant**, meaning that they are mutant at one or more loci.

MHC genes are inherited in most cases as an **MHC haplotype**, the set of genes in a haploid genome inherited from one parent. Thus, if the parents are ab or cd, then the offspring are most likely to be ac, ad, bc, or bd.

Antigen recognition by T cells is **MHC restricted**, which means that a given T cell will recognize an antigen only when the peptide is bound to a particular MHC molecule. Normally, as T cells are stimulated only in the presence of self MHC molecules, antigen is recognized only as peptides bound to self MHC molecules. However, experimental manipulations can produce T cells that recognize antigen only when its peptide fragments are bound to non-self MHC molecules. Thus, MHC restriction defines T-cell specificity both in terms of the antigen recognized and in terms of the MHC molecule that binds its peptide fragments.

MIC molecules are MHC class I-like molecules that are expressed in the gut under conditions of stress and are encoded within the class I region of the human MHC.

Microorganisms are microscopic organisms, unicellular except for some fungi, that include bacteria, yeasts and other fungi, and protozoa, all of which can cause human disease.

Anti-carbohydrate antibodies can bind either the ends or the middles of polysaccharide chains; the latter antibodies are called **middle-binders**.

Minor histocompatibility antigens (**minor H antigens**) are peptides of polymorphic cellular proteins bound to MHC molecules that can lead to graft rejection when they are recognized by T cells.

Minor lymphocyte stimulatory (**Mls**) **loci**: see **Mls antigens**.

Mitogen-activated protein kinases (**MAP kinases**) are kinases that become phosphorylated and activated on cellular stimulation by a variety of ligands, and lead to new gene expression by phosphorylating key transcription factors.

When lymphocytes from two unrelated individuals are cultured together in a **mixed lymphocyte culture**, the T cells proliferate in response to the allogeneic MHC molecules on the cells of the other donor. This **mixed lymphocyte reaction** is used in testing for histocompatibility.

Mls antigens are non-MHC antigens that provoke strong primary mixed lymphocyte responses. They are encoded by **minor lymphocyte stimulatory (Mls) loci,** which are endogenous mammary tumor viruses integrated in the mouse genome. Mls antigens are encoded in the 3′ (long terminal repeat of the integrated virus and act as superantigens. They stimulate a large number of T lymphocytes by binding to the V_β domain of many different T-cell receptors.

MMTV: see **mouse mammary tumor virus**.

It has been proposed that infectious agents could provoke autoimmunity by **molecular mimicry**, the induction of antibodies and T cells that react against the pathogen but also cross-react with self antigens.

Monoclonal antibodies are antibodies produced by a single clone of B lymphocytes. Monoclonal antibodies are usually produced by making hybrid antibody-forming cells from a fusion of myeloma cells with immune spleen cells.

Monocytes are white blood cells with a bean-shaped nucleus; they are precursors to macrophages.

Some antibodies recognize all allelic forms of a polymorphic molecule such as an MHC class I protein; these antibodies are thus said to recognize a **monomorphic** epitope.

An individual lymphocyte carries antigen receptors of a single antigen specificity and thus has the property of **monospecificity** in response to antigen.

Mouse mammary tumor virus (**MMTV**) is a retrovirus that encodes a viral superantigen; integrated copies of related viruses encode the endogenous superantigens known as Mls antigens.

Mucins are highly glycosylated cell-surface proteins. Mucin-like molecules are bound by L-selectin in lymphocyte homing.

The **mucosal-associated lymphoid tissue** (**MALT**) comprises all lymphoid cells in epithelia and in the lamina propria lying below the body's mucosal surfaces. The main sites of mucosal associated lymphoid tissues are the gut-associated lymphoid tissues (GALT), and the bronchial-associated lymphoid tissues (BALT).

Multiple myeloma is a tumor of plasma cells, almost always first detected as multiple foci in bone marrow. Myeloma cells produce a monoclonal immunoglobulin, called a myeloma protein, that is detectable in the patient's plasma.

Multiple sclerosis is a neurological disease characterized by focal demyelination in the central nervous system, lymphocytic infiltration in the brain, and a chronic progressive course. It is believed to be an autoimmune disease.

Myasthenia gravis is an autoimmune disease in which autoantibodies against the acetylcholine receptor on skeletal muscle cells cause a block in neuromuscular junctions, leading to progressive weakness and eventually death.

Myeloid progenitors are cells in bone marrow that give rise to the granulocytes and macrophages of the immune system.

Myeloma proteins are immunoglobulins secreted by myeloma tumors and are found in the patient's plasma.

Myelopoiesis is the production of monocytes and polymorphonuclear leukocytes in bone marrow.

Naive lymphocytes are lymphocytes that have never encountered their specific antigen and thus have never responded to it, as distinct from memory or effector lymphocytes. All lymphocytes leaving the central lymphoid organs are naive lymphocytes, those from the thymus being **naive T cells** and those from bone marrow being **naive B cells**.

Natural killer cells (**NK cells**) are large, usually granular, non-T, non-B lymphocytes, which kill certain tumor cells. NK cells are important in innate immunity to viruses and other intracellular pathogens, as well as in antibody-dependent cell-mediated cytotoxicity (ADCC).

Necrosis is the death of cells or tissues due to chemical or physical injury, as opposed to apoptosis, which is a biologically programmed form of cell death. Necrosis leaves extensive cellular debris that needs to be removed by phagocytes, whereas apoptosis does not.

During intrathymic development, thymocytes that recognize self are deleted from the repertoire, a process known as **negative selection**. Autoreactive B cells undergo a similar process in bone marrow.

Antibodies that can inhibit the infectivity of a virus or the toxicity of a toxin molecule are said to **neutralize** them. Such antibodies are known as **neutralizing antibodies** and the process of inactivation as **neutralization**.

Neutrophils, also known as **neutrophilic polymorphonuclear leukocytes**, are the major class of white blood cell in human peripheral blood. They have a multilobed nucleus and neutrophilic granules. Neutrophils are phagocytes and have an important role in engulfing and killing extracellular pathogens.

NK1.1 CD4 T cells are a small subset of T cells that express the NK1.1 marker, a molecule normally found on NK cells, as well as α:β T-cell receptors of limited variety and, usually, the co-receptor molecule CD4. They are the major producers of IL-4 early in the immune response.

NK cells: see **natural killer cells**.

Nodular sclerosis: see **Hodgkin's disease**.

Recombination signal sequences (RSS) flanking gene segments consist of a seven-nucleotide heptamer and a nine-nucleotide **nonamer** of conserved sequence, separated by 12 or 23 nucleotides. RSSs form the target for the site-specific recombinase that joins the gene segments in antigen receptor gene rearrangement.

When T- and B-cell receptor gene segments rearrange, they often form **non-productive rearrangements** that cannot encode a protein because the coding sequences are in the wrong translational reading frame.

N-regions are made up of nucleotides that are inserted into the junctions between gene segments of T-cell receptor and immunoglobulin heavy-chain V-region genes during gene segment joining. These **N-nucleotides** are not encoded in either gene segment, but are inserted by the enzyme terminal deoxynucleotidyl transferase (TdT). They markedly increase the diversity of these receptors.

The transcription factor called **Nuclear Factor of Activated T Cells (NFAT)** is a complex of a protein called NFATc, as it is held in the cytosol by serine/threonine phosphorylation, and the Fos/Jun dimer known as AP-1. It moves from the cytosol to the nucleus on cleavage of the phosphate residues by calcineurin, as serine/threonine protein phosphatase.

The ***nude*** mutation of mice produces hairlessness and defective formation of the thymic stroma, so that nude mice, which are homozygous for this mutation, have no mature T cells.

Oncogenes are genes involved in regulating cell growth. When these genes are defective in structure or expression, they can cause cells to grow continuously to form a tumor.

An **opportunistic pathogen** is a microorganism that causes disease only in individuals with compromised host defense mechanisms, as occurs in AIDS.

Opsonization is the alteration of the surface of a pathogen or other particle so that it can be ingested by phagocytes. Antibody and complement opsonize extracellular bacteria for destruction by neutrophils and macrophages.

Organ-specific autoimmune diseases are autoimmune diseases targeted at a particular organ, such as the thyroid in Graves' disease. They contrast with systemic autoimmune diseases, which do not show organ specificity.

Original antigenic sin describes the tendency of humans to make antibody responses to those epitopes shared between the original strain of a virus and subsequent related viruses, while ignoring other highly immunogenic epitopes on the second and subsequent viruses.

Lymphocyte subpopulations can be isolated by **panning** on petri dishes coated with monoclonal antibodies against cell-surface markers, to which the lymphocytes bind.

The **paracortical area**, or **paracortex**, is the T-cell area of lymph nodes, lying just below the follicular cortex that is primarily B cells.

Parasites are organisms that obtain sustenance from a live host. In medical practice, the term is restricted to worms and protozoa, the subject matter of parasitology.

Paroxysmal nocturnal hemoglobinuria (**PNH**) is a disease in which complement regulatory proteins are defective, so that complement activation leads to episodes of spontaneous hemolysis.

Partial agonist peptides, or altered peptide ligands, are able to stimulate a partial response from a cloned T-cell line, such as the induction of cytokine secretion but not proliferation.

Passive hemagglutination is a technique for detecting antibody in which red blood cells are coated with antigen and the antibody is detected by agglutination of the coated red blood cells.

The injection of antibody or immune serum into a naive recipient is called **passive immunization**. Cf. **active immunization**.

Pathogenic microorganisms, or **pathogens**, are microorganisms that can cause disease when they infect a host.

Pathology is the scientific study of disease. The term pathology is also used to describe detectable damage to tissues.

Pentadecacatechol is the chemical substance in the leaves of the poison ivy plant that causes the cell-mediated immunity associated with hypersensitivity to poison ivy.

Pentraxins are a family of acute-phase proteins formed of five identical subunits, to which C-reactive protein and serum amyloid protein belong.

Perforin is a protein that can polymerize to form the membrane pores that are an important part of the killing mechanism in cell-mediated cytotoxicity. Perforin is produced by cytotoxic T cells and NK cells and is stored in granules that are released by the cell when it contacts a specific target cell.

The **periarteriolar lymphoid sheath** (**PALS**) is part of the inner region of the white pulp of the spleen, and contains mainly T cells.

Peripheral blood mononuclear cells are lymphocytes and monocytes isolated from peripheral blood, usually by Ficoll Hypaque density centrifugation.

Peripheral lymphoid organs are the lymph nodes, spleen, and mucosal-associated lymphoid tissues, in which immune responses are induced, as opposed to the central lymphoid organs, in which lymphocytes develop.

Peripheral tolerance is tolerance acquired by mature lymphocytes in the peripheral tissues, as opposed to central tolerance that is acquired by immature lymphocytes during their development.

Peyer's patches are aggregates of lymphocytes along the small intestine, especially the ileum.

Antibody-like phage can be produced by cloning immunoglobulin V-region genes in filamentous phage, which thus express antigen-binding domains on their surfaces, forming a **phage display library**. Antigen-binding phage can be replicated in bacteria and used like antibodies. This technique is being used to develop novel antibodies of any specificity.

Phagocytosis is the internalization of particulate matter by cells. Usually, the **phagocytic cells** or **phagocytes** are macrophages or neutrophils, and the particles are bacteria that are taken up and destroyed. The ingested material is contained in a vesicle called a **phagosome**, which then fuses with one or more lysosomes to form a **phagolysosome**. The lysosomal enzymes are important in pathogen destruction and degradation to small molecules.

Phosphatidylinositol bisphosphate (**PIP_2**) is a membrane-associated phospholipid that is cleaved by phospholipase C-γ to give the signaling molecules diacylglycerol and inositol trisphosphate.

Phospholipase C-γ is a key enzyme in signal transduction. It is activated by protein tyrosine kinases that are themselves activated by receptor ligation, and activated phospholipase C-γ cleaves inositol phospholipid into inositol trisphosphate and diacylglycerol.

Plasma is the fluid component of blood containing water, electrolytes, and the plasma proteins.

A **plasmablast** is a B cell in a lymph node that already shows some features of a plasma cell.

Plasma cells are terminally differentiated B lymphocytes and are the main antibody-secreting cells of the body. They are found in the medulla of the lymph nodes, in splenic red pulp, and in bone marrow.

Platelets are small cell fragments found in the blood that are crucial for blood clotting. They are formed from megakaryocytes.

P-nucleotides are nucleotides found in junctions between gene segments of the rearranged V-region genes of antigen receptors. They are an inverse repeat of the sequence at the end of the adjacent gene segment, being generated from a hairpin intermediate during recombination, and hence are called palindromic or P-nucleotides.

Poison ivy is a plant whose leaves contain pentadecacatechol, a potent contact sensitizing agent and a frequent cause of contact hypersensitivity.

Antigen activates specific lymphocytes whereas all mitogens, by definition, activate most or all lymphocytes, a process known as **polyclonal activation** because it involves multiple clones of diverse specificity. Such mitogens are known as **polyclonal mitogens**.

The major histocompatibility complex is both **polygenic**, containing several loci encoding proteins of identical function, and polymorphic, having multiple alleles at each locus.

The **poly-Ig receptor** binds polymeric immunoglobulins, especially IgA, at the basolateral membrane of epithelia and transports them across the cell, where they are released from the apical surface. This transcytosis transfers IgA from its site of synthesis to its site of action at epithelial surfaces.

The **polymerase chain reaction** (**PCR**) is a technique for amplifying a specific sequence in DNA by repeated cycles of synthesis driven by pairs of reciprocally oriented primers.

Polymorphism literally means existing in a variety of different shapes. Genetic polymorphism is variability at a gene locus in which the variants occur at a frequency of greater than 1%. The major histocompatibility complex is the most polymorphic gene cluster known in humans.

Polymorphonuclear leukocytes are white blood cells with multi-lobed nuclei and cytoplasmic granules. There are three types of polymorphonuclear leukocyte: the neutrophils with granules that stain with neutral dyes, the eosinophils with granules that stain with eosin, and the basophils with granules that stain with basic dyes.

Some antibodies show **polyspecificity**, the ability to bind to many different antigens. This is also known as **polyreactivity**.

Only those developing T cells whose receptors can recognize antigens presented by self MHC molecules can mature in the thymus, a process known as **positive selection**. All other developing T cells die before reaching maturity.

During B-cell development, **pre-B cells** are cells that have rearranged their heavy-chain genes but not their light-chain genes.

Pre-B-cell receptor: see **surrogate light chain**.

The **precipitin reaction** was the first quantitative technique for measuring antibody production. The amount of antibody is determined from the amount of precipitate obtained with a fixed amount of antigen. The precipitin reaction also can be used to define antigen valence and zones of antibody or antigen excess in mixtures of antigen and antibody.

Prednisone is a synthetic steroid with potent anti-inflammatory and immunosuppressive activity used in treating acute graft rejection and autoimmune disease.

During T-dependent antibody responses, a **primary focus** (plural, **primary foci**) of B-cell activation forms in the vicinity of the margin between T and B cell areas of lymphoid tissue. Here, the T and B cells interact and B cells can differentiate directly into antibody-forming cells or migrate to lymphoid follicles for further proliferation and differentiation.

Lymphoid tissues contain lymphoid follicles made up of follicular dendritic cells and B lymphocytes. The **primary follicles** contain resting B lymphocytes and are the site at which germinal centers form when they are entered by activated B cells, forming secondary follicles.

The **primary immune response** is the adaptive immune response to an initial exposure to antigen. **Primary immunization**, also known as priming, generates both the primary immune response and immunological memory.

The binding of antibody molecules to antigen is called a **primary interaction**, as distinct from secondary interactions in which binding is detected by some associated change such as the precipitation of soluble antigen or agglutination of particulate antigen.

Priming of antigen-specific naive lymphocytes occurs when antigen is presented to them in an immunogenic form; the primed cells will either differentiate into armed effector cells or into memory cells that can respond in second and subsequent immune responses.

During B-cell development, **pro-B cells** are cells that have displayed B-cell surface marker proteins but have not yet completed heavy-chain gene rearrangement. They are divided into early pro-B cells and late pro-B cells.

Professional antigen-presenting cells (**APCs**) are cells that normally initiate the responses of naive T cells to antigen. To date, only dendritic cells, macrophages, and B cells have been shown to have this capacity. A professional antigen-presenting cell must be able to display peptide fragments of antigen on appropriate MHC molecules and also carry co-stimulatory molecules on its surface.

Progenitors are the more differentiated progeny of stem cells that give rise to distinct subsets of mature blood cells and lack the capacity for self-renewal posssed by true stem cells.

Programmed cell death, or apoptosis, is cell death triggered from within the dying cell. Apoptosis eliminates developing T cells that fail positive or negative selection, excess effector cells, and mature lymphocytes that do not encounter antigen. It is critical in maintaining the numbers of lymphocytes at appropriate levels.

Properdin, or factor P, is a positive regulatory component of the alternative pathway of complement activation. It acts by stabilizing the C3/C5 convertase of the alternative factor (comprising C3b,Bb) on the surface of bacterial cells.

Cytosolic proteins are degraded by a large catalytic multisubunit protease called a **proteasome**. It is thought that peptides that are presented by MHC class I molecules are generated by the action of proteasomes, and two subunits of some proteasomes are encoded in the MHC.

Protective immunity is the resistance to specific infection that follows infection or vaccination.

Protein A is a membrane component of *Staphylococcus aureus* that binds to the Fc region of IgG and is thought to protect the bacteria from IgG antibodies by inhibiting their interactions with complement and Fc receptors. It is useful for purifying IgG antibodies.

Protein kinase C is a serine/threonine kinase that is activated by diacylglycerol and calcium as a result of signaling via many different receptors.

Protein kinases add phosphate groups to proteins, and **protein phosphatases** remove these phosphate groups.

Enzymes that add phosphate groups to tyrosine residues are called **protein tyrosine kinases**. These enzymes have crucial roles in signal transduction and regulation of cell growth. Their activity is regulated by a second set of molecules called protein tyrosine phosphatases that remove the phosphate from the tyrosine residues.

Proto-oncogenes are cellular genes that regulate growth control. When mutated or aberrantly expressed, they can contribute to the malignant transformation of cells, leading to cancer.

A **provirus** is the DNA form of a retrovirus when it is integrated into the host cell genome, where it can remain transcriptionally inactive for a long period.

P-selectin: see **selectins**.

Purine nucleotide phosphorylase (**PNP**) **deficiency** is an enzyme defect that results in severe combined immunodeficiency. This enzyme is important in purine metabolism, and its deficiency causes the accumulation of purine nucleosides, which are toxic for developing T cells, causing the immune deficiency.

Bacteria with large capsules are difficult for phagocytes to ingest. Such encapsulated bacteria often produce pus at the site of infection, and are thus called **pyogenic bacteria**.

Radiation bone marrow chimeras are mice that have been heavily irradiated and then reconstituted with bone marrow cells of a different strain of mouse, so that the lymphocytes differ genetically from the environment in which they develop. Such chimeric mice have been important in studying lymphocyte development.

Antigen:antibody interaction can be studied by **radioimmunoassay** (**RIA**) in which antigen or antibody is labeled with radioactivity. An unlabeled antigen or antibody is attached to a solid support such as a plastic surface, and the fraction of the labeled antibody or antigen retained on the surface is determined in order to measure binding.

The recombination activating genes ***RAG-1*** and ***RAG-2*** encode the proteins **RAG-1** and **RAG-2,** which are critical to receptor gene rearrangement. Mice lacking either of these genes cannot form receptors and are severely immunodeficient.

Rapamycin is an immunosuppressive drug that blocks cytokine action.

Ras, Raf: see **G proteins**.

IgE antibodies responsible for immediate hypersensitivity reactions were originally called **reagins** or **reaginic antibodies**.

Antigen receptor expression requires gene segment **rearrangement** in developing lymphocytes. The expressed V-region sequences are composed of rearranged gene segments.

The distinguishing characteristic of lymphocytes is the expression of cell-surface **receptors** for antigen. Each lymphocyte bears a receptor of unique structure generated during lymphocyte development through the rearrangement of receptor gene segments to produce a complete gene.

The antigen receptors of lymphocytes are associated with **receptor-associated tyrosine kinases**, mainly of the Src family, which bind to receptor tails via their SH2 domains.

The replacement of a light chain of a self-reactive antigen receptor on immature B cells with a light chain that does not confer autoreactivity is known as **receptor editing**.

Receptor-mediated endocytosis is the internalization into endosomes of molecules bound to cell-surface receptors. Antigens bound to B-cell receptors are internalized by this process.

The totality of lymphocyte receptors for antigen is known as the lymphocyte **receptor repertoire**. It is made up of many millions of different receptors, with all the antigen receptors on a single lymphocyte being identical in structure.

A **recessive lethal gene** is a gene that is needed for the human or animal to develop to adulthood; when both copies are defective, the human or animal dies *in utero* or early after birth.

In any situation in which cells or tissues are transplanted, they come from a donor and are placed in a **recipient** or host.

Recombination activating genes: see **RAG-1** and **RAG-2**.

Strains of mice derived from intra-strain crosses that have undergone recombination within the MHC are called **recombinant inbred strains**.

Recombination signal sequences (**RSSs**) are short stretches of DNA that flank the gene segments that are rearranged to generate a V-region exon. They always consist of a conserved heptamer and nonamer separated by 12 or 23 base pairs. Gene segments are only joined if one is flanked by an RSS containing a 12 base pair spacer and the other is flanked by an RSS containing a 23 base pair spacer, the 12/23 rule of gene segment joining.

The non-lymphoid area of the spleen in which red blood cells are broken down is called the **red pulp**.

When neutrophils and macrophages take up opsonized particles, this triggers a metabolic change in the cell called the **respiratory burst**. It leads to the production of a number of mediators.

The virus known as **respiratory scyncytial virus** (**RSV**) is a human pathogen that is a common cause of severe chest infection in young children, often associated with wheezing.

The **Rev** protein is the product of the *rev* gene of the human immunodeficiency virus (HIV). The Rev protein promotes the passage of viral RNA from nucleus to cytoplasm during HIV replication.

The enzyme **reverse transcriptase** is an essential component of retroviruses, as it translates the RNA genome into DNA before integration into host cell DNA. Reverse transcriptase also enables RNA sequences to be converted into complementary DNA (cDNA), and so to be cloned, and thus is an essential reagent in molecular biology.

The **reverse transcriptase-polymerase chain reaction** (RT-PCR) is used to amplify RNA sequences. The enzyme reverse transcriptase is used to convert an RNA sequence into a cDNA sequence, which is then amplified by PCR.

The **Rhesus** (**Rh**) **blood group antigen** is a red cell membrane antigen that is also detectable on the cells of rhesus monkeys. Anti-Rh antibodies do not agglutinate human red blood cells, so antibody to Rh antigen must be detected by using a Coombs test.

Rheumatoid arthritis is a common inflammatory joint disease that is probably due to an autoimmune response. The disease is accompanied by the production of **rheumatoid factor**, an IgM anti-IgG antibody that can also be produced in normal immune responses.

The technique of **sandwich ELISA** uses antibody bound to a surface to trap a protein by binding to one of its epitopes. The trapped protein is then detected by an enzyme-linked antibody specific for a different epitope on the protein's surface. This gives the assay a high degree of specificity.

Scatchard analysis is a mathematical analysis of equilibrium binding that allows the affinity and valence of a receptor–ligand interaction to be determined.

SCID, ***scid***: see **severe combined immunodeficiency**.

SDS-PAGE is the common abbreviation for polyacrylamide gel electrophoresis (PAGE) of proteins dissolved in the detergent sodium dodecyl sulfate (SDS). This technique is widely used to characterize proteins, especially after labeling and immunoprecipitation.

A **secondary antibody response** is the antibody response induced by a second or booster injection of antigen—a **secondary immunization**. The secondary response starts sooner after antigen injection, reaches higher levels, is of higher affinity than the primary response, and is dominated by IgG antibodies.

Secondary interactions: see **primary interactions**.

When the recipient of a first tissue or organ graft has rejected that graft, a second graft from the same donor is rejected more rapidly and vigorously in what is called a **second set rejection**.

The co-stimulatory signal required for lymphocyte activation is often called a **second signal**, with the first signal coming from the binding of antigen by the antigen receptor. Both signals are required for the activation of most lymphocytes.

The **secretory component** attached to IgA antibodies in body secretions is a fragment of the poly-Ig receptor left attached to the IgA after transport across epithelial cells.

A cell is said to be **selected** by antigen when its receptors bind that antigen. If the cell starts to proliferate as a result, this is called clonal selection, and the cell founds a clone; if the cell is killed as a result of binding antigen, this is called negative selection or clonal deletion.

Selectins are a family of cell-surface adhesion molecules of leukocytes and endothelial cells that bind to sugar moieties on specific glycoproteins with mucin-like features.

Tolerance is the failure to respond to an antigen; when that antigen is borne by self tissues, tolerance is called **self tolerance**. See also **tolerance**.

Allergic reactions require prior immunization, called **sensitization**, by the allergen that elicits the acute response. Allergic reactions occur only in **sensitized** individuals.

Sepsis is infection of the bloodstream. This is a very serious and frequently fatal condition. Infection of the blood with Gram-negative bacteria triggers **septic shock** through the release of the cytokine TNF-α.

A **sequence motif** is a pattern of nucleotides or amino acids shared by different genes or proteins that often have related functions. Sequence motifs observed in peptides that bind a particular MHC glycoprotein are based on the requirements for particular amino acids to achieve binding to that MHC molecule.

Seroconversion is the phase of an infection when antibodies against the infecting agent are first detectable in the blood.

Serology is the use of antibodies to detect and measure antigens by using serological assays, so called because these assays were originally carried out with serum, the fluid component of clotted blood, from immunized individuals.

Serotonin is the principal vasoactive amine found in mast cell granules of rodents.

Serum is the fluid component of clotted blood.

Serum sickness occurs when foreign serum or serum proteins are injected into a person. It is caused by the formation of immune complexes between the injected protein and the antibodies formed against it. It is characterized by fever, arthralgias, and nephritis.

Severe combined immune deficiency (**SCID**) is an immune deficiency disease in which neither antibody nor T-cell responses are made. It is usually the result of T-cell deficiencies. The scid mutation in mice causes severe combined immune deficiency in mice.

SH1, SH2, SH3: see **Src-family tyrosine kinases**.

A **signal joint** is formed by the precise joining of recognition signal sequences in the process of somatic recombination that generates T-cell receptor and immunoglobulin genes.

Signal transduction describes the general process by which cells perceive changes in their environment. In lymphocytes, the most important changes are those occurring during infection that generate antigens that stimulate the cells of the immune system to bring about an adaptive immune response.

A **single-chain Fv** fragment, comprising a V region of a heavy chain linked by a stretch of synthetic peptide to a V region of a light chain, can be made by genetic engineering.

During T-cell maturation in the thymus, mature T cells are detected by the expression of either the CD4 or the CD8 co-receptor and are therefore called **single-positive thymocytes**.

Sirolimus is the drug name that has been adopted for the chemical rapamycin; the two terms are used interchangeably in literature.

Smallpox is an infectious disease, caused by the virus variola, that once killed at least 10% of infected people. It has now been eradicated by vaccination.

Small pre-B cells: see **large pre-B cells**.

Somatic gene therapy: see **gene therapy**.

During B-cell responses to antigen, the V-region DNA sequence undergoes **somatic hypermutation,** resulting in the generation of variant immunoglobulins, some of which bind antigen with a higher affinity. This allows the affinity of the antibody response to increase. These mutations affect only somatic cells, and are not inherited through germline transmission.

Somatic mutation theories of antibody diversity proposed that a single gene encoding all antibody molecules underwent mutation in somatic cells to generate the diversity of secreted antibodies. These are also known as somatic diversification theories.

During lymphocyte development, receptor gene segments undergo **somatic recombination** to generate intact V-region exons that encode the V region of each immunoglobulin and T-cell receptor chain. These events occur only in somatic cells; the changes are not inherited.

The **spleen** is an organ containing a red pulp, involved in removing senescent blood cells, and a white pulp of lymphoid cells that respond to antigens delivered to the spleen by the blood.

The **Src-family tyrosine kinases** are receptor-associated protein tyrosine kinases. They have several domains, called Src-homology 1, 2, and 3. The SH1 domain contains the active site of the kinase, the SH2 domain can bind to phosphotyrosine residues, and the SH3 domain is involved in interactions with proline-rich regions in other proteins.

Staphylococcal enterotoxins (**SEs**) cause food poisoning and also stimulate many T cells by binding to MHC class II molecules and the V_β domain of the T-cell receptor; the staphylococcal enterotoxins are thus superantigens.

STATs: see **Janus kinases**.

Stem-cell factor (**SCF**) is a transmembrane protein on bone marrow stromal cells that binds to Kit, a signaling receptor carried on developing B cells and other developing white blood cells.

Superantigens are molecules that stimulate a subset of T cells by binding to MHC class II molecules and V_β domains of T-cell receptors, stimulating the activation of T cells expressing particular V_β V gene segments.

Suppressor T cells are T cells that, when mixed with naive or effector T cells, suppress their activity. The precise nature of suppressor T cells and their modes of antigen recognition and activation remain mysterious.

The membrane-bound immunoglobulin that acts as the antigen receptor on B cells is often known as **surface immunoglobulin**.

During B-cell development, the shift from pro-B cells to large pre-B cells is accompanied by the expression of μ heavy chains on their surface in combination with the **surrogate light chain** made up of VpreB and λ5. Together, this is called the pre-B-cell receptor, and it includes Igα and Igβ.

When isotype switching occurs, the active heavy-chain V-region exon undergoes somatic recombination with a 3′ constant-region gene at a **switch region** of DNA. These DNA joints do not need to occur at precise sites, because they occur in intronic DNA. Thus, all switch recombinations are productive.

Syk: see **tyrosine kinase**.

When one eye is damaged, there is often an autoimmune response that damages the other eye, a syndrome known as **sympathetic ophthalmia**.

A **syngeneic graft** is a graft between two genetically identical individuals. It is accepted as self.

Systemic anaphylaxis is the most dangerous form of immediate hypersensitivity reaction. It involves antigen in the blood stream triggering mast cells all over the body. The activation of these mast cells causes widespread vasodilation, tissue fluid accumulation, epiglottal swelling, and often death.

Systemic autoimmunity or **systemic autoimmune disease** involves the production of antibodies to common self constituents. The major cause of pathology in systemic autoimmunity is the deposition of immune complexes. The classical example of a systemic autoimmune disease is **systemic lupus erythematosus**, in which autoantibodies against DNA, RNA, and proteins associated with nucleic acids form immune complexes that damage small blood vessels.

Tacrolimus: see **FK506**.

TAP-1 and **TAP-2** (transporters associated with antigen processing) are ATP-binding cassette proteins involved in transporting short peptides from the cytosol into the lumen of the endoplasmic reticulum, where they associate with MHC class I molecules.

Tapasin, or the **TAP-associated protein**, is a key molecule in the assembly of MHC class I molecules; a cell deficient in this protein has only unstable MHC class I molecules on the cell surface.

The functions of effector T cells are always assayed by the changes that they produce in antigen-bearing **target cells**. These cells can be B cells, which are activated to produce antibody; macrophages, which are activated to kill bacteria or tumor cells; or labeled cells that are killed by cytotoxic T cells.

The **Tat** protein is a product of the *tat* gene of HIV. It is produced when latently infected cells are activated, and it binds to a transcriptional enhancer in the long terminal repeat of the provirus, increasing the transcription of the proviral genome.

T-cell antigen receptor: see **T-cell receptor**.

T cells, or **T lymphocytes**, are a subset of lymphocytes defined by their development in the thymus and by heterodimeric receptors associated with the proteins of the CD3 complex. Most T cells have α:β heterodimeric receptors but γ:δ T cells have a γ:δ heterodimeric receptor.

A **T-cell clone** is derived from a single progenitor T cell. See also **cloned T-cell lines**.

T-cell hybrids are formed by fusing an antigen-specific, activated T cell with a T-cell lymphoma. The hybrid cells bear the receptor of the specific T-cell parent and grow in culture like the lymphoma.

T-cell lines are cultures of T cells grown by repeated cycles of stimulation, usually with antigen and antigen-presenting cells. When single T cells from these lines are propagated, they give rise to T-cell clones or cloned T-cell lines.

The **T-cell receptor** consists of a disulfide-linked heterodimer of the highly variable α and β chains expressed at the cell membrane as a complex with the invariant CD3 chains. T cells carrying this type of receptor are often called α:β T cells. An alternative receptor made up of variable γ and δ chains is expressed with CD3 on a subset of T cells.

The complement system can be activated directly or by antibody, but both pathways converge with the activation of the **terminal complement components,** which assemble to form the membrane-attack complex.

The enzyme **terminal deoxynucleotidyl transferase** (**TdT**) inserts non-templated or N-nucleotides into the junctions between gene segments in T-cell receptor and immunoglobulin heavy-chain V-region genes. The N-nucleotides contribute greatly to junctional diversity in V regions.

When the same antigen is injected a third time, the response elicited is called a **tertiary response** and the injection a **tertiary immunization**.

T_H1 cells are a subset of CD4 T cells that are characterized by the cytokines they produce. They are mainly involved in activating macrophages, and are sometimes called inflammatory CD4 T cells.

T_H2 cells are a subset of CD4 T cells that are characterized by the cytokines they produce. They are mainly involved in stimulating B cells to produce antibody, and are often called helper CD4 T cells.

The term **T_H3 cell** has been used to describe unique cells that produce mainly transforming growth factor-β in response to antigen; they develop predominantly in the mucosal immune response to antigens that are presented orally.

The lymph from most of the body, except the head, neck, and right arm, is gathered in a large lymphatic vessel, the **thoracic duct**, which runs parallel to the aorta through the thorax and drains into the left subclavian vein. The thoracic duct thus returns the lymphatic fluid and lymphocytes back into the peripheral blood circulation.

Surgical removal of the thymus is called **thymectomy**.

The **thymic anlage** is the tissue from which the thymic stroma develops during embryogenesis.

The **thymic cortex** is the outer region of each **thymic lobule** in which thymic progenitor cells proliferate, rearrange their T-cell receptor genes, and undergo thymic selection, especially positive selection on **thymic cortical epithelial cells**.

The **thymic stroma** consists of epithelial cells and connective tissue that form the essential microenvironment for T-cell development.

Thymocytes are lymphoid cells found in the thymus. They consist mainly of developing T cells, although a few thymocytes have achieved functional maturity.

The **thymus**, the site of T-cell development, is a lymphoepithelial organ in the upper part of the middle of the chest, just behind the breastbone.

Some antigens elicit responses only in individuals that have T cells; they are called **thymus-dependent antigens** or **TD antigens**. Other antigens can elicit antibody production in the absence of T cells and are called **thymus-independent antigens** or **TI antigens**. There are two types of TI antigen: the **TI-1 antigens,** which have intrinsic B-cell activating activity, and the **TI-2 antigens,** which seem to activate B cells by having multiple identical epitopes that crosslink the B-cell receptor.

T cell and T lymphocyte are short designations for **thymus-dependent lymphocyte**, the lymphocyte population that does not develop in the absence of a functioning thymus.

During the process of germinal center formation, cells called **tingible body macrophages** appear. These are phagocytic cells engulfing apoptotic B cells, which are produced in large numbers during the height of the germinal center response.

Almost all tissues have resident **tissue dendritic cells** that can take up antigen but achieve effective co-stimulatory activity only if they migrate to local lymphoid organs. Graft rejection is triggered by tissue dendritic cells that migrate from the graft to local lymph nodes to trigger an anti-graft response.

Transplantation of organ or **tissue grafts** such as skin grafts is used medically to repair organ or tissue deficits.

Some autoimmune diseases attack particular tissues, such as connective tissue, resulting in **tissue-specific autoimmune disease**.

The **titer** of an antiserum is a measure of its concentration of specific antibodies based on serial dilution to an end point, such as a certain level of color change in an ELISA assay.

T lymphocytes: see **T cells**.

TNF: see **lymphotoxin**; **tumor necrosis factor-α**.

Tolerance is the failure to respond to an antigen; the immune system is said to be tolerant to self antigens. Tolerance to self antigens is an essential feature of the immune system; when tolerance is lost, the immune system can destroy self tissues, as happens in autoimmune disease. See also: **central tolerance**; **peripheral tolerance**; **self tolerance**.

The palatine **tonsils** that lie on either side of the pharynx are large aggregates of lymphoid cells organized as part of the mucosal or gut-associated immune system.

The recombination activating genes, *RAG-1* and *RAG-2*, seem to be related to **topoisomerases**, enzymes involved in cleaving and sealing DNA molecules to allow DNA replication and repair.

Toxic shock syndrome is a systemic toxic reaction caused by the massive production of cytokines by CD4 T cells activated by the bacterial superantigen **toxic shock syndrome toxin-1** (**TSST-1**), which is secreted by *Staphylococcus aureus*.

Inactivated toxins called **toxoids** are no longer toxic but retain their immunogenicity so that they can be used for immunization.

The family of proteins known as TNF receptor associated factors, or **TRAFs**, consists of at least six members that bind to various TNF family receptors or TNFRs. They share a domain known as a TRAF domain, and have a crucial role as signal transducers between upstream members of the TNFR family and downstream transcription factors.

The active transport of molecules across epithelial cells is called **transcytosis**. Transcytosis of IgA molecules involves transport across intestinal epithelial cells in vesicles that originate on the baso-lateral surface and fuse with the apical surface in contact with the intestinal lumen.

The insertion of small pieces of DNA into cells is called **transfection**. If the DNA is expressed without integrating into host cell DNA, this is called a transient transfection; if the DNA integrates into host cell DNA, then it replicates whenever host cell DNA is replicated, producing a stable transfection.

Foreign genes can be placed in the mouse genome by **transgenesis**. This generates transgenic mice that are used to study the function of the inserted gene, the transgene, and the regulation of its expression.

Some cancers have chromosomal **translocation,** in which a piece of one chromosome is abnormally linked to another chromosome.

The grafting of organs or tissues from one individual to another is called **transplantation**. The **transplanted organs** or grafts can be rejected by the immune system unless the host is tolerant to the graft antigens or immunosuppressive drugs are used to prevent rejection.

Transporters associated with antigen processing: see **TAP-1** and **TAP-2**.

The **tuberculin test** is a clinical test in which a purified protein derivative (PPD) of *Mycobacterium tuberculosis*, the causative agent of tuberculosis, is injected subcutaneously. PPD elicits a delayed-type hypersensitivity reaction in individuals who have had tuberculosis or have been immunized against it.

Tuberculoid leprosy: see **leprosy**.

Tumor immunology is the study of host defenses against tumors, usually studied by tumor transplantation. Tumors transplanted into syngeneic recipients can grow progressively or can be rejected through T-cell recognition of **tumor-specific transplantation antigens** (**TSTA**) or **tumor rejection antigens**. TSTA are peptides of mutant or overexpressed cellular proteins bound to MHC class I molecules on the tumor cell surface.

Tumor necrosis factor-α (**TNF-α**) is a cytokine produced by macrophages and T cells that has multiple functions in the immune response. It is the defining member of the TNF family of cytokines. These cytokines function as cell-associated or secreted proteins that interact with receptors of the **tumor necrosis factor receptor** (**TNFR**) family, which in turn communicates with the interior of the cell via components known as TRAFs (tumor necrosis factor receptor-associated factors).

Tumor necrosis factor-β (**TNF-β**): see **lymphotoxin**.

The **TUNEL assay** (TdT-dependent d**U**TP–biotin **n**ick **e**nd **l**abeling) identifies apoptotic cells *in situ* by the characteristic fragmentation of their DNA. Biotin-tagged dUTP added to the free 3′ ends of the DNA fragments by the enzyme TdT can be detected by immunohistochemical staining with enzyme-linked streptavidin.

In **two-dimensional gel electrophoresis,** proteins are separated by isoelectric focusing in one dimension, followed by SDS-PAGE on a slab gel at right-angles to the first dimension. This can separate and identify large numbers of distinct proteins.

Hypersensitivity reactions are classified by mechanism: **type I hypersensitivity reactions** involve IgE antibody triggering of mast cells; **type II hypersensitivity reactions** involve IgG antibodies against cell surface or matrix antigens; **type III hypersensitivity reactions** involve antigen:antibody complexes; and **type IV hypersensitivity reactions** are T-cell mediated.

A **tyrosine kinase** is an enzyme that specifically phosphorylates tyrosine residues in proteins. They are critical in T- and B-cell activation. The kinases that are critical for B-cell activation are Blk, Fyn, Lyn, and Syk. The tyrosine kinases that are critical for T-cell activation are called Lck, Fyn, and ZAP-70.

Urticaria is the technical term for hives, which are red, itchy skin welts usually brought on by an allergic reaction.

Vaccination is the deliberate induction of adaptive immunity to a pathogen by injecting a **vaccine**, a dead or attenuated (non-pathogenic) form of the pathogen.

The first effective vaccine was **vaccinia**, a cowpox virus that causes a limited infection in humans that leads to immunity to the human smallpox virus.

The **valence** of an antibody or antigen is the number of different molecules that it can combine with at one time.

The **variability** of a protein is a measure of the difference between the amino acid sequences of different variants of that protein. The most variable proteins known are antibodies and T-cell receptors.

Variability plot: see **Wu and Kabat plot**.

Variable gene segments: see **V gene segments**.

The **variable region** (**V region**) of an immunoglobulin or T-cell receptor is formed of the amino-terminal domains of its component polypeptide chains. These are called the variable domains (V domains) and are the most variable parts of the molecule. They contain the antigen-binding sites.

Vascular addressins are molecules on endothelial cells to which leukocyte adhesion molecules bind. They have a key role in selective homing of leukocytes to particular sites in the body.

Vav: see **adaptor proteins**.

The enzyme that joins the gene segments of B-cell and T-cell receptor genes is called the **V(D)J recombinase**. It is made up of several enzymes, but most important are the products of the recombinase-activating genes *RAG-1* and *RAG-2* whose protein products RAG-1 and RAG-2 are expressed in developing lymphocytes and make up the only known lymphoid specific components of the V(D)J recombinase.

The **very late antigens** (**VLA**) are members of the β_1 family of integrins involved in cell–cell and cell–matrix interactions. Some VLAs are important in leukocyte and lymphocyte migration.

The variable region of the polypeptide chains of an immunoglobulin or T-cell receptor is composed of a single N-terminal **V domain**. Paired V domains form the antigen-binding sites of Immunoglobulins and T-cell receptors.

Vesicles are small membrane-bounded compartments within the cytosol.

The first 95 amino acids or so of immunoglobulin or T-cell receptor V domains are encoded in inherited **V gene segments**. There are multiple different V gene segments in the germline genome. To produce a complete exon encoding a V domain, one V gene segment must be rearranged to join up with a J or a rearranged DJ gene segment.

Many viruses that are produced by mammalian cells are enclosed in a **viral envelope** of host cell membrane lipid and proteins bound to the viral core by viral envelope proteins.

Virions are complete virus particles, the form in which viruses spread from cell to cell or from one individual to another.

Viruses are pathogens composed of a nucleic acid genome enclosed in a protein coat. They can replicate only in a living cell, as they do not possess the metabolic machinery for independent life.

VpreB: see **pre-B-cell receptor**.

Antibodies against the neutrophil granule proteinase-3 are formed in **Wegener's granulomatosis**, an autoimmune disease in which there is severe necrotizing vasculitis. The presence of anti-neutrophil cytoplasmic antigen, or ANCA, helps in the diagnosis of this disease.

Weibel–Palade bodies are granules within endothelial cells that contain P-selectin. Activation of the endothelial cell by mediators such as histamine and C5a leads to rapid translocation of P-selectin to the cell surface.

In **Western blotting**, a mixture of proteins is separated, usually by gel electrophoresis and transferred by blotting to a nitrocellulose membrane; labeled antibodies are then used as probes to detect specific proteins.

When small amounts of allergen are injected into the dermis of an allergic individual, a **wheal-and-flare reaction** is observed. This consists of a raised area of skin containing fluid and a spreading, red, itchy circular reaction.

The discrete areas of lymphoid tissue in the spleen are known as the **white pulp**.

The **Wiskott–Aldrich syndrome** is a congenital abnormality in which antibodies against polysaccharide antigens are defective. Patients with the Wiskott–Aldrich syndrome are susceptible to infection with pyogenic bacteria.

A **Wu and Kabat plot**, or variability plot, is generated from the amino acid sequences of related proteins by plotting the variability of the sequence against amino acid residue number. Variability is the number of different amino acids observed at a position divided by the frequency of the most common amino acid.

Animals of different species are **xenogeneic**.

The use of **xenografts**, organs from a different species, is being explored as a solution to the severe shortage of human organs for transplantation. The main problem with xenografting is the presence of natural antibodies against xenograft antigens; attempts are being made to modify these reactions by creating transgenic animals.

Mice with mutations in the *btk* gene have a deficiency in making antibodies, especially in primary responses. These mice are called **xid**, for X-linked immunodeficiency, the mouse equivalent of X-linked agammaglobulinemia, the human form of this disease.

X-linked agammaglobulinemia (**XLA**) is a genetic disorder in which B-cell development is arrested at the pre-B-cell stage and no mature B cells or antibodies are formed. The disease is due to a defect in the gene encoding the protein tyrosine kinase *btk*.

X-linked hyper IgM syndrome is a disease in which little or no IgG, IgE, or IgA antibody is produced and even IgM responses are deficient, but serum IgM levels are normal to high. It is due to a defect in the gene encoding the CD40 ligand.

X-linked severe combined immunodeficiency (**X-linked SCID**) is a disease in which T-cell development fails at an early intra-thymic stage and no production of mature T cells or T-cell dependent antibody occurs. It is due to a defect in a gene that encodes part of the receptors for several different cytokines.

ZAP-70: see **tyrosine kinase**.

INDEX